## ▼ Special Constants

$$\pi \approx 3.1416 \qquad e \approx 2.7183 \qquad \sqrt{2} \approx 1.4142 \qquad \sqrt{3} \approx 1.7321$$

## ▼ Special Products

$$(x + a)(x + b) = x^2 + (a + b)x + ab \qquad\qquad (a + b)(a - b) = a^2 - b^2$$
$$(a + b)^2 = a^2 + 2ab + b^2 \qquad\qquad (a + b)^3 = a^3 + 3a^2b + 3ab^2 + b^3$$
$$(a - b)^2 = a^2 - 2ab + b^2 \qquad\qquad (a - b)^3 = a^3 - 3a^2b + 3ab^2 - b^3$$

## ▼ Special Factorizations

$$x^2 + (a + b)x + ab = (x + a)(x + b) \qquad\qquad a^2 - b^2 = (a + b)(a - b)$$
$$a^2 + 2ab + b^2 = (a + b)^2 \qquad\qquad a^3 - b^3 = (a - b)(a^2 + ab + b^2)$$
$$a^2 - 2ab + b^2 = (a - b)^2 \qquad\qquad a^3 + b^3 = (a + b)(a^2 - ab + b^2)$$

## ▼ Formulas from Plane Geometry: $P \rightarrow$ perimeter, $C \rightarrow$ circumference, $A \rightarrow$ area

**Rectangle**
$$P = 2l + 2w$$
$$A = lw$$

**Square**
$$P = 4s$$
$$A = s^2$$

**Regular Polygon**
$$P = ns$$
$$A = \frac{ns}{2}a$$
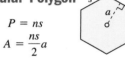

**Parallelogram**
$$A = bh$$

**Trapezoid**
$$A = \frac{a + b}{2} \cdot h$$

**Triangle**
$$A = \frac{1}{2}bh$$

**Triangle**
Sum of angles
$$A + B + C = 180°$$

**Right Triangle**
Pythagorean Theorem
$$a^2 + b^2 = c^2$$
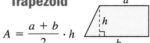

**Circle**
$$A = \pi r^2$$
$$C = 2\pi r = \pi d$$

**Ellipse**
$$A = \pi ab$$
$$P \approx 2\pi\sqrt{\frac{a^2 + b^2}{2}}$$
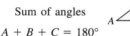

**Right Parabolic Segment**
$$A = \frac{4}{3}ab$$

## ▼ Formulas from Solid Geometry: $S \rightarrow$ surface area, $V \rightarrow$ volume

**Rectangular Solid**
$$V = LWH$$
$$S = 2(LW + LH + WH)$$
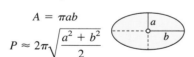

**Cube**
$$V = s^3$$
$$S = 6s^2$$

**Right Circular Cylinder**
$$V = \pi r^2 h$$
$$S = 2\pi r(r + h)$$

**Right Circular Cone**
$$V = \frac{1}{3}\pi r^2 h$$
$$S = \pi r(r + s)$$
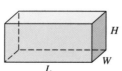

**Right Square Pyramid**
$$V = \frac{1}{3}b^2 h$$
$$S = b^2 + b\sqrt{b^2 + 4h^2}$$
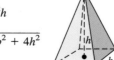

**Sphere**
$$V = \frac{4}{3}\pi r^3$$
$$S = 4\pi r^2$$

# ▼ Formulas from Analytical Geometry: $P_1 \to (x_1, y_1)$, $P_2 \to (x_2, y_2)$

### Distance between $P_1$ and $P_2$

$$d = \sqrt{(x_2 - x_1)^2 + (y_2 - y_1)^2}$$

### Slope of Line Containing $P_1$ and $P_2$

$$m = \frac{\Delta y}{\Delta x} = \frac{y_2 - y_1}{x_2 - x_1}$$

### Equation of Line Containing $P_1$ and $P_2$

Point-Slope Form

$$y - y_1 = m(x - x_1)$$

### Equation of Line Containing $P_1$ and $P_2$

Slope-Intercept Form (slope $m$, $y$-intercept $b$)

$$y = mx + b, \text{ where } b = y_1 - mx_1$$

### Parallel Lines

Slopes Are Equal: $m_1 = m_2$, $b_1 \neq b_2$

### Perpendicular Lines

Slopes Have a Product of $-1$: $m_1 m_2 = -1$

### Intersecting Lines

Slopes Are Unequal: $m_1 \neq m_2$

### Dependent (Coincident) Lines

Slopes and $y$-Intercepts Are Equal: $m_1 = m_2$, $b_1 = b_2$

# ▼ Logarithms and Logarithmic Properties

$$y = \log_b x \Leftrightarrow b^y = x \qquad \log_b b = 1 \qquad \log_b 1 = 0$$

$$\log_b b^x = x \qquad b^{\log_b x} = x \qquad \log_c x = \frac{\log_b x}{\log_b c}$$

$$\log_b(MN) = \log_b M + \log_b N \qquad \log_b\left(\frac{M}{N}\right) = \log_b M - \log_b N \qquad \log_b M^P = P \cdot \log_b M$$

# ▼ Applications of Exponentials and Logarithms

$A \to$ amount accumulated $\qquad P \to$ initial deposit $\qquad r \to$ interest rate per year

$n \to$ compounding periods per year $\qquad t \to$ time in years

$\mathcal{P} \to$ periodic payment $\qquad R \to$ interest rate per time period $\left(\dfrac{r}{n}\right)$ $\qquad N \to$ number of compounding periods $(nt)$

### Interest Compounded $n$ Times per Year

$$A = P\left(1 + \frac{r}{n}\right)^{nt}$$

### Interest Compounded Continuously

$$A = Pe^{rt}$$

### Accumulated Value of an Annuity

$$A = \frac{\mathcal{P}}{R}\left[(1 + R)^N - 1\right]$$

### Payments Required to Accumulate Amount $A$

$$\mathcal{P} = \frac{AR}{(1 + R)^N - 1}$$

# ▼ Sequences and Series

$a_1 \to$ 1st term $\qquad a_n \to$ nth term $\qquad S_n \to$ sum of $n$ terms $\qquad d \to$ common difference $\qquad r \to$ common ratio

### Arithmetic Sequences

$$a_1, a_2 = a_1 + d, a_3 = a_1 + 2d, \ldots, a_n = a_1 + (n - 1)d$$

$$S_n = \frac{n}{2}(a_1 + a_n)$$

$$S_n = \frac{n}{2}[2a_1 + (n - 1)d]$$

### Geometric Sequences

$$a_1, a_2 = a_1 r, a_3 = a_1 r^2, \ldots, a_n = a_1 r^{n-1}$$

$$S_n = \frac{a_1 - a_1 r^n}{1 - r}$$

$$S_\infty = \frac{a_1}{1 - r}; \ |r| < 1$$

# ▼ Binomial Theorem

$$(a + b)^n = \binom{n}{0}a^n b^0 + \binom{n}{1}a^{n-1}b^1 + \binom{n}{2}a^{n-2}b^2 + \cdots + \binom{n}{n-1}a^1 b^{n-1} + \binom{n}{n}a^0 b^n$$

$$n! = n(n - 1)(n - 2) \cdots (3)(2)(1) \qquad 0! = 1 \qquad \binom{n}{k} = \frac{n!}{k!(n - k)!}$$

# COLLEGE ALGEBRA

**JOHN W. COBURN**

*ST LOUIS COMMUNITY COLLEGE AT FLORISSANT VALLEY*

**JEREMY P. COFFELT**

*BLINN COLLEGE*

**THIRD EDITION**

Mc
Graw
Hill

*Connect
Learn
Succeed*™

The McGraw-Hill Companies

Connect
Learn
Succeed™

COLLEGE ALGEBRA, THIRD EDITION

Published by McGraw-Hill, a business unit of The McGraw-Hill Companies, Inc., 1221 Avenue of the Americas, New York, NY 10020. Copyright © 2014 by The McGraw-Hill Companies, Inc. All rights reserved. Printed in the United States of America. Previous editions © 2010 and 2007. No part of this publication may be reproduced or distributed in any form or by any means, or stored in a database or retrieval system, without the prior written consent of The McGraw-Hill Companies, Inc., including, but not limited to, in any network or other electronic storage or transmission, or broadcast for distance learning.

Some ancillaries, including electronic and print components, may not be available to customers outside the United States.

This book is printed on acid-free paper.

1 2 3 4 5 6 7 8 9 0 DOW/DOW 1 0 9 8 7 6 5 4 3

ISBN 978–0–07–351958–6
MHID 0–07–351958–8

ISBN 978–0–07–734087–2 (Annotated Instructor's Edition)
MHID 0–07–734087–6

Senior Vice President, Products & Markets: *Kurt L. Strand*
Vice President, General Manager, Products & Markets: *Marty Lange*
Vice President, Content Production & Technology Services: *Kimberly Meriwether David*
Director of Development: *Rose Koos*
Managing Director: *Ryan Blankenship*
Brand Manager: *Caroline Celano*
Director of Digital Content: *Emilie J. Berglund*
Development Editor: *Ashley Zellmer*
Marketing Manager: *Kevin M. Ernzen*
Senior Project Manager: *Vicki Krug*
Senior Buyer: *Laura Fuller*
Senior Media Project Manager: *Sandra M. Schnee*
Senior Designer: *Laurie B. Janssen*
Cover Designer: *Ron Bissell*
Cover Image: © *Michael & Patricia Fogden/CORBIS*
Content Licensing Specialist: *John C. Leland*
Compositor: *Aptara®, Inc.*
Typeface: *10.5/12 Times Roman*
Printer: *R. R. Donnelley*

All credits appearing on page or at the end of the book are considered to be an extension of the copyright page.

**Library of Congress Cataloging-in-Publication Data**

Coburn, John W.
    College algebra / John W. Coburn, St. Louis Community College at Florissant Valley, Jeremy Coffelt, Blinn College. – Third edition.
        pages cm
    Includes index.
    ISBN 978–0–07–351958–6 — ISBN 0–07–351958–8 (hard copy : alk. paper) — ISBN 978–0–07–734087–2— ISBN 0–07–734087–6 (annotated instructor's edition) (print)   1. Algebra–Textbooks.   I. Coffelt, Jeremy. II. Title.
    QA154.3.C594 2014
    512.9–dc23

                                    2012028108

www.mhhe.com

# Contents

# About the Authors

### John Coburn

John Coburn grew up in the Hawaiian Islands, the seventh of sixteen children. In 1977 he received his Associate of Arts Degree from Windward Community College, where he graduated with honors. In 1979 he earned a Bachelor's Degree in Education from the University of Hawai'i. After working in the business world for a number of years, he returned to teaching, accepting a position in high school mathematics, where he was recognized as Teacher of the Year in 1987. Soon afterward, a decision was made to seek a Master's Degree, which he received two years later from the University of Oklahoma. John is now a full professor at the Florissant Valley Campus of St. Louis Community College, where he has taught mathematics for the last twenty-one years. During this time he has received numerous nominations as an outstanding teacher by the local chapter of Phi Theta Kappa, earned recognition as a "Prime Time Teacher" by Eastern Illinois College in 2003, and was recognized as Post-Secondary Teacher of the Year in 2004 by Mathematics Educators of Greater St. Louis (MEGSL). John has made numerous presentations at local, state, and national conferences on a wide variety of topics, and maintains memberships in several mathematical organizations. Some of John's other interests include music, athletics, and the wild outdoors, as well as body surfing, snorkeling, and beach combing whenever he gets the chance. He is also an avid gamer, enjoying numerous board, card, and party games. John hopes that this love of life comes through in his writing, and helps to make the learning experience an interesting and engaging one for all students.

### Jeremy Coffelt

Jeremy Coffelt grew up in the small town of Archer City, Texas, made (in)famous as the inspiration and filming location of *The Last Picture Show*. After graduating from Archer City High School in 2000, he continued his education at Midwestern State University, where he graduated with a Bachelor's Degree in Mathematics in 2002. From there, he completed Master's Degrees in Mathematics from Kansas State University (2005) and in Civil Engineering from Texas A&M University (2008). During his graduate studies, Jeremy published several papers in topics ranging from analytic number theory to Bayesian regression and engineering systems reliability. In 2007, he joined the faculty at Blinn College, where he has since been nominated for several teaching awards. When not teaching or writing, Jeremy enjoys spending time with his ladies—his wife Vanessa, his Chihuahua Buttons, and his Catahoula Abby. His other interests include traveling and all things competitive, including cycling, pool, chess, poker, and tennis.

### Dedication

*I dedicate this work to each of my seven children, in hopes it will help them discover a love of mathematics from their father, as I discovered a love of mathematics from my own.     John Coburn*

*I dedicate my contributions to this text to my wife, Vanessa. For the times you left me alone to work, I thank you. For the times you interrupted my work, I love you.     Jeremy Coffelt*

# Making Connections

## A Focus on Applications

▶ **Chapter Openers** highlight Chapter Connections, an interesting application exercise from the chapter, and provide a list of other real-world connections to give context for students who wonder how math relates to them.

▶ **Application Exercises** at the end of each section are the hallmark of the Coburn-Coffelt series. Never contrived, always creative, and borne out of the authors' lives and experiences, each application tells a story and appeals to a variety of teaching styles, disciplines, backgrounds, and interests. The authors have ensured that the applications reflect the most common majors of college algebra students.

## Clear and Timely Examples

▶ **Examples** are designed with a direct focus on the skill at hand while linking to previous concepts and laying the groundwork for concepts to come. The examples provide students with a starting point for solving a variety of problems.

▶ **Caution Boxes** signal students to pause so that they avoid common errors.

▶ **Check Points** alert students when a specific learning objective has been covered to reinforce correct mathematical terms.

▶ **"Now Try"** boxes immediately following examples guide students to specific matched exercises at the end of the section and connect concepts to homework problems.

▶ **Graphical Examples** show students how the calculator can be used to enhance their understanding.

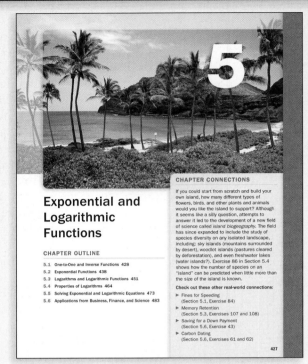

### Exponential and Logarithmic Functions

**CHAPTER OUTLINE**

5.1 One-to-One and Inverse Functions 428
5.2 Exponential Functions 438
5.3 Logarithms and Logarithmic Functions 451
5.4 Properties of Logarithms 464
5.5 Solving Exponential and Logarithmic Equations 473
5.6 Applications from Business, Finance, and Science 483

**CHAPTER CONNECTIONS**

If you could start from scratch and build your own island, how many different types of flowers, birds, and other plants and animals would you like the island to support? Although it seems like a silly question, attempts to answer it led to the development of a new field of science called *island biogeography*. The field has since expanded to include the study of species diversity on any isolated landscape, including: sky islands (mountains surrounded by desert), woodlot islands (pastures cleared by deforestation), and even freshwater lakes (water islands?). Exercise 86 in Section 5.4 shows how the number of species on an "island" can be predicted when little more than the size of the island is known.

Check out these other real-world connections:
▶ Fines for Speeding (Section 5.1, Exercise 84)
▶ Memory Retention (Section 5.3, Exercises 107 and 108)
▶ Saving for a Down Payment (Section 5.6, Exercise 43)
▶ Carbon Dating (Section 5.6, Exercises 61 and 62)

427

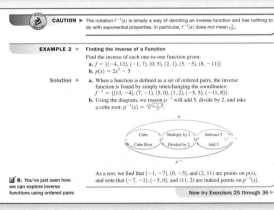

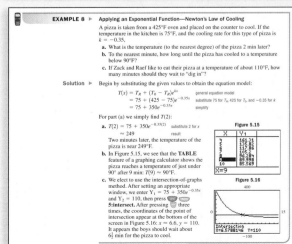

# Comprehensive Exercise Sets

▶ **Mid-Chapter Checks** provide students with a sampling of exercises to assess their knowledge before moving on to the second half of the chapter.

## End-of-Section Exercise Sets

▶ **Concepts and Vocabulary** exercises help students recall and articulate important terms.

▶ **Developing Your Skills** exercises provide students with practice of essential concepts with increasing levels of difficulty.

▶ **Working with Formulas** exercises demonstrate contextual applications of well-known formulas.

▶ **Extending the Concept** exercises challenge students to extend their knowledge and skills.

▶ **Maintaining Your Skills** exercises address skills from previous sections to help students retain knowledge after learning new concepts.

## End-of-Chapter Review Material

▶ **Making Connections** are matching exercises that help students interpret graphical and algebraic information.

▶ **Chapter Summary and Concept Reviews** present key concepts by section and are paired with corresponding exercises.

▶ **Practice Tests** enable students to check their knowledge and prepare for assessments.

▶ **Cumulative Reviews,** at the end of each chapter, revisit important concepts from earlier chapters so that students can retain their skills.

▶ **Graphing Calculator** icons appear next to exercises where concepts can be supported by graphing technology.

▶ **Homework Selection Guide** A list of suggested homework exercises has been provided for each section of the text (Annotated Instructor's Edition only). The guide provides preselected assignments at four levels: *Basic, Core, Standard,* and *Extended.*

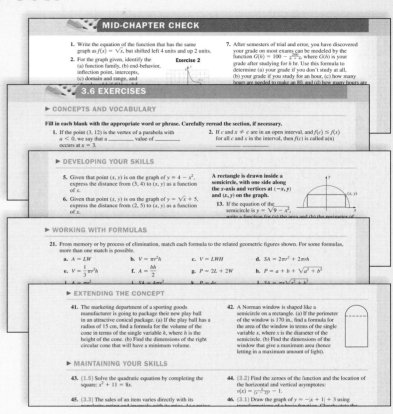

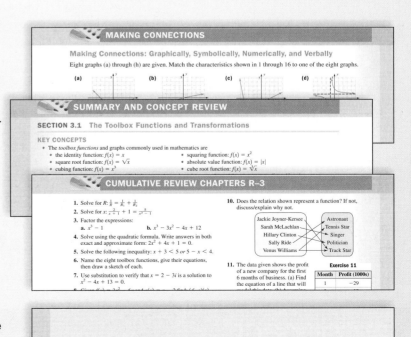

# Connect Math Hosted by ALEKS Corp.

## Built By Today's Educators, For Today's Students

**Fewer clicks means more time for you…**

Change assignment dates right from the home page.

Teaching multiple sections? Easily move from one to another.

Edit, print, and view assignments in just one click.

**…and your students.**

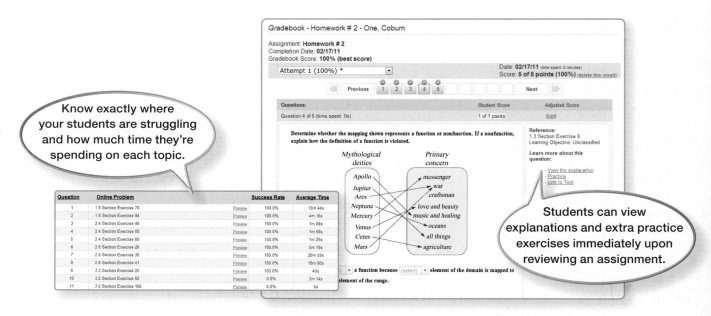

Know exactly where your students are struggling and how much time they're spending on each topic.

Students can view explanations and extra practice exercises immediately upon reviewing an assignment.

# Quality Content For Today's Online Learners

Online Exercises were carefully selected and developed to provide a seamless transition from textbook to technology.

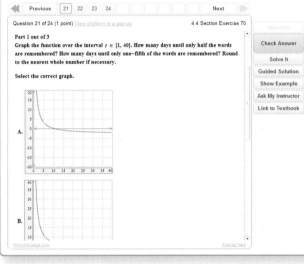

For consistency, the guided solutions match the style and voice of the original text, as though the author is guiding the students through the problems.

Multimedia eBook includes access to a variety of media assets and a place to highlight and keep track of class notes

ALEKS Corporation's experience with algorithm development ensures a commitment to accuracy and a meaningful experience for students to demonstrate their understanding with a focus towards online learning.

The ALEKS® Initial Assessment is an artificially intelligent (AI), diagnostic assessment that identifies precisely what a student knows. Instructors can then use this information to make more informed decisions on what topics to cover in more detail with the class.

ALEKS is a registered trademark of ALEKS Corporation.

# www.successinmath.com

McGraw Hill **connect** plus+ | MATH

Hosted by **ALEKS Corp.**

**ALEKS** is a unique, online program that significantly raises student proficiency and success rates in mathematics, while reducing faculty workload and office-hour lines. ALEKS uses artificial intelligence and adaptive questioning to assess precisely a student's knowledge, and deliver individualized learning tailored to the student's needs. With a comprehensive library of math courses, ALEKS delivers an unparalleled adaptive learning system that has helped millions of students achieve math success.

## ALEKS Delivers a Unique Math Experience:

- **Research-Based, Artificial Intelligence** precisely measures each student's knowledge
- **Individualized Learning** presents the exact topics each student is most **ready to learn**
- **Adaptive, Open-Response Environment** includes comprehensive tutorials and resources
- **Detailed, Automated Reports** track student and class progress toward course mastery
- **Course Management Tools** include textbook integration, custom features, and more

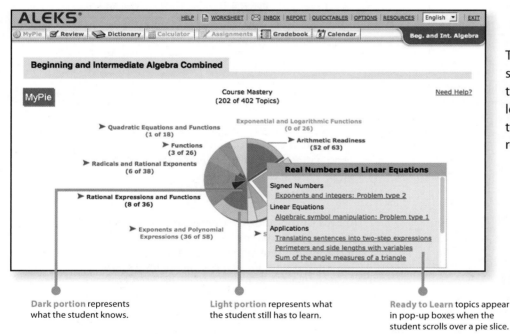

The ALEKS Pie summarizes a student's current knowledge, then delivers an individualized learning path with the exact topics the student is most ready to learn.

**Dark portion** represents what the student knows.

**Light portion** represents what the student still has to learn.

**Ready to Learn** topics appear in pop-up boxes when the student scrolls over a pie slice.

> ❝My experience with ALEKS has been effective, efficient, and eloquent. **Our students' pass rates improved from 49 percent to 82 percent with ALEKS.** We also saw student retention rates increase by 12% in the next course. Students feel empowered as they guide their own learning through ALEKS.❞
>
> —**Professor Eden Donahou,** *Seminole State College of Florida*

To learn more about ALEKS, please visit: **www.aleks.com/highered/math**

# ALEKS® Prep Products

**ALEKS Prep** products focus on prerequisite and introductory material, and can be used during the first six weeks of the term to ensure student success in math courses ranging from Beginning Algebra through Calculus. ALEKS Prep quickly fills gaps in prerequisite knowledge by assessing precisely each student's preparedness and delivering individualized instruction on the exact topics students are most ready to learn. As a result, instructors can focus on core course concepts and see improved student performance with fewer drops.

> "ALEKS is wonderful. It is a professional product that takes very little time as an instructor to administer. Many of our students have taken Calculus in high school, but they have forgotten important algebra skills. ALEKS gives our students an opportunity to review these important skills."
>
> —**Professor Edward E. Allen,** *Wake Forest University*

# A Total Course Solution

A cost-effective total course solution: fully integrated, interactive eBook combined with the power of ALEKS adaptive learning and assessment.

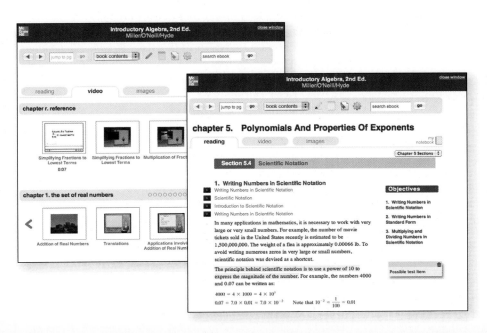

Students can easily access the full eBook content, multimedia resources, and their notes from within their ALEKS Student Accounts.

To learn more about ALEKS, please visit: **www.aleks.com/highered/math**

# Acknowledgments and Reviewers

We first want to express a deep appreciation for the guidance, comments, and suggestions offered by all reviewers. We have once again found their collegial exchange of ideas and experience very refreshing, instructive, and always helpful in creating a better learning tool for our students. Then there's Vicki Krug, who has continued to display an uncanny ability to bring innumerable pieces from all directions into a unified whole. With Patricia Steele's skill as a copy editor being as sharp as ever, her attention to detail continues to pay great dividends. For their useful suggestions, infinite patience, and tireless efforts in bringing this text to completion, we would also like to thank Ryan Blankenship, Caroline Celano, and Ashley Zellmer. We are especially grateful to Ashley for holding the enterprise together as the winds of change buffeted us once again. We must also thank Laurie Janssen and our magnificent design team, and Emilie Berglund as the Director of Digital Content, for helping shape a text that John Osgood, Kevin Ernzen, Tim Cote, Pete Vanaria, and our entire sales force can really get behind. In truth, our hats are off to all the fine people at McGraw-Hill for their continuing support and belief in this series. Additional gratitude must go to Hal Whipple for his attention to detail during the accuracy checks; Jon Booze and his team for their work on the test banks; J.D. Herdlick, Laura Taylor, Kara Allara, and Cathy Gonzalez for their contributions to the lecture and exercise video creation; with special thanks to Vanessa Coffelt, Stephen Toner, Amy Naughten, and Michael Lanstrum for working so hard to make Connect® Math Hosted by ALEKS the superb homework system it has become. A final word of thanks is due Rick Armstrong, Anne Marie Mosher, Rick Pescarino, Kelly Ballard, John Elliot, Jim Frost, Barb Kurt, Lillian Seese, Nate Wilson, and all of our colleagues at St. Louis Community College and Blinn College, whose friendship, encouragement, and love of mathematics makes going to work each day a joy.

## Digital Contributors

Vanessa Coffelt *Blinn College*
Stephen Toner *Victor Valley College*
Michael Lanstrum *Cuyahoga Community College*
Amy Naughten

## Chapter Reviews

Marwan Abu-Sawwa *Florida State College at Jacksonville*
Eldon Baldwin *Prince George's Community College*
Madelaine Bates *Bronx Community College*
Annette Benbow *Tarrant County College*
Adel Boules *University of North Florida*
Cheryl Cavaliero *Butler County Community College*
Ayona Chatterjee *University of West Georgia*
Katrina Cunningham *Southern University and A&M College–Baton Rouge*
Hugh Cornell *University of North Florida*
Kathy Cousins-Cooper *North Carolina A&T State University*
Nelson de la Rosa *Miami Dade College*
Shreyas Desai *Atlanta Metropolitan College*
Letitia Downen *Southern Illinois University–Edwardsville*
Mickey Dunlap *University of Tennessee at Martin*
Karen Estes *St. Petersburg College*
Patricia Foard *South Plains College*
William Forrest *Baton Rouge Community College*
Edna Greenwood *Tarrant County College–Northwest Campus*
Matt Henes *Pasadena City College*
Melissa Hardeman *University of Arkansas at Little Rock*
Sheyleah Harris-Plant *South Plains College*
Mary Beth Headlee *State College of Florida*
Esmarie Kennedy *San Antonio College*

Michael Kirby *Tidewater CC–Va. Beach Campus*
Debra Lackey *Odessa College*
Cynthia Landrigan *Erie Community College*
Rebecca Leefers *Michigan State University*
Nancy Liu *Miami Dade College*
John Lofberg *South Dakota School of Mines and Technology*
Lorraine Lopez *San Antonio College*
Sandra Maldonado *Florida Gulf Coast University*
Richard K. Maxwell *Penn State Fayette, The Eberly Campus*
Jennifer McNeilly *University of Illinois at Urbana-Champaign*
Mary Ann Moore *Florida Gulf Coast University*
Madhu Motha *Butler County Community College*
Tanya O'Keefe *Darton College*
Gilbert Perez *San Antonio College*
Mark W. Pierce *Macon State College*
Robert Plant, II *South Plains College*
David Platt *Front Range Community College*
Colleen Quinn *Erie Community College–South Campus*
Lynn Rickabaugh *Aiken Technical College*
Minnie M. Riley *Hinds Community College–Raymond Campus*
Christy Schmidt *Northwest Vista College*
Joy Shurley *Abraham Baldwin College*
Pavel Sikorskii *Michigan State University*
John Russell Taylor *University of North Carolina at Charlotte*
Jean Thornton *Western Kentucky University*
Diann H. Torrence *Delgado Community College*
Albert Urrechaga *Hillsborough Community College*
David J. Walker *Hinds Community College District–Raymond Campus*
Xubo Wang *Macon State College*

# Multimedia and Print Supplements

## Connect Math Hosted by ALEKS Corp.

Connect Math Hosted by ALEKS is an exciting, new assignment and assessment ehomework platform. Starting with an easily viewable, intuitive interface, students will be able to access key information, complete homework assignments, and utilize an integrated, media-rich eBook.

**ALEKS** (**A**ssessment and **LE**arning in **K**nowledge **S**paces) is a dynamic online learning system for mathematics education, available over the Web 24/7. ALEKS assesses students, accurately determines their knowledge, and then guides them to the material that they are most ready to learn. With a variety of reports, Textbook Integration Plus, quizzes, and homework assignment capabilities, ALEKS offers flexibility and ease of use for instructors.

- ALEKS uses artificial intelligence to determine exactly what each student knows and is ready to learn. ALEKS remediates student gaps and provides highly efficient learning and improved learning outcomes.
- ALEKS is a comprehensive curriculum that aligns with syllabi or specified textbooks. Used in conjunction with McGraw-Hill texts, students also receive links to text-specific videos, multimedia tutorials, and textbook pages.
- Textbook Integration with ALEKS 360 allows ALEKS to be automatically aligned with syllabi or specified McGraw-Hill textbooks with instructor chosen dates, chapter goals, homework, and quizzes.
- ALEKS with AI-2 gives instructors increased control over the scope and sequence of student learning. Students using ALEKS demonstrate a steadily increasing mastery of the content of the course.
- ALEKS offers a dynamic classroom management system that enables instructors to monitor and direct student progress toward mastery of course objectives.

## ALEKS Prep/Remediation

- Helps instructors meet the challenge of remediating under-prepared or improperly placed students.
- Assesses students on their prerequisite knowledge needed for the course they are entering and prescribes a unique and efficient learning path specifically to address their strengths and weaknesses.
- Students can address prerequisite knowledge gaps outside of class freeing the instructor to use class time pursuing course outcomes.

**McGraw-Hill Create™** With **McGraw-Hill Create™**, you can easily rearrange chapters, combine material from other content sources, and quickly upload content you have written, like your course syllabus or teaching notes. Find the content you need in Create by searching through thousands of leading McGraw-Hill textbooks. Arrange your book to fit your teaching style. Create even allows you to personalize your book's appearance by selecting the cover and adding your name, school, and course information. Order a Create book and you'll receive a complimentary print review copy in 3–5 business days or a complimentary electronic review copy (eComp) via e-mail in minutes. Go to www.mcgrawhillcreate.com today and register to experience how McGraw-Hill Create empowers you to teach *your* students *your* way. **www.mcgrawhillcreate.com**

**Student Solutions Manual** The *Student Solutions Manual* provides comprehensive, worked-out solutions to all of the odd-numbered exercises in the text.

**Lecture Notes** provide a lecture template for students to fill-in important topics, terms, and procedures so they spend less time creating notes and more time engaged in class.

**Computerized Test Bank (CTB) Online** The computerized test bank contains a variety of question types, including true-false, multiple-choice, short-answer. The Brownstone Diploma® system enables you to efficiently select, add, and organize questions by type or level of difficulty. It also allows for printing tests complete with answer keys, as well as editing existing questions or creating new ones. Printable tests and a print version of the test bank can also be found on the website.

**Lecture and Exercise Videos** Lecture videos introduce concepts, definitions, formulas, and problem-solving procedures to help students better comprehend the topic at hand. Exercise videos provide step-by-step instruction for the key exercises which students will most wish to see worked out. The videos are closed-captioned for the hearing-impaired, are subtitled in Spanish, and meet the Americans with Disabilities Act Standards for Accessible Design.

**Instructor Solutions Manual** The *Instructor Solutions Manual* provides comprehensive, worked-out solutions to all exercises. The steps shown in the solutions match the style of the textbook.

**Annotated Instructor's Edition** (instructors only) The Annotated Instructor's Edition contains answers to all exercises. The answers to most questions are printed in a second color next to each problem. Answers not appearing on the page can be found in the Answer Appendix at the end of the book.

# List of Changes to the Third Edition

▶ **More than a third of the examples** are new or revised and **nearly 1000 exercises** are new or revised in this edition.

▶ All chapters now have six sections or less—a tremendous aid to testing, review, coverage, and retention.

▶ Exercise instructions have been shortened and made easier to follow and understand.

▶ Exercises Sets have been reorganized according to difficulty in order to build student understanding.

▶ Most Applications have been carefully reviewed, improved, and updated.

▶ Most Multipart exercises have been broken down to clarify what's being asked and how a student should answer.

▶ Most Applications involving rates of interest have been modified to more nearly match interest rates of the day.

▶ **New Chapter Openers and *Chapter Connections*** have been added to all chapters to reflect modern issues and ideas.

▶ In **Chapter R, A Review of Basic Concepts,** sections R.4, R.5, and R.6 have been reordered so that radical expressions follow exponents.

▶ In **Chapter 2, Relations, Functions and Graphs,** coverage of topics including slopes, rates of change, and the difference quotient have been expanded.

▶ **New Section 2.6, Linear Models and Real Data,** is focused on real-world situations where data is best modeled by a linear function.

▶ **New Chapter 3, More on Functions,** builds on the concepts from Chapter 2, expanding a student's understanding of functions by introducing the algebra of functions and additional families of functions, including basic rational and power functions, and piecewise-defined functions.

▶ **Chapter 4, Polynomial and Rational Functions,** now boasts a more streamlined coverage of rational functions, with a greater focus on the most important characteristics of rational graphs (zeroes and asymptotic behavior). Coverage of complex solutions to polynomial equations has been regrouped to offer instructors better coverage options.

▶ **Chapter 5, Exponential and Logarithmic Functions,** has been restructured to include additional emphasis on logarithmic properties, the change of base formula, and real-world applications.

# Applications Index

# R

# A Review of Basic Concepts and Skills

## CHAPTER OUTLINE

## CHAPTER CONNECTIONS

Participation in many common recreational activities depends on the time of year, or even on the time of day. For instance, we expect that attendance at state parks will be greater in the spring than in the winter, and that a swimming pool will have more swimmers at 1:00 P.M. (in the heat of the day) than at 8:00 A.M. The equation $S = -h^2 + 10h$ can be used to estimate number of people in a swimming pool at any time of day, where $S$ is the number of swimmers and $h$ is the number of hours the pool has been open. This chapter reviews the skills required to estimate the number of swimmers in the pool at a given time of day, as well as other mathematical skills to be used throughout the course. This equation appears as Exercise 128 in Section R.3.

**Check out these other real-world connections:**

▶ Pediatric Dosages and Clark's Rule (Section R.1, Exercise 104)

▶ Maximizing Revenue of Video Game Sales (Section R.3, Exercise 129)

▶ Accident Investigation (Section R.4, Exercise 55)

▶ Growth of a New Stock Hitting the Market (Section R.6, Exercise 75)

1

## LEARNING OBJECTIVES

*In Section R.1 you will review how to:*

- ☐ **A.** Identify sets of numbers, graph real numbers, and use set notation
- ☐ **B.** Use inequality symbols and order relations
- ☐ **C.** Use the absolute value of a real number
- ☐ **D.** Apply the order of operations

The most fundamental requirement for learning algebra is mastering the words, symbols, and numbers used to express mathematical ideas. "Words are the symbols of knowledge, the keys to accurate learning" (Norman Lewis in *Word Power Made Easy*, Penguin Books).

## A. Sets of Numbers, Graphing Real Numbers, and Set Notation

To effectively use mathematics as a problem-solving tool, we must first be familiar with the **sets of numbers** used to quantify (give a numeric value to) the things we investigate. Only then can we make comparisons and develop equations that lead to informed decisions.

### Natural Numbers

The most basic numbers are those used to count physical objects: 1, 2, 3, 4, and so on. These are called **natural numbers** and are represented by the capital letter $\mathbb{N}$, often written in the special font shown. We use **set notation** to list or describe a set of numbers. Braces { } are used to group **members** or **elements** of the set, commas separate each member, and three dots (called an *ellipsis*) are used to indicate a pattern that continues indefinitely. The notation $\mathbb{N} = \{1, 2, 3, 4, 5, \ldots\}$ is read, "$\mathbb{N}$ is the set of numbers 1, 2, 3, 4, 5, and so on." To show membership in a set, the symbol $\in$ is used. It is read "is an element of" or "belongs to." The statements $6 \in \mathbb{N}$ (6 is an element of $\mathbb{N}$) and $0 \notin \mathbb{N}$ (0 is not an element of $\mathbb{N}$) are true statements. A set having no elements is called the **empty** or **null set,** and is designated by empty braces { } or the symbol $\varnothing$.

---

**EXAMPLE 1** ▶ **Writing Sets of Numbers Using Set Notation**

List the set of natural numbers that are

   **a.** greater than 100       **b.** negative

   **c.** greater than or equal to 5 and less than 12

Solution ▶   **a.** $\{101, 102, 103, 104, \ldots\}$

   **b.** { }; all natural numbers are positive.

   **c.** $\{5, 6, 7, 8, 9, 10, 11\}$

Now try Exercises 5 and 6 ▶

---

### Whole Numbers

Combining zero with the natural numbers produces a new set called the **whole numbers** $\mathbb{W} = \{0, 1, 2, 3, 4, \ldots\}$. We say that the natural numbers are a **proper subset** of the whole numbers, denoted $\mathbb{N} \subset \mathbb{W}$, since every natural number is also a whole number. The symbol $\subset$ means "is a proper subset of."

---

**EXAMPLE 2** ▶ **Determining Membership in a Set**

Given $A = \{1, 2, 3, 4, 5, 6\}$, $B = \{2, 4\}$, and $C = \{0, 1, 2, 3, 5, 8\}$, determine whether the following statements are true or false. Justify your response.

   **a.** $B \subset A$          **b.** $B \subset C$          **c.** $C \subset \mathbb{W}$

   **d.** $C \subset \mathbb{N}$          **e.** $104 \in \mathbb{W}$      **f.** $2 \notin \mathbb{W}$

---

**Solution** ▶   **a.** True: Every element of $B$ is in $A$.    **b.** False: $4 \notin C$, so $B \not\subseteq C$.
                 **c.** True: All elements are whole numbers.    **d.** False: $0 \notin \mathbb{N}$, so $C \not\subseteq \mathbb{N}$.
                 **e.** True: 104 is a whole number.            **f.** False: 2 *is* a whole number.

**Now try Exercises 7 through 12** ▶

### Integers

Numbers greater than zero are **positive numbers.** Every positive number has an *opposite* that is a **negative number** (a number less than zero). Combining zero with the natural numbers and their opposites produces the set of **integers** $\mathbb{Z} = \{\ldots, -3, -2, -1, 0, 1, 2, 3, \ldots\}$. We can illustrate the location and magnitude of a number (in relation to other numbers) using a **number line** (see Figure R.1).

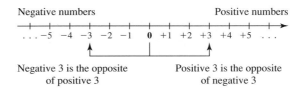

**Figure R.1**

The number that corresponds to a given point on the number line is called the **coordinate** of that point. When we want to note a specific location on the line, a bold dot "•" is used and we have then **graphed** the number. Since we need only one coordinate to denote a location on the number line, it is referred to as a **one-dimensional graph.**

### Rational Numbers

Fractions and mixed numbers are part of a set called the **rational numbers** $\mathbb{Q}$. A rational number is one that can be written as a fraction with an integer numerator and an integer denominator other than zero. In set notation we write $\mathbb{Q} = \{\frac{p}{q} | p, q \in \mathbb{Z}; q \neq 0\}$. The vertical bar "|" is read "such that" and indicates that a description follows. In words, we say, "$\mathbb{Q}$ is the set of numbers of the form $p$ over $q$, such that $p$ and $q$ are integers and $q$ is not equal to zero."

**EXAMPLE 3** ▶   **Graphing Rational Numbers**

Graph the fractions by converting to decimal form and estimating their location between two integers:

   **a.** $-2\frac{1}{3}$            **b.** $\frac{7}{2}$

**Solution** ▶   **a.** $-2\frac{1}{3} = -2.3333333\ldots$ or $-2.\overline{3}$       **b.** $\frac{7}{2} = 3.5$

**Now try Exercises 13 through 16** ▶

Since the division $\frac{7}{2}$ **terminated,** the result is called a **terminating decimal.** The decimal form of $-2\frac{1}{3}$ is called **repeating** and **nonterminating** (note that $-2.3 \neq -2.\overline{3}$). Recall that a repeating decimal is written with a horizontal bar over the first block of digit(s) that repeat. For instance $\frac{118}{55} = 2.1454545\ldots = 2.1\overline{45}$.

When using a calculator for computations involving repeating decimals, you must either use the rational form or "fill the display" with the digits that repeat. As an exploration, suppose that you are to inherit $\frac{1}{3} = 0.\overline{3}$ of a $90,000 estate. How many "repeating threes" (times 90,000) are needed until the calculator returns an answer of $30,000? **See Exercises 17 and 18.**

## Irrational Numbers

Although any fraction can be written in decimal form, not all decimal numbers can be written as a fraction. One example is the number represented by the Greek letter $\pi$ (pi), frequently seen in a study of circles. Although we often approximate $\pi$ using 3.14, its true value has a **nonrepeating** and *nonterminating* decimal form. Other numbers of this type include 2.101001000100001 ... (there is no block of digits that repeat—the number of zeroes between each "1" is increasing), and $\sqrt{5} \approx 2.2360679\ldots$ (the decimal form never terminates). Numbers with a nonrepeating and nonterminating decimal form belong to the set of irrational numbers $\mathbb{H}$.

**EXAMPLE 4** ▶ **Approximating Irrational Numbers**

Use a calculator as needed to approximate the value of each number given (round to 100ths), then graph them on the number line:

    **a.** $\sqrt{3}$         **b.** $\pi$         **c.** $\sqrt{19}$         **d.** $-\frac{\sqrt{2}}{2}$

**Solution** ▶   **a.** $\sqrt{3} \approx 1.73$   **b.** $\pi \approx 3.14$   **c.** $\sqrt{19} \approx 4.36$   **d.** $-\frac{\sqrt{2}}{2} \approx -0.71$

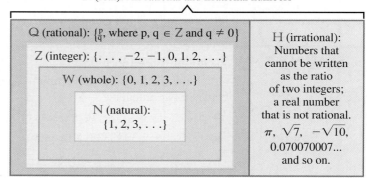

                                                      **Now try Exercises 19 through 22** ▶

## Real Numbers

The set of rational numbers combined with the set of irrational numbers produces the set of **real numbers** $\mathbb{R}$. Figure R.2 illustrates the relationship between the sets of numbers we've discussed so far. Notice how each subset appears "nested" in a larger set.

**Figure R.2**

**EXAMPLE 5** ▶ **Identifying Members of a Number Set**

List the numbers in set $A = \{-2, 0, 5, \sqrt{7}, 12, \frac{2}{3}, 4.5, \sqrt{21}, \pi, -0.75\}$ that belong to

    **a.** $\mathbb{Q}$         **b.** $\mathbb{H}$         **c.** $\mathbb{W}$         **d.** $\mathbb{Z}$

**Solution** ▶   **a.** $-2, 0, 5, 12, \frac{2}{3}, 4.5, -0.75 \in \mathbb{Q}$       **b.** $\sqrt{7}, \sqrt{21}, \pi \in \mathbb{H}$

           **c.** $0, 5, 12 \in \mathbb{W}$                      **d.** $-2, 0, 5, 12 \in \mathbb{Z}$

                                                      **Now try Exercises 23 through 26** ▶

**EXAMPLE 6** ▶ **Evaluating Statements about Sets of Numbers**

Determine whether the statements are true or false. Justify your response.

**a.** $\mathbb{N} \subset \mathbb{Q}$      **b.** $\mathbb{H} \subset \mathbb{Q}$      **c.** $\mathbb{W} \subset \mathbb{Z}$      **d.** $\mathbb{Z} \subset \mathbb{R}$

**Solution** ▶ **a.** True: All natural numbers can be written as fractions over 1.
**b.** False: No irrational number can be written in fraction form.
**c.** True: All whole numbers are integers.
**d.** True: Every integer is a real number.

☑ **A.** You've just reviewed how to identify sets of numbers, graph real numbers, and use set notation

**Now try Exercises 27 through 38** ▶

## B. Inequality Symbols and Order Relations

We compare numbers of different size using **inequality notation,** known as the **greater than** ($>$) and **less than** ($<$) symbols. Note that $-4 < 3$ is the same as saying $-4$ is to the left of 3 on the number line. In fact, on a number line, any given number is smaller than any number to the right of it (see Figure R.3).

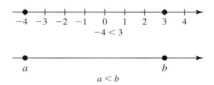

**Figure R.3**      $a < b$

| **Order Property of Real Numbers** |
|---|

Given any two real numbers $a$ and $b$.
     1. $a < b$ if $a$ is to the left of $b$ on the number line.
     2. $a > b$ if $a$ is to the right of $b$ on the number line.

Inequality notation is used with numbers and variables to write mathematical statements. A **variable** is a symbol, commonly a letter of the alphabet, used to represent an unknown quantity. Over the years $x$, $y$, and $n$ have become most common, although any letter (or symbol) can be used. Often we'll use variables that remind us of the quantities they represent, like $L$ for length, and $D$ for distance.

**EXAMPLE 7** ▶ **Writing Mathematical Models Using Inequalities**

Use a variable and an inequality symbol to represent the statement: "To hit a home run out of Jacobi Park, the ball must travel over three hundred twenty-five feet."

**Solution** ▶ Let $D$ represent distance: $D > 325$ ft.

**Now try Exercises 39 through 42** ▶

In Example 7, note the number 325 itself is not a possible value for $D$. If the ball traveled *exactly* 325 ft, it would hit the fence and stay in play. Numbers that mark the limit or boundary of an inequality are called **endpoints.** If the endpoint(s) are *not* included, the less than ($<$) or greater than ($>$) symbols are used. When the endpoints *are* included, the *less than or equal to symbol* ($\leq$) or the *greater than or equal to symbol* ($\geq$) is used. The decision to *include* or *exclude* an endpoint is often an important one, and many mathematical decisions (and real-life decisions) depend on a clear understanding of the distinction. **See Exercises 43 through 48.**

☑ **B.** You've just reviewed how to use inequality symbols and order relations

## C. The Absolute Value of a Real Number

Any nonzero real number "$n$" is either a positive number or a negative number. But in some applications, our primary interest is simply the *size* of $n$, rather than its sign. This is called the **absolute value** of $n$, denoted $|n|$, and can be thought of as its *distance from zero on the number line*, regardless of the direction (see Figure R.4). Since distance is always positive or zero, $|n| \geq 0$.

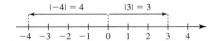

**Figure R.4**

---

**EXAMPLE 8** ▶

**Absolute Value Reading and Reasoning**

In the table shown, the absolute value of a number is given in column 1. Complete the remaining columns.

Solution ▶

| Column 1 (In Symbols) | Column 2 (Spoken) | Column 3 (Result) | Column 4 (Reason) |
|---|---|---|---|
| $\lvert 7.5 \rvert$ | "the absolute value of seven and five-tenths" | 7.5 | the distance between 7.5 and 0 is 7.5 units |
| $\lvert -2 \rvert$ | "the absolute value of negative two" | 2 | the distance between $-2$ and 0 is 2 units |
| $-\lvert -6 \rvert$ | "the opposite of the absolute value of negative six" | $-6$ | the distance between $-6$ and 0 is 6 units, the opposite of 6 is $-6$ |

**Now try Exercises 49 through 56** ▶

Example 8 illustrates that the absolute value of a positive number is the number itself, while the absolute value of a negative number is the *opposite of that number* (recall that $-n$ is positive if $n$ itself is negative). For this reason, the formal definition of absolute value is stated as follows.

**Absolute Value**

For any real number $n$,

$$|n| = \begin{cases} n & \text{if} \quad n \geq 0 \\ -n & \text{if} \quad n < 0 \end{cases}$$

The concept of absolute value can actually be used to find the distance between *any* two numbers on a number line. For instance, we know the distance between 2 and 8 is 6 (by counting). Using absolute values, we can write $|8 - 2| = |6| = 6$, or $|2 - 8| = |-6| = 6$. Generally, if $a$ and $b$ are two numbers on the real number line, the distance between them is $|a - b|$, which is identical to $|b - a|$.

---

**EXAMPLE 9** ▶

**Using Absolute Value to Find the Distance between Points**

Find the distance between $-5$ and 3 on the number line.

Solution ▶

Substituting $-5$ for $a$ and 3 for $b$ in the formula shown gives

$$|-5 - 3| = |-8| = 8 \quad \text{or} \quad |3 - (-5)| = |8| = 8.$$

☑ **C.** You've just reviewed how to use the absolute value of a real number

**Now try Exercises 57 through 64** ▶

## D. The Order of Operations

The operations of addition, subtraction, multiplication, and division are defined for the set of real numbers, and the concept of absolute value plays an important role. Prior to our study of the order of operations, we will review fundamental concepts related to division and zero, exponential notation, and square roots/cube roots.

### Division and Zero

The quotient $\frac{36}{9} = 4$ can be checked using the related multiplication: $4 \cdot 9 = 36\checkmark$. A similar check can be used to understand quotients involving zero.

---

**EXAMPLE 10** ▶ **Understanding Division with Zero by Writing the Related Product**

Rewrite each quotient *using the related product.*

     **a.** $0 \div 8 = p$       **b.** $\frac{16}{0} = q$       **c.** $\frac{0}{12} = n$

**Solution** ▶    **a.** $0 \div 8 = p$, if $p \cdot 8 = 0$.    **b.** $\frac{16}{0} = q$, if $q \cdot 0 = 16$.    **c.** $\frac{0}{12} = n$, if $n \cdot 12 = 0$.

       This shows $p = 0$.       There is no such number $q$.      This shows $n = 0$.

---

Now try Exercises 65 through 68 ▶

---

**WORTHY OF NOTE**

When a pizza is delivered to your home, it often has "8 parts to the whole," and in fraction form we have $\frac{8}{8}$. When all 8 pieces are eaten, 0 pieces remain and the fraction form becomes $\frac{0}{8} = 0$. However, the expression $\frac{8}{0}$ is meaningless (undefined), since it would indicate a pizza that has "0 parts to the whole (??)."

     In Example 10(a), a dividend of 0 and a divisor of 8 means we are going to divide zero into eight groups. The related multiplication shows there will be zero in each group. As in Example 10(b), an expression with a divisor of 0 *cannot be computed or checked.* Although it seems trivial, division by zero has many implications in a study of mathematics, so make an effort to know the facts: The quotient of zero and any nonzero number is zero ($\frac{0}{n} = 0$) but division *by* zero is undefined ($\frac{n}{0}$ is undefined). The special case of $\frac{0}{0}$ is said to be **indeterminate**, as $\frac{0}{0} = n$ *appears* to be true for all real numbers $n$ (since the check gives $n \cdot 0 = 0\checkmark$). The expression $\frac{0}{0}$ is studied in greater detail in more advanced classes.

---

**Division and Zero**

The quotient of zero and any nonzero number $n$ is zero ($n \neq 0$):

$$0 \div n = 0 \qquad \frac{0}{n} = 0.$$

The expressions $n \div 0$ and $\frac{n}{0}$ are undefined.

---

### Squares, Cubes, and Exponential Form

When a number is repeatedly multiplied by itself as in $(10)(10)(10)(10)$, we write it using **exponential notation** as $10^4$. The number used for repeated multiplication (in this case 10) is called the **base,** and the superscript number is called an **exponent.** The exponent tells how many times the base occurs as a factor, and we say $10^4$ is written in **exponential form.** Numbers that result from squaring an integer are called **perfect squares,** while numbers that result from cubing an integer are called **perfect cubes.** These are often collected into a table, such as Table R.1, and students

**Table R.1**

| Perfect Squares | | | | Perfect Cubes | |
|---|---|---|---|---|---|
| $N$ | $N^2$ | $N$ | $N^2$ | $N$ | $N^3$ |
| 1 | 1 | 7 | 49 | 1 | 1 |
| 2 | 4 | 8 | 64 | 2 | 8 |
| 3 | 9 | 9 | 81 | 3 | 27 |
| 4 | 16 | 10 | 100 | 4 | 64 |
| 5 | 25 | 11 | 121 | 5 | 125 |
| 6 | 36 | 12 | 144 | 6 | 216 |

are strongly encouraged to memorize these values to help complete many common calculations mentally. Only the square and cube of selected positive integers are shown.

**EXAMPLE 11** ▶ **Evaluating Numbers in Exponential Form**

Write each exponential in expanded form, then determine its value.

    **a.** $4^3$        **b.** $(-6)^2$        **c.** $-6^2$        **d.** $\left(\frac{2}{3}\right)^3$

Solution ▶     **a.** $4^3 = 4 \cdot 4 \cdot 4 = 64$          **b.** $(-6)^2 = (-6) \cdot (-6) = 36$

               **c.** $-6^2 = -(6 \cdot 6) = -36$        **d.** $\left(\frac{2}{3}\right)^3 = \frac{2}{3} \cdot \frac{2}{3} \cdot \frac{2}{3} = \frac{8}{27}$

**Now try Exercises 69 and 70** ▶

Examples 11(b) and 11(c) illustrate an important distinction. The expression $(-6)^2$ gives a single operation, "the square of negative six" and the negative sign is included in both factors. The expression $-6^2$ gives two operations, "six is squared, and the result is made negative." The square of six is calculated first, with the negative sign applied afterward.

### Square Roots and Cube Roots

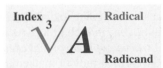

Index $\overset{3}{\phantom{x}}\!\!\sqrt{\phantom{A}}$ Radical

$\sqrt[3]{A}$

Radicand

For the square root operation, either the $\sqrt{\phantom{x}}$ or $\sqrt[\phantom{x}]{\phantom{x}}$ notation can be used. The $\sqrt{\phantom{x}}$ symbol is called a **radical,** the number under the radical is called the **radicand,** and the small number used is called the **index** (see figure). The index tells how many factors are needed to obtain the radicand. For example, $\sqrt{25} = 5$, since $5 \cdot 5 = 5^2 = 25$ (when the $\sqrt{\phantom{x}}$ symbol is used, the index is understood to be 2). In general, $\sqrt{a} = b$ only if $b^2 = a$. All numbers greater than zero have one positive and one negative square root. The *positive* or **principal square root** of 49 is 7 ($\sqrt{49} = 7$) since $7^2 = 49$. The *negative* square root of 49 is $-7$ ($-\sqrt{49} = -7$).

The cube root of a number has the form $\sqrt[3]{a} = b$, where $b^3 = a$. This means $\sqrt[3]{27} = 3$ since $3^3 = 27$, and $\sqrt[3]{-8} = -2$ since $(-2)^3 = -8$. The cube root of a real number has one unique real value. In general, we have the following:

**WORTHY OF NOTE**

It is helpful to note that both 0 and 1 are their own square root, cube root, and $n$th root. That is, $\sqrt{0} = 0$, $\sqrt[3]{0} = 0, \dots, \sqrt[n]{0} = 0$; and $\sqrt{1} = 1, \sqrt[3]{1} = 1, \dots, \sqrt[n]{1} = 1$.

| Square Roots | Cube Roots |
|---|---|
| For $a \geq 0$, $\sqrt{a} = b$ if $b^2 = a$. | For $a \in \mathbb{R}$, $\sqrt[3]{a} = b$ if $b^3 = a$. |
| This indicates that | This indicates that |
| $\sqrt{a} \cdot \sqrt{a} = a$ or $(\sqrt{a})^2 = a$ | $\sqrt[3]{a} \cdot \sqrt[3]{a} \cdot \sqrt[3]{a} = a$ or $(\sqrt[3]{a})^3 = a$ |

**EXAMPLE 12** ▶ **Evaluating Square Roots and Cube Roots**

Determine the value of each expression.

    **a.** $\sqrt{49}$      **b.** $\sqrt[3]{125}$      **c.** $\sqrt{\frac{9}{16}}$      **d.** $-\sqrt{16}$      **e.** $\sqrt{-25}$

Solution ▶     **a.** $\sqrt{49} = 7$ since $7 \cdot 7 = 49$        **b.** $\sqrt[3]{125} = 5$ since $5 \cdot 5 \cdot 5 = 125$

             **c.** $\sqrt{\frac{9}{16}} = \frac{3}{4}$ since $\frac{3}{4} \cdot \frac{3}{4} = \frac{9}{16}$        **d.** $-\sqrt{16} = -4$ since $\sqrt{16} = 4$

             **e.** $\sqrt{-25}$ is not a real number [note that $5 \cdot 5 = (-5)(-5) = 25$].

**Now try Exercises 71 through 76** ▶

For square roots, if the radicand is a perfect square or has perfect squares in both the numerator and denominator, the result is a rational number as in Examples 12(a) and 12(c). If the radicand is not a perfect square, the result is an irrational number. Similar statements can be made regarding cube roots [Example 12(b)].

## The Order of Operations

When basic operations are combined into a larger mathematical expression, we use a specified **priority** or **order of operations** to evaluate them.

> ### The Order of Operations
>
> 1. Simplify within grouping symbols (parentheses, brackets, braces, etc.). If there are "nested" symbols of grouping, begin with the innermost group. If a fraction bar is used, simplify the numerator and denominator separately.
> 2. Evaluate all exponents and roots.
> 3. Compute all multiplications or divisions *in the order they occur from left to right*.
> 4. Compute all additions or subtractions *in the order they occur from left to right*.

**WORTHY OF NOTE**

Sometimes the acronym **PEMDAS** is used as a more concise way to recall the order of operations: **P**arentheses, **E**xponents, **M**ultiplication, **D**ivision, **A**ddition, and **S**ubtraction. The idea has merit, so long as you remember that multiplication and division *have an equal rank,* as do addition and subtraction, and these must be computed in the order they occur (from left to right).

**EXAMPLE 13** ▶  **Evaluating Expressions Using the Order of Operations**

Simplify using the order of operations:

**a.** $5 + 2 \cdot 3$

**b.** $8 + 36 \div 4(12 - 3^2)$

**c.** $\dfrac{-4.5(8) - 3}{\sqrt[3]{125} + 2^3}$

**d.** $7500\left(1 + \dfrac{0.075}{12}\right)^{12 \cdot 15}$

**Solution** ▶

**a.** $5 + 2 \cdot 3 = 5 + 6$     multiplication before addition

$= 11$     result

**b.** $8 + 36 \div 4 \cdot (12 - 3^2)$

$= 8 + 36 \div 4 \cdot (12 - 9)$     simplify within parentheses

$= 8 + 36 \div 4 \cdot (3)$     $12 - 9 = 3$

$= 8 + 9(3)$     the division occurs first

$= 8 + 27$     multiply

$= 35$     result

**c.** $\dfrac{-4.5(8) - 3}{\sqrt[3]{125} + 2^3}$     original expression

$= \dfrac{-36 - 3}{5 + 8}$     simplify terms in the numerator and denominator

$= \dfrac{-39}{13}$     combine terms

$= -3$     result

**d.** $7500\left(1 + \dfrac{0.075}{12}\right)^{12 \cdot 15}$     original expression

$= 7500(1.00625)^{12 \cdot 15}$     simplify within the parenthesis (division before addition)

$= 7500(1.00625)^{180}$     simplify the exponent so it can be applied

$\approx 7500(3.069451727)$     exponents before multiplication

$\approx 23{,}020.88795$     result

**WORTHY OF NOTE**

Many common tendencies are hard to overcome. For instance, let's evaluate the expressions $3 + 4 \cdot 5$ and $24 \div 6 \cdot 2$. For the first, the correct result is 23 (multiplication before addition), though some will get 35 by adding first. For the second, the correct result is 8 (multiplication or division *in order*), though some will get 2 by multiplying first.

☑ **D.** You've just reviewed how to apply the order of operations

**Now try Exercises 77 through 102** ▶

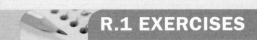

## R.1 EXERCISES

### ▶ CONCEPTS AND VOCABULARY

**Fill in each blank with the appropriate word or phrase. Carefully reread the section, if necessary.**

1. The decimal form of $\sqrt{7}$ contains an infinite number of non _____ and non _____ digits. This means that $\sqrt{7}$ is a(n) _____ number.

2. Discuss/Explain why the value of $12 \cdot \frac{1}{3} + \frac{2}{3}$ is $4\frac{2}{3}$ and not 12.

3. Every positive number has two square roots, one _____ and one _____. The two square roots of 49 are ____; and ____ $\sqrt{49}$ represents the _____ square root of 49.

4. Discuss/Explain (a) why $(-5)^2 = 25$, while $-5^2 = -25$; and (b) why $-5^3 = (-5)^3 = -125$.

### ▶ DEVELOPING YOUR SKILLS

5. List the natural numbers that are
   a. less than 6.
   b. less than 1.

6. List the natural numbers that are
   a. between 0 and 1.
   b. greater than 50.

**Identify each of the following statements as either true or false. If false, give an example that shows why.**

7. $\mathbb{N} \subset \mathbb{W}$                     8. $\mathbb{W} \not\subset \mathbb{N}$

9. $\{33, 35, 37, 39\} \subset \mathbb{W}$

10. $\{2.2, 2.3, 2.4, 2.5\} \subset \mathbb{W}$

11. $6 \in \{0, 1, 2, 3, \ldots\}$      12. $1297 \notin \{0, 1, 2, 3, \ldots\}$

**Convert to decimal form and graph by estimating the number's location between two integers.**

13. $\frac{4}{3}$       14. $-\frac{7}{8}$       15. $2\frac{5}{9}$       16. $-1\frac{5}{6}$

 17. A Texas rancher has 120,000 acres of range land, and wants to use two-thirds of it for cattle and the remaining $\frac{1}{3} = 0.\overline{3}$ for sheep. (a) Using the fraction $\frac{1}{3}$, how many acres will be set aside for sheep? (b) Using a calculator, determine the number of "repeating 3's" that are required (0.3, 0.33, 0.333, etc.) before the correct answer is returned.

18. An architect is reviewing the floor plan for a new office building that offers 36,000 ft² of office space on the first floor. On this floor, $\frac{13}{18} = 0.7\overline{2}$ is considered premium frontage space. (a) Using the fraction $\frac{13}{18}$, how many square feet is considered "premium?" (b) Using a calculator, determine the number of "repeating 2's" that are required (0.72, 0.722, 0.7222, etc.) before the correct answer is returned.

 **Use a calculator to approximate the value of each number (round to hundredths as needed). Then graph each number by estimating its location between two integers.**

19. $\sqrt{7}$      20. $\sqrt{\frac{75}{4}}$      21. $\sqrt{3}$      22. $\frac{25\pi}{12}$

**For the sets in Exercises 23 through 26:**
   a. List all numbers that are elements of (i) $\mathbb{N}$, (ii) $\mathbb{W}$, (iii) $\mathbb{Z}$, (iv) $\mathbb{Q}$, (v) $\mathbb{H}$, and (vi) $\mathbb{R}$.
   b. Reorder the elements of each set from smallest to largest.
   c. Graph the elements of each set on a number line.

23. $\{-1, 8, 0.75, \frac{9}{2}, 5.\overline{6}, 7, \frac{3}{5}, 6\}$

24. $\{-7, 2.\overline{1}, 5.73, -3\frac{5}{6}, 0, -1.12, \frac{7}{8}\}$

25. $\{-5, \sqrt{49}, 2, -3, 6, -1, \sqrt{3}, 0, 4, \pi\}$

26. $\{-8, 5, -2\frac{3}{5}, 1.75, -\sqrt{2}, -0.6, \pi, \frac{7}{2}, \sqrt{64}\}$

**State true or false. If false, state why.**

27. $\mathbb{R} \subset \mathbb{H}$                     28. $\mathbb{N} \subset \mathbb{R}$

29. $\mathbb{Q} \subset \mathbb{Z}$                     30. $\mathbb{Z} \subset \mathbb{Q}$

31. $\sqrt{25} \in \mathbb{H}$                     32. $\sqrt{19} \in \mathbb{H}$

**Match each set with its correct symbol and description/illustration.**

33. __ __ Irrational numbers      a. $\mathbb{R}$      I. $\{1, 2, 3, 4, \ldots\}$

34. __ __ Integers      b. $\mathbb{Q}$      II. $\{\frac{a}{b} \mid a, b \in \mathbb{Z}; b \neq 0\}$

35. __ __ Real numbers      c. $\mathbb{H}$      III. $\{0, 1, 2, 3, 4, \ldots\}$

36. __ __ Rational numbers      d. $\mathbb{W}$      IV. $\{\pi, \sqrt{7}, -\sqrt{13}, \text{etc.}\}$

37. __ __ Whole numbers      e. $\mathbb{N}$      V. $\{\ldots -3, -2, -1, 0, 1, 2, 3, \ldots\}$

38. __ __ Natural numbers      f. $\mathbb{Z}$      VI. $\mathbb{N}, \mathbb{W}, \mathbb{Z}, \mathbb{Q}, \mathbb{H}$

**Use a descriptive variable or the variable given with an inequality symbol ($<$, $>$, $\leq$, $\geq$) to write a model for each statement.**

**39.** To spend the night at a friend's house, Kylie must be at least 6 yr old.

**40.** Monty can spend at most $2500 on the purchase of a used automobile.

**41.** If Jerod gets no more than two words incorrect on his spelling test he can play in the soccer game this weekend.

**42.** Andy must weigh less than 112 lb to be allowed to wrestle in his weight class at the meet.

**43.** In order for the expression $\sqrt{2x - 3}$ to represent a real number, $x$ must be greater than or equal to $\frac{3}{2}$

**44.** In order for the expression $\sqrt{5 - 4x}$ to represent a real number, $x$ must be less than or equal to $\frac{5}{4}$.

**45.** In order for the expression $\dfrac{1}{\sqrt{2 - x}}$ to represent a real number, $x$ must be less than 2.

**46.** In order for the expression $\dfrac{1}{\sqrt{x - 7}}$ to represent a real number, $x$ must be greater than 7.

**47.** In order for a weight sensor to function properly, an item must weigh at least 5 g, but less than 32 g.

**48.** To warn against trespassers, a new motion detector is installed. The detector's range is from 2 m, to no more than 20 m.

**Evaluate/Simplify each expression.**

**49.** $|-2.75|$

**50.** $|-7.24|$

**51.** $-|4|$

**52.** $-|6|$

**53.** $\left|\dfrac{1}{2}\right|$

**54.** $\left|\dfrac{2}{5}\right|$

**55.** $\left|-\dfrac{3}{4}\right|$

**56.** $\left|-\dfrac{3}{7}\right|$

**Use the concept of absolute value to complete Exercises 57 through 64.**

**57.** Write the statement two ways, then simplify. "The distance between $-7.5$ and 2.5 is …"

**58.** Write the statement two ways, then simplify. "The distance between $13\frac{2}{5}$ and $-2\frac{3}{5}$ is …"

**59.** What two numbers on the number line are five units from negative three?

**60.** What two numbers on the number line are three units from two?

**61.** If $n$ is positive, then $-n$ is _____.

**62.** If $n$ is negative, then $-n$ is _____.

**63.** If $n < 0$, then $|n| = $ _____.

**64.** If $n > 0$, then $|n| = $ _____.

**Determine which expressions are equal to zero and which are undefined. Justify your responses by writing the related multiplication.**

**65.** $12 \div 0$

**66.** $0 \div 12$

**67.** $\dfrac{7}{0}$

**68.** $\dfrac{0}{7}$

**Without computing the actual answer, state whether the result will be positive or negative. Be careful to note what power is used and whether the negative sign is included in parentheses.**

**69. a.** $(-7)^2$      **b.** $-7^2$
     **c.** $(-7)^5$      **d.** $-7^5$

**70. a.** $(-7)^3$      **b.** $-7^3$
     **c.** $(-7)^4$      **d.** $-7^4$

**Evaluate without the aid of a calculator.**

**71.** $-\sqrt{\dfrac{121}{36}}$

**72.** $-\sqrt{\dfrac{25}{49}}$

**73.** $\sqrt[3]{-8}$

**74.** $\sqrt[3]{-64}$

**75.** What perfect square is closest to 78?

**76.** What perfect cube is closest to $-71$?

**Perform the operation indicated without the aid of a calculator.**

**77.** $-24 - (-31)$

**78.** $-45 - (-54)$

**79.** $7.045 - 9.23$

**80.** $0.0762 - 0.9034$

**81.** $4\frac{5}{6} + \left(-\frac{1}{2}\right)$

**82.** $1\frac{1}{8} + \left(-\frac{3}{4}\right)$

**83.** $\left(-\frac{2}{3}\right)\left(3\frac{5}{8}\right)$

**84.** $(-8)\left(2\frac{1}{4}\right)$

**85.** $(12)(-3)(0)$

**86.** $(-1)(0)(-5)$

**87.** $-60 \div 12$

**88.** $75 \div (-15)$

**89.** $\frac{4}{5} \div (-8)$

**90.** $-15 \div \frac{1}{2}$

**91.** $-\frac{2}{3} \div \frac{16}{21}$

**92.** $-\frac{3}{4} \div \frac{7}{8}$

**Evaluate without a calculator, using the order of operations.**

**93.** $12 - 10 \div 2 \times 5 + (-3)^2$

**94.** $(5 - 2)^2 - 16 \div 4 \cdot 2 - 1$

**95.** $\sqrt{\dfrac{9}{16}} - \dfrac{3}{5} \cdot \left(\dfrac{5}{3}\right)^2$

**96.** $\left(\dfrac{3}{2}\right)^2 \div \left(\dfrac{9}{4}\right) - \sqrt{\dfrac{25}{64}}$

**97.** $\dfrac{4(-7) - 6^2}{6 - \sqrt{49}}$

**98.** $\dfrac{5(-6) - 3^2}{9 - \sqrt{64}}$

**Evaluate using a calculator (round to hundredths).**

**99.** $2475\left(1 + \dfrac{0.06}{4}\right)^{4 \cdot 10}$

**100.** $5100\left(1 + \dfrac{0.078}{52}\right)^{52 \cdot 20}$

**101.** $\dfrac{-4 + \sqrt{(-4)^2 - 4(3)(-39)}}{2(3)}$

**102.** $\dfrac{-12 - \sqrt{(-12)^2 - 4(-2)(32)}}{2(-2)}$

## ▶ WORKING WITH FORMULAS

**103. Pitch diameter:** $D = \dfrac{d \cdot n}{n + 2}$

Mesh gears are used to transfer rotary motion and power from one shaft to another. The *pitch diameter D* of a drive gear is given by the formula shown, where $d$ is the outer diameter of the gear and $n$ is the number of teeth on the gear. Find the pitch diameter of a gear with 24 teeth and an outer diameter of 5 cm.

**104. Pediatric dosages and Clark's rule:** $D_C = \dfrac{D_A \cdot W}{150}$

The amount of medication prescribed for young children depends on their weight, height, age, body surface area, and other factors. **Clark's rule** is a formula that helps estimate the correct child's dose $D_C$ based on the adult dose $D_A$ and the weight $W$ of the child (an average adult weight of 150 lb is assumed). Compute a child's dose if the adult dose is 50 mg and the child weighs 30 lb.

## ▶ APPLICATIONS

*Use positive and negative numbers to model the situation, then compute.*

**105. Temperature changes:** At 6:00 P.M., the temperature was 50°F. A cold front moves through that causes the temperature to *drop* 3°F each hour until midnight. What is the temperature at midnight?

**106. Air conditioning:** Most air conditioning systems are designed to create a 2° *drop* in the air temperature each hour. How long would it take to reduce the air temperature from 86° to 71°?

**107. Record temperatures:** The state of California holds the record for the greatest temperature swing between a record high and a record low. The record high was 134°F and the record low was −45°F. How many degrees *difference* are there between the record high and the record low?

**108. Cold fronts:** In Juneau, Alaska, the temperature was 17°F early one morning. A cold front later moved in and the temperature *dropped* 32°F by lunchtime. What was the temperature at lunchtime?

 **Use a calculator and the rational/irrational numbers given to compute.**

**109. Distance traveled:** An insect crawls $15\sqrt{2}$ cm along the diagonal of a square, $\frac{31}{2}$ cm along the length of a line segment, and $10\pi$ cm around the circumference of a circle. What is the total distance traveled (to the nearest 100th of a cm)?

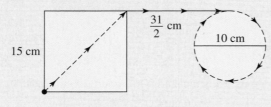

**110. Distance between points:** Find the distance between points A and B (rounded to the nearest 10th of a cm), given that the square and equilateral triangle have sides of length 10 cm, the circle has a *circumference* of 22 cm, and the line segment is $\frac{35}{2}$ cm long.

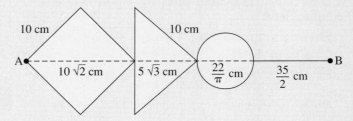

## ▶ EXTENDING THE CONCEPT

**111.** Here are some historical approximations for $\pi$. Which one is closest to the true value?

Archimedes: $3\frac{1}{7}$     Tsu Ch'ung-chih: $\frac{355}{113}$

Aryabhata: $\frac{62,832}{20,000}$     Brahmagupta: $\sqrt{10}$

**112.** If $A > 0$ and $B < 0$, is the product $A \cdot (-B)$ positive or negative?

**113.** If $A > 0$ and $B < 0$, is the quotient $-(A \div B)$ positive or negative?

## LEARNING OBJECTIVES

*In Section R.2 you will review how to:*

☐ **A.** Identify terms, coefficients, and expressions
☐ **B.** Create mathematical models
☐ **C.** Evaluate algebraic expressions
☐ **D.** Identify and use properties of real numbers
☐ **E.** Simplify algebraic expressions

To effectively use mathematics as a problem-solving tool, we must develop the ability to translate written or verbal information into a mathematical model. After obtaining a model, many applications require that you work effectively with algebraic terms and expressions. The basic ideas involved are reviewed here.

### A. Terms, Coefficients, and Algebraic Expressions

An **algebraic term** is a *collection of factors* that may include numbers, variables, or expressions within parentheses. Here are some examples:

(a) $3$      (b) $-6P$      (c) $5xy$      (d) $-8n^2$      (e) $n$      (f) $2(x + 3)$

If a term consists of a number only, it is called a **constant** term. In (a), 3 is a constant term. Any term that contains a variable is called a **variable term.** We call the constant factor of a term the **numerical coefficient** or simply the **coefficient.** The coefficients for (a), (b), (c), and (d) are 3, $-6$, 5, and $-8$ respectively. In (e), the coefficient of $n$ is 1, since $1 \cdot n = 1n = n$. The term in (f) has two factors as written, 2 and $(x + 3)$. The coefficient is 2.

An **algebraic expression** can be a single term or a sum or difference of terms. To avoid confusion when identifying the coefficient of each term, the expression can be rewritten using algebraic addition if desired: $A - B = A + (-B)$. For instance, $4 - 3x = 4 + (-3x)$ shows the coefficient of $x$ is $-3$. To identify the coefficient of a rational term, it sometimes helps to **decompose** the term, rewriting it using a fractional coefficient as in $\frac{n-2}{5} = \frac{1}{5}(n - 2)$ and $\frac{x}{2} = \frac{1}{2}x$.

---

**EXAMPLE 1** ▶ **Identifying Terms and Coefficients**

State the number of terms in each expression as given, then identify the coefficient of each term.

     **a.** $2x - 5y$      **b.** $\dfrac{x + 3}{7} - 2x$      **c.** $-(x - 12)$      **d.** $-2x^2 - x + 5$

**Solution** ▶ We can begin by rewriting each subtraction using algebraic addition.

| Rewritten: | **a.** $2x + (-5y)$ | **b.** $\frac{1}{7}(x + 3) + (-2x)$ | **c.** $-1(x - 12)$ | **d.** $-2x^2 + (-1x) + 5$ |
|---|---|---|---|---|
| Number of terms: | two | two | one | three |
| Coefficient(s): | 2 and $-5$ | $\frac{1}{7}$ and $-2$ | $-1$ | $-2$, $-1$, and 5 |

☑ **A.** You've just reviewed how to identify terms, coefficients, and expressions

**Now try Exercises 5 through 12** ▶

### B. Translating Written or Verbal Information into a Mathematical Model

The key to solving many applied problems is finding an algebraic expression that accurately models relationships described in context. First, we assign a variable to represent an unknown quantity, then build related expressions using words from the English language that suggest mathematical operations. Variables that remind us of what they represent are often used in the modeling process, such as $D = RT$ for Distance equals Rate times Time. These are often called **descriptive variables.** Capital letters are also used due to their widespread appearance in other fields.

**EXAMPLE 2** ▶   **Translating English Phrases into Algebraic Expressions**

Assign a variable to the unknown number, then translate each phrase into an algebraic expression.

    **a.** twice a number, increased by five

    **b.** eleven less than eight times the width

    **c.** ten less than triple the payment

    **d.** two hundred fifty dollars more than double the amount

Solution ▶   **a.** Let $n$ represent the number. Then $2n$ represents twice the number, and $2n + 5$ represents twice the number, increased by five.

    **b.** Let $W$ represent the width. Then $8W$ represents eight times the width, and $8W - 11$ represents 11 less than eight times the width.

    **c.** Let $p$ represent the payment. Then $3p$ represents triple the payment, and $3p - 10$ represents 10 less than triple the payment.

    **d.** Let $A$ represent the amount in dollars. Then $2A$ represents double the amount, and $2A + 250$ represents 250 dollars more than double the amount.

**Now try Exercises 13 through 24** ▶

Identifying and translating such phrases *when they occur in context* is an important problem-solving skill. Note how this is done in Example 3.

**EXAMPLE 3** ▶   **Creating a Mathematical Model**

*The cost for a rental car is $35 plus 15 cents per mile.* Express the cost of renting a car in terms of the number of miles driven.

Solution ▶   Let $m$ represent the number of miles driven. Then $0.15m$ represents the cost for each mile and $C = 35 + 0.15m$ represents the total cost for renting the car.

☑ **B.** You've just reviewed how to create mathematical models

**Now try Exercises 25 through 36** ▶

## C. Evaluating Algebraic Expressions

We often need to **evaluate** expressions to investigate patterns and note relationships.

| **Evaluating a Mathematical Expression** |
|---|
|     **1.** Replace each variable with open parentheses ( ). |
|     **2.** Substitute the values given for each variable. |
|     **3.** Simplify using the order of operations. |

In this process, it's best to use a **vertical format,** with the original expression written first, the substitutions shown next, followed by the simplified forms and the final result. The numbers substituted or "plugged into" the expression are often called the **input values,** with the result called the **output** value.

**EXAMPLE 4 ▶**   **Evaluating an Algebraic Expression**

Evaluate the expression $x^3 - 2x^2 + 5$ for $x = -3$.

**Solution ▶**   For $x = -3$:  $\begin{aligned} x^3 - 2x^2 + 5 &= (\quad)^3 - 2(\quad)^2 + 5 && \text{replace variables with open parentheses} \\ &= (-3)^3 - 2(-3)^2 + 5 && \text{substitute } -3 \text{ for } x \\ &= -27 - 2(9) + 5 && \text{simplify: } (-3)^3 = -27, (-3)^2 = 9 \\ &= -27 - 18 + 5 && \text{simplify: } 2(9) = 18 \\ &= -40 && \text{result} \end{aligned}$

When the input is $-3$, the output is $-40$.

**WORTHY OF NOTE**

In Example 4, note the importance of the first step in the evaluation process: *replace each variable with open parentheses.* Skipping this step could easily lead to confusion as we try to evaluate the squared term, since $-3^2 = -9$, while $(-3)^2 = 9$. **Also see Exercises 53 and 54.**

**Now try Exercises 37 through 52 ▶**

If the same expression is evaluated repeatedly, results are often collected and analyzed in a table of values, as shown in Example 5. As a practical matter, the substitutions and simplifications are often done mentally or on scratch paper, with the table showing only the input and output values.

**EXAMPLE 5 ▶**   **Evaluating an Algebraic Expression**

Evaluate $x^2 - 2x - 3$ to complete the table shown. Which input value(s) of $x$ cause the expression to have an output of 0?

**Solution ▶**

| Input $x$ | Output $x^2 - 2x - 3$ |
|:---:|:---:|
| $-2$ | $(-2)^2 - 2(-2) - 3 = 5$ |
| $-1$ | $0$ |
| $0$ | $-3$ |
| $1$ | $-4$ |
| $2$ | $-3$ |
| $3$ | $0$ |
| $4$ | $5$ |

**☑ C. You've just reviewed how to evaluate algebraic expressions**

The expression has an output of 0 when $x = -1$ and $x = 3$.

**Now try Exercises 53 through 58 ▶**

For exercises that combine the skills from Examples 3 through 5, **see Exercises 83 through 90.**

## D. Properties of Real Numbers

While the phrase, "an unknown number times five," is accurately modeled by the expression $n5$ for some number $n$, in algebra we prefer to have numerical coefficients precede variable factors. When we reorder the *factors* as $5n$, we are using the **commutative property of multiplication.** A reordering of *terms* involves the **commutative property of addition.**

### The Commutative Properties

Given that $a$ and $b$ represent real numbers:

ADDITION:  $a + b = b + a$       MULTIPLICATION:  $a \cdot b = b \cdot a$

Terms can be added in any order without changing the sum.       Factors can be multiplied in any order without changing the product.

Each property can be extended to include any number of terms or factors. While the commutative property implies a *reordering* or *movement* of terms (to commute implies back-and-forth movement), the **associative property** implies a *regrouping* or reassociation of terms. For example, the sum $(\frac{3}{4} + \frac{3}{5}) + \frac{2}{5}$ is easier to compute if we regroup the addends as $\frac{3}{4} + (\frac{3}{5} + \frac{2}{5})$. This illustrates the **associative property of addition.** Multiplication is also associative.

### The Associative Properties

Given that $a$, $b$, and $c$ represent real numbers:

ADDITION:                    MULTIPLICATION:

$$(a + b) + c = a + (b + c) \qquad (a \cdot b) \cdot c = a \cdot (b \cdot c)$$

Terms can be regrouped.       Factors can be regrouped.

**EXAMPLE 6** ▶ **Simplifying Expressions Using Properties of Real Numbers**

Use the commutative and associative properties to simplify each calculation.

**a.** $\frac{3}{8} - 19 + \frac{5}{8}$          **b.** $[-2.5 \cdot (-1.2)] \cdot 10$

Solution ▶ **a.** $\frac{3}{8} - 19 + \frac{5}{8} = -19 + \frac{3}{8} + \frac{5}{8}$          commutative property (order changes)

$\qquad\qquad\qquad = -19 + (\frac{3}{8} + \frac{5}{8})$          associative property (grouping changes)

$\qquad\qquad\qquad = -19 + 1$          simplify

$\qquad\qquad\qquad = -18$          result

**b.** $[-2.5 \cdot (-1.2)] \cdot 10 = -2.5 \cdot [(-1.2) \cdot 10]$          associative property (grouping changes)

$\qquad\qquad\qquad\qquad = -2.5 \cdot (-12)$          simplify

$\qquad\qquad\qquad\qquad = 30$          result

**Now try Exercises 59 and 60** ▶

**WORTHY OF NOTE**

Is subtraction commutative? Consider a situation involving money. If you had $100, you could easily buy an item costing $20: $100 − $20 leaves you with $80. But if you had $20, could you buy an item costing $100? Obviously $100 − $20 is not the same as $20 − $100. Subtraction is *not* commutative. Likewise, 100 ÷ 20 is not the same as 20 ÷ 100, and division is *not* commutative.

For any real number $x$, $x + 0 = x$. Since the original number was returned or "identified," 0 is called the **additive identity.** Similarly, 1 is called the **multiplicative identity** since $1 \cdot x = x$. The identity properties are used extensively in the process of solving equations.

### The Additive and Multiplicative Identities

Given that $x$ is a real number,

$$x + 0 = x \qquad\qquad\qquad 1 \cdot x = x$$

Zero is the identity for addition.       One is the identity for multiplication.

For any real number $x$, there is a real number $-x$ such that $x + (-x) = 0$. The number $-x$ is called the **additive inverse** of $x$, since their sum results in the additive identity. Similarly, the **multiplicative inverse** of any nonzero number $x$ is $\frac{1}{x}$, since $x \cdot \frac{1}{x} = 1$ (the multiplicative identity). This property can also be stated as $\frac{p}{q} \cdot \frac{q}{p} = 1$ $(p, q \neq 0)$ for any rational number $\frac{p}{q}$. Note that $\frac{p}{q}$ and $\frac{q}{p}$ are **reciprocals.**

### The Additive and Multiplicative Inverses

Given that $x$ is a real number $(x \neq 0)$:

$$x + (-x) = 0 \qquad\qquad x \cdot \frac{1}{x} = 1$$

$x$ and $-x$ are                    $x$ and $\frac{1}{x}$ are
additive inverses.              multiplicative inverses.

**EXAMPLE 7** ▶ **Determining Additive and Multiplicative Inverses**

Replace the box to create a true statement:

a. $\Box \cdot \dfrac{-3}{5}x = 1 \cdot x$           b. $x + 4.7 + \Box = x$

Solution ▶ a. $\Box = \dfrac{5}{-3}$, since $\dfrac{5}{-3} \cdot \dfrac{-3}{5} = 1$

b. $\Box = -4.7$, since $4.7 + (-4.7) = 0$

**Now try Exercises 61 and 62 ▶**

The **distributive property of multiplication over addition** is widely used in a study of algebra, because it enables us to rewrite a product as an equivalent sum and vice versa.

### The Distributive Property of Multiplication over Addition

Given that $a$, $b$, and $c$ represent real numbers:

$$a(b + c) = ab + ac \qquad\qquad ab + ac = a(b + c)$$

A factor outside a sum can be        A factor common to each addend
distributed to each addend in        in a sum can be "undistributed"
the sum.                             and written outside a group.

Recall that if no coefficient is indicated, it is assumed to be 1, as in $x = 1x$, $(x^2 + 3x) = 1(x^2 + 3x)$, and $-(x^3 - 5x^2) = -1(x^3 - 5x^2)$.

**EXAMPLE 8** ▶ **Simplifying Expressions Using the Distributive Property**

Apply the distributive property as appropriate. Simplify if possible.

a. $7(p + 5.2)$          b. $-(2.5 - x)$          c. $7x^3 - x^3$          d. $\dfrac{5}{2}n + \dfrac{1}{2}n$

Solution ▶ a. $7(p + 5.2) = 7p + 7(5.2)$           b. $-(2.5 - x) = -1(2.5 - x)$
$= 7p + 36.4$                          $= -1(2.5) - (-1)(x)$
$= -2.5 + x$

**WORTHY OF NOTE**

From Example 8(b) we learn that a negative sign outside a group changes the sign of all terms within the group: $-(2.5 - x) = -2.5 + x$.

c. $7x^3 - x^3 = 7x^3 - 1x^3$          d. $\dfrac{5}{2}n + \dfrac{1}{2}n = \left(\dfrac{5}{2} + \dfrac{1}{2}\right)n$
$= (7 - 1)x^3$                             $= \left(\dfrac{6}{2}\right)n$
$= 6x^3$                                   $= 3n$

☑ D. You've just reviewed how to identify and use properties of real numbers

**Now try Exercises 63 through 70 ▶**

## E. Simplifying Algebraic Expressions

Two terms are **like terms** only if they have the *same variable factors* (the coefficient is not used to identify like terms). For instance, $3x^2$ and $-\frac{1}{7}x^2$ are like terms, while $5x^3$ and $5x^2$ are not. We simplify expressions by **combining like terms** using the distributive property, along with the commutative and associative properties. Many times the distributive property is used to eliminate grouping symbols *and* combine like terms within the same expression.

**EXAMPLE 9** ▶    **Simplifying an Algebraic Expression**

Simplify the expression completely: $7(2p^2 + 1) - (p^2 + 3)$.

**Solution** ▶

| | |
|---|---|
| $7(2p^2 + 1) - 1(p^2 + 3)$ | original expression; note coefficient of $-1$ |
| $= 14p^2 + 7 - 1p^2 - 3$ | distributive property |
| $= (14p^2 - 1p^2) + (7 - 3)$ | commutative and associative properties (collect like terms) |
| $= (14 - 1)p^2 + 4$ | distributive property |
| $= 13p^2 + 4$ | result |

**Now try Exercises 71 through 80** ▶

The steps for simplifying an algebraic expression are summarized here:

### To Simplify an Expression

1. Eliminate parentheses by applying the distributive property.
2. Use the commutative and associative properties to group like terms.
3. Use the distributive property to combine like terms.

☑ E. You've just reviewed how to simplify algebraic expressions

As you practice with these ideas, many of the steps will become more automatic. At some point, the distributive property, the commutative and associative properties, as well as the use of algebraic addition will all be performed mentally.

## R.2 EXERCISES

### ▶ CONCEPTS AND VOCABULARY

**Fill in each blank with the appropriate word or phrase. Carefully reread the section, if necessary.**

1. The constant factor in a variable term is called the _____.

2. When $3 \cdot 14 \cdot \frac{2}{3}$ is written as $3 \cdot \frac{2}{3} \cdot 14$, the _____ property has been used.

3. Discuss/Explain why the additive inverse of $-5$ is 5, while the multiplicative inverse of $-5$ is $-\frac{1}{5}$.

4. Discuss/Explain how we can rewrite the sum $3x + 6y$ as a product, and the product $2(x + 7)$ as a sum.

## ▶ DEVELOPING YOUR SKILLS

**Identify the number of terms in each expression and the coefficient of each term.**

**5.** $3x - 5y$

**6.** $-2a - 3b$

**7.** $2x + \dfrac{x + 3}{4}$

**8.** $\dfrac{n - 5}{3} + 7n$

**9.** $-2x^2 + x - 5$

**10.** $3n^2 + n - 7$

**11.** $-(x + 5)$

**12.** $-(n - 3)$

**Translate each phrase into an algebraic expression.**

**13.** seven fewer than a number

**14.** a number decreased by six

**15.** the sum of a number and four

**16.** a number increased by nine

**17.** the difference between a number and five is squared

**18.** the sum of a number and two is cubed

**19.** thirteen less than twice a number

**20.** five less than double a number

**21.** five fewer than two-thirds of a number

**22.** fourteen more than one-half of a number

**23.** three times the sum of a number and five, decreased by seven

**24.** five times the difference of a number and two, increased by six

**Create a mathematical model using descriptive variables.**

**25.** The length of the rectangle is three meters less than twice the width.

**26.** The height of the triangle is six centimeters less than three times the base.

**27.** The speed of the car was fifteen miles per hour more than the speed of the bus.

**28.** It took Romulus three minutes more time than Remus to finish the race.

**29. Hovering altitude:** The helicopter was hovering 150 ft above the top of the building. Express the altitude of the helicopter in terms of the building's height.

**30. Stacks on a cargo ship:** The smoke stack of the cargo ship cleared the bridge by 25 ft as it passed beneath it. Express the height of the stack in terms of the bridge's height.

**31. Dimensions of a city park:** The length of a rectangular city park is 20 m more than twice its width. Express the length of the park in terms of the width.

**32. Dimensions of a parking lot:** In order to meet the city code while using the available space, a contractor planned to construct a parking lot with a length that was 50 ft less than three times its width. Express the length of the lot in terms of the width.

**33. Cost of milk:** In 2010, a gallon of milk cost two and one-half times what it did in 1990. Express the cost of a gallon of milk in 2010, in terms of the 1990 cost.

**34. Cost of gas:** In 2010, a gallon of gasoline cost two and one-half times what it did in 1990. Express the cost of a gallon of gas in 2010, in terms of the 1990 cost.

**35. Pest control:** In her pest control business, Judy charges $50 per call plus $12.50 per gallon of insecticide for the control of spiders and certain insects. Express the total charge in terms of the number of gallons of insecticide used.

**36. Computer repairs:** As his reputation and referral business grew, Keith began to charge $75 per service call plus an hourly rate of $50 for the repair and maintenance of home computers. Express the cost of a service call in terms of the number of hours spent on the call.

**Evaluate each expression given $x = 2$ and $y = -3$**

**37.** $4x - 2y$

**38.** $5x - 3y$

**39.** $-2x^2 + 3y^2$

**40.** $-5x^2 + 4y^2$

**41.** $2y^2 + 5y - 3$

**42.** $3x^2 + 2x - 5$

**43.** $-2(3y + 1)$

**44.** $-3(2y + 5)$

**45.** $(-3x)^2 - 4xy - y^2$

**46.** $(-2x)^2 - 5xy - y^2$

**47.** $\frac{1}{2}x - \frac{1}{3}y$

**48.** $\frac{2}{3}x - \frac{1}{2}y$

**49.** $\dfrac{-12y + 5}{-3x + 1}$

**50.** $\dfrac{12x + (-3)}{-3y + 1}$

**51.** $\sqrt{-12y} \cdot 4$

**52.** $7 \cdot \sqrt{-27y}$

**Evaluate each expression for integers from $-3$ to 3 inclusive. Identify input(s) that give an output of zero.**

**53.** $x^2 - 3x - 4$

**54.** $x^2 - 2x - 3$

**55.** $-3(1 - x) - 6$

**56.** $5(3 - x) - 10$

**57.** $x^3 - 6x + 4$

**58.** $x^3 + 5x + 18$

**Rewrite each expression using the given property and simplify if possible.**

**59.** Commutative property of addition

   **a.** $-5 + 7$

   **b.** $-2 + n$

   **c.** $-4.2 + a + 13.6$

   **d.** $7 + x - 7$

**60.** Associative property of multiplication

   **a.** $2 \cdot (3 \cdot 6)$

   **b.** $3 \cdot (4 \cdot b)$

   **c.** $-1.5 \cdot (6 \cdot a)$

   **d.** $-6 \cdot (-\frac{5}{6} \cdot x)$

**Replace the box so that a true statement results.**

**61. a.** $x + (-3.2) + \boxed{\phantom{x}} = x$

   **b.** $n - \frac{5}{6} + \boxed{\phantom{x}} = n$

**62. a.** $\boxed{\phantom{x}} \cdot \frac{2}{3}x = 1x$

   **b.** $\boxed{\phantom{x}} \cdot \dfrac{n}{-3} = 1n$

**Apply the distributive property and simplify if possible.**

**63.** $-5(x - 2.6)$          **64.** $-12(v - 3.2)$

**65.** $\frac{2}{3}(-\frac{1}{5}p + 9)$          **66.** $\frac{5}{6}(-\frac{2}{15}q + 24)$

**67.** $3a + (-5a)$          **68.** $13m + (-5m)$

**69.** $\frac{2}{3}x + \frac{3}{4}x$          **70.** $\frac{5}{12}y - \frac{3}{8}y$

**Simplify by removing all grouping symbols (as needed) and combining like terms.**

**71.** $3(a^2 + 3a) - (5a^2 + 7a)$

**72.** $2(b^2 + 5b) - (6b^2 + 9b)$

**73.** $x^2 - (3x - 5x^2)$

**74.** $n^2 - (5n - 4n^2)$

**75.** $(3a + 2b - 5c) - (a - b - 7c)$

**76.** $(x - 4y + 8z) - (8x - 5y - 2z)$

**77.** $\frac{3}{5}(5n - 4) + \frac{5}{8}(n + 16)$

**78.** $\frac{2}{3}(2x - 9) + \frac{3}{4}(x + 12)$

**79.** $(3a^2 - 5a + 7) + 2(2a^2 - 4a - 6)$

**80.** $2(3m^2 + 2m - 7) - (m^2 - 5m + 4)$

## ▶ WORKING WITH FORMULAS

**81. Electrical resistance:** $R = \dfrac{kL}{d^2}$

The electrical resistance in a wire depends on the length and diameter of the wire. This resistance can be modeled by the formula shown, where $R$ is the resistance in ohms, $L$ is the length in feet, and $d$ is the diameter of the wire in inches. Find the resistance if $k = 0.000025$, $d = 0.015$ in., and $L = 90$ ft

**82. Volume and pressure:** $P = \dfrac{k}{V}$

If temperature remains constant, the pressure of a gas held in a closed container is related to the volume of gas by the formula shown, where $P$ is the pressure in pounds per square inch, $V$ is the volume of gas in cubic inches, and $k$ is a constant that depends on given conditions. Find the pressure exerted by the gas if $k = 440,310$ and $V = 22,580$ in$^3$.

## ▶ APPLICATIONS

**Translate each key phrase into an algebraic expression, then evaluate as indicated.**

**83. Cruising speed:** A turbo-prop airliner has a cruising speed that is one-half the cruising speed of a 747 jet aircraft. (a) Express the speed of the turbo-prop in terms of the speed of the jet, and (b) determine the speed of the airliner if the cruising speed of the jet is 550 mph.

**84. Softball toss:** Macklyn can throw a softball two-thirds as far as her father. (a) Express the distance that Macklyn can throw a softball in terms of the distance her father can throw. (b) If her father can throw the ball 210 ft, how far can Macklyn throw the ball?

**85. Dimensions of a lawn:** The length of a rectangular lawn is 3 ft more than twice its width. (a) Express the length of the lawn in terms of the width. (b) If the width is 52 ft, what is the length?

**86. Pitch of a roof:** To obtain the proper pitch, the crossbeam for a roof truss must be 2 ft less than three-halves the rafter. (a) Express the length of the crossbeam in terms of the rafter. (b) If the rafter is 18 ft, how long is the crossbeam?

**87. Postage costs:** In 2011, a first-class stamp cost 31¢ more than it did in 1978. Express the cost of a 2011 stamp in terms of the 1978 cost. If a stamp cost 13¢ in 1978, what was the cost in 2011?

**88. Minimum wage:** In 2011, the federal minimum wage was $4.95 per hour more than it was in 1976. Express the 2011 wage in terms of the 1976 wage. If the hourly wage in 1976 was $2.30, what was it in 2009?

**89. Repair costs:** The TV repair shop charges a flat fee of $43.50 to come to your house and $25 per hour for labor. Express the cost of repairing a TV in terms of the time it takes to repair it. If the repair took 1.5 hr, what was the total cost?

**90. Repair costs:** At the local car dealership, shop charges are $79.50 to diagnose the problem and $85 per shop hour for labor. Express the cost of a repair in terms of the labor involved. If a repair takes 3.5 hr, how much will it cost?

▶ **EXTENDING THE CONCEPT**

**91.** If $C$ must be a positive odd integer and $D$ must be a negative even integer, then $C^2 + D^2$ must be a:
   **a.** positive odd integer.
   **b.** positive even integer.
   **c.** negative odd integer.
   **d.** negative even integer.
   **e.** cannot be determined.

**92.** Historically, several attempts have been made to create metric time using factors of 10, but our current system won out. If 1 day was 10 metric hours, 1 metric hour was 10 metric minutes, and 1 metric minute was 10 metric seconds, what time would it really be if a metric clock read 4:3:5? Assume that each new day starts at midnight.

---

## R.3   Exponents, Scientific Notation, and a Review of Polynomials

**LEARNING OBJECTIVES**

*In Section R.3 you will review how to:*

☐ **A.** Apply properties of exponents
☐ **B.** Perform operations in scientific notation
☐ **C.** Identify and classify polynomial expressions
☐ **D.** Add and subtract polynomials
☐ **E.** Compute the product of two polynomials
☐ **F.** Compute special products: binomial conjugates and binomial squares

In this section, we review basic exponential properties and operations on polynomials. Although there are five to eight exponential properties (depending on how you count them), all can be traced back to the basic definition involving repeated multiplication.

### A. The Properties of Exponents

An exponent is a superscript number or letter occurring to the upper right of a base number, and indicates how many times the base occurs as a factor. For $b \cdot b \cdot b = b^3$, we say $b^3$ is written in *exponential form*. In some cases, we may refer to $b^3$ as an **exponential term.**

**Exponential Notation**

For any positive integer $n$,

$$b^n = \underbrace{b \cdot b \cdot b \cdot \cdots \cdot b}_{n \text{ times}} \quad \text{and} \quad \underbrace{b \cdot b \cdot b \cdot \cdots \cdot b}_{n \text{ times}} = b^n$$

### The Product and Power Properties

There are two properties that follow immediately from this definition. When $b^3$ is multiplied by $b^2$, we have an uninterrupted string of five factors: $b^3 \cdot b^2 = (b \cdot b \cdot b) \cdot (b \cdot b)$, which can then be written as $b^5$. This is an example of the **product property of exponents.**

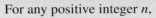

> **WORTHY OF NOTE**
>
> In this statement of the product property and the exponential properties that follow, it is assumed that for any expression of the form $0^m$, $m > 0$ (hence $0^m = 0$).

**Product Property of Exponents**

For any base $b$ and positive integers $m$ and $n$:
$$b^m \cdot b^n = b^{m+n}$$

In words the property says, *to multiply exponential terms with the* **same base,** *keep the common base and add the exponents.* A special application of the product property uses repeated factors of the *same* exponential term, as in $(x^2)^3$. Using the product property, we have $(x^2)(x^2)(x^2) = x^{2+2+2} = x^6$. Notice the same result can be found more quickly by multiplying the inner exponent by the outer exponent: $(x^2)^3 = x^{2\cdot3} = x^6$. We generalize this idea to state the **power property of exponents.** In words the property says, *to raise an exponential term to a power, keep the same base and multiply the exponents.*

**Power Property of Exponents**

For any base $b$ and positive integers $m$ and $n$:
$$(b^m)^n = b^{m\cdot n}$$

**EXAMPLE 1** ▶   **Multiplying Terms Using Exponential Properties**

Compute each product.

    **a.** $-4x^3 \cdot \frac{1}{2}x^2$     **b.** $(p^3)^2 \cdot (p^4)^5$

**Solution** ▶   **a.** $-4x^3 \cdot \frac{1}{2}x^2 = (-4 \cdot \frac{1}{2})(x^3 \cdot x^2)$    commutative and associative properties

                           $= (-2)(x^{3+2})$    simplify; product property

                           $= -2x^5$    result

    **b.** $(p^3)^2 \cdot (p^4)^5 = p^{3\cdot2} \cdot p^{4\cdot5}$    power property

                         $= p^6 \cdot p^{20}$    simplify

                         $= p^{6+20}$    product property

                         $= p^{26}$    result

**Now try Exercises 5 through 10** ▶

The power property can easily be extended to include more than one factor within the parentheses. This application of the power property is sometimes called the **product to a power property** and can be extended to include any number of factors. We can also raise a quotient of exponential terms to a power. The result is called the **quotient to a power property.** In words the properties say, to raise a product or quotient of exponential terms to a power, *multiply every exponent inside* the parentheses *by the exponent outside* the parentheses.

**Product to a Power Property**

For any bases $a$ and $b$, and positive integers $m$, $n$, and $p$:

$$(a^m b^n)^p = a^{mp} \cdot b^{np}$$

**Quotient to a Power Property**

For any bases $a$ and $b \neq 0$, and positive integers $m$, $n$, and $p$:

$$\left(\frac{a^m}{b^n}\right)^p = \frac{a^{mp}}{b^{np}}$$

**EXAMPLE 2** ▶   **Simplifying Terms Using the Power Properties**

Simplify using the power property (if possible):

    **a.** $(-3a)^2$     **b.** $-3a^2$     **c.** $\left(\dfrac{-5a^3}{2b}\right)^2$

**Solution** ▶   **a.** $(-3a)^2 = (-3)^2 \cdot (a^1)^2$     **b.** $-3a^2$ is in simplified form

                    $= 9a^2$

    **c.** $\left(\dfrac{-5a^3}{2b}\right)^2 = \dfrac{(-5)^2(a^3)^2}{2^2 b^2}$

                       $= \dfrac{25a^6}{4b^2}$

**WORTHY OF NOTE**

Regarding Examples 2(a) and 2(b), note the difference between the expressions $(-3a)^2 = (-3 \cdot a)^2$ and $-3a^2 = -3 \cdot a^2$. In the first, the exponent acts on both the negative 3 *and* the $a$; in the second, the exponent acts on only the $a$ and there is no "product to a power."

**Now try Exercises 11 through 20** ▶

Applications of exponents sometimes involve linking one exponential term with another using a substitution. The result is then simplified using exponential properties.

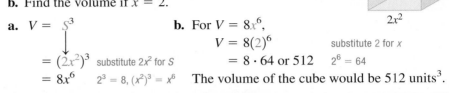

**EXAMPLE 3 ▶**   **Applying the Power Property after a Substitution**

The formula for the volume of a cube is $V = S^3$, where $S$ is the length of one edge. If the length of each edge is $2x^2$:

a. Find a formula for volume in terms of $x$.

b. Find the volume if $x = 2$.

**Solution ▶**   **a.** $V = S^3$

$= (2x^2)^3$   substitute $2x^2$ for $S$

$= 8x^6$    $2^3 = 8, (x^2)^3 = x^6$

**b.** For $V = 8x^6$,

$V = 8(2)^6$   substitute 2 for $x$

$= 8 \cdot 64$ or 512   $2^6 = 64$

The volume of the cube would be 512 units$^3$.

**Now try Exercises 21 and 22 ▶**

### The Quotient Property of Exponents

By combining exponential notation and the property $\dfrac{x}{x} = 1$ for $x \neq 0$, we note a pattern that helps to simplify a *quotient* of exponential terms. For $\dfrac{x^5}{x^2} = \dfrac{x \cdot x \cdot x \cdot x \cdot x}{x \cdot x} = x^3$, the exponent of the final result appears to be the *difference between the exponent in the numerator and the exponent in the denominator*. This seems reasonable since the subtraction would indicate a removal of the factors that reduce to 1. Regardless of how many factors are used, we can generalize the idea and state the **quotient property of exponents.** In words the property says, *to divide two exponential terms with the **same base**, keep the common base and subtract the exponent of the denominator* from *the exponent of the numerator.*

**Quotient Property of Exponents**

For any base $b \neq 0$ and positive integers $m$ and $n$:

$$\frac{b^m}{b^n} = b^{m-n}$$

### Zero and Negative Numbers as Exponents

If the exponent of the denominator is *greater* than the exponent in the numerator, the quotient property yields a negative exponent: $\dfrac{x^2}{x^5} = x^{2-5} = x^{-3}$. To help understand what a negative exponent *means,* let's look at the expanded form of the expression: $\dfrac{x^2}{x^5} = \dfrac{x \cdot x^1}{x \cdot x \cdot x \cdot x \cdot x} = \dfrac{1}{x^3}$. A negative exponent can literally be interpreted as "write the factors as a reciprocal." A good way to remember this is

$2^{-3}$   three factors of 2 written as a reciprocal    $2^{-3} = \dfrac{1}{2^3} = \dfrac{1}{8}$

Since the result would be similar regardless of the base used, we can generalize this idea and state the **property of negative exponents.**

### Property of Negative Exponents

For any base $b \neq 0$ and integer $n$:

$$b^{-n} = \frac{1}{b^n} \qquad \frac{1}{b^{-n}} = b^n \qquad \left(\frac{a}{b}\right)^{-n} = \left(\frac{b}{a}\right)^n ; a \neq 0$$

Finally, when we consider that $\dfrac{x^3}{x^3} = 1$ by division, and $\dfrac{x^3}{x^3} = x^{3-3} = x^0$ using the quotient property, we conclude that $x^0 = 1$ as long as $x \neq 0$. We can also generalize this observation and state the meaning of zero as an exponent. In words the property says, *any nonzero quantity raised to an exponent of zero is equal to 1.*

> **WORTHY OF NOTE**
>
> The use of zero as an exponent should not strike you as strange or odd; it's simply a way of saying that *no factors of the base remain*, since all terms have been reduced to 1.
>
> For $\dfrac{2^3}{2^3}$, we have $\dfrac{8}{8} = 1$, or
>
> $\dfrac{\overset{1}{\cancel{2}} \cdot \overset{1}{\cancel{2}} \cdot \overset{1}{\cancel{2}}}{2 \cdot 2 \cdot 2} = 1$, or $2^{3-3} = 2^0 = 1$.

### Zero Exponent Property

For any base $b \neq 0$:

$$b^0 = 1$$

---

**EXAMPLE 4** ▶  **Simplifying Expressions Using Exponential Properties**

Simplify using exponential properties. Answer using positive exponents only.

**a.** $\left(\dfrac{2a^3}{b^2}\right)^{-2}$ 
**b.** $(3hk^{-2})^3(6h^{-2}k^{-3})^{-2}$

**c.** $(3x)^0 + 3x^0 + 3^{-2}$ 
**d.** $\dfrac{(-2m^2n^3)^5}{(4mn^2)^3}$

**Solution** ▶ 
**a.** $\left(\dfrac{2a^3}{b^2}\right)^{-2} = \left(\dfrac{b^2}{2a^3}\right)^2$  property of negative exponents

$\qquad\qquad = \dfrac{(b^2)^2}{2^2(a^3)^2}$  power properties

$\qquad\qquad = \dfrac{b^4}{4a^6}$  result

**b.** $(3hk^{-2})^3(6h^{-2}k^{-3})^{-2} = 3^3h^3(k^{-2})^3 \cdot 6^{-2}(h^{-2})^{-2}(k^{-3})^{-2}$  power property

$\qquad\qquad = 3^3h^3k^{-6} \cdot 6^{-2}h^4k^6$  simplify

$\qquad\qquad = 3^3 \cdot 6^{-2} \cdot h^{3+4} \cdot k^{-6+6}$  product property

$\qquad\qquad = \dfrac{27h^7k^0}{36}$  simplify $\left(6^{-2} = \dfrac{1}{6^2} = \dfrac{1}{36}\right)$

$\qquad\qquad = \dfrac{3h^7}{4}$  result $(k^0 = 1)$

> **WORTHY OF NOTE**
>
> Notice in Example 4(c), we have $(3x)^0 = (3 \cdot x)^0 = 1$, while $3x^0 = 3 \cdot x^0 = 3(1) = 3$. This is another example of operations and grouping symbols working together: $(3x)^0 = 1$ because any *quantity* to the zero power is 1. However, for $3x^0$ there are no grouping symbols, so the exponent 0 acts only on the $x$ and not the 3: $3x^0 = 3 \cdot x^0 = 3(1) = 3$.

**c.** $(3x)^0 + 3x^0 + 3^{-2} = 1 + 3(1) + \dfrac{1}{3^2}$  zero exponent property; property of negative exponents

$\qquad\qquad = 4 + \dfrac{1}{9}$  simplify

$\qquad\qquad = 4\dfrac{1}{9} = \dfrac{37}{9}$  result

**d.** $\dfrac{(-2m^2n^3)^5}{(4mn^2)^3} = \dfrac{(-2)^5(m^2)^5(n^3)^5}{4^3m^3(n^2)^3}$     power property

$= \dfrac{-32m^{10}n^{15}}{64m^3n^6}$     simplify

$= \dfrac{-1m^7n^9}{2}$     quotient property

$= -\dfrac{m^7n^9}{2}$     result

**Now try Exercises 23 through 58 ▶**

### Summary of Exponential Properties

For real numbers $a$ and $b$, and integers $m$, $n$, $p$ (excluding 0 in any denominator)

Product property:     $b^m \cdot b^n = b^{m+n}$

Power property:     $(b^m)^n = b^{m \cdot n}$

Product to a power:     $(a^m b^n)^p = a^{mp} \cdot b^{np}$

Quotient to a power:     $\left(\dfrac{a^m}{b^n}\right)^p = \dfrac{a^{mp}}{b^{np}}$

Quotient property:     $\dfrac{b^m}{b^n} = b^{m-n}$

Zero exponents:     $b^0 = 1 \;(b \neq 0)$

Negative exponents:     $\dfrac{b^{-n}}{1} = \dfrac{1}{b^n}, \dfrac{1}{b^{-n}} = b^n, \left(\dfrac{a}{b}\right)^{-n} = \left(\dfrac{b}{a}\right)^n$

☑ **A.** You've just reviewed how to apply properties of exponents

## B. Exponents and Scientific Notation

In many technical and scientific applications, we encounter numbers that are either extremely large or very, very small. For example, the mass of the Moon is over 73 sextillion kilograms (73 followed by 21 zeroes), while the constant for universal gravitation contains 10 zeroes before the first nonzero digit. When computing with numbers of this magnitude, scientific notation has a distinct advantage over the common decimal notation (base-10 place values).

### Scientific Notation

A nonzero number written in scientific notation has the form

$$N \times 10^k$$

where $1 \leq |N| < 10$ and $k$ is an integer.

**WORTHY OF NOTE**

Recall that multiplying by 10's (or multiplying by $10^k$, $k > 0$) shifts the decimal point to the right $k$ places, making the number larger. Dividing by 10's (or multiplying by $10^{-k}$, $k > 0$) shifts the decimal point to the left $k$ places, making the number smaller.

To convert a number from decimal notation into scientific notation, we begin by placing the decimal point to the immediate right of the first nonzero digit (creating a number less than 10 but greater than or equal to 1) and multiplying by $10^k$. Then we determine the power of 10 (the value of $k$) needed to ensure that the two forms are equivalent. When writing large or small numbers in scientific notation, we often round the value of $N$ to two or three decimal places, since these numbers typically occur when exact measurements are impossible, and our concern primarily centers around their magnitude.

**EXAMPLE 5** ▶ **Converting from Decimal Notation to Scientific Notation**

The mass of the Moon is about 73,000,000,000,000,000,000,000 kg. Write this number in scientific notation.

Solution ▶ Place decimal to the right of first nonzero digit (7) and multiply by $10^k$.

$$73{,}000{,}000{,}000{,}000{,}000{,}000{,}000 = 7.3 \times 10^k$$

To return the decimal to its original position would require 22 shifts to the *right*, so $k$ must be *positive* 22.

$$73{,}000{,}000{,}000{,}000{,}000{,}000{,}000 = 7.3 \times 10^{22}$$

The mass of the Moon is $7.3 \times 10^{22}$ kg.

**Now try Exercises 59 and 60** ▶

Converting a number from scientific notation to decimal notation is simply an application of multiplication or division with powers of 10.

**EXAMPLE 6** ▶ **Converting from Scientific Notation to Decimal Notation**

The constant of gravitation is $6.67 \times 10^{-11}$. Write this number in common decimal form.

Solution ▶ Since the exponent is *negative* 11, shift the decimal 11 *places to the left,* using placeholder zeroes as needed to return the decimal to its original position:

$$6.67 \times 10^{-11} = 0.000\ 000\ 000\ 066\ 7$$

**Now try Exercises 61 and 62** ▶

Computations that involve scientific notation typically use real number properties and the properties of exponents.

**EXAMPLE 7** ▶ **Storage Space on a Hard Drive**

A typical 250-gigabyte portable hard drive can hold 250,000,000,000 bytes of information. A 2-hr DVD movie can take up as much as 8,000,000,000 bytes of storage space. Find the number of movies (to the nearest whole movie) that can be stored on this hard drive.

Solution ▶ Using the ideas from Example 5, the hard drive holds $2.5 \times 10^{11}$ bytes, while the DVD requires $8.0 \times 10^9$ bytes. Divide to find the number of DVDs the hard drive will hold.

$$\frac{2.5 \times 10^{11}}{8.0 \times 10^9} = \frac{2.5}{8.0} \times \frac{10^{11}}{10^9} \quad \text{rewrite the expression}$$
$$= 0.3125 \times 10^2 \quad \text{divide; subtract exponents}$$
$$= 31.25 \quad \text{result}$$

✓ **B.** You've just reviewed how to perform operations in scientific notation

The drive will hold approximately 31 DVD movies.

**Now try Exercises 63 and 64** ▶

## C. Identifying and Classifying Polynomial Expressions

A **monomial** is a term using *only whole number exponents* on variables, with no variables in the denominator. One important characteristic of a monomial is its **degree.** For a monomial in one variable, the degree is the same as the exponent *on the variable.* The degree of a monomial in two or more variables is the sum of exponents occurring on variable factors. A **polynomial** is a monomial or any sum or difference of monomial terms. For instance, $\frac{1}{2}x^2 - 5x + 6$ is a polynomial, while $3n^{-2} + 2n - 7$ is not (the exponent $-2$ is not a whole number). Identifying polynomials is an important skill because they represent a very different kind of real-world model than nonpolynomials. In addition, there are different **families of polynomials,** with each family having different characteristics. We classify polynomials according to their *degree* and *number of terms.* The **degree of a polynomial** in one variable is the largest exponent occurring on the variable. The degree of a polynomial in more than one variable is the largest sum of exponents in any one term. A polynomial with two terms is called a **binomial** (*bi* means two) and a polynomial with three terms is called a **trinomial** (*tri* means three). There are special names for polynomials with four or more terms, but for these, we simply use the general name *polynomial* (*poly* means many).

**EXAMPLE 8** ▶ **Classifying and Describing Polynomials**

For each expression:
   **a.** Classify as a monomial, binomial, trinomial, or polynomial.
   **b.** State the degree of the polynomial.
   **c.** Name the coefficient of each term.

Solution ▶

| Expression | Classification | Degree | Coefficients |
|---|---|---|---|
| $5x^2y - 2xy$ | binomial | three | $5, -2$ |
| $x^2 - 0.81$ | binomial | two | $1, -0.81$ |
| $z^3 - 3z^2 + 9z - 27$ | polynomial (four terms) | three | $1, -3, 9, -27$ |
| $\frac{-3}{4}x + 5$ | binomial | one | $\frac{-3}{4}, 5$ |
| $2x^2 + x - 3$ | trinomial | two | $2, 1, -3$ |

Now try Exercises 65 through 70 ▶

A polynomial expression is in **standard form** when the terms of the polynomial are written in *descending order of degree,* beginning with the highest-degree term. The coefficient of the highest-degree term is called the **leading coefficient.**

**EXAMPLE 9** ▶ **Writing Polynomials in Standard Form**

Write each polynomial in standard form, then identify the leading coefficient.

Solution ▶

| Polynomial | Standard Form | Leading Coefficient |
|---|---|---|
| $9 - x^2$ | $-x^2 + 9$ | $-1$ |
| $5z + 7z^2 + 3z^3 - 27$ | $3z^3 + 7z^2 + 5z - 27$ | $3$ |
| $2 + \left(\frac{-3}{4}\right)x$ | $\frac{-3}{4}x + 2$ | $\frac{-3}{4}$ |
| $-3 + 2x^2 + x$ | $2x^2 + x - 3$ | $2$ |

☑ **C.** You've just reviewed how to identify and classify polynomial expressions

Now try Exercises 71 through 76 ▶

## D. Adding and Subtracting Polynomials

Adding polynomials simply involves using the distributive, commutative, and associative properties to combine like terms (at this point, the properties are usually applied mentally). As with real numbers, the subtraction of polynomials involves adding the opposite of the second polynomial using algebraic addition. This can be viewed as distributing $-1$ to the second polynomial and combining like terms.

**EXAMPLE 10** ▶ **Adding and Subtracting Polynomials**

Perform the indicated operations:

$(0.7n^3 + 4n^2 + 8) + (0.5n^3 - n^2 - 6n) - (3n^2 + 7n - 10)$.

**Solution** ▶ $0.7n^3 + 4n^2 + 8 + 0.5n^3 - n^2 - 6n - 3n^2 - 7n + 10$    eliminate parentheses (distributive property)

$= 0.7n^3 + 0.5n^3 + 4n^2 - 1n^2 - 3n^2 - 6n - 7n + 8 + 10$    commutative property

$= 1.2n^3 - 13n + 18$    combine like terms

**Now try Exercises 77 through 82** ▶

Sometimes it's easier to add or subtract polynomials using a vertical format and aligning like terms. Note the use of a placeholder zero in Example 11.

**EXAMPLE 11** ▶ **Subtracting Polynomials Using a Vertical Format**

Compute the difference of $x^3 - 5x + 9$ and $x^3 + 3x^2 + 2x - 8$ using a vertical format.

**Solution** ▶

$$
\begin{array}{c}
x^3 + \mathbf{0}x^2 - 5x + 9 \\
-(x^3 + 3x^2 + 2x - 8)
\end{array}
\longrightarrow
\begin{array}{c}
x^3 + \mathbf{0}x^2 - 5x + 9 \\
-x^3 - 3x^2 - 2x + 8 \\
\hline
-3x^2 - 7x + 17
\end{array}
$$

☑ **D.** You've just reviewed how to add and subtract polynomials

The difference is $-3x^2 - 7x + 17$.

**Now try Exercises 83 and 84** ▶

## E. The Product of Two Polynomials

### Monomial Times Monomial

The simplest case of polynomial multiplication is the product of monomials shown in Example 1(a). These were computed using exponential properties and the properties of real numbers.

### Monomial Times Polynomial

To compute the product of a monomial and a polynomial, we use the distributive property.

**EXAMPLE 12** ▶ **Multiplying a Monomial by a Polynomial**

Find the product: $-2a^2(a^2 - 2a + 1)$.

**Solution** ▶ $-2a^2(a^2 - 2a + 1) = -2a^2(a^2) - (-2a^2)(2a^1) + (-2a^2)(1)$    distribute

$= -2a^4 + 4a^3 - 2a^2$    simplify

**Now try Exercises 85 and 86** ▶

### Binomial Times Polynomial

For products involving binomials, we still use a version of the distributive property—this time to distribute one polynomial to each term of the other polynomial factor. Note the distribution can be performed either from the left or from the right.

---

**EXAMPLE 13** ▶ Multiplying a Binomial by a Polynomial

Multiply as indicated:

    **a.** $(2z + 1)(z - 2)$                **b.** $(2v - 3)(4v^2 + 6v + 9)$

Solution ▶   **a.** $(2z + 1)(z - 2) = 2z(z - 2) + 1(z - 2)$    distribute to every term in the first binomial

                             $= 2z^2 - 4z + 1z - 2$     eliminate parentheses (distribute again)

                             $= 2z^2 - 3z - 2$        simplify

  **b.** $(2v - 3)(4v^2 + 6v + 9) = 2v(4v^2 + 6v + 9) - 3(4v^2 + 6v + 9)$    distribute

                           $= 8v^3 + 12v^2 + 18v - 12v^2 - 18v - 27$    simplify

                           $= 8v^3 - 27$           combine like terms

**Now try Exercises 87 through 92** ▶

---

### The F-O-I-L Method

By observing the product of two binomials in Example 13(a), we note a pattern that can make the process more efficient. The product of two binomials can quickly be computed using the **F**irst, **O**uter, **I**nner, **L**ast (**FOIL**) method, an acronym giving the respective position of each term in a product of binomials in relation to the other terms. We illustrate here using the product $(2x - 1)(3x + 2)$.

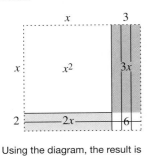
**The F-O-I-L Method for Multiplying Binomials**

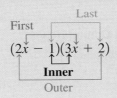

$6x^2 + 4x - \mathbf{3x} - 2$
First   Outer   Inner   Last

Combine like terms
$6x^2 + x - 2$

The first term of the result will always be the product of the first terms from each binomial, and the last term of the result is the product of their last terms. We also note that here, the middle term is found by adding the *outermost product* with the *innermost product* (see Worthy of Note). As you practice with the F-O-I-L process, much of the work can be done mentally and you can often compute the entire product without writing anything down except the answer.

---

**EXAMPLE 14** ▶ Multiplying Binomials Using F-O-I-L

Compute each product mentally:

    **a.** $(5n - 1)(n + 2)$

    **b.** $(2b + 3)(5b - 6)$

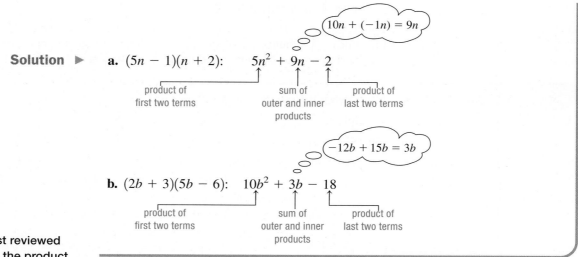

**Solution** ▶

**a.** $(5n - 1)(n + 2)$:     $5n^2 + 9n - 2$

$10n + (-1n) = 9n$

product of first two terms     sum of outer and inner products     product of last two terms

**b.** $(2b + 3)(5b - 6)$:     $10b^2 + 3b - 18$

$-12b + 15b = 3b$

product of first two terms     sum of outer and inner products     product of last two terms

☑ **E.** You've just reviewed how to compute the product of two polynomials

**Now try Exercises 93 through 104** ▶

## F. Special Polynomial Products

Certain polynomial products are considered "special" for two reasons: (1) the product follows a predictable pattern, and (2) the result can be used to simplify expressions, graph functions, solve equations, and/or develop other skills.

### Binomial Conjugates

Expressions like $x + 7$ and $x - 7$ are called **binomial conjugates.** For any given binomial, its conjugate is found by using the same two terms with the opposite sign between them. Example 15 shows that when we multiply a binomial and its conjugate, the "outers" and "inners" sum to zero and the result is a **difference of two squares.**

**EXAMPLE 15** ▶    **Multiplying Binomial Conjugates**

Compute each product mentally:

**a.** $(x + 7)(x - 7)$        **b.** $(2x - 5y)(2x + 5y)$        **c.** $\left(x + \dfrac{2}{5}\right)\left(x - \dfrac{2}{5}\right)$

$-7x + 7x = 0x$

**Solution** ▶    **a.** $(x + 7)(x - 7) = x^2 - 49$      difference of squares $x^2 - 7^2$

$10xy + (-10xy) = 0xy$

**b.** $(2x - 5y)(2x + 5y) = 4x^2 - 25y^2$    difference of squares: $(2x)^2 - (5y)^2$

$-\dfrac{2}{5}x + \dfrac{2}{5}x = 0$

**c.** $\left(x + \dfrac{2}{5}\right)\left(x - \dfrac{2}{5}\right) = x^2 - \dfrac{4}{25}$    difference of squares: $x^2 - \left(\dfrac{2}{5}\right)^2$

**Now try Exercises 105 through 112** ▶

In summary, we have the following.

### The Product of a Binomial and Its Conjugate

Given any expression that can be written in the form $A + B$, the conjugate of the expression is $A - B$ and their product is a difference of two squares:

$$(A + B)(A - B) = A^2 - B^2$$

### Binomial Squares

Expressions like $(x + 7)^2$ are called **binomial squares** and are useful for solving many equations and sketching a number of basic graphs. Note $(x + 7)^2 = (x + 7)(x + 7) = x^2 + 14x + 49$ using the F-O-I-L process. The expression $x^2 + 14x + 49$ is called a **perfect square trinomial** because it is the result of expanding a binomial square. If we write a binomial square in the more general form $(A + B)^2 = (A + B)(A + B)$ and compute the product, we notice a pattern that helps us write the expanded form more quickly.

$$(A + B)^2 = (A + B)(A + B) \quad \text{repeated multiplication}$$
$$= A^2 + AB + AB + B^2 \quad \text{F-O-I-L}$$
$$= A^2 + 2AB + B^2 \quad \text{simplify (perfect square trinomial)}$$

The first and last terms of the trinomial are squares of the terms $A$ and $B$. Also, the middle term of the trinomial is *twice the product of these two terms*: $AB + AB = 2AB$. The F-O-I-L process shows us why. Since the outer and inner products are identical, we always end up with two. A similar result holds for $(A - B)^2$ and the process can be summarized for both cases using the $\pm$ symbol.

### The Square of a Binomial

Given any expression that can be written in the form $(A \pm B)^2$,

1. $(A + B)^2 = A^2 + 2AB + B^2$
2. $(A - B)^2 = A^2 - 2AB + B^2$

**CAUTION** ▶ Note the square of a binomial always results in a trinomial (three terms). In particular, $(A + B)^2 \neq A^2 + B^2$.

**EXAMPLE 16** ▶ Find each binomial square without using F-O-I-L:

**a.** $(a + 9)^2$      **b.** $(3x - 5)^2$      **c.** $(3 + \sqrt{x})^2$

**Solution** ▶ **a.** $(a + 9)^2 = a^2 + 2(a \cdot 9) + 9^2$      $(A + B)^2 = A^2 + 2AB + B^2$
$\phantom{(a + 9)^2} = a^2 + 18a + 81$      simplify

**b.** $(3x - 5)^2 = (3x)^2 - 2(3x \cdot 5) + 5^2$      $(A - B)^2 = A^2 - 2AB + B^2$
$\phantom{(3x - 5)^2} = 9x^2 - 30x + 25$      simplify

**c.** $(3 + \sqrt{x})^2 = 9 + 2(3 \cdot \sqrt{x}) + (\sqrt{x})^2$      $(A + B)^2 = A^2 + 2AB + B^2$
$\phantom{(3 + \sqrt{x})^2} = 9 + 6\sqrt{x} + x$      simplify

☑ **F.** You've just reviewed how to compute special products: binomial conjugates and binomial squares

**Now try Exercises 113 through 122** ▶

With practice, you will be able to go directly from the binomial square to the resulting trinomial.

**R.3 EXERCISES**

▶ **CONCEPTS AND VOCABULARY**

**Fill in each blank with the appropriate word or phrase. Carefully reread the section, if necessary.**

1. The equation $(x^2)^3 = x^6$ is an example of the _____ property of exponents.

2. The expression $5x^2 - 3x - 10$ can be classified as a _____ of degree _____, with a leading coefficient of _____.

3. Discuss/Explain why one of the following expressions can be simplified further, while the other cannot: (a) $-7n^4 + 3n^2$; (b) $-7n^4 \cdot 3n^2$.

4. Discuss/Explain why the degree of $2x^2y^3$ is greater than the degree of $2x^2 + y^3$. Include additional examples for contrast and comparison.

▶ **DEVELOPING YOUR SKILLS**

**Determine each product using the product and/or power properties.**

5. $\dfrac{2}{3}n^2 \cdot 21n^5$

6. $24g^5 \cdot \dfrac{3}{8}g^9$

7. $(-6p^2q)(2p^3q^3)$

8. $(-1.5vy^2)(-8v^4y)$

9. $(a^2)^4 \cdot (a^3)^2 \cdot b^2 \cdot b^5$

10. $d^2 \cdot d^4 \cdot (c^5)^2 \cdot (c^3)^2$

**Simplify using the product to a power property.**

11. $(6pq^2)^3$

12. $(-3p^2q)^2$

13. $(3.2hk^2)^3$

14. $(-2.5h^5k)^2$

15. $\left(\dfrac{p}{2q}\right)^2$

16. $\left(\dfrac{b}{3a}\right)^3$

17. $(-0.7c^4)^2(10c^3d^2)^2$

18. $(-2.5a^3)^2(3a^2b^2)^3$

19. $(\tfrac{3}{4}x^3y)^2$

20. $(\tfrac{4}{5}x^3)^2$

21. **Volume of a cube:** The formula for the volume of a cube is $V = S^3$, where $S$ is the length of one edge. If the length of each edge is $3x^2$,

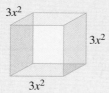

   a. Find a formula for volume in terms of the variable $x$.

   b. Find the volume of the cube if $x = 2$.

22. **Area of a circle:** The formula for the area of a circle is $A = \pi r^2$, where $r$ is the length of the radius. If the radius is given as $5x^3$,

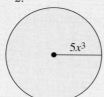

   a. Find a formula for area in terms of the variable $x$.

   b. Find the area of the circle if $x = 2$.

**Simplify using the quotient property or the property of negative exponents. Write answers using positive exponents only.**

23. $\dfrac{-6w^5}{-2w^2}$

24. $\dfrac{8z^7}{16z^5}$

25. $\dfrac{-12a^3b^5}{4a^2b^4}$

26. $\dfrac{5m^3n^5}{10mn^2}$

27. $\left(\dfrac{2}{3}\right)^{-3}$

28. $\left(\dfrac{5}{6}\right)^{-1}$

29. $\dfrac{2}{h^{-3}}$

30. $\dfrac{3}{m^{-2}}$

31. $(-2)^{-3}$

32. $(-4)^{-2}$

33. $\left(\dfrac{-1}{2}\right)^{-3}$

34. $\left(\dfrac{-2}{3}\right)^{-2}$

**Simplify each expression using the quotient to a power property.**

35. $\left(\dfrac{2p^4}{q^3}\right)^2$

36. $\left(\dfrac{-5v^4}{7w^3}\right)^2$

37. $\left(\dfrac{0.2x^2}{0.3y^3}\right)^3$

38. $\left(\dfrac{-0.5a^3}{0.4b^2}\right)^2$

39. $\left(\dfrac{5m^2n^3}{2r^4}\right)^2$

40. $\left(\dfrac{4p^3}{3x^2y}\right)^3$

**Use properties of exponents to simplify the following. Write the answer using positive exponents only.**

41. $\dfrac{9p^6q^4}{-12p^4q^6}$

42. $\dfrac{5m^5n^2}{10m^5n}$

43. $\dfrac{20h^{-2}}{12h^5}$

44. $\dfrac{5k^3}{20k^{-2}}$

45. $\dfrac{(a^2)^3}{a^4 \cdot a^5}$

46. $\dfrac{(5^3)^4}{5^9}$

47. $\left(\dfrac{a^{-3} \cdot b}{c^{-2}}\right)^{-4}$

48. $\dfrac{(p^{-4}q^8)^2}{p^5q^{-2}}$

49. $\dfrac{-6(2x^{-3})^2}{10x^{-2}}$

50. $\dfrac{18n^{-3}}{-8(3n^{-2})^3}$

51. $4^0 + 5^0$

52. $(-3)^0 + (-7)^0$

53. $2^{-1} + 5^{-1}$

54. $4^{-1} + 8^{-1}$

55. $3^0 + 3^{-1} + 3^{-2}$

56. $2^{-2} + 2^{-1} + 2^0$

57. $-5x^0 + (-5x)^0$

58. $-2n^0 + (-2n)^0$

**Convert the following numbers to scientific notation.**

**59.** In mid-2011, the U.S. Census Bureau estimated the world population at nearly 6,970,000,000 people.

**60.** The mass of a proton is generally given as 0.000 000 000 000 000 000 000 000 001 670 kg.

**Convert the following numbers to decimal notation.**

**61.** The smallest microprocessors in common use measure $6.5 \times 10^{-9}$ m across.

**62.** In 2011, the estimated net worth of Bill Gates, the founder of Microsoft, was $5.4 \times 10^{10}$ dollars.

**Compute using scientific notation. Show all work.**

**63.** The Metropolitan Water District of Chicago supplies roughly 1,000,000,000 gal of freshwater per day to a population of nearly 5,400,000 people. To the nearest whole number, what is the average number of gallons used per person, per day?

**64.** In fiscal terms, a nation's debt-per-capita is the ratio of its total debt to its total population. In the year 2011, the total U.S. debt was estimated at \$14,300,000,000,000, while the population was estimated at 311,000,000. What was the U.S. debt-per-capita ratio for 2011? Round to the nearest whole dollar.

**Identify each expression as a polynomial or nonpolynomial (if a nonpolynomial, state why); classify each as a monomial, binomial, trinomial, or none of these; and state the degree of the polynomial.**

**65.** $-35w^3 + 2w^2 + (-12w) + 14$

**66.** $-2x^3 + \frac{2}{3}x^2 - 12x + 1.2$

**67.** $5n^{-2} + 4n + \sqrt{17}$      **68.** $\dfrac{4}{r^3} + 2.7r^2 + r + 1$

**69.** $p^3 - \frac{2}{5}$          **70.** $q^3 + 2q^{-2} - 5q$

**Write each polynomial in standard form and name the leading coefficient.**

**71.** $7w + 8.2 - w^3 - 3w^2$

**72.** $-2k^2 - 12 - k$

**73.** $c^3 + 6 + 2c^2 - 3c$

**74.** $-3v^3 + 14 + 2v^2 + (-12v)$

**75.** $12 - \frac{2}{3}x^2$

**76.** $8 + 2n^2 + 7n$

**Find the indicated sum or difference.**

**77.** $(3p^3 - 4p^2 + 2p - 7) + (p^2 - 2p - 5)$

**78.** $(5q^2 - 3q + 4) + (-3q^2 + 3q - 4)$

**79.** $(5.75b^2 + 2.6b - 1.9) + (2.1b^2 - 3.2b)$

**80.** $(0.4n^2 + 5n - 0.5) + (0.3n^2 - 2n + 0.75)$

**81.** $(\frac{3}{4}x^2 - 5x + 2) - (\frac{1}{2}x^2 + 3x - 4)$

**82.** $(\frac{5}{9}n^2 + 4n - \frac{1}{2}) - (\frac{2}{3}n^2 - 2n + \frac{3}{4})$

**83.** Subtract $q^5 + 2q^4 + q^2 + 2q$ from $q^6 + 2q^5 + q^4 + 2q^3$ using a vertical format.

**84.** Find $x^4 + 2x^3 + x^2 + 2x$ decreased by $x^4 - 3x^3 + 4x^2 - 3x$ using a vertical format.

**Compute each product.**

**85.** $-3x(x^2 - x - 6)$        **86.** $-2v^2(v^2 + 2v - 15)$

**87.** $(3r - 5)(r - 2)$          **88.** $(s - 3)(5s + 4)$

**89.** $(x - 3)(x^2 + 3x + 9)$    **90.** $(z + 5)(z^2 - 5z + 25)$

**91.** $(b^2 - 3b - 28)(b + 2)$   **92.** $(2h^2 - 3h + 8)(h - 1)$

**93.** $(7v - 4)(3v - 5)$         **94.** $(6w - 1)(2w + 5)$

**95.** $(3 - m)(3 + m)$           **96.** $(5 + n)(5 - n)$

**97.** $(x + \frac{1}{2})(x + \frac{1}{4})$        **98.** $(z + \frac{1}{3})(z + \frac{5}{6})$

**99.** $(m + \frac{3}{4})(m - \frac{3}{4})$        **100.** $(n - \frac{2}{5})(n + \frac{2}{5})$

**101.** $(3x - 2y)(2x + 5y)$      **102.** $(6a + b)(a + 3b)$

**103.** $(4c + d)(3c + 5d)$       **104.** $(5x + 3y)(2x - 3y)$

**For each binomial, determine its conjugate and find the product of the binomial with its conjugate.**

**105.** $4m - 3$                  **106.** $6n + 5$

**107.** $7x - 10$                 **108.** $c + 3$

**109.** $6 + 5k$                  **110.** $11 - 3r$

**111.** $x + \sqrt{6}$            **112.** $p - \sqrt{2}$

**Find each binomial square.**

**113.** $(x + 4)^2$               **114.** $(a - 3)^2$

**115.** $(4g + 3)^2$              **116.** $(5x - 3)^2$

**117.** $(4 - \sqrt{x})^2$        **118.** $(\sqrt{x} + 7)^2$

**Compute each product.**

**119.** $(x - 3)(y + 2)$

**120.** $(a + 3)(b - 5)$

**121.** $(k - 5)(k + 6)(k + 2)$

**122.** $(a + 6)(a - 1)(a + 5)$

## ▶ WORKING WITH FORMULAS

**123.** **Medication in the bloodstream:** $M = 0.5t^4 + 3t^3 - 97t^2 + 348t$

If 400 mg of a pain medication are taken orally, the number of milligrams in the bloodstream is modeled by the formula shown, where $M$ is the number of milligrams and $t$ is the time in hours, $0 \leq t \leq 5$. Construct a table of values for $t = 1$ through 5, then answer the following.

**a.** How many milligrams are in the bloodstream after 2 hr? After 3 hr?

**b.** Based on part (a), would you expect the number of milligrams in the bloodstream after 4 hr to be less or more? Why?

**c.** Approximately how many hours until the medication wears off (the number of milligrams in the bloodstream is 0)?

**124. Amount of a mortgage payment:** $M = \dfrac{A\left(\dfrac{r}{12}\right)\left(1 + \dfrac{r}{12}\right)^n}{\left(1 + \dfrac{r}{12}\right)^n - 1}$

The monthly mortgage payment required to pay off (or amortize) a loan is given by the formula shown, where $M$ is the monthly payment, $A$ is the original amount of the loan, $r$ is the annual interest rate, and $n$ is the term of the loan in months. Find the monthly payment (to the nearest cent) required to purchase a $198,000 home, if the interest rate is 6.5% and the home is financed over 30 yr.

## ▶ APPLICATIONS

**125. Attraction between particles:** In electrical theory, the force of attraction between two point charges $P$ and $Q$ with opposite charges is modeled by $F = \dfrac{kPQ}{d^2}$, where $d$ is the distance between them and $k$ is a constant that depends on certain conditions. This is known as Coulomb's law. Rewrite the formula using a negative exponent.

**126. Intensity of light:** The intensity of illumination from a light source depends on the distance from the source according to $I = \dfrac{k}{d^2}$, where $I$ is the intensity measured in footcandles, $d$ is the distance from the source in feet, and $k$ is a constant that depends on the conditions. Rewrite the formula using a negative exponent.

**127. Rewriting an expression:** In advanced mathematics, negative exponents are widely used because they are easier to work with than rational expressions. Rewrite the expression $\dfrac{5}{x^3} + \dfrac{3}{x^2} + \dfrac{2}{x^1} + 4$ using negative exponents.

**128. Swimming pool hours:** A swimming pool opens at 8 A.M. and closes at 6 P.M. In summertime, the number

of people in the pool at any time can be approximated by the formula $S = -h^2 + 10h$, where $S$ is the number of swimmers and $h$ is the number of hours the pool has been open.

   **a.** How many swimmers are in the pool at 6 P.M.? Why?

   **b.** During which 2 hr would you expect the largest number of swimmers?

   **c.** Approximately how many swimmers are in the pool at 3 P.M.?

   **d.** Create a table of values for $t = 1, 2, 3, 4, \ldots$ and check your answer to part (b).

**129. Maximizing revenue:** A sporting goods store finds that if they price their video games at $20, they make 200 sales per day. For each decrease of $1, 20 additional video games are sold. This means the store's revenue can be modeled by the formula $R = (20 - 1x)(200 + 20x)$, where $x$ is the number of $1 decreases. Multiply out the binomials and use a table of values to determine what price will give the most revenue.

**130. Maximizing revenue:** Due to past experience, a jeweler knows that if they price jade rings at $60, they will sell 120 each day. For each decrease of $2, five additional sales will be made. This means the jeweler's revenue can be modeled by the formula $R = (60 - 2x)(120 + 5x)$, where $x$ is the number of $2 decreases. Multiply out the binomials and use a table of values to determine what price will give the most revenue.

## ▶ EXTENDING THE CONCEPT

**131.** If $(3x^2 + kx + 1) - (kx^2 + 5x - 7) + (2x^2 - 4x - k) = -x^2 - 3x + 2$, what is the value of $k$?

**132.** If $\left(2x + \dfrac{1}{2x}\right)^2 = 5$, then the expression $4x^2 + \dfrac{1}{4x^2}$ is equal to what number?

The square roots and cube roots seen in Section R.1 come from a much larger family called **radical expressions.** Expressions containing radicals can be found in virtually every field of mathematical study, and are an invaluable tool for modeling many real-world phenomena.

## A. Simplifying Radical Expressions of the Form $\sqrt[n]{a^n}$

In Section R.1, we noted that $\sqrt{a} = b$ only if $b^2 = a$. This definition results in a nonreal value for expressions like $\sqrt{-16}$, since there is no real number $b$ such that $b^2 = -16$. In other words, the expression $\sqrt{a}$ represents a real number only if $a \geq 0$. Of particular interest to us now is an inverse operation for $a^2$. What operation can be applied to $a^2$ to return $a$? Consider the following.

---

**EXAMPLE 1** ▶ **Evaluating a Radical Expression**

Evaluate $\sqrt{a^2}$ for the values given:

  **a.** $a = 3$      **b.** $a = 5$      **c.** $a = -6$

**Solution** ▶   **a.** $\sqrt{3^2} = \sqrt{9}$      **b.** $\sqrt{5^2} = \sqrt{25}$      **c.** $\sqrt{(-6)^2} = \sqrt{36}$

                           $= 3$                      $= 5$                        $= 6$

                                          **Now try Exercises 5 and 6** ▶

---

The pattern seemed to indicate that $\sqrt{a^2} = a$ and that our search for an inverse operation was complete—until Example 1(c), where we found that $\sqrt{(-6)^2} \neq -6$. Using the absolute value concept, we can "repair" this apparent discrepancy and state a general rule for simplifying these expressions: $\sqrt{a^2} = |a|$. For expressions like $\sqrt{49x^2}$ and $\sqrt{y^6}$, the radicands can be rewritten as perfect squares and simplified in the same manner: $\sqrt{49x^2} = \sqrt{(7x)^2} = 7|x|$ and $\sqrt{y^6} = \sqrt{(y^3)^2} = |y^3|$.

---

**The Square Root of $a^2$: $\sqrt{a^2}$**

For any real number $a$,
$$\sqrt{a^2} = |a|.$$

---

**EXAMPLE 2** ▶ **Simplifying Square Root Expressions**

Simplify each expression.

  **a.** $\sqrt{169x^2}$      **b.** $\sqrt{x^2 - 10x + 25}$

**Solution** ▶   **a.** $\sqrt{169x^2} = |13x|$  since $x$ could    **b.** $\sqrt{x^2 - 10x + 25} = \sqrt{(x-5)^2}$  since $x - 5$

                        $= 13|x|$  be negative                              $= |x - 5|$  could be

                                                                     negative

                                          **Now try Exercises 7 and 8** ▶

---

**CAUTION** ▶ In Section R.3, we noted that $(A + B)^2 \neq A^2 + B^2$, indicating that you cannot square the individual terms in a sum (the square of a binomial results in a perfect square trinomial). In a similar way, $\sqrt{A^2 + B^2} \neq A + B$, and you cannot take the square root of individual terms. There is a big difference between the expressions $\sqrt{A^2 + B^2}$ and $\sqrt{(A + B)^2} = |A + B|$. Try evaluating each when $A = 3$ and $B = 4$.

To investigate expressions like $\sqrt[3]{x^3}$, note the radicand in both $\sqrt[3]{8}$ and $\sqrt[3]{-64}$ can be written as a perfect cube. From our earlier definition of cube roots we know $\sqrt[3]{8} = \sqrt[3]{(2)^3} = 2$, $\sqrt[3]{-64} = \sqrt[3]{(-4)^3} = -4$, and that every real number has only one real cube root. For this reason, absolute value notation is not used or needed when taking cube roots.

> ### The Cube Root of $a^3$: $\sqrt[3]{a^3}$
>
> For any real number $a$,
> $$\sqrt[3]{a^3} = a.$$

**EXAMPLE 3** ▶ **Simplifying Cube Root Expressions**

Simplify each expression.

a. $\sqrt[3]{-27x^3}$      b. $\sqrt[3]{-64n^6}$

**Solution** ▶ a. $\sqrt[3]{-27x^3} = \sqrt[3]{(-3x)^3}$      b. $\sqrt[3]{-64n^6} = \sqrt[3]{(-4n^2)^3}$
$$= -3x \qquad\qquad\qquad\qquad = -4n^2$$

**Now try Exercises 9 and 10** ▶

We can extend these ideas to fourth roots, fifth roots, and so on. For example, the fifth root of $a$ is $b$ only if $b^5 = a$. In symbols, $\sqrt[5]{a} = b$ implies $b^5 = a$. Since an odd number of negative factors is always negative: $(-2)^5 = -32$, and an even number of negative factors is always positive: $(-2)^4 = 16$, we must take the index into account when evaluating expressions like $\sqrt[n]{a^n}$. If $n$ is even and the radicand is unknown, absolute value notation must be used.

> **WORTHY OF NOTE**
>
> Just as $\sqrt[2]{-16}$ is not a real number, $\sqrt[4]{-16}$ and $\sqrt[6]{-16}$ do not represent real numbers. An even number of repeated factors is always nonnegative!

> ### The $n$th Root of $a^n$: $\sqrt[n]{a^n}$
>
> For any real number $a$,
> 1. $\sqrt[n]{a^n} = |a|$ when $n$ is even.      2. $\sqrt[n]{a^n} = a$ when $n$ is odd.

**EXAMPLE 4** ▶ **Simplifying Radical Expressions**

Simplify each expression.

a. $\sqrt[4]{81}$      b. $\sqrt[4]{-81}$      c. $\sqrt[5]{32}$      d. $\sqrt[5]{-32}$
e. $\sqrt[4]{16m^4}$      f. $\sqrt[5]{32p^5}$      g. $\sqrt[6]{(m+5)^6}$      h. $\sqrt[7]{(x-2)^7}$

**Solution** ▶ a. $\sqrt[4]{81} = 3$      b. $\sqrt[4]{-81}$ is not a real number
c. $\sqrt[5]{32} = 2$      d. $\sqrt[5]{-32} = -2$
e. $\sqrt[4]{16m^4} = \sqrt[4]{(2m)^4}$      f. $\sqrt[5]{32p^5} = \sqrt[5]{(2p)^5}$
$$= |2m| \text{ or } 2|m| \qquad\qquad\qquad = 2p$$
g. $\sqrt[6]{(m+5)^6} = |m+5|$      h. $\sqrt[7]{(x-2)^7} = x-2$

☑ **A.** You've just reviewed how to simplify radical expressions of the form $\sqrt[n]{a^n}$

**Now try Exercises 11 and 12** ▶

## B. Radical Expressions and Rational Exponents

As an alternative to radical notation, a rational (fractional) exponent can be used, along with the power property of exponents. For $\sqrt[3]{a^3} = a$, notice that an exponent of one-third can replace the cube root notation and produce the same result: $\sqrt[3]{a^3} = (a^3)^{\frac{1}{3}} = a^{\frac{3}{3}} = a$. In the same way, an exponent of one-half can replace the square root notation: $\sqrt{a^2} = (a^2)^{\frac{1}{2}} = a^{\frac{2}{2}} = |a|$. In general, we have the following:

**Rational Exponents**

If $\sqrt[n]{a}$ represents a real number and $n \geq 2$ is an integer,

then $\sqrt[n]{a} = \sqrt[n]{a^1} = a^{\frac{1}{n}}$

**EXAMPLE 5 ▶ Simplifying Radical Expressions Using Rational Exponents**

Simplify by rewriting each radicand as a perfect $n$th power and converting to rational exponent notation.

**a.** $\sqrt[3]{-125}$ **b.** $-\sqrt[4]{16x^{20}}$ **c.** $\sqrt[4]{-81}$ **d.** $\sqrt[3]{\dfrac{8w^3}{27}}$

**Solution ▶**

**a.** $\sqrt[3]{-125} = \sqrt[3]{(-5)^3} = [(-5)^3]^{\frac{1}{3}} = (-5)^{\frac{3}{3}} = -5$

**b.** $-\sqrt[4]{16x^{20}} = -\sqrt[4]{(2x^5)^4} = -[(2x^5)^4]^{\frac{1}{4}} = -|2x^5| = -2|x|^5$

**c.** $\sqrt[4]{-81} = (-81)^{\frac{1}{4}}$ is not a real number

**d.** $\sqrt[3]{\dfrac{8w^3}{27}} = \sqrt[3]{\left(\dfrac{2w}{3}\right)^3} = \left[\left(\dfrac{2w}{3}\right)^3\right]^{\frac{1}{3}} = \dfrac{2w}{3}$

Now try Exercises 13 and 14 ▶

**Figure R.5**

When a rational exponent is used, as in $\sqrt[n]{a} = \sqrt[n]{a^1} = a^{\frac{1}{n}}$, the denominator of the exponent represents the index number, while the numerator of the exponent represents the original power on $a$. *This is true even when the exponent on a is something other than one!* In other words, the radical expression $\sqrt[4]{16^3}$ can be rewritten as $(16^3)^{\frac{1}{4}} = 16^{\frac{3}{1}\cdot\frac{1}{4}}$ or $16^{\frac{3}{4}}$. This is further illustrated in Figure R.5 where we see the rational exponent has the form, "power over root." To evaluate this expression without the aid of a calculator, we use the commutative property to rewrite $16^{\frac{3}{1}\cdot\frac{1}{4}}$ as $16^{\frac{1}{4}\cdot\frac{3}{1}}$ and begin with the fourth root of 16: $(16^{\frac{1}{4}})^{\frac{3}{1}} = 2^3 = 8$.

**WORTHY OF NOTE**

Any rational number can be decomposed into the product of a unit fraction and an integer: $\frac{m}{n} = \frac{1}{n} \cdot m$.

In general, if $m$ and $n$ have no common factors (other than 1) the expression $a^{\frac{m}{n}}$ can be interpreted in the following two ways.

**Rational Exponents**

If $\frac{m}{n}$ is a rational number expressed in lowest terms with $n \geq 2$, then

(1) $a^{\frac{m}{n}} = (\sqrt[n]{a})^m$ or (2) $a^{\frac{m}{n}} = \sqrt[n]{a^m}$

(compute $\sqrt[n]{a}$, then take the $m$th power), (compute $a^m$, then take the $n$th root),

provided $\sqrt[n]{a}$ represents a real number.

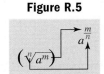

Expressions with rational exponents are generally easier to evaluate if we compute the root first, then apply the exponent. Computing the root first also helps us determine whether or not an expression represents a real number.

---

**EXAMPLE 6** ▶ **Simplifying Expressions with Rational Exponents**

Simplify each expression, if possible.

**a.** $-49^{\frac{3}{2}}$     **b.** $(-49)^{\frac{3}{2}}$     **c.** $(-8)^{\frac{2}{3}}$     **d.** $-8^{-\frac{2}{3}}$

**Solution** ▶

**a.** $\begin{aligned} -49^{\frac{3}{2}} &= -(49^{\frac{1}{2}})^3 \\ &= -(\sqrt{49})^3 \\ &= -(7)^3 \text{ or } -343 \end{aligned}$

**b.** $\begin{aligned} (-49)^{\frac{3}{2}} &= [(-49)^{\frac{1}{2}}]^3, \\ &= (\sqrt{-49})^3 \\ &\text{not a real number} \end{aligned}$

**c.** $\begin{aligned} (-8)^{\frac{2}{3}} &= [(-8)^{\frac{1}{3}}]^2 \\ &= (\sqrt[3]{-8})^2 \\ &= (-2)^2 \text{ or } 4 \end{aligned}$

**d.** $\begin{aligned} -8^{-\frac{2}{3}} &= -(8^{\frac{1}{3}})^{-2} \\ &= -(\sqrt[3]{8})^{-2} \\ &= -2^{-2} \text{ or } -\frac{1}{4} \end{aligned}$

Now try Exercises 15 through 20 ▶

> **WORTHY OF NOTE**
>
> While the expression $(-8)^{\frac{1}{3}} = \sqrt[3]{-8}$ represents the real number $-2$, the expression $(-8)^{\frac{2}{6}} = (\sqrt[6]{-8})^2$ is not a real number, even though $\frac{1}{3} = \frac{2}{6}$. Note that the second exponent is not in lowest terms.

 **B.** You've just reviewed how to rewrite and simplify radical expressions using rational exponents

## C. Using Properties of Radicals to Simplify Radical Expressions

The properties used to simplify radical expressions are closely connected to the properties of exponents. For instance, the product to a power property holds even when $n$ is a rational number. This means $(xy)^{\frac{1}{2}} = x^{\frac{1}{2}}y^{\frac{1}{2}}$ and $(4 \cdot 25)^{\frac{1}{2}} = 4^{\frac{1}{2}} \cdot 25^{\frac{1}{2}}$. When the second statement is expressed in radical form, we have $\sqrt{4 \cdot 25} = \sqrt{4} \cdot \sqrt{25}$, with both forms having a value of 10. This suggests the **product property of radicals,** which can be extended to include cube roots, fourth roots, and so on.

> **Product Property of Radicals**
>
> If $\sqrt[n]{A}$ and $\sqrt[n]{B}$ represent real-valued expressions, then
> $$\sqrt[n]{AB} = \sqrt[n]{A} \cdot \sqrt[n]{B} \quad \text{and} \quad \sqrt[n]{A} \cdot \sqrt[n]{B} = \sqrt[n]{AB}.$$

---

**CAUTION** ▶ Note that this property applies only to a *product* of two terms, not to a sum or difference. In other words, while $\sqrt{9x^2} = |3x|$, $\sqrt{9 + x^2} \neq |3 + x|$!

---

One application of the product property is to simplify radical expressions. In general, the expression $\sqrt[n]{a}$ is in simplified form if $a$ has no factors (other than 1) that are perfect $n$th roots.

---

**EXAMPLE 7** ▶ **Simplifying Radical Expressions**

Write each expression in simplest form using the product property.

**a.** $\sqrt{18}$     **b.** $5\sqrt[3]{125x^4}$     **c.** $\dfrac{-4 + \sqrt{20}}{2}$     **d.** $1.2\sqrt[3]{16n^4}\ \sqrt[3]{4n^5}$

**Solution** ▶

**a.** $\begin{aligned} \sqrt{18} &= \sqrt{9 \cdot 2} \\ &= \sqrt{9}\sqrt{2} \\ &= 3\sqrt{2} \end{aligned}$

**b.** $5\sqrt[3]{125x^4} = 5 \cdot \sqrt[3]{125 \cdot x^4}$

*These steps can be done mentally* $\begin{cases} = 5 \cdot \sqrt[3]{125} \cdot \sqrt[3]{x^3} \cdot \sqrt[3]{x^1} \\ = 5 \cdot 5 \cdot x \cdot \sqrt[3]{x} \\ = 25x\sqrt[3]{x} \end{cases}$

**c.** $\dfrac{-4 + \sqrt{20}}{2} = \dfrac{-4 + \sqrt{4 \cdot 5}}{2}$

$\qquad\qquad = \dfrac{-4 + 2\sqrt{5}}{2}$

$\qquad\qquad = \dfrac{\overset{1}{2}(-2 + \sqrt{5})}{\underset{1}{2}}$

$\qquad\qquad = -2 + \sqrt{5}$

**d.** $1.2\sqrt[3]{16n^4}\sqrt[3]{4n^5} = 1.2\sqrt[3]{16n^4 \cdot 4n^5}$

$\qquad\qquad = 1.2\sqrt[3]{64 \cdot n^9}$

$\qquad\qquad = 1.2\sqrt[3]{64}\sqrt[3]{n^9}$

$\qquad\qquad = 1.2\sqrt[3]{64}\sqrt[3]{(n^3)^3}$

$\qquad\qquad = 1.2(4)n^3$

$\qquad\qquad = 4.8n^3$

**Now try Exercises 21 through 24** ▶

When radicals are *combined* using the product property, the result may contain a perfect *n*th root, which should be simplified. Note that the *index numbers must be the same* in order to use this property.

The **quotient property of radicals** can also be established using exponential properties. The fact that $\dfrac{\sqrt{100}}{\sqrt{25}} = \sqrt{\dfrac{100}{25}} = 2$ suggests the following:

**Quotient Property of Radicals**

If $\sqrt[n]{A}$ and $\sqrt[n]{B}$ represent real-valued expressions with $B \neq 0$, then

$$\sqrt[n]{\dfrac{A}{B}} = \dfrac{\sqrt[n]{A}}{\sqrt[n]{B}} \quad \text{and} \quad \dfrac{\sqrt[n]{A}}{\sqrt[n]{B}} = \sqrt[n]{\dfrac{A}{B}}.$$

Many times the product and quotient properties must work together to simplify a radical expression, as shown in Example 8A.

**EXAMPLE 8A** ▶ **Simplifying Radical Expressions**

Simplify each expression:

**a.** $\dfrac{\sqrt{18a^5}}{\sqrt{2a}}$ 　　　　　　　　**b.** $\sqrt[3]{\dfrac{81}{125x^3}}$

**Solution** ▶ **a.** $\dfrac{\sqrt{18a^5}}{\sqrt{2a}} = \sqrt{\dfrac{18a^5}{2a}}$ 　　　**b.** $\sqrt[3]{\dfrac{81}{125x^3}} = \dfrac{\sqrt[3]{81}}{\sqrt[3]{125x^3}}$

$\qquad\qquad\quad = \sqrt{9a^4}$ 　　　　　　　　　　$= \dfrac{\sqrt[3]{27 \cdot 3}}{5x}$

$\qquad\qquad\quad = 3a^2$ 　　　　　　　　　　　　$= \dfrac{3\sqrt[3]{3}}{5x}$

Radical expressions can also be simplified using rational exponents.

**EXAMPLE 8B** ▶ **Using Rational Exponents to Simplify Radical Expressions**

Simplify using rational exponents:

**a.** $\sqrt{36p^4q^5}$ 　　　**b.** $v\sqrt[3]{v^4}$ 　　　**c.** $\sqrt[3]{m}\sqrt{m}$

**Solution** ▶ **a.** $\sqrt{36p^4q^5} = (36p^4q^5)^{\frac{1}{2}}$ 　　　**b.** $v\sqrt[3]{v^4} = v^1 \cdot v^{\frac{4}{3}}$ 　　　**c.** $\sqrt[3]{m}\sqrt{m} = m^{\frac{1}{3}}m^{\frac{1}{2}}$

$\qquad\qquad\qquad = 36^{\frac{1}{2}}p^{\frac{4}{2}}q^{\frac{5}{2}}$ 　　　　　　　$= v^{\frac{3}{3}} \cdot v^{\frac{4}{3}}$ 　　　　　　$= m^{\frac{1}{3}+\frac{1}{2}}$

$\qquad\qquad\qquad = 6p^2q^{\frac{4}{2}}q^{\frac{1}{2}}$ 　　　　　　　$= v^{\frac{7}{3}}$ 　　　　　　　　$= m^{\frac{5}{6}}$

$\qquad\qquad\qquad = 6p^2q^2q^{\frac{1}{2}}$ 　　　　　　　$= v^{\frac{6}{3}}v^{\frac{1}{3}}$ 　　　　　　$= \sqrt[6]{m^5}$

$\qquad\qquad\qquad = 6p^2q^2\sqrt{q}$ 　　　　　　　$= v^2\sqrt[3]{v}$

☑ **C.** You've just reviewed how to use properties of radicals to simplify radical expressions

**Now try Exercises 25 through 28** ▶

## D. Addition and Subtraction of Radical Expressions

Since $3x$ and $5x$ are like terms, we know $3x + 5x = 8x$. If $x = \sqrt[3]{7}$, the sum becomes $3\sqrt[3]{7} + 5\sqrt[3]{7} = 8\sqrt[3]{7}$, illustrating how *like* radical expressions can be combined. Like radicals are those that have *the same index and radicand.* In some cases, we can identify like radicals only after radical terms have been simplified.

**EXAMPLE 9** ▶ **Adding and Subtracting Radical Expressions**

Simplify and combine (if possible).

   **a.** $\sqrt{45} + 2\sqrt{20}$          **b.** $\sqrt[3]{16x^5} - x\sqrt[3]{54x^2}$

**Solution** ▶
$$
\begin{aligned}
\textbf{a. } \sqrt{45} + 2\sqrt{20} &= 3\sqrt{5} + 2(2\sqrt{5}) && \text{simplify radicals: } \sqrt{45} = \sqrt{9\cdot 5};\ \sqrt{20} = \sqrt{4\cdot 5}\\
&= 3\sqrt{5} + 4\sqrt{5} && \text{like radicals}\\
&= 7\sqrt{5} && \text{result}
\end{aligned}
$$

$$
\begin{aligned}
\textbf{b. } \sqrt[3]{16x^5} - x\sqrt[3]{54x^2} &= \sqrt[3]{8\cdot 2\cdot x^3\cdot x^2} - x\sqrt[3]{27\cdot 2\cdot x^2}\\
&= 2x\sqrt[3]{2x^2} - 3x\sqrt[3]{2x^2} && \text{simplify radicals}\\
&= -x\sqrt[3]{2x^2} && \text{result}
\end{aligned}
$$

☑ **D.** You've just reviewed how to add and subtract radical expressions

**Now try Exercises 29 through 32** ▶

## E. Multiplication and Division of Radical Expressions; Radical Expressions in Simplest Form

Multiplying radical expressions is simply an extension of our earlier work. The multiplication can take various forms, from the distributive property to any of the special products reviewed in Section R.3. For instance, $(A \pm B)^2 = A^2 \pm 2AB + B^2$, even if $A$ or $B$ is a radical term.

**EXAMPLE 10** ▶ **Multiplying Radical Expressions**

Compute each product and simplify.

   **a.** $5\sqrt{3}(\sqrt{6} - 4\sqrt{3})$          **b.** $(x + \sqrt{7})(x - \sqrt{7})$          **c.** $(3 - \sqrt{2})^2$

**Solution** ▶
$$
\begin{aligned}
\textbf{a. } 5\sqrt{3}(\sqrt{6} - 4\sqrt{3}) &= 5\sqrt{18} - 20(\sqrt{3})^2 && \text{distribute}\\
&= 5(3\sqrt{2}) - 20(3) && \text{simplify: } \sqrt{18} = 3\sqrt{2},\ (\sqrt{3})^2 = 3\\
&= 15\sqrt{2} - 60 && \text{result}
\end{aligned}
$$

**LOOKING AHEAD**

Notice that the answer for Example 10(b) contains no radical terms, since the outer and inner products sum to zero. This result will be used to simplify certain radical expressions in this section and later in Chapter 1.

$$
\begin{aligned}
\textbf{b. } (x + \sqrt{7})(x - \sqrt{7}) &= x^2 - (\sqrt{7})^2 && (A + B)(A - B) = A^2 - B^2\\
&= x^2 - 7 && \text{result}
\end{aligned}
$$

$$
\begin{aligned}
\textbf{c. } (3 - \sqrt{2})^2 &= (3)^2 - 2(3)(\sqrt{2}) + (\sqrt{2})^2 && (A - B)^2 = A^2 - 2AB + B^2\\
&= 9 - 6\sqrt{2} + 2 && \text{simplify each term}\\
&= 11 - 6\sqrt{2} && \text{result}
\end{aligned}
$$

**Now try Exercises 33 through 36** ▶

When we applied the quotient property in Example 8A, we obtained a denominator free of radicals. Sometimes the denominator is not automatically free of radicals, and the need to write radical expressions in *simplest form* comes into play. This process is called **rationalizing the denominator.**

### Radical Expressions in Simplest Form

A radical expression is in simplest form if:
1. The radicand has no perfect $n$th-root factors.
2. The radicand contains no fractions.
3. No radicals occur in a denominator.

As with other types of simplification, the desired form can be achieved in various ways. If the denominator is a single radical term, we multiply the numerator and denominator by the factors required to eliminate the radical in the denominator [see Examples 11(a) and 11(b)].

**EXAMPLE 11** ▶ **Simplifying Radical Expressions**

Simplify by rationalizing the denominator. Assume $a, x \neq 0$.

$$\textbf{a. } \frac{2}{5\sqrt{3}} \qquad \textbf{b. } \frac{-7}{\sqrt[3]{x}}$$

**Solution** ▶ 

$$\textbf{a. } \frac{2}{5\sqrt{3}} = \frac{2}{5\sqrt{3}} \cdot \frac{\sqrt{3}}{\sqrt{3}} \qquad \text{multiply numerator and denominator by } \sqrt{3}$$

$$= \frac{2\sqrt{3}}{5(\sqrt{3})^2} = \frac{2\sqrt{3}}{15} \qquad \text{simplify—denominator is now rational}$$

$$\textbf{b. } \frac{-7}{\sqrt[3]{x}} = \frac{-7(\sqrt[3]{x})(\sqrt[3]{x})}{\sqrt[3]{x}(\sqrt[3]{x})(\sqrt[3]{x})} \qquad \text{multiply using two additional factors of } \sqrt[3]{x}$$

$$= \frac{-7\sqrt[3]{x^2}}{\sqrt[3]{x^3}} \qquad \text{product property}$$

$$= \frac{-7\sqrt[3]{x^2}}{x} \qquad \sqrt[3]{x^3} = x$$

**Now try Exercises 37 through 42** ▶

In some applications, the denominator may be a sum or difference containing a radical term. In this case, the methods from Example 11 are ineffective, and instead we multiply by a conjugate since $(A + B)(A - B) = A^2 - B^2$. If either $A$ or $B$ is a square root, the result will be a denominator free of radicals.

**EXAMPLE 12** ▶ **Simplifying Radical Expressions Using a Conjugate**

Simplify the expression by rationalizing the denominator. Write the answer in exact form and approximate form rounded to three decimal places.  $\dfrac{2}{\sqrt{5} - 1}$

**Solution** ▶ 

$$\frac{2}{\sqrt{5} - 1} = \frac{2}{\sqrt{5} - 1} \cdot \frac{\sqrt{5} + 1}{\sqrt{5} + 1} \qquad \text{multiply by the conjugate of the denominator}$$

$$= \frac{2(\sqrt{5} + 1)}{(\sqrt{5})^2 - 1^2} \qquad \text{FOIL}$$
$$\qquad\qquad\qquad\qquad \text{difference of squares}$$

$$= \frac{2(\sqrt{5} + 1)}{4} \qquad \text{simplify}$$

$$= \frac{\sqrt{5} + 1}{2} \qquad \text{exact form}$$

$$\approx 1.618 \qquad \text{approximate form}$$

**Now try Exercises 43 through 46** ▶

☑ **E.** You've just reviewed how to multiply and divide radical expressions and write a radical expression in simplest form

## F. The Pythagorean Theorem

A right triangle is one that has a 90° angle (see Figure R.6). The longest side (opposite the right angle) is called the **hypotenuse,** while the other two sides are simply called "legs." The **Pythagorean theorem** is a formula that says if you add the square of each leg, the result will be equal to the square of the hypotenuse. Furthermore, we note the converse of this theorem is also true.

**Figure R.6**

Hypotenuse        Leg
        90°
        Leg

---

### Pythagorean Theorem

1. For any right triangle with legs $a$ and $b$ and hypotenuse $c$,

$$a^2 + b^2 = c^2$$

2. For any triangle with sides $a$, $b$, and longest side $c$, if $a^2 + b^2 = c^2$,
   the triangle is a right triangle.

---

A geometric interpretation of the theorem is given in Figure R.7, which shows $3^2 + 4^2 = 5^2$.

**Figure R.7**

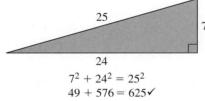

$5^2 + 12^2 = 13^2$
$25 + 144 = 169 ✓$

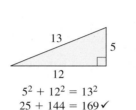

$7^2 + 24^2 = 25^2$
$49 + 576 = 625 ✓$

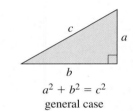

$a^2 + b^2 = c^2$
general case

---

**EXAMPLE 13** ▶  **Applying the Pythagorean Theorem**

An extension ladder is placed 9 ft from the base of a building in an effort to reach a third-story window that is 27 ft high. What is the minimum length of the ladder required? Answer in exact form using radicals, and in approximate form by rounding to one decimal place.

**Solution** ▶  We can assume the building makes a 90° angle with the ground, and use the Pythagorean theorem to find the required length. Let $c$ represent this length.

$c^2 = a^2 + b^2$     Pythagorean theorem
$c^2 = (9)^2 + (27)^2$     substitute 9 for $a$ and 27 for $b$
$c^2 = 81 + 729$     $9^2 = 81, 27^2 = 729$
$c^2 = 810$     add
$c = \sqrt{810}$     definition of square root; $c > 0$
$c = 9\sqrt{10}$     exact form: $\sqrt{810} = \sqrt{81 \cdot 10} = 9\sqrt{10}$
$c \approx 28.5$ ft     approximate form

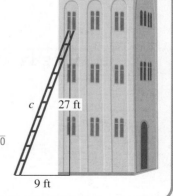

The ladder must be at least 28.5 ft tall.

☑ **F. You've just reviewed how to solve applications of the Pythagorean theorem using radicals**

**Now try Exercises 49 and 50** ▶

There are a number of additional applications of interest in the Exercise Set.

## R.4 EXERCISES

▶ **CONCEPTS AND VOCABULARY**

**Fill in each blank with the appropriate word or phrase. Carefully reread the section, if necessary.**

1. By decomposing the rational exponent, we can rewrite $16^{\frac{3}{4}}$ as $(16^{\frac{?}{?}})^?$.

2. $(x^{\frac{3}{2}})^{\frac{2}{3}} = x^{\frac{3}{2}\cdot\frac{2}{3}} = x^1$ is an example of the _____ property of exponents.

3. Discuss/Explain what it means when we say an expression like $\sqrt{A}$ has been written in simplest form.

4. Discuss/Explain why it would be easier to simplify the expression given using rational exponents rather than radicals.   $\dfrac{x^{\frac{1}{2}}}{x^{\frac{1}{3}}}$

▶ **DEVELOPING YOUR SKILLS**

**Evaluate the expression $\sqrt{x^2}$ for the values given.**

5. **a.** $x = 9$          **b.** $x = -10$

6. **a.** $x = 7$          **b.** $x = -8$

**Simplify each expression, assuming that variables can represent any real number.**

7. **a.** $\sqrt{49p^2}$      **b.** $\sqrt{(x-3)^2}$
   **c.** $\sqrt{81m^4}$      **d.** $\sqrt{x^2 - 6x + 9}$

8. **a.** $\sqrt{25n^2}$      **b.** $\sqrt{(y+2)^2}$
   **c.** $\sqrt{v^{10}}$      **d.** $\sqrt{4a^2 + 12a + 9}$

9. **a.** $\sqrt[3]{64}$      **b.** $\sqrt[3]{-216x^3}$
   **c.** $\sqrt[3]{216z^{12}}$      **d.** $\sqrt[3]{\dfrac{v^3}{-8}}$

10. **a.** $\sqrt[3]{-8}$      **b.** $\sqrt[3]{-125p^3}$
   **c.** $\sqrt[3]{27q^9}$      **d.** $\sqrt[3]{\dfrac{w^3}{-64}}$

11. **a.** $\sqrt[6]{64}$      **b.** $\sqrt[6]{-64}$
   **c.** $\sqrt[5]{243x^{10}}$      **d.** $\sqrt[5]{-243x^5}$
   **e.** $\sqrt[5]{(k-3)^5}$      **f.** $\sqrt[6]{(h+2)^6}$

12. **a.** $\sqrt[4]{81}$      **b.** $\sqrt[4]{-81}$
   **c.** $\sqrt[5]{1024z^{15}}$      **d.** $\sqrt[5]{-1024z^{20}}$
   **e.** $\sqrt[5]{(q-9)^5}$      **f.** $\sqrt[6]{(p+4)^6}$

13. **a.** $\sqrt[3]{-125}$      **b.** $-\sqrt[4]{81n^{12}}$
   **c.** $\sqrt{-36}$      **d.** $\sqrt{\dfrac{49v^{10}}{36}}$

14. **a.** $\sqrt[3]{-216}$      **b.** $-\sqrt[4]{16m^{24}}$
   **c.** $\sqrt{-121}$      **d.** $\sqrt{\dfrac{25x^6}{4}}$

15. **a.** $8^{\frac{2}{3}}$      **b.** $\left(\dfrac{16}{25}\right)^{\frac{3}{2}}$
   **c.** $\left(\dfrac{4}{25}\right)^{-\frac{3}{2}}$      **d.** $\left(\dfrac{-27p^6}{8q^3}\right)^{\frac{2}{3}}$

16. **a.** $9^{\frac{3}{2}}$      **b.** $\left(\dfrac{4}{9}\right)^{\frac{3}{2}}$
   **c.** $\left(\dfrac{16}{81}\right)^{-\frac{3}{4}}$      **d.** $\left(\dfrac{-125v^9}{27w^6}\right)^{\frac{2}{3}}$

17. **a.** $-144^{\frac{3}{2}}$      **b.** $\left(-\dfrac{4}{25}\right)^{\frac{3}{2}}$
   **c.** $(-27)^{-\frac{2}{3}}$      **d.** $-\left(\dfrac{27x^3}{64}\right)^{-\frac{4}{3}}$

18. **a.** $-100^{\frac{3}{2}}$      **b.** $\left(-\dfrac{49}{36}\right)^{\frac{3}{2}}$
   **c.** $(-125)^{-\frac{2}{3}}$      **d.** $-\left(\dfrac{x^9}{8}\right)^{-\frac{4}{3}}$

**Use properties of exponents to simplify. Assume all variables represent nonnegative real numbers. Answer without using negative exponents.**

19. **a.** $(2n^2p^{-\frac{2}{5}})^5$      **b.** $\left(\dfrac{8y^{\frac{3}{4}}}{64y^{\frac{3}{2}}}\right)^{\frac{1}{3}}$

20. **a.** $\left(\dfrac{24x^{\frac{3}{8}}}{4x^{\frac{1}{2}}}\right)^2$      **b.** $(2x^{-\frac{1}{4}}y^{\frac{3}{4}})^4$

**Simplify each expression. Assume all variables represent nonnegative real numbers.**

21. **a.** $\sqrt{18m^2}$      **b.** $-2\sqrt[3]{-125p^3q^7}$
   **c.** $\dfrac{3}{8}\sqrt[3]{64m^3n^5}$      **d.** $\sqrt{32p^3q^6}$
   **e.** $\dfrac{-6 + \sqrt{28}}{2}$      **f.** $\dfrac{27 - \sqrt{72}}{6}$

22. **a.** $\sqrt{8x^6}$      **b.** $3\sqrt[3]{128a^4b^2}$
   **c.** $\dfrac{2}{9}\sqrt[3]{27a^2b^6}$      **d.** $\sqrt{54m^6n^8}$
   **e.** $\dfrac{12 - \sqrt{48}}{8}$      **f.** $\dfrac{-20 + \sqrt{32}}{4}$

**23. a.** $2.5\sqrt{18a}\sqrt{2a^3}$   **b.** $-\dfrac{2}{3}\sqrt{3b}\sqrt{12b^2}$

 **c.** $\sqrt{\dfrac{x^3y}{3}}\sqrt{\dfrac{4x^5y}{12y}}$   **d.** $\sqrt[3]{9v^2u}\,\sqrt[3]{3u^5v^2}$

**24. a.** $5.1\sqrt{2p}\sqrt{32p^5}$   **b.** $-\dfrac{4}{5}\sqrt{5q}\sqrt{20q^3}$

 **c.** $\sqrt{\dfrac{ab^2}{3}}\sqrt{\dfrac{25ab^4}{27}}$   **d.** $\sqrt[3]{5cd^2}\,\sqrt[3]{25cd}$

**25. a.** $\dfrac{\sqrt{8m^5}}{\sqrt{2m}}$   **b.** $\dfrac{\sqrt[3]{108n^4}}{\sqrt[3]{4n}}$

 **c.** $\sqrt{\dfrac{45}{16x^2}}$   **d.** $12\sqrt[3]{\dfrac{81}{8z^9}}$

**26. a.** $\dfrac{\sqrt{27y^7}}{\sqrt{3y}}$   **b.** $\dfrac{\sqrt[3]{72b^5}}{\sqrt[3]{3b^2}}$

 **c.** $\sqrt{\dfrac{20}{4x^4}}$   **d.** $-9\sqrt[3]{\dfrac{125}{27x^6}}$

**27. a.** $\sqrt[5]{32x^{10}y^{15}}$   **b.** $x\sqrt[4]{x^5}$

 **c.** $\sqrt[4]{\sqrt[3]{b}}$   **d.** $\dfrac{\sqrt[3]{6}}{\sqrt{6}}$

 **e.** $\sqrt{b}\,\sqrt[4]{b}$

**28. a.** $\sqrt[4]{81a^{12}b^{16}}$   **b.** $a\sqrt[5]{a^6}$

 **c.** $\sqrt{\sqrt[4]{a}}$   **d.** $\dfrac{\sqrt[3]{3}}{\sqrt[4]{3}}$

 **e.** $\sqrt[3]{c}\,\sqrt[4]{c}$

**Simplify and add (if possible). Assume all variables represent positive quantities.**

**29. a.** $12\sqrt{72}-9\sqrt{98}$
 **b.** $8\sqrt{48}-3\sqrt{108}$
 **c.** $7\sqrt{18m}-\sqrt{50m}$
 **d.** $2\sqrt{28p}-3\sqrt{63p}$

**30. a.** $-3\sqrt{80}+2\sqrt{125}$
 **b.** $5\sqrt{12}+2\sqrt{27}$
 **c.** $3\sqrt{12x}-5\sqrt{75x}$
 **d.** $3\sqrt{40q}+9\sqrt{10q}$

**31. a.** $3x\sqrt[3]{54x}-5\sqrt[3]{16x^4}$
 **b.** $\sqrt{20}+\sqrt{3x}-\sqrt{12x}+\sqrt{45}$
 **c.** $\sqrt{28x^3}-\sqrt{75}+\sqrt{7x^3}-\sqrt{27}$

**32. a.** $5\sqrt[3]{54m^3}-2m\sqrt[3]{16m^3}$
 **b.** $\sqrt{10b}-\sqrt{250b}+\sqrt{80}-\sqrt{20}$
 **c.** $\sqrt{75r^3}+\sqrt{32}-\sqrt{27r^3}+\sqrt{200}$

**Compute each product and simplify the result.**

**33. a.** $(7\sqrt{2})^2$   **b.** $\sqrt{3}(\sqrt{5}+\sqrt{7})$
 **c.** $(n+\sqrt{5})(n-\sqrt{5})$   **d.** $(6-\sqrt{3})^2$

**34. a.** $(0.3\sqrt{5})^2$   **b.** $\sqrt{5}(\sqrt{6}-\sqrt{2})$
 **c.** $(4+\sqrt{3})(4-\sqrt{3})$   **d.** $(2+\sqrt{5})^2$

**35. a.** $(3+2\sqrt{7})(3-2\sqrt{7})$
 **b.** $(\sqrt{5}+\sqrt{20})(\sqrt{15}-\sqrt{10})$
 **c.** $(2\sqrt{2}+5\sqrt{6})(3\sqrt{2}+\sqrt{6})$

**36. a.** $(5+4\sqrt{10})(1-2\sqrt{10})$
 **b.** $(\sqrt{3}+\sqrt{2})(\sqrt{10}+\sqrt{15})$
 **c.** $(3\sqrt{5}+4\sqrt{2})(\sqrt{15}+\sqrt{6})$

**Use a substitution to verify the solutions to the quadratic equation given. Verify results using a calculator.**

**37.** $x^2-4x+1=0$
 **a.** $x=2+\sqrt{3}$   **b.** $x=2-\sqrt{3}$

**38.** $x^2-10x+18=0$
 **a.** $x=5-\sqrt{7}$   **b.** $x=5+\sqrt{7}$

**39.** $x^2+2x-9=0$
 **a.** $x=-1+\sqrt{10}$   **b.** $x=-1-\sqrt{10}$

**40.** $x^2-14x+29=0$
 **a.** $x=7-2\sqrt{5}$   **b.** $x=7+2\sqrt{5}$

**Rationalize each expression by building perfect $n$th root factors for each denominator. Assume all variables represent positive quantities.**

**41. a.** $\dfrac{3}{\sqrt{12}}$   **b.** $\sqrt{\dfrac{20}{27x^3}}$

 **c.** $\sqrt{\dfrac{27}{50b}}$   **d.** $\sqrt[3]{\dfrac{1}{4p}}$

 **e.** $\dfrac{5}{\sqrt[3]{a}}$

**42. a.** $\dfrac{-4}{\sqrt{20}}$   **b.** $\sqrt{\dfrac{125}{12n^3}}$

 **c.** $\sqrt{\dfrac{5}{12x}}$   **d.** $\sqrt[3]{\dfrac{3}{2m^2}}$

 **e.** $\dfrac{-8}{3\sqrt[3]{5}}$

**Simplify the following expressions by rationalizing the denominators. Where possible, state results in exact form and approximate form, rounded to hundredths.**

**43. a.** $\dfrac{8}{3+\sqrt{11}}$   **b.** $\dfrac{6}{\sqrt{x}-\sqrt{2}}$

**44. a.** $\dfrac{7}{\sqrt{7}+3}$   **b.** $\dfrac{12}{\sqrt{x}+\sqrt{3}}$

**45. a.** $\dfrac{\sqrt{6}-3}{\sqrt{3}+\sqrt{2}}$   **b.** $\dfrac{8+2\sqrt{2}}{3-3\sqrt{2}}$

**46. a.** $\dfrac{1+\sqrt{10}}{\sqrt{2}+\sqrt{5}}$   **b.** $\dfrac{1+\sqrt{6}}{5+2\sqrt{6}}$

## ▶ WORKING WITH FORMULAS

**47. Fish length to weight relationship:** $L = 1.13(W)^{\frac{1}{3}}$

The length to weight relationship of a female Pacific halibut can be approximated by the formula shown, where $W$ is the weight in pounds and $L$ is the length in feet. A fisherman lands a halibut that weighs 400 lb. Approximate the length of the fish (round to two decimal places).

**48. Timing a falling object:** $t = \dfrac{\sqrt{s}}{4}$

The time it takes an object to fall a certain distance is given by the formula shown, where $t$ is the time in seconds and $s$ is the distance the object has fallen. Approximate the time it takes an object to hit the ground, if it is dropped from the top of a building that is 80 ft in height (round to hundredths).

## ▶ APPLICATIONS

**49. Length of a cable:** A radio tower is secured by cables that are anchored in the ground 8 m from its base. If the cables are attached to the tower 24 m above the ground, what is the length of each cable? Answer in (a) exact form using radicals, and (b) approximate form by rounding to one decimal place.

**50. Height of a kite:** Benjamin Franklin is flying his kite in a storm once again. John Adams has walked to a position directly under the kite and is 75 ft from Ben. If the kite is 50 ft above John Adams' head and the string is taut, how much string $S$ has Ben let out? Answer in (a) exact form using radicals, and (b) approximate form by rounding to one decimal place.

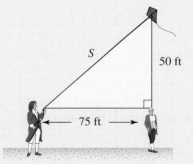

The time $T$ (in days) required for a planet to make one revolution around the sun is modeled by the equation $T = 0.407R^{\frac{3}{2}}$, where $R$ is the maximum radius of the planet's orbit (in millions of miles). This is known as Kepler's third law of planetary motion. Use the equation to approximate the number of days required for one complete orbit of each planet, given the maximum orbital radius shown.

**51. a.** Earth: 93 million mi
**b.** Mars: 142 million mi
**c.** Mercury: 36 million mi

**52. a.** Venus: 67 million mi
**b.** Jupiter: 480 million mi
**c.** Saturn: 890 million mi

**The Roche formula** $d = R\left(\dfrac{2D}{d}\right)^{\frac{1}{3}}$ **(Edward Roche, 1820–1883) gives the distance $d$ at which an orbiting satellite will begin to disintegrate due to the gravitational pull of the primary planet, where $R$ is the radius of the primary planet, $D$ is the density of the primary planet, and $d$ is the density of the satellite (distances are measured in meters, and densities in kilograms per cubic meter$^3$).**

**53.** To the nearest meter, find the distance at which the formula says the moon would begin to disintegrate due to the Earth's gravitational pull, given the radius of the Earth is 6,378,137 m, the density of the Earth is 5513 kg/m$^3$, and the density of the moon is 3346 kg/m$^3$.

**54.** To the nearest meter, find the distance at which the formula says the Earth would begin to disintegrate due to the Sun's gravitational pull, given the radius of the Sun is 696,000,000 m, the density of the Sun is 1408 kg/m$^3$, and the density of the Earth is 5513 kg/m$^3$.

**55. Accident investigation:** After an accident, police officers will try to determine the approximate velocity $V$ that a car was traveling using the formula $V = 2\sqrt{6L}$, where $L$ is the length of the skid marks in feet and $V$ is the velocity in miles per hour. (a) If the skid marks are 54 ft long, how fast was the car traveling? (b) Approximate the speed of the car if the skid marks were 90 ft long.

**56. Wind-powered energy:** If a wind-powered generator is delivering $P$ units of power, the velocity $V$ of the wind (in miles per hour) can be determined using $V = \sqrt[3]{\dfrac{P}{k}}$, where $k$ is a constant that depends on the size and efficiency of the generator. Rationalize the radical expression and use the new version to find the velocity of the wind if $k = 0.004$ and the generator is putting out 13.5 units of power.

**57. Surface area:** The lateral surface area (surface area excluding the base) $S$ of a cone is given by the formula $S = \pi r \sqrt{r^2 + h^2}$, where $r$ is the radius of the base and $h$ is the height of the cone. Find the lateral surface area of a cone that has a radius of 6 m and a height of 10 m. Answer in simplified radical form and in approximate form, rounded to hundredths.

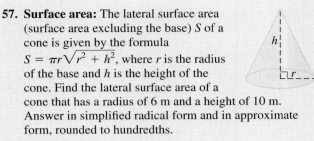

**58. Surface area:** The lateral surface area $S$ of a frustum (a truncated cone) is given by the formula $S = \pi(a + b)\sqrt{h^2 + (b - a)^2}$, where $a$ is the radius of the upper base, $b$ is the radius of the lower base, and $h$ is the height. Find the surface area of a frustum where $a = 6$ m, $b = 8$ m, and $h = 10$ m. Answer in simplified radical form and in approximate form, rounded to hundredths.

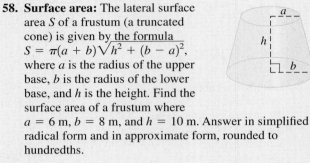

## ▶ EXTENDING THE CONCEPT

The expression $x^2 - 7$ is not factorable using *integer values*. But the expression *can be written* in the form $x^2 - (\sqrt{7})^2$, enabling us to factor it as a "binomial" and its conjugate: $(x + \sqrt{7})(x - \sqrt{7})$. Use this idea to factor the following expressions.

**59. a.** $x^2 - 5$
   **b.** $n^2 - 19$

**60. a.** $4v^2 - 11$
   **b.** $9w^2 - 17$

**61.** Find a quick way to simplify the expression without the aid of a calculator.

$$\left(\left(\left(\left(\left(\left(3^{\frac{5}{6}}\right)^{\frac{3}{2}}\right)^{\frac{4}{5}}\right)^{\frac{3}{4}}\right)^{\frac{2}{5}}\right)^{\frac{10}{3}}\right)$$

**62.** If $\left(x^{\frac{1}{2}} + x^{-\frac{1}{2}}\right)^2 = \dfrac{9}{2}$, find the value of $x^{\frac{1}{2}} + x^{-\frac{1}{2}}$.

 **63.** Rewrite by rationalizing the *numerator*:

$$\dfrac{\sqrt{x + h} - \sqrt{x}}{h}$$

Determine the values of $x$ for which each expression represents a real number.

**64.** $\dfrac{x^2 - 4}{\sqrt{x - 1}}$          **65.** $\dfrac{\sqrt{x - 1}}{x^2 - 4}$

**66.** Any three positive integers $(a, b, c)$ that satisfy the Pythagorean theorem are called Pythagorean triples. Two common triples were seen previously: $(3, 4, 5)$ and $(7, 24, 25)$. Others can be found using the following: $a = m^2 - n^2$, $b = 2mn$ and $c = m^2 + n^2$, for positive integers $m$ and $n$, where $m > n$. What triples are returned for (a) $m = 3$, $n = 2$? (b) For $m = 5$ and $n = 4$? Verify each result.

**67.** When working with the equation of an ellipse, radical expressions of the form $\sqrt{m + n} + \sqrt{m - n}$ may occur, and can be rewritten as the single radical $\sqrt{a + \sqrt{b}}$, where $a = 2m$ and $b = 4(m^2 - n^2)$. (a) Rewrite the radical expression $\sqrt{3 + \sqrt{2}} + \sqrt{3 - \sqrt{2}}$ as a single radical, and (b) use a calculator to verify the result is equivalent.

**68.** The equation used to determine the velocity necessary to escape the gravitational pull of a planet is given by $v = \sqrt{\dfrac{2GM}{R}}$, where $R$ is the radius of the planet (in meters), $M$ is the mass of the planet (in kilograms), and $G$ is the universal constant of gravitation ($6.67 \times 10^{-11}$). (a) Given the Earth's mass is about $5.98 \times 10^{24}$ kg with a radius of $6.38 \times 10^6$ m, at what velocity must the space shuttle travel to break free of Earth's gravity (the answer will be in meters per second)? (b) If the mass of Mars is $6.39 \times 10^{23}$, with a radius of $3.40 \times 10^6$ m, at what velocity must a spaceship travel to break free of Mars' gravity?

**LEARNING OBJECTIVES**

*In Section R.5 you will review how to:*

☐ **A.** Factor out the greatest common factor

☐ **B.** Factor out common binomial factors and factor by grouping

☐ **C.** Factor quadratic polynomials

☐ **D.** Factor special forms and quadratic forms

☐ **E.** Solve applications by factoring

It is often said that knowing which tool to use is just as important as knowing how to use the tool. In this section, we review the tools needed to factor an expression, an important part of solving polynomial equations. This section will also help us decide which factoring tool is appropriate when many different factorable expressions are presented.

## A. The Greatest Common Factor

To **factor** an expression means to *rewrite the expression as an equivalent product*. The distributive property is an example of factoring in action. To factor $2x^2 + 6x$, we might first rewrite each term using the common factor $2x$: $2x^2 + 6x = 2x \cdot x + 2x \cdot 3$, then apply the distributive property to obtain $2x(x + 3)$. We commonly say that we have *factored out* $2x$. The **greatest common factor** (or GCF) is the largest factor common to *all* terms in the polynomial.

---

**EXAMPLE 1** ▶ **Factoring Polynomials**

Factor each polynomial:

    **a.** $12x^2 + 18xy - 30y$          **b.** $x^5 + x^2$

**Solution** ▶   **a.** 6 is common to all three terms:

$$12x^2 + 18xy - 30y$$
$$= 6(2x^2 + 3xy - 5y)$$

mentally: $6 \cdot 2x^2 + 6 \cdot 3xy - 6 \cdot 5y$

  **b.** $x^2$ is common to both terms:

$$x^5 + x^2$$
$$= x^2(x^3 + 1)$$

mentally: $x^2 \cdot x^3 + x^2 \cdot 1$

☑ **A.** You've just reviewed how to factor out the greatest common factor

**Now try Exercises 5 and 6** ▶

---

## B. Common Binomial Factors and Factoring by Grouping

If the terms of a polynomial have a **common *binomial* factor,** it can also be factored out using the distributive property.

---

**EXAMPLE 2** ▶ **Factoring Out a Common Binomial Factor**

Factor:

    **a.** $(x + 3)x^2 + (x + 3)5$                    **b.** $x^2(x - 2) - 3(x - 2)$

**Solution** ▶   **a.** $(x + 3)x^2 + (x + 3)5$            **b.** $x^2(x - 2) - 3(x - 2)$

           $= (x + 3)(x^2 + 5)$                $= (x - 2)(x^2 - 3)$

**Now try Exercises 7 and 8** ▶

---

One application of removing a binomial factor involves **factoring by grouping.** At first glance, the expression $x^3 + 2x^2 + 3x + 6$ appears unfactorable. But by grouping the terms (applying the associative property), we can remove a monomial factor from each subgroup, which then reveals a common binomial factor.

$$\underline{x^3 + 2x^2} + \underline{3x + 6} = x^2(x + 2) + 3(x + 2)$$
$$= (x + 2)(x^2 + 3)$$

This grouping of terms must take into account any sign changes and common factors, as seen in Example 3. Also, it will be helpful to note that a general four-term polynomial $A + B + C + D$ is factorable by grouping only if $AD = BC$.

---

**EXAMPLE 3** ▶ **Factoring by Grouping**

Factor $3t^3 + 15t^2 - 6t - 30$.

**Solution** ▶ Notice that all four terms have a common factor of 3. Begin by factoring it out.

$3t^3 + 15t^2 - 6t - 30$     original polynomial

$= 3(t^3 + 5t^2 - 2t - 10)$     factor out 3

$= 3(\underline{t^3 + 5t^2} - \underline{2t - 10})$     group remaining terms

$= 3[t^2(t + 5) - 2(t + 5)]$     factor common *monomial*

$= 3(t + 5)(t^2 - 2)$     factor common *binomial*

☑ **B.** You've just reviewed how to factor out common binomial factors and factor by grouping

**Now try Exercises 9 and 10** ▶

---

When asked to factor an expression, first look for common factors. The resulting expression will be easier to work with and help ensure the final answer is written in **completely factored form.** If a four-term polynomial cannot be factored as written, try rearranging the terms to find a combination that enables factoring by grouping.

## C. Factoring Quadratic Polynomials

A quadratic polynomial is one that can be written in the form $ax^2 + bx + c$, where $a, b, c \in \mathbb{R}$ and $a \neq 0$. One common form of factoring involves quadratic trinomials such as $x^2 + 7x + 10$ and $2x^2 - 13x + 15$. While we know $(x + 5)(x + 2) = x^2 + 7x + 10$ and $(2x - 3)(x - 5) = 2x^2 - 13x + 15$ using F-O-I-L, how can we factor these trinomials without seeing the original expression in advance? First, it helps to place the trinomials in two families—those with a leading coefficient of 1 and those with a leading coefficient other than 1.

### $ax^2 + bx + c$, where $a = 1$

When $a = 1$, the only factor pair for $x^2$ (other than $1 \cdot x^2$) is $x \cdot x$ and the first term in each binomial will be $x$: $(x \quad )(x \quad )$. The following observation helps guide us to the complete factorization. Consider the product $(x + b)(x + a)$:

$$(x + b)(x + a) = x^2 + ax + bx + ab \quad \text{F-O-I-L}$$

$$= x^2 + (a + b)x + ab \quad \text{distributive property}$$

Note the last term is the product $ab$ (the *lasts*), while the coefficient of the middle term is $a + b$ (the sum of the *outers* and *inners*). Since the last term of $x^2 - 8x + 7$ is 7 and the coefficient of the middle term is $-8$, we are seeking two numbers with a product of positive 7 and a sum of negative 8. The numbers are $-7$ and $-1$, so the factored form is $(x - 7)(x - 1)$. It is also helpful to note that if the constant term is positive, the binomials will have *like* signs, since only *the product of like signs is positive.* If the constant term is negative, the binomials will have *unlike* signs, since only *the product of unlike signs is negative.* This means we can use the sign of the linear term (the term with degree 1) to guide our choice of factors.

> ### Factoring Trinomials with a Leading Coefficient of 1
>
> If the constant term is positive, the binomials will have *like* signs:
>
> $$(x + \quad)(x + \quad) \text{ or } (x - \quad)(x - \quad),$$
>
> to match the sign of the linear (middle) term.
>
> If the constant term is negative, the binomials will have *unlike* signs:
>
> $$(x + \quad)(x - \quad),$$
>
> with the larger factor placed in the binomial
> whose sign *matches* the linear (middle) term.

**EXAMPLE 4** ▶ **Factoring Trinomials**

Factor these expressions:

   **a.** $-x^2 + 11x - 24$        **b.** $x^2 - 10 - 3x$

**Solution** ▶ **a.** First rewrite the trinomial in standard form as $-1(x^2 - 11x + 24)$. For $x^2 - 11x + 24$, the constant term is positive so the binomials will have like signs. Since the linear term is negative,

$$-1(x^2 - 11x + 24) = -1(x - \quad)(x - \quad) \quad \text{like signs, both negative}$$
$$= -1(x - 8)(x - 3) \quad (-8)(-3) = 24; -8 + (-3) = -11$$

**b.** First rewrite the trinomial in standard form as $x^2 - 3x - 10$. The constant term is negative so the binomials will have unlike signs. Since the linear term is negative,

$$x^2 - 3x - 10 = (x + \quad)(x - \quad) \quad \text{unlike signs, one positive and one negative}$$
$$\phantom{x^2 - 3x - 10} \quad\quad\quad\quad\quad\quad\quad 5 > 2, 5 \text{ is placed in the second binomial;}$$
$$= (x + 2)(x - 5) \quad (2)(-5) = -10; 2 + (-5) = -3$$

**Now try Exercises 11 and 12** ▶

Sometimes we encounter **prime polynomials,** or polynomials that cannot be factored. For $x^2 + 9x + 15$, the factor pairs of 15 are $1 \cdot 15$ and $3 \cdot 5$, with neither pair having a sum of $+9$. We conclude that $x^2 + 9x + 15$ is prime.

### $ax^2 + bx + c$, where $a \neq 1$

If the leading coefficient is not one, the possible combinations of outers and inners are more numerous. Furthermore, the sum of the outer and inner products will change depending on the position of the possible factors. Note that $(2x + 3)(x + 9) = 2x^2 + 21x + 27$ and $(2x + 9)(x + 3) = 2x^2 + 15x + 27$ result in a different middle term, even though identical numbers were used.

To factor $2x^2 - 13x + 15$, note the constant term is positive so the binomials *must have like signs*. The negative linear term indicates these signs will be negative. We then list possible factors for the first and last terms of each binomial, then sum the outer and inner products.

| Possible First and Last Terms for $2x^2$ and 15 | Sum of Outers and Inners |
|---|---|
| 1. $(2x - 1)(x - 15)$ | $-30x - 1x = -31x$ |
| 2. $(2x - 15)(x - 1)$ | $-2x - 15x = -17x$ |
| 3. $(2x - 3)(x - 5)$ | $-10x - 3x = -13x$ ← |
| 4. $(2x - 5)(x - 3)$ | $-6x - 5x = -11x$ |

As you can see, only possibility 3 yields a linear term of $-13x$, and the correct factorization is then $(2x - 3)(x - 5)$. With practice, this **trial-and-error** process can be completed very quickly.

If the constant term is negative, the number of possibilities can be reduced by finding a factor pair with a sum *or* difference equal to the *absolute value* of the linear coefficient, as we can then arrange the sign in each binomial to obtain the needed result as shown in Example 5.

**EXAMPLE 5** ▶

**Factoring a Trinomial Using Trial and Error**

Factor $6z^2 - 11z - 35$.

**Solution** ▶

Note the constant term is negative (binomials will have unlike signs) and $|-11| = 11$. The factors of 35 are $1 \cdot 35$ and $5 \cdot 7$. Two possible first terms are: $(6z\ \ )(z\ \ )$ and $(3z\ \ )(2z\ \ )$, and we begin with 5 and 7 as factors of 35.

| $(6z\quad)(z\quad)$ | Outer and Inner Products | | $(3z\quad)(2z\quad)$ | Outer and Inner Products | |
| --- | --- | --- | --- | --- | --- |
| | **Sum** | **Difference** | | **Sum** | **Difference** |
| 1. $(6z\quad 5)(z\quad 7)$ | $42z + 5z$ <br> $47z$ | $42z - 5z$ <br> $37z$ | 3. $(3z\quad 5)(2z\quad 7)$ | $21z + 10z$ <br> $31z$ | $21z - 10z$ <br> $11z$ |
| 2. $(6z\quad 7)(z\quad 5)$ | $30z + 7z$ <br> $37z$ | $30z - 7z$ <br> $23z$ | 4. $(3z\quad 7)(2z\quad 5)$ | $15z + 14z$ <br> $29z$ | $15z - 14z$ <br> $1z$ |

Since possibility 3 yields a linear term of $11z$, we need not consider other factors of 35 and write the factored form as $6z^2 - 11z - 35 = (3z\ \ 5)(2z\ \ 7)$. The signs can then be arranged to obtain a middle term of $-11z$: $(3z + 5)(2z - 7)$, $-21z + 10z = -11z$ ✓.

☑ **C.** You've just reviewed how to factor quadratic polynomials

**Now try Exercises 13 and 14** ▶

## D. Factoring Special Forms and Quadratic Forms

Next we consider methods to factor each of the special products we encountered in Section R.2.

### The Difference of Two Squares

Multiplying and factoring are inverse processes. Since $(x - 7)(x + 7) = x^2 - 49$, we know that $x^2 - 49 = (x - 7)(x + 7)$. In words, *the difference of two squares will factor into a binomial and its conjugate.* To find the terms of the factored form, rewrite each term in the original expression as a square: $(\quad)^2$.

**Factoring the Difference of Two Perfect Squares**

Given any expression that can be written in the form $A^2 - B^2$,

$$A^2 - B^2 = (A + B)(A - B)$$

Note that the *sum* of two perfect squares $A^2 + B^2$ *cannot be factored* using real numbers (the expression is prime). As a reminder, always check for a common factor first and be sure to write all results in completely factored form. See Example 6(c).

---

**EXAMPLE 6** ▶ **Factoring the Difference of Two Perfect Squares**

Factor each expression completely.

**a.** $4w^2 - 81$     **b.** $v^2 + 49$     **c.** $-3n^2 + 48$     **d.** $z^4 - \frac{1}{81}$     **e.** $x^2 - 7$

**Solution** ▶

**a.** $4w^2 - 81 = (2w)^2 - 9^2$     write as a difference of squares
$\qquad = (2w + 9)(2w - 9)$     $A^2 - B^2 = (A + B)(A - B)$

**b.** $v^2 + 49$ is prime.

**c.** $-3n^2 + 48 = -3(n^2 - 16)$     factor out $-3$
$\qquad = -3(n^2 - 4^2)$     write as a difference of squares
$\qquad = -3(n + 4)(n - 4)$     $A^2 - B^2 = (A + B)(A - B)$

**d.** $z^4 - \frac{1}{81} = (z^2)^2 - \left(\frac{1}{9}\right)^2$     write as a difference of squares
$\qquad = \left(z^2 + \frac{1}{9}\right)\left(z^2 - \frac{1}{9}\right)$     $A^2 - B^2 = (A + B)(A - B)$
$\qquad = \left(z^2 + \frac{1}{9}\right)\left[z^2 - \left(\frac{1}{3}\right)^2\right]$     write as a difference of squares ($z^2 + \frac{1}{9}$ is prime)
$\qquad = \left(z^2 + \frac{1}{9}\right)\left(z + \frac{1}{3}\right)\left(z - \frac{1}{3}\right)$     result

**e.** $x^2 - 7 = x^2 - (\sqrt{7})^2$     write as a difference of squares
$\qquad = (x + \sqrt{7})(x - \sqrt{7})$     $A^2 - B^2 = (A + B)(A - B)$

**Now try Exercises 15 and 16** ▶

---

**CAUTION** ▶ In an attempt to factor a *sum* of two perfect squares, say $v^2 + 49$, let's list all possible binomial factors. These are (1) $(v + 7)(v + 7)$, (2) $(v - 7)(v - 7)$, and (3) $(v + 7)(v - 7)$. Note that (1) and (2) are the binomial squares $(v + 7)^2$ and $(v - 7)^2$, with each product resulting in a "middle" term, whereas (3) is a binomial times its conjugate, resulting in a *difference* of squares: $v^2 - 49$. With all possibilities exhausted, we conclude that *the sum of two squares is prime!*

## Perfect Square Trinomials

Since $(x + 7)^2 = x^2 + 14x + 49$, we know that $x^2 + 14x + 49 = (x + 7)^2$. In words, *a perfect square trinomial will factor into a binomial square.* To use this idea effectively, we must learn to *identify* perfect square trinomials. Note that the first and last terms of $x^2 + 14x + 49$ are *the squares* of $x$ and 7, and the middle term is *twice the product of these two terms*: $2(7x) = 14x$. These are the characteristics of a perfect square trinomial.

**Factoring Perfect Square Trinomials**

Given any expression that can be written in the form $A^2 \pm 2AB + B^2$,

**1.** $A^2 + 2AB + B^2 = (A + B)^2$
**2.** $A^2 - 2AB + B^2 = (A - B)^2$

**EXAMPLE 7** ▶  **Factoring a Perfect Square Trinomial**

Factor $12m^3 - 12m^2 + 3m$.

**Solution** ▶  $12m^3 - 12m^2 + 3m$          check for common factors: GCF = $3m$
$= 3m(4m^2 - 4m + 1)$     factor out $3m$

For the remaining trinomial $4m^2 - 4m + 1$ ...

1. Are the first and last terms perfect squares?
$$4m^2 = (2m)^2 \text{ and } 1 = (1)^2 \checkmark \text{ Yes.}$$

2. Is the linear term twice the product of $2m$ and 1?
$$2 \cdot 2m \cdot 1 = 4m \checkmark \text{ Yes.}$$

Factor as a binomial square: $4m^2 - 4m + 1 = (2m - 1)^2$
This shows $12m^3 - 12m^2 + 3m = 3m(2m - 1)^2$.

**Now try Exercises 17 and 18** ▶

**CAUTION** ▶  As shown in Example 7, be sure to include the GCF in your final answer. It is a common error to "leave the GCF behind," but it must be included at each step of the factoring process.

In actual practice, these calculations can be performed mentally, making the process much more efficient.

### Sum or Difference of Two Perfect Cubes

Recall that the *difference* of two perfect squares is factorable, but the *sum* of two perfect squares is prime. In contrast, *both the sum and difference of two perfect* **cubes** *are factorable*. For either $A^3 + B^3$ or $A^3 - B^3$ we have the following:

1. Each will factor into the product of a binomial and a trinomial:          (      )(          )
                                                                                                binomial     trinomial
2. The terms of the binomial are the quantities being cubed:          $(A \quad B)( \qquad )$
3. The terms of the trinomial are the square of $A$, the product $AB$, and the square of $B$ respectively:          $(A \quad B)(A^2 \quad AB \quad B^2)$
4. The binomial takes the same sign as the original expression          $(A \pm B)(A^2 \quad AB \quad B^2)$
5. The middle term of the trinomial takes the opposite sign of the original expression (the last term is always positive):          $(A \pm B)(A^2 \mp AB + B^2)$

---

**Factoring the Sum or Difference of Two Perfect Cubes**

Given any expression that can be written in the form $A^3 \pm B^3$

1. $A^3 + B^3 = (A + B)(A^2 - AB + B^2)$
2. $A^3 - B^3 = (A - B)(A^2 + AB + B^2)$

**EXAMPLE 8** ▶  **Factoring the Sum and Difference of Two Perfect Cubes**

Factor completely:

**a.** $x^3 + 125$      **b.** $-5m^3n + 40n^4$

Solution ▶  **a.**    $x^3 + 125 = x^3 + 5^3$                         write terms as perfect cubes

Use $A^3 + B^3 = (A + B)(A^2 - AB + B^2)$     factoring template

$x^3 + 5^3 = (x + 5)(x^2 - 5x + 25)$     $A \to x$ and $B \to 5$

**b.**    $-5m^3n + 40n^4 = -5n(m^3 - 8n^3)$                         check for common
factors (GCF $= -5n$)

$= -5n[m^3 - (2n)^3]$     write terms as perfect cubes

Use $A^3 - B^3 = (A - B)(A^2 + AB + B^2)$     factoring template

$m^3 - (2n)^3 = (m - 2n)[m^2 + m(2n) + (2n)^2]$     $A \to m$ and $B \to 2n$

$= (m - 2n)(m^2 + 2mn + 4n^2)$     simplify

$\Rightarrow -5m^3n + 40n^4 = -5n(m - 2n)(m^2 + 2mn + 4n^2).$     factored form

The results for parts (a) and (b) can be checked using multiplication.

**Now try Exercises 19 and 20** ▶

### Quadratic Forms and *u*-Substitution

For any quadratic expression $ax^2 + bx + c$ in standard form, the degree of the leading term is twice the degree of the middle term. Generally, a trinomial is in **quadratic form** if it can be written as $a(\_\_)^2 + b(\_\_) + c$, where the parentheses "hold" the same factors. The equation $x^4 - 13x^2 + 36 = 0$ is in quadratic form since $(x^2)^2 - 13(x^2) + 36 = 0$. In many cases, we can factor these expressions using a **placeholder substitution** that transforms them into a more recognizable form. In a study of algebra, the letter "*u*" often plays this role. If we let $u$ represent $x^2$, the expression $(x^2)^2 - 13(x^2) + 36$ becomes $u^2 - 13u + 36$, which can be factored into $(u - 9)(u - 4)$. After "unsubstituting" (replace $u$ with $x^2$), we have $(x^2 - 9)(x^2 - 4) = (x + 3)(x - 3)(x + 2)(x - 2)$.

**EXAMPLE 9** ▶  **Factoring a Quadratic Form**

Write in completely factored form: $(x^2 - 2x)^2 - 2(x^2 - 2x) - 3$.

Solution ▶  Expanding the binomials would produce a fourth-degree polynomial that would be very difficult to factor. Instead we note the expression is in *quadratic form*. Letting $u$ represent $x^2 - 2x$ (the variable part of the "middle" term), $(x^2 - 2x)^2 - 2(x^2 - 2x) - 3$ becomes $u^2 - 2u - 3$.

$u^2 - 2u - 3 = (u - 3)(u + 1)$     factor

To finish up, write the expression in terms of $x$, substituting $x^2 - 2x$ for $u$.

$= (x^2 - 2x - 3)(x^2 - 2x + 1)$     substitute $x^2 - 2x$ for $u$

The resulting trinomials can be further factored.

$= (x - 3)(x + 1)(x - 1)^2$     $x^2 - 2x + 1 = (x - 1)^2$

☑ **D.** You've just reviewed how to factor special forms and quadratic forms

**Now try Exercises 21 and 22** ▶

It is well known that information is retained longer and used more effectively when it's placed in an organized form. The "factoring flowchart" provided in Figure R.8 offers a streamlined and systematic approach to factoring and the concepts involved. However, with some practice the process tends to "flow" more naturally than following a chart, with many of the decisions becoming automatic.

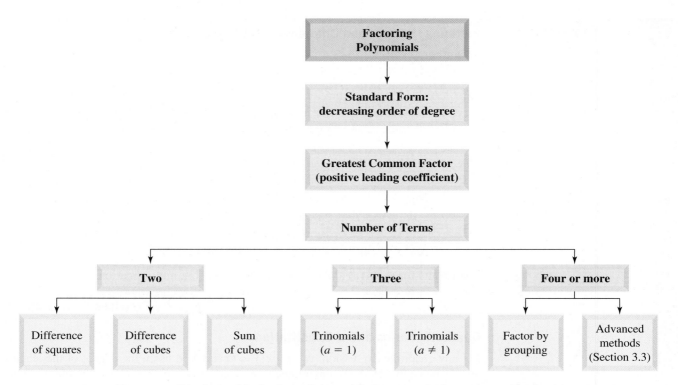

**Figure R.8**

• Can any result be factored further?     • Polynomials that cannot be factored are said to be *prime*.

For additional practice with these ideas, **see Exercises 23 through 50.**

### E. Applications of Factoring

The ability to solve linear and quadratic equations is the foundation on which a large percentage of our future studies are built. Both are closely linked to the solution of other equation types, as well as to the graphs and applications of these equations.

---

**EXAMPLE 10 ▶**     **The Surface Area of a Torus**

In common language, the geometric shape known as a torus is referred to as a mathematical donut. The surface area of this "donut" is given by the formula $S = \pi^2 R^2 - \pi^2 r^2$, where $R$ is the outer radius of the donut, and $r$ is the inner radius. (a) Write the formula in completely factored form. (b) Use the result to find the surface area of a donut given $R = 9$ cm and $r = 6$ cm.

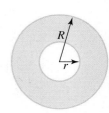

**Solution ▶**     **a.** Begin by factoring out the common factor $\pi^2$.

$$S = \pi^2 R^2 - \pi^2 r^2 \quad \text{given formula}$$
$$= \pi^2(R^2 - r^2) \quad \text{factor out } \pi^2$$

Finish by factoring the difference of two perfect squares

$$= \pi^2(R + r)(R - r) \quad \text{result}$$

**b.** Evaluate the result from part (a) using $R = 9$ and $r = 6$.

$$S = \pi^2(R + r)(R - r) \quad \text{factored form}$$
$$= \pi^2(9 + 6)(9 - 6) \quad \text{substitute 9 for } R \text{ and 6 for } r$$
$$= \pi^2(15)(3) \quad \text{simplify}$$
$$= 45\pi^2 \quad \text{result}$$

The surface area of this donut is $45\pi^2$ cm$^2$ or about 444.1 cm$^2$.

☑ **E.** You've just reviewed how to solve applications by factoring

**Now try Exercises 53 through 60 ▶**

## R.5 EXERCISES

### ▶ CONCEPTS AND VOCABULARY

**Fill in each blank with the appropriate word or phrase. Carefully reread the section, if necessary.**

1. The difference of two perfect squares always factors into the product of a(n) _____ and its _____.

2. The expression $x^2 + 6x + 9$ is said to be a(n) _____ _____ trinomial, since its factored form is a perfect (binomial) square.

3. Discuss/Explain why $4x^2 - 36 = (2x - 6)(2x + 6)$ is not written in completely factored form, then rewrite it so it is factored completely.

4. Discuss/Explain why $a^3 + b^3$ is factorable, but $a^2 + b^2$ is not. Demonstrate by writing $x^3 + 64$ in factored form, and by exhausting all possibilities for $x^2 + 64$ to show it is prime.

### ▶ DEVELOPING YOUR SKILLS

**Factor each expression using the method indicated.**

**Greatest Common Factor**

5. **a.** $-17x^2 + 51$    **b.** $21b^3 - 14b^2 + 56b$
   **c.** $-3a^4 + 9a^2 - 6a^3$

6. **a.** $-13n^2 - 52$    **b.** $9p^2 + 27p^3 - 18p^4$
   **c.** $-6g^5 + 12g^4 - 9g^3$

**Common Binomial Factor**

7. **a.** $2a(a + 2) + 3(a + 2)$
   **b.** $(b^2 + 3)3b + (b^2 + 3)2$
   **c.** $4m(n + 7) - 11(n + 7)$

8. **a.** $5x(x - 3) - 2(x - 3)$
   **b.** $(v - 5)2v + (v - 5)3$
   **c.** $3p(q^2 + 5) + 7(q^2 + 5)$

**Grouping**

9. **a.** $9q^3 + 6q^2 + 15q + 10$
   **b.** $h^5 - 12h^4 - 3h + 36$
   **c.** $k^5 - 7k^3 - 5k^2 + 35$

10. **a.** $6h^3 - 9h^2 - 2h + 3$
    **b.** $4k^3 + 6k^2 - 2k - 3$
    **c.** $3x^2 - xy - 6x + 2y$

**Trinomial Factoring where $|a| = 1$**

11. **a.** $-p^2 + 5p + 14$    **b.** $q^2 - 4q + 12$
    **c.** $n^2 + 20 - 9n$

12. **a.** $-m^2 + 13m - 42$    **b.** $x^2 + 12 + 13x$
    **c.** $v^2 + 10v + 15$

**Trinomial Factoring where $a \neq 1$**

13. **a.** $3p^2 - 13p - 10$    **b.** $4q^2 + 7q - 15$
    **c.** $10u^2 - 19u - 15$

14. **a.** $6v^2 + v - 35$    **b.** $20x^2 + 53x + 18$
    **c.** $15z^2 - 22z - 48$

**Difference of Perfect Squares**

15. **a.** $4s^2 - 25$    **b.** $9x^2 - 49$
    **c.** $50x^2 - 72$    **d.** $121h^2 - 144$
    **e.** $b^2 - 5$

16. **a.** $9v^2 - \frac{1}{25}$    **b.** $25w^2 - \frac{1}{49}$
    **c.** $v^4 - 1$    **d.** $16z^4 - 81$
    **e.** $x^2 - 17$

**Perfect Square Trinomials**

17. **a.** $a^2 - 6a + 9$    **b.** $b^2 + 10b + 25$
    **c.** $4m^2 - 20m + 25$    **d.** $9n^2 - 42n + 49$

18. **a.** $x^2 + 12x + 36$    **b.** $z^2 - 18z + 81$
    **c.** $25p^2 - 60p + 36$    **d.** $16q^2 + 40q + 25$

**Sum/Difference of Perfect Cubes**

19. **a.** $8p^3 - 27$    **b.** $m^3 + \frac{1}{8}$
    **c.** $g^3 - 0.027$    **d.** $-2t^4 + 54t$

20. **a.** $27q^3 - 125$    **b.** $n^3 + \frac{8}{27}$
    **c.** $b^3 - 0.125$    **d.** $3r^4 - 24r$

**$u$-Substitution**

21. **a.** $x^4 - 10x^2 + 9$    **b.** $x^4 + 13x^2 + 36$
    **c.** $x^6 - 7x^3 - 8$

**22. a.** $x^6 - 26x^3 - 27$
   **b.** $3(n + 5)^2 + 2(n + 5) - 21$
   **c.** $2(z + 3)^2 + 3(z + 3) - 54$

**23.** Completely factor each of the following (recall that "1" is its own perfect square and perfect cube).
   **a.** $n^2 - 1$     **b.** $n^3 - 1$
   **c.** $n^3 + 1$     **d.** $28x^3 - 7x$

**24.** Carefully factor each of the following trinomials, if possible. Note differences and similarities.
   **a.** $x^2 - x + 6$     **b.** $x^2 + x - 6$
   **c.** $x^2 + x + 6$     **d.** $x^2 - x - 6$
   **e.** $x^2 - 5x + 6$    **f.** $x^2 + 5x - 6$

**Factor each expression completely, if possible. Rewrite the expression in standard form (factor out "−1" if needed) and factor out the GCF if one exists. If you believe the expression will not factor, write "prime."**

**25.** $a^2 + 7a + 10$         **26.** $b^2 + 9b + 20$

**27.** $2x^2 - 24x + 40$       **28.** $10z^2 - 140z + 450$

**29.** $64 - 9m^2$            **30.** $25 - 16n^2$

**31.** $-9r + r^2 + 18$       **32.** $28 + s^2 - 11s$

**33.** $2h^2 + 7h + 6$        **34.** $3k^2 + 10k + 8$

**35.** $9k^2 - 24k + 16$      **36.** $4p^2 - 20p + 25$

**37.** $-6x^3 + 39x^2 - 63x$  **38.** $-28z^3 + 16z^2 + 80z$

**39.** $12m^2 - 40m + 4m^3$   **40.** $-30n - 4n^2 + 2n^3$

**41.** $a^2 - 7a - 60$        **42.** $b^2 - 9b - 36$

**43.** $8x^3 - 125$          **44.** $27r^3 + 64$

**45.** $m^2 + 9m - 24$        **46.** $n^2 - 14n - 36$

**47.** $x^3 - 5x^2 - 9x + 45$  **48.** $x^3 + 3x^2 - 4x - 12$

**49.** Match each expression with the description that fits *best*.
   \_\_\_\_ **a.** prime polynomial
   \_\_\_\_ **b.** standard trinomial $a = 1$
   \_\_\_\_ **c.** perfect square trinomial
   \_\_\_\_ **d.** difference of cubes
   \_\_\_\_ **e.** binomial square
   \_\_\_\_ **f.** sum of cubes
   \_\_\_\_ **g.** binomial conjugates
   \_\_\_\_ **h.** difference of squares
   \_\_\_\_ **i.** standard trinomial $a \neq 1$
   **A.** $x^3 + 27$              **B.** $(x + 3)^2$
   **C.** $x^2 - 10x + 25$        **D.** $x^2 - 144$
   **E.** $x^2 - 3x - 10$         **F.** $8s^3 - 125t^3$
   **G.** $2x^2 - x - 3$          **H.** $x^2 + 9$
   **I.** $(x - 7)$ and $(x + 7)$

**50.** Match each polynomial to its factored form. Two of them are prime.
   \_\_\_\_ **a.** $4x^2 - 9$
   \_\_\_\_ **b.** $4x^2 - 28x + 49$
   \_\_\_\_ **c.** $x^3 - 125$
   \_\_\_\_ **d.** $8x^3 + 27$
   \_\_\_\_ **e.** $x^2 - 3x - 10$
   \_\_\_\_ **f.** $x^2 + 3x + 10$
   \_\_\_\_ **g.** $2x^2 - x - 3$
   \_\_\_\_ **h.** $2x^2 + x - 3$
   \_\_\_\_ **i.** $x^2 + 25$
   **A.** $(x - 5)(x^2 + 5x + 25)$
   **B.** $(2x - 3)(x + 1)$     **C.** $(2x + 3)(2x - 3)$
   **D.** $(2x - 7)^2$          **E.** prime trinomial
   **F.** prime binomial        **G.** $(2x + 3)(x - 1)$
   **H.** $(2x + 3)(4x^2 - 6x + 9)$
   **I.** $(x - 5)(x + 2)$

► **WORKING WITH FORMULAS**

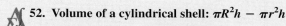

**51. Surface area of a cylinder: $2\pi r^2 + 2\pi rh$**

The surface area of a cylinder is given by the formula shown, where $h$ is the height of the cylinder and $r$ is the radius. Factor out the GCF and use the result to find the surface area of a cylinder where $r = 35$ cm and $h = 65$ cm. Answer in exact form and in approximate form rounded to the nearest whole number.

**52. Volume of a cylindrical shell: $\pi R^2 h - \pi r^2 h$**

The volume of a cylindrical shell (a larger cylinder with a smaller cylinder removed) can be found using the formula shown, where $R$ is the radius of the larger cylinder and $r$ is the radius of the smaller. Factor the expression completely and use the result to find the volume of a shell where

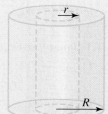

$R = 9$ cm, $r = 3$ cm, and $h = 10$ cm. Answer in exact form and in approximate form rounded to the nearest whole number.

▶ **APPLICATIONS**

**In many cases, factoring an expression can make it easier to evaluate as in the following applications.**

**53. Conical shells:** The volume of a conical shell (like the shell of an ice cream cone) is given by the formula $V = \frac{1}{3}\pi R^2 h - \frac{1}{3}\pi r^2 h$, where $R$ is the outer radius and $r$ is the inner radius of the cone. Write the formula in completely factored form, then find the volume of a shell when $R = 5.1$ cm, $r = 4.9$ cm, and $h = 9$ cm. Answer in exact form and in approximate form rounded to the nearest tenth.

**54. Spherical shells:** The volume of a spherical shell (like the outer shell of a cherry cordial) is given by the formula $V = \frac{4}{3}\pi R^3 - \frac{4}{3}\pi r^3$, where $R$ is the outer radius and $r$ is the inner radius of the shell. Write the right-hand side in completely factored form, then find the volume of a shell where $R = 1.8$ cm and $r = 1.5$ cm. Answer in exact form and in approximate form rounded to the nearest tenth.

**55. Volume of a box:** The volume of a rectangular box $x$ inches in height is given by the relationship $V = x^3 + 8x^2 + 15x$. Factor the right-hand side to determine: (a) The number of inches that the width exceeds the height, (b) the number of inches the length exceeds the height, and (c) the volume given the height is 2 ft.

**56. Shipping textbooks:** A publisher ships paperback books stacked $x$ copies high in a box. The total number of books shipped per box is given by the relationship $B = x^3 - 13x^2 + 42x$. Factor the right-hand side to determine (a) how many more or fewer books fit the width of the box (than the height), (b) how many more or fewer books fit the length of the box (than the height), and (c) the number of books shipped per box if they are stacked 10 high in the box.

**57. Space-Time relationships:** Due to the work of Albert Einstein and other physicists who labored on space-time relationships, it is known that the faster an object moves the shorter it appears to become. This phenomenon is modeled by the

**Lorentz transformation** $L = L_0\sqrt{1 - \left(\dfrac{v}{c}\right)^2}$,

where $L_0$ is the length of the object at rest, $L$ is the relative length when the object is moving at velocity $v$, and $c$ is the speed of light. Factor the radicand and use the result to determine the relative length of a 12-in. ruler if it is shot past a stationary observer at 0.75 times the speed of light ($v = 0.75c$).

**58. Tubular fluid flow:** As a fluid flows through a tube, it is flowing faster at the center of the tube than at the sides, where the tube exerts a backward drag. **Poiseuille's law** gives the velocity of the flow at any point of the cross section: $v = \dfrac{G}{4\eta}(R^2 - r^2)$, where $R$ is the inner radius of the tube, $r$ is the distance from the center of the tube to a point in the flow, $G$ represents what is called the pressure gradient, and $\eta$ is a constant that depends on the viscosity of the fluid. Factor the right-hand side and find $v$ given $R = 0.5$ cm, $r = 0.3$ cm, $G = 15$, and $\eta = 0.25$.

**Solve by factoring.**

**59. Packing and shipping:** To pack and ship fragile glass rods, a manufacturer designs a solid rectangular prism out of reprocessed cardboard, with a rectangular hollow (see diagram). (a) Find an expression representing the volume of cardboard used given $L = 4x$, $W = x$, and $H = x$, with $w = y$ and $h = y$. (b) Write the expression in completely factored form, and (c) find the volume of reprocessed cardboard used if $x = 4$ in. and $y = 1.5$ in.

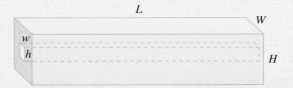

**60. Displaying artwork:** To display works of art, the design of the rectangular frame around the art is critical to the aesthetics of the art. (a) Find an expression that represents the area of the frame shown, given it has dimensions $x$ in. by $x$ in., with a width of $y$ in. (b) Write the expression in factored and simplified form, and (c) find the area of the frame given $x = 18$ in. and $y = 1.5$ in.

## ▶ EXTENDING THE CONCEPT

**61.** Factor out a constant that leaves integer coefficients for each term:

    **a.** $\frac{1}{2}x^4 + \frac{1}{8}x^3 - \frac{3}{4}x^2 + 4$

    **b.** $\frac{2}{3}b^5 - \frac{1}{6}b^3 + \frac{4}{9}b^2 - 1$

**62.** Factor each expression completely.

    **a.** $x^4 - 81$

    **b.** $16n^4 - 1$

    **c.** $q^4 - 28q^2 + 75$

    **d.** $a^4 - 18a^2 + 32$

**63.** If $x = 2$ is substituted into $2x^3 + hx + 8$, the result is zero. What is the value of $h$?

**64.** Factor the expression: $192x^3 - 164x^2 - 270x$.

**Identify the common factors in each expression, then factor completely.** [*Hint:* **For Exercises 67 through 70, note that** $kA^{\frac{1}{2}} + 3kA^{-\frac{1}{2}} = kA^{-\frac{1}{2}}(A + 3)$.]

**65.** $5x(x^2 - 4x)^2 - 25x(x^2 - 4x)$

**66.** $3x(x^3 - 18)^2 + 9x^2(x - 2)(x^3 - 18)$

**67.** $35c(c^2 - 5)^{\frac{1}{2}} - 7c^3(c^2 - 5)^{-\frac{1}{2}}$

**68.** $4d(5d - 6)(2d + 1)^{-\frac{1}{2}} - 8d(2d + 1)^{\frac{1}{2}}$

**69.** $4(b^2 - 1)^{\frac{1}{2}}(b - 8) + b(b - 8)^2(b^2 - 1)^{-\frac{1}{2}}$

**70.** $n(n - 3)^2(3n^2 - 1)^{-\frac{1}{2}} + (3n^2 - 1)^{\frac{1}{2}}(n - 3)$

**71.** As an alternative to evaluating polynomials by direct substitution, **nested factoring** can be used. The method has the advantage of using only products and sums—no powers. For $P = x^3 + 3x^2 + 1x + 5$, we begin by grouping all variable terms and factoring $x$: $P = [x^3 + 3x^2 + 1x] + 5 = x[x^2 + 3x + 1] + 5$. Then we group the inner terms with $x$ and factor again: $P = x[x^2 + 3x + 1] + 5 = x[x(x + 3) + 1] + 5$. The expression can now be evaluated using any input and the order of operations. If $x = 2$, we quickly find that $P = 27$. Use this method to evaluate $H = x^3 + 2x^2 + 5x - 9$ for $x = -3$.

---

## R.6 | Rational Expressions

### LEARNING OBJECTIVES

*In Section R.6 you will review how to:*

☐ **A.** Write a rational expression in simplest form

☐ **B.** Multiply and divide rational expressions

☐ **C.** Add and subtract rational expressions

☐ **D.** Simplify compound fractions

☐ **E.** Rewrite formulas and rational expressions

A rational number is one that can be written as the quotient of two integers. Similarly, a *rational expression* is one that can be written as the quotient of two polynomials. We can apply the skills developed in a study of fractions (how to reduce, add, subtract, multiply, and divide) to **rational expressions,** sometimes called **algebraic fractions.**

### A. Writing a Rational Expression in Simplest Form

A rational expression is in **simplest form** when the numerator and denominator have no common factors (other than 1). After factoring the numerator and denominator, we apply the **fundamental property of rational expressions.**

**Fundamental Property of Rational Expressions**

If $P$, $Q$, and $R$ are polynomials, with $Q, R \neq 0$,

$$(1) \ \frac{P \cdot R}{Q \cdot R} = \frac{P}{Q} \quad \text{and} \quad (2) \ \frac{P}{Q} = \frac{P \cdot R}{Q \cdot R}$$

In words, the property says (1) a rational expression can be simplified by canceling common factors in the numerator and denominator, and (2) an equivalent expression can be formed by multiplying numerator and denominator by the same nonzero polynomial.

**EXAMPLE 1** ▶ **Simplifying a Rational Expression**

Write the expression in simplest form: $\dfrac{x^2 - 1}{x^2 - 3x + 2}$.

**Solution** ▶

$$\frac{x^2 - 1}{x^2 - 3x + 2} = \frac{(x - 1)(x + 1)}{(x - 1)(x - 2)}$$   factor numerator and denominator

$$= \frac{\cancel{(x - 1)}(x + 1)}{\cancel{(x - 1)}(x - 2)}$$   common factors reduce to 1

$$= \frac{x + 1}{x - 2}$$   simplest form

Now try Exercises 5 through 8 ▶

Note that after simplifying the expression from Example 1, we are actually saying the resulting (simpler) expression is equivalent to the original expression for all values where both are defined. The first expression is not defined when $x = 1$ or $x = 2$, the second when $x = 2$ (since the denominators would be zero). The calculator screens shown in Figure R.9 help to illustrate this fact, and it appears that we would very much prefer to be working with the simpler expression!

**Figure R.9**

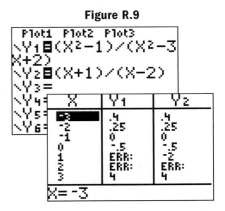

**CAUTION** ▶ When reducing rational numbers or expressions, only common *factors* can be reduced. It is incorrect to reduce (or divide out) individual terms: $\dfrac{x^2 - 1}{x^2 - 3x + 2} \neq \dfrac{-1}{-3x + 2}$.

**WORTHY OF NOTE**

If we view $a$ and $b$ as two points on the number line, we note that they are the same distance apart, regardless of the order they are subtracted. This tells us the numerator and denominator will have the same absolute value but be opposite in sign, giving a value of $-1$ (check using a few test values).

When simplifying rational expressions, we sometimes encounter expressions of the form $\dfrac{a - b}{b - a}$. If we factor $-1$ from the numerator, we see that $\dfrac{a - b}{b - a} = \dfrac{-1\cancel{(b - a)}}{\cancel{b - a}} = -1$.

**EXAMPLE 2** ▶  **Simplifying a Rational Expression**

Write the expression in simplest form: $\dfrac{6-2x}{x^2-9}$.

**Solution** ▶

$$\dfrac{6-2x}{x^2-9} = \dfrac{2\overset{-1}{\cancel{(3-x)}}}{\cancel{(x-3)}(x+3)} \qquad \text{factor numerator and denominator}$$

$$= \dfrac{(2)(-1)}{x+3} \qquad \text{reduce: } \dfrac{3-x}{x-3} = -1$$

$$= \dfrac{-2}{x+3} \qquad \text{simplest form}$$

☑ **A.** You've just reviewed
how to write a rational
expression in simplest form

Now try Exercises 9 through 14 ▶

## B. Multiplication and Division of Rational Expressions

Operations on rational expressions use the factoring skills reviewed earlier, along with much of what we know about rational numbers.

**Multiplying Rational Expressions**

Given that $P$, $Q$, $R$, and $S$ are polynomials with $Q, S \neq 0$,

$$\dfrac{P}{Q} \cdot \dfrac{R}{S} = \dfrac{PR}{QS}$$

1. Factor all numerators and denominators completely.
2. Reduce common factors.
3. Multiply numerator × numerator and denominator × denominator.

**EXAMPLE 3** ▶  **Multiplying Rational Expressions**

Compute the product: $\dfrac{2a+2}{3a-3a^2} \cdot \dfrac{3a^2-a-2}{9a^2-4}$.

**Solution** ▶

$$\dfrac{2a+2}{3a-3a^2} \cdot \dfrac{3a^2-a-2}{9a^2-4} = \dfrac{2(a+1)}{3a(1-a)} \cdot \dfrac{(3a+2)(a-1)}{(3a-2)(3a+2)} \qquad \text{factor}$$

$$= \dfrac{2(a+1)}{3a\cancel{(1-a)}} \cdot \dfrac{\overset{1}{\cancel{(3a+2)}}\overset{(-1)}{\cancel{(a-1)}}}{(3a-2)\cancel{(3a+2)}} \qquad \text{reduce: } \dfrac{a-1}{1-a} = -1$$

$$= \dfrac{-2(a+1)}{3a(3a-2)} \qquad \text{simplest form}$$

Now try Exercises 15 through 18 ▶

To divide fractions, we multiply the first expression by the *reciprocal of the second* (we sometimes say, "invert the divisor and multiply"). The quotient of two rational expressions is computed in the same way.

### Dividing Rational Expressions

Given that $P$, $Q$, $R$, and $S$ are polynomials with $Q, R, S \neq 0$,

$$\frac{P}{Q} \div \frac{R}{S} = \frac{P}{Q} \cdot \frac{S}{R} = \frac{PS}{QR}$$

Invert the divisor and multiply.

---

**EXAMPLE 4** ▶ **Dividing Rational Expressions**

Compute the quotient $\dfrac{4m^3 - 12m^2 + 9m}{m^2 - 49} \div \dfrac{10m^2 - 15m}{m^2 + 4m - 21}$.

**Solution** ▶
$$\frac{4m^3 - 12m^2 + 9m}{m^2 - 49} \div \frac{10m^2 - 15m}{m^2 + 4m - 21}$$

$$= \frac{4m^3 - 12m^2 + 9m}{m^2 - 49} \cdot \frac{m^2 + 4m - 21}{10m^2 - 15m} \qquad \text{invert and multiply}$$

$$= \frac{m(4m^2 - 12m + 9)}{(m + 7)(m - 7)} \cdot \frac{(m + 7)(m - 3)}{5m(2m - 3)} \qquad \text{factor}$$

$$= \frac{\overset{1}{\cancel{m}}(2m - 3)\overset{1}{\cancel{(2m - 3)}}}{\cancel{(m + 7)}(m - 7)} \cdot \frac{\overset{1}{\cancel{(m + 7)}}(m - 3)}{5\cancel{m}\cancel{(2m - 3)}} \qquad \text{factor and reduce}$$

$$= \frac{(2m - 3)(m - 3)}{5(m - 7)} \qquad \text{lowest terms}$$

 **B.** You've just reviewed how to multiply and divide rational expressions

Note that we sometimes refer to simplest form as *lowest terms*.

**Now try Exercises 19 through 38** ▶

---

**WATCH YOUR STEP**

**CAUTION** ▶ For products like $\dfrac{(w + 7)(w - 7)}{(w - 7)(w - 2)} \cdot \dfrac{w - 2}{w + 7}$, it is a common mistake to think that all factors "cancel," with "nothing left" and an answer of zero. Actually, all factors *reduce to 1,* and the result is a value of 1 for all inputs where the product is defined.

$$\frac{\overset{1}{\cancel{(w + 7)}}\overset{1}{\cancel{(w - 7)}}}{\cancel{(w - 7)}\cancel{(w - 2)}} \cdot \frac{\overset{1}{\cancel{w - 2}}}{\cancel{w + 7}} = 1$$

---

## C. Addition and Subtraction of Rational Expressions

Recall that the addition and subtraction of *fractions* requires finding the lowest common denominator (LCD) and building equivalent fractions. The sum or difference of the numerators is then placed over this denominator. The procedure for the addition and subtraction of *rational expressions* is very much the same. Note that the LCD can also be described as the least common multiple (LCM) of all denominators.

### Addition and Subtraction of Rational Expressions

1. Find the LCD of all rational expressions.
2. Build equivalent expressions using the LCD.
3. Add or subtract numerators as indicated.
4. Write the result in lowest terms.

**EXAMPLE 5** ▶ **Adding and Subtracting Rational Expressions**

Compute as indicated:

a. $\dfrac{7}{10x} + \dfrac{3}{25x^2}$     b. $\dfrac{10x}{x^2 - 9} - \dfrac{5}{x - 3}$

**Solution** ▶ a. The LCM for $10x$ and $25x^2$ is $50x^2$.   find the LCD

$$\dfrac{7}{10x} + \dfrac{3}{25x^2} = \dfrac{7}{10x} \cdot \dfrac{5x}{5x} + \dfrac{3}{25x^2} \cdot \dfrac{2}{2}$$   write equivalent expressions

$$= \dfrac{35x}{50x^2} + \dfrac{6}{50x^2}$$   simplify

$$= \dfrac{35x + 6}{50x^2}$$   add the numerators and write the result over the LCD

The result is in simplest form.

b. The LCM for $x^2 - 9$ and $x - 3$ is $(x - 3)(x + 3)$.   find the LCD

$$\dfrac{10x}{x^2 - 9} - \dfrac{5}{x - 3} = \dfrac{10x}{(x - 3)(x + 3)} - \dfrac{5}{x - 3} \cdot \dfrac{x + 3}{x + 3}$$   write equivalent expressions

$$= \dfrac{10x - 5(x + 3)}{(x - 3)(x + 3)}$$   subtract numerators, write the result over the LCD

$$= \dfrac{10x - 5x - 15}{(x - 3)(x + 3)}$$   distribute

$$= \dfrac{5x - 15}{(x - 3)(x + 3)}$$   combine like terms

$$= \dfrac{5\overset{1}{\cancel{(x - 3)}}}{\cancel{(x - 3)}(x + 3)} = \dfrac{5}{x + 3}$$   factor and reduce

**Now try Exercises 39 through 44** ▶

**EXAMPLE 6** ▶ **Adding and Subtracting Rational Expressions**

Perform the operations indicated:

a. $\dfrac{5}{n + 2} - \dfrac{n - 3}{n^2 - 4}$     b. $\dfrac{b^2}{4a^2} - \dfrac{c}{a}$

**Solution** ▶ a. The LCM for $n + 2$ and $n^2 - 4$ is $(n + 2)(n - 2)$.

$$\dfrac{5}{n + 2} - \dfrac{n - 3}{n^2 - 4} = \dfrac{5}{n + 2} \cdot \dfrac{n - 2}{n - 2} - \dfrac{n - 3}{(n + 2)(n - 2)}$$   write equivalent expressions

$$= \dfrac{5(n - 2) - (n - 3)}{(n + 2)(n - 2)}$$   subtract numerators, write the result over the LCD

$$= \dfrac{5n - 10 - n + 3}{(n + 2)(n - 2)}$$   distribute

$$= \dfrac{4n - 7}{(n + 2)(n - 2)}$$   result

**b.** The LCM for $a$ and $4a^2$ is $4a^2$:   $\dfrac{b^2}{4a^2} - \dfrac{c}{a} = \dfrac{b^2}{4a^2} - \dfrac{c}{a} \cdot \dfrac{4a}{4a}$   write equivalent expressions

$$= \dfrac{b^2}{4a^2} - \dfrac{4ac}{4a^2} \qquad \text{simplify}$$

$$= \dfrac{b^2 - 4ac}{4a^2} \qquad \text{subtract numerators, write the result over the LCD}$$

 **C.** You've just reviewed how to add and subtract rational expressions

**Now try Exercises 45 through 58** ▶

 **CAUTION** ▶ When the second term in a subtraction has a binomial numerator as in Example 6(b), be sure the subtraction *is applied to both terms*. It is a common error to write

$$\dfrac{5(n-2)}{(n+2)(n-2)} - \dfrac{n-3}{(n+2)(n-2)} = \dfrac{5n - 10 \;\widehat{-\; n - 3}}{(n+2)(n-2)} \; \text{✗}$$ in which the subtraction is applied

to the first term only. This is incorrect!

## D. Simplifying Compound Fractions

Rational expressions whose numerator or denominator contain a fraction are called **compound fractions.** The expression $\dfrac{\dfrac{2}{3m} - \dfrac{3}{2}}{\dfrac{3}{4m} - \dfrac{1}{3m^2}}$ is a compound fraction with a

numerator of $\dfrac{2}{3m} - \dfrac{3}{2}$ and a denominator of $\dfrac{3}{4m} - \dfrac{1}{3m^2}$. The two methods commonly used to simplify compound fractions are summarized in the following boxes.

---

**Simplifying Compound Fractions (Method I)**

1. Add/subtract fractions in the numerator, writing them as a single expression.
2. Add/subtract fractions in the denominator, also writing them as a single expression.
3. Multiply the numerator by the reciprocal of the denominator and simplify if possible.

---

**Simplifying Compound Fractions (Method II)**

1. Find the LCD of all fractions in the numerator and denominator.
2. Multiply the numerator and denominator by this LCD and simplify.
3. Simplify further if possible.

---

Method II is illustrated in Example 7.

**EXAMPLE 7** ▶ **Simplifying a Compound Fraction**

Simplify the compound fraction:

$$\dfrac{\dfrac{2}{3m} - \dfrac{3}{2}}{\dfrac{3}{4m} - \dfrac{1}{3m^2}}$$

**Solution** ▶  The LCD for all fractions is $12m^2$.

$$\dfrac{\dfrac{2}{3m} - \dfrac{3}{2}}{\dfrac{3}{4m} - \dfrac{1}{3m^2}} = \dfrac{\left(\dfrac{2}{3m} - \dfrac{3}{2}\right)\left(\dfrac{12m^2}{1}\right)}{\left(\dfrac{3}{4m} - \dfrac{1}{3m^2}\right)\left(\dfrac{12m^2}{1}\right)}$$

multiply numerator and
denominator by $12m^2 = \dfrac{12m^2}{1}$

$$= \dfrac{\left(\dfrac{2}{3m}\right)\left(\dfrac{12m^2}{1}\right) - \left(\dfrac{3}{2}\right)\left(\dfrac{12m^2}{1}\right)}{\left(\dfrac{3}{4m}\right)\left(\dfrac{12m^2}{1}\right) - \left(\dfrac{1}{3m^2}\right)\left(\dfrac{12m^2}{1}\right)}$$

distribute

$$= \dfrac{8m - 18m^2}{9m - 4}$$

simplify

$$= \dfrac{2m(4 \overset{-1}{\cancel{- 9m}})}{\cancel{9m - 4}} = -2m$$

factor and write in lowest terms

☑ **D.** You've just reviewed
how to simplify compound
fractions

Now try Exercises 59 through 66 ▶

## E.  Rewriting Formulas and Rational Expressions

In many fields of study, formulas and algebraic models involve rational expressions
and we often need to write them in an alternative form.

---

**EXAMPLE 8** ▶  **Rewriting a Formula**

In an electrical circuit with two resistors in parallel, the total resistance $R$ is related
to resistors $R_1$ and $R_2$ by the formula $\dfrac{1}{R} = \dfrac{1}{R_1} + \dfrac{1}{R_2}$. Rewrite the right-hand side as
a single term.

**Solution** ▶  $\dfrac{1}{R} = \dfrac{1}{R_1} + \dfrac{1}{R_2}$     LCD for the right-hand side is $R_1R_2$

$\quad\quad = \dfrac{R_2}{R_1R_2} + \dfrac{R_1}{R_1R_2}$     build equivalent expressions using LCD

$\quad\quad = \dfrac{R_2 + R_1}{R_1R_2}$     write as a single expression

Now try Exercises 67 and 68 ▶

---

**EXAMPLE 9** ▶  **Simplifying a Rational Expression**

When studying rational expressions and rates of change, we encounter the

expression $\dfrac{\dfrac{1}{x + h} - \dfrac{1}{x}}{h}$. Simplify the compound fraction.

**Solution** ▶ Using Method I gives:

$$\frac{\dfrac{1}{x+h}-\dfrac{1}{x}}{h}=\frac{\dfrac{x}{x(x+h)}-\dfrac{x+h}{x(x+h)}}{h} \qquad \text{LCD for the numerator is } x(x+h)$$

$$=\frac{\dfrac{x-(x+h)}{x(x+h)}}{h} \qquad \text{write numerator as a single expression}$$

$$=\frac{\dfrac{-h}{x(x+h)}}{h} \qquad \text{simplify}$$

$$=\frac{-h}{x(x+h)}\cdot\frac{1}{h} \qquad \text{invert and multiply}$$

$$=\frac{-1}{x(x+h)} \qquad \text{result}$$

 **E.** You've just reviewed how to rewrite formulas and rational expressions

**Now try Exercises 69 through 72** ▶

## R.6 EXERCISES

▶ **CONCEPTS AND VOCABULARY**

**Fill in each blank with the appropriate word or phrase. Carefully reread the section, if necessary.**

1. As with numeric fractions, algebraic fractions require a _____ _____ for addition and subtraction.

2. Since $x^2 + 9$ is prime, the expression $(x^2 + 9)/(x + 3)$ is already written in _____ _____.

**State T or F and discuss/explain your response.**

3. $\dfrac{x}{x+3} - \dfrac{x+1}{x+3} = \dfrac{1}{x+3}$

4. $\dfrac{\cancel{(x+3)}\cancel{(x-2)}}{\cancel{(x-2)}\cancel{(x+3)}} = 0$

▶ **DEVELOPING YOUR SKILLS**

**Reduce to lowest terms. Recall that $a - b = -1(b - a)$.**

5. a. $\dfrac{a-7}{3a-21}$  b. $\dfrac{2x+6}{4x^2-8x}$

6. a. $\dfrac{x-4}{7x-28}$  b. $\dfrac{3x-18}{6x^2-12x}$

7. a. $\dfrac{r^2+3r-10}{r^2+r-6}$  b. $\dfrac{m^2+3m-4}{m^2-4m}$

8. a. $\dfrac{x^2-5x-14}{x^2+6x-7}$  b. $\dfrac{a^2+3a-28}{a^2-49}$

9. a. $\dfrac{x-7}{7-x}$  b. $\dfrac{5-x}{x-5}$

10. a. $\dfrac{v^2-3v-28}{49-v^2}$  b. $\dfrac{u^2-10u+25}{25-u^2}$

11. a. $\dfrac{-12a^3b^5}{4a^2b^{-4}}$  b. $\dfrac{7x+21}{63}$

    c. $\dfrac{y^2-9}{3-y}$  d. $\dfrac{m^3n-m^3}{m^4-m^4n}$

12. a. $\dfrac{5m^{-3}n^5}{-10mn^2}$  b. $\dfrac{-5v+20}{25}$

    c. $\dfrac{n^2-4}{2-n}$  d. $\dfrac{w^4-w^4v}{w^3v-w^3}$

13. **a.** $\dfrac{2n^3 + n^2 - 3n}{n^3 - n^2}$  **b.** $\dfrac{6x^2 + x - 15}{4x^2 - 9}$

**c.** $\dfrac{x^3 + 8}{x^2 - 2x + 4}$  **d.** $\dfrac{mn^2 + n^2 - 4m - 4}{mn + n + 2m + 2}$

14. **a.** $\dfrac{x^3 + 4x^2 - 5x}{x^3 - x}$  **b.** $\dfrac{5p^2 - 14p - 3}{5p^2 + 11p + 2}$

**c.** $\dfrac{12y^2 - 13y + 3}{27y^3 - 1}$  **d.** $\dfrac{ax^2 - 5x^2 - 3a + 15}{ax - 5x + 5a - 25}$

**Compute as indicated. Write final results in lowest terms.**

15. $\dfrac{a^2 - 4a + 4}{a^2 - 9} \cdot \dfrac{a^2 - 2a - 3}{a^2 - 4}$

16. $\dfrac{b^2 + 5b - 24}{b^2 - 6b + 9} \cdot \dfrac{b}{b^2 - 64}$

17. $\dfrac{x^2 - 7x - 18}{x^2 - 6x - 27} \cdot \dfrac{2x^2 + 7x + 3}{2x^2 + 5x + 2}$

18. $\dfrac{6v^2 + 23v + 21}{4v^2 - 4v - 15} \cdot \dfrac{4v^2 - 25}{3v + 7}$

19. $\dfrac{p^3 - 64}{p^3 - p^2} \div \dfrac{p^2 + 4p + 16}{p^2 - 5p + 4}$

20. $\dfrac{a^2 + 3a - 28}{a^2 + 5a - 14} \div \dfrac{a^3 - 4a^2}{a^3 - 8}$

21. $\dfrac{3x - 9}{4x + 12} \div \dfrac{3 - x}{5x + 15}$

22. $\dfrac{5b - 10}{7b - 28} \div \dfrac{2 - b}{5b - 20}$

23. $\dfrac{a^2 + a}{a^2 - 3a} \cdot \dfrac{3a - 9}{2a + 2}$

24. $(m^2 - 16) \cdot \dfrac{m^2 - 5m}{m^2 - m - 20}$

25. $\dfrac{xy - 3x + 2y - 6}{x^2 - 3x - 10} \div \dfrac{xy - 3x}{xy - 5y}$

26. $\dfrac{2a - ab + 7b - 14}{b^2 - 14b + 49} \div \dfrac{ab - 2a}{ab - 7a}$

27. $\dfrac{m^2 + 2m - 8}{m^2 - 2m} \div \dfrac{m^2 - 16}{m^2}$

28. $\dfrac{18 - 6x}{x^2 - 25} \div \dfrac{2x^2 - 18}{x^3 - 2x^2 - 25x + 50}$

29. $\dfrac{y + 3}{3y^2 + 9y} \cdot \dfrac{y^2 + 7y + 12}{y^2 - 16} \div \dfrac{y^2 + 4y}{y^2 - 4y}$

30. $\dfrac{x^2 + 4x - 5}{x^2 - 5x - 14} \div \dfrac{x^2 - 1}{x^2 - 4} \cdot \dfrac{x + 1}{x + 5}$

31. $\dfrac{x^2 - 0.49}{x^2 + 0.5x - 0.14} \div \dfrac{x^2 - x + 0.21}{x^2 - 0.09}$

32. $\dfrac{x^2 - 0.25}{x^2 + 0.1x - 0.2} \div \dfrac{x^2 - 0.8x + 0.15}{x^2 - 0.16}$

33. $\dfrac{n^2 - \dfrac{4}{9}}{n^2 - \dfrac{13}{15}n + \dfrac{2}{15}} \div \dfrac{n^2 + \dfrac{4}{3}n + \dfrac{4}{9}}{n^2 - \dfrac{1}{25}}$

34. $\dfrac{q^2 - \dfrac{9}{25}}{q^2 - \dfrac{1}{10}q - \dfrac{3}{10}} \div \dfrac{q^2 + \dfrac{17}{20}q + \dfrac{3}{20}}{q^2 - \dfrac{1}{16}}$

35. $\dfrac{3a^3 - 24a^2 - 12a + 96}{a^2 - 11a + 24} \div \dfrac{6a^2 - 24}{3a^3 - 81}$

36. $\dfrac{p^3 + p^2 - 49p - 49}{p^2 + 6p - 7} \div \dfrac{p^2 + p + 1}{p^3 - 1}$

37. $\dfrac{4n^2 - 1}{12n^2 - 5n - 3} \cdot \dfrac{6n^2 + 5n + 1}{2n^2 + n} \cdot \dfrac{12n^2 - 17n + 6}{6n^2 - 7n + 2}$

38. $\dfrac{4x^2 - 25}{x^2 - 11x + 30} \div \dfrac{2x^2 - x - 15}{x^2 - 9x + 18} \cdot \dfrac{4x^2 + 25x - 21}{12x^2 - 5x - 3}$

**Compute as indicated. Write answers in lowest terms.**

39. $\dfrac{3}{8x^2} + \dfrac{5}{2x}$  40. $\dfrac{15}{16y} - \dfrac{7}{2y^2}$

41. $\dfrac{7}{4x^2y^3} - \dfrac{1}{8xy^4}$  42. $\dfrac{3}{6a^3b} + \dfrac{5}{9ab^3}$

43. $\dfrac{4p}{p^2 - 36} - \dfrac{2}{p - 6}$  44. $\dfrac{3q}{q^2 - 49} - \dfrac{3}{2q - 14}$

45. $\dfrac{m}{m^2 - 16} + \dfrac{4}{4 - m}$  46. $\dfrac{2}{4 - p^2} + \dfrac{p}{p - 2}$

47. $\dfrac{2}{m - 7} - 5$  48. $\dfrac{4}{x - 1} - 9$

49. $\dfrac{y + 1}{y^2 + y - 30} - \dfrac{2}{y + 6}$

50. $\dfrac{1}{a + 4} + \dfrac{a}{a^2 - a - 20}$

51. $\dfrac{2x - 1}{x^2 + 3x - 4} - \dfrac{x - 5}{x^2 + 3x - 4}$

52. $\dfrac{3y - 4}{y^2 + 2y + 1} - \dfrac{2y - 5}{y^2 + 2y + 1}$

53. $\dfrac{-2}{3a + 12} - \dfrac{7}{a^2 + 4a}$

54. $\dfrac{2}{m^2 - 9} + \dfrac{m - 5}{m^2 + 6m + 9}$

55. $\dfrac{y + 2}{5y^2 + 11y + 2} + \dfrac{5}{y^2 + y - 6}$

56. $\dfrac{m - 4}{3m^2 - 11m + 6} + \dfrac{m}{2m^2 - m - 15}$

**Write each term as a rational expression. Then compute the sum or difference indicated.**

**57. a.** $p^{-2} - 5p^{-1}$        **b.** $x^{-2} + 2x^{-3}$

**58. a.** $3a^{-1} + (2a)^{-1}$      **b.** $2y^{-1} - (3y)^{-1}$

**Simplify each compound fraction. Use either method.**

**59.** $\dfrac{\dfrac{5}{a} - \dfrac{1}{4}}{\dfrac{25}{a^2} - \dfrac{1}{16}}$      **60.** $\dfrac{\dfrac{8}{x^3} - \dfrac{1}{27}}{\dfrac{2}{x} - \dfrac{1}{3}}$

**61.** $\dfrac{p + \dfrac{1}{p-2}}{1 + \dfrac{1}{p-2}}$      **62.** $\dfrac{1 + \dfrac{3}{y-6}}{y + \dfrac{9}{y-6}}$

**63.** $\dfrac{\dfrac{2}{y^2 - y - 20}}{\dfrac{3}{y+4} - \dfrac{4}{y-5}}$      **64.** $\dfrac{\dfrac{2}{x^2 - 3x - 10}}{\dfrac{6}{x+2} - \dfrac{4}{x-5}}$

**Rewrite each expression as a compound fraction. Then simplify using either method.**

**65. a.** $\dfrac{1 + 3m^{-1}}{1 - 3m^{-1}}$      **b.** $\dfrac{1 + 2x^{-2}}{1 - 2x^{-2}}$

**66. a.** $\dfrac{4 - 9a^{-2}}{3a^{-2}}$      **b.** $\dfrac{3 + 2n^{-1}}{5n^{-2}}$

**Rewrite each expression as a single term.**

**67.** $\dfrac{1}{f_1} + \dfrac{1}{f_2}$      **68.** $\dfrac{1}{w} + \dfrac{1}{x} - \dfrac{1}{y}$

**69.** $\dfrac{\dfrac{a}{x+h} - \dfrac{a}{x}}{h}$      **70.** $\dfrac{\dfrac{a}{h-x} - \dfrac{a}{-x}}{h}$

**71.** $\dfrac{\dfrac{1}{2(x+h)^2} - \dfrac{1}{2x^2}}{h}$      **72.** $\dfrac{\dfrac{a}{(x+h)^2} - \dfrac{a}{x^2}}{h}$

## ▶ WORKING WITH FORMULAS

**73. Cost to seize illegal drugs:** $C = \dfrac{450P}{100 - P}$

The cost $C$, in millions of dollars, for a government to find and seize $P\%$ ($0 \leq P < 100$) of a certain illegal drug is modeled by the rational equation shown. Complete the table (round to the nearest dollar) and answer the following questions.

| $P$ | $\dfrac{450P}{100 - P}$ |
|-----|------|
| 40 | |
| 60 | |
| 80 | |
| 90 | |
| 93 | |
| 95 | |
| 98 | |
| 100 | |

  **a.** What is the cost of seizing 40% of the drugs? Estimate the cost at 85%.

  **b.** Why does cost increase dramatically the closer you get to 100%?

  **c.** Will 100% of the drugs ever be seized?

**74. Chemicals in the bloodstream:** $C = \dfrac{200H^2}{H^3 + 40}$

Rational equations are often used to model chemical concentrations in the bloodstream. The percent concentration $C$ of a certain drug $H$ hours after injection into muscle tissue can be modeled by the equation shown ($H \geq 0$). Complete the table (round to the nearest tenth of a percent) and answer the following questions.

| $H$ | $\dfrac{200H^2}{H^3 + 40}$ |
|-----|------|
| 0 | |
| 1 | |
| 2 | |
| 3 | |
| 4 | |
| 5 | |
| 6 | |
| 7 | |

  **a.** What is the percent concentration of the drug 3 hr after injection?

  **b.** Why is the concentration virtually equal at $H = 4$ and $H = 5$?

  **c.** Why does the concentration begin to decrease?

  **d.** How long will it take for the concentration to become less than 10%?

## ▶ APPLICATIONS

**75. Stock prices:** When a hot new stock hits the market, its price will often rise dramatically and then taper off over time. The equation $P = \dfrac{50(7d^2 + 10)}{d^3 + 50}$ models the price of stock XYZ $d$ days after it has "hit the market."

(a) Create a table of values showing the price of the stock for the first 10 days (rounded to the nearest cent) and comment on what you notice. (b) Find the opening price of the stock. (c) Does the stock ever return to its original price?

**76. Wildlife populations:** The Department of Wildlife introduces 60 elk into a new game reserve. It is projected that the size of the herd will grow according to the equation $N = \dfrac{10(6 + 3t)}{1 + 0.05t}$, where $N$ is the number of elk and $t$ is the time in years. Approximate the population of elk after 14 yr.

**77. Typing speed:** The number of words per minute that a beginner can type is approximated by the equation $N = \dfrac{60t - 120}{t}$, where $N$ is the number of words per minute after $t$ weeks, $3 < t < 12$. Use a table to determine how many weeks it takes for a student to be typing an average of forty-five words per minute.

**78. Memory retention:** A group of students is asked to memorize 50 Russian words that are unfamiliar to them. The number $N$ of these words that the average student remembers $D$ days later is modeled by the equation $N = \dfrac{5D + 35}{D}$ $(D \geq 1)$. How many words are remembered after (a) 1 day? (b) 5 days?

(c) 12 days? (d) 35 days? (e) 100 days? According to this model, is there a certain number of words that the average student never forgets? How many?

**79. Pollution removal:** For a steel mill, the cost $C$ (in millions of dollars) to remove toxins from the resulting sludge is given by $C = \dfrac{22P}{100 - P}$, where $P$ is the percent of the toxins removed. What is the cost to remove 75% of the toxins?

**80. Average speed:** For a commuter airline flying between two major cities, the average speed $s$ of the plane is given by the formula shown, where $d$ is the distance between the cities, $R$ is the outbound rate of the plane, and $r$ is the inbound rate. (a) Simplify the compound fraction. (b) Suppose the outbound rate is 250 mph, and the inbound rate is 200 mph (due to strong headwinds). To the nearest whole mph, what is the average speed of the plane for the entire trip? (c) Discuss/Explain why the average speed is not $(250 + 200)/2$.

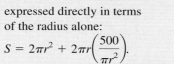
$$s = \dfrac{2d}{\dfrac{d}{R} + \dfrac{d}{r}}$$

## ▶ EXTENDING THE CONCEPT

**81.** One of these expressions is *not* equal to the others. Identify which and explain why.

    **a.** $\dfrac{20n}{10n}$          **b.** $20 \cdot n \div 10 \cdot n$

    **c.** $20n \cdot \dfrac{1}{10n}$      **d.** $\dfrac{20}{10} \cdot \dfrac{n}{n}$

**82.** Given the rational numbers $\dfrac{2}{5}$ and $\dfrac{3}{4}$, what is the reciprocal of the sum of their reciprocals? Given that $\dfrac{a}{b}$ and $\dfrac{c}{d}$ are *any* two numbers—what is the reciprocal of the sum of their reciprocals?

**83.** The average of $A$ and $B$ is $x$. The average of $C$, $D$, and $E$ is $y$. The average of $A$, $B$, $C$, $D$, and $E$ is:

    **a.** $\dfrac{3x + 2y}{5}$      **b.** $\dfrac{2x + 3y}{5}$

    **c.** $\dfrac{2(x + y)}{5}$    **d.** $\dfrac{3(x + y)}{5}$

**84.** The surface area of a cylinder is given by the formula $S = 2\pi r^2 + 2\pi rh$ (see diagram). If the cylinder has a volume of 500 cm³ $(V = \pi r^2 h)$, the surface area can be expressed directly in terms of the radius alone: $S = 2\pi r^2 + 2\pi r\left(\dfrac{500}{\pi r^2}\right)$.

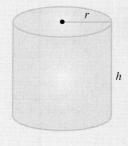

(a) Write the expression on the right-hand side as a single term. (b) Approximate the surface area if the cylinder has a radius of $r = 5$ cm.

## OVERVIEW OF CHAPTER R

## R.1 Prerequisite Definitions, Properties, Formulas, and Relationships

### Notation and Relations

| concept | notation | description | example |
|---|---|---|---|
| • Set notation: | $\{members\}$ | braces enclose the members of a set | set of even whole numbers $A = \{0, 2, 4, 6, 8, \ldots\}$ |
| • Is an element of | $\in$ | indicates membership in a set | $14 \in A$ |
| • Empty set | $\varnothing$ or $\{ \ \}$ | a set having no elements | odd numbers in $A$ |
| • Is a proper subset of | $\subset$ | indicates the elements of one set are entirely contained in another | $S = \{0, 6, 12, 18, 24, \ldots\}$ $S \subset A$ |
| • Defining a set | $\{x \mid x \ldots\}$ | the set of all $x$, *such that* $x \ldots$ | $S = \{x \mid x = 6n \text{ for } n \in \mathbb{W}\}$ |

### Sets of Numbers

- Natural: $\mathbb{N} = \{1, 2, 3, 4, \ldots\}$
- Integers: $\mathbb{Z} = \{\ldots, -3, -2, -1, 0, 1, 2, 3, \ldots\}$
- Irrational: $\mathbb{H} = \{$numbers with a nonterminating, nonrepeating decimal form$\}$
- Whole: $\mathbb{W} = \{0, 1, 2, 3, \ldots\}$
- Rational: $\mathbb{Q} = \left\{ \dfrac{p}{q}, \text{ where } p, q \in \mathbb{Z}; q \neq 0 \right\}$
- Real: $\mathbb{R} = \{$all rational and irrational numbers$\}$

### Absolute Value of a Number

- The distance between a number $n$ and zero (always positive)

$$|n| = \begin{cases} n & \text{if } n \geq 0 \\ -n & \text{if } n < 0 \end{cases}$$

### Distance between numbers $a$ and $b$ on the number line

$$|a - b|$$
$$\text{or}$$
$$|b - a|$$

## R.2 Properties of Real Numbers: For real numbers $a$, $b$, and $c$,

### Commutative Property
- Addition: $a + b = b + a$
- Multiplication: $a \cdot b = b \cdot a$

### Associative Property
- Addition: $(a + b) + c = a + (b + c)$
- Multiplication: $(a \cdot b) \cdot c = a \cdot (b \cdot c)$

### Identities
- Additive: $0 + a = a$
- Multiplicative: $1 \cdot a = a$

### Inverses
- Additive: $a + (-a) = 0$
- Multiplicative: $a \cdot \dfrac{1}{a} = 1; a \neq 0$

### Distributive Property (product to sum)
$$a(b + c) = ab + ac$$

### Distributive Property (sum to product)
$$ab + ac = a(b + c)$$

### Evaluating Algebraic Expressions
1. Replace each variable with empty/open parentheses.
2. Substitute the values given for each variable.
3. Simplify using the order of operations.

### Simplifying Algebraic Expressions
1. Eliminate parentheses by applying the distributive property.
2. Use the properties of real numbers to group like terms.
3. Use the distributive property to combine like terms.

## R.3  Properties of Exponents:
For real numbers $a$ and $b$, and integers $m$, $n$, and $p$ (excluding 0 in any denominator),

- Product property: $b^m \cdot b^n = b^{m+n}$
- Product to a power: $(a^m b^n)^p = a^{mp} \cdot b^{np}$
- Quotient property: $\dfrac{b^m}{b^n} = b^{m-n}$
- Negative exponents: $b^{-n} = \dfrac{1}{b^n}; \left(\dfrac{a}{b}\right)^{-n} = \left(\dfrac{b}{a}\right)^n$

- Power property: $(b^m)^n = b^{mn}$
- Quotient to a power: $\left(\dfrac{a^m}{b^n}\right)^p = \dfrac{a^{mp}}{b^{np}}$
- Zero exponents: $b^0 = 1 \ (b \neq 0)$
- Scientific notation: $N \times 10^k; \ 1 \le |N| < 10, \ k \in \mathbb{Z}$

## R.3  Polynomials

### Classification
- Polynomials are classified as a monomial, binomial, trinomial, or polynomial, depending on the number of terms.

### Degree
- The degree of a polynomial in one variable is the same as the largest exponent occurring on the variable in any term.

### Standard Form
- A polynomial expression is in standard form when written with the terms in descending order of degree.

### Leading Coefficient
- The leading coefficient of a polynomial is the numerical coefficient of the term with highest degree.

## R.3  Polynomial Products
- $(x + a)(x + b) = x^2 + (a + b)x + ab$
- $(A + B)^2 = A^2 + 2AB + B^2$
- $(A + B)(A^2 - AB + B^2) = A^3 + B^3$

- $(A + B)(A - B) = A^2 - B^2$
- $(A - B)^2 = A^2 - 2AB + B^2$
- $(A - B)(A^2 + AB + B^2) = A^3 - B^3$

## R.4  Properties of Radicals
- $\sqrt{a}$ is a real number only for $a \ge 0$
- $\sqrt[n]{a} = b$, only if $b^n = a$
- For any real number $a$, $\sqrt[n]{a^n} = |a|$ when $n$ is even
- If $\sqrt[n]{a}$ represents a real number and $n \ge 2$ is an integer, then $\sqrt[n]{a} = a^{\frac{1}{n}}$

- If $\sqrt[n]{A}$ and $\sqrt[n]{B}$ represent real numbers,
$$\sqrt[n]{AB} = \sqrt[n]{A} \cdot \sqrt[n]{B}$$

- A radical expression is in simplest form when:

- If $n$ is even, $\sqrt[n]{a}$ is a real number only if $a \ge 0$
- $\sqrt{a} = b$, only if $b^2 = a$
- For any real number $a$, $\sqrt[n]{a^n} = a$ when $n$ is odd
- If $\dfrac{m}{n}$ is a rational number written in lowest terms with $n \ge 2$, then $a^{\frac{m}{n}} = (\sqrt[n]{a})^m$ or $a^{\frac{m}{n}} = \sqrt[n]{a^m}$ provided $\sqrt[n]{a}$ represents a real number.
- If $\sqrt[n]{A}$ and $\sqrt[n]{B}$ represent real numbers and $B \neq 0$,
$$\sqrt[n]{\dfrac{A}{B}} = \dfrac{\sqrt[n]{A}}{\sqrt[n]{B}}$$

1. the radicand has no factors that are perfect $n$th roots,
2. the radicand contains no fractions, and
3. no radicals occur in a denominator.

## R.4  Pythagorean Theorem
- For any right triangle with legs $a$ and $b$ and hypotenuse $c$: $a^2 + b^2 = c^2$.

- For any triangle with sides $a$, $b$, and longest side $c$, if $a^2 + b^2 = c^2$, then the triangle is a right triangle.

## R.5  Factoring Polynomials
- Greatest common factor: $kx + ka = k(x + a)$
- Trinomial factoring $(a = 1)$:
$x^2 + (a + b)x + ab = (x + a)(x + b)$

- Common binomial factor:
$x(x + a) + b(x + a) = (x + a)(x + b)$

## R.5  Special Factorizations
- $A^2 - B^2 = (A + B)(A - B)$
- $A^3 + B^3 = (A + B)(A^2 - AB + B^2)$

- $A^2 \pm 2AB + B^2 = (A \pm B)^2$
- $A^3 - B^3 = (A - B)(A^2 + AB + B^2)$

## R.6 Rational Expressions: For polynomials $P$, $Q$, $R$, and $S$ with no denominator of zero,

- Lowest terms: $\dfrac{P \cdot R}{Q \cdot R} = \dfrac{P}{Q}$

- Multiplication: $\dfrac{P}{Q} \cdot \dfrac{R}{S} = \dfrac{P \cdot R}{Q \cdot S} = \dfrac{PR}{QS}$

- Addition: $\dfrac{P}{R} + \dfrac{Q}{R} = \dfrac{P + Q}{R}$

- Equivalence: $\dfrac{P}{Q} = \dfrac{P \cdot R}{Q \cdot R}$

- Division: $\dfrac{P}{Q} \div \dfrac{R}{S} = \dfrac{P}{Q} \cdot \dfrac{S}{R} = \dfrac{PS}{QR}$

- Subtraction: $\dfrac{P}{R} - \dfrac{Q}{R} = \dfrac{P - Q}{R}$

- Addition/subtraction with unlike denominators:
  1. Find the LCD of all rational expressions.
  2. Build equivalent expressions using LCD.
  3. Add/subtract numerators as indicated.
  4. Write the result in lowest terms.

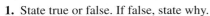

## PRACTICE TEST

1. State true or false. If false, state why.
   a. $\mathbb{H} \subset \mathbb{R}$
   b. $\mathbb{N} \subset \mathbb{Q}$
   c. $\sqrt{2} \in \mathbb{Q}$
   d. $\dfrac{1}{2} \notin \mathbb{W}$

2. State true or false. If false, state why.
   a. $(3 + 4)^2 = 25$
   b. $\dfrac{7}{0} = 0$
   c. $x - 3 = -3 + x$
   d. $-2(x - 3) = -2x - 3$

3. State the value of each expression.
   a. $\sqrt{121}$
   b. $\sqrt[3]{-125}$
   c. $\sqrt{-36}$
   d. $\sqrt{400}$

4. Evaluate each expression:
   a. $\dfrac{7}{8} - \left(-\dfrac{1}{4}\right)$
   b. $-\dfrac{1}{3} - \dfrac{5}{6}$
   c. $-0.7 + 1.2$
   d. $1.3 + (-5.9)$

5. Evaluate each expression:
   a. $(-4)\left(-2\dfrac{1}{3}\right)$
   b. $(-0.6)(-1.5)$
   c. $\dfrac{-2.8}{-0.7}$
   d. $4.2 \div (-0.6)$

6.  Evaluate using a calculator: $2000\left(1 + \dfrac{0.08}{12}\right)^{12 \cdot 10}$

7. State the value of each expression, if possible.
   a. $0 \div 6$
   b. $6 \div 0$

8. State the number of terms in each expression and identify the coefficient of each.
   a. $-2v^2 + 6v + 5$
   b. $\dfrac{c + 2}{3} + c$

9. Evaluate each expression given $x = -0.5$ and $y = -2$. Round to hundredths as needed.
   a. $2x - 3y^2$
   b. $\sqrt{2} - x(4 - x^2) + \dfrac{y}{x}$

10. Translate each phrase into an algebraic expression.
    a. Nine less than twice a number is subtracted from the number cubed.
    b. Three times the square of half a number is subtracted from twice the number.

11. Create a mathematical model using descriptive variables.
    a. The radius of the planet Jupiter is approximately 119 mi less than 11 times the radius of the Earth. Express the radius of Jupiter in terms of the Earth's radius.
    b. Last year, Video Venue Inc. earned $1.2 million more than four times what it earned this year. Express last year's earnings of Video Venue Inc. in terms of this year's earnings.

12. Simplify by combining like terms.
    a. $8v^2 + 4v - 7 + v^2 - v$
    b. $-4(3b - 2) + 5b$
    c. $4x - (x - 2x^2) + x(3 - x)$

13. Factor each expression completely.
    a. $9x^2 - 16$
    b. $4v^3 - 12v^2 + 9v$
    c. $x^3 + 5x^2 - 9x - 45$

14. Factor completely.
    a. $x^3 - 27$
    b. $24ac^3 + 81ab^3$

15. Simplify using the properties of exponents.
    a. $\dfrac{5}{b^{-3}}$
    b. $(-2a^3)^2(a^2b^4)^3$
    c. $\left(\dfrac{m^2}{2n}\right)^3$
    d. $\left(\dfrac{5p^2q^3r^4}{-2pq^2r^4}\right)^2$

16. Simplify using the properties of exponents.
    a. $\dfrac{-12a^3b^5}{3a^2b^4}$
    b. $(3.2 \times 10^{-17}) \times (2.0 \times 10^{15})$
    c. $\left(\dfrac{a^{-3} \cdot b}{c^{-2}}\right)^{-4}$
    d. $-7x^0 + (-7x)^0$

**17.** Compute each product.

  **a.** $(3x^2 + 5y)(3x^2 - 5y)$

  **b.** $(2a + 3b)^2$

**18.** Add or subtract as indicated.

  **a.** $(-5a^3 + 4a^2 - 3) + (7a^4 + 4a^2 - 3a - 15)$

  **b.** $(2x^2 + 4x - 9) - (7x^4 - 2x^2 - x - 9)$

**19.** Match the formula on the left to the figure on the right.

  **a.** \_\_\_\_\_ $V = LWH$     (i)    (ii)

  **b.** \_\_\_\_\_ $V = \dfrac{4}{3}\pi r^3$    (iii)

  **c.** \_\_\_\_\_ $V = s^3$     (iv)

  **d.** \_\_\_\_\_ $V = \dfrac{1}{3}\pi r^2 h$    (v)

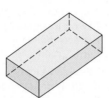

  **e.** \_\_\_\_\_ $V = \pi r^2 h$

Simplify or compute as indicated.

**20. a.** $\dfrac{x - 5}{5 - x}$       **b.** $\dfrac{4 - n^2}{n^2 - 4n + 4}$

  **c.** $\dfrac{x^3 - 27}{x^2 + 3x + 9}$     **d.** $\dfrac{3x^2 - 13x - 10}{9x^2 - 4}$

  **e.** $\dfrac{x^2 - 25}{3x^2 - 11x - 4} \div \dfrac{x^2 + x - 20}{x^2 - 8x + 16}$

  **f.** $\dfrac{m + 3}{m^2 + m - 12} - \dfrac{2}{5(m + 4)}$

**21. a.** $\sqrt{(x + 11)^2}$      **b.** $\sqrt[3]{\dfrac{-8}{27v^3}}$

  **c.** $\left(\dfrac{25}{16}\right)^{-\frac{3}{2}}$      **d.** $\dfrac{-4 + \sqrt{32}}{8}$

  **e.** $7\sqrt{40} - \sqrt{90}$    **f.** $(x + \sqrt{5})(x - \sqrt{5})$

  **g.** $\sqrt{\dfrac{2}{5x}}$        **h.** $\dfrac{8}{\sqrt{6} - \sqrt{2}}$

**22. Maximizing revenue:** Due to past experience, the manager of a video store knows that if a popular video game is priced at \$30, the store will sell 40 each day. For each decrease of \$0.50, one additional sale will be made. The formula for the store's revenue is then $R = (30 - 0.5x)(40 + x)$, where $x$ represents the number of times the price is decreased. Multiply the binomials and use a table of values to determine (a) the number of 50¢ decreases that will give the most revenue and (b) the maximum amount of revenue.

**23. Diagonal of a rectangular prism:** Use the Pythagorean theorem to determine the length of the diagonal of the rectangular prism shown in the figure. (*Hint:* First find the diagonal of the base.)

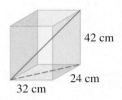

**24. Weight of a cloud:** Have you ever stopped to think about how much a cloud weighs? We get our rainwater from clouds, so they must weigh *something*. Depending on the density, 1 m³ of cloud can hold between $\frac{1}{10}$ of a gram to 5 g of water. Let's consider a typical, large and fluffy cloud, and conveniently "rearrange it" into a rectangular prism. For convenience, let's say our cloud holds 2 g of water per cubic meter. (a) Express the volume of the cloud in scientific notation, if it were 2000 m long, 1500 m wide, and 1200 m tall (an average-sized cloud). (b) Express the weight of the cloud in scientific notation (in grams and kilograms).

**25. NCIS training:** As part of an NCIS training exercise, a steel cable is secured between two telephone poles $D$ m apart, with recruits required to move hand-over-hand from one pole to the next. If the cable is depressed $d$ m when a recruit reaches the middle of the cable, $d = \dfrac{\sqrt{L^2 - D^2}}{2}$, where $L$ is the length of the cable after it has been stretched. (a) Factor the radicand, and (b) find the distance the cable is depressed if the poles are 12 m apart and the cable has been stretched to a length of 13 m.

# Equations and Inequalities

## CHAPTER OUTLINE

## CHAPTER CONNECTIONS

The more you understand equations, the better you can apply them in context. Concerned with rush hour congestion but limited by funds, traffic engineers must choose roadway improvements that balance safety at peak volume with "off hour" demands. To do this, they must determine how far from average the volume fluctuates throughout the day. The techniques illustrated in this chapter will assist you in answering questions of this nature. This application appears as Exercise 60 in Section 1.3.

**Check out these other real-world connections:**

► Forensic Analysis
  (Section 1.1, Exercise 55)
► Exam Averages
  (Section 1.2, Exercise 77)
► National Debt
  (Section 1.5, Exercise 133)
► Removing Environmental Toxins
  (Section 1.6, Exercise 91)

**LEARNING OBJECTIVES**

*In Section 1.1 you will see how we can:*

- ☐ **A.** Solve linear equations using properties of equality
- ☐ **B.** Recognize equations that are identities or contradictions
- ☐ **C.** Solve for a specified variable in a formula or literal equation
- ☐ **D.** Use the problem-solving guide to solve various problem types

In a study of algebra, you will encounter many **families of equations,** or groups of equations that share common characteristics. Of interest to us here is the family of **linear equations in one variable,** a study that lays the foundation for understanding more advanced families. This section will also lay the foundation for solving a formula for a specified variable, a practice widely used in science, business, industry, and research.

## A. Solving Linear Equations Using Properties of Equality

An **equation** is *a statement that two expressions are equal.* From the expressions $3(x - 1) + x$ and $-x + 7$, we can form the equation

$$3(x - 1) + x = -x + 7,$$

which is a **linear equation in one variable** (the exponent on any variable is a 1). To solve an equation, we attempt to find a specific input or $x$-value that will make the equation true, meaning the left-hand expression will be equal to the right. Using Table 1.1, we find that $3(x - 1) + x = -x + 7$ is a true equation when $x$ is replaced by 2, and is a false equation otherwise. Replacement values that make the equation true are called **solutions** or **roots** of the equation.

**Table 1.1**

| $x$ | $3(x - 1) + x$ | $-x + 7$ |
|-----|----------------|----------|
| $-2$ | $-11$ | 9 |
| $-1$ | $-7$ | 8 |
| 0 | $-3$ | 7 |
| 1 | 1 | 6 |
| **2** | **5** | **5** |
| 3 | 9 | 4 |
| 4 | 13 | 3 |

**CAUTION** ▶ From Section R.2, an **algebraic *expression*** is a sum or difference of algebraic terms. Algebraic expressions can be simplified, evaluated, or written in an equivalent form, but cannot be *"solved,"* since we're not seeking a specific value of the unknown.

Solving equations using a table is too time consuming to be practical. Instead we attempt to write a sequence of **equivalent equations,** each one simpler than the one before, until we reach a point where the solution is obvious. Equivalent equations are those that have the same solution set, and are obtained by using the distributive, additive, and multiplicative properties of equality to obtain an equation of the form *variable = number.*

| **Additive Property of Equality** | **Multiplicative Property of Equality** |
|---|---|
| If $A$, $B$, and $C$ represent algebraic expressions and $A = B$, then | If $A$, $B$, and $C$ represent algebraic expressions and $A = B$, then |
| $$A + C = B + C$$ and $$A - C = B - C$$ | $$AC = BC$$ and $$\frac{A}{C} = \frac{B}{C}(C \neq 0)$$ |

In words, the additive property says that like quantities, numbers, or terms can be added to both sides of an equation. A similar statement can be made for the multiplicative property. These properties are combined into a general guide for solving linear equations, which you've likely encountered in your previous studies. Note that not all steps in the guide are required to solve every equation.

> **Guide to Solving Linear Equations in One Variable**
> - Eliminate parentheses using the distributive property, then combine any like terms.
> - Use the additive property of equality to write the equation with all variable terms on one side, and all constants on the other. Simplify each side.
> - Use the multiplicative property of equality to obtain an equation of the form $x =$ constant.
> - For applications, answer in a complete sentence and include any units of measure indicated.

For our first example, we'll use the equation $3(x - 1) + x = -x + 7$ from our initial discussion.

**EXAMPLE 1** ▶ **Solving a Linear Equation Using Properties of Equality**

Solve for $x$: $3(x - 1) + x = -x + 7$.

**Solution** ▶

| | |
|---|---|
| $3(x - 1) + x = -x + 7$ | original equation |
| $3x - 3 + x = -x + 7$ | distributive property |
| $4x - 3 = -x + 7$ | combine like terms |
| $5x - 3 = 7$ | add $x$ to both sides (additive property of equality) |
| $5x = 10$ | add 3 to both sides (additive property of equality) |
| $x = 2$ | multiply both sides by $\frac{1}{5}$ or divide both sides by 5 (multiplicative property of equality) |

As we noted in Table 1.1, the solution is $x = 2$.

**Now try Exercises 5 through 8** ▶

To check a solution by substitution means we substitute the solution back into the original equation (this is sometimes called **back-substitution**), and verify the left-hand side is equal to the right. For Example 1 we have:

| | |
|---|---|
| $3(x - 1) + x = -x + 7$ | original equation |
| $3(2 - 1) + 2 = -2 + 7$ | substitute 2 for $x$ |
| $3(1) + 2 = 5$ | simplify |
| $5 = 5$✓ | solution checks |

If any coefficients in an equation are fractional, multiply both sides by the least common denominator (LCD) to *clear the fractions*. Since any decimal number can be written in fraction form, the same idea can be applied to decimal coefficients.

**EXAMPLE 2** ▶ **Solving a Linear Equation with Fractional Coefficients**

Solve for $n$: $\frac{1}{4}(n + 8) - 2 = \frac{1}{2}(n - 6)$.

**Solution** ▶

| | |
|---|---|
| $\frac{1}{4}(n + 8) - 2 = \frac{1}{2}(n - 6)$ | original equation |
| $\frac{1}{4}n + 2 - 2 = \frac{1}{2}n - 3$ | distributive property |
| $\frac{1}{4}n = \frac{1}{2}n - 3$ | combine like terms |
| $4(\frac{1}{4}n) = 4(\frac{1}{2}n - 3)$ | multiply both sides by LCD = 4 |
| $n = 2n - 12$ | distributive property |
| $-n = -12$ | subtract $2n$ |
| $n = 12$ | multiply by $-1$ |

☑ **A.** You've just seen how we can solve linear equations using properties of equality

Verify the solution is $n = 12$ using back-substitution.

**Now try Exercises 9 through 26** ▶

## B. Identities and Contradictions

Example 1 illustrates what is called a **conditional equation,** since the equation is true for $x = 2$, but false for all other values of $x$. The equation in Example 2 is also conditional. An **identity** is an equation that is *always true,* no matter what value is substituted for the variable. For instance, $2(x + 3) = 2x + 6$ is an identity with a solution set of all real numbers, written as $x \in \mathbb{R}$ in set notation, or $x \in (-\infty, \infty)$ in interval notation. **Contradictions** are equations that are *never true,* no matter what real number is substituted for the variable. The equations $x - 3 = x + 1$ and $-3 = 1$ are contradictions. To state the solution set for a contradiction, we use the symbol "$\varnothing$" (the null set) or "{ }" (the empty set). Recognizing these special equations will prevent some surprise and indecision in later chapters.

---

**EXAMPLE 3** ▶ Solving Equations (Special Cases)

Solve each equation and state the solution set.

   **a.** $2(x - 4) + 10x = 8 + 4(3x + 1)$     **b.** $8x - (6 - 10x) = 24 + 6(3x - 5)$

**Solution** ▶

**a.**
$$2(x - 4) + 10x = 8 + 4(3x + 1) \quad \text{original equation}$$
$$2x - 8 + 10x = 8 + 12x + 4 \quad \text{distributive property}$$
$$12x - 8 = 12x + 12 \quad \text{combine like terms}$$
$$-8 = 12 \quad \text{subtract } 12x; \text{ contradiction}$$

Since $-8$ is never equal to 12, the original equation is a contradiction. The solution set is empty: $x \in \{\ \}$.

**b.**
$$8x - (6 - 10x) = 24 + 6(3x - 5) \quad \text{original equation}$$
$$8x - 6 + 10x = 24 + 18x - 30 \quad \text{distributive property}$$
$$18x - 6 = 18x - 6 \quad \text{combine like terms}$$
$$-6 = -6 \quad \text{subtract } 18x; \text{ identity}$$

The result shows that the original equation is an identity, with an infinite number of solutions: $x \in \mathbb{R}$. You may recall this notation is read, "$x$ is an element of the set of real numbers" or simply "$x$ is a real number."

✓ **B.** You've just seen how we can recognize equations that are identities or contradictions

**Now try Exercises 27 through 32** ▶

---

In Example 3(a), our attempt to solve for $x$ ended with all variables being eliminated, leaving an equation that is *always false*—a contradiction ($-8$ is never equal to 12). There is nothing wrong with the solution process, the result is simply telling us the original equation has *no solution.* In Example 3(b), all variables were again eliminated but the end result was *always true*—an identity ($-6$ is always equal to $-6$). Once again we've done nothing wrong mathematically; the result is just telling us that the original equation will be true no matter what value of $x$ we use for an input.

## C. Solving for a Specified Variable in Literal Equations

A **literal equation** is an equation that has two or more variables. A **formula** is simply one that models a known relationship. For example, the formula $A = P + PRT$ models the growth of money in an account earning simple interest, where $A$ represents the total amount accumulated, $P$ is the initial deposit, $R$ is the annual interest rate, and $T$ is the number of years the money is left on deposit. To *describe* $A = P + PRT$, we might say the formula has been "solved for $A$" or that "$A$ is written in terms of $P$, $R$, and $T$." In some cases, before using a formula it may be convenient to solve for one of the other variables, say $P$. In this case, $P$ is called the **object variable.**

**EXAMPLE 4** ▶ **Solving for a Specified Variable**

Given $A = P + PRT$, write $P$ in terms of $A$, $R$, and $T$ (solve for $P$).

**Solution** ▶ Since the object variable occurs in more than one term, we first apply the distributive property.

$$A = \boldsymbol{P} + \boldsymbol{P}RT \qquad \text{focus on } \boldsymbol{P}\text{—the object variable}$$

$$A = \boldsymbol{P}(1 + RT) \qquad \text{factor out } \boldsymbol{P}$$

$$\frac{A}{1 + RT} = \frac{\boldsymbol{P}(1 + RT)}{(1 + RT)} \qquad \text{solve for } \boldsymbol{P} \text{ [divide by } (1 + RT)\text{]}$$

$$\frac{A}{1 + RT} = \boldsymbol{P} \qquad \text{result}$$

**Now try Exercises 33 through 42** ▶

In Example 4, our decision to solve for $P$ was arbitrary, and any of the other variables could have been selected instead. Verify that if we had solved for $T$, the result could have been written as either $T = \dfrac{A - P}{PR}$ or $T = \dfrac{A}{PR} - \dfrac{1}{R}$. As in this example, there is often more than one acceptable form of the correct answer. If you find your own answer seems to disagree with the one given, try manipulating either of the two to make them match.

We solve literal equations for a specified variable using the same methods we used for other equations and formulas. Remember that it's good practice to *focus on the object variable* to help guide you through the solution process, as again shown in Example 5.

**EXAMPLE 5** ▶ **Solving for a Specified Variable**

Given $2x + 3y = 15$, write $y$ in terms of $x$ (solve for $y$).

**Solution** ▶
$$2x + 3\boldsymbol{y} = 15 \qquad\qquad \text{focus on the object variable}$$
$$3\boldsymbol{y} = -2x + 15 \qquad \text{subtract } 2x \text{ (isolate } y\text{-term)}$$
$$\tfrac{1}{3}(3\boldsymbol{y}) = \tfrac{1}{3}(-2x + 15) \qquad \text{multiply by } \tfrac{1}{3} \text{ (solve for } y\text{)}$$
$$y = \tfrac{-2}{3}x + 5 \qquad\qquad \text{distribute and simplify}$$

**WORTHY OF NOTE**

In Example 5, notice that in the second step we wrote the subtraction of $2x$ as $-2x + 15$ instead of $15 - 2x$. For reasons that will become clearer as we continue our study, we generally write variable terms before constant terms.

**Now try Exercises 43 through 48** ▶

## Literal Equations and General Solutions

Solving literal equations for a specified variable can help us develop the general solution for an entire family of equations. This is demonstrated here for the family of linear equations written in the form $ax + b = c$. A side-by-side comparison with a specific linear equation demonstrates that identical ideas are used.

| **Specific Equation** | | **Literal Equation** |
|---|---|---|
| $2\boldsymbol{x} + 3 = 15$ | focus on object variable | $a\boldsymbol{x} + b = c$ |
| $2\boldsymbol{x} = 15 - 3$ | subtract constant | $a\boldsymbol{x} = c - b$ |
| $x = \dfrac{15 - 3}{2}$ | divide by coefficient | $x = \dfrac{c - b}{a}$ |

Of course the solution on the left would be written as $x = 6$ and checked in the original equation. On the right we now have a general formula for all equations of the form $ax + b = c$.

**EXAMPLE 6** ▶  **Solving Equations of the Form *ax + b = c* Using the General Formula**

Solve $6x - 1 = -25$ using the formula just developed, and check your solution in the original equation.

**Solution** ▶  For this equation, $a = 6$, $b = -1$, and $c = -25$, giving

$$x = \frac{c - b}{a}$$

$$= \frac{-25 - (-1)}{6}$$

$$= \frac{-24}{6}$$

$$= -4$$

→Check:    $6x - 1 = -25$

$6(-4) - 1 = -25$

$-24 - 1 = -25$

$-25 = -25$ ✓

☑ **C.** You've just seen how we can solve for a specified variable in a formula or literal equation

Now try Exercises 49 through 52 ▶

Developing a general solution for the linear equation $ax + b = c$ seems to have little practical use. But in Section 1.6 we'll use this idea to develop a general solution for *quadratic equations,* a result with much greater significance.

## D.  Using a Problem-Solving Guide

Becoming a good problem solver is an evolutionary process. Over time and with continued effort, your problem-solving skills grow, as will your ability to solve a wider range of applications. Most good problem solvers develop the following characteristics:

- A positive attitude
- A mastery of basic facts
- Strong mental arithmetic skills
- Good mental-visual skills
- Good estimation skills
- A willingness to persevere

These characteristics form a solid basis for applying what we call the **Problem-Solving Guide,** which simply organizes the basic elements of good problem solving. Using this guide will help save you from two common stumbling blocks—indecision and not knowing where to start.

**Problem-Solving Guide**

- **Gather and organize information.**
  Read the problem several times, forming a mental picture as you read. *Highlight key phrases.* List given information, including any related formulas. *Clearly identify what you are asked to find.*
- **Make the problem visual.**
  *Draw and label a diagram* or create a table of values, as appropriate. This will help you see how different parts of the problem fit together.
- **Develop an equation model.**
  *Assign a variable* to represent what you are asked to find and build any related expressions referred to in the problem. Write an equation model based on the relationships given in the problem. *Carefully reread the problem to double-check your equation model.*
- **Use the model and given information to solve the problem.**
  Substitute given values, then simplify and solve. State the answer in sentence form, and check that the answer is reasonable. Include any units of measure indicated.

### General Modeling Exercises

Translating word phrases into symbols is an important part of building equations from information given in paragraph form. Sometimes the variable *occurs more than once* in the equation, because two different items in the same exercise are related. If the relationship involves a comparison of size, we often use line segments or bar graphs to model the relative sizes.

---

**EXAMPLE 7 ▶  Solving an Application Using the Problem-Solving Guide**

The largest state in the United States is Alaska (AK), which covers an area that is 230 square miles ($mi^2$) more than 500 times that of the smallest state, Rhode Island (RI). If they have a combined area of 616,460 $mi^2$, how many square miles does each cover?

**Solution ▶**  Combined area is 616,460 $mi^2$,                    gather and organize information
AK covers 230 more than 500 times the area of RI.      highlight any key phrases

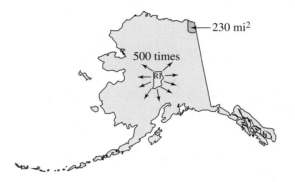

make the problem visual

Let $R$ represent the area of Rhode Island.            assign a variable
Then $500R + 230$ represents Alaska's area.            build related expressions

$$\text{Rhode Island's area} + \text{Alaska's area} = \text{Total}$$
$$R + (500R + 230) = 616{,}460 \quad \text{write the equation model}$$
$$501R = 616{,}230 \quad \text{combine like terms, subtract 230}$$
$$R = 1230 \quad \text{divide by 501}$$

Rhode Island covers an area of 1230 $mi^2$, while Alaska covers an area of $500(1230) + 230 = 615{,}230\ mi^2$.

**Now try Exercises 55 through 60 ▶**

---

### Exercises Involving Geometry

As your ability to solve equations grows, your ability to solve a wider range of more significant applications will grow as well. In many cases, the applications will involve some basic geometry, which we briefly review here (for a more complete review, see Appendix I).

Perimeter is a measure of the distance around a two-dimensional figure. Since this is a linear measure, results are stated in linear units like centimeters (cm), feet (ft), kilometers (km), miles (mi), and so on. If no unit is specified, simply write the result as *x units*. Area is a measure of the surface of a two-dimensional figure, with results stated in square units (as in *x units*$^2$). Volume is a measure of the amount of space occupied by a three-dimensional object and is measured in *cubic units*. Some of the most common formulas involving perimeter, area, and volume are given on the inside cover of this text.

If an exercise or application uses a formula, *begin by stating the formula.* Using the formula as a template for the values substituted will help to prevent many careless errors.

---

**EXAMPLE 8 ▶**   **Solving a Problem Involving Geometry**

Sand at a cement factory is being dumped from a conveyor belt into a pile shaped like a right circular cone atop a right circular cylinder. If the diameter of the cylinder is 10 ft and the cone is twice as tall as the cylinder, how tall is the cylinder when the volume is 625 ft$^3$?

**Solution ▶**   The diameter of the cylinder is 10 ft. The height       gather/organize information
of the cone is twice the height of the cylinder.       highlight any key phrases
The volume is 625 ft$^3$.

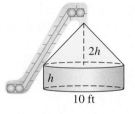

make the problem visual

If we let $h$ represent the height of the cylinder,       assign a variable
then $2h$ is the height of the cone.       build related expressions

Volume of cylinder + Volume of cone = Total volume

$$\pi r^2 h + \frac{1}{3}\pi r^2 H = 625 \qquad \text{write the equation model}$$

$$\pi(5)^2 h + \frac{1}{3}\pi(5)^2(2h) = 625 \qquad H = 2h,\ r = \frac{1}{2}(10) = 5$$

$$25\pi h + \frac{50}{3}\pi h = 625 \qquad \text{simplify terms}$$

$$\frac{125}{3}\pi h = 625 \qquad \text{combine terms}$$

$$h = \frac{15}{\pi} \qquad \text{multiply by } \frac{3}{125\pi};\ 625\left(\frac{3}{125\pi}\right) = \frac{15}{\pi}$$

$$\approx 4.77 \qquad \text{approximate}$$

The cylinder is approximately 4 ft 9 in. (4.77 ft) tall.

**Now try Exercises 61 through 64 ▶**

---

### Uniform Motion (Distance, Rate, Time) Exercises

Uniform motion problems have many variations, and it's important to draw a good diagram when you get started. Recall that if speed is constant, the distance traveled is equal to the rate (speed) multiplied by the time in motion: $D = RT$.

**EXAMPLE 9** ▶ **Solving an Application Involving Uniform Motion**

I live 260 mi from a popular mountain retreat. On my way there to do some mountain biking, my car had engine trouble—forcing me to bike the rest of the way. If I drove 2 hr longer than I biked and averaged 60 mi/hr driving and 10 mi/hr biking, how many hours did I spend pedaling to the resort?

**Solution** ▶ The sum of the two distances must be 260 mi.    gather/organize information
The **rates** are given, and the driving time is    highlight any key phrases
*2 hr more than biking* time.

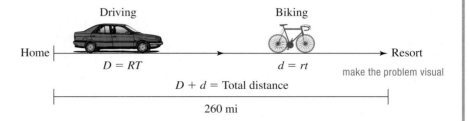

Driving                    Biking

Home |————————————————————————————▶| Resort
          $D = RT$                $d = rt$            make the problem visual
     |———————— $D + d$ = Total distance ————————|
                    260 mi

If we let $t$ represent the biking time,       assign a variable
then $T = t + 2$ represents time spent driving.   build related expressions

$$D + d = 260$$        write the equation model
$$RT + rt = 260$$      $RT = D$, $rt = d$
$$60(t + 2) + 10t = 260$$  substitute $t + 2$ for $T$, 60 for $R$, 10 for $r$
$$70t + 120 = 260$$    distribute and combine like terms
$$70t = 140$$          subtract 120
$$t = 2$$              divide by 70

I rode my bike for $t = 2$ hr, after driving $t + 2 = 4$ hr.

Now try Exercises 65 through 68 ▶

### Exercises Involving Mixtures

Mixture problems offer another opportunity to refine our problem-solving skills while using many elements from the problem-solving guide. They also lend themselves to a very useful mental-visual image and have many practical applications.

**EXAMPLE 10** ▶ **Solving an Application Involving Mixtures**

As a nasal decongestant, doctors sometimes prescribe saline solutions with a concentration between 6% and 20%. In "the old days," pharmacists had to create different mixtures, but only needed to stock these concentrations, since any percentage in between could be obtained using a mixture. An order comes in for a 15% solution. How many milliliters (mL) of the 20% solution must be mixed with 10 mL of the 6% solution to obtain the desired 15% solution?

**Solution** ▶

Only 6% and 20% concentrations are available; mix some 20% solution with 10 mL of the 6% solution. (See Figure 1.1.)

gather/organize information

highlight any key phrases

**WORTHY OF NOTE**

For mixture exercises, an estimate assuming equal amounts of each liquid can be helpful. For example, assume we use 10 mL of the 6% solution and 10 mL of the 20% solution. The final concentration would be halfway in between, $\frac{6 + 20}{2} = 13\%$. This is too low a concentration (we need a 15% solution), so we know that more than 10 mL of the stronger (20%) solution must be used.

**Figure 1.1**

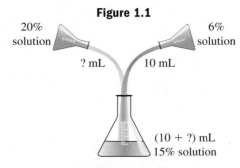

make the problem visual

If we let $x$ represent the amount of 20% solution, then $10 + x$ represents the total amount of 15% solution.

assign a variable

build related expressions

| 1st quantity times its concentration | 2nd quantity times its concentration | 1st + 2nd quantity times desired concentration | |
|---|---|---|---|
| $10(0.06)$ | $+ \quad x(0.2)$ | $= \quad (10 + x)(0.15)$ | write equation model |
| $0.6$ | $+ \quad 0.2x$ | $= \quad 1.5 + 0.15x$ | distribute/simplify |
| | | $0.2x = 0.9 + 0.15x$ | subtract 0.6 |
| | | $0.05x = 0.9$ | subtract 0.15x |
| | | $x = 18$ | divide by 0.05 |

To obtain a 15% solution, 18 mL of the 20% solution must be mixed with 10 mL of the 6% solution.

 **D.** You've just seen how we can use the problem-solving guide to solve various problem types

Now try Exercises 69 through 72 ▶

# 1.1 EXERCISES

▶ **CONCEPTS AND VOCABULARY**

**Fill in each blank with the appropriate word or phrase. Carefully reread the section, if necessary.**

1. A(n) _____ is an equation that is always true, regardless of the _____ value, while a(n) _____ is always false, regardless of the _____ value.

2. A(n) _____ equation is an equation having _____ or more unknowns. In $S = 2\pi r^2 + 2\pi rh$, which is an example of this type of equation, we say $S$ is written in terms of _____ and _____.

3. Discuss/Explain how to check the solution of a linear equation. Give examples in your discussion.

4. Discuss/Explain each of the four basic parts of the *problem-solving guide*. Include a solved example in your discussion.

▶ **DEVELOPING YOUR SKILLS**

**Solve each equation. Check your answer by substitution.**

5. $4x + 3(x - 2) = 18 - x$

6. $15 - 2x = -4(x + 1) + 9$

7. $8 - (3b + 5) = -5 + 2(b + 1)$

8. $2a + 4(a - 1) = 3 - (2a + 1)$

**Solve each equation.**

9. $\frac{1}{5}(b + 10) - 7 = \frac{1}{3}(b - 9)$

10. $\frac{1}{6}(n - 12) = \frac{1}{4}(n + 8) - 2$

11. $\frac{2}{3}(m + 6) = \frac{-1}{2}$

12. $\frac{4}{5}(n - 10) = \frac{-8}{9}$

13. $\frac{1}{2}x + 5 = \frac{1}{3}x + 7$　　　14. $-4 + \frac{2}{3}y = \frac{1}{2}y - 5$

15. $\frac{x + 3}{5} + \frac{x}{3} = 7$　　　16. $\frac{z - 4}{6} - 2 = \frac{z}{2}$

17. $15 = -6 - \frac{3p}{8}$　　　18. $-15 - \frac{2q}{9} = -21$

19. $0.2(24 - 7.5a) - 6.1 = 4.1$

20. $0.4(17 - 4.25b) - 3.15 = 4.16$

21. $6.2v - (2.1v - 5) = 1.1 - 3.7v$

22. $7.9 - 2.6w = 1.5w - (9.1 + 2.1w)$

23. $\frac{n}{2} + \frac{n}{5} = \frac{2}{3}$　　　24. $\frac{m}{3} - \frac{2}{5} = \frac{m}{4}$

25. $3p - \frac{p}{4} - 5 = \frac{p}{6} - 2p + 6$

26. $\frac{q}{6} + 1 - 3q = 2 - 4q + \frac{q}{8}$

**Identify the following equations as an identity, a contradiction, or a conditional equation, then state the solution.**

27. $-3(4z + 5) = -15z - 20 + 3z$

28. $5x - 9 - 2 = -5(2 - x) - 1$

29. $8 - 8(3n + 5) = -5 + 6(1 + n)$

30. $2a + 4(a - 1) = 1 + 3(2a + 1)$

31. $-4(4x + 5) = -6 - 2(8x + 7)$

32. $-(5x - 3) + 2x = 11 - 4(x + 2)$

**Solve for the specified variable in each formula or literal equation.**

33. $P = C + CM$ for $C$ (retail)

34. $S = P - PD$ for $P$ (retail)

35. $C = 2\pi r$ for $r$ (geometry)

36. $V = LWH$ for $W$ (geometry)

37. $V = \frac{1}{4}\pi d^2 L$ for $L$ (geometry)

38. $V = \frac{1}{3}\pi r^2 h$ for $h$ (geometry)

39. $S_n = n\left(\frac{a_1 + a_n}{2}\right)$ for $n$ (sequences)

40. $A = \frac{h(b_1 + b_2)}{2}$ for $h$ (geometry)

41. $S = B + \frac{1}{2}PS$ for $P$ (geometry)

42. $s = \frac{1}{2}gt^2 + vt$ for $g$ (physics)

43. $Ax + By = C$ for $y$

44. $\frac{x}{h} + \frac{y}{k} = 1$ for $y$

45. $\frac{5}{6}x + \frac{3}{8}y = 2$ for $y$

46. $\frac{2}{3}x - \frac{7}{9}y = 12$ for $y$

47. $y - 3 = \frac{-4}{5}(x + 10)$ for $y$

48. $y + 4 = \frac{-2}{15}(x + 10)$ for $y$

**The following equations are given in $ax + b = c$ form. Solve by identifying the values of $a$, $b$, and $c$, then using the formula $x = \frac{c - b}{a}$.**

49. $3x + 2 = -19$

50. $7x + 5 = 47$

51. $7x - 13 = -27$

52. $3x - 4 = -25$

## ▶ WORKING WITH FORMULAS

53. **Euler's Polyhedron Formula: $V + F - E = 2$**

Discovered by Leonhard Euler in 1752, this simple but powerful formula states that in any regular polyhedron, the number of vertices $V$ and faces $F$ is always two more than the number of edges $E$.
(a) Verify the formula for a simple cube. (b) Verify the formula for the octahedron shown in the figure. (c) Solve the formula for $V$. (d) Determine how many vertices a dodecahedron with 12 faces and 30 edges has.

54. **Surface area of a cylinder: $S = 2\pi r^2 + 2\pi rh$**

The surface area of a cylinder is given by the formula shown, where $h$ is the height of the cylinder and $r$ is the radius of the base. (a) Solve the formula for $h$. (b) Find the height of a cylinder that has a radius of 8 cm and a surface area of 1256 cm². Use $\pi \approx 3.14$.

▶ **APPLICATIONS**

**Solve by building an equation model and using the problem-solving guidelines as needed.**

**General Modeling Exercises**

**55. Forensic studies:** In forensic studies, skeletal remains are analyzed to determine the height, gender, race, age, and other characteristics of the decedent. For instance, the height of a male individual is approximated as 34 in. more than three and one-third times the length of the radial bone. If a live individual is 74 in. tall, how long is his radial bone?

**56. Mesopotamian rivers:** The area near the joining of the Tigris and Euphrates Rivers (in modern Iraq) has often been called the *Cradle of Civilization,* since the area has evidence of many ancient cultures. The length of the Euphrates River exceeds that of the Tigris by 620 mi. If they have a combined length of 2880 mi, how long is each river?

**57. Energy costs:** Through meticulous planning, careful design, timeless effort, and visionary leadership, MHHE was able to achieve an energy savings of nearly 39% in their new regional headquarters. If energy costs at the old building averaged $12,387/mo, what is their annual savings in energy costs at the new headquarters?

**58. Depreciating truck values:** Hi-Tech Home Improvements buys a fleet of identical trucks that cost $32,750 each. The company is allowed to depreciate the value of their trucks for tax purposes by $5250/yr. If company policies dictate that older trucks must be sold once their value declines to $6500, approximately how many years will they keep these trucks?

**59. Famous architecture:** The *Hall of Mirrors* is the central gallery of the Palace of Versailles and is one of the most famous rooms in the world. The length of this hall is 11 m less than 8 times the width. If the hall is 73 m long, what is its width?

**60. Interplanetary distances:** The Mars rover *Spirit* landed on January 3, 2004. Just over 1 yr later, on January 14, 2005, the *Huygens* probe landed on Titan (one of Saturn's moons). At their closest approach, the distance from the Earth to Saturn is 29 million mi more than 21 times the distance from the Earth to Mars. If the distance to Saturn is 743 million mi, what is the distance to Mars?

**Geometry Exercises**

**61.** U.S. postal regulations require that a package can have a maximum combined length and girth (distance around) of 108 in. A shipping carton is constructed so that it has a width of 14 in., a height of 12 in., and can be cut or folded to various lengths. What is the maximum length that can be used?

*Source:* www.USPS.com

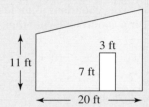

**62.** After the studs are up, the wall shown in the figure must be covered in drywall. If exactly seven 4-ft by 8-ft sheets of drywall are needed to cover the wall, how much taller is the right end of the wall than the left?

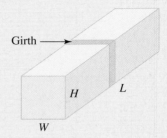

**63.** The base of a new trophy has the form of a cylinder sitting atop a rectangular solid. If 600 in$^3$ of aluminum are needed to form the base, how tall is the rectangular portion?

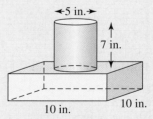

**64.** A grain silo has the form of a hemisphere atop a cylinder. If the maximum storage capacity of the silo is 500 m$^3$ or 95% of the total volume of the silo, how tall is its cylindrical base?

**Uniform Motion Exercises**

**65.** At 9:00 A.M., Linda leaves work on a business trip, gets on the interstate, and sets her cruise control at 60 mph. At 9:30 A.M., Bruce notices she's left her briefcase and cell phone, and immediately starts after her driving 75 mph. At what time will Bruce catch up with Linda?

**66.** A plane flying at 300 mph has a 3-hr head start on a "chase plane," which has a speed of 800 mph. How far from the airport will the chase plane overtake the first plane?

**67.** At a high-school cross-country meet, Jared jogged 8 mph for the first part of the race, then increased his speed to 12 mph for the second part. If the race was 21 mi long and Jared finished in 2 hr, how far did he jog at the faster pace?

**68.** Two friends decide to have a race. On "go," Jess takes off on her jet ski, headed upstream against a 10-mph current. 60 sec later, Deb gives chase with her dune buggy on a dirt track parallel to the river. If the top speed of the jet ski is 35 mph on still water and the dune buggy can hit 45 mph, how long will it take Deb to pass Jess?

**Mixture Exercises**

**69.** To help sell more of a lower grade meat, a butcher mixes some premium ground beef worth \$3.10/lb with 8 lb of lower grade ground beef worth \$2.05/lb. If the result was an intermediate grade of ground beef worth \$2.68/lb, how much premium ground beef was used?

**70.** Knowing that the camping/hiking season has arrived, a nutrition outlet is mixing GORP (Good Old Raisins and Peanuts) for the anticipated customers. How many pounds of peanuts worth \$1.29/lb should be mixed with 20 lb of deluxe raisins worth \$1.89/lb to obtain a mix that will sell for \$1.49/lb?

**71.** How many pounds of walnuts at 84¢/lb should be mixed with 20 lb of pecans at \$1.20/lb to give a mixture worth \$1.04/lb?

**72.** How many pounds of cheese worth 81¢/lb must be mixed with 10 lb of cheese worth \$1.29/lb to make a mixture worth \$1.11/lb?

## ▶ EXTENDING THE CONCEPT

**73.** Look up and read the following article. Then turn in a one page summary. "Don't Give Up!," William H. Kraus, *Mathematics Teacher*, Volume 86, Number 2, February 1993: pages 110–112.

**74.** A chemist has four solutions of a very rare and expensive chemical that are 15% acid (cost \$120/oz), 20% acid (cost \$180/oz), 35% acid (cost \$280/oz), and 45% acid (cost \$359/oz). She requires 200 oz of a 29% acid solution. Find the combination of any two of these concentrations that will minimize the total cost of the mix.

**75.** $P$, $Q$, $R$, $S$, $T$, and $U$ represent numbers. The arrows in the figure show the sum of the two or three numbers added in the indicated direction (Example: $Q + T = 23$). Find $P + Q + R + S + T + U$.

**76.** Given a sphere circumscribed by a cylinder, verify the volume of the sphere is $\frac{2}{3}$ that of the cylinder.

**Exercise 76**

**Exercise 75**

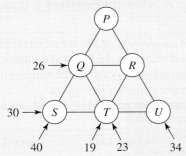

## ▶ MAINTAINING YOUR SKILLS

**77.** (R.1) Simplify the expression using the order of operations.

$$-2 - 6^2 \div 4 + 8$$

**78.** (R.2) Name the coefficient of each term in the expression:

$$-3v^3 + v^2 - \frac{v}{3} + 7$$

**79.** (R.5) Factor each expression:

    **a.** $4x^2 - 9$         **b.** $x^3 - 27$

**80.** (R.2) Identify the property illustrated:

$$\frac{6}{7} \cdot 5 \cdot 21 = \frac{6}{7} \cdot 21 \cdot 5$$

**EXAMPLE 2** ▶    **Solving an Inequality**

Solve the inequality, then graph the solution set and write it in interval notation: $\frac{-2}{3}x + \frac{1}{2} \leq \frac{5}{6}$.

**Solution** ▶

$$\frac{-2}{3}x + \frac{1}{2} \leq \frac{5}{6} \qquad \text{original inequality}$$

$$6\left(\frac{-2}{3}x + \frac{1}{2}\right) \leq 6\left(\frac{5}{6}\right) \qquad \text{clear fractions (multiply by LCD); sign remains the same (since } 6 > 0)$$

$$-4x + 3 \leq 5 \qquad \text{simplify}$$

$$-4x \leq 2 \qquad \text{subtract 3}$$

$$\frac{-4x}{-4} \geq \frac{2}{-4} \qquad \text{divide by } -4, \text{ reverse inequality sign}$$

$$x \geq -\frac{1}{2} \qquad \text{result}$$

> **WORTHY OF NOTE**
>
> As an alternative to multiplying or dividing by a negative value, the additive property of inequality can be used to ensure the variable term will be positive. From Example 2, the inequality $-4x \leq 2$ can be written as $-2 \leq 4x$ by adding $4x$ to both sides and subtracting 2 from both sides. This gives the solution $-\frac{1}{2} \leq x$, which is equivalent to $x \geq -\frac{1}{2}$.

- Graph:

- Interval notation: $x \in \left[-\frac{1}{2}, \infty\right)$

**Now try Exercises 21 through 26** ▶

To check a linear inequality, you often have an infinite number of choices—any number from the solution set/interval. If a test value from the solution interval results in a true inequality, all numbers in the interval are solutions. For Example 2, using $x = 0$ results in the true statement $\frac{1}{2} \leq \frac{5}{6}$ ✓.

Some inequalities have all real numbers as the solution set: $\mathbb{R}$, while other inequalities have no solutions, with the answer given as the empty set: { }.

**EXAMPLE 3** ▶    **Solving Inequalities (Special Cases)**

Solve the inequality and write the solution in set notation:

     **a.** $7 - (3x + 5) \geq 2(x - 4) - 5x$          **b.** $3(x + 4) - 5 < 2(x - 3) + x$

**Solution** ▶    **a.** $7 - (3x + 5) \geq 2(x - 4) - 5x$    original inequality

         $7 - 3x - 5 \geq 2x - 8 - 5x$    distributive property

         $-3x + 2 \geq -3x - 8$    combine like terms

         $2 \geq -8$    add $3x$

Since the resulting statement is always true, the original inequality is true for all real numbers. The solution is all real numbers: $x \in \mathbb{R}$.

   **b.** $3(x + 4) - 5 < 2(x - 3) + x$    original inequality

         $3x + 12 - 5 < 2x - 6 + x$    distributive property

         $3x + 7 < 3x - 6$    combine like terms

         $7 < -6$    subtract $3x$

Since the resulting statement is always false, the original inequality is false for all real numbers. The solution is $x \in$ { }.

☑ **B.** You've just seen how we can solve linear inequalities

**Now try Exercises 27 through 32** ▶

## C. Solving Compound Inequalities

In some applications of inequalities, we must consider more than one solution interval. These are called **compound inequalities,** and they require us to take a close look at the operations of **union** "∪" and **intersection** "∩". The intersection of two sets $A$ and $B$, written $A \cap B$, is the set of all elements *common to both sets*. The union of two sets $A$ and $B$, written $A \cup B$, is the set of all elements *that are in either set* (or both). When stating the union of two sets, repetitions are unnecessary.

**EXAMPLE 4** ▶   **Finding the Union and Intersection of Two Sets**

For set $A = \{-2, -1, 0, 1, 2, 3\}$ and set $B = \{1, 2, 3, 4, 5\}$, determine $A \cap B$ and $A \cup B$.

**Solution** ▶   $A \cap B$ is the set of all elements in *both A and B:*
$A \cap B = \{1, 2, 3\}$.
$A \cup B$ is the set of all elements in *either A or B:*
$A \cup B = \{-2, -1, 0, 1, 2, 3, 4, 5\}$.

**Now try Exercises 33 through 38** ▶

> **WORTHY OF NOTE**
>
> For the long term, it may help to rephrase the distinction as follows. The intersection is a *selection* of elements that are common to two sets, while the union is a *collection* of the elements from two sets.

Notice the intersection of two sets is described using the word "and," while the union of two sets is described using the word "or." When compound inequalities are formed using these words, the solution is modeled after the ideas from Example 4. (**See Exercises 39 through 44.**) If "and" is used, the solutions must satisfy *both* inequalities. If "or" is used, the solutions can satisfy *either* inequality.

**EXAMPLE 5** ▶   **Solving a Compound Inequality**

Solve the compound inequality, then write the solution in interval notation:
$-3x - 1 < -4$ **or** $4x + 3 < -6$.

**Solution** ▶   Begin by independently solving each simple inequality.

| | | | |
|---|---|---|---|
| $-3x - 1 < -4$ | or | $4x + 3 < -6$ | original statement |
| $-3x < -3$ | or | $4x < -9$ | isolate variable term |
| $x > 1$ | or | $x < -\dfrac{9}{4}$ | solve for $x$, reverse first inequality symbol |

The solution $x > 1$ **or** $x < -\frac{9}{4}$ is better understood by graphing each interval separately, *then selecting both intervals (the union).*

> **WORTHY OF NOTE**
>
> The graphs from Example 5 clearly show the solution consists of two disjoint (disconnected) intervals. This is reflected in the "or" statement: $x < -\frac{9}{4}$ or $x > 1$, and in the interval notation. Also, note the solution $x < -\frac{9}{4}$ or $x > 1$ is not equivalent to $-\frac{9}{4} > x > 1$, as there is no single number that is both greater than 1 and less than $-\frac{9}{4}$ at the same time.

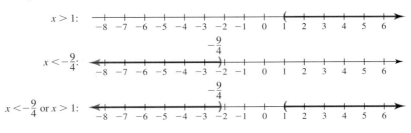

Interval notation: $x \in \left(-\infty, -\dfrac{9}{4}\right) \cup (1, \infty)$.

**Now try Exercises 45 and 46** ▶

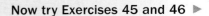

**EXAMPLE 6** ▶  **Solving a Compound Inequality**

Solve the compound inequality, then write the solution in interval notation: $3x + 5 > -13$ **and** $3x + 5 < -1$.

**Solution** ▶  Begin by independently solving each simple inequality.

| | | | |
|---|---|---|---|
| $3x + 5 > -13$ | and | $3x + 5 < -1$ | original statement |
| $3x > -18$ | and | $3x < -6$ | subtract 5 |
| $x > -6$ | and | $x < -2$ | divide by 3 |

The solution $x > -6$ **and** $x < -2$ can best be understood by graphing each interval separately, then *noting where they intersect.*

> **WORTHY OF NOTE**
>
> The inequality $a < b$ ($a$ is less than $b$) can equivalently be written as $b > a$ ($b$ is greater than $a$). In Example 6, the solution is read, "$x > -6$ and $x < -2$," but if we rewrite the first inequality as $-6 < x$ (with the "arrowhead" still pointing at $-6$), we have $-6 < x$ and $x < -2$ and can clearly see that $x$ must be in the single interval between $-6$ and $-2$.

$x > -6$:

$x < -2$:

$x > -6$ and $x < -2$:

Interval notation: $x \in (-6, -2)$.

**Now try Exercises 47 through 52** ▶

The solution from Example 6 consists of the single interval $(-6, -2)$, indicating the original inequality could actually be *joined* and written as $-6 < x < -2$, called a **joint inequality.** We solve joint inequalities in much the same way as simple inequalities, but must remember they *have three parts (left, middle, and right).* This means operations must be applied to *all three parts* in each step of the solution process, to obtain a solution form such as *smaller number $< x <$ larger number.* The same ideas apply when other inequality symbols are used.

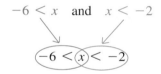

**EXAMPLE 7** ▶  **Solving a Joint Inequality**

Solve the joint inequality, then graph the solution set and write it in interval notation: $1 > \dfrac{2x + 5}{-3} \geq -6$.

**Solution** ▶

| | |
|---|---|
| $1 > \dfrac{2x + 5}{-3} \geq -6$ | original inequality |
| $-3 < 2x + 5 \leq 18$ | multiply all parts by $-3$; reverse the inequality symbols |
| $-8 < 2x \leq 13$ | subtract 5 from all parts |
| $-4 < x \leq \dfrac{13}{2}$ | divide all parts by 2 |

• Graph:

• Interval notation: $x \in \left(-4, \dfrac{13}{2}\right]$

✓ **C.** You've just seen how we can solve compound inequalities

**Now try Exercises 53 through 58** ▶

### D. Applications of Inequalities

**Allowable Values**

One application of inequalities involves the concept of allowable values, which are those values of the input that result in a real number output. Consider the expression $\frac{24}{x}$. As Table 1.2 suggests, we can evaluate this expression using any real number *other than zero,* since the expression $\frac{24}{0}$ is undefined. Using set notation the allowable values are written $\{x|x \neq 0\}$. To graph the solution we must be careful to exclude zero, as shown in Figure 1.2.

**Table 1.2**

| $x$ | $\frac{24}{x}$ |
|---|---|
| 6 | 4 |
| $-12$ | $-2$ |
| $\frac{1}{2}$ | 48 |
| 0 | error |

**Figure 1.2**

The graph gives us a snapshot of the solution using interval notation, which is written as a union of two **disjoint (disconnected) intervals** so as to exclude zero: $x \in (-\infty, 0) \cup (0, \infty)$. When the denominator of a fraction contains a variable expression, values that cause a denominator of zero are excluded.

**EXAMPLE 8** ▶ **Determining Allowable Values**

Determine the allowable values for $x$ in the expression $\dfrac{6}{x - 2}$. State the result in set notation, graphically, and using interval notation.

**Solution** ▶ Set the denominator equal to zero and solve: $x - 2 = 0$ yields $x = 2$. This means 2 *must be excluded.*

- Set notation: $\{x|x \neq 2\}$
- Graph: 
- Interval notation: $x \in (-\infty, 2) \cup (2, \infty)$

**Now try Exercises 59 through 66** ▶

A second area where allowable values are a concern involves the square root operation. Recall that $\sqrt{49} = 7$ since $7 \cdot 7 = 49$. However, $\sqrt{-49}$ cannot be written as the product of two real numbers since $(-7) \cdot (-7) = 49$ and $7 \cdot 7 = 49$. In other words, $\sqrt{X}$ represents a real number only if the radicand is positive or zero. If $X$ represents an algebraic expression, the set of allowable values is $\{X|X \geq 0\}$.

**EXAMPLE 9** ▶ **Determining Allowable Values**

Determine the allowable values for $x$ in $\sqrt{x + 3}$. State the allowable values in set notation, graphically, and in interval notation.

**Solution** ▶ The radicand must represent a nonnegative number. Solving $x + 3 \geq 0$ gives $x \geq -3$.

- Set notation: $\{x|x \geq -3\}$
- Graph: 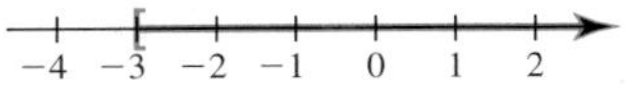
- Interval notation: $x \in [-3, \infty)$

**Now try Exercises 67 through 74** ▶

Inequalities are widely used to help gather information, and to make comparisons that will lead to informed decisions. Here, the problem-solving guide is once again a valuable tool.

**EXAMPLE 10** ▶ Using an Inequality to Compute Desired Test Scores

Justin earned scores of 78, 72, and 86 on the first three out of four exams. What score must he earn on the fourth exam to have an average of at least 80?

Solution ▶ **Gather and organize information; highlight any key phrases.**
First the scores: 78, 72, 86. An average of *at least* 80 means $A \geq 80$.
**Make the problem visual.**

| Test 1 | Test 2 | Test 3 | Test 4 | Computed Average | Minimum |
|--------|--------|--------|--------|------------------|---------|
| 78 | 72 | 86 | $x$ | $\dfrac{78 + 72 + 86 + x}{4}$ | 80 |

**Develop a model; assign a variable; build related expressions.**

Let $x$ represent Justin's score on the fourth exam, then $\dfrac{78 + 72 + 86 + x}{4}$ represents his average score.

$$\frac{78 + 72 + 86 + x}{4} \geq 80 \qquad \text{average must be greater than or equal to 80}$$

**Solve the problem; write the equation model and solve.**

$$78 + 72 + 86 + x \geq 320 \qquad \text{multiply by 4}$$
$$236 + x \geq 320 \qquad \text{simplify}$$
$$x \geq 84 \qquad \text{solve for } x \text{ (subtract 236)}$$

Justin must score at least an 84 on the last exam to earn an 80 average.

**Now try Exercises 77 through 82** ▶

As your problem-solving skills improve, the process outlined in the problem-solving guide naturally becomes less formal, as we work more directly toward the equation model. See Example 11.

**EXAMPLE 11** ▶ Using an Inequality to Make a Financial Decision

As Margaret starts her new job, her employer offers two salary options. Plan 1 is base pay of $2000/mo plus 3% of sales. Plan 2 is base pay of $500/mo plus 15% of sales. What level of monthly sales is needed for her to earn more under Plan 2?

Solution ▶ Let $x$ represent her monthly sales in dollars. The Plan 1 model would be $0.03x + 2000$; for Plan 2 we would have $0.15x + 500$. To find the sales volume needed for her to earn more under Plan 2, we solve the inequality

| Monthly Sales | Plan 1 Pay | Plan 2 Pay |
|---------------|------------|------------|
| 11,000 | 2330 | 2150 |
| 11,500 | 2345 | 2225 |
| 12,000 | 2360 | 2300 |
| 12,500 | 2375 | 2375 |
| 13,000 | 2390 | 2450 |
| 13,500 | 2405 | 2525 |
| 14,000 | 2420 | 2600 |

$$0.15x + 500 > 0.03x + 2000 \qquad \text{Plan 2 > Plan 1}$$
$$0.12x + 500 > 2000 \qquad \text{subtract } 0.03x$$
$$0.12x > 1500 \qquad \text{subtract 500}$$
$$x > 12{,}500 \qquad \text{divide by 0.12}$$

If Margaret can generate more than $12,500 in monthly sales, she will earn more under Plan 2.

☑ **D. You've just seen how we can solve applications of inequalities**

**Now try Exercises 83 through 86** ▶

## 1.2 EXERCISES

### ▶ CONCEPTS AND VOCABULARY

**Fill in each blank with the appropriate word or phrase. Carefully reread the section, if necessary.**

1. For inequalities, the three ways of writing a solution set are _____ notation, a(n) _____ _____ graph, and _____ notation.

2. The intersection of set $A$ with set $B$ is the set of elements in $A$ _____ $B$. The union of set $A$ with set $B$ is the set of elements in $A$ _____ $B$.

3. Discuss/Explain how the concept of allowable values relates to rational and radical expressions. Include a few examples.

4. Discuss/Explain the use of the words "and" and "or" in the statement of compound inequalities. Include a few examples to illustrate.

### ▶ DEVELOPING YOUR SKILLS

**Use an inequality to write a mathematical model for each statement.**

5. To qualify for a secretarial position, a person must type at least 45 words per minute.

6. The balance in a checking account must remain above $1000 or a fee is charged.

7. To bake properly, a turkey must be kept at temperatures above 250°, but below 450°.

8. To fly efficiently, the airliner must cruise at or between altitudes of 30,000 and 35,000 ft.

**Graph each inequality on a number line.**

9. $y < 3$

10. $x > -2$

11. $m \leq 5$

12. $n \geq -4$

13. $x \neq 1$

14. $x \neq -3$

15. $5 > x > 2$

16. $-3 < y \leq 4$

**Write the solution set illustrated on each graph in set notation and interval notation.**

17. 
18. 
19. 
20. 

**Solve the inequality and write the solution in set notation. Then graph the solution and write it in interval notation.**

21. $5a - 11 \geq 2a - 5$

22. $-8n + 5 > -2n - 12$

23. $2(n + 3) - 4 \leq 5n - 1$

24. $-5(x + 2) - 3 < 3x + 11$

25. $\dfrac{3x}{8} + \dfrac{x}{4} < -4$

26. $\dfrac{2y}{5} + \dfrac{y}{10} < -2$

**Solve each inequality and write the solution in set notation.**

27. $7 - 2(x + 3) \geq 4x - 6(x - 3)$

28. $-3 - 6(x - 5) \leq 2(7 - 3x) + 1$

29. $4(3x - 5) + 18 < 2(5x + 1) + 2x$

30. $8 - (6 + 5m) > -9m - (3 - 4m)$

31. $-6(p - 1) + 2p \leq -2(2p - 3)$

32. $9(w - 1) - 3w \geq -2(5 - 3w) + 1$

**Determine the intersection and union of sets $A$, $B$, $C$, and $D$ as indicated, given $A = \{-3, -2, -1, 0, 1, 2, 3\}$, $B = \{2, 4, 6, 8\}$, $C = \{-4, -2, 0, 2, 4\}$, and $D = \{4, 5, 6, 7\}$.**

33. $A \cap B$ and $A \cup B$

34. $A \cap C$ and $A \cup C$

35. $A \cap D$ and $A \cup D$

36. $B \cap C$ and $B \cup C$

37. $B \cap D$ and $B \cup D$

38. $C \cap D$ and $C \cup D$

**Express the compound inequalities graphically and in interval notation.**

39. $x < -2$ or $x > 1$

40. $x < -5$ or $x > 5$

41. $x < 5$ and $x \geq -2$

42. $x \geq -4$ and $x < 3$

43. $x \geq 3$ and $x \leq 1$

44. $x \leq -5$ or $x \geq -7$

**Solve the compound and joint inequalities, then graph the solution set.**

45. $4(x - 1) \leq 20$ or $x + 6 > 9$

46. $-3(x + 2) > 15$ or $x - 3 \leq -1$

47. $-2x - 7 \leq 3$ and $2x \leq 0$

48. $-3x + 5 \leq 17$ and $5x \leq 0$

49. $\frac{3}{5}x + \frac{1}{2} > \frac{3}{10}$ and $-4x > 1$

50. $\frac{2}{3}x - \frac{5}{6} \leq 0$ and $-3x < -2$

51. $\dfrac{3x}{8} + \dfrac{x}{4} < -3$ or $x + 1 > -5$

52. $\dfrac{2x}{5} + \dfrac{x}{10} < -2$ or $x - 3 > 2$

**53.** $-3 \le 2x + 5 < 7$     **54.** $2 < 3x - 4 \le 19$

**55.** $-0.5 \le 0.3 - x \le 1.7$

**56.** $-8.2 < 1.4 - x < -0.9$

**57.** $-7 < -\frac{3}{4}x - 1 \le 11$     **58.** $-21 \le -\frac{2}{3}x + 9 < 7$

Determine the allowable values in each expression. Write your answer in interval notation.

**59.** $\dfrac{12}{m}$     **60.** $\dfrac{-6}{n}$

**61.** $\dfrac{5}{y + 7}$     **62.** $\dfrac{4}{x - 3}$

**63.** $\dfrac{a + 5}{6a - 3}$     **64.** $\dfrac{m + 5}{8m + 4}$

**65.** $\dfrac{15}{3x - 12}$     **66.** $\dfrac{7}{2x + 6}$

**67.** $\sqrt{x - 2}$     **68.** $\sqrt{y + 7}$

**69.** $\sqrt{3n - 12}$     **70.** $\sqrt{2m + 5}$

**71.** $\sqrt{b - \frac{4}{3}}$     **72.** $\sqrt{a + \frac{3}{4}}$

**73.** $\sqrt{8 - 4y}$     **74.** $\sqrt{12 - 2x}$

## ▶ WORKING WITH FORMULAS

**75. Body mass index:** $B = \dfrac{704W}{H^2}$

The U.S. government publishes a body mass index formula to help people consider the risk of heart disease. An index $B$ of 27 or more means that a person is at risk. Here $W$ represents weight in pounds and $H$ represents height in inches. (a) Solve the formula for $W$. (b) While dominating as a competitive bodybuilder, Arnold Schwarzenegger typically weighed at least 235 lb. If he is 6′2″, what was his body mass index back then? (A person with an index above 30 is typically considered obese.)

*Source:* www.surgeongeneral.gov/topics

**76. Lift capacity:** $75S + 125B \le 750$

The capacity in pounds of the lift used by a roofing company to place roofing shingles and buckets of roofing nails on rooftops is modeled by the formula shown, where $S$ represents packs of shingles and $B$ represents buckets of nails. Solve the formula for (a) $S$ in terms of $B$ and (b) $B$ in terms of $S$. Then use the formula to find (c) the largest number of shingle packs that can be lifted, (d) the largest number of nail buckets that can be lifted, and (e) the largest number of shingle packs that can be lifted along with three nail buckets.

## ▶ APPLICATIONS

Write an inequality to model the given information and solve.

**77. Exam scores:** Jacques is going to college on an academic scholarship that requires him to maintain at least a 75% average in all of his classes. So far he has scored 82%, 76%, 65%, and 71% on four exams. What scores are possible on his last exam that will enable him to keep his scholarship?

**78. Timed trials:** In the first three trials of the 100-m butterfly, Johann had times of 50.2, 49.8, and 50.9 sec. How fast must he swim the final timed trial to have an average time of at most 50 sec?

**79. Checking account balance:** If the average daily balance in a certain checking account drops below $1000, the bank charges the customer a $7.50 service fee. The table gives the daily balance for one customer. What must the daily balance be for Friday to avoid a service charge?

| Weekday | Balance |
|---------|---------|
| Monday | $1125 |
| Tuesday | $850 |
| Wednesday | $625 |
| Thursday | $400 |

**80. Average weight:** In the National Football League, many consider an offensive line to be "small" if the average weight of the five down linemen is less than 325 lb. Using the table, what must the weight of the right tackle be so that the line will not be considered small?

| Lineman | Weight |
|---------|--------|
| Left tackle | 318 lb |
| Left guard | 322 lb |
| Center | 326 lb |
| Right guard | 315 lb |
| Right tackle | ? |

**81. Area of a rectangle:** Given the rectangle shown, what is the range of values for the width in order to keep the area less than 150 m²?

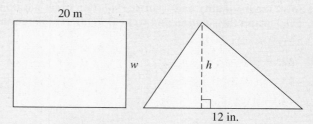

**82. Area of a triangle:** Using the triangle shown, find the height that will guarantee an area equal to or greater than 48 in².

**83. Heating and cooling subsidies:** As long as the outside temperature is over 45°F and less than 85°F ($45 < F < 85$), the city does not issue heating or cooling subsidies for low-income families. What is the corresponding range of Celsius temperatures $C$? Recall that $F = \frac{9}{5}C + 32$.

**84. U.S. and European shoe sizes:** To convert a European male shoe size "E" to an American male shoe size "A," the formula $A = 0.76E - 23$ can be used. Lillian has five sons in the U.S. military, with shoe sizes ranging from size 9 to size 14 ($9 \le A \le 14$). What is the corresponding range of European sizes? Round to the nearest half-size.

**85. Power tool rentals:** Sunshine Equipment Co. rents its power tools for a $20 fee, plus $4.50/hr. Kealoha's Rentals offers the same tools for an $11 fee plus $6.00/hr. How many hours $h$ must a tool be rented to make the cost at Sunshine a better deal?

**86. Moving van rentals:** Stringer Truck Rentals will rent a moving van for $15.75/day plus $0.35/mi. Bertz Van Rentals will rent the same van for $25/day plus $0.30/mi. How many miles $m$ must the van be driven to make the cost at Bertz a better deal on a one-day rental?

▶ **EXTENDING THE CONCEPT**

**87.** Jeff had an old '63 Chrysler that got 11 mpg back when gas was only $0.99/gal. He now owns a Honda Civic that gets 32 mpg, but today gas costs $3.50/gal. (a) If he commutes to work five days a week, 50 mi each way, what are the current weekly fuel costs in the Honda? (b) What were the weekly fuel costs when he was making the same commute in the Chrysler? (c) How much closer would he need to move to make the weekly commute cheaper now than it was then?

**88.** By transmitting data on the same lines that deliver cable to a TV, cable modems are able to download data at rates in excess of 54 Mbit/sec. Dial-up modems, which send and receive data through telephone lines, are typically limited to download speeds below 56 kbit/sec. (a) How long would it take to download an average-length (12-Gbit) movie using a cable modem? (b) How long would it take to download an average-quality (24-Mbit) photograph using dial-up? (c) In 1 hr, how many more movies can be downloaded via cable than photos via dial-up? (*Hint:* 1 Gbit = 1000 Mbit, 1 Mbit = 1000 kbit.)

**89.** Place the correct inequality symbol in the blank to make the statement true.
   **a.** If $m \le n$ and $p < 0$, then $mp$ _____ $np$.
   **b.** If $m > n$, then $-m$ _____ $-n$.
   **c.** If $m < n$, then $\dfrac{1}{m}$ _____ $\dfrac{1}{n}$.
   **d.** If $m < 0$ and $n > 0$, then $m^3$ _____ $n^2$.

**90.** What can be said about $x$, $y$, and $z$ if the following statements are assumed true?
   **a.** $xyz < 0$           **b.** $x^2y^3z^4 > 0$
   **c.** $x^2 + y^2 + z^2 \le 0$    **d.** $\dfrac{x^2}{yz} \ge 0$

**91.** The following argument leads to a false statement. Explain why the final statement is false and then find the error that led to it.

$$a < b \qquad \text{initial assumption}$$
$$a(a - b) < b(a - b) \qquad \text{multiply by } (a - b)$$
$$a^2 - ab < ab - b^2 \qquad \text{distribute}$$
$$a^2 - 2ab + b^2 < 0 \qquad \text{subtract } ab, \text{ add } b^2$$
$$(a - b)^2 < 0 \qquad \text{factor, false statement}$$

▶ **MAINTAINING YOUR SKILLS**

**92. (R.1)** Translate into an algebraic expression: eight subtracted from twice a number.

**94. (1.1)** Solve: $-4(x - 7) - 3 = 2x + 1$

**93. (R.2)** Simplify the algebraic expression:
$$2\left(\tfrac{5}{9}x - 1\right) - \left(\tfrac{1}{6}x + 3\right).$$

**95. (1.1)** Solve: $\tfrac{4}{5}m + \tfrac{2}{3} = \tfrac{1}{2}$

**LEARNING OBJECTIVES**

*In Section 1.3 you will see how we can:*

- [ ] **A.** Solve absolute value equations
- [ ] **B.** Solve "less than" absolute value inequalities
- [ ] **C.** Solve "greater than" absolute value inequalities
- [ ] **D.** Solve applications involving absolute value

While the equations $x + 1 = 5$ and $|x + 1| = 5$ are similar in many respects, note the first has only the solution $x = 4$, but both $x = 4$ and $x = -6$ satisfy the second. The fact there are two solutions shouldn't surprise us, as it's a natural result of how absolute value is defined.

## A. Solving Absolute Value Equations

The absolute value of a number $x$ can be thought of as its distance from zero on the number line, regardless of direction. This means $|x| = 4$ will have *two solutions*, since there are two numbers that are four units from zero: $x = -4$ and $x = 4$ (see Figure 1.3).

Distance from zero is exactly 4 ▼        Distance from zero is exactly 4

**Figure 1.3**

This basic idea can be extended to include situations where the quantity within absolute value bars *is an algebraic expression,* and suggests the following property.

> **Property of Absolute Value Equations**
>
> If $X$ represents an algebraic expression and $k$ is a positive real number,
> $$\text{then } |X| = k$$
> $$\text{implies } X = -k \text{ or } X = k.$$

**WORTHY OF NOTE**

Note if $k < 0$, the equation $|X| = k$ has no solutions since the absolute value of any quantity is always positive or zero. On a related note, we can verify that if $k = 0$, the equation $|X| = 0$ has only the solution $X = 0$.

As the statement of this property suggests, it can only be applied *after* the absolute value expression has been isolated on one side.

---

**EXAMPLE 1** ▶ **Solving an Absolute Value Equation**

Solve: $-5|x - 7| + 2 = -13$.

**Solution** ▶ Begin by isolating the absolute value expression.

$$-5|x - 7| + 2 = -13 \qquad \text{original equation}$$
$$-5|x - 7| = -15 \qquad \text{subtract 2}$$
$$|x - 7| = 3 \qquad \text{divide by } -5 \text{ (simplified form)}$$

Now consider $x - 7$ as the variable expression "$X$" in the property of absolute value equations, giving

$$x - 7 = -3 \quad \text{or} \quad x - 7 = 3 \qquad \text{apply the property of absolute value equations}$$
$$x = 4 \quad \text{or} \quad x = 10 \qquad \text{add 7}$$

Substituting into the original equation verifies the solution set is $\{4, 10\}$.

**Now try Exercises 5 through 16** ▶

---

**CAUTION** ▶ For equations like those in Example 1, be careful not to treat the absolute value bars as simple grouping symbols. The equation $-5(x - 7) + 2 = -13$ has only the solution $x = 10$, and "misses" the second solution since it yields $x - 7 = 3$ in simplified form. The equation $-5|x - 7| + 2 = -13$ simplifies to $|x - 7| = 3$ and there are actually *two* solutions. Also note that $-5|x - 7| \neq |-5x + 35|$!

---

**EXAMPLE 2** ▶ **Solving an Absolute Value Equation**

Solve: $\left|5 - \dfrac{2}{3}x\right| - 9 = 8$.

**Solution** ▶

$$\left|5 - \dfrac{2}{3}x\right| - 9 = 8 \qquad \text{original equation}$$

$$\left|5 - \dfrac{2}{3}x\right| = 17 \qquad \text{add 9}$$

$$5 - \dfrac{2}{3}x = -17 \quad \text{or} \quad 5 - \dfrac{2}{3}x = 17 \qquad \substack{\text{apply the property of absolute} \\ \text{value equations}}$$

$$-\dfrac{2}{3}x = -22 \quad \text{or} \quad -\dfrac{2}{3}x = 12 \qquad \text{subtract 5}$$

$$x = 33 \quad \text{or} \quad x = -18 \qquad \text{multiply by } -\dfrac{3}{2}$$

> **WORTHY OF NOTE**
>
> As illustrated in both Examples 1 and 2, the property we use to solve absolute value equations can be applied only *after* the absolute value term has been isolated. As you will see, the same is true for the properties used to solve absolute value inequalities.

**Check** ▶

For $x = 33$: $\left|5 - \dfrac{2}{3}(33)\right| - 9 = 8$ 　　For $x = -18$: $\left|5 - \dfrac{2}{3}(-18)\right| - 9 = 8$

$$|5 - 2(11)| - 9 = 8 \qquad\qquad |5 - 2(-6)| - 9 = 8$$

$$|5 - 22| - 9 = 8 \qquad\qquad |5 + 12| - 9 = 8$$

$$|-17| - 9 = 8 \qquad\qquad |17| - 9 = 8$$

$$17 - 9 = 8 \qquad\qquad 17 - 9 = 8$$

$$8 = 8 \checkmark \qquad\qquad 8 = 8 \checkmark$$

Both solutions check. The solution set is $\{-18, 33\}$.

**Now try Exercises 17 through 20** ▶

For some equations, it's helpful to apply the **multiplicative property of absolute value:**

---

### Multiplicative Property of Absolute Value

If $A$ and $B$ represent algebraic expressions,

$$\text{then } |AB| = |A||B|.$$

---

Note that if $A = -1$ the property says $|-1 \cdot B| = |-1||B| = |B|$, or more simply, $|-B| = |B|$. More generally the property is applied where $A$ is any constant.

---

**EXAMPLE 3** ▶ **Solving Equations Using the Multiplicative Property of Absolute Value**

Solve: $|-2x| + 5 = 13$.

**Solution** ▶

$$|-2x| + 5 = 13 \qquad \text{original equation}$$

$$|-2x| = 8 \qquad \text{subtract 5}$$

$$|-2||x| = 8 \qquad \text{apply multiplicative property of absolute value}$$

$$2|x| = 8 \qquad \text{simplify}$$

$$|x| = 4 \qquad \text{divide by 2}$$

$$x = -4 \quad \text{or} \quad x = 4 \qquad \text{apply property of absolute value equations}$$

Both solutions check. The solution set is $\{-4, 4\}$.

**Now try Exercises 21 and 22** ▶

In some instances, we have one absolute value quantity equal to another, as in $|A| = |B|$. From this equation, four possible solutions are immediately apparent:

$$(1)\ A = B \qquad (2)\ A = -B \qquad (3)\ -A = B \qquad (4)\ -A = -B$$

However, basic properties of equality show that equations (1) and (4) are equivalent, as are equations (2) and (3), meaning all solutions can be found using only equations (1) and (2).

---

**EXAMPLE 4** ▶ **Solving Absolute Value Equations with Two Absolute Value Expressions**

Solve the equation $|2x + 7| = |x - 1|$.

**Solution** ▶ This equation has the form $|A| = |B|$, where $A = 2x + 7$ and $B = x - 1$. From our previous discussion, all solutions can be found using $A = B$ and $A = -B$.

| | | | | |
|---|---|---|---|---|
| $A = B$ | solution template | | $A = -B$ | solution template |
| $2x + 7 = x - 1$ | substitute | | $2x + 7 = -(x - 1)$ | substitute |
| $2x = x - 8$ | subtract 7 | | $2x + 7 = -x + 1$ | distribute |
| $x = -8$ | subtract $x$ | | $3x = -6$ | add $x$, subtract 7 |
| | | | $x = -2$ | divide by 3 |

The solutions are $x = -8$ and $x = -2$. Verify the solutions by substituting them into the original equation.

☑ **A.** You've just seen how we can solve absolute value equations

**Now try Exercises 23 and 24** ▶

---

## B. Solving "Less Than" Absolute Value Inequalities

Absolute value *inequalities* can be solved using the basic concept underlying the property of absolute value equalities. Whereas the equation $|x| = 4$ asks for all numbers $x$ whose distance from zero is *equal* to 4, the inequality $|x| < 4$ asks for all numbers $x$ whose distance from zero is *less than* 4.

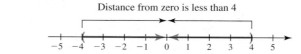

**Figure 1.4**

Distance from zero is less than 4

As Figure 1.4 illustrates, the solutions are $x > -4$ and $x < 4$, which can be written as the joint inequality $-4 < x < 4$. This idea can likewise be extended to include the absolute value of an algebraic expression $X$ as follows.

---

**Property I: Absolute Value Inequalities (Less Than)**

If $X$ represents an algebraic expression and $k$ is a positive real number,

$$\text{then } |X| < k$$

$$\text{implies } -k < X < k$$

---

Property I can also be applied when the "$\leq$" symbol is used. Also notice that if $k \leq 0$, the solution is the empty set since the absolute value of any quantity is always positive or zero.

**EXAMPLE 5** ▶ **Solving "Less Than" Absolute Value Inequalities**

Solve the inequalities:

**a.** $\dfrac{|3x + 2|}{4} \leq 1$      **b.** $|2x - 7| < -5$

**Solution** ▶     **a.**

$$\dfrac{|3x + 2|}{4} \leq 1 \qquad \text{original inequality}$$

$$|3x + 2| \leq 4 \qquad \text{multiply by 4}$$

$$-4 \leq 3x + 2 \leq 4 \qquad \text{apply Property I}$$

$$-6 \leq 3x \leq 2 \qquad \text{subtract 2 from all three parts}$$

$$-2 \leq x \leq \dfrac{2}{3} \qquad \text{divide all three parts by 3}$$

The solution interval is $\left[-2, \frac{2}{3}\right]$.

**b.** $|2x - 7| < -5$     original inequality

Since the absolute value of any quantity is always positive or zero, the solution for this inequality is the empty set: { }.

☑ **B.** You've just seen how we can solve "less than" absolute value inequalities

**Now try Exercises 25 through 30** ▶

As with the inequalities from Section 1.2, solutions to absolute value inequalities can be checked using a test value. For Example 5(a), substituting $x = 0$ from the solution interval yields:

$$\dfrac{1}{2} \leq 1 \checkmark$$

## C. Solving "Greater Than" Absolute Value Inequalities

For "greater than" inequalities, consider $|x| > 4$. Now we're asked to find all numbers $x$ whose distance from zero is *greater than* 4. As Figure 1.5 shows, solutions are found in the interval to the left of $-4$, or to the right of 4. The fact the intervals are disjoint (disconnected) is reflected in this graph, in the inequalities $x < -4$ **or** $x > 4$, as well as the interval notation $x \in (-\infty, -4) \cup (4, \infty)$.

Distance from zero is greater than 4                      Distance from zero is greater than 4

**Figure 1.5**     $-7 \quad -6 \quad -5 \quad -4 \quad -3 \quad -2 \quad -1 \quad 0 \quad 1 \quad 2 \quad 3 \quad 4 \quad 5 \quad 6 \quad 7$

As before, we can extend this idea to include algebraic expressions, as follows:

| **Property II: Absolute Value Inequalities (Greater Than)** |
| --- |
| If $X$ represents an algebraic expression and $k$ is a positive real number, |
| then $\|X\| > k$ |
| implies $X < -k$   or   $X > k$ |

**EXAMPLE 6** ▶ **Solving "Greater Than" Absolute Value Inequalities**

Solve the inequalities:

**a.** $-\dfrac{1}{3}\left|3 + \dfrac{x}{2}\right| < -2$     **b.** $|5x + 2| \geq -\dfrac{3}{2}$

**Solution** ▶ **a.** Note the exercise is given as a *less than* inequality, but as we multiply both sides by $-3$, we must *reverse the inequality symbol.*

$$-\frac{1}{3}\left|3 + \frac{x}{2}\right| < -2 \qquad \text{original inequality}$$

$$\left|3 + \frac{x}{2}\right| > 6 \qquad \text{multiply by } -3, \textit{ reverse the symbol}$$

$$3 + \frac{x}{2} < -6 \quad \text{or} \quad 3 + \frac{x}{2} > 6 \qquad \text{apply Property II}$$

$$\frac{x}{2} < -9 \quad \text{or} \quad \frac{x}{2} > 3 \qquad \text{subtract 3}$$

$$x < -18 \quad \text{or} \quad x > 6 \qquad \text{multiply by 2}$$

Property II yields the disjoint intervals $x \in (-\infty, -18) \cup (6, \infty)$ as the solution.

$$\xleftarrow{\quad}\ \overset{-30\ \ -24\ \ -18\ \ -12\ \ -6\ \ \ 0\ \ \ 6\ \ \ 12\ \ 18\ \ 24\ \ 30}{\longrightarrow}$$

**b.** $|5x + 2| \geq -\dfrac{3}{2}$  original inequality

Since the absolute value of any quantity is always positive or zero, the solution for this inequality is all real numbers: $x \in (-\infty, \infty)$.

☑ **C. You've just seen how we can solve "greater than" absolute value inequalities**

**Now try Exercises 31 through 48** ▶

Due to the nature of absolute value functions, there are times when an absolute value relation cannot be satisfied. For instance the equation $|x - 4| = -2$ has no solutions, as the left-hand expression will always represent a nonnegative value. The inequality $|2x + 3| < -1$ has no solutions for the same reason. On the other hand, the inequality $|9 - x| \geq 0$ is true for all real numbers, since any value substituted for $x$ will result in a nonnegative value. We can generalize many of these special cases as follows.

**Absolute Value Functions—Special Cases**

Given $k$ is a positive real number and $A$ represents an algebraic expression,

| $\mid A \mid = -k$ | $\mid A \mid < -k$ | $\mid A \mid > -k$ |
|---|---|---|
| has no solutions | has no solutions | is true for all real numbers |

**See Exercises 49 through 52.**

**CAUTION** ▶ Be sure you note the difference between the individual solutions of an absolute value equation, and the solution intervals that often result from solving absolute value inequalities. The solution $\{-2, 5\}$ indicates that both $x = -2$ and $x = 5$ are solutions, while the solution $[-2, 5)$ indicates that all numbers between $-2$ and $5$, including $-2$, are solutions.

## D. Applications Involving Absolute Value

Applications of absolute value often involve finding a range of values for which a given statement is true. Many times, the equation or inequality used must be modeled after a given description or from given information, as in Example 7.

**EXAMPLE 7** ▶ 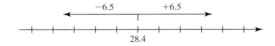 **Solving Applications Involving Absolute Value Inequalities**

For new cars, the number of miles per gallon (mpg) a car will get is heavily dependent on whether it is used mainly for short trips and city driving, or primarily on the highway for longer trips. For a certain car, the number of miles per gallon that a driver can expect varies by no more than 6.5 mpg above or below its field-tested average of 28.4 mpg. What range of mileage values can a driver expect for this car?

**Solution** ▶ Field tested average: 28.4 mpg                      gather information
mileage varies by no more than 6.5 mpg                      highlight key phrases

```
      ←  -6.5      +6.5  →
    ──┼──┼──┼──┼──┼──┼──┼──┼──┼──┼──┼──┼──→        make the problem visual
                28.4
```

Let $m$ represent the miles per gallon a driver can expect.    assign a variable
Then the difference between $m$ and 28.4 can be no more
than 6.5, or $|m - 28.4| \le 6.5$.                              write an equation model

$$|m - 28.4| \le 6.5$$                                          equation model
$$-6.5 \le m - 28.4 \le 6.5$$                                   apply Property I
$$21.9 \le m \le 34.9$$                                         add 28.4 to all three parts

The mileage that a driver can expect ranges from a low of 21.9 mpg to a high of 34.9 mpg.

 **D.** You've just seen how we can solve applications involving absolute value

**Now try Exercises 55 through 64** ▶

## 1.3 EXERCISES

### ▶ CONCEPTS AND VOCABULARY

**Fill in the blank with the appropriate word or phrase. Carefully reread the section, if needed.**

1. To write an absolute value equation or inequality in simplified form, we _____ the absolute value expression on one side.

2. The absolute value equation $|2x + 3| = 7$ is true when $2x + 3 =$ _____ or when $2x + 3 =$ _____.

**Describe the solution set for each inequality (assume $k > 0$). Justify your answer.**

3. $|ax + b| < -k$

4. $|ax + b| > -k$

### ▶ DEVELOPING YOUR SKILLS

**Solve each absolute value equation. Write the solution in set notation. When possible, verify solutions by substituting into the original equation.**

5. $2|m - 1| - 7 = 3$

6. $3|n - 5| - 14 = -2$

7. $2|4v + 5| - 6.5 = 10.3$

8. $7|2w + 5| + 6.3 = 11.2$

9. $-|7p - 3| + 6 = -5$

10. $-|3q + 4| + 3 = -5$

11. $-2|b| - 3 = -4$

12. $-3|c| - 5 = -6$

13. $-3|x + 5| + 6 = 6$

14. $-|y + 3| - 14 = -14$

15. $-2|3x| - 17 = -5$

16. $-5|2y| - 14 = 6$

17. $-3\left|\dfrac{w}{2} + 4\right| - 1 = -4$

18. $-2\left|3 - \dfrac{v}{3}\right| + 1 = -5$

19. $8.7|p - 7.5| - 26.6 = 8.2$

20. $5.3|q + 9.2| + 6.7 = 43.8$

21. $8.7|-2.5x| - 26.6 = 8.2$

22. $5.3|1.25n| + 6.7 = 43.8$

23. $|x - 2| = |3x + 4|$

24. $|2x - 1| = |x + 3|$

**Solve each absolute value inequality. Write solutions in interval notation. Check solutions by back-substitution.**

25. $3|p + 4| + 5 < 8$

26. $5|q - 2| - 7 \leq 8$

27. $-3|m| - 2 > 4$

28. $-2|n| + 3 > 7$

29. $|3b - 11| + 6 \leq 9$

30. $|2c + 3| - 5 < 1$

31. $|4 - 3z| + 12 > 7$

32. $|2 - 3u| - 4 \geq -5$

33. $\dfrac{|5v + 1|}{4} + 8 < 9$

34. $\dfrac{|3w - 2|}{2} + 6 < 8$

35. $\left|\dfrac{4x + 5}{3} - \dfrac{1}{2}\right| \leq \dfrac{7}{6}$

36. $\left|\dfrac{2y - 3}{4} - \dfrac{3}{8}\right| \leq \dfrac{15}{16}$

37. $|n + 3| > 7$

38. $|m - 1| > 5$

39. $-2|w| - 5 \leq -11$

40. $-5|v| - 3 \leq -23$

41. $\dfrac{|q|}{2} - \dfrac{5}{6} \geq \dfrac{1}{3}$

42. $\dfrac{|p|}{5} + \dfrac{3}{2} \geq \dfrac{9}{4}$

43. $3|5 - 7d| + 9 > 9$

44. $5|2c + 7| - 11 > -11$

45. $2 < \left|-3m + \dfrac{4}{5}\right| - \dfrac{1}{5}$

46. $4 \leq \left|\dfrac{5}{4} - 2n\right| - \dfrac{3}{4}$

47. $4|5 - 2h| - 9 \leq -9$

48. $3|7 + 2k| + 10 \leq 10$

49. $3.9|4q - 5| + 8.7 \leq -22.5$

50. $0.9|2p + 7| + 16.11 \leq 10.89$

51. $|4z - 9| + 6 \geq 4$

52. $|5u - 3| + 8 > 6$

## ▶ WORKING WITH FORMULAS

53. **Spring oscillation:** $|d - x| \leq L$

    A weight attached to a spring hangs at rest a distance of $x$ in. off the ground. If the weight is pulled down (stretched) a distance of $L$ inches and released, the weight begins to bounce and its distance $d$ off the ground must satisfy the indicated formula. (a) Solve for $x$ in terms of $L$ and $d$. (b) If $x$ equals 4 ft and the spring is stretched 3 in. and released, solve the inequality to find what distances from the ground the weight will oscillate between.

54. **A "fair" coin:** $\left|\dfrac{h - 50}{5}\right| < 1.645$

    If we flip a coin 100 times, we expect "heads" to come up about 50 times if the coin is "fair." In a study of probability, it can be shown that the number of heads $h$ that appears in such an experiment should satisfy the given inequality to be considered "fair." (a) Solve this inequality for $h$. (b) If you flipped a coin 100 times and obtained 40 heads, is the coin "fair"?

## ▶ APPLICATIONS

**Solve each application of absolute value.**

55. **Altitude of jet stream:** To take advantage of the jet stream, an airplane must fly at a height $h$ (in feet) that satisfies the inequality $|h - 35{,}050| \leq 2550$. Solve the inequality and determine if an altitude of 34,000 ft will place the plane in the jet stream.

56. **Quality control tests:** In order to satisfy quality control, the marble columns a company produces must earn a stress test score $S$ that satisfies the inequality $|S - 17{,}750| \leq 275$. Solve the inequality and determine if a score of 17,500 is in the passing range.

57. **Submarine depth:** The sonar operator on a submarine detects an old World War II submarine net and must decide to detour over or under the net. The computer gives him a depth model $|d - 394| - 20 > 164$, where $d$ is the depth in feet that represents safe passage. At what depth should the submarine travel to go under or over the net? Answer using simple inequalities.

58. **Optimal fishing depth:** When deep-sea fishing, the optimal depths $d$ (in feet) for catching a certain type of fish satisfy the inequality $28|d - 350| - 1400 < 0$. Find the range of depths that offer the best fishing. Answer using simple inequalities.

**For Exercises 59 through 62, (a) develop a model that uses an absolute value inequality, and (b) solve.**

59. **Stock value:** My stock in MMM Corporation fluctuated a great deal in 2009, but never by more than $3.35 from its current value. If the stock is worth $37.58 today, what was its range in 2009?

60. **Traffic studies:** On a given day, the volume of traffic at a busy intersection averages 726 cars per hour (cph). During rush hour the volume is much higher, during "off hours" much lower. Find the range of this volume if it never varies by more than 235 cph from the average.

61. **Car radio reception:** While on spring break, Paul decides to drive home to visit his family in Texas. His route takes him south on Interstate 35, a long straight stretch of freeway through Oklahoma City, home of his favorite radio station, 100.5 the KATT. At 12:00 noon, he is still 180 mi away, but knows his car radio will pick up the signal as long as he's within 90 mi of the city. If he cruises along this route at a steady 72 mph, at what times will he pick up and then lose the signal?

62. **Computer consultant salaries:** The national average salary for a computer consultant is $53,336. For a large computer firm, the salaries offered to their employees vary by no more than $11,994 from this national average. Find the range of salaries offered by this company.

63. **Tolerances for sport balls:** According to the official rules for golf, baseball, pool, and bowling, (a) golf balls must be within 0.03 mm of $d = 42.7$ mm, (b) baseballs must be within 1.01 mm of $d = 73.78$ mm, (c) billiard balls must be within 0.127 mm of $d = 57.150$ mm, and (d) bowling balls must be within 1.205 mm of $d = 217.105$ mm. Write each statement using an absolute value inequality, then (e) determine which sport gives the least tolerance $t$ $\left( t = \dfrac{\text{width of interval}}{\text{average value}} \right)$ for the diameter of the ball.

64. **Automated packaging:** The machines that fill boxes of breakfast cereal are programmed to fill each box within a certain tolerance. If the box is overfilled, the company loses money. If it is underfilled, it is considered unsuitable for sale. Suppose that boxes marked "14 ounces" must be filled to within 0.1 oz. Find the acceptable range of weights for this cereal.

▶ **EXTENDING THE CONCEPT**

65. Determine the value or values (if any) that will make the equation or inequality true.

  **a.** $|x| + x = 8$          **b.** $|x - 2| \le \dfrac{x}{2}$

  **c.** $x - |x| = x + |x|$     **d.** $|x + 3| \ge 6x$

  **e.** $|2x + 1| = x - 3$

66. In many cases, it can be helpful to view the solutions to absolute value equations and inequalities as follows. For any algebraic expression $X$ and positive constant $k$, the equation $|X| = k$ has solutions $X = k$ and $-X = k$, since the absolute value of either quantity on the left will indeed yield the positive constant $k$. Likewise, $|X| < k$ has solutions $X < k$ and $-X < k$. Note the inequality

symbol has not been reversed as yet, but will naturally be reversed as part of the solution process. Solve the following equations or inequalities using this idea.

  **a.** $|x - 3| = 5$           **b.** $|x - 7| > 4$

  **c.** $3|x + 2| \le 12$       **d.** $-3|x - 4| + 7 = -11$

67. You may have noticed that we discussed a multiplicative property of absolute value in this section, but not an additive property. (a) To see why no such property exists for addition, find values for $A$ and $B$ that make $|A + B| = |A| + |B|$ false. (b) Although typically $|A + B| \ne |A| + |B|$, there are exceptions to this rule. If $|A + B| = |A| + |B|$ happens to be true, what can be said about $A$ and $B$?

▶ **MAINTAINING YOUR SKILLS**

68. (1.1) Solve $V^2 = \dfrac{2W}{C\rho A}$ for $\rho$ (physics).

69. (R.5) Factor the expression completely:
$$18x^3 + 21x^2 - 60x.$$

70. (1.2) Solve the inequality, then write the solution set in interval notation:
$$-3(2x - 5) > 2(x + 1) - 7.$$

71. (R.4) Simplify $\dfrac{-1}{3 + \sqrt{3}}$ by rationalizing the denominator. State the result in exact form and approximate form (to hundredths).

## MID-CHAPTER CHECK

**1.** Solve each equation. If the equation is an identity or contradiction, so state and name the solution set.

**a.** $\dfrac{r}{3} + 5 = 2$

**b.** $5(2x - 1) + 4 = 9x - 7$

**c.** $m - 2(m + 3) = 1 - (m + 7)$

**d.** $\dfrac{1}{5}y + 3 = \dfrac{3}{2}y - 2$

**e.** $\dfrac{1}{2}(5j - 2) = \dfrac{3}{2}(j - 4) + j$

**f.** $0.6(x - 3) + 0.3 = 1.8$

Solve for the variable specified.

**2.** $H = -16t^2 + v_0 t$ for $v_0$

**3.** $S = 2\pi x^2 + \pi x^2 y$ for $y$

**4.** Solve each inequality and graph the solution set.

**a.** $-5x + 16 \le 11$ or $3x + 2 \le -4$

**b.** $\dfrac{1}{2} < \dfrac{1}{12}x - \dfrac{5}{6} \le \dfrac{3}{4}$

**5.** Determine the allowable values for each expression. Write your answer in interval notation.

**a.** $\dfrac{3x + 1}{2x - 5}$    **b.** $\sqrt{17 - 6x}$

**6.** Solve the following absolute value equations. Write the solution in set notation.

**a.** $\dfrac{2}{3}|d - 5| + 1 = 7$    **b.** $5 - |s + 3| = \dfrac{11}{2}$

**7.** Solve the following absolute value inequalities.

**a.** $3|q + 4| - 2 < 10$

**b.** $\left|\dfrac{x}{3} + 2\right| + 5 \le 5$

**8.** Solve the following absolute value inequalities. Write solutions in interval notation.

**a.** $3.1|d - 2| + 1.1 \ge 7.3$

**b.** $\dfrac{|1 - y|}{3} + 2 > \dfrac{11}{2}$

**c.** $-5|k - 2| + 3 < 4$

**9.** **Motocross:** An Enduro motocross motorcyclist averages 30 mph through the first part of a 115-mi course, and 50 mph though the second part. If the rider took 2 hr and 50 min to complete the course, how long was she on the first part?

**10.** **Kiteboarding:** With a correctly sized kite, a person can kiteboard when the wind is blowing at a speed $w$ (in mph) that satisfies the inequality $|w - 17| \le 9$. Solve the inequality and determine if a person can kiteboard with a windspeed of 9 mph.

---

## 1.4    Complex Numbers

**LEARNING OBJECTIVES**

*In Section 1.4 you will see how we can:*

☐ **A.** Identify and simplify imaginary and complex numbers

☐ **B.** Add and subtract complex numbers

☐ **C.** Multiply complex numbers and find powers of $i$

☐ **D.** Divide complex numbers

For centuries, even the most prominent mathematicians refused to work with equations like $x^2 + 1 = 0$. Using the principal of square roots gave the "solutions" $x = \sqrt{-1}$ and $x = -\sqrt{-1}$, which they found baffling and mysterious, since there is no real number whose square is $-1$. In this section, we'll see how this dilemma was finally resolved.

### A. Identifying and Simplifying Imaginary and Complex Numbers

The equation $x^2 = -1$ has no real solutions, since the square of any real number is positive or zero. But if we apply the principle of square roots we get $x = \sqrt{-1}$ and $x = -\sqrt{-1}$, which seem to check when substituted into the original equation:

$$x^2 + 1 = 0 \quad \text{original equation}$$

$$(1) \qquad (\sqrt{-1})^2 + 1 = 0 \quad \text{substitute } \sqrt{-1} \text{ for } x$$

$$-1 + 1 = 0 \checkmark \quad \text{answer "checks"}$$

$$(2) \qquad (-\sqrt{-1})^2 + 1 = 0 \quad \text{substitute } -\sqrt{-1} \text{ for } x$$

$$-1 + 1 = 0 \checkmark \quad \text{answer "checks"}$$

This observation likely played a part in prompting Renaissance mathematicians to study such numbers in greater depth, as they reasoned that while these were not *real number* solutions, they must be *solutions of a new and different kind*. Their study eventually resulted in the introduction of the set of **imaginary numbers** and the **imaginary unit *i*,** as follows.

### Imaginary Numbers and the Imaginary Unit

- Imaginary numbers are those of the form $\sqrt{-k}$, where $k$ is a positive real number.

- The imaginary unit $i$ is the number whose square is $-1$:
$$i = \sqrt{-1} \text{ and } i^2 = -1$$

**WORTHY OF NOTE**

It was René Descartes (in 1637) who first used the term *imaginary* to describe these numbers; Leonhard Euler (in 1777) who introduced the letter *i* to represent $\sqrt{-1}$; and Carl F. Gauss (in 1831) who first used the phrase *complex* number to describe solutions that had both a real part and an imaginary part. For more on complex numbers and their story, see www.mhhe.com/coburn.

As a convenience to understanding and working with imaginary numbers, we rewrite them in terms of $i$, allowing that the product property of radicals ($\sqrt{AB} = \sqrt{A}\sqrt{B}$) still applies if at most one of the radicands is negative. For $\sqrt{-3}$, we have $\sqrt{-1 \cdot 3} = \sqrt{-1}\sqrt{3} = i\sqrt{3}$. More generally, we simply make the following statement regarding imaginary numbers.

### Rewriting Imaginary Numbers

For any positive real number $k$,
$$\sqrt{-k} = i\sqrt{k}.$$

For $\sqrt{-16}$ and $\sqrt{-20}$ we have:

$$\sqrt{-16} = i\sqrt{16} \qquad\qquad \sqrt{-20} = i\sqrt{20}$$
$$= i(4) \qquad\qquad\qquad = i\sqrt{4 \cdot 5}$$
$$= 4i \qquad\qquad\qquad = 2i\sqrt{5},$$

and we say the expressions have been simplified and written in terms of $i$. Note that for $\sqrt{-20}$ we've written the result with the unit "$i$" *in front of the radical* to prevent it appearing as though it were *under the radical*. In symbols, $2i\sqrt{5} = 2\sqrt{5}i \neq 2\sqrt{5i}$.

The solutions to $x^2 = -1$ also serve to illustrate that for $k > 0$, there are two solutions to $x^2 = -k$, namely, $i\sqrt{k}$ and $-i\sqrt{k}$. In other words, like positive numbers, every negative number has two square roots. The first of these, $i\sqrt{k}$, is called the **principal square root** of $-k$.

**EXAMPLE 1** ▶  **Simplifying Imaginary Numbers**

Rewrite the imaginary numbers in terms of $i$ and simplify if possible.

**a.** $\sqrt{-7}$     **b.** $\sqrt{-81}$     **c.** $\sqrt{-24}$     **d.** $-3\sqrt{-16}$

**Solution** ▶

**a.** $\sqrt{-7} = i\sqrt{7}$          **b.** $\sqrt{-81} = i\sqrt{81}$
$$= 9i$$

**c.** $\sqrt{-24} = i\sqrt{24}$       **d.** $-3\sqrt{-16} = -3i\sqrt{16}$
$$= i\sqrt{4 \cdot 6} \qquad\qquad\qquad = -3i(4)$$
$$= 2i\sqrt{6} \qquad\qquad\qquad\quad = -12i$$

**Now try Exercises 5 through 10** ▶

**EXAMPLE 2** ▶ **Writing an Expression in Terms of *i***

The numbers $x = \dfrac{-6 + \sqrt{-16}}{2}$ and $x = \dfrac{-6 - \sqrt{-16}}{2}$ are not real, but are known

to be solutions of $x^2 + 6x + 13 = 0$. Simplify $\dfrac{-6 + \sqrt{-16}}{2}$.

**Solution** ▶ Using the *i* notation, we have

$$\frac{-6 + \sqrt{-16}}{2} = \frac{-6 + i\sqrt{16}}{2} \qquad \text{write } \sqrt{-16} \text{ in } i \text{ notation}$$

$$= \frac{-6 + 4i}{2} \qquad \text{simplify}$$

$$= \frac{\cancel{2}(-3 + 2i)}{\cancel{2}} \qquad \text{factor numerator and reduce}$$

$$= -3 + 2i \qquad \text{result}$$

**Now try Exercises 11 through 14** ▶

> **WORTHY OF NOTE**
>
> The expression $\dfrac{-6 + 4i}{2}$ from the solution of Example 2 can also be simplified by rewriting it as two separate terms, then simplifying each term:
> $$\frac{-6 + 4i}{2} = \frac{-6}{2} + \frac{4i}{2}$$
> $$= -3 + 2i.$$

The result in Example 2 contains both a **real part** ($-3$) and an **imaginary part** ($2i$). Numbers of this type are called **complex numbers.**

| **Complex Numbers** |
|---|
| Complex numbers are those that can be written in the form $a + bi$, where $a$ and $b$ are real numbers and $i = \sqrt{-1}$. |

The expression $a + bi$ is called the **standard form** of a complex number. From this definition we note that all real numbers are also complex numbers, since $a + 0i$ is complex with $b = 0$. In addition, all imaginary numbers are complex numbers, since $0 + bi$ is a complex number with $a = 0$ (see Figure 1.6).

**EXAMPLE 3** ▶ **Writing Complex Numbers in Standard Form**

Write each complex number in the form $a + bi$, and identify the values of $a$ and $b$.

**a.** $2 + \sqrt{-49}$     **b.** $\sqrt{-12}$     **c.** $7$     **d.** $\dfrac{4 + 3\sqrt{-25}}{20}$

**Solution** ▶ 

**a.** $2 + \sqrt{-49} = 2 + i\sqrt{49}$
$= 2 + 7i$
$a = 2, b = 7$

**b.** $\sqrt{-12} = 0 + i\sqrt{12}$
$= 0 + 2i\sqrt{3}$
$a = 0, b = 2\sqrt{3}$

**c.** $7 = 7 + 0i$
$a = 7, b = 0$

**d.** $\dfrac{4 + 3\sqrt{-25}}{20} = \dfrac{4 + 3i\sqrt{25}}{20}$
$= \dfrac{4 + 15i}{20}$
$= \dfrac{1}{5} + \dfrac{3}{4}i$
$a = \dfrac{1}{5}, b = \dfrac{3}{4}$

☑ **A.** You've just seen how we can identify and simplify imaginary and complex numbers

**Now try Exercises 15 through 22** ▶

Complex numbers complete the development of our "numerical landscape." Sets of numbers and their relationships are represented in Figure 1.6, which shows how some sets of numbers are nested within larger sets and highlights the fact that complex numbers consist of a real part (any number within the orange rectangle), and an imaginary number part (any number within the yellow rectangle).

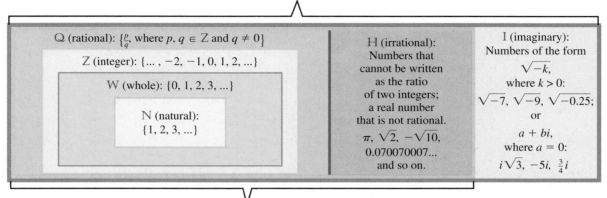

C (complex): Numbers of the form $a + bi$, where $a, b \in \mathbb{R}$ and $i = \sqrt{-1}$.

Q (rational): $\{\frac{p}{q}$, where $p, q \in \mathbb{Z}$ and $q \neq 0\}$

Z (integer): $\{... , -2, -1, 0, 1, 2, ...\}$

W (whole): $\{0, 1, 2, 3, ...\}$

N (natural): $\{1, 2, 3, ...\}$

H (irrational): Numbers that cannot be written as the ratio of two integers; a real number that is not rational. $\pi$, $\sqrt{2}$, $-\sqrt{10}$, 0.070070007... and so on.

I (imaginary): Numbers of the form $\sqrt{-k}$, where $k > 0$: $\sqrt{-7}$, $\sqrt{-9}$, $\sqrt{-0.25}$; or $a + bi$, where $a = 0$: $i\sqrt{3}$, $-5i$, $\frac{3}{4}i$

R (real): All rational and irrational numbers

**Figure 1.6**

## B. Adding and Subtracting Complex Numbers

The sum and difference of two polynomials is computed by identifying and combining like terms. The sum or difference of two complex numbers is computed in a similar way, by adding the real parts from each, and the imaginary parts from each. Notice in Example 4 that the commutative, associative, and distributive properties also apply to complex numbers.

**EXAMPLE 4** ▶ **Adding and Subtracting Complex Numbers**

Perform the indicated operation and write the result in $a + bi$ form, then identify the values of $a$ and $b$.

**a.** $(2 + 3i) + (-5 + 2i)$  **b.** $(-5 - 4i) - (-2 - 7i)$

Solution ▶

**a.** $(2 + 3i) + (-5 + 2i)$

$= 2 + 3i + (-5) + 2i$    distribute

$= 2 + (-5) + 3i + 2i$    commute terms

$= [2 + (-5)] + (3i + 2i)$    group like terms

$= -3 + 5i$    result

$a = -3, b = 5$

**b.** $(-5 - 4i) - (-2 - 7i)$

$= -5 - 4i + 2 + 7i$

$= -5 + 2 + (-4i) + 7i$

$= (-5 + 2) + (-4i + 7i)$

$= -3 + 3i$

$a = -3, b = 3$

☑ **B.** You've just seen how we can add and subtract complex numbers

Now try Exercises 23 through 28 ▶

## C. Multiplying Complex Numbers; Powers of $i$

The product of two binomials is computed using the distributive property and the F-O-I-L process, then combining like terms. The product of two complex numbers is computed in a similar manner. If any result gives a factor of $i^2$, remember that $i^2 = -1$.

**EXAMPLE 5** ▶ **Multiplying Complex Numbers**

Find the indicated product and write the answer in $a + bi$ form.

**a.** $\sqrt{-4}\sqrt{-9}$    **b.** $\sqrt{-6}\,(2 + \sqrt{-3})$

**c.** $(6 - 5i)(4 + i)$    **d.** $(2 + 3i)(2 - 3i)$

**Solution** ▶

**a.** $\sqrt{-4}\,\sqrt{-9}$

$= i\sqrt{4} \cdot i\sqrt{9}$    rewrite in terms of $i$

$= 2i \cdot 3i$    simplify

$= 6i^2$    multiply

$= 6(-1)$    $i^2 = -1$

$= -6 + 0i$    result

**b.** $\sqrt{-6}(2 + \sqrt{-3})$

$= i\sqrt{6}(2 + i\sqrt{3})$    rewrite in terms of $i$

$= 2i\sqrt{6} + i^2\sqrt{18}$    distribute

$= 2i\sqrt{6} + (-1)\sqrt{9}\sqrt{2}$    $i^2 = -1$

$= 2i\sqrt{6} - 3\sqrt{2}$    simplify

$= -3\sqrt{2} + 2i\sqrt{6}$    result

**c.** $(6 - 5i)(4 + i)$

$= 24 + 6i - 20i - 5i^2$    F-O-I-L

$= 24 + 6i - 20i - 5(-1)$    $i^2 = -1$

$= 24 + 6i - 20i + 5$    simplify

$= 29 - 14i$    result

**d.** $(2 + 3i)(2 - 3i)$

$= (2)^2 - (3i)^2$    $(A + B)(A - B) = A^2 - B^2$

$= 4 - 9i^2$    $(3i)^2 = 9i^2$

$= 4 - 9(-1)$    $i^2 = -1$

$= 13 + 0i$    result

Now try Exercises 29 through 38 ▶

**CAUTION** ▶ When computing with imaginary and complex numbers, always write the square root of a negative number in terms of $i$ before you begin, as shown in Examples 5(a) and 5(b). Otherwise we get conflicting results. If we multiply the radicands first, we get $\sqrt{-4}\,\sqrt{-9} = \sqrt{36} = 6$, which is an incorrect result because the original factors were imaginary.

Recall that expressions $2x + 5$ and $2x - 5$ are called binomial conjugates. In the same way, $a + bi$ and $a - bi$ are called **complex conjugates.** Note from Example 5(d) that the *product* of the complex number $a + bi$ with its complex conjugate $a - bi$ is *a real number.* This relationship is useful when rationalizing expressions with a complex number in the denominator, and we generalize the result as follows:

**Product of Complex Conjugates**

For a complex number $a + bi$ and its conjugate $a - bi$,
their product $(a + bi)(a - bi)$ is the real number $a^2 + b^2$;

$$(a + bi)(a - bi) = a^2 + b^2$$

**WORTHY OF NOTE**

Notice that the product of a complex number and its conjugate also gives us a method for *factoring the sum of two squares* using complex numbers! For the expression $x^2 + 4$, the factored form would be $(x + 2i)(x - 2i)$. For more on this idea, **see Exercise 72.**

Showing that $(a + bi)(a - bi) = a^2 + b^2$ is left as an exercise **(see Exercise 72),** but from here on, when asked to compute the product of complex conjugates, simply refer to the formula as illustrated here: $(-3 + 5i)(-3 - 5i) = (-3)^2 + 5^2$ or 34. **See Exercises 39 through 42.**

These operations on complex numbers enable us to verify complex solutions by substitution, in the same way we verify solutions for real numbers. In Example 2 we stated that $x = -3 + 2i$ was one solution to $x^2 + 6x + 13 = 0$. This is verified here.

**EXAMPLE 6 ▶**  **Checking a Complex Root by Substitution**

Verify that $x = -3 + 2i$ is a solution to $x^2 + 6x + 13 = 0$.

**Solution ▶**

$$x^2 + 6x + 13 = 0 \quad \text{original equation}$$
$$(-3 + 2i)^2 + 6(-3 + 2i) + 13 = 0 \quad \text{substitute } -3 + 2i \text{ for } x$$
$$(-3)^2 + 2(-3)(2i) + (2i)^2 - 18 + 12i + 13 = 0 \quad \text{square and distribute}$$
$$9 - 12i + 4i^2 + 12i - 5 = 0 \quad \begin{array}{l}\text{simplify; combine terms}\\ (-18 + 13 = -5)\end{array}$$
$$9 + (-4) - 5 = 0 \quad \text{combine terms } (12i - 12i = 0); i^2 = -1$$
$$0 = 0 \checkmark$$

**Now try Exercises 43 through 50 ▶**

**EXAMPLE 7 ▶**  **Checking a Complex Root by Substitution**

Show that $x = 2 - i\sqrt{3}$ is a solution of $x^2 - 4x = -7$.

**Solution ▶**

$$x^2 - 4x = -7 \quad \text{original equation}$$
$$(2 - i\sqrt{3})^2 - 4(2 - i\sqrt{3}) = -7 \quad \text{substitute } 2 - i\sqrt{3} \text{ for } x$$
$$4 - 4i\sqrt{3} + (i\sqrt{3})^2 - 8 + 4i\sqrt{3} = -7 \quad \text{square and distribute}$$
$$4 - 4i\sqrt{3} - 3 - 8 + 4i\sqrt{3} = -7 \quad (i\sqrt{3})^2 = -3$$
$$-7 = -7 \checkmark \quad \text{solution checks}$$

**Now try Exercises 51 and 52 ▶**

The imaginary unit $i$ has another interesting and useful property. Since $i = \sqrt{-1}$ and $i^2 = -1$, it follows that $i^3 = i^2 \cdot i = (-1)i = -i$ and $i^4 = (i^2)^2 = (-1)^2 = 1$. We can now simplify any *higher power of $i$* by rewriting the expression in terms of $i^4$, since $(i^4)^n = 1$ for any natural number $n$.

$$i = i \qquad\qquad i^5 = i^4 \cdot i = i$$
$$i^2 = -1 \qquad\qquad i^6 = i^4 \cdot i^2 = -1$$
$$i^3 = -i \qquad\qquad i^7 = i^4 \cdot i^3 = -i$$
$$i^4 = 1 \qquad\qquad i^8 = (i^4)^2 = 1$$

Notice the powers of $i$ "cycle through" the four values $i$, $-1$, $-i$ and $1$. In more advanced classes, powers of complex numbers play an important role, and next we learn to reduce higher powers using the power property of exponents and $i^4 = 1$. Essentially, we divide the exponent on $i$ by 4, then use the remainder to compute the value of the expression. For $i^{35}$, $35 \div 4 = 8$ remainder 3, showing $i^{35} = (i^4)^8 \cdot i^3 = (1)^8 \cdot (-i) = -i$.

**EXAMPLE 8 ▶**  **Simplifying Higher Powers of $i$**

Simplify:

**a.** $i^{22}$      **b.** $i^{28}$      **c.** $i^{57}$      **d.** $i^{75}$

**Solution ▶**

**a.** $22 \div 4 = 5$ remainder 2      **b.** $28 \div 4 = 7$ remainder 0
$$i^{22} = (i^4)^5 \cdot (i^2) \qquad\qquad i^{28} = (i^4)^7$$
$$= (1)^5(-1) \qquad\qquad\qquad = (1)^7$$
$$= -1 \qquad\qquad\qquad\qquad = 1$$

**c.** $57 \div 4 = 14$ remainder $1$

$$i^{57} = (i^4)^{14} \cdot i$$
$$= (1)^{14}i$$
$$= i$$

**d.** $75 \div 4 = 18$ remainder $3$

$$i^{75} = (i^4)^{18} \cdot (i^3)$$
$$= (1)^{18}(-i)$$
$$= -i$$

☑ **C.** You've just seen how we can multiply complex numbers and find powers of *i*

**Now try Exercises 53 and 54** ▶

## D. Division of Complex Numbers

Since $i = \sqrt{-1}$, expressions like $\dfrac{3 - i}{2 + i}$ actually have a radical in the denominator.

To divide complex numbers, we simply apply our earlier method of rationalizing denominators (Section R.4), but this time using a *complex* conjugate.

**EXAMPLE 9** ▶  **Dividing Complex Numbers**

Divide and write each result in $a + bi$ form.

**a.** $\dfrac{2}{5 - i}$          **b.** $\dfrac{3 - i}{2 + i}$          **c.** $\dfrac{6 + \sqrt{-36}}{3 + \sqrt{-9}}$

**Solution** ▶

**a.** $\dfrac{2}{5 - i} = \dfrac{2}{5 - i} \cdot \dfrac{5 + i}{5 + i}$

$$= \dfrac{2(5 + i)}{5^2 + 1^2}$$

$$= \dfrac{10 + 2i}{26}$$

$$= \dfrac{10}{26} + \dfrac{2}{26}i$$

$$= \dfrac{5}{13} + \dfrac{1}{13}i$$

**b.** $\dfrac{3 - i}{2 + i} = \dfrac{3 - i}{2 + i} \cdot \dfrac{2 - i}{2 - i}$

$$= \dfrac{6 - 3i - 2i + i^2}{2^2 + 1^2}$$

$$= \dfrac{6 - 5i + (-1)}{5}$$

$$= \dfrac{5 - 5i}{5} = \dfrac{5}{5} - \dfrac{5i}{5}$$

$$= 1 - i$$

**c.** $\dfrac{6 + \sqrt{-36}}{3 + \sqrt{-9}} = \dfrac{6 + i\sqrt{36}}{3 + i\sqrt{9}}$        convert to *i* notation

$$= \dfrac{6 + 6i}{3 + 3i}$$        simplify

The expression can be further simplified by reducing common factors.

$$= \dfrac{6(\cancel{1 + i})}{3(\cancel{1 + i})} = 2 + 0i$$        factor and reduce

**Now try Exercises 55 through 60** ▶

As mentioned, operations on complex numbers can be checked using inverse operations, just as we do for real numbers. To check the division from Example 9(b), we multiply $1 - i$ by the divisor $2 + i$:

$$(1 - i)(2 + i) = 2 + i - 2i - i^2$$
$$= 2 - i - (-1)$$
$$= 2 - i + 1$$
$$= 3 - i ✓$$

☑ **D.** You've just seen how we can divide complex numbers

Several checks are asked for in the Exercise Set.

## 1.4 EXERCISES

### ▶ CONCEPTS AND VOCABULARY

**Fill in each blank with the appropriate word or phrase. Carefully reread the section, if necessary.**

1. Given the complex number $3 + 2i$, its complex conjugate is _____.

2. For $i = \sqrt{-1}$, $i^2 =$ ___, $i^4 =$ ___, $i^6 =$ ___, and $i^8 =$ ___, $i^3 =$ ___, $i^5 =$ ___, $i^7 =$ ___, and $i^9 =$ ___.

3. Discuss/Explain which is correct:
   a. $\sqrt{-4} \cdot \sqrt{-9} = \sqrt{(-4)(-9)} = \sqrt{36} = 6$
   b. $\sqrt{-4} \cdot \sqrt{-9} = 2i \cdot 3i = 6i^2 = -6$

4. Compare/Contrast the product $(1 + \sqrt{2})(1 - \sqrt{3})$ with the product $(1 + i\sqrt{2})(1 - i\sqrt{3})$. What is the same? What is different?

### ▶ DEVELOPING YOUR SKILLS

**Simplify each radical (if possible). If imaginary, rewrite in terms of $i$ and simplify.**

5. a. $\sqrt{-144}$      b. $\sqrt{-49}$
   c. $\sqrt{27}$      d. $\sqrt{72}$

6. a. $\sqrt{-100}$      b. $\sqrt{-169}$
   c. $\sqrt{64}$      d. $\sqrt{98}$

7. a. $-\sqrt{-18}$      b. $-\sqrt{-50}$
   c. $3\sqrt{-25}$      d. $2\sqrt{-9}$

8. a. $-\sqrt{-32}$      b. $-\sqrt{-75}$
   c. $3\sqrt{-144}$      d. $2\sqrt{-81}$

9. a. $\sqrt{-19}$      b. $\sqrt{-31}$
   c. $\sqrt{\dfrac{-12}{25}}$      d. $\sqrt{\dfrac{-9}{32}}$

10. a. $\sqrt{-17}$      b. $\sqrt{-53}$
    c. $\sqrt{\dfrac{-45}{36}}$      d. $\sqrt{\dfrac{-49}{75}}$

**Simplify each expression, writing the result in terms of $i$.**

11. a. $\dfrac{2 + \sqrt{-4}}{2}$      b. $\dfrac{6 + \sqrt{-27}}{3}$

12. a. $\dfrac{16 - \sqrt{-8}}{2}$      b. $\dfrac{4 + 3\sqrt{-20}}{2}$

13. a. $\dfrac{8 + \sqrt{-16}}{2}$      b. $\dfrac{10 - \sqrt{-50}}{5}$

14. a. $\dfrac{6 - \sqrt{-72}}{4}$      b. $\dfrac{12 + \sqrt{-200}}{8}$

**Write each complex number in the standard form $a + bi$ and clearly identify the values of $a$ and $b$.**

15. a. $5$      b. $3i$

16. a. $-2$      b. $-4i$

17. a. $2\sqrt{-81}$      b. $\dfrac{\sqrt{-32}}{8}$

18. a. $-3\sqrt{-36}$      b. $\dfrac{\sqrt{-75}}{15}$

19. a. $4 + \sqrt{-50}$      b. $-5 + \sqrt{-27}$

20. a. $-2 + \sqrt{-48}$      b. $7 + \sqrt{-75}$

21. a. $\dfrac{14 + \sqrt{-98}}{8}$      b. $\dfrac{5 + \sqrt{-250}}{10}$

22. a. $\dfrac{21 + \sqrt{-63}}{12}$      b. $\dfrac{8 + \sqrt{-27}}{6}$

**Perform the addition or subtraction. Write the result in $a + bi$ form.**

23. a. $(12 - \sqrt{-4}) + (7 + \sqrt{-9})$
    b. $(3 + \sqrt{-25}) + (-1 - \sqrt{-81})$
    c. $(11 + \sqrt{-108}) - (2 - \sqrt{-48})$

24. a. $(-7 - \sqrt{-72}) + (8 + \sqrt{-50})$
    b. $(\sqrt{3} + \sqrt{-2}) - (\sqrt{12} + \sqrt{-8})$
    c. $(\sqrt{20} - \sqrt{-3}) + (\sqrt{5} - \sqrt{-12})$

25. a. $(2 + 3i) + (-5 - i)$
    b. $(5 - 2i) + (3 + 2i)$
    c. $(6 - 5i) - (4 + 3i)$

26. a. $(-2 + 5i) + (3 - i)$
    b. $(7 - 4i) - (2 - 3i)$
    c. $(2.5 - 3.1i) + (4.3 + 2.4i)$

27. a. $(3.7 + 6.1i) - (1 + 5.9i)$
    b. $\left(8 + \dfrac{3}{4}i\right) - \left(-7 + \dfrac{2}{3}i\right)$
    c. $\left(-6 - \dfrac{5}{8}i\right) + \left(4 + \dfrac{1}{2}i\right)$

28. a. $(9.4 - 8.7i) - (6.5 + 4.1i)$
    b. $\left(3 + \dfrac{3}{5}i\right) - \left(-11 + \dfrac{7}{15}i\right)$
    c. $\left(-4 - \dfrac{5}{6}i\right) + \left(13 + \dfrac{3}{8}i\right)$

**Multiply and write your answer in $a + bi$ form.**

29. a. $5i \cdot (-3i)$       b. $(4i)(-4i)$

30. a. $(-6i)(-3i)$       b. $-i \cdot 9i$

31. a. $-7(3 + 5i)$       b. $6i(-3 + 7i)$

32. a. $-7i(5 - 3i)$       b. $3(2 - 3i)$

33. a. $(-4 - 2i)(3 + 2i)$       b. $(2 - 3i)(-5 + i)$

34. a. $(5 + 2i)(-7 + 3i)$       b. $(4 - i)(7 + 2i)$

**Compute the special products.**

35. a. $(2 + 3i)^2$       b. $(3 - 4i)^2$

36. a. $(2 - i)^2$       b. $(3 - i)^2$

37. a. $(-2 + 5i)^2$       b. $(3 + i\sqrt{2})^2$

38. a. $(-2 - 5i)^2$       b. $(2 - i\sqrt{3})^2$

**For each complex number given, name the complex conjugate and compute the product.**

39. a. $4 + 5i$       b. $3 - i\sqrt{2}$

40. a. $2 - i$       b. $-1 + i\sqrt{5}$

41. a. $7i$       b. $\frac{1}{2} - \frac{2}{3}i$

42. a. $-5i$       b. $\frac{3}{4} + \frac{1}{5}i$

**Use substitution to determine if the value shown is a solution to the given equation. Show that the complex conjugate of each verified solution is also a solution.**

43. $x^2 + 36 = 0; x = -6$

44. $x^2 + 16 = 0; x = -4$

45. $x^2 + 49 = 0; x = -7i$

46. $x^2 + 25 = 0; x = -5i$

47. $(x - 3)^2 = -9; x = 3 - 3i$

48. $(x + 1)^2 = -4; x = -1 + 2i$

49. $x^2 - 2x - 5 = 0; x = 1 - 2i$

50. $x^2 + 6x + 11 = 0; x = -1 - 3i$

51. $x^2 - 4x + 9 = 0; x = 2 + i\sqrt{5}$

52. $x^2 - 2x + 4 = 0; x = 1 - i\sqrt{3}$

**Simplify the powers of $i$.**

53. a. $i^{48}$     b. $i^{26}$     c. $i^{39}$     d. $i^{53}$

54. a. $i^{36}$     b. $i^{50}$     c. $i^{19}$     d. $i^{65}$

**Divide and write your answer in $a + bi$ form. Check your answer using multiplication.**

55. a. $\dfrac{-2}{\sqrt{-49}}$       b. $\dfrac{2}{1 - \sqrt{-4}}$

56. a. $\dfrac{4}{\sqrt{-25}}$       b. $\dfrac{3}{2 + \sqrt{-9}}$

57. a. $\dfrac{3 + 4i}{4i}$       b. $\dfrac{3 - 2i}{-6 + 4i}$

58. a. $\dfrac{2 - 3i}{3i}$       b. $\dfrac{-4 + 8i}{2 - 4i}$

59. a. $\dfrac{7}{3 + 2i}$       b. $\dfrac{-5}{2 - 3i}$

60. a. $\dfrac{6}{1 + 3i}$       b. $\dfrac{7}{7 - 2i}$

## ▶ WORKING WITH FORMULAS

61. **Absolute value of a complex number:**

    $$|a + bi| = \sqrt{a^2 + b^2}$$

    The absolute value of any complex number $a + bi$ (sometimes called the *modulus* of the number) is computed by taking the square root of the sum of the squares of $a$ and $b$. Find the absolute value of the given complex numbers.

    a. $|2 + 3i|$     b. $|4 - 3i|$     c. $|3 + i\sqrt{2}|$

62. **Binomial cubes:**

    $$(A + B)^3 = A^3 + 3A^2B + 3AB^2 + B^3$$

    The cube of any binomial can be found using the formula shown, where $A$ and $B$ are the terms of the binomial. Use the formula to compute the cube of the given complex numbers.

    a. $1 - 2i$     b. $-1 + i\sqrt{3}$     c. $\dfrac{\sqrt{3}}{2} + \dfrac{1}{2}i$

## ▶ APPLICATIONS

63. **Dawn of imaginary numbers:** In a day when imaginary numbers were imperfectly understood, Girolamo Cardano (1501–1576) once posed the problem, "Find two numbers that have a sum of 10 and whose product is 40." In other words, $A + B = 10$ and $AB = 40$. Although the solution is routine today, at the time the problem posed an enormous challenge. Verify that $A = 5 + i\sqrt{15}$ and $B = 5 - i\sqrt{15}$ satisfy these conditions.

64. **Verifying calculations using $i$:** Suppose Cardano had said, "Find two numbers that have a sum of 4 and a product of 7" (see Exercise 63). Verify that $A = 2 + i\sqrt{3}$ and $B = 2 - i\sqrt{3}$ satisfy these conditions.

Although it may seem odd, complex numbers have several applications in the real world, including the study of *alternating current* (AC) circuits. Briefly, the components of an AC circuit are current *I* (in amperes), voltage *V* (in volts), and the impedance *Z* (in ohms).

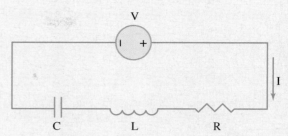

The impedance of an electrical circuit is a measure of the total opposition to the flow of current through the circuit and is calculated as $Z = R + (X_L - X_C)i$, where *R* represents a pure resistance, $X_C$ represents the capacitance, and $X_L$ represents the inductance. Each of these is also measured in ohms (symbolized by $\Omega$).

**65.** Find the impedance *Z* if $R = 7\ \Omega$, $X_L = 6\ \Omega$, and $X_C = 11\ \Omega$.

**66.** Find the impedance *Z* if $R = 9.2\ \Omega$, $X_L = 5.6\ \Omega$, and $X_C = 8.3\ \Omega$.

The voltage *V* (in volts) across any element in an AC circuit is calculated as a product of the current *I* and the impedance *Z*: $V = IZ$.

**67.** Find the voltage in a circuit with a current of $I = (3 - 2i)$ amperes and an impedance of $Z = (5 + 5i)\ \Omega$.

**68.** Find the voltage in a circuit with a current of $I = (2 - 3i)$ amperes and an impedance of $Z = (4 + 2i)\ \Omega$.

In an AC circuit, the total impedance (in ohms) is given by $Z = \dfrac{Z_1 Z_2}{Z_1 + Z_2}$, where *Z* represents the total impedance of a circuit that has $Z_1$ and $Z_2$ wired in parallel.

**69.** Find the total impedance *Z* if $Z_1 = (1 + 2i)\ \Omega$ and $Z_2 = (3 - 2i)\ \Omega$.

**70.** Find the total impedance *Z* if $Z_1 = (3 - i)\ \Omega$ and $Z_2 = (2 + i)\ \Omega$.

## ▶ EXTENDING THE CONCEPT

**71.** Simplify the expression $i^{17}(3 - 4i) - 3i^3(1 + 2i)^2$.

**72.** Up to this point, we've said that expressions like $x^2 - 9$ and $p^2 - 7$ are factorable:
$$x^2 - 9 = (x + 3)(x - 3) \text{ and } p^2 - 7 = (p + \sqrt{7})(p - \sqrt{7}),$$
while $x^2 + 9$ and $p^2 + 7$ are prime. More correctly, we should state that $x^2 + 9$ and $p^2 + 7$ are nonfactorable *using real numbers,* since they actually *can* be factored if complex numbers are used. Specifically,
$$(x + 3i)(x - 3i) = x^2 + 9 \text{ and } (p + i\sqrt{7})(p - i\sqrt{7}) = p^2 + 7.$$
  **a.** Verify that in general, $(a + bi)(a - bi) = a^2 + b^2$.

Use this idea to factor the following.
  **b.** $x^2 + 36$          **c.** $m^2 + 3$          **d.** $n^2 + 12$          **e.** $4x^2 + 49$

**73.** While it is a simple concept for real numbers, the square root of a complex number is much more involved due to the interplay between its real and imaginary parts. For $z = a + bi$ the square root of *z* can be found using the formula: $\sqrt{z} = \dfrac{\sqrt{2}}{2}\left(\sqrt{|z| + a} \pm i\sqrt{|z| - a}\right)$, where the sign is chosen to match the sign of *b* (see Exercise 61). Use the formula to find the square root of each complex number, then check by squaring.
  **a.** $z = -7 + 24i$          **b.** $z = 5 - 12i$          **c.** $z = 4 + 3i$

**74.** For $z = a + bi$ we have the conjugate $\bar{z} = a - bi$ and absolute value $|z| = \sqrt{a^2 + b^2}$ (see Exercise 61). These definitions allow us to concisely state several properties of complex numbers. For instance, we have $z\bar{z} = |z|^2$, which can be verified as follows: $z\bar{z} = (a + bi)(a - bi) = a^2 + b^2 = |z|^2$✓. Use this same idea to show the following:
  **a.** $\dfrac{1}{z} = \dfrac{1}{|z|^2} \cdot \bar{z}$          **b.** $a = \dfrac{1}{2}(z + \bar{z})$          **c.** $b = \dfrac{1}{2i}(z - \bar{z})$          **d.** $\overline{z_1 \cdot z_2} = \bar{z}_1 \cdot \bar{z}_2$

► MAINTAINING YOUR SKILLS

**75. (1.2)** Determine the allowable values in each expression. Write your answer in interval notation.

**a.** $\dfrac{1}{8 - w}$    **b.** $\sqrt{-t}$

**c.** $\sqrt{4a + 6}$    **d.** $\dfrac{1}{x^2 - 4}$

**76. (R.1)** Write the symbols in words and state True/False.

**a.** $6 \notin \mathbb{Q}$    **b.** $\mathbb{Q} \subset \mathbb{R}$

**c.** $103 \in \{3, 4, 5, \ldots\}$    **d.** $\mathbb{R} \not\subset \mathbb{C}$

**77. (1.1)** John can run 10 m/sec, while Rick can only run 9 m/sec. If Rick gets a 2-sec head start, who will hit the 200-m finish line first?

**78. (R.5)** Factor the following expressions completely.

**a.** $x^4 - 16$    **b.** $n^3 - 27$

**c.** $x^3 - x^2 - x + 1$    **d.** $4n^2m - 12nm^2 + 9m^3$

---

## 1.5  Solving Quadratic Equations

**LEARNING OBJECTIVES**

*In Section 1.5 you will see how we can:*

- **A.** Solve quadratic equations using the zero product property
- **B.** Solve quadratic equations using the square root property of equality
- **C.** Solve quadratic equations by completing the square
- **D.** Solve quadratic equations using the quadratic formula
- **E.** Use the discriminant to identify solutions
- **F.** Solve applications of quadratic equations

In Section 1.1 we solved the equation $ax + b = c$ for $x$ to establish a general solution for all linear equations of this form. In this section, we'll establish a general solution for the quadratic equation $ax^2 + bx + c = 0$ $(a \neq 0)$ using a process known as *completing the square*. Other applications of completing the square include the graphing of parabolas, circles, and other relations from the family of *conic sections*.

### A. Quadratic Equations and the Zero Product Property

A **quadratic equation** is one that can be written in the form $ax^2 + bx + c = 0$, where $a$, $b$, and $c$ are real numbers and $a \neq 0$. As shown, the equation is written in **standard form,** meaning the terms are in decreasing order of degree and the equation is set equal to zero.

> **Quadratic Equations**
>
> A quadratic equation can be written in the form
> $$ax^2 + bx + c = 0,$$
> with $a, b, c \in \mathbb{R}$, and $a \neq 0$.

Notice that $a$ is the leading coefficient, $b$ is the coefficient of the linear (first degree) term, and $c$ is a constant. All quadratic equations have degree two, but can have one, two, or three terms. The equation $n^2 - 81 = 0$ is a quadratic equation with two terms, where $a = 1$, $b = 0$, and $c = -81$.

**EXAMPLE 1** ► **Determining Whether an Equation Is Quadratic**

State whether the given equation is quadratic. If yes, identify coefficients $a$, $b$, and $c$.

**a.** $2x^2 - 18 = 0$    **b.** $z - 12 - 3z^2 = 0$    **c.** $\dfrac{-3}{4}x + 5 = 0$

**d.** $z^3 - 2z^2 + 7z = 8$    **e.** $0.8x^2 = 0$

**Solution** ▶

|   | **Standard Form** | **Quadratic** | **Coefficients** |
|---|---|---|---|
| **a.** | $2x^2 - 18 = 0$ | yes, deg 2 | $a = 2 \quad b = 0 \quad c = -18$ |
| **b.** | $-3z^2 + z - 12 = 0$ | yes, deg 2 | $a = -3 \quad b = 1 \quad c = -12$ |
| **c.** | $\dfrac{-3}{4}x + 5 = 0$ | no, deg 1 | (linear equation) |
| **d.** | $z^3 - 2z^2 + 7z - 8 = 0$ | no, deg 3 | (cubic equation) |
| **e.** | $0.8x^2 = 0$ | yes, deg 2 | $a = 0.8 \quad b = 0 \quad c = 0$ |

**Now try Exercises 5 through 16** ▶

With quadratic and other polynomial equations, we generally cannot isolate the variable on one side using only properties of equality, because the variable is raised to different powers. Instead we attempt to solve the equation by factoring and applying the **zero product property.**

**Zero Product Property**

If $A$ and $B$ represent real numbers or real-valued expressions

$$\text{and } A \cdot B = 0,$$

$$\text{then } A = 0 \text{ or } B = 0.$$

In words, the property says, *If the product of any two (or more) factors is equal to zero, then at least one of the factors must be equal to zero.* We can use this property to solve higher degree equations after rewriting them in terms of equations with lesser degree. As with linear equations, values that make the original equation true are called *solutions* or *roots* of the equation.

**EXAMPLE 2** ▶  **Solving Equations Using the Zero Product Property**

Solve by writing the equations in factored form and applying the zero product property.

**a.** $3x^2 = 5x$      **b.** $-5x + 2x^2 = 3$      **c.** $4x^2 = 12x - 9$

**Solution** ▶  **a.**

$$3x^2 = 5x \qquad \text{given equation}$$
$$3x^2 - 5x = 0 \qquad \text{standard form}$$
$$x(3x - 5) = 0 \qquad \text{factor}$$
$$x = 0 \quad \text{or} \quad 3x - 5 = 0 \qquad \text{set factors equal to zero (zero product property)}$$
$$x = 0 \quad \text{or} \qquad x = \frac{5}{3} \qquad \text{result}$$

**b.**

$$-5x + 2x^2 = 3 \qquad \text{given equation}$$
$$2x^2 - 5x - 3 = 0 \qquad \text{standard form}$$
$$(2x + 1)(x - 3) = 0 \qquad \text{factor}$$
$$2x + 1 = 0 \quad \text{or} \quad x - 3 = 0 \qquad \text{set factors equal to zero (zero product property)}$$
$$x = -\frac{1}{2} \quad \text{or} \qquad x = 3 \qquad \text{result}$$

**c.**
$$4x^2 = 12x - 9 \qquad \text{given equation}$$
$$4x^2 - 12x + 9 = 0 \qquad \text{standard form}$$
$$(2x - 3)(2x - 3) = 0 \qquad \text{factor}$$
$$2x - 3 = 0 \quad \text{or} \quad 2x - 3 = 0 \qquad \text{set factors equal to zero (zero product property)}$$
$$x = \frac{3}{2} \quad \text{or} \qquad x = \frac{3}{2} \qquad \text{result}$$

This equation has only the solution $x = \frac{3}{2}$, which we call a *repeated root*.

Now try Exercises 17 through 40 ▶

**CAUTION** ▶ Consider the equation $x^2 - 2x - 3 = 12$. While the left-hand side is factorable, the result is $(x - 3)(x + 1) = 12$ and finding a solution becomes a "guessing game" because the equation is not set equal to zero. If you *misapply* the zero product property and say that $x - 3 = 12$ or $x + 1 = 12$, the "solutions" are $x = 15$ or $x = 11$, which are both incorrect! After subtracting 12 from both sides, $x^2 - 2x - 3 = 12$ becomes $x^2 - 2x - 15 = 0$ giving $(x - 5)(x + 3) = 0$ with solutions $x = 5$ or $x = -3$.

☑ **A.** You've just seen how we can solve quadratic equations using the zero product property

## B. Quadratic Equations and the Square Root Property of Equality

The equation $x^2 = 9$ can be solved by factoring. In standard form we have $x^2 - 9 = 0$ (note $b = 0$), then $(x + 3)(x - 3) = 0$. The solutions are $x = -3$ or $x = 3$, which are simply the *positive and negative square roots of* 9. This result suggests an alternative method for solving equations of the form $X^2 = k$, known as the **square root property of equality.**

> **Square Root Property of Equality**
>
> If $X$ represents an algebraic expression and $X^2 = k$,
> $$\text{then } X = \sqrt{k} \text{ or } X = -\sqrt{k};$$
> $$\text{also written as } X = \pm\sqrt{k}.$$

**EXAMPLE 3** ▶ **Solving an Equation Using the Square Root Property of Equality**

Use the square root property of equality to solve each equation.

    **a.** $-4x^2 + 3 = -6$      **b.** $x^2 + 12 = 0$      **c.** $(x - 5)^2 = 7$

**Solution** ▶    **a.** $-4x^2 + 3 = -6$        original equation

$$x^2 = \frac{9}{4} \qquad \text{subtract 3, divide by } -4$$

$$x = \sqrt{\frac{9}{4}} \quad \text{or} \quad x = -\sqrt{\frac{9}{4}} \qquad \text{square root property of equality}$$

$$x = \frac{3}{2} \quad \text{or} \quad x = -\frac{3}{2} \qquad \text{simplify radicals}$$

This equation has two rational solutions.

**WORTHY OF NOTE**

In Section R.4 we noted that for any real number $a$, $\sqrt{a^2} = |a|$. From Example 3(a), solving the equation by taking the square root of both sides produces $\sqrt{x^2} = \sqrt{\frac{9}{4}}$. This is equivalent to $|x| = \sqrt{\frac{9}{4}}$, again showing this equation must have two solutions, $x = -\sqrt{\frac{9}{4}}$ and $x = \sqrt{\frac{9}{4}}$.

**b.** $\qquad x^2 + 12 = 0$         original equation

$\qquad\qquad x^2 = -12$        subtract 12

$\quad x = \sqrt{-12} \ \text{ or } \ x = -\sqrt{-12}$    square root property of equality

$\quad x = 2i\sqrt{3} \ \ \text{ or } \ x = -2i\sqrt{3}$    simplify radicals

This equation has two complex solutions.

**c.** $\qquad\qquad (x - 5)^2 = 7$        original equation

$\quad x - 5 = \sqrt{7} \qquad \text{ or } \ x - 5 = -\sqrt{7}$   square root property of equality

$\qquad x = 5 + \sqrt{7} \qquad\qquad x = 5 - \sqrt{7}$   solve for $x$

This equation has two irrational solutions.

> **Now try Exercises 41 through 56** ▶

**CAUTION** ▶   For equations of the form $(x + d)^2 = k$ as in Example 3(c), you should resist the temptation to expand the binomial square in an attempt to simplify the equation and solve by factoring—many times the result is nonfactorable. *Any* equation of the form $(x + d)^2 = k$ can quickly be solved using the square root property of equality.

Answers written using radicals are called **exact** or **closed form** solutions. Actually checking the exact solutions is a nice application of fundamental skills. Let's check $x = 5 + \sqrt{7}$ from Example 3(c).

**Check:** $\qquad\qquad (x - 5)^2 = 7$    original equation

$\qquad (5 + \sqrt{7} - 5)^2 = 7$    substitute $5 + \sqrt{7}$ for $x$

$\qquad\qquad\quad (\sqrt{7})^2 = 7$    simplify

$\qquad\qquad\qquad\quad 7 = 7 \ ✓$   $(\sqrt{7})^2 = 7$; result checks ($x = 5 - \sqrt{7}$ also checks)

☑ **B. You've just seen how we can solve quadratic equations using the square root property of equality**

## C. Solving Quadratic Equations by Completing the Square

Again consider $(x - 5)^2 = 7$ from Example 3(c). If we had first expanded the binomial square, we would have obtained $x^2 - 10x + 25 = 7$ then $x^2 - 10x + 18 = 0$ in standard form. Note that this equation *cannot be solved by factoring*. However, reversing this process does lead us to a strategy for solving nonfactorable quadratic equations, by creating a *perfect square trinomial* from the quadratic and linear terms. This process is known as **completing the square.** To transform $x^2 - 10x + 18 = 0$ back into $x^2 - 10x + 25 = 7$ [which we would then rewrite as $(x - 5)^2 = 7$ and solve], we subtract 18 from both sides, then add 25:

$$x^2 - 10x + 18 = 0$$
$$x^2 - 10x = -18 \qquad \text{subtract 18}$$
$$x^2 - 10x + 25 = -18 + 25 \qquad \text{add 25}$$
$$(x - 5)^2 = 7 \qquad \text{factor, simplify}$$

In general, after subtracting the constant term, the number that "completes the square" is computed as $\left[\frac{1}{2}(\text{coefficient of linear term})\right]^2$: $\left[\frac{1}{2}(10)\right]^2 = 25$. This is easily seen by comparing the square of a binomial with its product: $(x + p)^2 = x^2 + 2px + p^2$. Comparing $x^2 + 2px + p^2$ with $x^2 - 10x + p^2$ shows $2p$ must be equal to $-10$, so $p = -5$ and $p^2 = 25$. For additional practice finding this constant term, **see Exercises 57 through 62.**

**EXAMPLE 4** ▶ **Solving a Quadratic Equation by Completing the Square**

Solve by completing the square: $x^2 + 13 = 6x$.

Solution ▶

| | |
|---|---|
| $x^2 + 13 = 6x$ | original equation |
| $x^2 - 6x + 13 = 0$ | standard form |
| $x^2 - 6x + \underline{\quad} = -13 + \underline{\quad}$ | subtract 13 to make room for new constant |
| | compute $[\frac{1}{2}(\text{linear coefficient})]^2 = [\frac{1}{2}(-6)]^2 = 9$ |
| $x^2 - 6x + 9 = -13 + 9$ | add 9 to both sides (completing the square) |
| $(x - 3)^2 = -4$ | factor and simplify |
| $x - 3 = \sqrt{-4}$ or $x - 3 = -\sqrt{-4}$ | square root property of equality |
| $x = 3 + 2i$ or $x = 3 - 2i$ | simplify radicals and solve for $x$ |

**Now try Exercises 63 through 72** ▶

The process of completing the square can be applied to any quadratic equation with a leading coefficient of 1. If the leading coefficient is not 1, we simply divide through by $a$ before beginning, which brings us to this summary of the process.

**Completing the Square to Solve a Quadratic Equation**

To solve $ax^2 + bx + c = 0$ by completing the square:

1. Subtract the constant $c$ from both sides.
2. Divide both sides by the leading coefficient $a$.
3. Compute $\left[\frac{1}{2}\left(\frac{b}{a}\right)\right]^2 = \left(\frac{b}{2a}\right)^2$ and add the result to both sides.
4. Factor left-hand side as a binomial square; simplify right-hand side.
5. Solve using the square root property of equality.

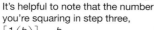
**WORTHY OF NOTE**

It's helpful to note that the number you're squaring in step three, $\left[\frac{1}{2}\left(\frac{b}{a}\right)\right] = \frac{b}{2a}$, turns out to be the constant term in the factored form. From Example 4, the number we squared was $\frac{1}{2}(-6) = -3$, and the binomial square was $(x - 3)^2$.

**EXAMPLE 5** ▶ **Solving a Quadratic Equation by Completing the Square**

Solve by completing the square: $-3x^2 + 1 = 4x$.

Solution ▶

| | |
|---|---|
| $-3x^2 + 1 = 4x$ | original equation |
| $-3x^2 - 4x + 1 = 0$ | standard form (nonfactorable) |
| $-3x^2 - 4x = -1$ | subtract 1 |
| $x^2 + \frac{4}{3}x = \frac{1}{3}$ | divide by $-3$ |
| $x^2 + \frac{4}{3}x + \frac{4}{9} = \frac{1}{3} + \frac{4}{9}$ | $\left[\frac{1}{2}\left(\frac{4}{3}\right)\right]^2 = \left(\frac{2}{3}\right)^2 = \frac{4}{9}$; add $\frac{4}{9}$ |
| $\left(x + \frac{2}{3}\right)^2 = \frac{7}{9}$ | factor and simplify $\left(\frac{1}{3} = \frac{3}{9}\right)$ |
| $x + \frac{2}{3} = \sqrt{\frac{7}{9}}$ or $x + \frac{2}{3} = -\sqrt{\frac{7}{9}}$ | square root property of equality |
| $x = -\frac{2}{3} + \frac{\sqrt{7}}{3}$ or $x = -\frac{2}{3} - \frac{\sqrt{7}}{3}$ | solve for $x$ and simplify (exact form) |
| $x \approx 0.22$ or $x \approx -1.55$ | approximate form (to hundredths) |

☑ **C.** You've just seen how we can solve quadratic equations by completing the square

**Now try Exercises 73 through 80** ▶

**CAUTION** ▶ For many of the skills/processes needed in a study of algebra, it's actually easier to work with the fractional form of a number, rather than the decimal form. For example, computing $\left(\frac{2}{3}\right)^2$ is easier than computing $(0.\overline{6})^2$, and finding $\sqrt{\frac{9}{16}}$ is much easier than finding $\sqrt{0.5625}$.

### D. Solving Quadratic Equations Using the Quadratic Formula

In Section 1.1 we found a general solution for the linear equation $ax + b = c$ by comparing it to $2x + 3 = 15$. Here we'll use a similar idea to find a general solution for quadratic equations. In a side-by-side format, we'll solve the equations $2x^2 + 5x + 3 = 0$ and $ax^2 + bx + c = 0$ by completing the square. Note the similarities.

|  |  |  |
|---|---|---|
| $2x^2 + 5x + 3 = 0$ | given equations | $ax^2 + bx + c = 0$ |
| $2x^2 + 5x + \phantom{ } = -3$ | subtract constant term | $ax^2 + bx + \phantom{ } = -c$ |
| $x^2 + \dfrac{5}{2}x + \underline{\phantom{xx}} = -\dfrac{3}{2}$ | divide by leading coefficient | $x^2 + \dfrac{b}{a}x + \underline{\phantom{xx}} = -\dfrac{c}{a}$ |
| $\left[\dfrac{1}{2}\left(\dfrac{5}{2}\right)\right]^2 = \dfrac{25}{16}$ | $\left[\dfrac{1}{2}(\text{linear coefficient})\right]^2$ | $\left[\dfrac{1}{2}\left(\dfrac{b}{a}\right)\right]^2 = \dfrac{b^2}{4a^2}$ |
| $x^2 + \dfrac{5}{2}x + \dfrac{25}{16} = \dfrac{25}{16} - \dfrac{3}{2}$ | add to both sides | $x^2 + \dfrac{b}{a}x + \dfrac{b^2}{4a^2} = \dfrac{b^2}{4a^2} - \dfrac{c}{a}$ |
| $\left(x + \dfrac{5}{4}\right)^2 = \dfrac{25}{16} - \dfrac{3}{2}$ | left side factors as a binomial square | $\left(x + \dfrac{b}{2a}\right)^2 = \dfrac{b^2}{4a^2} - \dfrac{c}{a}$ |
| $\left(x + \dfrac{5}{4}\right)^2 = \dfrac{25}{16} - \dfrac{24}{16}$ | determine LCDs | $\left(x + \dfrac{b}{2a}\right)^2 = \dfrac{b^2}{4a^2} - \dfrac{4ac}{4a^2}$ |
| $\left(x + \dfrac{5}{4}\right)^2 = \dfrac{1}{16}$ | simplify right side | $\left(x + \dfrac{b}{2a}\right)^2 = \dfrac{b^2 - 4ac}{4a^2}$ |
| $x + \dfrac{5}{4} = \pm\sqrt{\dfrac{1}{16}}$ | square root property of equality | $x + \dfrac{b}{2a} = \pm\sqrt{\dfrac{b^2 - 4ac}{4a^2}}$ |
| $x + \dfrac{5}{4} = \pm\dfrac{1}{4}$ | simplify radicals | $x + \dfrac{b}{2a} = \pm\dfrac{\sqrt{b^2 - 4ac}}{2a}$ |
| $x = -\dfrac{5}{4} \pm \dfrac{1}{4}$ | solve for $x$ | $x = -\dfrac{b}{2a} \pm \dfrac{\sqrt{b^2 - 4ac}}{2a}$ |
| $x = \dfrac{-5 \pm 1}{4}$ | combine terms | $x = \dfrac{-b \pm \sqrt{b^2 - 4ac}}{2a}$ |
| $x = \dfrac{-5 + 1}{4}$ or $x = \dfrac{-5 - 1}{4}$ | solutions | $x = \dfrac{-b + \sqrt{b^2 - 4ac}}{2a}$ or $x = \dfrac{-b - \sqrt{b^2 - 4ac}}{2a}$ |

On the left, our final solutions are $x = -1$ or $x = -\frac{3}{2}$. The general solution on the right is called the **quadratic formula,** which can be used to solve *any equation belonging to the quadratic family.*

> **Quadratic Formula**
>
> If $ax^2 + bx + c = 0$, with $a, b, c \in \mathbb{R}$ and $a \neq 0$, then
> $$x = \frac{-b + \sqrt{b^2 - 4ac}}{2a} \quad \text{or} \quad x = \frac{-b - \sqrt{b^2 - 4ac}}{2a};$$
> also written $x = \dfrac{-b \pm \sqrt{b^2 - 4ac}}{2a}.$

**CAUTION** ▶ It's very important to note the values of $a$, $b$, and $c$ come from an equation *written in standard form*. For $3x^2 - 5x = -7$, $a = 3$ and $b = -5$, but $c \neq -7$! In standard form we have $3x^2 - 5x + 7 = 0$, and note the value for use in the formula is actually $c = 7$.

**EXAMPLE 6** ▶ **Solving Quadratic Equations Using the Quadratic Formula**

Solve $4x^2 + 1 = 8x$ using the quadratic formula. State the solution(s) in both exact and approximate forms. Check one of the exact solutions in the original equation.

**Solution** ▶ Begin by writing the equation in standard form and identifying the values of $a$, $b$, and $c$.

$$4x^2 + 1 = 8x \qquad \text{original equation}$$

$$4x^2 - 8x + 1 = 0 \qquad \text{standard form } a = 4, b = -8, c = 1$$

$$x = \frac{-(-8) \pm \sqrt{(-8)^2 - 4(4)(1)}}{2(4)} \qquad \text{substitute 4 for } a, -8 \text{ for } b, \text{ and 1 for } c$$

$$x = \frac{8 \pm \sqrt{64 - 16}}{8} = \frac{8 \pm \sqrt{48}}{8} \qquad \text{simplify}$$

$$x = \frac{8 \pm 4\sqrt{3}}{8} = \frac{8}{8} \pm \frac{4\sqrt{3}}{8} = 1 \pm \frac{\sqrt{3}}{2} \qquad \text{simplify radical (see following CAUTION)}$$

$$x = 1 + \frac{\sqrt{3}}{2} \quad \text{or} \quad x = 1 - \frac{\sqrt{3}}{2} \qquad \text{exact solutions}$$

$$x \approx 1.87 \quad \text{or} \quad x \approx 0.13 \qquad \text{approximate solutions}$$

**Check** ▶
$$4x^2 + 1 = 8x \qquad \text{original equation}$$

$$4\left(1 + \frac{\sqrt{3}}{2}\right)^2 + 1 = 8\left(1 + \frac{\sqrt{3}}{2}\right) \qquad \text{substitute } 1 + \frac{\sqrt{3}}{2} \text{ for } x$$

$$4\left[1 + 2\left(\frac{\sqrt{3}}{2}\right) + \frac{3}{4}\right] + 1 = 8 + 4\sqrt{3} \qquad \text{square binomial; distribute}$$

$$4 + 4\sqrt{3} + 3 + 1 = 8 + 4\sqrt{3} \qquad \text{distribute}$$

$$8 + 4\sqrt{3} = 8 + 4\sqrt{3} ✓ \qquad \text{result checks}$$

☑ **D.** You've just seen how we can solve quadratic equations using the quadratic formula

**Now try Exercises 81 through 104** ▶

**CAUTION** ▶ For $\dfrac{8 \pm 4\sqrt{3}}{8}$, be careful not to incorrectly "cancel the eights" as in $\dfrac{\overset{1}{\cancel{8}} \pm 4\sqrt{3}}{\underset{1}{\cancel{8}}} \neq 1 \pm 4\sqrt{3}$.

*No!* Use a calculator to verify that the results are not equivalent. Both terms in the numerator are divided by 8 and we must either rewrite the expression as separate terms (as in Example 6) or factor the numerator to see if the expression simplifies further:

$$\frac{8 \pm 4\sqrt{3}}{8} = \frac{\overset{1}{\cancel{4}}(2 \pm \sqrt{3})}{\underset{2}{\cancel{8}}} = \frac{2 \pm \sqrt{3}}{2}, \text{ which is equivalent to } 1 \pm \frac{\sqrt{3}}{2}.$$

## E. The Discriminant of the Quadratic Formula

For any real-valued expression $X$, recall that $\sqrt{X}$ represents a real number only for $X \geq 0$. Since the quadratic formula contains the radical $\sqrt{b^2 - 4ac}$, the expression $b^2 - 4ac$, called the **discriminant,** will determine the nature (real or nonreal) and the number of solutions to a given quadratic equation.

| **The Discriminant of the Quadratic Formula** | | |
|---|---|---|
| For $ax^2 + bx + c$, where $a, b, c \in \mathbb{R}$ and $a \neq 0$, | | |
| **1.** If $b^2 - 4ac > 0$, there are two real roots | **2.** If $b^2 - 4ac = 0$, there is one real (repeated) root | **3.** If $b^2 - 4ac < 0$, there are two nonreal roots |

Further analysis of the discriminant reveals even more concerning the nature of quadratic solutions. Namely, if $a$, $b$, and $c$ are rational and the discriminant is:

1. zero, then the original equation is a perfect square trinomial.
2. positive and a perfect square, then there will be two rational roots which means the original equation can be solved by factoring.
3. positive and not a perfect square, then there will be two irrational roots.

---

**EXAMPLE 7** ▶ **Using the Discriminant to Analyze Solutions**

Use the discriminant to determine if the equation given has any real root(s). If so, state whether the roots are rational or irrational, and whether the quadratic expression is factorable.

    **a.** $2x^2 + 5x + 2 = 0$     **b.** $x^2 - 4x + 7 = 0$     **c.** $4x^2 - 20x + 25 = 0$

**Solution** ▶   **a.** $a = 2, b = 5, c = 2$     **b.** $a = 1, b = -4, c = 7$     **c.** $a = 4, b = -20, c = 25$

      $b^2 - 4ac = (5)^2 - 4(2)(2)$     $b^2 - 4ac = (-4)^2 - 4(1)(7)$     $b^2 - 4ac = (-20)^2 - 4(4)(25)$

          $= 9$                  $= -12$                   $= 0$

      Since $b^2 - 4ac > 0$,       Since $b^2 - 4ac < 0$,       Since $b^2 - 4ac = 0$,

      two rational roots,         two nonreal roots,        one rational root,

        factorable            nonfactorable           factorable

**Now try Exercises 105 through 116** ▶

In Example 7(b), $b^2 - 4ac = -12$ and the quadratic formula shows $x = \dfrac{4 \pm \sqrt{-12}}{2}$. After simplifying, we find the solutions are the complex conjugates $x = 2 + i\sqrt{3}$ or $x = 2 - i\sqrt{3}$. In general, when $b^2 - 4ac < 0$, the solutions *will be complex conjugates*.

### Complex Solutions

The complex solutions of a quadratic equation with real coefficients
must occur in conjugate pairs.
If $a + bi$ is a solution, then $a - bi$ is also a solution.

**EXAMPLE 8** ▶   **Solving Quadratic Equations Using the Quadratic Formula**

Solve: $2x^2 - 6x + 5 = 0$.

**Solution** ▶  With $a = 2$, $b = -6$, and $c = 5$, the discriminant becomes $(-6)^2 - 4(2)(5) = -4$, showing there will be two nonreal roots. The quadratic formula then yields

$$x = \frac{-b \pm \sqrt{b^2 - 4ac}}{2a} \qquad \text{quadratic formula}$$

$$x = \frac{-(-6) \pm \sqrt{-4}}{2(2)} \qquad b^2 - 4ac = -4 \text{, substitute 2 for } a \text{ and } -6 \text{ for } b$$

$$x = \frac{6 \pm 2i}{4} \qquad \text{simplify, write in } i \text{ form}$$

$$x = \frac{3}{2} \pm \frac{1}{2}i \qquad \text{solutions are complex conjugates}$$

☑ **E.** You've just seen how we can use the discriminant to identify solutions

**Now try Exercises 117 through 122** ▶

---

**WORTHY OF NOTE**

While it's possible to solve by completing the square if $\dfrac{b}{a}$ is a fraction or an odd number (see Example 5), the process is usually most efficient when $\dfrac{b}{a}$ is an even number. This is one observation you could use when selecting a solution method.

### Summary of Solution Methods for $ax^2 + bx + c = 0$

1. If $b = 0$, $ax^2 + c = 0$: isolate $x^2$ and use the square root property of equality.
2. If $c = 0$, $ax^2 + bx = 0$: factor out the GCF and use the zero product property.
3. If no coefficient is zero, you can attempt to solve by
   a. factoring   b. completing the square   c. using the quadratic formula

## F. Applications of Quadratic Equations

A projectile is any object that is thrown, shot, or *projected* upward with no sustaining source of propulsion. The height of the projectile at time $t$ is modeled by the equation $h = -16t^2 + vt + k$, where $h$ is the height of the object in feet, $t$ is the elapsed time in seconds, and $v$ is the initial velocity in feet per second. The constant $k$ represents the initial height of the object above ground level, as when a person releases an object 5 ft above the ground in a throwing motion ($k = 5$), or when a rocket runs out of fuel at an altitude of 240 ft ($k = 240$).

**EXAMPLE 9** ▶ **Solving an Application of Quadratic Equations—Rocketry**

A model rocketry club is testing a newly developed engine. A few seconds after liftoff, at a velocity of 160 ft/sec and a height of 240 ft, it runs out of fuel and becomes a projectile (see Figure 1.7).

    **a.** How high is the rocket 3 sec later?

    **b.** For how many seconds was the height of the rocket at least 496 ft?

    **c.** How many seconds until the rocket returns to the ground?

**Solution** ▶   **a.** Using the information given, the equation modeling the rocket's height in the projectile phase is $h = -16t^2 + 160t + 240$. For its height at $t = 3$ we have

$$h = -16(3)^2 + 160(3) + 240$$
$$= -16(9) + 480 + 240$$
$$= 576$$

Three seconds later, the rocket was at an altitude of 576 ft.

**b.** Since the height $h$ must be at least 496 ft, we use the model $h = -16t^2 + 160t + 240$ and write the equation $-16t^2 + 160t + 240 = 496$. In standard form, we obtain $-16t^2 + 160t - 256 = 0$ (subtract 496 from both sides), which we can now solve by factoring.

| | |
|---|---|
| $-16t^2 + 160t - 256 = 0$ | related equation |
| $t^2 - 10t + 16 = 0$ | divide by $-16$ |
| $(t - 2)(t - 8) = 0$ | factor |
| $t - 2 = 0 \quad \text{or} \quad t - 8 = 0$ | zero product property |
| $t = 2 \quad \text{or} \quad t = 8$ | result |

This shows the rocket is at exactly 496 ft after 2 sec (on its ascent) and after 8 sec (during its descent). We conclude the rocket's height was greater than 496 ft for $8 - 2 = 6$ sec.

**c.** When the rocket hits the ground, its height is $h = 0$. Substituting 0 for $h$ and solving gives

| | |
|---|---|
| $h = -16t^2 + 160t + 240$ | original function |
| $0 = -16t^2 + 160t + 240$ | substitute 0 for $h$ |
| $0 = t^2 - 10t - 15$ | divide by $-16$ |

The equation is nonfactorable, so we use the quadratic formula to solve, with $a = 1$, $b = -10$, and $c = -15$:

$$t = \frac{-b \pm \sqrt{b^2 - 4ac}}{2a} \qquad \text{quadratic formula}$$

$$= \frac{-(-10) \pm \sqrt{(-10)^2 - 4(1)(-15)}}{2(1)} \qquad \text{substitute 1 for } a, -10 \text{ for } b, -15 \text{ for } c$$

$$= \frac{10 \pm \sqrt{160}}{2} \qquad \text{simplify}$$

$$= \frac{10}{2} \pm \frac{4\sqrt{10}}{2} \qquad \sqrt{160} = 4\sqrt{10}$$

$$= 5 \pm 2\sqrt{10} \qquad \text{simplify}$$

Since we need the time $t$ in seconds, we use the approximate form of the answer, obtaining $t \approx -1.32$ and $t \approx 11.32$. The rocket will return to the ground in just over 11 sec (since $t$ represents time, the solution $t = -1.32$ does not apply).

**Figure 1.7**

Projectile phase

240 ft →

Power phase

**Now try Exercises 125 through 130** ▶

**EXAMPLE 10** ▶ **Solving an Application of Quadratic Equations—Mobile Broadband Usage**

For the years 2005 to 2009, the number $N$ of mobile broadband subscriptions (in millions) can be modeled by $N = 21.07x^2 + 63.01x + 74.34$, where $x = 0$ represents the year 2005 [*Source:* Data from the *2011 Statistical Abstract of the United States,* Table 1391, page 869]. If this trend continues, in what year will the number of subscribers reach or surpass 2 billion?

**Solution** ▶ We are essentially asked to solve the equation $N = 21.07x^2 + 63.01x + 74.34$, when $N = 2000$ (2 billion = 2000 million).

$$2000 = 21.07x^2 + 63.01x + 74.34 \qquad \text{given equation, } N = 2000$$
$$0 = 21.07x^2 + 63.01x - 1925.66 \qquad \text{subtract 2000}$$

For $a = 21.07$, $b = 63.01$, and $c = -1925.66$, the quadratic formula gives

$$x = \frac{-b \pm \sqrt{b^2 - 4ac}}{2a} \qquad \text{quadratic formula}$$

$$x = \frac{-(63.01) \pm \sqrt{(63.01)^2 - 4(21.07)(-1925.66)}}{2(21.07)} \qquad \text{substitute known values}$$

$$x \approx \frac{-63.01 \pm \sqrt{166{,}264.88}}{42.14} \qquad \text{simplify}$$

$$x \approx -11.17 \quad \text{or} \quad x \approx 8.18 \qquad \text{result}$$

We disregard the negative solution (since $x$ represents time), and find the total number of subscribers will reach or surpass 2 billion about 8 yr after 2005, or in the year 2013.

☑ **F. You've just seen how we can solve applications of quadratic equations**

**Now try Exercises 131 through 136** ▶

## 1.5 EXERCISES

▶ **CONCEPTS AND VOCABULARY**

**Fill in each blank with the appropriate word or phrase. Carefully reread the section, if necessary.**

1. A polynomial equation is in standard form when written in _____ order of degree and set equal to _____.

2. The solution $x = 2 + \sqrt{3}$ is called a(n) _____ form of the solution. Using a calculator, we find the _____ form is $x \approx 3.732$.

3. According to the summary on page 122, what method should be used to solve $4x^2 - 5x = 0$? What are the solutions?

4. Discuss/Explain why this version of the quadratic formula is incorrect:
$$x = -b \pm \frac{\sqrt{b^2 - 4ac}}{2a}$$

▶ **DEVELOPING YOUR SKILLS**

**Determine whether each equation is quadratic. If so, identify the coefficients $a$, $b$, and $c$. If not, discuss why.**

5. $2x - 15 - x^2 = 0$

6. $21 + x^2 - 4x = 0$

7. $\frac{2}{3}x - 7 = 0$

8. $12 - 4x = 9$

9. $\frac{1}{4}x^2 = 6x$

10. $0.5x = 0.25x^2$

11. $2x^2 + 7 = 0$

12. $5 = -4x^2$

13. $-3x^2 + 9x - 5 + 2x^3 = 0$

**14.** $z^2 - 6z + 9 - z^3 = 0$

**15.** $(x - 1)^2 + (x - 1) + 4 = 9$

**16.** $(x + 5)^2 - (x + 5) + 4 = 17$

**Solve using the zero product property. Be sure each equation is in standard form and factor out any common factors before attempting to solve. Check all answers in the original equation.**

**17.** $x^2 - 15 = 2x$      **18.** $z^2 - 10z = -21$

**19.** $m^2 = 8m - 16$      **20.** $-10n = n^2 + 25$

**21.** $5p^2 - 10p = 0$      **22.** $6q^2 - 18q = 0$

**23.** $-14h^2 = 7h$      **24.** $9w = -6w^2$

**25.** $a^2 - 17 = -8$      **26.** $b^2 + 8 = 12$

**27.** $g^2 + 18g + 70 = -11$      **28.** $h^2 + 14h - 2 = -51$

**29.** $m^3 + 5m^2 - 9m - 45 = 0$

**30.** $n^3 - 3n^2 - 4n + 12 = 0$    **31.** $(c - 12)c - 15 = 30$

**32.** $(d - 10)d + 10 = -6$    **33.** $9 + (r - 5)r = 33$

**34.** $7 + (s - 4)s = 28$    **35.** $(t + 4)(t + 7) = 54$

**36.** $(g + 17)(g - 2) = 20$    **37.** $2x^2 - 4x - 30 = 0$

**38.** $-3z^2 + 12z + 36 = 0$    **39.** $2w^2 - 5w = 3$

**40.** $-3v^2 = -v - 2$

**Solve using the square root property of equality. Write answers in exact form and approximate form rounded to hundredths. If there are no real solutions, so state.**

**41.** $m^2 = 16$      **42.** $p^2 = 49$

**43.** $y^2 - 28 = 0$      **44.** $m^2 - 20 = 0$

**45.** $p^2 + 36 = 0$      **46.** $n^2 + 5 = 0$

**47.** $x^2 = \frac{21}{16}$      **48.** $y^2 = \frac{13}{9}$

**49.** $(n - 3)^2 = 36$      **50.** $(p + 5)^2 = 49$

**51.** $(w + 5)^2 = 3$      **52.** $(m - 4)^2 = 5$

**53.** $(x - 3)^2 + 7 = 2$      **54.** $(m + 11)^2 + 5 = 3$

**55.** $(m - 2)^2 = \frac{18}{49}$      **56.** $(x - 5)^2 = \frac{12}{25}$

**Fill in the blank so the result is a perfect square trinomial, then factor into a binomial square.**

**57.** $x^2 + 6x + \_ = \_\_\_\_\_$      **58.** $y^2 + 10y + \_ = \_\_\_\_\_$

**59.** $n^2 + 3n + \_ = \_\_\_\_\_$      **60.** $x^2 - 5x + \_ = \_\_\_\_\_$

**61.** $p^2 + \frac{2}{3}p + \_ = \_\_\_\_\_$      **62.** $x^2 - \frac{3}{2}x + \_ = \_\_\_\_\_$

**Solve by completing the square. Write answers in exact form and approximate form rounded to hundredths. If there are no real solutions, so state.**

**63.** $x^2 + 6x = -5$      **64.** $m^2 + 8m = -12$

**65.** $p^2 - 6p + 3 = 0$      **66.** $n^2 = 4n + 10$

**67.** $p^2 + 6p = -4$      **68.** $x^2 - 8x - 1 = 0$

**69.** $m^2 + 3m = 1$      **70.** $n^2 + 5n - 2 = 0$

**71.** $n^2 = 5n + 5$      **72.** $w^2 - 7w + 3 = 0$

**73.** $2x^2 = -7x + 4$      **74.** $3w^2 - 8w + 4 = 0$

**75.** $2n^2 - 3n - 9 = 0$      **76.** $2p^2 - 5p = 1$

**77.** $4p^2 - 3p - 2 = 0$      **78.** $3x^2 + 5x - 6 = 0$

**79.** $m^2 = 7m - 4$      **80.** $a^2 - 15 = 4a$

**Solve each equation using the most efficient method: factoring, square root property of equality, completing the square, or the quadratic formula. Write answers in both exact and approximate form rounded to hundredths. Check one of the exact solutions in the original equation.**

**81.** $3p^2 + p = 0$      **82.** $8w^2 = 3w$

**83.** $w^2 + 6w - 1 = 0$      **84.** $n^2 + 4n - 8 = 0$

**85.** $6w^2 - w = 2$      **86.** $12x^2 - 1 = x$

**87.** $4a^2 - 4a = 1$      **88.** $4n^2 - 8n - 1 = 0$

**89.** $4m^2 - 25 = 0$      **90.** $9 - 49x^2 = 0$

**91.** $3n^2 - 2n - 3 = 0$      **92.** $8x^2 - 5x - 1 = 0$

**93.** $3m^2 - 7m - 6 = 0$      **94.** $4x^2 - x = 3$

**95.** $\frac{5}{9}x^2 - \frac{16}{15}x = \frac{3}{2}$      **96.** $\frac{5}{4}m^2 - \frac{8}{3}m + \frac{1}{6} = 0$

**97.** $0.2a^2 + 1.2a + 0.9 = 0$

**98.** $-5.4n^2 + 8.1n + 9 = 0$

**99.** $2x^2 - 4x + 5 = 0$

**100.** $2p^2 - 4p + 11 = 0$

**101.** $3a^2 - 5a + 6 = 0$

**102.** $4m^2 = 12m - 15$

**103.** $3a^2 - a + 2 = 0$

**104.** $2x^2 + x + 3 = 0$

**Use the discriminant to determine whether the given equation has irrational, rational, repeated, or nonreal roots. Also state whether the original equation is factorable using integers, but do not solve for x.**

**105.** $-3x^2 + 2x + 1 = 0$      **106.** $2x^2 - 5x - 3 = 0$

**107.** $-4x + x^2 + 13 = 0$      **108.** $-10x + x^2 + 41 = 0$

**109.** $15x^2 - x - 6 = 0$      **110.** $10x^2 - 11x - 35 = 0$

**111.** $-4x^2 + 6x - 5 = 0$      **112.** $-5x^2 - 3 = 2x$

**113.** $2x^2 + 8 = -9x$      **114.** $x^2 + 4 = -7x$

**115.** $4x^2 + 12x = -9$      **116.** $9x^2 + 4 = 12x$

**Solve the equations given. Simplify each result.**

**117.** $-6x + 2x^2 + 5 = 0$      **118.** $17 + 2x^2 = 10x$

**119.** $5x^2 + 5 = -5x$      **120.** $x^2 = -2x - 19$

**121.** $-2x^2 = -5x + 11$      **122.** $4x - 3 = 5x^2$

## ► WORKING WITH FORMULAS

**123. Height of a projectile:** $h = -16t^2 + vt$

If an object is projected vertically upward from ground level with no continuing source of propulsion, the height of the object (in feet) is modeled by the equation shown, where $v$ is the initial velocity in feet per second, and $t$ is the time in seconds. Use the quadratic formula to solve for $t$ in terms of $v$ and $h$. (*Hint:* Set the equation equal to zero and identify the coefficients as before.)

**124. Surface area of a cylinder:** $A = 2\pi r^2 + 2\pi rh$

The surface area of a cylinder is given by the formula shown, where $h$ is the height and $r$ is the radius of the base. The equation can be considered quadratic in the variable $r$. Use the quadratic formula to solve for $r$ in terms of $h$ and $A$. (*Hint:* Rewrite the equation in standard form and identify the coefficients as before.)

## ► APPLICATIONS

**125. Height of a projectile:** On the moon, the height of a golf ball hit with an initial velocity of 96 ft/sec is given by the equation $h = -2.7t^2 + 96t$, where $h$ represents the height of the ball in feet after $t$ sec. On Earth, the same shot would be modeled by the equation $h = -16t^2 + 96t$. How much longer will the flight time be on the moon than on the Earth?

**126. Height of a projectile:** The height of an object thrown upward from the floor of a canyon 106 ft deep, with an initial velocity of 120 ft/sec, is given by the equation $h = -16t^2 + 120t - 106$, where $h$ represents the height of the object in feet after $t$ sec. How long will it take the object to rise to the height of the canyon wall? Answer in exact form and approximate form rounded to hundredths.

**127. Ladder stability:** From past mishaps, Seth knows his 16-ft ladder should be leaned against his house in such a way that the top of the ladder is 6 ft farther above the ground than the bottom of the ladder is from the base of the outer wall. With this placement, how far up the side of the house will the top of the ladder reach?

**128. Walkway design:** To isolate her stunning antique roses from her husband's lawnmower, Vanessa decides to enclose her triangular flower garden with a stone walkway. The plot lies in the corner of the yard against the fence and forms a right triangle with legs 8 ft long. If 30 ft² of stone was needed, what is the length of the walkway along its outer edge?

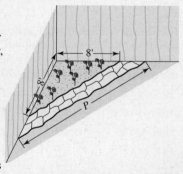

**129. Cost, revenue, and profit:** The revenue for a manufacturer of computer printers is given by the equation $R = x(30 - 0.4x)$, where revenue is in thousands of dollars and $x$ thousand printers are manufactured and sold. What is the minimum number of printers that must be sold to bring in a revenue of $440,000?

**130. Cost, revenue, and profit:** The cost to produce bottled spring water is given by the cost equation $C = 16x + 63$, where $x$ is the number of bottles in thousands. The total revenue from the sale of these bottles is given by the

equation $R = -x^2 + 326x - 18{,}463$. (a) Determine the profit equation (profit = revenue − cost). (b) After a bad flood contaminates the drinking water of a nearby community, the owners decide to bottle and donate as many bottles of water as they can, without taking a loss (i.e., they break even: profit or $P = 0$). How many bottles will they produce for the flood victims?

**131. Gutter design:** A rectangular sheet of aluminum was folded to form a rain gutter with a 31-in² rectangular cross section. If the cross section has a height that is 2 in. shorter than its width, how wide was the piece of aluminum before it was folded?

**132. Office organizer:** Always the crafter, Amanda decides to use a leftover sheet of poster board to make herself an office organizer. The rectangular sheet needs squares cut from each corner so it can then be folded to form an open box. If a 22-in. by 28-in. sheet is used, what two sizes can the squares be in order to create a box with volume 1100 in³?

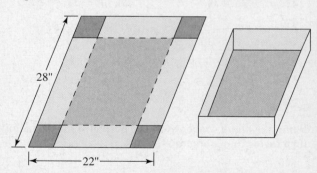

**133. National debt:** For the years 2005 to 2010, the national debt (in trillions of dollars) can be modeled by the equation $D = 0.195x^2 + 0.149x + 7.996$, where $x = 0$ represents the year 2005 [*Source:* www.treasurydirect.gov]. If this trend continues, in what year will the debt reach $25 trillion?

**134. U.S. international trade balance:** For the years 1995 to 2003, the international trade balance $B$ (in millions of dollars) can be approximated by the equation $B = -3.1x^2 + 4.5x - 19.9$, where $x = 0$ represents year 1995 [*Source:* Data from the *2005 Statistical Abstract of the United States*, Table 1278, page 799]. If this trend continues, in what year will the trade balance reach a deficit of $750 million dollars or more?

**135. Football field dimensions:** In a nearby park, a field has been marked off for the neighborhood Pop Warner football team. If the field has a perimeter of 310 yd and an area of 4950 yd$^2$, what are the dimensions of the field?

**136. Tennis court dimensions:** A regulation tennis court for a singles match is laid out so that its length is 3 ft less than three times its width. The area of the singles court is 2106 ft$^2$. What is the length and width of the singles court?

▶ **EXTENDING THE CONCEPT**

**137. Proof without words:** When solving $ax^2 + bx + c = 0$ by completing the square, we begin by subtracting the constant $c$ from both sides. With $ax^2 + bx$ now on the left, we divide both sides by the leading coefficient $a$ to get an equivalent equation with a leading coefficient of 1.

This results in $x^2 + \dfrac{b}{a}x$ remaining on the left, to which

we add $\left[\dfrac{1}{2}\left(\dfrac{b}{a}\right)\right]^2$ to complete the square algebraically.

Discuss how the diagram given (a) geometrically shows $\left[\dfrac{1}{2}\left(\dfrac{b}{a}\right)\right]^2$ literally completes the square and (b) justifies

the factored form $x^2 + \dfrac{b}{a}x + \left[\dfrac{1}{2}\left(\dfrac{b}{a}\right)\right]^2 = \left[x + \dfrac{1}{2}\left(\dfrac{b}{a}\right)\right]^2$.

**138. Discriminant:** For what values of $c$ will the equation $9x^2 - 12x + c = 0$ have

   **a.** no real roots?     **b.** one rational root?

   **c.** two real roots?     **d.** two integer roots?

**139. Complex polynomials:** Although the quadratic equations seen so far had only real coefficients, the techniques of this section can also be applied when the coefficients are complex. Use the quadratic formula to solve the following (carefully chosen) equations.

(Recall that $\dfrac{1}{i} = -i$.)

   **a.** $z^2 - 9iz = -22$

   **b.** $2iz^2 - 9z + 26i = 0$

   **c.** $0.5z^2 + (4 - 3i)z + (-9 - 12i) = 0$

**140. Conics:** In the study of conics, it is often useful to solve a quadratic equation involving two variables for one variable in terms of the other. Solve each of the following equations for $x$, and then for $y$.

   **a.** $\dfrac{x^2}{a^2} + \dfrac{y^2}{b^2} = 1$     **b.** $(x - h)^2 + (y - k)^2 = r^2$

   **c.** $ax^2 + bxy + cy^2 = d$

▶ **MAINTAINING YOUR SKILLS**

**141. (1.1)** State the formula for the perimeter and area of each figure illustrated.

  **a.**
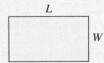
  **b.**

  **c.**

  **d.**

**142. (R.5)** Factor the following expressions:

   **a.** $3x^3 - 15x^2 - 108x$

   **b.** $4x^2 - 25$

   **c.** $x^3 + 6x^2 - 4x - 24$

**143. (1.1)** A total of 900 tickets were sold for a recent concert and $25,000 was collected. If good seats were $30 and cheap seats were $20, how many of each type were sold?

**144. (1.1)** Solve for $C$: $P = C + Ct$.

**Solving Other Types of Equations**

## LEARNING OBJECTIVES

*In Section 1.6 you will see how we can:*

- ☐ **A.** Solve polynomial equations of higher degree
- ☐ **B.** Solve rational equations
- ☐ **C.** Solve radical equations and equations with rational exponents
- ☐ **D.** Solve equations in quadratic form
- ☐ **E.** Solve applications of various equation types

The ability to solve linear and quadratic equations is the foundation on which a large percentage of our future studies are built. Both are closely linked to the solution of other equation types, as well as to the graphs of these equations. In this section, we get our first glimpse of these connections, as we learn to solve certain polynomial, rational, radical, and other equations.

### A. Polynomial Equations of Higher Degree

In standard form, linear and quadratic equations have a known number of terms, so we commonly represent their coefficients using the early letters of the alphabet, as in $ax^2 + bx + c = 0$. However, these equations belong to the larger family of **polynomial equations**. To write a general polynomial, where the number of terms is unknown, we often represent the coefficients using subscripts on a single letter such as $a_1$, $a_2$, $a_3$, and so on.

> **Polynomial Equations**
>
> A polynomial equation of degree $n$ is one of the form
> $$a_n x^n + a_{n-1} x^{n-1} + \cdots + a_1 x + a_0 = 0,$$
> where $a_n, a_{n-1}, \ldots, a_1, a_0$ are real numbers and $a_n \neq 0$.

Factorable polynomials of degree 3 and higher can also be solved using the zero product property and fundamental algebra skills.

---

**EXAMPLE 1** ▶ **Solving a Polynomial Equation by Factoring**

Solve by factoring: $4x^3 - 40x = 6x^2$.

**Solution** ▶

$$
\begin{aligned}
4x^3 - 40x &= 6x^2 && \text{given equation} \\
4x^3 - 6x^2 - 40x &= 0 && \text{standard form} \\
2x(2x^2 - 3x - 20) &= 0 && \text{common factor is } 2x \\
2x(2x + 5)(x - 4) &= 0 && \text{factored form}
\end{aligned}
$$

$2x = 0 \quad$ or $\quad 2x + 5 = 0 \quad$ or $\quad x - 4 = 0 \qquad$ zero product property

$\quad x = 0 \quad$ or $\qquad x = \dfrac{-5}{2} \quad$ or $\qquad x = 4 \qquad$ result—solve for $x$

Substituting these values into the original equation verifies they are solutions.

**Now try Exercises 5 through 10** ▶

---

**EXAMPLE 2** ▶ **Solving Polynomial Equations by Factoring**

Solve each equation:

**a.** $x^3 - 7x + 21 = 3x^2$      **b.** $x^4 - 16 = 0$

**Solution** ▶    **a.**

$$
\begin{aligned}
x^3 - 7x + 21 &= 3x^2 && \text{given equation} \\
x^3 - 3x^2 - 7x + 21 &= 0 && \text{standard form; factor by grouping} \\
x^2(x - 3) - 7(x - 3) &= 0 && \text{remove common factors from each group} \\
(x - 3)(x^2 - 7) &= 0 && \text{factored form}
\end{aligned}
$$

$x - 3 = 0 \quad$ or $\quad x^2 - 7 = 0 \qquad$ zero product property

$\qquad x = 3 \quad$ or $\qquad x^2 = 7 \qquad$ isolate variables

$\qquad\qquad\qquad\qquad\quad x = \pm\sqrt{7} \qquad$ square root property of equality

The solutions are $x = 3$, $x = \sqrt{7}$, and $x = -\sqrt{7}$.

**b.**
$$x^4 - 16 = 0 \qquad \text{given equation}$$
$$(x^2 + 4)(x^2 - 4) = 0 \qquad \text{factor as a difference of squares}$$
$$(x^2 + 4)(x + 2)(x - 2) = 0 \qquad \text{factor } x^2 - 4$$
$$x^2 + 4 = 0 \quad \text{or} \quad x + 2 = 0 \quad \text{or} \quad x - 2 = 0 \qquad \text{zero product property}$$
$$x^2 = -4 \quad \text{or} \qquad x = -2 \quad \text{or} \qquad x = 2 \qquad \text{isolate variables}$$
$$x = \pm\sqrt{-4} \qquad \text{square root property of equality}$$

Since $\pm\sqrt{-4} = \pm 2i$, the solutions are $x = 2i$, $x = -2i$, $x = 2$, and $x = -2$.

☑ **A.** You've just seen how we can solve polynomial equations of higher degree

**Now try Exercises 11 through 32** ▶

In Examples 1 and 2, we were able to solve higher degree equations by "breaking them down" into linear and quadratic equations. This basic idea can be applied to other kinds of equations as well, by rewriting them as equivalent linear and/or quadratic equations. For future use, it will be helpful to note that for a third-degree equation in the standard form $ax^3 + bx^2 + cx + d = 0$, a solution using factoring by grouping is always possible when $ad = bc$.

## B. Rational Equations

In Section 1.1 we solved linear equations using basic properties of equality. If any equation contained fractional terms, we "cleared the fractions" using the least common denominator (LCD). We can also use this idea to solve **rational equations,** or equations that contain rational *expressions*.

**Solving Rational Equations**

1. Identify and exclude any values that cause a zero denominator.
2. Multiply both sides by the LCD and simplify (this will eliminate all denominators).
3. Solve the resulting equation.
4. Check all solutions in the original equation.

**EXAMPLE 3** ▶ **Solving a Rational Equation**

Solve for $m$: $\dfrac{2}{m} - \dfrac{1}{m - 1} = \dfrac{4}{m^2 - m}$.

**Solution** ▶ Since $m^2 - m = m(m - 1)$, the LCD is $m(m - 1)$, where $m \neq 0$ and $m \neq 1$.

$$\frac{m(m - 1)}{1}\left(\frac{2}{m} - \frac{1}{m - 1}\right) = \frac{m(m - 1)}{1}\left[\frac{4}{m(m - 1)}\right] \qquad \text{multiply by LCD}$$

$$\frac{\cancel{m}(m - 1)}{1}\left(\frac{2}{\cancel{m}}\right) - \frac{m\cancel{(m - 1)}}{1}\left(\frac{1}{\cancel{m - 1}}\right) = \frac{\cancel{m(m - 1)}}{1}\left[\frac{4}{\cancel{m(m - 1)}}\right] \qquad \text{distribute and reduce}$$

$$2(m - 1) - m = 4 \qquad \text{simplify—denominators are eliminated}$$

$$2m - 2 - m = 4 \qquad \text{distribute}$$

$$m = 6 \qquad \text{solve for } m$$

Checking by substitution we have:

$$\frac{2}{m} - \frac{1}{m-1} = \frac{4}{m^2-m}$$ original equation

$$\frac{2}{6} - \frac{1}{6-1} = \frac{4}{6^2-6}$$ substitute 6 for $m$

$$\frac{1}{3} - \frac{1}{5} = \frac{4}{30}$$ simplify

$$\frac{5}{15} - \frac{3}{15} = \frac{2}{15}$$ common denominator

$$\frac{2}{15} = \frac{2}{15} \checkmark$$ result

Now try Exercises 33 through 38 ▶

Multiplying both sides of an equation by a variable sometimes introduces a solution that satisfies the *resulting equation,* but not the original equation—the one we're trying to solve. Such "solutions" are said to be **extraneous** and illustrate the need to check all apparent solutions in the original equation. In the case of rational equations, we are particularly aware that any value that causes a zero denominator is outside the set of allowable values and cannot be a solution.

**EXAMPLE 4** ▶ **Solving a Rational Equation**

Solve: $x + \dfrac{12}{x-3} = 1 + \dfrac{4x}{x-3}$.

**Solution** ▶ The LCD is $x - 3$, where $x \neq 3$.

$$(x-3)\left(x + \frac{12}{x-3}\right) = (x-3)\left(1 + \frac{4x}{x-3}\right)$$ multiply by LCD

$$(x-3)(x) + \frac{x-3}{1}\left(\frac{12}{x-3}\right) = (x-3)(1) + \frac{x-3}{1}\left(\frac{4x}{x-3}\right)$$ distribute and reduce

$$x^2 - 3x + 12 = x - 3 + 4x$$ simplify—denominators are eliminated

$$x^2 - 8x + 15 = 0$$ set equation equal to zero

$$(x-3)(x-5) = 0$$ factor

$$x = 3 \quad \text{or} \quad x = 5$$ zero product property

Checking shows $x = 3$ is extraneous, and $x = 5$ is the only valid solution.

Now try Exercises 39 through 44 ▶

In many fields of study, formulas involving rational expressions are used as equation models. Frequently, we need to solve these equations for one variable in terms of others, a skill closely related to our work in Section 1.1.

**EXAMPLE 5** ▶ **Solving for a Specified Variable in a Formula**

Solve for the indicated variable: $S = \dfrac{a}{1 - r}$ for $r$.

**Solution** ▶

$$S = \frac{a}{1 - r}$$   LCD is $1 - r$

$$(1 - r)(S) = (1 - r)\left(\frac{a}{1 - r}\right)$$   multiply by LCD

$$S - Sr = a$$   simplify—denominator is eliminated

$$-Sr = a - S$$   isolate term with $r$

$$r = \frac{a - S}{-S}$$   solve for $r$ (divide both sides by $-S$)

$$r = \frac{S - a}{S}; \; S \neq 0$$   multiply numerator/denominator by $-1$

> **WORTHY OF NOTE**
>
> Generally, we try to write rational answers with the fewest possible number of negative signs. Multiplying the numerator and denominator in Example 5 by $-1$ gave $r = \frac{S - a}{S}$, a more acceptable answer.

☑ **B.** You've just seen how we can solve rational equations

**Now try Exercises 45 through 52** ▶

## C. Radical Equations and Equations with Rational Exponents

A **radical equation** is any equation that contains terms with a variable in the radicand. To solve a radical equation, we attempt to isolate a radical term on one side, then apply the appropriate $n$th power to free up the radicand and solve for the unknown. This is an application of the **power property of equality.**

> **The Power Property of Equality**
>
> For any integer $n \geq 2$, if $\sqrt[n]{u}$ is a real-valued expression and $\sqrt[n]{u} = v$,
>
> then $(\sqrt[n]{u})^n = v^n$,
>
> and $u = v^n$.

Raising both sides of an equation to an *even* power can also introduce a false (extraneous) solution. Note that by inspection, the equation $x - 2 = \sqrt{x}$ has only the solution $x = 4$. But the equation $(x - 2)^2 = x$ (obtained by squaring both sides) has both $x = 4$ *and* $x = 1$ as solutions, yet $x = 1$ does not satisfy the original equation. This means we should *check all solutions of an equation where an even power is applied.*

**EXAMPLE 6** ▶ **Solving Radical Equations**

Solve each equation:

**a.** $\sqrt{3x - 2} + 12 = x + 10$       **b.** $2\sqrt[3]{x - 5} + 4 = 0$

**Solution** ▶    **a.** $\sqrt{3x - 2} + 12 = x + 10$       original equation

$\qquad\quad \sqrt{3x - 2} = x - 2$       isolate radical term (subtract 12)

$\qquad\quad (\sqrt{3x - 2})^2 = (x - 2)^2$       apply power property, power is even

$\qquad\qquad\quad 3x - 2 = x^2 - 4x + 4$       simplify; square binomial

$\qquad\qquad\qquad\quad 0 = x^2 - 7x + 6$       set equal to zero

$\qquad\qquad\qquad\quad 0 = (x - 6)(x - 1)$       factor

$x - 6 = 0 \quad$ or $\quad x - 1 = 0$       apply zero product property

$x = 6 \quad$ or $\quad x = 1$       result, check for extraneous roots

Check ▶  $x = 6$:     $\sqrt{3(6) - 2} + 12 = (6) + 10$
$$\sqrt{16} + 12 = 16$$
$$16 = 16\checkmark$$

Check ▶  $x = 1$:     $\sqrt{3(1) - 2} + 12 = (1) + 10$
$$\sqrt{1} + 12 = 11$$
$$13 = 11\,\textbf{x}$$

The only solution is $x = 6$; $x = 1$ is extraneous.

**b.**  $2\sqrt[3]{x - 5} + 4 = 0$     original equation

$\sqrt[3]{x - 5} = -2$     isolate radical term (subtract 4, divide by 2)

$(\sqrt[3]{x - 5})^3 = (-2)^3$     apply power property, power is odd

$x - 5 = -8$     simplify: $(\sqrt[3]{x - 5})^3 = x - 5$

$x = -3$     solve

Substituting $-3$ for $x$ in the original equation verifies it is a solution.

> **Now try Exercises 53 through 56** ▶

Sometimes squaring both sides of an equation still leaves an equation with a radical term, but often there is *one fewer* than before. In this case, we simply repeat the process, as indicated by the flowchart in Figure 1.8.

**Figure 1.8**

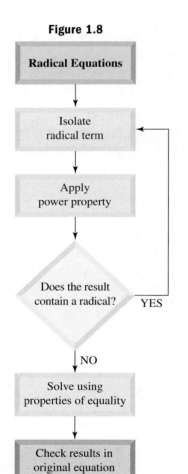

**EXAMPLE 7** ▶  **Solving a Radical Equation**
Solve: $\sqrt{x + 15} - \sqrt{x + 3} = 2$.

**Solution** ▶

$\sqrt{x + 15} - \sqrt{x + 3} = 2$     original equation

$\sqrt{x + 15} = \sqrt{x + 3} + 2$     isolate one radical

$(\sqrt{x + 15})^2 = (\sqrt{x + 3} + 2)^2$     power property

$x + 15 = (x + 3) + 4\sqrt{x + 3} + 4$     $(A + B)^2 = A^2 + 2AB + B^2$; $A = \sqrt{x+3}, B = 2$

$x + 15 = x + 4\sqrt{x + 3} + 7$     simplify

$8 = 4\sqrt{x + 3}$     isolate radical

$2 = \sqrt{x + 3}$     divide by 4

$4 = x + 3$     power property

$1 = x$     possible solution

**Check** ▶

$\sqrt{x + 15} - \sqrt{x + 3} = 2$     original equation

$\sqrt{(1) + 15} - \sqrt{(1) + 3} = 2$     substitute 1 for $x$

$\sqrt{16} - \sqrt{4} = 2$     simplify

$4 - 2 = 2$     solution checks

$2 = 2\checkmark$

> **Now try Exercises 57 and 58** ▶

Since rational exponents are so closely related to radicals, the solution process for each is very similar. The goal is still to "undo" the radical (rational exponent) and solve for the unknown.

**The Power Property of Equality**

For any positive integer $n$, if $u^{\frac{1}{n}}$ is a real-valued expression and $u^{\frac{1}{n}} = v$,

$$\text{then } (u^{\frac{1}{n}})^n = v^n,$$
$$\text{and } u = v^n.$$

The power property of equality basically says that if certain conditions are satisfied, both sides of an equation can be raised to any needed power.

**EXAMPLE 8** ▶ **Solving Equations with Rational Exponents**

Solve each equation:

    **a.** $3(x + 1)^{\frac{3}{4}} - 9 = 15$        **b.** $(x - 3)^{\frac{2}{3}} = 4$

**Solution** ▶   **a.**

| | |
|---|---|
| $3(x + 1)^{\frac{3}{4}} - 9 = 15$ | original equation |
| $(x + 1)^{\frac{3}{4}} = 8$ | isolate variable term (add 9, divide by 3) |
| $[(x + 1)^{\frac{3}{4}}]^4 = 8^4$ | apply power property |
| $(x + 1)^3 = 8^4$ | simplify: $\frac{3}{4}(4) = 3$ |
| $[(x + 1)^3]^{\frac{1}{3}} = (8^4)^{\frac{1}{3}}$ | take cube root of both sides |
| $x + 1 = 16$ | simplify: $(8^4)^{\frac{1}{3}} = (8^{\frac{1}{3}})^4 = 16$ |
| $x = 15$ | result |

**Check** ▶

| | |
|---|---|
| $3(15 + 1)^{\frac{3}{4}} - 9 = 15$ | substitute 15 for $x$ in the original equation |
| $3\left(16^{\frac{1}{4}}\right)^3 - 9 = 15$ | simplify, rewrite exponent |
| $3(2)^3 - 9 = 15$ | $\sqrt[4]{16} = 2$ |
| $3(8) - 9 = 15$ | $2^3 = 8$ |
| $15 = 15$ ✓ | solution checks |

**b.**

| | |
|---|---|
| $(x - 3)^{\frac{2}{3}} = 4$ | original equation |
| $[(x - 3)^{\frac{2}{3}}]^3 = 4^3$ | apply power property |
| $(x - 3)^2 = 4^3$ | simplify: $\frac{2}{3}(3) = 2$ |
| $x - 3 = \pm\sqrt{4^3}$ | square root property |
| $x - 3 = \pm 8$ | simplify: $\sqrt{4^3} = (\sqrt{4})^3 = 8$ |
| $x = 3 \pm 8$ | result |

The solutions are $3 + 8 = 11$ and $3 - 8 = -5$.
Verify by checking both in the original equation.

☑ **C.** You've just seen how we can solve radical equations and equations with rational exponents

Now try Exercises 59 through 64 ▶

**CAUTION** ▶ As you continue solving equations with radicals and rational exponents, be careful not to arbitrarily place the "$\pm$" sign in front of terms *given* in radical form. The expression $\sqrt{18}$ indicates the positive square root of 18, where $\sqrt{18} = 3\sqrt{2}$. The equation $x^2 = 18$ becomes $x = \pm\sqrt{18}$ after applying the power property, with solutions $x = \pm 3\sqrt{2}$ ($x = -3\sqrt{2}, x = 3\sqrt{2}$), since the square of either number produces 18.

## D. Equations in Quadratic Form

In Section R.5, we used a technique called *u-substitution* to factor expressions in quadratic form. The following equations are also in quadratic form since the degree of the leading term is twice the degree of the middle term: $x^{\frac{2}{3}} - 3x^{\frac{1}{3}} - 10 = 0$,

$(d^2 - d)^2 - 8(d^2 - d) + 12 = 0$ and $x - 3\sqrt{x + 4} + 4 = 0$ [*Note:* The last equation can be rewritten as $(x + 4) - 3(x + 4)^{\frac{1}{2}} = 0$]. A *u*-substitution will help to solve these equations by factoring. The first equation appears in Example 9; the other two are in **Exercises 70 and 74** respectively.

**EXAMPLE 9** ▶ **Solving Equations in Quadratic Form**

Solve using a *u*-substitution:

**a.** $x^{\frac{2}{3}} - 3x^{\frac{1}{3}} - 10 = 0$          **b.** $x^4 - 36 = 5x^2$

**Solution** ▶ **a.** This equation is in quadratic form since it can be rewritten as:

$\left(x^{\frac{1}{3}}\right)^2 - 3\left(x^{\frac{1}{3}}\right)^1 - 10 = 0$, where the degree of leading term is twice that of second term. If we let $u = x^{\frac{1}{3}}$, then $u^2 = x^{\frac{2}{3}}$ and the equation becomes $u^2 - 3u^1 - 10 = 0$, which is factorable.

$$(u - 5)(u + 2) = 0 \quad \text{factor}$$
$$u = 5 \quad \text{or} \quad u = -2 \quad \text{solution in terms of } u$$
$$x^{\frac{1}{3}} = 5 \quad \text{or} \quad x^{\frac{1}{3}} = -2 \quad \text{resubstitute } x^{\frac{1}{3}} \text{ for } u$$
$$\left(x^{\frac{1}{3}}\right)^3 = 5^3 \quad \text{or} \quad \left(x^{\frac{1}{3}}\right)^3 = (-2)^3 \quad \text{cube both sides: } \frac{1}{3}(3) = 1$$
$$x = 125 \quad \text{or} \quad x = -8 \quad \text{solve for } x$$

Both solutions check.

**b.** In the standard form $x^4 - 5x^2 - 36 = 0$, we note the equation is also in quadratic form, since it can be written as $(x^2)^2 - 5(x^2)^1 - 36 = 0$. If we let $u = x^2$, then $u^2 = x^4$ and the equation becomes $u^2 - 5u^1 - 36 = 0$, which is factorable.

$$(u - 9)(u + 4) = 0 \quad \text{factor}$$
$$u = 9 \quad \text{or} \quad u = -4 \quad \text{solution in terms of } u$$
$$x^2 = 9 \quad \text{or} \quad x^2 = -4 \quad \text{resubstitute } x^2 \text{ for } u$$
$$x = \pm\sqrt{9} \quad \text{or} \quad x = \pm\sqrt{-4} \quad \text{square root property}$$
$$x = \pm3 \quad \text{or} \quad x = \pm2i \quad \text{simplify}$$

The solutions are $x = -3$, $x = 3$, $x = -2i$, and $x = 2i$. Verify that all solutions check.

☑ **D.** You've just seen how we can solve equations in quadratic form

Now try Exercises 65 through 78 ▶

## E. Applications

Applications of the skills from this section come in many forms. **Number puzzles** and **consecutive integer** exercises help develop the ability to translate written information into algebraic forms **(see Exercises 81 through 84)**. Applications involving **geometry** or a stated relationship between two quantities often depend on these skills, and in many scientific fields, equation models involving radicals and rational exponents are commonplace **(see Exercises 93 and 94)**.

**EXAMPLE 10** ▶ **Solving a Geometry Application**

A hemispherical wash basin has a radius of 6 in. The volume of water in the basin can be modeled by $V = 6\pi h^2 - \frac{\pi}{3}h^3$, where $h$ is the height of the water (see diagram). At what height $h$ is the volume of water numerically equal to $15\pi$ times the height $h$?

**Solution** ▶   We are essentially asked to solve $V = 6\pi h^2 - \frac{\pi}{3}h^3$ when $V = 15\pi h$.

The equation becomes

$$15\pi h = 6\pi h^2 - \frac{\pi}{3}h^3 \qquad \text{original equation, substitute } 15\pi h \text{ for } V$$

$$\frac{\pi}{3}h^3 - 6\pi h^2 + 15\pi h = 0 \qquad \text{standard form}$$

$$h^3 - 18h^2 + 45h = 0 \qquad \text{multiply by } \frac{3}{\pi}$$

$$h(h^2 - 18h + 45) = 0 \qquad \text{factor out } h$$

$$h(h - 3)(h - 15) = 0 \qquad \text{factored form}$$

$$h = 0 \quad \text{or} \quad h = 3 \quad \text{or} \quad h = 15 \qquad \text{result}$$

The "solution" $h = 0$ can be discounted since there would be no water in the basin, and $h = 15$ is too large for this context (the radius is only 6 in.). The only solution that fits is $h = 3$.

**Check** ▶

$$15\pi h = 6\pi h^2 - \frac{\pi}{3}h^3 \qquad \text{resulting equation}$$

$$15\pi(3) = 6\pi(3)^2 - \frac{\pi}{3}(3)^3 \qquad \text{substitute 3 for } h$$

$$45\pi = 6\pi(9) - \frac{\pi}{3}(27) \qquad \text{apply exponents}$$

$$45\pi = 54\pi - 9\pi \qquad \text{simplify}$$

$$45\pi = 45\pi \checkmark \qquad \text{result checks}$$

**Now try Exercises 85 and 86** ▶

In this section, we noted that extraneous roots can occur when (1) both sides of an equation are multiplied by a variable term (as when solving rational equations) and (2) when both sides of an equation are raised to an even power (as when solving certain radical equations or equations with rational exponents). Example 10 illustrates a third way that extraneous roots can occur, as when a solution checks out fine algebraically, but does not fit the context or physical constraints of the situation.

Applications of rational equations can also take many forms. Work and uniform motion exercises help us develop important skills that can be used with more complex equation models. A uniform motion example follows here. For more on work, **see Exercises 87 and 88.**

**EXAMPLE 11** ▶   **Solving a Uniform Motion Application**

To celebrate reaching their weight loss goals, Jethro and Tony decide to go skydiving. At the determined altitude, the plane's door opens and they both take the leap. As instructed, Jethro maintains a spread-eagle position, but Tony recklessly takes the dive position, which increases his average free-fall speed to 120 fps (feet per second) more than Jethro's. If Jethro falls 3500 ft in the same time that Tony falls 6000 ft, what was the average speed of each during free fall?

**Solution** ▶ We begin by organizing the given information in a table using $D = RT$.

|        | Distance (ft) | Rate (fps) | Time (sec) |
|--------|---------------|------------|------------|
| Jethro | 3500          | $R$        | $T$        |
| Tony   | 6000          | $R + 120$  | $T$        |

Because the two free-fall times are the same, we rewrite the uniform motion equation in terms of time. To do this, we divide both sides of the equation by $R$ to get $T = \dfrac{D}{R}$. Since Jethro's time equals Tony's time, we have

$$\frac{3500}{R} = \frac{6000}{R + 120} \qquad \text{equal times; } T = \frac{D}{R}$$

$$\overset{1}{\cancel{R}}(R + 120)\frac{3500}{\cancel{R}} = R\overset{1}{\cancel{(R + 120)}}\frac{6000}{\cancel{(R + 120)}} \qquad \text{multiply by LCD}$$

$$3500(R + 120) = 6000R \qquad \text{simplify—denominators are eliminated}$$

$$3500R + 420{,}000 = 6000R \qquad \text{distribute}$$

$$420{,}000 = 2500R \qquad \text{subtract } 3500R$$

$$168 = R \qquad \text{solve for } R \text{ (divide by 2500)}$$

Jethro's average speed during free fall was 168 fps, while Tony's was $168 + 120 = 288$ fps.

**Now try Exercises 89 and 90** ▶

Here, we'll close the section using a rational equation to model the cost of removing industrial waste from drinking water.

**EXAMPLE 12** ▶ **Solving an Application Involving a Rational Equation**

In Verano City, the cost $C$ to remove industrial waste from drinking water is given by the equation $C = \dfrac{80P}{100 - P}$, where $P$ is the percent of total pollutants removed and $C$ is the cost in thousands of dollars. If the City Council budgets \$1,520,000 for the removal of these pollutants, what percentage of the waste will be removed?

**Solution** ▶

$$C = \frac{80P}{100 - P} \qquad \text{equation model}$$

$$1520 = \frac{80P}{100 - P} \qquad \text{substitute 1520 for } C$$

$$1520(100 - P) = 80P \qquad \text{multiply by LCD of } (100 - P)$$

$$152{,}000 = 1600P \qquad \text{distribute and simplify}$$

$$95 = P \qquad \text{result}$$

☑ **E. You've just seen how we can solve applications of various equation types**

On a budget of \$1,520,000, 95% of the pollutants will be removed.

**Now try Exercises 91 and 92** ▶

## 1.6 EXERCISES

▶ **CONCEPTS AND VOCABULARY**

**Fill in each blank with the appropriate word or phrase. Carefully reread the section, if necessary.**

1. Factorable polynomial equations can be solved using the _____ _____ property.

2. "False solutions" to a rational or radical equation are also called _____ roots.

3. Discuss/Explain the power property of equality as it relates to rational exponents and properties of reciprocals. Use the equation $(x - 2)^{\frac{2}{3}} = 9$ for your discussion.

4. One factored form of an equation is shown. Discuss/Explain why $x = -8$ and $x = 1$ are not solutions to the equation, and what must be done to find the actual solutions: $2(x + 8)(x - 1) = -16$.

▶ **DEVELOPING YOUR SKILLS**

**Solve using the zero product property. Be sure each equation is in standard form and factor out any common factors before attempting to solve. Check all answers in the original equation.**

5. $22x = x^3 - 9x^2$
6. $x^3 = 13x^2 - 42x$
7. $3x^3 = -7x^2 + 6x$
8. $7x^2 + 15x = 2x^3$
9. $2x^4 - 3x^3 = 9x^2$
10. $-7x^2 = 2x^4 - 9x^3$
11. $x^3 + 7x^2 - 63 = 9x$
12. $x^3 - 4x + 24 = 6x^2$
13. $x^3 - 25x = 2x^2 - 50$
14. $3x^2 + x = x^3 + 3$
15. $2x^4 - 16x = 0$
16. $x^4 + 64x = 0$
17. $x^3 - 4x = 5x^2 - 20$
18. $x^3 - 18 = 9x - 2x^2$
19. $4x - 12 = 3x^2 - x^3$
20. $x - 7 = 7x^2 - x^3$
21. $2x^3 - 12x^2 = 10x - 60$
22. $9x + 81 = 27x^2 + 3x^3$
23. $x^4 - 7x^3 + 4x^2 = 28x$
24. $x^4 + 3x^3 + 9x^2 = -27x$
25. $x^4 - 256 = 0$
26. $x^4 - 625 = 0$
27. $x^6 - 2x^4 - x^2 + 2 = 0$
28. $x^6 - 3x^4 - 16x^2 + 48 = 0$
29. $x^5 - x^3 - 8x^2 + 8 = 0$
30. $x^5 - 9x^3 - x^2 + 9 = 0$
31. $x^6 - 1 = 0$
32. $x^6 - 64 = 0$

**Solve each equation. Identify any extraneous roots.**

33. $\dfrac{2}{x} + \dfrac{1}{x + 1} = \dfrac{5}{x^2 + x}$

34. $\dfrac{3}{m + 3} - \dfrac{5}{m^2 + 3m} = \dfrac{1}{m}$

35. $\dfrac{21}{a + 2} = \dfrac{3}{a - 1}$

36. $\dfrac{4}{2y - 3} = \dfrac{7}{3y - 5}$

37. $\dfrac{1}{3y} - \dfrac{1}{4y} = \dfrac{1}{y^2}$

38. $\dfrac{3}{5x} - \dfrac{1}{2x} = \dfrac{1}{x^2}$

39. $x + \dfrac{14}{x - 7} = 1 + \dfrac{2x}{x - 7}$

40. $\dfrac{10}{x - 5} + x = 1 + \dfrac{2x}{x - 5}$

41. $\dfrac{6}{n + 3} + \dfrac{20}{n^2 + n - 6} = \dfrac{5}{n - 2}$

42. $\dfrac{7}{p + 2} - \dfrac{1}{p^2 + 5p + 6} = -\dfrac{2}{p + 3}$

43. $\dfrac{a}{2a + 1} - \dfrac{2a^2 + 5}{2a^2 - 5a - 3} = \dfrac{3}{a - 3}$

44. $\dfrac{-18}{6n^2 - n - 1} + \dfrac{3n}{2n - 1} = \dfrac{4n}{3n + 1}$

**Solve for the variable indicated.**

45. $\dfrac{1}{f} = \dfrac{1}{f_1} + \dfrac{1}{f_2}$ for $f$

46. $\dfrac{1}{x} - \dfrac{1}{y} = \dfrac{1}{z}$ for $z$

47. $I = \dfrac{E}{R + r}$ for $r$

48. $q = \dfrac{pf}{p - f}$ for $p$

49. $\dfrac{P_1 V_1}{T_1} = \dfrac{P_2 V_2}{T_2}$ for $T_2$

50. $\dfrac{C}{P_2} = \dfrac{P_1}{d^2}$ for $P_2$

51. $t = \dfrac{A - P}{Pr}$ for $r$

52. $d = \dfrac{S - a}{n - 1}$ for $n$

Solve each equation and check your solutions by *substitution*. Identify any extraneous roots.

**53. a.** $-3\sqrt{3x-5} = -9$    **b.** $x = \sqrt{3x+1} + 3$

**54. a.** $-2\sqrt{4x-1} = -10$    **b.** $-5 = \sqrt{5x-1} - x$

**55. a.** $2 = \sqrt[3]{3m-1}$    **b.** $2\sqrt[3]{7-3x} - 3 = -7$

   **c.** $\dfrac{\sqrt[3]{2m+3}}{-5} + 2 = 3$    **d.** $\sqrt[3]{2x-9} = \sqrt[3]{3x+7}$

**56. a.** $-3 = \sqrt[3]{5p+2}$    **b.** $3\sqrt[3]{3-4x} - 7 = -4$

   **c.** $\dfrac{\sqrt[3]{6x-7}}{4} - 5 = -6$

   **d.** $3\sqrt[3]{x+3} = 2\sqrt[3]{2x+17}$

**57. a.** $\sqrt{x-9} + \sqrt{x} = 9$

   **b.** $x = 3 + \sqrt{23-x}$

   **c.** $\sqrt{x-2} - \sqrt{2x} = -2$

   **d.** $\sqrt{12x+9} - \sqrt{24x} = -3$

**58. a.** $\sqrt{x+7} - \sqrt{x} = 1$

   **b.** $\sqrt{2x+31} + x = 2$

   **c.** $\sqrt{3x} = \sqrt{x-3} + 3$

   **d.** $\sqrt{3x+4} - \sqrt{7x} = -2$

Write the equation in simplified form, then solve. Check all answers by substitution.

**59.** $x^{\frac{3}{5}} + 17 = 9$

**60.** $-2x^{\frac{3}{4}} + 47 = -7$

**61.** $0.\overline{3}x^{\frac{5}{2}} - 39 = 42$

**62.** $0.\overline{5}x^{\frac{5}{3}} + 92 = -43$

**63.** $2(x+5)^{\frac{2}{3}} - 11 = 7$

**64.** $-3(x-2)^{\frac{4}{5}} + 29 = -19$

Solve each equation using a *u*-substitution. Check all answers.

**65.** $x^{\frac{2}{3}} - 2x^{\frac{1}{3}} - 15 = 0$    **66.** $x^3 - 9x^{\frac{3}{2}} = -8$

**67.** $x^4 - 29x^2 + 100 = 0$    **68.** $z^4 - 20z^2 + 64 = 0$

**69.** $(b^2 - 3b)^2 - 14(b^2 - 3b) + 40 = 0$

**70.** $(d^2 - d)^2 - 8(d^2 - d) + 12 = 0$

**71.** $x^{-2} - 3x^{-1} - 4 = 0$

**72.** $x^{-2} - 2x^{-1} - 35 = 0$

**73.** $x^{-4} - 13x^{-2} + 36 = 0$

Use a *u*-substitution to solve each radical equation.

**74.** $x + 4 - 3\sqrt{x+4} = 0$

**75.** $x + 4 = 7\sqrt{x+4}$

**76.** $2(x+1) = 5\sqrt{x+1} - 2$

**77.** $2\sqrt{x+10} + 8 = 3(x+10)$

**78.** $4\sqrt{x-3} = 3(x-3) - 4$

## ▶ WORKING WITH FORMULAS

 **79. Lateral surface area of a cone:** $S = \pi r\sqrt{r^2 + h^2}$

The lateral surface area (surface area excluding the base) $S$ of a cone is given by the formula shown, where $r$ is the radius of the base and $h$ is the height of the cone. (a) Solve the equation for $h$. (b) Find the surface area of a cone that has a radius of 6 m and a height of 10 m. Answer in simplest form.

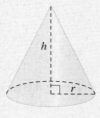

**80. Painted area on a canvas:** $A = \dfrac{4x^2 + 60x + 104}{x}$

A rectangular canvas is to contain a small painting with an area of 52 in², and requires 2-in. margins on the left and right, with 1-in. margins on the top and bottom for framing. The total area of such a canvas is given by the formula shown, where $x$ is the height of the *painted* area.

**a.** What is the area $A$ of the canvas if the height of the painting is $x = 10$ in.?

**b.** If the area of the canvas is $A = 120$ in², what are the dimensions of the painted area?

## ▶ APPLICATIONS

Find all real numbers that satisfy the following descriptions.

**81.** When the cube of a number is added to twice its square, the result is equal to 18 more than 9 times the number.

**82.** Four times a number decreased by 20 is equal to the cube of the number decreased by 5 times its square.

**83.** Find three consecutive even integers such that 4 times the largest plus the fourth power of the smallest is equal to the square of the remaining even integer increased by 24.

**84.** Find three consecutive integers such that the sum of twice the largest and the fourth power of the smallest is equal to the square of the remaining integer increased by 75.

85. **Composite figures—grain silos:** Grain silos can be described as a hemisphere sitting atop a cylinder. The interior volume $V$ of the silo can be modeled by $V = \frac{2}{3}\pi r^3 + \pi r^2 h$, where $h$ is the height of a cylinder with radius $r$. For a cylinder 6 m tall, what radius would give the silo a volume that is numerically equal to $24\pi$ times this radius?

86. **Composite figures—gelatin capsules:** The gelatin capsules manufactured for cold and flu medications are shaped like a cylinder with a hemisphere on each end. The interior volume $V$ of each capsule can be modeled by $V = \frac{4}{3}\pi r^3 + \pi r^2 h$, where $h$ is the height of the cylindrical portion and $r$ is its radius. If the cylindrical portion of the capsule is 8 mm long ($h = 8$ mm), what radius would give the capsule a volume that is numerically equal to $15\pi$ times this radius?

87. **Printing newspapers:** The editor of the school newspaper notes the college's new copier can complete the required print run 10 min quicker than the back-up copier. If it takes 12 min when both copiers are used, how long would it take the new copier to make a run on its own?

88. **Inflating a tire:** A foot pump can inflate a new bicycle tire in one-tenth the time it takes for a hole to deflate a fully inflated tire. If it takes 5 min to inflate a leaking tire, how long would it take to inflate a repaired tire?

89. **Triathlon competition:** As one part of a Mountain-Man triathlon, participants must row a canoe 5 mi down river (with the current), circle a buoy, and row 5 mi back up river (against the current) to the starting point. If the current is flowing at a steady rate of 4 mph and Tom Chaney made the round-trip in 3 hr,

how fast can he row in still water? (*Hint:* The time rowing down river and the time rowing up river must add up to 3 hr.)

90. **Flight time:** The flight distance from Cincinnati, Ohio, to Chicago, Illinois, is approximately 300 mi. On a recent round-trip between these cities, the plane encountered a steady 25-mph headwind on the way to Chicago, with a 25-mph tailwind on the return trip. If the total flying time came to exactly 5 hr, what was the flying time to Chicago? What was the flying time back to Cincinnati? (*Hint:* The flight time between the two cities must add up to 5 hr.)

91. **Pollution removal:** For a steel mill, the cost $C$ (in millions of dollars) to remove toxins from the resulting sludge is given by $C = \dfrac{22P}{100 - P}$, where $P$ is the percent of the toxins removed. What percent can be removed if the mill spends \$88,000,000 on the cleanup?

92. **Wildlife populations:** When the Department of Wildlife introduced 60 elk into a new game reserve, it was projected that the size of the herd will grow according to the equation $N = \dfrac{10(6 + 3t)}{1 + 0.05t}$, where $N$ is the number of elk and $t$ is the time in years. If recent counts find 300 elk, approximately how many years have passed?

93. **Planetary motion:** The time $T$ (in days) for a planet to make one revolution (elliptical orbit) around the sun is modeled by $T = 0.407R^{\frac{3}{2}}$, where $R$ is the maximum radius of the planet's orbit in millions of miles (*Kepler's third law of planetary motion*). Use the equation to approximate the maximum radius of each orbit, given the number of days it takes for one revolution.

    a. Mercury: 88 days

    b. Venus: 225 days

    c. Earth: 365 days

    d. Mars: 687 days

    e. Jupiter: 4333 days

    f. Saturn: 10,759 days

94. **Wind-powered energy:** If a wind-powered generator is delivering $P$ units of power, the velocity $V$ of the wind (in miles per hour) can be determined using $V = \sqrt[3]{\dfrac{P}{k}}$, where $k$ is a constant that depends on the size and efficiency of the generator. Given $k = 0.004$, approximately how many units of power are being delivered if the wind is blowing at 27 mi/hr?

► EXTENDING THE CONCEPT

**95.** The SMOG formula, $y = \sqrt{w} + 3$, approximates the reading level required to fully comprehend a given text. Here, $y$ is the number of years of education needed to understand a written work containing $w$ words of three or more syllables in a random selection of 30 sentences. Calculate the SMOG index of Shakespeare's *Hamlet* by picking 10 random sentences from Act I, 10 more from Act III, and another 10 from Act V. Then count the total number of words in your 30-sentence sample that contain at least three syllables. Use this number to compute $y$. Repeat this process to determine the reading level of Mark Twain's *The Adventures of Huckleberry Finn* and Darwin's *On the Origin of Species*. Which of these books demands the highest reading level? (The works listed above are available for free online at books.google.com.)

**96.** To solve the equation $3 - \dfrac{8}{x + 3} = \dfrac{1}{x}$, a student multiplied by the LCD $x(x + 3)$, simplified, and got this result: $3 - 8x = (x + 3)$. Identify and fix the mistake, then find the correct solution(s).

**97.** To solve $\sqrt{x - 5} = x - 7$, we would square both sides to get the quadratic equation $x - 5 = x^2 - 14x + 49$, which has solutions of $x = 6$ and $x = 9$. But from the original radical equation, we know $x - 7 \geq 0$, and so $x \geq 7$. This observation guarantees $x = 6$ is extraneous and makes checking by substitution unnecessary. Use the same reasoning to determine if any of the following solutions are *extraneous* in the given equations. Justify your choices.
  **a.** $\sqrt{x - 5} = 7 - x; x = 6, x = 9$
  **b.** $\sqrt{10 - x} = x - 8; x = 6, x = 9$
  **c.** $3\sqrt{x - 5} = x - 3; x = 6, x = 9$
  **d.** $3\sqrt{10 - x} = x - 12; x = 6, x = 9$

**98.** As an extension of our earlier work with absolute values, try solving the following exercises.
  Recall that for $|X| = k$, $X = -k$ or $X = k$.
  **a.** $|x^2 - 2x - 25| = 10$
  **b.** $|x^2 - 9| = -x + 3$
  **c.** $|x^2 - 7x| = -x + 7$
  **d.** $|x^2 - 5x - 2| = x + 5$

► MAINTAINING YOUR SKILLS

**99.** (1.1) Two jets take off on parallel runways going in opposite directions. The first travels at a rate of 250 mph and the second at 325 mph. How long until they are 980 mi apart?

**100.** (1.2) Graph the relation given:
$$2x - 3 < 7 \text{ and } x + 2 > 1$$

**101.** (R.3) Simplify using properties of exponents:
$$2^{-1} + (2x)^0 + 2x^0$$

**102.** (R.4) Find the missing side.

---

## MAKING CONNECTIONS

## Making Connections: Graphically, Symbolically, Numerically, and Verbally

Eight relations (a) through (h) are given. Match the characteristics or descriptions shown in 1 through 16 to one of the eight relations.

**(a)** $|x| = 5$     **(b)** $|x| \leq 5$     **(c)** $2x^2 + 13 = -5$     **(d)** $\dfrac{2x - 13}{3} = -5$

**(e)** $2x^2 - 13 = -5$     **(f)** $x = -5i^6$     **(g)** $2x - 13 \geq -5$     **(h)** $2\sqrt{x} - 13 = -5$

**1.** _____ less than or equal to
**2.** _____ $x = \pm 2$
**3.** _____ solutions are complex
**4.** _____ $x = \pm 5$
**5.** _____ $x \in [-5, 5]$

**6.** _____ $x = -1$
**7.** _____ rational equation
**8.** _____ $x = 3$ is a solution
**9.** _____ linear inequality
**10.** _____ $x \in [4, \infty)$

**11.** _____ $x = 16$
**12.** _____ $x = 5$
**13.** _____ $x = \pm 3i$
**14.** _____ radical equation
**15.** _____ quadratic equation
**16.** _____ absolute value equation

# SUMMARY AND CONCEPT REVIEW

## SECTION 1.1    Linear Equations, Formulas, and Problem Solving

### KEY CONCEPTS

- An equation is a statement that two expressions are equal.
- Replacement values that make an equation true are called solutions or roots.
- To solve an equation we use the distributive property and the properties of equality to write a sequence of simpler, equivalent equations until the solution is obvious. A guide for solving linear equations appears on page 75.
- An equation can be an identity (always true), a contradiction (never true), or conditional (true or false, depending on the input value[s]).
- If an equation contains fractions, multiplying both sides by the least common denominator of all fractions will "clear the denominators" and reduce the amount of work required to solve. Similarly, multiplying both sides of an equation by an appropriate power of 10 can be used to "clear the decimals."
- Solutions to an equation can be checked using back-substitution, by replacing the variable with the proposed solution and verifying the left-hand expression is equal to the right.
- The basic elements of good problem solving include:
  1. Gathering and organizing information
  2. Making the problem visual
  3. Developing an equation model
  4. Using the model to solve the application

  For a complete review, see the problem-solving guide on page 78.

### EXERCISES

1. Use substitution to determine if the indicated value is a solution to the equation given.

   **a.** $x - (2 - x) = 4(x - 5), x = -6$     **b.** $\frac{3}{4}b + 2 = \frac{5}{2}b + 16, b = -8$     **c.** $4d - 2 = -\frac{1}{2} + 3d, d = \frac{3}{2}$

Solve each equation.

2. $-2b + 7 = -5$

3. $3(2n - 6) + 1 = 3 - (20 - 6n)$

4. $4m - 5 = 11m + 2$

5. $\frac{1}{2}x + \frac{2}{3} = \frac{3}{4}$

6. $6p - (3p + 5) - 9 = 3(p - 3)$

7. $-\frac{g}{6} = 3 - \frac{1}{2} - \frac{5g}{12}$

Solve for the specified variable in each formula or literal equation.

8. $V = \pi r^2 h$ for $h$

9. $P = 2L + 2W$ for $L$

10. $ax + b = c$ for $x$

11. $2x - 3y = 6$ for $y$

Use the problem-solving guidelines (page 78) to solve the following applications.

12. At a large family reunion, two kegs of lemonade are available. One is 2% sugar (too sour) and the second is 7% sugar (too sweet). How many gallons of the 2% keg must be mixed with 12 gal of the 7% keg to get a 5% mix?

13. A rectangular window with a width of 3 ft and a height of 4 ft is topped by a semi-circular window. Find the total area of the window.

14. Two cyclists start from the same location and ride in opposite directions, one riding at 15 mph and the other at 18 mph. If their radio phones have a range of 22 mi, how many minutes will they be able to communicate?

## SECTION 1.2    Linear Inequalities in One Variable

### KEY CONCEPTS

- Inequalities are solved using properties similar to those used for solving equations. The one exception is when multiplying or dividing by a negative quantity, as the inequality symbol *must then be reversed* to maintain the truth of the resulting statement.
- Solutions to an inequality can be given using a simple inequality, graphed on a number line, stated in set notation, or stated using interval notation.
- Given two sets $A$ and $B$: $A$ intersect $B$ ($A \cap B$) is the set of elements *shared* by both $A$ **and** $B$ (elements common to both sets). $A$ union $B$ ($A \cup B$) is the set of elements *in either* $A$ **or** $B$ (elements are combined to form a larger set).

## EXERCISES

Use inequality symbols to write a mathematical model for each statement.

**15.** You must be 35 yr old or older to run for president of the United States.

**16.** A child must be under 2 yr of age to be admitted free.

**17.** The speed limit on many interstate highways is 65 mph.

**18.** Our caloric intake should not be less than 1200 calories/day.

Solve the inequality and write the solution using interval notation.

**19.** $7x > 35$

**20.** $-\dfrac{3}{5}m < 6$

**21.** $2(3m - 2) \le 8$

**22.** $-1 < \dfrac{1}{3}x + 2 \le 5$

**23.** $-4 < 2b + 8$ and $3b - 5 > -32$

**24.** $-5(x + 3) > -7$ or $x - 5.2 > -2.9$

**25.** Find the allowable values for each of the following expressions. Write your answer in interval notation.

   **a.** $\dfrac{7}{n - 3}$

   **b.** $\dfrac{5}{2x - 3}$

   **c.** $\sqrt{x + 5}$

   **d.** $\sqrt{-3n + 18}$

**26.** Latoya has earned grades of 72%, 95%, 83%, and 79% on her first four exams. What grade must she make on her fifth and last exam so that her average is 85% or more?

## SECTION 1.3    Absolute Value Equations and Inequalities

### KEY CONCEPTS

- To solve absolute value equations and inequalities, begin by writing the equation in simplified form, with the absolute value isolated on one side.
- If $X$ and $Y$ represent algebraic expressions and $k$ is a nonnegative constant:
  - Absolute value equations:    $|X| = k$ is equivalent to $X = -k$ or $X = k$
                                         $|X| = |Y|$ is equivalent to $X = Y$ or $X = -Y$
  - "Less than" inequalities:    $|X| < k$ is equivalent to $-k < X < k$
  - "Greater than" inequalities:    $|X| > k$ is equivalent to $X < -k$ or $X > k$
- If the absolute value quantity has been isolated on the left, the solution to a less-than inequality will be a single interval, while the solution to a greater-than inequality will consist of two disjoint intervals.

### EXERCISES

Solve each equation or inequality. Write solutions to inequalities in interval notation.

**27.** $7 = |x - 3|$

**28.** $-2|x + 2| = -10$

**29.** $|-2x + 3| = 13$

**30.** $\dfrac{|2x + 5|}{3} + 8 = 9$

**31.** $-3|x + 2| - 2 < -14$

**32.** $\left|\dfrac{x}{2} - 9\right| \le 7$

**33.** $|3x + 5| = -4$

**34.** $3|x + 1| < -9$

**35.** $2|x + 1| > -4$

**36.** $5|m - 2| - 12 \le 8$

**37.** $\dfrac{|3x - 2|}{2} + 6 \ge 10$

**38.** Monthly rainfall received in Omaha, Nebraska, rarely varies by more than 1.7 in. from an average of 2.5 in./month. Use this information to (a) write an absolute value inequality model, then (b) solve the inequality to find the highest and lowest amounts of monthly rainfall for this city.

## SECTION 1.4    Complex Numbers

### KEY CONCEPTS

- The italicized $i$ represents the number whose square is $-1$. This means $i^2 = -1$ and $i = \sqrt{-1}$.
- For $k > 0$, $\sqrt{-k} = i\sqrt{k}$ and we say the expression has been *written in terms of i*.
- The standard form of a *complex number* is $a + bi$, where $a$ is the *real part* and $bi$ is the *imaginary part*.
- To add or subtract complex numbers, combine the like terms.
- For any complex number $a + bi$, its *complex conjugate* is $a - bi$.
- Larger powers of $i$ can be simplified using $i^4 = 1$.
- To multiply complex numbers, use the F-O-I-L method and simplify.
- To find a *quotient* of complex numbers, multiply the numerator and denominator by the conjugate of the denominator.

## EXERCISES

Simplify each expression and write the result in standard form.

**39.** $\sqrt{-72}$

**40.** $6\sqrt{-48}$

**41.** $\dfrac{-10 + \sqrt{-50}}{5}$

**42.** $\sqrt{3}\sqrt{-6}$

**43.** $i^{57}$

Perform the operation indicated and write the result in standard form.

**44.** $(5 + 2i)^2$

**45.** $\dfrac{5i}{1 - 2i}$

**46.** $(-3 + 5i) - (2 - 2i)$

**47.** $(2 + 3i)(2 - 3i)$

**48.** $4i(-3 + 5i)$

Use substitution to show the given complex number and its conjugate are solutions to the equation shown.

**49.** $x^2 - 9 = -34; x = 5i$

**50.** $x^2 - 4x + 9 = 0; x = 2 + i\sqrt{5}$

## SECTION 1.5    Solving Quadratic Equations

### KEY CONCEPTS

- The standard form of a quadratic equation is $ax^2 + bx + c = 0$, where $a$, $b$, and $c$ are real numbers with $a \neq 0$. In words, we say the equation is written in decreasing order of degree and set equal to zero.
- Factorable quadratics can be solved using the zero product property, which states that if the product of two factors is zero, then one or both must equal zero. Symbolically, if $A \cdot B = 0$, then $A = 0$ or $B = 0$.
- The square root property of equality states that if $X^2 = k$, where $k \geq 0$, then $X = \sqrt{k}$ or $X = -\sqrt{k}$.
- Quadratic equations can also be solved by *completing the square,* or using the *quadratic formula.*
- If the discriminant $b^2 - 4ac = 0$, the equation has one real (repeated) root. If $b^2 - 4ac > 0$, the equation has two real roots; and if $b^2 - 4ac < 0$, the equation has two nonreal roots.

### EXERCISES

**51.** Determine whether the given equation is quadratic. If so, write the equation in standard form and identify the values of $a$, $b$, and $c$.

    **a.** $-3 = 2x^2$
    **b.** $7 = -2x + 11$
    **c.** $99 = x^2 - 8x$
    **d.** $20 = 4 - x^2$

**52.** Solve by factoring.

    **a.** $x^2 - 3x - 10 = 0$
    **b.** $2x^2 - 50 = 0$
    **c.** $x(6x - 1) = 2$
    **d.** $3x^2 - 15 = 4x$

**53.** Solve using the square root property of equality.

    **a.** $x^2 - 9 = 0$
    **b.** $2(x - 2)^2 + 1 = 11$
    **c.** $3x^2 + 15 = 0$
    **d.** $-2x^2 + 4 = -46$

**54.** Solve by completing the square. Give solutions in both exact and approximate form.

    **a.** $x^2 + 2x = 15$
    **b.** $x^2 + 6x = 16$
    **c.** $-4x + 2x^2 = 3$
    **d.** $3x^2 - 7x = -2$

**55.** Solve using the quadratic formula. Give solutions in both exact and approximate form.

    **a.** $4x^2 + 7 = 12x$
    **b.** $x(x + 8) = 5$
    **c.** $x^2 - 4x = -9$
    **d.** $2x^2 - 6x + 5 = 0$

Solve the following applications.

**56.** A batter has just popped up to the catcher, who catches the ball while standing on home plate. (a) If the batter made contact with the ball at a height of 4 ft and the ball left the bat with an initial velocity of 128 ft/sec, how long will it take the ball to reach a height of 116 ft? (b) How high is the ball 5 sec after contact? (c) If the catcher catches the ball at a height of 4 ft, how long was it airborne?

**57.** Letter size paper is $8\frac{1}{2}$ in. by 11 in. Legal size paper is $8\frac{1}{2}$ in. by 14 in. The next larger (common) size of paper is Ledger, which has an area of 187 in$^2$, with a length that is 6 in. longer than the width. What are the dimensions of the Ledger size paper?

## SECTION 1.6    Solving Other Types of Equations

### KEY CONCEPTS

- To solve rational equations, first clear denominators using the LCD, noting values that must be excluded.
- Multiplying an equation by a variable quantity sometimes introduces extraneous solutions. Check all results in the original equation.

- To solve radical equations, isolate the radical on one side, then apply the appropriate "$n$th power" to free up the radicand. Repeat the process if needed. See flowchart on page 132.
- For equations with a rational exponent $\frac{m}{n}$, isolate the variable term and raise both sides to the "$n$th power." Then carefully take the appropriate root of both sides, noting that even roots can result in two solutions.

## EXERCISES

Solve by factoring.

**58.** $x^3 - 7x^2 = 3x - 21$

**59.** $3x^3 + 5x^2 = 2x$

**60.** $x^4 - 8x = 0$

**61.** $x^4 - \frac{1}{16} = 0$

Solve each equation.

**62.** $\frac{3}{5x} + \frac{7}{10} = \frac{1}{4x}$

**63.** $\frac{3h}{h+3} - \frac{7}{h^2+3h} = \frac{1}{h}$

**64.** $\frac{2n}{n+2} - \frac{3}{n-4} = \frac{n^2+20}{n^2-2n-8}$

**65.** $\frac{\sqrt{x^2+7}}{2} + 3 = 5$

**66.** $3\sqrt{x+4} = x + 4$

**67.** $\sqrt{3x+4} = 2 - \sqrt{x+2}$

**68.** $3\left(x - \frac{1}{4}\right)^{-\frac{3}{2}} = \frac{8}{9}$

**69.** $-2(5x+2)^{\frac{2}{3}} + 17 = -1$

**70.** $(x^2 - 3x)^2 - 14(x^2 - 3x) + 40 = 0$

**71.** $x^4 - 7x^2 = 18$

**72.** In the evolution of locomotive travel, the new diesel locomotives could travel 20 mph faster than the steam locomotives they replaced. In a cross-country race from coast to coast (approximately 2880 mi), it would take the steam locomotives 12 hr longer to finish. How fast could the steam locomotive travel? How fast were the diesel locomotives of the age?

**73.** Buttons the Chihuahua cleans her bowl in a leisurely 80 sec. Abby the Catahoula, who eats without breathing, empties hers in 20 sec. If the two were willing to share, how long would it take to finish off one bowl together?

Exercise 73

## PRACTICE TEST

**1.** Solve each equation.

   **a.** $-\frac{2}{3}x - 5 = 7 - (x + 3)$

   **b.** $-5.7 + 3.1x = 14.5 - 4(x + 1.5)$

   **c.** $P = C + kC$ for $C$

   **d.** $2|2x + 5| - 17 = -11$

**2.** How much water that is 102°F must be mixed with 25 gal of water at 91°F, so that the resulting temperature of the water will be 97°F?

**3.** Solve each inequality.

   **a.** $-\frac{2}{5}x + 7 < 19$    **b.** $-1 < 3 - x \le 8$

   **c.** $\frac{1}{2}x + 3 < 9$  or  $\frac{2}{3}x - 1 \ge 3$

   **d.** $\frac{1}{2}|x - 3| + \frac{5}{4} \le \frac{7}{4}$    **e.** $-\frac{2}{3}|x + 1| - 5 < -7$

**4.** To make the bowling team, Jacques needs a three-game average of 160. If he bowled 141 and 162 for the first two games, what score $S$ must be obtained in the third game so that his average is at least 160?

Simplify each expression.

**5.** $\frac{-8 + \sqrt{-20}}{6}$    **6.** $i^{39}$

**7.** Given $x = \frac{1}{2} + \frac{\sqrt{3}}{2}i$ and $y = \frac{1}{2} - \frac{\sqrt{3}}{2}i$ find

   **a.** $x + y$    **b.** $x - y$    **c.** $xy$

**8.** Find the product. $(3i + 5)(2 - i)$

**9.** Compute the quotient: $\frac{3i}{1-i}$.

**10.** Show $x = 2 - 3i$ is a solution of $x^2 - 4x + 13 = 0$.

**11.** Solve by factoring.

   **a.** $z^2 - 7z - 30 = 0$    **b.** $3x^2 - 20x = -12$

**12.** Solve using the square root property of equality.

   **a.** $4x^2 - 3 = 97$    **b.** $(x - 1)^2 + 3 = 0$

**13.** Solve by completing the square.

   **a.** $2x^2 - 20x + 49 = 0$

   **b.** $2x^2 - 5x = -4$

**14.** Solve using the quadratic formula.

    **a.** $3x^2 + 2 = 6x$         **b.** $x^2 = 2x - 10$

**15.** The cost of raw materials to produce plastic toys is given by the cost equation $C = 2x + 35$, where $x$ is the number of toys in hundreds. The total income (revenue) from the sale of these toys is given by $R = -x^2 + 122x - 1965$. (a) Determine the profit equation (profit = revenue − cost). (b) During the Christmas season, the owners of the company decide to manufacture and donate as many toys as they can, without taking a loss (i.e., they break even: profit or $P = 0$). How many toys will they produce for charity?

**16.** Due to the seasonal nature of the business, the revenue of Wet Willey's Water World can be modeled by the equation $r = -3t^2 + 42t - 135$, where $t$ is the time in months ($t = 1$ corresponds to January) and $r$ is the dollar revenue in thousands. (a) What month does Wet Willey's open? (b) What month does Wet Willey's close? (c) Does Wet Willey's bring in more revenue in July or August? How much more?

Solve each equation.

**17.** $54x^3 = 6x$

**18.** $4x^3 + 8x^2 - 9x - 18 = 0$

**19.** $\dfrac{2}{x-3} + \dfrac{2x}{x+2} = \dfrac{x^2 + 16}{x^2 - x - 6}$

**20.** $\dfrac{4}{x-3} + 2 = \dfrac{5x}{x^2 - 9}$

**21.** $5 - 2\sqrt[3]{1-x} = 11$     **22.** $\sqrt{x} + 1 = \sqrt{2x - 7}$

**23.** $(x+3)^{-\frac{2}{3}} = \dfrac{1}{4}$     **24.** $x^4 + 16 = 17x^2$

**25.** Allometric studies tell us that the necessary food intake $F$ (in grams per day) of nonpasserine birds (birds other than songbirds and other small birds) can be modeled by the equation $F \approx 0.3W^{\frac{3}{4}}$, where $W$ is the bird's weight in grams. (a) If my green-winged macaw weighs 1296 g, what is her anticipated daily food intake? (b) If my blue-headed pionus consumes 19.2 g/day, what is his estimated weight?

## CALCULATOR EXPLORATION AND DISCOVERY

### I. Using a Graphing Calculator as an Investigative Tool

The mixture concept can be applied in a wide variety of ways, including mixing tin and copper to get bronze, different kinds of nuts for the holidays, diversifying investments, or mixing two acid solutions in order to get a desired concentration. Whether the value of each part in the mix is monetary or a percent of concentration, the general mixture equation has this form:

$$\text{Quantity 1} \cdot \text{Value I} + \text{Quantity 2} \cdot \text{Value II} = \text{Total quantity} \cdot \text{Desired value}$$

Graphing calculators are a great tool for exploring this relationship, because the TABLE feature enables us to test the result of various mixtures in an instant. Suppose 10 oz of an 80% glycerin solution are to be mixed with an unknown amount of a 40% solution. How much of the 40% solution is used if a 56% solution is needed? To begin, we might observe that using equal amounts of the 40% and 80% solutions would result in a 60% concentration (halfway between 40% and 80%). To illustrate, let $C$ represent the final concentration of the mix.

**Figure 1.9**

```
Plot1  Plot2  Plot3
\Y1◻.8(10)+.4X
\Y2◻.56(10+X)
\Y3=
\Y4=
\Y5=
\Y6=
\Y7=
```

$$
\begin{aligned}
10(0.8) + 10(0.4) &= (10 + 10)C \quad &\text{equal amounts} \\
8 + 4 &= 20C \quad &\text{simplify} \\
12 &= 20C \quad &\text{add} \\
0.6 &= C \quad &\text{divide by 20}
\end{aligned}
$$

Since this concentration is too high (a 56% = 0.56 solution is desired), we know more of the weaker solution should be used. To explore the relationship further, assume X oz of the 40% solution are added. Then enter $Y_1 = .8(10) + .4X$ on the (Y=) screen to compute the resulting volume of glycerin. To determine the volume required for a 56% mixture, enter $Y_2 = .56(10 + X)$ (see Figure 1.9). Next, set up a TABLE using (2nd) (WINDOW) (**TBLSET**) with **TblStart** = 10, **ΔTbl** = 1, and the calculator set in **Indpnt: AUTO** mode (see Figure 1.10). Finally, access the TABLE results using (2nd) (GRAPH) (**TABLE**). The resulting screen is shown in Figure 1.11, where we note that 15 oz of the 40% solution should be used (since the equation $Y_1 = Y_2$ is true when X is 15).

**Exercise 1:** Use this idea to solve Exercises 71 and 72 from Section 1.1.

**Figure 1.10**

```
TABLE SETUP
 TblStart=10
 ΔTbl=1
Indpnt: AUTO Ask
Depend: AUTO Ask
```

**Figure 1.11**

| X | Y1 | Y2 |
|---|-----|------|
| 10 | 12 | 11.2 |
| 11 | 12.4 | 11.76 |
| 12 | 12.8 | 12.32 |
| 13 | 13.2 | 12.88 |
| 14 | 13.6 | 13.44 |
| 15 | 14 | 14 |
| 16 | 14.4 | 14.56 |

X=10

## II. Absolute Value Equations and Inequalities

**Figure 1.12**

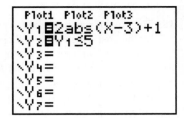

Graphing calculators can also be used in many different ways to explore and solve inequalities. Here we'll use a table of values and a *relational test*. To begin we'll consider the equation $2|x - 3| + 1 = 5$ by entering the left-hand side as $Y_1$ on the ⬚ screen. The calculator does not use absolute value bars the way they're written, and the equation is actually entered as $Y_1 = 2$ **abs** $(X - 3) + 1$ (see Figure 1.12). The **"abs("** notation is accessed by pressing **MATH** ▷ **(NUM)** ( 1 ) (this gives only the left parenthesis, you must supply the right). Preset the TABLE as shown earlier. By scrolling through the table (use the up ⌃ and down ⌄ arrows), we find $Y_1 = 5$ when $x = 1$ or $x = 5$ (see Figure 1.13).

Although we could also solve the *inequality* $2|x - 3| + 1 \leq 5$ using the table (the solution interval is $x \in [1, 5]$), a relational test can help. Relational tests have the calculator return a "1" if a given statement is true, and a "0" otherwise. Enter $Y_2 = Y_1 \leq 5$, by accessing $Y_1$ using **VARS** ▷ **(Y-VARS) 1:Function** ⬚, and the "≤" symbol using **2nd** **MATH** **(TEST)** (the "less than or equal to" symbol is option 6). Returning to the table shows $Y_1 \leq 5$ is true for $1 \leq x \leq 5$ (see Figure 1.13).

**Figure 1.13**

| X | Y₁ | Y₂ |
|---|----|----|
| 0 | 7 | 0 |
| 1 | 5 | 1 |
| 2 | 3 | 1 |
| 3 | 1 | 1 |
| 4 | 3 | 1 |
| 5 | 5 | 1 |
| 6 | 7 | 0 |

X=0

Use a table and a relational test to help solve the following inequalities. Verify the result algebraically.

**Exercise 2:** $3|x + 1| - 2 \geq 7$     **Exercise 3:** $-2|x + 2| + 5 \geq -1$

**Exercise 4:** $-1 \leq 4|x - 3| - 1$

---

## STRENGTHENING CORE SKILLS

## Using Distance to Understand Absolute Value Equations and Inequalities

In Section R.1 we noted that for any two numbers $a$ and $b$ on the number line, *the distance between $a$ and $b$ is written $|a - b|$* or $|b - a|$. In exactly the same way, the equation $|x - 3| = 4$ can be read, "the distance between 3 and an unknown number is equal to 4." The advantage of reading it in this way (instead of "the absolute value of $x$ minus 3 is 4") is that a much clearer *visualization* is formed, giving a constant reminder there are two solutions. In diagram form we have Figure 1.14.

**Figure 1.14**

From this we note the solution is $x = -1$ or $x = 7$.

In the case of an inequality such as $|x + 2| \leq 3$, we rewrite the inequality as $|x - (-2)| \leq 3$ and read it, "the distance between $-2$ and an unknown number is less than or equal to 3." With some practice, visualizing this relationship mentally enables a quick statement of the solution: $x \in [-5, 1]$. In diagram form we have Figure 1.15.

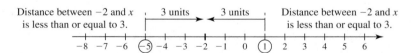

**Figure 1.15**

Equations and inequalities where the coefficient of $x$ is not 1 still lend themselves to this form of conceptual understanding. For $|2x - 1| \geq 3$ we read, "the distance between 1 and twice an unknown number is greater than or equal to 3." On the number line (Figure 1.16), the number 3 units to the left of 1 is $-2$, and the number 3 units to the right of 1 is 4.

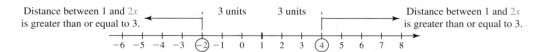

**Figure 1.16**

For $2x \leq -2$, $x \leq -1$, and for $2x \geq 4$, $x \geq 2$, so the solution is $x \in (-\infty, -1] \cup [2, \infty)$.

Attempt to solve the following equations and inequalities by visualizing a number line. Check all results algebraically.

**Exercise 1:** $|x - 2| = 5$     **Exercise 2:** $|x + 1| \leq 4$     **Exercise 3:** $|2x - 3| \geq 5$

# Relations, Functions, and Graphs

## CHAPTER OUTLINE

## CHAPTER CONNECTIONS

Viewing relations and functions in terms of an equation, a table of values, and the related graph, often brings a clearer understanding of the relationships involved. For instance, the magnitude 5.4 earthquake that shook the Midwest in 2008 was centered near West Salem, Illinois, but was felt up to 450 mi away in Topeka (Kansas), Oshkosh (Wisconsin), and Atlanta (Georgia). In areas where the density of the Earth is near constant, the effects of the quake emanate outward in a circular fashion. Using the concepts found in this chapter, we'll be able to determine what other American cities were affected by the quake. This application occurs as Exercise 92 in Section 2.1.

**Check out these other real-world connections:**

▶ Garbage Collection
  (Section 2.2, Exercise 43)
▶ Internet Use
  (Section 2.3, Exercise 103)
▶ Fuel Efficiency
  (Section 2.5, Exercise 112)
▶ Credit Card Use
  (Section 2.6, Exercise 31)

In everyday life, we encounter a large variety of relationships. For instance, the time it takes us to get to work is related to our average speed; the monthly cost of heating a home is related to the average outdoor temperature; and in many cases, the amount of our charitable giving is related to changes in the cost of living. In each case we say that a relation exists between the two quantities.

## A. Relations, Mapping Notation, and Ordered Pairs

In the most general sense, a **relation** is simply a correspondence between two sets. Relations can be represented in many different ways and may even be very "unmathematical," like the one shown in Figure 2.1 between a set of people and the set of their corresponding birthdays. If $P$ represents the set of people and $B$ represents the set of birthdays, we say that elements of $P$ correspond to elements of $B$, or the birthday relation maps elements of $P$ to elements of $B$. Using what is called **mapping notation,** we might simply write $P \rightarrow B$. From a purely practical standpoint, we note that while it is possible for two different people to share the same birthday, it is quite impossible for the same person to have two different birthdays. Later, this observation will help us mark the difference between a relation and special kind of relation called a function.

**Figure 2.1**

| $P$ | | $B$ |
|---|---|---|
| Missy | | April 12 |
| Jeff | | Nov 9 |
| Angie | | Sept 10 |
| Megan | | Nov 28 |
| Mackenzie | | May 7 |
| Michael | | April 14 |
| Mitchell | | |

The bar graph in Figure 2.2 is also an example of a relation. In the graph, each year is related to annual consumer spending per person on cable and satellite television. As an alternative to mapping or a bar graph, this relation could also be represented using ordered pairs. For example, the ordered pair (5, 234) would indicate that in 2005, spending per person on cable and satellite TV in the United States averaged $234.

Over a long period of time, we could collect many ordered pairs of the form $(t, s)$, where consumer spending $s$ *depends* on the time $t$. For this reason we often call the second coordinate of an ordered pair (in this case $s$) the **dependent variable,** with the first coordinate designated as the **independent variable.** The set of all first coordinates is called the **domain** of the relation. The set of all second coordinates is called the **range.**

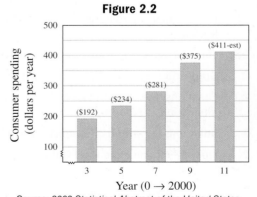

**Figure 2.2**

Consumer spending (dollars per year)

($192) ($234) ($281) ($375) ($411-est)

Year (0 → 2000)

*Source: 2009 Statistical Abstract of the United States, Table 1089 (some figures are estimates)*

**EXAMPLE 1** ▶ **Expressing a Relation as a Mapping and as a Pointwise-Defined Relation**

Represent the relation from Figure 2.2 in mapping notation and in ordered pair form, then state its domain and range.

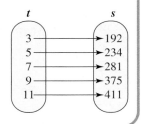

**Solution** ▶ Let $t$ represent the year and $s$ represent consumer spending. The mapping $t \rightarrow s$ gives the diagram shown. In ordered pair form we have {(3, 192), (5, 234), (7, 281), (9, 375), and (11, 411)}. The domain is the set {3, 5, 7, 9, 11}; the range is {192, 234, 281, 375, 411}.

☑ **A.** You've just seen how we can express a relation in mapping notation and ordered pair form

Now try Exercises 5 through 10 ▶

For more on this relation, **see Section 2.4, Exercise 95.**

## B. The Graph of a Relation

**Table 2.1** $y = x - 1$

| $x$ | $y$ |
|-----|-----|
| −4 | −5 |
| −2 | −3 |
| 0 | −1 |
| 2 | 1 |
| 4 | 3 |

**Table 2.2** $x = |y|$

| $x$ | $y$ |
|-----|-----|
| 2 | −2 |
| 1 | −1 |
| 0 | 0 |
| 1 | 1 |
| 2 | 2 |

Relations can also be stated in **equation form.** The equation $y = x - 1$ expresses a relation where each $y$-value is one less than the corresponding $x$-value (see Table 2.1). The equation $x = |y|$ expresses a relation where each $x$-value corresponds to the absolute value of $y$ (see Table 2.2). In each case, the relation is the set of all ordered pairs $(x, y)$ that create a true statement when substituted, and a few ordered pair solutions are shown in the tables for each equation.

Relations can be expressed graphically using a **rectangular coordinate system.** It consists of a horizontal number line (**the $x$-axis**) and a vertical number line (**the $y$-axis**) intersecting at their zero marks. The point of intersection is called the *origin*. The $x$- and $y$-axes create a flat, two-dimensional surface called the **$xy$-plane** and divide the plane into four regions called **quadrants.** These are labeled using a capital "Q" (for quadrant) and the Roman numerals I through IV, beginning in the upper right and moving counterclockwise (Figure 2.3). The **grid lines** shown denote the integer values on each axis and further divide the plane into a **coordinate grid,** where every point in the plane corresponds to an ordered pair. Since a point at the origin has not moved along either axis, it has coordinates (0, 0). To plot a point $(x, y)$ means we place a dot at its location in the $xy$-plane. A few of the ordered pairs from $y = x - 1$ are plotted in Figure 2.4, where a noticeable pattern emerges—the points seem to lie along a straight line.

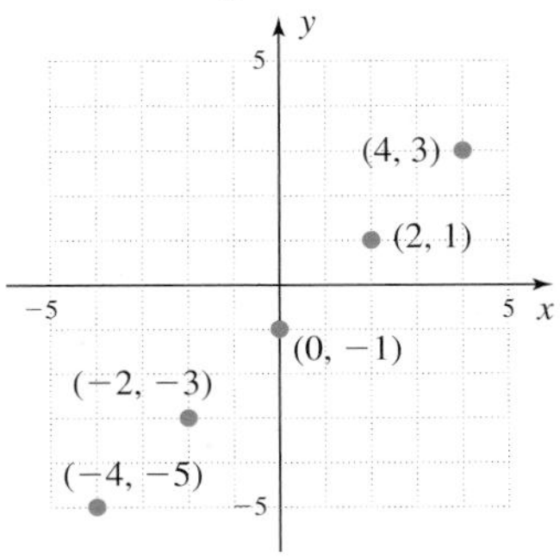

If a relation is pointwise-defined, the graph of the relation is simply the plotted points. The graph of a relation *in equation form,* such as $y = x - 1$, is the set of *all* ordered pairs $(x, y)$ that are solutions (make the equation true).

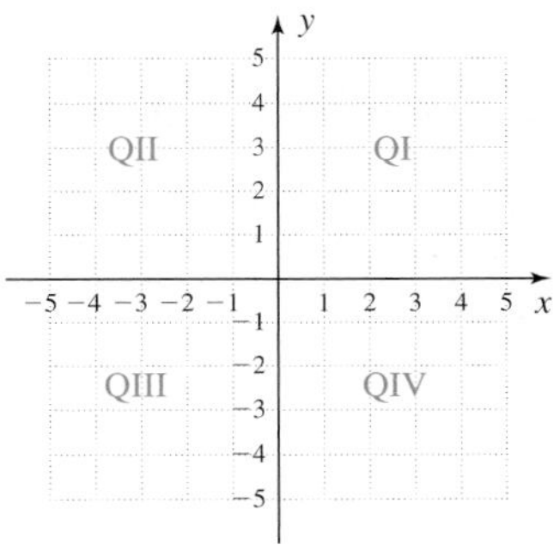

**Figure 2.3**

**Figure 2.4**

### Solutions to an Equation in Two Variables

1. If substituting $a$ for $x$ and $b$ for $y$ results in a true equation, the ordered pair $(a, b)$ is a solution and on the graph of the relation.

2. If the ordered pair $(a, b)$ is on the graph of a relation, it is a solution (substituting $a$ for $x$ and $b$ for $y$ will result in a true equation).

We generally use only a few select points to determine the shape of a graph, then draw a straight line or smooth curve through these points, as indicated by any patterns formed.

**EXAMPLE 2** ▶ **Graphing Relations**

Graph the relations $y = x - 1$ and $x = |y|$ using the ordered pairs given in Tables 2.1 and 2.2.

**Solution** ▶ For $y = x - 1$, we plot the points then connect them with a straight line (Figure 2.5). For $x = |y|$, the plotted points form a V-shaped graph made up of two half lines (Figure 2.6).

**Figure 2.5**

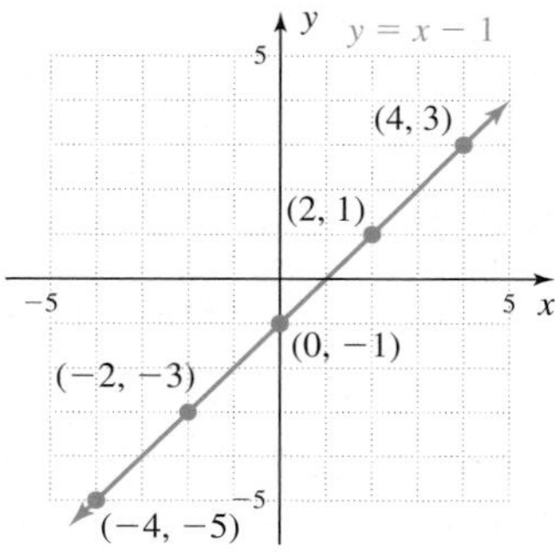

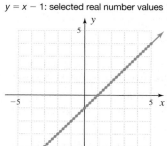

**Figure 2.6**

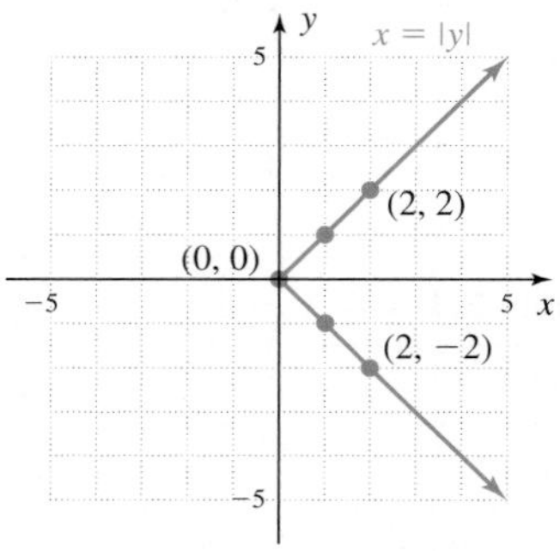

_(Figure 2.5 shows the line $y = x - 1$ with points $(4, 3)$, $(2, 1)$, $(0, -1)$, $(-2, -3)$, $(-4, -5)$.)_

_(Figure 2.6 shows $x = |y|$ with points $(0, 0)$, $(2, 2)$, $(2, -2)$.)_

**Now try Exercises 11 through 14** ▶

While we used only a few points to graph the relations in Example 2, they are actually made up of an *infinite number of ordered pairs* that satisfy each equation, including those that might be rational or irrational. This understanding is an important part of reading and interpreting graphs, and is illustrated for you in Figures 2.7 through 2.10.

**Figure 2.7**

$y = x - 1$: selected integer values

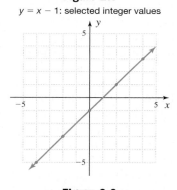

**Figure 2.8**

$y = x - 1$: selected rational values

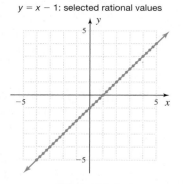

**Figure 2.9**

$y = x - 1$: selected real number values

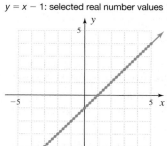

**Figure 2.10**

$y = x - 1$: all real number values

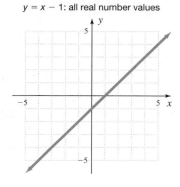

Since there are an infinite number of ordered pairs forming the graph of $y = x - 1$, the domain cannot be given in list form. Here we note $x$ can be any real number and write $D: x \in \mathbb{R}$. Likewise, $y$ can be any real number and for the range we have $R: y \in \mathbb{R}$. All of these points together make these graphs **continuous,** which for our purposes means you can draw the entire graph without lifting your pencil from the paper.

Actually, a majority of graphs cannot be drawn using only a straight line or directed line segments. In these cases, we rely on a "sufficient number" of points to outline the basic shape of the graph, then connect the points with a smooth curve. As your experience with graphing increases, this "sufficient number of points" tends to get smaller as you learn to anticipate what the graph of a given relation should look like. In particular, for the linear graph in Figure 2.5 we notice that both the $x$- and $y$-variables have an *implied exponent of 1*. This is in fact a characteristic of linear equations and graphs. In Example 3 we'll notice that if the exponent on one of the variables in the sum is 2 (either $x$ or $y$ is *squared*) while the other exponent is 1, the result is a graph called a **parabola.** If the $x$-term is squared [Example 3(a)] the parabola is oriented vertically, as in Figure 2.11, and its highest or lowest point is called the **vertex.** If the $y$-term is squared [Example 3(c)], the parabola is oriented horizontally, as in Figure 2.13 and the leftmost or rightmost point is the vertex. The graphs and equations of other relations likewise have certain identifying characteristics. **See Exercises 81 through 88.**

**EXAMPLE 3** ▶  **Graphing Relations**

Graph the following relations by completing the tables given. Then use the graph to state the domain and range of the relation.

    **a.** $y = x^2 - 2x$     **b.** $y = \sqrt{9 - x^2}$     **c.** $x = y^2$

**Solution** ▶  For each relation, we use each $x$-input in turn to determine the related $y$-output(s), if they exist. Results can be entered in a table and the ordered pairs used to assist in drawing a complete graph.

**a.**      $y = x^2 - 2x$

| $x$ | $y$ | $(x, y)$ Ordered Pairs |
|:---:|:---:|:---:|
| $-4$ | 24 | $(-4, 24)$ |
| $-3$ | 15 | $(-3, 15)$ |
| $-2$ | 8 | $(-2, 8)$ |
| $-1$ | 3 | $(-1, 3)$ |
| 0 | 0 | $(0, 0)$ |
| 1 | $-1$ | $(1, -1)$ |
| 2 | 0 | $(2, 0)$ |
| 3 | 3 | $(3, 3)$ |
| 4 | 8 | $(4, 8)$ |

**Figure 2.11**

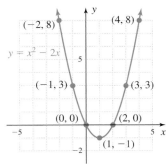

The resulting **vertical parabola** is shown in Figure 2.11. Although $(-4, 24)$ and $(-3, 15)$ were not plotted here, the arrowheads indicate an infinite extension of the graph, which will include these points. This "infinite extension" in the *upward* direction shows there is no largest $y$-value (the graph becomes infinitely "tall"). Since the smallest possible $y$-value is $-1$ [from the vertex $(1, -1)$], the range is $y \geq -1$. However, this extension also continues forever in the *outward* direction as well (the graph gets wider and wider). This means the $x$-value of all possible ordered pairs could vary from negative to positive infinity, and the domain is all real numbers. We then have $D: x \in \mathbb{R}$ and $R: y \geq -1$.

**b.**  $y = \sqrt{9 - x^2}$

| $x$ | $y$ | $(x, y)$ Ordered Pairs |
|-----|-----|------------------------|
| $-4$ | not real | — |
| $-3$ | 0 | $(-3, 0)$ |
| $-2$ | $\sqrt{5}$ | $(-2, \sqrt{5})$ |
| $-1$ | $2\sqrt{2}$ | $(-1, 2\sqrt{2})$ |
| 0 | 3 | $(0, 3)$ |
| 1 | $2\sqrt{2}$ | $(1, 2\sqrt{2})$ |
| 2 | $\sqrt{5}$ | $(2, \sqrt{5})$ |
| 3 | 0 | $(3, 0)$ |
| 4 | not real | — |

**Figure 2.12**

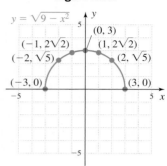

The result is the graph of a **semicircle** (Figure 2.12). The points with irrational coordinates were graphed by estimating their location. Note that when $x < -3$ or $x > 3$, the relation $y = \sqrt{9 - x^2}$ does not represent a real number and no points can be graphed. Also note that no arrowheads are used since the graph terminates at $(-3, 0)$ and $(3, 0)$. These observations and the graph itself show that for this relation, $D: -3 \leq x \leq 3$, and $R: 0 \leq x \leq 3$.

**c.** Similar to $x = |y|$, the relation $x = y^2$ is defined only for $x \geq 0$ since $y^2$ is always nonnegative ($-1 = y^2$ has no real solutions). In addition, we reason that each positive $x$-value will correspond to two $y$-values. For example, given $x = 4$, $(4, -2)$ and $(4, 2)$ are both solutions to $x = y^2$.

$x = y^2$

| $x$ | $y$ | $(x, y)$ Ordered Pairs |
|-----|-----|------------------------|
| $-2$ | not real | — |
| $-1$ | not real | — |
| 0 | 0 | $(0, 0)$ |
| 1 | $-1, 1$ | $(1, -1)$ and $(1, 1)$ |
| 2 | $-\sqrt{2}, \sqrt{2}$ | $(2, -\sqrt{2})$ and $(2, \sqrt{2})$ |
| 3 | $-\sqrt{3}, \sqrt{3}$ | $(3, -\sqrt{3})$ and $(3, \sqrt{3})$ |
| 4 | $-2, 2$ | $(4, -2)$ and $(4, 2)$ |

**Figure 2.13**

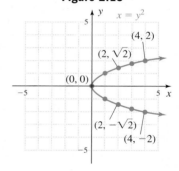

This relation is a horizontal parabola, with a vertex at $(0, 0)$ (Figure 2.13). The graph begins at $x = 0$ and extends infinitely to the right, showing the domain is $x \geq 0$. Similar to Example 3(a), this "infinite extension" also extends in both the *upward* and *downward* directions and the *y-value* of all possible ordered pairs could vary from negative to positive infinity. We then have $D: x \geq 0$ and $R: y \in \mathbb{R}$.

☑ **B.** You've just seen how we can graph relations

Now try Exercises 15 through 22 ▶

## C. The Equation and Graph of a Circle

Using the midpoint and distance formulas, we can develop the equation of another important relation, that of a circle. As the name suggests, the **midpoint of a line segment** is located halfway between the endpoints. On a standard number line, the midpoint of the line segment with endpoints 1 and 5 is 3, but more important, note that 3 is the

**average distance** between 1 and 5: $\frac{1+5}{2} = \frac{6}{2} = 3$. This observation can be extended to find the midpoint between any two points $(x_1, y_1)$ and $(x_2, y_2)$ in the $xy$-plane. We simply find the average distance between the $x$-coordinates and the average distance between the $y$-coordinates.

---

**The Midpoint Formula**

Given any line segment with endpoints $P_1 = (x_1, y_1)$ and $P_2 = (x_2, y_2)$, the midpoint $M$ is given by

$$M = \left( \frac{x_1 + x_2}{2}, \frac{y_1 + y_2}{2} \right)$$

---

The midpoint formula can be used in many different ways. Here we'll use it to find the coordinates of the center of a circle.

**EXAMPLE 4** ▶  **Using the Midpoint Formula**

The diameter of a circle has endpoints at $P_1 = (-3, -2)$ and $P_2 = (5, 4)$. Use the midpoint formula to find the coordinates of the center, then plot this point.

**Solution** ▶  Midpoint $= \left( \dfrac{x_1 + x_2}{2}, \dfrac{y_1 + y_2}{2} \right)$

$$M = \left( \frac{-3 + 5}{2}, \frac{-2 + 4}{2} \right)$$

$$M = \left( \frac{2}{2}, \frac{2}{2} \right) = (1, 1)$$

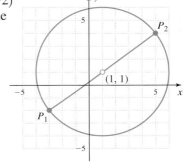

The center is at $(1, 1)$, which we graph directly on the diameter as shown.

**Now try Exercises 23 through 32** ▶

**Figure 2.14**

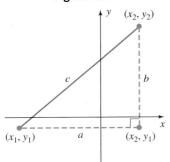

## The Distance Formula

In addition to a line segment's midpoint, we are often interested in the *length* of the segment. For any two points $(x_1, y_1)$ and $(x_2, y_2)$ not lying on a horizontal or vertical line, a right triangle can be formed as in Figure 2.14. Regardless of the triangle's orientation, side $a$ (the horizontal segment or base of the triangle) will have length $|x_2 - x_1|$ units, with side $b$ (the vertical segment or height) having length $|y_2 - y_1|$ units. From the Pythagorean theorem (Section R.4), we see that $c^2 = a^2 + b^2$ corresponds to $c^2 = (|x_2 - x_1|)^2 + (|y_2 - y_1|)^2$. By taking the square root of both sides we obtain the length of the hypotenuse, *which is identical to the distance between these two points*: $c = \sqrt{(x_2 - x_1)^2 + (y_2 - y_1)^2}$. The result is called the **distance formula,** although it's most often written using $d$ for distance, rather than $c$. Note the absolute value bars are dropped from the formula, since the square of any quantity is always nonnegative. This also means that *either* point can be used as the initial point in the computation.

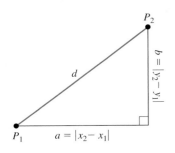

---

**The Distance Formula**

Given any two points $P_1 = (x_1, y_1)$ and $P_2 = (x_2, y_2)$, the distance $d$ between them is
$$d = \sqrt{(x_2 - x_1)^2 + (y_2 - y_1)^2}$$

**EXAMPLE 5** ▶    **Using the Distance Formula**

Use the distance formula to find the diameter of the circle from Example 4.

Solution ▶    For $(x_1, y_1) = (-3, -2)$ and $(x_2, y_2) = (5, 4)$, the distance formula gives

$$d = \sqrt{(x_2 - x_1)^2 + (y_2 - y_1)^2}$$
$$= \sqrt{[5 - (-3)]^2 + [4 - (-2)]^2}$$
$$= \sqrt{8^2 + 6^2}$$
$$= \sqrt{100} = 10$$

The diameter of the circle is 10 units long.

**Now try Exercises 33 through 42** ▶

A circle can be defined as the set of all points in a plane that are a *fixed distance* called the **radius,** from a *fixed point* called the **center.** Since the definition involves *distance,* we can construct the general equation of a circle using the distance formula. Assume the center has coordinates $(h, k)$, and let $(x, y)$ represent any point on the graph. The distance between these points is equal to the radius $r$, and the distance formula yields: $\sqrt{(x - h)^2 + (y - k)^2} = r$. Squaring both sides gives the equation of a circle in **standard form:** $(x - h)^2 + (y - k)^2 = r^2$.

### The Equation of a Circle

A circle of radius $r$ with center at $(h, k)$ has the equation

$$(x - h)^2 + (y - k)^2 = r^2$$

If $h = 0$ and $k = 0$, the circle is centered at $(0, 0)$ and the graph is a **central circle** with equation $x^2 + y^2 = r^2$. At other values for $h$ or $k$, the center is at $(h, k)$ with no change in the radius. Note that an open dot is used for the center, as it's actually a point of reference and not a part of the graph.

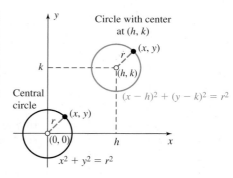

**EXAMPLE 6** ▶    **Finding the Equation of a Circle in Standard Form**

Find the equation of a circle with center $(0, -1)$ and radius 4.

Solution ▶    Since the center is at $(0, -1)$ we have $h = 0$, $k = -1$, and $r = 4$. Using the standard form $(x - h)^2 + (y - k)^2 = r^2$ we obtain

$$(x - 0)^2 + [y - (-1)]^2 = 4^2 \quad \text{substitute 0 for } h, -1 \text{ for } k, \text{ and 4 for } r$$
$$x^2 + (y + 1)^2 = 16 \quad \text{simplify}$$

The graph of $x^2 + (y + 1)^2 = 16$ is shown in the figure.

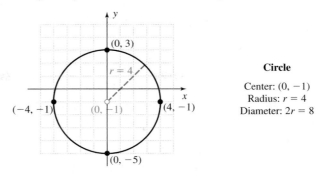

**Circle**

Center: $(0, -1)$
Radius: $r = 4$
Diameter: $2r = 8$

Now try Exercises 43 through 62 ▶

The graph of a circle can be obtained by first identifying the coordinates of the center and the length of the radius from the equation in standard form. After plotting the center point, we count a distance of $r$ units left and right of center (horizontally), and up and down from center (vertically), obtaining four points on the circle. Neatly graph a circle containing these four points.

**EXAMPLE 7** ▶    **Graphing a Circle**

Graph the circle represented by $(x - 2)^2 + (y + 3)^2 = 12$. Clearly label the center and radius.

**Solution** ▶    Comparing the given equation with the standard form, we find the center is at $(2, -3)$ and the radius is $r = 2\sqrt{3} \approx 3.5$.

$$(x - h)^2 + (y - k)^2 = r^2 \qquad \text{standard form}$$
$$(x - 2)^2 + (y + 3)^2 = 12 \qquad \text{given equation}$$
$$x - h = x - 2 \qquad y - k = y + 3 \qquad r^2 = 12 \qquad \text{corresponding parts}$$
$$h = 2 \qquad\qquad k = -3 \qquad\qquad r = \sqrt{12}$$
$$r \approx 3.5$$

Plot the center $(2, -3)$ and count approximately 3.5 units in the horizontal and vertical directions. Complete the circle by freehand drawing or using a compass. The graph shown is obtained.

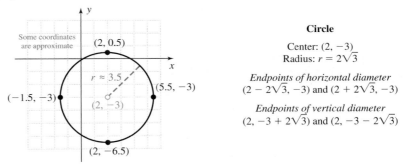

**Circle**

Center: $(2, -3)$
Radius: $r = 2\sqrt{3}$

*Endpoints of horizontal diameter*
$(2 - 2\sqrt{3}, -3)$ and $(2 + 2\sqrt{3}, -3)$

*Endpoints of vertical diameter*
$(2, -3 + 2\sqrt{3})$ and $(2, -3 - 2\sqrt{3})$

Now try Exercises 63 through 68 ▶

**WORTHY OF NOTE**

After writing the equation in standard form, it is possible to end up with a constant that is zero or negative. In the first case, the graph is a single point. In the second case, no graph is possible since roots of the equation will be complex numbers. These are called *degenerate cases*. **See Exercise 100.**

In Example 7, note the equation is composed of binomial squares in both $x$ and $y$. By expanding the binomials and collecting like terms, we can write the equation of the circle in **general form**:

$$(x - 2)^2 + (y + 3)^2 = 12 \quad \text{standard form}$$
$$x^2 - 4x + 4 + y^2 + 6y + 9 = 12 \quad \text{expand binomials}$$
$$x^2 + y^2 - 4x + 6y + 1 = 0 \quad \text{combine like terms—general form}$$

For future reference, observe the general form contains a *sum* of second-degree terms in $x$ and $y$, and that *both terms have the same coefficient* (in this case, "1").

Since this form of the equation was derived by squaring binomials, it seems reasonable to assume we can go back to the standard form by creating binomial squares in $x$ and $y$. This is accomplished by *completing the square*.

**EXAMPLE 8** ▶ **Finding the Center and Radius of a Circle**

Find the center and radius of the circle with equation $x^2 + y^2 + 2x - 4y - 4 = 0$. Then sketch its graph and label the center and radius.

**Solution** ▶ To find the center and radius, we complete the square in both $x$ and $y$.

$$x^2 + y^2 + 2x - 4y - 4 = 0 \qquad \text{given equation}$$
$$(x^2 + 2x + \underline{\phantom{1}}) + (y^2 - 4y + \underline{\phantom{1}}) = 4 \qquad \text{group } x\text{-terms and } y\text{-terms; add 4}$$
$$(x^2 + 2x + 1) + (y^2 - 4y + 4) = 4 + 1 + 4 \qquad \text{complete each binomial square}$$
$$\underbrace{\phantom{xxxx}}_{\text{adds 1 to left side}} \qquad \underbrace{\phantom{xxxx}}_{\text{adds 4 to left side}} \qquad \underbrace{\phantom{xxxx}}_{\text{add } 1 + 4 \text{ to right side}}$$
$$(x + 1)^2 + (y - 2)^2 = 9 \qquad \text{factor and simplify}$$

The center is at $(-1, 2)$ and the radius is $r = \sqrt{9} = 3$.

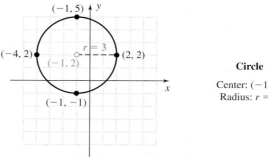

**Circle**

Center: $(-1, 2)$
Radius: $r = 3$

Now try Exercises 69 through 80 ▶

**EXAMPLE 9** ▶ **Applying the Equation of a Circle**

To aid in a study of nocturnal animals, some naturalists install a motion detector near a popular watering hole. The device has a range of 10 m in any direction. Assume the water hole has coordinates $(0, 0)$ and the device is placed at $(2, -1)$.

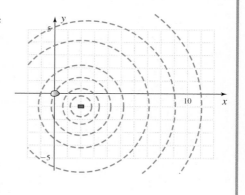

**a.** Write the equation of the circle that models the maximum effective range of the device.

**b.** Use the distance formula to determine if the device will detect a badger that is approaching the water and is now at coordinates $(11, -5)$.

**Solution** ▶    **a.** Since the device is at $(2, -1)$ and the radius (or reach) of detection is 10 m, any movement in the interior of the circle defined by $(x - 2)^2 + (y + 1)^2 = 10^2$ will be detected.

**b.** Using the points $(2, -1)$ and $(11, -5)$ in the distance formula yields:

$$d = \sqrt{(x_2 - x_1)^2 + (y_2 - y_1)^2} \quad \text{distance formula}$$
$$= \sqrt{(11 - 2)^2 + [-5 - (-1)]^2} \quad \text{substitute given values}$$
$$= \sqrt{9^2 + (-4)^2} \quad \text{simplify}$$
$$= \sqrt{81 + 16} \quad \text{compute squares}$$
$$= \sqrt{97} \approx 9.85 \quad \text{result}$$

Since $9.85 < 10$, the badger is within range of the device and will be detected.

 **C.** You've just seen how we can develop the equation and graph of a circle using the distance and midpoint formulas

**Now try Exercises 91 through 98** ▶

## 2.1 EXERCISES

▶ **CONCEPTS AND VOCABULARY**

**Fill in each blank with the appropriate word or phrase. Carefully reread the section, if necessary.**

**1.** If a relation is defined by a set of ordered pairs, the domain is the set of all _____ coordinates, the range is the set of all _____ coordinates.

**2.** A circle is defined as the set of all points that are an equal distance, called the _____, from a given point, called the _____.

**3.** For the equation $y = x + 5$ and the ordered pair $(x, y)$, $x$ is referred to as the input or _____ variable, while $y$ is called the _____ or dependent variable.

**4.** Discuss/Explain how to find the center and radius of the circle defined by the equation $x^2 + y^2 - 6x = 7$. How would this circle differ from the one defined by $x^2 + y^2 - 6y = 7$?

▶ **DEVELOPING YOUR SKILLS**

**Represent each relation in mapping notation, then state the domain and range.**

**5.**
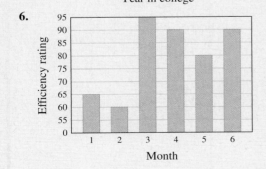

**State the domain and range of each pointwise-defined relation.**

**7.** $\{(1, 2), (3, 4), (5, 6), (7, 8), (9, 10)\}$

**8.** $\{(-2, 4), (-3, -5), (-1, 3), (4, -5), (2, -3)\}$

**9.** $\{(4, 0), (-1, 5), (2, 4), (4, 2), (-3, 3)\}$

**10.** $\{(-1, 1), (0, 4), (2, -5), (-3, 4), (2, 3)\}$

**Complete each table using the given equation. For some exercises, each input may correspond to two outputs. Use these points to graph the relation and state its domain and range.**

**11.** $y = -\dfrac{2}{3}x + 1$

| $x$ | $y$ |
|-----|-----|
| $-6$ | |
| $-3$ | |
| $0$ | |
| $3$ | |
| $6$ | |
| $8$ | |

**12.** $y = -\dfrac{5}{4}x + 3$

| $x$ | $y$ |
|-----|-----|
| $-8$ | |
| $-4$ | |
| $0$ | |
| $4$ | |
| $8$ | |
| $10$ | |

**6.**

**13.** $x + 2 = |y|$

| x | y |
|----|----|
| −2 | |
| 0 | |
| 1 | |
| 3 | |
| 6 | |
| 7 | |

**14.** $|y + 1| = x$

| x | y |
|----|----|
| 0 | |
| 1 | |
| 3 | |
| 5 | |
| 6 | |
| 7 | |

**15.** $y = x^2 - 1$

| x | y |
|----|----|
| −3 | |
| −2 | |
| 0 | |
| 2 | |
| 3 | |
| 4 | |

**16.** $y = -x^2 + 3$

| x | y |
|----|----|
| −2 | |
| −1 | |
| 0 | |
| 1 | |
| 2 | |
| 3 | |

**17.** $y = \sqrt{25 - x^2}$

| x | y |
|----|----|
| −4 | |
| −3 | |
| 0 | |
| 2 | |
| 3 | |
| 4 | |

**18.** $y = \sqrt{169 - x^2}$

| x | y |
|----|----|
| −12 | |
| −5 | |
| 0 | |
| 3 | |
| 5 | |
| 12 | |

**19.** $x - 1 = y^2$

| x | y |
|----|----|
| 10 | |
| 5 | |
| 4 | |
| 2 | |
| 1.25 | |
| 1 | |

**20.** $y^2 - 2 = x$

| x | y |
|----|----|
| −2 | |
| −1 | |
| 0 | |
| 1 | |
| 2 | |
| 7 | |

**21.** $y = \sqrt[3]{x + 1}$

| x | y |
|----|----|
| −9 | |
| −2 | |
| −1 | |
| 0 | |
| 4 | |
| 7 | |

**22.** $y = (x - 1)^3$

| x | y |
|----|----|
| −2 | |
| −1 | |
| 0 | |
| 1 | |
| 2 | |
| 3 | |

**Find the midpoint of each segment with the given endpoints.**

**23.** $(1, 8), (5, -6)$

**24.** $(5, 6), (6, -8)$

**25.** $(-4.5, 9.2), (3.1, -9.8)$

**26.** $(5.2, 7.1), (6.3, -7.1)$

**27.** $\left(\dfrac{1}{5}, -\dfrac{2}{3}\right), \left(-\dfrac{1}{10}, \dfrac{3}{4}\right)$

**28.** $\left(-\dfrac{3}{4}, -\dfrac{1}{3}\right), \left(\dfrac{3}{8}, \dfrac{5}{6}\right)$

**Find the midpoint of each segment.**

**29.**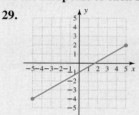

**30.**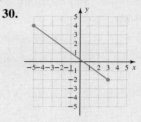

**Find the center of each circle with the diameter shown.**

**31.**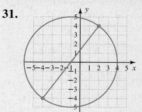

**32.**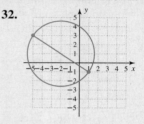

**33.** Use the distance formula to find the length of the line segment in Exercise 29.

**34.** Use the distance formula to find the length of the line segment in Exercise 30.

**35.** Use the distance formula to find the length of the diameter of the circle in Exercise 31.

**36.** Use the distance formula to find the length of the diameter of the circle in Exercise 32.

**In Exercises 37 through 42, three points that form the vertices of a triangle are given. Use the distance formula to determine if any of the triangles are right triangles (the three sides satisfy the Pythagorean theorem $a^2 + b^2 = c^2$).**

**37.** $(-3, 7), (2, 2), (5, 5)$

**38.** $(7, 0), (-1, 0), (7, 4)$

**39.** $(-4, 3), (-7, -1), (3, -2)$

**40.** $(5, 2), (0, -3), (4, -4)$

**41.** $(-3, 2), (-1, 5), (-6, 4)$

**42.** $(0, 0), (-5, 2), (2, -5)$

**Find the equation of a circle satisfying the conditions given, then sketch its graph.**

43. center $(0, 0)$, radius 3

44. center $(0, 0)$, radius 6

45. center $(5, 0)$, radius $\sqrt{3}$

46. center $(0, 4)$, radius $\sqrt{5}$

47. center $(4, -3)$, radius 2

48. center $(3, -8)$, radius 9

49. center $(-7, -4)$, radius $\sqrt{7}$

50. center $(-2, -5)$, radius $\sqrt{6}$

51. center $(1, -2)$, diameter 6

52. center $(-2, 3)$, diameter 10

53. center $(4, 5)$, diameter $4\sqrt{3}$

54. center $(5, 1)$, diameter $4\sqrt{5}$

55. center at $(7, 1)$, graph contains the point $(1, -7)$

56. center at $(-8, 3)$, graph contains the point $(-3, 15)$

57. center at $(3, 4)$, graph contains the point $(7, 9)$

58. center at $(-5, 2)$, graph contains the point $(-1, 3)$

59. diameter has endpoints $(5, 1)$ and $(5, 7)$

60. diameter has endpoints $(2, 3)$ and $(8, 3)$

61. diameter has endpoints $(-3, 4)$ and $(4, -3)$

62. diameter has endpoints $(5.4, 7.2)$ and $(-5.4, 7.2)$

**Identify the center and radius of each circle, then graph. Also state the domain and range of the relation.**

63. $(x - 2)^2 + (y - 3)^2 = 4$

64. $(x - 5)^2 + (y - 1)^2 = 9$

65. $(x + 1)^2 + (y - 2)^2 = 12$

66. $(x - 7)^2 + (y + 4)^2 = 20$

67. $(x + 4)^2 + y^2 = 81$

68. $x^2 + (y - 3)^2 = 49$

**Write each equation in standard form to find the center and radius of the circle. Then sketch the graph.**

69. $x^2 + y^2 - 10x - 12y + 4 = 0$

70. $x^2 + y^2 + 6x - 8y - 6 = 0$

71. $x^2 + y^2 - 10x + 4y + 4 = 0$

72. $x^2 + y^2 + 6x + 4y + 12 = 0$

73. $x^2 + y^2 + 6y - 5 = 0$

74. $x^2 + y^2 - 8x + 12 = 0$

75. $x^2 + y^2 + 4x + 10y + 18 = 0$

76. $x^2 + y^2 - 8x - 14y - 47 = 0$

77. $x^2 + y^2 + 14x + 12 = 0$

78. $x^2 + y^2 - 22y - 5 = 0$

79. $2x^2 + 2y^2 - 12x + 20y + 4 = 0$

80. $3x^2 + 3y^2 - 24x + 18y + 3 = 0$

**Match each of the eight graphs given with their corresponding equation (two equations given have no matching graph).**

**a.** $y = x^2 - 6x$          **b.** $x^2 + (y - 3)^2 = 36$

**c.** $x^2 + y = 9$          **d.** $3x - 4y = 12$

**e.** $y = \dfrac{-3}{2}x + 4$          **f.** $(x - 1)^2 + (y + 2)^2 = 49$

**g.** $(x - 3)^2 + y^2 = 16$          **h.** $(x - 1)^2 + (y + 2)^2 = 9$

**i.** $4x - 3y = 12$          **j.** $6x + y = x^2 + 9$

81.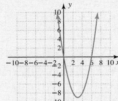

82.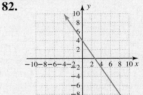

83.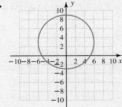

84.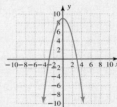

85.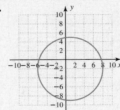

86.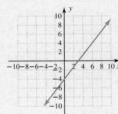

87.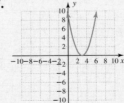

88.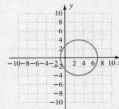

## ▶ WORKING WITH FORMULAS

**89. Pick's theorem:** $A = \frac{1}{2}B + I - 1$

Pick's theorem is an interesting yet little known formula for computing the area of a polygon drawn in the Cartesian coordinate system. The formula can be applied as long as the vertices of the polygon are lattice points (both $x$ and $y$ are integers). If $B$ represents the number of lattice points lying directly on the boundary of the polygon (including the vertices), and $I$ represents the number of points in the interior, the area of the polygon is given by the formula shown. Use some graph paper to carefully draw a triangle with vertices at $(-3, 1)$, $(3, 9)$, and $(7, 6)$, then use Pick's theorem to compute the triangle's area.

**90. Radius of a circumscribed circle:** $r = \sqrt{\dfrac{A}{2}}$

The radius $r$ of a circle circumscribed around a square is found by using the formula given, where $A$ is the area of the square. (a) Solve the formula for $A$ and use the result to find the area of the square shown. (b) Graph the relation from part (a).

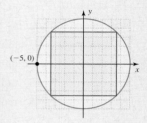

## ▶ APPLICATIONS

**91. Radar detection:** A luxury liner is located at map coordinates $(5, 12)$ and has a radar system with a range of 25 nautical miles in any direction. (a) Write the equation of the circle that models the range of the ship's radar, and (b) use the distance formula to determine if the radar can pick up the liner's sister ship located at coordinates $(15, 36)$.

**92. Earthquake range:** The epicenter (point of origin) of a large earthquake was located at map coordinates $(3, 7)$, with the quake being felt up to 12 mi away. (a) Write the equation of the circle that models the range of the earthquake's effect. (b) Use the distance formula to determine if a person living at coordinates $(13, 1)$ would have felt the quake.

**93. Inscribed circle:** Find the equation for both the red and blue circles, then find the area of the region shaded in blue.

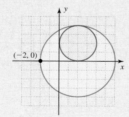

**94. Inscribed triangle:** The area of an equilateral triangle inscribed in a circle is given by the formula $A = \dfrac{3\sqrt{3}}{4}r^2$, where $r$ is the radius of the circle. Find the area of the equilateral triangle shown.

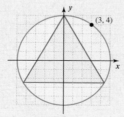

**95. Radio broadcast range:** Two radio stations may not use the same frequency if their broadcast areas *overlap*. Suppose station KXRQ has a broadcast area bounded by $x^2 + y^2 + 8x - 6y = 0$ and WLRT has a broadcast area bounded by $x^2 + y^2 - 10x + 4y = 0$. Graph the circle representing each broadcast area on the same grid to determine if both stations may broadcast on the same frequency.

**96. Emergency broadcasts:** The emergency broadcast system is designed to relay an emergency signal to all points of the country. A signal is sent from station ESMB with broadcast area bounded by $x^2 + y^2 = 2500$ ($x$ and $y$ in miles). The signal is relayed by a transmitter with range $(x - 20)^2 + (y - 30)^2 = 900$. Graph each circle on the same grid to determine the greatest distance from ESMB that this signal can be received.

**97. Nocturnal animals:** To alert naturalists to the presence of nocturnal animals, a motion detector is installed near a watering hole. With the water hole at center, the detector is placed at coordinates $(3, 5)$ with $x$ and $y$ in meters. If the detector has a range of 62 m, (a) find an equation modeling the detector's range. Suppose a raccoon and an opossum are approaching the watering hole and are at coordinates $(14, 65)$ and $(36, -51)$ respectively. (b) How far is each animal from the detector? (c) Which animal is detected first?

**98. Produce protection:** To protect his prized apples from any who would pluck them from the tree without permission, Mr. Grump installs a motion-activated alarm near the base of the tree. With the tree at the center, the motion detector is placed at coordinates $(2, -3)$ with $x$ and $y$ in feet. If the detector has a range of 32 ft, (a) find an equation modeling the detector's range. Suppose a neighborhood kid and a hungry passerby are approaching the tree and are at coordinates $(37, 9)$ and $(-19, 17)$ respectively. (b) How far is each person from the detector? (c) Which person sets off the alarm?

## ▶ EXTENDING THE CONCEPT

**99.** Although we use the word "domain" extensively in mathematics, it is also commonly seen in literature and heard in everyday conversation. Using a college-level dictionary, look up and write out the various meanings of the word, noting how closely the definitions given are related to its mathematical use.

**100.** When completing the square to find the center and radius of a circle, we sometimes encounter a value for $r^2$ that is negative or zero. These are called **degenerate cases.** If $r^2 < 0$, no circle is possible, while if $r^2 = 0$, the "graph" of the circle is simply the point $(h, k)$. Find the center and radius of the following circles (if possible).
   **a.** $x^2 + y^2 - 12x + 4y + 40 = 0$
   **b.** $x^2 + y^2 - 2x - 8y - 8 = 0$
   **c.** $x^2 + y^2 - 6x - 10y + 35 = 0$

**101.** Given the length of the diagonal of a square is equal to $\sqrt{2}$ times the length of one side, derive the formula from Exercise 90.

**102.** The area of a circular sector is given by $A = \frac{n}{360}\pi r^2$, where $n$ represents the degree measure of the sector and $r$ is the radius of each circle. Use this fact with your knowledge of triangles and circles to verify the region shown in purple has an area of $A = \frac{r^2}{6}(4\pi - 3\sqrt{3})$. Assume the circles have equal radii and each goes through the center of the other.

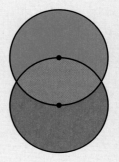

## ▶ MAINTAINING YOUR SKILLS

**103.** (R.3, R.4) Simplify the following expressions.
   **a.** $3^3 + 3^2 + 3^1 + 3^0 + 3^{-1}$
   **b.** $\dfrac{x^2 x^5}{x^3}$
   **c.** $125^{-\frac{1}{3}}$
   **d.** $27^{\frac{2}{3}}$
   **e.** $(2m^3 n)^2$
   **f.** $(5x)^0 + 5x^0$

**104.** (1.1) Solve the following equation.
$$\frac{x}{3} + \frac{1}{4} = \frac{5}{6}$$

**105.** (1.5) Solve $x^2 - 27 = 6x$ by factoring.

**106.** (1.6) Solve $1 - \sqrt{n + 3} = -n$ and check solutions by substitution. If a solution is extraneous, so state.

---

## 2.2  Linear Graphs and Rates of Change

**LEARNING OBJECTIVES**

*In Section 2.2 you will see how we can:*

☐ **A.** Graph linear equations using the intercept method

☐ **B.** Find the slope of a line and interpret it as a rate of change

☐ **C.** Graph horizontal and vertical lines

☐ **D.** Identify parallel and perpendicular lines

☐ **E.** Apply linear equations in context

In preparation for sketching graphs of other equations, we'll first look more closely at the characteristics of linear graphs. While linear graphs are fairly simple models, they have many substantive and meaningful applications. For instance, most of us are aware that satellite and cable TV have been increasing in popularity since they were first introduced. A closer look at consumer spending on these forms of entertainment (Figure 2.15) reveals an increase from $192 per person per year in 2003 to $281 in 2007 (also see page 148 and Exercise 95 on page 200). From an investor's or a producer's point of view, there is a very high interest in the questions, "How fast are sales increasing? Can this relationship be modeled mathematically to help predict sales in future years?" Answers to these and other questions are precisely what our study in this section is all about.

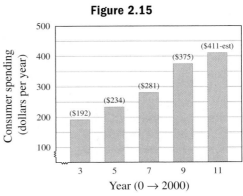

**Figure 2.15**

*Source: 2009 Statistical Abstract of the United States, Table 1089 (some figures are estimates)*

## A. The Graph of a Linear Equation

In general, a linear equation can be identified using these three tests:

1. the exponent on any variable is one,
2. no variable occurs in a denominator, and
3. no two variables are multiplied together.

The equation $3y = 9$ is a linear equation in one variable, while $2x + 3y = 12$ and $y = -\frac{2}{3}x + 4$ are linear equations in two variables. This leads us to the following definition:

### Linear Equations

A linear equation is one that can be written in the form

$$ax + by = c$$

where $a$, $b$, and $c$ are real numbers, with $a$ and $b$ not simultaneously equal to zero.

As in Section 2.1, the most basic method for graphing a line is to simply plot a few points, then draw a straight line through the points.

---

**EXAMPLE 1** ▶

**Solution** ▶

### WORTHY OF NOTE

If you cannot draw a straight line through the plotted points, a computational error has been made. All points satisfying a linear equation *lie on a straight line*.

### Graphing a Linear Equation in Two Variables

Graph the equation $3x + 2y = 4$ by plotting points.

Selecting $x = -2$, $x = 0$, $x = 1$, and $x = 4$ as inputs, we compute the related outputs and enter the ordered pairs in a table. The result is

| $x$ input | $y$ output | $(x, y)$ ordered pairs |
|:---:|:---:|:---:|
| $-2$ | $5$ | $(-2, 5)$ |
| $0$ | $2$ | $(0, 2)$ |
| $1$ | $\frac{1}{2}$ | $(1, \frac{1}{2})$ |
| $4$ | $-4$ | $(4, -4)$ |

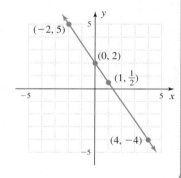

**Now try Exercises 5 through 10** ▶

---

Notice that the line in Example 1 crosses the $y$-axis at $(0, 2)$, and this point is called the **$y$-intercept** of the line. In general, $y$-intercepts have the form $(0, y)$. Although difficult to see graphically, substituting 0 for $y$ and solving for $x$ shows this line crosses the $x$-axis at $(\frac{4}{3}, 0)$ and this point is called the **$x$-intercept.** In general, $x$-intercepts have the form $(x, 0)$. The $x$- and $y$-intercepts are usually easier to calculate than other points (since $y = 0$ or $x = 0$ respectively) and we often graph linear equations using only these two points. This is called the **intercept method** for graphing linear equations.

### The Intercept Method

1. Substitute 0 for $x$ and solve for $y$. This will give the $y$-intercept $(0, y)$.
2. Substitute 0 for $y$ and solve for $x$. This will give the $x$-intercept $(x, 0)$.
3. Plot the intercepts and use them to graph a straight line.

**EXAMPLE 2** ▶ **Graphing Lines Using the Intercept Method**

Graph $3x + 2y = 9$ using the intercept method.

**Solution** ▶ Substitute 0 for $x$ ($y$-intercept):

$$3(0) + 2y = 9$$
$$2y = 9$$
$$y = \frac{9}{2}$$
$$\left(0, \frac{9}{2}\right)$$

Substitute 0 for $y$ ($x$-intercept):

$$3x + 2(0) = 9$$
$$3x = 9$$
$$x = 3$$
$$(3, 0)$$

Using a ruler or straightedge, drawing a straight line through these points results in the graph shown.

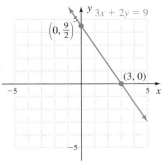

☑ **A.** You've just seen how we can graph linear equations using the intercept method

Now try Exercises 11 through 30 ▶

## B. The Slope of a Line and Rates of Change

After the $x$- and $y$-intercepts, we next consider the **slope of a line.** We see applications of this concept in many diverse areas, including the *grade* of a highway (trucking), the *pitch* of a roof (carpentry), the *climb* of an airplane (flying), the *drainage* of a field (landscaping), and the *slope* of a mountain (parks and recreation). While the general concept is an intuitive one, we seek to quantify the concept (assign it a numeric value) for purposes of comparison and decision making. In each of the preceding examples (grade, pitch, climb, etc.), slope is a measure of "steepness," as

**Figure 2.16**

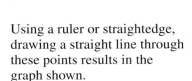

defined by the ratio $\frac{\text{vertical change}}{\text{horizontal change}}$. Using a line segment through arbitrary points $P_1 = (x_1, y_1)$ and $P_2 = (x_2, y_2)$, we can create the right triangle shown in Figure 2.16 to help us quantify this relationship. The figure illustrates that the **vertical change** or the **change in $y$** (also called the **rise**) is simply the difference in $y$-coordinates: $y_2 - y_1$. The **horizontal change** or **change in $x$** (also called the **run**) is the difference in $x$-coordinates: $x_2 - x_1$. In algebra, we typically use the letter "$m$" to represent slope, giving $m = \frac{y_2 - y_1}{x_2 - x_1}$ as the $\frac{\text{change in } y}{\text{change in } x}$. The result is called the **slope formula.**

**WORTHY OF NOTE**

While the original reason that "$m$" was chosen for slope is uncertain, some have speculated that it was because in French, the verb for "to climb" is *monter*. Others say it could be due to the "modulus of slope," the word *modulus* meaning a numeric measure of a given property, in this case the inclination of a line.

**The Slope Formula**

Given two points $P_1 = (x_1, y_1)$ and $P_2 = (x_2, y_2)$, the slope of the nonvertical line through $P_1$ and $P_2$ is

$$m = \frac{y_2 - y_1}{x_2 - x_1}$$

where $x_2 \neq x_1$.

Actually, the slope value does much more than quantify the slope of a line, it expresses a **rate of change** between the quantities measured along each axis. In applications of slope, the ratio $\frac{\text{change in } y}{\text{change in } x}$ is symbolized as $\frac{\Delta y}{\Delta x}$. The symbol $\Delta$ is the Greek letter **delta** and has come to represent a change in some quantity, and the notation $m = \frac{\Delta y}{\Delta x}$ is read, "slope is equal to the *change in y* over the *change in x*." Interpreting slope as a rate of change has many significant applications in college algebra and beyond.

---

**EXAMPLE 3** ▶ **Using the Slope Formula**

Find the slope of the line through the given points, then use $m = \frac{\Delta y}{\Delta x}$ to find an additional point on the line.

   **a.** $(-2, 6)$ and $(4, 2)$      **b.** $(2, 1)$ and $(8, 4)$

**Solution** ▶   **a.** For $P_1 = (-2, 6)$ and $P_2 = (4, 2)$,    **b.** For $P_1 = (2, 1)$ and $P_2 = (8, 4)$,

$$m = \frac{y_2 - y_1}{x_2 - x_1} \qquad\qquad m = \frac{y_2 - y_1}{x_2 - x_1}$$

$$= \frac{2 - 6}{4 - (-2)} \qquad\qquad = \frac{4 - 1}{8 - 2}$$

$$= \frac{-4}{6} = \frac{-2}{3} \qquad\qquad = \frac{3}{6} = \frac{1}{2}$$

The slope of this line is $\frac{-2}{3}$. Using $\frac{\Delta y}{\Delta x} = \frac{-2}{3}$, we note that $y$ decreases 2 units ($y$ is negative), as $x$ increases 3 units. Since $(4, 2)$ is known to be on the line, the point $(4 + 3, 2 - 2) = (7, 0)$ must also be on the line (see figure).

The slope of this line is $\frac{1}{2}$. Using $\frac{\Delta y}{\Delta x} = \frac{1}{2}$, we note that $y$ increases 1 unit ($y$ is positive), as $x$ increases 2 units. Since $(8, 4)$ is known to be on the line, the point $(8 + 2, 4 + 1) = (10, 5)$ must also be on the line.

**Now try Exercises 31 through 40** ▶

---

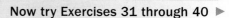

**CAUTION** ▶ When using the slope formula, try to avoid these common errors.

  **1.** The order that the $x$- and $y$-coordinates are subtracted must be consistent, since $\frac{y_2 - y_1}{x_2 - x_1} \neq \frac{y_2 - y_1}{x_1 - x_2}$.

  **2.** The vertical change (involving the $y$-values) always occurs in the numerator: $\frac{y_2 - y_1}{x_2 - x_1} \neq \frac{x_2 - x_1}{y_2 - y_1}$.

  **3.** When $x_1$ or $y_1$ is negative, use parentheses when substituting into the formula to prevent confusing the negative sign with the subtraction operation.

---

**EXAMPLE 4** ▶ **Interpreting the Slope Formula as a Rate of Change**

Jimmy works on the assembly line for an auto parts remanufacturing company. By 9:00 A.M. his group has assembled 29 carburetors. By 12:00 noon, they have completed 87 carburetors. Assuming the relationship is linear, find the slope of the line and discuss its meaning in this context.

**Solution** ▶

First write the information as ordered pairs using $c$ to represent the carburetors assembled and $t$ to represent time. This gives $(t_1, c_1) = (9, 29)$ and $(t_2, c_2) = (12, 87)$. The slope formula then gives:

$$\frac{\Delta c}{\Delta t} = \frac{c_2 - c_1}{t_2 - t_1} = \frac{87 - 29}{12 - 9}$$

$$= \frac{58}{3} \text{ or } 19.\overline{3}$$

Here the slope ratio measures $\frac{\text{carburetors assembled}}{\text{hours}}$, and we see that Jimmy's group can assemble 58 carburetors every 3 hr, or about 19 carburetors per hour.

**Now try Exercises 41 through 44** ▶

## Positive and Negative Slope

If you've ever traveled by air, you've likely heard the announcement, "Ladies and gentlemen, please return to your seats and fasten your seat belts as we begin our descent." For a time, the descent of the airplane follows a linear path, but the *slope of the line is negative* since the altitude of the plane is decreasing. Positive and negative slopes, as well as the rate of change they represent, are important characteristics of linear graphs. If $m < 0$ as in Example 3(a), the slope of the line is negative and the line slopes downward as you move left to right since $y$-values are decreasing. In Example 3(b), the slope was a positive number ($m > 0$) and the line will slope upward from left to right since the $y$-values are increasing.

$m > 0$, positive slope
$y$-values *increase* from left to right

$m < 0$, negative slope
$y$-values *decrease* from left to right

**EXAMPLE 5** ▶ **Applying Slope as a Rate of Change in Altitude**

At a horizontal distance of 10 mi from take-off, an airline pilot receives instructions to decrease altitude from their current level of 20,000 ft. A short time later, they are 17.5 mi from the airport at an altitude of 10,000 ft. Find the slope ratio for the descent of the plane and discuss its meaning in this context. Recall that 1 mi = 5280 ft.

**Solution** ▶

Let $a$ represent the altitude of the plane and $d$ its horizontal distance from the airport. Converting all measures to feet, we have $(d_1, a_1) = (52{,}800, 20{,}000)$ and $(d_2, a_2) = (92{,}400, 10{,}000)$, giving

$$\frac{\Delta a}{\Delta d} = \frac{a_2 - a_1}{d_2 - d_1} = \frac{10{,}000 - 20{,}000}{92{,}400 - 52{,}800}$$

$$= \frac{-10{,}000}{39{,}600} = \frac{-25}{99}$$

☑ **B.** You've just seen how we can find the slope of a line and interpret it as a rate of change

Since this slope ratio measures $\frac{\Delta \text{altitude}}{\Delta \text{distance}}$, we note the plane is decreasing 25 ft in altitude for every 99 ft it travels horizontally.

**Now try Exercises 45 through 48** ▶

## C. Horizontal Lines and Vertical Lines

Horizontal and vertical lines have a number of important applications, from finding the boundaries of a given graph (the domain and range), to performing certain tests on nonlinear graphs. To better understand them, consider that in *one dimension*, the graph of $x = 2$ is a single point (Figure 2.17), indicating a location on the number line 2 units from zero in the positive direction. In *two dimensions*, the equation $x = 2$ represents **all points** with an $x$-coordinate of 2. A few of these are graphed in Figure 2.18, but since there are an infinite number, we end up with a solid *vertical line* whose equation is $x = 2$ (Figure 2.19).

**Figure 2.17**

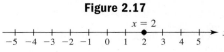

**Figure 2.18**

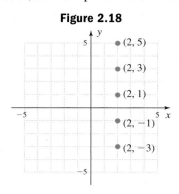

**Figure 2.19**

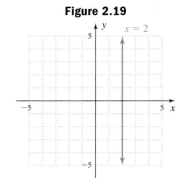

> **WORTHY OF NOTE**
>
> If we write the equation $x = 2$ in the form $ax + by = c$, the equation becomes $x + 0y = 2$, since the original equation has no $y$-variable. Notice that regardless of the value chosen for $y$, $x$ will always be 2 and we end up with the set of ordered pairs $(2, y)$, which gives us a vertical line.

The same idea can be applied to horizontal lines. In *two dimensions*, the equation $y = 4$ represents *all points* with a $y$-coordinate of positive 4, and there are an infinite number of these as well. The result is a solid horizontal line whose equation is $y = 4$. **See Exercises 49 through 58.**

| Horizontal Lines | Vertical Lines |
|---|---|
| The equation of a horizontal line is $$y = k$$ where $(0, k)$ is the $y$-intercept. | The equation of a vertical line is $$x = h$$ where $(h, 0)$ is the $x$-intercept. |

So far, the slope formula has only been applied to lines that were nonhorizontal or nonvertical. So what *is* the slope of a horizontal line? On an intuitive level, we expect that a perfectly level highway would have an incline or slope of zero. In general, for any two points on a horizontal line, $y_2 = y_1$ and $y_2 - y_1 = 0$, giving a slope of $m = \frac{0}{x_2 - x_1} = 0$. For any two points on a vertical line, $x_2 = x_1$ and $x_2 - x_1 = 0$, making the slope ratio undefined: $m = \frac{y_2 - y_1}{0}$ (see Figures 2.20 and 2.21).

**Figure 2.20**
**For any horizontal line, $y_2 = y_1$**

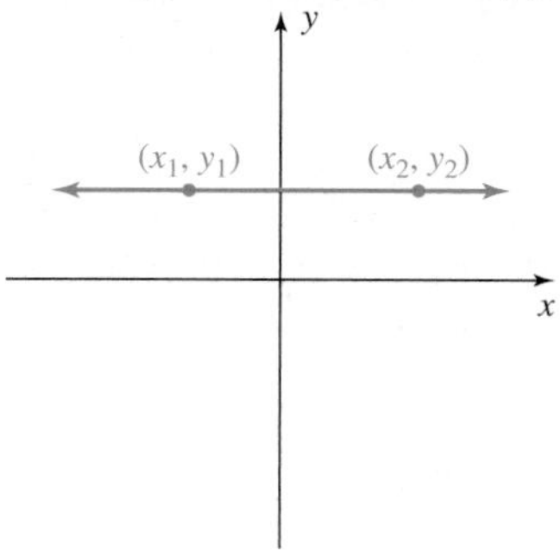

**Figure 2.21**
**For any vertical line, $x_2 = x_1$**

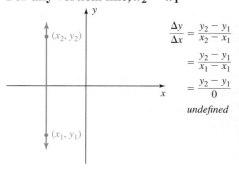

$$\frac{\Delta y}{\Delta x} = \frac{y_2 - y_1}{x_2 - x_1}$$
$$= \frac{y_1 - y_1}{x_2 - x_1}$$
$$= \frac{0}{x_2 - x_1}$$
$$= 0$$

$$\frac{\Delta y}{\Delta x} = \frac{y_2 - y_1}{x_2 - x_1}$$
$$= \frac{y_2 - y_1}{x_1 - x_1}$$
$$= \frac{y_2 - y_1}{0}$$
*undefined*

| The Slope of a Horizontal Line | The Slope of a Vertical Line |
|---|---|
| The slope of any horizontal line is zero. | The slope of any vertical line is undefined. |

**EXAMPLE 6** ▶ **Calculating Slopes**

The federal minimum wage remained constant from 1997 through 2006. However, the buying power (in 1996 dollars) of these wage earners fell each year due to inflation (see Table 2.3). This decrease in buying power is approximated by the red line shown.

**a.** Using the data or graph, find the slope of the line segment representing the minimum wage.

**b.** Select two points on the line representing buying power to approximate the slope of the line segment, and explain what it means in this context.

**Table 2.3**

| Time $t$ (years) | Minimum wage $w$ | Buying power $p$ |
|---|---|---|
| 1997 | 5.15 | 5.03 |
| 1998 | 5.15 | 4.96 |
| 1999 | 5.15 | 4.85 |
| 2000 | 5.15 | 4.69 |
| 2001 | 5.15 | 4.56 |
| 2002 | 5.15 | 4.49 |
| 2003 | 5.15 | 4.39 |
| 2004 | 5.15 | 4.28 |
| 2005 | 5.15 | 4.14 |
| 2006 | 5.15 | 4.04 |

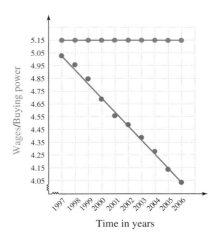

Time in years

**Solution** ▶ **a.** Since the minimum wage did not increase or decrease from 1997 to 2006, the line segment has slope $m = 0$.

**b.** The points (1997, 5.03) and (2006, 4.04) from the table appear to be on or close to the line drawn. For buying power $p$ and time $t$, the slope formula yields:

$$\frac{\Delta p}{\Delta t} = \frac{p_2 - p_1}{t_2 - t_1}$$

$$= \frac{4.04 - 5.03}{2006 - 1997}$$

$$= \frac{-0.99}{9} = \frac{-0.11}{1}$$

**WORTHY OF NOTE**

In the context of lines, try to avoid saying that a horizontal line has "no slope," since it's unclear whether a slope of zero or an undefined slope is intended.

The buying power of a minimum wage worker decreased by 11¢/yr during this time period.

☑ **C.** You've just seen how we can graph horizontal and vertical lines

Now try Exercises 59 and 60 ▶

## D. Parallel and Perpendicular Lines

Two lines in the same plane that never intersect are called **parallel lines.** When we place these lines on the coordinate grid, we find that "never intersect" is equivalent to saying "the lines have equal slopes but different $y$-intercepts." In Figure 2.22, notice the rise and run of each line is identical, and that by counting $\frac{\Delta y}{\Delta x}$ both lines have slope $m = \frac{3}{4}$.

**Figure 2.22**

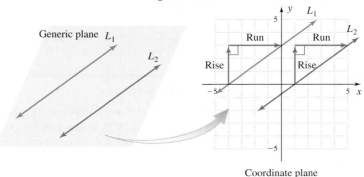

Coordinate plane

### Parallel Lines

Given $L_1$ and $L_2$ are distinct, nonvertical lines with slopes of $m_1$ and $m_2$ respectively.
1. If $m_1 = m_2$, then $L_1$ is parallel to $L_2$.
2. If $L_1$ is parallel to $L_2$, then $m_1 = m_2$.
In symbols, we write $L_1 \parallel L_2$.
*Any two vertical lines (undefined slope) are parallel.*

**EXAMPLE 7A** ▶ **Determining Whether Two Lines Are Parallel**

Teladango Park has been mapped out on a rectangular coordinate system, with a ranger station at $(0, 0)$. Brendan and Kapi are at coordinates $(-24, -18)$ and have set a direct course for the pond at $(11, 10)$. Caden and Kymani are at $(-27, 1)$ and are heading straight to the lookout tower at $(-2, 21)$. Are they hiking on parallel or nonparallel courses?

**Solution** ▶ To respond, we compute the slope of each trek across the park.

For Brendan and Kapi:    For Caden and Kymani:

$$m = \frac{y_2 - y_1}{x_2 - x_1} \qquad m = \frac{y_2 - y_1}{x_2 - x_1}$$

$$= \frac{10 - (-18)}{11 - (-24)} \qquad = \frac{21 - 1}{-2 - (-27)}$$

$$= \frac{28}{35} = \frac{4}{5} \qquad = \frac{20}{25} = \frac{4}{5}$$

Since the slopes are equal, the two groups are hiking on parallel courses.

Two lines in the same plane that intersect at right angles are called **perpendicular lines.** Using the coordinate grid, we note that *intersect at right angles* suggests that *their slopes are negative reciprocals*. While certainly not a proof, notice in Figure 2.23 the ratio $\frac{\text{rise}}{\text{run}}$ for $L_1$ is $\frac{4}{3}$ and the ratio $\frac{\text{rise}}{\text{run}}$ for $L_2$ is $\frac{-3}{4}$. Alternatively, we can say their **slopes have a product of −1,** since $m_1 \cdot m_2 = -1$ implies $m_1 = -\frac{1}{m_2}$. For a formal proof of this property, see Appendix IV.

**Figure 2.23**

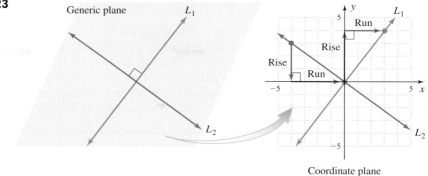

Generic plane

Coordinate plane

**Perpendicular Lines**

Given $L_1$ and $L_2$ are distinct, nonvertical lines with slopes of $m_1$ and $m_2$ respectively.
1. If $m_1 \cdot m_2 = -1$, then $L_1$ is perpendicular to $L_2$.
2. If $L_1$ is perpendicular to $L_2$, then $m_1 \cdot m_2 = -1$.
   In symbols we write $L_1 \perp L_2$.
   *Any vertical line (undefined slope) is perpendicular
   to any horizontal line (slope m = 0).*

We can easily find the slope of a line perpendicular to a second line whose slope is known or can be found—just find the reciprocal and make it negative. For a line with slope $m_1 = -\frac{3}{7}$, any line perpendicular to it will have a slope of $m_2 = \frac{7}{3}$. For $m_1 = -5$, the slope of any line perpendicular would be $m_2 = \frac{1}{5}$.

**EXAMPLE 7B** ▶ **Determining Whether Two Lines Are Perpendicular**

The three points $P_1 = (5, 1)$, $P_2 = (3, -2)$, and $P_3 = (-3, 2)$ form the vertices of a triangle. Use these points to draw the triangle, then use the slope formula to determine if they form a *right* triangle.

**Solution** ▶ For a right triangle to be formed, two of the lines through these points must be perpendicular (forming a right angle). From Figure 2.24, it *appears* a right triangle is formed, but we must *verify* that two of the sides are actually perpendicular. Using the slope formula, we have:

**Figure 2.24**

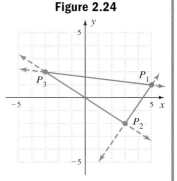

For $P_1$ and $P_2$

$$m_1 = \frac{-2 - 1}{3 - 5}$$

$$= \frac{-3}{-2} = \frac{3}{2}$$

For $P_1$ and $P_3$

$$m_2 = \frac{2 - 1}{-3 - 5}$$

$$= \frac{1}{-8}$$

For $P_2$ and $P_3$

$$m_3 = \frac{2 - (-2)}{-3 - 3}$$

$$= \frac{4}{-6} = \frac{2}{-3}$$

Since $m_1 \cdot m_3 = -1$, the triangle has a right angle and must be a right triangle.

☑ **D.** You've just seen how we can identify parallel and perpendicular lines

**Now try Exercises 61 through 72** ▶

## E. Applications of Linear Equations

The graph of a linear equation can be used to help solve many applied problems. If the numbers you're working with are either very small or very large, **scale the axes** appropriately. This can be done by letting each tick mark represent a smaller or larger unit so the data points given will fit on the grid. Also, many applications use only nonnegative values and although points with negative coordinates may be used to graph a line, only ordered pairs in QI can be meaningfully interpreted.

**EXAMPLE 8** ► **Applying a Linear Equation Model—Commission Sales**

Use the information given to create a linear equation model in two variables, then graph the line and answer the question posed:

*A salesperson gets a daily $20 meal allowance plus $7.50 for every item she sells. How many sales are needed for a daily income of $125?*

**Solution** ► Let $x$ represent sales and $y$ represent income. This gives

**verbal model:** Daily income ($y$) equals $7.50 per sale ($x$) + $20 for meals

**equation model:** $y = 7.5x + 20$

Using $x = 0$ and $x = 10$, we find (0, 20) and (10, 95) are points on this line and these are used to sketch the graph. From the graph, it appears that 14 sales are needed to generate a daily income of $125.00.

Since daily income is given as $125, we substitute 125 for $y$ and solve for $x$.

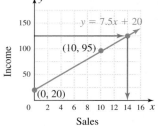

$$125 = 7.5x + 20 \quad \text{substitute 125 for } y$$
$$105 = 7.5x \quad \text{subtract 20}$$
$$14 = x \quad \text{divide by 7.5}$$

Now try Exercises 75 through 82 ►

**EXAMPLE 9** ► **Applying a Linear Equation Model—Modeling the Increase of SPAM**

The genre of e-mail known as SPAM is more formally known in computing circles as UBE: **unsolicited bulk e-mail.** While SPAM itself is not an acronym, many will say it stands for Stupid, Pointless, Annoying Messages. By any name, the use of SPAM is on the rise. The owner of a small company tracked the receipt of SPAM on a monthly basis from January of 2006 to December of 2011, a 72-month period. If the number of SPAM messages increased linearly from 1250 in December 2006 to 5738 in December of 2011, (a) find the slope of the line and explain its meaning in this context. (b) Use this rate of change to project the amount of SPAM that will be received in June of 2013, assuming no anti-SPAM software is put in place.

**Solution** ▶ **a.** Using ordered pairs of the form $(m, S)$ for (month, SPAM), we have $(12, 1250)$ for December of 2006 and $(72, 5738)$ for December of 2011. We then have

$$\frac{\Delta S}{\Delta m} = \frac{5738 - 1250}{72 - 12}$$

$$= \frac{4488}{60}$$

$$= \frac{74.8}{1},$$

showing that the amount of SPAM received is increasing at a rate of about 75 e-mail messages per month.

**b.** From December of 2011 to June of 2013 is a period of 18 months. This means the amount of SPAM would increase by $18 \times 74.8 \approx 1346$ messages. From $(72, 5738)$ we then have the ordered pair $(72 + 18, 5738 + 1346) = (90, 7084)$ for June of 2013, showing the company projects they will be receiving 7084 SPAM messages per month by June of 2013.

 **E. You've just seen how we can apply linear equations in context**

**Now try Exercises 83 through 86** ▶

## 2.2 EXERCISES

▶ **CONCEPTS AND VOCABULARY**

Fill in each blank with the appropriate word or phrase. Carefully reread the section, if necessary.

1. To find the $x$-intercept of a line, substitute _____ for $y$ and solve for $x$. To find the $y$-intercept, substitute _____ for $x$ and solve for $y$.

2. The slope of a horizontal line is _____, the slope of a vertical line is _____, and the slopes of two parallel lines are _____.

3. Discuss/Explain If $m_1 = 2.1$ and $m_2 = 2.01$, will the lines intersect? If $m_1 = \frac{2}{3}$ and $m_2 = -\frac{2}{3}$, are the lines perpendicular?

4. Discuss/Explain the relationship between the slope formula, the Pythagorean theorem, and the distance formula. Include several illustrations.

▶ **DEVELOPING YOUR SKILLS**

Create a table of values for each equation and sketch the graph.

5. $2x + 3y = 6$

| $x$ | $y$ |
|-----|-----|
|     |     |
|     |     |
|     |     |
|     |     |

6. $-3x + 5y = 10$

| $x$ | $y$ |
|-----|-----|
|     |     |
|     |     |
|     |     |
|     |     |

7. $y = \frac{3}{2}x + 4$

| $x$ | $y$ |
|-----|-----|
|     |     |
|     |     |
|     |     |
|     |     |

8. $y = \frac{5}{3}x - 3$

| $x$ | $y$ |
|-----|-----|
|     |     |
|     |     |
|     |     |
|     |     |

9. If you completed Exercise 7, verify that $(-3, -0.5)$ and $(\frac{1}{2}, \frac{19}{4})$ also satisfy the equation given. Do these points appear to be on the graph you sketched?

10. If you completed Exercise 8, verify that $(-1.5, -5.5)$ and $(\frac{11}{2}, \frac{37}{6})$ also satisfy the equation given. Do these points appear to be on the graph you sketched?

**Graph the following equations using the intercept method. If the line goes through the origin, select an additional point. Plot a third point as a check.**

**11.** $3x + y = 6$

**12.** $-2x + y = 12$

**13.** $5y - x = 5$

**14.** $-4y + x = 8$

**15.** $-5x + 2y = 6$

**16.** $3y + 4x = 9$

**17.** $2x - 5y = 4$

**18.** $-6x + 4y = 8$

**19.** $2x + 3y = -12$

**20.** $-3x - 2y = 6$

**21.** $y = -\dfrac{1}{2}x$

**22.** $y = \dfrac{2}{3}x$

**23.** $y - 25 = 50x$

**24.** $y + 30 = 60x$

**25.** $y = -\dfrac{2}{5}x - 2$

**26.** $y = \dfrac{3}{4}x + 2$

**27.** $2y - 3x = 0$

**28.** $y + 3x = 0$

**29.** $3y + 4x = 12$

**30.** $-2x + 5y = 8$

**Compute the slope of the line through the given points, then graph the line and use $m = \dfrac{\Delta y}{\Delta x}$ to find two additional points on the line. Answers may vary.**

**31.** $(3, 5), (4, 6)$

**32.** $(-2, 3), (5, 8)$

**33.** $(10, 3), (4, -5)$

**34.** $(-3, -1), (0, 7)$

**35.** $(1, -8), (-3, 7)$

**36.** $(-5, 5), (0, -5)$

**37.** $(-3, 6), (4, 2)$

**38.** $(-2, -4), (-3, -1)$

**39.** $\left(-\dfrac{5}{4}, \dfrac{5}{8}\right), \left(\dfrac{3}{4}, -\dfrac{1}{8}\right)$

**40.** $\left(-\dfrac{1}{2}, -\dfrac{3}{4}\right), \left(\dfrac{1}{4}, \dfrac{3}{8}\right)$

**41.** The graph shown models the relationship between the cost of a new home and the size of the home in square feet. (a) Determine the slope of the line and interpret what the slope ratio means in this context and (b) estimate the cost of a 3000 ft² home.

**Exercise 41**

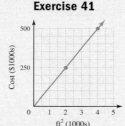

**Exercise 42**

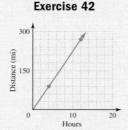

**42.** The graph shown models the relationship between the distance of an aircraft carrier from its home port and the number of hours since departure. (a) Determine the slope of the line and interpret what the slope ratio means in this context and (b) estimate the distance from port after 8.25 hr.

**43.** The graph shown models the relationship between the volume of garbage that is dumped in a landfill and the number of commercial garbage trucks that enter the site. (a) Determine the slope of the line and interpret what the slope ratio means in this context and (b) estimate the number of trucks entering the site daily if 1000 m³ of garbage is dumped per day.

**Exercise 43**

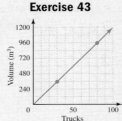

**Exercise 44**

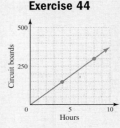

**44.** The graph shown models the relationship between the number of circuit boards that have been assembled at a factory and the number of hours since starting time. (a) Determine the slope of the line and interpret what the slope ratio means in this context and (b) estimate how many hours the factory has been running if 225 circuit boards have been assembled.

**45. Height and weight:** While there are many exceptions, numerous studies have shown a close relationship between an average height and average weight. Suppose a person 70 in. tall weighs 165 lb, while a person 64 in. tall weighs 142 lb. Assuming the relationship is linear, (a) find the slope of the line and discuss its meaning in this context and (b) determine how many pounds are added for each inch of height.

**46. Rate of climb:** Shortly after takeoff, a plane increases altitude at a constant (linear) rate. In 5 min the altitude is 10,000 ft. Fifteen minutes after takeoff, the plane has reached its cruising altitude of 32,000 ft. (a) Find the slope of the line and discuss its meaning in this context and (b) determine how long it takes the plane to climb from 12,200 ft to 25,400 ft.

**47. Sewer line slope:** Fascinated at how quickly the plumber was working, Ryan watched with great interest as the new sewer line was laid from the house to the main line, a distance of 48 ft. At the edge of the house, the sewer line was 6 in. under ground. If the plumber tied in to the main line at a depth of 18 in., what is the slope of the (sewer) line? What does this slope indicate?

**48. Slope (pitch) of a roof:** A contractor goes to a lumber yard to purchase some trusses (the triangular frames) for the roof of a house. Many sizes are available, so the contractor takes some measurements to ensure the roof will have the desired slope. In one case, the height of the truss (base to ridge) was 4 ft, with a width of 24 ft (eave to eave). Find the slope of the roof if these trusses are used. What does this slope indicate?

**Compute the slope of the line through the given points, then graph the line and use the slope to find two additional points. Answers may vary.**

**49.** $(6, -5), (-3, -5)$

**50.** $(3, -6), (3, 4)$

**51.** $(-5, 6), (-5, -2)$

**52.** $(-6, 4), (2, 4)$

**Graph each line using two or three ordered pairs that satisfy the equation.**

**53.** $x = -3$

**54.** $y = 4$

**55.** $x = 2$

**56.** $y = -2$

Write the equation for each line $L_1$ and $L_2$ shown. Specifically state their point of intersection.

**57.** 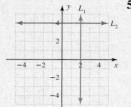  **58.**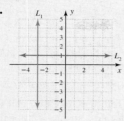

**59. Supreme Court justices:** The table shown gives the total number of justices $j$ sitting on the Supreme Court of the United States for selected time periods $t$ (in decades), along with the number of nonmale, nonwhite justices $n$ for the same years. (a) Use the data to graph the linear relationship between $t$ and $j$, then determine the slope of the line and discuss its meaning in this context. (b) Use the data to graph the linear relationship between $t$ and $n$, then determine the slope of the line and discuss its meaning.

**Exercise 59**

| Time $t$ (1960 → 0) | Justices $j$ | Nonwhite, nonmale $n$ |
|---|---|---|
| 0 | 9 | 0 |
| 10 | 9 | 1 |
| 20 | 9 | 2 |
| 30 | 9 | 3 |
| 40 | 9 | 4 |
| 50 | 9 | 5 |

**60. Boiling temperature:** The table shown gives the boiling temperature $t$ of water as related to the altitude $h$. Use the data to graph the linear relationship between $h$ and $t$, then determine the slope of the line and discuss its meaning in this context.

**Exercise 60**

| Altitude $h$ (ft) | Boiling Temperature $t$ (°F) |
|---|---|
| 0 | 212.0 |
| 1000 | 210.2 |
| 2000 | 208.4 |
| 3000 | 206.6 |
| 4000 | 204.8 |
| 5000 | 203.0 |
| 6000 | 201.2 |

Two points on $L_1$ and two points on $L_2$ are given. Use the slope formula to determine if lines $L_1$ and $L_2$ are parallel, perpendicular, or neither.

**61.** $L_1$: $(-2, 0)$ and $(0, 6)$
   $L_2$: $(1, 8)$ and $(0, 5)$

**62.** $L_1$: $(1, 10)$ and $(-1, 7)$
   $L_2$: $(0, 3)$ and $(1, 5)$

**63.** $L_1$: $(-3, -4)$ and $(0, 1)$
   $L_2$: $(0, 0)$ and $(-4, 4)$

**64.** $L_1$: $(6, 2)$ and $(8, -2)$
   $L_2$: $(5, 1)$ and $(3, 0)$

**65.** $L_1$: $(6, 3)$ and $(8, 7)$
   $L_2$: $(7, 2)$ and $(6, 0)$

**66.** $L_1$: $(-5, -1)$ and $(4, 4)$
   $L_2$: $(4, -7)$ and $(8, 10)$

In Exercises 67 through 72, three points that form the vertices of a triangle are given. Use the points to draw the triangle, then use the slope formula to determine if any of the triangles are right triangles. Also see Exercises 37 through 42 in Section 2.1.

**67.** $(-3, 7)$, $(2, 2)$, $(5, 5)$

**68.** $(5, 2)$, $(0, -3)$, $(4, -4)$

**69.** $(7, 0)$, $(-1, 0)$, $(7, 4)$

**70.** $(-3, 2)$, $(-1, 5)$, $(-6, 4)$

**71.** $(-4, 3)$, $(-7, -1)$, $(3, -2)$

**72.** $(0, 0)$, $(-5, 2)$, $(2, -5)$

▶ **WORKING WITH FORMULAS**

**73. Human life expectancy:** $L = 0.15T + 73.7$

In the United States, the average life expectancy has been steadily increasing over the years due to better living conditions and improved medical care. This relationship is modeled by the formula shown, where $L$ is the average life expectancy and $T$ is number of years since 1980. (a) What was the life expectancy in the year 2010? (b) Solve this formula for $T$. (c) In what year will average life expectancy reach 79 yr?

**74. Forensic studies:** $H = 1.88F + 32.01$

When partial skeletal remains of a human are found, forensic scientists attempt to estimate the height of the individual using known relationships between height and bone size. The formula shown gives the relationship between a male Caucasian's height (in inches) and the length of his femur bone. (a) If a 23-in. femur bone is found, how tall was the individual? (b) Solve the equation for $F$, and (c) if a male Caucasian is 68 in. tall, approximately how long is his femur?

► **APPLICATIONS**

**Use the information given to build a linear equation model, then use the equation to respond.**

**75. Business depreciation:** A business purchases a copier for $8500 and anticipates it will depreciate in value $1250/yr.

    **a.** What is the copier's value after 4 yr of use?

    **b.** How many years will it take for this copier's value to decrease to $2250?

**76. Baseball card value:** After purchasing an autographed baseball card for $85, its value increases by $1.50/yr.

    **a.** What is the card's value 7 yr after purchase?

    **b.** How many years will it take for this card's value to reach $100?

**77. Water level:** During a long drought, the water level in a local lake decreased at a rate of 3 in./month. The water level before the drought was 300 in.

    **a.** What was the water level after 9 months of drought?

    **b.** How many months will it take for the water level to decrease to 20 ft?

**78. Gas mileage:** When empty, a large dumptruck gets about 15 mpg. It is estimated that for each 3 tons of cargo it hauls, gas mileage decreases by $\frac{3}{4}$ mpg.

    **a.** If 10 tons of cargo is being carried, what is the truck's mileage?

    **b.** If the truck's mileage is down to 10 mpg, how much weight is it carrying?

**79. Cost of college:** For the years 2000 to 2008, the cost of tuition and fees per semester (in constant dollars) at a public 4-yr college can be approximated by the equation $y = 386x + 3500$, where $y$ represents the cost in dollars and $x = 0$ represents the year 2000. Use the equation to find (a) the cost of tuition and fees in 2010 and (b) the year this cost will exceed $9000.

*Source: The College Board*

**80. Decrease in smokers:** For the years 1990 to 2000, the percentage of the U.S. adult population who were smokers can be approximated by the equation $y = -0.52x + 28.7$, where $y$ represents the percentage of smokers (as a whole number) and $x = 0$ represents 1990. Use the equation to find (a) the percentage of adults who smoked in the year 2005 and (b) the year the percentage of smokers is projected to fall below 15%.

*Source: WebMD*

**81. Female physicians:** In 1960 only about 7% of physicians were female. Soon after, this percentage began to grow dramatically. For the years 1990 to 2000, the percentage of physicians that were female can be approximated by the equation $y = 0.6x + 18.1$, where $y$ represents the percentage and $x = 0$ represents the year 1990. Use the equation to find (a) the percentage of physicians that were female in 2000 and (b) the projected year this percentage would have exceeded 30%.

*Source: American Journal of Public Health*

**82. Temperature and cricket chirps:** Biologists have found a strong relationship between temperature and the number of times a cricket chirps. This is modeled by the equation $T = \frac{1}{4}N + 40$, where $N$ is the number of times the cricket chirps per minute and $T$ is the temperature in degrees Fahrenheit. Use the equation to find (a) the outdoor temperature if the cricket is chirping 48 times per minute and (b) the number of times a cricket chirps if the temperature is 70°.

**83. Parallel/nonparallel roads:** Aberville is 38 mi north and 12 mi west of Boschertown, with a straight "farm and machinery" road (FM 1960) connecting the two cities. In the next county, Crownsburg is 30 mi north and 9.5 mi west of Dower, and these cities are likewise connected by a straight road (FM 830). If the two roads continued indefinitely in both directions, would they intersect at some point?

**84. Perpendicular/nonperpendicular course headings:** Two shrimp trawlers depart Charleston Harbor at the same time. One heads for the shrimping grounds located 12 mi north and 3 mi east of the harbor. The other heads for a point 2 mi south and 8 mi east of the harbor. Are the routes of the trawlers perpendicular? If so, how far apart are the boats when they reach their destinations (to the nearest one-tenth mi)?

**85. Spellbound by technology:** In a 2005 report from the Kaiser Family Foundation, U.S. kids ages 8 to 18 spend nearly *44.5 hr/week* in front of a screen, largely due to the proliferation of TV, games, computers, and smartphones (according to the report, the only activity that takes more of their time is sleeping). A 1999 study by the American Academy of Pediatrics (AAP) found that children were spending an average of 6.5 hr/week with various media. Assuming the increase was linear, (a) write this information in ordered pair form, then find the slope of the line and explain its meaning in this context. (b) Use this rate of change to estimate the amount of time kids 8 to 18 might have spent in front of a screen in the year 2010.

**86.** The Human Development Index (HDI) developed by the United Nations attempts to measure the development of a country using indicators of increasing wealth and improvements in health and education. Ever since 1980, China has seen a steady increase in the HDI, with an index of 125 in 1990 and 180 in 2010. Assuming the index increased linearly over time, (a) write this information in ordered pair form, then find the slope of the line and explain its meaning in this context. Use this rate of change (b) to determine the value of the index in 2000 and (c) to predict the value of the index in 2014.

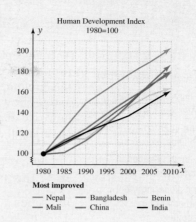

Human Development Index
1980=100

Most improved
— Nepal  — Bangladesh  — Benin
— Mali   — China       — India

## ▶ EXTENDING THE CONCEPT

**87.** If the lines $4y + 2x = -5$ and $3y + ax = -2$ are perpendicular, what is the value of $a$?

**88.** Let $m_1$, $m_2$, $m_3$, and $m_4$ be the slopes of lines $L_1$, $L_2$, $L_3$, and $L_4$ respectively. Which of the following statements is true?

   **a.** $m_4 < m_1 < m_3 < m_2$
   **b.** $m_3 < m_2 < m_4 < m_1$
   **c.** $m_3 < m_4 < m_2 < m_1$
   **d.** $m_1 < m_3 < m_4 < m_2$
   **e.** $m_1 < m_4 < m_3 < m_2$

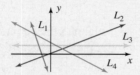

**89.** An *arithmetic sequence* is a sequence of numbers where each successive term is found by adding a fixed constant, called the common difference $d$, to the preceding term. For instance, 3, 7, 11, 15, . . . is an arithmetic sequence with $d = 4$. The formula for the "$n$th term" $t_n$ of an arithmetic sequence is a linear equation of the form $t_n = t_1 + (n - 1)d$, where $d$ is the common difference and $t_1$ is the first term of the sequence. Use the equation to find the term specified for each sequence.

   **a.** 2, 9, 16, 23, 30, . . . ; 21st term
   **b.** 7, 4, 1, −2, −5, . . . ; 31st term
   **c.** 5.10, 5.25, 5.40, 5.55, . . . ; 27th term
   **d.** $\frac{3}{2}, \frac{9}{4}, 3, \frac{15}{4}, \frac{9}{2}, \ldots$ ; 17th term

## ▶ MAINTAINING YOUR SKILLS

**90.** (2.1) Name the center and radius of the circle defined by $(x - 3)^2 + (y + 4)^2 = 169$.

**91.** (R.4) Compute the sum and product indicated:

   **a.** $\sqrt{20} + 3\sqrt{45} - \sqrt{5}$
   **b.** $(3 + \sqrt{5})(3 - \sqrt{5})$

**92.** (1.5) Solve the equation by factoring, then check the result(s) using substitution:

$12x^2 - 44x - 45 = 0$

**93.** (R.5) Factor the following polynomials completely:

   **a.** $x^3 - 3x^2 - 4x + 12$
   **b.** $x^2 - 23x - 24$
   **c.** $x^3 - 125$

# Graphs and Special Forms of Linear Equations

## LEARNING OBJECTIVES

*In Section 2.3 you will see how we can:*

☐ **A.** Write a linear equation in slope-intercept form

☐ **B.** Use slope-intercept form to graph linear equations

☐ **C.** Write a linear equation in point-slope form

☐ **D.** Apply the slope-intercept form and point-slope form in context

The concept of slope is an important part of mathematics, because it gives us a way to measure and compare change. The value of an automobile changes with time, the circumference of a circle increases as the radius increases, and the tension in a spring grows the more it is stretched. The real world is filled with examples of how one change affects another, and slope helps us understand how these changes are related.

## A. Linear Equations and Slope-Intercept Form

In Section 2.2, we learned that a linear equation is one that can be written in the form $ax + by = c$. Solving for $y$ in a linear equation offers distinct advantages to understanding linear graphs and their applications.

**EXAMPLE 1** ▶ **Solving for y in a Linear Equation**

Solve $2y - 6x = 4$ for $y$, then evaluate at $x = 4$, $x = 0$, and $x = -\frac{1}{3}$.

**Solution** ▶

$$2y - 6x = 4 \qquad \text{given equation}$$
$$2y = 6x + 4 \qquad \text{add } 6x$$
$$y = 3x + 2 \qquad \text{divide by 2}$$

Since the coefficients are integers, evaluate the equation mentally. Inputs are multiplied by 3, then increased by 2, yielding the ordered pairs $(4, 14)$, $(0, 2)$, and $\left(-\frac{1}{3}, 1\right)$.

**Now try Exercises 5 through 10** ▶

This form of the equation (where $y$ has been written in terms of $x$) enables us to quickly identify what operations are performed on $x$ in order to obtain $y$. For $y = 3x + 2$: *multiply inputs by 3, then add 2.* This makes evaluating equations more efficient, and offers additional advantages in graphing and applications.

**EXAMPLE 2** ▶ **Solving for y in a Linear Equation**

Solve the linear equation $3y - 2x = 6$ for $y$, then identify the new coefficient of $x$ and the constant term.

**Solution** ▶

$$3y - 2x = 6 \qquad \text{given equation}$$
$$3y = 2x + 6 \qquad \text{add } 2x$$
$$y = \frac{2}{3}x + 2 \qquad \text{divide by 3}$$

The coefficient of $x$ is $\frac{2}{3}$ and the constant term is 2.

**Now try Exercises 11 through 16** ▶

**WORTHY OF NOTE**

In Example 2, the final form can be written $y = \frac{2}{3}x + 2$ as shown (inputs are multiplied by two-thirds, then increased by 2), or written as $y = \frac{2x}{3} + 2$ (inputs are multiplied by two, the result divided by 3 and this amount increased by 2). The two forms are equivalent.

When the coefficient of $x$ is rational, it's helpful to select inputs that are multiples of the denominator if the context or application requires us to evaluate the equation. This enables us to perform most operations mentally. For $y = \frac{2}{3}x + 2$, possible inputs might be $x = -9, -6, 0, 3, 6$, and so on. **See Exercises 17 through 22.**

In Section 2.2, linear equations were graphed using the intercept method. When the equation is written with $y$ in terms of $x$, we notice a powerful connection between the graph and its equation—one that highlights the primary characteristics of a linear graph.

**EXAMPLE 3** ▶ **Noting Relationships between an Equation and Its Graph**

Find the intercepts of $4x + 5y = -20$ and use them to graph the line. Then,

**a.** Use the intercepts to calculate the slope of the line and identify the $y$-intercept.

**b.** Write the equation with $y$ in terms of $x$ and compare the calculated slope and $y$-intercept to the equation in this form. Comment on what you notice.

**Solution** ▶ Substituting 0 for $x$ in $4x + 5y = -20$, we find the $y$-intercept is $(0, -4)$. Substituting 0 for $y$ gives an $x$-intercept of $(-5, 0)$. The graph is displayed here.

**a.** The $y$-intercept is $(0, -4)$ and by calculation or counting $\frac{\Delta y}{\Delta x}$, the slope is $m = \frac{-4}{5}$. From the intercept $(-5, 0)$, $\Delta y = -4$ and $\Delta x = 5$, and we arrive at the intercept $(0, -4)$.

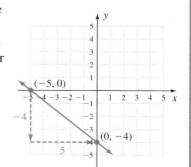

**b.** Solving for $y$:

$$4x + 5y = -20 \qquad \text{given equation}$$
$$5y = -4x - 20 \qquad \text{subtract } 4x$$
$$y = \frac{-4}{5}x - 4 \qquad \text{divide by 5}$$

The slope value seems to be the coefficient of $x$, while the $y$-intercept is indicated by the constant term.

**Now try Exercises 23 through 28** ▶

After solving a linear equation for $y$, an input of $x = 0$ causes the "$x$-term" to become zero, so the $y$-intercept is automatically the constant term. As Example 3 illustrates, we can also identify the slope of the line—it is the coefficient of $x$. In general, a linear equation of the form $y = mx + b$ is said to be in **slope-intercept form,** since the slope of the line is $m$ and the $y$-intercept is $(0, b)$.

**Slope-Intercept Form**

For a nonvertical line whose equation is $y = mx + b$,
the slope of the line is $m$ and the $y$-intercept is $(0, b)$.

**EXAMPLE 4** ▶ **Finding the Slope-Intercept Form of a Linear Equation**

Write each equation in slope-intercept form. Then identify the slope and $y$-intercept of the line.

**a.** $3x - 2y = 9$      **b.** $y + x = 5$      **c.** $2y = x$

**Solution** ▶

**a.** $3x - 2y = 9$

$$-2y = -3x + 9$$
$$y = \frac{3}{2}x - \frac{9}{2}$$
$$m = \frac{3}{2}, b = -\frac{9}{2}$$
$y$-intercept $\left(0, -\frac{9}{2}\right)$

**b.** $y + x = 5$

$$y = -x + 5$$
$$y = -1x + 5$$
$$m = -1, b = 5$$
$y$-intercept $(0, 5)$

**c.** $2y = x$

$$y = \frac{x}{2}$$
$$y = \frac{1}{2}x$$
$$m = \frac{1}{2}, b = 0$$
$y$-intercept $(0, 0)$

☑ **A.** You've just seen how we can write a linear equation in slope-intercept form

**Now try Exercises 29 through 36** ▶

Note that we can analytically develop the slope-intercept form of a line using the slope formula. Figure 2.25 shows the graph of a general line through the point $(x, y)$ with a $y$-intercept of $(0, b)$. Using these points in the slope formula, we have

**Figure 2.25**

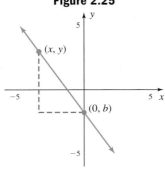

$$\frac{y_2 - y_1}{x_2 - x_1} = m \qquad \text{slope formula}$$

$$\frac{y - b}{x - 0} = m \qquad \text{substitute: } (0, b) \text{ for } (x_1, y_1), (x, y) \text{ for } (x_2, y_2)$$

$$\frac{y - b}{x} = m \qquad \text{simplify}$$

$$y - b = mx \qquad \text{multiply by } x$$

$$y = mx + b \qquad \text{add } b$$

This approach confirms the relationship between the graphical characteristics of a line and its slope-intercept form. Specifically, for any linear equation written in the form $y = mx + b$, the slope must be $m$ and the $y$-intercept is $(0, b)$.

## B. Slope-Intercept Form and the Graph of a Line

If the slope and $y$-intercept of a linear equation are known or can be found, we can construct its equation by substituting these values directly into the slope-intercept form $y = mx + b$.

**EXAMPLE 5** ▶ **Finding the Equation of a Line from Its Graph**

Find the slope-intercept equation of the line shown.

**Solution** ▶ Using $(-3, -2)$ and $(-1, 2)$ in the slope formula, or by simply counting $\frac{\Delta y}{\Delta x}$, the slope is $m = \frac{4}{2}$ or $\frac{2}{1}$. By inspection we see the $y$-intercept is $(0, 4)$. Substituting $\frac{2}{1}$ for $m$ and 4 for $b$ in the slope-intercept form we obtain the equation $y = 2x + 4$.

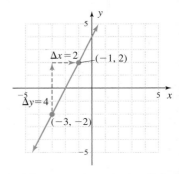

Now try Exercises 37 through 41 ▶

Actually, if the slope is known and we have *any* point $(x, y)$ on the line, we can still construct the equation since the given point *must satisfy the equation of the line*. In this case, we're treating $y = mx + b$ as a simple formula, solving for $b$ after substituting known values for $m$, $x$, and $y$.

**EXAMPLE 6** ▶ **Using $y = mx + b$ as a Formula**

Find the slope-intercept equation of a line that has slope $m = \frac{4}{5}$ and contains $(-5, 2)$.

**Solution** ▶ Use $y = mx + b$ as a "formula," with $m = \frac{4}{5}$, $x = -5$, and $y = 2$.

$$y = mx + b \qquad \text{slope-intercept form}$$

$$2 = \frac{4}{5}(-5) + b \qquad \text{substitute } \frac{4}{5} \text{ for } m, -5 \text{ for } x, \text{ and } 2 \text{ for } y$$

$$2 = -4 + b \qquad \text{simplify}$$

$$6 = b \qquad \text{solve for } b$$

The equation of the line is $y = \frac{4}{5}x + 6$.

Now try Exercises 42 through 46 ▶

Writing a linear equation in slope-intercept form enables us to draw its graph with a minimum of effort, since we can easily locate the $y$-intercept and a second point using the rate of change $\frac{\Delta y}{\Delta x}$. For instance, $\frac{\Delta y}{\Delta x} = \frac{5}{3}$ indicates that counting up 5 and right 3 from a known point will locate another point on this line.

**EXAMPLE 7** ▶ **Graphing a Line Using Slope-Intercept Form and the Rate of Change**

Write $3y - 5x = 9$ in slope-intercept form, then graph the line using the $y$-intercept and the rate of change (slope).

Solution ▶

$$3y - 5x = 9 \qquad \text{given equation}$$
$$3y = 5x + 9 \qquad \text{isolate } y \text{ term (add } 5x)$$
$$y = \tfrac{5}{3}x + 3 \qquad \text{divide by 3}$$

The slope is $m = \frac{5}{3}$ and the $y$-intercept is $(0, 3)$. Plot the $y$-intercept, then use $\frac{\Delta y}{\Delta x} = \frac{5}{3}$ (rise 5 and run 3—shown in blue) to find another point on the line (shown in red). Finish by drawing a line through these points.

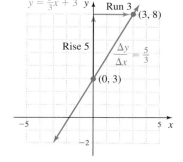

**WORTHY OF NOTE**

Noting the fraction $\frac{5}{3}$ is equal to $\frac{-5}{-3}$, we could also begin at $(0, 3)$ and count $\frac{\Delta y}{\Delta x} = \frac{-5}{-3}$ (down 5 and left 3) to find an additional point on the line: $(-3, -2)$. Also, for any negative slope $\frac{\Delta y}{\Delta x} = -\frac{a}{b}$, note $-\frac{a}{b} = \frac{-a}{b} = \frac{a}{-b}$.

**Now try Exercises 47 through 58** ▶

For a discussion of what graphing method might be most efficient for a given linear equation, **see Exercises 97 and 108.**

## Parallel and Perpendicular Lines

From Section 2.2 we know parallel lines have equal slopes: $m_1 = m_2$, and perpendicular lines have slopes with a product of $-1$: $m_1 \cdot m_2 = -1$ or $m_1 = -\frac{1}{m_2}$. In some applications, we need to find the equation of a second line parallel or perpendicular to a given line, through a given point. Using the slope-intercept form makes this a simple four-step process.

**Finding the Equation of a Line Parallel or Perpendicular to a Given Line**

1. Identify the slope $m_1$ of the given line.
2. Find slope $m_2$ of the new line using the parallel or perpendicular relationship.
3. Use $m_2$ with the point $(x, y)$ in the "formula" $y = mx + b$ and solve for $b$.
4. The desired equation will be $y = m_2 x + b$.

**EXAMPLE 8** ▶ **Finding the Equation of a Parallel Line**

Find the slope-intercept equation of a line that goes through $(-6, -1)$ and is parallel to $2x + 3y = 6$.

Solution ▶ Begin by writing the equation in slope-intercept form to identify the slope.

$$2x + 3y = 6 \qquad \text{given line}$$
$$3y = -2x + 6 \qquad \text{isolate } y\text{-term}$$
$$y = \tfrac{-2}{3}x + 2 \qquad \text{result}$$

The original line has slope $m_1 = \frac{-2}{3}$ and this will also be the slope of any line parallel to it. Using $m_2 = \frac{-2}{3}$ with $(x, y)$ corresponding to $(-6, -1)$ we have

$$y = mx + b \qquad \text{slope-intercept form}$$

$$-1 = \frac{-2}{3}(-6) + b \qquad \begin{array}{l}\text{substitute } \frac{-2}{3} \text{ for } m, \\ -6 \text{ for } x, \text{ and } -1 \text{ for } y\end{array}$$

$$-1 = 4 + b \qquad \text{simplify}$$

$$-5 = b \qquad \text{solve for } b$$

The equation of the new line is $y = \frac{-2}{3}x - 5$.

**WORTHY OF NOTE**

For $2x + 3y = 6$, any equation of the form $2x + 3y = k$ ($k$ a constant) will give the equation of a parallel line. Using the point $(-6, -1)$ from Example 8 gives $2(-6) + 3(-1) = k$, with $k = -15$. Note that $2x + 3y = -15$ is equivalent to the result from this example.

**Now try Exercises 59 through 72 ▶**

For any nonlinear graph, a straight line drawn through two points on the graph is called a **secant line.** The slope of a secant line, and lines parallel and perpendicular to this line, play fundamental roles in further development of the rate-of-change concept.

**EXAMPLE 9 ▶**   **Finding Equations for Parallel and Perpendicular Lines**

A secant line is drawn using the points $(-4, 0)$ and $(2, -2)$ on the graph of the relation shown. Find the equation of a line that is

**a.** parallel to the secant line through $(-1, -4)$.

**b.** perpendicular to the secant line through $(-1, -4)$.

**Solution ▶**   Either by using the slope formula or counting $\frac{\Delta y}{\Delta x}$, we find the secant line has slope $m = \frac{-2}{6} = \frac{-1}{3}$. Any line parallel to this secant line will have a slope of $-\frac{1}{3}$, and any line perpendicular will have a slope of 3.

**a.** For the parallel line through $(-1, -4)$, $m_2 = \frac{-1}{3}$.

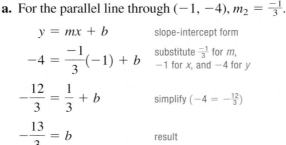

**WORTHY OF NOTE**

The word "secant" comes from the Latin word *secare,* meaning "to cut." Hence a secant line is one that cuts through a graph in two or more points.

$$y = mx + b \qquad \text{slope-intercept form}$$

$$-4 = \frac{-1}{3}(-1) + b \qquad \begin{array}{l}\text{substitute } \frac{-1}{3} \text{ for } m, \\ -1 \text{ for } x, \text{ and } -4 \text{ for } y\end{array}$$

$$-\frac{12}{3} = \frac{1}{3} + b \qquad \text{simplify } (-4 = -\frac{12}{3})$$

$$-\frac{13}{3} = b \qquad \text{result}$$

The equation of the parallel line (in blue) is $y = \frac{-1}{3}x - \frac{13}{3}$.

**b.** For the perpendicular line through $(-1, -4)$, $m_2 = 3$.

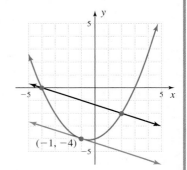

$$y = mx + b \qquad \text{slope-intercept form}$$

$$-4 = 3(-1) + b \qquad \text{substitute 3 for } m, -1 \text{ for } x, \text{ and } -4 \text{ for } y$$

$$-4 = -3 + b \qquad \text{simplify}$$

$$-1 = b \qquad \text{result}$$

The equation of the perpendicular line (in gold) is $y = 3x - 1$.

☑ **B.** You've just seen how we can use the slope-intercept form to graph linear equations

**Now try Exercises 73 through 76 ▶**

## C. Linear Equations in Point-Slope Form

As an alternative to using $y = mx + b$, we can find the equation of the line using the slope formula $\frac{y_2 - y_1}{x_2 - x_1} = m$, and the fact that *the slope of a line is constant*. For a given slope $m$, we can let $(x_1, y_1)$ represent a *given* point on the line and $(x, y)$ represent *any other point* on the line. The formula then becomes $\frac{y - y_1}{x - x_1} = m$. Isolating the "$y$" terms on one side gives a new form for the equation of a line, called the **point-slope form:**

$$\frac{y - y_1}{x - x_1} = m \qquad \text{slope formula}$$

$$\frac{(x - x_1)}{1}\left(\frac{y - y_1}{x - x_1}\right) = m(x - x_1) \qquad \text{multiply both sides by } (x - x_1)$$

$$y - y_1 = m(x - x_1) \qquad \text{simplify, point-slope form}$$

### The Point-Slope Form of a Linear Equation

For a nonvertical line with slope $m$ and containing the point $(x_1, y_1)$,
the equation of the line is $y - y_1 = m(x - x_1)$.

While using $y = mx + b$ (as in Example 6) may appear to be easier, both the slope-intercept form and point-slope form have their own advantages and it will help to be familiar with both.

**EXAMPLE 10** ▶ **Using $y - y_1 = m(x - x_1)$ as a Formula**

Find the equation of the line in point-slope form, if $m = \frac{2}{3}$ and $(-3, -3)$ is on the line. Then graph the line.

**Solution** ▶
$$y - y_1 = m(x - x_1) \qquad \text{point-slope form}$$

$$y - (-3) = \frac{2}{3}[x - (-3)] \qquad \begin{array}{l}\text{substitute } \frac{2}{3} \text{ for } m; (-3, -3) \\ \text{for } (x_1, y_1)\end{array}$$

$$y + 3 = \frac{2}{3}(x + 3) \qquad \text{simplify, point-slope form}$$

To graph the line, plot $(-3, -3)$ and use $\frac{\Delta y}{\Delta x} = \frac{2}{3}$ to find additional points on the line.

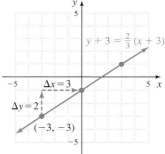

☑ **C.** You've just seen how we can write a linear equation in point-slope form

Now try Exercises 77 through 88 ▶

## D. Applications of Linear Equations and Rates of Change

As a mathematical tool, linear equations rank among the most common, powerful, and versatile. In all cases, it's important to remember that slope represents a *rate of change*. The notation $m = \frac{\Delta y}{\Delta x}$ literally means the quantity measured along the $y$-axis, is changing with respect to changes in the quantity measured along the $x$-axis.

**EXAMPLE 11** ▶ **Relating Temperature to Altitude**

In meteorological studies, it is well known that atmospheric temperature depends on the altitude according to the formula $T = -3.5h + 58.5$, where $T$ represents the approximate Fahrenheit temperature at height $h$ (in thousands of feet, $0 \leq h \leq 36$).

  **a.** Interpret the meaning of the slope in this context.

  **b.** Determine the temperature at an altitude of 12,000 ft.

  **c.** If the temperature is $-8°F$ what is the approximate altitude?

**Solution** ▶   **a.** Notice that $h$ is the input variable and $T$ is the output. This shows $\frac{\Delta T}{\Delta h} = \frac{-3.5}{1}$, meaning the temperature drops $3.5°F$ for every 1000-ft increase in altitude.

  **b.** Since height is in thousands, use $h = 12$.

$$
\begin{array}{ll}
T = -3.5h + 58.5 & \text{original formula} \\
T = -3.5(12) + 58.5 & \text{substitute 12 for } h \\
\phantom{T} = 16.5 & \text{result}
\end{array}
$$

At a height of 12,000 ft, the temperature is about $16.5°$.

  **c.** Replacing $T$ with $-8$ and solving gives

$$
\begin{array}{ll}
T = -3.5h + 58.5 & \text{original formula} \\
-8 = -3.5h + 58.5 & \text{substitute } -8 \text{ for } T \\
-66.5 = -3.5h & \text{subtract 58.5} \\
19 = h & \text{divide by } -3.5
\end{array}
$$

The temperature is about $-8°F$ at a height of $19 \times 1000 = 19,000$ ft.

**Now try Exercises 99 and 100** ▶

For additional help making connections between graphs and common activities, **see Exercises 89 through 96.**

In some applications, the relationship is known to be linear but only a few points on the line are given. In this case, we can use two of the known data points to calculate the slope, then the point-slope form to find an equation model. One such application is *linear depreciation,* as when a government allows businesses to depreciate vehicles and equipment over time (the less a piece of equipment is worth, the less you pay in taxes).

**EXAMPLE 12A** ▶ **Using Point-Slope Form to Find an Equation Model**

Five years after purchase, the auditor of a newspaper company estimates the value of their printing press is $60,000. Eight years after its purchase, the value $v$ of the press had depreciated to $42,000. Find a linear equation that models this depreciation and discuss the slope and $v$-intercept in context.

**Solution** ▶ Since the value of the press depends on time, the ordered pairs have the form (time, value) or $(t, v)$ where *time* is the input, and *value* is the output. This means the ordered pairs are (5, 60,000) and (8, 42,000).

$$
\begin{aligned}
m &= \frac{v_2 - v_1}{t_2 - t_1} && \text{slope formula} \\[6pt]
&= \frac{42,000 - 60,000}{8 - 5} && (t_1, v_1) = (5, 60,000); (t_2, v_2) = (8, 42,000) \\[6pt]
&= \frac{-18,000}{3} = \frac{-6000}{1} && \text{simplify and reduce}
\end{aligned}
$$

**WORTHY OF NOTE**

Actually, it doesn't matter which of the two points is used in Example 12A. Once the point (5, 60,000) is plotted, a constant slope of $m = -6000$ will "drive" the line through (8, 42,000). If we first graph (8, 42,000), the same slope would "drive" the line through (5, 60,000). Convince yourself by reworking the problem using the other point.

The slope of the line is $\frac{\Delta \text{value}}{\Delta \text{time}} = \frac{-6000}{1}$, indicating the printing press loses \$6000 in value with each passing year.

$$
\begin{aligned}
v - v_1 &= m(t - t_1) &\quad \text{point-slope form} \\
v - 60{,}000 &= -6000(t - 5) &\quad \text{substitute } -6000 \text{ for } m; (5, 60{,}000) \text{ for } (t_1, v_1) \\
v - 60{,}000 &= -6000t + 30{,}000 &\quad \text{simplify} \\
v &= -6000t + 90{,}000 &\quad \text{solve for } v
\end{aligned}
$$

The depreciation equation is $v = -6000t + 90{,}000$. The $v$-intercept $(0, 90{,}000)$ indicates the original value (cost) of the equipment was \$90,000.

Once the depreciation equation is found, it represents the (time, value) relationship for all future (and intermediate) ages of the press. In other words, we can now predict the value of the press for any given year. However, note that some equation models are valid for only a set period of time, and each model should be used with care.

**EXAMPLE 12B** ▶  **Using an Equation Model to Gather Information**

From Example 12A,
  **a.** How much will the press be worth after 11 yr?
  **b.** How many years until the value of the equipment is \$9000?
  **c.** Is this equation model valid for $t = 18$ yr (why or why not)?

**Solution** ▶  **a.** Find the value $v$ when $t = 11$:

$$
\begin{aligned}
v &= -6000t + 90{,}000 &\quad \text{equation model} \\
v &= -6000(11) + 90{,}000 &\quad \text{substitute 11 for } t \\
&= 24{,}000 &\quad \text{result } (11, 24{,}000)
\end{aligned}
$$

After 11 yr, the printing press will be worth only \$24,000.
  **b.** "… value is \$9000" means $v = 9000$:

$$
\begin{aligned}
v &= 9000 &\quad \text{value at time } t \\
-6000t + 90{,}000 &= 9000 &\quad \text{substitute } -6000t + 90{,}000 \text{ for } v \\
-6000t &= -81{,}000 &\quad \text{subtract 90,000} \\
t &= 13.5 &\quad \text{divide by } -6000
\end{aligned}
$$

After 13.5 yr, the printing press will be worth \$9000.
  **c.** Since substituting 18 for $t$ gives a negative quantity, the equation model is not valid for $t = 18$. In the current context, the model is only valid while $v \geq 0$ and solving $-6000t + 90{,}000 \geq 0$ shows the domain in this context is $t \in [0, 15]$.

☑ **D.** You've just seen how we can apply the slope-intercept form and point-slope form in context

**Now try Exercises 101 through 106** ▶

## 2.3 EXERCISES

### ▶ CONCEPTS AND VOCABULARY

**Fill in each blank with the appropriate word or phrase. Carefully reread the section, if necessary.**

1. For the equation $7x + 4y = 12$, the slope is _____ and the $y$-intercept is _____.

2. The equation $y - y_1 = m(x - x_1)$ is called the _____ form of a line.

3. Discuss/Explain how to graph a line using only the slope and a point on the line (no equations).

4. Discuss/Explain one method for finding the equation of a line parallel to $y = -2x + 1$, through the point $(0, 4)$.

### ▶ DEVELOPING YOUR SKILLS

**Solve each equation for $y$ and evaluate the result using $x = -5, x = -2, x = 0, x = 1,$ and $x = 3$.**

5. $4x + 5y = 10$

6. $3y - 2x = 9$

7. $-0.4x + 0.2y = 1.4$

8. $-0.2x + 0.7y = -2.1$

9. $\frac{1}{3}x + \frac{1}{5}y = -1$

10. $\frac{1}{7}y - \frac{1}{3}x = 2$

**For each equation, solve for $y$ and identify the new coefficient of $x$ and new constant term.**

11. $6x - 3y = 9$

12. $9y - 4x = 18$

13. $-0.5x - 0.3y = 2.1$

14. $-0.7x + 0.6y = -2.4$

15. $\frac{5}{6}x + \frac{1}{7}y = -\frac{4}{7}$

16. $\frac{7}{12}y - \frac{4}{15}x = \frac{7}{6}$

**Evaluate each equation by selecting three inputs that will result in integer values. Then graph each line.**

17. $y = -\frac{4}{3}x + 5$

18. $y = \frac{5}{4}x + 1$

19. $y = -\frac{3}{2}x - 2$

20. $y = \frac{2}{5}x - 3$

21. $y = -\frac{1}{6}x + 4$

22. $y = -\frac{1}{3}x + 3$

**Find the $x$- and $y$-intercepts for each line and use these two points to calculate the slope. Then solve for $y$ and compare the calculated slope and $y$-intercept to the resulting equation and comment.**

23. $3x + 4y = 12$

24. $3y - 2x = -6$

25. $2x - 5y = 10$

26. $2x + 3y = 9$

27. $4x - 5y = -15$

28. $5y + 6x = -25$

**Write each equation in slope-intercept form (solve for $y$), then identify the slope and $y$-intercept.**

29. $2x + 3y = 6$

30. $4y - 3x = 12$

31. $5x + 4y = 20$

32. $y + 2x = 4$

33. $x = 3y$

34. $2x = -5y$

35. $3x + 4y - 12 = 0$

36. $5y - 3x + 20 = 0$

**For Exercises 37 through 46, use the slope-intercept form to state the equation of each line.**

37.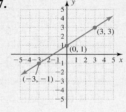

38.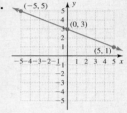

39. $m = -2$; $y$-intercept $(0, -3)$

40. $m = 3$; $y$-intercept $(0, 2)$

41. $m = \frac{-3}{2}$; $y$-intercept $(0, -4)$

42. $m = -4$; $(-3, 2)$ is on the line

43. $m = 2$; $(5, -3)$ is on the line

44. $m = \frac{-3}{2}$; $(-4, 7)$ is on the line

45.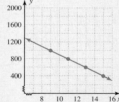

46.

**Graph each linear equation using the $y$-intercept and rate of change (slope) determined from each equation.**

47. $y = \frac{2}{3}x + 3$

48. $y = \frac{5}{2}x - 1$

49. $y = \frac{-1}{3}x + 2$

50. $y = \frac{-4}{5}x + 2$

51. $y = 2x - 5$

52. $y = -3x + 4$

53. $y = \frac{1}{2}x - 3$

54. $y = \frac{-3}{2}x + 2$

**Write each equation in slope-intercept form, then use the rate of change (slope) and $y$-intercept to graph the line.**

55. $3x + 5y = 20$

56. $2y - x = 4$

57. $2x - 3y = 15$

58. $-3x + 2y = 4$

**Find the equation of the line using the information given. Write answers in slope-intercept form.**

**59.** parallel to $2x - 5y = 10$, with $y$-intercept $(0, 4)$

**60.** parallel to $6x + 9y = 27$, with $y$-intercept $(0, -7)$

**61.** perpendicular to $5y - 3x = 9$, through the point $(6, -3)$

**62.** perpendicular to $x - 4y = 7$, through the point $(-5, 3)$

**63.** parallel to $12x + 5y = 65$, through the point $(-2, -1)$

**64.** parallel to $15y - 8x = 50$, through the point $(3, -4)$

**65.** parallel to $y = -3$, through the point $(2, 5)$

**66.** perpendicular to $y = -3$ through the point $(2, 5)$

**Write the equations in slope-intercept form and state whether the lines are parallel, perpendicular, or neither.**

**67.** $4y - 5x = 8$
$5y + 4x = -15$

**68.** $3y - 2x = 6$
$2x - 3y = 6$

**69.** $2x - 5y = 20$
$4x - 3y = 18$

**70.** $-4x + 6y = 12$
$2x + 3y = 6$

**71.** $3x + 4y = 12$
$6x + 8y = 2$

**72.** $5y = 11x + 135$
$11y + 5x = -77$

**For each graph given, find an equation of the line (a) parallel and (b) perpendicular to the secant line shown, and passing through the point indicated.**

**73.**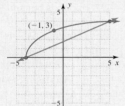

**74.**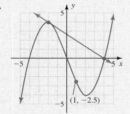

**75.**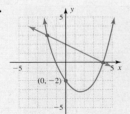

**76.**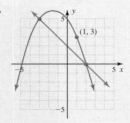

**Find the equation of the line in point-slope form, then graph the line.**

**77.** $m = 2; P_1 = (2, -5)$

**78.** $m = -1; P_1 = (2, -3)$

**79.** $P_1 = (3, -4), P_2 = (11, -1)$

**80.** $P_1 = (-1, 6), P_2 = (5, 1)$

**81.** $m = 0.5; P_1 = (1.8, -3.1)$

**82.** $m = 1.5; P_1 = (-0.75, -0.125)$

**Find the equation of the line in point-slope form, and state the meaning of the slope in context.**

**83.**

**84.**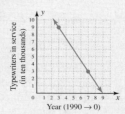

**85.**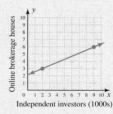

**86.**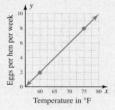

**87.**

**88.**

**Using the concept of slope, match each description with the graph that best illustrates it. Assume time is scaled on the horizontal axes.**

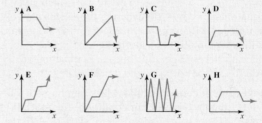

**89.** While driving today, I got stopped by a state trooper. After she warned me to slow down, I continued on my way.

**90.** After hitting the ball, I began trotting around the bases shouting, "Ooh, ooh, ooh!" When I saw it wasn't a home run, I began sprinting.

**91.** At first I ran at a steady pace, then I got tired and walked the rest of the way.

**92.** While on my daily walk, I had to run for a while when I was chased by a stray dog.

**93.** I climbed up a tree, then I jumped out.

**94.** I steadily swam laps at the pool yesterday.

**95.** I walked toward the candy machine, stared at it for a while then changed my mind and walked back.

**96.** For practice, the girls' track team did a series of 25-m sprints, with a brief rest in between.

## ▶ WORKING WITH FORMULAS

**97. General linear equation:** $ax + by = c$

The general equation of a line is shown here, where $a$, $b$, and $c$ are real numbers, with $a$ and $b$ not simultaneously zero. Solve the equation for $y$ and note the slope (coefficient of $x$) and $y$-intercept (constant term) are written in terms of $a$, $b$, and $c$. Use these to find the slope and $y$-intercept of the following lines, without solving for $y$ or computing points.

    **a.** $3x + 4y = 8$      **b.** $2x + 5y = -15$
    **c.** $5x - 6y = -12$     **d.** $3y - 5x = 9$

**98. Intercept-Intercept form:** $\dfrac{x}{h} + \dfrac{y}{k} = 1$

The $x$- and $y$-intercepts of a line can also be found by writing the equation in the form shown. The $x$-intercept will be $(h, 0)$ and the $y$-intercept will be $(0, k)$. Find the $x$- and $y$-intercepts of the following lines using this method. How is the slope of each line related to the values of $h$ and $k$?

    **a.** $2x + 5y = 10$      **b.** $3x - 4y = -12$
    **c.** $5x + 4y = 8$

## ▶ APPLICATIONS

**99. Speed of sound:** The speed of sound as it travels through the air depends on the temperature of the air according to the equation $V = \frac{3}{5}T + 331$, where $V$ represents the velocity of the sound waves in meters per second (m/sec), at a temperature of $T°$ Celsius. (a) Interpret the meaning of the slope and $y$-intercept in this context. (b) Determine the speed of sound at a temperature of 20°C. (c) If the speed of sound is measured at 361 m/sec, what is the temperature of the air?

**100. Automobile acceleration:** A driver going down a straight highway is traveling 60 ft/sec (about 41 mph) on cruise control, when he begins accelerating at a rate of 5.2 ft/sec². The velocity of the car is then given by the equation $V = \frac{26}{5}t + 60$, where $V$ represents the velocity at time $t$. (a) Interpret the meaning of the slope and $y$-intercept in this context. (b) Determine the velocity of the car after 9.4 sec. (c) If the car is traveling at 100 ft/sec, for how long did it accelerate?

**101. Investing in coins:** In 2004, Mark purchased a 1909-S VDB Lincoln Cent (in fair condition) for $190. By the year 2008, its value had grown to $210. (a) Use the relation (time since purchase, value) with $t = 0$ corresponding to 2004 to find a linear equation modeling the value of the coin. (b) Discuss what the slope and $y$-intercept indicate in this context. (c) How much will the penny be worth in 2012? (d) How many years after purchase will the penny's value be $270?

**102. Depreciation:** Once a piece of equipment is put into service, its value begins to depreciate. A business purchases some computer equipment for $18,500. At the end of a 2-yr period, the value of the equipment has decreased to $11,500. (a) Use the relation (time since purchase, value) to find a linear equation modeling the value of the equipment. (b) Discuss what the slope and $y$-intercept indicate in this context. (c) What is the equipment's value after 4 yr? (d) Generally, companies will sell used equipment while it still has value and use the funds to purchase new equipment. According to the equation how many years will it take this equipment to depreciate in value to $1000?

**103. Internet users:** According to InternetWorldStats.com, there were approximately 143 million Internet users in 2001. By 2008, this figure had grown to near 220 million. (a) Assuming the relationship is linear, use the relation (year, Internet users) with $t = 1$ corresponding to 2001 to find an equation model for the number of users each year. (b) Discuss what the slope indicates in this context. (c) According to this model, how many users were there in 2010? (d) In what year is the number of users projected to exceed 300 million?

**104. Pressure and depth:** Due to the increasing weight of the water above, the pressure on a diver increases as she dives deeper into a body of water. At a depth of 18 ft ($d = 18$), the pressure on a diver is about 22.7 pounds per square inch (psi). At a depth of 45 ft, the pressure would be 34.7 psi. (a) Given the relationship is linear, use the relation (depth in feet, pressure in psi) to find an equation model for the pressure at depth $d$. (b) Discuss what the slope and $y$-intercept indicate in this context. (c) According to this model, what would the pressure be at a depth of 75 ft? (d) If the pressure is 59.1 psi, how deep is the diver?

**105. Highway grade:** Upon exiting the Eisenhower Tunnel (through the Colorado Rockies) westbound on I-70, there is a large sign that says, "Steep Grade Next 8 Miles." The tunnel has an elevation of about 11,200 ft, and 4 mi later the elevation is 9740 ft. Use the relation (driving distance, elevation) to (a) find an equation model for the elevation after driving $d$ mi, and (b) discuss what the slope and $y$-intercept indicate in this context. (c) According to this model, what is the elevation when the grade levels out? (d) Given 1 mi = 5280 ft, what is the percent grade?

**106. Luxury taxes:** In professional baseball, the *luxury tax* is a tax on the amount a team's total payroll exceeds a set threshold established by the league. The tax is intended to prevent teams in larger and more profitable markets from dominating the sport by attracting a majority of the most talented (and hence highest paid) players. In 2004, this threshold was set at $120.5 million dollars. By 2009 the threshold had been raised to $162 million.

(a) Assuming the relationship is linear, use the relation (year, threshold) with $t = 4$ corresponding to 2004 to find an equation model for the yearly increase of the payroll threshold. (b) Discuss what the slope indicates in this context. (c) According to this model, what was the approximate threshold in 2011? (d) In what year is the threshold expected to exceed 196 million?

## ▶ EXTENDING THE CONCEPT

**107.** The technique discussed in the Worthy of Note found next to Example 8 can be generalized for any line in the form $ax + by = c$. A line parallel to this line will have the form $ax + by = k$, and any line perpendicular will have the form $bx - ay = k$, where the value of $k$ is determined by substituting a given point $(x, y)$. For the equation $3x + 2y = 12$, (a) write the equation in slope-intercept form, then use this technique to find the equation of the line (b) parallel and (c) perpendicular, through the point $(6, -7)$. Then write both results in slope-intercept form to verify the results.

**108.** The general form of a linear equation is $ax + by = c$, where $a$ and $b$ are not simultaneously zero. (a) Find the $x$- and $y$-intercepts using the general form (substitute 0 for $y$, then 0 for $x$). Based on what you see, when does the intercept method work most efficiently? (b) Find the slope and $y$-intercept using the general form (solve for $y$). Based on what you see, when does the slope-intercept method work most efficiently?

**109.** Match the correct graph to the conditions stated for $m$ and $b$. There are more choices than graphs.

**a.** $m < 0, b < 0$      **b.** $m > 0, b < 0$
**c.** $m < 0, b > 0$      **d.** $m > 0, b > 0$
**e.** $m = 0, b > 0$      **f.** $m < 0, b = 0$
**g.** $m > 0, b = 0$      **h.** $m = 0, b < 0$

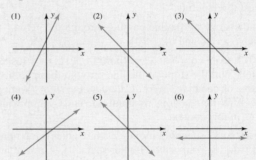

## ▶ MAINTAINING YOUR SKILLS

**110. (1.6)** Determine the allowable values:

    **a.** $y = \sqrt{2x - 5}$

    **b.** $y = \dfrac{5}{2x - 5}$

**111. (R.4)** Simplify without the use of a calculator.

    **a.** $27^{\frac{2}{3}}$         **b.** $\sqrt{81x^2}$

**112. (1.1)** Three equations follow. One is an identity, another is a contradiction, and a third has a unique solution. State which is which.

$2(x - 5) + 13 - 1 = 9 - 7 + 2x$

$2(x - 4) + 13 - 1 = 9 + 7 - 2x$

$2(x - 5) + 13 - 1 = 9 + 7 + 2x$

**113. (1.1)** Compute the area of the circular sidewalk shown here $(A = \pi r^2)$. Use your calculator's value of $\pi$ and round the answer (only) to hundredths.

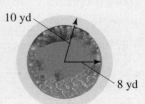

10 yd

8 yd

## MID-CHAPTER CHECK

1. Graph the relation $y = 2x^2 - 3$ by completing the table.

**Exercise 1**

| x | y |
|----|----|
| −6 |    |
| −4 |    |
| −2 |    |
| 0  |    |
| 2  |    |
| 4  |    |
| 6  |    |

2. While awaiting orders, two Coast Guard vessels sit idle at coordinates $(-6.4, 9.4)$ and $(6.8, -8.2)$. If a distress call comes from a boat located midway between the ships, what is the range and position of the boat in trouble? Assume that each unit is 1 mi.

3. Find the center and radius of the circle defined by $(x + 5)^2 + (y - 12)^2 = 169$, then sketch its graph.

4. Sketch the graph of the line $4x - 3y = 12$. Plot and label at least three points.

5. Find the slope of the line passing through the given points: $(-3, 8)$ and $(4, -10)$.

6. In 2011, Data.com lost $2 million. In 2012, they lost $0.5 million. Will the slope of the line through these points be positive or negative? Why? Calculate the slope. Were you correct? Write the slope as a unit rate and explain what it means in this context.

7. To earn some spending money, Sahara takes a job in a ski shop working primarily with her specialty—snowboards.

She is paid a monthly salary of $950 plus a commission of $7.50 for each snowboard she sells. (a) Write an equation that models her monthly earnings E. (b) Determine her income if she sells 20, 30, or 40 snowboards in one month. (c) Use the results of parts (a) and (b) to graph the line. (d) Determine the number of snowboards that must be sold for Sahara's monthly income to top $1300.

8. Find the equations and slopes of lines $L_1$ and $L_2$.

**Exercise 8**

9. Joe purchased his dream car in 2010 for $30,000. Just three yr later, the book value was only $21,000. (a) Find a linear equation modeling the value of the car. Then use the equation to (b) determine the car's value in 2015, and (c) determine the year in which the car will be worth only 10% of its purchase price.

10. (a) Find the intercepts of the lines $3x + 4y = 12$ and $4x - 3y = 12$. Using the intercepts, (b) find the slope of each line. Are the lines parallel, perpendicular, or neither?

---

## 2.4  Functions, Function Notation, and the Graph of a Function

**LEARNING OBJECTIVES**

*In Section 2.4 you will see how we can:*

☐ **A.** Identify the graph of a function

☐ **B.** Determine the domain and range of a function

☐ **C.** Use function notation and evaluate functions

☐ **D.** Read and interpret information given graphically

In this section, we introduce one of the most central ideas in mathematics—the concept of a function. Functions can model the cause-and-effect relationship that is so important to using mathematics as a decision-making tool. In addition, the study will help to unify and expand on many ideas that are already familiar.

### A. Functions and Relations

There is a special type of relation that merits further attention. A **function** is a relation where each element of the domain corresponds to exactly one element of the range. In other words, for each first coordinate or input value, there is only one possible second coordinate or output.

**Functions**

> A *function* is a relation that pairs each element from the *domain* with exactly one element from the *range*.

If the relation is defined by a mapping, we need only check that each element of the domain is mapped to exactly one element of the range. This is indeed the case for the mapping $P \rightarrow B$ from Figure 2.1 (page 148), where we saw that each person corresponded to only one birthday, and that it was impossible for one person to be born on two different days. For the relation $x = |y|$ shown in Figure 2.6 (page 150), each element of the domain except zero is paired with *more than one* element of the range. The relation $x = |y|$ is *not* a function.

---

**EXAMPLE 1** ▶    **Determining Whether a Relation Is a Function**

Three different relations are given in mapping notation below. Determine whether each relation is a function.

**a.**            **b.**            **c.**

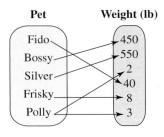
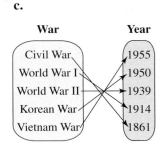

**Solution** ▶   Relation (a) is a function, since each person corresponds to exactly one room. This relation pairs math professors with their respective office numbers. Notice that while two people can be in one office, it is impossible for one person to physically be in two different offices.

Relation (b) is not a function, since we cannot tell whether Polly the Parrot weighs 2 lb or 3 lb (one element of the domain is mapped to two elements of the range).

Relation (c) is a function, where each major war is paired with the year it began.

Now try Exercises 5 through 10 ▶

---

If the relation is given as a set of ordered pairs or distinct plotted points, we need only check that no two points have the same first coordinate with a different second coordinate. This gives rise to an alternative definition for a function.

> **Functions (Alternate Definition)**
>
> A function is a set of ordered pairs $(x, y)$, in which each first coordinate
> is paired with only one second coordinate.

---

**EXAMPLE 2** ▶    **Identifying Functions**

Two relations named $f$ and $g$ are given; $f$ is stated as a set of ordered pairs, while $g$ is given as a set of plotted points. Determine whether each is a function.

$f$: $\{(-3, 0), (1, 4), (2, -5), (4, 2), (-3, -2), (3, 6), (0, -1), (4, -5), (6, 1)\}$

**Solution** ▶   The relation $f$ is not a function, since $-3$ is paired with two different outputs: $(-3, 0)$ and $(-3, -2)$.

The relation $g$ shown in the figure *is* a function. Each input corresponds to exactly one output, otherwise one point would be directly above another and have the same first coordinate.

Now try Exercises 11 through 18 ▶

**Figure 2.26**

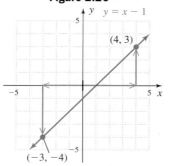

The graphs of $y = x - 1$ and $x = |y|$ from Section 2.1 offer additional insight into the definition of a function. Figure 2.26 shows the line $y = x - 1$ with emphasis on the plotted points $(4, 3)$ and $(-3, -4)$. The vertical movement shown from the $x$-axis to a point on the graph illustrates *the pairing of a given x-value with one related y-value*. Note the vertical line shows *only one related y-value* ($x = 4$ is paired with only $y = 3$). Figure 2.27 gives the graph of $x = |y|$, highlighting the points $(4, 4)$ and $(4, -4)$. The vertical movement shown here branches in two directions, associating one $x$-value with more than one $y$-value. This shows the relation $y = x - 1$ is a function, while the relation $x = |y|$ is not.

This "vertical connection" of a location on the $x$-axis to a point on the graph can be generalized into a **vertical line test** for functions.

**Figure 2.27**

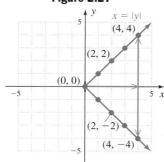

**Vertical Line Test**

A given graph is the graph of a function, if and only if every vertical line intersects the graph in at most one point.

Applying the test to the graph in Figure 2.26 helps to illustrate that the graph of any nonvertical line must be the graph of a function, as is the graph of any pointwise-defined relation where no $x$-coordinate is repeated. Compare the relations $f$ and $g$ from Example 2.

**EXAMPLE 3** ▶ **Using the Vertical Line Test**

Use the vertical line test to determine if any of the relations shown (from Section 2.1) are functions.

**Solution** ▶ Visualize a vertical line on each coordinate grid (shown in solid blue), then mentally shift the line to the left and right as shown in Figures 2.28, 2.29, and 2.30 (dashed lines). In Figures 2.28 and 2.29, every vertical line intersects the graph at most once, indicating both $y = x^2 - 2x$ and $y = \sqrt{9 - x^2}$ are functions. In Figure 2.30, a vertical line intersects the graph twice for any $x > 0$ [for instance, both $(4, 2)$ and $(4, -2)$ are on the graph]. The relation $x = y^2$ is *not* a function.

**Figure 2.28**

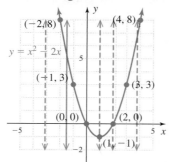

**Figure 2.29**

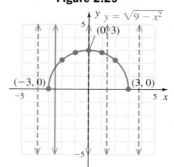

**Figure 2.30**

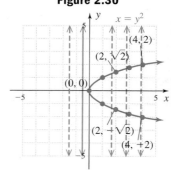

Now try Exercises 19 through 30 ▶

**EXAMPLE 4** ▶ **Using the Vertical Line Test**

Use a table of values to graph the relations defined by

**a.** $y = |x|$    **b.** $y = \sqrt{x}$,

then use the vertical line test to determine whether each relation is a function.

**Solution** ▶

**a.** For $y = |x|$, using input values from $x = -4$ to $x = 4$ produces the following table and graph (Figure 2.31). Note the result is a V-shaped graph that "opens upward." The point $(0, 0)$ of this absolute value graph is called the **vertex**. Since any vertical line will intersect the graph in at most one point, this is the graph of a function.

$y = |x|$

| $x$ | $y = |x|$ |
|-----|-----------|
| $-4$ | 4 |
| $-3$ | 3 |
| $-2$ | 2 |
| $-1$ | 1 |
| 0 | 0 |
| 1 | 1 |
| 2 | 2 |
| 3 | 3 |
| 4 | 4 |

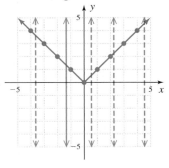

**Figure 2.31**

**b.** For $y = \sqrt{x}$, $x$-values less than zero do not produce a real number, so our graph actually begins at $(0, 0)$ (see Figure 2.32). Completing the table for nonnegative values produces the graph shown, which appears to rise to the right and remains in the first quadrant. Since any vertical line will intersect this graph in at most one place, $y = \sqrt{x}$ is also a function.

**Figure 2.32**

$y = \sqrt{x}$

| $x$ | $y = \sqrt{x}$ |
|-----|----------------|
| 0 | 0 |
| 1 | 1 |
| 2 | $\sqrt{2} \approx 1.4$ |
| 3 | $\sqrt{3} \approx 1.7$ |
| 4 | 2 |

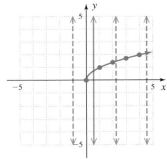

☑ **A.** You've just seen how we can identify the graph of a function

Now try Exercises 31 through 34 ▶

## B. The Domain and Range of a Function

### Vertical Boundary Lines and the Domain

In addition to its use as a graphical test for functions, a vertical line can help determine the domain of a function from its graph. For the graph of $y = \sqrt{x}$ (Figure 2.32), a vertical line will not intersect the graph until $x = 0$, and then will intersect the graph for all values $x \geq 0$ (showing the function is defined for these values). These **vertical boundary lines** indicate the domain is $x \in [0, \infty)$.

For the graph of $y = |x|$ (Figure 2.31), a vertical line will intersect the graph (or its infinite extension) for *all values* of $x$, and the domain is $x \in (-\infty, \infty)$. Using vertical lines in this way also affirms the domain of $y = x - 1$ (Figure 2.26) is $x \in (-\infty, \infty)$ while the domain of the relation $x = |y|$ (Figure 2.27) is $x \in [0, \infty)$.

### Range and Horizontal Boundary Lines

The range of a relation can be found using a **horizontal boundary line,** since it will associate a value on the y-axis with a point on the graph (if it exists). Simply visualize a horizontal line and move the line up or down until you determine the graph will always intersect the line, or will no longer intersect the line. This will give you the boundaries of the range. Mentally applying this idea to the graph of $y = \sqrt{x}$ (Figure 2.32) shows the range is $y \in [0, \infty)$. Although shaped very differently, a horizontal boundary line shows the range of $y = |x|$ (Figure 2.31) is also $y \in [0, \infty)$.

**EXAMPLE 5** ▶ **Determining the Domain and Range of a Function**

Use a table of values to graph the functions defined by

**a.** $y = x^2$    **b.** $y = \sqrt[3]{x}$

Then use boundary lines to determine the domain and range of each.

**Solution** ▶ **a.** For $y = x^2$, it seems convenient to use inputs from $x = -3$ to $x = 3$, producing the following table and graph. Note the result is a basic parabola that "opens upward" (both ends point in the positive y direction), with a vertex at (0, 0). Figure 2.33 shows a vertical line will intersect the graph or its extension anywhere it is placed. The domain is $x \in (-\infty, \infty)$. Figure 2.34 shows a horizontal line will intersect the graph only for values of $y$ that are greater than or equal to 0. The range is $y \in [0, \infty)$.

**Squaring Function**

| $x$ | $y = x^2$ |
|---|---|
| -3 | 9 |
| -2 | 4 |
| -1 | 1 |
| 0 | 0 |
| 1 | 1 |
| 2 | 4 |
| 3 | 9 |

Figure 2.33

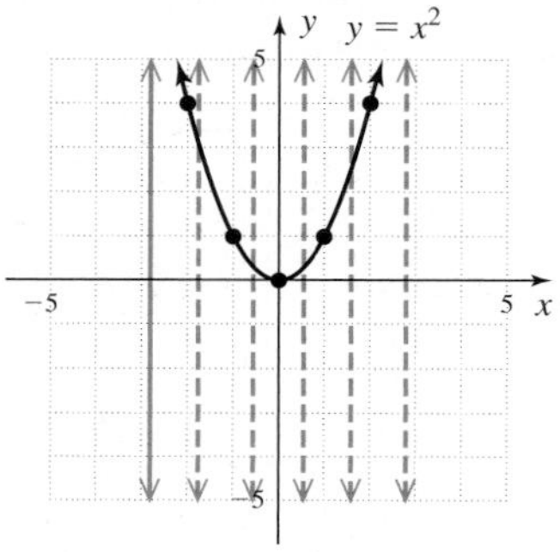

Figure 2.34

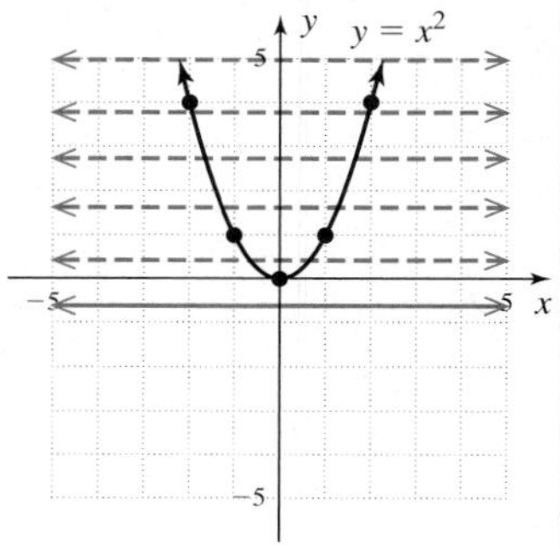

**b.** For $y = \sqrt[3]{x}$, we select points that are perfect cubes where possible, then a few others to round out the graph. The resulting table and graph are shown. Notice there is a "pivot point" at (0, 0) called a **point of inflection,** and the ends of the graph point in opposite directions. Figure 2.35 shows a vertical line will intersect the graph or its extension anywhere it is placed. Figure 2.36 shows any horizontal line will also intersect the graph. The domain is $x \in (-\infty, \infty)$, and the range is $y \in (-\infty, \infty)$.

**Cube Root Function**

| $x$ | $y = \sqrt[3]{x}$ |
|---|---|
| -8 | -2 |
| -4 | $\approx -1.6$ |
| -1 | -1 |
| 0 | 0 |
| 1 | 1 |
| 4 | $\approx 1.6$ |
| 8 | 2 |

Figure 2.35

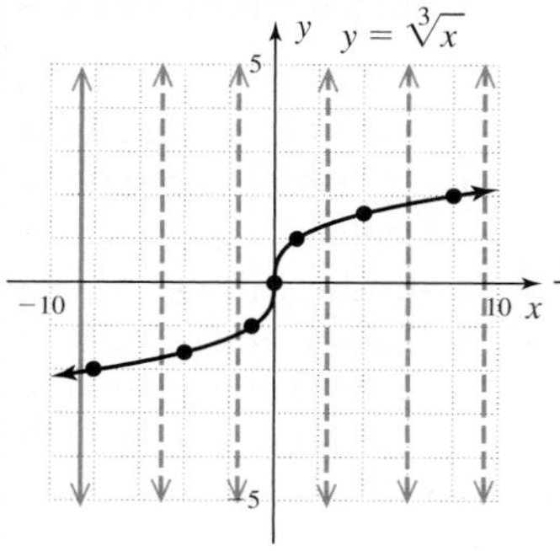

Figure 2.36

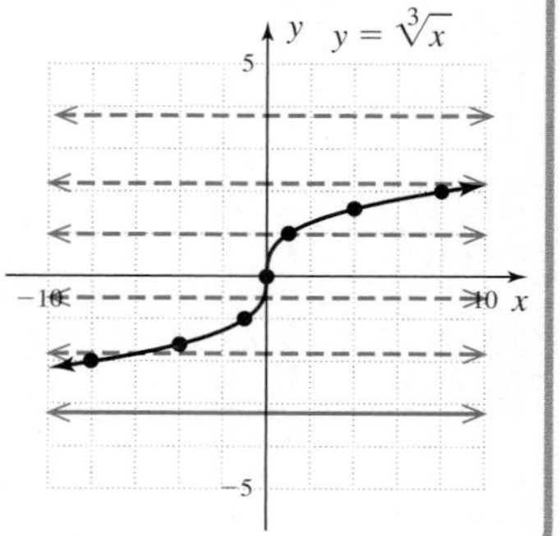

Now try Exercises 35 through 46 ▶

## Implied Domains

When stated in equation form, the domain of a function is often given implicitly by the expression used to define it, since the expression will dictate what input values are allowed. The **implied domain** is the set of all real numbers for which the function represents a real number. If the function involves a rational expression, the domain will exclude any input that causes a denominator of zero, since division by zero is undefined. If the function involves a radical expression with an even index ($\sqrt[n]{A}$ where $n$ is even), the domain will exclude inputs that create a negative radicand, since the expression will represent a real number only when $A \geq 0$.

**EXAMPLE 6** ▶ **Determining Implied Domains**

State the domain of each function using interval notation.

    **a.** $y = \dfrac{3}{x + 2}$      **b.** $y = \sqrt{2x + 3}$      **c.** $y = \dfrac{x - 1}{x^2 - 9}$      **d.** $y = x^2 - 5x + 7$

**Solution** ▶   **a.** Since $y = \frac{3}{x + 2}$ involves a rational expression, we must avoid division by zero. Setting the denominator equal to zero gives $x + 2 = 0$ with solution $x = -2$, showing $-2$ is not in the domain. The domain is $x \in (-\infty, -2) \cup (-2, \infty)$.

    **b.** Since $y = \sqrt{2x + 3}$ involves a radical expression with implied index 2 (even), the radicand must be nonnegative. Using the radicand and a "greater than or equal to" inequality gives $2x + 3 \geq 0$ with solution $x \geq -\frac{3}{2}$. The domain is $x \in \left[-\frac{3}{2}, \infty\right)$.

    **c.** Since $y = \frac{x - 1}{x^2 - 9}$ involves a rational expression, we must avoid division by zero. Setting the denominator equal to zero gives $x^2 - 9 = 0$ with solutions $x = \pm 3$, showing $-3$ and $3$ are not in the domain. The domain is $x \in (-\infty, -3) \cup (-3, 3) \cup (3, \infty)$. Note than $x = 1$ *is* in the domain, since $\frac{0}{-8} = 0$ is a real number.

    **d.** Since the squaring operation, multiplying by a constant, addition, and subtraction are defined for all real numbers, there are no restrictions for $y = x^2 - 5x + 7$ (also note that no rational or radical expressions are involved) and the domain is all real numbers: $x \in (-\infty, \infty)$.

**Now try Exercises 47 through 64** ▶

**EXAMPLE 7** ▶ **Determining Implied Domains**

Determine the domain of each function:

    **a.** $y = \sqrt{\dfrac{7}{x + 3}}$             **b.** $y = \dfrac{2x}{\sqrt{4x + 5}}$

**Solution** ▶   **a.** For $y = \sqrt{\frac{7}{x + 3}}$, we must have $\frac{7}{x + 3} \geq 0$ (for the radicand) **and** $x + 3 \neq 0$ (for the denominator). Since the numerator is *always* positive, we need $x + 3 > 0$, which gives $x > -3$. The domain is $x \in (-3, \infty)$.

    **b.** For $y = \frac{2x}{\sqrt{4x + 5}}$, we must have $4x + 5 \geq 0$ (for the radicand) **and** $\sqrt{4x + 5} \neq 0$ (for the denominator). This shows we need $4x + 5 > 0$, so $x > \frac{-5}{4}$. The domain is $x \in \left(\frac{-5}{4}, \infty\right)$.

☑ **B.** You've just seen how we can determine the domain and range of a function

**Now try Exercises 65 through 70** ▶

## C. Function Notation

In our study of functions, you've likely noticed that the relationship between input and output values is an important one. To highlight this fact, think of a function as a simple machine, which can *process inputs* using a stated sequence of operations, then deliver a single output. The inputs are *x*-values, a program we'll name *f* performs the operations on *x*, and *y* is the resulting output (see Figure 2.37). Once again we see that "the value of *y* depends on the value of *x*," or simply "*y* is a function of *x*." We write "*y* is a function of *x*" as $y = f(x)$ using **function notation.** You are already familiar with letting a variable represent a number. Here we do something quite different, as the letter *f* is used to represent *a sequence of operations to be performed on x.* Consider the function $y = \frac{x}{2} + 1$, which we'll now write as $f(x) = \frac{x}{2} + 1$ [since $y = f(x)$].

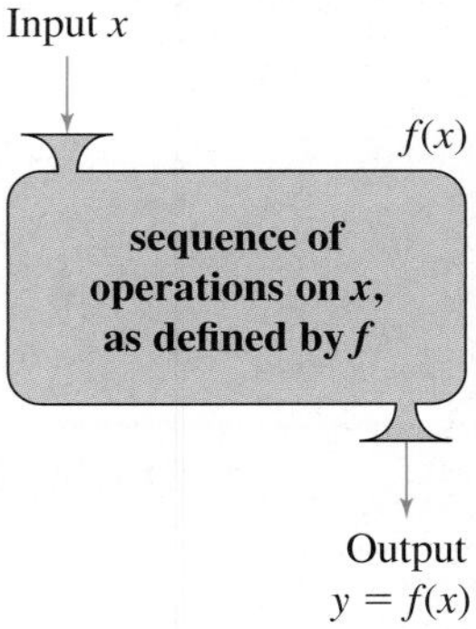

In words the function says, "divide inputs by 2, then add 1." To evaluate the function at $x = 4$ (Figure 2.38) we have:

**Figure 2.37**

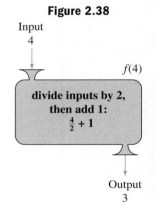

$$f(\overset{\text{input 4}}{x}) = \frac{\overset{\text{input 4}}{x}}{2} + 1$$

$$f(4) = \frac{4}{2} + 1$$

$$= 2 + 1$$

$$= 3$$

**Figure 2.38**

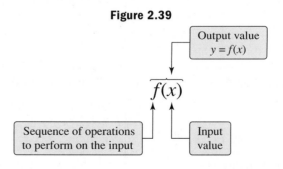

Function notation enables us to summarize the three most important aspects of a function using a single expression, as shown in Figure 2.39.

**Figure 2.39**

Output value
$y = f(x)$

$\overline{f(x)}$

Sequence of operations to perform on the input

Input value

Instead of saying, "... when $x = 4$, the value of the function is 3," we simply say "*f* of 4 is 3," or write $f(4) = 3$. Note that the ordered pair $(4, 3)$ is equivalent to $(4, f(4))$.

**CAUTION** ▶ Although *f(x)* is the favored notation for a "function of *x*," other letters can also be used. For example, *g(x)* and *h(x)* also denote functions of *x*, where *g* and *h* represent different sequences of operations on the *x*-inputs. It is also important to remember that these represent *function values* and not the product of two variables: $f(x) \neq f \cdot (x)$.

**EXAMPLE 8** ▶ **Evaluating a Function**

Given $f(x) = -2x^2 + 4x$, find

    **a.** $f(-2)$      **b.** $f\left(\dfrac{7}{2}\right)$      **c.** $f(2a)$      **d.** $f(a + 1)$

**Solution** ▶

**a.**
$$f(x) = -2x^2 + 4x$$
$$f(-2) = -2(-2)^2 + 4(-2)$$
$$= -2(4) + (-8)$$
$$= -8 - 8 = -16$$

**b.**
$$f(x) = -2x^2 + 4x$$
$$f\left(\frac{7}{2}\right) = -2\left(\frac{7}{2}\right)^2 + 4\left(\frac{7}{2}\right)$$
$$= \frac{-49}{2} + 14 = \frac{-21}{2}$$

**c.**
$$f(x) = -2x^2 + 4x$$
$$f(2a) = -2(2a)^2 + 4(2a)$$
$$= -2(4a^2) + 8a$$
$$= -8a^2 + 8a$$

**d.**
$$f(x) = -2x^2 + 4x$$
$$f(a + 1) = -2(a + 1)^2 + 4(a + 1)$$
$$= -2(a^2 + 2a + 1) + 4a + 4$$
$$= -2a^2 - 4a - 2 + 4a + 4$$
$$= -2a^2 + 2$$

☑ **C.** You've just seen how we can use function notation and evaluate functions

Now try Exercises 71 through 86 ▶

## D. Reading and Interpreting Information Given Graphically

Graphs are an important part of studying functions, and learning to read and interpret them correctly is a high priority. A graph highlights and emphasizes the all-important input/output relationship that defines a function. In this study, we hope to firmly establish that the following statements are synonymous:

    **1.** $f(-2) = 5$
    **2.** $(-2, f(-2)) = (-2, 5)$
    **3.** $(-2, 5)$ is on the graph of $f$, and
    **4.** when $x = -2$, $f(x) = 5$

**EXAMPLE 9A** ▶ **Reading a Graph**

For the functions $f$ and $g$ whose graphs are shown in Figures 2.40 and 2.41
    **a.** State the domain of the function.
    **b.** Evaluate the function at $x = 2$.
    **c.** Determine the value(s) of $x$ for which $y = 3$.
    **d.** State the range of the function.

**Figure 2.40**

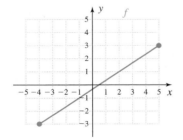

**Figure 2.41**

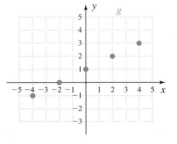

**Solution** ▶ **For $f$,**

a. The graph is a continuous line segment with endpoints at $(-4, -3)$ and $(5, 3)$, so we state the domain in interval notation. Using a vertical boundary line we note the smallest $x$-value is $-4$ and the largest is 5. The domain is $x \in [-4, 5]$.

b. The graph shows an input of $x = 2$ corresponds to an output of $y = 1$: $f(2) = 1$ since $(2, 1)$ is a point on the graph (see Figure 2.42).

c. Reading from the $y$-axis, $f(x) = 3$ (or $y = 3$) must correspond to $x = 5$, since $(5, 3)$ is a point on the graph (see Figure 2.43).

d. Using a horizontal boundary line, the smallest $y$-value is $-3$ and the largest is 3. The range is $y \in [-3, 3]$.

**Figure 2.42**

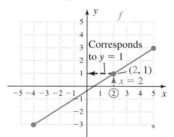

**$x = 2$ corresponds to $y = 1$**

**Figure 2.43**

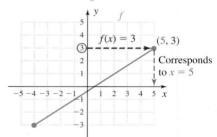

**$f(x) = 3$ corresponds to $x = 5$**

**For $g$,**

a. Since $g$ is given as a set of plotted points, we state the domain as the set of first coordinates: $D$: $\{-4, -2, 0, 2, 4\}$.

b. An input of $x = 2$ corresponds to an output of $y = 2$: $g(2) = 2$ since $(2, 2)$ is on the graph.

c. For $g(x) = 3$ (or $y = 3$) the input value must be $x = 4$, since $(4, 3)$ is a point on the graph.

d. The range is the set of all second coordinates: $R$: $\{-1, 0, 1, 2, 3\}$.

---

**EXAMPLE 9B** ▶ **Reading a Graph**

Use the graph of $f(x)$ given to answer the following questions:

a. What is the value of $f(-2)$?

b. What value(s) of $x$ satisfy $f(x) = 1$?

**Solution** ▶

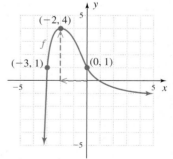

a. The notation $f(-2)$ says to find the value of the function $f$ when $x = -2$. Expressed graphically, we go to $x = -2$ and locate the corresponding point on the graph (blue arrows). Here we find that $f(-2) = 4$.

b. For $f(x) = 1$, we're looking for $x$-inputs that result in an output of $y = 1$ [since $y = f(x)$]. From the graph, we note there are two points with a $y$-coordinate of 1, namely, $(-3, 1)$ and $(0, 1)$. This shows $f(-3) = 1$, and $f(0) = 1$, so the required $x$-values are $x = -3$ and $x = 0$.

**Now try Exercises 87 through 92** ▶

In many applications involving functions, the domain and range can be determined by the context or situation given.

**EXAMPLE 10** ▶ **Determining the Domain and Range from the Context**

Paul's 2009 Voyager has a 20-gal tank and gets 18 mpg. The number of miles he can drive (his range) depends on how much gas is in the tank. As a function we have $M(g) = 18g$, where $M(g)$ represents the total distance in miles and $g$ represents the gallons of gas in the tank (see graph). Find the domain and range.

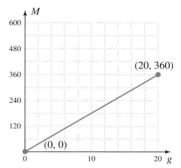

**Solution** ▶ Begin evaluating at $x = 0$, since the tank cannot hold less than zero gallons. With an empty tank, the (minimum) range is $M(0) = 18(0)$ or 0 mi. On a full tank, the maximum range is $M(20) = 18(20)$ or 360 mi. As shown in the graph, the domain is $g \in [0, 20]$ and the corresponding range is $M(g) \in [0, 360]$.

 **D. You've just seen how we can read and interpret information given graphically**

Now try Exercises 96 through 103 ▶

## 2.4 EXERCISES

▶ **CONCEPTS AND VOCABULARY**

**Fill in each blank with the appropriate word or phrase. Carefully reread the section, if necessary.**

1. If a relation is given in ordered pair form, we state the domain by listing all of the _____ coordinates in a set.

2. A relation is a function if each element of the _____ is paired with _____ _____ element of the range.

3. Discuss/Explain why the relation $y = x^2$ is a function, while the relation $x = y^2$ is not. Justify your response using graphs, ordered pairs, and so on.

4. Discuss/Explain the process of finding the domain and range of a function given its graph, using vertical and horizontal boundary lines. Include a few illustrative examples.

▶ **DEVELOPING YOUR SKILLS**

**Determine whether or not the mappings shown represent functions. If not, explain how the definition of a function is violated.**

5.
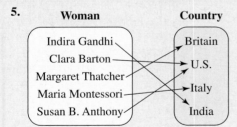

6.
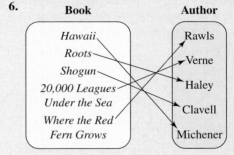

7.
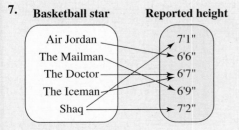

8.
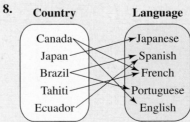

**For Exercises 9 and 10, state why the situations described are not "functional." In other words, explain how the definition of a function is violated in each case.**

9. In trying to remember the worst crop failure in recent history, one farmer said to another, "Yep—that was the year that summer temperatures hit 106°!"

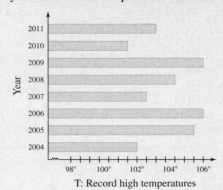

T: Record high temperatures

10. In an effort to increase the effectiveness of their marketing team, the Hypertext Corporation decides they are going to hire the person scoring highest on the National Association of Marketers Exam (NAME).

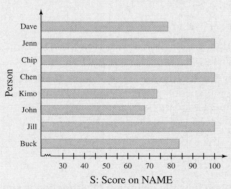

S: Score on NAME

**Determine whether or not the relations given represent functions. If not, explain how the definition of a function is violated.**

11. $\{(-2, 1), (0, 5), (3, -4), (4, 2), (-4, 7), (4, 7), (1, 0), (4, -5), (7, 2)\}$

12. $\{(-7, -5), (-5, 3), (4, 0), (-3, -5), (1, -6), (0, 9), (2, -8), (3, -2), (-5, 7)\}$

13. $\{(9, -10), (-7, 6), (6, -10), (4, -1), (2, -2), (1, 8), (0, -2), (-2, -7), (-6, 4)\}$

14. $\{(1, -81), (-2, 64), (-3, 49), (5, -36), (-8, 25), (13, -16), (-21, 9), (34, -4), (-55, 1)\}$

15.

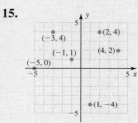

16.

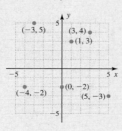

17.

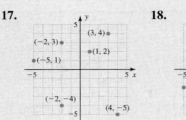

18.

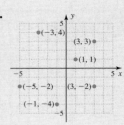

**Determine whether or not the relations shown represent functions. If not, give an example of how the definition of a function is violated.**

19.

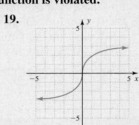

20.

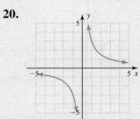

21.

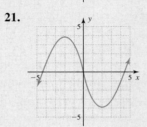

22.

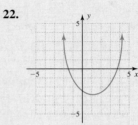

23.

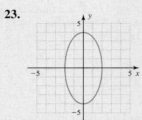

24.

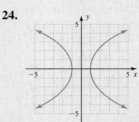

25.

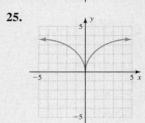

26.

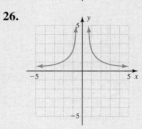

27.

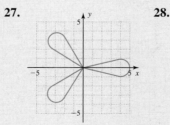

28.

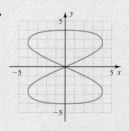

**29.**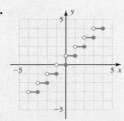

**30.**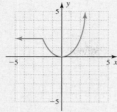

**45.**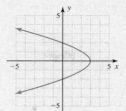

**46.**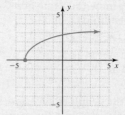

**Graph each relation using a table, then use the vertical line test to determine if the relation is a function.**

**31.** $y = x$

**32.** $y = \sqrt[3]{x}$

**33.** $y = (x + 2)^2$

**34.** $x = |y - 2|$

**Determine whether or not the relations shown represent functions, then determine the domain and range of each.**

**35.**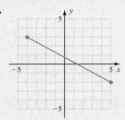

**36.**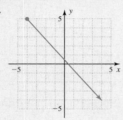

**37.**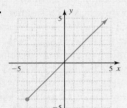

**38.**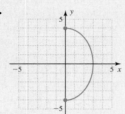

**39.**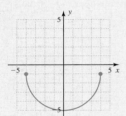

**40.**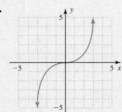

**41.**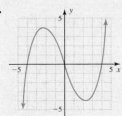

**42.**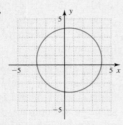

**43.**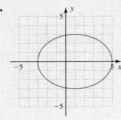

**44.**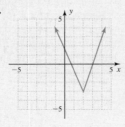

**Determine the domain of the following functions, and write your response in interval notation.**

**47.** $f(x) = \dfrac{3}{x - 5}$

**48.** $g(x) = \dfrac{-2}{3 + x}$

**49.** $h(a) = \sqrt{3a + 5}$

**50.** $p(a) = \sqrt{5a - 2}$

**51.** $v(x) = \dfrac{x + 2}{x^2 - 25}$

**52.** $w(x) = \dfrac{x - 4}{x^2 - 49}$

**53.** $u = \dfrac{v - 5}{v^2 - 18}$

**54.** $p = \dfrac{q + 7}{q^2 - 12}$

**55.** $y = \dfrac{17}{25}x + 123$

**56.** $y = \dfrac{11}{19}x - 89$

**57.** $m = n^2 - 3n - 10$

**58.** $s = t^2 - 3t - 10$

**59.** $y = 2|x| + 1$

**60.** $y = |x - 2| + 3$

**61.** $y_1 = \dfrac{x}{x^2 - 3x - 10}$

**62.** $y_2 = \dfrac{x - 4}{x^2 + 2x - 15}$

**63.** $y = \dfrac{\sqrt{x - 2}}{2x - 5}$

**64.** $y = \dfrac{\sqrt{x + 1}}{3x + 2}$

**65.** $h(x) = \dfrac{-2}{\sqrt{x + 4}}$

**66.** $f(x) = \sqrt{\dfrac{5}{x - 2}}$

**67.** $g(x) = \sqrt{\dfrac{-4}{3 - x}}$

**68.** $p(x) = \dfrac{-7}{\sqrt{5 - x}}$

**69.** $r(x) = \dfrac{2x - 1}{\sqrt{3x - 7}}$

**70.** $s(x) = \dfrac{x^2 - 4}{\sqrt{11 - 2x}}$

**Determine the value of $f(-6)$, $f(\frac{6}{5})$, $f(2c)$, and $f(c + 1)$, then simplify.**

**71.** $f(x) = \dfrac{1}{2}x + 3$

**72.** $f(x) = \dfrac{2}{3}x - 5$

**73.** $f(x) = 5x^2 - 4x$

**74.** $f(x) = 10x^2 + 3x$

**Determine the value of $h(3)$, $h(-\frac{2}{3})$, $h(3a)$, and $h(a - 2)$, then simplify.**

**75.** $h(x) = \dfrac{12}{x}$

**76.** $h(x) = \dfrac{36}{x^2}$

**77.** $h(x) = \dfrac{5|x|}{x}$

**78.** $h(x) = \dfrac{4|x|}{x}$

**Determine the value of $g(4)$, $g(\frac{3}{2})$, $g(2c)$, and $g(c + 3)$, then simplify.**

**79.** $g(r) = 2\pi r$

**80.** $g(r) = 2\pi rh$

**81.** $g(r) = \pi r^2$

**82.** $g(r) = \pi r^2 h$

Determine the value of $p(5)$, $p\left(\frac{3}{2}\right)$, $p(3a)$, and $p(a-1)$, then simplify.

**83.** $p(x) = \sqrt{2x - 1}$      **84.** $p(x) = \sqrt{4x + 3}$

**85.** $p(x) = \dfrac{3x^2 - 5}{x^2}$      **86.** $p(x) = \dfrac{2x^2 + 3}{x^2}$

Use the graph of each function given to (a) state the domain, (b) state the range, (c) evaluate $f(2)$, and (d) find the value(s) $x$ for which $f(x) = k$ ($k$ a constant).

**87.** $k = 4$            **88.** $k = 3$

**89.** $k = 1$            **90.** $k = -3$

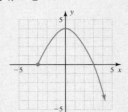 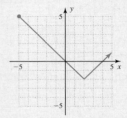

**91.** $k = 2$            **92.** $k = -1$

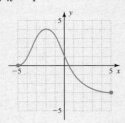 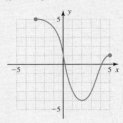

## ▶ WORKING WITH FORMULAS

**93. Ideal weight for males:** $W(H) = \frac{9}{2}H - 151$

The ideal weight for an adult male can be modeled by the function shown, where $W$ is his weight in pounds and $H$ is his height in inches. (a) Find the ideal weight for a male who is 75 in. tall. (b) If I am 72 in. tall and weigh 210 lb, how much weight should I lose?

**94. Fahrenheit to Celsius conversions:** $C = \frac{5}{9}(F - 32)$

The relationship between Fahrenheit and Celsius temperatures is modeled by the function shown. (a) What is the temperature in degrees Celsius if °F = 41? (b) Use the formula to solve for $F$ in terms of $C$, then substitute the result from part (a). What do you notice?

**95. Spending on cable and satellite TV:** $s = 29t + 96$

The data from Example 1, Section 2.1 is closely modeled by the formula shown, where $t$ represents the year ($t = 0$ corresponds to the year 2000) and $s$ represents the average amount spent per person, per year in the United States. (a) List five ordered pairs for this relation using $t = 3, 5, 7, 9, 11$. Does the model give a good approximation of the actual data? (b) According to the model, what will be the average amount spent on cable and satellite TV in the year 2013? (c) According to the model, in what year will annual spending surpass \$500?

## ▶ APPLICATIONS

**96. Vehicle range:** John's old '87 LeBaron has a 15-gal gas tank and gets 23 mpg. The number of miles he can drive is a function of how much gas is in the tank. (a) Write this relationship in function form and (b) determine the domain and range of the function in this context.

**97. Operating time:** Jackie has a gas-powered model boat with a 5-oz gas tank. The boat will run for 2.5 min on each ounce. The number of minutes she can operate the boat is a function of how much gas is in the tank. (a) Write this relationship in function form and (b) determine the domain and range of the function in this context.

 **98. Volume of a cube:** The volume of a cube depends on the length of the sides. In other words, volume is a function of the sides: $V(s) = s^3$. (a) In practical terms, what is the domain of this function? (b) Evaluate $V(6.25)$ and (c) evaluate the function for $s = 2x^2$.

 **99. Volume of a cylinder:** For a fixed radius of 10 cm, the volume of a cylinder depends on its height. In other words, volume is a function of height: $V(h) = 100\pi h$. (a) In practical terms, what is the domain of this function? (b) Evaluate $V(7.5)$ and (c) evaluate the function for $h = \frac{8}{\pi}$.

**100. Rental charges:** Temporary Transportation Inc. rents cars (local rentals only) for a flat fee of \$19.50 and an hourly charge of \$12.50. This means that cost is a function of the hours the car is rented. (a) Write this relationship in function form; (b) find the cost if the car is rented for 3.5 hr; (c) determine how long the car was rented if the bill came to \$119.75; and (d) determine the domain and range of the function in this context, if your budget limits you to a maximum of \$150 for the rental.

**101. Cost of a service call:** Paul's Plumbing charges a flat fee of $50 per service call plus an hourly rate of $42.50. This means that cost is a function of the hours the job takes to complete. (a) Write this relationship in function form; (b) find the cost of a service call that takes $2\frac{1}{2}$ hr; (c) find the number of hours the job took if the charge came to $262.50; and (d) determine the domain and range of the function in this context, if your insurance company has agreed to pay for all charges over $500 for the service call.

**102. Hair growth:** While influenced by age, health, genetics, and other factors, human hair grows at a rate of about 1.2 cm per month. If Charlene's hair is currently 35 cm long, (a) write a function that will determine the length of her hair after $t$ months. (b) Use the function to find the length of her hair in $7\frac{1}{2}$ months. (c) If she wants to grow her hair to a length of 50 cm, how many months will it take?

**103. Lawn care:** Most common grasses grow at a rate of about 2.1 cm per week, with the rate affected by temperature, humidity, time of year, and other factors. If the grass in Caden's yard is currently 1.9 cm tall, (a) write a function that will determine the height of the lawn after $t$ weeks. (b) Use the function to find the height of the lawn 3 weeks later. (c) How many weeks can Caden put off the mowing if the neighborhood association gives out warnings to anyone whose grass exceeds a height of 14.5 cm?

**104. Predicting tides:** The graph shown approximates the height of the tides at Fair Haven, New Brunswick, for a 12-hr period. (a) Is this the graph of a function? Why? (b) Approximately what time did high tide occur? (c) How high is the tide at 6 P.M.? (d) What time(s) will the tide be 2.5 m?

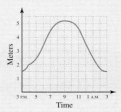

**105. Twitter account followers:** The graph shown models the growth in the number of people that follow my Twitter account. (a) Is this the graph of a function? Why? (b) What was the approximate number of followers at the start of 2010? (c) In what year did my number of followers surpass 120?

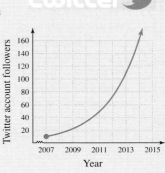

▶ **EXTENDING THE CONCEPT**

**106.** A father challenges his son to a 400-m race, depicted in the graph shown here.

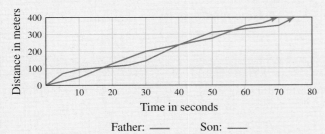

Father: —— Son: ——

a. Who won and what was the approximate winning time?

b. Approximately how many meters behind was the second place finisher?

c. Estimate the number of seconds the father was in the lead in this race.

d. How many times during the race were the father and son tied?

**107.** Sketch the graph of $f(x) = x^2 - 4$, then discuss how you could use this graph to obtain the graph of $F(x) = |x^2 - 4|$ without computing additional points.

Determine what the graph of $g(x) = \frac{|x^2 - 4|}{x^2 - 4}$ would look like.

**108.** If the equation of a function is given, the domain is implicitly defined by input values that generate real-valued outputs. But unless the graph is given or can be easily sketched, we must attempt to find the range analytically *by solving for x in terms of y.* We should note that sometimes this is an easy task, while at other times it is virtually impossible and we must rely on other methods. For the following functions, determine the implicit domain and find the range by solving for $x$ in terms of $y$.

a. $y = \frac{x-3}{x+2}$          b. $y = x^2 - 3$

▶ **MAINTAINING YOUR SKILLS**

**109.** (2.1) Find the equation of a circle whose center is $(4, -1)$ with a radius of 5. Then graph the circle.

**110.** (R.4) Compute the sum and product indicated:
   **a.** $\sqrt{24} + 6\sqrt{54} - \sqrt{6}$
   **b.** $(2 + \sqrt{3})(2 - \sqrt{3})$

**111.** (1.5) Solve the equation by factoring, then check the result(s) using substitution: $3x^2 - 4x = 7$.

**112.** (R.5) Factor the following polynomials completely:
   **a.** $x^3 - 3x^2 - 25x + 75$
   **b.** $2x^2 - 13x - 24$
   **c.** $8x^3 - 125$

---

## 2.5   Analyzing the Graph of a Function

**LEARNING OBJECTIVES**

*In Section 2.5 you will see how we can:*

☐ **A.** Determine whether a function is even, odd, or neither

☐ **B.** Determine intervals where a function is positive or negative

☐ **C.** Determine where a function is increasing or decreasing

☐ **D.** Identify the maximum and minimum values of a function

☐ **E.** Develop a formula to calculate rates of change for any function

In this section, we'll consolidate and refine many of the ideas we've encountered related to functions. When functions and graphs are applied as real-world models, we create numeric and visual representations that enable an informed response to questions involving *maximum* efficiency, *positive* returns, *increasing* costs, and other relationships that can have a great impact on our lives.

### A.  Graphs and Symmetry

While the domain and range of a function will remain dominant themes in our study, for the moment we turn our attention to other characteristics of a function's graph. We begin with the concept of symmetry.

#### Symmetry about the y-Axis

Consider the graph of $f(x) = x^4 - 4x^2$ shown in Figure 2.44, where the portion of the graph to the left of the $y$-axis appears to be a mirror image of the portion to the right. A function is **symmetric about the y-axis** if, given any point $(x, y)$ on the graph, the point $(-x, y)$ is also on the graph. We note that $(-1, -3)$ is on the graph, as is $(1, -3)$, and that $(-2, 0)$ is an $x$-intercept of the graph, as is $(2, 0)$. Functions that are symmetric about the $y$-axis are also known as **even functions** and in general we have:

**Figure 2.44**

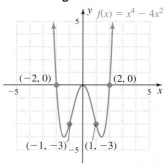

**Even Functions: y-Axis Symmetry**

A function $f$ is an *even function* if and only if, for each point $(x, y)$ on the graph of $f$, the point $(-x, y)$ is also on the graph. In function notation

$$f(-x) = f(x)$$

Symmetry can be a great help in graphing new functions, enabling us to plot fewer points and to complete the graph using properties of symmetry.

**EXAMPLE 1** ▶  **Graphing an Even Function Using Symmetry**

**a.** The function $g(x)$ in Figure 2.45 (shown in solid blue) is known to be even. Draw the complete graph.

**b.** Prove that $h(x) = x^{\frac{2}{3}}$ is an even function, then sketch the complete graph using $h(0)$, $h(1)$, $h(8)$, and $y$-axis symmetry.

**Figure 2.45**

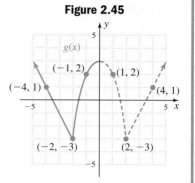

**Solution** ▶  **a.** To complete the graph of $g$ (see Figure 2.45) use the points $(-4, 1)$, $(-2, -3)$, $(-1, 2)$, and $y$-axis symmetry to find additional points. The corresponding ordered pairs are $(4, 1)$, $(2, -3)$, and $(1, 2)$, which we use to help draw a "mirror image" of the partial graph given.

**b.** To prove that $h(x) = x^{\frac{2}{3}}$ is an even function, we must show $h(-x) = h(x)$.

$$h(-x) \overset{?}{=} h(x) \qquad \text{first step of proof}$$

$$(-x)^{\frac{2}{3}} \overset{?}{=} (x)^{\frac{2}{3}} \qquad \text{evaluate } h(-x) \text{ and } h(x)$$

$$\left[(-x)^2\right]^{\frac{1}{3}} \overset{?}{=} \left[(x)^2\right]^{\frac{1}{3}} \qquad \text{rewrite; } x^{mn} = (x^m)^n$$

$$(x^2)^{\frac{1}{3}} = (x^2)^{\frac{1}{3}} \qquad \text{result; } (-x)^2 = x^2$$

**Figure 2.46**

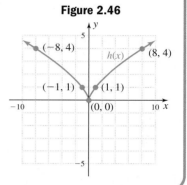

Using $h(0) = 0$, $h(1) = 1$, and $h(8) = 4$ with $y$-axis symmetry produces the graph shown in Figure 2.46.

**Now try Exercises 5 through 10** ▶

## Symmetry about the Origin

Another common form of symmetry is known as **symmetry about the origin.** As the name implies, the graph is somehow "centered" at $(0, 0)$. This form of symmetry is easy to see for closed figures with their center at $(0, 0)$, like certain polygons, circles, and ellipses (these will exhibit both $y$-axis symmetry *and* symmetry about the origin). Note the relation graphed in Figure 2.47 contains the points $(-3, 3)$ and $(3, -3)$, along with $(-1, -4)$ and $(1, 4)$. But the function $f(x)$ in Figure 2.48 also contains these points and is, in the same sense, symmetric about the origin (the paired points are on opposite sides of the $x$- and $y$-axes, and a like distance from the origin).

**Figure 2.47**

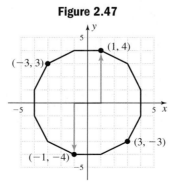

**Figure 2.48**

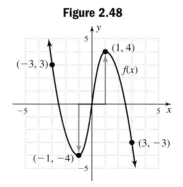

Functions symmetric about the origin are known as **odd functions** and in general we have:

> ### Odd Functions: Symmetry about the Origin
>
> A function $f$ is an *odd function* if and only if, for each point $(x, y)$ on the graph of $f$, the point $(-x, -y)$ is also on the graph. In function notation
>
> $$f(-x) = -f(x)$$

**EXAMPLE 2** ▶ **Graphing an Odd Function Using Symmetry**

    **a.** In Figure 2.49, the function $g(x)$ given (shown in solid blue) is known to be *odd*. Draw the complete graph.

    **b.** Prove that $h(x) = x^3 - 4x$ is an odd function, then sketch the graph using $h(-2)$, $h(-1)$, $h(0)$, and symmetry about the origin.

**Solution** ▶ **a.** To complete the graph of $g$, use the points $(-6, 3)$, $(-4, 0)$, and $(-2, 2)$ and symmetry about the origin to find additional points. The corresponding ordered pairs are $(6, -3)$, $(4, 0)$, and $(2, -2)$, which we use to help draw a "mirror image" of the partial graph given (see Figure 2.49).

**Figure 2.49**

**Figure 2.50**

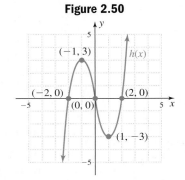

    **b.** To prove that $h(x) = x^3 - 4x$ is an odd function, we must show that $h(-x) = -h(x)$.

$$h(-x) \overset{?}{=} -h(x)$$
$$(-x)^3 - 4(-x) \overset{?}{=} -[x^3 - 4x]$$
$$-x^3 + 4x = -x^3 + 4x \checkmark$$

Using $h(-2) = 0$, $h(-1) = 3$, and $h(0) = 0$ with symmetry about the origin produces the graph shown in Figure 2.50.

☑ **A. You've just seen how we can determine whether a function is even, odd, or neither**

**Now try Exercises 11 through 24** ▶

Finally, some relations also exhibit a third form of symmetry, that of symmetry about the $x$-axis. If the graph of a circle is centered at the origin, the graph has both odd and even symmetry, and is also symmetric about the $x$-axis. Besides the one trivial exception $f(x) = 0$, any graph that exhibits $x$-axis symmetry cannot be the graph of a function since it fails the vertical line test.

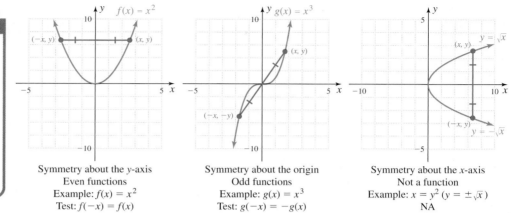

Symmetry about the $y$-axis      Symmetry about the origin      Symmetry about the $x$-axis

Even functions                 Odd functions                 Not a function

Example: $f(x) = x^2$        Example: $g(x) = x^3$        Example: $x = y^2$ $(y = \pm\sqrt{x})$

Test: $f(-x) = f(x)$        Test: $g(-x) = -g(x)$               NA

## B. Intervals Where a Function Is Positive or Negative

Consider the graph of $f(x) = x^2 - 4$ shown in Figure 2.51, which has $x$-intercepts at $(-2, 0)$ and $(2, 0)$. As in Section 2.2, the $x$-intercepts have the form $(x, 0)$ and are called the **zeroes** of the function (the $x$-input causes an output of 0). Just as zero on the number line separates negative numbers from positive numbers, the zeroes of a function (that crosses the $x$-axis) separate $x$-intervals where a function is negative from $x$-intervals where the function is positive. Noting that outputs ($y$-values) are positive in Quadrants I and II, $f(x) > 0$ in intervals where its graph is *above the x-axis*. Conversely, $f(x) < 0$ in $x$-intervals where its graph is *below the x-axis*. To illustrate, compare the graph of $f$ in Figure 2.51, with that of $g$ in Figure 2.52.

**Figure 2.51**                **Figure 2.52**

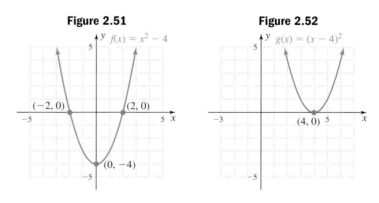

The graph of $f$ is a parabola, with $x$-intercepts of $(-2, 0)$ and $(2, 0)$. Using our previous observations, we note $f(x) \geq 0$ for $x \in (-\infty, -2] \cup [2, \infty)$ since the graph is on or above the $x$-axis, and $f(x) < 0$ for $x \in (-2, 2)$. The graph of $g$ is also a parabola, but is entirely on or above the $x$-axis, showing $g(x) \geq 0$ for $x \in \mathbb{R}$. The difference is that zeroes coming from factors of the form $(x - r)$ (with degree 1) allow the graph to cross the $x$-axis. The zeroes of $f$ came from $(x + 2)(x - 2) = 0$. Zeroes that come from factors of the form $(x - r)^2$ (with degree 2) cause the graph to "bounce" off the $x$-axis (intersect without crossing) since all outputs must be nonnegative. The zero of $g$ came from $(x - 4)^2 = 0$.

---

**EXAMPLE 3** ▶ **Solving an Inequality Using a Graph**

Use the graph of $g(x) = x^3 - 2x^2 - 4x + 8$ given to solve the inequalities
  **a.** $g(x) \geq 0$ **b.** $g(x) < 0$

**Solution** ▶ From the graph, the zeroes of $g$ ($x$-intercepts) occur at $(-2, 0)$ and $(2, 0)$.

  **a.** For $g(x) \geq 0$, the graph must be on or above the $x$-axis, meaning the solution is $x \in [-2, \infty)$.

  **b.** For $g(x) < 0$, the graph must be below the $x$-axis, and the solution is $x \in (-\infty, -2)$.

  As we might have anticipated from the graph, factoring by grouping gives $g(x) = (x + 2)(x - 2)^2$, with the graph crossing the $x$-axis at $-2$, and bouncing off the $x$-axis (intersecting without crossing) at $x = 2$.

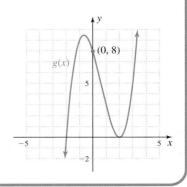

**Now try Exercises 25 through 30** ▶

Even if the function is not a polynomial, the zeroes can still be used to find $x$-intervals where the function is positive or negative.

---

**EXAMPLE 4** ▶ **Solving an Inequality Using a Graph**

For the graph of $r(x) = 2\sqrt{x + 6} - 4$ shown, solve
  **a.** $r(x) \leq 0$
  **b.** $r(x) > 0$

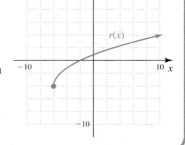

**Solution** ▶ **a.** The only zero of $r$ is at $(-2, 0)$. The graph is on or below the $x$-axis for $x \in [-6, -2]$, so $r(x) \leq 0$ in this interval.

  **b.** The graph is above the $x$-axis for $x \in (-2, \infty)$, and $r(x) > 0$ in this interval.

☑ **B.** You've just seen how we can determine intervals where a function is positive or negative

**Now try Exercises 31 and 32** ▶

This study of inequalities strengthens the foundation for the graphical solutions studied throughout college algebra.

## C. Intervals Where a Function Is Increasing or Decreasing

In our study of linear graphs, we said a graph was increasing if it "rose" when viewed from left to right. More generally, we say the graph of a function is increasing *on a given interval* if larger and larger $x$-values produce larger and larger $y$-values. This suggests the following tests for intervals where a function is increasing or decreasing.

### Increasing and Decreasing Functions

Given open interval $I$ that is a subset of the domain, with $x_1$ and $x_2$ in $I$ and $x_2 > x_1$,

1. A function is **increasing** on $I$ if $f(x_2) > f(x_1)$ for all $x_1$ and $x_2$ in $I$ (larger inputs produce larger outputs).
2. A function is **decreasing** on $I$ if $f(x_2) < f(x_1)$ for all $x_1$ and $x_2$ in $I$ (larger inputs produce smaller outputs).
3. A function is **constant** on $I$ if $f(x_2) = f(x_1)$ for all $x_1$ and $x_2$ in $I$ (larger inputs produce identical outputs).

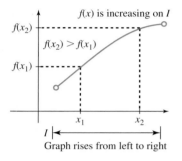

Graph rises from left to right

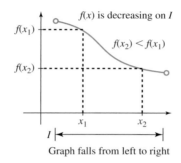

Graph falls from left to right

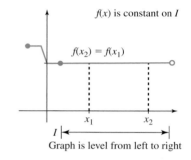

Graph is level from left to right

Consider the graph of $f(x) = -x^2 + 4x + 5$ given in Figure 2.53. Since the parabola opens downward with the vertex at $(2, 9)$, the function must increase until it reaches this peak at $x = 2$, and decrease thereafter. Notationally we'll write this as $f(x)\uparrow$ for $x \in (-\infty, 2)$ and $f(x)\downarrow$ for $x \in (2, \infty)$. In the interval $(-\infty, 2)$, we see that larger inputs will indeed produce larger outputs, and $f(x)\uparrow$ on the interval. For instance,

**Figure 2.53**

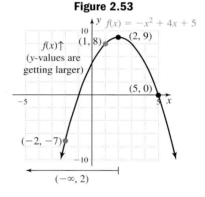

$$1 > -2 \qquad\qquad x_2 > x_1$$

$$\text{and} \qquad\qquad\qquad \text{and}$$

$$f(1) > f(-2) \qquad f(x_2) > f(x_1)$$
$$8 > -7$$

---

**EXAMPLE 5** ▶ **Finding Intervals Where a Function Is Increasing or Decreasing**

Use the graph of $v(x)$ given to name the interval(s) where $v$ is increasing, decreasing, or constant.

**Solution** ▶ From left to right, the graph of $v$ increases until leveling off at $(-2, 2)$, then it remains constant until reaching $(1, 2)$. The graph then increases once again until reaching a peak at $(3, 5)$ and decreases thereafter. The result is $v(x)\uparrow$ for $x \in (-\infty, -2) \cup (1, 3)$, $v(x)\downarrow$ for $x \in (3, \infty)$, and $v(x)$ is constant for $x \in (-2, 1)$.

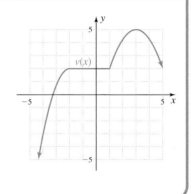

Now try Exercises 33 through 36 ▶

Notice the graph of *f* in Figure 2.53 and the graph of *v* in Example 5 have something in common. It appears that both the far left and far right branches of each graph point downward (in the negative *y*-direction). We say that the **end-behavior** of both graphs is identical, which is the term used to describe what happens to a graph as $|x|$ becomes very large. For $x < 0$, we say a graph is, "up on the left" or "down on the left," depending on the direction the "end" is pointing. For $x > 0$, we say the graph is "up on the right" or "down on the right," as the case may be.

**EXAMPLE 6** ▶

**Describing the End-Behavior of a Graph**

The graph of $f(x) = x^3 - 3x$ is shown. Use the graph to name intervals where *f* is increasing or decreasing, and comment on the end-behavior of the graph.

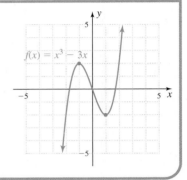

**Solution** ▶

From the graph we observe that $f(x)\uparrow$ for $x \in (-\infty, -1) \cup (1, \infty)$, and $f(x)\downarrow$ for $x \in (-1, 1)$. The end-behavior of the graph is down on the left, and up on the right (down/up).

☑ **C.** You've just seen how we can determine where a function is increasing or decreasing

Now try Exercises 37 through 40 ▶

## D. Maximum and Minimum Values

The *y*-coordinate of the vertex of a parabola that opens downward, and the *y*-coordinate of "peaks" from other graphs are called **maximum values.** A **global maximum** (also called an *absolute* maximum) names the largest *y*-value over the entire domain. A **local maximum** (also called a *relative* maximum) gives the largest range value in a specified interval; and an **endpoint maximum** can occur at an endpoint of the domain. The same can be said for any corresponding minimum values. See Figure 2.54.

More formally we define these **extreme values** as follows.

**Maximum and Minimum Values**

Given an open interval *I* that is a subset of the domain of *f*, with $c \in I$,

1.  $f(c)$ is a local maximum of *f* if $f(c) \geq f(x)$ for all *x* in *I* (no *nearby* input produces a larger output).
2.  $f(c)$ is a local minimum of *f* if $f(c) \leq f(x)$ for all *x* in *I* (no *nearby* input produces a smaller output).

If the inequality is satisfied for all *x* in the domain, $f(c)$ is called a *global* maximum or minimum.

If *c* is an endpoint of the domain, $f(c)$ is called an *endpoint* maximum or minimum.

**Figure 2.54**

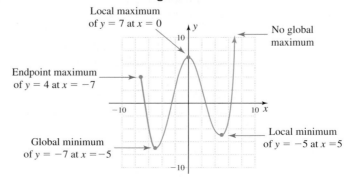

Local maximum of $y = 7$ at $x = 0$

No global maximum

Endpoint maximum of $y = 4$ at $x = -7$

Local minimum of $y = -5$ at $x = 5$

Global minimum of $y = -7$ at $x = -5$

We will soon develop the ability to locate maximum and minimum values for quadratic and other functions. In future courses, methods are developed to help locate maximum and minimum values for almost *any* function. For now, our work will rely chiefly on a function's graph.

**EXAMPLE 7** ▶  **Analyzing Characteristics of a Graph**

**Figure 2.55**

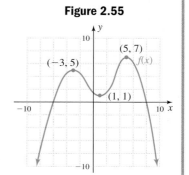

Analyze the graph of function *f* shown in Figure 2.55. Include specific mention of

  **a.** domain and range,

  **b.** intervals where *f* is increasing or decreasing,

  **c.** maximum (max) and minimum (min) values,

  **d.** intervals where $f(x) \geq 0$ and $f(x) < 0$, and

  **e.** whether the function is even, odd, or neither.

**Solution** ▶  **a.** Vertical and horizontal boundary lines show the domain is $x \in (-\infty, \infty)$, with a range of $y \in (-\infty, 7]$.

  **b.** $f(x){\uparrow}$ for $x \in (-\infty, -3) \cup (1, 5)$ shown in **blue** in Figure 2.56, and $f(x){\downarrow}$ for $x \in (-3, 1) \cup (5, \infty)$ as shown in **red**.

**Figure 2.56**

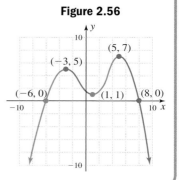

  **c.** From part (b) we find that $y = 5$ at $(-3, 5)$ and $y = 7$ at $(5, 7)$ are local maximums, with a local minimum of $y = 1$ at $(1, 1)$. Note that $y = 7$ is also a global maximum (there is no global minimum).

  **d.** $f(x) \geq 0$ for $x \in [-6, 8]$; $f(x) < 0$ for $x \in (-\infty, -6) \cup (8, \infty)$

☑ **D.** You've just seen how we can identify the maximum and minimum values of a function

  **e.** The function is neither even nor odd.

Now try Exercises 41 through 48 ▶

The ideas presented here can be applied to functions of all kinds, including rational functions, piecewise-defined functions, step functions, and so on. There are a number of interesting applications in the Exercise Set (**see Exercises 99 through 104**).

## E. Rates of Change and the Difference Quotient

As noted in Section 2.2 one of the defining characteristics of linear functions is that their rate of change is constant: $\frac{\Delta y}{\Delta x} = \frac{y_2 - y_1}{x_2 - x_1} = m$. For nonlinear functions the rate of change is not constant, but to aid in their study and application, we use a related concept called the **average rate of change,** given by *the slope of a secant line* through points $(x_1, y_1)$ and $(x_2, y_2)$ on the graph.

**EXAMPLE 8** ▶ **Calculating Average Rates of Change**

The graph shown displays the number of units shipped of vinyl records, cassette tapes, and CDs for the period 1980 to 2005.

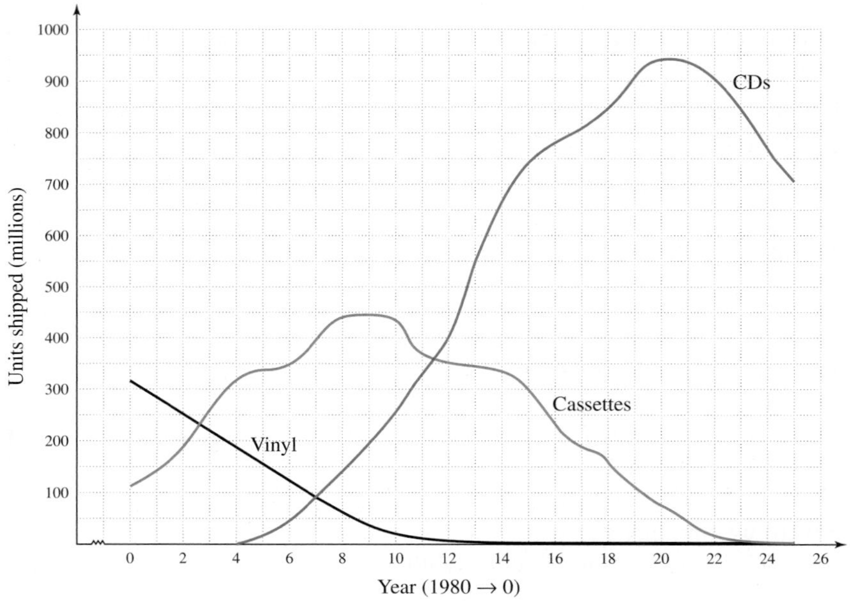

Source: Swivel.com

**Units shipped in millions**

| Year (1980 → 0) | Vinyl | Cassettes | CDs |
|---|---|---|---|
| 0 | 323 | 110 | 0 |
| 2 | 244 | 182 | 0 |
| 4 | 205 | 332 | 6 |
| 6 | 125 | 345 | 53 |
| 8 | 72 | 450 | 150 |
| 10 | 12 | 442 | 287 |
| 12 | 2 | 366 | 408 |
| 14 | 2 | 345 | 662 |
| 16 | 3 | 225 | 779 |
| 18 | 3 | 159 | 847 |
| 20 | 2 | 76 | 942 |
| 24 | 1 | 5 | 767 |
| 25 | 1 | 3 | 705 |

**a.** Find the average rate of change in CDs shipped and in cassettes shipped from 1994 to 1998. What do you notice?

**b.** Does it appear that the rate of increase in CDs shipped was greater from 1986 to 1992, or from 1992 to 1996? Compute the average rate of change for each period and comment on what you find.

**Solution** ▶ Using 1980 as year zero (1980 → 0), we have the following:

**a.**

| **CDs** | **Cassettes** |
|---|---|
| 1994: (14, 662), 1998: (18, 847) | 1994: (14, 345), 1998: (18, 159) |

$$\frac{\Delta y}{\Delta x} = \frac{847 - 662}{18 - 14} \qquad\qquad \frac{\Delta y}{\Delta x} = \frac{159 - 345}{18 - 14}$$

$$= \frac{185}{4} \qquad\qquad\qquad = -\frac{186}{4}$$

$$= 46.25 \qquad\qquad\qquad = -46.5$$

The average rate of decrease in the number of cassettes shipped was roughly equal to the average rate of increase in the number of CDs shipped (about 46,000,000 per year).

**b.** From the graph, the secant line for 1992 to 1996 appears to have a greater slope.

| **1986–1992 CDs** | **1992–1996 CDs** |
|---|---|
| 1986: (6, 53), 1992: (12, 408) | 1992: (12, 408), 1996: (16, 779) |

$$\frac{\Delta y}{\Delta x} = \frac{408 - 53}{12 - 6} \qquad\qquad \frac{\Delta y}{\Delta x} = \frac{779 - 408}{16 - 12}$$

$$= \frac{355}{6} \qquad\qquad\qquad = \frac{371}{4}$$

$$= 59.1\overline{6} \qquad\qquad\qquad = 92.75$$

For the years 1986 to 1992, the average rate of change for CD sales was about 59 million per year. For the years 1992 to 1996, the average rate of change was significantly higher at almost 93 million per year.

**Now try Exercises 49 through 56, 105 and 106** ▶

The importance of the rate of change concept would be hard to overstate. In many business, scientific, and economic applications, it is this attribute of a function that draws the most attention. In Section 2.3 we computed average rates of change by selecting two points from a graph, and computing the slope of the secant line: $m = \frac{\Delta y}{\Delta x} = \frac{y_2 - y_1}{x_2 - x_1}$. With a simple change of notation, we can use the function's *equation* rather than relying on a graph. Note that $y_2$ corresponds to the function evaluated at $x_2$: $y_2 = f(x_2)$. Likewise, $y_1 = f(x_1)$. Substituting these into the slope formula yields $\frac{\Delta y}{\Delta x} = \frac{f(x_2) - f(x_1)}{x_2 - x_1}$, giving the average rate of change between $x_1$ and $x_2$ *for any function f.*

> ### Average Rate of Change
>
> For a function $f$ and $[x_1, x_2]$ a subset of the domain,
> the average rate of change between $x_1$ and $x_2$ is
>
> $$\frac{\Delta y}{\Delta x} = \frac{f(x_2) - f(x_1)}{x_2 - x_1}, x_1 \neq x_2$$

### Average Rates of Change Applied to Projectile Velocity

A projectile is any object that is thrown, shot, or cast upward, with no continuing source of propulsion. The object's height (in feet) after $t$ sec is modeled by the function $h(t) = -16t^2 + vt + k$, where $v$ is the initial velocity of the projectile, and $k$ is the height of the object at $t = 0$. For instance, if a soccer ball is kicked vertically upward from ground level ($k = 0$) with an initial speed of 64 ft/sec, the height of the ball $t$ sec later is $h(t) = -16t^2 + 64t$. Evaluating the function for $t = 0$ to 4 produces Table 2.4 and the parabolic graph shown in Figure 2.57. Experience tells us the ball is traveling faster immediately after being kicked, as compared to when it nears its maximum height where the effects of gravity cause it to momentarily stop, then begin its descent. In other words, the rate of change $\frac{\Delta \text{height}}{\Delta \text{time}}$ has a larger value at any time prior to reaching its maximum height. To quantify this, in Example 9 we'll compute the average rate of change between (a) $t = 0.5$ and $t = 1$, and compare it to the average rates of change between, (b) $t = 1$ and $t = 1.5$, and (c) $t = 1.5$ and $t = 2$.

**WORTHY OF NOTE**

Keep in mind the graph of $h$ represents the relationship between the soccer ball's height in feet and the elapsed time $t$. It does not model the actual path of the ball.

**Table 2.4**

| Time in seconds | Height in feet |
|:---:|:---:|
| 0 | 0 |
| 1 | 48 |
| 2 | 64 |
| 3 | 48 |
| 4 | 0 |

**Figure 2.57**

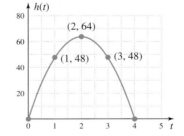

**EXAMPLE 9** ▶  **Average Rates of Change Applied to Projectiles**

For the projectile function $h(t) = -16t^2 + 64t$, find the average rate of change for

**a.** $t \in [0.5, 1]$.        **b.** $t \in [1, 1.5]$.        **c.** $t \in [1.5, 2.0]$.

Then graph the secant lines representing these average rates of change and comment.

**Solution** ▶  Using the given intervals in the formula $\dfrac{\Delta h}{\Delta t} = \dfrac{h(t_2) - h(t_1)}{t_2 - t_1}$ yields

**a.** $\dfrac{\Delta h}{\Delta t} = \dfrac{h(1) - h(0.5)}{1 - (0.5)}$        **b.** $\dfrac{\Delta h}{\Delta t} = \dfrac{h(1.5) - h(1)}{1.5 - 1}$        **c.** $\dfrac{\Delta h}{\Delta t} = \dfrac{h(2) - h(1.5)}{2 - 1.5}$

$= \dfrac{48 - 28}{0.5}$                          $= \dfrac{60 - 48}{0.5}$                          $= \dfrac{64 - 60}{0.5}$

$= 40$                                  $= 24$                                  $= 8$

For $t \in [0.5, 1]$, the average rate of change is $\frac{40}{1}$, meaning the height of the ball is increasing at an average rate of 40 ft/sec. For $t \in [1, 1.5]$, the average rate of change has slowed to $\frac{24}{1}$, and the soccer ball's height is increasing at only 24 ft/sec. In the interval $[1.5, 2]$, the average rate of change has slowed even further to a mere 8 ft/sec. The secant lines representing these rates of change are shown in the figure, where we note the line from the first interval (in **red**), has a much steeper slope than the line from the third interval (in **gold**).

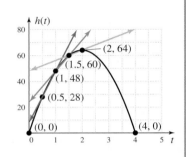

**Now try Exercises 57 through 64, 107 and 108** ▶

The calculation for average rates of change can be applied to any function $y = f(x)$ and will yield valuable information—particularly in an applied context. For practice with other functions, **see Exercises 65 through 72.**

You may have had the experience of riding in the external elevator of a modern building, with a superb view of the surrounding area as you rise from the bottom floor. For the first few floors, you note you can see much farther than from ground level. As you ride to the higher floors, you can see still farther, but not that much farther due to the curvature of the Earth. This is another example of a nonconstant rate of change.

**EXAMPLE 10** ▶  **Average Rates of Change Applied to Viewing Distance**

The distance a person can see depends on their elevation above level ground. On a clear day, this viewing distance can be approximated by the function $d(h) = 1.2\sqrt{h}$, where $d(h)$ represents the viewing distance (in miles) at height $h$ (in feet) above level ground. Find the average rate of change to the nearest 100th, for

**a.** $h \in [9, 16]$
**b.** $h \in [196, 225]$
**c.** Graph the function along with the lines representing the average rate of change and comment on what you notice.

**Solution** ▶  Use the points given in the formula: $\dfrac{\Delta d}{\Delta h} = \dfrac{d(h_2) - d(h_1)}{h_2 - h_1}$

**a.** $\dfrac{\Delta d}{\Delta h} = \dfrac{d(16) - d(9)}{16 - 9}$ $\qquad$ **b.** $\dfrac{\Delta d}{\Delta h} = \dfrac{d(225) - d(196)}{225 - 196}$

$\qquad\quad = \dfrac{4.8 - 3.6}{7}$ $\qquad\qquad\qquad\qquad = \dfrac{18 - 16.8}{29}$

$\qquad\quad \approx 0.17$ $\qquad\qquad\qquad\qquad\quad\; \approx 0.04$

**c.** For $h \in [9, 16]$, $\frac{\Delta d}{\Delta h} \approx \frac{0.17}{1}$, meaning the viewing distance is increasing at an average rate of 0.17 mi (about 898 ft) for each 1 ft increase in elevation. For $h \in [196, 225]$, $\frac{\Delta d}{\Delta h} \approx \frac{0.04}{1}$ and the viewing distance is increasing at a rate of only 0.04 mi (about 211 ft) for each increase of 1 ft. We'll sketch the graph using the points (0, 0), (9, 3.6), (16, 4.8), (196, 16.8) and (225, 18), along with (100, 12) and (169, 15.6) to help round out the graph (Figure 2.58).

**Figure 2.58**

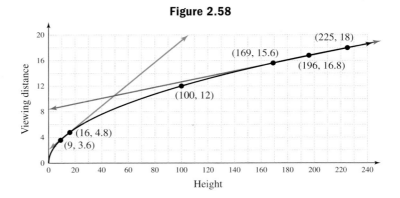

Note the slope of the secant line through the points (9, 3.6) and (16, 4.8), has a much steeper slope than the line through (196, 16.8) and (225, 18), indicating that initially, small changes in height result in larger increases in the viewing distance.

**Now try Exercises 109 and 110** ▶

We now continue our discussion of average rates of change, seeking to make the calculation more efficient due to (1) the large number of times the concept is applied, and (2) its value to virtually all fields of study. Our current calculation $\frac{\Delta y}{\Delta x} = \frac{f(x_2) - f(x_1)}{x_2 - x_1}$ works very well, but requires us to recalculate $\frac{\Delta y}{\Delta x}$ for each new interval chosen. Using a slightly different approach, we can develop a *formula* for a given function, and use it to calculate various rates of change with greater efficiency. This is done by selecting a point $x = x_1$ in the domain, and a point $x_2 = x + h$, where $h$ is assumed to be some very small, arbitrary number. Making these substitutions in our current formula gives

$$\dfrac{\Delta y}{\Delta x} = \dfrac{f(x_2) - f(x_1)}{x_2 - x_1} \qquad \text{average rate of change}$$

$$= \dfrac{f(x + h) - f(x)}{(x + h) - x} \qquad \text{substitute } x + h \text{ for } x_2, x \text{ for } x_1$$

$$= \dfrac{f(x + h) - f(x)}{h} \qquad \text{simplify}$$

The advantage of this new formula, called the **difference quotient,** is that the result is a new function that can be evaluated repeatedly for any interval.

### The Difference Quotient

For a given function $f(x)$ and constant $h \neq 0$,

$$\frac{f(x + h) - f(x)}{h}$$

is the difference quotient for $f$.

Note that the formula has three parts: (1) the function evaluated at $x + h$: $f(x + h)$, (2) the function $f$ itself: $f(x)$, and (3) the constant $h$. The expression for $f(x + h)$ can be evaluated and simplified prior to its use in the difference quotient.

$$\overset{\overset{(1)}{} \qquad \overset{(2)}{}}{\frac{f(x + h) - f(x)}{\underset{(3)}{h}}}$$

**CAUTION** ▶ Be very careful when computing $f(x + h)$ in the difference quotient. One common mistake is to replace $f(x + h)$ with $f(x) + h$ or $f(x) + f(h)$. To see how these three expressions compare, consider the function $f(x) = x^2$. In this case, $f(x) + h = x^2 + h$ and $f(x) + f(h) = x^2 + h^2$, but the correct expression is $f(x + h) = (x + h)^2 = x^2 + 2xh + h^2$. **Also see Exercise 116.**

**EXAMPLE 11A** ▶ **Computing a Difference Quotient and Average Rates of Change**

Use the difference quotient to find an *average rate of change* formula for the reciprocal function $f(x) = \frac{1}{x}$. Then use the formula to find the average rate of change on the intervals [0.5, 0.51] and [3, 3.01]. Note $h = 0.01$ for both intervals.

**Solution** ▶ For $f(x) = \frac{1}{x}$ we have $f(x + h) = \frac{1}{x + h}$ and we compute as follows:

$$\frac{\Delta y}{\Delta x} = \frac{f(x + h) - f(x)}{h} \qquad \text{difference quotient}$$

$$= \frac{\dfrac{1}{x + h} - \dfrac{1}{x}}{h} \qquad \text{substitute } \frac{1}{x + h} \text{ for } f(x + h) \text{ and } \frac{1}{x} \text{ for } f(x)$$

$$= \frac{\dfrac{x - (x + h)}{(x + h)x}}{h} \qquad \text{common denominator}$$

$$= \frac{\dfrac{-h}{(x + h)x}}{h} \qquad \text{simplify numerator}$$

$$= \frac{-1}{(x + h)x} \qquad \text{invert and multiply}$$

This is the formula for computing rates of change given $f(x) = \frac{1}{x}$. To find $\frac{\Delta y}{\Delta x}$ on the interval $[0.5, 0.51]$ we have:

$$\frac{\Delta y}{\Delta x} = \frac{-1}{(0.5 + 0.01)(0.5)} \qquad \text{substitute 0.5 for } x, 0.01 \text{ for } h$$

$$\approx \frac{-4}{1} \qquad \text{result (approximate)}$$

On this interval $y$ is decreasing by about 4 units for every 1 unit $x$ is increasing. The slope of the secant line (in blue) through these points is $m \approx -4$.

For the interval $[3, 3.01]$ we have

$$\frac{\Delta y}{\Delta x} = \frac{-1}{(3 + 0.01)(3)} \qquad \text{substitute 3 for } x, 0.01 \text{ for } h$$

$$\approx -\frac{1}{9} \qquad \text{result (approximate)}$$

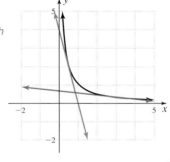

On this interval, $y$ is decreasing 1 unit for every 9 units $x$ is increasing. The slope of the secant line (in red) through these points is $m \approx -\frac{1}{9}$.

In context, suppose the graph of $f$ were modeling the declining area of the rainforest over time. The difference quotient would give us information on just how fast the destruction was taking place (and give us plenty of cause for alarm).

When applying the difference quotient to various functions, we attempt to simplify the resulting expression in a way that makes it useful for college algebra, and for further study in more advanced classes. In Example 11A, this involved simplifying a complex fraction. In Example 11B, the simplification involves rationalizing a numerator, and leaving radicals in the denominator (see Worthy of Note at Example 11B).

**EXAMPLE 11B** ▶ **Computing a Difference Quotient and Average Rates of Change**

Use the difference quotient to find an average rate of change formula for $g(x) = 5 - \sqrt{x}$.

Solution ▶ From $g(x) = 5 - \sqrt{x}$ we have $g(x + h) = 5 - \sqrt{x + h}$ and

$$\frac{\Delta y}{\Delta x} = \frac{g(x + h) - g(x)}{h} \qquad \text{difference quotient}$$

$$= \frac{(5 - \sqrt{x + h}) - (5 - \sqrt{x})}{h} \qquad \begin{array}{l}\text{substitute } 5 - \sqrt{x + h} \text{ for } g(x + h) \\ \text{and } 5 - \sqrt{x} \text{ for } g(x)\end{array}$$

$$= \frac{\sqrt{x} - \sqrt{x + h}}{h} \qquad \text{simplify: } 5 - 5 = 0$$

$$= \frac{\sqrt{x} - \sqrt{x + h}}{h} \cdot \frac{\sqrt{x} + \sqrt{x + h}}{\sqrt{x} + \sqrt{x + h}} \qquad \text{rationalize numerator}$$

$$= \frac{x - (x + h)}{h(\sqrt{x} + \sqrt{x + h})} \qquad \text{multiply: } (A - B)(A + B) = A^2 - B^2$$

$$= \frac{-h}{h(\sqrt{x} + \sqrt{x + h})} \qquad \text{simplify numerator}$$

$$= \frac{-1}{\sqrt{x} + \sqrt{x + h}} \qquad \text{result}$$

**WORTHY OF NOTE**

In a calculus class, you'll investigate what happens to the difference quotient as $h$ becomes very close to zero. This makes it helpful to manipulate the resulting expression here, so that we avoid cases where the denominator itself is becoming very close to zero.

**Now try Exercises 73 through 96** ▶

It is important that you see these calculations as much more than just an algebraic exercise. If the functions were modeling the height of a rocket after the retro-rockets fired, the difference quotient could provide us with valuable information regarding the decreasing velocity of the rocket, and even help pinpoint the moment the velocity was zero. In Example 12, we'll use the difference quotient to analyze the velocity of an object falling under the influence of gravity. Neglecting air resistance, the distance an object falls is modeled by $d(t) = 16t^2$, where $d(t)$ represents the distance fallen after $t$ sec.

Due to the effects of gravity, the velocity of the object increases as it falls. We can analyze this rate of change using the difference quotient.

**EXAMPLE 12** ▶ **Applying the Difference Quotient in Context**

A construction worker drops a heavy wrench from atop a girder of a new skyscraper. Use the function $d(t) = 16t^2$ to

  **a.** Compute the distance the wrench has fallen after 2 sec and after 7 sec.

  **b.** Find a formula for the velocity of the wrench (average rate of change in distance per unit time).

  **c.** Use the formula to find the rate of change in the intervals [2, 2.01] and [7, 7.01].

  **d.** Graph the function and the secant lines representing the average rate of change. Comment on what you notice.

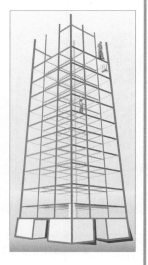

**Solution** ▶   **a.** Substituting $t = 2$ and $t = 7$ in this function yields

$$d(2) = 16(2)^2 \qquad d(7) = 16(7)^2 \qquad \text{evaluate } d(t) = 16t^2$$
$$\quad\;\; = 16(4) \qquad\qquad\;\; = 16(49) \qquad \text{square input}$$
$$\quad\;\; = 64 \qquad\qquad\qquad = 784 \qquad\;\; \text{multiply}$$

After 2 sec, the wrench has fallen 64 ft; after 7 sec, the wrench has fallen 784 ft.

  **b.** For $d(t) = 16t^2$, $d(t + h) = 16(t + h)^2$, which we compute separately.

$$d(t + h) = 16(t + h)^2 \qquad\qquad\qquad \text{substitute } t + h \text{ for } t$$
$$\qquad\quad\;\; = 16(t^2 + 2th + h^2) \qquad\;\; \text{square binomial}$$
$$\qquad\quad\;\; = 16t^2 + 32th + 16h^2 \qquad \text{distribute 16}$$

Using this result in the difference quotient yields

$$\frac{d(t + h) - d(t)}{h} = \frac{(16t^2 + 32th + 16h^2) - 16t^2}{h} \qquad \text{substitute into the difference quotient}$$

$$= \frac{16t^2 + 32th + 16h^2 - 16t^2}{h} \qquad \text{eliminate parentheses}$$

$$= \frac{32th + 16h^2}{h} \qquad \text{combine like terms}$$

$$= \frac{h(32t + 16h)}{h} \qquad \text{factor out } h \text{ and simplify}$$

$$= 32t + 16h \qquad \text{result}$$

For any number of seconds $t$ and small increment of time $h$ thereafter, the velocity of the wrench is modeled by $\frac{\Delta \text{distance}}{\Delta \text{time}} = 32t + 16h$

**c.** For the interval $[t, t + h] = [2, 2.01]$, $t = 2$ and $h = 0.01$:

$$\frac{\Delta \text{distance}}{\Delta \text{time}} = 32(2) + 16(0.01) \qquad \text{substitute 2 for } t \text{ and 0.01 for } h$$

$$= 64 + 0.16 = 64.16$$

Two seconds after being dropped, the velocity of the wrench is close to 64.16 ft/sec (44 mph). For the interval $[t, t + h] = [7, 7.01]$, $t = 7$ and $h = 0.01$:

$$\frac{\Delta \text{distance}}{\Delta \text{time}} = 32(7) + 16(0.01) \qquad \text{substitute 7 for } t \text{ and 0.01 for } h$$

$$= 224 + 0.16 = 224.16$$

Seven seconds after being dropped, the velocity of the wrench is approximately 224.16 ft/sec (about 153 mph).

**d.**

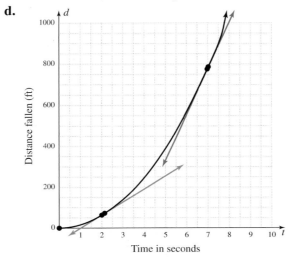

The velocity increases with time, as indicated by the steepness of each secant line.

☑ **D.** You've just seen how we can develop a formula to calculate rates of change for any function

**Now try Exercises 111 and 112** ▶

## 2.5 EXERCISES

▶ **CONCEPTS AND VOCABULARY**

**Fill in each blank with the appropriate word or phrase. Carefully reread the section, if necessary.**

**1.** If $f(-x) = f(x)$ for all $x$ in the domain, we say that $f$ is an _____ function and symmetric about the _____. If $f(-x) = -f(x)$, the function is _____ and is symmetric about the _____.

**2.** The _____ rate of change of a function can be found by calculating the _____ of the line that passes through two points on the graph of the function. For nonlinear functions, this is called a _____ _____.

**3.** Without referring to notes or textbook, list as many features/attributes as you can that are related to analyzing the graph of a function. Include details on how to locate or determine each attribute.

**4.** Discuss/Explain how and why using the difference quotient differs from using the average rate of change formula.

▶ **DEVELOPING YOUR SKILLS**

**The following functions are known to be even. Complete each graph using symmetry.**

**5.**
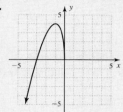

**6.**
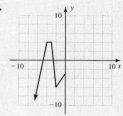

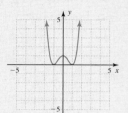

**27.** $g(x) > 0$

**28.** $g(x) < 0$

$g(x) = x^4 - 2x^2 + 1$

**Determine whether the following functions are even:**
$f(-x) = f(x)$.

**7.** $f(x) = -7|x| + 3x^2 + 5$    **8.** $p(x) = 2x^4 - 6x + 1$

**9.** $g(x) = \dfrac{1}{3}x^4 - 5x^2 + 1$    **10.** $q(x) = \dfrac{1}{x^2} - |x|$

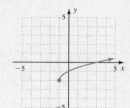

**29.** $p(x) \geq 0$

**30.** $p(x) \leq 0$

$p(x) = x^3 + 2x^2 - 4x - 8$

**The following functions are known to be odd. Complete each graph using symmetry.**

**11.**
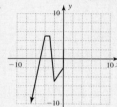

**12.**
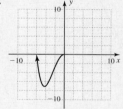

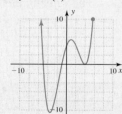

**31.** $q(x) > 0$

**32.** $q(x) < 0$

$q(x) = \sqrt{x + 1} - 2$

**Determine whether the following functions are odd:**
$f(-x) = -f(x)$.

**13.** $f(x) = 4\sqrt[3]{x} - x$    **14.** $g(x) = \dfrac{1}{2}x^3 - 6x$

**15.** $p(x) = 3x^3 - 5x^2 + 1$    **16.** $q(x) = \dfrac{1}{x} - x$

**Determine whether the following functions are even, odd, or neither.**

**17.** $w(x) = x^3 - x^2$    **18.** $q(x) = \dfrac{3}{4}x^2 + 3|x|$

**19.** $p(x) = 2\sqrt[3]{x} - \dfrac{1}{4}x^3$    **20.** $g(x) = x^3 + 7x$

**21.** $v(x) = x^3 + 3|x|$    **22.** $f(x) = x^4 + 7x^2 - 30$

**23.** $h(x) = 4x\sqrt{x^2 - 6}$    **24.** $n(x) = \dfrac{5x^2}{7x^4 + 3}$

**Name the interval(s) where the following functions are increasing, decreasing, or constant. Write answers using interval notation. Assume all endpoints have integer values.**

**33.** $y = V(x)$    **34.** $y = H(x)$

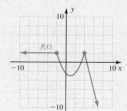

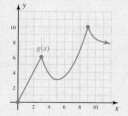

**Use the graphs given to solve the inequalities indicated. Write all answers in interval notation.**

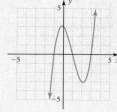

**25.** $f(x) \geq 0$

**26.** $f(x) < 0$

**35.** $y = f(x)$    **36.** $y = g(x)$

$f(x) = x^3 - 3x^2 - x + 3$

For Exercises 37 through 40, (a) name the interval(s) where the function is increasing, decreasing, or constant, and (b) comment on the end-behavior.

**37.** $p(x) = 0.5(x + 2)^3$

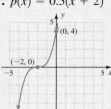

**38.** $q(x) = -\sqrt[3]{x + 1}$

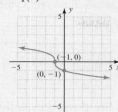

**39.** $y = f(x)$

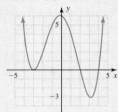

**40.** $y = g(x)$

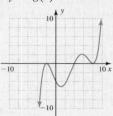

For each function given, determine the following (answer in interval notation as appropriate): (a) the domain and range, (b) the zeroes of the function, (c) where the function is positive or negative, (d) interval(s) where the function is increasing, decreasing, or constant, and (e) the location of any local max or min value(s).

**41.** $y = H(x)$

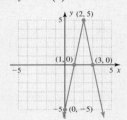

**42.** $y = f(x)$

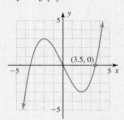

**43.** $y = g(x)$

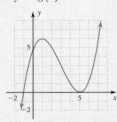

**44.** $y = h(x)$

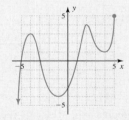

**45.** $y = Y_1$

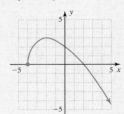

**46.** $y = Y_2$

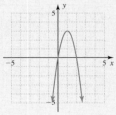

**47.** $p(x) = (x + 3)^3 + 1$

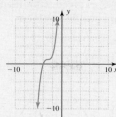

**48.** $q(x) = |x - 5| + 3$

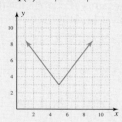

Using the graphs shown, find the average rate of change of $f$ and $g$ for the intervals specified.

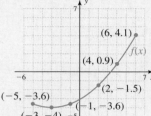

**49.** $f(x)$ for $x \in [-5, -1]$

**50.** $f(x)$ for $x \in [2, 6]$

**51.** $f(x)$ for $x \in [-3, 2]$

**52.** $f(x)$ for $x \in [-5, -3]$

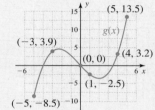

**53.** $g(x)$ for $x \in [-3, 1]$

**54.** $g(x)$ for $x \in [0, 1]$

**55.** $g(x)$ for $x \in [-3, 4]$

**56.** $g(x)$ for $x \in [0, 5]$

The functions shown are projectile equations where $h(t)$ is the height of the projectile after $t$ sec. Calculate $\frac{\Delta h}{\Delta t}$ over the intervals indicated. Include units of measurement in your answer.

**57.** $h(t) = -16t^2 + 160t$, $h$ in feet
    **a.** $[2, 5]$             **b.** $[3, 5]$
    **c.** $[4, 5]$             **d.** $[5, 7]$

**58.** $h(t) = -16t^2 + 32t$, $h$ in feet
    **a.** $[0.25, 1]$      **b.** $[0.5, 1]$
    **c.** $[0.75, 1]$      **d.** $[1, 1.5]$

**59.** $h(t) = -4.9t^2 + 34.3t + 2.6$, $h$ in meters
    **a.** $[1, 6]$             **b.** $[2, 5]$
    **c.** $[2.5, 3.5]$      **d.** $[3.5, 4.5]$

**60.** $h(t) = -4.9t^2 + 14.7t + 4.1$, $h$ in meters
    **a.** $[1, 2]$             **b.** $[0.2, 2.8]$
    **b.** $[0.5, 1.5]$      **c.** $[1.5, 2.5]$

**Graph the function and secant lines representing the average rates of change for the exercises given. Comment on what you notice in terms of the projectile's velocity.**

**61.** Exercise 57      **62.** Exercise 58

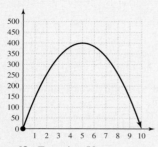 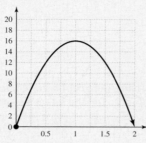

**63.** Exercise 59      **64.** Exercise 60

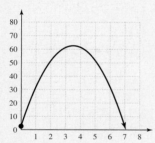 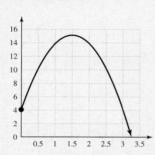

 **Find the average rate of change for the interval specified. Round to hundredths as needed.**

**65.** $y = x^3 - 8; [2, 5]$

**66.** $y = \sqrt[3]{x + 5}; [-5, 3]$

**67.** $y = 2|x + 3|; [-4, 0]$

**68.** $y = |3x + 1| - 2; [-2, 2]$

**69.** $F = 9.8m; [70, 100]$

**70.** $\rho = 0.2m; [1.3, 1.5]$

**71.** $A = \pi r^2; [5, 7]$

**72.** $V = \dfrac{4}{3}\pi r^3; [5, 7]$

**Use the difference quotient to find (a) an average rate of change formula for the functions given and (b)/(c) calculate the rate of change in the intervals shown. Then (d) sketch the graph of each function along with the secant lines and comment on what you notice.**

**73.** $g(x) = x^2 + 2x$      **74.** $j(x) = x^2 - 6x$

   $[-3.0, -2.9], [0.50, 0.51]$    $[1.9, 2.0], [5.0, 5.01]$

**75.** $g(x) = x^3 + 1$      **76.** $v(x) = \sqrt{x}$

   $[-2.1, -2.0], [0.40, 0.41]$    $[1.0, 1.1], [4.0, 4.1]$

**Use the difference quotient to find (a) an average rate of change formula for the functions given and (b)/(c) calculate the rate of change for the intervals indicated. Then (d) comment on how the rate of change in each interval corresponds to the graph of the function.**

**77.** $j(x) = \dfrac{1}{x^2}$      **78.** $f(x) = x^2 - 4x$

   $[0.50, 0.51], [1.50, 1.51]$    $[0.00, 0.01], [3.00, 3.01]$

**79.** $g(x) = x^3 + 1$      **80.** $r(x) = \sqrt{x}$

   $[-2.01, -2.00], [0.40, 0.41]$    $[1.00, 1.01], [4.00, 4.01]$

**Compute and simplify the difference quotient $\dfrac{f(x + h) - f(x)}{h}$ for each function given.**

**81.** $F(x) = 4$      **82.** $f(x) = 2x - 3$

**83.** $j(x) = x^2 + 3$      **84.** $q(x) = 4x^2 + 2x - 3$

**85.** $u(x) = x^3 - 1$      **86.** $f(x) = \dfrac{2}{x}$

**87.** $F(x) = \dfrac{1}{x^2} + 3$      **88.** $p(x) = 2\sqrt{x}$

**89.** $G(x) = \dfrac{2}{3}$      **90.** $g(x) = 4x + 1$

**91.** $p(x) = x^2 - 2$      **92.** $r(x) = 3x^2 - 5x + 2$

**93.** $v(x) = 2x^3$      **94.** $g(x) = 3 - \dfrac{1}{x}$

**95.** $G(x) = \dfrac{-3}{x^2}$      **96.** $q(x) = \sqrt{3x}$

▶ **WORKING WITH FORMULAS**

**97. Trigonometric graphs: $y = \sin(x)$ and $y = \cos(x)$**

While trigonometric functions are not covered in detail until future courses, we've already developed a number of tools that will help us understand these relations and their graphs. Here $y = \sin x$ and $y = \cos x$ are given, graphed over the interval $x \in [-360°, 360°]$. Use them to find (a) the range of the functions, (b) the zeroes of the functions, (c) interval(s) where $y$ is increasing/ decreasing, (d) location of minimum/maximum values, and (e) whether each relation is even, odd, or neither.

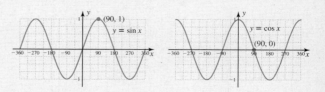

**EXAMPLE 1B** ▶  **Drawing a Scatterplot and Observing Associations**

A cup of coffee is placed on a table and allowed to cool. The temperature of the coffee is measured every 10 min and the data are shown in the table. Draw the scatterplot and state whether the association is positive, negative, or cannot be determined.

| Elapsed Time (minutes) | Temperature (°F) |
|:---:|:---:|
| 0 | 110 |
| 10 | 89 |
| 20 | 76 |
| 30 | 72 |
| 40 | 71 |

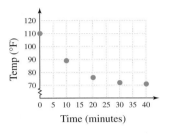

**Solution** ▶  Since temperature depends on cooling time, *time* is the input *x* and temperature is the output *y*. Scale the *x*-axis from 0 to 40 and the *y*-axis from 70 to 120 to comfortably fit the data. As you see in the figure, there is a negative association between the variables, meaning the temperature *decreases* over time.

☑ **A.** You've just seen how we can draw a scatterplot and identify positive and negative associations

**Now try Exercises 5 and 6** ▶

## B. Scatterplots and Linear/Nonlinear Associations

The data in Example 1A had a positive association, while the association in Example 1B was negative. But the data from these examples differ in another important way. In Example 1A, the data seem to cluster about an imaginary line. This indicates a linear equation model might be a good approximation for the data, and we say there is a **linear association** between the variables. The data in Example 1B could not accurately be modeled using a straight line, and we say the variables *time* and cooling *temperature* exhibit a **nonlinear association.**

**EXAMPLE 2** ▶  **Drawing a Scatterplot and Observing Associations**

A corporate intern tracked her annual salary from 2002 to 2011 and the data are shown in the table. Draw the scatterplot and determine if there is a linear or nonlinear association between the variables. Also state whether the association is positive, negative, or cannot be determined.

| Year | Salary ($1000s) |
|:---:|:---:|
| 2002 | 29.1 |
| 2003 | 29.5 |
| 2004 | 30.3 |
| 2005 | 31.5 |
| 2006 | 33.2 |
| 2007 | 36.1 |
| 2008 | 40.2 |
| 2009 | 46.3 |
| 2010 | 55.8 |
| 2011 | 70.2 |

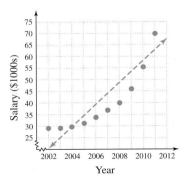

**Solution** ▶

Since salary earned depends on a given year, *year* is the input *x* and *salary* is the output *y*. Scale the *x*-axis from 2002 to 2012, and the *y*-axis from 25 to 75 to comfortably fit the data. A line doesn't seem to model the data very well, and the association appears to be nonlinear. The data rises from left to right, indicating a positive association between the variables. This makes good sense, since we expect our salaries to increase over time.

☑ **B.** You've just seen how we can use a scatterplot to identify linear and nonlinear associations

Now try Exercises 7 and 8 ▶

## C. Identifying Strong and Weak Correlations

Using Figures 2.59 and 2.60 shown, we can make one additional observation regarding the data in a scatterplot. While both associations shown appear linear, the data in Figure 2.59 seems to cluster more tightly about an imaginary straight line than the data in Figure 2.60.

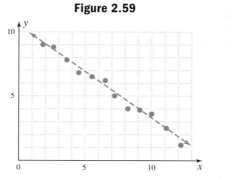

**Figure 2.59**

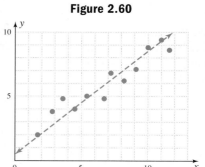

**Figure 2.60**

We refer to this "clustering" as the "goodness of fit," or in statistical terms, the strength of the correlation. If the data points form a perfectly straight line, we say the strength of the correlation is either −1 or 1(−100% or 100%), depending on the association. If the data points appear clustered about the line, but are scattered on either side of it, the strength of the correlation falls somewhere between −1 and 1, depending on whether the association is positive or negative and how tightly/loosely they're scattered. This is summarized in Figure 2.61.

**Figure 2.61**

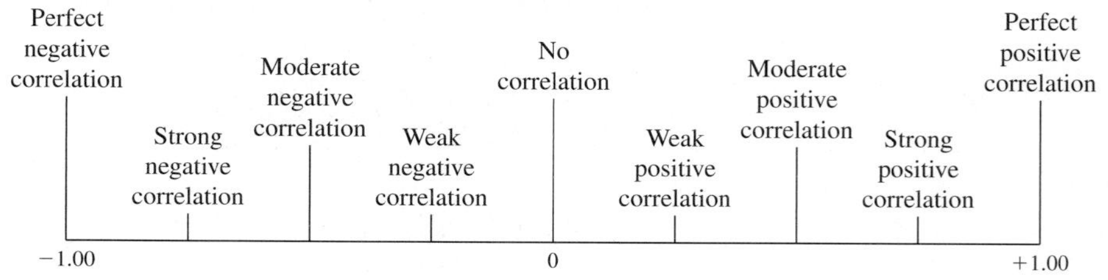

Perfect negative correlation
Strong negative correlation
Moderate negative correlation
Weak negative correlation
No correlation
Weak positive correlation
Moderate positive correlation
Strong positive correlation
Perfect positive correlation

−1.00                                            0                                            +1.00

The following scatterplots help to further illustrate this idea. Figure 2.62 shows a linear and negative association between the value of a car and the age of a car, with a strong correlation. Figure 2.63 shows there is no apparent association between family income and the number of children, and Figure 2.64 appears to show a linear and positive association between a man's height and weight, with a weak correlation.

**Figure 2.62**

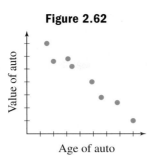

**Figure 2.63**

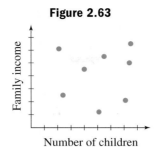

**Figure 2.64**

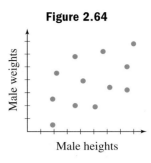

Until we develop a more accurate method of calculating a numeric value for this correlation, the best we can do are these broad generalizations: weak correlation, strong correlation, or no correlation.

---

**EXAMPLE 3A** ▶ **High School and College GPAs**

Many colleges use a student's high school GPA as a possible indication of their future college GPA. Use the data from Table 2.5 (high school GPA, college GPA) to draw a scatterplot. Then

  **a.** Sketch a line that seems to approximate the data, meaning it has the same general direction while appearing to pass through the observed "center" of the data.

  **b.** State whether the association is positive, negative, or cannot be determined.

  **c.** Decide whether the correlation is weak or strong.

**Table 2.5**

| High School GPA | College GPA |
|:---:|:---:|
| 1.8 | 1.8 |
| 2.2 | 2.3 |
| 2.8 | 2.5 |
| 3.2 | 2.9 |
| 3.4 | 3.6 |
| 3.8 | 3.9 |

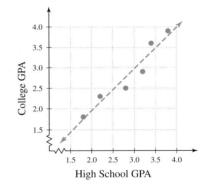

**Solution** ▶   **a.** A line approximating the data set as a whole is shown in the figure.

  **b.** Since the line has positive slope, there is a positive association between a student's high school GPA and their GPA in college.

  **c.** The correlation appears strong.

---

**EXAMPLE 3B** ▶ **Natural Gas Consumption**

The amount of natural gas consumed by homes and offices varies with the season, with the highest consumption occurring in the winter months. Use the data from Table 2.6 (outdoor temperature, gas consumed) to draw a scatterplot. Then

  **a.** Sketch an estimated line of best fit.

  **b.** State whether the association is positive, negative, or cannot be determined.

  **c.** Decide whether the correlation is weak or strong.

**Table 2.6**

| Outdoor Temperature (F°) | Gas Consumed (cubic feet) |
|---|---|
| 30 | 800 |
| 40 | 620 |
| 50 | 570 |
| 60 | 400 |
| 70 | 290 |
| 80 | 220 |

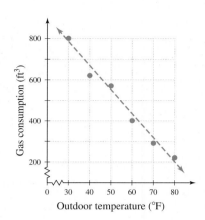

Solution ▶

**a.** We again use appropriate scales and sketch a line that seems to best model the data (see figure).

**b.** There is a negative association between temperature and the amount of natural gas consumed.

✓ **C.** You've just seen how we can use a scatterplot to identify strong and weak correlations

**c.** The correlation appears to be strong.

Now try Exercises 9 and 10 ▶

## D. Linear Functions That Model Relationships Observed in a Set of Data

Finding a **linear function model** for a set of data often involves visually estimating and sketching a line that appears to best "fit" the data. This means answers may vary slightly, but a good, usable model can often be obtained. To find the function, we select two points on this imaginary line and use either the slope-intercept form or the point-slope formula to construct the function. Points on this estimated line but not actually in the data set *can still be used* to help determine the equation of the line.

EXAMPLE 4 ▶   **Finding a Linear Function to Model the Relationship between GPAs**

Use the scatterplot from Example 3A to find a function model for the line a college might use to project an incoming student's future GPA.

Solution ▶   Any two points on or near the estimated best-fit line can be used to help determine the linear function (see the figure in Example 3A). In general, it's best to pick two points that are a good distance apart, as this tends to improve the accuracy of the model. It appears (1.8, 1.8) and (3.8, 3.9) are both near the line, giving

$$m = \frac{y_2 - y_1}{x_2 - x_1} \qquad \text{slope formula}$$

$$= \frac{3.9 - 1.8}{3.8 - 1.8} \qquad \text{substitute (3.8, 3.9) for } (x_2, y_2), (1.8, 1.8) \text{ for } (x_1, y_1)$$

$$= 1.05 \qquad \text{slope}$$

$$y - y_1 = m(x - x_1) \qquad \text{point-slope form}$$

$$y - 1.8 = 1.05(x - 1.8) \qquad \text{substitute 1.05 for } m, (1.8, 1.8) \text{ for } (x_1, y_1)$$

$$y - 1.8 = 1.05x - 1.89 \qquad \text{distribute}$$

$$y = 1.05x - 0.09 \qquad \text{add 1.8 (solve for } y)$$

One possible function model for this data is $f(x) = 1.05x - 0.09$. Slightly different functions may be obtained, depending on the points chosen.

Now try Exercises 11 through 20 ▶

The function from Example 4 predicts that a student with a high school GPA of 3.2 will have a college GPA of almost 3.3: $f(3.2) = 1.05(3.2) - 0.09 \approx 3.3$, yet the data gives an observed value of only 2.9. When working with data and function models, we should expect some variation when the two are compared, especially if the correlation is weak.

Applications of data analysis can be found in virtually all fields of study. In Example 5 we apply these ideas to an Olympic swimming event.

**EXAMPLE 5** ▶ **Finding a Linear Function to Model the Relationship (Year, Gold Medal Times)**

The men's 400-m freestyle times (gold medal times—to the nearest second) for the 1976 through 2008 Olympics are given in Table 2.7 (1900→0). Let the year be the input $x$, and winning race time be the output $y$. Based on the data, draw a scatterplot and answer the following questions.

**a.** Does the association appear linear or nonlinear?

**b.** Is the association positive or negative?

**c.** Classify the correlation as weak or strong.

**d.** Find a function model that approximates the data, then use it to predict the winning time for the 2012 Olympics.

**Table 2.7**

| Year ($x$) (1900→0) | Time ($y$) (sec) |
|---|---|
| 76 | 232 |
| 80 | 231 |
| 84 | 231 |
| 88 | 227 |
| 92 | 225 |
| 96 | 228 |
| 100 | 221 |
| 104 | 223 |
| 108 | 223 |

**Solution** ▶ Begin by choosing appropriate scales for the axes. The $x$-axis (year) could be scaled from 76 to 116, and the $y$-axis (swim time) from 210 to 246. This will allow for a "frame" around the data. After plotting the points, we obtain the scatterplot shown in the figure.

**a.** The association appears to be linear.

**b.** The association is negative, showing that winning times tend to decrease over time.

**c.** There is a moderate to strong correlation.

**d.** The points (76, 232) and (104, 223) appear to be on a line approximating the data, and we select these to develop our equation model.

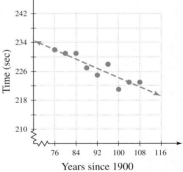

$$m = \frac{y_2 - y_1}{x_2 - x_1} \qquad \text{slope formula}$$

$$= \frac{223 - 232}{104 - 76} \qquad (x_1, y_1) \to (76, 232), (x_2, y_2) \to (104, 223)$$

$$\approx -0.32 \qquad \text{slope (rounded to tenths)}$$

$$y - 232 = -0.32(x - 76) \qquad \text{point-slope form}$$

$$y - 232 = -0.32x + 24.32 \qquad \text{distribute}$$

$$y = -0.32x + 256.32 \qquad \text{add 232 (solve for } y)$$

One model for this data is $f(x) = -0.32x + 256.32$. To find the winning time for the 2012 Olympics, we substitute 112 for $t$.

$$f(x) = -0.32x + 256.32 \qquad \text{function model}$$

$$f(112) = -0.32(112) + 256.32 \qquad \text{substitute 112 for } x \text{ (2012)}$$

$$= 220.48 \qquad \text{result}$$

In 2012 the winning time is projected to be about 220.5 sec.

☑ **D.** You've just seen how we can find a linear function that models relationships observed in a set of data

**Now try Exercises 21 and 22** ▶

As a reminder, great care should be taken when using equation models obtained from real data. It would be foolish to assume that in the year 2700, swim times for the 400-m freestyle would be near 0 sec—even though that's what the model predicts for $x = 800$. Most function models are limited by numerous constraining factors, and data collected over a much longer period of time might even be better approximated using a nonlinear model.

## E.  Linear Regression and the Line of Best Fit

There is actually a sophisticated method for calculating the equation of a line that best fits a data set, called the **regression line.** The method minimizes the vertical distance (called the deviation) between all data points and the line itself, making it the unique **line of best fit.** Most graphing calculators have the ability to perform this calculation quickly. The process involves these steps: (1) clearing old data, (2) entering new data, (3) displaying the data, (4) calculating the regression line, and (5) displaying and using the regression line. We'll illustrate by finding the regression line for the data shown in Table 2.7 in Example 5, which gives the men's 400-m freestyle gold medal times (in seconds) for the 1976 through the 2008 Olympics, with 1900→0.

### Step 1: Clear Old Data

To prepare for the new data, we first clear out any old data. Press the ⬛STAT key and select option **4:ClrList.** This places the **ClrList** command on the home screen. We tell the calculator which lists to clear by pressing ⬛2nd 1 to indicate List1 (L1), then enter a comma using the ⬛, key, and continue entering other lists we want to clear: ⬛2nd 2 ⬛, ⬛2nd 3 ⬛ENTER will clear List1 (L1), List2 (L2), and List3 (L3).

### Step 2: Enter New Data

**Figure 2.65**

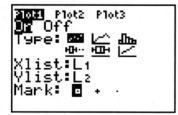

| L1 | L2 | L3 | 2 |
|----|----|----|---|
| 88 | 227 | | |
| 92 | 225 | | |
| 96 | 228 | | |
| 100 | 221 | | |
| 104 | 223 | | |
| 108 | 223 | | |
| ------ | ------ | | |

L2(9) =223

Press the ⬛STAT key and select option **1:Edit.** Move the cursor to the first position of List1, and simply enter the data from the first column of Table 2.7 in order: 76 ⬛ENTER 80 ⬛ENTER 84 ⬛ENTER, and so on. Then use the right arrow ▷ to navigate to List2, and enter the data from the second column: 232 ⬛ENTER 231 ⬛ENTER 231 ⬛ENTER, and so on. When finished, you should obtain the screen shown in Figure 2.65.

### Step 3: Display the Data

**Figure 2.66**

```
Plot1  Plot2  Plot3
On  Off
Type: ▨  ⬚  ⬚
      ⬚  ⬚  ⬚
Xlist:L1
Ylist:L2
Mark:  ▫  +  ·
```

With the data held in these lists, we can now display the related ordered pairs on the coordinate grid. First press the ⬛Y= key and ⬛CLEAR any existing equations. Then press ⬛2nd ⬛Y= to access the "**STATPLOTS**" screen. With the cursor on **1:Plot1,** press ⬛ENTER and be sure the options shown in Figure 2.66 are highlighted. If you need to make any changes, navigate the cursor to the desired option, input the change, and press ⬛ENTER. Note the data in L1 ranges from 76 to 108, while the data in L2 ranges from 221 to 232. This means an appropriate viewing window might be [70, 120] for the x-values, and [210, 240] for the y-values. Press the ⬛WINDOW key and set up the window accordingly. After you're finished, pressing the ⬛GRAPH key should produce the graph shown in Figure 2.67 (be sure that "Plot1" on the ⬛Y= screen is highlighted).

**Figure 2.67**

```
       240
P1:L1,L2
  ⬚ ⬚ ⬚
70          ⬚   ⬚        120
              ⬚  ⬚ ⬚
           ⬚
X=76        Y=232
       210
```

### Step 4: Calculate the Regression Equation

To have the calculator compute the regression equation, press the ⬛STAT and ▷ keys to move the cursor over to the **CALC** options (see Figure 2.68). Since it appears the data is best modeled by a linear equation, we choose option **4:LinReg(ax + b).** Pressing the number 4 places this option on the home screen, and pressing ⬛ENTER computes the

**Figure 2.68**

```
EDIT CALC TESTS
1:1-Var Stats
2:2-Var Stats
3:Med-Med
4▮LinReg(ax+b)
5:QuadReg
6:CubicReg
7↓QuartReg
```

**Figure 2.69**

```
LinReg
 y=ax+b
 a=-.3291666667
 b=257.0611111
```

values of a and b (the calculator automatically uses the values in L1 and L2 unless instructed otherwise). Rounded to hundredths, the linear regression model is $y = -0.33x + 257.06$ (Figure 2.69).

### Step 5: Display and Use the Results

Although graphing calculators have the ability to paste the regression equation directly into $Y_1$ on the  screen, for now we'll enter $Y_1 = -0.33x + 257.06$ by hand. Afterward, pressing the GRAPH key will plot the data points (if Plot1 is still active) and graph the line. Your display screen should now look like the one in Figure 2.70. The regression line is the best estimator for the set of data as a whole, but there will still be some difference between the values it generates and the values from the set of raw data (the output in Figure 2.70 shows the estimated time for the 2000 Olympics was about 224 sec, when actually it was the year Ian Thorpe of Australia set a world record of 221 sec).

**Figure 2.70**

In Example 6, we consider data for the average salary of a social worker in the United States. While social work can be a very exciting and rewarding profession, it can also be very demanding and often requires a strong dedication to help cure social ills, promote social change, and assist the empowerment of helpless or downtrodden individuals. Example 6 shows the average salary of a social worker has been rising steadily over time.

**EXAMPLE 6 ▶**

### Modeling the Average Annual Salary of a Social Worker

The data given in the table show the average annual salary for a social worker from 2000 to 2010, according to the U.S. Department of Health and Human Services, the U.S. Department of Labor, and other sources. (a) Find the regression equation for the data, then (b) use it to project the annual salary for a social worker in 2013. (c) In what year is this salary projected to exceed $55,000?

| Year (2000 → 0) | Annual Salary |
|---|---|
| 0 | 33.5 |
| 1 | 35.0 |
| 2 | 35.5 |
| 3 | 36.5 |
| 4 | 37.6 |
| 5 | 38.5 |
| 6 | 40.0 |
| 7 | 42.1 |
| 8 | 43.6 |
| 9 | 45.0 |
| 10 | 46.3 |

**Solution ▶**

a. Since the salary $S$ depends on the year $t$, following the prescribed sequence (steps 1 through 5) produces the function $S(t) \approx 1.29t + 33.0$, with a strong correlation.

b. For $t = 13$, we obtain $S(13) \approx 49.8$, or an annual salary of about $49,800.

c. For $S(t) = 55$, we have $55 = 1.29t + 33.0$, which gives $t \approx 17.0$. The average salary of a social worker is projected to reach $55,000 in the year 2017.

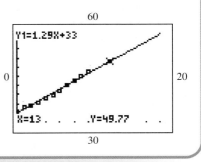

✓ **E.** You've just seen how we can use a linear regression to find the line of best fit

Now try Exercises 25 through 32 ▶

## 2.6 EXERCISES

### ▶ CONCEPTS AND VOCABULARY

**Fill in each blank with the appropriate word or phrase. Carefully reread the section, if necessary.**

**1.** If the data points seems to cluster along an imaginary line, the data is said to have a _____ association.

**2.** Compare/Contrast: One scatterplot is linear, with a weak and positive association. Another is linear, with a strong and negative association. Draw a sample and give a written description of each.

**3.** If the pattern of data points seems to increase as they are viewed left to right, the data is said to have a _____ association.

**4.** Discuss/Explain how this is possible: Working from the same scatterplot, Demetrius obtained the equation $y = -0.64x + 44$ for his equation model, while Jessie got the equation $y = -0.59x + 42$.

### ▶ DEVELOPING YOUR SKILLS

**5. First class postage rates:** The data given in the table show the cost of sending a first class letter for selected years. (a) Draw a scatterplot of the data, then (b) decide if the association is positive, negative, or cannot be determined.

*Source: 2009 Statistical Abstract of the United States*

| Year (2000→0) | Cost in Cents |
|---|---|
| 0 | 33 |
| 1 | 34 |
| 2 | 37 |
| 6 | 39 |
| 7 | 41 |
| 8 | 42 |
| 10 | 44 |

**6. Smoking in the United States:** The Surgeon General's warning in 1964 had a profound effect on cigarette consumption, as advertising was banned and public awareness of the dangers grew. The percentage of U.S. adults who smoke is shown in the table for selected years. (a) Draw a scatterplot of the data, then (b) decide if the association is positive, negative, or cannot be determined.

*Source: Centers for Disease Control and Prevention*

| Year (1965→0) | Percentage of Population |
|---|---|
| 0 | 42.4 |
| 5 | 37.1 |
| 10 | 36.9 |
| 15 | 32.5 |
| 20 | 29.9 |
| 25 | 25.3 |
| 30 | 24.6 |
| 35 | 23.1 |
| 40 | 21.2 |
| 45 | 19.8 |

**7. Women in politics:** Since the 1980s women have made tremendous gains in the political arena, with more and more female candidates winning seats in state legislatures across the country. The number of women serving is

| Year (1980→0) | Women in State Legislatures |
|---|---|
| 5 | 1103 |
| 10 | 1270 |
| 15 | 1532 |
| 20 | 1670 |
| 25 | 1674 |
| 30 | 1809 |

shown in the table for selected years. (a) Draw a scatterplot of the data, (b) decide if the association is linear or nonlinear, and (c) if the association is positive, negative, or cannot be determined.

*Source: Center for American Women and Politics*

**8. Shares traded:** The number of shares traded on the New York Stock Exchange experienced dramatic change in the 1990s as more and more individual investors gained access to the stock

| Year (1990→0) | Shares Traded (billions) |
|---|---|
| 5 | 88 |
| 7 | 134 |
| 11 | 311 |
| 15 | 523 |
| 16 | 598 |
| 19 | 738 |

market via the Internet and online brokerage houses. The volume is shown in the table for selected years (in billions of shares). (a) Draw a scatterplot of the data, (b) decide if the association is linear or nonlinear, and (c) if the association is positive, negative, or cannot be determined.

*Source: Statistical Abstract of the United States (various years)*

**The data sets in Exercises 9 and 10 are known to be linear.**

**9. Gross domestic product (GDP):** The GDP per capita (in $1000s) for the United States is shown in the table for selected years. (a) Draw a scatterplot and sketch an estimated line of best fit, (b) decide if the association is positive or negative, then (c) decide whether the correlation is weak or strong.

*Source: 2009 Statistical Abstract of the United States, Table 1306*

| Year (1970→0) | GDP per Capita ($1000s) |
|---|---|
| 0 | 5.1 |
| 5 | 7.6 |
| 10 | 12.3 |
| 15 | 17.7 |
| 20 | 23.3 |
| 25 | 27.7 |
| 30 | 35.0 |
| 33 | 37.8 |
| 36 | 39.7 |

10. **Home sales:** Real estate brokers carefully track sales of new homes looking for trends in location, price, size, and other factors. The table relates the average selling price (in thousands) to the number of new homes sold by Homestead Realty in 2012. (a) Draw a scatterplot and sketch an estimated line of best fit,

| Price | Sales |
|-------|-------|
| 240 | 123 |
| 250 | 103 |
| 260 | 95 |
| 270 | 75 |
| 280 | 50 |
| 290 | 44 |
| 300 | 30 |

(b) decide if the association is positive or negative, then (c) decide whether the correlation is weak or strong.

**For the scatterplots given: (a) Arrange them in order from the weakest to the strongest correlation, (b) sketch a line that seems to approximate the data, (c) state whether the association is positive, negative, or cannot be determined, and (d) choose two points on (or near) the line and use them to approximate its slope (rounded to tenths).**

11.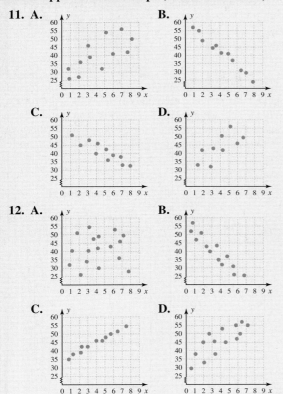

12. A.   B.   C.   D.

**For the scatterplots given, (a) determine whether a linear or nonlinear model is more appropriate. (b) Determine if the association is positive or negative. (c) If linear, classify the correlation as weak or strong and (d) sketch a line that seems to approximate the data and choose two points on (or near) the line and use them to approximate its slope (rounded to tenths).**

13.   14.

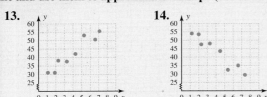

15. 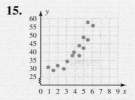   16.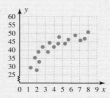

17. 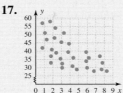   18.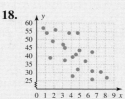

19. **Law enforcement:** The number of law enforcement officers employed by the federal government and having the authority to carry firearms and make arrests is shown in the table for selected years.
(a) Draw a scatterplot and sketch an estimated line of best fit, then (b) decide if the association is positive or negative. (c) Choose two points on or near this line, and use them to find a function model for the data, then predict the number of officers in 1995 and 2011. Answers may vary.

| x (1990→0) | y (1000s) |
|------------|-----------|
| 3 | 68.8 |
| 6 | 74.5 |
| 8 | 83.1 |
| 10 | 88.5 |
| 14 | 93.4 |

*Source:* U.S. Bureau of Justice, Statistics at www.ojp.usdoj.gov/bjs/fedle.htm

20. **Boiling temperature:** Using special equipment designed to duplicate atmospheric pressure, a lab experiment is set up to study the boiling temperature for altitudes up to 8000 ft. The data collected are shown in the table, with the boiling temperature y in degrees Fahrenheit and the altitude x in feet. (a) Draw a scatterplot and sketch an estimated line of best fit, then (b) decide if the association is positive or negative. (c) Choose two points on or near this line,

| x | y |
|------|-------|
| −1000 | 213.8 |
| 0 | 212.0 |
| 1000 | 210.2 |
| 2000 | 208.4 |
| 3000 | 206.5 |
| 4000 | 204.7 |
| 5000 | 202.9 |
| 6000 | 201.0 |
| 7000 | 199.2 |
| 8000 | 197.4 |

and use them to find a function model and predict the boiling point on the summit of Mt. Hood in Washington State (11,239 ft), and along the shore of the Dead Sea (1312 ft below sea level). Answers may vary.

21. **Modeling gross domestic product:** For the data given in Exercise 9 (GDP per capita), choose two points on or near the line you sketched and use them to find a function model for the data. Based on this model, what was the GDP per capita for the year 2010?

22. **Modeling home sales:** For the data given in Exercise 10 (home sales), choose two points on or near the line you sketched and use them to find a function model for the data. Based on this model, how many sales can be expected for homes costing $275,000? $317,000?

## ▶ WORKING WITH FORMULAS

23. **Circumference of a circle: $C = 2\pi r$:** The formula for the circumference of a circle can be written as a function of $C$ in terms of $r$: $C(r) = 2\pi r$. (a) Set up a table of values for $r = 1$ through 6 and draw a scatterplot of the data. (b) Is the association positive or negative? Why? (c) What can you say about the strength of the correlation? (d) Sketch a line that "approximates" the data. What can you say about the slope of this line?

24. **Volume of a cylinder: $V = \pi r^2 h$:** As part of a project, students cut a long piece of PVC pipe with a diameter of 10 cm into sections that are 5, 10, 15, 20, and 25 cm long. The bottom of each is then made watertight and

each section is filled to the brim with water. The volume is then measured with the results collected into the table shown. (a) Draw a scatterplot of the data. (b) Is the association positive or negative? Why? (c) What can you say about the strength of the correlation? (d) Would the correlation here be stronger or weaker than in Exercise 23? Why?

| Height (cm) | Volume (cm³) |
|---|---|
| 5 | 380 |
| 10 | 800 |
| 15 | 1190 |
| 20 | 1550 |
| 25 | 1955 |

## ▶ APPLICATIONS

 **Use the regression capabilities of a graphing calculator to complete Exercises 25 through 32.**

25. **Height versus wingspan:** Leonardo da Vinci's famous diagram is an illustration of how the human body comes in predictable proportions. Careful measurements of height versus wingspan were taken on eight students and the set of data is shown here. Using the data, (a) draw the scatterplot; (b) determine whether the association is linear or nonlinear; (c) state whether the association is positive or negative; and (d) find the regression equation and use it to predict the wingspan of a student with a height of 65 in.

| Height | Wingspan |
|---|---|
| 61 | 60.5 |
| 61.5 | 62.5 |
| 54.5 | 54.5 |
| 73 | 71.5 |
| 67.5 | 66 |
| 51 | 50.75 |
| 57.5 | 54 |
| 52 | 51.5 |
| 64 | 63 |

26. **Height versus shoe size:** The data in the table show the height (in inches) compared to the shoe size worn for a random sample of 12 male students. Using the data, (a) draw the scatterplot, (b) determine whether the association is linear or nonlinear, (c) state whether the association is positive or negative, and (d) find the regression equation and use it to predict the shoe size of a man 80 in.

| Height | Shoe Size |
|---|---|
| 69 | 10 |
| 72 | 9 |
| 75 | 14 |
| 74 | 12 |
| 73 | 10.5 |
| 71 | 10 |
| 69.5 | 11.5 |
| 66.5 | 8.5 |
| 65.5 | 9 |

27. **Starbucks coffee:** While the growth in the number of Starbucks stores has slowed in the new decade, the stores continue to enjoy immense popularity. The number of stores nationwide is given in the table for selected years. Using the data, (a) draw the scatterplot, (b) determine whether the association is linear or nonlinear, (c) state whether the association is positive or negative, and (d) find the regression equation and use it to predict the number of stores in 2015.

*Source:* starbucks.com

### Exercise 27

| Year (2000→0) | Stores (1000s) |
|---|---|
| 0 | 3.5 |
| 1 | 4.7 |
| 3 | 7.2 |
| 5 | 9.0 |
| 7 | 15.0 |
| 9 | 16.6 |
| 10 | 16.9 |

### Exercise 28

| Quarter (2009 QI→1) | Total Sales (millions) |
|---|---|
| 1 | 18.3 |
| 2 | 22.1 |
| 3 | 27.1 |
| 4 | 34.6 |
| 5 | 43.2 |
| 6 | 51.8 |
| 7 | 54.2 |
| 8 | 65.5 |

28. **iPhone sales:** The Apple iPhone has proven to be one of the most successful technologies launched in recent years. The cumulative sales total for the iPhone is shown in the table beginning with the first quarter of 2009. Using the data, (a) draw the scatterplot, (b) determine whether the association is linear or nonlinear, (c) state whether the association is positive or negative, and (d) find the regression equation and use it to predict the cumulative sales for the fourth quarter of 2013.

*Source:* poweredbysteam.com, zdnet.com, other sources

**29. High jump records:** The winning height at the summer Olympics has steadily increased over time, as shown in the table for selected years. Using the data, (a) draw the scatterplot, (b) determine whether the association is linear or nonlinear, (c) state whether the association is positive or negative, and (d) find the regression equation and predict the winning height for the 2016 Olympics.

*Source: The World Almanac*

| Year (1900→0) | Height (in.) |
| --- | --- |
| 12 | 76.0 |
| 24 | 78.0 |
| 36 | 79.9 |
| 60 | 85.0 |
| 72 | 87.8 |
| 84 | 92.5 |
| 96 | 94.1 |
| 108 | 92.9 |

**30. Pole vault records:** The winning height for the men's pole vault at the summer Olympics (to the nearest unit) has steadily increased over time, as shown in the table for selected years. Using the data, (a) draw the scatterplot, (b) determine whether the association is linear or nonlinear, (c) state whether the association is positive or negative, and (d) find the regression equation and use it to predict the winning height for the 2016 Olympics.

*Source: The World Almanac*

| Year (1900→0) | Height (feet) |
| --- | --- |
| 12 | 13.0 |
| 20 | 13.4 |
| 36 | 14.3 |
| 52 | 14.9 |
| 60 | 15.4 |
| 64 | 16.7 |
| 72 | 18.1 |
| 84 | 18.9 |
| 96 | 19.4 |
| 108 | 19.6 |

**31. Plastic money:** The total amount of business transacted using credit cards has grown rapidly in recent years. The total volume (in billions of dollars) is shown in the table for selected years. (a) Use a graphing calculator to draw a scatterplot of the data and decide if the association is linear or nonlinear. (b) If linear, calculate a regression equation for the data and display the scatterplot and graph on the same screen. (c) If current trends continue, how many billions of dollars will be transacted in 2015? (d) In what year will the volume exceed 4000 billion?

*Source: 2009 Statistical Abstract of the United States, various years*

| Year (2000→0) | Volume (billions) |
| --- | --- |
| 0 | 1458 |
| 2 | 1638 |
| 4 | 1882 |
| 6 | 2237 |
| 7 | 2414 |
| 8 | 2604 |
| 9 | 2742 |
| 10 | 2859 |

**32. ACH payments:** Due to the convenience and ease of use, the number of ACH (Automated Clearing House) payments is increasing rapidly. The number of ACH transactions is shown in the table for selected years. (a) Use a graphing calculator to draw a scatterplot of the data and decide if the association is linear or nonlinear. (b) If linear, calculate a regression equation for the data and display the scatterplot and graph on the same screen. (c) If current trends continue, how many ACH transactions will there be in 2013? (d) In what year will the number of transactions surpass 30 billion?

*Source: NACHA Reports*

| Year (2000→0) | Transactions (billions) |
| --- | --- |
| 1 | 6.8 |
| 2 | 7.9 |
| 3 | 8.8 |
| 4 | 12.1 |
| 5 | 14.0 |
| 6 | 14.6 |
| 7 | 17.0 |
| 8 | 18.2 |
| 9 | 18.7 |
| 10 | 19.4 |

▶ **EXTENDING THE CONCEPT**

**33.** It can be very misleading to rely on the correlation coefficient alone when selecting a regression model. To illustrate, (a) run a linear regression on the data set given (without doing a scatterplot), and note the strength of the correlation [be sure the **"DiagnosticOn"** feature is active: (2nd) (0) (CATALOG)]. (b) Now run a quadratic regression ( (STAT) **CALC 5:QuadReg**) and note the strength of the correlation. (c) What do you notice? What factors other than the correlation coefficient must be taken into account when choosing a form of regression?

**34.** In his book *Gulliver's Travels,* Jonathan Swift describes how the Lilliputians were able to measure Gulliver for new clothes, even though he was a giant compared to them. According to the text, "Then they measured my right thumb, and desired no more … for by mathematical computation, once around the thumb is twice around the wrist, and so on to the neck and waist." Is it true that once around the neck is twice around the waist? Find at least 10 willing subjects and take measurements of their necks and waists in millimeters. Arrange the data in ordered pair form (circumference of neck, circumference of waist). Draw the scatterplot for this data. Does the association appear to be linear? Find the equation of the best fit line for this set of data. What is the slope of this line? Is the slope near $m = 2$?

**Exercise 33**

| x | y |
| --- | --- |
| 50 | 5 |
| 100 | 125 |
| 150 | 250 |
| 200 | 275 |
| 250 | 370 |
| 300 | 500 |
| 350 | 600 |

▶ **MAINTAINING YOUR SKILLS**

**35.** (2.4) Is the graph shown here the graph of a function? Discuss why or why not.

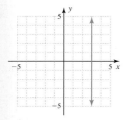

**36.** (1.1) Determine the area of the figure shown $(A = LW, A = \pi r^2)$.

**Exercise 36**

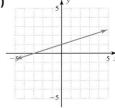

18 cm

24 cm

**37.** (1.1) Solve for $r$:
$A = P + Prt$

**38.** (1.1) Solve for $w$ (if possible):
$-2(6w^2 + 5) - 1 = 7w - 4(3w^2 + 1)$

## MAKING CONNECTIONS

### Making Connections: Graphically, Symbolically, Numerically, and Verbally

Match the characteristics shown in 1 through 16 to one of the eight graphs given in (a) through (h).

**(a)**
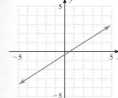

**(b)**
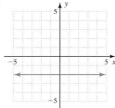

**(c)**
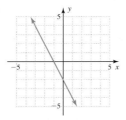

**(d)**
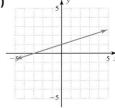

**(e)**
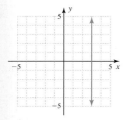

**(f)**
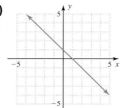

**(g)**
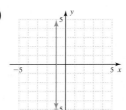

**(h)**
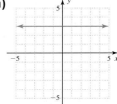

**1.** ____ $y = \frac{1}{3}x + 1$

**2.** ____ $y = -x + 1$

**3.** ____ $m > 0, b < 0$

**4.** ____ $x = -1$

**5.** ____ $y = -2$

**6.** ____ $m < 0, b < 0$

**7.** ____ $m = -2$

**8.** ____ $m = \frac{2}{3}$

**9.** ____ $f(-3) = 4, f(1) = 0$

**10.** ____ $f(-4) = 3, f(4) = 3$

**11.** ____ $f(x) \geq 0$ for $x \in [-3, \infty)$

**12.** ____ $x = 3$

**13.** ____ $f(x) \leq 0$ for $x \in [1, \infty)$

**14.** ____ $m = 0, b > 0$

**15.** ____ function is increasing, $y$-intercept is negative

**16.** ____ function is decreasing, $y$-intercept is negative

## SUMMARY AND CONCEPT REVIEW

### SECTION 2.1   Rectangular Coordinates; Graphing Circles and Other Relations

**KEY CONCEPTS**

- A relation is a collection of ordered pairs $(x, y)$ and can be represented as a set, mapping, graph, or equation.
- The domain of a relation is the set of all first coordinates, and the range is the set of all corresponding second coordinates.
- The graph of a relation in equation form is the set of all ordered pairs $(x, y)$ that satisfy the equation. We plot a sufficient number of points and connect them with a straight line or smooth curve, depending on the pattern formed.
- The midpoint of a line segment with endpoints $(x_1, y_1)$ and $(x_2, y_2)$ is $\left(\dfrac{x_1 + x_2}{2}, \dfrac{y_1 + y_2}{2}\right)$.
- The distance between the points $(x_1, y_1)$ and $(x_2, y_2)$ is $d = \sqrt{(x_2 - x_1)^2 + (y_2 - y_1)^2}$.
- The equation of a circle centered at $(h, k)$ with radius $r$ is $(x - h)^2 + (y - k)^2 = r^2$.

## EXERCISES

1. Represent the relation in mapping notation, then state the domain and range.

   $\{(-7, 3), (-4, -2), (5, 1), (-7, 0), (3, -2), (0, 8)\}$

2. Graph the relation $y = \sqrt{25 - x^2}$ by completing the table, then state the domain and range of the relation.

   | x | −5 | −4 | −2 | 0 | 2 | 4 | 5 |
   |---|----|----|----|---|---|---|---|
   | y |    |    |    |   |   |   |   |

3. Graph the relation $5x + 3y = -15$. Then determine the value of $y$ when $x = 0$, and the value(s) of $x$ when $y = 0$. Write the results in ordered pair form.

Mr. Northeast and Mr. Southwest live in Coordinate County and are good friends. Mr. Northeast lives at *19 East and 25 North* or (19, 25), while Mr. Southwest lives at *14 West and 31 South* or $(-14, -31)$. If the streets in Coordinate County are laid out in 1-mi squares,

4. Use the distance formula to find how far apart they live.

5. If they agree to meet halfway between their homes, what are the coordinates of their meeting place?

6. Sketch the graph of $x^2 + y^2 = 16$.

7. Sketch the graph of $x^2 + y^2 + 6x + 4y + 9 = 0$.

8. Find an equation of the circle whose diameter has the endpoints $(-3, 0)$ and $(0, 4)$.

## SECTION 2.2    Linear Graphs and Rates of Change

### KEY CONCEPTS

- A linear equation can be written in the form $ax + by = c$, where $a$ and $b$ are simultaneously equal to 0.
- The slope of the line through $(x_1, y_1)$ and $(x_2, y_2)$ is $m = \dfrac{y_2 - y_1}{x_2 - x_1}$, where $x_1 \neq x_2$.
- Other designations for slope are $m = \dfrac{\text{rise}}{\text{run}} = \dfrac{\text{change in } y}{\text{change in } x} = \dfrac{\Delta y}{\Delta x} = \dfrac{\text{vertical change}}{\text{horizontal change}}$.
- Lines with positive slope $(m > 0)$ rise from left to right; lines with negative slope $(m < 0)$ fall from left to right.
- The equation of a horizontal line is $y = k$; the slope is $m = 0$.
- The equation of a vertical line is $x = h$; the slope is undefined.
- Lines can be graphed using the intercept method. First determine $(x, 0)$ (substitute 0 for $y$ and solve for $x$), then $(0, y)$ (substitute 0 for $x$ and solve for $y$). Then draw a straight line through these points.
- Parallel lines have equal slopes $(m_1 = m_2)$; perpendicular lines have slopes that are negative reciprocals $\left(m_1 = -\dfrac{1}{m_2}\right)$.

### EXERCISES

9. Plot the points and determine the slope. Then use the ratio $m = \dfrac{\Delta y}{\Delta x}$ to find an additional point on the line:

   **a.** $(-4, 3)$ and $(5, -2)$     **b.** $(3, 4)$ and $(-6, 1)$.

10. Use the slope formula to determine if lines $L_1$ and $L_2$ are parallel, perpendicular, or neither:

    **a.** $L_1$: $(-2, 0)$ and $(0, 6)$; $L_2$: $(1, 8)$ and $(0, 5)$

    **b.** $L_1$: $(1, 10)$ and $(-1, 7)$: $L_2$: $(-2, -1)$ and $(1, -3)$

11. Graph each equation by plotting points: (a) $y = 3x - 2$ (b) $y = -\dfrac{3}{2}x + 1$.

12. Find the intercepts for each line and sketch the graph: (a) $2x + 3y = 6$     (b) $y = \dfrac{4}{3}x - 2$.

13. Identify each line as either horizontal, vertical, or neither, and graph each line.

    **a.** $x = 5$     **b.** $y = -4$     **c.** $2y + x = 5$

14. Find the slope and $y$-intercept of the line shown and discuss the slope in this context.

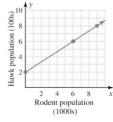

### SECTION 2.3    Graphs and Special Forms of Linear Equations

### KEY CONCEPTS

- The equation of a nonvertical line in slope-intercept form is $y = mx + b$. The slope of the line is $m$ and the $y$-intercept is $(0, b)$.
- To graph a line given its equation in slope-intercept form, plot the $y$-intercept, use the slope ratio $m = \dfrac{\Delta y}{\Delta x}$ to find a second point, and then draw a line through these points.
- If the slope $m$ and a point $(x_1, y_1)$ on the line are known, the equation of the line can be written in point-slope form: $y - y_1 = m(x - x_1)$.
- A secant line is the straight line drawn through two points on a graph.
- The notation $m = \dfrac{\Delta y}{\Delta x}$ indicates the rate at which $y$ is changing with respect to changes in $x$.

## EXERCISES

**15.** Write each equation in slope-intercept form, then identify slope and y-intercept.

    **a.** $4x + 3y - 12 = 0$                        **b.** $5x - 3y = 15$

**16.** Graph each equation using the slope and y-intercept.

    **a.** $f(x) = -\frac{2}{3}x + 1$       **b.** $h(x) = \frac{5}{2}x - 3$

**17.** Find the equation of each line with the given slope through the given point.

    **a.** $m = \frac{2}{3}$; $(1, 4)$         **b.** $m = -\frac{1}{2}$; $(-2, 3)$

**18.** What are the equations of the horizontal and vertical lines passing through $(-2, 5)$? Which line is the point $(7, 5)$ on?

**19.** Find the equation of the line passing through $(1, 2)$ and $(-3, 5)$. Write your final answer in slope-intercept form.

**20.** Find the equation of the line that is parallel to $4x - 3y = 12$ and passes through the point $(3, 4)$. Write your final answer in slope-intercept form.

**21.** (a) Determine the slope and y-intercept of the line shown. Then (b) write the equation of the line in slope-intercept form and (c) interpret the slope $m = \frac{\Delta W}{\Delta R}$ in this context.

**22.** For the graph given, (a) find the equation of the line in point-slope form, (b) write the equation in slope-intercept form, (c) use the equation to predict the x- and y-intercepts, and (d) find y when $x = 20$, and the value of x for which $y = 15$.

**Exercise 21**

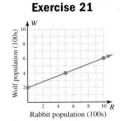

Rabbit population (100s)

**Exercise 22**

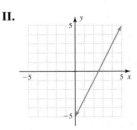

---

## SECTION 2.4  Functions, Function Notation, and the Graph of a Function

### KEY CONCEPTS

- A function is a relation that pairs each element from the domain with exactly one element of the range.
- The vertical line test says that if every vertical line crosses the graph of a relation in at most one point, the relation is a function.
- On a graph, vertical lines can be used to identify the domain, or the set of "allowable inputs" for a function.
- On a graph, horizontal lines can be used to identify the range, or the set of outputs generated by the function.
- x-values that cause a denominator of zero or that cause a negative radicand in a radical expression with an even index must be excluded from the domain.
- The phrase "y is a function of x," is written as $y = f(x)$.

### EXERCISES

**23.** State the implied domain of each function:

    **a.** $f(x) = \sqrt{4x + 5}$

    **b.** $g(x) = \dfrac{x - 4}{x^2 - x - 6}$

**24.** Determine $h(-2)$, $h(-\frac{2}{3})$, and $h(3a)$ for $h(x) = 2x^2 - 3x$.

**25.** Determine if the mapping given represents a function. If not, explain how the definition of a function is violated.

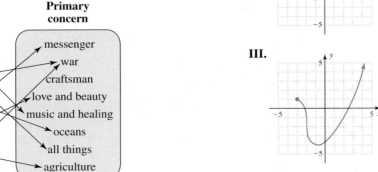

**26.** For each function shown, (a) state the domain and range, (b) find the value of $f(2)$, and (c) determine the value(s) of x for which $f(x) = 1$.

**I.**

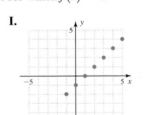

**II.**

**III.**

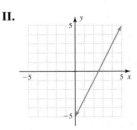

## SECTION 2.5    Analyzing the Graph of a Function

### KEY CONCEPTS

- A function $f$ is even (symmetric about the $y$-axis), if and only if when a point $(x, y)$ is on the graph, then $(-x, y)$ is also on the graph. In function notation: $f(-x) = f(x)$.
- A function $f$ is odd (symmetric about the origin), if and only if when a point $(x, y)$ is on the graph, then $(-x, -y)$ is also on the graph. In function notation: $f(-x) = -f(x)$.

*Intuitive descriptions of the characteristics of a graph are given here. Formal definitions can be found within Section 2.5.*

- A function is *increasing* in an interval if the graph rises from left to right (larger inputs produce larger outputs).
- A function is *decreasing* in an interval if the graph falls from left to right (larger inputs produce smaller outputs).
- A function is *positive* in an interval if the graph is above the $x$-axis in that interval.
- A function is *negative* in an interval if the graph is below the $x$-axis in that interval.
- A function is *constant* in an interval if the graph is parallel to the $x$-axis in that interval.
- A function value is a local maximum if it is the largest value in a designated interval. If it is also the largest value over the entire domain, then it is a global maximum. Similar statements can be made for minimum values.

### EXERCISES

State the domain and range for each function $f(x)$ given. Then state the intervals where $f$ is increasing or decreasing and intervals where $f$ is positive or negative. Assume all endpoints have integer values.

**27.**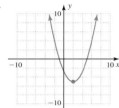

**28.**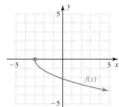

**29.**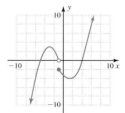

**30.** Determine whether the following are even, odd, or neither.

    **a.** $f(x) = 2x^5 - \sqrt[3]{x}$     **b.** $g(x) = x^4 - \dfrac{\sqrt[3]{x}}{x}$     **c.** $p(x) = |3x| - x^3$     **d.** $q(x) = \dfrac{x^2 - |x|}{x}$

**31.** Draw the function $f$ that has all of the following characteristics, then name the zeroes of the function and the location of all local maximum and minimum values.

    **a.** Domain: $x \in [-6, 10)$         **b.** Range: $y \in [-8, 6)$         **c.** $f(0) = 0$

    **d.** $f(x)\downarrow$ for $x \in (-6, -3) \cup (3, 7.5)$     **e.** $f(x)\uparrow$ for $x \in (-3, 3) \cup (7.5, 10)$     **f.** $f(x) < 0$ for $x \in (-6, 0) \cup (6, 9)$

    **g.** $f(x) > 0$ for $x \in (0, 6) \cup (9, 10)$

## SECTION 2.6    Linear Functions and Real Data

### KEY CONCEPTS

- A scatterplot is the graph of all the ordered pairs in a real data set.
- If larger inputs tend to produce larger output values, we say there is a positive association.
- If larger inputs tend to produce smaller output values, we say there is a negative association.
- If the data seem to cluster around an imaginary line, we say there is a linear association between the variables.
- If the data clearly cannot be approximated by a straight line, we say the variables exhibit a nonlinear association (or sometimes no association).
- Correlation is a measure of how tightly a set of data points cluster about an imaginary curve. The stronger the correlation, the tighter the cluster.
- A regression line minimizes the vertical distance between all data points and the graph itself, making it the unique line of best fit.

## EXERCISES

| t | g |
|----|----|
| 45 | 70 |
| 30 | 63 |
| 10 | 59 |
| 20 | 67 |
| 60 | 73 |
| 70 | 85 |
| 90 | 82 |
| 75 | 90 |

**32.** To determine the value of doing homework, a student in college algebra collects data on the time $t$ spent by classmates on their homework and the grade $g$ they receive on a quiz. Her data is entered in the table shown. (a) Draw a scatterplot of the data. (b) Does the association appear linear or nonlinear? (c) Is the association positive or negative?

**33.** If the association in Exercise 32 is linear, (a) use a graphing calculator to find a linear function that models the relation (study time, grade), then (b) graph the data and the line on the same screen. (c) Does the correlation appear weak or strong?

**34.** According to the function model from Exercise 33, (a) what grade can I expect if I study for an hour? (b) How long would I have to study to reasonably expect a 90 on the quiz?

## PRACTICE TEST

**1.** Two relations here are functions and two are not. Identify the nonfunctions (justify your response).

  **a.** $x = y^2 + 2y$    **b.** $y = \sqrt{5 - 2x}$
  **c.** $|y| + 1 = x$    **d.** $y = x^2 + 2x$

**2.** Determine whether the lines are parallel, perpendicular, or neither:

  $L_1: 2x + 5y = -15$ and $L_2: y = \frac{2}{5}x + 7$.

**3.** Find the slope and $y$-intercept of the line defined by $x + 4y = 8$, then sketch its graph.

**4.** Find the center and radius of the circle defined by $x^2 + y^2 - 4x + 6y - 3 = 0$, then sketch its graph.

**5.** After 2 sec, a car is traveling 20 mph. After 5 sec, its velocity is 40 mph. Assuming the relationship is linear, find the velocity equation and use it to determine the speed of the car after 9 sec.

**6.** Find the equation of the line parallel to $6x + 5y = 3$, containing the point $(2, -2)$. Answer in slope-intercept form.

**7.** My partner and I are at coordinates $(-20, 15)$ on a map. If our destination is at coordinates $(35, -12)$, (a) what are the coordinates of the rest station located halfway to our destination? (b) How far away is our destination? Assume that each unit is 1 mi.

**8.** Write the equations for lines $L_1$ and $L_2$ shown.

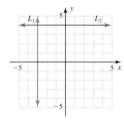

**9.** State the domain and range of the relations shown in graphs 9(a) and 9(b).

**Exercise 9(a)**

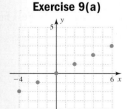

**Exercise 9(b)**

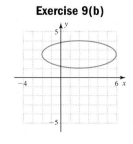

**10.** For the linear function shown,

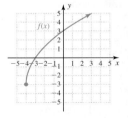

  **a.** Determine the value of $W(24)$ from the graph.
  **b.** What input $h$ will give an output of $W(h) = 375$?
  **c.** Find a linear function for the graph.
  **d.** What does the slope indicate in this context?
  **e.** State the domain and range of $h$.

**11.** Given $f(x) = \dfrac{2 - x^2}{x^2}$, evaluate and simplify:

  **a.** $f(\frac{2}{3})$    **b.** $f(a + 3)$

**12.** Determine the following from the graph shown.

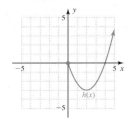

  **a.** the domain and range
  **b.** estimate the value of $f(-1)$
  **c.** interval(s) where $f(x)$ is negative or positive
  **d.** interval(s) where $f(x)$ is increasing, decreasing, or constant

For the function $h(x)$ whose partial graph is given,

**13.** Complete the graph if $h$ is known to be even.

**14.** Complete the graph if $h$ is known to be odd.

**15.** Graph the function $f(x) = 3\sqrt{4 - x}$ by completing the table.

**Exercises 13–14**

**Exercise 15**

| x | f(x) |
|----|------|
| -5 | |
| -3 | |
| -1 | |
| 0 | |
| 1 | |
| 3 | |
| 5 | |

**16.** For the function graphed here, determine the following: (a) domain and range, (b) end-behavior, and (c) any maximum or minimum values and where they occur.

(0, 50)

(57, 33)

f(x)

(23, 12)

**17.** A rabbit population is growing according to the model $r(t) = 2 + 8t^2$, where $r(t)$ is the number of rabbits $t$ yr after the first pair was first spotted in the vegetable garden. (a) Use the difference quotient to find an average rate of change formula for the rabbit population. Then (b) use the formula to find the average rate of change for $t = 0$ to $t = 3$. What does the number represent in this context?

**18.** In 2007, there were 3.3 million iPhones sold worldwide. By 2011, this figure had jumped to a total of 65.5 million [*Source*: zdnet.com, poweredbysteam.com, others]. Assume that for a time, this growth could be modeled by a linear function. (a) Determine the rate of change $\frac{\Delta \text{sales}}{\Delta \text{time}}$, and (b) interpret it in this context. Then use the rate of change to (c) approximate the number of sales by 2009, and what the projected sales would be by 2012 and 2015.

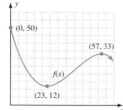

**19.** To study how annual rainfall affects the ability to attain certain levels of livestock production, a local university collects data on the average annual rainfall for a particular area and compares this to the average number of free-ranging cattle per acre for ranchers in that area. The data collected are shown in the table. (a) Draw a scatterplot of the data. (b) Does the association appear linear or nonlinear? (c) Is the association positive or negative?

**Exercise 19**

| Rainfall (in.) | Cattle per Acre |
|---|---|
| 12 | 2 |
| 16 | 3 |
| 19 | 7 |
| 23 | 9 |
| 28 | 11 |
| 32 | 22 |
| 37 | 23 |
| 40 | 26 |

**20.** If the association in Exercise 19 is linear, (a) use a graphing calculator to find a linear function that models the relation (rainfall, cattle per acre), (b) use the function to find the number of cattle per acre that might be possible for an area receiving 50 in. of rainfall per year, and (c) state whether the correlation is weak or strong.

## CALCULATOR EXPLORATION AND DISCOVERY

### I. Linear Equations, Window Size, and Friendly Windows

To graph linear equations on the TI-84 Plus, we (1) solve the equation for the variable $y$, (2) enter the equation on the [Y=] screen, and (3) [GRAPH] the equation and adjust the [WINDOW] if necessary.

**1. Solve the equation for $y$.**

$$2x - 3y = -3 \qquad \text{given equation}$$
$$-3y = -2x - 3 \qquad \text{subtract } 2x \text{ from each side}$$
$$y = \frac{2}{3}x + 1 \qquad \text{divide both sides by } -3$$

**2. Enter the equation on the [Y=] screen.**

On the [Y=] screen, enter $\frac{2}{3}x + 1$. Note that for some calculators parentheses are needed to group $(2 \div 3)x$, to prevent the calculator from interpreting this term as $2 \div (3x)$.

**3. [GRAPH] the equation, adjust the [WINDOW].**

Since much of our work is centered at (0, 0) on the coordinate grid, the calculator's default settings have a domain of $x \in [-10, 10]$ and a range of $y \in [-10, 10]$, as shown in Figure 2.71. This is referred to as the [WINDOW] size. To graph the line in this window, it is easiest to use the [ZOOM] key and select **6:ZStandard**, which resets the window to these default settings. The graph is shown in Figure 2.72. The **Xscl** and **Yscl** entries give the scale used on each axis, indicating that each "tick mark" represents 1 unit. Graphing calculators have many features that enable us to find ordered pairs on a line. One is the [2nd] [GRAPH] (**TABLE**) feature we have seen previously. We can also use the calculator's [TRACE] feature. As the name implies, this feature enables us to trace along the line by moving a blinking cursor using the left [◁] and right [▷] arrow keys. The calculator simultaneously displays the coordinates of the current location of the cursor. After pressing the [TRACE] button, the

**Figure 2.71**

```
WINDOW
 Xmin=-10
 Xmax=10
 Xscl=1
 Ymin=-10
 Ymax=10
 Yscl=1
 Xres=1
```

**Figure 2.72**

Y1=(2/3)X+1

X=3.4042553   Y=3.2695035

cursor appears automatically—usually at the $y$-intercept. Moving the cursor left and right, note the coordinates changing at the bottom of the screen. The point (3.4042553, 3.2695035) is on the line and satisfies the equation of the line. The calculator is displaying decimal values because the screen is exactly 95 pixels wide, 47 pixels to the left of the $y$-axis, and 47 pixels to the right. This means that each time you press the left or right arrow, the $x$-value changes by 1/47—which is *not* a nice round number. To TRACE through "friendlier" values, we can use the ZOOM **4:ZDecimal** feature, which sets Xmin = −4.7 and Xmax = 4.7, or **8:Zinteger,** which sets Xmin = −47 and Xmax = 47. Press ZOOM **4:ZDecimal** and the calculator will automatically regraph the line. Now when you TRACE the line, "friendly" decimal values are displayed.

**Exercise 1:** Use the ZOOM **4:ZDecimal** and **TRACE** features to identify the $x$- and $y$-intercepts for $Y_1 = \frac{2}{3}x + 1$.

**Exercise 2:** Use the ZOOM **8:Zinteger** and **TRACE** features to graph $79x - 55y = 869$, then identify the $x$- and $y$-intercepts.

## II. Locating Zeroes, Maximums, and Minimums

Graphically, the **zeroes** of a function appear as $x$-intercepts with coordinates $(x, 0)$. An estimate for these zeroes can easily be found using a graphing calculator. To illustrate, enter the function $y = x^2 - 8x + 9$ on the Y= screen and graph it using the standard window (ZOOM 6). We access the option for finding zeroes by pressing 2nd TRACE (**CALC**), which displays the screen shown in Figure 2.73. Pressing the number "2" selects **2:zero** and returns you to the graph, where you're asked to enter a "Left Bound." The calculator is asking you to narrow the area it has to search. Select any number conveniently to the left of the $x$-intercept you're interested in. For this graph, we entered a left bound of "0" (press ENTER). The calculator marks this choice with a "▶" marker (pointing to the right), then asks you to enter a "Right Bound." Select any value to the right of the $x$-intercept, but be sure the value you enter *bounds only one intercept* (see Figure 2.74). For this graph, a choice of 10 would include both $x$-intercepts, while a choice of 3 would bound only the intercept on the left. After entering 3, the calculator asks for a "Guess." This option is used when there is more than one zero in the interval, and most of the time we'll bypass this option by pressing ENTER again. The calculator then finds the zero in the selected interval (if it exists), with the coordinates displayed at the bottom of the screen (Figure 2.75).

The maximum and minimum values of a function are located in the same way. Enter $y = x^3 - 3x - 2$ on the Y= screen and graph the function. As seen in Figure 2.76, it appears a local maximum occurs near $x = -1$. To check, we access the **CALC** **4:maximum** option, which returns you to the graph and asks you for a *Left Bound,* a *Right Bound,* and a *Guess* as before. After entering a left bound of "−3" and a right bound of "0," and bypassing the Guess option (note the "▶" and "◀" markers), the calculator locates the maximum you selected, and again displays the coordinates. Due to the algorithm used by the calculator to find these values, a decimal number is sometimes displayed, even if the actual value is an integer (see Figure 2.77).

**Figure 2.73**

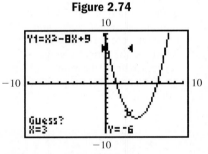

**Figure 2.74**

**Figure 2.75**

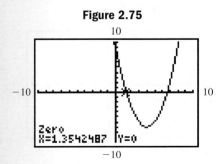

**Figure 2.76**

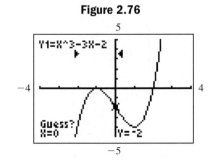

**Figure 2.77**

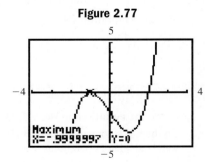

**Use a calculator to find all zeroes and to locate the local maximum and minimum values. Round to the nearest hundredth as needed.**

**Exercise 3:** $y = 2x^2 + 4x - 5$

**Exercise 4:** $y = x^3 - 3x + 1$

**Exercise 5:** $y = x^2 - 8x + 9$

**Exercise 6:** $y = x^3 - 2x^2 - 4x + 8$

**Exercise 7:** $y = x^4 - 5x^2 - 2x$

**Exercise 8:** $y = x\sqrt{x + 4}$

# STRENGTHENING CORE SKILLS

## The Various Forms of a Linear Equation

Learning mathematics is very much like the construction of a skyscraper. The final height of the skyscraper ultimately depends on the strength of the foundation and quality of the frame supporting each new floor as it is built. Our previous work with linear functions and their graphs, while important in its own right, is really the foundation on which much of our future work is built. For this reason, it's important you gain a certain fluency with linear functions and relationships—even to a point where things come to you effortlessly and automatically. As noted mathematician Henri Lebesque once said, "An idea reaches its maximum level of usefulness only when you understand it so well that it seems like you have always known it. You then become incapable of seeing the idea as anything but a trivial and immediate result." These formulas and concepts, while simple, have an endless number of significant and substantial applications.

### Forms and Formulas

| slope formula | point-slope form | slope-intercept form | standard form |
|---|---|---|---|
| $m = \dfrac{y_2 - y_1}{x_2 - x_1}$ | $y - y_1 = m(x - x_1)$ | $y = mx + b$ | $Ax + By = C$ |
| given any two points on the line | given slope $m$ and any point $(x_1, y_1)$ | given slope $m$ and $y$-intercept $(0, b)$ | $A$, $B$, and $C$ are integers (used in linear systems) |

### Characteristics of Lines

| $y$-intercept | $x$-intercept | increasing | decreasing |
|---|---|---|---|
| $(0, y)$ | $(x, 0)$ | $m > 0$ | $m < 0$ |
| let $x = 0$, solve for $y$ | let $y = 0$, solve for $x$ | line slants upward from left to right | line slants downward from left to right |

### Relationships between Lines

| parallel | perpendicular | intersecting | dependent |
|---|---|---|---|
| $m_1 = m_2$, $b_1 \neq b_2$ | $m_1 m_2 = -1$ | $m_1 \neq m_2$ | $m_1 = m_2$, $b_1 = b_2$ |
| lines do not intersect | lines intersect at right angles | lines intersect at one point | lines intersect at all points |

### Special Lines

| horizontal | vertical | identity |
|---|---|---|
| $y = k$ | $x = h$ | $y = x$ |
| passes through $(0, k)$ slope $m = 0$ | passes through $(h, 0)$ slope $m$ is undefined | the input value *identifies* the output |

**Use the formulas and concepts reviewed here to complete the following exercises.**

For the two points given, if possible: (a) compute the slope of the line through the points and state whether the line is increasing, decreasing, or constant, (b) find the equation of the line in point-slope form, then slope-intercept form, and (c) find the and intercepts and use them to graph the line.

**Exercise 1:** $(0, 5)$, $(6, 7)$     **Exercise 2:** $(3, 2)$, $(0, 9)$     **Exercise 3:** $(3, 2)$, $(9, 5)$     **Exercise 4:** $(-5, -4)$, $(3, 2)$

**Exercise 5:** $(-2, 5)$, $(6, -1)$     **Exercise 6:** $(2, -7)$, $(-8, -2)$     **Exercise 7:** $(-3, 4)$, $(-3, 2)$     **Exercise 8:** $(2, 8)$, $(-6, 8)$

## CUMULATIVE REVIEW CHAPTERS R–2

1. Perform the division by factoring the numerator: $(x^3 - 5x^2 + 2x - 10) \div (x - 5)$.

2. Find the solution interval for: $|x - 3| + 4 < 9$

3. The area of a circle is 69 cm². Find the circumference of the same circle.

4. The surface area of a cylinder is $A = 2\pi r^2 + 2\pi rh$. Write $r$ in terms of $A$ and $h$ (solve for $r$).

5. Solve for $x$: $-2(3 - x) + 5x = 4(x + 1) - 7$.

6. Evaluate without using a calculator: $\left(\dfrac{27}{8}\right)^{\frac{-2}{3}}$.

7. Find the slope of each line:
   a. through the points: $(-4, 7)$ and $(2, 5)$.
   b. a line with equation $3x - 5y = 20$.

8. Graph using a table of values.
   a. $f(x) = \sqrt{x - 2} + 3$.
   b. $f(x) = -|x + 2| - 3$.

9. Graph the line passing through $(-3, 2)$ with a slope of $m = \frac{1}{2}$, then state its equation.

10. Show that $x = 1 + 5i$ is a solution to $x^2 - 2x + 26 = 0$.

11. Given $f(x) = 3x^2 - 6x$, find the average rate of change in the interval $[-1, 2]$.

12. Graph by plotting the $y$-intercept, then counting $m = \frac{\Delta y}{\Delta x}$ to find additional points: $y = \frac{1}{3}x - 2$.

13. For the graph of the function given, determine the following:
   a. the domain and range
   b. the value of $f(-3), f(-1), f(1)$, and $f(3)$
   c. the zeroes of the function
   d. interval(s) where $f(x)$ is negative/positive
   e. interval(s) where $f(x)$ is increasing/decreasing
   f. location of any max/min values

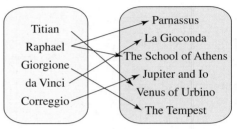

14. Given $f(x) = x^2$ and $g(x) = x^3$, use the formula for average rate of change to determine which of these functions is increasing faster in the intervals:
   a. $[0.5, 0.6]$    b. $[1.5, 1.6]$.

15. Add the rational expressions:
   a. $\dfrac{-2}{x^2 - 3x - 10} + \dfrac{1}{x + 2}$
   b. $\dfrac{b^2}{4a^2} - \dfrac{c}{a}$

16. Simplify the radical expressions:
   a. $\dfrac{-10 + \sqrt{72}}{4}$
   b. $\dfrac{1}{\sqrt{2}}$

17. Determine the equation of the function whose graph is given.

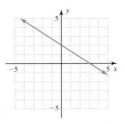

18. Determine if the following relation is a function. If not, how is the definition of a function violated?

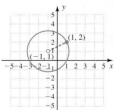

| Titian |
| Raphael |
| Giorgione |
| da Vinci |
| Correggio |

Parnassus
La Gioconda
The School of Athens
Jupiter and Io
Venus of Urbino
The Tempest

19. Solve by completing the square. Answer in both exact and approximate form: $2x^2 + 49 = -20x$

20. Solve using the quadratic formula. If solutions are complex, write them in $a + bi$ form. $2x^2 + 20x = -51$

21. The *National Geographic Atlas of the World* is a very large, rectangular book with an almost inexhaustible panoply of information about the world we live in. The length of the front cover is 16 cm more than its width, and the area of the cover is 1457 cm². Use this information to write an equation model, then use the quadratic formula to determine the length and width of the Atlas.

22. Compute as indicated:
   a. $(2 + 5i)^2$    b. $\dfrac{1 - 2i}{1 + 2i}$

23. Solve by factoring:
   a. $6x^2 - 7x = 20$
   b. $x^3 + 5x^2 - 15 = 3x$

24. A theorem from elementary geometry states, *"A line tangent to a circle is perpendicular to the radius at the point of tangency."* Find the equation of the tangent line for the circle and radius shown.

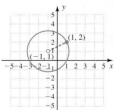

25. A triangle has its vertices at $(-4, 5), (4, -1)$, and $(0, 8)$. Find the perimeter of the triangle and determine whether or not it is a *right* triangle.

# More on Functions

## CHAPTER CONNECTIONS

Viewing a function in terms of an equation, table of values, and the related graph often brings a clearer understanding of the relationships involved. This is particularly important today, as environmental issues and an increased awareness of the fragile "balance of nature" have taken center stage in the national consciousness of many countries. In some oceans, sharks have been hunted to near extinction, causing a huge imbalance in the population of other aquatic species. Mathematical models can be developed to help us better understand these relationships. This application appears as Exercise 64 in Section 3.2.

**Check out these other real-world connections:**

▶ Wind Power Generation
   (Section 3.1, Exercise 83)
▶ Race Car Aerodynamics
   (Section 3.3, Exercise 53)
▶ Overtime Wages
   (Section 3.4, Exercise 50)
▶ Crime Scene Protection
   (Section 3.6, Exercise 36)

## LEARNING OBJECTIVES

*In Section 3.1 you will see how we can:*

- ☐ **A.** Identify basic characteristics of the toolbox functions
- ☐ **B.** Apply vertical/horizontal shifts of a basic graph
- ☐ **C.** Apply vertical/horizontal reflections of a basic graph
- ☐ **D.** Apply vertical stretches and compressions of a basic graph
- ☐ **E.** Apply transformations on a general function

Many applications of mathematics require that we select a function known to fit the context, or build a function model from the information supplied. So far we've looked at linear functions. Here we'll introduce the absolute value, squaring, square root, cubing, and cube root functions. Together these form the **toolbox functions,** so called because they give us a variety of "tools" to model the real world (see Section 3.3). In the same way a study of arithmetic depends heavily on the multiplication table, a study of algebra and mathematical modeling depends (in large part) on a solid working knowledge of these functions.

## A. The Toolbox Functions

While we can accurately graph a line using only two points, most functions require more points to show all of the graph's important features. However, our work is greatly simplified in that most functions belong to a **function family,** in which all graphs from the family share the characteristics of one basic graph, called the **parent function.** This means the number of points required for graphing will quickly decrease as we start anticipating what the graph of a given function should look like. The parent functions and their identifying characteristics are summarized here.

### The Toolbox Functions

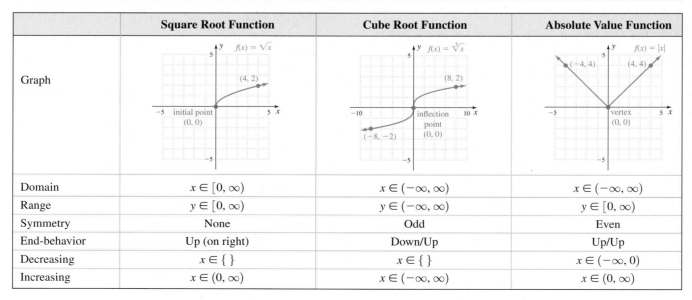

|  | **Constant Function** | **Identity Function** | **Squaring Function** | **Cubing Function** |
|---|---|---|---|---|
| Graph | $f(x) = k$ | $f(x) = x$ | $f(x) = x^2$ | $f(x) = x^3$ |
| Domain | $x \in (-\infty, \infty)$ | $x \in (-\infty, \infty)$ | $x \in (-\infty, \infty)$ | $x \in (-\infty, \infty)$ |
| Range | $y = k$ | $y \in (-\infty, \infty)$ | $y \in [0, \infty)$ | $y \in (-\infty, \infty)$ |
| Symmetry | Even | Odd | Even | Odd |
| End-behavior | $y = k$ | Down/Up | Up/Up | Down/Up |
| Decreasing | $x \in \{\,\}$ | $x \in \{\,\}$ | $x \in (-\infty, 0)$ | $x \in \{\,\}$ |
| Increasing | $x \in \{\,\}$ | $x \in (-\infty, \infty)$ | $x \in (0, \infty)$ | $x \in (-\infty, \infty)$ |

|  | **Square Root Function** | **Cube Root Function** | **Absolute Value Function** |
|---|---|---|---|
| Graph | $f(x) = \sqrt{x}$ | $f(x) = \sqrt[3]{x}$ | $f(x) = |x|$ |
| Domain | $x \in [0, \infty)$ | $x \in (-\infty, \infty)$ | $x \in (-\infty, \infty)$ |
| Range | $y \in [0, \infty)$ | $y \in (-\infty, \infty)$ | $y \in [0, \infty)$ |
| Symmetry | None | Odd | Even |
| End-behavior | Up (on right) | Down/Up | Up/Up |
| Decreasing | $x \in \{\,\}$ | $x \in \{\,\}$ | $x \in (-\infty, 0)$ |
| Increasing | $x \in (0, \infty)$ | $x \in (-\infty, \infty)$ | $x \in (0, \infty)$ |

In applications of the toolbox functions, the parent graph may be "morphed" and/or shifted from its original position, yet the graph will still retain its basic shape and features. The result is called a **transformation** of the parent graph.

---

**EXAMPLE 1** ▶  **Identifying the Characteristics of a Transformed Graph**

The graph of $f(x) = x^2 - 2x - 3$ is given. Use the graph to identify each of the features or characteristics indicated.

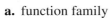

  **a.** function family
  **b.** domain and range
  **c.** vertex
  **d.** intervals where $f$ is increasing or decreasing
  **e.** max or min value(s)
  **f.** end-behavior
  **g.** $x$- and $y$-intercept(s)

**Solution** ▶  **a.** The graph is a parabola, from the squaring function family.
  **b.** domain: $x \in (-\infty, \infty)$; range: $y \in [-4, \infty)$
  **c.** vertex: $(1, -4)$
  **d.** decreasing: $x \in (-\infty, 1)$, increasing: $x \in (1, \infty)$
  **e.** minimum value $y = -4$ at $(1, -4)$
  **f.** end-behavior: up/up
  **g.** $y$-intercept: $(0, -3)$; $x$-intercepts: $(-1, 0)$ and $(3, 0)$

 **A.** You've just seen how we can identify basic characteristics of the toolbox functions

**Now try Exercises 5 through 14** ▶

---

Note that for Example 1(g), we can algebraically verify the $x$-intercepts by substituting 0 for $f(x)$ and solving the equation by factoring. This gives $0 = (x + 1)(x - 3)$, with solutions $x = -1$ and $x = 3$. It's also worth noting that while the parabola is no longer symmetric to the $y$-axis, it *is* symmetric to the vertical line $x = 1$. This line is called the **axis of symmetry** for the parabola, and for a vertical parabola, it will always be a vertical line that goes through the vertex.

## B. Vertical and Horizontal Shifts

As we study specific transformations of a graph, try to develop a *global view* as the transformations can be applied to *any* function. When these are applied to the toolbox functions, we rely on characteristic features of the parent function to assist in completing the transformed graph.

### Vertical Translations

We'll first investigate vertical translations or vertical shifts of the toolbox functions, using the absolute value function to illustrate.

---

**EXAMPLE 2** ▶  **Graphing Vertical Translations**

Construct a table of values for $f(x) = |x|$, $g(x) = |x| + 1$, and $h(x) = |x| - 3$ and graph the functions on the same coordinate grid. Then discuss what you observe.

**Solution** ▶  A table of values for all three functions is given, with the corresponding graphs shown in the figure.

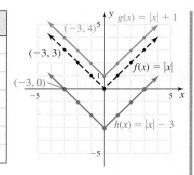

| $x$ | $f(x) = |x|$ | $g(x) = |x| + 1$ | $h(x) = |x| - 3$ |
|---|---|---|---|
| $-3$ | 3 | 4 | 0 |
| $-2$ | 2 | 3 | $-1$ |
| $-1$ | 1 | 2 | $-2$ |
| 0 | 0 | 1 | $-3$ |
| 1 | 1 | 2 | $-2$ |
| 2 | 2 | 3 | $-1$ |
| 3 | 3 | 4 | 0 |

Note that outputs of $g(x)$ are one more than the outputs of $f(x)$, and that each point on the graph of $f$ has been shifted *upward 1 unit* to form the graph of $g$. Similarly, each point on the graph of $f$ has been shifted *downward 3 units* to form the graph of $h$, since $h(x) = f(x) - 3$.

**Now try Exercises 15 through 18 ▶**

We describe the transformations in Example 2 as a **vertical shift** or **vertical translation** of a basic graph. The graph of $g$ is the graph of $f$ *shifted up 1 unit,* and the graph of $h$ is the graph of $f$ *shifted down 3 units.* In general, we have the following:

### Vertical Translations of a Basic Graph

Given $k > 0$ and any function whose graph is determined by $y = f(x)$,

1. The graph of $y = f(x) + k$ is the graph of $f(x)$ shifted upward $k$ units.
2. The graph of $y = f(x) - k$ is the graph of $f(x)$ shifted downward $k$ units.

### Horizontal Translations

The graph of a parent function can also be shifted left or right. This happens when we *alter the inputs to the basic function,* as opposed to adding to or subtracting from the function itself. For $Y_1 = x^2 + 2$ note that we first square inputs, then add 2, which results in a vertical shift. For $Y_2 = (x + 2)^2$, we add 2 to $x$ *prior to squaring* and since the input values are affected, we might anticipate the graph will shift along the $x$-axis—horizontally.

**EXAMPLE 3 ▶** **Graphing Horizontal Translations**

Construct a table of values for $f(x) = x^2$ and $g(x) = (x + 2)^2$, then graph the functions on the same grid and discuss what you observe.

**Solution ▶** Both $f$ and $g$ belong to the squaring family and their graphs are parabolas. A table of values is shown along with the corresponding graphs.

| $x$ | $f(x) = x^2$ | $g(x) = (x + 2)^2$ |
|---|---|---|
| $-3$ | 9 | 1 |
| $-2$ | 4 | 0 |
| $-1$ | 1 | 1 |
| 0 | 0 | 4 |
| 1 | 1 | 9 |
| 2 | 4 | 16 |
| 3 | 9 | 25 |

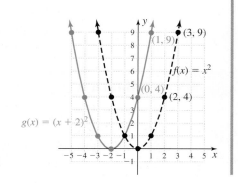

It is apparent the graphs of $g$ and $f$ are identical, except that the graph of $g$ has been shifted horizontally 2 units left.

Now try Exercises 19 and 20 ▶

We describe the transformation in Example 3 as a **horizontal shift** or **horizontal translation** of a basic graph. The graph of $g$ is the graph of $f$ *shifted 2 units to the left*. Once again it seems reasonable that since *input* values were altered, the shift must be horizontal rather than vertical. From this example, we also learn the direction of the shift is **opposite the sign:** $y = (x + 2)^2$ is 2 units *to the left* of $y = x^2$. Although it may seem counterintuitive, the shift *opposite the sign* can be "seen" by locating the new $x$-intercept, which in this case is also the vertex. Substituting 0 for $y$ gives $0 = (x + 2)^2$ with $x = -2$, as shown in the graph. In general, we have

### Horizontal Translations of a Basic Graph

Given $h > 0$ and any function whose graph is determined by $y = f(x)$,
1. The graph of $y = f(x + h)$ is the graph of $f(x)$ shifted *left h* units.
2. The graph of $y = f(x - h)$ is the graph of $f(x)$ shifted *right h* units.

**EXAMPLE 4 ▶    Graphing Horizontal Translations**

Sketch the graphs of $g(x) = |x - 2|$ and $h(x) = \sqrt{x + 3}$ using a horizontal shift of the parent function and a few characteristic points (not a table of values).

**Solution ▶**    The graph of $g(x) = |x - 2|$ (Figure 3.1) is the absolute value function shifted 2 units to the right (shift the vertex and two other points from $y = |x|$). The graph of $h(x) = \sqrt{x + 3}$ (Figure 3.2) is a square root function, shifted 3 units to the left (shift the initial point and one or two points from $y = \sqrt{x}$).

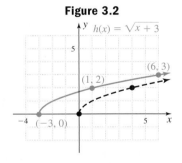

Figure 3.1

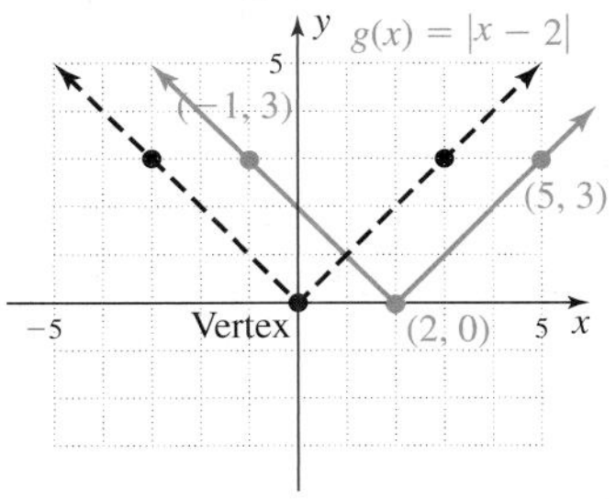

Figure 3.2

☑ **B.** You've just seen how we can perform vertical/horizontal shifts of a basic graph

Now try Exercises 21 and 22 ▶

**CAUTION ▶**    When performing horizontal translations, remember the direction of the shift is always **opposite the sign.** For $h > 0$, shift left $h$ units to get the graph of $y = f(x + h)$ and right $h$ units to get the graph of $y = f(x - h)$.

## C. Vertical and Horizontal Reflections

The next transformation we investigate is called a **vertical reflection,** in which we compare the function $Y_1 = f(x)$ with the negative of the function: $Y_2 = -f(x)$.

### Vertical Reflections

**EXAMPLE 5** ▶ **Graphing Vertical Reflections**

Construct a table of values for $Y_1 = x^2$ and $Y_2 = -x^2$, then graph the functions on the same grid and discuss what you observe.

**Solution** ▶ A table of values is given for both functions, along with the corresponding graphs.

| $x$ | $Y_1 = x^2$ | $Y_2 = -x^2$ |
|-----|-----------|-----------|
| $-2$ | 4 | $-4$ |
| $-1$ | 1 | $-1$ |
| 0 | 0 | 0 |
| 1 | 1 | $-1$ |
| 2 | 4 | $-4$ |

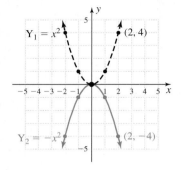

As you might have anticipated, the outputs for $f$ and $g$ differ only in sign. Each output is a **reflection** of the other, being an equal distance from the $x$-axis but on opposite sides.

**Now try Exercises 23 and 24** ▶

The vertical reflection in Example 5 is called a **reflection across the $x$-axis**. In general,

---
**Vertical Reflections of a Basic Graph**

For any function $y = f(x)$, the graph of $y = -f(x)$
is the graph of $f(x)$ reflected across the $x$-axis.

---

### Horizontal Reflections

It's also possible for a graph to be reflected horizontally *across the y-axis*. Just as we noted that $f(x)$ versus $-f(x)$ resulted in a vertical reflection, $f(x)$ versus $f(-x)$ results in a horizontal reflection.

**EXAMPLE 6** ▶ **Graphing a Horizontal Reflection**

Construct a table of values for $f(x) = \sqrt{x}$ and $g(x) = \sqrt{-x}$, then graph the functions on the same grid and discuss what you observe.

**Solution** ▶ A table of values is given here, along with the corresponding graphs.

| $x$ | $f(x) = \sqrt{x}$ | $g(x) = \sqrt{-x}$ |
|-----|-----------------|------------------|
| $-4$ | not real | 2 |
| $-2$ | not real | $\sqrt{2} \approx 1.41$ |
| $-1$ | not real | 1 |
| 0 | 0 | 0 |
| 1 | 1 | not real |
| 2 | $\sqrt{2} \approx 1.41$ | not real |
| 4 | 2 | not real |

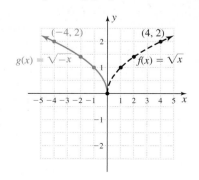

The graph of *g* is the same as the graph of *f*, but it has been reflected across the *y*-axis. A study of the domain shows why—*f* represents a real number only for nonnegative inputs, so its graph occurs to the right of the *y*-axis, while *g* represents a real number for nonpositive inputs, so its graph occurs to the left.

☑ **C. You've just seen how we can apply vertical/horizontal reflections of a basic graph**

**Now try Exercises 25 and 26** ▶

The transformation in Example 6 is called a **horizontal reflection** of a basic graph. In general,

> **Horizontal Reflections of a Basic Graph**
>
> For any function $y = f(x)$, the graph of $y = f(-x)$
> is the graph of $f(x)$ reflected across the *y*-axis.

## D. Vertically Stretching/Compressing a Basic Graph

As the words "stretching" and "compressing" imply, the graph of a basic function can also become elongated or flattened after certain transformations are applied. However, even these transformations preserve the key characteristics of the graph.

**EXAMPLE 7** ▶ **Stretching and Compressing a Basic Graph**

Construct a table of values for $f(x) = x^2$, $g(x) = 3x^2$, and $h(x) = \frac{1}{3}x^2$, then graph the functions on the same grid and discuss what you observe.

**Solution** ▶ A table of values is given for all three functions, along with the corresponding graphs.

| $x$ | $f(x) = x^2$ | $g(x) = 3x^2$ | $h(x) = \frac{1}{3}x^2$ |
|-----|------|------|------|
| $-3$ | 9 | 27 | 3 |
| $-2$ | 4 | 12 | $\frac{4}{3}$ |
| $-1$ | 1 | 3 | $\frac{1}{3}$ |
| 0 | 0 | 0 | 0 |
| 1 | 1 | 3 | $\frac{1}{3}$ |
| 2 | 4 | 12 | $\frac{4}{3}$ |
| 3 | 9 | 27 | 3 |

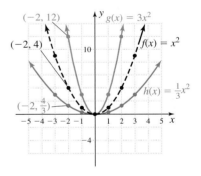

The outputs of *g* are triple those of *f*, making these outputs farther from the *x*-axis and *stretching g* upward (making the graph more narrow). The outputs of *h* are one-third those of *f*, and the graph of *h* is *compressed* downward, with its outputs closer to the *x*-axis (making the graph wider).

**Now try Exercises 27 through 30** ▶

**WORTHY OF NOTE**

In a study of trigonometry, you'll find that a basic graph may need to be stretched or compressed horizontally, a phenomenon known as *frequency variations*. Until then, we can generally avoid the issue by simply rewriting the function so that an equivalent *vertical* stretch or compression can be applied. **See Exercise 89.**

The transformations in Example 7 are called **vertical stretches** or **compressions** of a basic graph. Notice that while the outputs are increased or decreased by a constant factor (making the graph appear narrower or wider), the domain of the function remains unchanged. In general,

☑ **D. You've just seen how we can apply vertical stretches and compressions of a basic graph**

> **Stretches and Compressions of a Basic Graph**
>
> For any function $y = f(x)$, the graph of $y = af(x)$ is
> 1. the graph of $f(x)$ stretched vertically if $|a| > 1$,
> 2. the graph of $f(x)$ compressed vertically if $0 < |a| < 1$.

## E. Transformations of a General Function

If more than one transformation is applied to a basic graph, it's helpful to use the following sequence for graphing the new function.

> **General Transformations of a Basic Graph**
>
> Given a function $y = f(x)$, the graph of $y = af(x \pm h) \pm k$ can be obtained by applying the following sequence of transformations:
> **1.** horizontal shifts  **2.** reflections  **3.** stretches/compressions  **4.** vertical shifts

We generally use a few characteristic points to track the transformations involved, then draw the transformed graph through the new location of these points.

**EXAMPLE 8 ▶ Graphing Functions Using Transformations**

Use transformations of a parent function to sketch the graphs of
**a.** $g(x) = -(x + 2)^2 + 3$    **b.** $h(x) = 2\sqrt[3]{x - 2} - 1$

**Solution ▶ a.** The graph of $g$ is a parabola, (1) shifted left 2 units, (2) reflected across the $x$-axis, and (3) shifted up 3 units. This sequence of transformations is shown in Figures 3.3 through 3.5. Note that since the graph has been shifted 2 units left and 3 units up, the vertex of the parabola has likewise shifted from $(0, 0)$ to $(-2, 3)$.

**Figure 3.3**

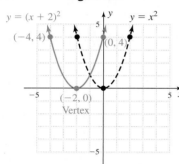

(1) Shifted left 2

**Figure 3.4**

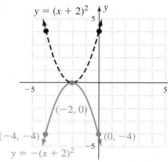

(2) Reflected across the $x$-axis

**Figure 3.5**

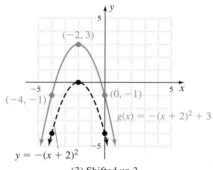

(3) Shifted up 3

**b.** The graph of $h$ is a cube root function, (1) shifted right 2, (2) stretched vertically by a factor of 2, then (3) shifted down 1. This sequence is shown in Figures 3.6 through 3.8 and illustrate how the inflection point has shifted from $(0, 0)$ to $(2, -1)$.

**Figure 3.6**

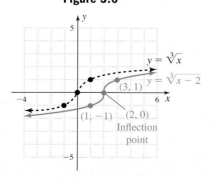

(1) Shifted right 2

**Figure 3.7**

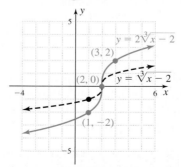

(2) Stretched vertically by a factor of 2

**Figure 3.8**

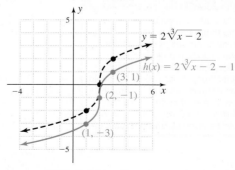

(3) Shifted down 1

**Now try Exercises 31 through 62 ▶**

It's important to note that the transformations can actually be applied to *any function,* even those that are new and unfamiliar. Consider the following pattern:

| **Parent Function** | **Transformation of Parent Function** |
|---|---|
| quadratic:  $y = x^2$ | $y = -2(x - 3)^2 + 1$ |
| absolute value:  $y = |x|$ | $y = -2|x - 3| + 1$ |
| cube root:  $y = \sqrt[3]{x}$ | $y = -2\sqrt[3]{x - 3} + 1$ |
| general:  $y = f(x)$ | $y = -2f(x - 3) + 1$ |

In each case, the transformation involves a horizontal shift right 3, a vertical reflection, a vertical stretch by a factor of 2, and a vertical shift up 1. Since the shifts are the same regardless of the initial function, we can generalize the results to any function $f(x)$.

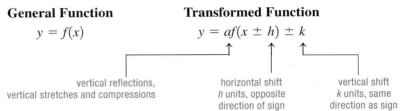

**General Function**          **Transformed Function**

$$y = f(x) \qquad\qquad y = af(x \pm h) \pm k$$

vertical reflections,           horizontal shift          vertical shift
vertical stretches and compressions      *h* units, opposite        *k* units, same
                              direction of sign         direction as sign

Also bear in mind that the graph will be reflected across the *y*-axis (horizontally) if *x* is replaced with $-x$. This process is illustrated in Example 9 for selected transformations.

**EXAMPLE 9** ▶  **Graphing Transformations of a General Function**

Given the graph of $f(x)$ shown in Figure 3.9, graph $g(x) = -f(x + 1) - 2$.

**Solution** ▶  For *g*, the graph of *f* is (1) shifted horizontally 1 unit left (Figure 3.10), (2) reflected across the *x*-axis (Figure 3.11), and (3) shifted vertically 2 units down (Figure 3.12). The final result is that in Figure 3.12.

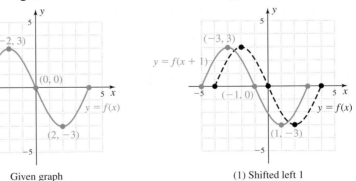

**Figure 3.9**                    **Figure 3.10**

Given graph                       (1) Shifted left 1

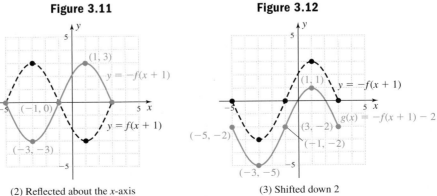

**Figure 3.11**                   **Figure 3.12**

(2) Reflected about the *x*-axis   (3) Shifted down 2

**Now try Exercises 63 through 72** ▶

As noted in Example 9, these shifts and transformation are often combined—particularly when the toolbox functions are used as real-world models (Section 3.3).

Using the general equation $y = af(x \pm h) \pm k$, we can identify the vertex, initial point, or inflection point of any toolbox function and sketch its graph. Given the *graph* of a toolbox function, we can likewise identify these points and reconstruct its equation. We first identify the function family and the location $(h, k)$ of any characteristic point. By selecting one other point $(x, y)$ on the graph, we then use the general equation as a formula (substituting $h$, $k$, and the $x$- and $y$-values of the second point) to solve for $a$ and complete the equation.

**EXAMPLE 10** ▶ **Writing the Equation of a Function from a Description**

As a fan oscillates $180°$ around a room, the direction the fan blows depends on time. Using the diagram from Figure 3.13 and the table of values shown,

  **a.** Plot the relationship (time, position).

  **b.** Find a function that models this relationship.

**Figure 3.13**

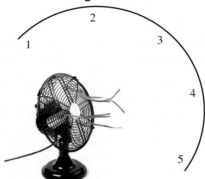

| Time (sec) | Position | Time (sec) | Position |
|---|---|---|---|
| 5 | 5 | 30 | 2 |
| 10 | 4 | 35 | 3 |
| 15 | 3 | 40 | 4 |
| 20 | 2 | 45 | 5 |
| 25 | 1 | | |

**Solution** ▶

**a.** The graph of the relationship is shown. Notice the graph passes the vertical line test, and is that of a function.

**b.** To find a formula for the function, we first note that the graph is that of an absolute value function with vertex $(h, k)$ at $(25, 1)$. For an additional point, choose $(45, 5)$ and work as follows:

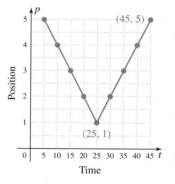

$$p(t) = a|t - h| + k \qquad \text{general equation (function is shifted right and up)}$$
$$5 = a|45 - 25| + 1 \qquad \text{substitute 25 for } h \text{ and 1 for } k, \text{ along with 45 for } t \text{ and 5 for } p(t)$$
$$5 = 20a + 1 \qquad \text{simplify}$$
$$4 = 20a \qquad \text{subtract 1}$$
$$\frac{1}{5} = a \qquad \text{solve for } a$$

The function is $p(t) = \frac{1}{5}|t - 25| + 1$ with domain $[5, 45]$.

☑ **E.** You've just seen how we can apply transformations on a general function

**Now try Exercises 73 through 78** ▶

## 3.1 EXERCISES

▶ **CONCEPTS AND VOCABULARY**

**Fill in each blank with the appropriate word or phrase. Carefully reread the section, if necessary.**

1. The vertex of $h(x) = 3(x + 5)^2 - 9$ is at _____ and the graph opens _____.

2. The inflection point of $f(x) = -2(x - 4)^3 + 11$ is at _____ and the end-behavior is _____, _____.

3. Discuss/Explain how the graph of $F(x) = -2f(x + 1) - 3$ can be obtained from the graph of $f(x)$. If $(0, 5)$, $(6, 7)$, and $(-9, -4)$ are on the graph of $f$, where do they end up on the graph of $F$?

4. Discuss/Explain why the shift of $f(x) = x^2 + 3$ is a *vertical shift* of 3 units in the *positive* direction, while the shift of $g(x) = (x + 3)^2$ is a *horizontal shift* 3 units in the *negative* direction. Include several examples along with a table of values for each.

▶ **DEVELOPING YOUR SKILLS**

**By carefully inspecting each graph given, identify the (a) function family; (b) vertex, inflection point, or initial point; (c) end-behavior; (d) intercepts; (e) maximum or minimum values and where they occur; (f) domain and range; (g) interval(s) where positive and negative; and (h) interval(s) where increasing and decreasing. Assume required features have integer values.**

5. $f(x) = x^2 + 6x + 5$

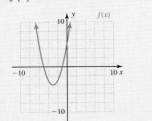

6. $g(x) = -3\sqrt{4 - x} + 3$

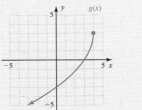

7. $h(x) = -3|x - 2| + 3$

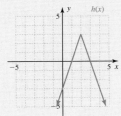

8. $p(x) = \sqrt[3]{x - 1} - 1$

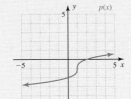

9. $q(x) = -\sqrt[3]{x} + 1$

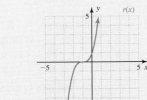

10. $r(x) = (x + 1)^3$

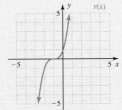

11.

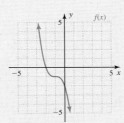

12.

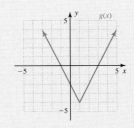

13.

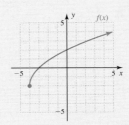

14.

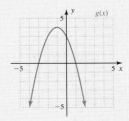

**Sketch each graph by hand using transformations of a parent function (without a table of values).**

15. $f(x) = x^3 - 2$

16. $g(x) = \sqrt{x} - 4$

17. $h(x) = x^2 + 3$

18. $t(x) = |x| - 3$

19. $p(x) = (x - 3)^2$

20. $q(x) = \sqrt{x - 1}$

21. $h(x) = |x + 3|$

22. $f(x) = \sqrt[3]{x} + 2$

23. $g(x) = -|x|$

24. $j(x) = -\sqrt{x}$

25. $f(x) = \sqrt[3]{-x}$

26. $g(x) = (-x)^3$

27. $f(x) = 4\sqrt[3]{x}$

28. $g(x) = -2|x|$

29. $p(x) = \frac{1}{3}x^3$

30. $q(x) = \frac{3}{4}\sqrt{x}$

Use the characteristics of each function family to match a given function to its corresponding graph. The graphs are not scaled—make your selection based on a careful comparison.

**31.** $f(x) = \frac{1}{2}x^3$

**32.** $f(x) = \frac{-2}{3}x + 2$

**33.** $f(x) = -(x - 3)^2 + 2$

**34.** $f(x) = -\sqrt[3]{x - 1} - 1$

**35.** $f(x) = |x + 4| + 1$

**36.** $f(x) = -\sqrt{x + 6}$

**37.** $f(x) = -\sqrt{x + 6} - 1$

**38.** $f(x) = x + 1$

**39.** $f(x) = (x - 4)^2 - 3$

**40.** $f(x) = |x - 2| - 5$

**41.** $f(x) = \sqrt{x + 3} - 1$

**42.** $f(x) = -(x + 3)^2 + 5$

**a.**

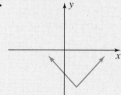

**b.**

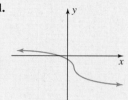

**c.**

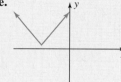

**d.**

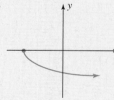

**e.**

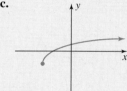

**f.**

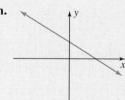

**g.**

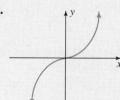

**h.**

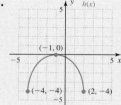

**i.**

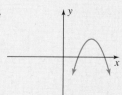

**j.**

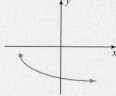

**k.**

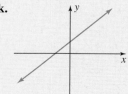

**l.**

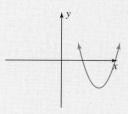

Graph each function using transformations of a parent function and a few characteristic points. *Clearly state and indicate the transformations used* and identify the location of all vertices, initial points, and/or inflection points.

**43.** $f(x) = \sqrt{x + 2} - 1$

**44.** $g(x) = \sqrt{x - 3} + 2$

**45.** $h(x) = -(x + 3)^2 - 2$

**46.** $H(x) = -(x - 2)^2 + 5$

**47.** $p(x) = (x + 3)^3 - 1$

**48.** $q(x) = (x - 2)^3 + 1$

**49.** $s(x) = \sqrt[3]{x + 1} - 2$

**50.** $t(x) = \sqrt[3]{x - 3} + 1$

**51.** $f(x) = -|x + 3| - 2$

**52.** $g(x) = -|x - 4| - 2$

**53.** $h(x) = -2(x + 1)^2 - 3$

**54.** $h(x) = \frac{1}{5}(x - 3)^2 + 1$

**55.** $p(x) = -\frac{1}{3}(x + 2)^3 - 1$

**56.** $P(x) = 4(x - 3)^3 + 1$

**57.** $Q(x) = \frac{1}{2}\sqrt[3]{-x + 3} - 1$

**58.** $q(x) = 4\sqrt[3]{x + 1} + 2$

**59.** $u(x) = -2\sqrt{-x - 1} + 3$

**60.** $v(x) = 3\sqrt{-x + 2} - 1$

**61.** $H(x) = \frac{1}{2}|x + 2| - 3$

**62.** $H(x) = -2|x - 3| + 4$

Apply the transformations indicated for the graph of the general function given.

**63.**

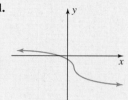

**a.** $f(x - 2)$
**b.** $-f(x) - 3$
**c.** $\frac{1}{2}f(x + 1)$
**d.** $f(-x) + 1$

**64.**

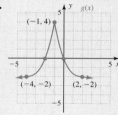

**a.** $g(x) - 2$
**b.** $-g(x) + 3$
**c.** $2g(x + 1)$
**d.** $\frac{1}{2}g(x - 1) + 2$

**65.**

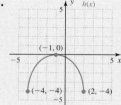

**a.** $h(x) + 3$
**b.** $-h(x - 2)$
**c.** $h(x - 2) - 1$
**d.** $\frac{1}{4}h(x) + 5$

**66.**

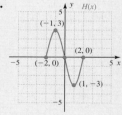

**a.** $H(x - 3)$
**b.** $-H(x) + 1$
**c.** $2H(x - 3)$
**d.** $\frac{1}{3}H(x - 2) + 1$

**67.** Suppose the intercepts of $f(x)$ are $(-3, 0)$, $(0, -4)$ and $(1, 0)$. Find the new location of these points for each of the following transformations: (a) $-2f(x)$, (b) $3f(-x)$, (c) $-4f(x + 2)$, and (d) $f(x - 1) + 5$.

**68.** Suppose the intercepts of $g(x)$ are $(0, 2)$, $(6, 0)$, and $(8, 0)$. Find the new location of these points for each of the following transformations: (a) $-\frac{1}{2}g(x)$, (b) $g(-x) + 4$, (c) $2g(x + 5)$, and (d) $g(x - 7) - 9$.

69. Suppose the domain of $f(x)$ is $x \in [2, \infty)$, while the range is $y \in (-\infty, 3)$. Find the domain and range of (a) $4f(x)$, (b) $-f(-x)$, (c) $2f(x + 1)$, and (d) $f(x - 2) + 3$.

70. Suppose the domain of $g(x)$ is $x \in (-3, 4]$, while the range is $y \in [10, \infty)$. Find the domain and range of (a) $-2g(x)$, (b) $g(-x) - 1$, (c) $\frac{1}{2}g(x - 3)$, and (d) $g(x + 1) - 2$.

71. Suppose $f(x)$ is a continuous function that has a down/up end-behavior, increases for $x \in (-\infty, -2) \cup (4, \infty)$, decreases for $x \in (-2, 4)$, has a local maximum at $(-2, 3)$, and has a local minimum at $(4, -3)$. For the function $2f(x - 3) - 4$, find the following: (a) end-behavior; (b) interval(s) where increasing and decreasing; and (c) location of any maximum or minimum values.

72. Suppose $g(x)$ is a continuous function that has an up/up end-behavior, increases for $x \in (-2, 0) \cup (3, \infty)$, decreases for $x \in (-\infty, -2) \cup (0, 3)$, has local minima at $(-2, 4)$ and $(3, 5)$, and has a local maximum at $(0, 6)$. For the function $-3g(-x) + 1$, find the following: (a) end-behavior; (b) interval(s) where increasing and decreasing; and (c) location of any maximum or minimum values.

Use the graph given and the points indicated to determine the equation of the function shown using the general form $y = af(x \pm h) \pm k$.

73.
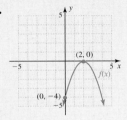

74.
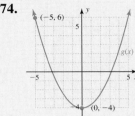

75.
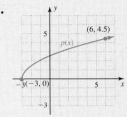

76.
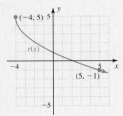

77.
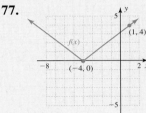

78.
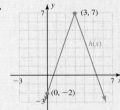

## ▶ WORKING WITH FORMULAS

79. **Volume of a sphere:** $V(r) = \frac{4}{3}\pi r^3$

The volume of a sphere is given by the function shown, where $V(r)$ is the volume in cubic units and $r$ is the radius. Note this function belongs to the *cubic family* of functions. (a) Approximate the value of $\frac{4}{3}\pi$ to one decimal place, then graph the function on the interval $[0, 3]$. (b) From your *graph,* estimate the volume of a sphere with radius 2.5 in., then compute the actual volume. Are the results close? (c) Solve for $r$ in terms of $V$. (d) If the volume is $288\pi$ in$^3$, what is the radius?

80. **Fluid motion:** $V(h) = -4\sqrt{h} + 20$

The velocity of a fluid flowing from an open tank is given by the function shown, where $V(h)$ is the velocity (in ft/sec) at water height $h$ (in ft). Note this function belongs to the *square root family* of functions. An open tank 25 ft tall is filled with fluid. (a) Use a table to graph the function on the interval $[0, 25]$. (b) From your graph, estimate the velocity of the fluid when $h = 7$ ft, then find the actual velocity. Are the answers close? (c) Solve for $h$ in terms of $V$. (d) If the velocity is 5 ft/sec, how high is the water in the tank?

► **APPLICATIONS**

81. **Gravity, distance, time:** After being released, the time it takes an object to fall $x$ ft is given by the function $T(x) = \frac{1}{4}\sqrt{x}$, where $T(x)$ is in seconds. (a) Describe the transformation applied to obtain the graph of $T$ from the graph of $y = \sqrt{x}$, then sketch the graph of $T$ for $x \in [0, 100]$. (b) How long would it take an object to hit the ground if it were dropped from a height of 81 ft? Is this point on your graph?

82. **Stopping distance:** In certain weather conditions, accident investigators will use the function $v(x) = 4.9\sqrt{x}$ to estimate the speed of a car (in miles per hour) that has been involved in an accident, based on the length $x$ of the skid marks (in feet). (a) Describe the transformation applied to obtain the graph of $v$ from the graph of $y = \sqrt{x}$, then sketch the graph of $v$ for $x \in [0, 400]$. (b) If the skid marks were 225 ft long, how fast was the car traveling? Is this point on your graph?

83. **Wind power:** The power $P$ generated by a certain wind turbine is given by the function $P(v) = \frac{8}{125}v^3$ where $P(v)$ is the power in watts at wind velocity $v$ (in miles per hour). (a) Describe the transformation applied to obtain the graph of $P$ from the graph of $y = v^3$, then sketch the graph of $P$ for $v \in [0, 25]$ (scale the axes appropriately). (b) How much power is being generated when the wind is blowing at 15 mph? Is this point on your graph?

84. **Wind power:** If the power $P$ (in watts) being generated by a wind turbine is known, the velocity of the wind (in mph) can be determined using the function $v(P) = \frac{5}{2}\sqrt[3]{P}$. (a) Describe the transformation applied to obtain the graph of $v$ from the graph of $y = \sqrt[3]{P}$, then sketch the graph of $v$ for $P \in [0, 512]$ (scale the axes appropriately). (b) How fast is the wind blowing if 343 W of power is being generated? Is this point on your graph?

85. **Distance rolled due to gravity:** The *distance* a ball rolls down an inclined plane is given by the function $d(t) = 2t^2$, where $d(t)$ represents the distance in feet after $t$ sec. (a) Describe the transformation applied to obtain the graph of $d$ from the graph of $y = t^2$, then sketch the graph of $d$ for $t \in [0, 3]$. (b) How far has the ball rolled after 2.5 sec? Is this point on your graph?

86. **Velocity due to gravity:** The *velocity* of a steel ball bearing as it rolls down an inclined plane is given by the function $v(t) = 4t$, where $v(t)$ represents the velocity in feet per second after $t$ sec. (a) Describe the transformation applied to obtain the graph of $v$ from the graph of $y = t$, then sketch the graph of $v$ for $t \in [0, 3]$. (b) What is the velocity of the ball bearing after 2.5 sec? Is this point on your graph?

► **EXTENDING THE CONCEPT**

87. (a) Graph $y = x^2$, $y = x^4$, and $y = x^6$, then identify and use the shared features to draw a sketch for *any* function of the form $y = x^{2n}$. (b) Repeat part (a) for the family of functions with odd powers: $y = x^3$, $y = x^5$, $y = x^7$, and $y = x^{2n+1}$. Then repeat parts (a) and (b) for the root functions: (c) $y = \sqrt[2n]{x} = x^{\frac{1}{2n}}$ and (d) $y = \sqrt[2n+1]{x} = x^{\frac{1}{2n+1}}$.

88. Carefully graph the functions $f(x) = |x|$ and $g(x) = 2\sqrt{x}$ on the same coordinate grid. From the graph, in what interval is the graph of $g(x)$ *above* the graph of $f(x)$? Pick a number (call it $a$) from this interval and substitute it in both functions. Is $g(a) > f(a)$? In what interval is the graph of $g(x)$ below the graph of $f(x)$? Pick a number from this interval (call it $b$) and substitute it in both functions. Is $g(b) < f(b)$?

89. For any function $y = f(x)$, the graph of $y = f(bx)$ is the graph of $y = f(x)$ compressed horizontally by a factor of $b$ if $|b| > 1$. For instance, the graph of $y = \sqrt{64x}$ is the same as the graph of $y = \sqrt{x}$, but compressed horizontally by a factor of 64. But $y = \sqrt{64x} = \sqrt{64}\sqrt{x} = 8\sqrt{x}$, showing this horizontal compression is equivalent to a vertical stretch by a factor of 8. For each of the following functions, state the factor of the horizontal compression and the equivalent vertical stretch.
   a. $f(x) = (3x)^2$        b. $g(x) = (2x)^3$
   c. $h(x) = |5x|$          d. $p(x) = \sqrt[3]{27x}$

90. (a) Find the equation of the function obtained by shifting the identity function $x_1$ units to the right, vertically stretching by a factor of $m$, and then shifting up $y_1$ units. Does this expression look familiar? (b) What is the slope of this new line? (c) Where has the point $(0, 0)$ on $y = x$ been moved to under this transformation?

**91.** For $f(x) = x^2$, graph (a) $y = f(x + 3)$, (b) $y = f(x) + 3$, and (c) $y = f(x) + f(3)$. How do the three graphs compare to the graph of $f(x)$? Describe each graph in terms of a transformation of $y = f(x)$. Then (d) explain in terms of transformations, why $f(x + h) \neq f(x) + h$ and $f(x + h) \neq f(x) + f(h)$.

**92.** Sketch the graph of $f(x) = -2|x - 3| + 8$ using transformations of the parent function, then determine the area of the region in quadrant I that is beneath the graph and bounded by the vertical line $x = 6$.

**93.** Sketch the graph of $f(x) = x^2 - 4$, then sketch the graph of $F(x) = |x^2 - 4|$ using your intuition and the meaning of absolute value (not a table of values). What happens to the graph?

▶ **MAINTAINING YOUR SKILLS**

**94.** **(2.1)** Find the distance between the points $(-13, 9)$ and $(7, -12)$, and the slope of the line containing these points.

**95.** **(1.1)** Solve for $x$: $\frac{2}{3}x + \frac{1}{4} = \frac{1}{2}x - \frac{7}{12}$.

**96.** **(1.1)** Find the perimeter and area of the figure shown.

**97.** **(2.5)** Without graphing, state intervals where $f(x)\uparrow$ and $f(x)\downarrow$ for $f(x) = (x - 4)^2 + 3$.

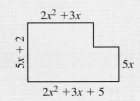

---

| 3.2 | **Basic Rational Functions and Power Functions** |

## LEARNING OBJECTIVES

*In Section 3.2 you will see how we can:*

☐ **A.** Graph basic rational functions, identify vertical and horizontal asymptotes, and describe end-behavior

☐ **B.** Use asymptotes and transformations to graph basic rational functions and write the equation for a given graph

☐ **C.** Graph basic power functions and state their domains

☐ **D.** Solve applications involving basic rational and power functions

In this section, we introduce two new kinds of relations, **rational functions** and **power functions**. While we've already studied a variety of functions, we still lack the ability to model a large number of important situations. For example, functions that model the cost of removing environmental pollutants, the relationship between time and medication remaining in the bloodstream, and the equations modeling planetary motion come from these two families.

## A. Rational Functions and Asymptotes

Just as a rational number is the ratio of two integers, a **rational function** is the ratio of two polynomials. In general,

> **Rational Functions**
>
> A rational function $V(x)$ is one of the form
> $$V(x) = \frac{p(x)}{d(x)},$$
> where $p$ and $d$ are polynomials and $d(x) \neq 0$.
> The domain of $V(x)$ is all real numbers, *except the zeroes of d.*

The simplest rational functions are the reciprocal function $y = \frac{1}{x}$ and the reciprocal square function $y = \frac{1}{x^2}$, as both have a constant numerator and a single term in the denominator. Since division by zero is undefined, the domain of both *excludes $x = 0$*. A preliminary study of these two functions will provide a strong foundation for our study of general rational functions in Chapter 4.

### The Reciprocal Function: $y = \dfrac{1}{x}$

The reciprocal function takes any input (other than zero) and gives its reciprocal as the output. This means large inputs produce small outputs and vice versa. A table of values (Table 3.1) and the resulting graph (Figure 3.14) are shown.

**Table 3.1**

| x | y |
|---|---|
| -100 | -0.01 |
| -10 | -0.1 |
| -1 | -1 |
| -0.1 | -10 |
| -0.01 | -100 |
| 0 | undefined |
| 0.01 | 100 |
| 0.1 | 10 |
| 1 | 1 |
| 10 | 0.1 |
| 100 | 0.01 |

**Figure 3.14**

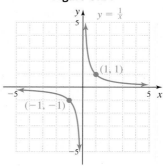

**WORTHY OF NOTE**

The notation used for graphical behavior always begins by describing what is happening to the *x*-values, and is followed by the resulting effect on the *y*-values. Using Figure 3.15, visualize that for a point (*x*, *y*) on the graph of $y = \frac{1}{x}$, as *x* gets larger, *y* must become smaller, particularly since their product must always be 1 ($y = \frac{1}{x} \Rightarrow xy = 1$).

**Figure 3.15**

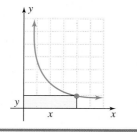

Table 3.1 and Figure 3.14 reveal some interesting features. First, the graph passes the vertical line test, verifying $y = \frac{1}{x}$ is indeed a function. Second, since division by zero is undefined, there can be no corresponding point on the graph, *creating a break at* $x = 0$. In line with our definition of rational functions, the domain is $x \in (-\infty, 0) \cup (0, \infty)$. Third, this is an odd function, with a "branch" of the graph in the first quadrant and one in the third quadrant, as the reciprocal of any input maintains its sign. Finally, we note that as *x* becomes an infinitely large positive number, *y* becomes an infinitely small positive number (closer and closer to zero). It seems convenient to symbolize this end-behavior using the following notation:

$$\text{as } x \to \infty, \qquad\qquad y \to 0$$

as *x* becomes an infinitely          *y* approaches 0
large positive number

Graphically, the curve becomes very close to, or *approaches the x-axis.*

We also note that as *x* approaches zero from the right, *y* becomes an infinitely large positive number: as $x \to 0^+$, $y \to \infty$. Note a superscript + or − sign is used to indicate the *direction of the approach,* meaning *from the positive side* (right) or *from the negative side* (left).

---

**EXAMPLE 1** ▶ **Describing the End-Behavior of Rational Functions**

For $y = \frac{1}{x}$ in QIII (Figure 3.14),

  **a.** Describe the end-behavior of the graph.

  **b.** Describe what happens as *x* approaches zero.

**Solution** ▶ Similar to the graph's behavior in QI, we have

  **a.** In words: As *x* becomes an infinitely large negative number, *y* approaches zero. In notation: As $x \to -\infty$, $y \to 0$.

  **b.** In words: As *x* approaches zero from the left, *y* becomes an infinitely large negative number. In notation: As $x \to 0^-$, $y \to -\infty$.

Now try Exercises 5 and 6 ▶

### The Reciprocal Square Function: $y = \dfrac{1}{x^2}$

From our previous work, we anticipate this graph will also have a break at $x = 0$. But since the square of any negative number is positive, the branches of the **reciprocal square function** are both *above the x-axis*. Note the result is the graph of an even function. See Table 3.2 and Figure 3.16.

**Table 3.2**

| $x$ | $y$ |
|---|---|
| $-100$ | $0.0001$ |
| $-10$ | $0.01$ |
| $-1$ | $1$ |
| $-0.1$ | $100$ |
| $-0.01$ | $10,000$ |
| $0$ | undefined |
| $0.01$ | $10,000$ |
| $0.1$ | $100$ |
| $1$ | $1$ |
| $10$ | $0.01$ |
| $100$ | $0.0001$ |

**Figure 3.16**

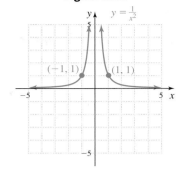

Similar to $y = \frac{1}{x}$, large positive inputs generate small, positive outputs: as $x \to \infty$, $y \to 0$. This is one indication of **asymptotic behavior** in the *horizontal* direction, and we say the line $y = 0$ (the $x$-axis) is a **horizontal asymptote** for the reciprocal and reciprocal square functions. In general,

**Horizontal Asymptotes**

Given a constant $k$, the line $y = k$ is a horizontal asymptote for $V$ if, as $x$ increases or decreases without bound, $V(x)$ approaches $k$:

$$\text{as } x \to -\infty, V(x) \to k \qquad \text{or} \qquad \text{as } x \to \infty, V(x) \to k$$

As shown in Figures 3.17 and 3.18, asymptotes are represented graphically as dashed lines that seem to "guide" the branches of the graph. Figure 3.17 shows a horizontal asymptote at $y = 1$, which suggests the graph of $f(x)$ is the graph of $y = \frac{1}{x}$ shifted up 1 unit. Figure 3.18 shows a horizontal asymptote at $y = -2$, which suggests the graph of $g(x)$ is the graph of $y = \frac{1}{x^2}$ shifted down 2 units.

**Figure 3.17**

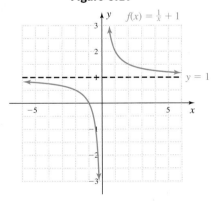

**Figure 3.18**

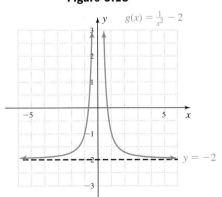

---

**EXAMPLE 2 ▶ Describing the End-Behavior of Rational Functions**

For the graph in Figure 3.18, use mathematical notation to
  **a.** Describe the end-behavior of the graph and name the horizontal asymptote.
  **b.** Describe what happens as $x$ approaches zero.

**Solution ▶**    **a.** as $x \to -\infty$, $g(x) \to -2$,          **b.** as $x \to 0^-$, $g(x) \to \infty$,
                   as $x \to \infty$, $g(x) \to -2$,              as $x \to 0^+$, $g(x) \to \infty$

$y = -2$ is a horizontal asymptote

**Now try Exercises 7 and 8 ▶**

While the graphical view of Example 2(a) (Figure 3.18) makes these concepts believable, a numerical view of this end-behavior can be even more compelling. Table 3.3 shows that as $|x|$ becomes very large, $g(x)$ becomes closer and closer to $-2$, but will never be equal to $-2$.

**Table 3.3**

| $x$ | $g(x)$ |
|---|---|
| 1 | $-1$ |
| 5 | $-1.96$ |
| 10 | $-1.99$ |
| 50 | $-1.9996$ |
| 100 | $-1.9999$ |
| 500 | $-1.999996$ |
| 1000 | $-1.999999$ |

**Table 3.4**

| $x$ | $g(x)$ |
|---|---|
| $-0.1$ | 98 |
| $-0.01$ | 9998 |
| $-0.001$ | 999,998 |
| 0 | undefined |
| 0.001 | 999,998 |
| 0.01 | 9998 |
| 0.1 | 98 |

From Example 2(b) and Table 3.4, we note that as $x$ *approaches 0*, $g$ becomes very large and *increases without bound*. This is one indication of asymptotic behavior in the *vertical* direction, and we say the line $x = 0$ (the $y$-axis) is a **vertical asymptote** for $g$ ($x = 0$ is also a vertical asymptote for $f$ in Figure 3.17). In general,

**Vertical Asymptotes**

Given a constant $h$, the vertical line $x = h$ is a vertical asymptote for a function $V$ if, as $x$ approaches $h$, $V(x)$ increases or decreases without bound:

as $x \to h^-$, $|V(x)| \to \infty$          or          as $x \to h^+$, $|V(x)| \to \infty$

Here is a brief summary to add to our toolbox functions:

**Reciprocal Function**

$$f(x) = \frac{1}{x}$$

Domain: $x \in (-\infty, 0) \cup (0, \infty)$
Range: $y \in (-\infty, 0) \cup (0, \infty)$
Horizontal asymptote: $y = 0$
Vertical asymptote: $x = 0$

**Reciprocal Quadratic Function**

$$g(x) = \frac{1}{x^2}$$

Domain: $x \in (-\infty, 0) \cup (0, \infty)$
Range: $y \in (0, \infty)$
Horizontal asymptote: $y = 0$
Vertical asymptote: $x = 0$

☑ **A.** You've just seen how we can graph basic rational functions, identify vertical and horizontal asymptotes, and describe end-behavior

For additional practice locating asymptotes from a given graph, **see Exercises 9 through 14.**

## B. Using Asymptotes to Graph Basic Rational Functions

Identifying these asymptotes is useful because the graphs of $y = \frac{1}{x}$ and $y = \frac{1}{x^2}$ can be transformed *in exactly the same way as the other toolbox functions*. When their graphs shift, the vertical and horizontal asymptotes shift with them and can be used as guides to redraw the graph. In shifted form,

$$f(x) = \frac{a}{x \pm h} \pm k \text{ for the reciprocal function, and}$$
$$g(x) = \frac{a}{(x \pm h)^2} \pm k \text{ for the reciprocal square function.}$$

When horizontal and/or vertical shifts are applied to simple rational functions, we first apply them to the asymptotes, then calculate the $x$- and $y$-intercepts as before. An additional point or two can be computed as needed to round out the graph.

**EXAMPLE 3** ▶ **Graphing Transformations of the Reciprocal Function**

Sketch the graph of $g(x) = \frac{1}{x-2} + 1$ using transformations of the parent function.

**Solution** ▶ The graph of $g$ is the same as that of $y = \frac{1}{x}$, but shifted 2 units right and 1 unit upward. This means the vertical asymptote is also shifted 2 units right, and the horizontal asymptote is shifted 1 unit up. The $y$-intercept is $g(0) = \frac{1}{2}$. For the $x$-intercept:

$$0 = \frac{1}{x-2} + 1 \qquad \text{substitute 0 for } g(x)$$

$$-1 = \frac{1}{x-2} \qquad \text{subtract 1}$$

$$-1(x-2) = 1 \qquad \text{multiply by } (x-2)$$

$$x = 1 \qquad \text{solve}$$

The $x$-intercept is $(1, 0)$. Knowing the graph is from the reciprocal function family, and shifting the asymptotes and intercepts yields the graph shown.

**Now try Exercises 15 through 30** ▶

These ideas can be "used in reverse" to determine the equation of a basic rational function from its given graph, as in Example 4.

**EXAMPLE 4** ▶ **Writing the Equation of a Basic Rational Function Given Its Graph**

Identify the function family for the graph given, then use the graph to write the equation of the function in "shifted form." Assume $|a| = 1$.

**Solution** ▶ The graph appears to be from the reciprocal square family, and has been shifted 2 units right (the vertical asymptote is at $x = 2$), and 1 unit down (the horizontal asymptote is at $y = -1$). From $y = \frac{1}{x^2}$, we obtain $f(x) = \frac{1}{(x-2)^2} - 1$ as the shifted form.

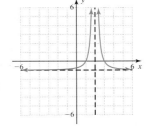

☑ **B.** You've just seen how we can use asymptotes and transformations to graph basic rational functions and write the equation for a given graph

**Now try Exercises 31 through 36** ▶

Using the definition of negative exponents, the basic reciprocal and reciprocal square functions can be written as $y = x^{-1}$ and $y = x^{-2}$ respectively. In this form, we note that these functions also belong to a family of functions known as the *power functions* **(see Exercise 78).**

## C. Graphs of Basic Power Functions

Italian physicist and astronomer Galileo Galilei (1564–1642) made numerous contributions to astronomy, physics, and other fields. But perhaps he is best known for his experiments with gravity, in which he dropped objects of different weights from the Leaning Tower of Pisa. Due in large part to his work, we know that the velocity of an object after it has fallen a certain distance is $v = \sqrt{2gs}$, where $g$ is the acceleration due to gravity (32 ft/sec$^2$), $s$ is the distance in feet the object has fallen, and $v$ is the

velocity of the object in feet per second **(see Exercise 69).** As you will see, this is an example of a formula that uses a power function.

From previous coursework or a review of radicals and rational exponents (Section R.4), we know that $\sqrt{x}$ can be written as $x^{\frac{1}{2}}$, and $\sqrt[3]{x}$ as $x^{\frac{1}{3}}$, enabling us to write these functions in *rational exponent form:* $f(x) = x^{\frac{1}{2}}$ and $g(x) = x^{\frac{1}{3}}$. In this form, we see that these actually belong to a larger family of functions, where $x$ is raised to some power, called the **power functions.**

### Power Functions and Root Functions

For any constant real number $p$ and variable $x$, functions of the form
$$f(x) = x^p$$
are called *power functions* in $x$.
If $p$ is of the form $\frac{1}{n}$ for integers $n \geq 2$, the functions
$$f(x) = x^{\frac{1}{n}} \Leftrightarrow f(x) = \sqrt[n]{x}$$
are called *root functions* in $x$.

The functions $y = x^2$, $y = x^{\frac{1}{4}}$, $y = x^3$, $y = \sqrt[5]{x}$, and $y = x^{\frac{3}{2}}$ are all power functions, but only $y = x^{\frac{1}{4}}$ and $y = \sqrt[5]{x}$ are also root functions. Initially we will focus on power functions where $p > 0$.

**EXAMPLE 5** ▶    **Comparing the Graphs of Power Functions**

Use a graphing calculator to graph the power functions $f(x) = x^{\frac{1}{4}}$, $g(x) = x^{\frac{2}{3}}$, $h(x) = x^1$, $p(x) = x^{\frac{3}{2}}$, and $q(x) = x^2$ in the standard ($x \in [-10, 10]$, $y \in [-10, 10]$) viewing window. Make an observation in QI regarding the effect of the exponent on each function, then discuss what the graphs of $y = x^{\frac{1}{6}}$ and $y = x^{\frac{7}{2}}$ would look like.

**Solution** ▶    Using a graphing calculator (or a very carefully chosen and computed table of values), we can produce graphs like those shown in Figure 3.19. Narrowing the window to focus on QI (Figure 3.20: $x \in [0, 5]$, $y \in [0, 5]$), we quickly see that for $x \geq 1$, larger values of $p$ cause the graph of $y = x^p$ to increase at a faster rate, and smaller values at a slower rate. In other words (for $x \geq 1$), since $\frac{1}{6} < \frac{1}{4}$, the graph of $y = x^{\frac{1}{6}}$ would increase slower and appear to be "under" the graph of $f(x) = x^{\frac{1}{4}}$.

Since $\frac{7}{2} > 2$, the graph of $y = x^{\frac{7}{2}}$ would increase faster and appear to be "more narrow" than the graph of $q(x) = x^2$ (verify this).

**Figure 3.19**

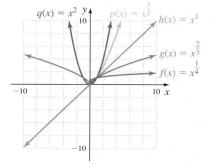

**Figure 3.20**

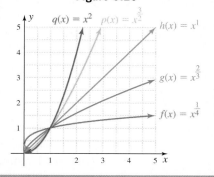

Now try Exercises 37 through 46 ▶

### The Domain of a Power Function

In addition to the observations made in Example 5, we can make other important notes, particularly regarding the *domains* of power functions. When the exponent on a power function is a positive rational number $\frac{m}{n}$ in simplest form, it appears the domain is all real numbers if $n$ is odd, as seen in the graphs of $g(x) = x^{\frac{2}{3}}$, $h(x) = x^1 = x^{\frac{1}{1}}$, and $q(x) = x^2 = x^{\frac{2}{1}}$. If $n$ is an even number, the domain is all nonnegative real numbers as seen in the graphs of $f(x) = x^{\frac{1}{4}}$ and $p(x) = x^{\frac{3}{2}}$. Further exploration will show that if $p$ is irrational, as in $y = x^{\sqrt{2}}$ or $y = x^{\pi}$, the domain is also all nonnegative real numbers and we have the following:

> ### The Domain of a Power Function
>
> Given a power function $f(x) = x^p$ with $p > 0$,
>
> 1. If $p = \frac{m}{n}$ is a rational number in simplest form,
>    a. the domain of $f$ is all real numbers if $n$ is odd: $x \in (-\infty, \infty)$,
>    b. the domain of $f$ is all nonnegative real numbers if $n$ is even: $x \in [0, \infty)$.
> 2. If $p$ is irrational, the domain of $f$ is all nonnegative real numbers: $x \in [0, \infty)$.

Further confirmation of statement 1 can be found by recalling the graphs of $y = \sqrt{x} = x^{\frac{1}{2}}$ and $y = \sqrt[3]{x} = x^{\frac{1}{3}}$ from Section 3.1 (Figures 3.21 and 3.22).

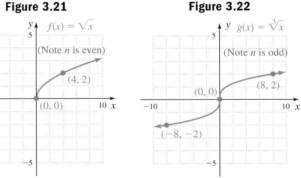

**Figure 3.21**

$f(x) = \sqrt{x}$

(Note $n$ is even)

(4, 2)

(0, 0)

Domain: $x \in [0, \infty)$
Range: $y \in [0, \infty)$

**Figure 3.22**

$g(x) = \sqrt[3]{x}$

(Note $n$ is odd)

(0, 0)

(8, 2)

(−8, −2)

Domain: $x \in (-\infty, \infty)$
Range: $y \in (-\infty, \infty)$

---

**EXAMPLE 6** ▶ **Determining the Domains of Power Functions**

State the domain of the following power functions, and identify whether each is also a root function.

a. $f(x) = x^{\frac{4}{5}}$          b. $g(x) = x^{\frac{1}{10}}$          c. $h(x) = \sqrt[8]{x}$

d. $q(x) = x^{\frac{2}{3}}$          e. $r(x) = x^{\sqrt{5}}$

**Solution** ▶

a. Since $n$ is odd, the domain of $f$ is all real numbers; $f$ is not a root function.
b. Since $n$ is even, the domain of $g$ is $x \in [0, \infty)$; $g$ is a root function.
c. In rational exponent form, $h(x) = x^{\frac{1}{8}}$. Since $n$ is even, the domain of $h$ is $x \in [0, \infty)$; $h$ is a root function.
d. Since $n$ is odd, the domain of $q$ is all real numbers; $q$ is not a root function.
e. Since $p$ is irrational, the domain of $r$ is $x \in [0, \infty)$; $r$ is not a root function.

**Now try Exercises 47 through 56** ▶

### Transformations of Power and Root Functions

As we saw in Section 3.1 (Toolbox Functions and Transformations), the graphs of the root functions $y = \sqrt{x}$ and $y = \sqrt[3]{x}$ can be transformed using shifts, stretches, reflections, and so on. In Example 8(b) (Section 3.1) we noted the graph of $h(x) = 2\sqrt[3]{x - 2} - 1$ was the graph of $y = \sqrt[3]{x}$ shifted 2 units right, stretched vertically by a factor of 2, and shifted 1 unit down. Graphs of other power functions can be transformed in exactly the same way.

**EXAMPLE 7** ▶ **Graphing Transformations of Power Functions**

Based on our previous observations,

    **a.** Determine the domain of $f(x) = x^{\frac{2}{3}}$ and $g(x) = x^{\frac{3}{2}}$, then verify by graphing.

    **b.** Next, discuss what the graphs of $F(x) = (x - 2)^{\frac{2}{3}} - 3$ and $G(x) = -x^{\frac{3}{2}} + 2$ will look like, then graph each to verify.

**Solution** ▶   **a.** Both $f$ and $g$ are power functions of the form $y = x^{\frac{m}{n}}$. For $f$, $n$ is odd so its domain is all real numbers. For $g$, $n$ is even and the domain is $x \in [0, \infty)$. Their graphs, which can be found with a graphing calculator or a table of values, support this conclusion (Figures 3.23 and 3.24).

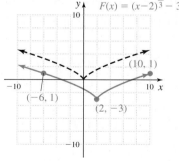

**Figure 3.23**

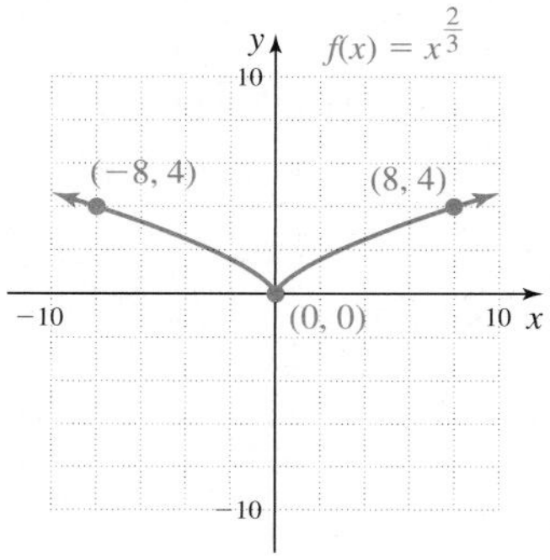

**Figure 3.24**

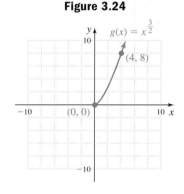

    **b.** The graph of $F$ will be the same as the graph of $f$, but shifted two units right and three units down, moving the vertex to $(2, -3)$. The graph of $G$ will be the same as the graph of $g$, but reflected across the $x$-axis, and shifted 2 units up (Figures 3.25 and 3.26).

**Figure 3.25**

**Figure 3.26**

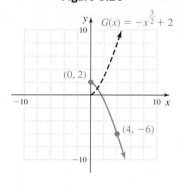

☑ **C.** You've just seen how we can graph basic power functions and state their domains

Now try Exercises 57 through 60 ▶

## D. Applications of Rational and Power Functions

These new functions have a variety of interesting and significant applications in the real world. Examples 8 through 10 provide a small sample, and there are a number of additional applications in the Exercise Set. In many applications, the coefficients may be rather large, and the axes should be scaled accordingly.

**EXAMPLE 8** ▶ **Modeling the Cost to Remove Waste**

For a large urban-centered county, the cost to remove chemical waste and other pollutants from a local river is given by the function $C(p) = \frac{18{,}000}{100 - p} - 180$, where $C(p)$ represents the cost (in thousands of dollars) to remove $p$ percent of the pollutants.

   **a.** Find the cost to remove 25%, 50%, and 75% of the pollutants and comment on the results.

   **b.** Graph the function using an appropriate scale.

   **c.** Use mathematical notation to state what happens as the county attempts to remove 100% of the pollutants.

**Solution** ▶    **a.** We evaluate the function as indicated, finding that $C(25) = 60$, $C(50) = 180$, and $C(75) = 540$. The cost is escalating rapidly. The change from 25% to 50% brought a $120,000 increase, but the change from 50% to 75% brought *a $360,000 increase!*

   **b.** From the context, we need only graph the portion from $0 \leq p < 100$. For the $C$-intercept we substitute $p = 0$ and find $C(0) = 0$, which seems reasonable as 0% would be removed if $0 were spent. We also note there must be a vertical asymptote at $x = 100$, since this $x$-value causes a denominator of 0. Using this information and the points from part (a) produces the graph shown.

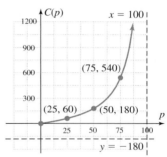

   **c.** As the percentage of pollutants removed approaches 100%, the cost of the cleanup skyrockets. Using notation: as $p \to 100^{-}$, $C \to \infty$.

**Now try Exercises 63 through 68** ▶

While not obvious at first, the function $C(p)$ in Example 8 is from the family of reciprocal functions $y = \frac{1}{x}$. A closer inspection shows it can be written in the form $y = -\frac{a}{x - h} - k = -\frac{18{,}000}{x - 100} - 180$, showing the graph of $y = \frac{1}{x}$ is shifted right 100 units, reflected across the $x$-axis, stretched by a factor of 18,000 and shifted 180 units down (the horizontal asymptote is $y = -180$). As sometimes occurs in real-world applications, portions of the graph were ignored due to the context. To see the full graph, we reason that the second branch occurs on the opposite side of the vertical and horizontal asymptotes. After adjusting the window to frame these key features, we get the graph shown in Figure 3.27.

**Figure 3.27**

$$C(p) = \frac{18{,}000}{100 - p} - 180$$

Next, we'll use a root function to model the distance to the horizon from a given height.

**EXAMPLE 9** ▶ **The Distance to the Horizon**

On a clear day, the distance a person can see from a certain height (the distance to the horizon) is closely approximated by the root function $d(h) = 3.57\sqrt{h}$, where $d(h)$ represents the viewing distance (in kilometers) from a height of $h$ meters above sea level.

**a.** To the nearest kilometer, how far can a person see when standing on the observation level of the John Hancock building in Chicago, Illinois, about 335 m high?

**b.** To the nearest meter, how high is the observer's eyes, if the viewing distance is 130 km?

**Solution** ▶ **a.** Substituting 335 for $h$ we have

$$d(h) = 3.57\sqrt{h} \qquad \text{original function}$$
$$d(335) = 3.57\sqrt{335} \qquad \text{substitute 335 for } h$$
$$\approx 65.34 \qquad \text{result}$$

On a clear day, a person can see about 65 kilometers.

**b.** We substitute 130 for $d(h)$:

$$d(h) = 3.57\sqrt{h} \qquad \text{original function}$$
$$130 = 3.57\sqrt{h} \qquad \text{substitute 130 for } d(h)$$
$$36.415 \approx \sqrt{h} \qquad \text{divide by 3.57}$$
$$1326.052 \approx h \qquad \text{square both sides}$$

If the distance to the horizon is 130 km, the observer's eyes are at a height of approximately 1326 m.

**Now try Exercises 69 through 72** ▶

One area where power functions and modeling with regression are used extensively is **allometric studies.** This area of inquiry studies the relative growth of a part of an animal in relation to the growth of the whole, like the wingspan of a bird compared to its weight, or the daily food intake of a mammal compared to its size.

**EXAMPLE 10** ▶ **Modeling the Food Requirements of Certain Bird Species**

To study the relationship between the weight of a nonpasserine bird and its daily food intake, the data shown in the table was collected (nonpasserine: nonsinging, nonperching birds).

**a.** On a graphing calculator, enter the data in L1 and L2, then set an appropriate window to view a scatterplot of the data. Does a power regression seem appropriate?

| Bird | Average Weight (g) | Daily Food Intake (g) |
|---|---|---|
| Common pigeon | 350 | 25 |
| Ring-necked duck | 725 | 50 |
| Ring-necked pheasant | 1400 | 70 |
| Canadian goose | 4525 | 165 |
| White swan | 9075 | 240 |

**b.** Use a graphing calculator to find an equation model using a power regression on the data (round values to three decimal places).

**c.** Use the equation to estimate the daily food intake required by a barn owl (470 g), and a gray-headed albatross (6800 g).

**d.** Find the weight of a Great Spotted Kiwi, given the daily food requirement is 130 g.

**Solution** ▶

**a.** After entering the weights in L1 and food intake in L2, we set a window that will comfortably fit the data. Using $x \in [0, 10{,}000]$ and $y \in [-30, 300]$ produces the scatterplot shown (Figure 3.28). The data does not appear linear, and based on our work in Example 5, a power function seems appropriate.

**Figure 3.28**

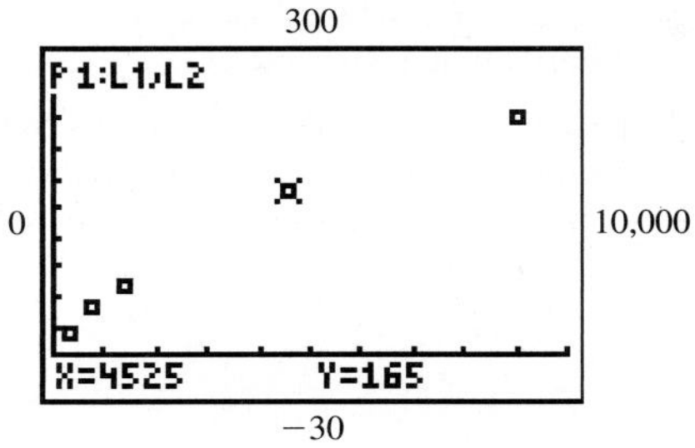

**b.** To access the power regression option, use **STAT** ▶ (**CALC**) **A:PwrReg.** To three decimal places, the equation for $Y_1$ would be $y = 0.493x^{0.685}$ (Figure 3.29).

**Figure 3.29**

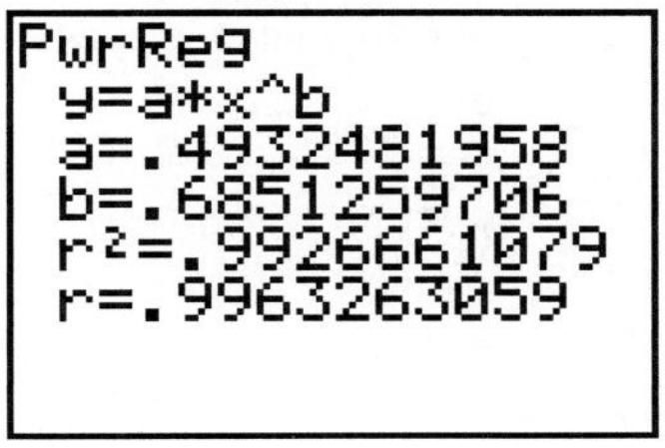

PwrReg
y=a*x^b
a=.4932481958
b=.6851259706
r²=.9926661079
r=.9963263059

**c.** For the barn owl, $x = 470$ and we find the estimated food requirement is about 33 g/day (Figure 3.30). For the gray-headed albatross $x = 6800$ and the model estimates about 208 g of food is required daily.

**d.** Here we're given the food intake of the Great Spotted Kiwi (the output value), and want to know what weight (input value) was used. Entering $Y_2 = 130$, we'll attempt to find where the graphs of $Y_1$ and $Y_2$ intersect (it will help to deactivate **Plot1** on the screen, so that only the graphs of $Y_1$ and $Y_2$ appear). Using **2nd** **TRACE** (**CALC**) **5:intersect** shows the graphs intersect at about (3423, 130) (Figure 3.31), indicating the average weight of a Great Spotted Kiwi is near 3423 g (about 7.5 lb).

**Figure 3.30**

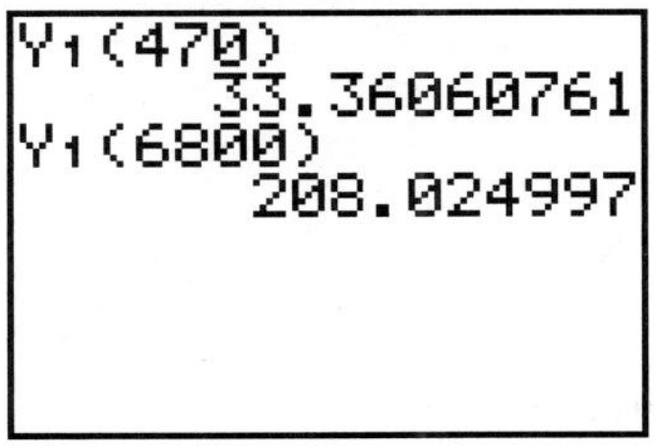

Y₁(470)
　　　33.36060761
Y₁(6800)
　　　208.024997

**Figure 3.31**

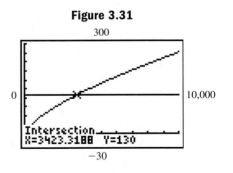

☑ **D.** You've just seen how we can solve applications involving basic rational and power functions

Now try Exercises 73 through 76 ▶

## 3.2 EXERCISES

► CONCEPTS AND VOCABULARY

**Fill in each blank with the appropriate word or phrase. Carefully reread the section, if necessary.**

1. Given the function $g(x) = \frac{1}{(x-3)^2} + 2$, a _____ asymptote occurs at $x = 3$ and a horizontal asymptote at _____.

2. The domains of $f(x) = x^{0.25}$ and $g(x) = x^{-0.\overline{3}}$ are _____ and _____ respectively.

3. Discuss/Explain how and why the range of the reciprocal function differs from the range of the reciprocal square function even though the domains are the same.

4. Discuss/Explain which of the functions $y_1 = x^{\frac{2}{3}}$, $y_2 = x^{\frac{3}{4}}$, $y_3 = x^{\frac{4}{3}}$, and $y_4 = x^{\frac{3}{2}}$ will grow fastest for (a) small inputs ($0 < x < 1$) and (b) large inputs ($x > 1$).

► DEVELOPING YOUR SKILLS

**For each graph given, use mathematical notation to (a) describe the end-behavior of each graph and (b) describe what happens as $x$ approaches 1.**

5. $V(x) = \dfrac{1}{x-1} + 2$

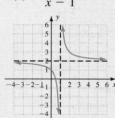

6. $v(x) = \dfrac{1}{x-1} - 2$

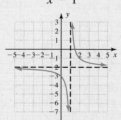

**For each graph given, use mathematical notation to (a) describe the end-behavior of each graph, (b) name the horizontal asymptote, and (c) describe what happens as $x$ approaches $-2$.**

7. $Q(x) = \dfrac{1}{(x+2)^2} + 1$

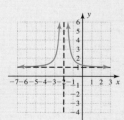

8. $q(x) = \dfrac{-1}{(x+2)^2} + 2$

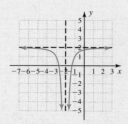

**Use the graph shown to complete each statement using the direction/approach notation.**

9. As $x \to -\infty$, $y$ _____.

10. As $x \to \infty$, $y$ _____.

11. As $x \to -1^+$, $y$ _____.

12. As $x \to -1^-$, $y$ _____.          **Exercises 9 through 14**

13. The line $x = -1$ is a vertical asymptote, since: as $x \to$ _____, $y \to$ _____.

14. The line $y = -2$ is a horizontal asymptote, since: as $x \to$ _____, $y \to$ _____.

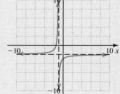

**Sketch the graph of each function using transformations of the parent function (not by plotting points). Clearly state the transformations used, and label the horizontal and vertical asymptotes as well as the $x$- and $y$-intercepts (if they exist). Also state the domain and range of each function.**

15. $f(x) = \dfrac{1}{x} - 1$

16. $g(x) = \dfrac{1}{x} + 2$

17. $h(x) = \dfrac{1}{x+2}$

18. $f(x) = \dfrac{1}{x-3}$

19. $g(x) = \dfrac{-1}{x-2}$

20. $h(x) = \dfrac{-1}{x} - 2$

21. $f(x) = \dfrac{1}{x+2} - 1$

22. $g(x) = \dfrac{1}{x-3} + 2$

23. $h(x) = \dfrac{1}{(x-1)^2}$

24. $f(x) = \dfrac{1}{(x+5)^2}$

25. $g(x) = \dfrac{-1}{(x+2)^2}$

26. $h(x) = \dfrac{-1}{x^2} - 2$

27. $f(x) = \dfrac{1}{x^2} - 2$

28. $g(x) = \dfrac{1}{x^2} + 3$

29. $h(x) = 1 + \dfrac{1}{(x+2)^2}$

30. $g(x) = -2 + \dfrac{1}{(x-1)^2}$

Identify the parent function for each graph given, then use the graph to construct the equation of the function in shifted form. Assume $|a| = 1$.

**31.**
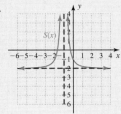

**32.**
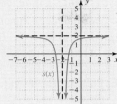

**33.**
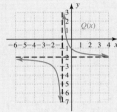

**34.**
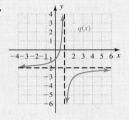

**35.**
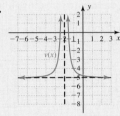

**36.**
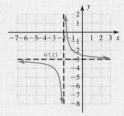

For each pair of functions given, state which function increases faster for $x > 1$, then use a graphing calculator to find where (a) $f(x) = g(x)$, (b) $f(x) > g(x)$, and (c) $f(x) < g(x)$.

**37.** $f(x) = x^2$, $g(x) = x^3$    **38.** $f(x) = x^4$, $g(x) = x^5$

**39.** $f(x) = x^4$, $g(x) = x^2$    **40.** $f(x) = x^3$, $g(x) = x^5$

**41.** $f(x) = x^{\frac{2}{3}}$, $g(x) = x^{\frac{4}{5}}$    **42.** $f(x) = x^{\frac{7}{4}}$, $g(x) = x^{\frac{3}{2}}$

**43.** $f(x) = \sqrt[6]{x}$, $g(x) = \sqrt[3]{x}$    **44.** $f(x) = \sqrt[5]{x}$, $g(x) = \sqrt[4]{x}$

**45.** $f(x) = \sqrt[3]{x^2}$, $g(x) = x^{\frac{5}{4}}$    **46.** $f(x) = x^{\frac{3}{2}}$, $g(x) = \sqrt[4]{x^3}$

**State the domain of the following functions.**

**47.** $f(x) = x^{\frac{7}{8}}$    **48.** $g(x) = x^{\frac{6}{7}}$

**49.** $h(x) = x^{\frac{6}{5}}$    **50.** $q(x) = x^{\frac{5}{6}}$

**51.** $r(x) = \sqrt[7]{x}$    **52.** $s(x) = x^{\frac{1}{6}}$

**Using the functions from Exercises 47 through 52, identify which of the following are defined and which are not. Do not use a calculator or evaluate.**

**53. a.** $f(-2)$    **b.** $f(2)$    **c.** $g(-2)$    **d.** $g(2)$

**54. a.** $h(0.3)$    **b.** $h(-0.3)$    **c.** $q(0.3)$    **d.** $q(-0.3)$

**55. a.** $h(-1.2)$    **b.** $r(-7)$    **c.** $s(-\pi)$    **d.** $s(0)$

**56. a.** $f\left(-\dfrac{7}{8}\right)$    **b.** $g\left(-\dfrac{8}{7}\right)$    **c.** $q(-1.9)$    **d.** $q(0)$

Compare and discuss the graphs of the following functions. Verify your answer by graphing both on a graphing calculator.

**57.** $f(x) = x^{\frac{7}{8}}$; $F(x) = (x + 1)^{\frac{7}{8}} - 2$

**58.** $g(x) = x^{\frac{8}{7}}$; $G(x) = (x - 3)^{\frac{8}{7}} + 2$

**59.** $p(x) = x^{\frac{6}{5}}$; $P(x) = -(x - 2)^{\frac{6}{5}}$

**60.** $q(x) = x^{\frac{5}{6}}$; $Q(x) = 2x^{\frac{5}{6}} - 5$

▶ **WORKING WITH FORMULAS**

**61. Gravitational attraction:** $F = \dfrac{km_1m_2}{d^2}$

The gravitational force $F$ between two objects with masses $m_1$ and $m_2$ depends on the distance $d$ between them and some constant $k$. (a) If the masses of the two objects are constant while the distance between them gets larger and larger, what happens to $F$? (b) Let $m_1$ and $m_2$ equal 1 mass unit each and suppose $k = 1$ as well. Investigate $F$ using a table of values. What family does this function belong to? (c) Solve for $m_2$ in terms of $k$, $m_1$, $d$ and $F$.

**62. Velocity of a bullet:** $v = \dfrac{m + M}{m}\sqrt{2gh}$

The velocity $v$ of a bullet of mass $m$ can be found using a device called a **ballistic pendulum.** The bullet is fired into a stationary block of wood of mass $M$, suspended from the end of a pendulum. The height $h$ the pendulum swings after impact is measured, and the velocity of the bullet is then estimated using $g = 9.8$ m/sec$^2$ (acceleration due to gravity). When a .22-caliber bullet of mass 2.6 g is fired into a wood block of mass 400 g, their combined mass swings to a height of 0.23 m. Find the velocity of the bullet.

▶ **APPLICATIONS**

63. **Deer and predators:** By banding deer over a period of 10 yr, a capture-and-release project determines the number of deer per square mile in the Mark Twain National Forest can be modeled by the function $D(p) = \frac{75}{p}$, where $p$ is the number of predators present and $D$ is the number of deer. Use this model to answer the following.

   a. As the number of predators increases, what will happen to the population of deer? Evaluate the function at $D(1)$, $D(3)$, and $D(5)$ to verify.

   b. What happens to the deer population if the number of predators becomes very large?

   c. Graph the function using an appropriate scale. Judging from the graph, use mathematical notation to describe what happens to the deer population if the number of predators becomes very small.

64. **Balance of nature:** A marine biology research group finds that in a certain reef area, the number of fish present depends on the number of sharks in the area. The relationship can be modeled by the function $F(s) = \frac{20,000}{s}$, where $F(s)$ is the fish population when $s$ sharks are present.

   a. As the number of sharks increases, what will happen to the population of fish? Evaluate the function at $F(10)$, $F(50)$, and $F(200)$ to verify.

   b. What happens to the fish population if the number of sharks becomes very large?

   c. Graph the function using an appropriate scale. Judging from the graph, use mathematical notation to describe what happens to the fish population if the number of sharks becomes very small.

65. **Intensity of light:** The intensity $I$ of a light source depends on the distance of the observer from the source. If the intensity is 100 W/m² at a distance of 5 m, the relationship can be modeled by the function $I(d) = \frac{2500}{d^2}$. Use the model to answer the following.

   a. As the distance from the lightbulb increases, what happens to the intensity of the light? Evaluate the function at $I(5)$, $I(10)$, and $I(15)$ to verify.

   b. If the intensity is increasing, is the observer moving closer to or farther from the light source?

   c. Graph the function using an appropriate scale. Judging from the graph, use mathematical notation to describe what happens to the intensity if the distance from the lightbulb becomes very small.

66. **Electrical resistance:** The resistance $R$ (in ohms) to the flow of electricity in a wire is related to the length of the wire and its gauge (diameter in fractions of an inch). For a certain wire with fixed length, this relationship can be modeled by the function $R(d) = \frac{0.2}{d^2}$, where $R(d)$ represents the resistance in a wire with diameter $d$.

   a. As the diameter of the wire increases, what happens to the resistance? Evaluate the function at $R(0.05)$, $R(0.25)$, and $R(0.5)$ to verify.

   b. If the resistance is increasing, is the diameter of the wire getting larger or smaller?

   c. Graph the function using an appropriate scale. Judging from the graph, use mathematical notation to describe what happens to the resistance in the wire as the diameter gets larger and larger.

67. **Pollutant removal:** For a certain coal-burning power plant, the cost to remove pollutants from plant emissions can be modeled by $C(p) = \frac{8000}{100 - p} - 80$, where $C(p)$ represents the cost (in thousands of dollars) to remove $p$ percent of the pollutants. (a) Find the cost to remove 20%, 50%, and 80% of the pollutants, then comment on the results; (b) graph the function using an appropriate scale; and (c) use mathematical notation to state what happens if the power company attempts to remove 100% of the pollutants.

68. **City-wide recycling:** A large city has initiated a new recycling effort and wants to distribute recycling bins for use in separating various recyclable materials. City planners anticipate the cost of the program can be modeled by the function $C(p) = \frac{22,000}{100 - p} - 220$, where $C(p)$ represents the cost (in $10,000) to distribute the bins to $p$ percent of the population. (a) Find the cost to distribute bins to 25%, 50%, and 75% of the population, then comment on the results; (b) graph the function using an appropriate scale; and (c) use mathematical notation to state what happens if the city attempts to give recycling bins to 100% of the population.

69. **Hot air ballooning:** If air resistance is neglected, the velocity (in ft/sec) of a falling object can be closely approximated by the function $V(s) = 8\sqrt{s}$, where $s$ is the distance the object has fallen (in feet). A balloonist suddenly finds it necessary to release some ballast in order to quickly gain altitude. (a) If she were flying at an altitude of 1000 ft, with what velocity will the ballast strike the ground? (b) If the ballast strikes the ground with a velocity of 225 ft/sec, what was the altitude of the balloon?

70. **River velocities:** The ability of a river or stream to move sand, dirt, or other particles depends on the size of the particle and the velocity of the river. This relationship can be used to approximate the velocity (in mph) of the river using the function $V(d) = 1.77\sqrt{d}$, where $d$ is the diameter (in inches) of the particle being moved. (a) If a creek can move a particle of diameter 0.095 in., how fast is it moving? (b) What is the largest particle that can be moved by a stream flowing 1.1 mph?

71. **Shoe sizes:** Although there may be some notable exceptions, the size of shoe worn by the average man is related to his height. This relationship is modeled by the function $S(h) = 0.75h^{1.5}$, where $h$ is the person's height in feet and $S$ is the U.S. shoe size. (a) Approximate Spud Webb's shoe size given he is 5 ft, 7 in. tall. (b) Approximate Shaquille O'Neal's height given his shoe size is 14.

**72. Whale weight:** For a certain species of whale, the relationship between its length and weight can be modeled by the function $W(l) = 0.03l^{2.45}$, where $l$ is the length of the whale in meters and $W$ is its weight in metric tons (1 metric ton ≈ 2205 pounds). (a) Estimate the weight of a newborn calf that is 6 m long. (b) At 106 metric tons, how long is an average adult?

**73. Gestation periods:** The data shown in the table can be used to study the relationship between the weight of a mammal and its length of pregnancy. (a) Graph a scatterplot of the data and (b) find an equation model using a power regression (round to three decimal places). Use the equation to estimate (c) the length of pregnancy of a raccoon (15.5 kg) and (d) the weight of a fox, given the length of pregnancy is 52 days.

| Mammal | Average Weight (kg) | Gestation (days) |
|---|---|---|
| Rat | 0.4 | 24 |
| Rabbit | 3.5 | 50 |
| Armadillo | 6.0 | 51 |
| Coyote | 13.1 | 62 |
| Dog | 24.0 | 64 |

**74. Bird wingspans:** The data in the table explores the relationship between a bird's weight and its wingspan. (a) Graph a scatterplot of the data and (b) find an equation model using a power regression (round to three decimal places). Use the equation to estimate (c) the wingspan of a Bald Eagle (16 lb) and (d) the weight of a Bobwhite Quail with a wingspan of 0.9 ft.

| Bird | Weight (lb) | Wingspan (ft) |
|---|---|---|
| Golden Eagle | 10.5 | 6.5 |
| Horned Owl | 3.1 | 2.6 |
| Peregrine Falcon | 3.3 | 4.0 |
| Whooping Crane | 17.0 | 7.5 |
| Raven | 1.5 | 2.0 |

**75. Species-area relationship:** To study the relationship between the number of species of birds on islands in the Caribbean, the data shown in the table was collected. (a) Graph a scatterplot of the data and (b) find an equation model using a power regression (round to three decimal places). Use the equation to estimate (c) the number of species of birds on Andros (2300 mi²) and (d) the area of Cuba, given there are 98 such species.

| Island | Area (mi²) | Species |
|---|---|---|
| Great Inagua | 600 | 16 |
| Trinidad | 2000 | 41 |
| Puerto Rico | 3400 | 47 |
| Jamaica | 4500 | 38 |
| Hispaniola | 30,000 | 82 |

**76. Planetary orbits:** The table shown gives the time required for the first five planets to make one complete revolution around the Sun (in years), along with the average orbital radius of the planet in astronomical units (1 AU ≈ 93 million miles). (a) Graph a scatterplot of the data and (b) find an equation model using a power regression (round to four decimal places). Use the equation to estimate (c) the average orbital radius of Saturn, given it orbits the Sun every 29.46 yr, and (d) estimate how many years it takes Uranus to orbit the Sun, given it has an average orbital radius of 19.2 AU.

| Planet | Years | Radius |
|---|---|---|
| Mercury | 0.24 | 0.39 |
| Venus | 0.62 | 0.72 |
| Earth | 1.00 | 1.00 |
| Mars | 1.88 | 1.52 |
| Jupiter | 11.86 | 5.20 |

▶ **EXTENDING THE CONCEPT**

**77.** Consider the graph of $f(x) = \frac{1}{x}$ once again, and the $x$ by $f(x)$ rectangles mentioned in the Worthy of Note on page 260. Calculate the area of each rectangle formed for $x \in \{1, 2, 3, 4, 5, 6\}$. What do you notice? Repeat the exercise for $g(x) = \frac{1}{x^2}$ and the $x$ by $g(x)$ rectangles. Can you detect the pattern formed here?

**78.** All of the power functions presented in this section had positive exponents, but the definition of these types of functions does allow for negative exponents as well. In addition to the reciprocal and reciprocal square functions ($y = x^{-1}$ and $y = x^{-2}$), these types of power functions have significant applications. For example, the

temperature of ocean water depends on several factors, including salinity, latitude, depth, and density. However, between depths of 125 m and 2000 m, ocean temperatures are relatively predictable, as indicated by the data shown for tropical oceans in the table. Use a graphing calculator to find the power regression model and use it to estimate the water temperature at a depth of 2850 m.

| Depth (meters) | Temp (°C) |
|---|---|
| 125 | 13.0 |
| 250 | 9.0 |
| 500 | 6.0 |
| 750 | 5.0 |
| 1000 | 4.4 |
| 1250 | 3.8 |
| 1500 | 3.1 |
| 1750 | 2.8 |
| 2000 | 2.5 |

► **MAINTAINING YOUR SKILLS**

**79.** (2.3) Solve the equation for $y$, then sketch its graph using the slope/intercept method: $2x + 3y = 15$.

**80.** (2.4) Charlie gave the following pairings: {(corndog, mustard), (hamburger, mustard), (taco, hot sauce), (nachos, queso), (hot dog, ketchup), (hamburger, mayo)}. Is the relation (entrée, condiment) as given also a function? State why or why not.

**81.** (1.6) Solve for $c$: $E = mc^2$.

**82.** (2.5) Compute the difference quotient of $f(x) = -2x^2 + 3x - 1$.

---

## 3.3  Variation: The Toolbox Functions in Action

**LEARNING OBJECTIVES**

*In Section 3.3 you will see how we can:*

☐ **A.** Solve direct variations
☐ **B.** Solve inverse variations
☐ **C.** Solve joint and combined variations

A study of direct and inverse variation offers perhaps our clearest view of how mathematics is used to model real-world phenomena. While the basis of our study is elementary, involving only the toolbox functions, the applications are at the same time elegant, powerful, and far reaching. In addition, these applications unite some of the most important ideas in algebra, including functions, transformations, rates of change, and graphical analysis, to name a few.

### A.  Toolbox Functions and Direct Variation

**Table 3.5**

| $g$ | $d$ |
|-----|-----|
| 1 | 24 |
| 2 | 48 |
| 3 | 72 |
| 4 | 96 |

If a car gets 24 miles per gallon (mpg), we could express the distance $d$ it could travel as $d = 24g$. Table 3.5 verifies the distance traveled by the car changes in *direct proportion* to the number of gallons used, and here we say, "distance traveled *varies directly* with gallons used." The equation $d = 24g$ is called a **direct variation,** and the coefficient 24 is called the **constant of variation.** Using the rate of change notation, $\frac{\Delta \text{distance}}{\Delta \text{gallons}} = \frac{\Delta d}{\Delta g} = \frac{24}{1}$, and we note this is actually a *linear equation* with slope $m = 24$. When working with variations, the constant $k$ is preferred over $m$, and in general we have the following:

> **Direct Variation**
>
> *y varies directly with x*, or *y is directly proportional to x*, if there is a nonzero constant $k$ such that
>
> $$y = kx.$$
>
> $k$ is called the *constant of variation* or *proportionality constant*.

---

**EXAMPLE 1** ►   **Writing a Variation Equation**

Write the variation equation for these statements:

  **a.** Wages earned vary directly with the number of hours worked.

  **b.** The value of an office machine varies directly with time.

  **c.** The circumference of a circle is directly proportional to the length of the diameter.

**Solution** ►   **a.** Wages vary directly with **h**ours worked: $W = k\boldsymbol{h}$

  **b.** The **V**alue of an office machine varies directly with **t**ime: $V = k\boldsymbol{t}$

  **c.** The **C**ircumference is directly proportional to the **d**iameter: $C = k\boldsymbol{d}$

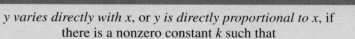
**Now try Exercises 5 through 8** ►

Once we determine the relationship between two variables is a direct variation, we try to find the value of $k$ and develop an equation model that can more generally be applied. *Note that "varies directly" indicates that one value is a constant multiple of the other.* In Example 1, you may have realized that if any one relationship between the variables is known, we can solve for $k$ by substitution. For instance, if the circumference of a circle is 314 cm when the diameter is 100 cm, $C = kd$ becomes $314 = k(100)$ and division shows $k = 3.14$ (our estimate for $\pi$). The result is a formula for the circumference of *any* circle. This suggests the following procedure:

> ### Solving Applications of Variation
>
> 1. Translate the information given into an equation model, using $k$ as the constant of variation.
> 2. Substitute the first relationship (pair of values) given and solve for $k$.
> 3. Substitute this value for $k$ in the original model to obtain the variation equation.
> 4. Use the variation equation to complete the application.

**EXAMPLE 2** ▶ **Solving an Application of Direct Variation**

The weight of an astronaut on the surface of another planet **varies directly** with their weight on Earth. An astronaut weighing 140 lb on Earth weighs only 53.2 lb on Mars. How much would a 170-lb astronaut weigh on Mars?

**Solution** ▶

1.  $M = kE$        "**M**ars weight varies directly with **E**arth weight"
2.  $53.2 = k(140)$   substitute 53.2 for $M$ and 140 for $E$
    $k = 0.38$      solve for $k$ (constant of variation)

Substitute this value of $k$ in the original equation to obtain the variation equation, then find the weight of a 170-lb astronaut on Mars.

3.  $M = 0.38E$       variation equation
4.     $= 0.38(170)$   substitute 170 for $E$
       $= 64.6$      result

An astronaut weighing 170 lb on Earth weighs only 64.6 lb on Mars.

**Now try Exercises 9 through 12** ▶

The toolbox function from Example 2 was a line with slope $k = 0.38$, or $k = \frac{19}{50}$ as a fraction in simplest form. As a rate of change, $k = \frac{\Delta M}{\Delta E} = \frac{19}{50}$, and we see that for every 50 additional pounds on Earth, the weight of an astronaut would increase by only 19 lb on Mars.

**EXAMPLE 3** ▶ **Making Estimates from the Graph of a Variation**

The scientists at NASA are planning to send additional probes to the red planet (Mars) that will weigh from 250 to 450 lb. Graph the variation equation from Example 2, then *use the graph* to estimate the corresponding range of weights on Mars. Check your estimate using the variation equation.

**Solution** ▶  After selecting an appropriate scale, begin at (0, 0) and count off the slope $k = \frac{\Delta M}{\Delta E} = \frac{19}{50}$. This gives the points (50, 19), (100, 38), (200, 76), and so on. From the graph (see dashed arrows), it appears the weights corresponding to 250 lb and 450 lb on Earth are near 95 lb and 170 lb on Mars. Using the equation gives

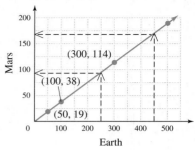

$M = 0.38E$    variation equation

$\quad = 0.38(250)$    substitute 250 for $E$

$\quad = 95,$

$M = 0.38E$    variation equation

$\quad = 0.38(450)$    substitute 450 for $E$

$\quad = 171,$ very close to our estimate from the graph.

> **Now try Exercises 13 through 20** ▶

When toolbox functions are used to model variations, our knowledge of their graphs and defining characteristics strengthens a contextual understanding of the application. Consider Examples 4 and 5, where the squaring function is used.

**EXAMPLE 4** ▶  **Writing Variation Equations**

Write the variation equation for these statements:

a. In free fall, the distance traveled by an object is directly proportional to the square of the time.

b. The area of a circle varies directly with the square of its radius.

**Solution** ▶  a. **D**istance varies directly with the square of the **t**ime: $D = kt^2$.

b. **A**rea varies directly with the square of the **r**adius: $A = kr^2$.

> **Now try Exercises 21 through 24** ▶

Both variations in Example 4 use the squaring function, where $k$ determines the amount of stretch or compression applied, and whether the graph will open upward or downward. However, regardless of the function used, the four-step solution process remains the same.

**EXAMPLE 5** ▶  **Solving an Application of Direct Variation**

The range of a projectile varies directly with the square of its initial velocity. As part of a circus act, Bailey the Human Bullet is shot out of a cannon with an initial velocity of 80 feet per second (ft/sec), into a net 200 ft away.

a. Find the constant of variation and write the variation equation.

b. Graph the equation and *use the graph* to estimate how far away the net should be placed if initial velocity is increased to 95 ft/sec.

c. Determine the accuracy of the estimate from (b) using the variation equation.

**Solution** ▶  a. 1.  $R = kv^2$          "Range varies directly with the square of the **v**elocity"

2.  $200 = k(80)^2$    substitute 200 for $R$ and 80 for $v$

$\quad k = 0.03125$    solve for $k$ (constant of variation)

3.  $R = 0.03125v^2$   variation equation (substitute 0.03125 for $k$)

**b.** Since velocity and distance are positive, we again use only QI. The graph is a parabola that opens upward, with the vertex at (0, 0). Selecting velocities from 50 to 100 ft/sec, we have:

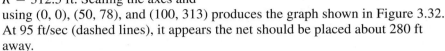

$$R = 0.03125v^2 \qquad \text{variation equation}$$
$$= 0.03125(50)^2 \qquad \text{substitute 50 for } v$$
$$= 78.125 \qquad \text{result}$$

**Figure 3.32**

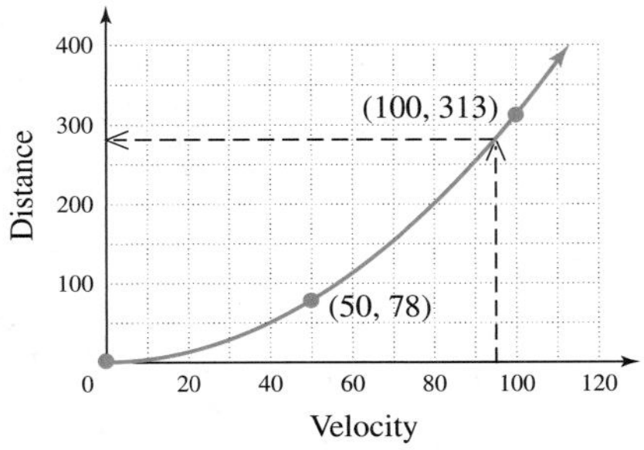

Likewise substituting 100 for $v$ gives $R = 312.5$ ft. Scaling the axes and using (0, 0), (50, 78), and (100, 313) produces the graph shown in Figure 3.32. At 95 ft/sec (dashed lines), it appears the net should be placed about 280 ft away.

**c.** Using the variation equation gives:

  **4.** $R = 0.03125v^2 \qquad \text{variation equation}$
$$= 0.03125(95)^2 \qquad \text{substitute 95 for } v$$
$$= 282.03125 \qquad \text{result}$$

Our estimate was off by about 2 ft. The net should be placed about 282 ft away.

☑ **A.** You've just seen how we can solve direct variations

**Now try Exercises 25 through 30 ▶**

We now have a complete picture of this relationship, in which the required information can be presented graphically (Figure 3.32), numerically (Table 3.6), verbally, and in equation form. This enables the people requiring the information, that is, Bailey himself (for obvious reasons) and the Circus Master who is responsible, to make more informed (and safe) decisions.

*Note:* For Examples 7 and 8, the four steps of the solution process will be used in sequence, but not numbered.

**Table 3.6**

| Velocity, $v$ (ft/sec) | Range, $R$ (ft) |
|:---:|:---:|
| 40 | 50 |
| 50 | 78 |
| 60 | 113 |
| 70 | 153 |
| 80 | 200 |
| 90 | 253 |
| 100 | 313 |

## B. Inverse Variation

Numerous studies have been done that relate the price of a commodity to the demand—the willingness of a consumer to pay that price. For instance, if there is a sudden increase in the price of a popular tool, hardware stores know there will be a corresponding decrease in the demand for that tool. The question remains, "What is this rate of decrease?" Can it be modeled by a linear function with a negative slope? A parabola that opens downward? Some other function? Table 3.7 shows some (simulated) data regarding price versus demand. It appears that a linear function is not appropriate because the rate of change in the number of tools sold is not constant. Likewise a quadratic model seems inappropriate, since we don't expect demand to suddenly start rising again as the price continues to increase. This phenomenon is actually an example of **inverse variation,** modeled by a transformation of the reciprocal function $y = \frac{k}{x}$. We will often rewrite the equation as $y = k(\frac{1}{x})$ to clearly see the inverse relationship. In the case at hand, we might write $D = k(\frac{1}{P})$, where $k$ is the constant of variation, $D$ represents the demand for the product, and $P$ the price of the product. In words, we say that "demand *varies inversely* with the price." In other applications of inverse variation, one quantity may vary inversely with the *square* of another [Example 6(b)], and in general we have

**Table 3.7**

| Price (dollars) | Demand (1000s) |
|:---:|:---:|
| 8 | 288 |
| 9 | 144 |
| 10 | 96 |
| 11 | 72 |
| 12 | 57.6 |

**Inverse Variation**

> *y varies inversely with x,* or *y is inversely proportional to x,* if
> there is a nonzero constant $k$ such that
>
> $$y = k\left(\frac{1}{x}\right).$$
>
> $k$ is called the *constant of variation* or *proportionality constant.*

**EXAMPLE 6 ▶  Writing Inverse Variation Equations**

Write the variation equation for these statements:

**a.** The time it takes to complete a road trip is inversely proportional to the average speed.

**b.** The intensity of light varies inversely with the square of the distance from the source.

**Solution ▶**   **a.** **T**ime is inversely proportional to **s**peed: $t = k\left(\frac{1}{s}\right)$.

**b.** **I**ntensity of light varies inversely with the square of the **d**istance: $I = k\left(\frac{1}{d^2}\right)$.

**Now try Exercises 31 through 34 ▶**

**EXAMPLE 7 ▶  Solving an Application of Inverse Variation**

Boyle's law tells us that in a closed container with constant temperature, the volume of a gas varies inversely with the pressure applied (see illustration). Suppose the air pressure in a closed cylinder is 50 pounds per square inch (psi) when the volume of the cylinder is 60 in³.

**a.** Find the constant of variation and write the variation equation.

**b.** Use the equation to find the volume, if the pressure is increased to 150 psi.

**Illustration of Boyle's Law**

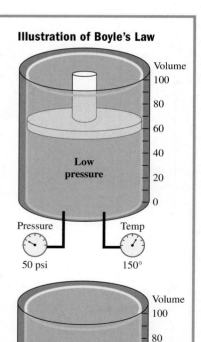

**Solution ▶**   **a.**  $V = k\left(\dfrac{1}{P}\right)$    "**V**olume varies inversely with the **P**ressure"

$60 = k\left(\dfrac{1}{50}\right)$    substitute 60 for *V* and 50 for *P*.

$k = 3000$    constant of variation

$V = 3000\left(\dfrac{1}{P}\right)$    variation equation (substitute 3000 for *k*)

**b.** Using the variation equation we have:

$V = 3000\left(\dfrac{1}{P}\right)$    variation equation

$= 3000\left(\dfrac{1}{150}\right)$    substitute 150 for *P*

$= 20$    result

When the pressure is increased to 150 psi, the volume decreases to 20 in³.

☑ **B.** You've just seen how we can solve inverse variations

**Now try Exercises 35 through 38 ▶**

As an application of the reciprocal function, the relationship in Example 7 is easily graphed as a transformation of $y = \frac{1}{x}$. Using an appropriate scale and values in QI, only a vertical stretch of 3000 is required and the result is shown in Figure 3.33. As noted, when the pressure increases the volume decreases, or in notation: as $P \to \infty$, $V \to 0$. Applications of this sort can be as sophisticated as the manufacturing of industrial pumps and synthetic materials, or as simple as cooking a homemade dinner. Simply based on the equation, how much pressure is required to reduce the volume of gas to 1 in³?

**Figure 3.33**

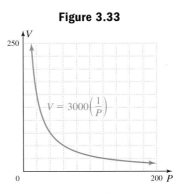

## C. Joint or Combined Variations

Just as some decisions might be based on many considerations, often several variables are needed to model a mathematical relationship. Imagine a wooden plank laid across the banks of a stream for hikers to cross the streambed (see Figure 3.34). The amount of weight the plank will support depends on the type of wood, the width and height of the plank's cross section, and the distance between the supported ends **(see Exercises 67 and 68).** This is an example of a **combined variation,** which involves both direct and indirect variation. Another possibility is **joint variation** in which one quantity is directly proportional to several others. A couple typical examples are: (1) *y varies jointly with x and p: y = kxp;* and (2) *y varies jointly with x and p, but inversely with the square of q: $y = k(\frac{xp}{q^2})$.* For practice writing joint and combined variations as equation models, **see Exercises 39 through 44.**

**Figure 3.34**

**EXAMPLE 8** ▶ **Solving an Application of Joint Variation**

The amount of fuel used by a certain ship traveling at a uniform speed varies jointly with the distance it travels and the square of the velocity. If 200 barrels of fuel are used to travel 10 mi at 20 knots, how far does the ship travel on 500 barrels of fuel at 30 knots?

**Solution** ▶

$$F = kdv^2$$    "Fuel use *varies jointly* with **d**istance and **v**elocity squared"

$$200 = k(10)(20)^2$$    substitute 200 for *F*, 10 for *d*, and 20 for *v*

$$200 = 4000k$$    simplify and solve for *k*

$$0.05 = k$$    constant of variation

$$F = 0.05dv^2$$    equation of variation

To find the distance traveled at 30 knots using 500 barrels of fuel, substitute 500 for *F* and 30 for *v*:

$$F = 0.05dv^2$$    equation of variation

$$500 = 0.05d(30)^2$$    substitute 500 for *F* and 30 for *v*

$$500 = 45d$$    simplify

$$11.\overline{1} = d$$    result

If 500 barrels of fuel are consumed while traveling 30 knots, the ship covers a distance of just over 11 mi.

☑ **B.** You've just seen how we can solve joint and combined variations

**Now try Exercises 45 through 52** ▶

It's interesting to note that the ship covers just over one additional mile, but consumes 2.5 times the fuel. The additional speed dramatically reduces fuel efficiency.

There is a variety of additional applications in the Exercise Set. **See Exercises 55 through 63.**

# 3.3 EXERCISES

## ▶ CONCEPTS AND VOCABULARY

**Fill in each blank with the appropriate word or phrase. Carefully reread the section, if necessary.**

1. If $y$ varies directly with $x$, and $x$ decreases, then $y$ _____.

2. The statement "$y$ varies inversely with the square of $x$" is written _____.

3. Discuss/Explain the general procedure for solving applications of variation. Include references to keywords, and illustrate using an example.

4. The basic percent formula is *amount equals percent times base,* or $A = PB$. In words, write this out as a direct variation with $B$ as the constant of variation, then as an inverse variation with $A$ as the constant of variation.

## ▶ DEVELOPING YOUR SKILLS

**Write the variation equation for each statement.**

5. distance traveled varies directly with time

6. cost varies directly with quantity

7. force varies directly with acceleration

8. length of a spring is directly proportional to attached weight

**For Exercises 9 and 10, find the variation equation, and use it to complete the table.**

9. $y$ varies directly with $x$; $y = 0.6$ when $x = 24$.

| $x$ | $y$ |
|-----|-----|
| 500 |     |
|     | 16.25 |
| 750 |     |

10. $w$ varies directly with $v$; $w = \frac{1}{3}$ when $v = 5$.

| $v$ | $w$ |
|-----|-----|
| 291 |     |
|     | $\frac{109}{5}$ |
| 339 |     |

11. **Water pressure and depth:** The hydrostatic pressure exerted on a scuba diver is directly proportional to the diver's depth. At the deep end of the training pool, the gauges displayed a depth of 9 ft and a pressure of 4 psi. Write the variation equation and estimate the pressure at a shipwreck lying 60 ft below the surface. What does the value of $k$ represent?

12. **Car value and loan payment:** The amount of the monthly payments on an auto loan varies directly with the selling price of the vehicle. Based on your credit history, one dealership is willing to offer you a 2013 Hyundai Sonata for $19,795 cash, or $385/month. Write the variation equation and estimate the monthly payment for a 2013 Shelby Mustang convertible, which is valued at $54,470. What does the value of $k$ represent?

13. **Building height and number of stairs:** The number of stairs in the stairwells of tall buildings and other structures varies directly with the height of the structure. The base and pedestal for the Statue of Liberty are 47 m tall, with 192 stairs from ground level to the observation deck at the top of the pedestal (at the statue's feet). (a) Find the variation equation, (b) use the graph of the variation equation to estimate the number of stairs from ground level to the observation deck in the statue's crown 81 m above ground level, and (c) use the equation to check this estimate. Was it close?

14. **Projected images:** The height of a projected image varies directly with the distance of the projector from the screen. At a distance of 48 in., the image on the screen is 16 in. high. (a) Find the variation equation, (b) use the graph of the variation equation to estimate the height of the image if the projector is placed at a distance of 5 ft 3 in., and (c) use the equation to check this estimate. Was it close?

Unit conversions offer a particularly important application of direct variation. When converting miles to feet, we use [feet] = $k$[miles] with the conversion factor $k = \frac{5280 \text{ ft}}{1 \text{ mi}}$. Use the conversion factors and quantities given to convert to the units indicated.

15. $\frac{1}{4}$ gallon to ounces; $k = \frac{128 \text{ oz.}}{1 \text{ gal}}$

16. 1 moment to minutes; $k = \frac{60 \text{ min}}{40 \text{ moments}}$

17. 3 pinches to tablespoons; $k = \frac{1 \text{ T}}{48 \text{ pinches}}$

18. 492 degrees Rankine to Kelvin; $k = \frac{5 \text{ K}}{9 \text{ R}}$

19. 1000 kilograms per cubic meter to pounds per cubic foot; $k \approx \frac{1 \text{ lb/ft}^3}{16 \text{ kg/m}^3}$

20. 60 miles per hour to feet per second; $k = \frac{1.46 \text{ fps}}{1 \text{ mph}}$

**Write the variation equation for each statement.**

21. Volume of a cube varies directly with the cube of a side.

22. Potential energy in a spring varies directly with the square of the distance the spring is compressed.

23. Electric power is directly proportional to the square of the current.

24. Manufacturing cost varies directly with the square root of the number of items made.

**For Exercises 25 and 26, find the variation equation, and use it to complete the table.**

25. $p$ varies directly with the fourth power of $q$; $p = 800$ when $q = 5$

| q | p |
|---|---|
| | 6.48 |
| 2.5 | |
| | 12,800 |

26. $n$ varies directly with $m$ squared; $n = 24.75$ when $m = 30$

| m | n |
|---|---|
| 40 | |
| | 99 |
| 88 | |

**For Exercises 27 through 30, describe the relationship indicated (a) in words, (b) in equation form, (c) graphically, and (d) in table form, then (e) solve the application.**

27. **The Borg Collective:** The surface area of a cube varies directly with the square of one side. A cube with sides of $14\sqrt{3}$ cm has a surface area of 3528 cm$^2$. Find the surface area of the spaceships used by the Borg Collective in *Star Trek—The Next Generation,* which were cubical spacecraft with sides of 3036 m.

28. **Geometry and geography:** The area of an equilateral triangle varies directly with the square of one side. A triangle with sides of 50 yd has an area of 1082.5 yd$^2$. Find the area of the region bounded by straight lines connecting the cities of Cincinnati, Ohio; Washington, D.C.; and Columbia, South Carolina, which are all approximately 400 mi apart. Assume the curvature of the earth has a negligible impact on this estimate.

29. **Galileo and gravity:** The distance an object falls varies directly with the square of the time it has been falling. The cannonballs dropped by Galileo from the Leaning Tower of Pisa fell 169 ft in about 3.25 sec. How long would it take a hammer, accidentally dropped from a height of 196 ft by a bridge repair crew, to splash into the water below? According to the equation, if a camera accidentally fell out of the *News 4 Eye-in-the-Sky* helicopter and hit the ground in 2.75 sec, how high was the helicopter?

30. **Soap bubble surface area:** When a child blows small soap bubbles, they come out in the form of a sphere because the surface tension in the soap seeks to minimize the surface area. The surface area of any sphere varies directly with the square of its radius. A soap bubble with a 0.75-in. radius has a surface area of approximately 7.07 in$^2$. What is the radius of a seventeenth-century cannonball that has a surface area of 113.1 in$^2$? What is the surface area of an orange with a radius of 1.5 in.?

**Write the variation equation for each statement.**

31. The force of gravity varies inversely with the square of the distance between objects.

32. Pressure varies inversely with the area over which it is applied.

33. The safe load of a beam supported at both ends varies inversely with its length.

34. The intensity of sound is inversely proportional to the square of its distance from the source.

**For Exercises 35 through 38, find the variation equation, and use it to complete the table or solve the application.**

35. $Y$ varies inversely with the square of $Z$; $Y = 1369$ when $Z = 3$

| Z | Y |
|---|---|
| 37 | |
| | 2.25 |
| 111 | |

36. $A$ varies directly with $B^{1.25}$; $A = 200$ when $B = 16$

| B | A |
|---|---|
| 0.4096 | |
| | 6.25 |
| 81 | |

**37. Gravitational force:** The effect of Earth's gravity on an object (its weight) varies inversely with the square of its distance from the center of the planet (assume the Earth's radius is 6400 km). If the weight of an astronaut is 75 kg on Earth (when $r = 6400$), what would this weight be at an altitude of 1600 km *above the surface* of the Earth?

**38. Popular running shoes:** The demand for a popular new running shoe varies inversely with the price of the shoes. When the wholesale price is set at $45, the manufacturer ships 5500 orders per week to retail outlets. Based on this information, how many orders would be shipped per week if the wholesale price rose to $55?

**Write the variation equation for each statement.**

**39.** Interest earned varies jointly with the rate of interest and the length of time on deposit.

**40.** Horsepower varies jointly with the number of cylinders in the engine and the square of the cylinder's diameter.

**41.** The area of a trapezoid varies jointly with its height and the sum of the bases.

**42.** The volume of metal in a circular coin varies directly with the thickness of the coin and the square of its radius.

**43.** The electrical resistance in a wire varies directly with its length and inversely with its cross-sectional area.

**44.** For an ideal gas, volume varies directly with temperature, but inversely with pressure.

**For Exercises 45 through 48, find the constant of variation and write the related variation equation. Then use the equation to complete the table or solve the application.**

**45.** *C* varies directly with *R* and inversely with *S* squared, and $C = 21$ when $R = 7$ and $S = 1.5$.

| R | S | C |
|---|---|---|
| 120 | | 22.5 |
| 200 | 12.5 | |
| | 15 | 10.5 |

**46.** *J* varies directly with *P* and inversely with the square root of *Q*, and $J = 19$ when $P = 4$ and $Q = 25$.

| P | Q | J |
|---|---|---|
| 47.5 | | 118.75 |
| 112 | 31.36 | |
| | 44.89 | 66.5 |

**47. Kinetic energy:** Kinetic energy (energy attributed to motion) varies jointly with the mass of the object and the square of its velocity. An object with a mass of 1 kg and velocity of 20 m per sec (m/sec) has kinetic energy of 200 J. How much energy is produced if the velocity is doubled?

**48. Safe load:** The load that a horizontal beam can support varies directly with the width of the beam and the square of its height, but inversely with the length of the beam. A beam 4 in. wide and 8 in. tall can safely support a load of 1 ton when the beam has a length of 12 ft. How much could a similar beam 10 in. tall safely support?

**Match the variation relationship described to its corresponding graph.**

**49.** *y* varies directly with *x*

**50.** *y* varies directly with the cube of *x*

**51.** *y* is directly proportional to the cube root of *x*

**52.** *y* is inversely proportional to *x*

a.
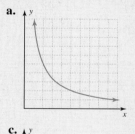

b.
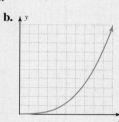

c.
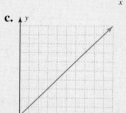

d.
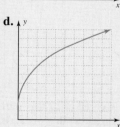

► **WORKING WITH FORMULAS**

**53. Drag force on a race car:** $F = kv^2A$

Whenever an object moves through a fluid (gas or liquid), the fluid exerts a force on the object opposing its movement. This force, called drag, is given by the formula shown, where *F* is the force (in pounds—lb), *v* is the velocity of the object (in feet per sec—fps), *A* is its cross-sectional area (in square feet—ft²), and *k* is a drag coefficient that depends on the aerodynamics of the object and the density of the fluid. (a) Write the variation equation in words. (b) Solve for *k* and use the formula to find the drag coefficient of a race car given $F = 175$ lb, $v = 176$ fps, and $A = 14$ ft². (c) Find the drag on a replica of the car described in (b) that has $\frac{1}{4}$ the cross-sectional area and travels at $\frac{1}{4}$ the speed.

**54. Force between charged particles:** $F = k\dfrac{Q_1 Q_2}{d^2}$

The force between two charged particles is given by the formula shown, where $F$ is the force (in joules—J), $Q_1$ and $Q_2$ represent the electrical charge on each particle (in coulombs—C), and $d$ is the distance between them (in meters). If the particles have a like charge, the force is repulsive; if the charges are unlike, the force is attractive. (a) Write the variation equation in words. (b) Solve for $k$ and use the formula to find the electrical constant $k$, given $F = 0.36$ J, $Q_1 = 2 \times 10^{-6}$ C, $Q_2 = 4 \times 10^{-6}$ C, and $d = 0.2$ m. Express the result in scientific notation.

## ▶ APPLICATIONS

**Find the variation equation, and use it to solve.**

**55. Cleanup time:** The time required to pick up the trash along a stretch of highway varies inversely with the number of volunteers who are working. If 12 volunteers can do the cleanup in 4 hr, how many volunteers are needed to complete the cleanup in just 1.5 hr?

**56. Wind power:** The wind farms in southern California contain wind generators whose power production varies directly with the cube of the wind's speed. If one such generator produces 1000 W of power in a 25 mph wind, find the power it generates in a 35 mph wind.

**57. Pull of gravity:** The weight of an object on the moon varies directly with the weight of the object on Earth. A 96-kg object on Earth would weigh only 16 kg on the moon. How much would a fully suited 250-kg astronaut weigh on the moon?

**58. Period of a pendulum:** The time that it takes for a simple pendulum to complete one period (swing over and back) varies directly with the square root of its length. If a pendulum 20 ft long has a period of 5 sec, find the period of a pendulum 30 ft long.

**59. Stopping distance:** The stopping distance of an automobile varies directly with the square root of its speed when the brakes are applied. If a car requires 108 ft to stop from a speed of 25 mph, estimate the stopping distance from a speed of 45 mph.

**60. Supply and demand:** A chain of hardware stores finds that the demand for a special power tool varies inversely with the advertised price of the tool. If the price is advertised at $85, there is a monthly demand for 10,000 units at all participating stores. Find the projected demand if the price were lowered to $70.83.

**61. Cost of copper tubing:** The cost of copper tubing varies jointly with the length and diameter of the tube. If a 36-ft spool of $\frac{1}{4}$-in.-diameter tubing costs $76.50, how much does a 24-ft spool of $\frac{3}{8}$-in.-diameter tubing cost?

**62. Electrical resistance:** The electrical resistance of a copper wire varies directly with its length and inversely with the square of the diameter of the wire. If a wire 30 m long with a diameter of 3 mm has a resistance of 25 $\Omega$, find the resistance of a wire 40 m long with a diameter of 3.5 mm.

**63. Volume of phone calls:** The number of phone calls per day between two cities varies directly with the product of their populations and inversely with the square of the distance between them. The city of Tampa, Florida (pop. 340,000), is 430 mi from the city of Atlanta, Georgia (pop. 420,000). Telecommunications experts estimate there are about 300 calls per day between the two cities. Use this information to estimate the number of daily phone calls between Amarillo, Texas (pop. 190,000), and Denver, Colorado (pop. 600,000), which are also separated by a distance of about 430 mi. Note: Population figures are for the year 2010 and rounded to the nearest ten-thousand.

*Source: 2010 U.S. Census*

**64. Internet commerce:** The likelihood of an eBay® item being sold for its "Buy it Now®" price $P$, varies directly with the feedback rating of the seller, and inversely with the cube of $\frac{P}{\text{MSRP}}$, where MSRP represents the manufacturer's suggested retail price. A power eBay® seller with a feedback rating of 99.6% knows she has a 60% likelihood of selling an item at 90% of the MSRP. What is the likelihood a seller with a 95.3% feedback rating can sell the same item at 95% of the MSRP?

**65. Volume of an egg:** The volume of an egg laid by an average chicken varies jointly with its length and the square of its width. An egg measuring 2.50 cm wide and 3.75 cm long has a volume of 12.27 cm³. A Barret's Blue Ribbon hen can lay an egg measuring 3.10 cm wide and 4.65 cm long. (a) What is the volume of this egg? (b) As a percentage, how much greater is this volume than that of an average chicken's egg?

**66. Athletic performance:** Researchers have estimated that a sprinter's time in the 100-m dash varies directly with the square root of her age and inversely with the number of hours spent training each week. At 20 yr old, Gail trains 10 hr per week (hr/wk) and has an average time of 11 sec. (a) Assuming she continues to train 10 hr/wk, what will her average time be at 30 yr old? (b) If she wants to keep her average time at 11 sec, how many hours per week should she train?

**67. Maximum safe load:** The maximum safe load $M$ that can be placed on a uniform horizontal beam supported at both ends varies directly with the width $w$ and the square of the height $h$ of the beam's cross section, and inversely with its length $L$. (a) Write the variation equation. (b) If a beam 18 in. wide, 2 in. high, and 8 ft long can safely support 270 lb, what is the safe load for a beam of like dimensions with a length of 12 ft?

**68. Maximum safe load:** Suppose a 10-ft wooden beam with dimensions 4 in. by 6 in. is made from the same material as the beam in Exercise 67 (the same $k$ value can be used). (a) What is the maximum safe load if the beam is turned so that the width is 6 in. and height is 4 in.? (b) What is the maximum safe load if the beam is turned sideways so that the width is 4 in. and height is 6 in.?

**69. Speed of a racer:** A car's speed varies directly with the radius of its tires and output (in rpm) of its engine, but inversely with the product of the transmission and differential gear ratios. If a Spec Ford Racer has a top speed of 130 mph at 5000 rpm in top gear (transmission ratio of 0.7), what is its speed in first gear (transmission ratio of 1.8) at 3000 rpm?

**70. Celestial satellites:** According to Kepler's third law of planetary motion, the mass of a planet is directly proportional to the cube of the average distance between it and an orbiting satellite, but inversely proportional to the square of the satellite's orbital period. The Earth has a mass of $5.97 \times 10^{24}$ kg and is at an average distance of $3.84 \times 10^5$ km from the moon, which completes an orbit every 27.3 days. If the Earth orbits the Sun every 365.3 days at an average distance of $1.496 \times 10^8$ km, what is the mass of the Sun?

## ▶ EXTENDING THE CONCEPT

**71.** The gravitational force $F$ in newtons (N) between two celestial bodies varies directly with the product of their masses and inversely with the square of the distance $d$ between them. The relationship is modeled by Newton's law of universal gravitation: $F = k \frac{m_1 m_2}{d^2}$. Given that $k = 6.67 \times 10^{-11}$, (a) what is the gravitational force exerted by a 1000-kg sphere on another identical sphere that is 10 m away? (b) What is the force if 10-kg spheres are placed 1000 m away?

**72.** The intensity of light and sound both vary inversely with the square of their distance from the source.

    **a.** Suppose you're relaxing one evening with a copy of *Twelfth Night* (Shakespeare), and the reading light is placed 5 ft from the surface of the book. At what distance would the intensity of the light be twice as great?

    **b.** *Tamino's Aria* (*The Magic Flute*—Mozart) is playing in the background, with the speakers 12 ft away. At what distance from the speakers would the intensity of sound be three times as great?

**73.** It's possible to solve variation problems like the one in Example 2 without first finding $k$. To see how this is done, we first solve $y = kx$ for $k$ to get $k = \frac{y}{x}$. We then substitute $(x_1, y_1)$ and $(x_2, y_2)$ into this formula, giving $k = \frac{y_1}{x_1}$ and $k = \frac{y_2}{x_2}$. Since $k$ is constant, we can conclude $\frac{y_1}{x_1} = \frac{y_2}{x_2}$. (a) Solve this equation for $y_1$ and (b) use this as a formula to compute the answer in Example 2 directly.

## ▶ MAINTAINING YOUR SKILLS

**74. (R.3)** Simplify: $\left(\frac{2x^4}{3x^3 y}\right)^{-2}$

**75. (1.6)** Solve: $x^3 + 6x^2 + 8x = 0$.

**76. (2.4)** State the domains of $f$ and $g$ given here:
    **a.** $f(x) = \frac{x - 3}{x^2 - 16}$
    **b.** $g(x) = \frac{x - 3}{\sqrt{x^2 - 16}}$

**77. (3.1)** Graph by using transformations of the parent function and plotting a minimum number of points: $f(x) = -2|x - 3| + 5$.

## MID-CHAPTER CHECK

1. Write the equation of the function that has the same graph as $f(x) = \sqrt{x}$, but shifted left 4 units and up 2 units.

2. For the graph given, identify the (a) function family, (b) end-behavior, inflection point, intercepts, (c) domain and range, and (d) value of $k$ if $f(k) = 2.5$. Assume required features have integer values.

**Exercise 2**

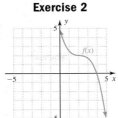

3. Use transformations to graph the given functions in the same window.
$$p(x) = (x - 3)^2, \qquad q(x) = -(x - 3)^2, \qquad r(x) = -\tfrac{1}{2}(x - 3)^2$$

4. Given the graph of $f(x)$ shown, graph $g(x) = -f(x + 2) + 6$.

**Exercise 4**

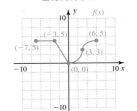

**Exercise 5**

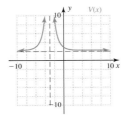

5. Use the graph shown to complete each statement using the direction/approach notation.

   a. The line $x = -3$ is a vertical asymptote, since: as $x \to -3$, $V(x) \to$ _____.

   b. The line $y = 2$ is a horizontal asymptote, since: as $|x| \to \infty$, $V(x) \to$ _____.

6. State the domain and range of the following power functions.

   a. $f(x) = x^{\frac{1}{2}}$        b. $g(x) = x^{\frac{1}{3}}$

   c. $p(x) = x^{-\frac{1}{2}}$        d. $q(x) = x^{-\frac{1}{3}}$

7. After semesters of trial and error, you have discovered your grade on most exams can be modeled by the function $G(h) = 100 - \frac{200}{h + 4}$, where $G(h)$ is your grade after studying for $h$ hr. Use this formula to determine (a) your grade if you don't study at all, (b) your grade if you study for an hour, (c) how many hours are needed to make an 80, and (d) how many hours are needed to make a 90. (e) If the model holds, can you ever make a 100?

8. According to Hooke's law, the distance a spring (or "springy" object) is stretched is directly proportional to the force applied. When raising your wrists from thighs to shoulders during a set of bicep curls, you notice the yoga band offers the same resistance at the "top" of the exercise as a 10-lb barbell. If you use the same band for shoulder presses, how "heavy" will it feel at the top of each press if the band is stretched twice as far as it was for curls?

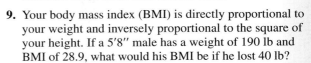

9. Your body mass index (BMI) is directly proportional to your weight and inversely proportional to the square of your height. If a 5'8" male has a weight of 190 lb and BMI of 28.9, what would his BMI be if he lost 40 lb?

10. Suppose $W$ varies jointly with the cube of $x$ and the cube root of $y$. If $W = 32$ when $x = 2$ and $y = 1$, find the variation equation and use it to complete the table.

| $W$ | $x$ | $y$ |
|-----|-----|-----|
| 8   | 1   |     |
|     | 3   | 27  |
| 128 |     | 64  |

## 3.4 Piecewise-Defined Functions

### LEARNING OBJECTIVES

*In Section 3.4 you will see how we can:*

☐ A. State the equation, domain, and range of a piecewise-defined function from its graph

☐ B. Graph functions that are piecewise-defined

☐ C. Solve applications involving piecewise-defined functions

Most of the functions we've studied thus far have been smooth and continuous. Although "smooth" and "continuous" are defined more formally in advanced courses, for our purposes *smooth* simply means the graph has no sharp turns or jagged edges, and *continuous* means you can draw the entire graph without lifting your pencil. In this section, we study a special class of functions, called **piecewise-defined functions,** whose graphs may be various combinations of smooth/not smooth and continuous/not continuous. The absolute value function is one example **(see Exercise 37).** Such functions have a tremendous number of applications in the real world.

## A. The Domain of a Piecewise-Defined Function

For the years 1990 to 2000, the American bald eagle remained on the nation's endangered species list, although the number of breeding pairs was growing slowly. After 2000, the population of eagles grew at a much faster rate, and they were removed from the list soon afterward. From Table 3.8 and plotted points modeling this growth (see Figure 3.35), we observe that a linear model would fit the period from 1992 to 2000 very well, but a line with greater slope would be needed for the years 2000 to 2006 and (perhaps) beyond.

**Table 3.8**

| Year (1990 → 0) | Bald Eagle Breeding Pairs |
|---|---|
| 2 | 3700 |
| 4 | 4400 |
| 6 | 5100 |
| 8 | 5700 |
| 10 | 6500 |
| 12 | 7600 |
| 14 | 8700 |
| 16 | 9800 |

*Source:* www.fws.gov/midwest/eagle/population

**Figure 3.35**

*Bald eagle breeding pairs* vs *Years since 1990*

The combination of these two lines would be a single function that modeled the population of breeding pairs from 1990 to 2006, but it would be *defined in two pieces.* This is an example of a **piecewise-defined function.**

The notation for these functions is a large "left brace" indicating the equations it groups are part of a single function. Using selected data points and techniques from Section 2.3, we find equations that could represent each piece are $p(t) = 350t + 3000$ for $2 \leq t \leq 10$ and $p(t) = 550t + 1000$ for $t > 10$, where $p(t)$ is the number of breeding pairs in year $t$. The complete function is then written:

function name       function pieces       domain of each piece

$$p(t) = \begin{cases} 350t + 3000, & 2 \leq t \leq 10 \\ 550t + 1000, & t > 10 \end{cases}$$

In Figure 3.35, note that we indicated the exclusion of $t = 10$ from the second piece of the function using an open half-circle.

**EXAMPLE 1** ▶ **Writing the Equation and Domain of a Piecewise-Defined Function**

The linear piece of the function shown has an equation
of $y = -2x + 10$. The equation of the quadratic piece is $y = -x^2 + 9x - 14$.

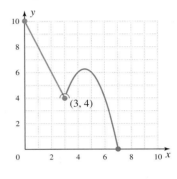

a. Use the correct notation to write them as a
single piecewise-defined function and state the
domain of each piece by inspecting the graph.

b. State the range of the function.

**Solution** ▶ a. From the graph we note the linear portion is
defined between 0 and 3, with these endpoints
included as indicated by the closed dots. The
domain here is $0 \le x \le 3$. The quadratic
portion begins at $x = 3$ *but does not include 3*,
as indicated by the half-circle notation. The
equation is

$$\underset{\text{function name}}{f(x)} = \begin{cases} \underset{\text{function pieces}}{-2x + 10,} & \underset{\text{domain}}{0 \le x \le 3} \\ -x^2 + 9x - 14, & 3 < x \le 7 \end{cases}$$

☑ **A.** You've just seen how
we can state the equation,
domain, and range of a
piecewise-defined function
from its graph

b. The largest $y$-value is 10 and the smallest is zero. The range is $y \in [0, 10]$.

Now try Exercises 5 through 8 ▶

Piecewise-defined functions can be composed of more than two pieces, and can
involve functions of many kinds.

## B. Graphing Piecewise-Defined Functions

As with other functions, piecewise-defined functions can be graphed by simply plotting points. Careful attention must be paid to the domain of each piece, both to evaluate the function correctly and to consider the inclusion/exclusion of endpoints. In
addition, try to keep the transformations of a basic function in mind, as this will often
help graph the function more efficiently.

**EXAMPLE 2** ▶ **Graphing a Piecewise-Defined Function**

Evaluate the piecewise-defined function by noting the effective domain of each piece,
then graph by plotting these points and using your knowledge of basic functions.

$$h(x) = \begin{cases} -x - 2, & -5 \le x < -1 \\ 2\sqrt{x + 1} - 1, & x \ge -1 \end{cases}$$

**Solution** ▶ The first piece of $h$ is a line with negative slope, while the second is a transformed
square root function. Using the endpoints of each piece's domain and a few
additional points, we obtain the following:

For $h(x) = -x - 2$, $-5 \le x < -1$,                For $h(x) = 2\sqrt{x + 1} - 1$, $x \ge -1$,

| $x$ | $h(x)$ |
|-----|--------|
| $-5$ | $3$ |
| $-3$ | $1$ |
| $-1$ | $-1$ |

| $x$ | $h(x)$ |
|-----|--------|
| $-1$ | $-1$ |
| $0$ | $1$ |
| $3$ | $3$ |

After plotting the points from the first piece, we connect them with a line segment noting the left endpoint is included, while the right endpoint is not (indicated using a semicircle around the point). Then we plot the points from the second piece and draw a square root graph, noting the left endpoint here *is* included, and the graph rises to the right. From the graph we note the complete domain of $h$ is $x \in [-5, \infty)$, and the range is $y \in [-1, \infty)$.

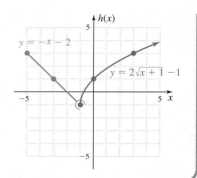

**Now try Exercises 9 through 12** ▶

As an alternative to plotting points, we can graph each piece of the function using transformations of a basic graph, then erase those parts that are outside of the corresponding domain. Repeat this procedure for each piece of the function. One interesting and highly instructive aspect of these functions is the opportunity to investigate restrictions on their domain and the ranges that result.

### Piecewise and Continuous Functions

**EXAMPLE 3** ▶  **Graphing a Continuous Piecewise-Defined Function**

Graph the function and state its domain and range:

$$f(x) = \begin{cases} -(x-3)^2 + 12, & 0 < x \le 6 \\ 3, & x > 6 \end{cases}$$

**Solution** ▶  The first piece of $f$ is a basic parabola, shifted three units right, reflected across the $x$-axis (opening downward), and shifted 12 units up. The vertex is at (3, 12) and the axis of symmetry is $x = 3$, producing the following graphs.

1. Graph first piece of $f$ (Figure 3.36).

2. Erase portion outside domain (Figure 3.37).

3. Partial graph (Figure 3.38).

| **Figure 3.36** | **Figure 3.37** | **Figure 3.38** |
|---|---|---|

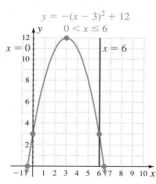
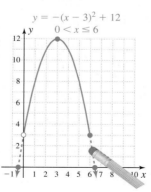
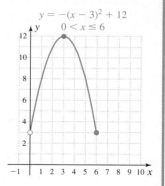

The second piece is simply a horizontal line through (0, 3).

**Solution** ▶  The 88¢ charge applies to letters weighing between 0 and 1 oz. Zero is not included since we have to mail *something,* but 1 is included since a large envelope and its contents weighing exactly 1 oz still costs 88¢. The graph will be a horizontal line segment.

The function is defined for all weights between 0 and 13 oz, excluding zero and including 13: $x \in (0, 13]$. The range consists of discrete outputs that are 20¢ apart: $R \in \{88, 108, 128, \ldots, 308, 328\}$.

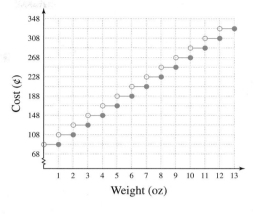

a. The cost of mailing a 7.5-oz report is $2.28 (228¢).

b. The cost of mailing an 8.0-oz report is still $2.28 (228¢).

c. The cost of mailing an 8.1-oz report is $228 + 20 = 248$¢, or $2.48, since this brings you up to the next step.

☑ **C.** You've just seen how we can solve applications involving piecewise-defined functions

**Now try Exercises 51 through 54** ▶

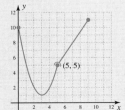

## 3.4 EXERCISES

▶ **CONCEPTS AND VOCABULARY**

**Fill in each blank with the appropriate word or phrase. Carefully reread the section, if necessary.**

1. A graph is called _____ if it has no sharp turns or jagged edges.

2. When graphing $2x + 3$ over a domain of $x > 0$, we leave a(n) _____ dot at $(0, 3)$.

3. Discuss/Explain how to determine if a piecewise-defined function is continuous, without having to graph the function. Illustrate with an example.

4. Discuss/Explain how it is possible for the domain of a function to be defined for all real numbers, but have a range that is split across more than one interval. Construct an illustrative example.

▶ **DEVELOPING YOUR SKILLS**

**(a) Determine the equation of each piecewise-defined function shown, then (b) state the domain and range of the function.**

5. $Y_1 = X^2 - 6x + 10$; $Y_2 = \frac{3}{2}X - \frac{5}{2}$

6. $Y_1 = -1.5|X - 5| + 10$; $Y_2 = -\sqrt{X - 7} + 5$

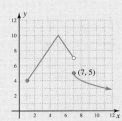

**Determine the domain and range of the piecewise-defined function shown.**

**7.** $f(x) = \begin{cases} 12 - \frac{1}{2}x^2, & 2 < x \le 4 \\ 5 - \frac{1}{3}x, & x > 6 \end{cases}$

**8.** $g(x) = \begin{cases} \sqrt{6 - x} + 8, & x \le 6 \\ |x - 10| + 2, & x > 8 \end{cases}$

**Evaluate each piecewise-defined function as indicated (if possible).**

**9.** $h(x) = \begin{cases} -2, & x < -2 \\ |x|, & -2 \le x < 3 \\ 5, & x \ge 3 \end{cases}$

$h(-5)$, $h(-2)$, $h(-\frac{1}{2})$, $h(0)$, $h(2.999)$, and $h(3)$

**10.** $H(x) = \begin{cases} 2x + 3, & x < 0 \\ x^2 + 1, & 0 \le x < 2 \\ 5, & x > 2 \end{cases}$

$H(-3)$, $H(-\frac{3}{2})$, $H(-0.001)$, $H(1)$, $H(2)$, and $H(3)$

**11.** $p(x) = \begin{cases} 5, & x < -3 \\ x^2 - 4, & -3 \le x \le 3 \\ 2x + 1, & x > 3 \end{cases}$

$p(-5)$, $p(-3)$, $p(-2)$, $p(0)$, $p(3)$, and $p(5)$

**12.** $q(x) = \begin{cases} -x - 3, & x < -1 \\ 2, & -1 \le x < 2 \\ -\frac{1}{2}x^2 + 3x - 3, & x \ge 2 \end{cases}$

$q(-3)$, $q(-1)$, $q(0)$, $q(1.999)$, $q(2)$, and $q(4)$

**Graph each piecewise-defined function and state its domain and range. Use transformations of the toolbox functions where possible.**

**13.** $g(x) = \begin{cases} -(x - 1)^2 + 5, & -2 \le x \le 4 \\ 2x - 12, & x > 4 \end{cases}$

**14.** $h(x) = \begin{cases} \frac{1}{2}x + 1, & x \le 0 \\ (x - 2)^2 - 3, & 0 < x \le 5 \end{cases}$

**15.** $H(x) = \begin{cases} -x + 3, & x < 1 \\ -|x - 5| + 6, & 1 \le x < 9 \end{cases}$

**16.** $w(x) = \begin{cases} \sqrt[3]{x - 1}, & x < 2 \\ (x - 3)^2, & 2 \le x \le 6 \end{cases}$

**17.** $f(x) = \begin{cases} -x - 3, & x < -3 \\ 9 - x^2, & -3 \le x < 2 \\ 4, & x \ge 2 \end{cases}$

**18.** $h(x) = \begin{cases} -\frac{1}{2}x - 1, & x < -3 \\ -|x| + 5, & -3 \le x \le 5 \\ 3\sqrt{x - 5}, & x > 5 \end{cases}$

**19.** $p(x) = \begin{cases} \frac{1}{2}x + 1, & x \ne 4 \\ 2, & x = 4 \end{cases}$

**20.** $q(x) = \begin{cases} \frac{1}{2}(x - 1)^3 - 1, & x \ne 3 \\ -2, & x = 3 \end{cases}$

**Graph the first piece of each function, and find the point $(a, f(a))$ where the two pieces would connect if the function were continuous. Use this point to find the value of $c$ so that a continuous function results. For Exercises 25 through 28, verify that your value of $c$ is correct by using it to draw a complete graph.**

**21.** $f(x) = \begin{cases} \dfrac{x^2 - 9}{x + 3}, & x \ne -3 \\ c, & x = -3 \end{cases}$

**22.** $f(x) = \begin{cases} \dfrac{x^2 - 3x - 10}{x - 5}, & x \ne 5 \\ c, & x = 5 \end{cases}$

**23.** $f(x) = \begin{cases} \dfrac{x^3 - 1}{x - 1}, & x \ne 1 \\ c, & x = 1 \end{cases}$

**24.** $f(x) = \begin{cases} \dfrac{4x - x^3}{x + 2}, & x \ne -2 \\ c, & x = -2 \end{cases}$

**25.** $f(x) = \begin{cases} 3x - 1, & x \le 4 \\ cx + 2, & x > 4 \end{cases}$

**26.** $f(x) = \begin{cases} 2x + 17, & x < -3 \\ -2x + c, & x \ge -3 \end{cases}$

**27.** $f(x) = \begin{cases} -x^3 + 5, & x < 0 \\ |x - 3| + c, & x \ge 0 \end{cases}$

**28.** $f(x) = \begin{cases} x^2 - 4, & x \le 1 \\ c\sqrt{x}, & x > 1 \end{cases}$

**Determine the equation of each piecewise-defined function shown. Assume all pieces are toolbox functions.**

29.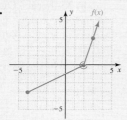

30.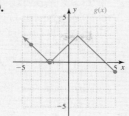

33.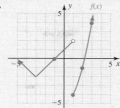

34.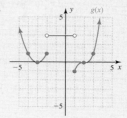

31.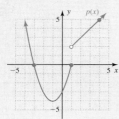

32.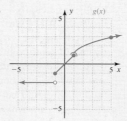

35.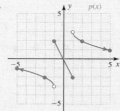

36.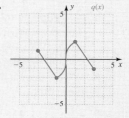

## ▶ WORKING WITH FORMULAS

37. **Definition of absolute value:** $|x| = \begin{cases} -x, & x < 0 \\ x, & x \geq 0 \end{cases}$

The absolute value function can be stated as a piecewise-defined function, a technique that is sometimes useful in graphing variations of the function or solving absolute value equations and inequalities. How does this definition ensure that the absolute value of a number is always positive? Use this definition to help sketch the graph of $f(x) = \frac{|x|}{x}$. Discuss what you notice.

38. **Heaviside step function:** $H(x) = \begin{cases} 0, & x < 0 \\ 1, & x \geq 0 \end{cases}$

The function given is one of many piecewise-defined functions with widespread applications. This particular function effectively acts like an on/off switch, returning a 1 for nonnegative inputs and a 0 otherwise. Because of this behavior, it has many uses in electronics, statistics, and engineering. (a) Graph the function and explain why the function is "on" when $x \geq 0$ and "off" when $x < 0$. (b) Graph $y = H(x + 3)$ and describe the on/off behavior. (c) Repeat part (b) for $y = H(2 - x)$.

## ▶ APPLICATIONS

**For Exercises 39 and 40, (a) write the information given as a piecewise-defined function. (b) Give the domain and range of each.**

39. **Results from advertising:**
Due to heavy advertising, initial sales of the Lynx Digital Camera grew very rapidly, but started to decline once the advertising blitz was over. During the advertising campaign, sales were modeled by the function $S(t) = -t^2 + 6t$, where $S(t)$ represents hundreds of sales in month $t$. However, as Lynx Inc. had hoped, the new product secured a foothold in the market and sales leveled out at a steady 500 sales per month.

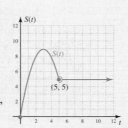

40. **Decline of newspaper publishing:** From the turn of the twentieth century, the number of newspapers sold (per thousand in population) grew rapidly until the 1930s, when the growth slowed down and then declined. The years 1940 to 1946 saw a "spike" in growth, but the years 1947 to 1954 saw an almost equal decline. Since 1954 the number has continued to decline, but at a slower rate. The number $N$ of papers sold per thousand in population for each period respectively can be approximated by

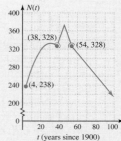

$N_1(t) = -0.13t^2 + 8.1t + 208 \quad N_3(t) = -2.45t + 460$

$N_2(t) = -5.75|t - 46| + 374$

*Source:* Data from the *Statistical Abstract of the United States*, various years; data from *The First Measured Century, The AEI Press*, Caplow, Hicks, and Wattenberg, 2001.

**41. Families that own stocks:** The percentage of American households that own publicly traded stocks began rising in the early 1950s, peaked in 1970, then began to decline until 1980 when there was a dramatic increase due to easy access over the Internet, an improved economy, and other factors. This phenomenon is modeled by the function $P(t)$, where $P(t)$ represents the percentage of households owning stock in year $t$, with 1950 corresponding to year 0.

$$P(t) = \begin{cases} -0.03t^2 + 1.28t + 1.68, & 0 \le t \le 30 \\ 1.89t - 43.5, & 30 < t < 50 \\ 51.2, & t \ge 50 \end{cases}$$

a. According to this model, what percentage of American households held stock in the years 1955, 1965, 1975, 1985, 1995, and 2005? If this pattern continues, what percent will hold stock in 2015?

b. Why is there a discrepancy in the outputs of each piece of the function for the year 1980 ($t = 30$)? According to how the function is defined, which output should be used?

*Source: 2011 Statistical Abstract of the United States, Table 1210; various other years.*

**42. Dependence on foreign oil:** America's dependency on foreign oil has always been a "hot" political topic, with the amount of imported oil fluctuating over the years due to political climate, public awareness, the economy, and other factors. The amount of crude oil imported can be approximated by the function given, where $A(t)$ represents the number of barrels (in billions) imported in year $t$, with 1980 corresponding to year 0.

$$A(t) = \begin{cases} 0.047t^2 - 0.38t + 1.9, & 0 \le t < 8 \\ -0.075t^2 + 1.495t - 5.265, & 8 \le t \le 11 \\ -0.011t^2 + 0.541t - 3.047, & t > 11 \end{cases}$$

a. Use $A(t)$ to estimate the number of barrels imported in the years 1983, 1989, 1995, and 2005. If this trend continues, how many barrels will be imported in 2015?

b. Use a table of values to determine the year in which the number of barrels imported was minimized.

*Source: 2011 Statistical Abstract of the United States, Table 932; various other years.*

**43. Energy rationing:** In certain areas of the United States, power blackouts have forced some counties to ration electricity. Suppose the cost is $0.09 per kilowatt–hour (kW·h) for the first 1000 kW·h a household uses. After 1000 kW·h, the cost increases to $0.18 per kW·h. (a) Write these charges for electricity in the form of a piecewise-defined function $C(h)$, where $C(h)$ is the cost for $h$ kilowatt–hours. Then (b) sketch the graph and determine the cost for 1200 kW·h.

**44. Water rationing:** Many southwestern states have a limited water supply, and some state governments try to control consumption by manipulating the cost of water usage. Suppose for the first 5000 gal a household uses per month, the charge is $0.05/gal. Once 5000 gal is used the charge doubles to $0.10/gal. (a) Write these charges for water usage in the form of a piecewise-defined function $C(w)$, where $C(w)$ is the cost for $w$ gal of water. Then (b) sketch the graph and determine the cost to a household that used 9500 gal of water during a very hot summer month.

**45. Pricing for natural gas:** A local gas company charges $0.75 per therm for natural gas, up to 25 therms. Once the 25 therms has been exceeded, the charge doubles to $1.50 per therm due to limited supply and great demand. (a) Write these charges for natural gas consumption in the form of a piecewise-defined function $C(t)$, where $C(t)$ is the charge for $t$ therms. Then (b) sketch the graph and determine the cost to a household that used 45 therms during a very cold winter month.

**46. Multiple births:** The number of multiple births has steadily increased in the United States during the twentieth century and beyond. Between 1985 and 1995 the number of twin births could be modeled by the function $T(x) = -0.21x^2 + 6.1x + 52$, where $x$ is the number of years since 1980 and $T$ is in thousands. After 1995, the incidence of twins becomes more linear, with $T(x) = 3.49x + 47.2$ serving as a better model. (a) Write the piecewise-defined function modeling the incidence of twins for these years. Then (b) sketch the graph and use the function to estimate the incidence of twins in 1985, 1995, and 2005. If this trend continues, how many sets of twins will be born in 2015?

*Source: National Vital Statistics Report, Vol. 58, No. 24, August 9, 2010*

**47. U.S. military expenditures:** Except for the year 1991 when military spending was cut drastically, the amount spent by the U.S. government on national defense and veterans' benefits rose steadily from 1980 to 1992. These expenditures can be modeled by the function $S(t) = -1.35t^2 + 31.9t + 152$, where $S(t)$ is in billions of dollars and 1980 corresponds to $t = 0$. From 1992 to 1996 this spending declined, then began to rise in the following years. From 1992 to 2002, military-related spending can be modeled by $S(t) = 2.6t^2 - 81.5t + 939$. (a) Write $S(t)$ as a single piecewise-defined function. Then (b) sketch the graph and use the function to estimate the amount spent by the United States in 2005, 2010, and 2015 if this trend continues.

*Source: 2011 Statistical Abstract of the United States, Table 501*

**48. Amusement arcades:** At a local amusement center, the owner has the SkeeBall machines programmed to reward very high scores. For scores of 200 or less, the function $T(x) = \frac{x}{10}$ models the number of tickets awarded (rounded to the nearest whole). For scores over 200, the number of tickets is modeled by $T(x) = 0.001x^2 - 0.3x + 40$. (a) Write $T(x)$ as a single piecewise-defined function. Then (b) sketch the graph and find the number of tickets awarded to a person who scores 390 points.

**49. Phone service charges:** When it comes to phone service, a large number of calling plans are available. Under one plan, the first 30 min of any phone call costs only 3.3¢ per minute. The charge increases to 7¢ per minute thereafter. (a) Write this information in the form of a piecewise-defined function. Then (b) sketch the graph and find the cost of a 46-min phone call.

**50. Overtime wages:** Tara works on an assembly line, putting together computer monitors. She is paid $9.50/hr for regular time (0, 40 hr], $14.25 for overtime (40, 48 hr], and when demand for computers is high, $19.00 for double-overtime (48, 84 hr]. (a) Write this information in the form of a simplified piecewise-defined function. Then (b) sketch the graph and find the gross amount of Tara's check for the week she put in 54 hr.

**51. Admission prices:** At Wet Willy's Water World, infants (under 2) are free, then admission is charged according to age. Children (2 and older, but less than 13) pay $2, teenagers (13 and older, but less than 20) pay $5, adults (20 and older, but less than 65) pay $7, and senior citizens (65 and older) get in at the teenage rate. (a) Write this information in the form of a piecewise-defined function. Then (b) sketch the graph and find the cost of admission for a family of nine which includes: one grandparent (70), two adults (44/45), 3 teenagers, 2 children, and one infant.

**52. Demographics:** One common use of the floor function $y = \lfloor x \rfloor$ is the reporting of ages. As of 2007, the record for longest living human is 122 yr, 164 days for the life of Jeanne Calment, formerly of France. While she actually lived $x = 122\frac{164}{365}$ yr, ages are normally reported using the floor function, or the greatest integer number of years less than or equal to the actual age: $\lfloor 122\frac{164}{365} \rfloor = 122$ yr. (a) Write a function $A(t)$ that gives a person's age, where $A(t)$ is the reported age at time $t$. (b) State the domain of the function (be sure to consider Madame Calment's record). Report the age of a person who has been living for (c) 36 yr; (d) 36 yr, 364 days; (e) 37 yr; and (f) 37 yr, 1 day.

**53. Postage rates:** The postal charge function from Example 8 is simply a transformation of the basic ceiling function $y = \lceil x \rceil$. Using the ideas from Section 3.1, (a) write the postal charges as a step function $C(w)$, where $C(w)$ is the cost of mailing a large envelope weighing $w$ oz, and (b) state the domain of the function. Then use the function to find the cost of mailing reports weighing: (c) 0.7 oz, (d) 5.1 oz, (e) 5.9 oz, (f) 6 oz, and (g) 6.1 oz.

**54. Cell phone charges:** A national cell phone company advertises that calls of 1 min or less do not count toward monthly usage. Calls lasting longer than 1 min are calculated normally using a ceiling function, meaning a call of 1 min, 1 sec will be counted as a 2-min call. Using the ideas from Section 3.1, (a) write the cell phone charges as a piecewise-defined function $C(m)$, where $C(m)$ is the number of "minutes" charged for a call lasting $m$ minutes, and include the domain of the function. Then (b) graph the function, and (c) use the graph or function to determine if a cell phone subscriber has exceeded the 30 free "minutes" granted by her calling plan for calls lasting 2 min 3 sec, 13 min 46 sec, 1 min 5 sec, 3 min 59 sec, and 8 min 2 sec. (d) What was the actual usage in minutes and seconds?

**55. Combined absolute value graphs:** (a) Carefully graph the function $h(x) = |x - 2| - |x + 3|$ using a table of values over the interval $x \in [-5, 5]$. Is the function continuous? (b) Rewrite $h(x)$ as a piecewise-defined function.

**56. Combined absolute value graphs:** (a) Carefully graph the function $H(x) = |x - 2| + |x + 3|$ using a table of values over the interval $x \in [-5, 5]$. Is the function continuous? (b) Rewrite $H(x)$ as a piecewise-defined function.

▶ **EXTENDING THE CONCEPT**

**57.** You've heard it said, "*any number divided by itself is one.*" If this were true, the function $f(x) = \frac{x + 2}{x + 2}$ would always return a value of one. Does it? Where is the function discontinuous? Graph $f(x) = \frac{x + 2}{x + 2}$ and $g(x) = \frac{|x + 2|}{x + 2}$ and compare the results.

**58.** Find a linear function $h(x)$ that will make the function shown a *continuous* function. Be sure to include its domain.

$$f(x) = \begin{cases} x^2, & x < 1 \\ h(x), & ??? \\ 2x + 3, & x > 3 \end{cases}$$

**59.** (a) Graph the function $f(x)$ given, and then (b) graph $g(x) = 3f(x + 2) + 1$. (c) Using the techniques of Section 3.1, rewrite $g(x)$ as a piecewise-defined function.

$$f(x) = \begin{cases} x^2 - 4, & -2 \le x \le 1 \\ 3, & 1 < x < 3 \\ -2x + 10, & x \ge 3 \end{cases}$$

**60.** The floor function $F(x) = \lfloor x \rfloor$ and the ceiling function $C(x) = \lceil x \rceil$ can produce some interesting average rates of change. The average rate of change of these two functions over any interval 1 unit or longer must lie within what range of values?

▶ **MAINTAINING YOUR SKILLS**

**61. (1.6)** Solve: $\frac{3}{x - 2} + 1 = \frac{30}{x^2 - 4}$.

**62. (R.6)** Compute the following and write the result in lowest terms:

$$\frac{x^3 + 3x^2 - 4x - 12}{x - 3} \cdot \frac{2x - 6}{x^2 + 5x + 6} \div (3x - 6)$$

**63. (2.3)** Find an equation of the line perpendicular to $3x + 4y = 8$ that passes through the point $(0, -2)$. Write the result in slope-intercept form.

**64. (R.4/1.1)** For the figure shown, (a) use the Pythagorean theorem to find the length of the missing side, then compute (b) the area of the triangular face and (c) the volume of the triangular prism.

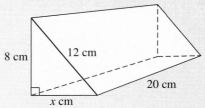

**The Algebra and Composition of Functions**

## LEARNING OBJECTIVES

*In Section 3.5 you will see how we can:*

- ☐ **A.** Compute a sum or difference of functions and determine the domain of the result
- ☐ **B.** Compute a product or quotient of functions and determine the domain
- ☐ **C.** Compose two functions and determine the domain; decompose a function
- ☐ **D.** Interpret operations on functions graphically
- ☐ **E.** Apply the algebra and composition of functions in context

In Section 3.1, we created new functions *graphically* by applying transformations to basic functions. In this section, we'll use two (or more) functions to create new functions *algebraically*. Previous courses often contain material on the sum, difference, product, and quotient of polynomials. Here we'll combine functions with the basic operations, noting the result is also a function that can be evaluated, graphed, and analyzed. We call these basic operations on functions the **algebra of functions.**

## A. Sums and Differences of Functions

This section introduces the notation used for basic operations on functions. Here we'll note the result is a new function whose domain depends on the original functions. In general, if $f$ and $g$ are functions *with overlapping domains*, $f(x) + g(x) = (f + g)(x)$ and $f(x) - g(x) = (f - g)(x)$.

| **Sums and Differences of Functions** | |
|---|---|
| For functions $f$ and $g$ with domains $D_f$ and $D_g$ respectively, the sum and difference of $f$ and $g$ are defined by: | |
| | **Domain of result** |
| $(f + g)(x) = f(x) + g(x)$ | $D_f \cap D_g$ |
| $(f - g)(x) = f(x) - g(x)$ | $D_f \cap D_g$ |

**EXAMPLE 1A** ▶ **Evaluating a Difference of Functions Using the Definition**

Given $f(x) = x^2 - 5x$ and $g(x) = 2x - 9$,
  **a.** Determine the domain of $h(x) = (f - g)(x)$.   **b.** Find $h(3)$ using the definition.

**Solution** ▶   **a.** Since the domain of both $f$ and $g$ is $\mathbb{R}$, their intersection and the domain of $h$ is also $\mathbb{R}$.

  **b.** $\begin{aligned} h(3) &= (f - g)(3) & \text{given difference} \\ &= f(3) - g(3) & \text{by definition} \\ &= [(3)^2 - 5(3)] - [2(3) - 9] & \text{evaluate} \\ &= (9 - 15) - (6 - 9) & \text{multiply} \\ &= (-6) - (-3) & \text{subtract} \\ &= -3 & \text{result} \end{aligned}$

**Now try Exercises 5 and 6** ▶

If the function $h$ is to be graphed or evaluated numerous times, it helps to compute a *new function rule* for $h$, rather than repeatedly apply the definition.

**EXAMPLE 1B** ▶ **Evaluate a Difference of Functions Using a New Function Rule**

For the functions $f$, $g$, and $h$, as defined in Example 1A,
  **a.** Find a new function rule for $h$.   **b.** Use the result to find $h(3)$.

**Solution** ▶   **a.** $\begin{aligned} h(x) &= (f - g)(x) & \text{given difference} \\ &= f(x) - g(x) & \text{by definition} \\ &= (x^2 - 5x) - (2x - 9) & \text{replace } f(x) \text{ with } (x^2 - 5x) \text{ and } g(x) \text{ with } (2x - 9) \\ &= x^2 - 7x + 9 & \text{distribute and combine like terms} \end{aligned}$

**b.** $h(3) = (3)^2 - 7(3) + 9$     substitute 3 for $x$

$\quad\quad = 9 - 21 + 9$     multiply

$\quad\quad = -3$     result

Notice the result from part (b) is identical to that in Example 1A.

**Now try Exercises 7 and 8** ▶

**CAUTION** ▶ From Example 1, note the importance of using grouping symbols with the algebra of functions. Without them, we could easily confuse the signs of $g$ when computing the difference. Also, note that any operation applied to the functions $f$ and $g$ simply results in an *expression* representing a new function rule for $h$, and is not an *equation* that needs to be factored or solved.

**EXAMPLE 2** ▶ **Computing a Sum of Functions**

For $f(x) = \sqrt{x - 2}$, and $g(x) = x^2$

**a.** Determine the domain of $h(x) = (g + f)(x)$.

**b.** Find a new function rule for $h$.

**c.** Evaluate $h(3)$.

**d.** Evaluate $h(-1)$.

**Solution** ▶ **a.** The domain of $f$ is $x \in [2, \infty)$, while the domain of $g$ is $\mathbb{R}$. Since their intersection is $[2, \infty)$, this is the domain of the new function $h$.

**b.** $h(x) = (g + f)(x)$     given sum

$\quad\quad = g(x) + f(x)$     by definition

$\quad\quad = x^2 + \sqrt{x - 2}$     substitute $x^2$ for $g(x)$ and $\sqrt{x - 2}$ for $f(x)$ (no other simplifications possible)

**c.** $h(3) = (3)^2 + \sqrt{3 - 2}$     substitute 3 for $x$

$\quad\quad = 10$     result

**d.** $x = -1$ is outside the domain of $h$ so $h(-1)$ is undefined.

**Now try Exercises 9 through 12** ▶

**WORTHY OF NOTE**

If we *did* try to evaluate $h(-1)$, the result would be $1 + \sqrt{-3}$, which is not a real number. While it's true we could write $1 + \sqrt{-3}$ as $1 + i\sqrt{3}$ and consider it an "answer," our study here focuses on real numbers and the graphs of functions in a coordinate system where $x$ and $y$ are both real.

This "intersection of domains" is illustrated in Figure 3.50.

**Figure 3.50**

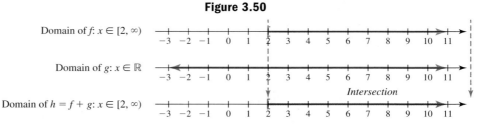

Domain of $f$: $x \in [2, \infty)$

Domain of $g$: $x \in \mathbb{R}$

Domain of $h = f + g$: $x \in [2, \infty)$

☑ **A.** You've just seen how we can compute a sum or difference of functions and determine the domain of the result

## B. Products and Quotients of Functions

The product and quotient of two functions is defined in a manner similar to that for sums and differences. For example, if $f$ and $g$ are functions *with overlapping domains*, $(f \cdot g)(x) = f(x) \cdot g(x)$ and $\left(\frac{f}{g}\right)(x) = \frac{f(x)}{g(x)}$. As you might expect, for quotients we must stipulate $g(x) \neq 0$.

### Products and Quotients of Functions

For functions $f$ and $g$ with domains $D_f$ and $D_g$ respectively,
the product and quotient of $f$ and $g$ are defined by:

**Domain of result**

$$(f \cdot g)(x) = f(x) \cdot g(x) \qquad\qquad D_f \cap D_g$$

$$\left(\tfrac{f}{g}\right)(x) = \tfrac{f(x)}{g(x)} \qquad\qquad D_f \cap D_g, \text{ for all } g(x) \neq 0$$

**EXAMPLE 3** ▶ **Computing a Product of Functions**

Given $f(x) = \sqrt{1 + x}$ and $g(x) = \sqrt{3 - x}$,

**a.** Determine the domain of $h(x) = (f \cdot g)(x)$.

**b.** Find a new function rule for $h$.

**c.** Use the result from part (b) to evaluate $h(2)$ and $h(4)$.

**Solution** ▶ **a.** The domain of $f$ is $x \in [-1, \infty)$ and the domain of $g$ is $x \in (-\infty, 3]$. The intersection of these domains gives $x \in [-1, 3]$, which is the domain for $h$.

**b.**
$$
\begin{aligned}
h(x) &= (f \cdot g)(x) & &\text{given product}\\
&= f(x) \cdot g(x) & &\text{by definition}\\
&= \sqrt{1 + x} \cdot \sqrt{3 - x} & &\text{substitute } \sqrt{1+x} \text{ for } f \text{ and } \sqrt{3-x} \text{ for } g\\
&= \sqrt{3 + 2x - x^2} & &\text{combine using properties of radicals}
\end{aligned}
$$

**c.**
$$
\begin{aligned}
h(2) &= \sqrt{3 + 2(2) - (2)^2} & &\text{substitute 2 for } x\\
&= \sqrt{3} \approx 1.732 & &\text{result}\\
h(4) &= \sqrt{3 + 2(4) - (4)^2} & &\text{substitute 4 for } x\\
&= \sqrt{-5} & &\textit{not a real number}
\end{aligned}
$$

The second result of part (c) is not surprising, since $x = 4$ is not in the domain of $h$ [meaning $h(4)$ is not defined for this function].

**Now try Exercises 13 through 16** ▶

In future sections, we use polynomial division as a tool for factoring, as an aid to graphing, and to determine whether two expressions are equivalent. Understanding the notation and domain issues related to division will strengthen our ability in these areas.

**EXAMPLE 4** ▶ **Computing a Quotient of Functions**

Given $f(x) = x - 3$ and $g(x) = x^3 - 3x^2 + 2x - 6$,

**a.** Determine the domain of $h(x) = \left(\tfrac{g}{f}\right)(x)$.

**b.** Find a new function rule for $h$.

**c.** Use the result from part (b) to evaluate $h(0)$.

**Solution** ▶ **a.** While the domain of both $f$ and $g$ is $\mathbb{R}$ and their intersection is also $\mathbb{R}$, we know from the definition (and past experience) *that $f(x)$ cannot be zero*. The domain of $h$ is $x \in (-\infty, 3) \cup (3, \infty)$.

**b.**
$$
\begin{aligned}
h(x) &= \left(\frac{g}{f}\right)(x) & &\text{given quotient}\\[4pt]
&= \frac{g(x)}{f(x)} & &\text{by definition}\\[4pt]
&= \frac{x^3 - 3x^2 + 2x - 6}{x - 3} & &\text{replace } g \text{ with } x^3 - 3x^2 + 2x - 6 \text{ and } f \text{ with } x - 3
\end{aligned}
$$

Using techniques from Section R.6, the expression that defines $h$ can be simplified: $\frac{x^3 - 3x^2 + 2x - 6}{x - 3} = \frac{x^2(x - 3) + 2(x - 3)}{x - 3} = \frac{(x^2 + 2)(x - 3)}{x - 3} = x^2 + 2$. But from the original expression, we know $h$ is not defined if $x = 3 [f(x) = 0]$, *even if the result for h is.* In this case, we write the simplified form as $h(x) = x^2 + 2, x \neq 3$.

**c.** For $h(0)$ we have:

$$h(0) = (0)^2 + 2 \quad \text{replace } x \text{ with } 0$$
$$h(0) = 2$$

**Now try Exercises 17 through 32 ▶**

☑ **B. You've just seen how we can compute a product or quotient of functions and determine the domain**

It's important to note that operations on functions mimic the operations on real numbers, and as such subtraction and division are not commutative. For additional practice with the algebra of functions, **see Exercises 33 through 40.**

## C. The Composition of Functions

The composition of functions gives us an efficient way to study how relationships are "linked." For example, the number of wolves $w$ in a countywide area depends on the human population $x$, and a simple model might be $w(x) = 600 - 0.02x$. But the number of rodents $r$ then depends on the number of wolves, say $r(w) = 10{,}000 - 9.5w$, so the human population also has a measurable effect on the rodent population. In this section, we'll show that this effect is modeled by the function $r(x) = 4300 + 0.19x$.

The composition of functions is best understood by studying the "input/output" nature of a function. Consider $g(x) = x^2 - 3$. For $g(x)$ we might say, "inputs are squared, then decreased by three." In diagram form we have:

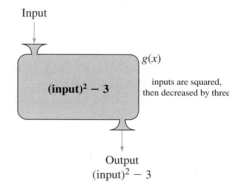

Input

**(input)² − 3**

$g(x)$

inputs are squared, then decreased by three

Output
(input)² − 3

In many respects, a function box can be regarded as a very simple machine, running a simple program. It doesn't matter what the input is, this machine is going to *square the input then subtract three.*

**EXAMPLE 5 ▶** **Evaluating a Function**

For $g(x) = x^2 - 3$, find
**a.** $g(-5)$      **b.** $g(5t)$      **c.** $g(t - 4)$

**Solution** ▶

**a.**
$$g(x) = x^2 - 3 \qquad \text{original function}$$
input $-5$
$$g(-5) = (-5)^2 - 3 \qquad \text{square input, then subtract 3}$$
$$= 25 - 3 \qquad \text{simplify}$$
$$= 22 \qquad \text{result}$$

**b.**
$$g(x) = x^2 - 3 \qquad \text{original function}$$
input $5t$
$$g(5t) = (5t)^2 - 3 \qquad \text{square input, then subtract 3}$$
$$= 25t^2 - 3 \qquad \text{result}$$

**c.**
$$g(x) = x^2 - 3 \qquad \text{original function}$$
input $t - 4$
$$g(t - 4) = (t - 4)^2 - 3 \qquad \text{square input, then subtract 3}$$
$$= t^2 - 8t + 16 - 3 \qquad \text{expand binomial}$$
$$= t^2 - 8t + 13 \qquad \text{result}$$

**Now try Exercises 41 and 42** ▶

> **WORTHY OF NOTE**
>
> It's important to note that $t$ and $t - 4$ are two different, distinct values—the number represented by $t$, and a number four less than $t$. Examples would be 7 and 3, 12 and 8, as well as $-10$ and $-14$. There should be nothing awkward or unusual about evaluating $g(t)$ versus evaluating $g(t - 4)$ as in Example 5(c).

**Figure 3.51**

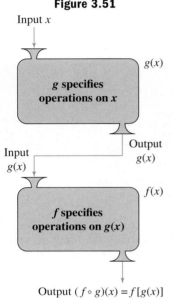

Input $x$

$g$ specifies operations on $x$    $g(x)$

Input $g(x)$    Output $g(x)$

$f$ specifies operations on $g(x)$    $f(x)$

Output $(f \circ g)(x) = f[g(x)]$

When the input value is itself a function (rather than a single number or variable), the result is a **composition of functions.** The evaluation method is exactly the same, we are simply using a function input. Using a general function $g(x)$ and a function diagram as before, we illustrate the process in Figure 3.51.

The notation used for the composition of $f$ with $g$ is an open dot " $\circ$ " placed between them, and is read, "$f$ composed with $g$." The notation $(f \circ g)(x)$ indicates that $g(x)$ is an input for $f$: $(f \circ g)(x) = f[g(x)]$. If the order is reversed, as in $(g \circ f)(x)$, $f(x)$ becomes the input for $g$: $(g \circ f)(x) = g[f(x)]$. Figure 3.51 also helps us determine the domain of a composite function, in that the first function $g$ can operate only if $x$ is a valid input for $g$, and the second function $f$ can operate only if $g(x)$ is a valid input for $f$. In other words, $(f \circ g)(x)$ is defined for *all x in the domain of g, such that g(x) is in the domain of f.*

**CAUTION** ▶ Try not to confuse the new "open dot" notation for the *composition* of functions, with the multiplication dot used to indicate the *product* of two functions: $(f \cdot g)(x) = f(x) \cdot g(x)$ while $(f \circ g)(x) = f[g(x)]$.

---

### The Composition of Functions

Given two functions $f$ and $g$, the composition of $f$ with $g$ is defined by
$$(f \circ g)(x) = f[g(x)]$$
The domain of the composition is all $x$ in the domain of $g$ for which $g(x)$ is in the domain of $f$.

---

**EXAMPLE 6** ▶ **Finding a Composition of Functions**

Given $f(x) = \sqrt{x - 4}$ and $g(x) = 3x + 2$, find
**a.** $(f \circ g)(x)$
**b.** $(g \circ f)(x)$
Also determine the domain for each.

**Solution** ▶    **a.** $f(x) = \sqrt{x - 4}$ says "decrease inputs by 4, and take the square root of the result."

$$
\begin{aligned}
(f \circ g)(x) &= f[g(x)] &&\text{$g(x)$ is an input for $f$}\\
&= \sqrt{g(x) - 4} &&\text{decrease input by 4, and take the square root of the result}\\
&= \sqrt{(3x + 2) - 4} &&\text{substitute $3x + 2$ for $g(x)$}\\
&= \sqrt{3x - 2} &&\text{result}
\end{aligned}
$$

While $g$ is defined for all real numbers, $f(x)$ represents a real number only when $x \geq 4$. For $f[g(x)]$, this means we need $g(x) \geq 4$, giving $3x + 2 \geq 4$, $x \geq \frac{2}{3}$. In interval notation, the domain of $(f \circ g)(x)$ is $x \in [\frac{2}{3}, \infty)$.

**b.** The function $g$ says "multiply inputs by 3, then increase by 2."

$$
\begin{aligned}
(g \circ f)(x) &= g[f(x)] &&\text{$f(x)$ is an input for $g$}\\
&= 3f(x) + 2 &&\text{multiply input by 3, then increase by 2}\\
&= 3\sqrt{x - 4} + 2 &&\text{substitute $\sqrt{x - 4}$ for $f(x)$}
\end{aligned}
$$

For $g[f(x)]$, $g$ can accept any real number input, but $f$ supplies only those when $x \geq 4$. The domain of $(g \circ f)(x)$ is $x \in [4, \infty)$.

Now try Exercises 43 through 52 ▶

Example 6 shows that $(f \circ g)(x)$ is generally not equal to $(g \circ f)(x)$. On those occasions when they *are* equal, the functions have a unique relationship that we'll study in Section 5.1.

---

**EXAMPLE 7** ▶    **Finding a Composition of Functions**

For $f(x) = \frac{3x}{x - 1}$ and $g(x) = \frac{2}{x}$, analyze the domain of

    **a.** $(f \circ g)(x)$     **b.** $(g \circ f)(x)$     **c.** Find the actual compositions and comment.

**Solution** ▶    **a.** $(f \circ g)(x)$: For $g$ to be defined, $x \neq 0$ must be our first restriction. Once $g(x)$ is used as the input, we have $f[g(x)] = \frac{3g(x)}{g(x) - 1}$, and additionally note that $g(x)$ cannot equal 1. This means $\frac{2}{x} \neq 1$, and so $x \neq 2$. The domain of $(f \circ g)(x)$ is all real numbers *except* $x = 0$ and $x = 2$.

**b.** $(g \circ f)(x)$: For $f$ to be defined, $x \neq 1$ must be our first restriction. Once $f(x)$ is used as the input, we have $g[f(x)] = \frac{2}{f(x)}$, and additionally note that $f(x)$ cannot be 0. This means $\frac{3x}{x - 1} \neq 0$, and so $x \neq 0$. The domain of $(g \circ f)(x)$ is all real numbers *except* $x = 0$ and $x = 1$.

**c.** For $(f \circ g)(x)$:

$$
\begin{aligned}
f[g(x)] &= \frac{3g(x)}{g(x) - 1} &&\text{composition of $f$ with $g$}\\[2mm]
&= \frac{3\left(\dfrac{2}{x}\right)}{\left(\dfrac{2}{x}\right) - 1} &&\text{substitute $\frac{2}{x}$ for $g(x)$}\\[2mm]
&= \frac{\dfrac{6}{x}}{\dfrac{2 - x}{x}} = \frac{6}{x} \cdot \frac{x}{2 - x} &&\text{simplify denominator; invert and multiply}\\[2mm]
&= \frac{6}{2 - x} &&\text{result}
\end{aligned}
$$

Notice the function rule for $(f \circ g)(x)$ implies $x = 2$ is not in the domain of the function, but does not show that $g$ (the inner function) is undefined when $x = 0$ [see part (a)]. The domain of $(f \circ g)(x)$ is actually $\{x | x \neq 0, 2\}$. For $(g \circ f)(x)$ we have:

$$g[f(x)] = \frac{2}{f(x)} \qquad \text{composition of } g \text{ with } f$$

$$= \frac{2}{\dfrac{3x}{x-1}} \qquad \text{substitute } \frac{3x}{x-1} \text{ for } f(x)$$

$$= \frac{2}{1} \cdot \frac{x-1}{3x} \qquad \text{invert and multiply}$$

$$= \frac{2(x-1)}{3x} \qquad \text{result}$$

Similarly, the function rule for $(g \circ f)(x)$ has an implied domain of $x \neq 0$, but does not show that $f$ (the inner function) is undefined when $x = 1$ [see part (b)]. The domain of $(g \circ f)(x)$ is actually all real numbers except $x = 0$ and $x = 1$.

**Now try Exercises 53 through 58** ▶

As Example 7 illustrates, the domain of $h(x) = (f \circ g)(x)$ *cannot simply be taken from the new function rule for h*. It *must* be determined from the functions composed to obtain $h$. To further explore concepts related to the domain of a composition, **see Exercise 95.**

### Decomposing a Composite Function

**Figure 3.52**

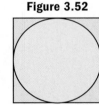

Based on Figure 3.52, would you say that the circle is inside the square or the square is inside the circle? The decomposition of a composite function is related to a similar question, as we ask ourselves what function (of the composition) is on the "inside"—the input quantity—and what function is on the "outside." For instance, consider $h(x) = \sqrt{x-4}$, where we see that $x - 4$ is "inside" the radical. Letting $f(x) = \sqrt{x}$, and $g(x) = x - 4$ we have $h(x) = (f \circ g)(x)$ or $f[g(x)]$.

**WORTHY OF NOTE**

The decomposition of a function is not unique and can often be done in many different ways.

**EXAMPLE 8** ▶ Decomposing a Composite Function

Given $h(x) = (\sqrt[3]{x} + 1)^2 - 3$, identify functions $f$ and $g$ so that $(f \circ g)(x) = h(x)$, then check by composing the functions to obtain $h(x)$.

**Solution** ▶ Noting that $\sqrt[3]{x} + 1$ is inside the squaring function, we assign $g(x)$ as this inner function: $g(x) = \sqrt[3]{x} + 1$. The outer function is the squaring function decreased by 3, so $f(x) = x^2 - 3$.

Check: $(f \circ g)(x) = f[g(x)] \qquad g(x) \text{ is an input for } f$

$\qquad\qquad = [g(x)]^2 - 3 \qquad f \text{ squares inputs, then decreases the result by 3}$

$\qquad\qquad = [\sqrt[3]{x} + 1]^2 - 3 \qquad \text{substitute } \sqrt[3]{x} + 1 \text{ for } g(x)$

$\qquad\qquad = h(x) ✓$

☑ **C.** You've just seen how we can compose two functions and determine the domain, and decompose a function

**Now try Exercises 59 through 68** ▶

## D. Graphical Views of Operations on Functions

The algebra of functions also has an instructive *graphical interpretation,* in which values for $f(k)$ and $g(k)$ are read from a graph ($k$ is a given constant), with operations like $(f + g)(k) = f(k) + g(k)$ then computed using the standard definition. Once the value of $g(k)$ is known, $(f \circ g)(x) = f[g(k)]$ is likewise interpreted and computed (also **see Exercise 94**).

**EXAMPLE 9** ▶ **Interpreting Operations on Functions Graphically**

Use the graph given to find the value of each expression:

a. $(f + g)(-2)$

b. $(f \circ g)(7)$

c. $(g - f)(6)$

d. $\left(\dfrac{g}{f}\right)(8)$

e. $(f \cdot g)(4)$

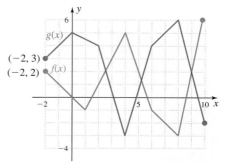

**Solution** ▶ Since the needed input values for this example are $x = -2, 4, 6, 7,$ and 8, we begin by reading the value of $f(x)$ and $g(x)$ at each point. From the graph, we note that $f(-2) = 2$ and $g(-2) = 3$. The other values are likewise found and appear in the table. For $(f + g)(-2)$ we have:

| $x$ | $f(x)$ | $g(x)$ |
|-----|--------|--------|
| $-2$ | 2 | 3 |
| 4 | 5 | $-3$ |
| 6 | $-1$ | 4 |
| 7 | $-2$ | 5 |
| 8 | $-3$ | 6 |

a. $(f + g)(-2) = f(-2) + g(-2)$    definition

  $= 2 + 3$         substitute 2 for $f(-2)$ and 3 for $g(-2)$

  $= 5$           result

b. $(f \circ g)(7) = f[g(7)]$       definition

  $= f(5)$         substitute 5 for $g(7)$

  $= 2$           result read from graph: $f(5) = 2$

With some practice, the computations can be done mentally and we have

c. $(g - f)(6) = g(6) - f(6)$

  $= 4 - (-1) = 5$

d. $\left(\dfrac{g}{f}\right)(8) = \dfrac{g(8)}{f(8)}$

  $= \dfrac{6}{-3} = -2$

e. $(f \cdot g)(4) = f(4) \cdot g(4)$

  $= 5(-3) = -15$

✓ **D.** You've just seen how we can interpret operations of functions graphically

**Now try Exercises 70 through 78** ▶

## E. Applications of the Algebra and Composition of Functions

The algebra of functions plays an important role in the business world. For example, the cost to manufacture an item, the revenue a company brings in, and the profit a company earns are all functions of the number of items made and sold. Further, we know a company "breaks even" (making $0 profit) when the difference between their revenue $R$ and cost $C$ is zero.

**EXAMPLE 10** ▶    **Applying Operations on Functions in Context**

The fixed costs to publish *Relativity Made Simple* (by N.O. Way) is $2500, while the marginal cost is $4.50 per book. Marketing studies indicate the best selling price for the book is $9.50 per copy.

**a.** Find the cost, revenue, and profit functions for this book.

**b.** Determine how many copies must be sold for the company to break even.

**Solution** ▶    **a.** Let $x$ represent the number of books published and sold. The cost of publishing is $4.50 per copy, plus fixed costs (labor, storage, etc.) of $2500. The cost function is $C(x) = 4.50x + 2500$. If the company charges $9.50 per book, the revenue function will be $R(x) = 9.50x$. Since profit equals revenue minus costs,

$$
\begin{aligned}
P(x) &= (R - C)(x) & &\text{profit equals revenue minus cost} \\
&= R(x) - C(x) & &\text{by definition} \\
&= 9.50x - (4.50x + 2500) & &\text{substitute } 9.50x \text{ for } R \text{ and } 4.50x + 2500 \text{ for } C \\
&= 9.50x - 4.50x - 2500 & &\text{distribute} \\
&= 5x - 2500 & &\text{result}
\end{aligned}
$$

The profit function is $P(x) = 5x - 2500$.

**b.** When a company "breaks even," the profit is zero: $P(x) = 0$.

$$
\begin{aligned}
P(x) &= 5x - 2500 & &\text{profit function} \\
0 &= 5x - 2500 & &\text{substitute 0 for } P(x) \\
2500 &= 5x & &\text{add 2500} \\
500 &= x & &\text{divide by 5}
\end{aligned}
$$

In order for the company to break even, 500 copies must be sold.

**Now try Exercises 81 through 86** ▶

Suppose that due to a collision, an oil tanker is spewing oil into the open ocean. The oil is spreading outward in a shape that is roughly circular, with the radius of the circle modeled by the function $r(t) = 2\sqrt{t}$, where $t$ is the time in minutes and $r$ is measured in feet. How could we determine the *area* of the oil slick in terms of $t$? As you can see, the radius depends on the time and the area depends on the radius. In diagram form we have:

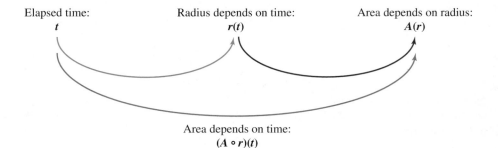

Elapsed time:
$t$

Radius depends on time:
$r(t)$

Area depends on radius:
$A(r)$

Area depends on time:
$(A \circ r)(t)$

It is possible to create a direct relationship between the elapsed time and the area of the circular spill using a composition of functions.

**EXAMPLE 11 ▶**   **Composition and the Area of an Oil Spill**

Given $r(t) = 2\sqrt{t}$ and $A(r) = \pi r^2$,

a. Express the area directly as a function of time by computing $(A \circ r)(t)$.

b. Find the area of the oil spill after 30 min.

**Solution ▶**   a. The function $A$ squares inputs, then multiplies by $\pi$.

$$
\begin{aligned}
(A \circ r)(t) &= A[r(t)] && \text{$r(t)$ is the input for $A$} \\
&= [r(t)]^2 \cdot \pi && \text{square input, multiply by $\pi$} \\
&= [2\sqrt{t}]^2 \cdot \pi && \text{substitute $2\sqrt{t}$ for $r(t)$} \\
&= 4\pi t && \text{result}
\end{aligned}
$$

Since the result contains no variable $r$, we can now compute the area of the spill directly at a time $t$ (in minutes): $(A \circ r)(t) = 4\pi t$.

b. To find the area after 30 min, use $t = 30$.

$$
\begin{aligned}
(A \circ r)(t) &= 4\pi t && \text{composite function} \\
(A \circ r)(30) &= 4\pi(30) && \text{substitute 30 for $t$} \\
&= 120\pi && \text{simplify} \\
&\approx 377 && \text{result (rounded to the nearest unit)}
\end{aligned}
$$

After 30 min, the area of the spill is approximately 377 ft$^2$.

**Now try Exercises 87 through 90 ▶**

**EXAMPLE 12 ▶**   **Composition and Related Populations**

Using a statistical study, environmentalists find a current population of 600 wolves in the county, and believe the wolf population $w$ will decrease as the human population $x$ grows, according to the formula $w(x) = 900 - 0.02x$. Further, the population of rodents $r$ depends on the number of wolves $w$, with the rodent population estimated by $r(w) = 10{,}000 - 9.5w$.

a. Use the models to estimate the number of wolves and rodents present in the county before humans arrived.

b. Find the current human and rodent populations, then use a composition to find a function modeling how the human population relates directly to the number of rodents.

c. Use the function to estimate the number of rodents in the county, if the human population grows by 35,000.

**Solution** ▶ **a.** To estimate the number of wolves and rodents in the county before humans arrived (when $x = 0$), we first compute $w(0) = 900$. Knowing there were 900 wolves, we then find $r(900) = 1450$. Initially, a population of 900 wolves kept the rodent population in check at 1450.

**b.** To determine the current size of the human population, we substitute 600 for $w(x)$ and solve for $x$.

$$w(x) = 900 - 0.02x \qquad \text{wolf population}$$
$$600 = 900 - 0.02x \qquad \text{substitute 600 for } w(x)$$
$$-300 = -0.02x \qquad \text{subtract 900}$$
$$15{,}000 = x \qquad \text{divide by } -0.02$$

There are currently 15,000 humans and $r(600) = 4300$ rodents in the county.
 To find a function directly relating $x$ to $r$, we use the composition $(r \circ w)(x) = r[w(x)]$.

$$r(w) = 10{,}000 - 9.5w \qquad r \text{ depends on } w$$
$$r[w(x)] = 10{,}000 - 9.5[900 - 0.02x] \qquad \text{compose } r \text{ with } w(x)$$
$$= 10{,}000 - 8550 + 0.19x \qquad \text{distribute}$$
$$(r \circ w)(x) = 1450 + 0.19x \qquad \text{simplify; } r \text{ depends on } x$$

**c.** If the human population grows by 35,000, the number of humans in the county will reach 50,000. Evaluating $(r \circ w)(50{,}000)$ gives 10,950 rodents, showing a decline in the wolf population will eventually cause the rodent population to flourish.
 Notice that the two function $w(x)$ and $(r \circ w)(x)$ are both linear, with their slopes quantifying the sensitivity of the animal populations to the human population. In the wolf model, $m = -0.02$, or $\frac{-20}{1000}$, suggesting 20 wolves are lost for every increase of 1000 in the human population. In the rodent model, $m = 0.19 \approx \frac{1}{5}$ indicating the rodent population is growing at a rate of almost one rodent for every five humans!

 **E. You've just seen how we can apply the algebra and composition of functions in context**

**Now try Exercises 91 and 92 ▶**

## 3.5 EXERCISES

▶ **CONCEPTS AND VOCABULARY**

**Fill in each blank with the appropriate word or phrase. Carefully reread the section, if necessary.**

1. For the product $h(x) = f(x) \cdot g(x)$, $h(5)$ can be found by evaluating $f$ and $g$ then multiplying the result, or multiplying $f \cdot g$ and evaluating the result. Notationally these are written _____ and _____.

2. The notation $(f \circ g)(x)$ indicates that $g(x)$ is the input value for $f(x)$, which is written _____.

3. Discuss/Explain the domain of $h(x) = (f \cdot g)(x)$, given $f(x) = \sqrt{x - 3}$ and $g(x) = \sqrt{2 - x}$.

4. Discuss/Explain how the domain of $(f \circ g)(x)$ is determined, given $f(x) = \sqrt{2x + 7}$ and $g(x) = \frac{2x}{x - 1}$.

## ▶ DEVELOPING YOUR SKILLS

**5.** Given $f(x) = 2x^2 - x - 3$ and $g(x) = x^2 + 5x$,
(a) determine the domain for $h(x) = f(x) - g(x)$ and
(b) find $h(-2)$ using the definition.

**6.** Given $f(x) = 2x^2 - 18$ and $g(x) = -3x - 7$,
(a) determine the domain for $h(x) = f(x) + g(x)$ and
(b) find $h(5)$ using the definition.

**7.** For the functions $f$, $g$, and $h$, as defined in Exercise 5,
(a) find a new function rule for $h$, and (b) use the result
to find $h(-2)$. (c) How does the result compare to that
of Exercise 5?

**8.** For the functions $f$, $g$, and $h$ as defined in Exercise 6,
(a) find a new function rule for $h$, and (b) use the result
to find $h(5)$. (c) How does the result compare to that in
Exercise 6?

**9.** For $f(x) = \sqrt{x - 3}$ and $g(x) = 2x^3 - 54$, (a) determine
the domain of $h(x) = (f + g)(x)$, (b) find a new function
rule for $h$, and (c) evaluate $h(4)$ and $h(2)$, if possible.

**10.** For $f(x) = 4x^2 - 2x + 3$ and $g(x) = \sqrt{2x - 5}$,
(a) determine the domain of $h(x) = (f - g)(x)$, (b) find
a new function rule for $h$, and (c) evaluate $h(7)$ and $h(2)$,
if possible.

**11.** For $p(x) = \sqrt{x + 5}$ and $q(x) = \sqrt{3 - x}$, (a) determine
the domain of $r(x) = (p + q)(x)$, (b) find a new function
rule for $r$, and (c) evaluate $r(2)$ and $r(4)$, if possible.

**12.** For $p(x) = \sqrt{6 - x}$ and $q(x) = \sqrt{x + 2}$, (a) determine
the domain of $r(x) = (p - q)(x)$, (b) find a new function
rule for $r$, and (c) evaluate $r(-3)$ and $r(2)$, if possible.

**13.** For $f(x) = \sqrt{x + 4}$ and $g(x) = 2x + 3$, (a) determine
the domain of $h(x) = (f \cdot g)(x)$, (b) find a new function
rule for $h$, and (c) evaluate $h(-4)$ and $h(21)$, if possible.

**14.** For $f(x) = -3x + 5$ and $g(x) = \sqrt{x - 7}$, (a) determine
the domain of $h(x) = (f \cdot g)(x)$, (b) find a new function
rule for $h$, and (c) evaluate $h(8)$ and $h(11)$, if possible.

**15.** For $p(x) = \sqrt{x + 1}$ and $q(x) = \sqrt{7 - x}$, (a) determine
the domain of $r(x) = (p \cdot q)(x)$, (b) find a new function
rule for $r$, and (c) evaluate $r(15)$ and $r(3)$, if possible.

**16.** For $p(x) = \sqrt{4 - x}$ and $q(x) = \sqrt{x + 4}$, (a) determine
the domain of $r(x) = (p \cdot q)(x)$, (b) find a new function
rule for $r$, and (c) evaluate $r(-5)$ and $r(-3)$, if possible.

**For the functions $f$ and $g$ given, (a) determine the domain
of $h(x) = \left(\frac{f}{g}\right)(x)$ and (b) find a new function rule for $h$ in
simplified form (if possible), noting the domain restrictions
alongside.**

**17.** $f(x) = x + 1$ and $g(x) = x - 5$

**18.** $f(x) = x + 3$ and $g(x) = x - 7$

**19.** $f(x) = x^2 - 16$ and $g(x) = x + 4$

**20.** $f(x) = x^2 - 49$ and $g(x) = x - 7$

**21.** $f(x) = x^3 + 4x^2 - 2x - 8$ and $g(x) = x + 4$

**22.** $f(x) = x^3 - 5x^2 + 2x - 10$ and $g(x) = x - 5$

**23.** $f(x) = x^3 - 7x^2 + 6x$ and $g(x) = x - 1$

**24.** $f(x) = x^3 - 1$ and $g(x) = x - 1$

**For the functions $p$ and $q$ given, (a) determine the domain
of $r(x) = \left(\frac{p}{q}\right)(x)$, (b) find a new function rule for $r$, and
(c) use it to evaluate $r(6)$ and $r(-6)$, if possible.**

**25.** $p(x) = 1 - x$ and $q(x) = \sqrt{3 - x}$

**26.** $p(x) = x + 2$ and $q(x) = \sqrt{x + 3}$

**27.** $p(x) = x^2 - 36$ and $q(x) = \sqrt{2x + 13}$

**28.** $p(x) = x^2 - 6x$ and $q(x) = \sqrt{7 + 3x}$

**For the functions $f$ and $g$ given, (a) find a new function rule
for $h(x) = \left(\frac{f}{g}\right)(x)$ in simplified form. (b) If $h(x)$ were the
original function, what would be its domain? (c) Since we
know $h(x) = \left(\frac{f}{g}\right)(x) = \frac{f(x)}{g(x)}$, what additional restrictions exist
for the domain of $h$?**

**29.** $f(x) = \dfrac{6x}{x - 3}$ and $g(x) = \dfrac{3x}{x + 2}$

**30.** $f(x) = \dfrac{4x}{x + 1}$ and $g(x) = \dfrac{2x}{x - 2}$

**31.** $f(x) = \dfrac{x^2 - 5x - 6}{x^2 - 5x + 6}$ and $g(x) = \dfrac{x^2 - 1}{x^2 - 4}$

**32.** $f(x) = \dfrac{6x^2 - x - 2}{4x^2 + 7x - 2}$ and $g(x) = \dfrac{6x^2 + 5x + 1}{4x^2 + 9x + 2}$

**For each pair of functions $f$ and $g$ given, determine the
sum, difference, product, and quotient of $f$ and $g$, and
determine the domain in each case.**

**33.** $f(x) = 2x + 3$ and $g(x) = x - 2$

**34.** $f(x) = x - 5$ and $g(x) = 2x - 3$

**35.** $f(x) = x^2 + 2x - 3$ and $g(x) = x - 1$

**36.** $f(x) = x^2 - 2x - 15$ and $g(x) = x + 3$

**37.** $f(x) = x + 2$ and $g(x) = \sqrt{x + 6}$

**38.** $f(x) = x^2 + 2$ and $g(x) = \sqrt{x - 5}$

**39.** $f(x) = \dfrac{2}{x - 3}$ and $g(x) = \dfrac{5}{x + 2}$

**40.** $f(x) = \dfrac{4}{x - 3}$ and $g(x) = \dfrac{1}{x + 5}$

**41.** Given $f(x) = x^2 - 5x - 14$, find $f(-2)$, $f(7)$, $f(2a)$, and
$f(a - 2)$.

**42.** Given $g(x) = x^3 - 9x$, find $g(-3)$, $g(2)$, $g(3t)$, and
$g(t + 1)$.

**For each pair of functions below, find (a) $h(x) = (f \circ g)(x)$
and (b) $H(x) = (g \circ f)(x)$, and (c) determine the domain of
each result.**

**43.** $f(x) = \sqrt{x + 3}$ and $g(x) = 2x - 5$

**44.** $f(x) = x + 3$ and $g(x) = \sqrt{9 - x}$

**45.** $f(x) = \sqrt{x - 3}$ and $g(x) = 3x + 4$

**46.** $f(x) = \sqrt{x + 5}$ and $g(x) = 4x - 1$

**47.** $f(x) = x^2 - 3x$ and $g(x) = x + 2$

**48.** $f(x) = 2x^2 - 1$ and $g(x) = 3x + 2$

**49.** $f(x) = x^2 + x - 4$ and $g(x) = x + 3$

**50.** $f(x) = x^2 - 4x + 2$ and $g(x) = x - 2$

**51.** $f(x) = |x| - 5$ and $g(x) = -3x + 1$

**52.** $f(x) = |x - 2|$ and $g(x) = 3x - 5$

**For the functions $f(x)$ and $g(x)$ given, analyze the domain of (a) $(f \circ g)(x)$ and (b) $(g \circ f)(x)$, then (c) find the actual compositions and comment.**

**53.** $f(x) = \dfrac{2x}{x + 3}$ and $g(x) = \dfrac{5}{x}$

**54.** $f(x) = \dfrac{-3}{x}$ and $g(x) = \dfrac{x}{x - 2}$

**55.** $f(x) = \dfrac{4}{x}$ and $g(x) = \dfrac{1}{x - 5}$

**56.** $f(x) = \dfrac{3}{x}$ and $g(x) = \dfrac{1}{x - 2}$

**57.** For $f(x) = x^2 - 8$, $g(x) = x + 2$, and $h(x) = (f \circ g)(x)$, find $h(5)$ in two ways:

    **a.** $(f \circ g)(5)$         **b.** $f[g(5)]$

**58.** For $p(x) = x^2 - 8$, $q(x) = x + 2$, and $R(x) = (p \circ q)(x)$, find $R(-2)$ in two ways:

    **a.** $(p \circ q)(-2)$      **b.** $p[q(-2)]$

**59.** For $h(x) = 5(x^3 + 7)^8$ and $g(x) = x^3 + 7$, find the function $f$ such that $h(x) = (f \circ g)(x)$.

**60.** For $r(x) = 2\sqrt[3]{3x - 1}$ and $q(x) = 3x - 1$, find the function $p$ such that $r(x) = (p \circ q)(x)$.

**61.** For $h(x) = \sqrt{2x + 3} - 5$ and $f(x) = \sqrt{x} - 5$, find the function $g$ such that $h(x) = (f \circ g)(x)$.

**62.** For $r(x) = \dfrac{4}{1 - x^2} - 3$ and $p(x) = \dfrac{4}{x} - 3$, find the function $q$ such that $r(x) = (p \circ q)(x)$.

**63.** For $h(x) = (\sqrt{x - 2} + 1)^3 - 5$, find two functions $f$ and $g$ such that $(f \circ g)(x) = h(x)$.

**64.** For $R(x) = \sqrt[3]{x^2 - 5} + 2$, find two functions $p$ and $q$ such that $(p \circ q)(x) = R(x)$.

**65.** Given $f(x) = 2x - 1$, $g(x) = x^2 - 1$, and $h(x) = x + 4$, find $j(x) = f\{g[h(x)]\}$ and $k(x) = g\{f[h(x)]\}$.

**66.** Given $p(x) = 3x^2$, $q(x) = 1 - 2x$, and $r(x) = 3 - x$, find $s(x) = r\{q[p(x)]\}$ and $t(x) = p\{q[r(x)]\}$.

**67.** Given $f(x) = 2x + 3$ and $g(x) = \frac{x - 3}{2}$, find (a) $(f \circ f)(x)$, (b) $(g \circ g)(x)$, (c) $(f \circ g)(x)$, and (d) $(g \circ f)(x)$.

**68.** Given $p(x) = \frac{2}{4 + x}$ and $q(x) = \frac{3}{1 - x}$, find (a) $(p \circ p)(x)$, (b) $(q \circ q)(x)$, (c) $(p \circ q)(x)$, and (d) $(q \circ p)(x)$.

**69. Reading a graph—Used vehicle sales:** The graph given shows the number of cars $C(t)$ and trucks $T(t)$ sold by Ullery Used Autos for the years 2000 to 2010. Use the graph to estimate the number of

    **a.** cars sold in 2005: $C(5)$

    **b.** trucks sold in 2008: $T(8)$

**Exercise 69**

    **c.** vehicles sold in 2009: $C(9) + T(9)$

    **d.** In function notation, how would you determine how many more cars than trucks were sold in 2009? What was the actual number?

**70. Reading a graph— Government investment:** The graph given shows a government's investment in its military $M(t)$ and public works $P(t)$ over time, in millions of dollars. Use the graph to estimate the investment in

**Exercise 70**

    **a.** the military in 2002: $M(2)$

    **b.** public works in 2005: $P(5)$

    **c.** public works and the military in 2009: $M(9) + P(9)$

    **d.** In function notation, how would you determine how much more will be invested in public works than the military in 2010? What is the actual number?

**71. Reading a graph—Space travel:** The graph given shows the revenue $R(t)$ and operating costs $C(t)$ of Space Travel Resources (STR), for the years 2000 to 2010. Use the graph to find the

**Exercise 71**

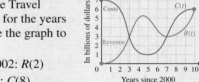

    **a.** revenue in 2002: $R(2)$

    **b.** costs in 2008: $C(8)$

    **c.** years STR broke even: $R(t) = C(t)$

    **d.** years costs exceeded revenue: $C(t) > R(t)$

    **e.** years STR made a profit: $R(t) > C(t)$

    **f.** For the year 2005, use function notation to write the profit equation for STR. What was their profit?

**72. Reading a graph— Corporate expenditures:** The graph given shows a large corporation's investment in research and development $R(t)$ over time, and the amount paid to investors as dividends $D(t)$. Use the graph to find the

**Exercise 72**

    **a.** dividend payments in 2002: $D(2)$

    **b.** investment in 2006: $R(6)$

    **c.** years where $R(t) = D(t)$

    **d.** years where $R(t) > D(t)$

    **e.** years where $R(t) < D(t)$

    **f.** Use function notation to write an equation for the total expenditures of the corporation in year $t$. What was the total for 2010?

**73. Reading a graph— Operations on functions:** Use the given graphs to find the result of the operations indicated.

**Exercise 73**

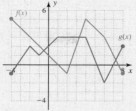

Note $f(-4) = 5$, $g(-4) = -1$, and so on.

a. $(f + g)(-4)$
b. $(f \cdot g)(1)$
c. $(f - g)(4)$
d. $(f + g)(0)$
e. $\left(\dfrac{f}{g}\right)(2)$
f. $(f \cdot g)(-2)$
g. $(g \cdot f)(2)$
h. $(g - f)(-1)$
i. $(g + f)(8)$
j. $\left(\dfrac{g}{f}\right)(7)$
k. $(g \circ f)(4)$
l. $(f \circ g)(4)$

**74. Reading a graph— Operations on functions:** Use the given graphs to find the result of the operations indicated.

**Exercise 74**

Note $p(-1) = -2$, $q(5) = 6$, and so on.

a. $(p + q)(-4)$
b. $(p \cdot q)(1)$
c. $(p - q)(4)$
d. $(p + q)(0)$

e. $\left(\dfrac{p}{q}\right)(5)$
f. $(p \cdot q)(-2)$
g. $(q \cdot p)(2)$
h. $(q - p)(-1)$
i. $(q + p)(7)$
j. $\left(\dfrac{q}{p}\right)(6)$
k. $(q \circ p)(4)$
l. $(p \circ q)(-1)$

**Find a function rule in simplified form for the vertical distance $h(x)$ between the graphs of $f$ and $g$ shown over the interval indicated.**

**75.** $x \in [0, 6]$

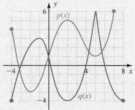

**76.** $x \in [1, 7]$

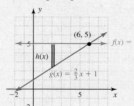

**77.** $x \in [0, 4]$

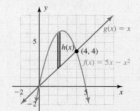

**78.** $x \in [0, 5]$

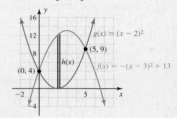

► **WORKING WITH FORMULAS**

**79. Surface area of a cylinder:** $A = 2\pi rh + 2\pi r^2$

If the height of a cylinder is fixed at 20 cm, the formula becomes $A = 40\pi r + 2\pi r^2$. (a) Write this formula in factored form and find functions $f(r)$ and $g(r)$ such that $A(r) = (f \cdot g)(r)$. (b) Then find $A(5)$ by direct calculation and also by computing the product of $f(5)$ and $g(5)$, then comment on the results.

**80. Compound annual growth:** $A(r) = P(1 + r)^t$

The amount of money $A$ in a savings account $t$ yr after an initial investment of $P$ dollars depends on the interest rate $r$. If \$1000 is invested for 5 yr, find functions $f(r)$ and $g(r)$ such that $A(r) = (f \circ g)(r)$.

► **APPLICATIONS**

**81. Boat manufacturing:** Giaro Boats manufactures a popular recreational vessel, the *Revolution*. To plan for expanded production and increased labor costs, the company carefully tracks current costs and income. The fixed cost to produce this boat is \$108,000 and the marginal costs are \$28,000 per boat. If the *Revolution* sells for \$40,000, (a) find the cost, revenue, and profit function; determine (b) how many boats must be sold to break even; and (c) the profit if 15 boats are sold.

**82. Nonprofit publications:** Adobe Hope, a nonprofit agency, publishes the weekly newsletter *Community Options*. The fixed cost for publishing the newsletter is \$900 per week, with a marginal cost of \$0.25 per newsletter. If the newsletter is sold for \$1.50 per copy, (a) find the cost, revenue, and profit function for the newsletter; (b) determine how many newsletters must be sold to break even; and (c) determine how much money will be returned to the community if 1000 newsletters are sold.

**83. Cost, revenue, and profit:** Suppose the total cost of manufacturing a certain computer component can be modeled by the function $C(n) = 0.1n^2$, where $C(n)$ is the cost in dollars for making $n$ components. If each component is sold at a price of $11.45, complete the following.

    **a.** Find the function that represents the total profit made from sales of the components.

    **b.** How much profit is earned if 12 components are made and sold?

    **c.** How much profit is earned if 60 components are made and sold?

    **d.** Explain why the company is making a "negative profit" after the 114th component is made and sold.

**84. Cost, revenue, and profit:** For a certain manufacturer, revenue has been increasing but so has the cost of materials and employee benefits. Suppose revenue can be modeled by $R(t) = 10\sqrt{t}$, the cost of materials by $M(t) = 2t + 1$, and the cost of benefits by $B(t) = 0.1t^2 + 2$, where outputs are in thousands of dollars and $t$ is the time in months since operations began. Use this information to complete the following.

    **a.** Find the function that represents the total manufacturing costs.

    **b.** Find the function that represents how much more the cost of benefits are than the cost of materials.

    **c.** What was the cost of benefits in the 10th month after operations began?

    **d.** How much less were the benefits costs than the materials cost in the 10th month?

    **e.** Find the function that represents the profit earned by this company.

    **f.** Find the amount of profit earned in the 5th month and 10th month. Discuss each result.

**85. Measuring the depth of a well:** To estimate the depth of a well, a large stone is dropped into the opening and a stopwatch is used to measure the number of seconds until a splash is heard. By solving the freefall equation $d = 16t^2$ for $t$, we find the function $t_1(d) = \frac{\sqrt{d}}{4}$ models the time it takes for the stone to hit the water. Since the speed of sound is known to be 1116 ft/sec, we use the distance = rate × time equation to obtain $t_2(d) = \frac{d}{1116}$ as a function modeling the time it takes the sound of the splash to reach our ears. (a) Find a formula for the total time $T(d)$. (b) If the well is 230 ft deep, what is the total time? (c) If it takes 6 sec for the sound to be heard, how deep is the well?

**86. Predator-prey concentrations:** Suppose the monthly whitetip reef shark population off the coast of Manuel Antonio National Park (Costa Rica) can be modeled by the function $s(t) = -40t^2 + 500t + 1300$ (where $t = 1$ corresponds to January). One of its favorite foods is the spiny lobster, whose monthly population in these waters might be modeled by the function $l(t) = 110t^2 - 1500t + 13,400$.

(a) Find a formula for the number of lobsters per shark in any given month. (b) How many lobsters per shark are there in February? (c) When is the number of lobsters per shark minimized at 2.9 lobsters per shark?

**87. International shoe sizes:** Peering inside her athletic shoes, Morgan notes the following shoe sizes: US 8.5, UK 6, EUR 40. The function that relates the U.S. sizes to the European (EUR) sizes is $g(x) = 2x + 23$, where $x$ represents the U.S. size and $g(x)$ represents the EUR size. The function that relates European sizes to sizes in the United Kingdom (UK) is $f(x) = 0.5x - 14$, where $x$ represents the EUR size and $f(x)$ represents the UK size. (a) Find the function $h(x)$ that relates the U.S. measurement directly to the UK measurement by finding $h(x) = (f \circ g)(x)$. Then find (b) the UK size for a U.S. size 13 and (c) the U.S. size for a UK 13.5.

**88. Currency conversion:** On a trip to Europe, Megan had to convert American dollars to euros using the function $E(x) = 1.12x$, where $x$ represents the number of dollars and $E(x)$ is the equivalent number of euros. Later, she converts her euros to Japanese yen using the function $Y(x) = 1061x$, where $x$ represents the number of euros and $Y(x)$ represents the equivalent number of yen. (a) Convert 100 U.S. dollars to euros. (b) Convert the answer from part (a) into Japanese yen. (c) Express yen as a function of dollars by finding $M(x) = (Y \circ E)(x)$, then use $M(x)$ to convert $100 directly to yen. Do parts (b) and (c) agree?

*Source: 2005 World Almanac, p. 231*

**89. Spread of a fire:** Due to a lightning strike, a forest fire begins to burn and is spreading outward in a shape that is roughly circular. The radius of the circle is modeled by the function $r(t) = 2t$, where $t$ is the time in minutes and $r$ is measured in meters. (a) Write a function for the area burned by the fire directly as a function of $t$ by computing $(A \circ r)(t)$. (b) Find the area of the circular burn after 60 min.

**90. Expanding supernova:** The surface area of a star goes through an expansion phase prior to going *supernova*. As the star begins expanding, the radius becomes a function of time. Suppose this function is $r(t) = 1.05t$, where $t$ is in days and $r(t)$ is in gigameters (Gm). (a) Find the radius of the star two days after the expansion phase begins. (b) Find the surface area after two days. (c) Express the surface area as a function of time by finding $f(t) = (S \circ r)(t)$, then use $f(t)$ to compute the surface area after two days directly. Do the answers agree?

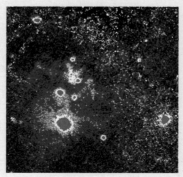

**91. Composition and dependent relationships:** In the wild, the balance of nature is often very fragile, with any sudden changes having dramatic and unforeseen consequences. With a huge increase in population and tourism near an African wildlife preserve, the number of lions is decreasing due to loss of habitat and a disruption in normal daily movements. This is causing a related increase in the hyena population, as the lion is one of the hyena's only natural predators. If this increase remains unchecked, animals lower in the food chain will suffer. If the lion population $L$ depends on the increase in human population $x$ according to the formula $L(x) = 500 - 0.015x$, and the hyena population depends on the lion population as modeled by the formula $H(L) = 650 - 0.5L$, (a) what is the current lion population and hyena population if no humans are present? (b) Use a composition to find a function modeling how the hyena population relates directly to the number of humans, and use the function to estimate the number of hyenas in the area if the human population grows to 16,000. (c) If the administrators of the preserve consider a population of 625 hyenas as "extremely detrimental," at what point should the human population be capped?

**92. Composition and dependent relationships:** The recent opening of a landfill in the area has caused the raccoon population to flourish, with an adverse effect on the number of purple martins. Wildlife specialists believe the population of martins $p$ will decrease as the raccoon population $r$ grows. Further, since mosquitoes are the primary diet of purple martins, the mosquito population $m$ is likewise affected. If the first relationship is modeled by the function $p(r) = 750 - 3.75r$ and the second by $m(p) = 50,000 - 45p$, (a) what are the current number of purple martins and mosquitoes if no raccoons are present? (b) Use a composition to find a function modeling how the raccoon population relates directly to the number of mosquitoes, and use the function to estimate the number of mosquitoes in the area if the raccoon population grows to 50. (c) If the health department considers 36,500 mosquitoes to be a "dangerous level," what increase in the raccoon population will bring this about?

▶ **EXTENDING THE CONCEPT**

**93.** Given $f(x) = \sqrt{1 - x}$ and $g(x) = \sqrt{x - 2}$, what can you say about the domain of $(f + g)(x)$? Enter the functions as $Y_1$ and $Y_2$ on a graphing calculator, then enter $Y_3 = Y_1 + Y_2$. See if you can determine why the calculator gives an error message for $Y_3$, regardless of the input.

**94.** Instead of calculating the result of an operation on two functions at a *specific point* as in Exercises 69–74, we can actually *graph the function* that results from the

operation. This skill, called the **addition of ordinates,** is widely applied in numerous areas, including the study of tides. For $f(x) = (x - 3)^2 + 2$ and $g(x) = 4|x - 3| - 5$, complete a table of values like the one shown for $x \in [-2, 8]$. Use the relation $(f - g)(x) = f(x) - g(x)$ to complete the last column. Finally, use the ordered pairs to graph the new function. Is the new function smooth? Is the new function continuous?

**95.** Suppose $f(x) = \frac{1}{x^2 - 3}$, $g(x) = \sqrt{x^2 - 1}$, and $h(x) = \frac{1}{x^2 - 4}$. (a) Determine the domain of $h$ and complete the column for $h$ in the table. Then (b) show $h$ offers a new function rule for $(f \circ g)(x)$. (c) Determine the domain of $f \circ g$ and complete the column for $f \circ g$ in the table. Why are the last two columns different?

| $x$ | $f(x)$ | $g(x)$ | $(f - g)(x)$ |
|---|---|---|---|
| −2 | | | |
| −1 | | | |
| 0 | | | |
| 1 | | | |
| 2 | | | |
| 3 | | | |
| 4 | | | |
| 5 | | | |
| 6 | | | |
| 7 | | | |
| 8 | | | |

| $x$ | $h(x)$ | $(f \circ g)(x)$ |
|---|---|---|
| −3 | | |
| −2 | | |
| −1 | | |
| 0 | | |
| 1 | | |
| 2 | | |
| 3 | | |

▶ **MAINTAINING YOUR SKILLS**

**96.** (1.4) Find the sum and product of the complex numbers $2 + 3i$ and $2 - 3i$.

**97.** (3.1) Draw a sketch of the functions *from memory*.
  **a.** $f(x) = \sqrt{x}$     **b.** $g(x) = \sqrt[3]{x}$
  **c.** $h(x) = |x|$

**98.** (1.5) Use the quadratic formula to solve $2x^2 - 3x + 4 = 0$.

**99.** (2.3) Find an equation of the line perpendicular to $-2x + 3y = 9$ that goes through the origin.

---

## 3.6 Another Look at Formulas, Functions, and Problem Solving

**LEARNING OBJECTIVES**

*In Section 3.6 you will see how we can:*

☐ **A.** Create functions using relationships defined on the coordinate plane
☐ **B.** Create functions using geometric relationships
☐ **C.** Create functions by combining basic formulas

In Section 1.1, we were introduced to the basic fundamentals of working with formulas, using algebraic models, and making substitutions in order to solve problems. In Chapter 2, we analyzed the graph of a function to locate specific features and characteristics. In this section, we attempt to refine and extend these skills by applying them in new ways, and in ways that will lead to more significant applications. In many cases, this will involve looking at how two different functions or formulas are related, and combining them to obtain a function in a single variable.

### A. Functions and Relationships Defined on the Coordinate Plane

From your previous coursework, recall that if two quantities are equal, one can be substituted for the other without changing the solution set. This simple idea is used throughout mathematics. Note how it's used here as we study relationships in the *xy*-plane.

**EXAMPLE 1** ▶ **Geometric Recreations—A Rectangle within a Parabola**

A rectangle is inscribed within a parabola in Quadrant I (QI) as shown, with its base on the *x*-axis, one side along the *y*-axis, and one corner on the graph of the parabola. If the parabola is defined by $y = 18 - 0.5x^2$,

  **a.** Find a function of a single variable that models the area of all such rectangles. What is the domain of this function?

  **b.** Find the area of the rectangle if $x = 5$.

  **c.** Find the value of *x* that will give the largest rectangle possible.

**Solution** ▶ **a.** The formula for the area of a rectangle is $A = LW$, with the area expressed in terms of the two variables *L* and *W*. From the diagram, we note that for any point $(x, y)$, the width of the rectangle can be represented by the *x*-value, and the length (or height) by the *y*-value, giving $A = xy$. Since the point $(x, y)$ must always be on the graph defined by $y = 18 - 0.5x^2$, we can substitute $18 - 0.5x^2$ for *y* in the area formula, giving a function for the area in terms of *x* alone.

$$A = xy \qquad \text{formula for area}$$
$$A(x) = x(18 - 0.5x^2) \qquad \text{substitute } 18 - 0.5x^2 \text{ for } y$$
$$= 18x - 0.5x^3 \qquad \text{function in one variable}$$

Because the rectangle must be in Quadrant I, we are restricted to values of *x* where $0 < x < 6$.

**b.** To find the area of the rectangle when $x = 5$, we substitute 5 for $x$ and obtain

$$A(5) = 18(5) - 0.5(5)^3 \quad \text{substitute 5 for } x$$
$$= 90 - 62.5 \quad \text{simplify}$$
$$= 27.5 \quad \text{result}$$

When $x = 5$ the rectangle has an area of 27.5 units$^2$.

**c.** For the maximum area, we must find where $A(x) = 18x - 0.5x^3$ reaches a maximum value for $0 < x < 6$. After setting a viewing window accordingly (Figure 3.53), we can use the graph of $A$ to find the maximum value in the specified interval. Using the [2nd] [TRACE] **(CALC) 4:maximum** feature, we find that a maximum area of $A \approx 41.57$ units$^2$ occurs when $x \approx 3.46$ (Figure 3.54).

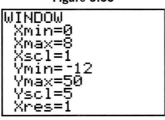

**Figure 3.53**

```
WINDOW
 Xmin=0
 Xmax=8
 Xscl=1
 Ymin=-12
 Ymax=50
 Yscl=5
 Xres=1
```

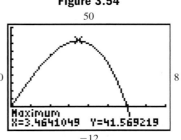

**Figure 3.54**

Maximum
X=3.4641049  Y=41.569219

Now try Exercises 25 and 26 ▶

Substitutions like that shown in Example 1 appear in a variety of ways, often making the resulting function easier to work with, or providing the means to answer questions that would be virtually impossible to answer otherwise. In Example 2, we'll try to minimize the distance between two points. Here it appears as an algebraic exercise. In the real world, minimizing distances can save energy, time, fuel, and/or money.

**EXAMPLE 2** ▶ **Minimizing Distances**

The graph of the parabola defined by $f(x) = 9 - x^2$ is shown.

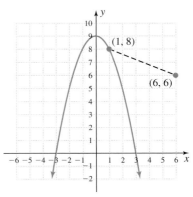

**a.** Find a function of a single variable that models the distance between the point $(6, 6)$ and a point $(x, y)$ on this parabola.

**b.** Use the function to find the distance between $(6, 6)$ and the point $(1, 8)$ on the parabola.

**c.** Use the function to find the point on the parabola that is the *minimum distance* from $(6, 6)$.

**Solution** ▶ **a.** The formula for the distance between two points is $d = \sqrt{(x_2 - x_1)^2 + (y_2 - y_1)^2}$. The distance between $(6, 6)$ and an arbitrary point $(x, y)$ is then given by $d = \sqrt{(x - 6)^2 + (y - 6)^2}$, with the distance expressed in terms of the two variables $x$ and $y$. But here, the point $(x, y)$ must always be on the graph defined by $y = 9 - x^2$, and we can substitute $9 - x^2$ for $y$ in the formula, giving a function for the distance in terms of $x$ alone.

$$d = \sqrt{(x - 6)^2 + (y - 6)^2} \quad \text{distance formula}$$
$$d(x) = \sqrt{(x - 6)^2 + [(9 - x^2) - 6]^2} \quad \text{substitute } 9 - x^2 \text{ for } y$$
$$= \sqrt{(x - 6)^2 + (3 - x^2)^2} \quad \text{function in one variable}$$

**b.** To find the distance between the point (6, 6) and the point (1, 8) on the parabola, we simply substitute $x = 1$ (the $x$-coordinate of the point) into $d(x)$, *since the function is now written in terms of x alone.*

$$
\begin{aligned}
d(1) &= \sqrt{(1 - 6)^2 + (3 - 1^2)^2} && \text{substitute 1 for } x \\
&= \sqrt{25 + 4} && \text{simplify} \\
&= \sqrt{29} && \text{result}
\end{aligned}
$$

The distance between the points (1, 8) and (6, 6) is $\sqrt{29} \approx 5.4$ units.

**c.** For the point on the parabola that is a *minimum* distance from (6, 6), we must find where $d(x) = \sqrt{(x - 6)^2 + (3 - x^2)^2}$ reaches a minimum value. From the graph of $f$ it appears this will occur when $0 < x < 3$, and we set a viewing window accordingly (Figure 3.55). We then use the graph of $d$ to find the minimum value using the (2nd) (TRACE) **(CALC) 3:minimum** feature. The result shows a minimum distance of $d \approx 4.1$ units occurs when $x = 2$ (Figure 3.56). For $x = 2$, the point on $f(x) = 9 - x^2$ is (2, 5), and this is the point that is a minimum distance from (6, 6).

Figure 3.55

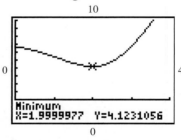

Figure 3.56

☑ **A.** You've just seen how we can create functions using relationships defined on the coordinate plane

Now try Exercises 27 and 28 ▶

## B. Functions and Geometric Relationships

The techniques used to write related functions in terms of a single variable can also be applied to geometric formulas. The resulting function will often provide needed information.

**EXAMPLE 3** ▶ **Finding the Volume of Water in a Conical Tank**

The water tank for a small village is simply a large inverted circular cone, with base radius $r = 2$ m and height $h = 5$ m. Finding the volume of water when the tank is full poses no problem, as $r = 2$ and $h = 5$ are simply substituted into the formula $V = \frac{1}{3}\pi r^2 h$. During the dry season, finding the volume is more problematic, as the radius at the water's surface cannot be measured directly even though the depth of the water is easily found.

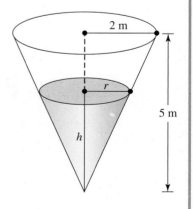

**a.** Find a function that will model the volume of water for any height $h$.

**b.** Find the volume when $h = 3.25$ m.

**c.** On emergency rations, the village consumes 0.3 m³ of water per day. How many days will the water last if it is not replenished?

**Solution** ▶   **a.** Similar to our earlier work, we try to find a relationship between the two objects involved, in this case the cone when it's full and the smaller cone formed as the water is depleted. Here we note the triangle formed by the height, radius, and side of the tank is similar to the triangle formed by the height and radius of the water. Using a proportional relationship, we can solve for $r$ in terms of $h$ and make a substitution.

$$\frac{\text{depleted radius}}{\text{full radius}} = \frac{\text{depleted height}}{\text{full height}} \qquad \text{similar triangles (sides are proportional)}$$

$$\frac{r}{2} = \frac{h}{5} \qquad \text{substitute known values}$$

$$r = \frac{2h}{5} \qquad \text{solve for } r \text{ (multiply by 2)}$$

Substituting for $r$ in the volume formula gives

$$V = \frac{1}{3}\pi r^2 h \qquad \text{volume formula}$$

$$V(h) = \frac{1}{3}\pi\left(\frac{2h}{5}\right)^2 h \qquad \text{substitute } \frac{2h}{5} \text{ for } r$$

$$= \frac{4\pi h^3}{75} \qquad \text{function in one variable}$$

**b.** To find the volume when $h = 3.25$ m, we substitute 3.25 for $h$.

$$V(3.25) = \frac{4\pi(3.25)^3}{75} \qquad \text{substitute 1 for } x$$

$$\approx 5.75 \qquad \text{result}$$

When the height of the water is 3.25 m, there is about 5.75 m$^3$ of water left in the tank.

**c.** With 5.75 m$^3$ of water available and a consumption rate of 0.3 m$^3$/day, the water will only last for $\frac{5.75}{0.3} \approx 19$ days.

**Now try Exercises 29 through 32** ▶

Once basic ideas are in place and the goal is clear, looking for the relationships needed to make an appropriate substitution becomes an easier task. Note how the same idea is applied in Example 4, but in a very different context.

**EXAMPLE 4** ▶   **Maximizing the Volume of a Cylinder as Part of a Marketing Strategy**

The marketing department of a large company wants to package a toy telescope by placing it in a cylinder, and displaying the cylinder inside a spherical package.

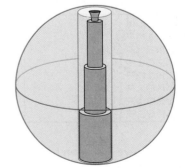

   **a.** Find a function that models the volume of any cylinder that will tightly fit into a sphere with fixed radius $R$.

   **b.** If $R = 15$ cm, what is the volume of a cylinder with height $h = 10$ cm?

   **c.** What is the maximum volume for a cylinder that will fit snugly within the sphere?

**Solution ▶**

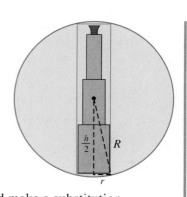

**a.** Here we attempt to find a relationship between the volume of a sphere, $V = \frac{4}{3}\pi R^3$ (with fixed radius $R$), and the volume of a cylinder, $V = \pi r^2 h$. Using the observations gleaned from the previous examples, we attempt to find a connection between $R$ and $r$ so that a substitution can be made. Note that if we draw a radius of the sphere to any "corner" of the cylinder, it acts as the hypotenuse of a right triangle with a height of $\frac{h}{2}$ and *the radius of the cylinder as the base.* Using the Pythagorean theorem, we can solve for $r^2$ in terms of $R^2$ and make a substitution.

$$a^2 + b^2 = c^2 \qquad \text{Pythagorean theorem}$$

$$r^2 + \left(\frac{h}{2}\right)^2 = R^2 \qquad \text{substitute } r \text{ for } a, \frac{h}{2} \text{ for } b, \text{ and } R \text{ for } c$$

$$r^2 = R^2 - \left(\frac{h}{2}\right)^2 \qquad \text{solve for } r^2$$

Prior to substituting $R^2 - \left(\frac{h}{2}\right)^2$ for $r^2$ in the formula for the cylinder, it will

help to write it as a single term: $R^2 - \dfrac{h^2}{4} = \dfrac{4R^2 - h^2}{4}$.

$$V = \pi r^2 h \qquad \text{volume formula}$$

$$= \pi\left(\frac{4R^2 - h^2}{4}\right)h \qquad \text{substitute } \frac{4R^2 - h^2}{4} \text{ for } r^2$$

**b.** Since we know $R = 15$ cm, the formula can be simplified further and written as a function of a single variable.

$$V(h) = \pi\left[\frac{4(15)^2 - h^2}{4}\right]h \qquad \text{substitute 15 for } R$$

$$= \frac{\pi}{4}(900h - h^3) \qquad \text{simplify}$$

To find the volume when $h = 10$ cm, we substitute 10 for $h$.

$$V(10) = \frac{\pi}{4}\left[900(10) - (10)^3\right] \qquad \text{substitute 10 for } h$$

$$\approx 6283 \qquad \text{result}$$

When the cylinder's height is 10 cm, it has a volume of about 6283 cm$^3$.

**c.** To find the maximum volume, we must find where $V(h) = \frac{\pi}{4}(900h - h^3)$ reaches a maximum value. Since $h$ can't exceed $2R = 2(15) = 30$, this value must occur for $0 < h < 30$. After setting a viewing window accordingly (Figure 3.57), we can use the graph of $V$ to find the maximum value in the specified interval. Using the **2nd** **TRACE** (CALC) **4:maximum** feature, we find that a maximum volume of $V \approx 8162$ cm$^3$ occurs when the height of the cylinder is $h \approx 17.3$ cm (Figure 3.58). Substituting 8162 for $V$ and 17.3 for $h$ in $V = \pi r^2 h$ shows the radius will be about 12.2 cm.

**Figure 3.57**

```
WINDOW
 Xmin=0
 Xmax=30
 Xscl=1
 Ymin=-2500
 Ymax=10000
 Yscl=1000
 Xres=1
```

**Figure 3.58**

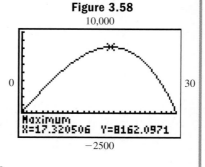

☑ **B.** You've just seen how we can create functions using geometric relationships

Now try Exercises 33 and 34 ▶

## C. Functions and Basic Formulas

The final two examples show how functions of a single variable can be constructed by combining very basic concepts and formulas. In the first example, the formulas for the area of a circle and the area of a square are used ($A = \pi r^2$ and $A = s^2$ respectively).

**EXAMPLE 5** ▶ **Making Jewelry for an Arts and Crafts Show**

Julia wants to fashion some new, asymmetric earrings for an upcoming crafts show. She intends to cut a 12-cm length of colored wire into two pieces, using one piece to make a circular earring and the other to make a square earring.

**a.** Find a function of $x$ that will model the area for each.

**b.** Where should she make the cut if she wants the areas of the earrings to be equal?

**c.** Where should she make the cut so that the total area of both earrings is a minimum?

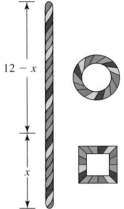

12 − x

x

**Solution** ▶ **a.** Begin by making the problem visual: "… a 12-cm length of wire will be cut into two pieces." If we let $x$ represent the distance we cut from one end, the remaining length of wire is $12 - x$. If we form the square from the first piece, each side will have length $\frac{x}{4}$ and the formula for its area will be $A_1 = \left(\frac{x}{4}\right)^2$. There are $(12 - x)$ cm left for the circle, and since this length will form the circumference we have $12 - x = 2\pi r$. To write the area function in terms of $x$ alone, we solve for $r$ and substitute.

$$12 - x = 2\pi r \qquad \text{circumference formula}$$

$$\frac{12 - x}{2\pi} = r \qquad \text{divide by } 2\pi$$

Substituting this result into the area formula gives

$$A_2 = \pi r^2 \qquad \text{volume formula}$$

$$= \pi \left(\frac{12 - x}{2\pi}\right)^2 \qquad \text{substitute } \tfrac{12 - x}{2\pi} \text{ for } r$$

**b.** To find where the cut should be made for equal areas, we set the area formulas equal to one another, and solve using the intersection of graphs method.

$$A_1 = A_2 \qquad \text{equal areas}$$

$$\left(\frac{x}{4}\right)^2 = \pi\left(\frac{12 - x}{2\pi}\right)^2 \qquad \text{equal areas}$$

Noting that $0 < x < 12$ is a constraint, we set a viewing window accordingly (Figure 3.59). Using the 2nd TRACE (**CALC**) **5:intersect** feature, we find the areas will be equal if the cut is made about 6.4 cm from one end and this piece is used to make the square earring (Figure 3.60).

**Figure 3.59**

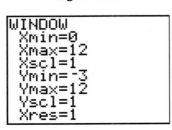

**Figure 3.60**

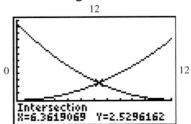

**c.** To find the value of $x$ that will minimize the total area, we note the total is $A = A_1 + A_2$, giving $A = \pi(\frac{12 - x}{2\pi})^2 + (\frac{x}{4})^2$, a function of the single variable $x$.

$$A(x) = \pi\left(\frac{12 - x}{2\pi}\right)^2 + \left(\frac{x}{4}\right)^2 \quad \text{total area}$$

Using $\boxed{\text{2nd}}$ $\boxed{\text{TRACE}}$ **(CALC) 3:minimum** feature (and the earlier constraints repeated in Figure 3.61), we find a minimum area of $A \approx 5.0 \text{ cm}^2$ occurs when the cut is made $x \approx 6.7$ cm from one end, and the square earring is made from this piece (Figure 3.62).

**Figure 3.61**

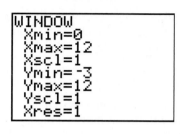

**Figure 3.62**

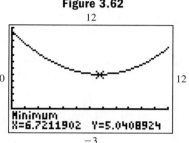

**Now try Exercises 35 through 38 ▶**

In Example 6, a combination of the Pythagorean theorem and the formula for uniform motion $(D = RT)$ is used to model the time it would take to respond to a crisis.

**EXAMPLE 6 ▶    Time Required to Respond to a Distress Call**

Mack is swimming 20 yd out from a straight shoreline when he hears cries for help coming from a campsite 100 yd down shore. Anxious to help, he swims at an angle toward the shore, in the general direction of the campsite. If he can swim 2 yd/sec and run 5 yd/sec,

**a.** Find a function that will model the time it takes for him to arrive at the campsite to help.

**b.** Find the time it will take him to get there if there are 52 yd left for him to run after he makes it to shore.

 **c.** How far down shore should he leave the water in order to reach the campsite in a *minimum* amount of time?

**Solution ▶    a.** We begin by making the problem visual (Figure 3.63). We know that (1) Mack is 20 yd from a straight shoreline, (2) the campsite is 100 yd down shore, (3) he is swimming to shore at an angle toward the campsite, and (4) he will run the rest of the way. If we let $x$ represent the distance down shore where Mack leaves the water, the distance to run must be $100 - x$.

**Figure 3.63**

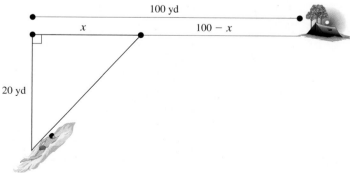

Using the Pythagorean theorem and the right triangle formed, we can find the distance that Mack covered on his swim to shore.

$$c^2 = a^2 + b^2$$         Pythagorean theorem

$$c^2 = 20^2 + x^2$$        substitute 20 for $a$ and $x$ for $b$

$$c = \sqrt{400 + x^2}$$        solve for $c$ ($c > 0$)

From the formula $T = \frac{D}{R}$ and Mack's swimming speed of 2 yd/sec, the time swimming is $T_S = \frac{\sqrt{400 + x^2}}{2}$. Likewise, a running speed of 5 yd/sec results in a running time of $T_R = \frac{100 - x}{5}$. The function representing the total time it would take him to arrive at the campsite is $T(x) = T_S + T_R = \frac{\sqrt{400 + x^2}}{2} + \frac{100 - x}{5}$.

**b.** If Mack has 52 yd left to run, he must have landed at a point 48 yd down shore, so $x = 48$.

$$T(48) = \frac{\sqrt{400 + 48^2}}{2} + \frac{100 - 48}{5}$$        evaluate $T(x)$ at $x = 48$

$$= \frac{\sqrt{2704}}{2} + \frac{52}{5}$$        simplify

$$= \frac{52}{2} + 10.4$$        simplify

$$= 26 + 10.4$$        divide

$$= 36.4$$        result

It will take Mack just over 36 sec to reach the campsite.

**c.** To find the distance $x$ down shore that will minimize the time needed to reach the campsite, we must find where $T$ reaches a minimum value. Since Mack can run much faster than he can swim, it seems reasonable that he should cover most of the distance running, so we'll use $0 < x < 50$ to set a viewing window (Figure 3.64). We then use the graph of $T$ to find the minimum value using the (2nd) (TRACE) (CALC) **3:minimum** feature. The result shows a minimum time of $T \approx 29$ sec occurs when $x \approx 8.7$ (Figure 3.65). Mack should swim to a point on shore that leaves about 91 yd left to run.

**Figure 3.64**

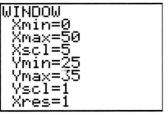

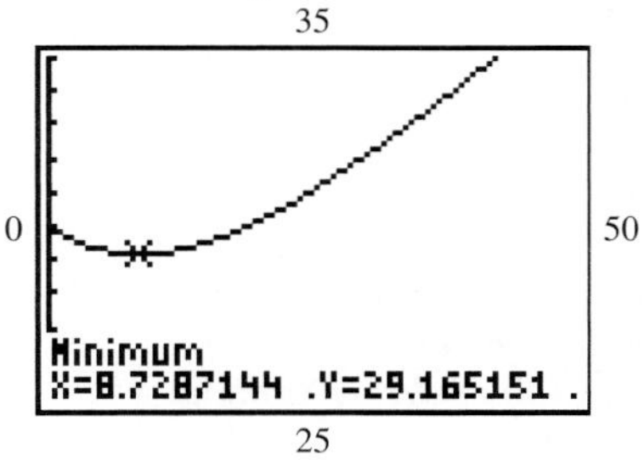

☑ **C.** You've just seen how we can create functions by combining basic formulas

**Figure 3.65**

Minimum
X=8.7287144 .Y=29.165151 .

Now try Exercises 39 and 40 ▶

## 3.6 EXERCISES

### ▶ CONCEPTS AND VOCABULARY

**Fill in each blank with the appropriate word or phrase. Carefully reread the section, if necessary.**

**1.** If the point (3, 12) is the vertex of a parabola with $a < 0$, we say that a _____ value of _____ occurs at $x = 3$.

**2.** If $c$ and $x \neq c$ are in an open interval, and $f(c) \leq f(x)$ for all $c$ and $x$ in the interval, then $f(c)$ is called a(n) _____ _____.

**3.** Discuss/Explain how $y = 2$ can be a minimum value, while $y = 1$ is a maximum value (see figure).

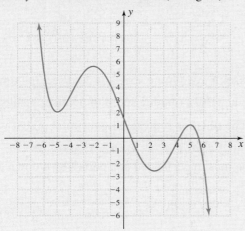

**4.** A square with sides of $x$ cm is inscribed in a circle. Discuss/Explain how you would express the area of the circle in terms of $x$.

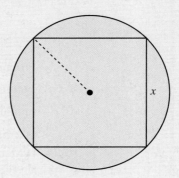

## ▶ DEVELOPING YOUR SKILLS

**5.** Given that point $(x, y)$ is on the graph of $y = 4 - x^2$, express the distance from $(3, 4)$ to $(x, y)$ as a function of $x$.

**6.** Given that point $(x, y)$ is on the graph of $y = \sqrt{x} + 5$, express the distance from $(2, 5)$ to $(x, y)$ as a function of $x$.

**7.** The area of a rectangle is 50 cm$^2$. Express the perimeter of the rectangle as a function of a single variable.

**8.** The area of a right triangle is 15 cm$^2$. Express the perimeter of the triangle as a function of a single variable.

**A triangle is drawn in the first quadrant, with a vertex at $(0, 0)$, its base along the $y$-axis and another vertex at a point $(x, y)$ on the graph of the polynomial shown (see figure).**

**9.** If the polynomial is defined by the equation $y = x^4$, write a function for the area of the triangle in terms of $x$.

**10.** If the polynomial is defined by the equation $y = 2x^6$, write a function for the area of the triangle in terms of $x$.

**A triangle is drawn in Quadrant I, with a vertex at $(0, 0)$, its base along the $x$-axis and another vertex at a point $(x, y)$ on the graph of the parabola (see figure).**

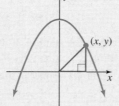

**11.** If the parabola is defined by the equation $y = 8 - 2x^2$, write a function for the area of the triangle in terms of $x$.

**12.** If the parabola is defined by the equation $y = 9 - 4x^2$, write a function for the area of the triangle in terms of $x$.

**A rectangle is drawn inside a semicircle, with one side along the $x$-axis and vertices at $(-x, y)$ and $(x, y)$ on the graph.**

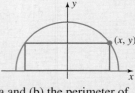

**13.** If the equation of the semicircle is $y = \sqrt{9 - x^2}$, write a function for (a) the area and (b) the perimeter of the rectangle in terms of $x$.

**14.** If the equation of the semicircle is $y = \sqrt{7 - x^2}$, write a function for (a) the area and (b) the perimeter of the rectangle in terms of $x$.

**A right triangle is drawn in Quadrant I with legs along the $x$- and $y$-axes. A rectangle is drawn inside the triangle with one vertex on the hypotenuse of the triangle (see figure).**

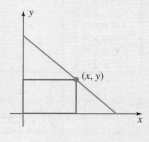

**15.** Write a function for the area of the rectangle if the legs of the triangle are 5 units ($y$) and 12 units ($x$) in length. What is the domain of this area function?

**16.** Write a function for the area of the rectangle if the legs of the triangle are 7 units ($y$) and 24 units ($x$) in length. What is the domain of this area function?

**17.** A rectangle is inscribed in a circle of radius 5 cm, as shown. Write a function for the area of the rectangle in terms of a single variable $x$, where $(x, y)$ is the first quadrant corner point.

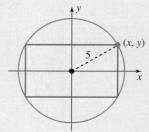

**18.** An isosceles triangle is inscribed in a circle of radius 7 cm, as shown. Write a function for the area of the triangle in terms of the single variable $x$, where $(x, y)$ is the first quadrant corner point.

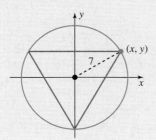

**19.** A cylinder is inscribed in a right circular cone with height of 24 cm and a base radius of 8 cm. Write a function for the volume of the cylinder in terms of the single variable $r$, where $r$ is the radius of the cylinder.

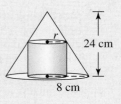

**20.** A cone is inverted and inscribed in another (larger) cone. The larger cone has a height of 36 cm and a radius of 9 cm. Write a function for the volume of the smaller cone in terms of the single variable $r$, where $r$ is the radius of the smaller cone.

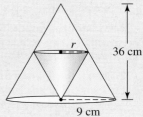

## ▶ WORKING WITH FORMULAS

**21.** From memory or by process of elimination, match each formula to the related geometric figures shown. For some formulas, more than one match is possible.

a. $A = LW$  
b. $V = \pi r^2 h$  
c. $V = LWH$  
d. $SA = 2\pi r^2 + 2\pi rh$  

e. $V = \dfrac{1}{3}\pi r^2 h$  
f. $A = \dfrac{bh}{2}$  
g. $P = 2L + 2W$  
h. $P = a + b + \sqrt{a^2 + b^2}$  

i. $A = \pi r^2$  
j. $SA = 4\pi r^2$  
k. $P = 4s$  
l. $SA = \pi r\sqrt{r^2 + h^2}$  

m. $a^2 + b^2 = c^2$  
n. $C = 2\pi r$  
o. $A = s^2$  
p. $V = \dfrac{4}{3}\pi r^3$  

q. $A = \dfrac{h}{2}(b_1 + b_2)$  
r. $A = \dfrac{\sqrt{3}}{4}s^2$  
s. $d = 2r$  
t. $SA = 2(LW + LH + WH)$  

**1.**            **2.**            **3.**            **4.**            **5.**

**6.**            **7.**            **8.**            **9.**            **10.**

**22.** Solve for $h$.  
a. $V = lwh$  
b. $V = \pi r^2 h$  
c. $SA = 2\pi r^2 + 2\pi rh$  
d. $A = \frac{h}{2}(b_1 + b_2)$

**23.** Solve for $h$.  
a. $A = \frac{bh}{2}$  
b. $V = \frac{1}{3}\pi r^2 h$  
c. $SA = \pi r\sqrt{r^2 + h^2}$  
d. $SA = 2(lw + lh + wh)$

**24.** Solve for $r$.  
a. $C = 2\pi r$  
b. $A = \pi r^2$  
c. $V = \frac{1}{3}\pi r^2 h$  
d. $V = \frac{4}{3}\pi r^3$

## ▶ APPLICATIONS

**25. Maximizing area:** Find the dimensions of the largest rectangle that can be inscribed in the semicircle defined by $y = \sqrt{9 - x^2}$.

**26. Maximizing area:** A rectangle has one side along the $x$-axis, and two vertices on the graph of $y = 15 - x^2$ (in Quadrants I and II). Find the dimensions of the rectangle with the largest area possible.

**27. Minimizing distance:** Find the point $(x, y)$ on the graph of $y = 4 - x^2$ that is a minimum distance from the point $(-7, 6)$.

**28. Minimizing distance:** Find the point $(x, y)$ on the graph of $y = \sqrt{x} + 5$ that is a minimum distance from the point $(3, 11)$.

**29. Maximizing area:** A rectangle is inscribed in a right triangle that has legs 8 cm and 15 cm in length. Find the dimensions of the rectangle with the largest area possible, given the length and width of the rectangle lie on the legs of the triangle.

**30. Maximizing area:** Find the dimensions of a rectangle with the largest area possible that can be inscribed within a semicircle of radius 12 in.

**31. Maximizing area:** Find the dimensions of a rectangle with the largest area possible that can be inscribed within a circle of radius 12 in.

**32. Maximizing area:** Find the dimensions of an isosceles triangle with the largest area possible that can be inscribed in a circle of radius $r = 10$ cm.

**33. Maximizing volume:** Determine the maximum volume of a cylinder that can be inscribed within a right circular cone, if the cone is 18 cm high with a base radius of $r = 6$ cm.

**34. Maximizing volume:** Determine the maximum volume of a right circular cone that can be inscribed within a larger cone if the smaller cone is inverted and the larger cone has a height of 24 cm and a base radius of $r = 8$ cm.

**35. Jewelry for pirates:** Captain Jack Sparrow has commissioned Julia to create some asymmetric earrings for his next movie. He instructs her to cut a 15-cm length of colored wire into two pieces, using one piece to make an equilateral triangular earring and the other to make a square earring. (a) If Jack wants the area of each earring to be the same, where should the cut be made? (b) If Jack wants the total area of both earrings to be a minimum, where should the cut be made?

**36. Protecting a crime scene:** Jethro Gibbs needs to mark out a crime scene using  yellow police tape. He has 100 yd of tape left, and needs to mark a circular area around where the body was found, and a large square area around where the bullet casings were found. (a) Where should the tape be cut if each enclosed area must be equal? (b) If Jethro decides the total area enclosed by both should be a minimum, where should the tape be cut?

**37. Barrels of oil:** In the construction of the cylindrical metal drums used to store crude oil, two circular metal pieces are used for the top and bottom, and a rectangular sheet is rolled into a cylinder and welded to form the sides. If the circular pieces have radius $r$, and the rectangular piece has height $h$, (a) write the formula for the surface area of a barrel. (b) In the United States these barrels are referred to as 55-gal drums (the volume is 55 gal). Use this information to write a function for the surface area of a barrel in terms of a single variable. (*Hint:* 1 gal = 231 in$^3$.) (c) Find the dimensions of a barrel that will minimize the amount of metal used to make the barrel.

**38. Storing agricultural products:** While the silos popularized in American agriculture are used to store grain, silos are also commonly used for bulk storage of coal, cement, food products, wood products, and other items. In many cases these silos are large, domed cylinders. Assuming the dome is a hemisphere with radius $r$, (a) write a formula for the total surface area of this silo (including the base). (b) If the volume of the cylinder alone is 150,000 ft$^3$, write a function for the total surface area in terms of a single variable. (c) Find the dimensions of a silo that will minimize the amount of metal used in its construction.

**39. Minimizing orienteering time:** Suzanne is orienteering in the field and is going for the next marker (see figure). She could run along a clear path and follow the road to the marker, or save time by cutting through the woods and underbrush to reach the marker. If she can run 7 yd/sec in the open but only 4 yd/sec through the underbrush, (a) find a function that will model the time it takes for her to arrive at the marker. (b) At what point should she leave the woods to minimize the time required to reach the marker? How many yards are left to run?

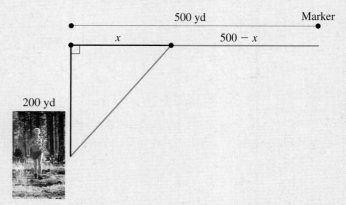

**40. Minimizing mission time:** On a reconnaissance mission, the team leader realizes they can travel along the roadways to reach the pickup location, or cut through the town. The reconnaissance team can move 5 mph on the roads, but is slowed to 3 mph when going through town in order to avoid detection. (a) Find a function that will model the time it takes the team to arrive at the pickup location. (b) At what point should the team leave the town to minimize the time required to reach the pickup location? How many miles are they from this location?

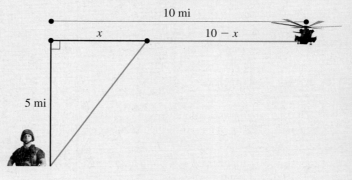

▶ **EXTENDING THE CONCEPT**

**41.** The marketing department of a sporting goods manufacturer is going to package their new play ball in an attractive conical package. (a) If the play ball has a radius of 15 cm, find a formula for the volume of the cone in terms of the single variable $h$, where $h$ is the height of the cone. (b) Find the dimensions of the right circular cone that will have a minimum volume.

**42.** A Norman window is shaped like a semicircle on a rectangle. (a) If the perimeter of the window is 170 in., find a formula for the area of the window in terms of the single variable $x$, where $x$ is the diameter of the semicircle. (b) Find the dimensions of the window that give a maximum area (hence letting in a maximum amount of light).

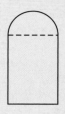

▶ **MAINTAINING YOUR SKILLS**

**43.** (1.5) Solve the quadratic equation by completing the square: $x^2 + 11 = 8x$.

**44.** (3.2) Find the zeroes of the function and the location of the horizontal and vertical asymptotes:
$v(x) = \frac{1}{(x - 2)^2} - 1$.

**45.** (3.3) The sales of an item varies directly with its popularity rating and inversely with its price. At a rating of 7 (out of 10) and a cost of $58, the SolarRay Phone sold 49 thousand units. How many phones would be sold if the rating rose to 9 but the cost rose to $84?

**46.** (3.1) Draw the graph of $y = -|x + 1| + 3$ using transformations of a basic function. Clearly state the transformations used.

---

## MAKING CONNECTIONS

## Making Connections: Graphically, Symbolically, Numerically, and Verbally

Eight graphs (a) through (h) are given. Match the characteristics shown in 1 through 16 to one of the eight graphs.

**(a)** 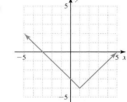   **(b)** 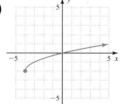   **(c)** 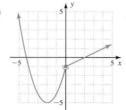   **(d)**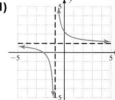

**(e)** 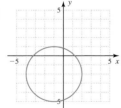   **(f)** 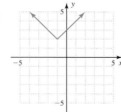   **(g)** 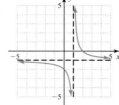   **(h)**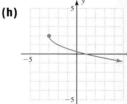

**1.** _____ domain: $x \in (-\infty, 1) \cup (1, \infty)$   **2.** _____ $y = \sqrt{x + 4} - 2$   **3.** _____ $f(x)\uparrow$ for $x \in (1, \infty)$

**4.** _____ horizontal asymptote at $y = -1$   **5.** _____ $y = \frac{1}{x + 1} + 1$   **6.** _____ domain: $x \in [-4, 2]$

**7.** _____ $y = |x - 1| - 4$   **8.** _____ $f(x) \leq 0$ for $x \in [1, \infty)$   **9.** _____ $f(-3) = -4, f(2) = 0$

**10.** _____ domain: $x \in [-4, \infty)$   **11.** _____ as $x \to \infty, y \to 1$   **12.** _____ $f(x) > 0$ for $x \in (-\infty, \infty)$

**13.** _____ basic function is shifted 1 unit left, 2 units up   **14.** _____ $y = \begin{cases} (x + 2)^2 - 5, & x < 0 \\ \frac{1}{2}x - 1, & x \geq 0 \end{cases}$   **15.** _____ $f(-3) = -1, f(5) = 1$

**16.** _____ basic function is shifted 3 units left, reflected across $x$-axis, then shifted up 2 units

## SUMMARY AND CONCEPT REVIEW

### SECTION 3.1    The Toolbox Functions and Transformations

#### KEY CONCEPTS

- The *toolbox functions* and graphs commonly used in mathematics are
  - the identity function: $f(x) = x$
  - square root function: $f(x) = \sqrt{x}$
  - cubing function: $f(x) = x^3$
  - squaring function: $f(x) = x^2$
  - absolute value function: $f(x) = |x|$
  - cube root function: $f(x) = \sqrt[3]{x}$
- For a basic or parent function $y = f(x)$, the general equation of the transformed function is $y = af(x \pm h) \pm k$. For any function $y = f(x)$ and $h, k > 0$,
  - the graph of $y = f(x) + k$ is the graph of $y = f(x)$ shifted upward $k$ units
  - the graph of $y = f(x + h)$ is the graph of $y = f(x)$ shifted left $h$ units
  - the graph of $y = -f(x)$ is the graph of $y = f(x)$ reflected across the $x$-axis
  - $y = af(x)$ results in a vertical stretch when $a > 1$
  - the graph of $y = f(x) - k$ is the graph of $y = f(x)$ shifted downward $k$ units
  - the graph of $y = f(x - h)$ is the graph of $y = f(x)$ shifted right $h$ units
  - the graph of $y = f(-x)$ is the graph of $y = f(x)$ reflected across the $y$-axis
  - $y = af(x)$ results in a vertical compression when $0 < a < 1$
- Transformations are applied in the following order: (1) horizontal shifts, (2) reflections, (3) stretches or compressions, and (4) vertical shifts.

#### EXERCISES

Identify the function family for each graph given, then identify the (a) end-behavior; (b) intercepts; (c) vertex, initial point, or point of inflection (as applicable); and (d) domain and range.

**1.**
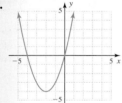

**2.**
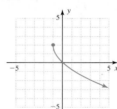

**3.**
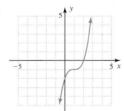

**4.**
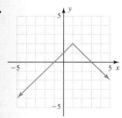

**5.**
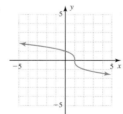

Identify the function family, then sketch the graph using shifts of a parent function and a few characteristic points.

**6.** $f(x) = -(x + 2)^2 - 5$       **7.** $f(x) = 2|x + 3|$       **8.** $f(x) = x^3 - 1$

**9.** $f(x) = \sqrt{x - 5} + 2$       **10.** $f(x) = \sqrt[3]{x + 2}$

**11.** Apply the transformations indicated for the graph of $f(x)$ given.

**a.** $f(x - 2)$
**b.** $-f(x) + 4$
**c.** $\frac{1}{2}f(x)$

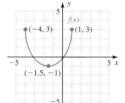

## SECTION 3.2   Basic Rational Functions and Power Functions

### KEY CONCEPTS

- A rational function is one of the form $V(x) = \frac{p(x)}{d(x)}$, where $p$ and $d$ are polynomials and $d(x) \neq 0$.
- The most basic rational functions are the reciprocal function $f(x) = \frac{1}{x}$ and the reciprocal square function $g(x) = \frac{1}{x^2}$.
- The line $y = k$ is a horizontal asymptote of $V$ if, as $|x|$ increases without bound, $V(x)$ approaches $k$: as $|x| \to \infty$, $V(x) \to k$.
- The line $x = h$ is a vertical asymptote of $V$ if, as $x$ approaches $h$, $V(x)$ increases/decreases without bound: as $x \to h$, $|V(x)| \to \infty$.
- A power function can be written in the form $f(x) = x^p$ where $p$ is a constant real number and $x$ is a variable. If $p = \frac{1}{n}$, where $n$ is a natural number, $f(x) = x^{\frac{1}{n}} = \sqrt[n]{x}$ is called a root function in $x$.

### EXERCISES

Sketch the graph of each function using shifts of the parent function (not by using a table of values). Find and label the $x$- and $y$-intercepts (if they exist) and redraw the asymptotes.

**12.** $f(x) = \dfrac{1}{x + 2} - 1$

**13.** $h(x) = \dfrac{-1}{(x - 2)^2} - 3$

**14.** In a certain county, the cost to keep public roads free of trash is given by $C(p) = \frac{-7500}{p - 100} - 75$, where $C(p)$ represents the cost (in thousands of dollars) to keep $p$ percent of the trash picked up. (a) Find the cost to pick up 30%, 50%, 70%, and 90% of the trash, and comment on the results. (b) Sketch the graph using transformations of a toolbox function. (c) Use mathematical notation to describe what happens if the county tries to keep 100% of the trash picked up.

**15.** Use a graphing calculator to graph the functions $f(x) = x^{\frac{5}{3}}$, $g(x) = x^{\frac{1}{2}}$, and $h(x) = x^{\pi}$ in the same viewing window. What is the domain of each function?

**16.** The expression $T = \frac{2\pi}{37,840} r^{\frac{3}{2}}$ models the time $T$ (in hr) it takes for a satellite to complete one revolution around the Earth, where $r$ represents the radius (in km) of the orbit measured from the center of the Earth. If the Earth has a radius of 6370 km, (a) how long does it takes for a satellite at a height of 200 km to complete one orbit? (b) What is the orbital height of a satellite that completes one revolution in 4 days (96 hr)?

## SECTION 3.3   Variation: The Toolbox Functions in Action

### KEY CONCEPTS

- *Direct variation:* If there is a nonzero constant $k$ such that $y = kx$, we say, "$y$ varies directly with $x$" or "$y$ is directly proportional to $x$" ($k$ is called the constant of variation).
- *Inverse variation:* If there is a nonzero constant $k$ such that $y = k\left(\frac{1}{x}\right)$ we say, "$y$ varies inversely with $x$" or "$y$ is inversely proportional to $x$."
- The process for solving variation equations can be found on page 275.

### EXERCISES

Find the equation model, and use it to complete the table.

**17.** $y$ varies directly with the cube root of $x$;
$y = 52.5$ when $x = 27$.

**18.** $z$ varies directly with $v$ and inversely with the square of $w$; $z = 1.62$ when $w = 8$ and $v = 144$.

| $x$ | $y$ |
|-----|-----|
| 216 | |
| | 12.25 |
| 729 | |

| $v$ | $w$ | $z$ |
|-----|-----|-----|
| 196 | 7 | |
| | 1.25 | 17.856 |
| 24 | | 48 |

**19.** Suppose $t$ varies jointly with $u$ and $v$, and inversely with $w$. If $t = 30$ when $u = 2$, $v = 3$, and $w = 5$, find $t$ when $u = 8$, $v = 12$, and $w = 15$.

**20.** The time that it takes for a simple pendulum to complete one period (swing over and back) is directly proportional to the square root of its length. If a pendulum 16 ft long has a period of 3 sec, find the time it takes for a 36-ft pendulum to complete one period.

## SECTION 3.4   Piecewise-Defined Functions

### KEY CONCEPTS

- To evaluate a piecewise-defined function, identify the domain interval containing the input value, then use the piece of the function corresponding to this interval.
- To graph a piecewise-defined function you can plot points, or graph each piece in its entirety, then erase portions of the graph outside the domain indicated for each piece.
- If the graph of a function can be drawn without lifting your pencil from the paper, the function is continuous.
- A discontinuity is said to be removable if we can redefine the function to "fill the hole."

### EXERCISES

**21.** For the graph and functions given, (a) use the correct notation to write the relation as a single piecewise-defined function and (b) state the range of the function: $Y_1 = 5$, $Y_2 = -X + 1$, $Y_3 = 3\sqrt{X - 3} - 1$.

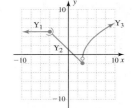

**22.** Use a table of values as needed to graph $h(x)$, then state its domain and range. If the function has a removable discontinuity, state how the second piece could be redefined so that a continuous function results.

$$h(x) = \begin{cases} \dfrac{x^2 - 2x - 15}{x + 3}, & x \neq -3 \\ -6, & x = -3 \end{cases}$$

**23.** Evaluate the piecewise-defined function $p(x)$: $p(-4)$, $p(-2)$, $p(2.5)$, $p(2.99)$, $p(3)$, and $p(3.5)$

$$p(x) = \begin{cases} -4, & x < -2 \\ -|x| - 2, & -2 \leq x < 3 \\ 3\sqrt{x} - 9, & x \geq 3 \end{cases}$$

**24.** Sketch the graph of the function and state its domain and range. Use transformations of the toolbox functions where possible.

$$q(x) = \begin{cases} 2\sqrt{-x - 3} - 4, & x \leq -3 \\ -2|x| + 2, & -3 < x < 3 \\ 2\sqrt{x - 3} - 4, & x \geq 3 \end{cases}$$

**25.** Many home improvement outlets now rent flatbed trucks in support of customers who purchase large items. The cost is $20/hr for the first 2 hr, $30/hr for the next 2 hr, then $40 for each hour afterward. Write this information as a piecewise-defined function, then sketch its graph. What is the total cost to rent this truck for 5 hr?

## SECTION 3.5   The Algebra and Composition of Functions

### KEY CONCEPTS

- The notation used to represent the basic operations on two functions is
  - $(f + g)(x) = f(x) + g(x)$
  - $(f - g)(x) = f(x) - g(x)$
  - $(f \cdot g)(x) = f(x) \cdot g(x)$
  - $\left(\dfrac{f}{g}\right)(x) = \dfrac{f(x)}{g(x)}$; $g(x) \neq 0$
- The result of these operations is a new function $h(x)$. The domain of $h$ is the intersection of domains for $f$ and $g$, excluding values that make $g(x) = 0$ in $h(x) = \left(\dfrac{f}{g}\right)(x)$.
- The composition of two functions is written $(f \circ g)(x) = f[g(x)]$ ($g$ is an input for $f$).
- The domain of $f \circ g$ is all $x$ in the domain of $g$, such that $g(x)$ is in the domain of $f$.

### EXERCISES

For $f(x) = x^2 + 4x$ and $g(x) = 3x - 2$, find the following:

**26.** $(f + g)(a)$

**27.** $(f \cdot g)(3)$

**28.** the domain of $\left(\dfrac{f}{g}\right)(x)$

Given $p(x) = 4x - 3$, $q(x) = x^2 + 2x$, and $r(x) = \frac{x + 3}{4}$ find:

**29.** $(p \circ q)(x)$

**30.** $(q \circ p)(3)$

**31.** $(p \circ r)(x)$ and $(r \circ p)(x)$

For each function here, find functions $f(x)$ and $g(x)$ such that $h(x) = f[g(x)]$:

**32.** $h(x) = \sqrt{3x - 2} + 1$

**33.** $h(x) = x^{\frac{2}{3}} - 3x^{\frac{1}{3}} - 10$

**34.** Use the graph given to find the value of each expression:

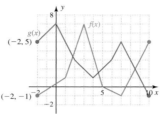

    **a.** $(f + g)(-2)$         **d.** $(f \cdot g)(3)$

    **b.** $(g - f)(7)$         **e.** $(f \circ g)(-2)$

    **c.** $\left(\dfrac{g}{f}\right)(10)$       **f.** $(g \circ f)(5)$

**35.** As the availability of free, public Wi-Fi has increased, so has the number of devices that can utilize this protocol. A new company has just released a Wi-Fi phone that provides free phone service anywhere it has access to a wireless network. The total cost for manufacturing these phones can be modeled by the function $C(n) = -0.002n^2 + 20n + 30,000$, where $n$ is the number of phones made and $C(n)$ is in dollars. If each phone is sold at a price of $84.95, the revenue is modeled by $R(n) = 84.95n$.

    **a.** Find the function that represents the total profit made from sales of the phones.

    **b.** How much profit is earned if 400 phones are sold?

    **c.** How much profit is earned if 5000 phones (the production limit) are sold?

    **d.** How many phone sales are necessary for the company to break even?

**36.** A stone is thrown into a pond causing a circular ripple to move outward from the point of entry. The radius of the circle is modeled by $r(t) = 2t + 3$, where $t$ is the time in seconds. Find a function that will give the area of the circle directly as a function of time. In other words, find $A(t)$.

## SECTION 3.6  Another Look at Formulas, Functions, and Problem Solving

### KEY CONCEPTS

- Many applications of algebra require that we determine how two different functions or formulas are related and combine them into a function of a single variable.
- If two quantities are equal, one can be substituted for the other without changing the solution set.
- As with other forms of problem solving, gathering all formulas and facts and making the problem visual using charts or diagrams are important parts of the process.

### EXERCISES

**37.** A rancher has 600 ft of fencing and wants to enclose a rectangular area for his goats. (a) Write a function that will model the area he can enclose in terms of the single variable $W$ (the width of the rectangle). (b) What dimensions will enclose a maximum area? (c) What is this maximum area?

**38.** A pyramid with square base $b$ sits atop a cube with dimensions $b \times b \times b$. (a) Write a formula for the total volume. (b) If the height of the pyramid is three times that of the base, write a function for the total volume in terms of a single variable. (c) Find the dimensions of the pyramid and the cube if a volume of 18,522 cm$^3$ is required.

## PRACTICE TEST

**1.** Each function graphed here is from a toolbox function family. For each graph, identify the (a) function family, (b) domain and range, (c) intercepts, (d) end-behavior, and (e) solve the inequalities $f(x) > 0$ and $f(x) < 0$.

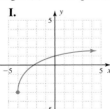
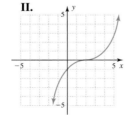
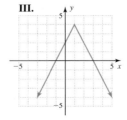
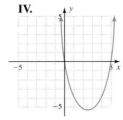

Sketch each graph using transformations.

    **2.** $f(x) = |x - 2| + 3$         **3.** $g(x) = -(x + 3)^2 - 2$

4. Sketch the graph of $f(x) = \frac{2}{x+3}$. Find and label the $x$- and $y$-intercepts, if they exist, along with all asymptotes.

5. Find the domains of the following functions.
   **a.** $f(x) = 2.09x^{\frac{2}{3}}$       **b.** $g(x) = -4.22x^{\frac{5}{2}}$
   **c.** $h(t) = 4.5t^{\pi}$

6. Identify the vertical and horizontal asymptotes of $g(x) = \frac{3}{(x+2)^2} - 1$.

7. Using time-lapse photography, the spread of a liquid is tracked as a small amount of it is dropped on a piece of fabric. (a) Graph a scatterplot of the data and (b) find an equation model using a power regression (round to two decimal places). Use the equation to estimate (c) the size of the stain after 0.5 sec and (d) how long it will take the stain to reach a size of 15 mm.

| Time (sec) | Size (mm) |
|---|---|
| 0.2 | 0.39 |
| 0.4 | 1.27 |
| 0.6 | 3.90 |
| 0.8 | 10.60 |
| 1.0 | 21.50 |

8. The following function has two removable discontinuities. Find the values of $a$ and $b$ so that a continuous function results.

$$g(x) = \begin{cases} \dfrac{x^3 + x^2 - 4x - 4}{x^2 - x - 2}, & x \neq -1, 2 \\ a, & x = -1 \\ b, & x = 2 \end{cases}$$

9. Given $h(x) = \begin{cases} 4, & x < -2 \\ 2x, & -2 \leq x \leq 2 \\ x^2, & x > 2 \end{cases}$
   **a.** Find $h(-3)$, $h(-2)$, and $h(\frac{5}{2})$.
   **b.** Sketch the graph of $h$. Label important points.

10. After starting the turkey-sandwich diet, the weight initially melted off Jay, then began to trickle off, and eventually stopped altogether. Use the graph shown to (a) model Jay's weight $W(t)$ after $t$ weeks on the diet by a piecewise-defined function. Then determine his (b) starting weight and (c) weight 6 weeks in.

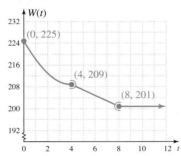

11. The output of a wind turbine varies jointly with the square of the blade diameter and the cube of the average wind speed. If a 10-ft-diameter turbine in 12-mph average winds produces 2300 kW·h/yr, how much will a 6-ft-diameter turbine produce in 15-mph average winds? Round to the nearest kW·h/yr.

12. According to Kepler's third law of planetary motion, the mass of a large celestial body is directly proportional to the cube of the average distance between it and an orbiting satellite, but inversely proportional to the square of the satellite's orbital period. Write the variation equation. Using the mean Earth/Sun distance of $1.496 \times 10^8$ km and the Earth's orbital period of 1 yr, the mass of the Sun has been calculated to be $1.98892 \times 10^{30}$ kg. Given the orbital period of Mars is 1.88 yr, find its mean distance from the Sun.

13. The maximum load that can be supported by a rectangular beam varies jointly with its width and the square of its height, but inversely with its length. If a beam 10 ft long, 3 in. wide, and 4 in. high can support 624 lb, how many pounds could a 12-ft-long beam with the same cross-section support?

14. A company's revenue and cost (in dollars) are given by the functions $R(x) = 300x$ and $C(x) = 20,000 + 100x$, where $x$ is the number of items produced and sold. Find a new function rule for the (a) profit function $P(x) = (R - C)(x)$ and (b) average profit function $\overline{P}(x) = \frac{P(x)}{x}$. Use the function in (b) to determine the average profit for (c) the first 50 items and (d) the first 200 items. Explain the meaning of these values. (e) What happens to the average profit as more and more items are produced and sold?

15. Given $f(x) = \sqrt{2 - x}$ and $g(x) = x - 1$, determine a new function rule for (a) $(f \cdot g)(x)$, (b) $(\frac{f}{g})(x)$, and (c) $(\frac{g}{f})(x)$. Specify the domain of new function.

16. Given $f(x) = x^2 + 2$ and $g(x) = \sqrt{3x - 1}$, determine $(f \circ g)(x)$ and its domain.

17. A snowball increases in size as it rolls downhill. The snowball is roughly spherical with a radius that can be modeled by the function $r(t) = \sqrt{t}$, where $t$ is time in seconds and $r$ is measured in inches. The volume of the snowball is given by the function $V(r) = \frac{4}{3}\pi r^3$. Use a composition to (a) express the volume directly as a function of time and (b) find the volume of the snowball after 9 sec.

18. Use a graphing calculator to find the maximum and minimum values of $f(x) = |x^2 + 4x - 11| - 7$. Round answers to the nearest hundredth when necessary.

19. Many of the silos in the Midwest are actually shaped like a cone on a cylinder. If the cylinder has height $H$ and the cone has height $h$, (a) find a formula for the volume of the silo in terms of $h$ alone, given $r = 12$ ft and the height of the cylinder is three times the height of the cone. (b) What is the volume of the silo if the cone is 15 feet tall?

20. Use the graphs given to compute (a) $(f + g)(2)$, (b) $(g - f)(-2)$, (c) $(g \div f)(1)$, and (d) $(f \circ g)(4)$.

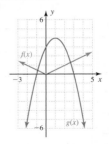

## CALCULATOR EXPLORATION AND DISCOVERY

### I. Function Families

Graphing calculators are able to display a number of graphs simultaneously, making them a wonderful tool for studying families of functions. Let's begin by entering the function $y = |x|$ [actually $y = abs(x)$ (MATH) ▶ **1:abs(** ] as $Y_1$ on the (Y=) screen. Next, we enter different variations of the function, but always in terms of its variable name "$Y_1$." This enables us to easily change the parent function, and observe how the changes affect the graph. Recall that to access the function name $Y_1$ press (VARS) ▶ (to access the Y-VARS menu) (ENTER) (to access the function variables menu) and (ENTER) (to select $Y_1$). Enter the functions $Y_2 = Y_1 + 3$ and $Y_3 = Y_1 - 6$ (see Figure 3.66). Graph all three functions in the (ZOOM) **6:ZStandard** window. The calculator draws each graph in the order they were entered and you can always identify the functions by pressing the (TRACE) key and then the up arrow (▲) or down arrow (▼) keys. In the upper left corner of the window shown in Figure 3.67, the calculator identifies which function the cursor is currently on. Most importantly, note that all functions in this family maintain the same "V" shape.

Next, change $Y_1$ to $Y_1 = abs(x - 3)$, leaving $Y_2$ and $Y_3$ as is. What do you notice when these are graphed again?

**Exercise 1:** Change $Y_1$ to $Y_1 = \sqrt{x}$ and graph, then enter $Y_1 = \sqrt{x - 3}$ and graph once again. What do you observe? What comparisons can be made with the translations of $Y_1 = abs(x)$?

**Exercise 2:** Change $Y_1$ to $Y_1 = x^2$ and graph, then enter $Y_1 = (x - 3)^2$ and graph once again. What do you observe? What comparisons can be made with the translations of $Y_1 = abs(x)$ and $Y_1 = \sqrt{x}$?

**Figure 3.66**

**Figure 3.67**

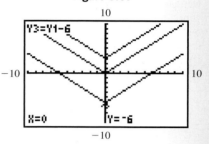

### II. Piecewise-Defined Functions

Most graphing calculators are able to graph piecewise-defined functions. Consider the function $f$ shown here:

$$f(x) = \begin{cases} x + 2, & x < 2 \\ (x - 4)^2 + 3, & x \geq 2 \end{cases}$$

Both "pieces" are well known—the first is a line with slope $m = 1$ and $y$-intercept $(0, 2)$. The second is a parabola that opens upward, shifted 4 units to the right and 3 units up. If we attempt to graph $f(x)$ using $Y_1 = x + 2$ and $Y_2 = (x - 4)^2 + 3$ as they stand, the resulting graph may be difficult to analyze because the pieces intersect and overlap (Figure 3.68). To graph the functions correctly we must indicate the domain for each piece, separated by a slash and enclosed in parentheses. For instance, for the first piece we enter $Y_1 = x + 2/(x < 2)$, and for the second, $Y_2 = (x - 4)^2 + 3/(x \geq 2)$ (Figure 3.69). The slash is the division symbol, but in this context, it has the effect of separating the function from the domain. The inequality symbols are accessed using the (2nd) (MATH) (**TEST**) keys. The graph is shown in Figure 3.70, where we see the function is linear for $x \in (-\infty, 2)$ and quadratic for $x \in [2, \infty)$. How does the calculator remind us the function is defined only for $x = 2$ on the second piece? Using the (2nd) (GRAPH) (**TABLE**) feature reveals the calculator will give an **ERR:** (ERROR) message for inputs outside of its domain (Figure 3.71).

We can also use the calculator to investigate endpoints of the domain. For instance, we know that $Y_1 = x + 2$ is not defined for $x = 2$, but what about numbers very close to 2?

**Figure 3.68**

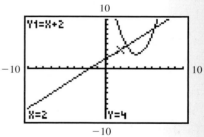

**Figure 3.69**

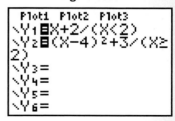

**Figure 3.70**

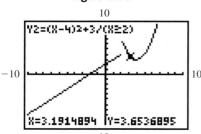

Go to  (**TBLSET**) and place the calculator in the **Indpnt: Auto ASK** mode. With both $Y_1$ and $Y_2$ enabled, use the ⟨2nd⟩ ⟨GRAPH⟩ (**TABLE**) feature to evaluate the functions at numbers very near 2. Use $x = 1.9, 1.99, 1.999$, and so on.

**Exercise 3:** What appears to be happening to the output values for $Y_1$? What about $Y_2$?

**Exercise 4:** What do you notice about the output values when 1.99999 is entered? Use the right arrow key ⟨▷⟩ to move the cursor into columns $Y_1$ and $Y_2$. Comment on what you think the calculator is doing. Will $Y_1$ ever really have an output equal to 4?

## STRENGTHENING CORE SKILLS

### Finding the Domain and Range of a Relation from Its Graph

The concepts of domain and range are an important and fundamental part of working with relations and functions. In Chapter 2, we learned to determine the domain of any relation from its graph using a "vertical boundary line," and the range by using a "horizontal boundary line." These approaches to finding the domain and range can be combined into a single step by envisioning a rectangle drawn around or about the graph. If the entire graph can be "bounded" within the rectangle, the domain and range can be based on the rectangle's length and width. If it's impossible to bound the graph in a particular direction, the related $x$- or $y$-values continue infinitely. Consider the graph in Figure 3.72. This is the graph of an ellipse (Section 8.2), and a rectangle that bounds the graph in all directions is shown in Figure 3.73.

**Figure 3.72**

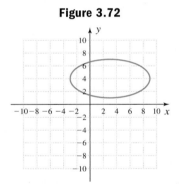

**Figure 3.73**

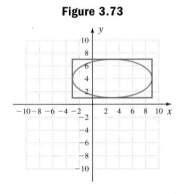

The rectangle extends from $x = -3$ to $x = 9$ in the horizontal direction, and from $y = 1$ to $y = 7$ in the vertical direction. The domain of this relation is $x \in [-3, 9]$ and the range is $y \in [1, 7]$.

The graph in Figure 3.74 is a parabola, and no matter how large we draw the rectangle, an infinite extension of the graph will extend beyond its boundaries in the left and right directions, and in the upward direction (Figure 3.75).

**Figure 3.74**

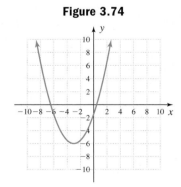

**Figure 3.75**

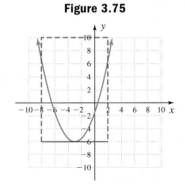

The domain of this relation is $x \in (-\infty, \infty)$ and the range is $y \in [-6, \infty)$.

Finally, the graph in Figure 3.76 is the graph of a square root function, and a rectangle can be drawn that bounds the graph below and to the left, but not above or to the right (Figure 3.77).

**Figure 3.76**

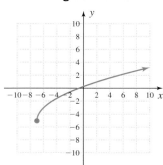

**Figure 3.77**

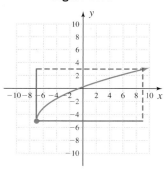

The domain of this relation is $x \in [-7, \infty)$ and the range is $y \in [-5, \infty)$.

Use this approach to find the domain and range of the following relations and functions.

**Exercise 1:**

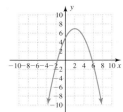

**Exercise 2:**

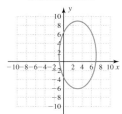

**Exercise 3:**

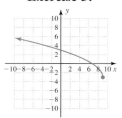

**Exercise 4:**

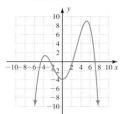

## CUMULATIVE REVIEW CHAPTERS R–3

**1.** Solve for $R$: $\frac{1}{R} = \frac{1}{R_1} + \frac{1}{R_2}$

**2.** Solve for $x$: $\frac{2}{x+1} + 1 = \frac{5}{x^2 - 1}$

**3.** Factor the expressions:
   **a.** $x^3 - 1$           **b.** $x^3 - 3x^2 - 4x + 12$

**4.** Solve using the quadratic formula. Write answers in both exact and approximate form: $2x^2 + 4x + 1 = 0$.

**5.** Solve the following inequality: $x + 3 < 5 \ or \ 5 - x < 4$.

**6.** Name the eight toolbox functions, give their equations, then draw a sketch of each.

**7.** Use substitution to verify that $x = 2 - 3i$ is a solution to $x^2 - 4x + 13 = 0$.

**8.** Given $f(x) = 3x^2 - 6x$ and $g(x) = x - 2$ find: $(f \cdot g)(x)$, $(f \div g)(x)$, and $(g \circ f)(-2)$.

**9.** As part of a study on traffic conditions, the mayor of a small city tracks her driving time to work each day for 6 months and finds a linear and increasing relationship. On day 1, her drive time was 17 min. By day 61 the drive time had increased to 28 min. Find a linear function that models the drive time and use it to estimate the drive time on day 121, if the trend continues. Explain what the slope of the line means in this context.

**10.** Does the relation shown represent a function? If not, discuss/explain why not.

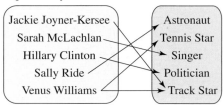

**11.** The data given shows the profit of a new company for the first 6 months of business. (a) Find the equation of a line that will model this data. (b) Assuming this trend continues, use the equation to find the first month a profit will be earned.

**Exercise 11**

| Month | Profit (1000s) |
|-------|---------------|
| 1 | −29 |
| 2 | −27 |
| 3 | −21 |
| 4 | −22 |
| 5 | −16 |
| 6 | −14 |

**12.** Graph the function $g(x) = \frac{-1}{(x+2)^2} + 3$ using transformations of a basic function.

**13.** Given $f(x) = (x-3)^3 + 2$ and $g(x) = (x-2)^{\frac{1}{3}} + 3$, find (a) $(f \circ g)(x)$ and (b) $(g \circ f)(x)$

**14.** Graph $f(x) = (x-2)^2 + 3$ then state intervals where:
  **a.** $f(x) \geq 0$        **b.** $f(x)\uparrow$

**15.** Given the graph of the general function $f(x)$ shown, graph $F(x) = -f(x+1) + 2$.

**Exercise 15**

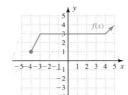

**16.** Graph the piecewise-defined function given:
$$f(x) = \begin{cases} -3, & x < -1 \\ x, & -1 \leq x \leq 1 \\ 3x, & x > 1 \end{cases}$$

**17.** $Y$ varies directly with $X$ and inversely with the square of $Z$. If $Y = 10$ when $X = 32$ and $Z = 4$, find $X$ when $Z = 15$ and $Y = 1.4$.

**18.** Compute as indicated:
  **a.** $(2+5i)^2$       **b.** $\dfrac{1-2i}{1+2i}$

**19.** Use the graphs of $f(x)$ and $g(x)$ as shown to determine the value of each expression.
  **a.** $f(4)$, $g(2)$, $(f \circ g)(2)$
  **b.** $g(4)$, $f(8)$, and $(g \circ f)(8)$
  **c.** $(f \cdot g)(0)$ and $\left(\frac{g}{f}\right)(0)$
  **d.** $(f + g)(1)$ and $(g - f)(9)$

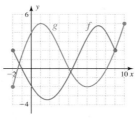

**20.** Solve using the quadratic formula. If solutions are nonreal, write them in $a + bi$ form.
$$2x^2 + 20x = -51$$

**21.** State the domain of each power function: (a) $f(x) = x^{\frac{2}{5}}$ (b) $g(x) = x^{\frac{5}{2}}$.

**22.** For $f(x) = -x^2 - 3x$, find (a) $f(2)$, (b) $-f(2)$, (c) $f(-2)$, and (d) $-f(-2)$

**23.** Solve the inequality and write your answer in interval notation: $-6 < 2x + 5 < 13$

**24.** Simplify:
  **a.** $5^2 + 5^1 + 5^0 + 5^{-1}$
  **b.** $3(0) + 3^0 + 0^3 + \frac{0}{3}$

**25.** State whether each equation is true or false. If false, give the correct result:
  **a.** $(x+3)^2 = x^2 + 9$    **b.** $(5x^2y^3)^2 = 5x^4y^6$
  **c.** $-3^2 = 9$              **d.** $\frac{x}{0} = 0$
  **e.** $3^{-2} = -9$         **f.** $2 + 3 \cdot 5 = 25$

# Polynomial and Rational Functions

## CHAPTER OUTLINE

## CHAPTER CONNECTIONS

The concept of supply and demand plays an important role in many national economies. If an item is scarce or in high demand, its price tends to be higher. If an item is plentiful or demand is low, manufacturers, wholesalers, and retailers tend to lower prices to keep factories operating and store doors open. This interplay between selling price and the revenue generated is a delicate balance, because while more product may be sold at a lower price, total revenue could decrease if less money is coming in. This application appears as Exercise 49 in Section 4.1.

**Check out these other real-world connections:**

► Tourist Population of a Resort Town (Section 4.2, Exercise 73)
► County Deficits (Section 4.3, Exercise 93)
► Volume of Traffic (Section 4.4, Exercise 73)
► Average Speed for a Round-Trip (Section 4.6, Exercise 85)

## LEARNING OBJECTIVES

*In Section 4.1 you will see how we can:*

☐ **A.** Graph quadratic functions by completing the square

☐ **B.** Graph quadratic functions using the vertex formula

☐ **C.** Find the equation of a quadratic function from its graph

☐ **D.** Solve applications involving extreme values

As our knowledge of functions grows, our ability to apply mathematics in new ways likewise grows. In this section, we'll build on the foundation laid in Section 3.1 and previous chapters, as we introduce additional function families and the tools needed to apply them effectively. We begin with the family of quadratic functions, noting that the squaring function $f(x) = x^2$ is actually a member of this (larger) family.

### Quadratic Functions

A quadratic function is one that can be written in the form
$$f(x) = ax^2 + bx + c,$$
where $a$, $b$, and $c$ are real numbers and $a \neq 0$.

## A. Graphing Quadratic Functions by Completing the Square

**Figure 4.1** $f(x) = ax^2 + bx + c$

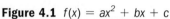
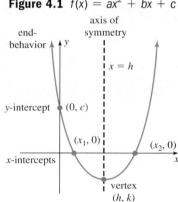

Our earlier work in Chapter 3 suggests the graph of *any* quadratic equation will be a parabola. Figure 4.1 provides a summary of a parabola's characteristic features. As pictured,

1. The parabola opens upward ($y = ax^2 + bx + c, a > 0$).
2. The vertex is at $(h, k)$ and $y = k$ is a global minimum.
3. The vertex is below the $x$-axis, so there are two $x$-intercepts.
4. The axis of symmetry contains the vertex, with equation $x = h$.
5. The $y$-intercept is $(0, c)$, since $f(0) = c$.

In Section 3.1, we graphed transformations of $f(x) = x^2$, using $y = a(x \pm h)^2 \pm k$. Here, we'll show that by completing the square, we can graph *any* quadratic function as a transformation of this basic graph.

When completing the square on a quadratic *equation* (Section 1.5), we applied the standard properties of equality to both sides of the equation. When completing the square on a **quadratic function**, the process is altered slightly, so that we operate on only one side.

**EXAMPLE 1** ▶ **Graphing a Quadratic Function by Completing the Square**

Given $g(x) = x^2 - 6x + 5$, complete the square to rewrite $g$ as a transformation of $f(x) = x^2$, then graph the function.

**Solution** ▶ To begin we note the leading coefficient is $a = 1$.

$$\begin{aligned}
g(x) &= x^2 - 6x + 5 && \text{given function} \\
&= 1(x^2 - 6x + \underline{\phantom{9}}) + 5 && \text{group variable terms} \\
&= 1(\underbrace{x^2 - 6x + 9}_{\text{adds } 1 \cdot 9 = 9}) \underbrace{- 9}_{\text{subtract 9}} + 5 && [\tfrac{1}{2}(-6)]^2 = 9 \\
&= (x - 3)^2 - 4 && \text{factor and simplify}
\end{aligned}$$

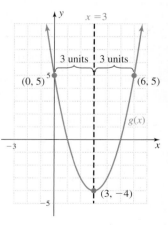

The graph of $g$ is the graph of $f$ shifted 3 units right, and 4 units down. The graph opens upward ($a > 0$) with the vertex at $(3, -4)$, and axis of symmetry $x = 3$. From the original equation we find $g(0) = 5$, giving a $y$-intercept of $(0, 5)$. The point $(6, 5)$ was obtained using the symmetry of the graph. The graph is shown in the figure.

**Now try Exercises 5 through 8** ▶

Note that by **adding 9** and simultaneously **subtracting 9** (essentially adding "0"), we changed only the *form* of the function, not its value. In other words, the resulting expression is equivalent to the original. If the leading coefficient is not 1, we factor it out from the variable terms, but take it into account when we add the constant needed to maintain an equivalent expression (Steps 2 and 3). The basic ideas are summarized here.

---

**Graphing $f(x) = ax^2 + bx + c$ by Completing the Square**

1. Group the variable terms apart from the constant $c$.
2. Factor out the leading coefficient $a$ from this group.
3. Compute $\left[\frac{1}{2}\left(\frac{b}{a}\right)\right]^2$ and add the result to the variable terms, then subtract $a \cdot \left[\frac{1}{2}\left(\frac{b}{a}\right)\right]^2$ from $c$ to maintain an equivalent expression.
4. Factor the grouped terms as a binomial square and combine constant terms.
5. Graph using transformations of $f(x) = x^2$.

---

**EXAMPLE 2** ▶    **Graphing a Quadratic Function by Completing the Square**

Given $p(x) = -2x^2 - 8x - 3$, complete the square to rewrite $p$ as a transformation of $f(x) = x^2$, then graph the function.

**Solution** ▶

$$p(x) = -2x^2 - 8x - 3 \qquad \text{given function}$$
$$= (-2x^2 - 8x + \underline{\quad}) - 3 \qquad \text{group variable terms}$$
$$= -2(x^2 + 4x + \underline{\quad}) - 3 \qquad \text{factor out } a = -2 \text{ (notice sign change)}$$
$$= \underbrace{-2(x^2 + 4x + 4)}_{\text{adds } -2 \cdot 4 = -8} \underbrace{- (-8)}_{\text{subtract } -8} - 3 \qquad \left[\frac{1}{2}(4)\right]^2 = 4$$
$$= -2(x + 2)^2 + 8 - 3 \qquad \begin{array}{l}\text{factor trinomial,}\\\text{simplify}\end{array}$$
$$= -2(x + 2)^2 + 5 \qquad \text{result}$$

The graph of $p$ is a parabola, shifted 2 units left, stretched vertically by a factor of 2, reflected across the $x$-axis (opens downward), and shifted up 5 units. The vertex is $(-2, 5)$, and the axis of symmetry is $x = -2$. From the original function, the $y$-intercept is $(0, -3)$. The point $(-4, -3)$ was obtained using the symmetry of the graph. The graph is shown in the figure.

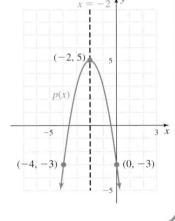

**WORTHY OF NOTE**

In cases like $f(x) = 3x^2 - 10x + 5$, where the leading coefficient is not a factor of $b$, we factor out 3 and *simultaneously divide the linear coefficient by 3*. This yields $h(x) = 3(x^2 - \frac{10}{3}x + \underline{\quad}) + 5$, and the process continues as before: $\left[\frac{1}{2}(\frac{10}{3})\right]^2 = (\frac{5}{3})^2 = \frac{25}{9}$, and so on. For more on this idea, **see Exercises 15 and 16**.

☑ **A.** You've just seen how we can graph quadratic functions by completing the square

**Now try Exercises 9 through 14** ▶

In Example 2, note that by adding 4 to the variable terms within parentheses, we actually added $-2 \cdot 4 = -8$ to the value of the function due to the distributive property. To adjust for this we subtracted $-8$ $[-(-8) = +8]$.

## B. Graphing Quadratic Functions Using the Vertex Formula

When the process of completing the square is applied to $f(x) = ax^2 + bx + c$, we obtain a very useful result. Notice the close similarities to Example 2.

$$f(x) = ax^2 + bx + c \qquad \text{quadratic function}$$

$$= (ax^2 + bx + \underline{\phantom{--}}) + c \qquad \text{group variable terms apart from the constant } c$$

$$= a\left(x^2 + \frac{b}{a}x + \underline{\phantom{--}}\right) + c \qquad \text{factor out } a$$

$$= a\left(x^2 + \frac{b}{a}x + \frac{b^2}{4a^2}\right) - a\left(\frac{b^2}{4a^2}\right) + c \qquad [\tfrac{1}{2}(\tfrac{b}{a})]^2 = \tfrac{b^2}{4a^2}, \text{ add within group, subtract } a\left(\tfrac{b^2}{4a^2}\right)$$

$$= a\left(x + \frac{b}{2a}\right)^2 - \frac{b^2}{4a} + c \qquad \text{factor the trinomial, simplify}$$

$$= a\left(x + \frac{b}{2a}\right)^2 + \frac{4ac - b^2}{4a} \qquad \text{result}$$

By comparing this result with previous transformations, we note the $x$-coordinate of the vertex is $h = \frac{-b}{2a}$ (since the graph shifts horizontally "opposite the sign"). While we could use the expression $\frac{4ac - b^2}{4a}$ for $k$, we find it easier to substitute $\frac{-b}{2a}$ back into the function: $k = f\left(\frac{-b}{2a}\right)$. The result is called the **vertex formula.**

### Vertex Formula

For the quadratic function $f(x) = ax^2 + bx + c$, the coordinates of the vertex are

$$(h, k) = \left(\frac{-b}{2a}, f\left(\frac{-b}{2a}\right)\right)$$

Since all characteristic features of the graph (end-behavior, vertex, axis of symmetry, $x$-intercepts, and $y$-intercept) can now be determined using the original equation, we'll rely on these features to sketch quadratic graphs, rather than having to complete the square.

---

**EXAMPLE 3 ▶**    **Graphing a Quadratic Function Using the Vertex Formula**

Graph $f(x) = 2x^2 + 8x + 3$ using the vertex formula and other features of a quadratic graph.

**Solution ▶**    The graph will open upward since $a > 0$.
The $y$-intercept is $(0, 3)$.
The vertex formula gives

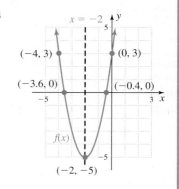

$$h = \frac{-b}{2a} \qquad \text{x-coordinate of vertex}$$

$$= \frac{-8}{2(2)} \qquad \text{substitute 2 for } a \text{ and 8 for } b$$

$$= -2 \qquad \text{simplify}$$

Computing $f(-2)$ to find the $y$-coordinate of the vertex yields

$$f(-2) = 2(-2)^2 + 8(-2) + 3 \qquad \text{substitute } -2 \text{ for } x$$
$$= 2(4) - 16 + 3 \qquad \text{multiply}$$
$$= 8 - 16 + 3 \qquad \text{simplify}$$
$$= -5 \qquad \text{result}$$

✓ **B.** You've just seen how we can graph quadratic functions using the vertex formula

The vertex is $(-2, -5)$. Using the quadratic formula we find the $x$-intercepts are approximately $(-3.6, 0)$ and $(-0.4, 0)$. The graph is shown in the figure, with the point $(-4, 3)$ obtained using symmetry.

Now try Exercises 17 through 24 ▶

## C. Finding the Equation of a Quadratic Function from Its Graph

While most of our emphasis so far has centered on graphing quadratic functions, it would be hard to overstate the importance of the reverse process—determining the equation of the function from its graph. This reverse process, which began with our study of lines, will be a continuing theme each time we consider a new function.

**EXAMPLE 4 ▶**  **Finding the Equation of a Quadratic Function**

The graph shown is a transformation of $f(x) = x^2$. What function defines this graph?

**Solution ▶**  Compared to the graph of $f(x) = x^2$, the vertex has been shifted left 1 and up 2, so the function will have the form $y = a(x + 1)^2 + 2$. Since the graph opens downward, we know $a$ will be negative. As before, we select one additional point on the graph and substitute to find the value of $a$. With $(x, y)$ corresponding to $(1, 0)$ we obtain

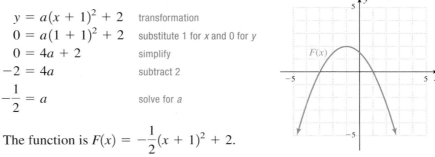

> **WORTHY OF NOTE**
>
> It helps to remember that any point $(x, y)$ on the parabola can be used. To verify this, try the calculation again using $(-3, 0)$.

$$y = a(x + 1)^2 + 2 \quad \text{transformation}$$
$$0 = a(1 + 1)^2 + 2 \quad \text{substitute 1 for } x \text{ and 0 for } y$$
$$0 = 4a + 2 \quad \text{simplify}$$
$$-2 = 4a \quad \text{subtract 2}$$
$$-\frac{1}{2} = a \quad \text{solve for } a$$

✓ **C.** You've just seen how we can find the equation of a quadratic function from its graph

The function is $F(x) = -\dfrac{1}{2}(x + 1)^2 + 2$.

Now try Exercises 25 through 30 ▶

## D. Quadratic Functions and Extreme Values

If $a > 0$, the parabola opens upward, and the $y$-coordinate of the vertex is a global minimum, the smallest value attained by the function anywhere in its domain. Conversely, if $a < 0$ the parabola opens downward and the vertex yields a global maximum. As noted in Section 2.5, these greatest and least points are known as **extreme values** and have a number of significant applications.

**EXAMPLE 5 ▶**  **Applying a Quadratic Model to Manufacturing**

An airplane manufacturer has the capacity to produce up to 15 planes per month. Suppose that due to economies of scale, the profit made from the sale of these planes is modeled by $P(x) = -0.2x^2 + 4x - 3$, where $P(x)$ is the profit in hundred-thousands of dollars per month, and $x$ is the number of planes sold. Based on this model,

 **a.** Find the $y$-intercept and explain what it means in this context.
 **b.** How many planes should be made and sold to maximize profit?
 **c.** What is the maximum profit?

**Solution** ▶   **a.** $P(0) = -3$, which means the manufacturer loses $300,000 each month if the company produces no planes.

**b.** Since $a < 0$, we know the graph opens downward and has a maximum value. To find the required number of sales needed to "maximize profit," we use the vertex formula with $a = -0.2$ and $b = 4$:

$$x = \frac{-b}{2a} \qquad \text{vertex formula}$$

$$= \frac{-4}{2(-0.2)} \qquad \text{substitute } -0.2 \text{ for } a \text{ and } 4 \text{ for } b$$

$$= 10 \qquad \text{result}$$

The result shows 10 planes should be sold each month for maximum profit.

**c.** Evaluating $P(10)$ we find that a maximum profit of 17 "hundred-thousand dollars" will be earned ($1,700,000).

Now try Exercises 33 through 38 ▶

Recall that if the leading coefficient is positive and the vertex is below the $x$-axis ($k < 0$), the graph will have two $x$-intercepts (see Figure 4.2). If $a > 0$ and the vertex is above the $x$-axis ($k > 0$), the graph will not cross the $x$-axis (Figure 4.3). Similar statements can be made for the case where $a$ is negative. See Figure 4.4 for the case where $k = 0$.

**Figure 4.2**

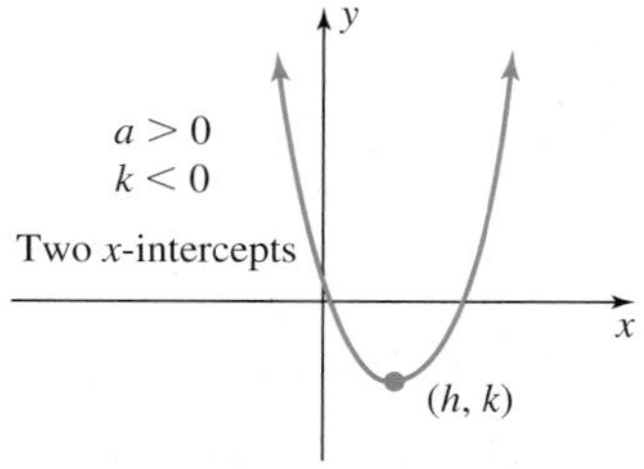

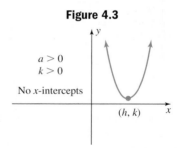

$a > 0$
$k < 0$

Two $x$-intercepts

$(h, k)$

**Figure 4.3**

$a > 0$
$k > 0$

No $x$-intercepts

$(h, k)$

In some applications of quadratic functions, our interest includes the $x$-intercepts of the graph. Drawing on our previous work, we note that the following statements are equivalent, meaning if any one statement is true, then all four statements are true.

**Figure 4.4**

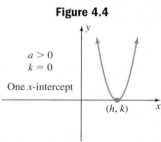

$a > 0$
$k = 0$

One $x$-intercept

$(h, k)$

- $(r, 0)$ is an **$x$-intercept** of the graph of $f(x)$.
- $x = r$ is a **solution** or **root** of the equation $f(x) = 0$.
- $(x - r)$ is a **factor** of $f(x)$.
- $r$ is a **zero** of $f(x)$.

**EXAMPLE 6** ▶   **Modeling the Height of a Projectile—Football**

In the 1976 Pro Bowl, NFL punter Ray Guy of the Oakland Raiders kicked the ball so high it hit the scoreboard hanging from the roof of the New Orleans Superdome (forcing officials to raise the scoreboard from about 90 ft to 200 ft). If we assume the ball made contact with the scoreboard near the vertex of the kick, the function $h(t) = -16t^2 + 76t + 1$ is one possible model for the height of the ball, where $h(t)$ represents the height (in feet) after $t$ sec.

**a.** What does the $y$-intercept of this function represent?

**b.** After how many seconds did the football reach its maximum height?

**c.** What was the maximum height of this kick?

**d.** To the nearest hundredth of a second, how long until the ball returned to the ground (what was the hang time)?

**Solution** ▶   **a.** $h(0) = 1$, meaning the ball was 1 ft off the ground when Ray Guy kicked it.

**b.** Since $a < 0$, we know the graph opens downward and has a maximum value. To find the time needed to reach the maximum height, we use the vertex formula with $a = -16$ and $b = 76$:

$$t = \frac{-b}{2a} \qquad \text{vertex formula}$$

$$= \frac{-76}{2(-16)} \qquad \text{substitute } -16 \text{ for } a \text{ and } 76 \text{ for } b$$

$$= 2.375 \qquad \text{result}$$

The ball reached its maximum height after 2.375 sec.

**c.** To find the maximum height, we substitute 2.375 for $t$ [evaluate $h(2.375)$]:

$$h(t) = -16t^2 + 76t + 1 \qquad \text{given function}$$

$$h(2.375) = -16(2.375)^2 + 76(2.375) + 1 \qquad \text{substitute 2.375 for } t$$

$$= 91.25 \qquad \text{result}$$

The ball reached a maximum height of 91.25 ft.

**d.** When the ball returns to the ground it has a height of 0 ft. Substituting 0 for $h(t)$ gives $0 = -16t^2 + 76t + 1$, which we solve using the quadratic formula.

$$t = \frac{-b \pm \sqrt{b^2 - 4ac}}{2a} \qquad \text{quadratic formula}$$

$$= \frac{-76 \pm \sqrt{76^2 - 4(-16)(1)}}{2(-16)} \qquad \text{substitute } -16 \text{ for } a, 76 \text{ for } b, \text{ and } 1 \text{ for } c$$

$$= \frac{-76 \pm \sqrt{5840}}{-32} \qquad \text{simplify}$$

$$t \approx -0.013 \quad \text{or} \quad t \approx 4.763$$

The punt had a hang time of just under 4.8 sec.

Now try Exercises 39 through 44 ▶

While the calculator is a wonderful tool for removing some computational drudgery, it does not (cannot) replace the analytical thought required to be an effective problem solver. Most real-world applications still require an attentive analysis of the context and the question asked, as well as a careful development of the equation model to be used. This is particularly important in applications like those in Example 7, where the original equation has more than one independent variable, and similar to our work in Section 3.6, a given or known relationship must be used to eliminate one of them.

**EXAMPLE 7** ▶   **Determining Maximum Area from a Fixed Perimeter**

Due to an increase in consumer demand, a local nursery is building new chain-link pens for holding various kinds of mulch. What is the maximum total area that can be fenced off, if five open-front pens are needed (see Figure 4.5) and 108 ft of fencing is available?

**Figure 4.5**

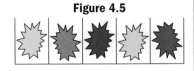

**Solution** ▶   We are asked to maximize the total area, an impossibility right now since the formula for area has *two independent variables*: Area = $LW$. Knowing that only 108 ft of fencing will be used, we observe that the finished pens will have *1 length and 6 identical widths,* giving the equation $108 = L + 6W$. This enables us to write $L$ in terms of $W$ (solve for $L$), and *substitute the result for L in A = LW,* giving an equation model with the single independent variable $W$.

$$108 = L + 6W \qquad \text{known relationship}$$
$$108 - 6W = L \qquad \text{solve for } L$$

Using $L = 108 - 6W$, we substitute for $L$ in the formula Area $= LW$.

$$A = LW \qquad \text{area formula}$$
$$= (108 - 6W)W \qquad \text{substitute } 108 - 6W \text{ for } L$$
$$A(W) = -6W^2 + 108W \qquad \text{result: a function in one variable}$$

Noting the result is a quadratic function, the maximum value will occur at $W = -\frac{b}{2a}$, and substituting 108 for $b$ and $-6$ for $a$ shows this value is $W = 9$. Evaluating $A(W)$ for $W = 9$ shows that a maximum area of 486 ft will be enclosed $[A(9) = 486]$ if a width of 9 ft is used. Note this gives a length of $L = 108 - 6(9) = 54$ ft.

> **Now try Exercises 45 through 48** ▶

In a free-market economy, it is well known that if the price of an item is decreased, more people are likely to buy it. This is why stores have sales and bargain days. But if the item is sold too cheaply, revenue starts to decline because less money is coming in—even though more sales are being made. This phenomenon is analyzed in Example 8 using the standard formula for revenue: *revenue = price · number of sales* or $R = P \cdot S$.

**EXAMPLE 8** ▶ **Supply and Demand: Analyzing Retail Sales Revenue**

When a popular running shoe is priced at $80, The Shoe Warehouse will sell an average of 96 pairs per week. Based on sales of similar shoes, the company believes that for each decrease of $2.50 in price, four additional pairs will be sold.

**a.** Construct a function that models the store's revenue at various prices. What is the maximum possible revenue under these conditions? What price should be charged to obtain this maximum revenue?

**b.** What is the cheapest price the manager could charge, while still bringing in a monthly revenue of at least $7000?

**Solution** ▶ **a.** Based on the given price and sales figures, the store has a current revenue of ($80)(96) = $7680. If we let $x$ represent the number of times the price is decreased by $2.50, the model for the resulting price would be $80 - 2.50x$ and the model for resulting sales would be $96 + 4x$.

$$(80 - 2.5x)(96 + 4x) = R(x) \qquad \begin{array}{l} \text{for each price decrease of \$2.50,} \\ \text{sales increase by 4} \end{array}$$

In standard form, we have

$$7680 + 320x - 240x - 10x^2 = R(x) \qquad \text{multiply binomials}$$
$$-10x^2 + 80x + 7680 = R(x) \qquad \text{simplify}$$

We know the maximum value will occur at $x = \frac{-b}{2a}$, and substituting 80 for $b$ and $-10$ for $a$ gives $x = \frac{-80}{2(-10)} = 4$. Evaluating $R(x)$ for $x = 4$ gives a maximum revenue of $7840 $[R(4) = 7840]$ after four price decreases of $2.50 each. This shows the selling price will be $80 - 4(2.50) = $70.

**b.** Since the desired revenue level is $7000, we substitute 7000 for $R(x)$ and solve.

$$-10x^2 + 80x + 7680 = 7000 \qquad \text{substitute 7000 for } R(x)$$
$$x^2 - 8x - 768 = -700 \qquad \text{divide by } -10$$
$$x^2 - 8x - 68 = 0 \qquad \text{add 700}$$

Since the trinomial will not easily factor, we use the quadratic formula to solve for $x$. With $a = 1$, $b = -8$, and $c = -68$ we have

$$x = \frac{-b \pm \sqrt{b^2 - 4ac}}{2a}$$　　　　quadratic formula

$$= \frac{-(-8) \pm \sqrt{(-8)^2 - 4(1)(-68)}}{2(1)}$$　　　　substitute 1 for $a$, $-8$ for $b$, $-68$ for $c$

$$= \frac{8 \pm \sqrt{336}}{2}$$　　　　simplify

$$x \approx -5.2, \quad x \approx 13.2$$　　　　result

The result indicates that to keep a revenue of at least $7000 per week, the price must decrease by no more than $13.2(\$2.50) = \$33$ (the negative solution does not fit the context). The selling price should be kept near $80 - 33 = \$47$.

**☑ D. You've just seen how we can solve applications involving extreme values**

**Now try Exercises 49 and 50 ▶**

Previously we've seen real data sets that were best modeled by linear functions (Section 2.6) and power functions (Section 3.2). These models enabled us to make meaningful decisions and comparisons, as well as reasonable projections for future occurrences. Here, we explore data relationships that are best modeled by quadratic functions.

After a set of data is collected and organized, any patterns or relationships that exist may not be easily seen. The regression model chosen for the data will depend on the context of the data and any patterns noted in the scatterplot, along with a careful assessment of the correlation coefficient. Regardless of the form of regression used, the steps for inputting the data, setting a window size, viewing the scatterplot, and so on, are identical.

**EXAMPLE 9 ▶** **Growth of Online Sales of Pet Supplies**

Since the year 2000, there has been a tremendous increase in the online sales of pet food, pet medications, and other pet supplies. The data given in the table show the amount spent (in billions of dollars) for the years indicated.

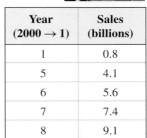

*Source: 2009 Statistical Abstract of the United States, Page 646, Table 1016.*

a. Input the data into a graphing calculator, set an appropriate window and view the scatterplot, then use the context and the scatterplot to decide on an appropriate form of regression.

b. Determine the regression equation, and use it to find the amount of online sales projected for the year 2012.

c. If the predicted trends continue, in what year will online sales of pet supplies surpass $25 billion dollars?

| Year (2000 → 1) | Sales (billions) |
|---|---|
| 1 | 0.8 |
| 5 | 4.1 |
| 6 | 5.6 |
| 7 | 7.4 |
| 8 | 9.1 |

**Solution ▶**

a. Begin by entering the years in L1 and the dollar amounts in L2. From the data given, a viewing window of Xmin = 0, Xmax = 15, Ymin = -2, and Ymax = 15 seems appropriate. The scatterplot in Figure 4.6 shows a definite increasing, nonlinear pattern and we opt for a quadratic regression.

**Figure 4.6**

```
         15
P1:L1,L2
                        □
                    □
              □
     □
 0 ━━━━━━━━━━━━━━━━━━━━━ 15
   □
  X=5          Y=4.1
        -2
```

**Figure 4.7**

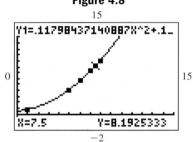

```
QuadReg Y₁
```

b. Using **STAT** ▶ (**CALC**) **5:QuadReg** places this option on the home screen. Pressing ⊙ at this point will give us the quadratic coefficients, but we can also have the calculator paste the equation itself directly into $Y_1$ on the ⊙ screen, simply by appending $Y_1$ to the **QuadReg** command (Figure 4.7). The resulting equation is $y \approx 0.118x^2 + 0.136x + 0.537$ (to three decimal places), and the graph is shown in Figure 4.8. Using the home screen, we find that $Y_1(12) \approx 19.2$ (Figure 4.9), indicating that in 2012, about $19.2 billion is projected to be spent online for pet supplies.

c. To find the year when spending will surpass $25 billion, we set $Y_2 = 25$ and use the intersection-of-graphs method (be sure to increase Ymax so this graph can be seen: Ymax = 30). The result is shown in Figure 4.10 and indicates that spending will likely surpass $25 billion late in the year 2013.

**Figure 4.8**

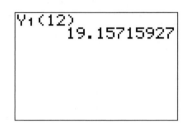

**Figure 4.9**

**Figure 4.10**

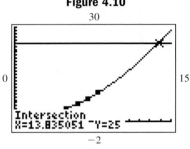

**Now try Exercises 51 through 54** ▶

## 4.1 EXERCISES

▶ **CONCEPTS AND VOCABULARY**

**Fill in each blank with the appropriate word or phrase. Carefully reread the section, if necessary.**

1. Maximum and minimum values are called _____ values. For quadratic functions, these can be found using the _____ formula.

2. If the leading coefficient is positive and the vertex $(h, k)$ is in Quadrant IV, the graph will have _____ $x$-intercepts.

3. Compare/Contrast how to complete the square on an *equation,* versus how to complete the square on a function. Use the equation $2x^2 + 6x - 3 = 0$ and the function $f(x) = 2x^2 + 6x - 3$ to illustrate.

4. Discuss/Explain why the graph of a quadratic function has no $x$-intercepts if $a$ and $k$ [vertex $(h, k)$] have like signs. Under what conditions will the function have a single real root?

▶ **DEVELOPING YOUR SKILLS**

**Graph each function by completing the square and using transformations of the basic function. Label the vertex and all intercepts (if they exist).**

5. $f(x) = x^2 + 4x - 5$
6. $g(x) = x^2 - 6x - 7$
7. $h(x) = -x^2 + 2x + 3$
8. $H(x) = -x^2 + 8x - 7$
9. $u(x) = 3x^2 + 6x - 5$
10. $v(x) = 4x^2 - 24x + 15$
11. $f(x) = -2x^2 + 8x + 7$
12. $g(x) = -3x^2 + 12x - 7$
13. $p(x) = x^2 - 5x + 2$
14. $q(x) = x^2 + 7x + 4$
15. $q(x) = 4x^2 - 9x + 2$
16. $g(x) = -2x^2 + 9x - 7$

**Graph each function using the vertex formula and other features of a quadratic graph. Label important features.**

**17.** $f(x) = x^2 + 2x - 6$        **18.** $g(x) = x^2 + 8x + 11$

**19.** $h(x) = -x^2 + 4x + 2$        **20.** $H(x) = -x^2 + 10x - 19$

**21.** $f(x) = 4x^2 - 12x + 3$        **22.** $g(x) = 3x^2 + 12x + 5$

**23.** $p(x) = \frac{1}{2}x^2 + 3x - 5$        **24.** $q(x) = \frac{1}{3}x^2 - 2x - 4$

**State the equation of the function whose graph is shown.**

**25.** 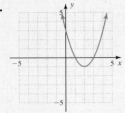                **26.**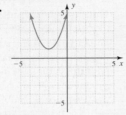

**27.** 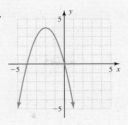                **28.**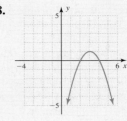

**29.** 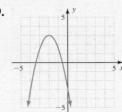                **30.**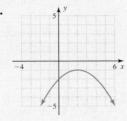

## ▶ WORKING WITH FORMULAS

**31. Vertex/intercept formula:** $x = h \pm \sqrt{-\dfrac{k}{a}}$

As an alternative to using the quadratic formula *prior* to completing the square, the $x$-intercepts can easily be found using the vertex/intercept formula *after* completing the square, when the coordinates of the vertex are known. (a) Beginning with the shifted form $y = a(x - h)^2 + k$, substitute 0 for $y$ and solve for $x$ to derive the formula. (b) Use the formula to find all zeroes, real or complex, of the following functions.

  **i.** $y = (x + 3)^2 - 5$        **ii.** $y = -(x - 4)^2 + 3$        **iii.** $y = 2(x + 4)^2 - 7$

  **iv.** $y = -3(x - 2)^2 + 6$        **v.** $s(t) = 0.2(t + 0.7)^2 - 0.8$        **vi.** $r(t) = -0.5(t - 0.6)^2 + 2$

**32. Surface area of a rectangular box with square ends:** $S = 2h^2 + 4Lh$

The surface area of a rectangular box with square ends is given by the formula shown, where $h$ is the height and width of the square ends, and $L$ is the length of the box. (a) If $L$ is 3 ft and the box must have a surface area of 32 ft$^2$, find the dimensions of the square ends. (b) Solve for $L$, then find the length if the height is 1.5 ft and surface area is 22.5 ft$^2$.

## ▶ APPLICATIONS

**33. Maximum profit:** An automobile manufacturer can produce up to 300 cars per day. The profit made from the sale of these vehicles can be modeled by the function $P(x) = -10x^2 + 3500x - 66{,}000$, where $P(x)$ is the profit in dollars and $x$ is the number of automobiles made and sold. Based on this model:

  **a.** Find the $y$-intercept and explain what it means in this context.

  **b.** Find the $x$-intercepts and explain what they mean in this context.

  **c.** How many cars should be made and sold to maximize profit?

  **d.** What is the maximum profit?

**34. Maximum profit:** The profit for a manufacturer of collectible grandfather clocks is given by the function $P(x) = -1.6x^2 + 240x - 375$, where $P(x)$ is the profit in dollars and $x$ is the number of clocks made and sold.

  **a.** Find the $y$-intercept and explain what it means in this context.

  **b.** Find the $x$-intercepts and explain what they mean in this context.

  **c.** How many clocks should be made and sold to maximize profit?

  **d.** What is the maximum profit?

**35. Optimal pricing strategy:** The director of the Ferguson Valley drama club must decide what to charge for a ticket to the club's performance of *The Music Man*. If the price is set too low, the club will lose money; and if the price is too high, people won't come. From past experience she estimates that the profit $P$ from sales (in hundreds) can be approximated by $P(x) = -x^2 + 46x - 88$, where $x$ is the cost of a ticket and $0 \le x \le 50$.

  **a.** Find the lowest cost of a ticket that would enable the club to break even.

  **b.** What is the highest cost that the club can charge to break even?

  **c.** If the theater were to close down before any tickets are sold, how much money would the club lose?

  **d.** How much should the club charge to maximize their profits? What is the maximum profit?

**36. Maximum profit:** A kitchen appliance manufacturer can produce up to 200 appliances per day. The profit made from the sale of these machines can be modeled by the function $P(x) = -0.5x^2 + 175x - 3300$, where $P(x)$ is the profit in dollars, and $x$ is the number of appliances made and sold. Based on this model,

   **a.** Find the $y$-intercept and explain what it means in this context.

   **b.** Find the $x$-intercepts and explain what they mean in this context.

   **c.** Determine the domain of the function and explain its significance.

   **d.** How many should be sold to maximize profit? What is the maximum profit?

**37. Cost of production:** The cost of producing a plastic toy is given by the function $C(x) = 2x + 35$, where $x$ is the number of hundreds of toys. The revenue from toy sales is given by $R(x) = -x^2 + 122x - 365$. Since profit = revenue − cost, the profit function must be $P(x) = -x^2 + 120x - 400$ (verify). How many toys sold will produce the maximum profit? What is the maximum profit?

**38. Cost of production:** The cost to produce bottled spring water is given by $C(x) = 16x + 63$, where $x$ is the number of thousands of bottles. The total income (revenue) from the sale of these bottles is given by the function $R(x) = -x^2 + 326x - 7337$. Since profit = revenue − cost, the profit function must be $P(x) = -x^2 + 310x - 7400$ (verify). How many bottles sold will produce the maximum profit? What is the maximum profit?

**The projectile function: $h(t) = -16t^2 + vt + k$ applies to any object projected upward with an initial velocity $v$, from a height $k$, but not to an object under propulsion. Consider this situation and answer the questions that follow.**

**39. Model rocketry:** A member of the local rocketry club launches her latest rocket from a large field. At the moment its fuel is exhausted, the rocket has a velocity of 240 ft/sec and an altitude of 544 ft ($t$ is in seconds).

   **a.** Write the function that models the height of the rocket.

   **b.** How high is the rocket at $t = 0$? If it took off from the ground, why is it this high at $t = 0$?

   **c.** How high is the rocket 5 sec after the fuel is exhausted?

   **d.** How high is the rocket 10 sec after the fuel is exhausted?

   **e.** How could the rocket be at the same height at $t = 5$ and at $t = 10$?

   **f.** What is the maximum height attained by the rocket?

   **g.** How many seconds was the rocket airborne *after* its fuel was exhausted?

**40. Hunting slings:** A stone is slung upward with an initial velocity of 176 ft/sec. After $t$ sec, its height $h(t)$ above the ground is given by the function $h(t) = -16t^2 + 176t$.

   **a.** Find the stone's height above the ground after 2 sec.

   **b.** Sketch the graph modeling the stone's height.

   **c.** What is the stone's maximum height? What is the value of $t$ at this height?

   **d.** How many seconds after it is slung will the stone strike the ground?

**41. Hollywood catapult:** In the movie *The Court Jester* (1956; Danny Kaye, Basil Rathbone, Angela Lansbury, and Glynis Johns), a catapult is used to toss the nefarious adviser to the king into a river. Suppose the path flown by the king's adviser is modeled by the function $h(d) = -0.02d^2 + 1.64d + 14.4$, where $h(d)$ is the height of the adviser in feet at a distance of $d$ ft from the base of the catapult.

   **a.** How high was the release point of this catapult?

   **b.** How far from the catapult did the adviser reach a maximum altitude?

   **c.** What was this maximum altitude attained by the adviser?

   **d.** How far from the catapult did the adviser splash into the river?

**42. Blanket toss competition:** The fraternities at Steele Head University are participating in a blanket toss competition, an activity borrowed from the whaling villages of the Inuit Eskimos. If the person being tossed is traveling at 32 ft/sec as he is projected into the air, and the Frat members are holding the canvas blanket at a height of 5 ft,

   **a.** Write the function that models the height at time $t$ of the person being tossed.

   **b.** How high is the person when (i) $t = 0.5$, (ii) $t = 1.5$?

   **c.** From part (b) what do you know about *when* the maximum height is reached?

   **d.** To the nearest tenth of a second, when is the maximum height reached?

   **e.** To the nearest one-half foot, what was the maximum height?

   **f.** To the nearest tenth of a second, how long was this person airborne?

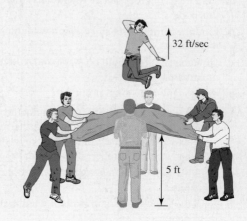

32 ft/sec

5 ft

**43. Motorcycle jumps:** On December 31, 2007, Australian freestyle motocross legend Robbie Maddison set a world record by jumping his motorcycle 322 ft, $7\frac{1}{2}$ in. During practice the day before, the prevailing wind conditions actually enabled him to jump farther. Suppose his height on one such jump is given by the equation $h = -16t^2 + 52t + 25$, where $h$ represents his height (in feet) above ground level $t$ sec after takeoff.

   **a.** How high is the top of the takeoff ramp?

   **b.** If he touched down on the landing ramp 15 ft above ground level, how long was "Maddo" in the air?

   **c.** What would have been the daredevil's maximum height?

**44. SuperPipe Finals:** In the Winter X Games, one of the most exciting events to watch is the SuperPipe Final. The height of a professional snowboarder during one particularly huge jump is given by the function $h = -16t^2 + 35t + 16$, where $h$ represents her height (in feet) above the pipe base $t$ sec after leaving the upper edge, or lip.

   **a.** How high is the lip of the superpipe?

   **b.** If the snowboarder lands her trick 11 ft above the base of the pipe, how long was she in the air?

   **c.** What was the athlete's maximum height above the base of the pipe?

**45. Fencing a backyard:** Tina and Imai have just purchased a purebred German Shepherd, and need to fence in their backyard so the dog can run. What is the maximum rectangular area they can enclose with 200 ft of fencing, if (a) they use fencing material along all four sides? What are the dimensions of the rectangle? (b) What is the maximum area if they are able to use the house as one of the sides? What are the dimensions of *this* rectangle?

**46. Building sheep pens:** It's time to drench the sheep again, so Chance and Chelsea-Lou are fencing off a large rectangular area to build some temporary holding pens. To prep the males, females, and lambs, they are separated into three smaller and equal-size pens partitioned within the large rectangle. If 384 ft of fencing is available and the maximum area is desired, what will be (a) the dimensions of the larger, outer rectangle? (b) the dimensions of the smaller holding pens?

**47. Building windows:** Window World, Inc., is responsible for designing new windows for the expansion of the campus chapel. The current design is shown in the figure. The metal trim used to secure the perimeter of the frame is 126 in. long. If the maximum window area is desired (to let in the most sunlight), what will be (a) the dimensions of the rectangular portion of each window? (b) the total area of each window?

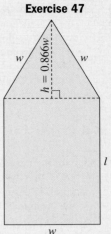

**Exercise 47**

**48. Building windows:** Before construction on the chapel expansion begins (see Exercise 47), a new design is submitted. The new design is very different (see figure), but must still have a perimeter of 126 in. If the maximum window area is still desired, what will be (a) the dimensions of the rectangular portion of each window? (b) the total area of each window?

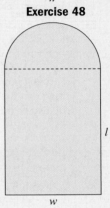

**Exercise 48**

**49. Maximizing soft drink revenue:** A convenience store owner sells 20-oz soft drinks for $1.50 each, and sells an average of 500 per week. Using a market survey, she believes that for each $0.05 decrease in price, an additional 25 soft drinks will be sold. (a) What price should be charged to maximize revenue? What is the maximum revenue? (b) What is the lowest price the owner could charge and still bring in at least $700 per week in revenue?

**50. Maximizing restaurant revenue:** At Figaro's Pizzeria, large pizzas usually sell for $12.50 and at this price an average of 320 are sold each weekend. Using a survey and years of experience, the manager of Figaro's believes that for each $0.25 decrease in price, 16 additional pizzas will be sold. (a) What price should be charged to maximize revenue? What is the maximum revenue? (b) What selling price(s) will generate exactly $4800 in revenue?

**51. Registration for 5-km race:** A local community hosts a 5-km race to raise money for cancer research. Due to legal restrictions, only the first 5000 registrants will be allowed to compete. The table shows the cumulative number of registered participants at the end of the day, for the first 5 days. (a) Find a quadratic regression equation that models the data. Use this equation to estimate (b) the number of participants after 1 week of registration, (c) the number of days it will take for the race to fill up, and (d) the maximum number of participants that would have signed up had there been no limit.

| Day | Registration Total |
|-----|--------------------|
| 1   | 791                |
| 2   | 1688               |
| 3   | 2407               |
| 4   | 3067               |
| 5   | 3692               |

**52. Concert tickets:** The *Javier Mendoza Band* has scheduled a concert at Candlestick Park. Once the tickets go on sale, the band is sure to sell out this 70,000 person venue. The table shows the cumulative number of tickets sold each week, for the first 4 weeks. (a) Find a quadratic regression equation that models the data. Use this equation to estimate (b) the number of tickets sold after 5 weeks, (c) the number of weeks it will take for the concert to sell out, and (d) the number of fans that won't get to attend the show.

| Week | Ticket Sales Total |
|------|--------------------|
| 1 | 17,751 |
| 2 | 31,266 |
| 3 | 45,311 |
| 4 | 54,986 |

▶ **EXTENDING THE CONCEPT**

**53.** Use the general solutions from the quadratic formula to show that the average value of the x-intercepts is $\frac{-b}{2a}$. Explain/Discuss why the result is valid even if the roots are complex.

**54.** Write the equation of a quadratic function whose x-intercepts are given by $x = 2 \pm 3i$.

**55.** For any quadratic function $f(x) = ax^2 + bx + c$, we know $a$ controls vertical reflections and whether the graph will be wider ($|a| < 1$) or more narrow ($|a| > 1$). This fact can be used to quickly graph additional points without evaluating the function, by noting that for a vertex at $(h, k)$, $(h + \Delta h, k + a|\Delta h|^2)$ will also be on the graph. (a) Use the vertex formula to verify that for $f(x) = x^2 - 4x$, the vertex is at $(2, -4)$. (b) Noting that for this equation $a = 1$ and using integer values for $\Delta h$, show the points $(2 \pm 1; -4 + 1^2)$, $(2 \pm 2; -4 + 2^2)$, and $(2 \pm 3; -4 + 3^2)$ are also on the graph. Note this process is made very easy by simply counting units on the grid, rather than using a computation.

**56.** Referring to Exercise 31, discuss the nature (real or complex, rational or irrational) and number of zeroes (0, 1, or 2) given by the vertex/intercept formula if (a) $a$ and $k$ have like signs, (b) $a$ and $k$ have unlike signs, (c) $k$ is zero, (d) the ratio $-\frac{k}{a}$ is positive and a perfect square, and (e) the ratio $-\frac{k}{a}$ is positive and not a perfect square.

▶ **MAINTAINING YOUR SKILLS**

**57. (2.3)** Identify the slope and y-intercept for $-4x + 3y = 9$. Do not graph.

**58. (R.6)** Multiply: $\frac{x^2 - 4x + 4}{x^2 + 3x - 10} \cdot \frac{x^2 - 25}{x^2 - 10x + 25}$

**59. (3.1)** Given $f(x) = \sqrt[3]{x + 3}$, find the equation of the function whose graph is that of $f(x)$, shifted right 2 units, reflected across the x-axis, and then down 3 units.

**60. (2.5)** Given $f(x) = 3x^2 + 7x - 6$, solve $f(x) \leq 0$ using the x-intercepts and end-behavior of $f$.

---

| 4.2 | **Synthetic Division; the Remainder and Factor Theorems** |

**LEARNING OBJECTIVES**

*In Section 4.2 you will see how we can:*

☐ **A.** Divide polynomials using long division and synthetic division

☐ **B.** Use the remainder theorem to evaluate polynomials

☐ **C.** Use the factor theorem to factor and build polynomials

☐ **D.** Solve applications using the remainder theorem

To find the zero of a linear function, we can use properties of equality to isolate $x$. To find the zeroes of a quadratic function, we can factor or use the quadratic formula. To find the zeroes of higher degree polynomials, we must first develop additional tools, including synthetic division and the remainder and factor theorems. These will help us write a higher degree polynomial in terms of linear and quadratic polynomials, whose zeroes can easily be found.

## A. Long Division and Synthetic Division

To help understand **synthetic division** and its use as a mathematical tool, we first review the process of **long division.**

### Long Division

Polynomial long division closely resembles the division of whole numbers, with the main difference being that *we group each partial product* in parentheses to prevent errors in subtraction.

**EXAMPLE 1 ▶** **Dividing Polynomials Using Long Division**

Divide $x^3 - 4x^2 + x + 6$ by $x - 1$.

**Solution ▶** The divisor is $(x - 1)$ and the dividend is $(x^3 - 4x^2 + x + 6)$. To find the first multiplier, we compute *the ratio of leading terms* from each expression. Here the ratio $\frac{x^3 \text{ from dividend}}{x \text{ from divisor}}$ shows our first multiplier will be "$x^2$," with $x^2(x - 1) = x^3 - x^2$.

$$\begin{array}{r} x^2 \phantom{- 4x^2 + x + 6} \\ x - 1 \overline{)\, x^3 - 4x^2 + x + 6} \\ \underline{-(x^3 - x^2)} \phantom{+ x + 6} \end{array} \text{ subtraction } \longrightarrow \begin{array}{r} x^2 \phantom{- 4x^2 + x + 6} \\ x - 1 \overline{)\, x^3 - 4x^2 + x + 6} \\ \underline{\mathbf{-x^3 + x^2}} \phantom{+ x + 6} \\ -3x^2 + x \phantom{+ 6} \end{array} \text{ algebraic addition}$$

At each stage, after writing the subtraction as algebraic addition (distributing the negative) we compute the sum in each column and "bring down" the next term. Each following multiplier is found as before, using the ratio $\frac{ax^k \text{ next leading term}}{x \text{ from divisor}}$.

$$\begin{array}{lrl}
 & x^2 - 3x - 2 \phantom{+ 6} & \\
 & x - 1 \overline{)\, x^3 - 4x^2 + x + 6} & \\
 & \underline{-(x^3 - x^2)} \phantom{+ x + 6} & \\
\text{next multiplier: } \frac{-3x^2}{x} = -3x & -3x^2 + x \phantom{+ 6} & \\
 & \underline{-(-3x^2 + 3x)} \phantom{+ 6} & -3x(x - 1) = -3x^2 + 3x\text{, subtract } -3x^2 + 3x \\
\textbf{next multiplier: } \frac{-2x}{x} = -2 & -2x + 6 & \text{algebraic addition, bring down next term} \\
 & \underline{\mathbf{-(-2x + 2)}} & -2(x - 1) = -2x + 2\text{, subtract } -2x + 2 \\
 & 4 & \text{algebraic addition, remainder is 4}
\end{array}$$

The result shows $\frac{x^3 - 4x^2 + x + 6}{x - 1} = x^2 - 3x - 2 + \frac{4}{x - 1}$, or after multiplying both sides by $x - 1$, $x^3 - 4x^2 + x + 6 = (x - 1)(x^2 - 3x - 2) + 4$.

**Now try Exercises 5 through 10 ▶**

The process illustrated is called the **division algorithm,** and like the division of whole numbers, the final result can be checked by multiplication.

$$\begin{array}{rll}
\text{check: } \overset{\text{dividend}}{x^3 - 4x^2 + x + 6} & = \overset{\text{divisor}}{(x - 1)}\overset{\text{quotient}}{(x^2 - 3x - 2)} + \overset{\text{remainder}}{4} & \\
 & = (x^3 - 3x^2 - 2x - x^2 + 3x + 2) + 4 & \text{multiply} \\
 & = (x^3 - 4x^2 + x + 2) + 4 & \text{combine like terms} \\
 & = x^3 - 4x^2 + x + 6 \checkmark & \text{add remainder}
\end{array}$$

In general, the division algorithm for polynomials says:

**Division of Polynomials**

Given polynomials $p(x)$ and $d(x) \neq 0$, there exist unique polynomials $q(x)$ and $r(x)$ such that

$$p(x) = d(x)q(x) + r(x),$$

where $r(x) = 0$ or the degree of $r(x)$ is less than the degree of $d(x)$.
Here, $d(x)$ is called the *divisor,* $q(x)$ is the *quotient,* and $r(x)$ is the *remainder.*

In other words, "a polynomial of greater degree can be divided by a polynomial of equal or lesser degree to obtain a quotient and a remainder." As with whole numbers, if the remainder is zero, the divisor is a factor of the dividend.

## Synthetic Division

As the word "synthetic" implies, synthetic division not only *simulates* the long division process, but also condenses it and makes it more efficient *when the divisor is linear.* The process works by capitalizing on the repetition found in the division algorithm. First, the polynomials involved are written in decreasing order of degree, so the variable part of each term is unnecessary as we can let the *position of each coefficient* indicate the degree of the term. For the dividend from Example 1, $1 \quad -4 \quad 1 \quad 6$ would represent the polynomial $1x^3 - 4x^2 + 1x + 6$. Also, each stage of the algorithm involves a product of the divisor with the next multiplier, followed by a subtraction. These can likewise be computed using the coefficients only, as the degree of each term is still determined by its position. Here is the division from Example 1 in the synthetic division format. Note that we must use the *zero of the divisor* (as in $x = \frac{3}{2}$ for a divisor of $2x - 3$, or in this case, "1" from $x - 1 = 0$) and the coefficients of the dividend in the following format:

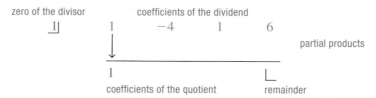

As this template indicates, the quotient and remainder will be read from the last row.

The arrow indicates we begin by "dropping the leading coefficient into place." We then multiply this coefficient by the "divisor," then place the result in the next column and add. Note that using the zero of the divisor enables us to *add in each column directly,* rather than subtracting then changing to algebraic addition as before.

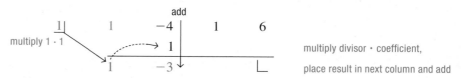

In a sense, we "multiply in the diagonal direction," and "add in the vertical direction." Repeat the process until the division is complete.

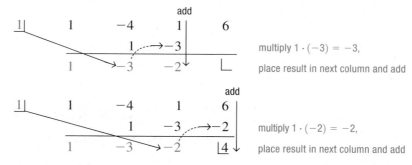

The quotient is read from the last row by noting the remainder is 4, leaving the coefficients $1 \quad -3 \quad -2$, which translate back into the polynomial $1x^2 - x - 2$. The final result is identical to that in Example 1, but the new process is more efficient, since all stages are actually computed on a single template as shown here:

$$\begin{array}{r}
\text{zero of the divisor} \qquad\qquad\qquad \text{coefficients of the dividend} \\
\left. 1\right| \qquad 1 \qquad -4 \qquad 1 \qquad 6 \\
\downarrow \qquad\quad 1 \qquad -3 \qquad -2 \\
\hline
1 \qquad -3 \qquad -2 \qquad \left| 4 \right. \\
\text{coefficients of the quotient} \qquad \text{remainder}
\end{array}$$

**EXAMPLE 2 ▶**  **Dividing Polynomials Using Synthetic Division**

Compute the quotient of $(x^3 + 3x^2 - 4x - 12)$ and $(x + 2)$, then check your answer.

**Solution ▶**  Using $-2$ as our "divisor" (from $x + 2 = 0$), we set up the synthetic division template and begin.

use $-2$ as a "divisor"  $\underline{-2}|$

$$\begin{array}{rrrr}
1 & 3 & -4 & -12 \\
\downarrow & -2 & -2 & 12 \\
\hline
1 & 1 & -6 & \left| 0 \right.
\end{array}$$

drop lead coefficient into place;
multiply by divisor, place result
in next column and add

The result shows $\frac{x^3 + 3x^2 - 4x - 12}{x + 2} = x^2 + x - 6$, with no remainder.

**Check ▶**
$$(x + 2)(x^2 + x - 6)$$
$$(x^3 + x^2 - 6x + 2x^2 + 2x - 12)$$
$$x^3 + 3x^2 - 4x - 12 \checkmark$$

**Now try Exercises 11 through 18 ▶**

Note that in synthetic division, the degree of $q(x)$ will always be one less than $p(x)$, since the process requires a linear divisor (degree 1).

Since the division process is so dependent on the place value (degree) of each term, polynomials such as $2x^3 + 3x + 7$, which has no term of degree 2, must be written using a zero *placeholder:* $2x^3 + \mathbf{0}x^2 + 3x + 7$. This ensures that like place values "line up" as we carry out the division.

**EXAMPLE 3 ▶**  **Dividing Polynomials Using a Zero Placeholder**

Compute the quotient $\frac{2x^3 + 3x + 7}{x - 3}$ and check your answer.

**Solution ▶**  use 3 as a "divisor"  $\underline{3}|$

$$\begin{array}{rrrr}
2 & 0 & 3 & 7 \\
\downarrow & 6 & 18 & 63 \\
\hline
2 & 6 & 21 & \left| 70 \right.
\end{array}$$

note placeholder **0** for "$x^2$" term

The result shows $\frac{2x^3 + 3x + 7}{x - 3} = 2x^2 + 6x + 21 + \frac{70}{x - 3}$. Multiplying by $x - 3$ gives $2x^3 + 3x + 7 = (x - 3)(2x^2 + 6x + 21) + 70$

**Check ▶**
$$(x - 3)(2x^2 + 6x + 21) + 70$$
$$(2x^3 + 6x^2 + 21x - 6x^2 - 18x - 63) + 70$$
$$2x^3 + 3x + 7 \checkmark$$

**Now try Exercises 19 through 28 ▶**

As noted earlier, for synthetic division the divisor must be a linear polynomial and the zero of this divisor is used. This means for the quotient $\frac{3x^4 - 2x^3 - 21x^2 + 32x - 12}{3x - 2}$ we have $3x - 2 = 0$, and $x = \frac{2}{3}$ would be used for synthetic division. **See Example 6(c).** Finally, if the divisor is nonlinear, long division must be used.

**EXAMPLE 4** ▶ **Division with a Nonlinear Divisor**

Compute the quotient: $\frac{2x^4 + x^3 - 7x^2 + 3}{x^2 - 2}$.

**Solution** ▶ Write the dividend as $2x^4 + x^3 - 7x^2 + \mathbf{0}x + 3$, and the divisor as $x^2 + \mathbf{0}x - 2$.

The quotient of leading terms gives $\dfrac{\overset{\text{from dividend}}{2x^4}}{\underset{\text{from divisor}}{x^2}} = 2x^2$ as our first multiplier.

$$
\begin{array}{r}
2x^2 + x - \mathbf{3} \\
x^2 + 0x - 2 \overline{\smash{)}2x^4 + x^3 - 7x^2 + 0x + 3} \\
\end{array}
$$

Multiply $2x^2(x^2 + 0x - 2)$   $-(2x^4 + 0x^3 - 4x^2)$   subtract (algebraic addition)

$\overline{\phantom{xx}x^3 - 3x^2 + 0x}$   bring down next term

Multiply $x(x^2 + 0x - 2)$   $-(x^3 + 0x^2 - 2x)$   subtract (algebraic addition)

$\overline{\phantom{xx}-3x^2 + 2x + 3}$   bring down next term

**Multiply $-3(x^2 + 0x - 2)$**   $-(-3x^2 + \mathbf{0}x + \mathbf{6})$   subtract (algebraic addition)

$\overline{\phantom{xxxx}2x - 3}$   remainder is $2x - 3$

Since the degree of $2x - 3$ (degree 1) is less than the degree of the divisor (degree 2), the process is complete.

$$\frac{2x^4 + x^3 - 7x^2 + 3}{x^2 - 2} = 2x^2 + x - 3 + \frac{2x - 3}{x^2 - 2}$$

☑ **A.** You've just seen how we can divide polynomials using long division and synthetic division

**Now try Exercises 29 through 32** ▶

Note that we elected to keep the solution to Example 4 in the form $\frac{p(x)}{d(x)} = q(x) + \frac{r(x)}{d(x)}$, instead of multiplying both sides by $d(x)$.

## B. The Remainder Theorem

In Example 2, we saw that $(x^3 + 3x^2 - 4x - 12) \div (x + 2) = x^2 + x - 6$, with remainder zero. Similar to whole number division, this means $x + 2$ must be a factor of $x^3 + 3x^2 - 4x - 12$, a fact made clear when we checked our answer: $(x + 2)(x^2 + x - 6) = x^3 + 3x^2 - 4x - 12$. Now consider the functions $p(x) = x^3 + 5x^2 + 2x - 8$, $d(x) = x + 3$, and their quotient $\frac{p(x)}{d(x)} = \frac{x^3 + 5x^2 + 2x - 8}{x + 3}$. Using $-3$ as the divisor in synthetic division gives

$$
\text{use } -3 \text{ as a "divisor"} \quad \underline{-3} \begin{array}{|rrrr}
1 & 5 & 2 & -8 \\
\downarrow & -3 & -6 & 12 \\
\hline
1 & 2 & -4 & \underline{|4}
\end{array}
$$

This shows $x + 3$ is *not* a factor of $p(x)$, since it didn't divide evenly (the remainder is not zero). However, from the result $p(x) = (x + 3)(x^2 + 2x - 4) + 4$, we make a remarkable observation—if we evaluate $p(-3)$, *the quotient portion becomes zero*, showing $p(-3) = 4$ (the remainder).

$$p(-3) = (-3 + 3)[(-3)^2 + 2(-3) - 4] + 4$$
$$= (\mathbf{0})(-1) + 4$$
$$= 4$$

This result can be verified by evaluating $p(-3)$ in its original form:

$$p(x) = x^3 + 5x^2 + 2x - 8$$
$$p(-3) = (-3)^3 + 5(-3)^2 + 2(-3) - 8$$
$$= -27 + 45 + (-6) - 8$$
$$= 4$$

The result is no coincidence, and illustrates the conclusion of the **remainder theorem.**

### The Remainder Theorem

If a polynomial $p(x)$ is divided by $x - c$ using synthetic division, the remainder is equal to $p(c)$.

To see why, we begin with the division algorithm $p(x) = (x - c)d(x) + r(x)$. Since $d(x)$ is linear, $r(x)$ must be constant. Evaluating at $x = c$ gives

$$p(c) = (c - c)d(c) + r$$
$$p(c) = (0)d(c) + r$$
$$p(c) = r$$

This gives us a powerful tool for evaluating polynomials. Where a direct evaluation involves powers of numbers and a long series of calculations, synthetic division reduces the process to simple products and sums.

**EXAMPLE 5** ▶ **Using the Remainder Theorem to Evaluate Polynomials**

Use the remainder theorem to find $p(-5)$ for $p(x) = x^4 + 3x^3 - 8x^2 + 5x - 6$. Verify the result using a substitution.

**Solution** ▶

use $-5$ as a "divisor"   $-5\rfloor$   $\begin{array}{rrrrr} 1 & 3 & -8 & 5 & -6 \\ & -5 & 10 & -10 & 25 \\ \hline 1 & -2 & 2 & -5 & \underline{19} \end{array}$

The result shows $p(-5) = 19$.

**Check** ▶

$$p(-5) = (-5)^4 + 3(-5)^3 - 8(-5)^2 + 5(-5) - 6$$
$$= 625 - 375 - 200 - 25 - 6$$
$$= 625 - 606$$
$$= 19$$

☑ **B.** You've just seen how we can use the remainder theorem to evaluate polynomials

Now try Exercises 33 through 46 ▶

Since $p(-5) = 19$, we know $(-5, 19)$ must be a point of the graph of $p(x)$. The ability to quickly evaluate polynomial functions using the remainder theorem will be used extensively in the sections that follow.

## C. The Factor Theorem

As a consequence of the remainder theorem, when $p(x)$ is divided by $x - c$ and the remainder is 0, $p(c) = 0$ and $c$ is a zero of the polynomial. The relationship between $x - c$, $c$, and $p(c) = 0$ are given in the **factor theorem.**

## The Factor Theorem

For a polynomial $p(x)$,

1. If $c$ is a zero of $p$, then $x - c$ is a factor of $p(x)$.
2. If $x - c$ is a factor of $p(x)$, then $c$ is a zero of $p$: $p(c) = 0$.

The remainder and factor theorems often work together to help us find factors of higher degree polynomials.

**EXAMPLE 6** ▶ **Using the Factor Theorem to Find Factors of a Polynomial**

Use the factor theorem to determine if

   **a.** $x - 2$     **b.** $x + 1$      **c.** $3x - 2$

are factors of $p(x) = 3x^4 - 2x^3 - 21x^2 + 32x - 12$.

**Solution** ▶ **a.** If $x - 2$ is a factor, then $p(2)$ must be 0. Using the remainder theorem we have

$$
\begin{array}{r|rrrrr}
2 & 3 & -2 & -21 & 32 & -12 \\
  &   & 6  & 8   & -26 & 12 \\
\hline
  & 3 & 4  & -13 & 6   & \underline{|0}
\end{array}
$$

Since the remainder is zero, we know $p(2) = 0$ (remainder theorem) and $x - 2$ is a factor (factor theorem).

**b.** Similarly, if $x + 1$ is a factor, then $p(-1)$ must be 0.

$$
\begin{array}{r|rrrrr}
-1 & 3 & -2 & -21 & 32 & -12 \\
   &   & -3 & 5   & 16 & -48 \\
\hline
   & 3 & -5 & -16 & 48 & \underline{|-60}
\end{array}
$$

Since the remainder is not zero, $x + 1$ is not a factor of $p$.

**c.** The zero of the divisor $3x - 2 = 0$ is $x = \frac{2}{3}$, and this value is used in the synthetic division.

$$
\begin{array}{r|rrrrr}
\frac{2}{3} & 3 & -2 & -21 & 32 & -12 \\
   &   & 2 & 0   & -14 & 12 \\
\hline
   & 3 & 0 & -21 & 18 & \underline{|0}
\end{array}
$$

Since the remainder is zero, $p\left(\frac{2}{3}\right) = 0$ (remainder theorem) and $x - \frac{2}{3}$ is the related factor (factor theorem). The original factor $3x - 2$ is recovered by noting that the quotient polynomial $q(x) = 3x^3 - 21x + 18$ has a common factor of 3, which will be factored out and applied to $x - \frac{2}{3}$. Starting with the partially factored form we have

$$
\begin{aligned}
p(x) &= \left(x - \frac{2}{3}\right)(3x^3 - 21x + 18) &&\text{partially factored form} \\
     &= \left(x - \frac{2}{3}\right)(3)(x^3 - 7x + 6) &&\text{factor out 3} \\
     &= (3x - 2)(x^3 - 7x + 6) &&\text{multiply } 3\left(x - \frac{2}{3}\right)
\end{aligned}
$$

This form of simplification will always take place when the zero found using synthetic division is a fraction and the coefficients of the polynomial are integers.

**Now try Exercises 47 through 52** ▶

As a final note on Example 6, if the given polynomial has integer coefficients, there should be no hesitation to use fractions in the synthetic division process. The fraction will be a zero only if all values in the quotient line are integers (i.e., all of the products and sums must be integers).

**EXAMPLE 7** ▶   **Building a Polynomial Using the Factor Theorem**

A polynomial $p(x)$ has three zeroes at $x = 3$, $\sqrt{2}$, and $-\sqrt{2}$. Use the factor theorem to find such a polynomial.

Solution ▶   Using the factor theorem, the factors of $p(x)$ must be $(x - 3)$, $(x - \sqrt{2})$, and $(x + \sqrt{2})$. Computing the product will yield the polynomial.

$$p(x) = (x - 3)(x - \sqrt{2})(x + \sqrt{2})$$
$$= (x - 3)(x^2 - 2)$$
$$= x^3 - 3x^2 - 2x + 6$$

Now try Exercises 53 through 60 ▶

Actually, the result obtained in Example 7 is not unique, since any polynomial of the form $a(x^3 - 3x^2 - 2x + 6)$ will also have the same three zeroes for $a \in \mathbb{R}$. Figure 4.11 shows the graph of $Y_1 = p(x)$, as well as graph of $Y_2 = 2p(x)$. Although they differ by a vertical stretch, the zeroes remain the same. Likewise, the graph of $-1p(x)$ would be a vertical reflection, *but still with the same zeroes.*

**Figure 4.11**

**EXAMPLE 8** ▶   **Finding Zeroes Using the Factor Theorem**

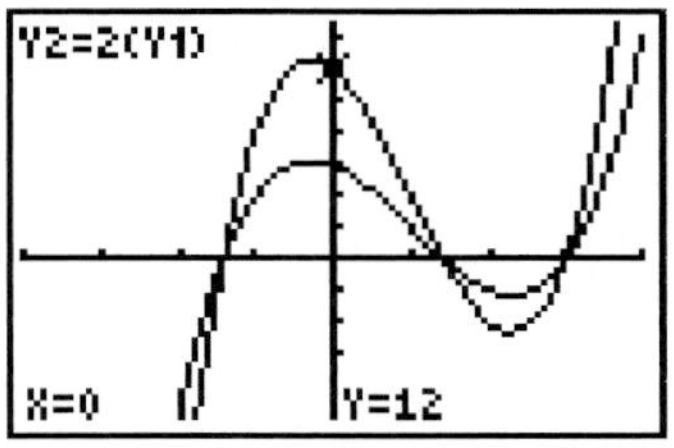

Given that 2 is a zero of $p(x) = x^4 + x^3 - 10x^2 - 4x + 24$, use the factor theorem to help find all other zeroes.

Solution ▶   Using synthetic division gives:

$$
\begin{array}{r|rrrrr}
\text{use 2 as a "divisor"} \quad 2 & 1 & 1 & -10 & -4 & 24 \\
& \downarrow & 2 & 6 & -8 & -24 \\
\hline
& 1 & 3 & -4 & -12 & \underline{|0}
\end{array}
$$

> **WORTHY OF NOTE**
>
> In Section R.5 we noted a third-degree polynomial $ax^3 + bx^2 + cx + d$ can be factored by grouping if $ad = bc$. In Example 8, $1(-12) = 3(-4)$ and the polynomial is factorable.

Since the remainder is zero, $x - 2$ is a factor and $p$ can be written:

$$x^4 + x^3 - 10x^2 - 4x + 24 = (x - 2)(x^3 + 3x^2 - 4x - 12)$$

Note the quotient polynomial can be factored by grouping to find the remaining factors of $p$.

$$
\begin{aligned}
x^4 + x^3 - 10x^2 - 4x + 24 &= (x - 2)(x^3 + 3x^2 - 4x - 12) && \text{group terms (in color)} \\
&= (x - 2)[x^2(x + 3) - 4(x + 3)] && \text{remove common factors from each group} \\
&= (x - 2)[(x + 3)(x^2 - 4)] && \text{factor common binomial} \\
&= (x - 2)(x + 3)(x + 2)(x - 2) && \text{factor difference of squares} \\
&= (x + 3)(x + 2)(x - 2)^2 && \text{completely factored form}
\end{aligned}
$$

☑ **C. You've just seen how we can use the factor theorem to factor and build polynomials**

The final result shows $x - 2$ is actually a repeated factor, and the remaining zeroes of $p$ are $-3$ and $-2$.

Now try Exercises 61 through 70 ▶

## D. Applications

While the factor and remainder theorems are valuable tools for factoring higher degree polynomials, each has applications that extend beyond this use.

**EXAMPLE 9** ▶    Using the Remainder Theorem to Solve a Discharge Rate Application

The *discharge rate* of a river is a measure of the river's water flow as it empties into a lake, sea, or ocean. The rate depends on many factors, but is primarily influenced by the precipitation in the surrounding area and is often seasonal. Suppose the discharge rate of the Shimote River was modeled by $D(m) = -m^4 + 22m^3 - 147m^2 + 317m + 150$, where $D(m)$ represents the discharge rate in thousands of cubic meters of water per second in month $m$ ($m = 1 \rightarrow$ Jan).

   **a.** What was the discharge rate in June (summer heat)?

   **b.** Is the discharge rate higher in February (winter runoff) or October (fall rains)?

Solution ▶    **a.** To find the discharge rate in June, we evaluate $D$ at $m = 6$.
Using the remainder theorem gives

$$
\begin{array}{r|rrrrr}
6 & -1 & 22 & -147 & 317 & 150 \\
  & \downarrow & -6 & 96 & -306 & 66 \\
\hline
  & -1 & 16 & -51 & 11 & \underline{|216}
\end{array}
$$

In June, the discharge rate is about 216,000 m³/sec.

   **b.** For the discharge rates in February ($m = 2$) and October ($m = 10$), we have

$$
\begin{array}{r|rrrrr}
2 & -1 & 22 & -147 & 317 & 150 \\
  & \downarrow & -2 & 40 & -214 & 206 \\
\hline
  & -1 & 20 & -107 & 103 & \underline{|356}
\end{array}
\qquad
\begin{array}{r|rrrrr}
10 & -1 & 22 & -147 & 317 & 150 \\
   & \downarrow & -10 & 120 & -270 & 470 \\
\hline
   & -1 & 12 & -27 & 47 & \underline{|620}
\end{array}
$$

The discharge rate during the fall rains in October is much higher: $620 > 356$.

☑ **D.** You've just seen how we can solve applications using the remainder theorem

Now try Exercises 73 through 78 ▶

---

## 4.2 EXERCISES

▶ **CONCEPTS AND VOCABULARY**

**Fill in each blank with the appropriate word or phrase. Carefully reread the section, if necessary.**

1. If polynomial $P(x)$ is divided by a linear divisor of the form $x - c$, the remainder is identical to _____. This is a statement of the _____ theorem.

2. If $P(c) = 0$, then _____ must be a factor of $P(x)$. Conversely, if _____ is a factor of $P(x)$, then $P(c) = 0$. These are statements from the _____ theorem.

3. Discuss/Explain how to write the quotient and remainder using the last line from a synthetic division.

4. Discuss/Explain why $(a, b)$ is a point on the graph of $P$, given $b$ was the remainder after $P$ was divided by $x - a$ using synthetic division.

## ▶ DEVELOPING YOUR SKILLS

**Divide using long division. Write the result as dividend = (divisor)(quotient) + remainder.**

5. $\dfrac{x^3 - 5x^2 - 4x + 23}{x - 2}$

6. $\dfrac{x^3 + 5x^2 - 17x - 26}{x + 7}$

7. $(2x^3 + 5x^2 + 4x + 17) \div (x + 3)$

8. $(3x^3 + 14x^2 - 2x - 37) \div (x + 4)$

9. $(x^3 - 8x^2 + 11x + 20) \div (x - 5)$

10. $(x^3 - 5x^2 - 22x - 16) \div (x + 2)$

**Divide using synthetic division. Write answers in two ways: (a) $\frac{\text{dividend}}{\text{divisor}} = \text{quotient} + \frac{\text{remainder}}{\text{divisor}}$, and (b) dividend = (divisor)(quotient) + remainder.**

11. $\dfrac{2x^2 - 5x - 3}{x - 3}$    12. $\dfrac{3x^2 + 13x - 10}{x + 5}$

13. $(x^3 - 3x^2 - 14x - 8) \div (x + 2)$

14. $(x^3 - 6x^2 - 24x - 17) \div (x + 1)$

15. $\dfrac{x^3 - 5x^2 - 4x + 23}{x - 2}$    16. $\dfrac{x^3 + 12x^2 + 34x - 9}{x + 7}$

17. $(2x^3 - 5x^2 - 11x - 17) \div (x - 4)$

18. $(3x^3 - x^2 - 7x + 27) \div (x - 1)$

**Divide using synthetic division. Write answers as dividend = (divisor)(quotient) + remainder.**

19. $(x^3 + 5x^2 + 7) \div (x + 1)$

20. $(x^3 - 3x^2 - 37) \div (x - 5)$

21. $(x^3 - 13x - 12) \div (x - 4)$

22. $(x^3 - 7x + 6) \div (x + 3)$

23. $\dfrac{3x^3 - 8x + 12}{x - 1}$    24. $\dfrac{2x^3 + 7x - 81}{x - 3}$

25. $(n^3 + 27) \div (n + 3)$

26. $(m^3 - 8) \div (m - 2)$

27. $(x^4 + 3x^3 - 16x - 8) \div (x - 2)$

28. $(x^4 + 3x^3 + 29x - 21) \div (x + 3)$

**Compute each indicated quotient. Write answers in the form $\frac{\text{dividend}}{\text{divisor}} = \text{quotient} + \frac{\text{remainder}}{\text{divisor}}$.**

29. $\dfrac{2x^3 + 7x^2 - x + 26}{x^2 + 3}$    30. $\dfrac{x^4 + 3x^3 + 2x^2 - x - 5}{x^2 - 2}$

31. $\dfrac{x^4 - 5x^2 - 4x + 7}{x^2 - 1}$    32. $\dfrac{x^4 + 2x^3 - 8x - 16}{x^2 + 5}$

**Use the remainder theorem to evaluate $P(x)$ as given.**

33. $P(x) = x^3 - 6x^2 + 5x + 12$
    **a.** $P(-2)$          **b.** $P(5)$

34. $Q(x) = x^3 + 4x^2 - 8x - 15$
    **a.** $Q(-2)$          **b.** $Q(3)$

35. $p(x) = x^4 - 4x^2 + x + 1$
    **a.** $p(-2)$          **b.** $p(2)$

36. $q(x) = x^4 + 3x^3 - 2x - 4$
    **a.** $q(-2)$          **b.** $q(2)$

37. $f(x) = 2x^3 - 7x + 33$
    **a.** $f(-2)$          **b.** $f(-3)$

38. $g(x) = -2x^3 + 9x^2 - 11$
    **a.** $g(-2)$          **b.** $g(-1)$

39. $h(x) = 2x^3 + 3x^2 - 9x - 10$
    **a.** $h(\frac{3}{2})$          **b.** $h(-\frac{5}{2})$

40. $s(x) = 3x^3 + 11x^2 + 2x - 16$
    **a.** $s(\frac{1}{3})$          **b.** $s(-\frac{8}{3})$

**Use the remainder theorem to show the given value is a zero.**

41. $P(x) = x^3 + 2x^2 - 5x - 6; x = -3$

42. $Q(x) = x^3 + 3x^2 - 16x + 12; x = -6$

43. $f(x) = x^3 - 7x + 6; x = 2$

44. $g(x) = x^3 - 13x + 12; x = -4$

45. $h(x) = 9x^3 + 18x^2 - 4x - 8; x = \dfrac{2}{3}$

46. $t(x) = 5x^3 + 13x^2 - 9x - 9; x = -\dfrac{3}{5}$

**Use the factor theorem to determine if the expressions given are factors of the polynomial shown.**

47. $f(x) = x^3 - 3x^2 - 13x + 15$
    **a.** $(x + 3)$          **b.** $(x - 5)$

48. $g(x) = x^3 + 2x^2 - 11x - 12$
    **a.** $(x + 4)$          **b.** $(x - 3)$

49. $h(x) = x^3 - 6x^2 + 3x + 10$
    **a.** $(x + 2)$          **b.** $(x - 5)$

50. $p(x) = x^3 + 2x^2 - 5x - 6$
    **a.** $(x - 2)$          **b.** $(x + 4)$

51. $q(x) = -2x^3 - x^2 + 12x - 9$
    **a.** $(x + 3)$          **b.** $(2x - 3)$

52. $r(x) = 3x^3 - 19x^2 + 30x - 8$
    **a.** $(3x - 1)$          **b.** $(x + 4)$

**A polynomial $P$ with integer coefficients has the zeroes and degree indicated. Use the factor theorem to write the function in factored form and standard form. Assume $a = 1$.**

53. $-2, 3, -5$; degree 3

54. $1, -4, 2$; degree 3

55. $-2, \sqrt{3}, -\sqrt{3}$; degree 3

56. $\sqrt{5}, -\sqrt{5}, 4$; degree 3

57. $-5, 2\sqrt{3}, -2\sqrt{3}$; degree 3

58. $4, 3\sqrt{2}, -3\sqrt{2}$; degree 3

59. $1, -2, \sqrt{10}, -\sqrt{10}$; degree 4

60. $\sqrt{7}, -\sqrt{7}, 3, -1$; degree 4

In Exercises 61 through 66, a known zero of the polynomial is given. Use the factor theorem to write the polynomial in completely factored form.

**61.** $P(x) = x^3 - 5x^2 - 2x + 24; x = -2$

**62.** $Q(x) = x^3 - 7x^2 + 7x + 15; x = 3$

**63.** $p(x) = x^4 + 2x^3 - 12x^2 - 18x + 27; x = -3$

**64.** $q(x) = x^4 + 4x^3 - 6x^2 - 4x + 5; x = 1$

**65.** $f(x) = 2x^3 + 11x^2 - x - 30; x = \frac{3}{2}$

**66.** $g(x) = 3x^3 + 2x^2 - 75x - 50; x = -\frac{2}{3}$

If $p(x)$ is a polynomial with rational coefficients and a leading coefficient of $a = 1$, the rational zeroes of $p$ (if they exist) *must be factors of the constant term*. Use this property to factor each polynomial completely.

**67.** $p(x) = x^3 - 3x^2 - 9x + 27$

**68.** $p(x) = x^3 - 4x^2 - 16x + 64$

**69.** $p(x) = x^3 - 6x^2 + 12x - 8$

**70.** $p(x) = x^3 - 15x^2 + 75x - 125$

## ▶ WORKING WITH FORMULAS

**Volume of an open box:** $V(x) = 4x^3 - 84x^2 + 432x$

An open box is constructed by cutting square corners from a 24 in. by 18 in. sheet of cardboard and folding up the sides. Its volume is given by the formula shown, where $x$ represents the length of the square cuts.

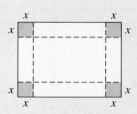

**71.** Find a function $V$ that models the volume of the box for any $x$ by $x$ squares cut. Write the result in factored form and standard form.

**72.** Given the volume is 640 in$^3$, use synthetic division and the remainder theorem to determine if the squares were 2-, 3-, 4-, or 5-in. squares and state the dimensions of the box.

## ▶ APPLICATIONS

**73. Tourist population:** During the 12 weeks of summer, the population of tourists at a popular beach resort is modeled by the polynomial $P(w) = -0.1w^4 + 2w^3 - 14w^2 + 52w + 5$, where $P(w)$ is the tourist population (in 1000s) during week $w$. Use the remainder theorem to help answer the following questions.

   **a.** Were there more tourists at the resort in week 5 ($w = 5$) or week 10? How many more tourists?

   **b.** Were more tourists at the resort one week after opening ($w = 1$) or 1 week before closing ($w = 11$). How many more tourists?

   **c.** The tourist population peaked (reached its highest) between weeks 7 and 10. Use the remainder theorem to determine the peak week.

**74. Debt load:** Due to a fluctuation in tax revenues, a county government is projecting a deficit for the next 12 months, followed by a quick recovery and the repayment of all debt near the end of this period. The projected balance can be modeled by the polynomial $D(m) = 0.1m^4 - 2m^3 + 15m^2 - 64m - 3$, where $D(m)$ represents the amount of debt (in millions of dollars) in month $m$. Use the remainder theorem to help answer the following questions.

   **a.** Was the debt higher in month 5 ($m = 5$) or month 10 of this period? How much higher?

   **b.** Was the debt higher in the first month of this period (1 month into the deficit) or after the 11th month (1 month before the expected recovery)? How much higher?

   **c.** The total debt reached its maximum between months 7 and 10. Use the remainder theorem to determine which month.

**75. Volume of water:** The volume of water in a rectangular, inground, swimming pool is given by $V(x) = x^3 + 11x^2 + 24x$, where $V(x)$ is the volume in cubic feet when the water is $x$ ft high. (a) Use the remainder theorem to find the volume when $x = 3$ ft. (b) If the volume is 100 ft$^3$ of water, what is the height $x$? (c) If the maximum capacity of the pool is 1000 ft$^3$, what is the maximum depth (to the nearest integer)?

**76. Amusement park attendance:** Attendance at an amusement park depends on the weather. After opening in spring, attendance rises quickly, slows during the summer, soars in the fall, then quickly falls with the approach of winter when the park closes. The model for attendance is given by $A(m) = -\frac{1}{4}m^4 + 6m^3 - 52m^2 + 196m - 260$, where $A(m)$ represents the number of people attending in month $m$ (in thousands). (a) Use the remainder theorem to determine whether more people go to the park in April ($m = 4$) or June ($m = 6$). (b) If maximum attendance was 28 thousand, in what month did the maximum occur? (c) When did the park close?

**77. Heating and cooling costs:** The average cost of heating and cooling a home is modeled by the polynomial $C(m) = -0.2m^4 + 5m^3 - 40.2m^2 + 109.2m + 80.6$, where $C(m)$ represents the cost in month $m$, and $1 \le m \le 12$. Use synthetic division to (a) find the cost in February ($m = 2$), and (b) determine if the cost in July ($m = 7$) is higher or lower than the cost in February. Discuss possible reasons why.

**78. Music and heart rates:** It's a well-known fact that music has a physiological effect on the human body. As an experiment, the heart rate of a subject was monitored as

he listened closely to all four movements of the William Tell Overture: (1) the sunrise, (2) the storm, (3) the call to the dairy cows, and (4) the (Lone Ranger) finale. His heart rate during this time was modeled by the polynomial $h(t) = -0.2t^4 + 5.8t^3 - 58t^2 + 234t - 218$, where $h(t)$ represents his heart rate in beats per minute after $t$ min, with $2 \le t \le 13$. Use synthetic division to (a) find his heart rate during the storm ($t = 4$) and (b) determine if his heart rate during the storm was higher or lower than near the end of the finale ($t = 10$). Discuss possible reasons why.

**In these applications, synthetic division is applied in the usual way, treating $k$ as an unknown constant.**

**79.** Find a value of $k$ that will make $-2$ a zero of $f(x) = x^3 - 3x^2 - 5x + k$.

**80.** Find a value of $k$ that will make $3$ a zero of $g(x) = x^3 + 2x^2 - 7x + k$.

**81.** For what value(s) of $k$ will $x - 2$ be a factor of $p(x) = x^3 - 3x^2 + kx + 10$?

**82.** For what value(s) of $k$ will $x + 5$ be a factor of $q(x) = x^3 + 6x^2 + kx + 50$?

▶ **EXTENDING THE CONCEPT**

**83.** Since we use a base-10 number system, numbers like 1196 can be written in polynomial form as $p(x) = 1x^3 + 1x^2 + 9x + 6$, where $x = 10$. Divide $p(x)$ by $x + 3$ using synthetic division and write your answer as $\frac{x^3 + x^2 + 9x + 6}{x + 3} = $ quotient $+ \frac{\text{remainder}}{\text{divisor}}$. For $x = 10$, what is the value of quotient $+ \frac{\text{remainder}}{\text{divisor}}$? What is the result of dividing 1196 by $10 + 3 = 13$? What can you conclude?

**84.** To investigate whether the remainder and factor theorems can be applied when the coefficients or zeroes of a polynomial are complex, try using the factor theorem to find a polynomial with degree 3, whose zeroes are $x = 2i$, $x = -2i$, and $x = 3$. Then see if the

result can be verified using the remainder theorem and these zeroes. What does the result suggest? Also see Exercise 85.

**85.** Though not a direct focus of this course, the remainder and factor theorems, as well as synthetic division, *can also be applied using complex numbers*. Use the remainder theorem to show the value given is a zero of $P(x)$.

   **a.** $P(x) = x^3 - 4x^2 + 9x - 36$; $x = 3i$
   **b.** $P(x) = x^4 + x^3 + 2x^2 + 4x - 8$; $x = -2i$
   **c.** $P(x) = -x^3 + x^2 - 3x - 5$; $x = 1 + 2i$
   **d.** $P(x) = -x^3 + x^2 - 8x - 10$; $x = 1 + 3i$

▶ **MAINTAINING YOUR SKILLS**

**86. (1.1)** John and Rick are out orienteering. Rick finds the last marker first and is heading for the finish line, 1275 yd away. John is just seconds behind, and after locating the last marker tries to overtake Rick, who by now has a 250-yd lead. If Rick runs at 4 yd/sec and John runs at 5 yd/sec, will John catch Rick before they reach the finish line?

**87. (1.5)** Solve for $w$: $-2(3w^2 + 5) + 3 = -7w + w^2 - 7$

**88. (2.2)** The profit of a small business increased linearly from $5000 in 2005 to $12,000 in 2010. Find a linear function $G(t)$ modeling the growth of the company's profit (let $t = 0$ correspond to 2005).

**89. (2.5)** Given $f(x) = x^2 - 4x$, use the average rate of change formula to find $\frac{\Delta y}{\Delta x}$ in the interval $x \in [1.0, 1.1]$.

## LEARNING OBJECTIVES

*In Section 4.3 you will see how we can:*

☐ **A.** Use the intermediate value theorem to identify intervals containing a real zero

☐ **B.** Find rational zeros of a real polynomial function using the rational zeroes theorem

☐ **C.** Obtain information on the zeroes of real polynomials using Descartes' rule of signs and the upper/lower bounds theorem

☐ **D.** Apply the fundamental theorem of algebra and the linear factorization theorem

☐ **E.** Solve applications of polynomial functions

This section represents one of the highlights in the college algebra curriculum, because it offers a look at what many call *the big picture*. The ideas presented are the result of a cumulative knowledge base developed over a long period of time, and give a fairly comprehensive view of the study of polynomial functions.

## A. Real Polynomials and the Intermediate Value Theorem

In Section 4.2, we used synthetic division along with the remainder and factor theorems to evaluate and factor certain polynomials. These were important steps toward the more general task of locating the real zeroes of a polynomial, which is the focus of this section. Because polynomial graphs are continuous (there are no holes or breaks in the graph), one useful tool in this effort is called the **intermediate value theorem (IVT)**.

### The Intermediate Value Theorem

Given $P$ is a polynomial with real coefficients, if $P(a)$ and $P(b)$ have opposite signs, then there is *at least* one number $c$ between $a$ and $b$ such that $P(c) = 0$.

**Figure 4.12**

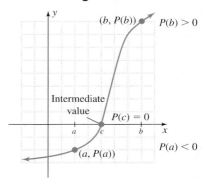

---

### WORTHY OF NOTE

You might recall a similar idea was used in Section 2.5, where we noted the graph of $P(x)$ crosses the $x$-axis at the zeroes determined by linear factors, with a corresponding change of sign in the function values.

While not a formal proof, the graph in Figure 4.12 offers an intuitive understanding of the theorem. If a continuous function $P(x)$ changes sign from positive to negative, or from negative to positive, it must cross the $x$-axis at least once along the way. If it crosses at $x = c$, then $P(c) = 0$.

---

**EXAMPLE 1** ▶ **Finding Zeroes Using the Intermediate Value Theorem**

Use the intermediate value theorem to show $P(x) = x^3 - 9x + 6$ has at least one zero in the interval given:

    **a.** $[-4, 3]$      **b.** $[-4, -3]$      **c.** $[0, 1]$

**Solution** ▶    **a.** Begin by evaluating $P$ at $x = -4$ and $x = 3$.

$$P(-4) = (-4)^3 - 9(-4) + 6 \qquad P(3) = (3)^3 - 9(3) + 6$$
$$= -64 + 36 + 6 \qquad\qquad\qquad = 27 - 27 + 6$$
$$= -22 \qquad\qquad\qquad\qquad\qquad = 6$$

Since $P(-4) < 0$ and $P(3) > 0$, there must be at least one number $c$ between $-4$ and $3$ where $P(c) = 0$.

**b.** In part (a) we found $P(-4) < 0$. Evaluating $P(-3)$ gives

$$P(-3) = (-3)^3 - 9(-3) + 6$$
$$= -27 + 27 + 6$$
$$= 6$$

Since $P(-4) < 0$ and $P(-3) > 0$, there must be at least one number $c_1$ between $-4$ and $-3$ where $P(c_1) = 0$. The graph must cross the $x$-axis at least once in this interval.

**c.** Evaluate $P$ at $x = 0$ and $x = 1$.

$$P(0) = (0)^3 - 9(0) + 6 \qquad\qquad P(1) = (1)^3 - 9(1) + 6$$
$$= 0 - 0 + 6 \qquad\qquad\qquad = 1 - 9 + 6$$
$$= 6 \qquad\qquad\qquad\qquad = -2$$

Since $P(0) > 0$ and $P(1) < 0$, there must be at least one number $c_2$ between 0 and 1 where $P(c_2) = 0$. The graph of $P$ shown in Figure 4.13 illustrates why the intermediate value theorem uses the phrase, "there is *at least* one number $c$." In the interval $[-4, 3]$ from part (a), polynomial $P$ actually has three zeores, which include the two located in parts (b) and (c).

**Figure 4.13**

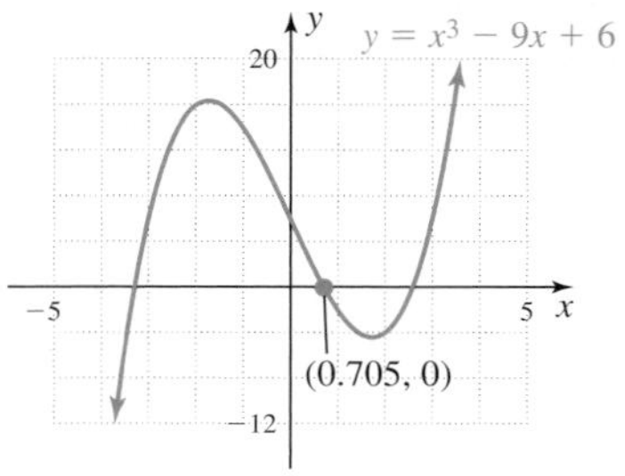

**A.** You've just seen how we can use the intermediate value theorem to identify intervals containing a real zero

**Now try Exercises 5 through 8** ▶

## B. The Rational Zeroes Theorem

The intermediate value theorem tells us whether or not a given interval contains a zero. Our next theorem gives us the information we need to actually *find* certain zeroes of a polynomial. Recall from Section 4.2 that if $c$ is a zero of $P$, when $P(x)$ is divided by $x - c$ using synthetic division, the remainder is zero (from the remainder and factor theorems). To find *divisors that give a remainder of zero,* we make the following observations. To solve $3x^2 - 11x - 20 = 0$ by factoring, a beginner might write out all possible binomial pairs where the **F**irst term in the F-O-I-L process multiplies to $3x^2$ and the **L**ast term multiplies to 20. The six possibilities are shown here:

$$(3x \quad 1)(x \quad 20) \qquad (3x \quad 2)(x \quad 10) \qquad (3x \quad 4)(x \quad 5)$$
$$(3x \quad 20)(x \quad 1) \qquad (3x \quad 10)(x \quad 2) \qquad (3x \quad 5)(x \quad 4)$$

If $3x^2 - 11x - 20$ is factorable using integers, the factors *must be somewhere in this list*. Also note the first coefficient in each binomial must be a factor of the leading coefficient, and the second coefficient must be a factor of the constant term. This means that regardless of which factored form is correct, the solution will be a rational number whose numerator comes from the factors of 20, and whose denominator comes from the factors of 3. The correct factored form is shown here, along with the solution:

$$3x^2 - 11x - 20 = 0$$
$$(3x + 4)(x - 5) = 0$$
$$3x + 4 = 0 \qquad x - 5 = 0$$
$$x = \frac{-4}{3} \begin{array}{l} \leftarrow \text{from the factors of 20} \\ \leftarrow \text{from the factors of 3} \end{array} \qquad x = \frac{5}{1} \begin{array}{l} \leftarrow \text{from the factors of 20} \\ \leftarrow \text{from the factors of 3} \end{array}$$

This same principle also applies to polynomials of higher degree, and these observations suggest the following theorem.

### The Rational Zeroes Theorem

Given $P$ is a polynomial with integer coefficients and $\frac{p}{q}$ is a rational number in lowest terms. If $\frac{p}{q}$ is a zero of $P$, then $p$ must be a factor of the constant term, and $q$ must be a factor of the leading coefficient.

It's helpful to note the theorem implies that if the leading coefficient is 1, the possible rational zeroes are limited to factors of the constant term: $\frac{p}{1} = p$. If the leading coefficient is not 1 and the constant term has a large number of factors, the set of possible rational zeroes becomes rather large. To list these possibilities, it helps to begin with all factor *pairs* of the constant term, then divide each of these by factors of the leading coefficient as shown in Example 2.

**EXAMPLE 2** ▶ **Identifying the Possible Rational Zeroes of a Polynomial**

List all possible rational zeroes of $P(x) = 3x^4 + 14x^3 - x^2 - 42x - 24$.

**Solution** ▶ All rational zeroes must be of the form $\frac{p}{q}$, where $p$ is a factor of $-24$ and $q$ is a factor of 3. The factor pairs of $-24$ are: $\pm1$, $\pm24$, $\pm2$, $\pm12$, $\pm3$, $\pm8$, $\pm4$ and $\pm6$. Dividing each by $\pm1$ and $\pm3$ (the factor pairs of 3), we note division by $\pm1$ will not change any of the previous values, while division by $\pm3$ gives $\pm\frac{1}{3}$, $\pm\frac{2}{3}$, $\pm\frac{8}{3}$, $\pm\frac{4}{3}$ as additional possibilities. Any rational zeroes must be from the set

$$\{\pm1,\ \pm24,\ \pm2,\ \pm12,\ \pm3,\ \pm8,\ \pm4,\ \pm6,\ \pm\tfrac{1}{3},\ \pm\tfrac{2}{3},\ \pm\tfrac{8}{3},\ \pm\tfrac{4}{3}\}.$$

**Now try Exercises 9 through 16** ▶

**Figure 4.14**

The actual zeroes of the function in Example 2 are $x = \sqrt{3}$, $x = -\sqrt{3}$, $x = -\frac{2}{3}$, and $x = -4$ and the graph of $P(x) = 3x^4 + 14x^3 - x^2 - 42x - 24$ is shown in Figure 4.14. Although the *rational* zeroes are indeed in the set noted, it's apparent we need a way to narrow down the number of possibilities (we don't want to try all 24 possible zeroes). If we're able to find even one zero easily, we can rewrite the polynomial using the related factor and the quotient polynomial, then perhaps use trinomial factoring or factoring by grouping to factor further. Many times testing to see if 1 or $-1$ are zeroes will help.

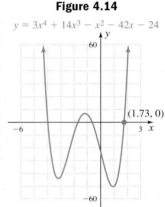

$y = 3x^4 + 14x^3 - x^2 - 42x - 24$

$(1.73, 0)$

**Tests to Determine if 1 or $-1$ Is a Zero of $P$**

For any polynomial $P$,

1. If the sum of the coefficients is zero, then 1 is a root and $x - 1$ is a factor.
2. After changing the sign of all terms with odd degree, if the sum of the coefficients is zero, then $-1$ is a root and $x + 1$ is a factor.

**EXAMPLE 3** ▶ **Finding the Rational Zeroes of a Polynomial**

Find all rational zeroes of $P(x) = 3x^4 - x^3 - 14x^2 + 4x + 8$, and use them to write the function in completely factored form. Then use the factored form to name all zeroes of $P$.

**Solution** ▶

$y = 3x^4 - x^3 - 14x^2 + 4x + 8$

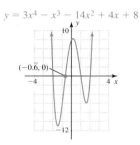

Instead of listing all possibilities using the rational zeroes theorem, we first test for 1 and $-1$, then see if we're able to complete the factorization using other means. The sum of the coefficients is: $3 - 1 - 14 + 4 + 8 = 0$, which means 1 is a zero and $x - 1$ is a factor. By changing the sign on terms of odd degree, we have $3x^4 + x^3 - 14x^2 - 4x + 8$ and $3 + 1 - 14 - 4 + 8 = -6$, showing $-1$ is *not* a zero. Using $x = 1$ and the factor theorem, we have

$$\text{use 1 as a “divisor”} \quad 1 \underline{\rvert} \begin{array}{rrrrr} 3 & -1 & -14 & 4 & 8 \\ & 3 & 2 & -12 & -8 \\ \hline 3 & 2 & -12 & -8 & \underline{\rvert 0} \end{array}$$

and we write $P$ as $P(x) = (x - 1)(3x^3 + 2x^2 - 12x - 8)$. Noting the quotient polynomial can be factored by grouping ($ad = bc$), we need not continue with synthetic division or the factor theorem.

$$\begin{aligned}
P(x) &= (x - 1)(\underline{3x^3 + 2x^2} - \underline{12x - 8}) &\text{group terms} \\
&= (x - 1)[x^2(3x + 2) - 4(3x + 2)] &\text{factor common terms} \\
&= (x - 1)(3x + 2)(x^2 - 4) &\text{factor common binomial} \\
&= (x - 1)(3x + 2)(x + 2)(x - 2) &\text{completely factored form}
\end{aligned}$$

The zeroes of $P$ are 1, $\frac{-2}{3}$, and $\pm 2$. The graph of $P$ is shown in the figure.

**Now try Exercises 17 through 32** ▶

In cases where the quotient polynomial is not easily factored, we continue with synthetic division and other possible zeroes, until the remaining zeroes can be determined.

**EXAMPLE 4** ▶

**Finding the Real Zeroes of a Polynomial**

Find all zeroes of $P(x) = x^5 - 3x^4 - 3x^3 + 13x^2 - 12$.

**Solution** ▶

The test for 1 shows 1 is not a zero. After changing the signs of all terms with odd degree, we have $-1 - 3 + 3 + 13 - 12 = 0$, and find $-1$ is a zero. Using $-1$ with the factor theorem, we continue our search for additional zeroes (note the placeholder zero for the "missing" linear term).

$$\text{use } -1 \text{ as a “divisor”} \quad -1 \underline{\rvert} \begin{array}{rrrrrr} 1 & -3 & -3 & 13 & 0 & -12 \\ & -1 & 4 & -1 & -12 & 12 \\ \hline 1 & -4 & 1 & 12 & -12 & \underline{\rvert 0} \end{array} \quad \begin{array}{l}\text{coefficients of } P \\ \\ \text{coefficients of } q_1(x)\end{array}$$

Here the quotient polynomial is not easily factored, and testing again for $-1$ shows it is not a repeated root. Using the rational zeroes theorem on the quotient polynomial, the possibilities are $\{\pm 1, \pm 12, \pm 2, \pm 6, \pm 3, \pm 4\}$, so we then try 2 (the next simplest possibility), *using the quotient polynomial*:

$$\text{use 2 as a “divisor”} \quad 2 \underline{\rvert} \begin{array}{rrrrr} 1 & -4 & 1 & 12 & -12 \\ & 2 & -4 & -6 & 12 \\ \hline 1 & -2 & -3 & 6 & \underline{\rvert 0} \end{array} \quad \begin{array}{l}\text{coefficients of } q_1(x) \\ \\ \text{coefficients of } q_2(x)\end{array}$$

If you miss the fact that $q_2(x)$ is actually factorable by grouping ($ad = bc$), the process could continue using $-2$ and the current quotient.

$$\text{use } -2 \text{ as a “divisor”} \quad -2 \underline{\rvert} \begin{array}{rrrr} 1 & -2 & -3 & 6 \\ & -2 & 8 & -10 \\ \hline 1 & -4 & 5 & \underline{\rvert -4} \end{array} \quad \text{coefficients of } q_2(x)$$

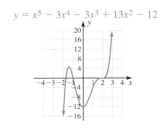

$y = x^5 - 3x^4 - 3x^3 + 13x^2 - 12$

We find $-2$ is not a zero, and in fact, trying all other possibilities shows that *none of them are zeroes*. This reminds us of two important ideas.

1. This process can only find rational zeros (the remaining zeroes may be irrational or nonreal), and
2. Some of the zeroes may be repeated roots.

Testing 2 for a second time using $q_2(x)$ gives

use 2 as a "divisor"

$$\begin{array}{r|rrrr} 2 & 1 & -2 & -3 & 6 \\ & & 2 & 0 & -6 \\ \hline & 1 & 0 & -3 & \underline{|0} \end{array}$$

coefficients of $q_2(x)$

2 is a repeated zero

and we see that 2 is actually a repeated zero and the final quotient is the quadratic factor $x^2 - 3$. Using this information produces the factored form $P(x) = (x + 1)(x - 2)^2(x^2 - 3)$ and the real zeros of $P$ are $-1$, 2 (as a repeated root), and $\pm\sqrt{3}$. The graph of $P$ is shown in the figure.

**Now try Exercises 33 through 36** ▶

☑ **B.** You've just seen how we can find rational zeroes of a real polynomial function using the rational zeroes theorem

## C. Descartes' Rule of Signs and Upper/Lower Bounds

Testing $x = 1$ and $x = -1$ is one way to reduce the number of possible rational zeroes, but unless we're very lucky, factoring the polynomial can still be a challenge. **Descartes' rule of signs** and the **upper and lower bounds property** offer additional assistance.

> **Descartes' Rule of Signs**
>
> Given the real polynomial equation $P(x) = 0$,
>
> 1. The number of positive real zeroes is either equal to the number of variations in sign for $P(x)$, or is an even number less.
> 2. The number of negative real zeroes is either equal to the number of variations in sign for $P(-x)$, or is an even number less.
>
> Note how Descartes' rule is employed in Example 5.

**EXAMPLE 5** ▶ **Using Descartes' Rule to Help Find Real Zeroes of a Polynomial**

For $P(x) = x^4 + 7x^3 + 12x^2 - 4x - 16$,

    **a.** Use the rational zeroes theorem to list all possible rational zeroes.
    **b.** Apply Descartes' rule to count the number of possible positive real zeroes, and the number of possible negative real zeroes.
    **c.** Use this information and the tools of this section to find all real zeroes of $P$.

**Solution** ▶     **a.** The factor pairs of the constant term are $\pm 1$, $\pm 16$; $\pm 2$, $\pm 8$; and $\pm 4$. Since the lead coefficient is 1, the possible rational zeroes are $\{\pm 1, \pm 2, \pm 4, \pm 8, \pm 16\}$.
    **b.** For this illustration, positive coefficients are in blue, with negative coefficients in red. Checking for the number of positive roots:

$$P(x) = x^4 + 7x^3 + 12x^2 - 4x - 16$$

The terms change sign only once (from positive to negative), showing there is *exactly* one positive root. To check for the number of negative real roots we use $P(-x)$, which simply changes the sign of all odd degree terms:

$$P(-x) = x^4 - 7x^3 + 12x^2 + 4x - 16$$

Here there are three changes in sign, meaning there are either three negative roots or one negative root.

**c.** The test for 1 shows it is a zero, and using 1 with synthetic division gives

use 1 as a "divisor"    $1\rfloor$   1   7   12   $-4$   $-16$    coefficients of $P$

               1   8   20   16

             1   8   20   16   $\lfloor 0$    coefficients of $q_1(x)$

Since there can be only one positive zero, we ignore the remaining positive values and continue our search using $-2$ and $q_1(x)$.

use $-2$ as a "divisor"    $-2\rfloor$   1    8    20    16    coefficients of $P$

                    $-2$   $-12$   $-16$

                 1    6    8    $\lfloor 0$    coefficients of $q_2(x)$

The result is $P(x) = (x - 1)(x + 2)(x^2 + 6x + 8) = (x - 1)(x + 2)^2(x + 4)$. The rational roots are $x = 1$, $x = -2$ as a repeated root, and $x = -4$. We note there is indeed one positive root and three negative roots—with one of these being a repeated root.

**Now try Exercises 37 through 44** ▶

One final idea that helps reduce the number of possible zeroes is the **upper and lower bounds property.** A number $b$ is an **upper bound** on the positive zeroes of a function if no positive zero is greater than $b$. In the same way, a number $a$ is a **lower bound** on the negative zeroes if no negative zero is less than $a$.

### Upper and Lower Bounds Property

Given $P(x)$ is a polynomial with real coefficients.

**1.** If $P(x)$ is divided by $x - b$ ($b > 0$) using synthetic division and all coefficients in the quotient row are either positive or zero, then $b$ is an upper bound on the zeroes of $P$.

**2.** If $P(x)$ is divided by $x - a$ ($a < 0$) using synthetic division and all coefficients in the quotient row alternate in sign, then $a$ is a lower bound on the zeroes of $P$.

For both 1 and 2, zero coefficients may be considered either positive or negative.

✓ **C.** You just seen how we can obtain information on the zeroes of real polynomials using Descartes' rule of signs and upper/lower bounds property

While this test certainly helps narrow the possibilities, we gain the additional benefit of knowing the property actually places boundaries on *all* real zeroes of the polynomial, both rational and irrational. In part (c) of Example 5, the quotient row of the first division is all positive, showing $x = 1$ is both a zero and an upper bound on the real zeroes of $P$. For more on the upper and lower bounds property, **see Exercise 97.**

### D. The Fundamental Theorem of Algebra

From Section 1.4, we know the set of real numbers is a subset of the complex numbers. Because complex numbers are the "larger" set (containing all other number sets), properties and theorems about complex numbers are more powerful and far reaching than theorems about real numbers. In the same way, real polynomials are a subset of the complex polynomials, and the same principle applies.

### Complex Polynomial Functions

A complex polynomial of degree $n$ has the form
$$P(x) = a_n x^n + a_{n-1} x^{n-1} + \cdots + a_1 x^1 + a_0,$$
where $a_n, a_{n-1}, \ldots, a_1, a_0$ are complex numbers and $a_n \neq 0$.

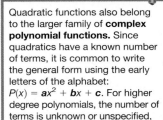
Notice that real polynomials have the same form, but here $a_n, a_{n-1}, \ldots, a_1, a_0$ *represent complex numbers.* In 1799, Carl Friedrich Gauss (1777–1855) proved that *all* polynomial functions have zeroes, and that the number of zeroes is equal to the degree of the polynomial. The proof of this statement is based on a theorem that is the bedrock for a complete study of polynomial functions, and has come to be known as the **fundamental theorem of algebra.**

### The Fundamental Theorem of Algebra

Every complex polynomial has at least one complex zero.

Although the statement may seem trivial, it enables us to draw two important conclusions. The first is that our search for a solution will not be fruitless or wasted—zeroes for *all* polynomial equations exist. Second, the fundamental theorem combined with the factor theorem enables us to state the **linear factorization theorem.**

### The Linear Factorization Theorem

If $P(x)$ is a polynomial function of degree $n \geq 1$, then $p$ has exactly $n$ linear factors and can be written in the form,
$$P(x) = a(x - c_1)(x - c_2) \cdot \cdots \cdot (x - c_n)$$
where $a \neq 0$ and $c_1, c_2, \ldots, c_n$ are (not necessarily distinct) complex numbers.

In other words, every complex polynomial of degree $n$ can be rewritten as the product of a nonzero constant and exactly $n$ linear factors (for a proof of this theorem, see Appendix IV).

**EXAMPLE 6** ▶ **Writing a Polynomial as a Product of Linear Factors**

Rewrite $P(x) = x^4 - 8x^2 - 9$ as a product of linear factors, and find its zeroes.

**Solution** ▶ From its given form, we know $a = 1$. Since $P$ has degree 4, the factored form must be $P(x) = (x - c_1)(x - c_2)(x - c_3)(x - c_4)$. Noting that $P$ is in quadratic form, we substitute $u$ for $x^2$ and $u^2$ for $x^4$ and attempt to factor:

$$\begin{aligned} x^4 - 8x^2 - 9 &\rightarrow u^2 - 8u - 9 \qquad \text{substitute } u \text{ for } x^2;\ u^2 \text{ for } x^4 \\ &= (u - 9)(u + 1) \qquad \text{factor in terms of } u \\ &= (x^2 - 9)(x^2 + 1) \qquad \text{rewrite in terms of } x \text{ (substitute } x^2 \text{ for } u) \end{aligned}$$

We know $x^2 - 9$ will factor since it is a difference of squares. From our work with complex numbers (Section 1.4), we know $(a + bi)(a - bi) = a^2 + b^2$, and the factored form of $x^2 + 1$ must be $(x + i)(x - i)$. The completely factored form is
$$P(x) = (x + 3)(x - 3)(x + i)(x - i),$$
and the zeroes of $P$ are $-3$, $3$, $-i$, and $i$.

**Now try Exercises 45 through 48** ▶

**EXAMPLE 7** ▶ **Writing a Polynomial as a Product of Linear Factors**

Rewrite $P(x) = x^3 + 2x^2 - 4x - 8$ as a product of linear factors and find its zeroes.

**Solution** ▶ We observe that $a = 1$ and $P$ has degree 3, so the factored form must be $P(x) = (x - c_1)(x - c_2)(x - c_3)$. Noting that $ad = bc$, we start with factoring by grouping.

$$\begin{aligned} P(x) &= x^3 + 2x^2 - 4x - 8 && \text{group terms (in color)} \\ &= x^2(x + 2) - 4(x + 2) && \text{remove common factors (note sign change)} \\ &= (x + 2)(x^2 - 4) && \text{factor common binomial} \\ &= (x + 2)(x + 2)(x - 2) && \text{factor difference of squares} \end{aligned}$$

The zeroes of $P$ are $-2$, $-2$, and $2$.

Now try Exercises 49 through 52 ▶

Note the polynomial in Example 7 has three zeroes, but the zero $-2$ was repeated two times. In this case we say $-2$ is a zero of multiplicity two, and a zero of **even multiplicity.** It is also possible for a zero to be repeated three or more times, with those repeated an odd number of times called zeroes of **odd multiplicity** [the factor $(x - 2) = (x - 2)^1$ also gives a zero of odd multiplicity]. In general, repeated factors are written in exponential form and we have

**Multiplicity of Zeroes**

If $P$ is a polynomial function and $(x - c)$ occurs as a factor of $P$ exactly $m$ times, then $c$ is a zero of multiplicity $m$.

**EXAMPLE 8** ▶ **Identifying the Multiplicity of a Zero**

Factor the given function completely, writing repeated factors in exponential form. Then state the multiplicity of each zero: $P(x) = (x^2 + 8x + 16)(x^2 - x - 20)(x - 5)$

**Solution** ▶
$$\begin{aligned} P(x) &= (x^2 + 8x + 16)(x^2 - x - 20)(x - 5) && \text{given polynomial} \\ &= (x + 4)(x + 4)(x - 5)(x + 4)(x - 5) && \text{trinomial factoring} \\ &= (x + 4)^3(x - 5)^2 && \text{exponential form} \end{aligned}$$

For function $P$, $-4$ is a zero of multiplicity 3 (odd multiplicity), and 5 is a zero of multiplicity 2 (even multiplicity).

Now try Exercises 53 through 56 ▶

These examples help illustrate three important consequences of the linear factorization theorem. As seen in Example 6, if the coefficients of $P$ are real, the polynomial can be factored into linear and quadratic factors using real numbers only $[(x + 3)(x - 3)(x^2 + 1)]$, where the quadratic factors have no real zeroes. Quadratic factors of this type are said to be **irreducible.**

**Corollary I: Irreducible Quadratic Factors**

If $P$ is a polynomial with real coefficients, $P$ can be factored into a product of linear factors (which are not necessarily distinct) and irreducible quadratic factors having real coefficients.

Closely related to this corollary and our previous study of quadratic functions, complex zeroes of the irreducible factors must occur in conjugate pairs.

### Corollary II: Complex Conjugates

If $P$ is a polynomial with real coefficients, complex zeroes must occur in conjugate pairs. If $a + bi$, $b \neq 0$ is a zero, then $a - bi$ will also be a zero.

Finally, the polynomial in Example 6 has degree 4 with 4 zeroes (two real, two nonreal), and the polynomial in Example 7 has degree 3 with 3 zeroes (three real, one of these is a repeated root). While not shown explicitly, the polynomial in Example 8 has degree 5, and there were 5 zeroes (one repeated twice, one repeated three times). This suggests our final corollary.

### Corollary III: Number of Zeroes

If $P$ is a polynomial with degree $n \geq 1$, then $P$ has exactly $n$ zeroes (real or nonreal), where zeroes of multiplicity $m$ are counted $m$ times.

These corollaries help us gain valuable information about a polynomial, when only partial information is given or known.

**EXAMPLE 9** ▶ **Building a Polynomial from Its Zeroes**

A real polynomial $P$ of degree 3 has zeroes of $-1$ and $2 + i\sqrt{3}$. Find the polynomial (assume $a = 1$).

**Solution** ▶ Using the factor theorem, two of the factors are $(x + 1)$ and $x - (2 + i\sqrt{3})$. From Corollary II, $2 - i\sqrt{3}$ must also be a zero and $x - (2 - i\sqrt{3})$ is also a factor of $P$. This gives

$$
\begin{aligned}
P(x) &= (x + 1)[x - (2 + i\sqrt{3})][x - (2 - i\sqrt{3})] \\
&= (x + 1)[(x - 2) - i\sqrt{3}][(x - 2) + i\sqrt{3}] \quad \text{associative property} \\
&= (x + 1)[(x^2 - 4x + 4) + 3] \quad (a + bi)(a - bi) = a^2 + b^2 \\
&= (x + 1)(x^2 - 4x + 7) \quad \text{simplify} \\
&= x^3 - 3x^2 + 3x + 7 \quad \text{result}
\end{aligned}
$$

The polynomial is $P(x) = x^3 - 3x^2 + 3x + 7$, which can be verified using the remainder theorem and any of the original zeroes.

**WORTHY OF NOTE**

When reconstructing a polynomial $P$ having complex zeroes, it is often more efficient to determine the irreducible quadratic factors of $P$ separately, as shown here. For the zeroes $2 \pm i\sqrt{3}$ we have

$$
\begin{aligned}
x &= 2 \pm i\sqrt{3} \\
x - 2 &= \pm i\sqrt{3} \\
(x - 2)^2 &= (\pm i\sqrt{3})^2 \\
x^2 - 4x + 4 &= -3 \\
x^2 - 4x + 7 &= 0.
\end{aligned}
$$

The quadratic factor is $(x^2 - 4x + 7)$.

**Now try Exercises 57 through 64** ▶

Example 10 helps to summarize the various ideas related to the search for polynomial zeroes.

**EXAMPLE 10** ▶ **Finding all Zeroes of a Polynomial**

For $P(x) = 2x^5 - 5x^4 + x^3 + x^2 - x + 6$,

**a.** Use the rational zeroes theorem to list all possible rational zeroes.

**b.** Apply Descartes' rule to count the number of possible positive, negative, and complex zeroes.

**c.** Use this information and the tools of this section to find all zeroes of $P$.

**Solution** ▶ **a.** The factors of 2 are $\{\pm 1, \pm 2\}$ and the factors of 6 are $\{\pm 1, \pm 6, \pm 2, \pm 3\}$. The possible rational zeroes for $P$ are $\{\pm 1, \pm 6, \pm 2, \pm 3, \pm \frac{1}{2}, \pm \frac{3}{2}\}$.

**b.** For Descartes' rule, we organize our work in a table. Since $P$ has degree 5, there must be a total of five zeroes. For this illustration, positive coefficients are in **blue** and negative coefficients in **red**: $P(x) = 2x^5 - 5x^4 + x^3 + x^2 - x + 6$. The terms change sign a total of four times, meaning there are four, two, or zero positive roots. For the negative roots, recall that $P(-x)$ will change the sign of *all odd-degree terms,* giving $P(-x) = -2x^5 - 5x^4 - x^3 + x^2 + x + 6$. This time there is only one sign change (from negative to positive) showing there will be exactly one negative root, a fact that is highlighted in the following table. Since there must be 5 zeroes, the number of possible nonreal zeroes is none, two, or four, as shown.

| possible positive zeroes | known negative zeroes | possible nonreal zeroes | total number must be 5 |
|---|---|---|---|
| 4 | 1 | 0 | 5 |
| 2 | 1 | 2 | 5 |
| 0 | 1 | 4 | 5 |

**c.** Testing 1 and $-1$ shows only $x = -1$ is a root, and using $-1$ in synthetic division gives:

use $-1$ as a "divisor"   $\underline{-1|}$ $\quad$ 2 $\quad$ $-5$ $\quad$ 1 $\quad$ 1 $\quad$ $-1$ $\quad$ 6 $\qquad$ coefficients of $P(x)$

$\qquad\qquad\qquad\qquad\qquad\qquad\quad$ $-2$ $\quad$ 7 $\quad$ $-8$ $\quad$ 7 $\quad$ $-6$

$\qquad\qquad\qquad\qquad\qquad\qquad$ $\overline{2 \quad -7 \quad 8 \quad -7 \quad 6 \quad \underline{|0}}$ $\quad$ $q_1(x)$ is not easily factored

Since there is *only one* negative root, we need only check the remaining positive zeroes. The quotient $q_1(x)$ is not easily factored, so we continue with synthetic division using the next larger positive root, $x = 2$.

use 2 as a "divisor" $\qquad\qquad$ $\underline{2|}$ 2 $\quad$ $-7$ $\quad$ 8 $\quad$ $-7$ $\quad$ 6 $\qquad$ coefficients of $q_1(x)$

$\qquad\qquad\qquad\qquad\qquad\qquad\qquad$ 4 $\quad$ $-6$ $\quad$ 4 $\quad$ $-6$

$\qquad\qquad\qquad\qquad\qquad\qquad$ $\overline{2 \quad -3 \quad 2 \quad -3 \quad \underline{|0}}$ $\quad$ $q_2(x)$

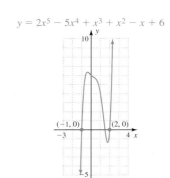

$y = 2x^5 - 5x^4 + x^3 + x^2 - x + 6$

$(-1, 0)$ $\quad$ $(2, 0)$

The partially factored form is $P(x) = (x + 1)(x - 2)(2x^3 - 3x^2 + 2x - 3)$. Graphing $P(x)$ at this point (see the figure) verifies that $-1$ and 2 are zeroes, but also indicates there is an additional zero between 1 and 2. If it's a *rational* zero, it must be $x = \frac{3}{2}$ from our list of possible zeroes. Checking $\frac{3}{2}$ using synthetic division and $q_2(x)$ we have

$$\frac{3}{2}\Big| \quad 2 \quad -3 \quad 2 \quad -3$$
$$\downarrow \quad\quad 3 \quad\; 0 \quad\; 3$$
$$\overline{2 \quad\;\; 0 \quad 2 \quad \underline{|0}}$$

Since the remainder is zero, $P\left(\frac{3}{2}\right) = 0$ (remainder theorem) and $x - \frac{3}{2}$ is a factor (factor theorem). Since the zero is a fraction, we'll use the ideas discussed in Section 4.2 to help write $P(x)$ in factored form:

$$P(x) = (x + 1)(x - 2)\left(x - \frac{3}{2}\right)(2x^2 + 2) \qquad \text{partially factored form}$$

$$= (x + 1)(x - 2)\left(x - \frac{3}{2}\right)(2)(x^2 + 1) \qquad \text{factor out 2}$$

$$= (x + 1)(x - 2)(2x - 3)(x^2 + 1) \qquad \text{multiply } 2\left(x - \frac{3}{2}\right)$$

$$= (x + 1)(x - 2)(2x - 3)(x + i)(x - i) \qquad \text{factored form}$$

The zeroes of $P$ are $-1, 2, \frac{3}{2}, -i$ and $i$, with one negative, two positive, and two nonreal zeroes (row two of the table).

☑ **D.** You've just seen how we can apply the fundamental theorem of algebra and the linear factorization theorem

**Now try Exercises 65 through 82 ▶**

### E. Applications of Polynomial Functions

Polynomial functions can be very accurate models of real-world phenomena, though we often must restrict their domain, as illustrated in Example 11.

---

**EXAMPLE 11** ▶    **Using the Remainder Theorem to Solve an Oceanography Application**

As part of an environmental study, scientists use radar to map the ocean floor from the coastline to a distance 12 mi from shore. In this study, ocean trenches appear as negative values and underwater mountains as positive values, as measured from the surrounding ocean floor. The terrain due west of a particular island can be modeled by $h(x) = x^4 - 25x^3 + 200x^2 - 560x + 384$, where $h(x)$ represents the height in feet, $x$ mi from shore $(0 < x \le 12)$.

   **a.** Use the remainder theorem to find the "height of the ocean floor" 10 mi out.

   **b.** Use the tools developed in this section to find the number of times the ocean floor has height $h(x) = 0$ in this interval, given this occurs 12 mi out.

**Solution** ▶    **a.** For part (a) we simply evaluate $h(10)$ using the remainder theorem.

$$
\begin{array}{r|rrrrr}
\text{use 10 as a "divisor"} \quad 10 & 1 & -25 & 200 & -560 & 384 \quad \text{coefficients of } h(x)\\
 & & 10 & -150 & 500 & -600 \\
\hline
 & 1 & -15 & 50 & -60 & \underline{|-216} \quad \text{remainder is } -216
\end{array}
$$

Ten miles from shore, there is an ocean trench 216 ft deep.

   **b.** For part (b), we're given 12 is a zero, so we again use the remainder theorem and work with the quotient polynomial.

$$
\begin{array}{r|rrrrr}
\text{use 12 as a "divisor"} \quad 12 & 1 & -25 & 200 & -560 & 384 \quad \text{coefficients of } h(x)\\
 & & 12 & -156 & 528 & -384 \\
\hline
 & 1 & -13 & 44 & -32 & \underline{|0} \quad q_1(x)
\end{array}
$$

The quotient is $q_1(x) = x^3 - 13x^2 + 44x - 32$. Since $a = 1$, we know the remaining rational zeroes must be factors of $-32$: $\{\pm 1, \pm 32, \pm 2, \pm 16, \pm 4, \pm 8\}$. Testing $x = 1$ shows it is a zero and synthetic division yields

$$
\begin{array}{r|rrrr}
\text{use 1 as a "divisor"} \quad 1 & 1 & -13 & 44 & -32 \quad \text{coefficients of } q_1(x)\\
 & & 1 & -12 & 32 \\
\hline
 & 1 & -12 & 32 & \underline{|0} \quad q_2(x)
\end{array}
$$

The function can now be written as $h(x) = (x - 12)(x - 1)(x^2 - 12x + 32)$ and in completely factored form $h(x) = (x - 12)(x - 1)(x - 4)(x - 8)$. The ocean floor has height zero at distances of 1, 4, 8, and 12 mi from shore.

The graph of $h(x)$ is shown in the figure. The graph shows a great deal of variation in the ocean floor, but the zeroes occurring at 1, 4, 8, and 12 mi out are clearly evident.

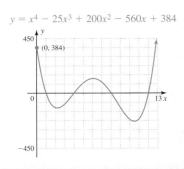

$y = x^4 - 25x^3 + 200x^2 - 560x + 384$

☑ **E.** You've just seen how we can solve an application of polynomial functions

**Now try Exercises 85 through 96** ▶

## 4.3 EXERCISES

▶ **CONCEPTS AND VOCABULARY**

**Fill in each blank with the appropriate word or phrase. Carefully reread the section, if necessary.**

1. A polynomial function of degree $n$ will have exactly _____ zeroes, real or _____, where zeroes of multiplicity $m$ are counted $m$ times.

2. According to Descartes' rule of signs, there are as many _____ real roots as changes in sign from term to term, or a(n) _____ number less.

3. Discuss/Explain how to determine which of the following values is *not* a possible root of $f(x) = 6x^3 - 2x^2 + 5x - 12$:
   **a.** $x = \frac{4}{3}$   **b.** $x = \frac{3}{4}$   **c.** $x = \frac{1}{2}$
   Discuss/Explain why.

4. Discuss/Explain each of the following: (a) irreducible quadratic factors, (b) zeroes that are complex conjugates, (c) zeroes of multiplicity $m$, (d) upper bounds on zeroes of a polynomial.

▶ **DEVELOPING YOUR SKILLS**

**Use the intermediate value theorem to verify the given polynomial has at least one zero "$c_i$" in the intervals specified. Do not find the zeroes.**

5. $f(x) = x^3 + 2x^2 - 8x - 5$
   **a.** $[-4, -3]$   **b.** $[2, 3]$

6. $g(x) = x^4 - 2x^2 + 6x - 3$
   **a.** $[-3, -2]$   **b.** $[0, 1]$

**For Exercises 7 and 8, create a table of values for integer inputs from $x = -4$ to $x = 4$, then use the intermediate value theorem to locate unit intervals that contain a zero of the function.**

7. $f(x) = x^3 - 3x + 1$      8. $g(x) = x^4 - 10x^2 + 15$

**List all possible rational zeroes for the polynomials given, but do not find the zeroes.**

9. $f(x) = 4x^3 - 19x - 15$   10. $g(x) = 3x^3 - 2x + 20$

11. $h(x) = 2x^3 - 5x^2 - 28x + 15$

12. $H(x) = 2x^3 - 19x^2 + 37x - 14$

13. $p(x) = 6x^4 - 2x^3 + 5x^2 - 28$

14. $q(x) = 7x^4 + 6x^3 - 49x^2 + 36$

15. $h(t) = 32t^3 - 52t^2 + 17t + 3$

16. $s(t) = 24t^3 + 17t^2 - 13t - 6$

**Use the rational zeroes theorem to write each function in factored form and find all zeroes.**

17. $f(x) = x^3 - 13x + 12$

18. $g(x) = x^3 - 21x + 20$

19. $h(x) = x^3 - 19x - 30$

20. $H(x) = x^3 - 28x - 48$

21. $p(x) = x^3 - 2x^2 - 11x + 12$

22. $q(x) = x^3 - 4x^2 - 7x + 10$

23. $r(x) = x^3 - 6x^2 - x + 30$

24. $t(x) = x^3 - 4x^2 - 20x + 48$

25. $f(x) = x^4 + 7x^3 - 7x^2 - 55x - 42$

26. $g(x) = x^4 + 4x^3 - 17x^2 - 24x + 36$

27. $f(x) = 4x^3 - 7x + 3$

28. $g(x) = 9x^3 - 7x - 2$

29. $h(x) = 4x^3 + 8x^2 - 3x - 9$

30. $H(x) = 9x^3 + 3x^2 - 8x - 4$

31. $Y_1 = 2x^3 - 3x^2 - 9x + 10$

32. $Y_2 = 3x^3 - 14x^2 + 17x - 6$

33. $p(x) = 2x^4 + 3x^3 - 9x^2 - 15x - 5$

34. $q(x) = 3x^4 + x^3 - 11x^2 - 3x + 6$

35. $r(x) = 3x^4 + 4x^3 - 10x^2 - 8x + 8$

36. $s(x) = 2x^4 - 7x^3 - 18x^2 + 49x + 28$

**For each polynomial given, (a) use the rational zeroes theorem to list all possible rational zeroes, (b) apply Descartes' rule of signs to count the number of possible positive and negative zeroes, and (c) use this information and other tools from this section to find all real zeroes.**

37. $f(x) = x^3 - 2x^2 - 11x + 12$

38. $g(x) = x^3 - 6x^2 + 3x + 10$

39. $h(x) = x^4 + 3x^3 - 2x^2 - 12x - 8$

40. $H(x) = x^4 - 4x^3 - 6x^2 + 36x - 27$

41. $p(x) = x^4 - 2x^3 - 19x^2 + 8x + 60$

42. $q(x) = x^4 + 2x^3 - 17x^2 - 18x + 72$

43. $r(x) = 2x^4 + 5x^3 - 24x^2 - 28x + 80$

44. $t(x) = 3x^4 + 28x^3 + 79x^2 + 42x - 72$

Rewrite each polynomial as a product of linear factors, and find the zeroes of the polynomial.

**45.** $P(x) = x^4 + 5x^2 - 36$

**46.** $Q(x) = x^4 + 21x^2 - 100$

**47.** $Q(x) = x^4 - 16$

**48.** $P(x) = x^4 - 81$

**49.** $P(x) = x^3 + x^2 - x - 1$

**50.** $Q(x) = x^3 - 3x^2 - 9x + 27$

**51.** $Q(x) = x^3 - 5x^2 - 25x + 125$

**52.** $P(x) = x^3 + 4x^2 - 16x - 64$

Factor each polynomial completely. Write any repeated factors in exponential form, then name all zeroes and their multiplicity.

**53.** $p(x) = (x^2 - 10x + 25)(x^2 + 4x - 45)(x + 9)$

**54.** $q(x) = (x^2 + 12x + 36)(x^2 + 2x - 24)(x - 4)$

**55.** $P(x) = (x^2 - 5x - 14)(x^2 - 49)(x + 2)$

**56.** $Q(x) = (x^2 - 9x + 18)(x^2 - 36)(x - 3)$

Find a polynomial $P(x)$ having real coefficients, with the degree and zeroes indicated. All real zeroes are given. Assume the lead coefficient is 1. Recall $(a + bi)(a - bi) = a^2 + b^2$.

**57.** degree 3, $x = 3$, $x = 2i$

**58.** degree 3, $x = -5$, $x = -3i$

**59.** degree 4, $x = -1$, $x = 2$, $x = i$

**60.** degree 4, $x = -1$, $x = 3$, $x = -2i$

**61.** degree 4, $x = 3$, $x = 2i$

**62.** degree 4, $x = -2$, $x = -3i$

**63.** degree 4, $x = -1$, $x = 1 + 2i$

**64.** degree 4, $x = -2$, $x = 1 + i\sqrt{3}$

Apply Descartes' rule of signs to determine the possible combinations of real and complex zeroes for each polynomial. Then graph the function using a graphing calculator. With all real zeroes in clear view, use a list of possible rational zeroes to factor the polynomial and find all zeroes.

**65.** $f(x) = 4x^3 - 16x^2 - 9x + 36$

**66.** $g(x) = 6x^3 - 41x^2 + 26x + 24$

**67.** $p(x) = 4x^4 + 40x^3 - 93x^2 + 30x - 72$

**68.** $q(x) = 4x^4 - 42x^3 - 70x^2 - 21x - 36$

**69.** $r(x) = x^4 - 5x^3 + 20x - 16$

**70.** $t(x) = x^4 - 10x^3 + 90x - 81$

Find the zeroes of the polynomials given using any combination of the rational zeroes theorem, testing for 1 and −1, and/or the remainder and factor theorems.

**71.** $f(x) = 2x^4 - 9x^3 + 4x^2 + 21x - 18$

**72.** $g(x) = 3x^4 + 4x^3 - 21x^2 - 10x + 24$

**73.** $p(x) = 2x^4 + 3x^3 - 24x^2 - 68x - 48$

**74.** $q(x) = 3x^4 - 19x^3 + 6x^2 + 96x - 32$

**75.** $r(x) = 3x^4 - 20x^3 + 34x^2 + 12x - 45$

**76.** $s(x) = 4x^4 - 15x^3 + 9x^2 + 16x - 12$

**77.** $r(x) = x^4 - x^3 - 14x^2 + 2x + 24$

**78.** $s(x) = x^4 - 3x^3 - 13x^2 + 9x + 30$

**79.** $r(x) = x^5 + 6x^2 - 49x + 42$

**80.** $t(x) = x^5 + 2x^2 - 9x + 6$

**81.** $p(x) = 2x^4 - x^3 + 3x^2 - 3x - 9$

**82.** $q(x) = 3x^4 + x^3 + 13x^2 + 5x - 10$

▶ **WORKING WITH FORMULAS**

**83. Extent of the Arctic sea ice:**
$A(t) = 0.062x^3 - 1.117x^2 + 4.782x + 7.912$

For the years 2002 to 2009, the average area of the Arctic sea ice can be modeled by the formula shown, where $A(t)$ represents the total area of the sea ice (in millions of square kilometers) during month $t$ of the year ($t = 1$ corresponds to January). (a) Use the remainder theorem to determine how many square kilometers of sea ice exist in January, and in October. (b) Use a graphing calculator to graph the function, and determine the month where the area of the sea ice is smallest. What is this area?

**84. Parental IQ versus age of child:**
$I(a) = -0.003a^4 + 0.186a^3 - 3.044a^2 + 4.234a + 180.5$

Like his father, Michael also became a math professor, and one day decided to make his father smile. After considering his childhood, formative years, young adulthood, and then his own family responsibilities, Michael came up with the function shown. Here, $I(a)$ is the IQ of a parent, as perceived by a child at age $a$ ($0 < a < 30$). (a) Use the remainder theorem to determine the perceived IQ of a parent if the child is 1 yr old, and 10 yr old. (b) Use a graphing calculator to graph the function, and determine the age at which parents are perceived to have the least intelligence. (c) According to this function, at what age does a son or daughter believe their parents actually have an IQ over 130?

## ▶ APPLICATIONS

**85. Altitude and terrain contours:** To avoid detection, a drone is flying just above the contours of the terrain, in the region of the Twin Peaks. The polynomial $A(d) = -d^4 + 28d^3 - 229d^2 + 462d + 720$ models the altitude of the drone (in meters) *above or below an altitude of 2000 m*, at a distance of $d$ mi ($0 < d < 16$). (a) Use the remainder theorem to find the altitude at $d = 1$, $d = 7$, and $d = 13$ mi. (b) Use the tools developed in this section to find the number of times the altitude was 2000 m during this period, and for what values of $d$.

**86. Engine trouble at NASCAR:** In the final race to the finish, the car driven by Steve Hannan developed engine trouble and began sputtering, causing the car to decelerate and accelerate wildly. The polynomial $V(t) = t^4 - 19t^3 + 108t^2 - 180t$ models the velocity of his car in mph *above or below his average speed* for the race (200 mph), $t$ sec after the trouble began ($0 < t < 10.3$). (a) Use the remainder theorem to find his speed at $t = 1$, $t = 5$, and $t = 8$. (b) Use the tools developed in this section to find the number of times his speed hit 200 mph during this period, and for what values of $t$.

 **87. Maximum and minimum values:** To locate the maximum and minimum values of the polynomial $F(x) = x^4 - 4x^3 - 12x^2 + 32x + 15$ requires finding the zeroes of $f(x) = 4x^3 - 12x^2 - 24x + 32$. Use the rational zeroes theorem and synthetic division to find the zeroes of $f$, then graph $F(x)$ on a calculator and see if the graph tends to support your calculations—do the maximum and minimum values occur at the zeroes of $f$?

 **88. Maximum and minimum values:** To locate the maximum and minimum values of the polynomial $G(x) = x^4 - 6x^3 + x^2 + 24x - 20$ requires finding the zeroes of $g(x) = 4x^3 - 18x^2 + 2x + 24$. Use the rational zeroes theorem and synthetic division to find the zeroes of $g$, then graph $G(x)$ on a calculator and see if the graph tends to support your calculations—do the maximum and minimum values occur at the zeroes of $g$?

**Geometry:** The volume of a cube is $V = x \cdot x \cdot x = x^3$, where $x$ represents the length of the edges. If a slice 1 unit thick is removed from the cube, the remaining volume is $v = x \cdot x \cdot (x - 1) = x^3 - x^2$. Use this information for Exercises 89 and 90.

**89.** A slice 1 unit in thickness is removed from one side of a cube. Use the rational zeroes theorem and synthetic division to find the original dimensions of the cube, if the remaining volume is (a) 48 cm³ and (b) 100 cm³.

**90.** A slice 1 unit in thickness is removed from one side of a cube, then a second slice of the same thickness is removed from a different side (not the opposite side). Use the rational zeroes theorem and synthetic division to find the original dimensions of the cube, if the remaining volume is (a) 36 cm³ and (b) 80 cm³.

**Geometry:** The volume of a rectangular box is $V = LWH$. For the box to satisfy certain requirements, its length must be twice the width, and its height must be two inches less than the width. Use this information for Exercises 91 and 92.

**91.** Use the rational zeroes theorem and synthetic division to find the dimensions of the box if it must have a volume of 150 in³.

**92.** Suppose the box must have a volume of 64 in³. Use the rational zeroes theorem and synthetic division to find the dimensions required.

**Government deficits:** Over a 14-yr period, the balance of payments (deficit versus surplus) for a certain county government was modeled by the function $f(x) = \frac{1}{4}x^4 - 6x^3 + 42x^2 - 72x - 64$, where $x = 0$ corresponds to 1990 and $f(x)$ is the deficit or surplus in tens of thousands of dollars. Use this information for Exercises 93 and 94.

**93.** Use the rational zeroes theorem and synthetic division to find the years when the county "broke even" (debt = surplus = 0) from 1990 to 2004. How many years did the county run a surplus during this period?

**94.** The deficit was at the $84,000 level $[f(x) = -84]$, four times from 1990 to 2004. Given this occurred in 1992 and 2000 ($x = 2$ and $x = 10$), use the rational zeroes theorem, synthetic division, and the remainder theorem to find the other two years the deficit was at $84,000.

 **95. Drag resistance on a boat:** In a study on the effects of drag against the hull of a sculling boat, if all other factors are held constant and we assume a flat, calm water surface, length could be made the sole variable. For a fixed speed of 5.5 knots, the relationship between drag and length can be modeled by $f(x) = -0.4192x^4 + 18.9663x^3 - 319.9714x^2 + 2384.2x - 6615.8$, where $f(x)$ is the efficiency rating of a boat with length $x$ ($8.7 < x < 13.6$). Here, $f(x) = 0$ represents an *average* rating. (a) Under these conditions, what lengths (to the nearest hundredth) will give the boat an average rating? (b) What length will maximize the efficiency of the boat? What is this rating?

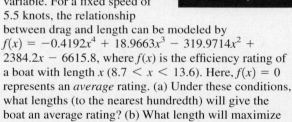

 **96. Comparing densities:** Why do some objects sink, while other objects float when thrown into water? It all depends on the density of the object compared to the density of water ($d = 1$). For uniformity, we'll consider spherical objects with radius 5 cm. When placed into water, the depth the sphere will sink (while still floating) is modeled by $p(x) = \frac{\pi}{3}x^3 - 5\pi x^2 + \frac{500\pi}{3}d$, where $d$ is the density of the object and the smallest positive zero of $p$ is the depth of the sphere below the surface. How submerged is a sphere of (a) balsa wood, $d = 0.17$; (b) pine wood, $d = 0.55$; (c) ebony wood, $d = 1.12$; (d) a large bobber made of lightweight plastic, $d = 0.05$ (see figure)?

## ▶ EXTENDING THE CONCEPT

**97.** In the figure, $P(x) = 0.02x^3 - 0.24x^2 - 1.04x + 2.68$ is graphed on the standard screen ($-10 \leq x \leq 10$), which shows two real zeroes. Since $P$ has degree 3, there must be one more real zero but is it negative or positive? Use the upper/lower bounds property (a) to see if $-10$ is a lower bound and (b) to see if 10 is an upper bound. (c) Then use your calculator to find the remaining zero.

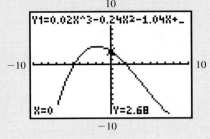

**98.** While the difference of two squares is factorable, it is often said the sum of two squares is prime. To be 100% correct, we should say the sum of two squares cannot be factored *using real numbers* since we now know that $a^2 + b^2 = (a + bi)(a - bi)$. Use this idea to factor the following binomials.

    **a.** $p(x) = x^2 + 25$     **b.** $q(x) = x^2 + 9$
    **c.** $r(x) = x^2 + 7$

**99.** While $x^2 - 16$ is factorable as a difference of squares, it is often said $x^2 - 17$ is not. To be 100% correct, we should say that $x^2 - 17$ is not factorable *using integers*.

Since $(\sqrt{17})^2 = 17$, it actually can be factored: $x^2 - 17 = (x + \sqrt{17})(x - \sqrt{17})$. Use this idea to solve the following equations.

    **a.** $x^2 - 7 = 0$     **b.** $x^2 - 12 = 0$
    **c.** $x^2 - 18 = 0$

**100.** Every general cubic equation $aw^3 + bw^2 + cw + d = 0$ can be written in the form $x^3 + px + q = 0$ (where the squared term has been "depressed"), using the transformation $w = x - \frac{b}{3a}$. Use this transformation to solve the following equations.

    **a.** $w^3 - 3w^2 + 6w - 4 = 0$
    **b.** $w^3 - 6w^2 + 21w - 26 = 0$

Note: For a complete treatment of cubic equations and their solutions, visit our website at www.mhhe.com/coburn.

**101.** For each of the following complex polynomials, one of its zeroes is given. Use this zero to help write the polynomial in completely factored form.

    **a.** $C(z) = z^3 + (1 - 4i)z^2 + (-6 - 4i)z + 24i$; $z = 4i$
    **b.** $C(z) = z^3 + (5 - 9i)z^2 + (4 - 45i)z - 36i$; $z = 9i$
    **c.** $C(z) = z^3 + (-2 - i)z^2 + (5 + 4i)z + (-6 + 3i)$; $z = 2 - i$
    **d.** $C(z) = z^3 - 2z^2 + (19 + 6i)z + (-20 + 30i)$; $z = 2 - 3i$

## ▶ MAINTAINING YOUR SKILLS

**102.** (3.4) Graph the piecewise-defined function and find the values of $f(-3), f(2),$ and $f(5)$.

$$f(x) = \begin{cases} 2, & x \leq -1 \\ |x - 1|, & -1 < x < 5 \\ 4, & x \geq 5 \end{cases}$$

**103.** (4.1) For a county fair, officials need to fence off a large rectangular area, then subdivide it into three equal (rectangular) areas. If the county provides 1200 ft of fencing, (a) what dimensions will maximize the area of the larger (outer) rectangle? (b) What is the area of each smaller rectangle?

**104.** (2.5) Use the graph given to (a) state intervals where $f(x) \geq 0$, (b) locate local maximum and minimum values, and (c) state intervals where $f(x)\uparrow$ and $f(x)\downarrow$.

**105.** (3.1) Write the equation of the function shown.

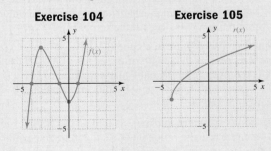

**Exercise 104**        **Exercise 105**

## MID-CHAPTER CHECK

**1.** Compute $(x^3 + 8x^2 + 7x - 14) \div (x + 2)$ using long division and write the result in two ways:

    (a) dividend = (quotient)(divisor) + remainder and
    (b) $\frac{\text{dividend}}{\text{divisor}} = (\text{quotient}) + \frac{\text{remainder}}{\text{divisor}}$.

**2.** Given that $x - 2$ is a factor of $f(x) = 2x^4 - x^3 - 8x^2 + x + 6$, use synthetic division to help write $f(x)$ in completely factored form.

**3.** Use the remainder theorem to evaluate $f(-2)$, given $f(x) = -3x^4 + 7x^2 - 8x + 11$.

**4.** Use the factor theorem to find a third-degree polynomial having $x = -2$ and $x = 1 + i$ as roots.

**5.** Use the intermediate value theorem to show that $g(x) = x^3 - 6x - 4$ has a root in the interval $(2, 3)$.

**6.** Use the rational zeroes theorem, tests for $-1$ and 1, synthetic division, and the remainder theorem to write $f(x) = x^4 + 5x^3 - 20x - 16$ in completely factored form.

**7.** Find all the zeroes of $h$, real and nonreal: $h(x) = x^4 + 3x^3 + 10x^2 + 6x - 20$.

**8.** Given $f(x) = 3x^2 - 24x + 37$, (a) does this function have a maximum or minimum value? (b) Use the vertex formula to find and correctly state this value.

**9.** A motorcycle shop finds that if their X-Terra model is priced at $2500, they will sell 17 bikes each month. For each decrease of $200 in price, they will sell two additional bikes. (a) What price should be charged to maximize revenue? (b) What is this maximum revenue?

**10.** When fighter pilots train for dogfighting, a "hard-deck" is usually established below which no competitive activity can take place. The polynomial graph given shows Maverick's altitude above and below this hard-deck during a 5-sec interval.

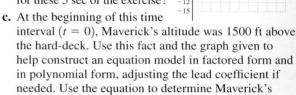

**a.** What is the minimum possible degree polynomial that could form this graph? Why?

**b.** How many seconds (total) was Maverick below the hard-deck for these 5 sec of the exercise?

**c.** At the beginning of this time interval ($t = 0$), Maverick's altitude was 1500 ft above the hard-deck. Use this fact and the graph given to help construct an equation model in factored form and in polynomial form, adjusting the lead coefficient if needed. Use the equation to determine Maverick's altitude in relation to the hard-deck at $t = 2$ and $t = 4$.

---

## 4.4   Graphing Polynomial Functions

**LEARNING OBJECTIVES**

*In Section 4.4 you will see how we can:*

- ☐ **A.** Identify the graph of a polynomial function and determine its degree
- ☐ **B.** Describe the end-behavior of a polynomial graph
- ☐ **C.** Discuss the attributes of a polynomial graph with zeroes of multiplicity
- ☐ **D.** Graph polynomial functions in standard form
- ☐ **E.** Solve applications of polynomials and polynomial modeling

As with linear and quadratic functions, understanding graphs of *polynomial* functions will help us apply them more effectively as mathematical models. Since all real polynomials can be written in terms of their linear and quadratic factors (Section 4.3), these functions provide the basis for our continuing study.

### A. Identifying the Graph of a Polynomial Function

Consider the graphs of $f(x) = x + 2$ and $g(x) = (x - 1)^2$, which we know are smooth, continuous curves. The graph of $f$ is a straight line with positive slope, that crosses the $x$-axis at $-2$. The graph of $g$ is a parabola, opening upward, shifted 1 unit to the right, and touching the $x$-axis at $x = 1$. When $f$ and $g$ are "combined" into the single function $P(x) = (x + 2)(x - 1)^2$, the behavior of the graph at these zeroes is still evident. In Figure 4.15, the graph of $P$ crosses the $x$-axis at $x = -2$, "bounces" off the $x$-axis at $x = 1$, and is still a smooth, continuous curve. This observation could be extended to include additional linear or quadratic factors, and helps affirm that the graph of a polynomial function is a *smooth, continuous curve*.

**Figure 4.15**

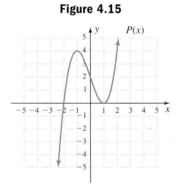

Further, after the graph of $P$ crosses the axis at $x = -2$, it must "turn around" at some point to reach the zero at $x = 1$, then turn again as it touches the $x$-axis without crossing. By combining this observation with our work in Section 4.3, we can state the following:

---

**Polynomial Graphs and Turning Points**

1. If $P(x)$ is a polynomial function of degree $n$, then the graph of $P$ has at most $n - 1$ turning points.
2. If the graph of a function $P$ has $n - 1$ turning points, then the degree of $P(x)$ is at least $n$.

While defined more precisely in a future course, we will take "smooth" to mean the graph has no sharp turns or jagged edges, and "continuous" to mean the entire graph can be drawn without lifting your pencil (Figure 4.16). In other words, a polynomial graph has none of the attributes shown in Figure 4.17.

**Figure 4.16**

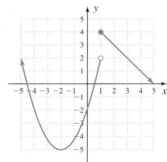

polynomial

**Figure 4.17**

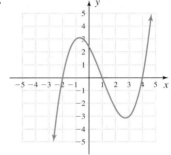

nonpolynomial

**EXAMPLE 1** ▶    Identifying Polynomial Graphs

Determine whether each graph could be the graph of a polynomial. If not, discuss why. If so, use the number of turning points and zeroes to identify the least possible degree of the function.

a.

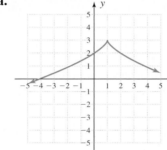

b.

c.

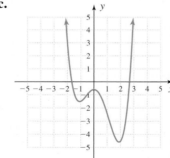

d.

**Solution** ▶    a. This is not a polynomial graph, as it has a cusp at (1, 3). A polynomial graph is always smooth.

b. This graph is smooth and continuous, and could be that of a polynomial. With two turning points and three zeroes, the function is at least degree 3.

c. This graph is smooth and continuous, and could be that of a polynomial. With three turning points and two zeroes, the function is at least degree 4.

d. This is not a polynomial graph, as it has a gap (discontinuity) at $x = 1$. A polynomial graph is always continuous.

☑ **A.** You've just seen how we can identify the graph of a polynomial function and determine its degree

**Now try Exercises 5 through 10** ▶

## B. The End-Behavior of a Polynomial Graph

Once the graph of a function has crossed or touched its last real zero and "made its last turn," it will continue to increase or decrease without bound as $|x|$ becomes large. As before, we refer to this as the **end-behavior** of the graph. In previous sections, we noted that quadratic functions (degree 2) with a positive leading coefficient ($a > 0$) had the end-behavior "up on the left" and "up on the right" (up/up). If the leading coefficient was negative ($a < 0$), end-behavior was "down on the left" and "down on the right" (down/down). These descriptions were also applied to the graph of a linear function $y = mx + b$ (degree 1). A positive leading coefficient ($m > 0$) indicates the graph will be down on the left, up on the right (down/up), and so on. All polynomial graphs exhibit some form of end-behavior, which can be likewise described.

**EXAMPLE 2** ▶ **Identifying the End-Behavior of a Graph**

State the end-behavior of each graph shown:

**a.** $f(x) = x^3 - 4x + 1$    **b.** $g(x) = -2x^5 + 7x^3 - 4x$    **c.** $h(x) = -2x^4 + 5x^2 + x - 1$

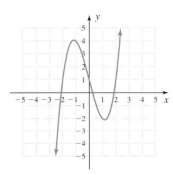 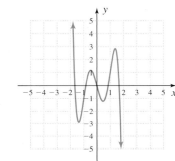 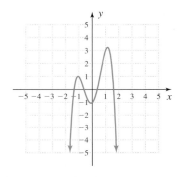

**Solution** ▶    **a.** down on the left, up on the right      **b.** up on the left, down on the right

     **c.** down on the left, down on the right

Now try Exercises 11 through 14 ▶

The leading term $ax^n$ of a polynomial function is said to be the **dominant term,** because for large values of $|x|$, the value of $ax^n$ is much larger than all other terms combined. Figure 4.18 shows a table of values for $Y_1 = 0.4x^5 - 3x^4 - 9x^3 - 15x^2 - 30$, where the leading coefficient is positive and very small, with all other coefficients negative and much larger. Initially, all outputs are negative as these terms overpower the leading term. But eventually (in this case for any integer greater than 10), the leading term will dominate all others since it becomes much more "powerful" for larger values. See Figure 4.19.

**Figure 4.18**

$Y_1 = 0.4x^5 - 3x^4 - 9x^3 - 15x^2 - 30$

| X | Y1 | |
|---|-----|---|
| 0 | -30 | |
| 1 | -56.6 | |
| 2 | -197.2 | |
| 3 | -553.8 | |
| 4 | -1204 | |
| 5 | -2155 | |
| 6 | -3292 | |

X=0

**Figure 4.19**

| X | Y1 | |
|---|-----|---|
| 7 | -4332 | |
| 8 | -4779 | |
| 9 | -3869 | |
| 10 | -530 | |
| 11 | 6673.4 | |
| 12 | 19583 | |
| 13 | 40496 | |

X=13

This means that like linear and quadratic graphs, polynomial end-behavior can be predicted in advance by analyzing this term alone (also **see Exercise 79**).

1. For $ax^n$ when $n$ is even, any nonzero number raised to an even power is positive, so the ends of the graph must point in the same direction. If $a > 0$, both point upward. If $a < 0$, both point downward.

2. For $ax^n$ when $n$ is odd, any number raised to an odd power has the same sign as the input value, so the ends of the graph must point in opposite directions. If $a > 0$, end-behavior is down on the left, up on the right. If $a < 0$, end-behavior is up on the left, down on the right.

From this we find that end-behavior depends on two things: *the degree of the function* (even or odd) and the *sign of the leading coefficient* (positive or negative). In more formal terms, this is described in terms of how the graph "behaves" for large values of $|x|$. For end-behavior that is "up on the right," we mean that as $x$ becomes a large positive number, $y$ becomes a large positive number. This is indicated using the notation: as $x \to \infty$, $y \to \infty$. Similar notation is used for the other possibilities. These facts are summarized in Table 4.1. The interior portion of each graph is dashed since the actual number of turning points may vary, although a polynomial of odd degree will have an even number of turning points, and a polynomial of even degree will have an odd number of turning points.

**Table 4.1**
**Polynomial End-Behavior**

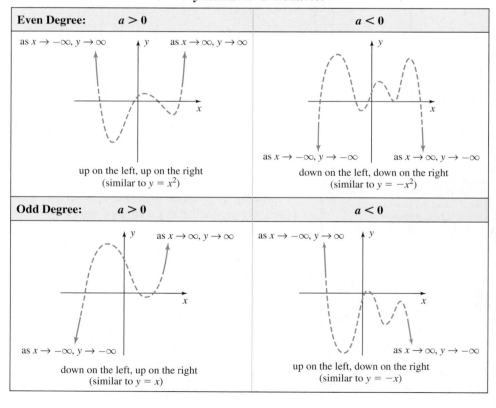

Note the end-behavior of $y = mx$ can be used as a representative of all odd degree functions, and the end-behavior of $y = ax^2$ as a representative of all even degree functions.

### The End-Behavior of a Polynomial Graph

Given a polynomial $P(x)$ with leading term $ax^n$ and $n \geq 1$.
If $n$ is **even,** ends will point in the **same direction,**

1. for $a > 0$: up on the left, up on the right (*as with $y = x^2$*);

    as $x \to -\infty$, $y \to \infty$;     as $x \to \infty$, $y \to \infty$

2. for $a < 0$: down on the left, down on the right (*as with $y = -x^2$*);

    as $x \to -\infty$, $y \to -\infty$;     as $x \to \infty$, $y \to -\infty$

If $n$ is **odd,** the ends will point in **opposite directions,**

1. for $a > 0$: down on the left, up on the right (*as with $y = x$*);

    as $x \to -\infty$, $y \to -\infty$;     as $x \to \infty$, $y \to \infty$

2. for $a < 0$: up on the left, down on the right (*as with $y = -x$*);

    as $x \to -\infty$, $y \to \infty$;     as $x \to \infty$, $y \to -\infty$

---

**EXAMPLE 3** ▶  **Identifying the End-Behavior of a Function**

State the end-behavior of each function.

    **a.** $f(x) = 0.8x^4 - 3x^3 + 0.5x^2 + 4x - 1$    **b.** $g(x) = -2x^5 + 6x^3 - 1$

**Solution** ▶  **a.** The function has degree 4 (even), so the ends will point in the same direction. The leading coefficient is positive, so end-behavior is up/up. See Figure 4.20.

    **b.** The function has degree 5 (odd), so the ends will point in opposite directions. The leading coefficient is negative, so the end-behavior is up/down. See Figure 4.21.

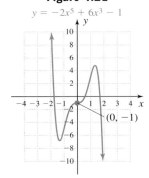

**Figure 4.20**

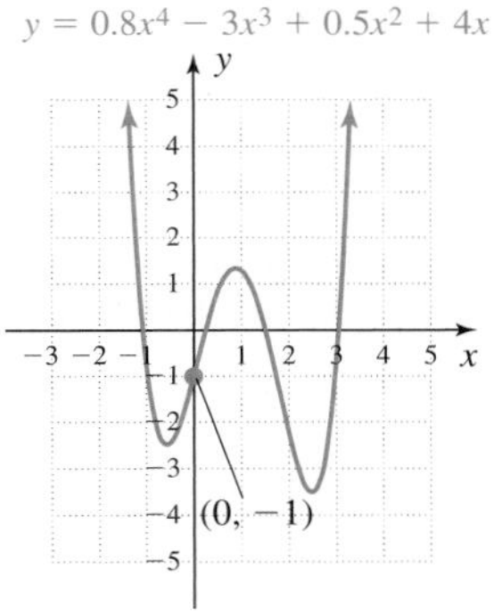

$y = 0.8x^4 - 3x^3 + 0.5x^2 + 4x - 1$

**Figure 4.21**

$y = -2x^5 + 6x^3 - 1$

☑ **B.** You've just seen how we can describe the end-behavior of a polynomial graph

Now try Exercises 15 through 20 ▶

## C. Attributes of Polynomial Graphs with Zeroes of Multiplicity

Another important aspect of polynomial functions is the behavior of a graph near its zeroes. In the simplest case, consider the functions $f(x) = x$ and $g(x) = x^3$ in Figure 4.22. Both have odd degree, like end-behavior (down/up), and a zero at $x = 0$. But the zero of $f$ has multiplicity 1, while the zero from $g$ has multiplicity 3. Notice the graph of $g$ is vertically compressed near $x = 0$ and flattens out on its approach and departure from this zero.

**Figure 4.22**

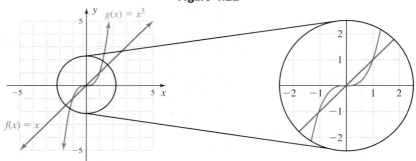

This behavior can be explained by noting that for $x = -1$ and $1, f(x) = g(x)$. But for $|x| < 1$, the graph of $g$ *will be closer to the x-axis* ($g$ increases slower than $f$) since the cube of a fractional number is smaller than the fraction itself. We further note that for $|x| > 1$, $g$ increases much faster than $f$, and $g$ will be farther from the x-axis. Similar observations can be made regarding $f(x) = x^2$ and $g(x) = x^4$ in Figure 4.23. Both functions have even degree, a zero at $x = 0$, and $f(x) = g(x)$ for $x = -1$ and 1. But for $|x| < 1$, the function with higher degree is once again closer to the x-axis.

**Figure 4.23**

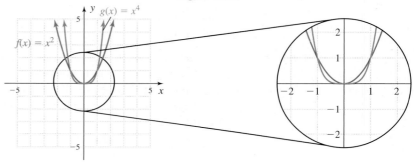

These observations can be generalized and applied to all real zeroes of a function.

### Polynomial Graphs and Zeroes of Multiplicity

Given $P(x)$ is a polynomial with factors of the form $(x - c)^m$, with $c$ a real number,

• If $m$ is odd, the graph will cross through the x-axis at $x = c$.
• If $m$ is even, the graph will "bounce" off the x-axis at $x = c$ (touch at just one point).

In each case, the graph will be more compressed (flatter) near $c$ for larger values of $m$.

**Figure 4.24**

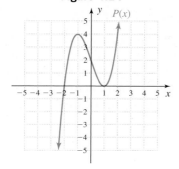

**Figure 4.25**

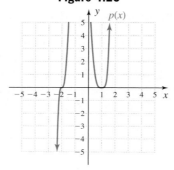

To illustrate, compare the graph of $P(x)$ in Figure 4.24, with the graph of $p(x)$ in Figure 4.25, noting the increased multiplicity of each zero:

$$P(x) = (x + 2)(x - 1)^2 \qquad p(x) = (x + 2)^3(x - 1)^4$$

Both graphs show the expected zeroes at $x = -2$ and $x = 1$, but the graph of $p(x)$ is flatter near $x = -2$ and $x = 1$, due to the increased multiplicity of each zero. We also lose sight of the graph of $p(x)$ between $x = -2$ and $x = 0$, since the increased multiplicities produce larger values than the original grid could display.

**EXAMPLE 4** ▶ **Identifying Attributes of a Function from Its Graph**

The graph of a polynomial $f(x)$ is shown.

a. State whether the degree of $f$ is even or odd.
b. Use the graph to name the zeroes of $f$, then state whether their multiplicity is even or odd.
c. State the minimum possible degree of $f$.
d. State the domain and range of $f$.

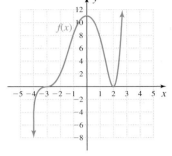

**Solution** ▶ a. Since the ends of the graph point in opposite directions, the degree of the function must be odd.

b. The graph crosses the $x$-axis at $x = -3$ and is compressed near $-3$, meaning it must have odd multiplicity with $m \geq 3$. The graph "bounces" off the $x$-axis at $x = 2$ and 2 must be a zero with multiplicity $m \geq 2$.

c. The minimum possible degree of $f$ is 5, as in $f(x) = a(x - 2)^2(x + 3)^3$ with $a > 0$.

d. $x \in \mathbb{R}, y \in \mathbb{R}$.

Now try Exercises 21 through 26 ▶

To find the degree of a polynomial from its factored form, add the exponents on all linear factors, then add 2 for each irreducible quadratic factor (the degree of any quadratic factor is 2). The sum gives the degree of the polynomial, from which end-behavior can be determined. To find the $y$-intercept, substitute 0 for $x$ as before, noting this is equivalent to applying the exponent to the constant from each factor.

**EXAMPLE 5** ▶ **Identifying Attributes of a Function from Its Factored Form**

State the degree of each function, then describe the end-behavior and name the $y$-intercept of each graph.

a. $f(x) = (x + 2)^3(x - 3)$     b. $g(x) = -(x + 2)^2(x^2 + 5)(x - 5)$

**Solution** ▶ a. The degree of $f$ is $3 + 1 = 4$. With even degree and positive leading coefficient, end-behavior is up/up. For $f(0) = (2)^3(-3) = -24$, the $y$-intercept is $(0, -24)$. See Figure 4.26.

b. The degree of $g$ is $2 + 2 + 1 = 5$. With odd degree and negative leading coefficient, end-behavior is up/down. For $g(0) = -1(2)^2(5)(-5) = 100$, the $y$-intercept is $(0, 100)$. See Figure 4.27.

**Figure 4.26**

$y = (x + 2)^3(x - 3)$

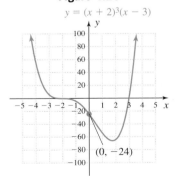

**Figure 4.27**

$y = -(x + 2)^2(x^2 + 5)(x - 5)$

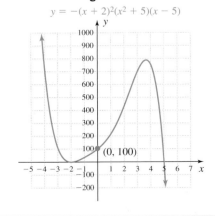

Now try Exercises 27 through 34 ▶

**EXAMPLE 6** ▶  **Matching Graphs to Functions Using Zeroes of Multiplicity**

The following functions all have zeroes at $x = -2, -1$, and 1. Match each function to the corresponding graph *using its degree and the multiplicity of each zero.*

   **a.** $y = (x + 2)(x + 1)^2(x - 1)^3$      **b.** $y = (x + 2)(x + 1)(x - 1)^3$

   **c.** $y = (x + 2)^2(x + 1)^2(x - 1)^3$      **d.** $y = (x + 2)^2(x + 1)(x - 1)^3$

**Figure 4.28**

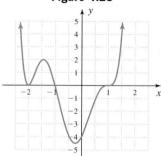

**Figure 4.29**

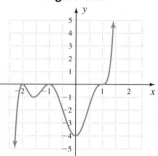

**Figure 4.30**

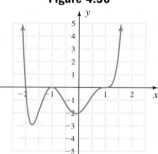

**Figure 4.31**

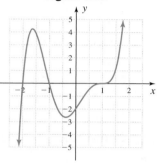

**Solution** ▶  The functions in Figures 4.28 and 4.30 must have even degree due to end-behavior, so each corresponds to (a) or (d). At $x = -1$ the graph in Figure 4.28 "crosses" and must have odd multiplicity, while the graph in Figure 4.30 "bounces" and must have even multiplicity, This indicates Figure 4.28 matches equation (d), while Figure 4.30 matches equation (a).

The graphs in Figures 4.29 and 4.31 must have odd degree due to end-behavior, so each corresponds to (b) or (c). Here, one graph "bounces" at $x = -2$, while the other "crosses." The graph in Figure 4.29 matches equation (c), the graph in Figure 4.31 matches equation (b).

**Now try Exercises 35 through 40** ▶

Using the ideas from Examples 5 and 6, we're able to draw a fairly accurate graph given the factored form of a polynomial. Convenient values between two zeroes, called **midinterval points,** should be used to help complete the graph.

**EXAMPLE 7** ▶ **Graphing a Function Given in Factored Form**

Sketch the graph of $f(x) = (x - 2)(x - 1)^2(x + 1)^3$ using end-behavior; the $x$- and $y$-intercepts, and zeroes of multiplicity.

**Solution** ▶ Adding the exponents of each factor, we find that $f$ is a function of degree 6 with a positive lead coefficient, so end-behavior will be up/up. Since $f(0) = -2$, the $y$-intercept is $(0, -2)$. The graph will bounce off the $x$-axis at $x = 1$ (even multiplicity), and cross the axis at $x = -1$ and 2 (odd multiplicities). The graph will "flatten out" near $x = -1$ because of its higher multiplicity. To help "round-out" the graph we evaluate $f$ at $x = 1.5$, giving $(-0.5)(0.5)^2(2.5)^3 \approx -1.95$ (note scaling of the $x$- and $y$-axes).

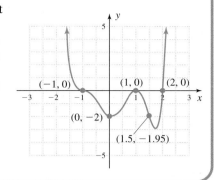

☑ **C.** You've just seen how we can discuss the attributes of a polynomial graph with zeroes of multiplicity

Now try Exercises 41 through 52 ▶

## D. The Graph of a Polynomial Function

Using the cumulative observations from this and previous sections, a general strategy emerges for the graphing of polynomial functions.

---

**Guidelines for Graphing Polynomial Functions**

1. Determine the end-behavior of the graph.
2. Find the $y$-intercept $(0, a_0)$
3. Find the zeroes using any combination of factoring, tests for 1 and $-1$, the factor and remainder theorems, the quadratic formula, and the rational zeroes theorem.
4. Use the $y$-intercept, end-behavior, the multiplicity of each zero, and midinterval points as needed to sketch a smooth, continuous curve.

   *Additional tools* include (a) number of zeroes corollary, (b) complex conjugates corollary, (c) number of turning points, (d) Descartes' rule of signs, (e) upper and lower bounds, and (f) symmetry.

---

**WORTHY OF NOTE**

Although of somewhat limited value, symmetry (item f in the guidelines) can sometimes aid in the graphing of polynomial functions. If all terms of the function have even degree, the graph will be symmetric to the $y$-axis (even). If all terms have odd degree, the graph will be symmetric to the origin. Recall that a constant term has degree zero, an even number.

---

**EXAMPLE 8** ▶ **Graphing a Polynomial Function**

Sketch the graph of $g(x) = -x^4 + 9x^2 - 4x - 12$.

**Solution** ▶
1. End-behavior: The function has degree 4 (even) with a negative leading coefficient, so end-behavior is *down on the left, down on the right*.
2. Since $g(0) = -12$, the $y$-intercept is $(0, -12)$.
3. Zeroes: Using the test for $x = 1$ gives $-1 + 9 - 4 - 12 = -8$, showing $x = 1$ is not a zero but $(1, -8)$ is a point on the graph. Using the test for $x = -1$ gives $-1 + 9 + 4 - 12 = 0$, so $-1$ is a zero and $x + 1$ is a factor. Using $x = -1$ with the factor theorem yields

$$
\begin{array}{r|rrrrr}
-1 & -1 & 0 & 9 & -4 & -12 \\
   &    & 1 & -1 & -8 & 12 \\
\hline
   & -1 & 1 & 8 & -12 & |\underline{0}
\end{array}
$$

The quotient polynomial is not easily factorable so we continue with synthetic division. Using the rational zeroes theorem, the possible rational zeroes are $\{\pm 1, \pm 12, \pm 2, \pm 6, \pm 3, \pm 4\}$, so we try $x = 2$.

use 2 as a "divisor" on *the quotient polynomial*

$$
\begin{array}{r|rrrr}
2 & -1 & 1 & 8 & -12 \\
  &    & -2 & -2 & 12 \\
\hline
  & -1 & -1 & 6 & \underline{|0}
\end{array}
$$

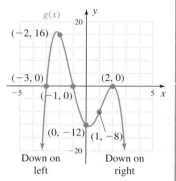

This shows $x = 2$ is a zero, $x - 2$ is a factor, and the function can now be written as

$$g(x) = (x + 1)(x - 2)(-x^2 - x + 6).$$

Factoring $-1$ from the trinomial gives

$$
\begin{aligned}
g(x) &= -1(x + 1)(x - 2)(x^2 + x - 6) \\
&= -1(x + 1)(x - 2)(x + 3)(x - 2) \\
&= -1(x + 1)(x - 2)^2(x + 3)
\end{aligned}
$$

The zeroes of $g$ are $x = -1$ and $-3$, both with multiplicity 1, and $x = 2$ with multiplicity 2.

4. To help "round-out" the graph we evaluate the midinterval point $x = -2$ using the remainder theorem or the factored form of $g(x)$, which shows that $(-2, 16)$ is also a point on the graph.

use $-2$ as a "divisor"

$$
\begin{array}{r|rrrrr}
-2 & -1 & 0 & 9 & -4 & -12 \\
   &    & 2 & -4 & -10 & 28 \\
\hline
   & -1 & 2 & 5 & -14 & \underline{|16}
\end{array}
$$

The final result is the graph shown.

**Now try Exercises 53 through 64** ▶

**CAUTION** ▶ Sometimes using a midinterval point to help draw a graph will give the impression that a maximum or minimum value has been located. This is rarely the case, as demonstrated in the figure in Example 8, where the maximum value in Quadrant II is actually closer to $(-2.22, 16.95)$. Determining the exact location of these extreme values requires the tools of calculus.

**EXAMPLE 9** ▶ **Using the Guidelines to Sketch a Polynomial Graph**

Sketch the graph of $h(x) = x^7 - 4x^6 + 7x^5 - 12x^4 + 12x^3$.

**Solution** ▶ 1. End-behavior: The function has degree 7 (odd) so the ends will point in opposite directions. The leading coefficient is positive and the end-behavior will be *down on the left* and *up on the right*.

2. $y$-intercept: Since $h(0) = 0$, the $y$-intercept is $(0, 0)$.

3. Zeroes: Testing 1 and $-1$ shows neither are zeroes but $(1, 4)$ and $(-1, -36)$ are points on the graph. Factoring out $x^3$ produces $h(x) = x^3(x^4 - 4x^3 + 7x^2 - 12x + 12)$, and we see that $x = 0$ is a zero of multiplicity 3. We next use synthetic division with $x = 2$ on the fourth-degree polynomial:

use 2 as a "divisor"

$$
\begin{array}{r|rrrrr}
2 & 1 & -4 & 7 & -12 & 12 \\
  &   & 2 & -4 & 6 & -12 \\
\hline
  & 1 & -2 & 3 & -6 & \underline{|0}
\end{array}
$$

This shows $x = 2$ is a zero and $x - 2$ is a factor. At this stage, it appears the quotient can be factored by grouping. From $h(x) = x^3(x - 2)(x^3 - 2x^2 + 3x - 6)$, we obtain $h(x) = x^3(x - 2)(x^2 + 3)(x - 2)$ after factoring and

$$h(x) = x^3(x - 2)^2(x^2 + 3)$$

as the completely factored form. We find that $x = 2$ is a zero of multiplicity 2, and the remaining two zeroes are complex.

4. Using this information produces the graph shown in the figure.

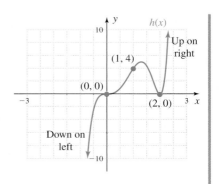

**☑ D.** You've just seen how we can graph polynomial functions in standard form

Now try Exercises 65 through 68 ▶

Similar to our work in previous sections, **Exercises 69 and 70** ask you to reconstruct the complete equation of a polynomial from its given graph.

## E. Applications of Polynomials and Polynomial Modeling

**EXAMPLE 10** ▶ **Modeling the Value of an Investment**

In the year 2000, Marc and his wife Maria decided to invest some money in precious metals. As expected, the value of the investment fluctuated over the years, sometimes being worth more than they paid, other times less. Suppose that through 2010, the gain or loss on the investment was modeled by $v(t) = t^4 - 11t^3 + 38t^2 - 40t$, where $v(t)$ represents the gain or loss (in hundreds of dollars) in year $t$ ($t = 0 \rightarrow 2000$).

 **a.** Use the rational zeroes theorem to find the years when their gain/loss was zero.

 **b.** Sketch the graph of the function.

 **c.** In what years was the investment worth less than they paid?

 **d.** What was their gain or loss in 2010?

**Solution** ▶ **a.** Writing the function as $v(t) = t(t^3 - 11t^2 + 38t - 40)$, we note $t = 0$ shows no gain or loss on purchase, and attempt to find the remaining zeroes. Testing for 1 and $-1$ shows neither is a zero, but $(1, -12)$ and $(-1, 90)$ are points on the graph of $v$. Next we try $t = 2$ with the factor theorem and the cubic polynomial.

$$
\begin{array}{r|rrrr}
2 & 1 & -11 & 38 & -40 \\
  &   & 2 & -18 & 40 \\
\hline
  & 1 & -9 & 20 & \underline{|0}
\end{array}
$$

> **WORTHY OF NOTE**
>
> Due to the context, the domain of $v(t)$ in Example 10 actually begins at $t = 0$, which we could designate with a point at (0, 0). In addition, note there are three sign changes in the terms of $v(t)$, indicating there will be 3 or 1 positive roots (we found 3).

We find that 2 is a zero and write $v(t) = t(t - 2)(t^2 - 9t + 20)$, then factor the quotient polynomial to obtain $v(t) = t(t - 2)(t - 4)(t - 5)$. Since $v(t) = 0$ for $t = 0, 2, 4,$ and 5, they "broke even" in years 2000, 2002, 2004, and 2005.

**Figure 4.32**

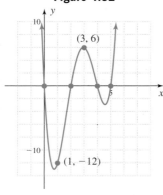

**b.** With even degree and a positive leading coefficient, the end-behavior is up/up. All zeroes have multiplicity 1. As an additional midinterval point we find $v(3) = 6$:

$$
\begin{array}{r|rrrrr}
3 & 1 & -11 & 38 & -40 & 0 \\
  &   & 3   & -24 & 42  & 6 \\
\hline
  & 1 & -8  & 14  & 2   & \underline{6}
\end{array}
$$

The graph is shown in Figure 4.32.

**c.** The investment was worth less than what they paid (outputs are negative) from 2000 to 2002 and 2004 to 2005.

**d.** In 2010, they were "sitting pretty," as their investment had gained $2,400.

$$
\begin{array}{r|rrrrr}
10 & 1 & -11 & 38  & -40 & 0 \\
   &   & 10  & -10 & 280 & 2400 \\
\hline
   & 1 & -1  & 28  & 240 & \underline{2400}
\end{array}
$$

**Now try Exercises 73 through 76 ▶**

As with linear and quadratic regression models, applications of other polynomial models begins with a scatterplot and a decision as to which form of regression might be appropriate. This can depend on a number of factors, such as any end-behavior that is evident, the number of apparent turning points, any anticipated behavior, and so on. However, due to the end-behavior of polynomial models, great care must be exercised when these models are used to make projections beyond the limits of the given data.

**EXAMPLE 11 ▶** The Earth's atmosphere consists of several layers that are defined in terms of altitude and the characteristics of the air in each layer. In order, these are the Troposphere (0–12 km), Stratosphere (12–50 km), Mesosphere (50–80 km), and Thermosphere (80–100 km). Due to their chemical and physical characteristics, the air temperature within each layer and from layer to layer varies a great deal. The data in the table give the temperature in °C at an altitude of $h$ kilometers (km). Use the data to:

**a.** Draw a scatterplot and decide on an appropriate form of regression, then find the regression equation.

**b.** Use the regression equation to find the temperature at altitudes of 32.6 km and 63.6 km.

**c.** As the space shuttle rockets into orbit, a temperature reading of $-75°C$ is taken. What are the possible altitudes for the shuttle at this point?

| Altitude (km) | Temperature (°C) |
|---|---|
| 0 | 20 |
| 4 | −20 |
| 8 | −45 |
| 12 | −55 |
| 20 | −57 |
| 30 | −43 |
| 40 | −16 |
| 50 | −2 |
| 60 | −14 |
| 70 | −54 |
| 80 | −91 |
| 90 | −93 |
| 100 | −45 |

**Solution** ▶   **a.** The scatterplot is shown in Figure 4.33. Using the characteristics exhibited (end-behavior, three turning points), it appears a quartic regression (degree 4) is appropriate and the equation is shown in Figure 4.34.

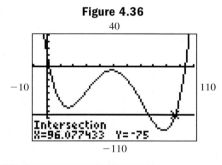

**b.** At altitudes of 32.6 km and 63.6 km, the temperature is very near $-30.0°C$ (Figure 4.35).

**c.** Setting $Y_2 = -75$, we note the line intersects the graph in two places, one indicating an altitude of about 76.4 km, and the other an altitude of about 96.1 km (Figure 4.36).

☑ **E.** You've just seen how to solve applications of polynomials and polynomial modeling

Now try Exercises 77 and 78 ▶

## 4.4 EXERCISES

### ▶ CONCEPTS AND VOCABULARY

**Fill in each blank with the appropriate word or phrase. Carefully reread the section, if necessary.**

1. For a polynomial with factors of the form $(x - c)^m$, $c$ is called a(n) _____ of multiplicity _____.

2. A polynomial function of degree $n$ has _____ zeroes and at most _____ "turning points."

3. Discuss/Explain how to find the degree and $y$-intercept of a function that is given in factored form. Use $f(x) = (x + 1)^3(x - 2)(x + 4)^2$ to illustrate.

4. Name all of the "tools" at your disposal that play a role in the graphing of polynomial functions. Which tools are indispensable and always used? Which tools are used only as the situation merits?

## ▶ DEVELOPING YOUR SKILLS

Determine whether each graph is the graph of a polynomial function. If yes, state the least possible degree of the function. If no, state why.

**5.**
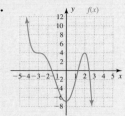

**6.**
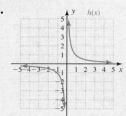

**7.**
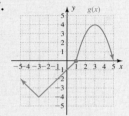

**8.**
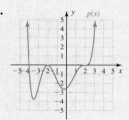

**9.**
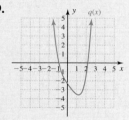

**10.**
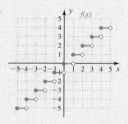

State the end-behavior of the functions shown.

**11.** $f(x)$
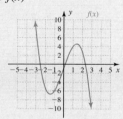

**12.** $g(x)$
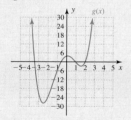

**13.** $H(x)$
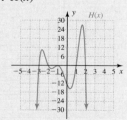

**14.** $h(x)$
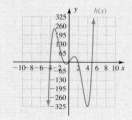

State the end-behavior and $y$-intercept of the functions given. Do not graph.

**15.** $f(x) = x^3 + 6x^2 - 5x - 2$

**16.** $g(x) = x^4 - 4x^3 - 2x^2 + 16x - 12$

**17.** $p(x) = -2x^4 + x^3 + 7x^2 - x - 6$

**18.** $q(x) = -2x^3 - 18x^2 + 7x + 3$

**19.** $Y_1 = -3x^5 + x^3 + 7x^2 - 6$

**20.** $Y_2 = -x^6 - 4x^5 + 4x^3 + 16x - 12$

For each graph given, (a) state whether the degree is even or odd; (b) name the zeroes and state their multiplicity; (c) state the minimum possible degree of $f$ and write $f$ in factored form; and (d) estimate the domain and range. Assume all zeroes are real.

**21.**
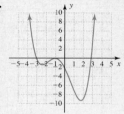

**22.**
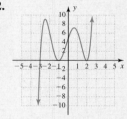

**23.**
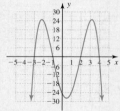

**24.**
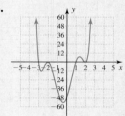

**25.**
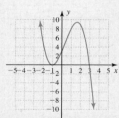

**26.**
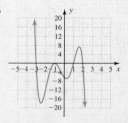

State the degree of each function, the end-behavior, and $y$-intercept of its graph.

**27.** $f(x) = (x - 3)(x + 1)^3(x - 2)^2$

**28.** $g(x) = (x + 2)^2(x - 4)(x + 1)$

**29.** $p(x) = -(x + 1)^2(x - 2)(2x - 3)(x + 4)$

**30.** $P(x) = -(x + 1)(x - 2)^3(5x - 3)$

**31.** $r(x) = (x^2 + 3)(x + 4)^3(x - 1)$

**32.** $s(x) = (x + 2)^2(x - 1)^2(x^2 + 5)$

**33.** $h(x) = (x^2 + 2)(x - 1)^2(1 - x)$

**34.** $H(x) = (x + 2)^2(2 - x)(x^2 + 4)$

**Every function in Exercises 35 through 40 has the zeroes** $x = -1, x = -3,$ **and** $x = 2.$ **Match each to its corresponding graph using degree, end-behavior, and the multiplicity of each zero.**

**35.** $f(x) = (x + 1)^2(x + 3)(x - 2)$

**36.** $F(x) = (x + 1)(x + 3)^2(x - 2)$

**37.** $g(x) = (x + 1)(x + 3)(x - 2)^3$

**38.** $G(x) = (x + 1)^3(x + 3)(x - 2)$

**39.** $h(x) = (x + 1)^2(x + 3)(x - 2)^2$

**40.** $H(x) = (x + 1)^3(x + 3)(x - 2)^2$

**a.**
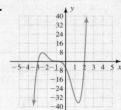

**b.**
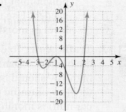

**c.**
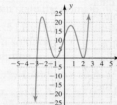

**d.**
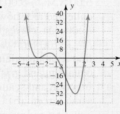

**e.**
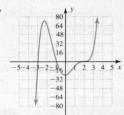

**f.**
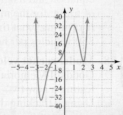

**Sketch the graph of each function using the degree, end-behavior, x- and y-intercepts, multiplicity of zeroes, and a few midinterval points to round-out the graph. Connect all points with a smooth, continuous curve.**

**41.** $f(x) = (x + 3)(x + 1)(x - 2)$

**42.** $g(x) = (x + 2)(x - 4)(x - 1)$

**43.** $p(x) = -(x + 1)^2(x - 3)$

**44.** $q(x) = -(x + 2)(x - 2)^2$

**45.** $Y_1 = (x + 1)^2(3x - 2)(x + 3)$

**46.** $Y_2 = (x + 2)(x - 1)^2(5x - 2)$

**47.** $r(x) = -(x + 1)^2(x - 2)^2(x - 1)$

**48.** $s(x) = -(x - 3)(x - 1)^2(x + 1)^2$

**49.** $f(x) = (2x + 3)(x - 1)^3$

**50.** $g(x) = (3x - 4)(x + 1)^3$

**51.** $h(x) = (x + 1)^3(x - 3)(x - 2)$

**52.** $Y_4 = (x - 3)(x - 1)^3(x + 1)^2$

**Use the Guidelines for Graphing Polynomial Functions to graph the polynomials.**

**53.** $y = x^3 + 3x^2 - 4$

**54.** $y = x^3 - 13x + 12$

**55.** $f(x) = x^3 - 3x^2 - 6x + 8$

**56.** $g(x) = x^3 + 2x^2 - 5x - 6$

**57.** $h(x) = -x^3 - x^2 + 5x - 3$

**58.** $H(x) = -x^3 - x^2 + 8x + 12$

**59.** $p(x) = -x^4 + 10x^2 - 9$

**60.** $q(x) = -x^4 + 13x^2 - 36$

**61.** $r(x) = x^4 - 6x^3 + 8x^2 + 6x - 9$

**62.** $s(x) = x^4 - 4x^3 - 3x^2 + 10x + 8$

**63.** $F(x) = 2x^4 + 3x^3 - 9x^2$

**64.** $G(x) = 3x^4 + 2x^3 - 8x^2$

**65.** $f(x) = x^5 + 4x^4 - 16x^2 - 16x$

**66.** $g(x) = x^5 - 3x^4 + x^3 - 3x^2$

**67.** $h(x) = x^6 - 2x^5 - 4x^4 + 8x^3$

**68.** $H(x) = x^6 + 3x^5 - 4x^4$

**Use the graph of each function to construct its equation in factored form and in polynomial form. Be sure to check the y-intercept and adjust the lead coefficient if necessary.**

**69.**
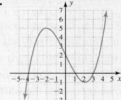

**70.**
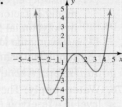

## ▶ WORKING WITH FORMULAS

**71. Roots tests for** $ax^4 + bx^3 + cx^2 + dx + e$:
$$(r_1)^2 + (r_2)^2 + (r_3)^2 + (r_4)^2 = \frac{b^2 - 2ac}{a^2}$$

Given $a = 1$ with $r_1, r_2, r_3,$ and $r_4$ as roots of the polynomial, the sum of the squares of the roots must be equal to $b^2 - 2c$. (a) Given that $x = -3, -1, 2,$ and $4$ are the roots of $x^4 - 2x^3 - 13x^2 + 14x + 24 = 0$, show these roots satisfy the test and (b) use the roots and the factored form to write the equation in polynomial form to confirm results.

**72.** It is worth noting that the root test in Exercise 71 still applies when the roots are irrational and/or complex. Given that $x = -\sqrt{3}, \sqrt{3}, 1 + 2i,$ and $1 - 2i$ are the solutions to $x^4 - 2x^3 + 2x^2 + 6x - 15 = 0$, show these roots satisfy the test and use these solutions and the factored form to write the equation in polynomial form to confirm results.

**Solution** ▶ Begin by writing $f$ in factored form: $f(x) = \frac{(x + 2)(x - 3)}{(x + 3)(x - 2)}$.

1. **$y$-intercept:** $f(0) = \frac{(2)(-3)}{(3)(-2)} = 1$, so the $y$-intercept is $(0, 1)$.
2. **$x$-intercepts:** Setting the numerator equal to zero gives $(x + 2)(x - 3) = 0$, showing the $x$-intercepts will be $(-2, 0)$ and $(3, 0)$.
3. **Vertical asymptote(s):** Setting the denominator equal to zero gives $(x + 3)(x - 2) = 0$, showing there will be vertical asymptotes at $x = -3$ and $x = 2$.
4. **Horizontal asymptote:** Since the degree of the numerator and the degree of the denominator are equal, $y = \frac{x^2}{x^2} = 1$ is a horizontal asymptote.
5. Solving
$$\frac{x^2 - x - 6}{x^2 + x - 6} = 1 \qquad f(x) = 1 \rightarrow \text{horizontal asymptote}$$
$$x^2 - x - 6 = x^2 + x - 6 \quad \text{multiply by } x^2 + x - 6$$
$$-2x = 0 \qquad \text{simplify}$$
$$x = 0 \qquad \text{solve}$$

The graph will cross the horizontal asymptote at $(0, 1)$.

The information from Steps 1 through 5 is shown in Figure 4.47, and indicates we have little information about the graph in the interval $(-\infty, -3)$. Since rational functions are defined for all real numbers except the zeroes of $d$, we know there must be a "piece" of the graph in this interval.

6. Selecting $x = -4$ to compute one additional point, we find $f(-4) = \frac{(-2)(-7)}{(-1)(-6)} = \frac{14}{6} = \frac{7}{3}$. The point is $(-4, \frac{7}{3})$.

All factors of $f$ are linear, so function values will alternate sign in the intervals created by $x$-intercepts and vertical asymptotes. The $y$-intercept $(0, 1)$ shows $f(x)$ is positive in the interval containing 0 (Figure 4.48). To meet all necessary conditions, we complete the graph as shown in Figure 4.49.

> **WORTHY OF NOTE**
>
> It's useful to note that the number of "pieces" forming a rational graph will always be one more than the number of vertical asymptotes. The graph of $f(x) = \frac{3x}{x^2 + 2}$ (Figure 4.44) has no vertical asymptotes and one piece, $y = \frac{1}{x}$ has one vertical asymptote and two pieces, $g(x) = \frac{x^2 - 4}{x^2 - 1}$ (Figure 4.45) has two vertical asymptotes and three pieces, and so on.

**Figure 4.47**          **Figure 4.48**          **Figure 4.49**

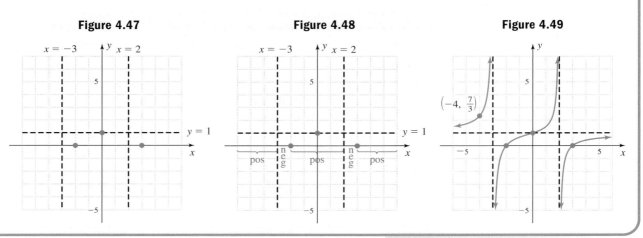

**Now try Exercises 27 through 46** ▶

Examples 2 and 3 demonstrate that graphs of rational functions come in a large variety. Once the components of the graph have been found, completing the graph presents an intriguing and puzzle-like challenge as we attempt to sketch a graph that meets all conditions. As we've done with other functions, can you reverse this process? That is, given the *graph* of a rational function, can you construct its equation?

**EXAMPLE 4** ▶ **Finding the Equation of a Rational Function from Its Graph**

Use the graph of $f(x)$ shown to construct its equation.

Solution ▶ The $x$-intercepts are $(-1, 0)$ and $(4, 0)$, so the numerator must contain the factors $(x + 1)$ and $(x - 4)$. The vertical asymptotes are $x = -2$ and $x = 3$, so the denominator must have the factors $(x + 2)$ and $(x - 3)$. So far we have:

$$f(x) = \frac{a(x + 1)(x - 4)}{(x + 2)(x - 3)}$$

Since $(2, 3)$ is on the graph, we substitute 2 for $x$ and 3 for $f(x)$ to solve for $a$:

$$3 = \frac{a(2 + 1)(2 - 4)}{(2 + 2)(2 - 3)} \qquad \text{substitute 3 for } f(x) \text{ and 2 for } x$$

$$3 = \frac{3a}{2} \qquad \text{simplify}$$

$$2 = a \qquad \text{solve}$$

☑ **C.** You've just seen how we can graph general rational functions with vertical asymptotes

The result is $f(x) = \frac{2(x + 1)(x - 4)}{(x + 2)(x - 3)} = \frac{2x^2 - 6x - 8}{x^2 - x - 6}$, with a horizontal asymptote at $y = 2$ and a $y$-intercept of $(0, \frac{4}{3})$, which fit the graph very well.

Now try Exercises 47 through 50 ▶

## D. Rational Functions with Oblique or Nonlinear Asymptotes

So far, we've found that for $p(x)$ with leading term $ax^n$ and $d(x)$ with leading term $bx^m$,

- If $n < m$, the line $y = 0$ is a horizontal asymptote.
- If $n = m$, the line $y = \frac{a}{b}$ is a horizontal asymptote.

But what happens if the degree of the numerator is *greater than* the degree of the denominator? To investigate, consider the functions $f$, $g$, and $h$ in Figures 4.50 through 4.52, whose only difference is the degree of the numerator.

**Figure 4.50**

$$f(x) = \frac{2x}{x^2 + 1}$$

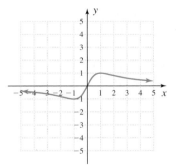

**Figure 4.51**

$$g(x) = \frac{2x^2}{x^2 + 1}$$

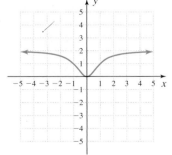

**Figure 4.52**

$$h(x) = \frac{2x^3}{x^2 + 1}$$

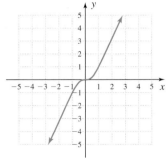

The graph of $f$ has a horizontal asymptote at $y = 0$ since the denominator is of larger degree (as $|x| \to \infty$, $y \to 0$). As we might have anticipated, the horizontal asymptote for $g$ is $y = 2$, the ratio of leading coefficients (as $|x| \to \infty$, $y \to 2$). The graph of $h$ has no horizontal asymptote, yet appears to be asymptotic to some slanted line. To see why, note the function $h(x) = \frac{2x^3}{x^2 + 1}$ can be considered an "improper fraction," similar to how we apply this designation to the fraction $\frac{3}{2}$. To write $h$ in "proper" form, we use long division, writing the dividend as $2x^3 + 0x^2 + 0x + 0$, and the divisor as $x^2 + 0x + 1$. The ratio $\frac{2x^3}{x^2}$ $\overset{\text{from dividend}}{\underset{\text{from divisor}}{}}$ shows $\mathbf{2x}$ will be our first multiplier.

$$
\begin{array}{r}
2x \phantom{0000000000} \\
x^2 + 0x + 1 \overline{)2x^3 + 0x^2 + 0x + 0} \\
\underline{-(2x^3 + 0x^2 + 2x)} \quad \text{\scriptsize multiply } 2x(x^2 + 0x + 1) \\
-2x \quad \text{\scriptsize subtract, next term is 0}
\end{array}
$$

The result shows $h(x) = 2x + \frac{-2x}{x^2 + 1}$. Note as $|x| \to \infty$, the term $\frac{-2x}{x^2 + 1}$ becomes very small and closer to zero (since the degree of the numerator is less than the degree of the denominator), so $h(x) \approx 2x$ for large $x$. This is an example of an **oblique (slanted) asymptote.** In general,

---

### Oblique and Nonlinear Asymptotes

Given $V(x) = \frac{p(x)}{d(x)}$ is a rational function in simplest form, where the degree of $p$ is greater than the degree of $d$, the graph will have an oblique or nonlinear asymptote as determined by $q(x)$, where $q(x)$ is the quotient polynomial after division.

---

If the denominator is a monomial, term-by-term division is the most efficient means of computing the quotient. If the denominator is not a monomial, either synthetic division or long division must be used. From our earlier work, $q(x)$ will give the equation of the asymptote, and the zeroes of the remainder (if they exist) will indicate where the function crosses the asymptote.

We conclude that an oblique or slant asymptote occurs when the degree of the numerator is one more than the degree of the denominator, as that indicates $q(x)$ will be linear. A nonlinear asymptote will occur when the degree of the numerator is larger by two or more.

**EXAMPLE 5** ▶    **Graphing a Rational Function with an Oblique Asymptote**

Graph the function: $h(x) = \frac{x^2}{x - 1}$

**Solution** ▶    The function is already in "factored form."

1. **$y$-intercept:** Since $h(0) = 0$, the $y$-intercept is $(0, 0)$.
2. **$x$-intercept:** $(0, 0)$; From, $x^2 = 0$, we have $x = 0$ with multiplicity two. The $x$-intercept is $(0, 0)$ and the function will not change sign here.
3. **Vertical asymptote:** Solving $x - 1 = 0$ gives $x = 1$ with multiplicity one. There is a vertical asymptote at $x = 1$ and the function will change sign here.
4. **Horizontal/oblique asymptote:** Since the degree of numerator $>$ the degree of denominator, we rewrite $h$ using division. The denominator is linear so we use synthetic division:

$$
\begin{array}{c}
\text{\scriptsize use 1 as a "divisor"} \quad \underline{1|} \quad
\begin{array}{r r r}
1 & 0 & 0 \quad \text{\scriptsize coefficients of dividend} \\
\downarrow & 1 & 1 \\
\hline
1 & 1 & \underline{|1} \quad \text{\scriptsize quotient and remainder}
\end{array}
\end{array}
$$

Since $q(x) = x + 1$ the graph has an oblique asymptote of $y = x + 1$.

5. The remainder $r(x) = 1$ is never zero and the graph will not cross the slant asymptote.

The information gathered in steps 1 through 5 is shown Figure 4.53, and is actually sufficient to complete a reasonable sketch of the graph. If you feel a little unsure about how to "puzzle" out the graph, find additional points in the first and third quadrants: $h(2) = 4$ and $h(-2) = -\frac{4}{3}$. Since the graph will "bounce" at $x = 0$ and output values must change sign at $x = 1$, all conditions are met with the graph shown in Figure 4.54.

**Figure 4.53**

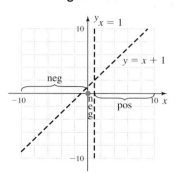

**Figure 4.54**

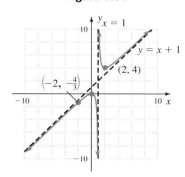

**Now try Exercises 51 through 56** ▶

Finally, it would be a mistake to think that all asymptotes are linear. In fact, when the degree of the numerator is two more than the degree of the denominator, a parabolic asymptote results. Functions of this type often occur in applications of rational functions, and are used to minimize cost, materials, distances, or other considerations of great importance to business and industry. For $f(x) = \frac{x^4 + 1}{x^2}$, term-by-term division gives $x^2 + \frac{1}{x^2}$ and the quotient $q(x) = x^2$ is a nonlinear, parabolic asymptote (see Figure 4.55). For more on nonlinear asymptotes, **see Exercises 57 through 60.**

**Figure 4.55**

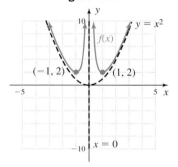

## E. Applications of Rational Functions

In many applications of rational functions, the coefficients can be rather large and the graph should be scaled appropriately.

**EXAMPLE 6** ▶ **Modeling the Rides at an Amusement Park**

A popular amusement park wants to add new rides and asks various contractors to submit ideas. Suppose one ride engineer offers plans for a ride that begins with a near vertical drop into a dark tunnel, quickly turns and becomes more horizontal, pops out the tunnel's end, then coasts up to the exit platform, braking 20 m from the release point. The height of a rider above ground is modeled by the function $h(x) = \frac{39x^2 - 507x + 468}{3x^2 + 23x + 20}$, where $h(x)$ is the height in meters at a horizontal distance of $x$ meters from the release point.

**a.** Graph the function for $x \in [-1, 20]$.

**b.** How high is the release point for this ride?

**c.** What is the distance along the ground from entrance to exit?

**d.** What is the height of the exit platform?

**Solution** ▶  **a.** Here we begin by noting the $y$-intercept is $h(0) = \frac{468}{20}$ or 23.4 m. Also, since the degree of the numerator is equal to that of the denominator, the ratio of leading terms $\frac{39x^2}{3x^2}$ indicates a horizontal asymptote is $y = 13$. Writing $h(x)$ in factored form gives $h(x) = \frac{39(x - 1)(x - 12)}{(3x + 20)(x + 1)}$, showing the $x$-intercepts will be (1, 0) and (12, 0), with vertical asymptotes at $x = -6.\overline{6}$ and $x = -1$. Computing midinterval points of $x = 4, 8$, and 16 gives (4, −5.85), (8, −2.76), and (16, 2.02). Graphing the function over the specified interval produces the graph shown in Figure 4.56.

**Figure 4.56**

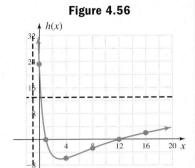

**b.** From context, the release point is $h(0) = 23.4$ m.

**c.** The ride enters the tunnel at $x = 1$ and exits at $x = 12$, making the tunnel 11 m long.

**d.** Since the ride begins braking at a distance of 20 m, the platform must be $h(20) \approx 3.5$ m high. See Figure 4.57.

**Figure 4.57**

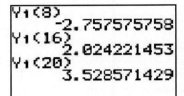

*Now try Exercises 63 through 72* ▶

As a final note, the ride proposed in Example 6 was never approved due to excessive $g$-forces on the riders. However, Example 6 helps to illustrate that when it comes to applications of rational functions, portions of the graph may be ignored due to the context. In addition, some applications may focus on a specific attribute of the graph, such as the horizontal asymptotes in **Exercises 63, 64,** and elsewhere, or the vertical asymptotes in **Exercises 65 and 66.**

**EXAMPLE 7** ▶  **Solving an Application of Rational Functions**

Suppose the cost (in dollars) of manufacturing $x$ thousand of a molded plastic toy is modeled by the function $C(x) = x^2 + 4x + 3$. The *average cost* of each item would then be expressed by

$$A(x) = \frac{\text{total cost}}{\text{number of items}} = \frac{x^2 + 4x + 3}{x}$$

**a.** Graph the function $A(x)$.

**b.** Find how many thousand items are manufactured when the average cost is $8.

**c.** Determine how many thousand items should be manufactured to minimize the average cost (use the graph to estimate this minimum average cost).

**Solution** ▶  **a.** The function is already in simplest form.

**1.** *y*-intercept: none [$A(0)$ is undefined]

**2.** *x*-intercept(s): After factoring we obtain $(x + 3)(x + 1) = 0$, and the zeroes of the numerator are $x = -1$ and $x = -3$, both with multiplicity one. The graph will cross the $x$-axis at each intercept.

**3.** **Vertical asymptote:** $x = 0$, multiplicity one; the function will change sign at $x = 0$.

4. **Horizontal/oblique asymptote:** The degree of numerator $>$ the degree of denominator, so we divide using term-by-term division:

$$\frac{x^2 + 4x + 3}{x} = \frac{x^2}{x} + \frac{4x}{x} + \frac{3}{x}$$

$$= x + 4 + \frac{3}{x}$$

The line $q(x) = x + 4$ is an oblique asymptote.

5. The remainder $\frac{3}{x}$ has no real zeroes, so the graph will not cross the slant asymptote.

The function changes sign at both $x$-intercepts and at the asymptote $x = 0$. The information from Steps 1 through 5 is shown in Figure 4.58 and perhaps an additional point in Quadrant I would help to complete the graph: $A(1) = 8$. The point $(1, 8)$ is on the graph, showing $A$ is positive in the interval containing 1 (since $y = 8$ is positive). Since output values will alternate in sign as stipulated above, all conditions are met with the graph shown in Figure 4.59.

**Figure 4.58**

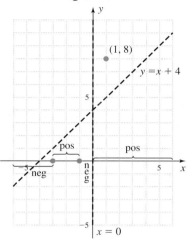

**Figure 4.59**

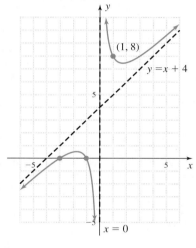

b. To find the number of items manufactured when the average cost is \$8, we replace $A(x)$ with 8 and solve: $\frac{x^2 + 4x + 3}{x} = 8$:

$$x^2 + 4x + 3 = 8x$$
$$x^2 - 4x + 3 = 0$$
$$(x - 1)(x - 3) = 0$$
$$x = 1 \quad \text{or} \quad x = 3$$

The average cost will be \$8 whenever 1,000 or 3,000 molded toys are made.

c. While the domain of this function is all real numbers except 0, the context limits us to positive values. From the graph in Figure 4.60, it appears that the minimum average cost is close to \$7.50, when approximately 1500 to 1800 items are manufactured. Using a graphing calculator, we find that the minimum average cost is actually closer to \$7.46, when about 1732 items are manufactured (Figure 4.60).

**Figure 4.60**

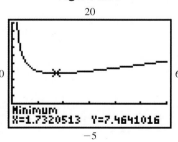

☑ **E.** You've just seen how we can solve applications of rational functions

**Now try Exercises 73 through 76** ▶

## 4.5 EXERCISES

### ▶ CONCEPTS AND VOCABULARY

**Fill in each blank with the appropriate word or phrase. Carefully reread the section, if necessary.**

1. Write the following in direction/approach notation. *As x becomes an infinitely large negative number, y approaches 2.* _____

2. Vertical asymptotes are found by setting the _____ equal to zero. The x-intercepts are found by setting the _____ equal to zero.

3. For any constant $k$, the notation as $|x| \to +\infty$, $y \to k$ is an indication of a(n) _____ asymptote, while $x \to k$, $|y| \to +\infty$ indicates a(n) _____ asymptote. Discuss/Explain why, and give examples.

4. If the degree of the numerator is equal to the degree of the denominator, a horizontal asymptote occurs at $y = \frac{a}{b}$, where $\frac{a}{b}$ represents the ratio of the _____ _____. Discuss/Explain why, and give examples.

### ▶ DEVELOPING YOUR SKILLS

**Give the location of the vertical asymptote(s) if they exist, and state the function's domain.**

5. $f(x) = \dfrac{x + 2}{x - 3}$

6. $F(x) = \dfrac{4x}{2x - 3}$

7. $h(x) = \dfrac{x^2 - 1}{2x^2 + 3x - 5}$

8. $H(x) = \dfrac{x - 5}{2x^2 - x - 3}$

9. $p(x) = \dfrac{2x + 3}{x^2 + x + 1}$

10. $q(x) = \dfrac{2x^3}{x^2 + 4}$

**Give the location of the vertical asymptote(s) if they exist, and state whether function values will change sign (positive to negative or negative to positive) from one side of the asymptote to the other.**

11. $Y_1 = \dfrac{x + 1}{x^2 - x - 6}$

12. $Y_2 = \dfrac{2x + 3}{x^2 - x - 20}$

13. $r(x) = \dfrac{x^2 + 3x - 10}{x^2 - 6x + 9}$

14. $R(x) = \dfrac{x^2 - 2x - 15}{x^2 - 4x + 4}$

15. $Y_1 = \dfrac{x}{x^3 + 2x^2 - 4x - 8}$

16. $Y_2 = \dfrac{-2x}{x^3 + x^2 - x - 1}$

**For the functions given, (a) determine if a horizontal asymptote exists and (b) determine if the graph will cross the asymptote, and if so, where it crosses.**

17. $Y_1 = \dfrac{2x - 3}{x^2 + 1}$

18. $Y_2 = \dfrac{4x + 3}{2x^2 + 5}$

19. $r(x) = \dfrac{4x^2 - 9}{x^2 - 3x - 18}$

20. $R(x) = \dfrac{2x^2 - x - 10}{x^2 + 5}$

21. $p(x) = \dfrac{3x^2 - 5}{x^2 - 1}$

22. $P(x) = \dfrac{3x^2 - 5x - 2}{x^2 - 4}$

**Apply long division to find the quotient and remainder for each function. Use this information to determine the equation of the horizontal asymptote, and whether the graph will cross this asymptote.**

23. $v(x) = \dfrac{8x}{x^2 + 1}$

24. $f(x) = \dfrac{4x + 8}{x^2 + 1}$

25. $g(x) = \dfrac{2x^2 - 8x}{x^2 - 4}$

26. $h(x) = \dfrac{x^2 - x - 6}{x^2 - 1}$

**Give the location of the x- and y-intercepts (if they exist), and discuss the behavior of the function at each x-intercept.**

27. $f(x) = \dfrac{x^2 - 3x}{x^2 - 5}$

28. $F(x) = \dfrac{2x - x^2}{x^2 + 2x - 3}$

29. $g(x) = \dfrac{x^2 + 3x - 4}{x^2 - 1}$

30. $G(x) = \dfrac{x^2 + 7x + 6}{x^2 - 2}$

31. $h(x) = \dfrac{x^3 - 6x^2 + 9x}{4 - x^2}$

32. $H(x) = \dfrac{4x + 4x^2 + x^3}{x^2 - 1}$

**Use the *Guidelines for Graphing Rational Functions* to graph the functions given.**

33. $f(x) = \dfrac{x + 3}{x - 1}$

34. $g(x) = \dfrac{x - 4}{x + 2}$

35. $F(x) = \dfrac{8x}{x^2 + 4}$

36. $G(x) = \dfrac{-12x}{x^2 + 3}$

37. $p(x) = \dfrac{-2x^2}{x^2 - 4}$

38. $P(x) = \dfrac{3x^2}{x^2 - 9}$

39. $q(x) = \dfrac{2x - x^2}{x^2 + 4x - 5}$

40. $Q(x) = \dfrac{x^2 + 3x}{x^2 - 2x - 3}$

41. $h(x) = \dfrac{-3x}{x^2 - 6x + 9}$

42. $H(x) = \dfrac{2x}{x^2 - 2x + 1}$

**43.** $r(x) = \dfrac{x - 1}{x^2 - 3x - 4}$  **44.** $R(x) = \dfrac{1 - x}{x^2 - 2x}$

**45.** $s(x) = \dfrac{x^2 - 4}{x^2 - 1}$  **46.** $S(x) = \dfrac{x^2 - x - 12}{x^2 + x - 12}$

**Construct an equation that corresponds to each graph.**

**47.**

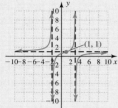

**48.**

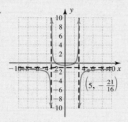

**49.**

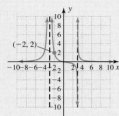

**50.**

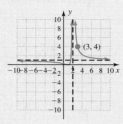

Graph the following using the *Guidelines for Graphing Rational Functions*, which is simply modified to include the possibility of oblique and nonlinear asymptotes.

**51.** $v(x) = \dfrac{x^2 - 4}{x}$  **52.** $f(x) = \dfrac{9 - x^2}{x - 1}$

**53.** $g(x) = \dfrac{x^2}{x - 1}$  **54.** $h(x) = \dfrac{1 - x^2}{x + 2}$

**55.** $g(x) = \dfrac{x^2 + 4x + 4}{x + 3}$  **56.** $G(x) = \dfrac{x^2 - 2x + 1}{x - 2}$

**57.** $p(x) = \dfrac{x^4 + 4}{x^2 + 1}$  **58.** $P(x) = \dfrac{x^4 - 5x^2 + 4}{x^2 + 2}$

**59.** $q(x) = \dfrac{10 + 9x^2 - x^4}{x^2 + 5}$  **60.** $Q(x) = \dfrac{x^4 - 2x^2 + 3}{x^2}$

▶ **WORKING WITH FORMULAS**

**61. Cost of removing pollutants:** $C(x) = \dfrac{kx}{100 - x}$

Some industries resist cleaner air standards because the cost of removing pollutants rises dramatically as higher standards are set. This phenomenon can be modeled by the formula given, where $C(x)$ is the cost (in thousands of dollars) of removing $x\%$ of the pollutant and $k$ is a constant that depends on the type of pollutant and other factors. Graph the function for $k = 250$ over the interval $x \in [0, 100]$, and then use the graph to answer the following questions.

  **a.** What is the significance of the *vertical asymptote* (what does it mean in this context)?

  **b.** If new laws are passed that require 80% of a pollutant to be removed, while the existing law requires only 75%, how much will the new legislation cost the company? Compare the cost of the 5% increase from 75% to 80% with the cost of the 1% increase from 90% to 91%.

  **c.** What percent of the pollutants can be removed if the company budgets $2250 thousand dollars?

**62. Surface area of a cylinder with fixed volume:**

$$S = \dfrac{2\pi r^3 + 2V}{r}$$

It's possible to construct many different cylinders that will hold a specified volume, by changing the radius and height. This is critically important to producers who want to minimize the cost of packing canned goods and marketers who want to present an attractive product. The surface area of the cylinder can be found using the formula shown, where the radius is $r$ and the volume $V = \pi r^2 h$ is known. Assume the fixed volume is 750 cm$^3$.

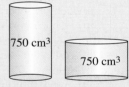

  **a.** Find the equation of the vertical asymptote. How would you describe the nonlinear asymptote?

  **b.** If the radius of the cylinder is 2 cm, what is its surface area?

  **c.** Use a graphing calculator to graph the function in an appropriate window, and use it to find the value of $r$ that minimizes $S(r)$. That is, find the radius that results in a cylinder with the smallest possible surface area, while still holding a volume of 750 cm$^3$.

▶ **APPLICATIONS**

**63. Medication in the blood-stream:** The concentration $C$ of a certain medicine in the bloodstream $h$ hours after being injected into the shoulder is given by the function: $C(h) = \frac{2h^2 + h}{h^3 + 70}$. Use the given graph of the function to answer the following questions.

*(graph: C(h) with vertical axis marked 0.1, 0.2, 0.3, 0.4 and horizontal axis h marked 2, 6, 10, 14, 18, 22, 26)*

a. Approximately how many hours after injection did the maximum concentration occur? What was the maximum concentration?

b. Use $C(h)$ to *compute* the average rate of change for the intervals $h = 8$ to $h = 10$ and $h = 20$ to $h = 22$. What do you notice?

c. Use mathematical notation to state what happens to the concentration $C$ as the number of hours becomes infinitely large. What role does the $h$-axis play for this function?

**64. Supply and demand:** In response to certain market demands, manufacturers will quickly get a product out on the market to take advantage of consumer interest. Once the product is released, it is not uncommon for sales to initially skyrocket, taper off, and then gradually decrease as consumer interest wanes. For a certain product, sales can be modeled by the function $S(t) = \frac{250t}{t^2 + 150}$, where $S(t)$ represents the daily sales (in $10,000) $t$ days after the product has debuted. Use the given graph of the function to answer the following questions.

*(graph: S(t) with vertical axis marked 2.5, 5.0, 7.5, 10.0 and horizontal axis t marked 10, 20, 30, 40, 50, 60, 70)*

a. Approximately how many days after the product came out did sales reach a maximum? What was the maximum sales?

b. Use $S(t)$ to compute the average rate of change for the intervals $t = 7$ to $t = 8$ and $t = 60$ to $t = 62$. What do you notice?

c. Use mathematical notation to state what happens to the daily sales $S$ as the number of days becomes infinitely large. What role does the $t$-axis play for this function?

**65. Cost to remove pollutants:** For a certain coal-burning power plant, the cost to remove pollutants from plant emissions can be modeled by $C(p) = \frac{80p}{100 - p}$, where $C(p)$ represents the cost (in thousands of dollars) to remove $p$ percent of the pollutants. (a) Find the cost to remove 20%, 50%, and 80% of the pollutants, then comment on the results; (b) graph the function using an appropriate scale; and (c) use mathematical notation to state what happens if the power company attempts to remove 100% of the pollutants.

**66. Costs of recycling:** A large city has initiated a new recycling effort, and wants to distribute recycling bins for use in separating various recyclable materials. City planners anticipate the cost of the program can be

modeled by the function $C(p) = \frac{220p}{100 - p}$, where $C(p)$ represents the cost (in $10,000) to distribute the bins to $p$ percent of the population. (a) Find the cost to distribute bins to 25%, 50%, and 75% of the population, then comment on the results; (b) graph the function using an appropriate scale; and (c) use mathematical notation to state what happens if the city attempts to give recycling bins to 100% of the population.

**Memory retention:** Due to their asymptotic behavior, rational functions are often used to model the mind's ability to retain information over a long period of time—the "use it or lose it" phenomenon.

**67. Language retention:** A large group of students is asked to memorize a list of 50 Italian words, a language that is unfamiliar to them. The group is then tested regularly to see how many of the words are retained over a period of time. The average number of words retained is modeled by the function $W(t) = \frac{6t + 40}{t}$, where $W(t)$ represents the number of words remembered after $t$ days.

a. Graph the function over the interval $t \in [1, 40]$. How many days pass until only half the words are remembered? How many days pass until only one-fifth of the words are remembered?

b. After 5 days, what is the average number of words retained? How many days pass until only 8 words can be recalled?

c. What is the significance of the horizontal asymptote (what does it mean in this context)?

**68. Language retention:** A similar study asked students to memorize 50 Hawaiian words, a language that is both unfamiliar and phonetically foreign to them (see Exercise 67). The average number of words retained is modeled by the function $W(t) = \frac{4t + 20}{t}$, where $W(t)$ represents the number of words after $t$ days.

a. Graph the function over the interval $t \in [1, 40]$. How many days pass until only half the words are remembered? How does this compare to Exercise 67? How many days pass until only one-fifth of the words are remembered?

b. After 7 days, what is the average number of words retained? How many days pass until only 5 words can be recalled?

c. What is the significance of the horizontal asymptote (what does it mean in this context)?

**Concentration and dilution:** When antifreeze is mixed with water, it becomes diluted—less than 100% antifreeze. The more water added, the less concentrated the antifreeze becomes, with this process continuing until a desired concentration is met. This application and many similar to it can be modeled by rational functions.

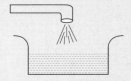

**69. Concentration of antifreeze:** A 400-gal tank currently holds 40 gal of a 25% antifreeze solution. To raise the concentration of the antifreeze in the tank, $x$ gal of a 75% antifreeze solution is pumped in.

**a.** Show the formula for the resulting concentration is $C(x) = \frac{40 + 3x}{160 + 4x}$ after simplifying, and graph the function over the interval $x \in [0, 360]$.

**b.** What is the concentration of the antifreeze in the tank after 10 gal of the new solution are added? After 120 gal have been added? How much liquid is now in the tank?

**c.** If the concentration level is now at 65%, how many gallons of the 75% solution have been added? How many gallons of liquid are in the tank now?

**d.** What is the maximum antifreeze concentration that can be attained in a tank of this size? What is the maximum concentration that can be attained in a tank of "unlimited" size?

**70. Concentration of sodium chloride:** A sodium chloride solution has a concentration of 0.2 oz (weight) per gallon. The solution is pumped into an 800-gal tank currently holding 40 gal of pure water, at a rate of 10 gal/min.

**a.** Find a function $A(t)$ modeling the amount of liquid in the tank after $t$ min, and a function $S(t)$ for the amount of sodium chloride in the tank after $t$ min.

**b.** The concentration $C(t)$ in ounces per gallon is measured by the ratio $\frac{S(t)}{A(t)}$, a rational function. Graph the function on the interval $t \in [0, 100]$. What is the concentration level (in ounces per gallon) after 6 min? After 28 min? How many gallons of liquid are in the tank each time?

**c.** If the concentration level is now 0.184 oz/gal, how long have the pumps been running? How many gallons of liquid are in the tank now?

**d.** What is the maximum concentration that can be attained in a tank of this size? What is the maximum concentration that can be attained in a tank of "unlimited" size?

**Test averages and grade point averages:** To calculate a test average we sum all test points $P$ and divide by the number of tests $N$: $\frac{P}{N}$. To compute the score or scores needed on future tests to raise the average grade to a

desired grade $G$, we add the number of additional tests $n$ to the denominator, and the number of additional tests times the projected grade $g$ on each test to the numerator: $G(n) = \frac{P + ng}{N + n}$. The result is a rational function with some "eye-opening" results.

**71. Computing an average grade:** After four tests, Bobby Lou's test average was an 84. [*Hint:* $P = 4(84) = 336$.]

**a.** Assume that she gets a 95 on all remaining tests ($g = 95$). Graph the resulting function on a calculator using the window $n \in [0, 20]$ and $G(n) \in [80 \text{ to } 100]$. Use the calculator to determine how many tests are required to lift her grade to a 90 under these conditions.

**b.** At some colleges, the range for an "A" grade is 93–100. How many tests would Bobby Lou have to score a 95 on, to raise her average to higher than 93? Were you surprised?

**c.** Describe the significance of the horizontal asymptote of the average grade function. Is a test average of 95 possible if she earns a 95 on all remaining exams?

**d.** Assume now that Bobby Lou scores 100 on all remaining tests ($g = 100$). Approximately how many more tests are required to lift her grade average to higher than 93?

**72. Computing a GPA:** At most colleges, $A \rightarrow 4$ grade points, $B \rightarrow 3$, $C \rightarrow 2$, and $D \rightarrow 1$. After taking 56 credit hours, Aurelio's GPA is 2.5. [*Hint:* In the formula given, $P = 2.5(56) = 140$.]

**a.** Assume Aurelio is determined to get A's (4 grade points or $g = 4$), for all remaining credit hours. Graph the resulting function on a calculator using the window $n \in [0, 60]$ and $G(n) \in [2, 4]$. Use the calculator to determine the number of credit hours required to lift his GPA to over 2.75 under these conditions.

**b.** At some colleges, scholarship money is available only to students with a 3.0 average or higher. How many (perfect 4.0) credit hours would Aurelio have to earn, to raise his GPA to 3.0 or higher? Were you surprised?

**c.** Describe the significance of the horizontal asymptote of the GPA function. Is a GPA of 4.0 possible for him if he earns an A in all remaining classes?

**Average cost of manufacturing an item:** The cost $C$ to manufacture an item depends on the relatively fixed costs $K$ for remaining in business (utilities, maintenance, transportation, etc.) and the actual cost $c$ of manufacturing the item (labor and materials). For $x$ items the cost is $C(x) = K + cx$. The average cost $A$ of manufacturing an item is then $A(x) = \frac{C(x)}{x}$.

**73. Manufacturing water heaters:** A company that manufactures water heaters finds their fixed costs are normally $50,000/month, while the cost to manufacture each heater is $125. Due to factory size and the current equipment, the company can produce a maximum of 5000 water heaters per month during a good month.

**a.** Use the average cost function to find the average cost if 500 water heaters are manufactured each month. What is the average cost if 1000 heaters are made?

**b.** What level of production will bring the average cost down to $150 per water heater?

**c.** If the average cost is currently $137.50, how many water heaters are being produced that month?

**d.** What's the significance of the horizontal asymptote for the average cost function (what does it mean in this context)? Will the company ever break the $130 average cost level? Why or why not?

**74. Producing biodegradable disposable diapers:** An enterprising company has finally developed a better disposable diaper that is biodegradable. The brand becomes wildly popular and production is soaring. The fixed cost of production is \$20,000/month, while the cost of manufacturing is \$6.00 per case (48 diapers). Even while working three shifts around-the-clock, the maximum production level is 16,000 cases per month. The company figures it will be profitable if it can bring costs down to an average of \$7 per case.

    **a.** Use the average cost function to find the average cost if 2000 cases are produced each month. What is the average cost if 4000 cases are made?

    **b.** What level of production will bring the average cost down to \$8 per case?

    **c.** If the average cost is currently \$10 per case, how many cases are being produced?

    **d.** What's the significance of the horizontal asymptote for the average cost function (what does it mean in this context)? Will the company ever reach its goal of \$7/case? What level of production would help them meet their goal?

**75. Manufacturing storage sheds:** Assume the monthly cost of manufacturing custom-crafted storage sheds is modeled by the function $C(x) = 4x^2 + 53x + 250$.

    **a.** Write the average cost function and state the equation of the vertical and oblique asymptotes.

    **b.** Find the cost and average cost of making 1, 2, and 3 sheds.

    **c.** Graph the average cost function and its asymptotes.

    **d.** Use the graph to estimate how many sheds should be made each month to minimize costs. What is the minimum cost? Verify results using a graphing calculator.

**76. Manufacturing playground equipment:** Assume the monthly cost of manufacturing playground equipment is modeled by the function $C(x) = 5x^2 + 94x + 576$. The company has projected that they will be profitable if they can bring their average cost down to \$200 per set of playground equipment.

    **a.** Write the average cost function and state the equation of the vertical and oblique asymptotes.

    **b.** Find the cost and average cost of making 1, 2, and 3 playground equipment combinations. Why would the average cost fall so dramatically early on?

    **c.** Graph the average cost function and its asymptotes.

    **d.** Use the graph to estimate how many sets of equipment should be made each month to minimize costs? What is the minimum cost? Will the company be profitable under these conditions? Verify results using a graphing calculator.

▶ **EXTENDING THE CONCEPT**

**77.** Referring to Exercise 62, suppose that instead of a closed cylinder, with both a top and bottom, we needed to manufacture *open cylinders,* like tennis ball cans that use a lid made from a different material.

    **a.** Derive the formula that will minimize the surface area of an open cylinder.

    **b.** Use it to find the dimensions of a cylinder with minimum surface area that will hold 90 in$^3$ of material.

**78.** Consider the function $f(x) = \frac{ax^2 + c}{bx^2 + d}$, where $a, b, c,$ and $d$ are constants and $a, b > 0$.

    **a.** What can you say about asymptotes and intercepts of this function if $c, d > 0$?

    **b.** Now assume $c < 0$ and $d > 0$. How does this affect the asymptotes? The intercepts?

    **c.** How is the graph affected if $c > 0$ and $d < 0$?

    **d.** Find values of $a, b, c,$ and $d$ that create a function with a horizontal asymptote at $y = \frac{3}{2}$, $x$-intercepts at $(-2, 0)$ and $(2, 0)$, a $y$-intercept of $(0, -4)$, and no vertical asymptotes.

▶ **MAINTAINING YOUR SKILLS**

**79. (2.3)** Find the equation of a line that is perpendicular to $3x - 4y = 12$ and contains the point $(2, -3)$.

**80. (4.2)** Use the remainder theorem to find the value of $f(4), f\left(\frac{3}{2}\right),$ and $f(2)$: $f(x) = 2x^3 - 7x^2 + 5x + 3$.

**81. (1.5)** Solve the following equation using the quadratic formula, then write the equation in factored form: $12x^2 + 55x - 48 = 0$.

**82. (R.1/1.4)** Describe/Define each set of numbers: complex $\mathbb{C}$, rational $\mathbb{Q}$, and integers $\mathbb{Z}$.

## A. Quadratic Inequalities

The study of quadratic inequalities is simply an extension of our earlier work in analyzing functions (Section 2.5). While we've developed the ability to graph a variety of new functions, the solution set for an inequality will still be determined by analyzing the behavior of the function at its zeroes. The key idea is to recognize the following statements are synonymous:

**1.** $f(x) > 0$.   **2.** Outputs are positive.   **3.** The graph is *above the x-axis*.

Similar statements can be made using the other inequality symbols.

Solving a quadratic inequality only requires that we (a) locate any real zeroes of the function and (b) determine whether the graph opens upward or downward. If there are no *x*-intercepts, the graph is entirely above the *x*-axis (output values are positive), or entirely below the *x*-axis (output values are negative), making the solution either all real numbers or the empty set.

**EXAMPLE 1** ▶ **Solving a Quadratic Inequality**

For $f(x) = x^2 + x - 6$, solve $f(x) > 0$.

**Solution** ▶ The graph of $f$ will open upward since $a > 0$. Factoring gives $f(x) = (x + 3)(x - 2)$, with zeroes at $-3$ and $2$. Using the *x*-axis alone (since graphing the function is not our focus), we plot $(-3, 0)$ and $(2, 0)$ and visualize a parabola opening upward through these points (Figure 4.61).

**Figure 4.61**

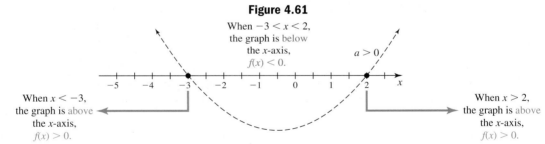

When $-3 < x < 2$, the graph is below the *x*-axis, $f(x) < 0$.

$a > 0$

When $x < -3$, the graph is above the *x*-axis, $f(x) > 0$.

When $x > 2$, the graph is above the *x*-axis, $f(x) > 0$.

The diagram clearly shows the graph is *above* the *x*-axis (outputs are positive) when $x < -3$ or when $x > 2$. The solution is $x \in (-\infty, -3) \cup (2, \infty)$. The actual graph is shown in Figure 4.62.

**Figure 4.62**

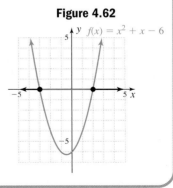

$y \quad f(x) = x^2 + x - 6$

**Now try Exercises 5 through 14** ▶

When solving general inequalities, zeroes of multiplicity continue to play a role. In Example 1, the zeroes of $f$ were both of multiplicity 1, and the graph crossed the *x*-axis at these points. In other cases, the zeroes may have even multiplicity.

**EXAMPLE 2** ▶ **Solving a Quadratic Inequality**

Solve the inequality $-x^2 + 6x \le 9$.

**Solution** ▶ Begin by writing the inequality in standard form: $-x^2 + 6x - 9 \le 0$. Note this is equivalent to $g(x) \le 0$ for $g(x) = -x^2 + 6x - 9$. Since $a < 0$, the graph of $g$ will open downward. The factored form is $g(x) = -(x - 3)^2$, showing 3 is a zero and a repeated root. Using the $x$-axis, we plot the point $(3, 0)$ and visualize a parabola opening downward through this point.

Figure 4.63 shows the graph is *below* the $x$-axis (outputs are negative) for *all values* of $x$ except $x = 3$. But since this is a less than *or equal to* inequality, the solution is $x \in \mathbb{R}$. The complete graph of $g$ shown in Figure 4.64 confirms the analytical solution.

> **WORTHY OF NOTE**
>
> Since $x = 3$ was a zero of multiplicity 2, the graph "bounced off" the $x$-axis at this point, with no change of sign for $g$. The graph is entirely below the $x$-axis, except at the vertex $(3, 0)$.

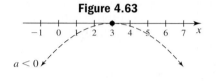

**Figure 4.63**

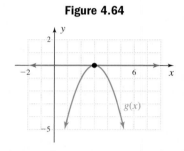

**Figure 4.64**

Now try Exercises 15 through 22 ▶

As an alternative to the zeroes method, the **interval test method** can be used to solve quadratic inequalities. Using the fact that all polynomials are continuous, test values are selected from certain intervals of the domain and substituted into the original function.

> **Interval Test Method for Solving Inequalities**
>
> 1. Find all real roots of the related equation (if they exist) and plot them on the $x$-axis.
> 2. Select any convenient test value from each interval created by the zeroes, and substitute these into the function.
> 3. The sign of the function at these test values will be the sign of the function for all values of $x$ in this interval.

**EXAMPLE 3** ▶ **Solving a Quadratic Inequality**

Solve the inequality $-2x^2 - 3x + 20 \le 0$ using interval tests.

**Solution** ▶ To begin, we find the zeroes of $f(x) = -2x^2 - 3x + 20$ by factoring.

$$-2x^2 - 3x + 20 = 0 \quad \text{related equation}$$
$$2x^2 + 3x - 20 = 0 \quad \text{multiply by } -1$$
$$(2x - 5)(x + 4) = 0 \quad \text{factored form}$$
$$2x - 5 = 0 \text{ or } x + 4 = 0 \quad \text{zero factor property}$$
$$x = \frac{5}{2} \text{ or } x = -4 \quad \text{solutions}$$

Plotting these intercepts creates three intervals on the $x$-axis (Figure 4.65).

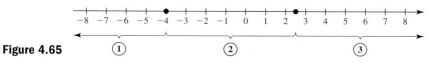

**Figure 4.65**

**Figure 4.67**

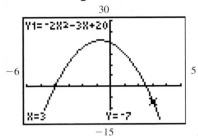

**A.** You've just seen how we can solve quadratic inequalities

Selecting a test value from each interval (in red) gives Figure 4.66:

**Figure 4.66**

| ① | ② | ③ |
|---|---|---|
| $x = -5$ | $x = 0$ | $x = 5$ |
| $f(-5) = -15$ | $f(0) = 20$ | $f(5) = -45$ |
| $f(x) < 0$ in ① | $f(x) > 0$ in ② | $f(x) < 0$ in ③ |

The interval tests show $-2x^2 - 3x + 20 \leq 0$ for $x \in (-\infty, -4) \cup (\frac{5}{2}, \infty)$, which is supported by the graph shown in Figure 4.67.

Now try Exercises 23 through 36 ▶

---

**WORTHY OF NOTE**

When evaluating a function using the interval test method, it's usually easier to use the factored form instead of the polynomial form, since all you really need is whether the result will be positive or negative. For instance, you could likely tell
$f(-5) = -[2(-5) - 5](-5 + 4)$ is going to be negative, more quickly than $f(-5) = -2(-5)^2 - 3(-5) + 20$.

## B. Polynomial Inequalities

Our work with quadratic inequalities transfers seamlessly to inequalities involving higher degree polynomials and the same two methods can be employed. The first involves drawing a quick sketch of the function, and using the concepts of multiplicity and end-behavior. The second involves the use of multiple interval tests, to check on the sign of the function in each interval.

### Solving Polynomial Inequalities Graphically

After writing the polynomial in standard form, find the zeroes, plot them on the $x$-axis, and determine the solution set using end-behavior and the behavior at each zero (cross—sign change; or bounce—no change in sign). In this process, any irreducible quadratic factors can be ignored, as they have no effect on the solution set. In summary,

---

**Solving Polynomial Inequalities**

Given $f(x)$ is a polynomial in standard form,

1. Write $f$ in completely factored form.
2. Plot real zeroes on the $x$-axis, noting their multiplicity.
   - If the multiplicity is odd the function will **change** sign.
   - If the multiplicity is even, there will be **no change** in sign.
3. Use the end-behavior to determine the sign of $f$ in the outermost intervals, then label the other intervals as $f(x) < 0$ or $f(x) > 0$ by analyzing the multiplicity of neighboring zeroes.
4. State the solution in interval notation.

---

**EXAMPLE 4** ▶  **Solving a Polynomial Inequality**

Solve the inequality $x^3 - 18 < -4x^2 + 3x$.

**Solution** ▶  In standard form we have $x^3 + 4x^2 - 3x - 18 < 0$, which is equivalent to $f(x) < 0$ where $f(x) = x^3 + 4x^2 - 3x - 18$. The polynomial cannot be factored by grouping and testing 1 and $-1$ shows neither is a zero. Using $x = 2$ and synthetic division gives

use 2 as a "divisor" $\underline{2|}$ $\quad 1 \quad\quad 4 \quad\quad -3 \quad\quad -18$
$$\quad\quad\quad\quad\quad\quad\quad\quad\quad\quad\quad\quad 2 \quad\quad 12 \quad\quad 18$$
$$\overline{\quad\quad\quad\quad\quad\quad\quad\quad 1 \quad\quad 6 \quad\quad\quad 9 \quad\quad |0}$$

with a quotient of $x^2 + 6x + 9$ and a remainder of zero.

**1.** The factored form is $f(x) = (x - 2)(x^2 + 6x + 9) = (x - 2)(x + 3)^2$.

**2.** The graph will bounce off the $x$-axis at $x = -3$ ($f$ will not change sign), and cross the $x$-axis at $x = 2$ ($f$ will change sign). This is illustrated in Figure 4.68, which uses open dots due to the strict inequality.

**Figure 4.68**

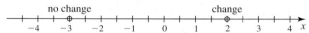

**3.** The polynomial has odd degree with a positive lead coefficient, so end-behavior is down/up, which we note in the outermost intervals. Working from the left, $f$ will not change sign at $x = -3$, showing $f(x) < 0$ in the left and middle intervals. This is supported by the $y$-intercept $(0, -18)$. See Figure 4.69.

**Figure 4.69**

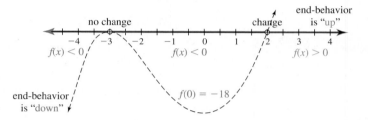

**4.** From the diagram, we see that $f(x) < 0$ for $x \in (-\infty, -3) \cup (-3, 2)$, which must also be the solution interval for $x^3 - 18 < -4x^2 + 3x$.

> **Now try Exercises 37 through 50 ▶**

**EXAMPLE 5 ▶  Solving a Polynomial Inequality**

Solve the inequality $x^4 + 4x \le 9x^2 - 12$.

**Solution ▶**  Writing the inequality in the form $f(x) < 0$ gives $x^4 - 9x^2 + 4x + 12 \le 0$. Testing 1 and $-1$ shows $x = 1$ is not a zero, but $x = -1$ is. Using synthetic division with $x = -1$ gives

$$
\begin{array}{r|rrrrr}
\text{use } -1 \text{ as a "divisor"} \quad -1 & 1 & 0 & -9 & 4 & 12 \\
& \downarrow & -1 & 1 & 8 & -12 \\
\hline
& 1 & -1 & -8 & 12 & \underline{|0} \\
\end{array}
$$

with a quotient of $q_1(x) = x^3 - x^2 - 8x + 12$ and a remainder of zero. As $q_1(x)$ is not easily factored, we continue with synthetic division using $x = 2$.

$$
\begin{array}{r|rrrr}
\text{use 2 as a "divisor"} \quad 2 & 1 & -1 & -8 & 12 \\
& \downarrow & 2 & 2 & -12 \\
\hline
& 1 & 1 & -6 & \underline{|0} \\
\end{array}
$$

The result is $q_2(x) = x^2 + x - 6$ with a remainder of zero.

**1.** The factored form is
$$f(x) = (x + 1)(x - 2)(x^2 + x - 6) = (x + 1)(x - 2)^2(x + 3).$$

2. The graph will "cross" at $x = -1$ and $-3$, and $f$ will change sign. The graph will "bounce" at $x = 2$ and $f$ will not change sign. This is illustrated in Figure 4.70 which uses closed dots since $f(x)$ *can* be equal to zero.

**Figure 4.70**

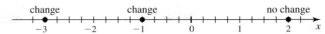

3. With even degree and positive lead coefficient, the end-behavior is up/up. Working from the leftmost interval, $f(x) > 0$, the function must change sign at $x = -3$ (going below the $x$-axis), and again at $x = -1$ (going above the $x$-axis). This is supported by the $y$-intercept $(0, 12)$. The graph then "bounces" at $x = 2$, remaining above the $x$-axis (no sign change). This produces the sketch shown in Figure 4.71.

**Figure 4.71**

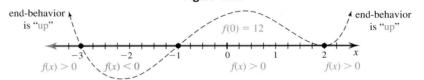

4. From the diagram, we see that $f(x) \leq 0$ for $x \in [-3, -1]$, and at the single point $x = 2$. This shows the solution for $x^4 + 4x \leq 9x^2 - 12$ is $x \in [-3, -1] \cup \{2\}$.

> **Now try Exercises 51 through 54 ▶**

### Solving Function Inequalities Using Interval Tests

As an alternative to graphical analysis an **interval test method** can be employed to solve polynomial (and rational) inequalities. The $x$-intercepts (and vertical asymptotes in the case of rational functions) are noted on the $x$-axis, then a test number is selected from each interval. Since polynomial and rational functions are continuous over their entire domain, the sign of the function at these test values will be the sign of the function for all values of $x$ in the chosen interval.

**EXAMPLE 6** ▶ **Solving a Polynomial Inequality**

Solve the inequality $x^3 + 2x + 8 \leq 5x^2$.

**Solution** ▶ Writing the relationship in function form gives $p(x) = x^3 - 5x^2 + 2x + 8$, with solutions needed to $p(x) \leq 0$. The tests for 1 and $-1$ show $x = -1$ is a root, and using $-1$ with synthetic division gives

$$\text{use } -1 \text{ as a "divisor" } \underline{-1|} \quad \begin{array}{rrrr} 1 & -5 & 2 & 8 \\ \downarrow & -1 & 6 & -8 \\ \hline 1 & -6 & 8 & |\underline{0} \end{array}$$

The quotient is $q(x) = x^2 - 6x + 8$, with a remainder of 0.

The factored form is $p(x) = (x + 1)(x^2 - 6x + 8) = (x + 1)(x - 2)(x - 4)$. The $x$-intercepts are $(-1, 0)$, $(2, 0)$, and $(4, 0)$. Plotting these intercepts creates four intervals on the $x$-axis (Figure 4.72).

**Figure 4.72**

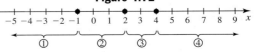

☑ **B.** You've just seen how we can solve polynomial inequalities

Selecting a test value from each interval gives the information shown in Figure 4.73.

**Figure 4.73**

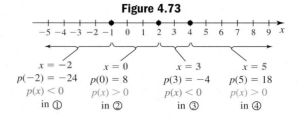

The interval tests show $x^3 + 2x - 8 \le 5x^2$ for $x \in (-\infty, -1] \cup [2, 4]$.

**Now try Exercises 55 and 56** ▶

## C. Rational Inequalities

In general, the solution process for polynomial and rational inequalities is virtually identical, once we recognize that vertical asymptotes also break the $x$-axis into intervals where function values may change sign. However, for rational functions it's more efficient to begin the analysis using the $y$-intercept or a test point, rather than end-behavior, although either will do.

**EXAMPLE 7** ▶  **Solving a Rational Inequality by Analysis**

Solve $\dfrac{x^2 - 9}{x^3 - x^2 - x + 1} \le 0$.

**Solution** ▶  In function form, $v(x) = \dfrac{x^2 - 9}{x^3 - x^2 - x + 1}$ and we want the solution for $v(x) \le 0$. The numerator and denominator are in standard form. The numerator factors easily, and the denominator can be factored by grouping.

1. The factored form is $v(x) = \dfrac{(x - 3)(x + 3)}{(x - 1)^2(x + 1)}$.

2. $v(x)$ will change sign at $x = 3, -3$, and $-1$ as all have odd multiplicity, but will not change sign at $x = 1$ (even multiplicity). Note that zeroes of the denominator will always be indicated by open dots (Figure 4.74) as they are excluded from any solution set.

**Figure 4.74**

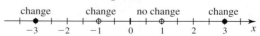

3. The $y$-intercept is $(0, -9)$, indicating that function values will be negative in the interval containing zero. Working outward from this interval using the "change/no change" approach, gives the solution indicated in Figure 4.75.

**Figure 4.75**

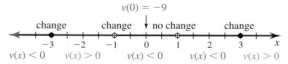

4. For $v(x) \le 0$, the solution is $x \in (-\infty, -3] \cup (-1, 1) \cup (1, 3]$.

**Now try Exercises 57 through 66** ▶

If the rational inequality is not given in function form or is composed of more than one term, start by writing the inequality with zero on one side, then combine terms into a single expression. This is the most efficient way of determining the zeroes of the function and the location of any vertical asymptotes.

**EXAMPLE 8** ▶ **Solving a Rational Inequality Using Interval Tests**

Solve the inequality $\frac{x-2}{x-3} \le \frac{1}{x+3}$.

**Solution** ▶ Rewrite the inequality with zero on one side: $\frac{x-2}{x-3} - \frac{1}{x+3} \le 0$. This is equivalent to $v(x) \le 0$, where $v(x) = \frac{x-2}{x-3} - \frac{1}{x+3}$. Combining the expressions on the right, we have

$$v(x) = \frac{(x-2)(x+3) - 1(x-3)}{(x+3)(x-3)} \qquad \text{LCD is } (x+3)(x-3)$$

$$= \frac{x^2 + x - 6 - x + 3}{(x+3)(x-3)} \qquad \text{multiply}$$

$$= \frac{x^2 - 3}{(x+3)(x-3)} \qquad \text{simplify}$$

1. The factored form is $v(x) = \frac{(x+\sqrt{3})(x-\sqrt{3})}{(x+3)(x-3)}$.  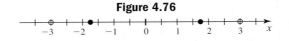 $x^2 - k = (x + \sqrt{k})(x - \sqrt{k})$
2. The two zeroes and two asymptotes will occur at $x = -\sqrt{3}, \sqrt{3}, -3$, and $3$ respectively, which we plot on the $x$-axis, creating five intervals (Figure 4.76).

**Figure 4.76**

3. From left to right, we select one test value from each interval created by the zeroes and vertical asymptotes. We chose: $x = -4, -2, 0, 2$, and $4$ (see Figure 4.77). The results are shown in Figure 4.78, with the conclusion noted beneath each interval.

**Figure 4.77**

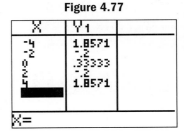

**Figure 4.78**

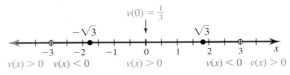

4. The solution for $\frac{x-2}{x-3} \le \frac{1}{x+3}$ is $x \in (-3, -\sqrt{3}] \cup [\sqrt{3}, 3)$.

The graph of $v(x) = \frac{x-2}{x-3} - \frac{1}{x+3}$ is shown in Figure 4.79 and strongly supports our solution.

**Figure 4.79**

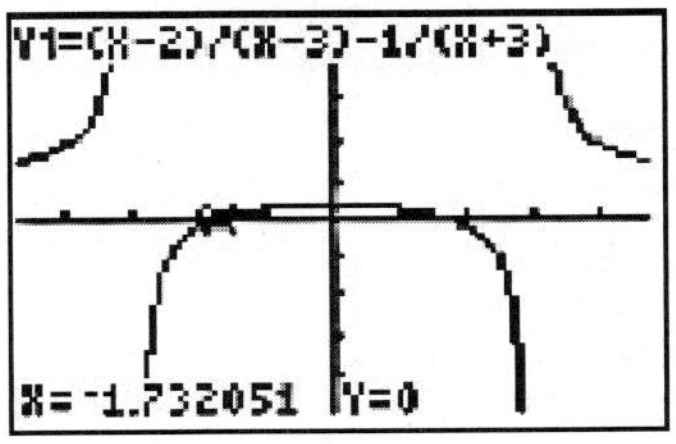

☑ **C.** You've just seen how we can solve rational inequalities

Now try Exercises 67 through 76 ▶

For additional practice with the concepts involved in solving polynomial and rational inequalities graphically, **see Exercises 77 through 80.**

## D. Applications of Inequalities

Applications of inequalities come in many varieties. In addition to stating the solution algebraically, these exercises often compel us to consider the *context of each application* as we state the solution set.

**EXAMPLE 9** ▶   **Solving Applications of Inequalities**

By adjusting his position and limbs, Shane causes himself to rise and fall in a vertical wind tunnel designed to train skydivers. His velocity (in feet per second) as he floats through the turbulence is given by $V(t) = t^5 - 10t^4 + 35t^3 - 50t^2 + 24t$, where $t$ is the time in seconds and $0 < t < 4.5$. During what intervals of time is he moving upward or in the positive direction $[V(t) > 0]$?

**Solution** ▶   Begin by writing $V$ in factored form. Factoring out $t$ gives
$V(t) = t(t^4 - 10t^3 + 35t^2 - 50t + 24)$, and testing 1 and $-1$ shows $t = 1$ is a root.

$$\text{use 1 as a “divisor”} \quad \underline{1|} \begin{array}{rrrrr} 1 & -10 & 35 & -50 & 24 \\ \downarrow & 1 & -9 & 26 & -24 \\ \hline 1 & -9 & 26 & -24 & \underline{|0} \end{array}$$

The quotient is $q_1(t) = t^3 - 9t^2 + 26t - 24$. Using $t = 2$, we continue with the division on $q_1(t)$ which gives:

$$\text{use 2 as a “divisor”} \quad \underline{2|} \begin{array}{rrrr} 1 & -9 & 26 & -24 \\ \downarrow & 2 & -14 & 24 \\ \hline 1 & -7 & 12 & \underline{|0} \end{array}$$

This shows $V(t) = t(t - 1)(t - 2)(t^2 - 7t + 12)$.

**1.** The completely factored form is $V(t) = t(t - 1)(t - 2)(t - 3)(t - 4)$.
**2.** All zeroes have odd multiplicity and function values will change sign.
**3.** With odd degree and a positive leading coefficient, end-behavior is down/up.

Function values will be negative in the far left interval and alternate in sign thereafter. The solution diagram is shown in the figure (the parentheses shown on the number line indicate the domain given).

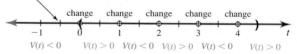

Since end-behavior is down/up, function values are negative in this interval, and will alternate thereafter.

**4.** For $V(t) > 0$, the solution is $t \in (0, 1) \cup (2, 3) \cup (4, 4.5)$. Shane is moving in the positive direction in these time intervals.

☑ **D. You've just seen how we can solve applications of inequalities**

Now try Exercises 83 through 90 ▶

## 4.6 EXERCISES

▶ **CONCEPTS AND VOCABULARY**

**Fill in each blank with the appropriate word or phrase. Carefully reread the section, if necessary.**

1. To solve a polynomial or rational inequality, begin by plotting the location of all zeroes and _____ asymptotes (if they exist), then consider the _____ of each.

2. For strict inequalities, the zeroes are _____ from the solution set. For nonstrict inequalities, zeroes are _____. The values at which vertical asymptotes occur are always _____.

3. Compare/Contrast the process for solving $x^2 - 3x - 4 \geq 0$ with $\frac{1}{x^2 - 3x - 4} \geq 0$. Are there similarities? What are the differences?

4. Compare/Contrast the process for solving $(x + 1)(x - 3)(x^2 + 1) > 0$ with $(x + 1)(x - 3) > 0$. Are there similarities? What are the differences?

▶ **DEVELOPING YOUR SKILLS**

**Solve each inequality by locating any x-intercept(s) and noting the end-behavior of the graph. Begin by writing the inequality in function form as needed.**

5. $f(x) = -x^2 + 4x; f(x) > 0$

6. $g(x) = x^2 - 5x; g(x) < 0$

7. $h(x) = x^2 + 4x - 5; h(x) \geq 0$

8. $p(x) = -x^2 + 3x + 10; p(x) \leq 0$

9. $q(x) = 2x^2 - 5x - 7; q(x) < 0$

10. $r(x) = -2x^2 - 3x + 5; r(x) > 0$

11. $7 \geq x^2$        12. $x^2 \leq 13$

13. $3x^2 \geq -2x + 5$        14. $4x^2 \geq 3x + 7$

15. $s(x) = x^2 - 8x + 16; s(x) \geq 0$

16. $t(x) = x^2 - 6x + 9; t(x) \geq 0$

17. $r(x) = 4x^2 + 12x + 9; r(x) < 0$

18. $f(x) = 9x^2 - 6x + 1; f(x) < 0$

19. $g(x) = -x^2 + 10x - 25; g(x) < 0$

20. $h(x) = -x^2 + 14x - 49; h(x) < 0$

21. $-x^2 > 2$        22. $x^2 < -4$

**Solve each inequality using the interval test method. Write your answer in interval notation when possible.**

23. $(x + 3)(x - 1) < 0$        24. $(x + 2)(x - 5) < 0$

25. $2x^2 - x - 6 \geq 0$        26. $-3x^2 + x + 4 \leq 0$

27. $x^2 + 14x + 49 \geq 0$        28. $-x^2 + 6x - 9 \leq 0$

**Determine the domain of each radical function.**

29. $h(x) = \sqrt{x^2 - 25}$

30. $p(x) = \sqrt{25 - x^2}$

31. $q(x) = \sqrt{x^2 - 5x}$

32. $r(x) = \sqrt{6x - x^2}$

**Match the correct solution with the inequality and graph given.**

33. $f(x) > 0$
   a. $x \in (-\infty, -2) \cup (1, \infty)$
   b. $x \in (-\infty, -2] \cup [1, \infty)$
   c. $(-2, 1)$
   d. $[-2, 1]$
   e. none of these

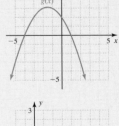

34. $g(x) \geq 0$
   a. $x \in (-\infty, -4) \cup (1, \infty)$
   b. $x \in (-\infty, -4] \cup [1, \infty)$
   c. $(-4, 1)$
   d. $[-4, 1]$
   e. none of these

35. $r(x) \leq 0$
   a. $x \in (-\infty, 3) \cup (3, \infty)$
   b. $x \in (-\infty, \infty)$
   c. $\{3\}$
   d. $\{\ \}$
   e. none of these

36. $s(x) < 0$
   a. $x \in (-\infty, 0) \cup (0, \infty)$
   b. $x \in (-\infty, \infty)$
   c. $\{0\}$
   d. $\{\ \}$
   e. none of these

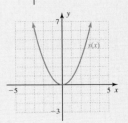

Solve each polynomial inequality indicated using a number line and the behavior of the graph at each zero. Write all answers in interval notation.

**37.** $(x + 3)(x - 5) < 0$      **38.** $(x - 2)(x + 7) < 0$

**39.** $(x + 1)^2(x - 4) \geq 0$     **40.** $(x + 6)(x - 1)^2 \leq 0$

**41.** $(x + 2)^3(x - 2)^2(x - 4) \geq 0$

**42.** $(x - 1)^3(x + 2)^2(x - 3) \leq 0$

**43.** $x^2 + 4x + 1 < 0$     **44.** $x^2 - 6x + 4 > 0$

**45.** $x^2 - 2x \geq 15$     **46.** $x^2 + 3x \geq 18$

**47.** $x^3 \geq 9x$     **48.** $x^3 \leq 4x$

**49.** $x^3 - 7x + 6 > 0$     **50.** $x^3 - 13x + 12 > 0$

**51.** $x^4 - 10x^2 > -9$     **52.** $x^4 + 36 < 13x^2$

**53.** $x^4 - 9x^2 > 4x - 12$     **54.** $x^4 - 16 > 5x^3 - 20x$

Solve each inequality using the interval test method.

**55.** $-4x + 12 < -x^3 + 3x^2$     **56.** $x^3 + 8 < 5x^2 - 2x$

Solve each rational inequality indicated using a number line and the behavior of the graph at each zero. Write all answers in interval notation.

**57.** $f(x) = \dfrac{x + 3}{x - 2}; f(x) \leq 0$

**58.** $F(x) = \dfrac{x - 4}{x + 1}; F(x) \geq 0$

**59.** $g(x) = \dfrac{x + 1}{x^2 + 4x + 4}; g(x) < 0$

**60.** $G(x) = \dfrac{x - 3}{x^2 - 2x + 1}; G(x) > 0$

**61.** $\dfrac{2 - x}{x^2 - x - 6} \geq 0$     **62.** $\dfrac{1 - x}{x^2 - 2x - 8} \leq 0$

**63.** $\dfrac{2x - x^2}{x^2 + 4x - 5} < 0$     **64.** $\dfrac{x^2 + 3x}{x^2 - 2x - 3} > 0$

**65.** $\dfrac{x^2 - 4}{x^3 - 13x + 12} \geq 0$     **66.** $\dfrac{x^2 + x - 6}{x^3 - 7x + 6} \leq 0$

**67.** $\dfrac{2}{x - 2} \leq \dfrac{1}{x}$     **68.** $\dfrac{5}{x + 3} \geq \dfrac{3}{x}$

**69.** $\dfrac{x - 3}{x + 17} > \dfrac{1}{x - 1}$     **70.** $\dfrac{1}{x + 5} < \dfrac{x - 2}{x - 7}$

**71.** $\dfrac{x + 1}{x - 2} \geq \dfrac{x + 2}{x + 3}$     **72.** $\dfrac{x - 3}{x - 6} \leq \dfrac{x + 1}{x + 4}$

**73.** $\dfrac{x + 2}{x^2 + 9} > 0$     **74.** $\dfrac{x^2 + 4}{x - 3} < 0$

**75.** $\dfrac{x^4 - 5x^2 - 36}{x^2 - 2x + 1} > 0$     **76.** $\dfrac{x^4 - 3x^2 - 4}{x^2 - x - 20} < 0$

**Match the correct solution with the inequality and graph given.**

**77.** $R(x) \leq 0$

  **a.** $x \in (-\infty, -1) \cup (0, 2)$

  **b.** $x \in [0, 1] \cup (2, \infty)$

  **c.** $x \in [-5, -1] \cup [2, 5]$

  **d.** $x \in (-\infty, -1) \cup [0, 2)$

  **e.** none of these

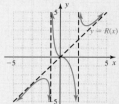

**78.** $g(x) \geq 0$

  **a.** $x \in (-4, -0.5) \cup (4, \infty)$

  **b.** $x \in [-0.5, 4] \cup [4, 5]$

  **c.** $x \in (-\infty, -4) \cup (-0.5, 4)$

  **d.** $x \in [-4, -0.5] \cup [4, \infty)$

  **e.** none of these

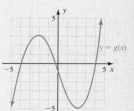

**79.** $f(x) < 0$

  **a.** $x \in (-5, -2) \cup (3, 5)$

  **b.** $x \in (-\infty, -2) \cup (-2, 1)$
     $\cup (3, \infty)$

  **c.** $x \in (-\infty, -2) \cup (3, \infty)$

  **d.** $x \in (-\infty, -2) \cup (-2, 1]$
     $\cup [3, \infty)$

  **e.** none of these

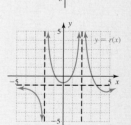

**80.** $r(x) \geq 0$

  **a.** $x \in (-\infty, -2) \cup [-1, 1]$
     $\cup [3, \infty)$

  **b.** $x \in (-2, -1] \cup [1, 2)$
     $\cup (2, 3]$

  **c.** $x \in (-\infty, -2) \cup (2, \infty)$

  **d.** $x \in (-2, -1) \cup (1, 2)$
     $\cup (2, 3]$

  **e.** none of these

## ▶ WORKING WITH FORMULAS

**81. The slope formula:** $m = \dfrac{y_2 - y_1}{x_2 - x_1}$

If we consider the slope of a line through the points $(x, 5)$ and $(3, 7)$, the slope formula becomes a rational function of the form $m(x) = \frac{2}{3 - x}$. Using a graphical analysis, (a) what is the domain of this function? (b) Where is $m(x) > 0$? (c) Where is $m(x) < 0$? (d) If the slope of the line must be less than $\frac{4}{3}$, what values of $x$ are possible?

**82. The pricing function:** $p = c + mc$

The price $p$ of an item depends on the retailers cost $c$ and the markup percentage $m$ (expressed as a decimal). To be competitive, the retailer knows it must charge \$36 for a particular item, but there are many manufacturers who can supply the item and at varying costs. (a) Solve the formula for $m$, and write the result as a function of $c$. (b) If the retailer must have a markup of 25% or more, what range of costs can be considered? (c) If the retailer wanted to undercut the competition and sell the item for \$30, what cost would be needed in order to maintain a 25% markup?

## ▶ APPLICATIONS

**Deflection of a beam:** The amount of deflection in a rectangular wooden beam of length $L$ ft can be approximated by $d(x) = k(x^3 - 3L^2x + 2L^3)$, where $k$ is a constant that depends on the characteristics of the wood and the force applied, and $x$ is the *distance from the unsupported end* of the beam $(x < L)$.

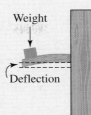

Weight

Deflection

**83.** Find the equation for a beam 8 ft long and use it for the following:

    **a.** For what distances $x$ is the quantity $\frac{d(x)}{k}$ less than 189 units?

    **b.** What is the amount of deflection 4 ft from the unsupported end $(x = 4)$?

    **c.** For what distances $x$ is the quantity $\frac{d(x)}{k}$ greater than 475 units?

    **d.** If safety concerns prohibit a deflection of more than $648k$ units, what is the shortest distance from the end of the beam that the force can be applied?

**84.** Find the equation for a beam 9 ft long and use it for the following:

    **a.** For what distances $x$ is the quantity $\frac{d(x)}{k}$ less than 216 units?

    **b.** What is the amount of deflection 4 ft from the unsupported end $(x = 4)$?

    **c.** For what distances $x$ is the quantity $\frac{d(x)}{k}$ greater than $550k$ units?

     **d.** Compare the answer to 83(b) with the answer to 84(b). What can you conclude?

**Average speed for a round-trip:** Surprisingly, the average speed of a round-trip is *not* the sum of the average speed in each direction divided by two. For a fixed distance $D$, consider rate $r_1$ in time $t_1$ for one direction, and rate $r_2$ in time $t_2$ for the other, giving $r_1 = \frac{D}{t_1}$ and $r_2 = \frac{D}{t_2}$. The average speed for the round-trip is $R = \frac{2D}{t_1 + t_2}$.

**85.** The distance from St. Louis, Missouri, to Springfield, Illinois, is approximately 80 mi. Suppose that Sione, due to the age of his vehicle, made the round-trip with an average speed of 40 mph.

    **a.** Use the relationships stated to verify that $r_2 = \frac{20r_1}{r_1 - 20}$.

**b.** Discuss the meaning of the horizontal and vertical asymptotes in this context.

**c.** Verify algebraically the speed returning would be greater than the speed going for $20 < r_1 < 40$. In other words, solve the inequality $\frac{20r_1}{r_1 - 20} > r_1$ using the ideas from this section.

**86.** The distance from Boston, Massachusetts, to Hartford, Connecticut, is approximately 100 mi. Suppose that Stella, due to excellent driving conditions, made the round-trip with an average speed of 60 mph.

    **a.** Use the relationships above to verify that $r_2 = \frac{30r_1}{r_1 - 30}$.

    **b.** Discuss the meaning of the horizontal and vertical asymptotes in this context.

    **c.** Verify algebraically the speed returning would be greater than the speed going for $30 < r_1 < 60$. In other words, solve the inequality $\frac{30r_1}{r_1 - 30} > r_1$ using the ideas from this section.

**Electrical resistance and temperature:** The amount of electrical resistance $R$ in a medium depends on the temperature, and for certain materials can be modeled by the equation $R(t) = 0.01t^2 + 0.1t + k$, where $R(t)$ is the resistance (in ohms $\Omega$) at temperature $t$ $(t \geq 0°)$ in degrees Celsius, and $k$ is the resistance at $t = 0°C$.

**87.** Suppose $k = 30$ for a certain medium. Write the resistance equation and use it to answer the following questions.

    **a.** For what temperatures is the resistance less than 42 $\Omega$?

    **b.** For what temperatures is the resistance greater than 36 $\Omega$?

    **c.** If it becomes uneconomical to run electricity through the medium for resistances greater than 60 $\Omega$, for what temperatures should the electricity generator be shut down?

**88.** Suppose $k = 20$. Write the resistance equation and answer the following questions.

    **a.** For what temperatures is the resistance less than 26 $\Omega$?

    **b.** For what temperatures is the resistance greater than 40 $\Omega$?

    **c.** If it becomes uneconomical to run electricity through the medium for resistances greater than 50 $\Omega$, for what temperatures should the electricity generator be shut down?

## ▶ EXTENDING THE CONCEPT

 **89.** Using the tools of calculus, it can be shown that $f(x) = x^4 - 4x^3 - 12x^2 + 32x + 39$ is increasing in the intervals where $F(x) = x^3 - 3x^2 - 6x + 8$ is positive. Solve the inequality $F(x) > 0$ using the ideas from this section.

**90.** As in our earlier studies, if $n$ is an even number, the expression $\sqrt[n]{A}$ represents a real number only if $A \geq 0$. Use this idea to find the domain of the following functions.

    **a.** $g(x) = \sqrt[4]{2x^3 + x^2 - 22x + 24}$

    **b.** $p(x) = \sqrt[4]{\dfrac{x + 2}{x^2 - 2x - 35}}$

**91.** Find one polynomial inequality and one rational inequality that have the solution $x \in (-\infty, -2) \cup (0, 1) \cup (1, \infty)$.

**92.** Using the tools of calculus, it can be shown that $r(x) = \frac{x^2 - 3x - 4}{x - 8}$ is decreasing in the intervals where $R(x) = \frac{x^2 - 16x + 28}{(x - 8)^2}$ is negative. Solve the inequality $R(x) < 0$ using the ideas from this section.

## ▶ MAINTAINING YOUR SKILLS

**93.** **(3.1)** Use the graph of $f(x)$ given to sketch the graph of $y = f(x + 2) - 3$.

**Exercise 93**

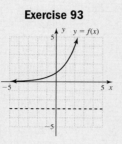

**94.** **(1.3)** Solve the absolute value inequality: $-2|x - 3| + 5 \geq -7$

**95.** **(3.4/4.5)** Graph the function $f(x) = \frac{x^2 + 2x - 8}{x + 4}$. If there is a removable discontinuity, repair the break using an appropriate piecewise-defined function.

**96.** **(R.4)** Solve the equation $\frac{1}{2}\sqrt{16 - x} - \frac{x}{2} = 2$. Check solutions in the original equation.

## MAKING CONNECTIONS

## Making Connections: Graphically, Symbolically, Numerically, and Verbally

Match the characteristics shown in 1 through 16 to one or more of the eight graphs given in (a) through (h).

**(a)**

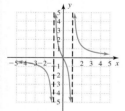

**(b)**

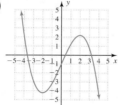

**(c)**

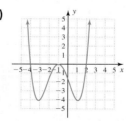

**(d)**

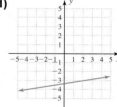

**(e)**

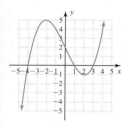

**(f)**

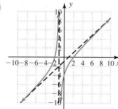

**(g)**

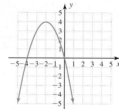

**(h)**

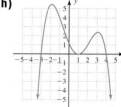

**1.** _____ $\frac{1}{6}(x + 4)(x - 1)(x - 3)$

**2.** _____ $y = -(x + 2)^2 + 4$

**3.** _____ min degree 3, y-int: $(0, -1)$

**4.** _____ min degree 4, one repeated root, relative max $y = 0$ at $x = -1$

**5.** _____ as $|x| \to \infty, y \to 0$

**6.** _____ $f(x) > 0$ only for $x \in (-4, 0)$

**7.** _____ $f(-1) = 4, f(2) = 1$

**8.** _____ $f(x)\uparrow$ only for $x \in (-2, 2)$

**9.** _____ $(y + 4) = \frac{1}{6}(x + 4)$

**10.** _____ $f(x) > 0$ for $x \in (-2, -1) \cup (2, \infty)$

**11.** _____ $f(x)\uparrow$ for $x \in D$

**12.** _____ min degree 4, one repeated root, lead coefficient $a < 0$

**13.** _____ axis of symmetry $x = -2, f(-3) = 3$

**14.** _____ $f(x) > 0$ for $x \in (-4, 1) \cup (3, \infty)$

**15.** _____ $f(x)\uparrow$ only for $x \in (-3, -1) \cup (1, \infty)$

**16.** _____ $f(x) = 4.2$ for $x = -2$ and $x = 4$

# SUMMARY AND CONCEPT REVIEW

## SECTION 4.1    Quadratic Functions and Applications

### KEY CONCEPTS

- A quadratic function is one of the form $f(x) = ax^2 + bx + c$; $a \neq 0$. The simplest quadratic is the squaring function $f(x) = x^2$, where $a = 1$ and $b, c = 0$.
- The graph of a quadratic function is a parabola. Parabolas have three distinctive features: (1) like end-behavior on the left and right, (2) an axis of symmetry, (3) a highest or lowest point called the vertex.
- For a quadratic function in the standard form $y = ax^2 + bx + c$,
  - End-behavior: graph opens upward if $a > 0$, opens downward if $a < 0$
  - Zeroes/$x$-intercepts (if they exist): substitute 0 for $y$ and solve for $x$
  - $y$-intercept: substitute 0 for $x$ to obtain $(0, c)$
  - Vertex: $(h, k)$, where $h = \frac{-b}{2a}$, $k = f\left(\frac{-b}{2a}\right)$
  - Maximum value: If the parabola opens downward, $y = k$ is the maximum value of $f$.
  - Minimum value: If the parabola opens upward, $y = k$ is the minimum value of $f$.
  - Symmetry: $x = h$ is the axis of symmetry [if $(h + d, y)$ is on the graph, then $(h - d, y)$ is also on the graph].

### EXERCISES

Graph $f(x)$ by completing the square and using transformations of the parent function. Graph $g(x)$ and $h(x)$ using the vertex formula and $y$-intercept. Find the $x$-intercepts (if they exist) for all functions.

**1.** $f(x) = x^2 + 8x + 15$          **2.** $g(x) = -x^2 + 4x - 5$          **3.** $h(x) = 4x^2 - 12x + 3$

**4. Height of a superball:** A teenager tries to see how high she can bounce her superball by throwing it downward on her driveway. The height of the ball (in feet) at time $t$ (in seconds) is given by $h(t) = -16t^2 + 96t$. (a) How high is the ball at $t = 0$? (b) How high is the ball after 1.5 sec? (c) How long until the ball is 135 ft high? (d) What is the maximum height attained by the ball? At what time $t$ did this occur?

## SECTION 4.2    Synthetic Division; the Remainder and Factor Theorems

### KEY CONCEPTS

- Synthetic division is an abbreviated form of long division. Only the coefficients of the dividend are used, since "standard form" ensures like place values are aligned. Zero placeholders are used for "missing" terms. The "divisor" must be linear with leading coefficient 1.
- To divide a polynomial by $x - c$, use $c$ in the synthetic division; to divide by $x + c$, use $-c$.
- After setting up the synthetic division template, drop the leading coefficient of the dividend into place, then multiply in the diagonal direction, place the product in the next column, and add in the vertical direction, continuing to the last column.
- The final sum is the remainder $r$; the numbers preceding it are the coefficients of the quotient $q(x)$.
- Remainder theorem: If $p(x)$ is divided by $x - c$, the remainder is equal to $p(c)$. The theorem can be used to evaluate polynomials at $x = c$.
- Factor theorem: If $p(c) = 0$, then $c$ is a zero of $p$ and $x - c$ is a factor. Conversely, if $x - c$ is a factor of $p$, then $p(c) = 0$. The theorem can be used to factor a polynomial or build a polynomial from its zeroes.
- The remainder and factor theorems also apply when $c$ is a complex number.

### EXERCISES

Divide using long division and clearly identify the quotient and remainder:

**5.** $\dfrac{x^3 + 4x^2 - 5x - 6}{x - 2}$          **6.** $\dfrac{x^3 + 2x - 4}{x^2 - x}$

**7.** Use the factor and remainder theorems to help factor $p(x) = x^3 + 2x^2 - 11x - 12$ completely.

Using the remainder theorem:

**8.** Find $h(-7)$ given $h(x) = x^3 + 9x^2 + 13x - 10$.

**9.** Show $x = \frac{1}{2}$ is a zero of $V$: $V(x) = 4x^3 + 8x^2 - 3x - 1$.

**10.** Show $x = 3i$ is a zero of $W$: $W(x) = x^3 - 2x^2 + 9x - 18$.

Using the factor theorem,

**11.** Find a degree 3 polynomial in standard form with zeroes $x = 1$, $x = -\sqrt{5}$, and $x = \sqrt{5}$.

**12.** Use synthetic division and the remainder theorem to answer: At a busy shopping mall, customers are constantly coming and going. One summer afternoon during the hours from 12 o'clock noon to 6 in the evening, the number of customers in the mall could be modeled by $C(t) = 3t^3 - 28t^2 + 66t + 35$, where $C(t)$ is the number of customers (in tens), $t$ hr after 12 noon. (a) How many customers were in the mall at noon? (b) Were more customers in the mall at 2:00 or at 3:00 P.M.? How many more? (c) Was the mall busier at 1:00 P.M. (after lunch) or 6:00 P.M. (around dinner time)?

## SECTION 4.3    The Zeroes of Polynomial Functions

### KEY CONCEPTS

- Fundamental theorem of algebra: Every complex polynomial has at least one complex zero.
- Linear factorization theorem: Every complex polynomial of degree $n \geq 1$ has exactly $n$ linear factors, and can be written in the form $p(x) = a(x - c_1)(x - c_2)\ldots(x - c_n)$, where $a \neq 0$ and $c_1, c_2, \ldots, c_n$ are (not necessarily distinct) complex numbers.
- For a polynomial $p$ in factored form with repeated factors $(x - c)^m$, $c$ is a zero of multiplicity $m$. If $m$ is odd, $c$ is a zero of odd multiplicity; if $m$ is even, $c$ is a zero of even multiplicity.
- Corollaries to the linear factorization theorem:
  - I. If $p$ is a polynomial with real coefficients, $p$ can be factored into linear factors (not necessarily distinct) and irreducible quadratic factors having real coefficients.
  - II. If $p$ is a polynomial with real coefficients, the complex zeroes of $p$ must occur in conjugate pairs. If $a + bi$ ($b \neq 0$) is a zero, then $a - bi$ is also a zero.
  - III. If $p$ is a polynomial with degree $n \geq 1$, then $p$ will have exactly $n$ zeroes (real or nonreal), where zeroes of multiplicity $m$ are counted $m$ times.
- Intermediate value theorem: If $p$ is a polynomial with real coefficients where $p(a)$ and $p(b)$ have opposite signs, then there is at least one $c$ between $a$ and $b$ such that $p(c) = 0$.
- Rational zeroes theorem: If a real polynomial has integer coefficients, rational zeroes must be of the form $\frac{p}{q}$, where $p$ is a factor of the constant term and $q$ is a factor of the leading coefficient.
- Descartes' rule of signs, upper and lower bounds property, tests for $-1$ and 1, and graphing technology can all be used with the rational zeroes theorem to factor, solve, and graph polynomial functions.

### EXERCISES

Using the tools from this section,

**13.** List all possible rational zeroes of $p(x) = 4x^3 - 16x^2 + 11x + 10$.

**14.** Find all rational zeroes of $p(x) = 4x^3 - 16x^2 + 11x + 10$.

**15.** Write $P(x) = 2x^3 - 3x^2 - 17x - 12$ in completely factored form.

**16.** Prove that $h(x) = x^4 - 7x^2 - 2x + 3$ has no rational zeroes.

**17.** Identify two intervals (of those given) that contain a zero of $Px = x^4 - 3x^3 - 8x^2 + 12x + 6$: $[-2, -1]$, $[1, 2]$, $[2, 3]$, $[4, 5]$.

**18.** Discuss the number of possible positive, negative, and nonreal zeroes for $g(x) = x^4 + 3x^3 - 2x^2 - x - 30$. Then identify which combination is correct.

## SECTION 4.4    Graphing Polynomial Functions

### KEY CONCEPTS

- All polynomial graphs are smooth, continuous curves.
- A polynomial of degree $n$ has *at most* $n - 1$ turning points.
- If the degree of a polynomial is odd, the ends of its graph will point in opposite directions (like $y = mx$). If the degree is even, the ends will point in the same direction (like $y = ax^2$). The sign of the lead coefficient determines the actual behavior.
- The "behavior" of a polynomial graph near its zeroes is determined by the multiplicity of the zero. For any factor $(x - c)^m$, the graph will "cross through" the $x$-axis if $m$ is odd and "bounce off" the $x$-axis (touching at just one point) if $m$ is even. The larger the value of $m$, the flatter (more compressed) the graph will be near $c$.
- To "round-out" a graph, additional *midinterval points* can be found between known zeroes.
- These ideas help to establish the *Guidelines for Graphing Polynomial Functions*. See page 385.

## EXERCISES

State the degree, end-behavior, and *y*-intercept, but do not graph.

**19.** $f(x) = -3x^5 + 2x^4 + 9x - 4$

**20.** $g(x) = (x - 1)(x + 2)^2(x - 2)$

Graph using the *Guidelines for Graphing Polynomials.*

**21.** $p(x) = (x + 1)^3(x - 2)^2$

**22.** $q(x) = 2x^3 - 3x^2 - 9x + 10$

**23.** $h(x) = x^4 - 6x^3 + 8x^2 + 6x - 9$

**24.** For the graph of $P(x)$ shown, (a) state whether the degree of $P$ is even or odd, (b) use the graph to locate the zeroes of $P$ and state whether their multiplicity is even or odd, and (c) find the minimum possible degree of $P$ and write it in factored form. Assume all zeroes are real.

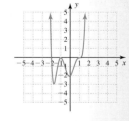

## SECTION 4.5 Graphing Rational Functions

### KEY CONCEPTS

- A rational function is one of the form $V(x) = \frac{p(x)}{d(x)}$, where $p$ and $d$ are polynomials and $d(x) \neq 0$.
- The domain of $V$ is all real numbers, except the zeroes of $d$.
- If zero is in the domain of $V$, substitute 0 for $x$ to find the $y$-intercept.
- The zeroes of $V$ (if they exist) are solutions to $p(x) = 0$.
- If $\frac{p(x)}{d(x)}$ is in simplest form, vertical asymptotes will occur at the zeroes of $d$.
- If the degree of $p <$ degree of $d$, $y = 0$ (the $x$-axis) is a horizontal asymptote. If the degree of $p =$ degree of $d$, $y = \frac{a}{b}$ is a horizontal asymptote, where $a$ is the leading coefficient of $p$, and $b$ is the leading coefficient of $d$.
- If $V = \frac{p(x)}{d(x)}$ is in simplest form, and the degree of $p$ is greater than the degree of $d$, the graph will have an oblique or nonlinear asymptote, as determined by the quotient polynomial after division. If the degree of $p$ is greater by 1, the result is a slant (oblique) asymptote. If the degree of $p$ is greater by 2, the result is a parabolic asymptote.
- The *Guidelines for Graphing Rational Functions* can be found on page 397.

### EXERCISES

**25.** For the function $V(x) = \frac{x^2 - 9}{x^2 - 3x - 4}$, state the following but do not graph: (a) domain (in set notation), (b) equations of the horizontal and vertical asymptotes, (c) the $x$- and $y$-intercept(s), and (d) the value of $V(1)$.

**26.** For $v(x) = \frac{(x + 1)^2}{x + 2}$, will the function change sign at $x = -1$? Will the function change sign at $x = -2$? Justify your responses.

Graph using the *Guidelines for Graphing Rational Functions.*

**27.** $v(x) = \dfrac{x^2 - 4x}{x^2 - 4}$

**28.** $t(x) = \dfrac{2x^2}{x^2 - 5}$

**29.** $h(x) = \dfrac{x^2 - 2x}{x - 3}$

**30.** $t(x) = \dfrac{x^3 - 7x + 6}{x^2}$

**31.** Use the vertical asymptotes, $x$-intercepts, and their multiplicities to construct an equation that corresponds to the given graph. Be sure the $y$-intercept on the graph matches the value given by your equation. Assume these features are integer-valued.

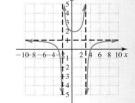

**32.** The average cost of producing a popular board game is given by the function $A(x) = \frac{5000 + 15x}{x}$; $x \geq 1000$. (a) Identify the horizontal asymptote of the function and explain its meaning in this context. (b) To be profitable, management believes the average cost must be below \$17.50. What levels of production will make the company profitable?

**33.** The cost to make $x$ thousand party favors is given by $C(x) = x^2 - 2x + 6$, where $x \geq 1$ and $C$ is in thousands of dollars. For the average cost of production $A(x) = \frac{x^2 - 2x + 6}{x}$, (a) graph the function, (b) use the graph to estimate the level of production that will make average cost a minimum, and (c) state the average cost of a single party favor at this level of production.

## SECTION 4.6    Polynomial and Rational Inequalities

### KEY CONCEPTS

- To solve polynomial inequalities, write $P(x)$ in factored form and note the multiplicity of each real zero.
- Plot real zeroes on a number line. The graph will cross the $x$-axis at zeroes of odd multiplicity ($P$ will change sign), and bounce off the axis at zeroes of even multiplicity ($P$ will not change sign).
- Use the end-behavior, $y$-intercept, or a test point to determine the sign of $P$ in a given interval, then label all other intervals as $P(x) > 0$ or $P(x) < 0$ by analyzing the multiplicity of neighboring zeroes. Use the resulting diagram to state the solution.
- The solution process for rational inequalities and polynomial inequalities is virtually identical, considering that vertical asymptotes also create intervals where function values may change sign, depending on their multiplicity.
- Polynomial and rational inequalities can also be solved using an interval test method. Since polynomials and rational functions are continuous on their domains, the sign of the function at any one point in an interval will be the same as for all other points in that interval.

### EXERCISES

Solve each inequality indicated using a number line and the behavior of the graph at each zero.

**34.** $x^3 + x^2 > 10x - 8$

**35.** $\dfrac{x^2 - 3x - 10}{x - 2} \geq 0$

**36.** $\dfrac{x}{x - 2} \leq \dfrac{-1}{x}$

### PRACTICE TEST

**1.** Complete the square to write each function as a transformation. Then graph each function and label the vertex and all intercepts (if they exist).

    **a.** $f(x) = -x^2 + 10x - 16$    **b.** $g(x) = \dfrac{1}{2}x^2 + 4x + 16$

**2.** The graph of a quadratic function has a vertex of $(-1, -2)$, and passes through the origin. Find the other intercept, and the equation of the graph in standard form.

**3.** Suppose the function $d(t) = t^2 - 14t$ models the depth of a scuba diver at time $t$, as she dives underwater from a steep shoreline, reaches a certain depth, and swims back to the surface.

    **a.** What is her depth after 4 sec? After 6 sec?

    **b.** What was the maximum depth of the dive?

    **c.** How many seconds was the diver beneath the surface?

**4.** Compute the quotient using long division: $\dfrac{x^3 - 3x^2 + 5x - 2}{x^2 + 2x + 1}$.

**5.** Find the quotient and remainder using synthetic division: $\dfrac{x^3 + 4x^2 - 5x - 20}{x + 2}$.

**6.** Use the factor and remainder theorems to show $x + 3$ is a factor of $x^4 - 15x^2 - 10x + 24$.

**7.** Given $f(x) = 2x^3 + 4x^2 - 5x + 2$, find the value of $f(-3)$ using the remainder theorem.

**8.** Given $x = 2$ and $x = 3i$ are two zeroes of a real polynomial $P(x)$ with degree 3. Use the factor theorem to find $P(x)$ given $(1, -20)$ is on the graph.

**9.** Factor the polynomial and state the multiplicity of each zero: $Q(x) = (x^2 - 3x + 2)(x^3 - 2x^2 - x + 2)$.

**10.** Given $C(x) = x^4 + x^3 + 7x^2 + 9x - 18$, (a) use the rational zeroes theorem to list all possible rational zeroes; (b) apply Descartes' rule of signs to count the number of possible positive, negative, and nonreal zeroes; and (c) use this information along with the tests for 1 and $-1$, synthetic division, and the factor theorem to factor $C$ completely.

**11.** Over a 10-yr period, the balance of payments (deficit versus surplus) for a small county was modeled by the function $f(x) = \dfrac{1}{2}x^3 - 7x^2 + 28x - 32$, where $x = 0$ corresponds to 2000 and $f(x)$ is the deficit or surplus in millions of dollars. (a) Use the rational roots theorem and synthetic division to find the years the county "broke even" (debt = surplus = 0) from 2000 to 2010. (b) How many years did the county run a surplus during this period? (c) What was the surplus/deficit in 2007?

**12.** Sketch the graph of $f(x) = (x - 3)(x + 1)^3(x + 2)^2$ using the degree, end-behavior, $x$- and $y$-intercepts, zeroes of multiplicity, and a few "midinterval" points.

**13.** Use the *Guidelines for Graphing Polynomials* to graph $g(x) = x^4 - 9x^2 - 4x + 12$.

**14.** Use the *Guidelines for Graphing Rational Functions* to graph $h(x) = \dfrac{x - 2}{x^2 - 3x - 4}$.

**15.** Suppose the cost of cleaning contaminated soil from a dump site is modeled by $C(x) = \dfrac{300x}{100 - x}$, where $C(x)$ is the cost (in \$1000s) to remove $x\%$ of the contaminants. Graph using $x \in [0, 100]$, and use the graph to answer the following questions.

    **a.** What is the significance of the *vertical asymptote* (what does it mean in this context)?

**b.** If EPA regulations are changed so that 85% of the contaminants must be removed, instead of the 80% previously required, how much additional cost will the new regulations add? Compare the cost of the 5% increase from 80% to 85% with the cost of the 5% increase from 90% to 95%. What do you notice?

**c.** What percent of the pollutants can be removed if the company budgets $2,200,000?

**16.** Graph using the *Guidelines for Graphing Rational Functions.*

**a.** $r(x) = \dfrac{x^3 - x^2 - 9x + 9}{x^2}$  **b.** $R(x) = \dfrac{x^3 + 7x - 6}{x^2 - 4}$

**17.** Find the level of production that will minimize the average cost of an item, if production costs are modeled by $C(x) = 2x^2 + 25x + 128$, where $C(x)$ is the cost to manufacture $x$ hundred items.

**18.** Solve each inequality

**a.** $x^3 - 13x \le 12$   **b.** $\dfrac{3}{x-2} < \dfrac{2}{x}$

**19.** Suppose the concentration of a chemical in the bloodstream of a large animal $h$ hr after injection into muscle tissue is modeled by the formula

$$C(h) = \frac{2h^2 + 5h}{h^3 + 55}.$$

**a.** Sketch a graph of the function for the intervals $x \in [-5, 20]$, $y \in [0, 1]$.

**b.** Where is the vertical asymptote? Does it play a role in this context?

**c.** What is the concentration after 2 hr? After 8 hr?

**d.** How long does it take the concentration to fall below 20% $[C(h) < 0.2]$?

**e.** When does the maximum concentration of the chemical occur? What is this maximum?

**f.** Describe the significance of the horizontal asymptote in this context.

**20.** Use the vertical asymptotes, $x$-intercepts, and their multiplicities to construct an equation that corresponds to the given graph. Be sure the $y$-intercept on the graph matches the value given by your equation. Assume these features are integer-valued.

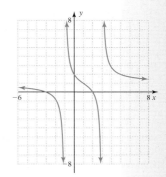

## CALCULATOR EXPLORATION AND DISCOVERY

### I. Complex Zeroes, Repeated Zeroes, and Inequalities

This *Calculator Exploration and Discovery* will explore the relationship between the solution of a polynomial (or rational) inequality and the complex zeroes and repeated zeroes of the related function. After all, if nonreal zeroes can never create an $x$-intercept, how do they affect the function? And if a function never crosses the $x$-axis at a zero of even multiplicity (always bounces), can it still affect a nonstrict (*less than or equal to* or *greater than or equal to*) inequality? These are interesting and important questions, with numerous avenues of exploration. To begin, consider the function $Y_1 = (x + 3)^2(x^3 - 1)$ or $Y_1 = (x + 3)^2(x - 1)(x^2 + x + 1)$ in completely factored form. This is a polynomial function of degree 5 with two real zeroes (one repeated), two complex zeroes (the quadratic factor is irreducible), and after viewing the graph on Figure 4.80, four turning points. From the graph (or by analysis), we have $Y_1 \le 0$ for $x \le 1$. Now let's consider $Y_2 = (x + 3)^2(x - 1)$, the same function as $Y_1$, less the quadratic factor. Since a function will never "cross the $x$-axis" at nonreal zeroes anyway, the removal of this factor *cannot affect the solution set of the inequality!* But how does it affect the function? $Y_2$ is now a function of degree three, with three real zeroes (one repeated) and only two turning points (Figure 4.81). But even so, the solution to $Y_2 \le 0$ is the same as for $Y_1 \le 0$: $x \le 1$. Finally, let's look at $Y_3 = x - 1$, the same function as $Y_2$ but with the repeated zero removed. The key here is to notice that since $(x - 3)^2$ will be nonnegative for any value of $x$, it too does not change the solution set of the "less than or equal to inequality," only the shape of the graph. $Y_3$ is a function of degree 1, with one real zero and no turning points, *but the solution interval for $Y_3 \le 0$ is the same solution interval as $Y_2$ and $Y_1$: $x \le 1$* (see Figure 4.82).

**Figure 4.80**

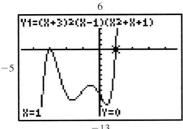

**Figure 4.81**

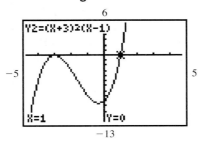

**Figure 4.82**

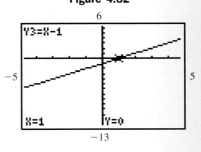

Explore these relationships further using the exercises in (A) and (B) given, using a "greater than or equal to" inequality. Begin by writing $Y_1$ in completely factored form.

**(A)** $Y_1 = (x^3 - 6x^2 + 32)(x^2 + 1)$
$Y_2 = x^3 - 6x^2 + 32$
$Y_3 = x + 2$

**(B)** $Y_1 = (x + 3)^2(x^3 - 2x^2 + x - 2)$
$Y_2 = (x + 3)^2(x - 2)$
$Y_3 = x - 2$

**Exercise 1:** Based on what you've noticed, comment on how the irreducible quadratic factors of a polynomial affect its graph. What role do they play in the solution of inequalities?

**Exercise 2:** How do zeroes of even multiplicity affect the solution set of nonstrict inequalities (less/greater than or equal to)?

## II. Removable Discontinuities

Graphing calculators offer both numerical and visual representations of removable discontinuities. For instance, enter the function $r(x) = \frac{x^2 - 4x + 3}{x - 1}$ on the Y= screen, then use the **(TBLSET)** feature to set up the table as shown in Figure 4.83. Pressing 2nd GRAPH displays the expected table, which shows the function cannot be evaluated at $x = 1$ (see Figure 4.84). Now change the **(TBLSET)** screen so that $\mathbf{\Delta Tbl = 0.01}$. Note again that the function is defined for all values except $x = 1$. Reset the table to $\mathbf{\Delta Tbl = 0.01}$ and investigate further.

We can actually see the gap or hole in the graph using a "friendly window." Since the screen of the TI-84 Plus is 95 pixels wide and 63 pixels high, multiples of 4.7 for Xmin and Xmax, and multiples of 3.1 for Ymin and Ymax, display what happens at integer (and other) values (see Figure 4.85). Pressing GRAPH gives Figure 4.86, which shows a noticeable gap at $(1, -2)$. With the TRACE feature, move the cursor over to the gap and notice what happens.

Use these ideas to view the discontinuities in the following rational functions. State the ordered pair location of each discontinuity.

**Figure 4.83**

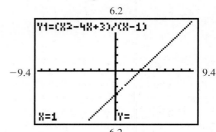

```
TABLE SETUP
 TblStart=.5
 ΔTbl=.1
Indpnt: Auto Ask
Depend: Auto Ask
```

**Figure 4.84**

| X | Y1 |
|---|---|
| .5 | -2.5 |
| .6 | -2.4 |
| .7 | -2.3 |
| .8 | -2.2 |
| .9 | -2.1 |
| 1 | ERROR |
| 1.1 | -1.9 |

X=.5

**Figure 4.85**

```
WINDOW
 Xmin=-9.4
 Xmax=9.4
 Xscl=1
 Ymin=-6.2
 Ymax=6.2
 Yscl=1
 Xres=1
```

**Figure 4.86**

**Exercise 3:** $r(x) = \dfrac{x^2 - 4}{x + 2}$

**Exercise 4:** $f(x) = \dfrac{x^2 - 2x - 3}{x + 1}$

**Exercise 5:** $r(x) = \dfrac{x^3 + 1}{x + 1}$

**Exercise 6:** $f(x) = \dfrac{x^3 - 7x + 6}{x^2 + x - 6}$

# STRENGTHENING CORE SKILLS

## Solving Inequalities Using the Push Principle

The most common method for solving polynomial inequalities involves finding the zeroes of the function and checking the sign of the function in the intervals between these zeroes. In Section 4.6, we relied on the end-behavior of the graph, the sign of the function at the $y$-intercept, and the multiplicity of the zeroes to determine the solution. There is a third method that is more conceptual in nature, but in many cases highly efficient. It is based on two very simple ideas, the first involving only order relations and the number line:

**A.** Given any number $x$ and constant $k > 0$: $x > x - k$ and $x < x + k$.

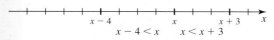

This statement simply reinforces the idea that if $a$ is left of $b$ on the number line, then $a < b$. As shown in the diagram, $x - 4 < x$ and $x < x + 3$, from which we get $x - 4 < x + 3$ for any $x$.

**B.** The second idea reiterates well-known ideas regarding the multiplication of signed numbers. For any number of factors:

*if there is an even number of negative factors, the result is positive;*
*if there is an odd number of negative factors, the result is negative.*

These two ideas work together to solve inequalities using what we'll call the *push principle*. Consider the inequality $x^2 - x - 12 > 0$. The factored form is $(x - 4)(x + 3) > 0$ and we want the product of these two factors to be positive. From (A), both factors will be positive if $x - 4 > 0$, since it's the smaller of the two; and both factors will be negative if $x + 3 < 0$, since it's the larger. The solution set is found by solving these two simple inequalities, giving the solution $x > 4$ or $x < -3$. If the inequality were $(x - 4)(x + 3) < 0$ instead, we require one negative factor and one positive factor. Due to order relations and the number line, the larger factor must be the positive one: $x + 3 > 0$ so $x > -3$. The smaller factor must be the negative one: $x - 4 < 0$ giving $x < 4$. This gives the solution $-3 < x < 4$ as can be verified using any alternative method. Solutions to all other polynomial and rational inequalities are an extension of these two cases.

**Illustration 1** ▶ Solve $x^3 - 7x + 6 < 0$ using the push principle.

**Solution** ▶ The polynomial can be factored using the tests for 1 and $-1$ and synthetic division. The factored form is $(x - 2)(x - 1)(x + 3) < 0$, which we've conveniently written with the factors in increasing order. For the product of three factors to be negative we require: (1) three negative factors or (2) one negative and two positive factors. The first condition is met by simply making the largest factor negative, as it will ensure the smaller factors are also negative: $x + 3 < 0$ so $x < -3$. The second condition is met by making the smaller factor negative and the "middle" factor positive (ensuring the larger factor is also positive): $x - 2 < 0$ *and* $x - 1 > 0$. The second solution interval is $x < 2$ and $x > 1$, or $1 < x < 2$.

Note the push principle does not require the testing of intervals between the zeroes, nor the "cross/bounce" analysis at the zeroes and vertical asymptotes (of rational functions). In addition, irreducible quadratic factors can still be ignored as they contribute nothing to the solution of real inequalities, and factors of even multiplicity can be overlooked precisely because there is no sign change at these roots.

**Illustration 2** ▶ Solve $(x^2 + 1)(x - 2)^2(x + 3) \geq 0$ using the push principle.

**Solution** ▶ Since the factor $x^2 + 1$ does not affect the solution set, this inequality will have the same solution as $(x - 2)^2(x + 3) \geq 0$. Further, since $(x - 2)^2$ will be nonnegative for all $x$, the original inequality *has the same solution set as* $x + 3 \geq 0$! The solution is $x \geq -3$.

With some practice, the push principle can be a very effective tool. Use it to solve the following exercises. Check all solutions by graphing the function on a graphing calculator.

**Exercise 1:** $x^3 - 3x - 18 \leq 0$

**Exercise 2:** $\dfrac{x + 1}{x^2 - 4} > 0$

**Exercise 3:** $x^3 - 13x + 12 < 0$

**Exercise 4:** $x^3 - 3x + 2 \geq 0$

**Exercise 5:** $x^4 - x^2 - 12 > 0$

**Exercise 6:** $(x^2 + 5)(x^2 - 9)(x + 2)^2(x - 1) \geq 0$

# CUMULATIVE REVIEW CHAPTERS R–4

**1.** Solve for $h$: $A = 2\pi r^2 + 2\pi rh$

**2.** Solve for $x$: $a(x - h)^2 + k = 0$

**3.** Factor each expression completely:
   **a.** $4x^2 - 12x + 9$   **b.** $x^3 - 3x + 2$

**4.** Solve using the quadratic formula. Write answers in both exact and approximate form: $x^2 - 6x + 2 = 0$.

**5.** Solve the following inequality: $x + 3 < 5$ and $5 - x < 1$.

**6.** State the domain of $f$ and $g$, given $f(x) = x^{\frac{2}{3}}$ and $g(x) = x^{\frac{3}{2}}$.

**7.** Use substitution to verify that $x = \frac{3}{2}$ is a solution to $4x^2 + 8x - 21 = 0$.

**8.** Solve the rational inequality: $\dfrac{x + 4}{x - 2} < 3$.

**9.** When he started working out on day 1, Ben could bench press 125 lb. On day 31 he could bench 140 lb. Assuming the growth was linear, (a) find a linear function that models his increasing strength and (b) use it to estimate how much he could bench on day 60. (c) If he can now bench 200 lb, how many days has it been since he started lifting weights?

**10.** State the domain of the function $y = \sqrt{9 - x^2}$.

**11.** The data given show the increasing height of a skyscraper for the first 6 months after the foundation and base were complete. (a) Use the data to find a linear function that models this data, then (b) use the function to estimate the building's height after 15 months. (c) If the completed height of the building will be 1635 ft, how many months will it take to complete?

**Exercise 11**

| Month | Height (ft) |
|-------|-------------|
| 0 | 10 |
| 1 | 52 |
| 2 | 90 |
| 3 | 155 |
| 4 | 180 |
| 5 | 235 |
| 6 | 280 |

**12.** Graph the function $g(x) = -|x + 2| + 3$ using transformations of a basic function.

**13.** For $f(x) = \sqrt[3]{2x - 3}$, find $x$ if $f(x) = -4$.

**14.** Graph $f(x) = x^3 - 12x$, then state intervals where:
  **a.** $f(x) \geq 0$
  **b.** $f(x)\uparrow$

**15.** Given $g(x) = x^5 + x^3 + x + 3$. (a) How many zeroes will this function have? (b) According to Descartes' rule of signs, how many positive roots are possible? (c) How many negative roots are possible? (d) What are the possible rational roots? (e) What possibilities are eliminated by parts (b) and (c)? (f) Find the only real root for this function.

**16.** An ice cream sandwich is a delicacy made up of a rectangular block of ice cream placed between two rectangular graham crackers. If the crackers are 0.2 cm × 6 cm × 12 cm, and the total sandwich measures 2 cm × 6 cm × 12 cm, how much ice cream is in an ice cream sandwich?

**17.** For $f(x) = x^2 - 5x + 4$, find the average rate of change in the following intervals: (a) [0, 1] and (b) [4, 5].

**18.** Use the rational zeroes theorem and synthetic division to find all zeroes (real and complex) of $f(x) = x^4 - 2x^2 + 16x - 15$.

**19.** Sketch the graph of $f(x) = x^3 - 3x^2 - 6x + 8$.

**20.** Sketch the graph of $h(x) = \frac{x - 1}{x^2 - 4}$ and use the zeroes and vertical asymptotes to solve $h(x) \geq 0$.

**21.** Given $g(x) = 1.47x^3 - 0.51x^2 + 1.9x$, use your calculator to evaluate $g(0.1)$ and $g(1)$.

**22.** Find the (a) $x$-intercept(s) and (b) $y$-intercept, and the equations of any (c) vertical asymptotes or (d) horizontal asymptotes for $r(x) = \frac{12 - 3x^2}{1.5x^2 + 2}$.

**23.** Perform the operation indicated.

  **a.** $(2 + 3i)^2$          **b.** $\dfrac{9 + i}{4 + 5i}$

  **c.** $(5 - 4i)(5 + 4i)$          **d.** $i^{15}$

**24.** The function $g(t) = 0.25t^2$ models the growth in sales of a company, for the 12 months of 2012. Use the difference quotient to find their average rate of growth between May ($t = 5$) and June ($t = 6$).

**25.** Using the calculator screen shown, identify which function ($Y_2$ through $Y_7$) performs the indicated transformation of $Y_1$.

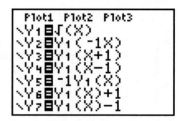

  **a.** $Y_1$ shifted up 1 unit
  **b.** $Y_1$ shifted down 1 unit
  **c.** $Y_1$ shifted right 1 unit
  **d.** $Y_1$ shifted left 1 unit
  **e.** $Y_1$ reflected across $x$-axis
  **f.** $Y_1$ reflected across $y$-axis

# 5

# Exponential and Logarithmic Functions

## CHAPTER CONNECTIONS

If you could start from scratch and build your own island, how many different types of flowers, birds, and other plants and animals would you like the island to support? Although it seems like a silly question, attempts to answer it led to the development of a new field of science called *island biogeography*. The field has since expanded to include the study of species diversity on any isolated landscape, including: sky islands (mountains surrounded by desert), woodlot islands (pastures cleared by deforestation), and even freshwater lakes (water islands?). Exercise 86 in Section 5.4 shows how the number of species on an "island" can be predicted when little more than the size of the island is known.

**Check out these other real-world connections:**

▶ Fines for Speeding
   (Section 5.1, Exercise 84)
▶ Memory Retention
   (Section 5.3, Exercises 107 and 108)
▶ Saving for a Down Payment
   (Section 5.6, Exercise 43)
▶ Carbon Dating
   (Section 5.6, Exercises 61 and 62)

**LEARNING OBJECTIVES**

*In Section 5.1 you will see how we can:*

☐ **A.** Identify one-to-one functions
☐ **B.** Explore inverse functions using ordered pairs
☐ **C.** Find inverse functions using an algebraic method
☐ **D.** Graph a function and its inverse
☐ **E.** Solve applications of inverse functions

Consider the function $f(x) = 2x - 3$. If $f(x) = 7$, the equation becomes $2x - 3 = 7$, and the corresponding value of $x$ can be found using *inverse operations*. In this section, we introduce the concept of an *inverse function*, which can be viewed as a formula for finding $x$-values that correspond to *any* given value of $f(x)$.

## A. Identifying One-to-One Functions

The graphs of $y = 2x$ and $y = x^2$ are shown in Figures 5.1 and 5.2. The dashed, vertical lines clearly indicate both are functions, with each $x$-value corresponding to only one $y$. But the points on $y = 2x$ have one characteristic those from $y = x^2$ do not—*each y-value also corresponds to only one x* (for $y = x^2$, 4 corresponds to both $-2$ and 2). If each element from the range of a function corresponds to only one element of the domain, the function is said to be **one-to-one.**

**Figure 5.1**

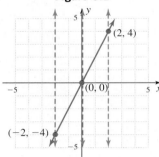

**Figure 5.2**

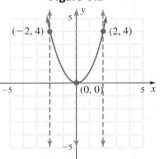

**One-to-One Functions**

A function $f$ is one-to-one if every element in the range corresponds to only one element of the domain.

In symbols, if $f(x_1) = f(x_2)$, then $x_1 = x_2$,
or if $x_1 \neq x_2$, then $f(x_1) \neq f(x_2)$.

From this definition, we note the graph of a one-to-one function must not only pass a vertical line test (to show each $x$ corresponds to only one $y$), but also pass a **horizontal line test** (to show each $y$ corresponds to only one $x$).

**Horizontal Line Test**

If every horizontal line intersects the graph of a function in at most one point, the function is one-to-one.

Notice the graph of $y = 2x$ (Figure 5.3) passes the horizontal line test, while the graph of $y = x^2$ (Figure 5.4) does not.

**Figure 5.3**

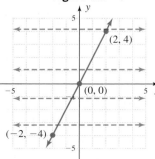

**Figure 5.4**

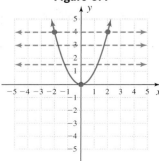

If the function is given in ordered pair form, we simply check to see that no second coordinate is paired with more than one first coordinate.

**EXAMPLE 1** ▶   Identifying One-to-One Functions

Determine whether each graph or relation shown depicts a function. If so, determine whether the function is one-to-one.

**a.**

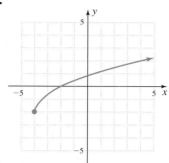

**b.**

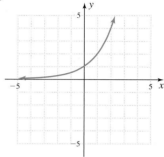

**c.**

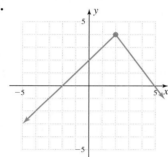

**d.**

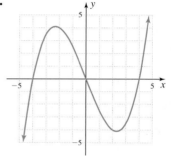

**e.**

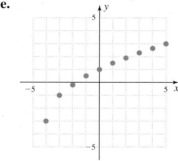

**f.** $\{(-4, 1), (-3, 5), (-1, 1),$
$(-2, 3), (-3, -2), (4, 4)\}$

Solution ▶   A careful inspection shows all five graphs depict functions, since each passes the vertical line test [the relation in (f) is not a function since the input $-3$ corresponds to two outputs]. Only (a), (b), and (e) pass the horizontal line test and are *one-to-one* functions.

☑ **A.** You've just seen how we can identify one-to-one functions

Now try Exercises 5 through 24 ▶

## B. Inverse Functions and Ordered Pairs

Consider the function $f(x) = 2x - 3$ and the values shown in Table 5.1. Figure 5.5 shows this function in diagram form (in blue), and illustrates that for each element of the domain, we *multiply by 2, then subtract 3*. An **inverse function** for $f$ is one that takes the result of these operations (elements of the range), and returns the original

**Table 5.1**

| x  | f(x) |
|----|------|
| −3 | −9   |
| 0  | −3   |
| 2  | 1    |
| 5  | 7    |
| 8  | 13   |

**Table 5.2**

| x  | F(x) |
|----|------|
| −9 | −3   |
| −3 | 0    |
| 1  | 2    |
| 7  | 5    |
| 13 | 8    |

domain element. Figures 5.5 and 5.6 show that function $F$ achieves this by "undoing" the operations in reverse order: *add 3, then divide by 2* (in red). A table of values for $F(x)$ is shown (Table 5.2).

**Figure 5.5**

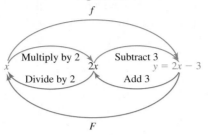

**Figure 5.6**

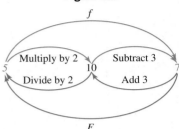

From this illustration we make the following observations regarding an inverse function, which we actually denote as $f^{-1}(x)$.

**Inverse Functions**

If $f$ is a one-to-one function with ordered pairs $(a, b)$,

1. $f^{-1}$ is a one-to-one function with ordered pairs $(b, a)$.
2. The range of $f$ will be the domain of $f^{-1}$.
3. The domain of $f$ will be the range of $f^{-1}$.

**WATCH YOUR STEP**

**CAUTION** ▶ The notation $f^{-1}(x)$ is simply a way of denoting an inverse function and has nothing to do with exponential properties. In particular, $f^{-1}(x)$ does *not* mean $\frac{1}{f(x)}$.

**EXAMPLE 2** ▶  **Finding the Inverse of a Function**

Find the inverse of each one-to-one function given:

**a.** $f = \{(-4, 13), (-1, 7), (0, 5), (2, 1), (5, -5), (8, -11)\}$
**b.** $p(x) = 2x^3 - 5$

Solution ▶  **a.** When a function is defined as a set of ordered pairs, the inverse function is found by simply interchanging the coordinates:
$f^{-1} = \{(13, -4), (7, -1), (5, 0), (1, 2), (-5, 5), (-11, 8)\}$.

**b.** Using the diagram, we reason $p^{-1}$ will add 5, divide by 2, and take a cube root: $p^{-1}(x) = \sqrt[3]{\frac{x + 5}{2}}$.

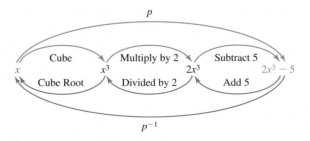

As a test, we find that $(-1, -7)$, $(0, -5)$, and $(2, 11)$ are points on $p(x)$, and note that $(-7, -1)$, $(-5, 0)$, and $(11, 2)$ are indeed points on $p^{-1}(x)$.

☑ **B.** You've just seen how we can explore inverse functions using ordered pairs

**Now try Exercises 25 through 36** ▶

## C. Finding Inverse Functions Using an Algebraic Method

The fact that interchanging $x$- and $y$-values helps determine an inverse function can be generalized to develop an **algebraic method** for finding inverses. Instead of interchanging *specific x*- and *y*-values, we actually interchange the *x*- and *y-variables,* then solve the equation for $y$. The process is summarized here.

> **Finding an Inverse Function**
>
> 1. Replace $f(x)$ with $y$.
> 2. Interchange $x$ and $y$.
> 3. Solve the new equation for $y$.
> 4. The result gives the inverse function: Replace $y$ with $f^{-1}(x)$.

If a function is *not* one-to-one, no inverse function exists. To see why, consider the points $(-2, 4)$ and $(2, 4)$ from $y = x^2$, which both lie on the horizontal line $y = 4$. When the coordinates are interchanged, the points $(4, -2)$ and $(4, 2)$ result. Since both lie on the vertical line $x = 4$, the relation fails the vertical line test and cannot be a function.

---

**EXAMPLE 3** ▶ **Finding Inverse Functions Algebraically**

State the domain and range of each function given, then use the algebraic method to find the inverse function, and state *its* domain and range.

**a.** $f(x) = \sqrt[3]{x + 5}$       **b.** $g(x) = \dfrac{2x}{x + 1}$

**Solution** ▶ **a.**  $f(x) = \sqrt[3]{x + 5}, x \in \mathbb{R}, y \in \mathbb{R}$

$\qquad y = \sqrt[3]{x + 5}$      use $y$ instead of $f(x)$ (original relationship)

$\qquad x = \sqrt[3]{y + 5}$      interchange $x$ and $y$ (inverse relationship)

$\qquad x^3 = y + 5$      cube both sides

$\qquad x^3 - 5 = y$      solve for $y$

$\qquad f^{-1}(x) = x^3 - 5, x \in \mathbb{R}, y \in \mathbb{R}$

**b.**  $g(x) = \dfrac{2x}{x + 1}, x \neq -1, y \neq 2$

$\qquad y = \dfrac{2x}{x + 1}$      use $y$ instead of $f(x)$ (original relationship)

$\qquad x = \dfrac{2y}{y + 1}$      interchange $x$ and $y$ (inverse relationship)

$\qquad xy + x = 2y$      multiply by $y + 1$ and distribute

$\qquad x = 2y - xy$      gather terms with $y$

$\qquad x = y(2 - x)$      factor

$\qquad \dfrac{x}{2 - x} = y$      solve for $y$

$\qquad g^{-1}(x) = \dfrac{x}{2 - x}, x \neq 2, y \neq -1$

**Now try Exercises 37 through 44** ▶

---

In cases where a given function is *not* one-to-one, we can sometimes restrict the domain to create a function that *is,* and then determine an inverse. The restriction we use is somewhat arbitrary, and only requires that the result produce all possible range values. Most often, we simply choose a limited domain that seems convenient or reasonable.

**EXAMPLE 4** ▶  **Restricting the Domain to Create a One-to-One Function**

Given $f(x) = (x - 4)^2$, restrict the domain to create a one-to-one function, then find $f^{-1}(x)$. State the domain and range of both resulting functions.

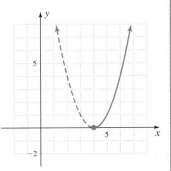

**Solution** ▶  The graph of $f$ is a parabola, opening upward with the vertex at $(4, 0)$. Restricting the domain to $x \geq 4$ (see figure) leaves only the "right branch" of the parabola, creating a one-to-one function without affecting the range, $y \in [0, \infty)$. For $f(x) = (x - 4)^2$ with restricted domain $x \geq 4$, we have

$$f(x) = (x - 4)^2 \quad \text{given function}$$
$$y = (x - 4)^2 \quad \text{use } y \text{ instead of } f(x) \text{ (original relationship)}$$
$$x = (y - 4)^2 \quad \text{interchange } x \text{ and } y \text{ (inverse relationship)}$$
$$\pm\sqrt{x} = y - 4 \quad \text{take square roots}$$
$$\sqrt{x} + 4 = y \quad \text{solve for } y, \text{ use } \sqrt{x} \text{ since } x \geq 4$$

The result shows $f^{-1}(x) = \sqrt{x} + 4$, with domain $x \in [0, \infty)$ and range $y \in [4, \infty)$ (the domain of $f$ becomes the range of $f^{-1}$, and the range of $f$ becomes the domain of $f^{-1}$).

**Now try Exercises 45 through 50** ▶

While we now have the ability to *find* the inverse of a function, we still lack a definitive method of *verifying* the inverse is correct. Actually, the diagrams in Figures 5.5 and 5.6 suggest just such a method. If we use $f(x)$ as an input for $f^{-1}$, or $f^{-1}(x)$ as an input for $f$, the end result should simply be $x$, as each function "undoes" the operations of the other. From Section 3.5 this is called a composition of functions and using the notation for composition we have,

**Verifying Inverse Functions**

If $f$ is a one-to-one function, then the function $f^{-1}$ exists and satisfies

$$(f \circ f^{-1})(x) = x \qquad \text{and} \qquad (f^{-1} \circ f)(x) = x$$

**EXAMPLE 5** ▶  **Finding and Verifying an Inverse Function**

Use the algebraic method to find the inverse function for $f(x) = \sqrt{x + 2}$. Then verify the inverse you found is correct.

**Solution** ▶  Since the graph of $f$ is the graph of $y = \sqrt{x}$ shifted 2 units left, we know $f$ is one-to-one with domain $x \in [-2, \infty)$ and range $y \in [0, \infty)$. This is important since the *domain and range values will be interchanged for the inverse function*. The domain of $f^{-1}$ will be $x \in [0, \infty)$ and its range $y \in [-2, \infty)$.

$$f(x) = \sqrt{x + 2} \quad \text{given function; } x \geq -2$$
$$y = \sqrt{x + 2} \quad \text{use } y \text{ instead of } f(x) \text{ (original relationship)}$$
$$x = \sqrt{y + 2} \quad \text{interchange } x \text{ and } y \text{ (inverse relationship)}$$
$$x^2 = y + 2 \quad \text{solve for } y \text{ (square both sides)}$$
$$x^2 - 2 = y \quad \text{subtract 2}$$
$$f^{-1}(x) = x^2 - 2 \quad \text{the result is } f^{-1}(x); \text{ } D{:} \text{ } x \in [0, \infty), \text{ } R{:} \text{ } y \in [-2, \infty)$$

**Verify ▶**

$$(f \circ f^{-1})(x) = f[f^{-1}(x)]$$     $f^{-1}(x)$ is an input for $f$

$$= \sqrt{f^{-1}(x) + 2}$$     $f$ adds 2 to inputs, then takes the square root

$$= \sqrt{(x^2 - 2) + 2}$$     substitute $x^2 - 2$ for $f^{-1}(x)$

$$= \sqrt{x^2}$$     simplify

$$= x \checkmark$$     since the domain of $f^{-1}(x)$ is $x \in [0, \infty)$

$$(f^{-1} \circ f)(x) = f^{-1}[f(x)]$$     $f(x)$ is an input for $f^{-1}$

$$= [f(x)]^2 - 2$$     $f^{-1}$ squares inputs, then subtracts 2

$$= [\sqrt{x + 2}]^2 - 2$$     substitute $\sqrt{x + 2}$ for $f(x)$

$$= x + 2 - 2$$     simplify

$$= x \checkmark$$     result

☑ **C.** You've just seen how we can find inverse functions using an algebraic method

**Now try Exercises 51 through 70 ▶**

## D. The Graph of a Function and Its Inverse

Graphing a function and its inverse on the same axes reveals an interesting and useful relationship—the graphs are reflections across the line $y = x$ (the identity function). Consider the function $f(x) = 2x + 3$, and its inverse $f^{-1}(x) = \frac{x - 3}{2} = \frac{1}{2}x - \frac{3}{2}$. In Figure 5.7, the points $(1, 5)$, $(0, 3)$, $(-\frac{3}{2}, 0)$, and $(-4, -5)$ from $f$ (see Table 5.3) are graphed in blue, with the points $(5, 1)$, $(3, 0)$, $(0, -\frac{3}{2})$, and $(-5, -4)$ (see Table 5.4) from $f^{-1}$ graphed in red (note the $x$- and $y$-values are reversed). Graphing both lines illustrates this symmetry (Figure 5.8).

**Table 5.3**

| $x$ | $f(x)$ |
|---|---|
| 1 | 5 |
| 0 | 3 |
| $-\frac{3}{2}$ | 0 |
| $-4$ | $-5$ |

**Figure 5.7**

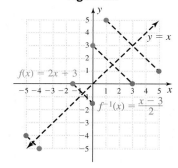

**Figure 5.8**

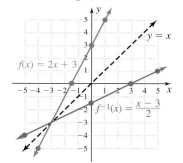

**Table 5.4**

| $x$ | $f^{-1}(x)$ |
|---|---|
| 5 | 1 |
| 3 | 0 |
| 0 | $-\frac{3}{2}$ |
| $-5$ | $-4$ |

**EXAMPLE 6 ▶** **Graphing a Function and Its Inverse**

Given the graph shown in Figure 5.9, draw a graph of the inverse function.

**Figure 5.9**

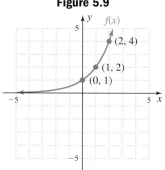

**Solution** ▶ From the graph, the domain of $f$ appears to be $x \in \mathbb{R}$ and the range is $y \in (0, \infty)$. This means the domain of $f^{-1}$ will be $x \in (0, \infty)$ and the range will be $y \in \mathbb{R}$. To sketch $f^{-1}$, draw the line $y = x$, interchange the $x$- and $y$-coordinates of the selected points, then plot these points and draw a smooth curve using the domain and range boundaries as a guide. The result is shown in Figure 5.10.

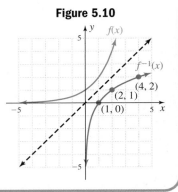
**Figure 5.10**

☑ **D. You've just seen how we can graph a function and its inverse**

Now try Exercises 71 through 80 ▶

A summary of important concepts is provided here.

---

### Functions and Inverse Functions

1. If the graph of a function passes the horizontal line test, the function is one-to-one.
2. If a function $f$ is one-to-one, the function $f^{-1}$ exists.
3. The domain of $f$ is the range of $f^{-1}$, and the range of $f$ is the domain of $f^{-1}$.
4. For a function $f$ and its inverse $f^{-1}$, $(f \circ f^{-1})(x) = x$ and $(f^{-1} \circ f)(x) = x$.
5. The graphs of $f$ and $f^{-1}$ are symmetric about the line $y = x$.

---

### E. Applications of Inverse Functions

Our final example illustrates one of the many ways that inverse functions can be applied.

**EXAMPLE 7** ▶ **Using Volume to Understand Inverse Functions**

The volume of an equipoise cylinder (height equal to diameter) is given by $v(x) = 2\pi x^3$ (since $h = d = 2x$), where $v(x)$ represents the volume and $x$ represents the radius of the cylinder.

**a.** Find the volume of such a cylinder if $x = 10$ ft.

**b.** Find $v^{-1}(x)$, and discuss what the input and output variables represent.

**Solution** ▶ **a.**  $v(x) = 2\pi x^3$     given function

$v(10) = 2\pi(10)^3$     substitute 10 for $x$

$= 2000\pi$     $10^3 = 1000$, exact form

With a radius of 10 ft, the volume of the cylinder would be $2000\pi$ ft$^3$.

**b.**  $v(x) = 2\pi x^3$     given function

$y = 2\pi x^3$     use $y$ instead of $v(x)$ (original relationship)

$x = 2\pi y^3$     interchange $x$ and $y$ (inverse relationship)

$\dfrac{x}{2\pi} = y^3$     solve for $y$

$\sqrt[3]{\dfrac{x}{2\pi}} = y$     result

The inverse function is $v^{-1}(x) = \sqrt[3]{\dfrac{x}{2\pi}}$. Here, the input $x$ is a given volume, and the output $v^{-1}(x)$ is the radius of an equipoise cylinder that will hold this volume.

☑ **E. You've just seen how we can solve applications of inverse functions**

Now try Exercises 83 through 88 ▶

## 5.1 EXERCISES

### ▶ CONCEPTS AND VOCABULARY

**Fill in each blank with the appropriate word or phrase. Carefully reread the section, if necessary.**

1. A function is one-to-one if each _____ coordinate corresponds to exactly_____ first coordinate.

2. To find $f^{-1}$ using the algebraic method, we (1) replace $f(x)$ with ____, (2) _____ $x$ and $y$, (3) _____ for $y$, and replace $y$ with $f^{-1}(x)$.

3. State true or false and explain why: *To show that g is the inverse of f, simply show that $(f \circ g)(x) = x$.* Include an example in your response.

4. Discuss/Explain why no inverse function exists for $f(x) = (x + 3)^2$ and $g(x) = \sqrt{4 - x^2}$. How would the domain of each function have to be restricted to allow for an inverse function?

### ▶ DEVELOPING YOUR SKILLS

**Determine whether each graph given is the graph of a one-to-one function. If not, give examples of how the definition of one-to-oneness is violated.**

5.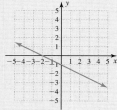

6.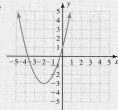

7.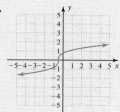

8.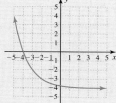

9.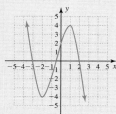

10.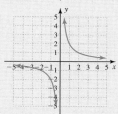

11.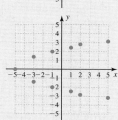

12.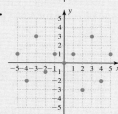

**Determine whether the functions given are one-to-one. If not, state why.**

13. $\{(-7, 4), (-1, 9), (0, 5), (-2, 1), (5, -5)\}$

14. $\{(9, 1), (-2, 7), (7, 4), (3, 9), (2, 7)\}$

15. $\{(-6, 1), (4, -9), (0, 11), (-2, 7), (-4, 5), (8, 1)\}$

16. $\{(-6, 2), (-3, 7), (8, 0), (12, -1), (2, -3), (1, 3)\}$

**Determine if the functions given are one-to-one by noting the function family to which each belongs and mentally picturing the shape of the graph. If a function is not one-to-one, discuss how the definition of one-to-oneness is violated.**

17. $f(x) = 3x - 5$

18. $g(x) = (x + 2)^3 - 1$

19. $h(x) = -|x - 4| + 3$

20. $p(t) = 3t^2 + 5$

21. $s(t) = \sqrt{2t - 1} + 5$

22. $r(t) = \sqrt[3]{t + 1} - 2$

23. $y = 3$

24. $y = -2x$

**For Exercises 25 through 28, find the inverse function of the one-to-one functions given.**

25. $f = \{(-2, 1), (-1, 4), (0, 5), (2, 9), (5, 15)\}$

26. $g = \{(-2, 30), (-1, 11), (0, 4), (1, 3), (2, 2)\}$

27. $v(x)$ is defined by the ordered pairs shown.

| $x$ | $v(x)$ |
|-----|--------|
| $-4$ | 3 |
| $-3$ | 2 |
| 0 | 1 |
| 5 | 0 |
| 12 | $-1$ |
| 21 | $-2$ |
| 32 | $-3$ |

28. $w(x)$ is defined by the ordered pairs shown.

| $x$ | $w(x)$ |
|-----|--------|
| $-6$ | 4 |
| $-5$ | 2.5 |
| $-2$ | $-2$ |
| 0 | $-5$ |
| 3 | $-9.5$ |
| 4 | $-11$ |
| 7 | $-15.5$ |

**Find the inverse function using diagrams similar to that illustrated in Example 2. Check the result using three test points.**

**29.** $f(x) = x + 5$

**30.** $g(x) = x - 4$

**31.** $p(x) = -\dfrac{4}{5}x$

**32.** $r(x) = \dfrac{3x}{4}$

**33.** $f(x) = 4x + 3$

**34.** $g(x) = 5x - 2$

**35.** $t(x) = \sqrt[3]{x - 4}$

**36.** $s(x) = \sqrt[3]{x + 2}$

**State the domain and range of $f(x)$, then use the algebraic method to find the inverse function and state its domain and range. Finally, find any three ordered pairs $(a, b)$ on the graph of $f$, and verify the ordered pairs $(b, a)$ are on the graph of $f^{-1}$.**

**37.** $f(x) = \sqrt[3]{x - 2}$

**38.** $f(x) = \sqrt[3]{x + 3}$

**39.** $f(x) = x^3 + 1$

**40.** $f(x) = x^3 - 2$

**41.** $f(x) = \dfrac{8}{x + 2}$

**42.** $f(x) = \dfrac{12}{x - 1}$

**43.** $f(x) = \dfrac{x}{x + 1}$

**44.** $f(x) = \dfrac{x + 2}{1 - x}$

**The functions given in Exercises 45 through 50 are not one-to-one. (a) Determine a domain restriction that preserves all range values and creates a one-to-one function, then state the new domain and range. (b) State the domain and range of the inverse function and find its equation.**

**45.** $f(x) = (x + 5)^2$

**46.** $g(x) = x^2 + 3$

**47.** $v(x) = \dfrac{8}{(x - 3)^2}$

**48.** $V(x) = \dfrac{4}{x^2} + 2$

**49.** $p(x) = (x + 4)^2 - 2$

**50.** $q(x) = \dfrac{4}{(x - 2)^2} + 1$

**For each function $f(x)$ given, prove (using a composition) that $g(x) = f^{-1}(x)$.**

**51.** $f(x) = -2x + 5, \; g(x) = \dfrac{x - 5}{-2}$

**52.** $f(x) = 3x - 4, \; g(x) = \dfrac{x + 4}{3}$

**53.** $f(x) = \sqrt[3]{x + 5}, \; g(x) = x^3 - 5$

**54.** $f(x) = \sqrt[3]{x - 4}, \; g(x) = x^3 + 4$

**55.** $f(x) = x^2 - 3, x \geq 0; \; g(x) = \sqrt{x + 3}$

**56.** $f(x) = x^2 + 8, x \geq 0; \; g(x) = \sqrt{x - 8}$

**Find the inverse of each function $f(x)$ given, then prove (by composition) your inverse function is correct. Note the domain and range of $f$ in each case is all real numbers.**

**57.** $f(x) = \dfrac{x - 5}{2}$

**58.** $f(x) = \dfrac{x + 4}{3}$

**59.** $f(x) = \frac{1}{2}x - 3$

**60.** $f(x) = \frac{2}{3}x + 1$

**61.** $f(x) = \sqrt[3]{2x + 1}$

**62.** $f(x) = \sqrt[3]{3x - 2}$

**63.** $f(x) = \dfrac{(x - 1)^3}{8}$

**64.** $f(x) = \dfrac{(x + 3)^3}{-27}$

**The functions given in Exercises 65 through 70 are one-to-one. State the implied domain of each function given, and use these to state the domain and range of the inverse function. Then find the inverse and prove by composition that your inverse is correct.**

**65.** $f(x) = \sqrt{3x + 2}$

**66.** $g(x) = \sqrt{2x - 5}$

**67.** $p(x) = 2\sqrt{x - 3}$

**68.** $q(x) = 4\sqrt{x + 1}$

**69.** $v(x) = x^2 + 3, x \geq 0$

**70.** $w(x) = x^2 - 1, x \geq 0$

**Determine the domain and range for each one-to-one function whose graph is given, and use this information to state the domain and range of the inverse function. Then sketch in the line $y = x$, estimate the location of two or more points on the graph, and use this information to graph the inverse function on the same grid.**

**71.**

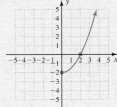

**72.**

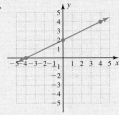

**73.**

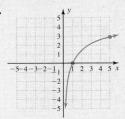

**74.**

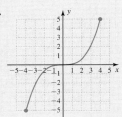

**75.**

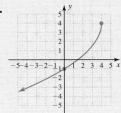

**76.**

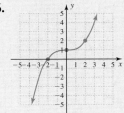

**For the functions given, (a) find the inverse function, then verify they are inverses by (b) using ordered pairs, (c) composing the functions, and (d) showing their graphs are symmetric about $y = x$.**

**77.** $f(x) = 2x + 1$

**78.** $g(x) = x^2 + 1, x \geq 0$

**79.** $h(x) = \dfrac{x}{x + 1}$

**80.** $j(x) = 2\sqrt{x + 9} - 6$

## ▶ WORKING WITH FORMULAS

**81. The height of a projected image: $h(x) = \dfrac{1}{2}x - 8.5$**

The height of an image projected on a screen is given by the formula shown, where $h(x)$ represents the projected height of the image (in centimeters) and $x$ is the distance of the projector from the screen (in centimeters). (a) When the projector is 80 cm from the screen, how large is the image? (b) Show that the inverse function is $h^{-1}(x) = 2x + 17$, then input your answer from part (a) and comment on the result. What information does the inverse function give?

**82. Degrees Fahrenheit to degrees Celsius: $C(x) = \dfrac{5}{9}(x - 32)$**

The formula shown is used to convert a temperature $x$ in degrees Fahrenheit to an equivalent temperature $C(x)$ in degrees Celsius. (a) If the living room thermostat reads 72°F, what is the temperature in degrees Celsius? (b) Show that the inverse function is $C^{-1}(x) = \frac{9}{5}x + 32$, then input your answer from part (a) and comment on the result. What information does the inverse give?

## ▶ APPLICATIONS

**83. Temperature and altitude:** The temperature at a given altitude can be approximated by the function $T(x) = -\frac{7}{2}x + 59$, where $T(x)$ represents the temperature in degrees Fahrenheit and $x$ represents the altitude in thousands of feet. (a) What is the approximate temperature at an altitude of 35,000 ft (normal cruising altitude for commercial airliners)? (b) Find $T^{-1}(x)$, and state what the independent and dependent variables represent. (c) If the temperature outside a weather balloon is $-18°$F, what is the approximate altitude of the balloon?

**84. Fines for speeding:** In some localities, there is a set formula to determine the amount of a fine for exceeding posted speed limits. Suppose the amount of the fine for exceeding a 50 mph speed limit was given by the function $f(x) = 12x - 560 \ (x > 50)$ where $f(x)$ represents the fine in dollars for a speed of $x$ mph. (a) What is the fine for traveling 65 mph through this speed zone? (b) Find $f^{-1}(x)$, and state what the independent and dependent variables represent. (c) If a fine of $172 was assessed, how fast was the driver going through this speed zone?

**85. Effect of gravity:** Due to the effect of gravity, the distance an object has fallen after being dropped is given by the function $d(x) = 16x^2 \ (x \geq 0)$, where $d(x)$ represents the distance in feet after $x$ sec. (a) How far has the object fallen 3 sec after it has been dropped? (b) Find $d^{-1}(x)$, and state what the independent and dependent variables represent. (c) If the object is dropped from a height of 784 ft, how many seconds until it hits the ground (stops falling)?

**86. Area and radius:** The area of a circle is given by $A(x) = \pi x^2$, where $A(x)$ represents the area in square units for a circle with radius $x$ units. (a) A pet dog is tethered to a stake in the backyard. If the tether is 10 ft long, how much area does the dog have to roam (use $\pi \approx 3.14$)? (b) Find $A^{-1}(x)$, and state what the independent and dependent variables represent. (c) If the owners want to allow the dog 1256 ft$^2$ of area to live and roam, how long a tether should be used?

**87. Volume of a cone:** The volume of an equipoise cone (height equal to diameter) is given by $V(x) = \frac{1}{12}\pi x^3$, where $V(x)$ represents volume and $x$ represents the height of the cone. (a) Find the volume of such a cone if its height is 60 ft (use $\pi \approx 3.14$). (b) Find $V^{-1}(x)$, and state what the independent and dependent variables represent. (c) If the volume of water in the cone is 1526.04 ft$^3$, how deep is the water at its deepest point?

**88. Wind power:** The power delivered by a certain wind-powered generator can be modeled by the function $P(x) = \frac{x^3}{2500}$, where $P(x)$ is the horsepower (hp) delivered by the generator and $x$ represents the speed of the wind in miles per hour. (a) Use the model to determine how much horsepower is generated by a 30 mph wind. (b) Find $P^{-1}(x)$ and state what the independent and dependent variables represent. (c) If gauges show 25.6 hp is being generated, how fast is the wind blowing?

## ▶ EXTENDING THE CONCEPT

**89.** (a) Find a formula for the inverse of $f(x) = \frac{ax + b}{cx + d}$, then (b) use this formula to find the inverse of $f(x) = \frac{2x}{x + 1}$. Compare your answer to the one found in Example 3(b).

**90.** After asking the gas station attendant for directions, you decide to follow his advice and take a back-roads shortcut to the interstate. Once the tank is topped off, you turn right and head down Old Possum Lane. Three miles later you cross Widowmaker bridge, where you take a left at Desperation Drive. After driving 10 more mi (5 past where the interstate was supposed to be), you decide you've had enough of this "shortcut." Since your cell phone doubles as your GPS but has no reception, you decide to head back. Use a diagram similar to that used in Example 2 to describe the inverse route back to the gas station.

**91.** By inspection, which of the following is the inverse of $f(x) = \frac{2}{3}(x - \frac{1}{2})^5 + \frac{4}{5}$?

**a.** $f^{-1}(x) = \sqrt[5]{\frac{1}{2}\left(x - \frac{2}{3}\right)} - \frac{4}{5}$

**b.** $f^{-1}(x) = \frac{3}{2}\sqrt[5]{x - 2} - \frac{5}{4}$

**c.** $f^{-1}(x) = \frac{3}{2}\sqrt[5]{x + \frac{1}{2}} - \frac{5}{4}$

**d.** $f^{-1}(x) = \sqrt[5]{\frac{3}{2}\left(x - \frac{4}{5}\right)} + \frac{1}{2}$

## ▶ MAINTAINING YOUR SKILLS

**92.** (4.6) Given $f(x) = x^2 - x - 2$, solve the inequality $f(x) \leq 0$ using the $x$-intercepts and end-behavior of the graph.

**93.** (1.1) Write as many of the following formulas as you can from memory:

    **a.** perimeter of a rectangle     **b.** area of a circle

    **c.** volume of a cylinder     **d.** volume of a cone

    **e.** circumference of a circle     **f.** area of a triangle

    **g.** area of a trapezoid     **h.** volume of a sphere

    **i.** Pythagorean theorem

**94.** (1.6) Solve the following cubic equations by factoring:

    **a.** $x^3 - 5x = 0$

    **b.** $x^3 - 7x^2 - 4x + 28 = 0$

    **c.** $x^3 - 3x^2 = 0$

    **d.** $x^3 - 3x^2 - 4x = 0$

**95.** (2.5) For the function $y = 2\sqrt{x + 3}$, find the average rate of change between $x = 1$ and $x = 2$, and between $x = 4$ and $x = 5$. Which is greater? Why?

---

## 5.2 Exponential Functions

### LEARNING OBJECTIVES

*In Section 5.2 you will see how we can:*

☐ **A.** Evaluate an exponential function

☐ **B.** Graph general exponential functions

☐ **C.** Graph base-$e$ exponential functions

☐ **D.** Solve exponential equations and applications

*Demographics* is the statistical study of human populations. In this section, we introduce the family of *exponential functions,* which are widely used to model population growth or decline with additional applications in science, engineering, and many other fields. As with other functions, we begin with a study of the graph and its characteristics.

### A. Evaluating Exponential Functions

In the boomtowns of the old west, it was not uncommon for a town to double in size every year (at least for a time) as the lure of gold drew more and more people westward. When this type of growth is modeled using mathematics, exponents play a lead role. Suppose the town of Goldsboro had 1000 residents when gold was first discovered. After 1 yr the population doubled to **2000** residents. The next year it doubled again to **4000**, then

again to **8000**, then to **16,000** and so on. You probably recognize the digits in blue as powers of two (indicating the population is *doubling*), with each one multiplied by 1000 (the initial population). This suggests we can model the relationship using

$$P(x) = 1000 \cdot 2^x$$

where $P(x)$ is the population after $x$ yr. Further, we can evaluate this function, called an **exponential function,** for *fractional parts of a year* using rational exponents. The population of Goldsboro one-and-a-half years after the gold rush was

$$P\left(\frac{3}{2}\right) = 1000 \cdot 2^{\frac{3}{2}}$$
$$= 1000 \cdot (\sqrt{2})^3$$
$$\approx 2828 \text{ people}$$

In general, exponential functions are defined as follows.

> **WORTHY OF NOTE**
>
> To properly understand the exponential function and its graph requires that we evaluate $f(x) = 2^x$ even when $x$ is *irrational*. For example, what does $2^{\sqrt{5}}$ mean? While the technical details require calculus, it can be shown that successive approximations of $2^{\sqrt{5}}$ as in $2^{2.2360}, 2^{2.23606}, 2^{2.236067}, \ldots$ approach a unique real number, and that $f(x) = 2^x$ exists for all real numbers $x$.

### Exponential Functions

For $b > 0$, $b \neq 1$, and all real numbers $x$,
$$f(x) = b^x$$
defines the base-$b$ exponential function.

Limiting $b$ to positive values ensures that outputs will be real numbers, and the restriction $b \neq 1$ is needed since $y = 1^x$ is a constant function (1 raised to *any* power is still 1). Specifically note the domain of an exponential function is *all real numbers,* and that all of the familiar properties of exponents still hold. A summary of these properties follows. For a complete review, see Section R.3.

### Exponential Properties

For real numbers $a$, $b$, $m$, and $n$, with $a, b > 0$,

$$b^m \cdot b^n = b^{m+n} \qquad \frac{b^m}{b^n} = b^{m-n} \qquad (b^m)^n = b^{mn}$$

$$(ab)^n = a^n \cdot b^n \qquad b^{-n} = \frac{1}{b^n} \qquad \left(\frac{b}{a}\right)^{-n} = \left(\frac{a}{b}\right)^n$$

**EXAMPLE 1** ▶  **Evaluating Exponential Functions**

Evaluate each exponential function for $x = 2$, $x = -1$, $x = \frac{1}{2}$, and $x = \pi$. Use a calculator for $x = \pi$, rounding to five decimal places.

   **a.** $f(x) = 4^x$         **b.** $g(x) = \left(\frac{4}{9}\right)^x$

**Solution** ▶   **a.** For $f(x) = 4^x$,                **b.** For $g(x) = \left(\frac{4}{9}\right)^x$,

$$f(2) = 4^2 = 16 \qquad\qquad g(2) = \left(\frac{4}{9}\right)^2 = \frac{16}{81}$$

$$f(-1) = 4^{-1} = \frac{1}{4} \qquad\qquad g(-1) = \left(\frac{4}{9}\right)^{-1} = \frac{9}{4}$$

$$f\left(\frac{1}{2}\right) = 4^{\frac{1}{2}} = \sqrt{4} = 2 \qquad g\left(\frac{1}{2}\right) = \left(\frac{4}{9}\right)^{\frac{1}{2}} = \sqrt{\frac{4}{9}} = \frac{2}{3}$$

$$f(\pi) = 4^{\pi} \approx 77.88023 \qquad g(\pi) = \left(\frac{4}{9}\right)^{\pi} \approx 0.07827$$

☑ **A.** You've just seen how we can evaluate an exponential function

Now try Exercises 5 through 8 ▶

## B. Graphing Exponential Functions

To gain a better understanding of exponential functions, we'll graph examples of $y = b^x$ and note some of the characteristic features. Since $b \neq 1$, it seems reasonable that we graph one exponential function where $b > 1$ and one where $0 < b < 1$.

**EXAMPLE 2** ▶ **Graphing Exponential Functions with $b > 1$**

Graph $y = 2^x$ using a table of values.

**Solution** ▶ To get an idea of the graph's shape we'll use integer values from $-3$ to $3$ in our table, then draw the graph as a continuous curve, since the function is defined for all real numbers.

| $x$ | $y = 2^x$ |
|-----|-----------|
| $-3$ | $2^{-3} = \frac{1}{8}$ |
| $-2$ | $2^{-2} = \frac{1}{4}$ |
| $-1$ | $2^{-1} = \frac{1}{2}$ |
| $0$ | $2^0 = 1$ |
| $1$ | $2^1 = 2$ |
| $2$ | $2^2 = 4$ |
| $3$ | $2^3 = 8$ |

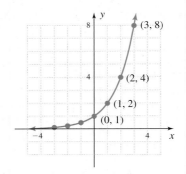

**Now try Exercises 9 and 10** ▶

**WORTHY OF NOTE**

As in Example 2, functions that are increasing for all $x \in D$ are said to be **monotonically increasing** or simply **monotonic functions**. The function in Example 3 is monotonically decreasing.

Several important observations can now be made. First note the $x$-axis (the line $y = 0$) is a horizontal asymptote for the function, because as $x \to -\infty$, $y \to 0$. Second, the function is increasing over its entire domain, giving the function a range of $y \in (0, \infty)$.

**EXAMPLE 3** ▶ **Graphing Exponential Functions with $0 < b < 1$**

Graph $y = \left(\frac{1}{2}\right)^x$ using a table of values.

**Solution** ▶ Using properties of exponents, we can write $\left(\frac{1}{2}\right)^x$ as $\left(\frac{2}{1}\right)^{-x} = 2^{-x}$. Again using integers from $-3$ to $3$, we plot the ordered pairs and draw a continuous curve.

| $x$ | $y = 2^{-x}$ |
|-----|--------------|
| $-3$ | $2^{-(-3)} = 2^3 = 8$ |
| $-2$ | $2^{-(-2)} = 2^2 = 4$ |
| $-1$ | $2^{-(-1)} = 2^1 = 2$ |
| $0$ | $2^0 = 1$ |
| $1$ | $2^{-1} = \frac{1}{2}$ |
| $2$ | $2^{-2} = \frac{1}{4}$ |
| $3$ | $2^{-3} = \frac{1}{8}$ |

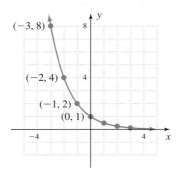

**Now try Exercises 11 and 12** ▶

We note this graph is also asymptotic to the $x$-axis, but *decreasing on its domain*. In addition, both $y = 2^x$ and $y = \left(\frac{1}{2}\right)^x = 2^{-x}$ have a $y$-intercept of $(0, 1)$ and both are one-to-one, which suggests that an inverse function can be found. Finally, observe that

$y = b^{-x}$ is *a reflection of* $y = b^x$ *across the y-axis,* a property that indicates these basic graphs might also be transformed in other ways, as were the other toolbox functions. The characteristics of exponential functions are summarized here:

---

### $f(x) = b^x, b > 0$ and $b \neq 1$

- one-to-one function
- domain: $x \in \mathbb{R}$
- increasing if $b > 1$
- $y$-intercept $(0, 1)$
- range: $y \in (0, \infty)$
- decreasing if $0 < b < 1$
- asymptotic to the $x$-axis (the line $y = 0$)

**Figure 5.11**

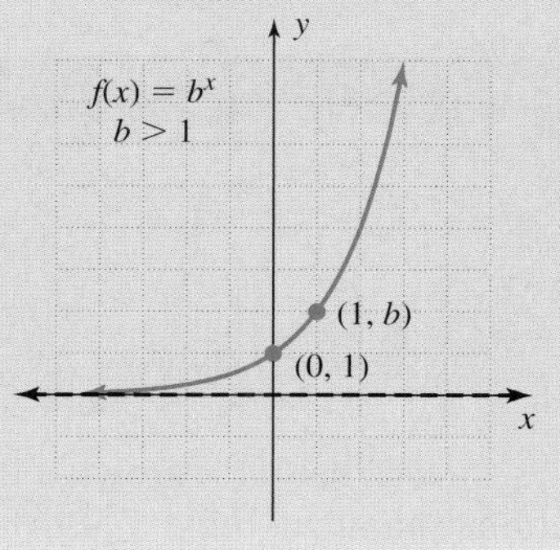

$f(x) = b^x$
$b > 1$

$(1, b)$
$(0, 1)$

**Figure 5.12**

$f(x) = b^x$
$0 < b < 1$

$(0, 1)$    $(1, b)$

---

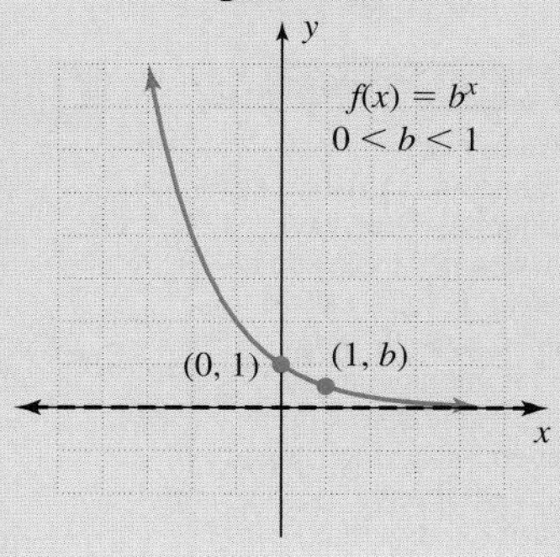

**WORTHY OF NOTE**

When an exponential function is increasing, it can be referred to as a "growth function." When decreasing, it is often called a "decay function." Each of the graphs shown in Figures 5.11 and 5.12 should now be added to your repertoire of basic functions, to be sketched from memory and analyzed or used as needed.

Just as the graph of a quadratic function maintains its parabolic shape regardless of the transformations applied, exponential functions will also maintain their general shape and features. Any sum or difference applied to the basic function ($y = b^x \pm k$ vs. $y = b^x$) will cause a vertical shift in the same direction as the sign, and any change to input values ($y = b^{x \pm h}$ vs. $y = b^x$) will cause a horizontal shift in a direction opposite the sign. For cases where multiple transformations are to be applied, refer to the sequence outlined on page 252.

---

**EXAMPLE 4** ▶ **Graphing Exponential Functions Using Transformations**

Graph $F(x) = 2^{x-1} + 2$ using transformations of the basic function $f(x) = 2^x$ (not by simply plotting points). Clearly state what transformations are applied.

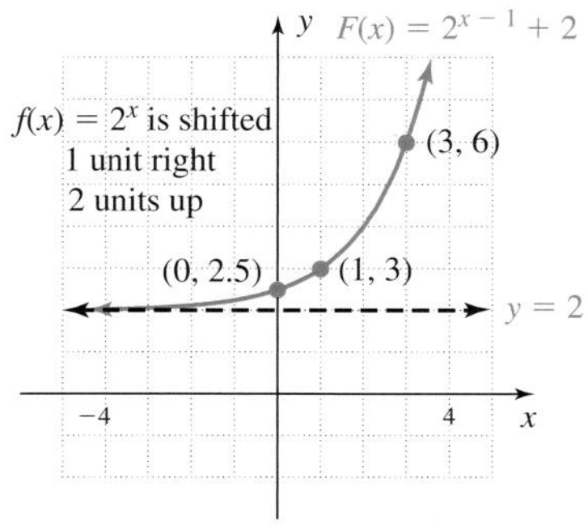

$y$  $F(x) = 2^{x-1} + 2$

$f(x) = 2^x$ is shifted 1 unit right 2 units up

$(3, 6)$

$(0, 2.5)$  $(1, 3)$

$y = 2$

$-4$    $4$    $x$

**Solution** ▶ Using the four-step process for graphing transformations (Section 3.1), we note that the graph of $F$ is that of the basic function $f(x) = 2^x$ with a horizontal shift 1 unit right and a vertical shift 2 units up. With this in mind the horizontal asymptote shifts from $y = 0$ to $y = 2$ and the $y$-intercept $(0, 1)$ from the basic graph becomes $(0 + 1, 1 + 2) = (1, 3)$. The $y$-intercept of $F$ is at $(0, 2.5)$:

$$F(0) = 2^{0-1} + 2$$
$$= 2^{-1} + 2$$
$$= \frac{1}{2} + 2$$
$$= 2.5$$

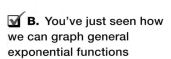 **B.** You've just seen how we can graph general exponential functions

To help sketch a more accurate graph, the point $(3, 6)$ can be used: $F(3) = 6$.

**Now try Exercises 13 through 28** ▶

## C. The Base-*e* Exponential Function: $f(x) = e^x$

In nature, exponential growth occurs when the instantaneous rate of change in a population is proportional to the current size of the population. Using the rate of change notation, $\frac{\Delta P}{\Delta t} = kP$, where $k$ is a constant and $\Delta t \to 0$. For the city of Goldsboro, we know the population at time $t$ is given by $P(t) = 1000 \cdot 2^t$, but have no information on this value of $k$ **(see Exercise 96)**. We can actually rewrite this function, and other exponential functions, using a base that gives the value of $k$ directly and without having to apply the difference quotient. This new base is an irrational number, symbolized by the letter $e$. In Section 5.6 we'll develop the number $e$ in the context of compound interest, while making numerous references to our discussion here, where we define $e$ as follows.

**WORTHY OF NOTE**

Just as the ratio of a circle's circumference to its diameter is an irrational number symbolized by $\pi$, the irrational number that results from $(1 + \frac{1}{x})^x$ for infinitely large $x$ is symbolized by $e$. Writing exponential functions in terms of $e$ simplifies many calculations in advanced courses, and offers additional advantages in applications of exponential functions.

### The Number *e*

$$\text{As } x \to \infty, \left(1 + \frac{1}{x}\right)^x \to e$$

In words, $e$ is the number that $(1 + \frac{1}{x})^x$ approaches as $x$ becomes infinitely large.

It can be shown that as $x$ grows without bound, $(1 + \frac{1}{x})^x$ indeed approaches a unique, irrational number. Table 5.5 gives approximate values of the expression for selected values of $x$, and shows $e \approx 2.71828$ to five decimal places.

This leads to the base-*e* **exponential function:** $f(x) = e^x$, also called the **natural exponential function.** Instead of having to enter a decimal approximation when computing with $e$, most calculators have an "$e^x$" key, usually as the (2nd) function for the key marked (LN). To find the value of $e^2$, use the keystrokes (2nd) (LN) 2 (⟩) (ENTER), and the calculator display should read 7.389056099. Note the calculator supplies the left parenthesis for the exponent, and you must supply the right. See Figure 5.13.

**Table 5.5**

| $x$ | Approximate Value $(1 + \frac{1}{x})^x$ |
|---|---|
| 10 | 2.59 |
| 100 | 2.705 |
| 1000 | 2.7169 |
| 10,000 | 2.71815 |
| 100,000 | 2.718268 |
| 1,000,000 | 2.7182805 |
| 10,000,000 | 2.71828169 |

**Figure 5.13**

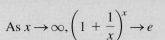

```
e^(2)
        7.389056099
```

---

**EXAMPLE 5** ▶  **Evaluating the Natural Exponential Function**

Use a calculator to evaluate $f(x) = e^x$ for the values of $x$ given. Round to six decimal places as needed.

  **a.** $f(3)$      **b.** $f(1)$      **c.** $f(0)$      **d.** $f(\frac{1}{2})$

**Solution** ▶  **a.** $f(3) = e^3 \approx 20.085537$    **b.** $f(1) = e^1 \approx 2.718282$
  **c.** $f(0) = e^0 = 1$ (exactly)    **d.** $f(\frac{1}{2}) = e^{\frac{1}{2}} \approx 1.648721$

**Now try Exercises 29 through 34** ▶

Although $e$ is an irrational number, the graph of $y = e^x$ has the same characteristics as other exponential graphs. Figure 5.14 shows this graph on the same grid as $y = 2^x$ and $y = 3^x$. As we might expect, all three graphs are increasing, have an asymptote at $y = 0$, and contain the point $(0, 1)$, with the graph of $y = e^x$ "between" the other two. The domain for all three functions, as with all basic exponential functions, is $x \in (-\infty, \infty)$ with range $y \in (0, \infty)$. The same transformations applied earlier can also be applied to the graph of $y = e^x$. **See Exercises 35 through 44.**

**Figure 5.14**

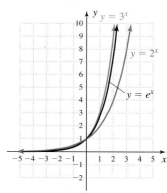

☑ **C.** You've just seen how we can graph base-$e$ exponential functions

## D. Solving Exponential Equations Using the Uniqueness Property

Since exponential functions are one-to-one, we can solve equations where each side is an exponential term with the identical base. This is because one-to-oneness guarantees a unique solution to the equation.

> **WORTHY OF NOTE**
>
> Exponential functions are very different from the power functions studied earlier. For power functions, the base is variable and the exponent is constant: $y = x^b$, while for exponential functions the *exponent is a variable* and the *base is constant*: $y = b^x$.

### Exponential Equations and the Uniqueness Property

For all real numbers $m$, $n$, and $b$, where $b > 0$ and $b \neq 1$,

**1.** If $b^m = b^n$, then $m = n$.

**2.** If $m \neq n$, then $b^m \neq b^n$

Equal bases imply equal exponents.

The equation $2^x = 32$ can be rewritten as $2^x = 2^5$, where we note $x = 5$ is a solution. Although $3^x = 32$ can be written as $3^x = 2^5$, the bases are not alike and the solution to this equation must wait until additional tools are developed in Section 5.5.

**EXAMPLE 6** ▶ **Solving Exponential Equations**

Solve the exponential equations using the uniqueness property.

**a.** $3^{2x-1} = 81$     **b.** $\left(\frac{1}{6}\right)^{-3x-2} = 36^{x+1}$     **c.** $e^x e^2 = \dfrac{e^4}{e^{x+1}}$

**Solution** ▶ **a.**

$$3^{2x-1} = 81 \qquad \text{given}$$
$$3^{2x-1} = 3^4 \qquad \text{rewrite using base 3}$$
$$\Rightarrow 2x - 1 = 4 \qquad \text{uniqueness property}$$
$$x = \frac{5}{2} \qquad \text{solve for } x$$

**Check** ▶

$$3^{2x-1} = 81 \qquad \text{given}$$
$$3^{2\left(\frac{5}{2}\right)-1} = 81 \qquad \text{substitute } \tfrac{5}{2} \text{ for } x$$
$$3^{5-1} = 81 \qquad \text{simplify}$$
$$3^4 = 81 \qquad \text{result checks}$$
$$81 = 81 \checkmark$$

The remaining checks are left to the student.

**b.** $\left(\dfrac{1}{6}\right)^{-3x-2} = 36^{x+1}$    given

$\left(6^{-1}\right)^{-3x-2} = \left(6^2\right)^{x+1}$    rewrite using base 6

$6^{3x+2} = 6^{2x+2}$    power property of exponents

$\Rightarrow 3x + 2 = 2x + 2$    uniqueness property

$x = 0$    solve for $x$

**c.** $e^x e^2 = \dfrac{e^4}{e^{x+1}}$    given

$e^{x+2} = e^{4-(x+1)}$    product property; quotient property

$e^{x+2} = e^{3-x}$    simplify

$\Rightarrow x + 2 = 3 - x$    uniqueness property

$2x = 1$    add $x$, subtract 2

$x = \dfrac{1}{2}$    solve for $x$

**Now try Exercises 45 through 64** ▶

Two common applications of exponential functions involve appreciation (as when collectibles grow in value over time), and depreciation (as when equipment decreases in value over time).

**EXAMPLE 7** ▶  **Applying an Exponential Function—Depreciation**

As part of their new buyback program, a major electronics retailer estimates that electronic devices lose 20% of their value every 6 months. The current buyback value can then be modeled by the function $V(t) = V_0(0.8)^{\frac{t}{6}}$, where $V_0$ is the initial value and $V(t)$ represents the value after $t$ months. How long will it take a tablet that cost $625 new to depreciate to $256?

**Solution** ▶  For this exercise, $V_0 = \$625$ and $V(t) = \$256$. The formula yields

$V(t) = V_0(0.8)^{\frac{t}{6}}$    given

$256 = 625\left(\dfrac{4}{5}\right)^{\frac{t}{6}}$    substitute known values

$\dfrac{256}{625} = \left(\dfrac{4}{5}\right)^{\frac{t}{6}}$    divide by 625

$\left(\dfrac{4}{5}\right)^4 = \left(\dfrac{4}{5}\right)^{\frac{t}{6}}$    equate bases; $\frac{256}{625} = \left(\frac{4}{5}\right)^4$

$\Rightarrow 4 = \dfrac{t}{6}$    Uniqueness Property

$24 = t$    multiply by 6; result

After 2 yr (24 months), the tablet's value will have dropped to $256.

**Now try Exercises 71 through 76** ▶

Another very practical application of the natural exponential function involves **Newton's law of cooling.** This law or formula models the temperature of an object as it cools down, as when a pizza is removed from the oven and placed on the kitchen counter. The function model is

$$T(x) = T_R + (T_0 - T_R)e^{kx}, \, k < 0$$

where $T_0$ represents the initial temperature of the object, $T_R$ represents the temperature of the room or surrounding medium, $T(x)$ is the temperature of the object $x$ min later, and $k$ is the cooling rate as determined by the physical properties of the object.

**EXAMPLE 8 ▶**     **Applying an Exponential Function—Newton's Law of Cooling**

A pizza is taken from a 425°F oven and placed on the counter to cool. If the temperature in the kitchen is 75°F, and the cooling rate for this type of pizza is $k = -0.35$,

   **a.** What is the temperature (to the nearest degree) of the pizza 2 min later?

   **b.** To the nearest minute, how long until the pizza has cooled to a temperature below 90°F?

   **c.** If Zack and Raef like to eat their pizza at a temperature of about 110°F, how many minutes should they wait to "dig in"?

**Solution ▶**     Begin by substituting the given values to obtain the equation model:

$$
\begin{aligned}
T(x) &= T_R + (T_0 - T_R)e^{kx} && \text{general equation model} \\
&= 75 + (425 - 75)e^{-0.35x} && \text{substitute 75 for } T_R, \text{425 for } T_0, \text{ and } -0.35 \text{ for } k \\
&= 75 + 350e^{-0.35x} && \text{simplify}
\end{aligned}
$$

For part (a) we simply find $T(2)$:

**a.** $T(2) = 75 + 350e^{-0.35(2)}$      substitute 2 for $x$

$\approx 249$      result

Two minutes later, the temperature of the pizza is near 249°F.

**b.** In Figure 5.15, we see that the **TABLE** feature of a graphing calculator shows the pizza reaches a temperature of just under 90° after 9 min: $T(9) \approx 90°F$.

**Figure 5.15**

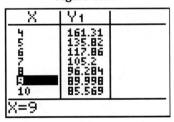

**c.** We elect to use the intersection-of-graphs method. After setting an appropriate window, we enter $Y_1 = 75 + 350e^{-0.35x}$ and $Y_2 = 110$, then press **2nd** **TRACE** **5:intersect.** After pressing [ENTER] three times, the coordinates of the point of intersection appear at the bottom of the screen in Figure 5.16: $x \approx 6.6, y = 110$. It appears the boys should wait about $6\frac{1}{2}$ min for the pizza to cool.

**Figure 5.16**

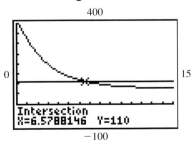

**Now try Exercises 77 and 78 ▶**

We now focus our attention on regression models that involve exponential functions. Recall the process of developing a regression equation involves these five stages: (1) clearing old data, (2) entering new data, (3) displaying the data, (4) calculating the regression equation, and (5) displaying and using the regression graph and equation.

**EXAMPLE 9** ▶ **Calculating an Exponential Regression Model**

The number of centenarians (people who are 100 yr of age or older) has been climbing steadily over the last half century. The table shows the number of centenarians (per million population) for selected years. Use the data and a graphing calculator to draw the scatterplot, then use the scatterplot and context to decide on an appropriate form of regression.

Source: Data from 2004 *Statistical Abstract of the United States,* Table 14; various other years

| Year (1950 → 0) | Number (per million) |
|---|---|
| 0 | 16 |
| 10 | 18 |
| 20 | 25 |
| 30 | 74 |
| 40 | 115 |
| 50 | 262 |

**Solution** ▶ After clearing any existing data in the data lists, enter the input values (years since 1950) in L1 and the output values (number of centenarians per million population) in L2 (Figure 5.17). For the viewing window, scale the x-axis (years since 1950) from −10 to 70 and the y-axis (number per million) from −50 to 300 to comfortably fit the data and allow room for the coordinates to be shown at the bottom of the screen (Figure 5.18). The scatterplot rules out a linear model. While a quadratic model may fit the data, we expect that the correct model should exhibit asymptotic behavior since extremely few people lived to be 100 yr of age prior to dramatic advances in hygiene, diet, and medical care. This would lead us toward an exponential equation model. The keystrokes **STAT** ▷ brings up the **CALC** menu, with **ExpReg** (exponential regression) being option "0."

**Figure 5.17**

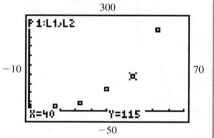

**Figure 5.18**

The option can be selected by simply pressing "0," or by using the up arrow △ or down arrow ▽ to scroll to **0:ExpReg** then pressing ◯. The exponential model seems to fit the data very well (Figures 5.19 and 5.20). To four decimal places the equation model is $y = (11.5090)1.0607^x$.

**Figure 5.19**

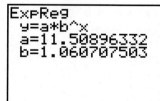

**Figure 5.20**

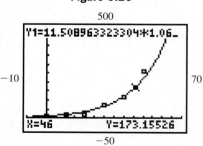

☑ **D.** You've just seen how we can solve exponential equations and applications

Now try Exercises 87 through 90 ▶

## 5.2 EXERCISES

### ▶ CONCEPTS AND VOCABULARY

**Fill in each blank with the appropriate word or phrase. Carefully reread the section, if necessary.**

1. The domain of $y = b^x$ is all _____ _____, and the range is $y \in$ _____. Further, as $x \to -\infty$, $y$ _____.

2. If each side of an equation can be written as an exponential term with the same base, the equation can be solved using the _____ _____.

3. State true or false and explain why: $y = b^x$ is always increasing if $0 < b < 1$.

4. Discuss/Explain the statement, "For $k > 0$, the $y$-intercept of $y = ab^x + k$ is $(0, a + k)$."

### ▶ DEVELOPING YOUR SKILLS

**Evaluate each function as indicated. Use a calculator only as needed, rounding results to thousandths.**

5. $P(t) = 4^t$;
   $t = 2, t = \frac{1}{2}$,
   $t = \frac{3}{2}, t = \sqrt{3}$

6. $Q(t) = 8^t$;
   $t = 2, t = \frac{1}{3}$,
   $t = \frac{5}{3}, t = \sqrt{5}$

7. $V(n) = \left(\frac{1}{8}\right)^n$;
   $n = 0, n = 2$,
   $n = \frac{2}{3}, n = -2$

8. $W(m) = \left(\frac{4}{9}\right)^m$;
   $m = 0, m = 3$,
   $m = \frac{3}{2}, m = -2$

**Graph each function using a table of values and integer inputs between $-3$ and $3$. Clearly label the $y$-intercept and one additional point, then state whether the function is increasing or decreasing.**

9. $f(x) = 3^x$

10. $g(x) = 4^x$

11. $p(x) = \left(\frac{1}{3}\right)^x$

12. $q(x) = \left(\frac{1}{4}\right)^x$

**Graph each function *using transformations* of $y = b^x$, sketching the asymptote, and strategically plotting a few points to round out the graph. Clearly state the transformations applied.**

13. $y = 3^x + 2$

14. $y = 3^x - 3$

15. $y = 3^{x+3}$

16. $y = 3^{x-2}$

17. $y = 3^{-x}$

18. $y = 3^{-x} - 2$

19. $y = \left(\frac{1}{3}\right)^x + 1$

20. $y = \left(\frac{1}{3}\right)^x - 4$

21. $y = \left(\frac{1}{3}\right)^{x-2}$

22. $y = \left(\frac{1}{3}\right)^{x+2}$

**Match each exponential equation to the correct graph.**

23. $y = 5^{-x}$

24. $y = 4^{-x}$

25. $y = 3^{-x+1}$

26. $y = 3^{-x} + 1$

27. $y = 2^{x+1} - 2$

28. $y = 2^{x+2} - 1$

a.
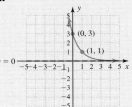

b.
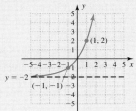

c.
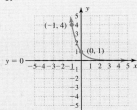

d.
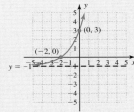

e.
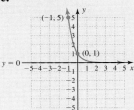

f.
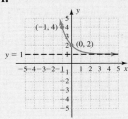

**Use a calculator to evaluate each expression, rounded to six decimal places.**

29. $e^1$

30. $e^0$

31. $e^2$

32. $e^{-3.2}$

33. $e^{\sqrt{2}}$

34. $e^{\pi}$

**Graph each exponential function.**

35. $F(x) = e^x - 2$

36. $G(x) = e^x + 1$

37. $m(t) = e^{t+4}$

38. $n(t) = e^{t-3}$

39. $f(x) = e^{x+3} - 2$

40. $g(x) = e^{x-2} + 1$

41. $r(t) = -e^t + 2$

42. $s(t) = -e^{t+2}$

43. $p(x) = e^{-x+2} - 1$

44. $q(x) = e^{-x-1} + 2$

**Solve each exponential equation and check your answer by substituting into the original equation.**

**45.** $10^x = 1000$       **46.** $144 = 12^x$

**47.** $25^x = 125$       **48.** $81 = 27^x$

**49.** $8^{x+2} = 32$       **50.** $9^{x-1} = 27$

**51.** $32^x = 16^{x+1}$       **52.** $100^{x+2} = 1000^x$

**53.** $(\frac{1}{5})^x = 125$       **54.** $(\frac{1}{4})^x = 64$

**55.** $(\frac{1}{3})^{2x} = 9^{x-6}$       **56.** $(\frac{1}{2})^{3x} = 8^{x-2}$

**57.** $(\frac{1}{9})^{x-5} = 3^{3x}$       **58.** $2^{-2x} = (\frac{1}{32})^{x-3}$

**59.** $25^{3x} = 125^{x-2}$       **60.** $27^{2x+4} = 9^{4x}$

**61.** $\dfrac{e^4}{e^{2-x}} = e^3 e$       **62.** $e^x(e^x + e) = \dfrac{e^x + e^{3x}}{e^{-x}}$

**63.** $(e^{2x-4})^3 = \dfrac{e^{x+5}}{e^2}$       **64.** $e^x e^{x+3} = (e^{x+2})^3$

**Estimate the answer to the following equations by bounding it between two integers.**

**65.** $3^x = 22$       **66.** $2^x = 2.125$

**67.** $e^{x-1} = 9$       **68.** $e^{0.5x} = 8$

## ▶ WORKING WITH FORMULAS

**69. The growth of bacteria:** $P(t) = 1000 \cdot 3^t$

If the initial population of a common bacterium is 1000 and the population triples every day, its population is given by the formula shown, where $P(t)$ is the total population after $t$ days. (a) Find the total population 12 hr, 1 day, $1\frac{1}{2}$ days, and 2 days later. (b) Do the outputs show the population is tripling every 24 hr (1 day)? (c) Explain why this is an increasing function. (d) Graph the function using an appropriate scale.

**70. Spinners with numbers 1 through 4:** $P(x) = (\frac{1}{4})^x$

Games that involve moving pieces around a board using a fair spinner are fairly common. If the spinner has the numbers 1 through 4, the probability that any one number is spun repeatedly is given by the formula shown, where $x$ represents the number of spins and $P(x)$ represents the probability the same number results $x$ times. (a) What is the probability that the first player spins a 2? (b) What is the probability that all four players spin a 2? (c) Explain why this is a decreasing function.

## ▶ APPLICATIONS

**71. Construction equipment:** The financial analyst for a large construction firm estimates that its heavy equipment loses one-fifth of its value each year. The  current value of the equipment is then modeled by the function $V(t) = V_0(\frac{4}{5})^t$, where $V_0$ represents the initial value, $t$ is in years, and $V(t)$ represents the value after $t$ yr. (a) How much is a large earthmover worth after 1 yr if it cost $125 thousand new? (b) How many years does it take for the earthmover to depreciate to a value of $64 thousand?

**72. Office equipment:** Photocopiers have become a critical part of the operation of many businesses, and due to their heavy use they can depreciate in value very quickly. If a copier loses three-eighths of its value each year, the current value of the copier can be modeled by the function $V(t) = V_0(\frac{5}{8})^t$, where $V_0$ represents the initial value, $t$ is in years, and $V(t)$ represents the value after $t$ yr. (a) How much is this copier worth after 1 yr if it cost $64 thousand new? (b) How many years does it take for the copier to depreciate to a value of $25 thousand?

**73. Medical equipment:** Margaret Madison, DDS, estimates that her dental equipment loses one-sixth of its value each year. (a) Determine the value of an x-ray machine after 5 yr if it cost $216 thousand new, and (b) determine how long until the machine is worth less than $125 thousand.

**74. Field deterioration:** The groundskeeper of a local high school estimates that due to heavy usage by the baseball and softball teams, the pitcher's mound loses one-fifth of its height every month.  Use this information to find an equation that models this information and then determine: (a) If the mound was 25 cm to begin, how long until the height becomes less than 16 cm high (meaning it must be entirely rebuilt)? (b) If the mound were allowed to deteriorate further, find the height after 3 months.

**75. Business revenue:** Similar to a small town doubling in size after a discovery of gold, a business that develops a product in high demand has the potential for doubling its revenue each year for a number of years. The revenue would be modeled by the function $R(t) = R_0 2^t$, where $R_0$ represents the initial revenue, and $R(t)$ represents the revenue after $t$ yr. (a) How much revenue is being generated after 4 yr, if the company's initial revenue was $2.5 million? (b) How many years does it take for the business to be generating $320 million in revenue?

**76. Business revenue:** If a company's revenue grows at a rate of 150% per year (rather than doubling as in Exercise 75), the revenue would be modeled by the function $R(t) = R_0(\frac{3}{2})^t$, where $R_0$ represents the initial revenue, and $R(t)$ represents the revenue after $t$ yr. (a) How much revenue is being generated after 3 yr, if the company's initial revenue was $256 thousand? (b) How long until the business is generating $1944 thousand in revenue? (*Hint:* Reduce the fraction.)

Use Newton's law of cooling to complete Exercises 77 and 78: $T(x) = T_R + (T_0 - T_R)e^{kx}$.

**77. Cold party drinks:** Janae was late getting ready for the party, and the liters of soft drinks she bought were still at room temperature (73°F) with guests due to arrive in 15 min. If she puts these in her freezer at −10°F, will the drinks be cold enough (35°F) for her guests? Assume $k \approx -0.031$.

**78. Warm party drinks:** Newton's law of cooling applies equally well if the "cooling is negative," meaning the object is taken from a colder medium and placed in a warmer one. If a can of soft drink is taken from a 35°F cooler and placed in a room where the temperature is 75°F, how long will it take the drink to warm to 65°F? Assume $k \approx -0.031$.

**Photochromatic sunglasses:** Sunglasses that darken in sunlight (photochromatic sunglasses) contain millions of molecules of a substance known as *silver halide*. The molecules are transparent indoors in the absence of ultraviolet (UV) light. Outdoors, UV light from the sun causes the molecules to change shape, darkening the lenses in response to the intensity of the UV light. For certain lenses, the function $T(x) = 0.85^x$ models the transparency of the lenses (as a percentage) based on a UV index $x$. Find the transparency (to the nearest percent), if the lenses are exposed to

**79.** sunlight with a UV index of 7 (a high exposure).

**80.** sunlight with a UV index of 5.5 (a moderate exposure).

**81.** Given that a UV index of 11 is very high and most individuals should stay indoors, what is the minimum transparency percentage for these lenses?

**82.** Use a trial-and-error process and a graphing calculator to determine the UV index when the lenses are 50% transparent.

**Modeling inflation:** Assuming the rate of inflation is 5% per year, the predicted price of an item can be modeled by the function $P(t) = P_0(1.05)^t$, where $P_0$ represents the initial price of the item and $t$ is in years. Use this information to solve Exercises 83 and 84.

**83.** What will the price of a new car be in the year 2015, if it cost $20,000 in the year 2010?

**84.** What will the price of a gallon of milk be in the year 2015, if it cost $3.95 in the year 2010? Round to the nearest cent.

**Modeling radioactive decay:** The half-life of a radioactive substance is the time required for half an initial amount of the substance to disappear through decay. The amount of the substance remaining is given by the formula $Q(t) = Q_0(\frac{1}{2})^{\frac{t}{h}}$, where $h$ is the half-life, $t$ represents the elapsed time, and $Q(t)$ represents the amount that remains ($t$ and $h$ must have the same unit of time). Use this information to solve Exercises 85 and 86.

**85.** Some isotopes of the substance known as thorium have a half-life of only 8 min. (a) If 64 grams are initially present, how many grams (g) of the substance remain after 24 min? (b) How many minutes until only 1 g of the substance remains?

**86.** Some isotopes of sodium have a half-life of about 16 hr. (a) If 128 g are initially present, how many grams of the substance remain after 2 days (48 hr)? (b) How many hours until only 1 g of the substance remains?

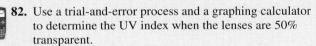 **Answer the questions using the given data and the related regression equation. All extrapolations assume the mathematical model will continue to represent future trends.**

**87. Milk production:** Since 1980, the number of family farms with milk cows for commercial production has been decreasing. Use the data from the table given to draw a scatterplot, then use the context and scatterplot to find the regression equation.

*Source: Statistical Abstract of the United States, 2011.*

**a.** What was the approximate number of farms with milk cows in 2003?

**b.** Approximately how many farms will have milk cows in 2015?

**c.** In what year will the number of farms drop below 30 thousand?

| **Exercise 87** | |
|---|---|
| **Year (1980 → 0)** | **Number (in 1000s)** |
| 0 | 334 |
| 5 | 269 |
| 10 | 193 |
| 15 | 140 |
| 20 | 105 |
| 25 | 78 |

| **Exercise 88** | |
|---|---|
| **Time (sec)** | **Height of Froth (in.)** |
| 0 | 0.90 |
| 2 | 0.65 |
| 4 | 0.40 |
| 6 | 0.21 |
| 8 | 0.15 |
| 10 | 0.12 |
| 12 | 0.08 |

**88. Froth height—carbonated beverages:** The height of the froth on carbonated drinks and other beverages can be manipulated by the ingredients used in making the beverage. The data in the table given show the froth height of a certain beverage as a function of time, after the froth has reached a maximum height. Use the data to draw a scatterplot, then use the context and scatterplot to find the regression equation.

**a.** What was the approximate height of the froth after 6.5 sec?

**b.** How long does it take for the height of the froth to reach one-half of its maximum height?

**c.** According to the model, how many seconds until the froth height is 0.02 in.?

**89. Musical notes:** The table shown gives the frequency (vibrations per second for each of the twelve notes in a selected octave) from the standard chromatic scale. Use the data to draw a scatterplot, then use the context and scatterplot to find the regression equation.

| # | Note | Frequency |
|---|------|-----------|
| 1 | A | 110.00 |
| 2 | A# | 116.54 |
| 3 | B | 123.48 |
| 4 | C | 130.82 |
| 5 | C# | 138.60 |
| 6 | D | 146.84 |
| 7 | D# | 155.56 |
| 8 | E | 164.82 |
| 9 | F | 174.62 |
| 10 | F# | 185.00 |
| 11 | G | 196.00 |
| 12 | G# | 207.66 |

  **a.** What is the frequency of the "A" note that is an octave higher than the one shown? [*Hint:* The names repeat every 12 notes (one octave), so this would be the 13th note in this sequence.]

  **b.** If the frequency is 370.00 what note is being played?

  **c.** What pattern do you notice for the F#'s in each octave (the 10th, 22nd, 34th, and 46th notes in sequence)? Does the pattern hold for all notes?

**90. Cost of cable service:** The average monthly cost of cable TV has been rising steadily since it became very popular in the early 1980s. The data given show the average monthly rate for selected years (1980 → 0). Use the data to draw a scatterplot, then use the context and scatterplot to find the regression equation. According to the model, what will be the cost of cable service in 2010? 2015?

*Source:* 2004–2005 *Statistical Abstract of the United States*, Table 1138.

| Year (1980 → 0) | Monthly Charge |
|-----------------|----------------|
| 0 | $7.69 |
| 5 | $9.73 |
| 10 | $16.78 |
| 20 | $23.07 |
| 25 | $30.70 |

## ▶ EXTENDING THE CONCEPT

**91.** If $10^{2x} = 25$, what is the value of $10^{-x}$?

**92.** If $5^{3x} = 27$, what is the value of $5^{2x}$?

**93.** If $3^{0.5x} = 5$, what is the value of $3^{x+1}$?

**94.** If $\left(\frac{1}{2}\right)^{x+1} = \frac{1}{3}$, what is the value of $\left(\frac{1}{2}\right)^{-x}$?

**95.** The formula $f(x) = \left(\frac{1}{2}\right)^x$ gives the probability that "$x$" number of flips result in heads (or tails). First determine the probability that 20 flips results in *20 heads in a row*. Then use the Internet or some other resource to determine the probability of winning a state lottery (expressed as a decimal). Which has the greater probability? Were you surprised?

 **The growth rate constant that governs an exponential function was introduced on page 442.**

**96.** In later sections, we will easily be able to find the growth constant $k$ for Goldsboro, where $P(t) = 1000 \cdot 2^t$. For now we'll approximate its value using the rate of change formula on a very small interval of the domain. From the definition of an exponential function, it can be shown that $\frac{\Delta P}{\Delta t} = kP(t)$ as $\Delta t \to 0$. Since $k$ is constant, we can choose any value of $t$, say $t = 4$. For $h = 0.0001$, we have $\frac{P(4 + 0.0001) - P(4)}{0.0001} = k \cdot P(4)$. (a) Use the equation shown to solve for $k$ (round to thousandths). (b) Show that $k$ appears constant by completing the same exercise for $t = 2$ and $t = 6$. (c) Verify that $P(t) = 1000 \cdot 2^t$ and $P(t) = 1000e^{kt}$ give approximately the same results.

## ▶ MAINTAINING YOUR SKILLS

**97.** (2.4) Given $f(x) = 2x^2 - 3x$, determine: $f(-1)$, $f(\frac{1}{3})$, $f(a)$, $f(a + h)$

**98.** (4.6) Determine the domain of each function.

  **a.** $p(x) = \sqrt{2x + 3}$

  **b.** $q(x) = \sqrt[4]{x^2 - 2x}$

  **c.** $r(x) = \left(\frac{3 - x}{x + 3}\right)^{\frac{1}{2}}$

**99.** (1.6) Solve the following equations:

  **a.** $-2\sqrt{x - 3} + 7 = 21$

  **b.** $\frac{9}{x + 3} + 3 = \frac{12}{x - 3}$

**100.** (1.1) Identify each formula:

  **a.** $\frac{4}{3}\pi r^3$

  **b.** $\frac{1}{2}bh$

  **c.** $lwh$

  **d.** $a^2 + b^2 = c^2$

A **transcendental function** is one whose solutions are beyond or *transcend* the methods applied to polynomial functions. The exponential function and its inverse, called the logarithmic function, are transcendental functions. In this section, we'll use the concept of an inverse to develop an understanding of the logarithmic function, which has numerous applications that include measuring pH levels, sound and earthquake intensities, barometric pressure, and other natural phenomena.

## A. Exponential Equations and Logarithmic Form

While exponential functions have a large number of significant applications, we can't appreciate their full value until we develop the inverse function. Without it, we're unable to solve all but the simplest equations, of the type encountered in Section 5.2. Using the fact that $f(x) = b^x$ is one-to-one, we have the following:

1. The function $f^{-1}(x)$ must exist.
2. We can graph $f^{-1}(x)$ by interchanging the $x$- and $y$-coordinates of points from $f(x)$.
3. The domain of $f(x)$ will become the range of $f^{-1}(x)$.
4. The range of $f(x)$ will become the domain of $f^{-1}(x)$.
5. The graph of $f^{-1}(x)$ will be a reflection of $f(x)$ across the line $y = x$.

Table 5.6 contains selected values for $f(x) = 2^x$. The values for $f^{-1}(x)$ in Table 5.7 were found by interchanging $x$- and $y$-coordinates. Both functions were then graphed using these values.

**Table 5.6**
$f(x) = 2^x$

| $x$ | $y = f(x)$ |
|---|---|
| $-3$ | $\frac{1}{8}$ |
| $-2$ | $\frac{1}{4}$ |
| $-1$ | $\frac{1}{2}$ |
| $0$ | $1$ |
| $1$ | $2$ |
| $2$ | $4$ |
| $3$ | $8$ |

**Table 5.7**
$f^{-1}(x) = ?$

| $x$ | $y = f^{-1}(x)$ |
|---|---|
| $\frac{1}{8}$ | $-3$ |
| $\frac{1}{4}$ | $-2$ |
| $\frac{1}{2}$ | $-1$ |
| $1$ | $0$ |
| $2$ | $1$ |
| $4$ | $2$ |
| $8$ | $3$ |

**Figure 5.21**

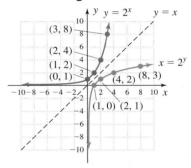

The interchange of $x$ and $y$ and the graphs in Figure 5.21 show that $f^{-1}(x)$ has an $x$-intercept of $(1, 0)$, a vertical asymptote at $x = 0$, a domain of $x \in (0, \infty)$, and a range of $y \in (-\infty, \infty)$. To find *an equation* for $f^{-1}(x)$, we'll attempt to use the algebraic approach employed previously. For $f(x) = 2^x$,

1. replace $f(x)$ with $y$: $y = 2^x$.   2. interchange $x$ and $y$: $x = 2^y$.

At this point we have an *implicit* equation for the inverse function, but no algebraic operations that enable us to solve *explicitly* for $y$ in terms of $x$. Instead, we write $x = 2^y$ in function form by noting that "$y$ is the exponent that goes on base 2 to obtain $x$." In the language of mathematics, this phrase is represented by $y = \log_2 x$ and is called a **logarithmic function** with base 2. For example, from Table 5.7 we have: $-3$ is the exponent that goes on base 2 to get $\frac{1}{8}$, and this is written $-3 = \log_2 \frac{1}{8}$ since $2^{-3} = \frac{1}{8}$ ✓. For $y = b^x$, $y = \log_b x$ is the inverse function, and is read, "$y$ is the logarithm base $b$ of $x$." For this new function, we must always keep in mind what $y$ *represents*— $y$ is an exponent. In fact, $y$ *is the exponent that goes on base $b$ to obtain $x$*: $y = \log_b x$.

**WORTHY OF NOTE**

The word *logarithm* was coined by John Napier in 1614, and loosely translated from its Greek origins means "to reason with numbers."

## Logarithmic Functions

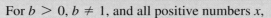

For $b > 0$, $b \neq 1$, and all positive numbers $x$,

$$y = \log_b x \text{ if and only if } x = b^y$$

The function $f(x) = \log_b x$ is a logarithmic function with base $b$. The expression $\log_b x$ is simply called a logarithm, and represents the exponent on $b$ that yields $x$. Finally, note the equations $x = b^y$ and $y = \log_b x$ are equivalent. We say that $x = b^y$ is the **exponential form** of the equation, whereas $y = \log_b x$ is written in **logarithmic form.** Of all possible bases for $\log_b x$, the most common are base 10 (likely due to our base-10 number system), and base $e$ (due to the advantages it offers in advanced courses). The expression $\log_{10} x$ is called a **common logarithm,** and we simply write $\log x$ for $\log_{10} x$. The expression $\log_e x$ is called a **natural logarithm,** and is written in abbreviated form as $\ln x$. Note specifically this shows that

$$y = \log x \text{ is equivalent to } 10^y = x, \text{ and}$$

$$y = \ln x \text{ is equivalent to } e^y = x.$$

### WORTHY OF NOTE

Since base-10 logarithms occur so frequently, we usually use only log $x$ to represent $\log_{10} x$. We do something similar with square roots. Technically, the "square root of $x$" should be written $\sqrt[2]{x}$. However, square roots are so common we often leave off the two, assuming that if no index is written, an index of two is intended.

---

**EXAMPLE 1** ▶ **Converting from Logarithmic Form to Exponential Form**

Write each equation in words, then in exponential form.

    **a.** $3 = \log_2 8$      **b.** $1 = \log 10$      **c.** $0 = \ln 1$      **d.** $-2 = \log_3 \frac{1}{9}$

Solution ▶    **a.**   $3 = \log_2 8 \Rightarrow$   3 is the exponent that goes on base 2 to obtain 8:   $2^3 = 8$.

         **b.**   $1 = \log 10 \Rightarrow$   1 is the exponent that goes on base 10 to obtain 10: $10^1 = 10$.

         **c.**   $0 = \ln 1 \Rightarrow$   0 is the exponent that goes on base $e$ to obtain 1:   $e^0 = 1$.

         **d.**   $-2 = \log_3 \frac{1}{9} \Rightarrow$ $-2$ is the exponent that goes on base 3 to obtain $\frac{1}{9}$: $3^{-2} = \frac{1}{9}$.

**Now try Exercises 5 through 20** ▶

To convert from exponential form to logarithmic form, note the exponent on the base and read from there. For $5^3 = 125$, "3 is the exponent that goes on base 5 to obtain 125," or *3 is the logarithm base 5 of 125*: $3 = \log_5 125$.

---

**EXAMPLE 2** ▶ **Converting from Exponential Form to Logarithmic Form**

Write each equation in words, then in logarithmic form.

    **a.** $10^3 = 1000$      **b.** $2^{-1} = \frac{1}{2}$      **c.** $e^2 \approx 7.389$      **d.** $9^{\frac{3}{2}} = 27$

Solution ▶    **a.**   $10^3 = 1000 \Rightarrow$   3 is the exponent that goes on base 10 to obtain 1000, or 3 is the logarithm base 10 of 1000: $3 = \log 1000$.

         **b.**   $2^{-1} = \frac{1}{2}$    $\Rightarrow -1$ is the exponent that goes on base 2 to obtain $\frac{1}{2}$, or $-1$ is the logarithm base 2 of $\frac{1}{2}$: $-1 = \log_2 \frac{1}{2}$.

         **c.**   $e^2 \approx 7.389 \Rightarrow$   2 is the exponent that goes on base $e$ to obtain 7.389, or 2 is the logarithm base $e$ of 7.389: $2 \approx \ln 7.389$.

         **d.**   $9^{\frac{3}{2}} = 27$    $\Rightarrow$   $\frac{3}{2}$ is the exponent that goes on base 9 to obtain 27, or $\frac{3}{2}$ is the logarithm base 9 of 27: $\frac{3}{2} = \log_9 27$.

☑ **A.** You've just seen how we can write exponential equations in logarithmic form

**Now try Exercises 21 through 36** ▶

### B. Finding Common Logarithms and Natural Logarithms

Some logarithms are easy to evaluate. For example, $\log 100 = 2$ since $10^2 = 100$, and $\log \frac{1}{100} = -2$ since $10^{-2} = \frac{1}{100}$. But what about the expressions $\log 850$ and $\ln 4$? Because logarithmic functions are continuous on their domains, a value exists for $\log 850$ and the equation $10^x = 850$ must have a solution. Further, the inequalities

$$\log 100 < \log 850 < \log 1000$$

$$2 < \log 850 < 3$$

tell us that $\log 850$ must be between 2 and 3. Fortunately, modern calculators can compute base-10 and base-$e$ logarithms instantly, often with nine-decimal-place accuracy. For $\log 850$, press $\boxed{\text{LOG}}$, then input 850 and press $\boxed{\phantom{)}}$. The display should read 2.929418926. We can also use the calculator to verify $10^{2.929418926} \approx 850$ (see Figure 5.22). For $\ln 4$, press the $\boxed{\text{LN}}$ key, then input 4 and press $\boxed{\phantom{)}}$ to obtain 1.386294361. Figure 5.23 verifies that $e^{1.386294361} \approx 4$.

**WORTHY OF NOTE**

For functions like log(X), e^(X), and $\sqrt{}$(X) the calculator supplies the left parenthesis automatically— you need only enter the argument and close the group with a right parenthesis. This is especially important for times when the argument contains a sum.

**Figure 5.22**

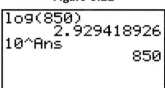

```
log(850)
         2.929418926
10^Ans
                  850
```

**Figure 5.23**

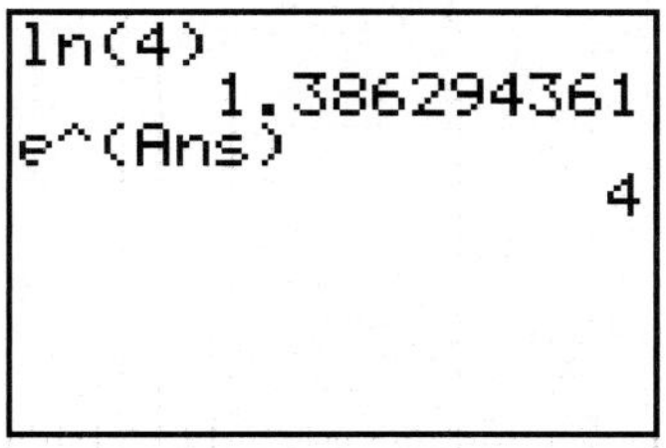

```
ln(4)
         1.386294361
e^(Ans)
                   4
```

---

**EXAMPLE 3** ▶ **Finding the Value of a Logarithm**

Determine the value of each logarithm without using a calculator:

   **a.** $\log_2 8$     **b.** $\log_5 \frac{1}{25}$     **c.** $\ln e$     **d.** $\log \sqrt{10}$

**Solution** ▶   **a.** $\log_2 8$ represents the exponent that goes on 2 to obtain 8: $\log_2 8 = 3$, since $2^3 = 8$.

    **b.** $\log_5 \frac{1}{25}$ represents the exponent that goes on 5 to obtain $\frac{1}{25}$: $\log_5 \frac{1}{25} = -2$, since $5^{-2} = \frac{1}{25}$.

    **c.** $\ln e$ represents the exponent that goes on $e$ to obtain $e$: $\ln e = 1$, since $e^1 = e$.

    **d.** $\log \sqrt{10}$ represents the exponent that goes on 10 to obtain $\sqrt{10}$: $\log \sqrt{10} = \frac{1}{2}$, since $10^{\frac{1}{2}} = \sqrt{10}$.

**Now try Exercises 37 through 50** ▶

---

**EXAMPLE 4** ▶ **Using a Calculator to Find Logarithms**

Use a calculator to evaluate each logarithmic expression. Verify the result.

   **a.** $\log 1857$     **b.** $\log 0.258$     **c.** $\ln 3.592$

**Solution** ▶   **a.** $\log 1857 \approx 3.268811904$,       **c.** $\ln 3.592 \approx 1.27870915$
      $10^{3.268811904} \approx 1857$ ✓        $e^{1.27870915} \approx 3.592$ ✓

    **b.** $\log 0.258 \approx -0.588380294$,
      $10^{-0.588380294} \approx 0.258$ ✓

☑ **B.** You've just seen how we can find common logarithms and natural logarithms

**Now try Exercises 51 through 58** ▶

Finally, note that if $x > b$, the value of $\log_b x$ is greater than 1, but if $0 < x < b$ the value of $\log_b x$ is less than 1. Also, if $x < 0$, the expression $\log_b x$ does not represent a real number (the domain of $y = \log_b x$ does not include negative numbers). See Figures 5.24 and 5.25.

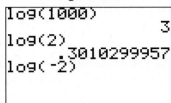

**Figure 5.24**

**Figure 5.25**

## C. Graphing Logarithmic Functions

For convenience and ease of calculation, our first examples of logarithmic graphs are done using base-2 logarithms. However, the basic shape of a logarithmic graph remains unchanged regardless of the base used, and transformations can be applied to $y = \log_b(x)$ for any value of $b$. For $y = a \log(x \pm h) \pm k$, $a$ continues to govern stretches, compressions, and vertical reflections, the graph will shift horizontally $h$ units opposite the sign, and shift $k$ units vertically in the same direction as the sign. Our earlier graph of $y = \log_2 x$ was completed using $x = 2^y$ as the inverse function for $y = 2^x$ (Figure 5.21). For reference, the graph is repeated in Figure 5.26.

**Figure 5.26**

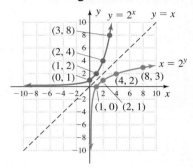

> **WORTHY OF NOTE**
>
> As with the basic graphs we studied in Section 3.1, logarithmic graphs maintain the same characteristics when transformations are applied, and these graphs *should be added to your collection of basic functions,* ready for recall or analysis as the situation requires.

**EXAMPLE 5** ▶ **Graphing Logarithmic Functions Using Transformations**

Graph $f(x) = \log_2(x - 3) + 1$ using transformations of $y = \log_2 x$ (not by simply plotting points). Clearly state what transformations are applied.

**Solution** ▶ Using the four-step process for graphing transformations (Section 3.1), we note that the graph of $f$ is the same as that of $y = \log_2 x$, but shifted 3 units right and 1 unit up. The vertical asymptote will be at $x = 3$ and the $x$-intercept $(1, 0)$ from the basic graph becomes $(1 + 3, 0 + 1) = (4, 1)$. Knowing the graph's basic shape, we compute one additional point using $x = 7$:

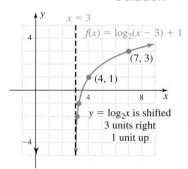

$$f(7) = \log_2(7 - 3) + 1$$
$$= \log_2 4 + 1$$
$$= 2 + 1$$
$$= 3$$

The point $(7, 3)$ is on the graph, shown in the figure.

**Now try Exercises 59 through 74** ▶

As with the exponential functions, much can be learned from graphs of logarithmic functions and a summary of important characteristics is given here.

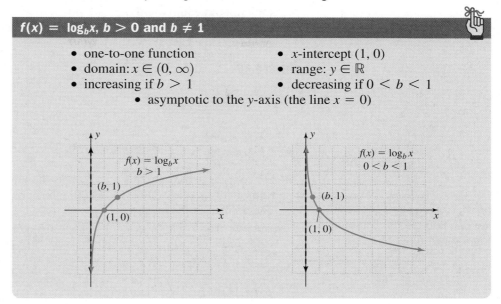

### $f(x) = \log_b x$, $b > 0$ and $b \neq 1$

- one-to-one function
- domain: $x \in (0, \infty)$
- increasing if $b > 1$
- $x$-intercept $(1, 0)$
- range: $y \in \mathbb{R}$
- decreasing if $0 < b < 1$
- asymptotic to the $y$-axis (the line $x = 0$)

**C.** You've just seen how we can graph logarithmic functions

## D. Finding the Domain of a Logarithmic Function

Examples 5 and 6 illustrate how the domain of a logarithmic function can change when certain transformations are applied. Since the domain of a nonshifted logarithmic function consists of *positive* real numbers, the argument of a logarithm must be greater than zero. This means finding the domain often consists of solving various inequalities, which can be done using the skills acquired in Section 4.6.

**EXAMPLE 6** ▶ **Finding the Domain of a Logarithmic Function**

Determine the domain of each function.

**a.** $p(x) = \log_2(2x + 3)$        **b.** $q(x) = \log_5(x^2 - 2x)$

**c.** $r(x) = \log\left(\dfrac{3 - x}{x + 3}\right)$        **d.** $f(x) = \ln|x - 2|$

**Solution** ▶ Begin by writing the argument of each function as a "greater than" inequality.

**a.** Solving $2x + 3 > 0$ for $x$ gives $x > -\frac{3}{2}$, and the domain of $p$ is $x \in (-\frac{3}{2}, \infty)$.

**b.** For $x^2 - 2x > 0$, we note $y = x^2 - 2x$ is a parabola, opening upward, with zeroes at $x = 0$ and $x = 2$ (see Figure 5.27). This means $x^2 - 2x$ will be positive for $x < 0$ and $x > 2$. The domain of $q$ is $x \in (-\infty, 0) \cup (2, \infty)$.

**Figure 5.27**

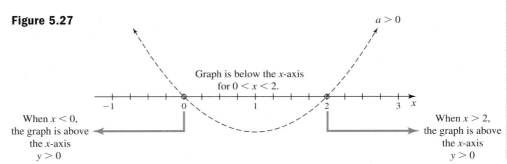

$a > 0$

Graph is below the $x$-axis
for $0 < x < 2$.

When $x < 0$,
the graph is above
the $x$-axis
$y > 0$

When $x > 2$,
the graph is above
the $x$-axis
$y > 0$

**c.** For $\frac{3-x}{x+3} > 0$, we note $y = \frac{3-x}{x+3}$ has a zero at $x = 3$, with a vertical asymptote at $x = -3$ and here we opt to use the interval test method to solve the inequality. Outputs are positive when $x = 0$ (see Figure 5.28), so $y$ is positive in the interval $(-3, 3)$ and negative elsewhere. The domain of $r$ is $x \in (-3, 3)$.

**Figure 5.28**

When −3 < x < 3, y > 0
(interval test)
−4
−3
−2
−1
0
1
2
3
4
x
When x < −3, y < 0
When x > 3, y < 0

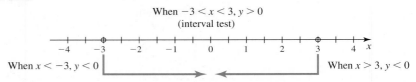

**d.** For $|x - 2| > 0$, we note $y = |x - 2|$ is the graph of $y = |x|$ shifted 2 units right, with its vertex at (2, 0). The graph is positive for all $x$, except at $x = 2$. The domain of $f$ is $x \in (-\infty, 2) \cup (2, \infty)$.

**D.** You've just seen how we can find the domain of a logarithmic function

**Now try Exercises 75 through 82** ▶

## E. Applications of Logarithms

The use of logarithmic scales as a tool of measurement is primarily due to the range of values for the phenomenon being measured. For instance, time is generally measured on a linear scale, and for short periods a linear scale is appropriate. For the time line in Figure 5.29, each tick-mark *represents 1 unit,* and the time line can display a period of 10 yr. However, the scale would be useless in a study of geology or the age of the universe. If we scale the number line logarithmically, each tick-mark *represents a power of 10* (Figure 5.30) and a scale of the same length can now display a time period of 10 billion years.

**Figure 5.29**

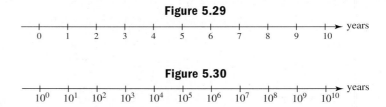

0
1
2
3
4
5
6
7
8
9
10
years

**Figure 5.30**

$10^0$
$10^1$
$10^2$
$10^3$
$10^4$
$10^5$
$10^6$
$10^7$
$10^8$
$10^9$
$10^{10}$
years

**WORTHY OF NOTE**

The **decibel** (dB) is the reference unit for sound, and is based on the faintest sound a person can hear, called the **threshold of audibility.** It is a base-10 logarithmic scale, meaning a sound 10 times more intense is 1 bel louder.

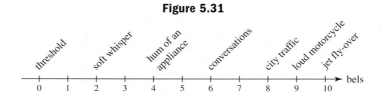

In much the same way, logarithmic measures are needed in a study of sound and earthquake intensity, as the scream of a jet engine is over 1 billion times more intense than the threshold of hearing, and the most destructive earthquakes are billions of times stronger than the slightest earth movement that can be felt. Similar ranges exist in the measurement of light, acidity, and voltage. Figures 5.31 and 5.32 show logarithmic scales for measuring sound in decibels (1 bel = 10 decibels) and earthquake intensity in Richter values (or magnitudes).

**Figure 5.31**

threshold
soft whisper
hum of an appliance
conversations
city traffic
loud motorcycle
jet fly-over
0
1
2
3
4
5
6
7
8
9
10
bels

**Figure 5.32**

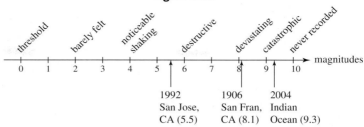

The slightest earth movement perceptible is called the **reference intensity** $I_0$, with the intensity $I$ of stronger earthquakes expressed as a multiple of $I_0$. The earthquake that struck Haiti in January of 2010 was measured at over 10,500,000 times this reference intensity, or $I = 10,5000,000I_0$. To find the Richter value (magnitude) of this earthquake, we simply take the base-10 logarithm of the ratio $\frac{I}{I_0}$ to express these values on a logarithmic scale. In function form, $M(I) = \log\left(\frac{I}{I_0}\right)$, and we find that the Haitian earthquake had a magnitude of just over 7.0: $\log\left(\frac{10,500,000I_0}{I_0}\right) = \log(10,500,000) \approx 7.0$.

---

**EXAMPLE 7A** ▶ **Finding the Magnitude of an Earthquake**

Find the magnitude of the earthquakes (rounded to tenths) with the intensities given.

**a.** Eureka earthquake; January 9, 2010, near Humboldt county, California: $I = 3,162,000I_0$.

**b.** Sumatra-Andaman earthquake; December 26, 2004, near the west coast of Sumatra, Indonesia: $I = 1,995,260,000I_0$.

**Solution** ▶

**a.**
$$M(I) = \log\left(\frac{I}{I_0}\right) \qquad \text{magnitude equation}$$

$$M(3,162,000I_0) = \log\left(\frac{3,162,000I_0}{I_0}\right) \qquad \text{substitute } 3,162,000I_0 \text{ for } I$$

$$= \log 3,162,000 \qquad \text{simplify}$$

$$\approx 6.5 \qquad \text{result}$$

The earthquake had a magnitude of about 6.5.

**b.**
$$M(I) = \log\left(\frac{I}{I_0}\right) \qquad \text{magnitude equation}$$

$$M(1,995,260,000I_0) = \log\left(\frac{1,995,260,000I_0}{I_0}\right) \qquad \text{substitute } 1,995,260,000I_0 \text{ for } I$$

$$= \log 1,995,260,000 \qquad \text{simplify}$$

$$\approx 9.3 \qquad \text{result}$$

The earthquake had a magnitude of about 9.3.

---

**EXAMPLE 7B** ▶ **Comparing Earthquake Intensity to the Reference Intensity**

How many times more intense than the reference intensity $I_0$ was the Peruvian earthquake of June 23, 2001, with magnitude 8.4?

Solution ▶

$$M(I) = \log\left(\frac{I}{I_0}\right) \quad \text{magnitude equation}$$

$$8.4 = \log\left(\frac{I}{I_0}\right) \quad \text{substitute 8.4 for } M(I)$$

$$10^{8.4} = \frac{I}{I_0} \quad \text{exponential form}$$

$$I = 10^{8.4}I_0 \quad \text{solve for } I$$

$$I \approx 251{,}188{,}643 I_0 \quad 10^{8.4} \approx 251{,}188{,}643$$

The earthquake was over 251 million times more intense than the reference intensity.

**EXAMPLE 7C** ▶ **Comparing Earthquake Intensities**

The 1906 earthquake that leveled much of San Francisco had a magnitude of 8.1. How many times more intense was the Japanese earthquake of March 11, 2011, with a magnitude of 9.0?

Solution ▶ The Japanese quake had a Richter value of 9.0, with an intensity of $10^{9.0}I_0$. Similarly, the San Francisco earthquake was estimated at 8.1 on the Richter scale, with an intensity of $10^{8.1}I_0$. Using these intensities, we find the Japanese quake was $\frac{10^{9.0}I_0}{10^{8.1}I_0} = 10^{9.0-8.1} = 10^{0.9}$ or about 8 times more intense than the San Francisco quake.

Now try Exercises 85 through 98 ▶

A second application of logarithmic functions involves the relationship between altitude and barometric pressure. The altitude or height above sea level can be determined by the formula $H = (30T + 8000)\ln\left(\frac{P_0}{P}\right)$, where $H$ is the altitude in meters for a temperature $T$ in degrees Celsius and barometric pressure $P$ in units called **centimeters of mercury** (cmHg). Here, $P_0$ is the barometric pressure at sea level: 76 cmHg.

**EXAMPLE 8** ▶ **Using Logarithms to Determine Altitude**

Hikers at the summit of Mt. Shasta in northern California take a pressure reading of 45.1 cmHg at a temperature of 9°C. How high is Mt. Shasta?

Solution ▶ For this exercise, $P_0 = 76$, $P = 45.1$, and $T = 9$. The formula yields

$$H = (30T + 8000)\ln\left(\frac{P_0}{P}\right) \quad \text{given formula}$$

$$= [30(9) + 8000]\ln\left(\frac{76}{45.1}\right) \quad \text{substitute given values}$$

$$= 8270\ln\left(\frac{76}{45.1}\right) \quad \text{simplify}$$

$$\approx 4316 \quad \text{result}$$

Mt. Shasta is about 4316 m high.

Now try Exercises 99 through 102 ▶

Our final application shows the versatility of logarithmic functions, and their value as a real-world model. Large advertising agencies are well aware that after a new ad campaign, sales will increase rapidly as more people become aware of the product. Continued advertising will give the new product additional market share, but once the "newness" wears off and the competition begins responding, sales tend to taper off—regardless of any additional amount spent on ads. This phenomenon can be modeled by the function

$$S(d) = k + a \ln d$$

where $S(d)$ is the number of expected sales after $d$ dollars are spent, and $a$ and $k$ are constants related to product type and market size.

---

**EXAMPLE 9** ▶  **Using Logarithms for Marketing Strategies**

Market research has shown that MusicMaster subscriptions, a new service for downloading and playing music, can be approximated by the equation $S(d) = 2500 + 250 \ln d$, where $S(d)$ is the number of subscriptions after $d$ thousand dollars is spent on advertising. The graph of $y = S(d)$ is shown.

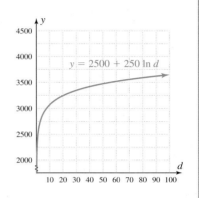

a. What sales volume is expected if the advertising budget is $40,000?

b. If the company needs 3500 subscriptions to begin making a profit, how much should be spent on advertising?

**Solution** ▶  a. For sales volume, we simply evaluate the function for $d = 40$ ($d$ in thousands):

$$
\begin{aligned}
S(d) &= 2500 + 250 \ln d & &\text{given equation} \\
S(40) &= 2500 + 250 \ln 40 & &\text{substitute 40 for } d \\
&\approx 2500 + 922 & &250 \ln 40 \approx 922 \\
&= 3422
\end{aligned}
$$

Spending $40,000 on advertising will generate approximately 3422 subscriptions.

b. To find the advertising budget needed, we substitute number of sales and solve for $d$.

$$
\begin{aligned}
S(d) &= 2500 + 250 \ln d & &\text{given equation} \\
3500 &= 2500 + 250 \ln d & &\text{substitute 3500 for } S(d) \\
1000 &= 250 \ln d & &\text{subtract 2500} \\
4 &= \ln d & &\text{divide by 250} \\
e^4 &= d & &\text{exponential form} \\
54.598 &\approx d & &e^4 \approx 54.598
\end{aligned}
$$

About $54,600 should be spent in order to sell 3500 subscriptions.

☑ **E. You've just seen how we can solve applications of logarithmic functions**

**Now try Exercises 103 and 104** ▶

## 5.3 EXERCISES

### ▶ CONCEPTS AND VOCABULARY

**Fill in each blank with the appropriate word or phrase. Carefully reread the section, if necessary.**

**1.** The range of $y = \log_b x$ is all _____ _____, and the domain is $x \in$ _____. Further, as $x \to 0$, $y \to$ _____.

**2.** The function $y = \log_b x$ is an increasing function if _____, and a decreasing function if _____.

**3.** What number does the expression $\log_2 32$ represent? Discuss/Explain how $\log_2 32 = \log_2 2^5$ justifies this fact.

**4.** Explain how the graph of $Y = \log_b(x - a)$ can be obtained from $y = \log_b x$. Where is the "new" $x$-intercept? Where is the new asymptote?

### ▶ DEVELOPING YOUR SKILLS

**Write each equation in exponential form.**

**5.** $3 = \log_2 8$

**6.** $2 = \log_3 9$

**7.** $x = \log_7 \dfrac{1}{7}$

**8.** $y = \ln \dfrac{1}{e^3}$

**9.** $0 = \log_9 1$

**10.** $0 = \ln 1$

**11.** $\dfrac{1}{3} = \log_8 y$

**12.** $\dfrac{1}{2} = \log_{81} x$

**13.** $1 = \log_2 2$

**14.** $1 = \ln e$

**15.** $\log_x 49 = 2$

**16.** $\log_y 16 = 2$

**17.** $\log 100 = 2$

**18.** $\log 10{,}000 = 4$

**19.** $\ln 54.598 \approx y$

**20.** $\log 0.001 = x$

**Write each equation in logarithmic form.**

**21.** $4^3 = 64$

**22.** $e^3 \approx 20.086$

**23.** $3^x = \dfrac{1}{9}$

**24.** $2^y = \dfrac{1}{8}$

**25.** $e^0 = 1$

**26.** $8^0 = 1$

**27.** $\left(\dfrac{1}{3}\right)^{-3} = y$

**28.** $\left(\dfrac{1}{5}\right)^{-2} = x$

**29.** $10^3 = 1000$

**30.** $e^1 = e$

**31.** $10^x = \dfrac{1}{100}$

**32.** $10^y = \dfrac{1}{10{,}000}$

**33.** $4^{\frac{3}{2}} = 8$

**34.** $e^{\frac{3}{2}} \approx 2.117$

**35.** $y^{\frac{-3}{2}} = \dfrac{1}{8}$

**36.** $x^{\frac{-2}{3}} = \dfrac{1}{9}$

**Determine the value of each logarithm without using a calculator.**

**37.** $\log_4 4$

**38.** $\log_9 9$

**39.** $\log 1000$

**40.** $\log 1{,}000{,}000$

**41.** $\ln e$

**42.** $\ln e^2$

**43.** $\log_4 2$

**44.** $\log_{81} 9$

**45.** $\log_7 \dfrac{1}{49}$

**46.** $\log_9 \dfrac{1}{81}$

**47.** $\ln \dfrac{1}{e^2}$

**48.** $\ln \dfrac{1}{\sqrt{e}}$

**49.** $\log \sqrt[3]{100}$

**50.** $\log 0.01$

**Use a calculator to evaluate each expression, rounded to four decimal places.**

**51.** $\log 50$

**52.** $\log 47$

**53.** $\ln 1.6$

**54.** $\ln 0.75$

**55.** $\ln 225$

**56.** $\ln 381$

**57.** $\log \sqrt{37}$

**58.** $\log 4\pi$

**Graph each function *using transformations* of $y = \log_b x$, sketching the asymptote, and strategically plotting a few points to round out the graph. Clearly state the transformations applied.**

**59.** $f(x) = \log_2 x + 3$

**60.** $g(x) = \log_3 x - 2$

**61.** $q(x) = \ln(x + 1)$

**62.** $r(x) = \ln(x + 1) - 2$

**63.** $Y_1 = -\ln(x + 1)$

**64.** $Y_2 = -\ln x + 2$

**65.** $p(x) = 3 \log(x - 2)$

**66.** $q(x) = 2 \log(x + 1)$

**67.** $F(t) = -\log t + 2$

**68.** $G(t) = \log(-t) - 3$

**Match each logarithmic equation to the correct graph. Assume $b > 1$.**

**69.** $y = \log_b(x + 2)$

**70.** $y = 2 \log_b x$

**71.** $f(x) = 1 - \log_b x$

**72.** $g(x) = \log_b x - 1$

**73.** $y = \log_b x + 2$

**74.** $y = -\log_b x$

I.
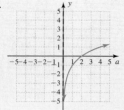

II.
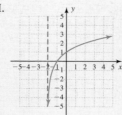

III.
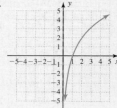

IV.
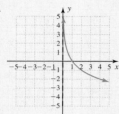

V.
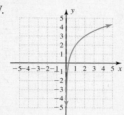

VI.
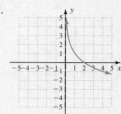

**Determine the domain of the following functions.**

**75.** $y = \log(x - 3)$  **76.** $y = \log(2 - x)$  **77.** $y = \log_6\left(\dfrac{x + 1}{x - 3}\right)$  **78.** $y = \ln\left(\dfrac{x - 2}{x + 3}\right)$

**79.** $f(x) = \log_5\sqrt{2x - 3}$  **80.** $g(x) = \ln\sqrt{5 - 3x}$  **81.** $y = \log(9 - x^2)$  **82.** $y = \ln(9x - x^2)$

## ▶ WORKING THE FORMULAS

**83. pH level: pH $= -\log[\text{H}^+]$**

The pH level of a solution indicates the concentration of hydrogen ions ($\text{H}^+$) in a unit called *moles per liter*. The pH level is given by the formula shown where $[\text{H}^+]$ is the ion concentration. A solution with pH $< 7$ is called an acid (lemon juice: pH $\approx 2$), and a solution with pH $> 7$ is called a base (household ammonia: pH $\approx 11$). Use the formula to determine the pH level of tomato juice if $[\text{H}^+] = 7.94 \times 10^{-5}$ moles per liter. Is this an acid or base solution?

**84. Time required for an investment to double:**
$$T(r) = \frac{\log 2}{\log(1 + r)}$$

The time $T(r)$ required for an investment to double in value is given by the formula shown, where $r$ represents the interest rate (expressed as a decimal). How long would it take an investment to double if the interest rate were (a) 5%, (b) 8%, (c) 12%?

## ▶ APPLICATIONS

**Earthquake intensity:** Use the information provided in Example 7 to answer the following.

**85.** Find the magnitude $M(I)$ of the earthquake given
  **a.** $I = 50,000I_0$  **b.** $I = 75,000,000I_0$.

**86.** Find the intensity $I$ of the earthquake given
  **a.** $M(I) = 3.2$  **b.** $M(I) = 8.1$.

**Determine how many times more intense the first quake was compared to the second.**

**87.** Great Chilean quake (1960): magnitude 9.5
Kobe, Japan, quake (1995): magnitude 6.9

**88.** Northern Sumatra (2004): magnitude 9.1
Southern Greece (2008): magnitude 4.5

**89. Earthquake intensity:** On June 25, 1989, an earthquake with magnitude 6.2 shook the southeast side of the Island of Hawaii (near Kalapana), causing some $1,000,000 in damage. On October 15, 2006, an earthquake measuring 6.7 on the Richter scale shook the northwest side of the island, causing over $100,000,000 in damage. How much more intense was the 2006 quake?

**90. Earthquake intensity:** The most intense earthquake of the modern era occurred in Chile on May 22, 1960, and measured 9.5 on the Richter scale. How many times more intense was this earthquake, than the quake that hit Northern Sumatra (Indonesia) on March 28, 2005, and measured 8.7?

**Brightness of a star:** The brightness or intensity $I$ of a star as perceived by the naked eye is measured in units called *magnitudes*. The brightest stars have magnitude 1 $[M(I) = 1]$ and the dimmest have magnitude 6 $[M(I) = 6]$. The magnitude of a star is given by the equation $M(I) = 6 - 2.5 \log\left(\frac{I}{I_0}\right)$, where $I$ is the actual intensity of light from the star and $I_0$ is the faintest light visible to the human eye, called the reference intensity. The intensity is often given as a multiple of $I_0$.

**91.** Find the brightness $M(I)$ of the star given
  **a.** $I = 27I_0$  **b.** $I = 85I_0$.

**92.** Find the intensity $I$ of a star given
  **a.** $M(I) = 1.6$  **b.** $M(I) = 5.2$.

**Intensity of sound:** The intensity of sound as perceived by the human ear is measured in units called decibels (dB). The loudest sounds that can be withstood without damage to the eardrum are in the 120- to 130-dB range, while a whisper may measure in the 15- to 20-dB range. Decibel measure is given by the equation $D(I) = 10 \log\left(\frac{I}{I_0}\right)$, where $I$ is the actual intensity of the sound and $I_0$ is the faintest sound perceptible by the human ear—called the reference intensity. The intensity is often given as a multiple of $I_0$, where $10^{-16}$ (watts per cm$^2$; W/cm$^2$) is used as the threshold of audibility.

93. Find the loudness $D(I)$ of the sound given
    **a.** $I = 10^{-14}$          **b.** $I = 10^{-4}$.

94. Find the intensity $I$ of the sound given
    **a.** $D(I) = 83$          **b.** $D(I) = 125$.

**Determine how many times more intense the first sound is compared to the second.**

95. pneumatic hammer: 11.2 bels
    heavy lawn mower: 8.5 bels

96. train horn: 7.5 bels
    soft music: 3.4 bels

97. **Sound intensity of a hair dryer:** Every morning (it seems), Jose is awakened by the mind-jarring, ear-jamming sound of his daughter's hair dryer (75 dB). He knew he was exaggerating, but told her (many times) of how it reminded him of his railroad days, when the air compressor for the pneumatic tools was running (110 dB). In fact, how many times more intense was the sound of the air compressor compared to the sound of the hair dryer?

98. **Sound intensity of a busy street:** The decibel level of noisy, downtown traffic has been estimated at 87 dB, while the laughter and banter at a loud party might be in the 60-dB range. How many times more intense is the sound of the downtown traffic?

The *barometric equation* $H = (30T + 8000)\ln\left(\frac{P_0}{P}\right)$ was discussed in Example 8.

99. **Temperature and atmospheric pressure:** Determine the height of Mount McKinley (Alaska), if the temperature at the summit is $-10°C$, with a barometric reading of 34 cmHg.

100. **Temperature and atmospheric pressure:** A large passenger plane is flying cross-country. The instruments on board show an air temperature of 3°C, with a barometric pressure of 22 cmHg. What is the altitude of the plane?

101. **Altitude and atmospheric pressure:** By definition, a mountain pass is a low point between two mountains. Passes may be very short with steep slopes, or as large as a valley between two peaks. Perhaps the highest drivable pass in the world is the Semo La pass in central Tibet. At its highest elevation, a temperature reading of 8°C was taken, along with a barometer reading of 39.3 cmHg. (a) Approximately how high is the Semo La pass? (b) While traveling up to this pass, an elevation marker is seen. If the barometer reading was 47.1 cmHg at a temperature of 12°C, what height did the marker give?

102. **Altitude and atmospheric pressure:** Hikers on Mt. Everest take successive readings of 35 cmHg at 5°C and 30 cmHg at $-10°C$. (a) How far up the mountain are they at each reading? (b) Approximate the height of Mt. Everest if the temperature at the summit is $-27°C$ and the barometric pressure is 22.2 cmHg.

103. **Marketing budgets:** An advertising agency has determined the number of items sold by a certain client is modeled by the equation $N(A) = 1500 + 315\ln A$, where $N(A)$ represents the number of sales after spending $A$ thousands of dollars on advertising. Determine the approximate number of items sold on an advertising budget of (a) \$10,000; (b) \$50,000. (c) Use the TABLE feature of a calculator to estimate how large a budget is needed (to the nearest \$500 dollars) to sell 3000 items.

104. **Sports promotions:** The accountants for a major boxing promoter have determined that the number of pay-per-view subscriptions sold to their championship bouts can be modeled by the function $N(d) = 15,000 + 5850\ln d$, where $N(d)$ represents the number of subscriptions sold after spending $d$ thousand dollars on promotional activities. Determine the number of subscriptions sold if (a) \$50,000 and (b) \$100,000 is spent. (c) Determine how much should be spent (to the nearest \$1000 dollars) to sell over 50,000 subscriptions by simplifying the logarithmic equation and writing the result in exponential form.

105. **Home ventilation:** If too little outdoor air enters a home, pollutants can sometimes accumulate to levels that pose a health risk. For a three-bedroom home, ventilation requirements can be modeled by the function $C(x) = 42\ln x - 270$, where $C(x)$ represents the number of cubic feet of air per minute (cfm) that should be exchanged with outside air in a home with floor area $x$ (in square feet). (a) How many cfm of exchanged air are needed for a three-bedroom home with a floor area of 2500 ft$^2$? (b) If a three-bedroom home is being mechanically ventilated by a system with 40 cfm capacity, what is the maximum square footage of the home, assuming it is built to code?

106. **Runway takeoff distance:** The minimum required length of a runway depends on the maximum allowable takeoff weight (mtw) of a specific plane. This relationship can be approximated by the function $L(x) = 2085\ln x - 14,900$, where $L(x)$ represents the required length of a runway in feet, for a plane with $x$ mtw in pounds.

a. The Airbus-320 has a 169,750-lb mtw. What minimum runway length is required for takeoff?

b. By simplifying the logarithmic equation that results and writing the equation in exponential form, determine the mtw of a Learjet 30, which requires a runway of 5550 ft to take off safely.

**Memory retention:** Under certain conditions, a person's retention of random facts can be modeled by the equation $P(x) = 95 - 14\log_2 x$, where $P(x)$ is the percentage of those facts retained after $x$ days. Find the percentage of facts a person might retain after $x$ days for the values given. Note that the values given are powers of 2.

107. **a.** 1 day           **b.** 4 days

108. **a.** 32 days         **b.** 64 days

## ▶ EXTENDING THE CONCEPT

109. Many texts and reference books give estimates of the noise level (in decibels dB) of common sounds. Through reading and research, try to locate or approximate where the following sounds would fall along this scale. In addition, determine at what point pain or ear damage begins to occur.

     **a.** threshold of audibility       **b.** lawn mower

     **c.** whisper                     **d.** loud rock concert

     **e.** lively party                **f.** jet engine

110. Determine the value of $x$ that makes the equation true: $\log_3[\log_3(\log_3 x)] = 0$.

111. Find the value of each expression without using a calculator.

     **a.** $\log_{64}\frac{1}{16}$      **b.** $\log_{\frac{4}{9}}\frac{27}{8}$      **c.** $\log_{0.25}32$

## ▶ MAINTAINING YOUR SKILLS

112. **(R.5)** Factor the following expressions:

     **a.** $x^3 - 8$               **b.** $a^2 - 49$

     **c.** $n^2 - 10n + 25$      **d.** $2b^2 - 7b + 6$

114. **(2.2)** A function $f(x)$ is defined by the ordered pairs shown in the table. Is the function (a) linear? (b) increasing? Justify your answers.

| $x$ | $f(x)$ |
|-----|--------|
| $-10$ | $0$ |
| $-9$ | $-2$ |
| $-8$ | $-8$ |
| $-6$ | $-18$ |
| $-5$ | $-50$ |
| $-4$ | $-72$ |

113. **(3.1)** Graph $g(x) = \sqrt[3]{x + 2} - 1$ by shifting the parent function. Then state the domain and range of $g$.

115. **(4.6)** For the graph shown, write the solution set for $f(x) < 0$. Then write the equation of the graph in factored form and in polynomial form.

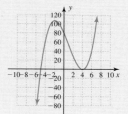

## MID-CHAPTER CHECK

1. Write the following in logarithmic form.

     **a.** $27^{\frac{2}{3}} = 9$              **b.** $81^{\frac{5}{4}} = 243$

2. Write the following in exponential form.

     **a.** $\log_8 32 = \frac{5}{3}$         **b.** $\log_{1296}6 = 0.25$

3. Solve each equation for the unknown:

     **a.** $4^{2x} = 32^{x-1}$        **b.** $\left(\frac{1}{3}\right)^{4b} = 9^{2b-5}$

4. Solve each equation for the unknown:

     **a.** $\log_{27}x = \frac{1}{3}$        **b.** $\log_b 125 = 3$

 5. The homes in a popular neighborhood are growing in value according to the formula $V(t) = V_0\left(\frac{9}{8}\right)^t$, where $t$ is the time in years, $V_0$ is the purchase price of the home, and $V(t)$ is the current value of the home. (a) In 3 yr, how much will a $50,000 home be worth? (b) Use the TABLE feature of your calculator to estimate how many years (to the nearest year) until the home doubles in value.

6. The graph of the function $f(x) = 5^x$ has been shifted right 3 units and up 2 units, then stretched by a factor of 4. What is the equation of the resulting function?

7. State the domain and range for $f(x) = \sqrt{x-3} + 1$, then find $f^{-1}(x)$ and state its domain and range. Verify the inverse relationship using composition.

8. Write the following equations in logarithmic form, then verify the result on a calculator.
   **a.** $2 \approx 10^{0.301}$    **b.** $e^4 \approx 54.598$

9. Write the following equations in exponential form, then verify the result on a calculator.
   **a.** $2.845 \approx \log 700$    **b.** $1.4 \approx \ln 4.0552$

10. On August 15, 2007, an earthquake measuring 8.0 on the Richter scale struck coastal Peru. On October 17, 1989, right before Game 3 of the World Series between the Oakland A's and the San Francisco Giants, the Loma Prieta earthquake, measuring 7.1 on the Richter scale, struck the San Francisco Bay area. How much more intense was the Peruvian earthquake?

---

## 5.4 | Properties of Logarithms

### LEARNING OBJECTIVES

*In Section 5.4 you will see how we can:*

- ☐ **A.** Solve logarithmic equations using the fundamental properties of logarithms
- ☐ **B.** Apply the product, quotient, and power properties of logarithms
- ☐ **C.** Apply the change-of-base formula
- ☐ **D.** Solve applications using properties of logarithms

Logarithmic and exponential expressions have several fundamental properties that enable us to solve some basic equations. In this section, we'll learn how to use these relationships effectively, and introduce additional properties that enable us to simplify more complex equations before relying on the same fundamental properties to complete the solution.

### A. Solving Equations Using the Fundamental Properties of Logarithms

In Section 5.3, we converted expressions from exponential form to logarithmic form using the basic definition: $x = b^y \Leftrightarrow y = \log_b x$. This relationship reveals the following fundamental properties:

> **Fundamental Properties of Logarithms**
>
> For any base $b > 0$, $b \neq 1$,
>
> **I.** $\log_b b = 1$, since $b^1 = b$     **III.** $\log_b b^x = x$, since $b^x = b^x$
>
> **II.** $\log_b 1 = 0$, since $b^0 = 1$     **IV.** $b^{\log_b x} = x \, (x > 0)$, since $\log_b x = \log_b x$

Although Properties III and IV follow directly from the inverse nature of $y = b^x$ and $y = \log_b x$, they can also be verified algebraically. Note that for $x = b^y$, $\log_b x = y$ is the equivalent logarithmic form, and substituting $b^y$ for $x$ yields $\log_b b^y = y$. Similarly, for $y = \log_b x$, rewriting in exponential form gives $b^y = x$ and then $b^{\log_b x} = x$. Alternatively, we can use compositions to verify Properties III and IV just as in Example 5 from Section 5.1. For $f(x) = b^x$, $f^{-1}(x) = \log_b x$ and we have

$$b^{\log_b x} = b^{f^{-1}(x)} \qquad \log_b b^x = \log_b f(x)$$
$$= f[f^{-1}(x)] \qquad\qquad = f^{-1}[f(x)]$$
$$= x \checkmark \qquad\qquad\qquad = x \checkmark$$

In common language, Properties III and IV show "a base-$b$ logarithm *undoes* a base-$b$ exponential" and "a base-$b$ exponential *undoes* a base-$b$ logarithm." These properties and others can be used to solve basic equations involving logarithms and exponentials. In particular, from the uniqueness property of exponents (page 443), note that if $\log_b x = k$, then $b^{\log_b x} = b^k$.

**EXAMPLE 1** ▶ **Solving Basic Logarithmic Equations**

Solve each equation by applying fundamental properties. Answer in exact form and approximate form using a calculator (round to thousandths).

    **a.** $\ln x = 2$     **b.** $-0.52 = \log x$

Solution ▶

| | | |
|---|---|---|
| **a.** $\ln x = 2$ | given | |
| $e^{\ln x} = e^2$ | exponentiate both sides | |
| $x = e^2$ | Property IV, exact form | |
| $\approx 7.389$ | approximate form | |

| | | |
|---|---|---|
| **b.** $-0.52 = \log x$ | given | |
| $10^{-0.52} = 10^{\log x}$ | exponentiate both sides | |
| $10^{-0.52} = x$ | Property IV, exact form | |
| $0.302 \approx x$ | approximate form | |

*Now try Exercises 5 through 8* ▶

Note that checking the exact solutions by substitution is a direct application of Property III (Figure 5.33).

    Also, we observe that exponentiating both sides of the equation produces the same result as simply writing the original equation in exponential form, and the process can be viewed in terms of either approach.

**Figure 5.33**

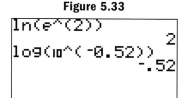

**EXAMPLE 2** ▶ **Solving Basic Exponential Equations**

Solve each equation by applying fundamental properties. Answer in exact form and approximate form using a calculator (round to thousandths).

    **a.** $e^x = 167$     **b.** $10^x = 8.223$

Solution ▶

| | | |
|---|---|---|
| **a.** $e^x = 167$ | given | |
| $\ln e^x = \ln 167$ | use natural log | |
| $x = \ln 167$ | Property III, exact form | |
| $\approx 5.118$ | approximate form | |

| | | |
|---|---|---|
| **b.** $10^x = 8.223$ | given | |
| $\log 10^x = \log 8.223$ | use common log | |
| $x = \log 8.223$ | Property III, exact form | |
| $\approx 0.915$ | approximate form | |

*Now try Exercises 9 through 12* ▶

Similar to our observations from Example 1, taking the logarithm of both sides produced the same result as writing the equation in logarithmic form, and the process can be viewed in terms of either approach. Also note that here, checking the exact solution by substitution is a direct application of Property IV (Figure 5.34).

    If an equation has a single logarithmic or exponential term, the equation can be solved by isolating this term and applying one of the fundamental properties as in Examples 1 and 2.

**Figure 5.34**

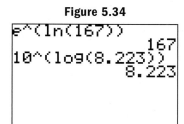

**EXAMPLE 3** ▶ **Solving Exponential Equations**

Solve each equation. Write answers in exact form and approximate form to four decimal places.

    **a.** $10^x - 29 = 51$     **b.** $3e^{x+1} - 5 = 7$

**Solution** ▶ **a.** $10^x - 29 = 51$      given

$10^x = 80$      add 29

Since the exponential is base 10, we apply the common logarithm.

$\log 10^x = \log 80$      take the common log of both sides

$x = \log 80$      Property III (exact form)

$\approx 1.9031$      approximate form

**b.** $3e^{x+1} - 5 = 7$      given

$3e^{x+1} = 12$      add 5

$e^{x+1} = 4$      divide by 3

Since the exponential is base $e$, we apply the natural logarithm.

$\ln e^{x+1} = \ln 4$      take the natural log of both sides

$x + 1 = \ln 4$      Property III

$x = \ln 4 - 1$      solve for $x$ (exact form)

$\approx 0.3863$      approximate form

Now try Exercises 13 through 18 ▶

---

**EXAMPLE 4** ▶ **Solving Logarithmic Equations**

Solve each equation. Write answers in exact form and approximate form to four decimal places.

**a.** $2 \log(7x) + 1 = 4$      **b.** $-4 \ln(x + 1) - 5 = 7$

**Solution** ▶ **a.** $2 \log(7x) + 1 = 4$      given

$2 \log(7x) = 3$      subtract 1

$\log(7x) = \dfrac{3}{2}$      divide by 2

$7x = 10^{\frac{3}{2}}$      exponential form

$x = \dfrac{10^{\frac{3}{2}}}{7}$      divide by 7 (exact form)

$\approx 4.5175$      approximate form

**b.** $-4 \ln(x + 1) - 5 = 7$      given

$-4 \ln(x + 1) = 12$      add 5

$\ln(x + 1) = -3$      divide by $-4$

$x + 1 = e^{-3}$      exponential form

$x = e^{-3} - 1$      subtract 1 (exact form)

$\approx -0.9502$      approximate form

☑ **A. You've just seen how we can solve logarithmic equations using the fundamental properties of logarithms**

Now try Exercises 19 through 24 ▶

## B. The Product, Quotient, and Power Properties of Logarithms

Generally speaking, equation solving involves simplifying the equation, isolating a variable term on one side, and applying an inverse to solve for the unknown. For logarithmic equations such as $\log x + \log(x + 3) = 1$, we must find a way to combine the terms on the left, before we can work toward a solution. This requires a further exploration of logarithmic properties.

Due to the close connection between exponents and logarithms, their properties are very similar. To illustrate, we'll use factors that can all be written in the form $2^x$, and write the equations $8 \cdot 4 = 32$, $\frac{8}{4} = 2$, and $8^2 = 64$ in both exponential form and logarithmic form.

*The exponents from a*    exponential form:       $2^3 \cdot 2^2 = 2^{3+2}$
*product are added:*    logarithmic form:       $\log_2(8 \cdot 4) = \log_2 8 + \log_2 4$

*The exponents from a*    exponential form:       $\dfrac{2^3}{2^2} = 2^{3-2}$
*quotient are subtracted:*

   logarithmic form:       $\log_2\left(\dfrac{8}{4}\right) = \log_2 8 - \log_2 4$

*The exponents from a*    exponential form:       $(2^3)^2 = 2^{3 \cdot 2}$
*power are multiplied:*    logarithmic form:       $\log_2 8^2 = 2 \cdot \log_2 8$

Each illustration can be generalized and applied with any base $b$.

**WORTHY OF NOTE**

For a more detailed verification of these properties, see Appendix IV.

**Properties of Logarithms**

Given *positive* numbers $M$, $N$, and $b \neq 1$, and *any* real number $p$.

**Product Property**
$\log_b(MN) = \log_b M + \log_b N$

The log of a product is a sum of logs.

**Quotient Property**
$\log_b\left(\frac{M}{N}\right) = \log_b M - \log_b N$

The log of a quotient is a difference of logs.

**Power Property**
$\log_b M^p = p \log_b M$

The log of a quantity to a power is the power times the log of the quantity.

**CAUTION** ▶ It's very important that you read and understand these properties correctly. In particular: (1) $\log(M + N) \neq \log M + \log N$, (2) $\log\left(\frac{M}{N}\right) \neq \frac{\log M}{\log N}$, and (3) $\frac{\log M}{\log N} \neq \log M - \log N$. For $M = 100$ and $N = 10$, statement (1) would indicate $\log 110 \neq 2 + 1$, (2) would indicate that $1 \neq \frac{2}{1}$, and (3) would indicate that $\frac{2}{1} \neq 2 - 1$.

In the statement of these properties, it's worth reminding ourselves that the equal sign "works both ways," and we have $\log_b M + \log_b N = \log_b(MN)$. These properties are often used to write a sum or difference of logarithmic terms as a single term.

**EXAMPLE 5** ▶ **Rewriting Expressions Using Logarithmic Properties**

Use the properties of logarithms to write each expression as a single term.
   **a.** $\log_2 7 + \log_2 5$    **b.** $2 \ln x + \ln(x + 6)$    **c.** $\log(x + 2) - \log x$

**Solution** ▶    **a.** $\log_2 7 + \log_2 5 = \log_2(7 \cdot 5)$      product property
                $= \log_2 35$           simplify

   **b.** $2 \ln x + \ln(x + 6) = \ln x^2 + \ln(x + 6)$      power property
                     $= \ln[x^2(x + 6)]$      product property
                     $= \ln(x^3 + 6x^2)$      simplify

   **c.** $\log(x + 2) - \log x = \log\left(\dfrac{x + 2}{x}\right)$      quotient property

**Now try Exercises 25 through 50** ▶

**EXAMPLE 6** ▶  **Rewriting Logarithmic Expressions Using the Power Property**

Use the power property of logarithms to rewrite each term as a product.

a. $\ln 5^x$    b. $\log 32^{x+2}$    c. $\log \sqrt{x}$

Solution ▶  a. $\ln 5^x = x \ln 5$                          power property

b. $\log 32^{x+2} = (x + 2) \log 32$        power property

c. $\log \sqrt{x} = \log x^{\frac{1}{2}}$                  write radical using a rational exponent

$= \dfrac{1}{2} \log x$                    power property

**Now try Exercises 51 through 58** ▶

For examples of how these properties are used in context, **see Exercises 85 and 86.**

**CAUTION** ▶  Note from Example 6(b) that parentheses *must be used* whenever the exponent is a sum or difference. There is a huge difference between $(x + 2) \log 32$ and $x + 2 \log 32$.

Examples 5 and 6 illustrate how the properties of logarithms are used to *consolidate* logarithmic terms, primarily in preparation for equation solving. In other cases, the properties are used to rewrite or *expand* logarithmic expressions, so that certain other procedures can be applied more easily. Example 7 actually lays the foundation for more advanced mathematical work.

**EXAMPLE 7** ▶  **Rewriting Expressions Using Logarithmic Properties**

Use the properties of logarithms to write the following expressions as sums or differences of simple logarithmic terms.

a. $\log(x^2 z)$    b. $\ln \sqrt{\dfrac{x}{x+5}}$

Solution ▶  a. $\log(x^2 z) = \log x^2 + \log z$              product property

$= 2 \log x + \log z$              power property

b. $\ln \sqrt{\dfrac{x}{x+5}} = \ln\left(\dfrac{x}{x+5}\right)^{\frac{1}{2}}$          write radical using a rational exponent

$= \dfrac{1}{2} \ln\left(\dfrac{x}{x+5}\right)$          power property

$= \dfrac{1}{2}[\ln x - \ln(x + 5)]$      quotient property

$= \dfrac{1}{2} \ln x - \dfrac{1}{2} \ln(x + 5)$      simplify

☑ **B.** You've just seen how we can apply the product, quotient, and power properties of logarithms

**Now try Exercises 59 through 70** ▶

As you begin working with applications of logarithmic properties, it may help to have them written on a separate note card. This will enable you to compare each step and property as they are applied, remembering that "$M$" and "$N$" can represent any positive number or real-valued expression. For Example 7(a) we then have

$$\log(x^2 z) = \log x^2 + \log z$$

## C. The Change-of-Base Formula

Although base-10 and base-$e$ logarithms dominate the mathematical landscape, there are many practical applications of other bases. Fortunately, a formula exists that will convert any given base into either base 10 or base $e$. It's called the **change-of-base formula.**

### Change-of-Base Formula

For positive numbers $M$, $a$, and $b$, with $a, b \neq 1$,

$$\log_b M = \frac{\ln M}{\ln b} \qquad\qquad \log_b M = \frac{\log M}{\log b} \qquad\qquad \log_b M = \frac{\log_a M}{\log_a b}$$

base $e$                       base 10                       arbitrary base $a$

### Proof of the Change-of-Base Formula

For $y = \log_b M$, we have $b^y = M$ in exponential form. It follows that

$$\log_a(b^y) = \log_a M \qquad \text{take base-}a\text{ logarithm of both sides}$$

$$y \log_a b = \log_a M \qquad \text{power property of logarithms}$$

$$y = \frac{\log_a M}{\log_a b} \qquad \text{divide by } \log_a b$$

$$\log_b M = \frac{\log_a M}{\log_a b} \qquad \text{substitute } \log_b M \text{ for } y$$

**EXAMPLE 8** ▶ **Using the Change-of-Base Formula to Evaluate Expressions**

Find the value of each expression using the change-of-base formula. Answer in exact form and approximate form using nine digits, then *verify the result* using the original base. Note that either base 10 or base $e$ can be used.

**a.** $\log_3 29$      **b.** $\log_5 3.6$

Solution ▶    **a.** $\log_3 29 = \dfrac{\log 29}{\log 3}$            **b.** $\log_5 3.6 = \dfrac{\ln 3.6}{\ln 5}$

$\approx 3.065044752$           $\approx 0.795888947$

☑ **C. You've just seen how we can apply the change-of-base formula**

**Check:** $3^{3.065044752} \approx 29$ ✓        **Check:** $5^{0.795888947} \approx 3.6$ ✓

Now try Exercises 71 through 78 ▶

The change-of-base formula can also be used to study and graph logarithmic functions of *any* base. For $y = \log_b x$, the right-hand expression is simply rewritten using the formula and the equivalent function is $y = \frac{\log x}{\log b}$. The new function can then be evaluated as in Example 8, or used to study the graph of $y = \log_b x$ for any base $b$. **See Exercises 79 through 82.**

## D. Solving Applications of Logarithms

We end this section with one additional application of logarithms. For all living things, the concentration of hydrogen ions in a solution plays an important role as their presence or absence alters the environment of other molecules in the solution. This can dramatically affect the functionality of the solution, or the ability of an organism to survive. The concentration of hydrogen ions (in moles per liter) is commonly expressed in terms of what is called the **pH scale.** A low pH number corresponds to high hydrogen ion concentration, and the solution is said to be acidic. A high pH corresponds to low hydrogen ion concentration, and the solution is said to be basic. Since the range of values is so large, the pH scale is logarithmic, meaning each unit change in the pH scale represents a 10-fold increase or decrease in the concentration of hydrogen ions. For example, tomato juice, with a pH level of 2, is 10 times more acidic than orange juice, with a pH level of 3, and 100 times more acidic than grape juice, with a pH level of 4. The pH values range from 0 to 14, with pure water at pH = 7 being deemed "neutral" (neither basic nor acidic). Measuring pH levels plays an important role in biology, chemistry, food science, environmental science, medicine, oceanography, personal care products, and many other areas. The concentration of hydrogen atoms is usually represented by the term $[H^+]$, with the pH defined as $pH = -\log[H^+]$.

**EXAMPLE 9A ▶**

### The Concentration of Hydrogen Atoms in Ocean Water

Ocean water has a pH near 7.9. What is the concentration of hydrogen ions? Write the result in scientific notation.

**Solution ▶**

Begin with the basic formula and work from there.

$$pH = -\log[H^+] \qquad \text{pH formula}$$
$$7.9 = -\log[H^+] \qquad \text{substitute 7.9 for pH}$$
$$-7.9 = \log[H^+] \qquad \text{multiply by } -1$$
$$10^{-7.9} = [H^+] \qquad \text{exponential form}$$
$$1.26 \times 10^{-8} \approx [H^+] \qquad \text{result}$$

The hydrogen ion concentration in ocean water is about $1.26 \times 10^{-8}$ moles/liter.

**EXAMPLE 9B ▶**

### Finding the pH Level of an Apple

The concentration of hydrogen ions in an everyday apple is very near $7.94 \times 10^{-4}$. What is the pH level of an apple?

**Solution ▶**

$$pH = -\log[H^+] \qquad \text{pH formula}$$
$$= -\log[7.94 \times 10^{-4}] \qquad \text{substitute } 7.94 \times 10^{-4} \text{ for } [H^+]$$
$$\approx 3.1 \qquad \text{result}$$

☑ **D. You've just seen how we can solve applications of logarithms**

An apple has a pH level near 3.1.

Now try Exercises 87 through 90 ▶

## 5.4 EXERCISES

▶ **CONCEPTS AND VOCABULARY**

**Fill in each blank with the appropriate word or phrase. Carefully reread the section, if necessary.**

1. To solve $\ln 2x - \ln(x + 3) = 0$, we can combine terms using the _____ property, or add $\ln(x + 3)$ to both sides and use the _____ property.

2. The statement $\log_e 10 = \frac{\log 10}{\log e}$ is an example of the _____ -of- _____ formula.

3. Use all factor pairs of 36 to illustrate the product property of logarithms. For example, since $36 = 4 \cdot 9$, is $\log(4 \cdot 9) = \log 4 + \log 9$?

4. Use integer divisors of 24 to illustrate the quotient property of logarithms. For example, since $12 = \frac{24}{2}$, is $\log 12 = \log\left(\frac{24}{2}\right) = \log 24 - \log 2$?

## ▶ DEVELOPING YOUR SKILLS

Solve each equation by applying fundamental properties. Answer in exact form and approximate form rounded to thousandths.

**5.** $3.4 = \ln x$          **6.** $\ln x = \frac{1}{2}$

**7.** $\log x = \frac{1}{4}$        **8.** $1.6 = \log x$

**9.** $9.025 = e^x$        **10.** $e^x = 0.343$

**11.** $10^x = 18.197$      **12.** $0.024 = 10^x$

Solve each exponential equation. Write answers in exact form and in approximate form rounded to four decimal places.

**13.** $4e^{x-2} + 5 = 70$      **14.** $2 - 3e^{0.4x} = -7$

**15.** $10^{x+5} - 228 = -150$    **16.** $10^{2x} + 27 = 190$

**17.** $-150 = 290.8 - 190e^{-0.75x}$

**18.** $250e^{0.05x+1} + 175 = 1175$

Solve each logarithmic equation. Write answers in exact form and in approximate form rounded to four decimal places.

**19.** $3\ln(x + 4) - 5 = 3$     **20.** $-15 = -8\ln(3x) + 7$

**21.** $-1.5 = 2\log(5 - x) - 4$

**22.** $-4\log(2x) + 9 = 3.6$

**23.** $\frac{1}{2}\ln(2x + 5) + 3 = 3.2$

**24.** $\frac{3}{4}\ln(4x) - 6.9 = -5.1$

Use properties of logarithms to write each expression as a single term.

**25.** $\log_2 7 + \log_2 6$      **26.** $\log_9 2 + \log_9 15$

**27.** $\ln(2x) + \ln(x - 7)$    **28.** $\ln(x + 2) + \ln(3x)$

**29.** $\log(x + 1) + \log(x - 1)$   **30.** $\log(x - 3) + \log(x + 3)$

**31.** $\log_3 28 - \log_3 7$      **32.** $\log_6 30 - \log_6 10$

**33.** $\log x - \log(x + 1)$     **34.** $\log(x - 2) - \log x$

**35.** $\ln(x - 5) - \ln x$      **36.** $\ln(x + 3) - \ln(x - 1)$

**37.** $\ln(x^2 - 4) - \ln(x + 2)$

**38.** $\ln(x^2 - 25) - \ln(x + 5)$

**39.** $\log_5(x^2 - 2x) + \log_5 x^{-1}$

**40.** $\log_3(3x^2 + 5x) - \log_3 x$

**41.** $\log_3(x + 2) + \log_3(x - 3) - \log_3(x + 6)$

**42.** $\log x^2 - \log(2x + 1) - \log(2x - 1)$

**43.** $\ln 3 - \ln y^2 + \ln z$

**44.** $-\log_2\sqrt{p} + \log_2 5 + \log_2\sqrt[3]{q}$

**45.** $3\log x + \log(x - 1)$

**46.** $\log_7(2 - x) - 4\log_7(3 + x)$

**47.** $2\log x - \frac{1}{2}\log(x^2 + 1)$

**48.** $\frac{1}{2}[\ln(x - 1) - 3\ln x]$

**49.** $2\log_3 p - 4(\log_3 2 + \log_3 q)$

**50.** $\frac{1}{2}\ln x - 3(\ln y + 2\ln z^2)$

Use the power property of logarithms to rewrite each term as a product.

**51.** $\log 8^{x+2}$         **52.** $\log 15^{x-3}$

**53.** $\ln 5^{2x-1}$         **54.** $\ln 10^{3x+2}$

**55.** $\log\sqrt{22}$         **56.** $\log\sqrt[3]{34}$

**57.** $\log_5 81$          **58.** $\log_7 121$

Use the properties of logarithms to write the following expressions as sums or differences of simple logarithmic terms.

**59.** $\log(a^3 b)$         **60.** $\log(m^2 n)$

**61.** $\ln(x\sqrt[4]{y})$        **62.** $\ln(\sqrt[3]{pq})$

**63.** $\ln\left(\dfrac{x^2}{y}\right)$       **64.** $\ln\left(\dfrac{m^2}{n^3}\right)$

**65.** $\log\sqrt{\dfrac{x - 2}{x}}$     **66.** $\log\sqrt[3]{\dfrac{3 - v}{2v}}$

**67.** $\log_2(3x^2 y^4\sqrt{z})$    **68.** $\log_5\left(\dfrac{x^4}{2yz^2}\right)$

**69.** $\log_3\sqrt{\dfrac{x + 2}{(x - 2)(x + 3)}}$   **70.** $\log_7\left[\dfrac{x^3(x + 1)}{(x + 3)^2}\right]^2$

Evaluate each expression using the change-of-base formula and either base 10 or base $e$. Answer in exact form and in approximate form using nine decimal places, then verify the result using the original base.

**71.** $\log_7 60$         **72.** $\log_8 92$

**73.** $\log_5 152$        **74.** $\log_6 200$

**75.** $\log_3 1.73205$     **76.** $\log_2 1.41421$

**77.** $\log_{0.5} 0.125$      **78.** $\log_{0.2} 0.008$

Use the change-of-base formula to write an equivalent function, then evaluate the function as indicated (round to six decimal places). Investigate and discuss any patterns you notice in the output values, then determine the next input that will continue the pattern.

**79.** $f(x) = \log_3 x; f(5), f(15), f(45)$

**80.** $g(x) = \log_2 x; g(5), g(10), g(20)$

**81.** $h(x) = \log_9 x; h(2), h(4), h(8)$

**82.** $H(x) = \log_\pi x; H(\sqrt{2}), H(2), H(2\sqrt{2})$

## ▶ WORKING WITH FORMULAS

**83.** $\log_b M = \dfrac{1}{\log_M b}$

Use the change-of-base formula to verify the "formula" shown.

**84.** $\log_B A \cdot \log_C B \cdot \log_D C = \log_D A$

Use the change-of-base formula to verify the "formula" shown.

## ▶ APPLICATIONS

**85. Pareto's 80/20 principle:** After observing that 80% of the land in his native Italy was owned by 20% of the population, Italian economist Vilfredo Pareto (1848–1923) noted this disparity in many other areas (20% of the workers produce 80% of the output, 20% of the customers create 80% of the revenue, etc.) and developed a mathematical model for this phenomenon, called **Pareto's law.** If $N$ represents the number of people with incomes greater than $X$, then $\log N = \log A - m \log X$, where $A$ and $m$ are predetermined constants. (a) Solve the equation for $N$ and (b) given $m = 1.5$ and $A = 9900$, find the number of people earning over \$200,000. Assume $X$ is in hundreds of thousands of dollars.

**86. The species/area relationship:** In the study of *island biogeography,* the relationship between island area and the number of species present could be  modeled by the equation $\log S = \log C + k \log A$, where $S$ represents the total number of species, $A$ represents the area of the island, while $C$ and $k$ are predetermined constants that depend on numerous factors, including the size and proximity of other land masses. This makes it possible to predict the number of species on an island, when little other information is available. (a) Solve the equation for $S$ and (b) given $k = 0.81$ and $C = 8$, find the predicted number of species on an island with area of $A = 2000 \text{ km}^2$.

**87. Blood plasma pH levels:** To be safe and usable, the blood plasma held by blood banks must have a pH level between 7.35 and 7.45. Blood outside of this normal range can cause disorientation, behavioral changes, or even death. Using ion-sensitive electrodes, a  sample of blood plasma is known to have a concentration of $[\text{H}^+] = 4.786 \times 10^{-8}$. Is the plasma usable?

**88. Fresh milk:** As milk begins to sour, there is a corresponding decrease in pH level. Fresh milk has a pH level of near 6.5. After transport from farm to market, a sample of milk is tested using ion-sensitive electrodes and is found to have a concentration of $[\text{H}^+] = 3.981 \times 10^{-5}$. Is this shipment of milk still suitable for market?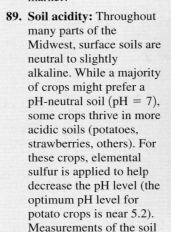

**89. Soil acidity:** Throughout many parts of the Midwest, surface soils are neutral to slightly alkaline. While a majority of crops might prefer a pH-neutral soil (pH = 7), some crops thrive in more acidic soils (potatoes, strawberries, others). For these crops, elemental sulfur is applied to help decrease the pH level (the optimum pH level for potato crops is near 5.2). Measurements of the soil  on a certain midwestern farm indicate a hydrogen ion concentration of $[\text{H}^+] = 1.259 \times 10^{-6}$. Is the soil ready for a potato crop to be planted?

**90. Acidity of gastric juices:** The normal pH value of human gastric juice can vary from 1 to 3, depending on genetics, diet, and other factors. The acidity is designed to control various harmful microorganisms that a person may ingest as they eat. Drinking large quantities of water before a meal can have a dramatic effect on this pH value, sometimes raising it as high as 4 or 5, making it possible for some harmful bacteria to survive. If a hospital patient's stomach fluid has a hydrogen ion concentration of $[\text{H}^+] = 3.981 \times 10^{-3}$, is the pH level within a normal range?

## ▶ EXTENDING THE CONCEPT

**91.** Logarithmic properties can also be used to compare the magnitude of very large and very small numbers, numbers too large or small for a handheld calculator to manage. Use the power property of logarithms to compare the numbers $600^{601}$ and $601^{600}$. Which is larger? Of the two numbers $\frac{1}{99^{100}}$ and $\frac{1}{100^{99}}$, which is smaller?

**92.** Suppose you and I represent two different numbers. Is the following cryptogram true or false? *The log of me base me is one and the log of you base you is one, but the log of you base me is equal to the log of me base you turned upside down.*

**93.** (a) Show that for the linear function $f(x) = mx + b$, $f(x + 1) = m + f(x)$. Then (b) show that for the exponential function $g(x) = a \cdot b^x$, $g(x + 1) = b \cdot g(x)$. (c) Use parts (a) and (b) to describe what increasing the input by one does to the output of each type of function.

**94.** (a) Show that any exponential function of the form $f(x) = a \cdot b^x$ can be rewritten in the form $f(x) = a \cdot e^{kx}$ by using $k = \ln b$. (b) Use the formula in part (a), to rewrite $V(t) = V_0 \cdot \left(\frac{4}{5}\right)^t$ and $A(t) = A_0 \cdot 2^t$ in terms of the natural exponential.

## ▶ MAINTAINING YOUR SKILLS

**95.** (4.5) State the zeroes of $f$ and the equation of any horizontal or vertical asymptotes given $f(x) = \frac{x^2 - x - 6}{x^2 - 1}$.

**96.** (1.5) Find all values of $x$ that make the equation true: $2x(x - 3) + 4 = (3x + 5)(x - 1)$.

**97.** (2.3) A sports shop can stock 36 cans of tennis balls on shelves that are 9 in. deep, and 54 cans on 12-in. shelves. Assuming the relationship is linear, (a) find the equation relating shelf size to number of cans, and (b) use it to determine what size shelf should be used to stock a full shipment of 72 cans.

**98.** (4.1) Determine the equation of the function shown in (a) shifted form and (b) standard form

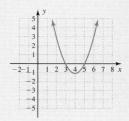

---

## 5.5  Solving Exponential and Logarithmic Equations

**LEARNING OBJECTIVES**

*In Section 5.5 you will see how we can:*

❑ **A.** Solve general logarithmic and exponential equations

❑ **B.** Solve applications involving logistic, exponential, and logarithmic functions

In this section, we'll develop the ability to solve more general logarithmic and exponential equations. A logarithmic equation has at least one term that involves the logarithm of a variable. Likewise, an exponential equation is one that involves a variable exponent on some base. In the same way that we might square both sides or divide both sides of an equation in the solution process, we'll show that we can also exponentiate both sides or take logarithms of both sides to help obtain a solution.

### A. Solving Logarithmic and Exponential Equations

One of the most common mistakes in solving exponential and logarithmic equations is to apply the inverse function too early—before the equation has been simplified. Just as we would naturally try to combine like terms for the equation $2\sqrt{x} + 7\sqrt{x} = 36$ (prior to squaring both sides), the logarithmic terms in $\log x + \log(x + 3) = 1$ must be combined prior to applying the exponential form. In addition, since the domain of $y = \log_b x$ is $x > 0$, logarithmic equations can sometimes produce **extraneous roots,** and checking all answers is a good practice. We'll illustrate by solving the equation $\log x + \log(x + 3) = 1$.

**EXAMPLE 1** ▶  **Solving a Logarithmic Equation**

Solve for $x$ and check your answer: $\log x + \log(x + 3) = 1$.

Solution ▶

$$
\begin{aligned}
\log x + \log(x + 3) &= 1 && \text{original equation} \\
\log[x(x + 3)] &= 1 && \text{product property} \\
x^2 + 3x &= 10^1 && \text{exponential form, distribute } x \\
x^2 + 3x - 10 &= 0 && \text{set equal to 0} \\
(x + 5)(x - 2) &= 0 && \text{factor} \\
x = -5 \text{ or } x &= 2 && \text{result}
\end{aligned}
$$

**Check:** The "solution" $x = -5$ is outside the domain and is extraneous. For $x = 2$,

$$
\begin{aligned}
\log x + \log(x + 3) &= 1 && \text{original equation} \\
\log 2 + \log(2 + 3) &= 1 && \text{substitute 2 for } x \\
\log 2 + \log 5 &= 1 && \text{simplify} \\
\log(2 \cdot 5) &= 1 && \text{product property} \\
\log 10 &= 1 && \text{Property I}
\end{aligned}
$$

You could also use a calculator to verify $\log 2 + \log 5 = 1$ directly.

**Now try Exercises 5 through 10** ▶

If the simplified form of an equation yields a logarithmic term on both sides, the **uniqueness property of logarithms** provides an efficient way to work toward a solution. Since logarithmic functions are one-to-one, we have

**The Uniqueness Property of Logarithms**

For positive real numbers $m$, $n$, and $b \neq 1$,

1. If $\log_b m = \log_b n$,     2.     If $m \neq n$,
   then $m = n$                           then $\log_b m \neq \log_b n$

Equal bases imply equal arguments.

**EXAMPLE 2** ▶  **Solving Logarithmic Equations Using the Uniqueness Property**

Solve each equation using the uniqueness property.

**a.** $\log(x + 2) = \log 7 + \log x$     **b.** $\ln 87 - \ln x = \ln 29$

Solution ▶

**a.**
$$
\begin{aligned}
\log(x + 2) &= \log 7 + \log x && \text{original equation} \\
\log(x + 2) &= \log 7x && \text{product property} \\
x + 2 &= 7x && \text{uniqueness property} \\
2 &= 6x && \text{solve for } x \\
\frac{1}{3} &= x && \text{result}
\end{aligned}
$$

**b.**
$$
\begin{aligned}
\ln 87 - \ln x &= \ln 29 && \text{original equation} \\
\ln\left(\frac{87}{x}\right) &= \ln 29 && \text{quotient property} \\
\frac{87}{x} &= 29 && \text{uniqueness property} \\
87 &= 29x && \text{clear denominator} \\
3 &= x && \text{result}
\end{aligned}
$$

**Now try Exercises 11 through 16** ▶

Often the solution may depend on using a variety of algebraic skills in addition to logarithmic or exponential properties.

**EXAMPLE 3 ▶ Solving Logarithmic Equations**

Solve the equation and check your answer.

$\log(x + 12) - \log x = \log(x + 9)$

**Solution ▶**

$$\log(x + 12) - \log x = \log(x + 9) \qquad \text{given equation}$$

$$\log\left(\frac{x + 12}{x}\right) = \log(x + 9) \qquad \text{quotient property}$$

$$\frac{x + 12}{x} = x + 9 \qquad \text{uniqueness property}$$

$$x + 12 = x^2 + 9x \qquad \text{clear denominator}$$

$$0 = x^2 + 8x - 12 \qquad \text{set equal to 0}$$

The equation is not factorable, and the quadratic formula must be used.

$$x = \frac{-b \pm \sqrt{b^2 - 4ac}}{2a} \qquad \text{quadratic formula}$$

$$= \frac{-8 \pm \sqrt{8^2 - 4(1)(-12)}}{2(1)} \qquad \begin{array}{l}\text{substitute 1 for } a, 8 \\ \text{for } b, -12 \text{ for } c\end{array}$$

$$= \frac{-8 \pm \sqrt{112}}{2} = \frac{-8 \pm 4\sqrt{7}}{2} \qquad \text{simplify}$$

$$= -4 \pm 2\sqrt{7} \qquad \text{result}$$

Substitution shows $x = -4 + 2\sqrt{7} \approx 1.29150$ checks, but substituting $-4 - 2\sqrt{7} \approx -9.2915$ for $x$ gives $\log(2.7085) - \log(-9.2915) = \log(-0.2915)$ and since the domain of a logarithm requires positive arguments, $x = -4 - 2\sqrt{7}$ must be an extraneous root.

**Now try Exercises 17 through 26 ▶**

**CAUTION ▶** Be careful not to dismiss or discard a possible solution simply because it's negative. For the equation $\log(-6 - x) = 1$, $x = -16$ is the solution (the domain here allows negative $x$-inputs: $-6 - x > 0$ yields $x < -6$ as the domain). In general, when a logarithmic equation has multiple solutions, all solutions should be checked.

Solving an exponential equation likewise involves isolating an exponential term on one side, or writing the equation where exponential terms of like base occur on each side. The latter case can be solved using the uniqueness property. If the exponential base is neither 10 nor $e$, logarithms of base $b$ can be used along with the change-of-base formula to solve the equation.

**EXAMPLE 4 ▶ Solving an Exponential Equation Using Base $b$**

Solve the exponential equation. Answer in both exact form, and approximate form to four decimal places: $4^{3x} - 1 = 8$

**Solution ▶**

$4^{3x} - 1 = 8 \qquad \text{given equation}$

$4^{3x} = 9 \qquad \text{add 1}$

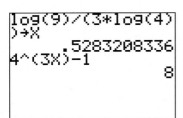

The left-hand side is neither base 10 or base $e$, so here we choose base 4 to solve.

$$\log_4 4^{3x} = \log_4 9 \qquad \text{logarithms base 4}$$

$$3x = \frac{\log 9}{\log 4} \qquad \text{Property III; change-of-base property}$$

$$x = \frac{\log 9}{3 \log 4} \qquad \text{multiply by } \tfrac{1}{3} \text{ (exact form)}$$

$$\approx 0.5283 \qquad \text{approximate form}$$

A calculator check is shown here.

**Now try Exercises 27 through 30** ▶

In some cases, two exponential terms with *unlike* bases may be involved. In this case, either common logs or natural logs can be used, but be sure to distinguish between constant terms like *ln 5* and variable terms like *x ln 5*. As with all equations, the goal is to isolate the *variable terms* on one side.

**EXAMPLE 5** ▶ **Solving an Exponential Equation with Unlike Bases**

Solve the exponential equation $5^{x+1} = 6^{2x}$.

**Solution** ▶ Begin by taking the natural log of both sides:

$$\ln(5^{x+1}) = \ln(6^{2x}) \qquad \text{apply base-}e \text{ logarithms}$$

$$(x+1)\ln 5 = 2x \ln 6 \qquad \text{power property}$$

$$x \ln 5 + \ln 5 = 2x \ln 6 \qquad \text{distribute}$$

$$\ln 5 = 2x \ln 6 - x \ln 5 \qquad \text{variable terms to one side}$$

$$\ln 5 = x(2 \ln 6 - \ln 5) \qquad \text{factor out } x$$

$$\frac{\ln 5}{2 \ln 6 - \ln 5} = x \qquad \text{solve for } x \text{ (exact form)}$$

$$0.8153 \approx x \qquad \text{approximate form}$$

**Now try Exercises 31 through 36** ▶

As an alternative to taking the natural log of both sides directly, we can use the properties of exponents to simplify and combine the exponential terms as follows.

$$5^{x+1} = 6^{2x} \qquad \text{original equation}$$

$$5^x 5^1 = (6^2)^x \qquad \text{product and power properties}$$

$$5 \cdot 5^x = 36^x \qquad \text{rewrite factors; simplify}$$

$$5 = \frac{36^x}{5^x} = \left(\frac{36}{5}\right)^x \qquad \text{divide by } 5^x \text{, apply power property}$$

$$\ln 5 = x \ln 7.2 \qquad \text{take natural logs } (\tfrac{36}{5} = 7.2)$$

$$\frac{\ln 5}{\ln 7.2} = x \qquad \text{solve for } x$$

The result is equivalent to our original solution: $x \approx 0.81528463$.

Logarithmic equations come in many different forms, and the following ideas summarize the basic approaches used in solving them. For this summary, recall that for any positive number $k$, $\log_b k$ *is a constant*. Assume $M$, $N$, and $X$ represent algebraic expressions in $x$.

1.  If the equation can be written in the form $\log_b M$ = constant, use the exponential form and algebra to solve: $M = b^{\text{constant}}$.
2.  If the equation can be written in the form $\log_b M = \log_b N$, use the uniqueness property and algebra to solve: $M = N$.
3.  If the equation has an additional constant term as in $\log_b M = \log_b N$ + constant, move all logarithmic terms to one side and consolidate using the product or quotient properties, then use the exponential form and algebra to solve:

$$\log_b M = \log_b N + \text{constant}$$
$$\log_b M - \log_b N = \text{constant}$$
$$\log_b\left(\frac{M}{N}\right) = \text{constant}$$
$$\frac{M}{N} = b^{\text{constant}}$$

4.  If the equation has multiple logarithmic terms as in $\log_b X = \log_b M + \log_b N$, consolidate logarithmic terms using the product or quotient properties, then use the uniqueness property and algebra to solve:

$$\log_b X = \log_b M + \log_b N$$
$$\log_b X = \log_b(MN)$$
$$X = MN$$

Many other forms and varieties are possible.

In advanced applications, the equations used are sometimes impossible to solve using inverse functions. This is often the case when logarithmic or exponential functions are mixed with other functions (polynomial, radical, rational, etc.). In these cases graphing and calculating technologies become indispensable tools, and the emphasis in working towards a solution shifts more to an understanding of the domain, the graphical attributes of the functions involved, and setting an appropriate window.

**EXAMPLE 6** ▶ **Solving Equations Using Technology**

Find all solutions to $e^{\sqrt[3]{x-8}} = \sqrt{x+9}$.

Solution ▶ Begin by noting that the domain of the function on the left, call it $Y_1$, is all real numbers, since this is the domain of both $y = e^x$ and $y = \sqrt[3]{x}$. However, the domain of the function on the right (call it $Y_2$) is $x \geq -9$, and we expect that any solution(s) to the equation must occur to the right of $-9$, so we opt for a standard viewing window to begin (Figure 5.35). At first it appears the graphs do not intersect to the left, but our knowledge of the domain

**Figure 5.35**

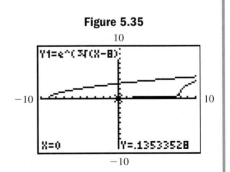

indicates they must intersect near $x = -9$, since $Y_1$ and $Y_2$ are both positive. Using the intersection-of-graphs method, we find one solution is $x \approx -8.994$. To the right, no point of intersection is initially visible, so we extend our window to explore whether the graphs intersect again. Using Xmax $= 20$ we note the graphs indeed intersect (Figure 5.36) and locate a second solution at $x \approx 11.433$.

**Figure 5.36**

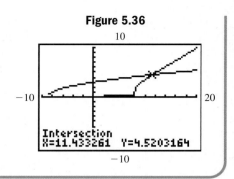

☑ **A.** You've just seen how we can solve general logarithmic and exponential equations

**Now try Exercises 37 through 40** ▶

## B. Applications of Logistic, Exponential, and Logarithmic Functions

Applications of exponential and logarithmic functions take many different forms and it would be impossible to illustrate them all. As you work through the exercises, try to adopt a "big picture" approach, applying the general principles illustrated here to other applications.

In applications involving the **logistic growth** of animal populations, the initial stage of growth is virtually exponential, but due to limitations on food, space, or other resources, growth slows and at some point it reaches a limit. In business, the same principle applies to the logistic growth of sales or profits, due to market saturation. In these cases, the exponential term appears in the denominator of a quotient, and we "clear denominators" to begin the solution process.

**EXAMPLE 7** ▶    **Solving a Logistic Equation**

A small business makes a new discovery and begins an aggressive advertising campaign, confident they can capture 66% of the market in a short period of time. They anticipate their market share will be modeled by the function $M(t) = \frac{66}{1 + 10e^{-0.05t}}$, where $M(t)$ represents the percentage after $t$ days. Use this function to answer the following.

**a.** What was the company's initial market share ($t = 0$)? What was their market share 30 days later?

**b.** How long will it take the company to reach a 60% market share?

**Solution** ▶    **a.** $M(t) = \dfrac{66}{1 + 10e^{-0.05t}}$    given

$M(0) = \dfrac{66}{1 + 10e^{-0.05(0)}}$    substitute 0 for $t$

$= \dfrac{66}{1 + 10}$    simplify ($e^0 = 1$)

$= 6$    result

The company originally had only a 6% market share.

$M(30) = \dfrac{66}{1 + 10e^{-0.05(30)}}$    substitute 30 for $t$

$= \dfrac{66}{1 + 10e^{-1.5}}$    simplify

$\approx 20.4$    result

After 30 days, they held a 20.4% market share.

**b.** For part (b), we replace $M(t)$ with 60 and solve for $t$.

$$60 = \frac{66}{1 + 10e^{-0.05t}} \qquad \text{given}$$

$$60(1 + 10e^{-0.05t}) = 66 \qquad \text{multiply by } 1 + 10e^{-0.05t}$$

$$1 + 10e^{-0.05t} = 1.1 \qquad \text{divide by 60}$$

$$10e^{-0.05t} = 0.1 \qquad \text{subtract 1}$$

$$e^{-0.05t} = 0.01 \qquad \text{divide by 10}$$

$$\ln e^{-0.05t} = \ln 0.01 \qquad \text{apply base-}e \text{ logarithms}$$

$$-0.05t = \ln 0.01 \qquad \text{Property III}$$

$$t = \frac{\ln 0.01}{-0.05} \qquad \text{solve for } t \text{ (exact form)}$$

$$\approx 92 \qquad \text{approximate form}$$

The company will reach a 60% market share in about 92 days.

Now try Exercises 43 and 44 ▶

Earlier in our study (page 458) we used the formula $H = (30T + 8000) \ln\left(\frac{P_0}{P}\right)$ to find an altitude $H$, given a temperature in degrees Celsius and the atmospheric (barometric) pressure in centimeters of mercury (cmHg). Using the tools from this section, we are now able to find the atmospheric pressure for a given altitude and temperature.

**EXAMPLE 8** ▶ **Using Logarithms to Determine Atmospheric Pressure**

Suppose a group of climbers has just scaled Mt. Rainier, the highest mountain of the Cascade Range in western Washington State. If the mountain is about 4395 m high and the temperature at the summit is $-22.5°C$, what is the atmospheric pressure at this altitude? The pressure at sea level is $P_0 = 76$ cmHg.

Solution ▶

$$H = (30T + 8000) \ln\left(\frac{P_0}{P}\right) \qquad \text{given}$$

$$4395 = [30(-22.5) + 8000] \ln\left(\frac{76}{P}\right) \qquad \begin{array}{l}\text{substitute 4395}\\\text{for } H, \text{ 76 for } P_0,\\\text{and } -22.5 \text{ for } T\end{array}$$

$$4395 = 7325 \ln\left(\frac{76}{P}\right) \qquad \text{simplify}$$

$$0.6 = \ln\left(\frac{76}{P}\right) \qquad \text{divide by 7325}$$

$$e^{0.6} = \frac{76}{P} \qquad \text{exponential form}$$

$$Pe^{0.6} = 76 \qquad \text{multiply by } P$$

$$P = \frac{76}{e^{0.6}} \qquad \begin{array}{l}\text{divide by } e^{0.6}\\\text{(exact form)}\end{array}$$

$$\approx 41.7 \text{ cmHg} \qquad \text{approximate form}$$

☑ **B.** You've just seen how we can solve applications involving logistic, exponential, and logarithmic functions

Now try Exercises 45 and 46 ▶

Additional applications involving appreciation/depreciation, Newton's law of cooling, spaceship velocities and more, can be found in the Exercise Set. **See Exercises 47 through 56.**

## 5.5 EXERCISES

### ▶ CONCEPTS AND VOCABULARY

**Fill in each blank with the appropriate word or phrase. Carefully reread the section, if necessary.**

1. To solve the equation $\ln(x + 3) - \ln x = 7$, we _____ like terms using logarithmic properties, prior to writing the equation in _____ form.

2. Since the domain of $y = \log_b x$ is _____, solving logarithmic equations will sometimes produce _____ roots. Checking all solutions to logarithmic equations is a necessary step.

3. Answer true or false and explain your response:
$$\log_b(M + N) = \log_b(M) + \log_b(N)$$

4. Answer true or false and explain your response:
$$\log_b\left(\frac{M}{N}\right) = \frac{\log_b M}{\log_b N}$$

### ▶ DEVELOPING YOUR SKILLS

**Solve each equation and check your answers.**

5. $\log 4 + \log(x - 7) = 2$

6. $\log 5 + \log(x - 9) = 1$

7. $\log(x - 15) - 2 = -\log x$

8. $\log x - 1 = -\log(x - 9)$

9. $\log(2x + 1) = 1 - \log x$

10. $\log(3x - 13) = 2 - \log x$

**Solve each equation using the uniqueness property.**

11. $\log(5x + 2) = \log 2$

12. $\log(2x - 3) = \log 3$

13. $\log_4(x + 2) - \log_4 3 = \log_4(x - 1)$

14. $\log_3(x + 6) - \log_3 x = \log_3 5$

15. $\ln(8x - 4) = \ln 2 + \ln x$

16. $\ln(x - 1) + \ln 6 = \ln(3x)$

**Solve each equation using any appropriate method. State solutions in both exact form and in approximate form rounded to four decimal places. Clearly identify any extraneous roots. If there are no solutions, so state.**

17. $\log_2 9 + \log_2(x + 3) = 3$

18. $\log_3(x - 4) + \log 7 = 2$

19. $\ln(x + 7) + \ln 9 = 2$

20. $\ln 5 + \ln(x - 2) = 1$

21. $\log(x + 8) + \log x = \log(x + 18)$

22. $\log(x + 14) - \log x = \log(x + 6)$

23. $\log(-x - 1) = \log(5x) + \log x$

24. $\log(1 - x) + \log x = \log(x + 4)$

25. $\log(x - 1) - \log x = \log(x - 3)$

26. $\ln x + \ln(x - 2) = \ln 4$

27. $7^x = 231$

28. $6^x = 3589$

29. $5^{3x} - 2 = 128{,}965$

30. $9^{3x} - 3 = 78{,}462$

31. $2^{x+1} = 3^x$

32. $7^x = 4^{2x-1}$

33. $5^{2x+1} = 9^{x+1}$

34. $\left(\frac{1}{5}\right)^{x-1} = \left(\frac{1}{2}\right)^{3-x}$

35. $\dfrac{250}{1 + 4e^{-0.06x}} = 200$

36. $\dfrac{80}{1 + 15e^{-0.06x}} = 50$

**Solve each equation using the zeroes method or the intersection-of-graphs method. Round approximate solutions to three decimal places.**

37. $\sqrt[3]{x} = \ln(x + 5)$

38. $\dfrac{x^2 - 25}{x^2 - 9} = -\ln(x + 9) + 6$

39. $2^{x^2 - x - 6} = x^2 + x - 6$

40. $x^3 - 9x = \frac{1}{2}e^x$

### ▶ WORKING WITH FORMULAS

41. **Logistic growth:** $P(t) = \dfrac{C}{1 + ae^{-kt}}$

For populations that exhibit logistic growth, the population at time $t$ is modeled by the function shown, where $C$ is the carrying capacity of the population (the maximum population that can be supported over a long period of time), $k$ is the growth constant, and $a = \frac{C - P(0)}{P(0)}$. Solve the formula for $t$, then use the result to find the value of $t$ given $P(0) = 50$, $C = 450$, $P = 400$, and $k = 0.075$.

**42. Estimating time of death:** $h = -3.9 \cdot \ln\left(\frac{T - T_R}{T_0 - T_R}\right)$

Using the formula shown, a forensic expert can compute the approximate time of death for a person found recently expired, where $T$ is the body temperature when it was found, $T_R$ is the (constant) temperature of the room, $T_0$ is the body temperature at the time of death ($T_0 = 98.6°F$), and $h$ is the number of hours since death. If the body was discovered at 9:00 A.M. with a temperature of 86.2°F, in a room at 73°F, at approximately what time did the person expire? (Note this formula is a version of Newton's law of cooling.)

## ▶ APPLICATIONS

**43. Stocking a lake:** A farmer wants to stock a private lake on his property with catfish. A specialist studies the area and depth of the lake, along with other factors, and determines it can support a maximum population of around 750 fish, with growth modeled by the function $P(t) = \frac{750}{1 + 24e^{-0.075t}}$, where $P(t)$ gives the current population after $t$ months. (a) How many catfish did the farmer initially put in the lake? (b) How many months until the population reaches 300 fish?

**44. Increasing sales:** After expanding their area of operations, a manufacturer of small storage buildings believes the larger area can support sales of 40 units per month. After increasing the advertising budget and enlarging the sales force, sales are expected to grow according to the model $S(t) = \frac{40}{1 + 1.5e^{-0.08t}}$, where $S(t)$ is the expected number of sales after $t$ months. (a) How many sales were being made each month, prior to the expansion? (b) How many months until sales reach 25 units per month?

Use the *barometric equation* $H = (30T + 8000) \ln\left(\frac{P_0}{P}\right)$ for Exercises 45 and 46. Recall that $P_0 = 76$ cmHg.

**45. Altitude and temperature:** A sophisticated spy plane is cruising at an altitude of 18,250 m. If the temperature at this altitude is −75°C, what is the barometric pressure?

**46. Altitude and temperature:** A large weather balloon is released and takes altitude, pressure, and temperature readings as it climbs, and radios the information back to Earth. What is the pressure reading at an altitude of 5000 m, given the temperature is −18°C?

Use *Newton's law of cooling* $T = T_R + (T_0 - T_R)e^{kh}$ to complete Exercises 47 and 48. Use $k = -0.012$ and refer to Section 5.2, page 445 as needed.

**47. Making popsicles:** On a hot summer day, Sean and his friends mix some Kool-Aid® and decide to freeze it in an ice tray to make popsicles. If the water used for the Kool-Aid® was 75°F and the freezer has a temperature of −20°F, how long will they have to wait to enjoy the treat?

**48. Freezing time:**
Suppose the current temperature in Esconabe, Michigan, was 47°F when a 5°F arctic cold front moved over the state. How long would it take the wet roads to ice over and form dangerous "black ice"?

**Depreciation/appreciation: As time passes, the value of certain items decrease (appliances, automobiles, etc.), while the value of other items increase (collectibles, real estate, etc.). The time $T$ in years for an item to reach a future value can be modeled by the formula $T = k \ln\left(\frac{V_n}{V_f}\right)$, where $V_n$ is the purchase price when new, $V_f$ is its future value, and $k$ is a constant that depends on the item.**

**49. Automobile depreciation:** If a new car is purchased for $28,500, find its value 3 yr later if $k = 5$.

**50. Home appreciation:** If a new home in an "upscale" neighborhood is purchased for $130,000, find its value 12 yr later if $k = -16$.

**Drug absorption: The time required for a certain percentage of a drug to be *absorbed* by the body after injection depends on the drug's absorption rate. This can be modeled by the function $T(p) = \frac{-\ln p}{k}$, where $p$ represents the percent of the drug that *remains unabsorbed* (expressed as a decimal), $k$ is the absorption rate of the drug, and $T(p)$ represents the elapsed time.**

**51.** For a drug with an absorption rate of 7.2% per hour, (a) find the time required (to the nearest hour) for the body to *absorb* 35% of the drug, and (b) find the percent of this drug (to the nearest half percent) that remains unabsorbed after 24 hr.

**52.** For a drug with an absorption rate of 5.7% per hour, (a) find the time required (to the nearest hour) for the body to *absorb* 50% of the drug, and (b) find the percent of this drug (to the nearest half percent) that remains unabsorbed after 24 hr.

**Spaceship velocity: In space travel, the change in the velocity of a spaceship $V_s$ (in km/sec) depends on the mass of the ship $M_s$ (in tons), the mass of the fuel which has been burned $M_f$ (in tons) and the escape velocity of the exhaust $V_e$ (in km/sec). Disregarding frictional forces, these are related by the equation $V_s = V_e \ln\left(\frac{M_s}{M_s - M_f}\right)$.**

**53.** For the Jupiter VII rocket, find the mass of the fuel $M_f$ that has been burned if $V_s = 6$ km/sec when $V_e = 8$ km/sec, and the ship's mass is 100 tons.

**54.** For the Neptune X satellite booster, find the mass of the ship $M_s$ if $M_f = 75$ tons of fuel has been burned when $V_s = 8$ km/sec and $V_e = 10$ km/sec.

**Learning curve: The job performance of a new employee when learning a repetitive task (as on an assembly line) improves very quickly at first, then grows more slowly over time. This can be modeled by the function $P(t) = a + b \ln t$, where $a$ and $b$ are constants that depend on the type of task and the training of the employee.**

55. The number of toy planes an employee can assemble is modeled by $P(t) = 5.9 + 12.6 \ln t$, where $P(t)$ is the number of planes assembled daily after working $t$ days. (a) How many planes is an employee making after 5 days on the job? (b) How many days until the employee is able to assemble 34 planes per day?

56. The number of circuit boards an associate can assemble from its component parts is modeled by $B(t) = 1 + 2.3 \ln t$, where $B(t)$ is the number of boards assembled daily after working $t$ days. (a) How many boards is an employee completing after 9 days on the job? (b) How long will it take until the employee is able to complete 10 boards per day?

## ▶ EXTENDING THE CONCEPT

**In Exercises 57 and 58, solve the given equation. Note that each equation is in quadratic form.**

57. $2e^{2x} - 7e^x = 15$

58. $3e^{2x} - 4e^x - 7 = -3$

59. Find $f^{-1}(x)$ for the functions given, then state the domain and range of each inverse.
    a. $f(x) = \log(x - 2) + 1$     b. $f(x) = e^{x+5} - 2$

60. In his *Triangular Theory of Love*, psychologist Robert Sternberg studied the ways in which commitment, intimacy, and passion varied over the life of a relationship. Suppose the following functions model these three components of love as a function of time $t$ in years:

    Commitment: $C(x) = \dfrac{9.8}{1 + 400e^{-1.5x}}$

    Intimacy: $I(x) = \dfrac{7.3}{1 + 34e^{-0.93x}}$

    Passion: $P(x) = \dfrac{215x}{x^3 + 44}$

    a. If a proposal is in order once commitment reaches an intensity of 5, how many years should a couple date before popping the question?

    b. If "the honeymoon's over" once passion drops below an intensity of 8, how long will the honeymoon last? Compare your answer to the one found in part (a).

    c. Although passion may fade over time, intimacy and commitment generally don't. According to this model, what are the maximum sustainable intensities for intimacy and commitment?

61. Match each equation with the most appropriate solution strategy, and justify/discuss why.
    a. $e^{x+1} = 25$     _____     apply base-10 logarithm to both sides
    b. $\log(2x + 3) = \log 53$     _____     rewrite and apply uniqueness property for exponentials
    c. $\log(x^2 - 3x) = 2$     _____     apply uniqueness property for logarithms
    d. $10^{2x} = 97$     _____     apply either base-10 or base-$e$ logarithm
    e. $2^{5x-3} = 32$     _____     apply base-$e$ logarithm
    f. $7^{x+2} = 23$     _____     write in exponential form

## ▶ MAINTAINING YOUR SKILLS

62. (4.1) Match the graph shown with its correct equation, without actually graphing the function.
    a. $y = x^2 + 4x - 5$
    b. $y = -x^2 - 4x + 5$
    c. $y = -x^2 + 4x + 5$
    d. $y = x^2 - 4x - 5$

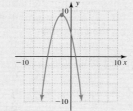

63. (2.4) State the domain and range of the functions.
    a. $y = \sqrt{2x + 3}$     b. $y = |x + 2| - 3$

64. (4.5) Graph the function $r(x) = \frac{x^2 - 4}{x - 1}$. Label all intercepts and asymptotes.

65. (3.3) Suppose the maximum load (in tons) that can be supported by a cylindrical post varies directly with its diameter raised to the fourth power and inversely with the square of its height. A post 8 ft high and 2 ft in diameter can support 6 tons. How many tons can be supported by a post 12 ft high and 3 ft in diameter?

# 5.6 Applications from Business, Finance, and Science

## LEARNING OBJECTIVES

*In Section 5.6 you will see how we can:*

☐ **A.** Calculate simple interest and compound interest

☐ **B.** Calculate interest compounded continuously

☐ **C.** Solve applications of annuities and amortization

☐ **D.** Solve applications of exponential growth and decay

Would you pay $750,000 for a home worth only $250,000? Surprisingly, when a conventional mortgage is repaid over 30 yr, this is not at all rare. Over time, the accumulated interest on the mortgage is easily more than two or three times the original value of the house. In this section, we explore how interest is paid or charged, and look at other applications of exponential and logarithmic functions from business, finance, as well as the physical and social sciences.

## A. Simple and Compound Interest

**Simple interest** is an amount of interest that is computed only once during the lifetime of an investment (or loan). In the world of finance, the initial deposit or base amount is referred to as the **principal $P$,** the **interest rate $r$** is given as a percentage and stated as an annual rate, with the term of the investment or loan most often given as *time $t$* in years. Simple interest is merely an application of the basic percent equation, with the additional element of time coming into play: *interest = principal × rate × time,* or $I = Prt$. To find the total amount $A$ that has accumulated (for deposits) or is due (for loans) after $t$ yr, we merely add the accumulated interest to the initial principal: $A = P + Prt$.

> **WORTHY OF NOTE**
>
> If a loan is kept for only a certain number of months, weeks, or days, the time $t$ should be stated as a fractional part of a year so the units for the rate and time (both in years) match.

### Simple Interest Formula

If principal $P$ is deposited or borrowed at interest rate $r$ for a period of $t$ yr, the simple interest on this account will be

$$I = Prt$$

The total amount $A$ accumulated or due after this period will be

$$A = P + Prt \quad \text{or} \quad A = P(1 + rt)$$

---

**EXAMPLE 1** ▶  **Solving an Application of Simple Interest**

Many finance companies offer what have become known as payday loans—a small $50 loan to help people get by until payday, usually no longer than 2 weeks. If the cost of this service is $12.50, determine the annual rate of interest charged by these companies.

**Solution** ▶  The interest charge is $12.50, the principal is $50.00, and the time period is 2 weeks or $\frac{2}{52} = \frac{1}{26}$ of a year. The simple interest formula yields

$$I = Prt \qquad \text{simple interest formula}$$

$$12.50 = 50r\left(\frac{1}{26}\right) \qquad \text{substitute \$12.50 for } I, \text{\$50.00 for } P, \text{ and } \tfrac{1}{26} \text{ for } t$$

$$6.5 = r \qquad \text{solve for } r$$

The annual interest rate on these loans is a whopping 650%!

**Now try Exercises 5 through 14** ▶

## Compound Interest

Many financial institutions pay **compound interest** on deposits they receive, which is interest paid on previously accumulated interest. The most common compounding frequencies are yearly, semiannually (two times per year), quarterly, monthly, and daily. For convenience, consider $1000 in principal, deposited at 8% for 3 yr. The simple interest calculation shows $240 in interest is earned and there will be $1240 in the account: $A = 1000[1 + (0.08)(3)] = \$1240$. If the interest is *compounded each year* instead of once at the end of the 3-yr period, the interest calculation shows

$A_1 = 1000(1 + 0.08) = 1080$ in the account at the end of year 1,

$A_2 = 1080(1 + 0.08) = 1166.40$ in the account at the end of year 2,

$A_3 = 1166.40(1 + 0.08) \approx 1259.71$ in the account at the end of year 3.

The account has earned an additional $1259.71 - \$1240 = \$19.71$ interest. More importantly, notice that we're multiplying by $(1 + 0.08)$ each compounding period, meaning results can be computed more efficiently by simply applying the factor $(1 + 0.08)^t$ to the initial principal $P$. For example,

$$A_3 = 1000(1 + 0.08)^3 \approx \$1259.71.$$

In general, for interest compounded yearly the **accumulated value** is $A = P(1 + r)^t$. Notice that solving this equation for $P$ will tell us the amount we need to deposit *now*, in order to accumulate $A$ dollars in $t$ yr: $P = \frac{A}{(1 + r)^t}$. This is called the **present value equation.**

### Interest Compounded Annually

If a principal $P$ is deposited at interest rate $r$ compounded yearly for a period of $t$ yr, the *accumulated value* is

$$A = P(1 + r)^t$$

If an accumulated value $A$ is desired after $t$ yr, and the money is deposited at interest rate $r$ compounded yearly, the *present value* is

$$P = \frac{A}{(1 + r)^t}$$

---

**EXAMPLE 2** ▶ **Finding the Doubling Time for Interest Compounded Yearly**

After being promised an 8% return, compounded yearly, Steve decides to invest $1000 in his friend's startup company. How long will it take for the money to double?

**Solution** ▶ Using the formula for interest compounded yearly we have

| | |
|---|---|
| $A = P(1 + r)^t$ | given |
| $2000 = 1000(1 + 0.08)^t$ | substitute 2000 for $A$, 1000 for $P$, and 0.08 for $r$ |
| $2 = 1.08^t$ | isolate variable term |
| $\ln 2 = t \ln 1.08$ | apply base-$e$ logarithms; power property |
| $\dfrac{\ln 2}{\ln 1.08} = t$ | solve for $t$ |
| $9.0 \approx t$ | approximate form |

The money will double in just over 9 yr.

**Now try Exercises 15 through 20** ▶

If interest is compounded monthly, the bank will divide the yearly interest rate by 12 to compute the monthly interest rate, but then pay you interest 12 times per year. The net effect is an increased gain in the interest you earn, and the final compound interest formula takes this form:

$$\text{accumulated amount} = \text{principal}\underbrace{\left(1 + \underbrace{\frac{\text{interest rate}}{\text{compoundings per year}}}_{\substack{\text{interest rate per}\\\text{compounding period}}}\right)^{\overbrace{(\text{compoundings per year} \times \text{years})}^{\substack{\text{total number of}\\\text{compounding periods}}}}}$$

### Compounded Interest Formula

If a principal $P$ is deposited at interest rate $r$ compounded $n$ times per year for a period of $t$ yr, the *accumulated value* will be:

$$A = P\left(1 + \frac{r}{n}\right)^{nt}$$

---

**EXAMPLE 3** ▶ **Solving an Application of Compound Interest**

Macalyn won $150,000 in the Missouri lottery and decides to invest the money for retirement in 20 yr. Of all the options available here, which one will produce the most money for retirement?

  **a.** A certificate of deposit paying 5.4% compounded yearly.
  **b.** A money market certificate paying 5.35% compounded semiannually.
  **c.** A bank account paying 5.25% compounded quarterly.
  **d.** A bond issue paying 5.2% compounded daily.

**Solution** ▶
**a.** $A = \$150,000\left(1 + \dfrac{0.054}{1}\right)^{(20 \times 1)}$      **c.** $A = \$150,000\left(1 + \dfrac{0.0525}{4}\right)^{(20 \times 4)}$

    $\approx \$429,440.97$                   $\approx \$425,729.59$

**b.** $A = \$150,000\left(1 + \dfrac{0.0535}{2}\right)^{(20 \times 2)}$      **d.** $A = \$150,000\left(1 + \dfrac{0.052}{365}\right)^{(20 \times 365)}$

    $\approx \$431,200.96$                   $\approx \$424,351.12$

☑ **A.** You've just seen how we can calculate simple interest and compound interest

The best choice is (b), semiannual compounding at 5.35% for 20 yr.

**Now try Exercises 21 through 28** ▶

---

## B. Interest Compounded Continuously

It seems natural to wonder what happens to the interest accumulation as $n$ (the number of compounding periods per year) becomes very large. It appears the interest rate becomes very small (because we're dividing it by $n$), but the exponent becomes very large (since we're multiplying it by $n$). To see the result of this interplay more clearly, it will help to rewrite the compound interest formula $A = P\left(1 + \frac{r}{n}\right)^{nt}$ using the substitution $n = xr$. This gives $\frac{r}{n} = \frac{1}{x}$, and by direct substitution ($xr$ for $n$ and $\frac{1}{x}$ for $\frac{r}{n}$) we obtain the form

$$A = P\left[\left(1 + \frac{1}{x}\right)^{x}\right]^{rt}$$

by regrouping. This allows for a more careful study of the "denominator versus exponent" relationship using the expression $\left(1 + \frac{1}{x}\right)^{x}$, *the same expression we used in*

*Section 5.2 to define the number e:* as $x \to \infty$, $(1 + \frac{1}{x})^x \to e$. From $n = xr$, $n$ becoming infinitely large implies $x$ becomes infinitely large, and we're able to develop a formula for **interest compounded continuously** by replacing $(1 + \frac{1}{x})^x$ with the number $e$ in the formula for compound interest:

$$A = P\left[\left(1 + \frac{1}{x}\right)^x\right]^{rt} \to Pe^{rt}$$

---

### Interest Compounded Continuously

If a principal $P$ is deposited at interest rate $r$ compounded continuously for a period of $t$ yr, the *accumulated value* will be

$$A = Pe^{rt}$$

---

**EXAMPLE 4** ▶ **Solving an Application of Interest Compounded Continuously**

Jaimin has $10,000 to invest and wants to have at least $25,000 in the account in 10 yr for his daughter's college education fund. If the account pays interest compounded continuously, what interest rate is required?

**Solution** ▶ In this case, $P = \$10{,}000$, $A = \$25{,}000$, and $t = 10$.

| | |
|---|---|
| $A = Pe^{rt}$ | given |
| $25{,}000 = 10{,}000e^{10r}$ | substitute given values |
| $2.5 = e^{10r}$ | isolate variable term |
| $\ln 2.5 = 10r \ln e$ | use natural logs; power property |
| $\dfrac{\ln 2.5}{10} = r$ | solve for $r$ ($\ln e = 1$) |
| $0.092 \approx r$ | approximate form |

☑ **B.** You've just seen how we can calculate interest compounded continuously

Jaimin will need an interest rate of about 9.2% to meet his goal.

**Now try Exercises 29 through 38** ▶

## C. Applications Involving Annuities and Amortization

Our previous calculations for simple and compound interest involved a single (lump) deposit (the principal) that accumulated interest over time. Many savings and investment plans involve a regular schedule of deposits (monthly, quarterly, or annual deposits) over the life of the investment. Such an investment plan is called an **annuity.**

Suppose that for 4 yr, $100 is deposited annually into an account paying 8% compounded yearly. Using the compound interest formula we can track the accumulated value $A$ in the account:

$$A = \underset{\substack{\text{accumulated} \\ \text{value of} \\ \text{4th payment}}}{100} + \underset{\substack{\text{accumulated} \\ \text{value of} \\ \text{3rd payment}}}{100(1.08)^1} + \underset{\substack{\text{accumulated} \\ \text{value of} \\ \text{2nd payment}}}{100(1.08)^2} + \underset{\substack{\text{accumulated} \\ \text{value of} \\ \text{1st payment}}}{100(1.08)^3}$$

To develop an annuity formula, we multiply the annuity equation by 1.08, then subtract the original equation. This leaves only the first and last terms, since the other (interior) terms cancel:

$$1.08A = 100(1.08) + 100(1.08)^2 + 100(1.08)^3 + 100(1.08)^4 \qquad \text{multiply by 1.08}$$
$$\underline{-A = -[100 + 100(1.08)^1 + 100(1.08)^2 + 100(1.08)^3]} \qquad \text{original equation}$$
$$1.08A - A = 100(1.08)^4 - 100 \qquad \text{subtract ("interior terms" cancel)}$$
$$0.08A = 100[(1.08)^4 - 1] \qquad \text{factor out 100}$$
$$A = \frac{100[(1.08)^4 - 1]}{0.08} \qquad \text{solve for } A$$

This result can be generalized for any periodic payment $\mathcal{P}$, interest rate $r$, number of yearly compounding periods $n$, and number of years $t$. This would give

$$A = \frac{\mathcal{P}\left[\left(1 + \dfrac{r}{n}\right)^{nt} - 1\right]}{\dfrac{r}{n}}$$

The formula can be made less formidable using $R = \frac{r}{n}$ and $N = nt$, where $R$ is the interest rate per compounding period and $N$ is the total number of compounding periods.

**Accumulated Value of an Annuity**

If a periodic payment $\mathcal{P}$ is deposited $n$ times per year at an *annual interest rate* $r$ with interest compounded $n$ times per year for $t$ yr, the accumulated value is given by

$$A = \frac{\mathcal{P}}{R}[(1 + R)^N - 1], \text{ where } R = \frac{r}{n} \text{ and } N = nt.$$

This is also referred to as the **future value** of the account.

**EXAMPLE 5** ▶   **Solving an Application of Annuities**

Since he was a young child, Fitisemanu's parents have been depositing $50 each month into an annuity that pays 6% annually and is compounded monthly. If the account is now worth $9875, how long has it been open?

**Solution** ▶   In this case, $A = 9875$, $\mathcal{P} = 50$, $r = 0.06$, $n = 12$, $R = \frac{r}{n} = 0.005$, and $N = nt = 12t$.

$$A = \frac{\mathcal{P}}{R}[(1 + R)^N - 1] \qquad \text{future value formula}$$

$$9875 = \frac{50}{0.005}[(1.005)^{12t} - 1] \qquad \text{substitute given values}$$

$$1.9875 = 1.005^{12t} \qquad \text{simplify and isolate variable term}$$

$$\ln 1.9875 = 12t \ln 1.005 \qquad \text{apply base-}e \text{ logarithms; power property}$$

$$\frac{\ln 1.9875}{12 \ln 1.005} = t \qquad \text{solve for } t \text{ (exact form)}$$

$$11.5 \approx t \qquad \text{approximate form}$$

The account has been open approximately 11.5 yr.

**Now try Exercises 39 through 42** ▶

The periodic payment required to meet a future goal or obligation can be computed by solving for $\mathcal{P}$ in the future value formula: $\mathcal{P} = \frac{AR}{(1 + R)^N - 1}$. In this form, $\mathcal{P}$ is referred to as a **sinking fund.**

---

**EXAMPLE 6** ▶    Solving an Application of Sinking Funds

Sheila is determined to stay out of debt and decides to save \$20,000 to pay cash for a new car in 4 yr. The best investment vehicle she can find pays 9% compounded monthly. If \$300 is the most she can invest each month, can she meet her 4-yr goal?

Solution ▶    Here we have $\mathcal{P} = 300$, $A = 20,000$, $r = 0.09$, $n = 12$, $R = \frac{r}{n} = 0.0075$, and $N = nt = 12t$. The sinking fund formula gives

$$\mathcal{P} = \frac{AR}{(1 + R)^N - 1}$$    sinking fund

$$300 = \frac{(20,000)(0.0075)}{(1.0075)^{12t} - 1}$$    substitute 300 for $\mathcal{P}$, 20,000 for $A$, 0.0075 for $R$, and $12t$ for $N$

$$300(1.0075^{12t} - 1) = 150$$    multiply in numerator; clear denominators

$$1.0075^{12t} = 1.5$$    isolate variable term

$$12t \ln 1.0075 = \ln 1.5$$    apply base-$e$ logarithms; power property

$$t = \frac{\ln 1.5}{12 \ln 1.0075}$$    solve for $t$ (exact form)

$$\approx 4.5$$    approximate form

☑ **C.** You've just seen how we can solve applications of annuities and amortization

No. She is close, but misses her original 4-yr goal.

**Now try Exercises 43 and 44** ▶

For Example 6, we could have found $\mathcal{P}$ directly by substituting 4 for $t$ and 20,000 for $A$. You can verify the result would be $\mathcal{P} \approx \$347.70$, which is what Sheila would need to invest to meet her 4-yr goal exactly.

For additional practice with the formulas for interest earned or paid, the *Working with Formulas* portion of this Exercise Set has been expanded. **See Exercises 45 through 52.**

**WORTHY OF NOTE**

Notice the formula for exponential growth is virtually identical to the formula for interest compounded continuously. In fact, both are based on the same principles. If we let $A(t)$ represent the amount in an account after $t$ yr and $A_0$ represent the initial deposit (instead of $P$), we have: $A(t) = A_0e^{rt}$ versus $Q(t) = Q_0e^{rt}$ and the two cannot be distinguished.

## D. Applications Involving Exponential Growth and Decay

Closely related to interest compounded continuously are applications of **exponential growth** and **exponential decay.** If $Q$ (quantity) and $t$ (time) are variables, then $Q$ grows exponentially as a function of $t$ if $Q(t) = Q_0e^{rt}$ for positive constants $Q_0$ and $r$. Careful studies have shown that population growth, whether it be humans, bats, or bacteria, can be modeled by these "base-$e$" exponential growth functions. If $Q(t) = Q_0e^{-rt}$, then we say $Q$ decreases or **decays exponentially** over time. The constant $r$ determines how rapidly a quantity grows or decays and is known as the **growth rate** or **decay rate** constant.

---

**EXAMPLE 7A** ▶    Solving an Application of Exponential Growth

Because fruit flies multiply very quickly, they are often used in studies of genetics. Given the necessary space and food supply, a certain population of fruit flies is known to double every 12 days. If there were 100 flies to begin, find (a) the growth rate $r$ and (b) the number of days until the population reaches 2000 flies.

## 5.6 EXERCISES

▶ **CONCEPTS AND VOCABULARY**

**Fill in each blank with the appropriate word or phrase. Carefully reread the section, if necessary.**

1. _____ interest is interest paid to you on previously accumulated interest.

2. Investment plans calling for regularly scheduled deposits are called _____. The annuity formula gives the _____ value of the account.

3. Explain/Describe the difference between the future value and present value of an annuity. Include an example.

4. Explain/Describe the difference between linear and exponential growth. Include an example.

▶ **DEVELOPING YOUR SKILLS**

**For simple interest accounts, the interest accumulated or due depends on the principal $P$, interest rate $r$, and the time $t$ in years according to the formula $I = Prt$.**

5. Find $P$ given $I = \$229.50$, $r = 6.25\%$, and $t = 9$ months.

6. Find $r$ given $I = \$1928.75$, $P = \$8500$, and $t = 3.75$ yr.

7. **Simple interest:** Larry came up a little short one month at bill-paying time and had to take out a title loan on his car at Check Casher's, Inc. He borrowed \$260, and 3 weeks later he paid off the note for \$297.50. What was the annual simple interest rate on this title loan?

8. **Simple interest:** Angela has \$750 in a passbook savings account that pays 2.5% simple interest. How long will it take the account balance to hit the \$1000 mark, if she makes no further deposits?

**For simple interest accounts, the amount $A$ accumulated or due depends on the principal $P$, interest rate $r$, and the time $t$ in years according to the formula $A = P(1 + rt)$.**

9. Find $P$ given $A = \$2500$, $r = 6.25\%$, and $t = 31$ months.

10. Find $r$ given $A = \$15,800$, $P = \$10,000$, and $t = 3.75$ yr.

11. **Simple interest:** Olivette Custom Auto Service borrowed \$120,000 at 4.75% simple interest to expand their facility from three service bays to four. If they repaid \$149,925, what was the term of the loan?

12. **Simple interest:** Healthy U borrows \$50,000 to expand their line of nutritional supplements. When the note is due 3 yr later, they repay the lender \$62,500. If it was a simple interest note, what was the annual interest rate?

13. **Simple interest:** The owner of Paul's Pawn Shop loans Larry \$200.00 using his Toro riding mower as collateral. Thirteen weeks later Larry comes back to get his mower out of pawn and pays Paul \$240.00. What was the annual simple interest rate on this loan?

14. **Simple interest:** To open business in a new strip mall, Laurie's Custom Card Shoppe borrows \$50,000 from a group of investors at 4.55% simple interest. Business booms and blossoms, enabling Laurie to repay the loan fairly quickly. If Laurie repays \$62,500, how long did it take?

**For accounts where interest is compounded annually, the amount $A$ accumulated or due depends on the principal $P$, interest rate $r$, and the time $t$ in years according to the formula $A = P(1 + r)^t$.**

15. Find $t$ given $A = \$48,428$, $P = \$38,000$, and $r = 6.25\%$.

16. Find $P$ given $A = \$30,146$, $r = 5.3\%$, and $t = 7$ yr.

17. How long would it take \$1525 to triple if invested at 7.1%?

18. What interest rate will ensure a \$747.26 deposit will be worth \$1000 in 5 yr?

**For accounts where interest is compounded annually, the principal $P$ needed to ensure an amount $A$ has been accumulated in the time period $t$ when deposited at interest rate $r$ is given by the formula $P = \dfrac{A}{(1 + r)^t}$.**

19. **Compound interest:** The Stringers need to make a \$10,000 balloon payment in 5 yr. How much should be invested now at 5.75% compounded yearly so that the money will be available?

20. **Compound interest:** Morgan is 8 yr old. If her mother wants to have \$25,000 for Morgan's first year of college (in 10 yr), how much should be invested now if the account pays a 6.375% fixed rate compounded yearly?

For compound interest accounts, the amount $A$ accumulated or due depends on the principal $P$, interest rate $r$, number of compoundings per year $n$, and the time $t$ in years according to the formula $A = P\left(1 + \frac{r}{n}\right)^{nt}$.

21. Find $t$ given $A = \$129,500$, $P = \$90,000$, and $r = 7.125\%$ compounded weekly.

22. Find $r$ given $A = \$95,375$, $P = \$65,750$, and $t = 15$ yr with interest compounded monthly.

23. How long would it take a $5000 deposit to double, if invested at a 9.25% rate and compounded daily?

24. What principal should be deposited at 8.375% compounded monthly to ensure the account will be worth $20,000 in 10 yr?

25. **Compound interest:** As a curiosity, David decides to invest $10 in an account paying 10% interest compounded 10 times per year for 10 yr. Is that enough time for the $10 to triple in value?

26. **Compound interest:** As a follow-up experiment (see Exercise 25), David invests $10 in an account paying 12% interest compounded 10 times per year for 10 yr, and another $10 in an account paying 10% interest compounded 12 times per year for 10 yr. Which produces the better investment—more compounding periods or a higher interest rate?

27. **Compound interest:** Due to demand, Donovan's Dairy (Wisconsin, USA) plans to double its size in 4 yr and will need $250,000 to begin development. If they invest $175,000 in an account that pays 8.75% compounded semiannually, (a) will there be sufficient funds to break ground in 4 yr? (b) If not, find the *minimum interest rate* that will enable the dairy to meet its 4-yr goal.

28. **Compound interest:** To celebrate the birth of a new daughter, Helyn invests 6000 Swiss francs in a college savings plan to pay for her daughter's first year of college in 18 yr. She estimates that 25,000 francs will be needed. If the account pays 7.2% compounded daily, (a) will she meet her investment goal? (b) If not, find the *minimum interest rate* that will enable her to meet this 18-yr goal.

For accounts where interest is compounded continuously, the amount $A$ accumulated or due depends on the principal $P$, interest rate $r$, and the time $t$ in years according to the formula $A = Pe^{rt}$.

29. Find $t$ given $A = \$2500$, $p = \$1750$, and $r = 4.5\%$.

30. Find $r$ given $A = \$325,000$, $p = \$250,000$, and $t = 10$ yr.

31. How long would it take $5000 to double if it is invested at 9.25%? Compare the result to Exercise 23.

32. What principal should be deposited at 8.375% to ensure the account will be worth $20,000 in 10 yr? Compare the result to Exercise 24.

33. **Interest compounded continuously:** Valance wants to build an addition to his home outside Madrid (Spain) so he can watch over and help his parents in their old age. He hopes to have 20,000 euros put aside for this purpose

within 5 yr. If he invests 12,500 euros in an account paying 8.6% interest compounded continuously, (a) will he meet his investment goal? (b) If not, find the *minimum interest rate* that will enable him to meet this 5-yr goal.

34. **Interest compounded continuously:** Minh-Ho just inherited her father's farm near Mito (Japan), which badly needs a new barn. The estimated cost of the barn is 8,465,000 yen and she would like to begin construction in 4 yr. If she invests 6,250,000 yen in an account paying 6.5% interest compounded continuously, (a) will she meet her investment goal? (b) If not, find the *minimum interest rate* that will enable her to meet this 4-yr goal.

35. **Interest compounded continuously:** William and Mary buy a small cottage in Dovershire (England), where they hope to move after retiring in 7 yr. The cottage needs about 20,000 euros worth of improvements to make it the retirement home they desire. If they invest 12,000 euros in an account paying 5.5% interest compounded continuously, (a) will they have enough to make the repairs? (b) If not, find the *minimum amount they need to deposit* that will enable them to meet this goal in 7 yr.

36. **Interest compounded continuously:** After living in Oslo (Norway) for 20 yr, Kjell and Torill decide to move inland to help operate the family ski resort. They hope to make the move in 6 yr, after they have put aside 140,000 kroner. If they invest 85,000 kroner in an account paying 6.9% interest compounded continuously, (a) will they meet their 140,000 kroner goal? (b) If not, find the *minimum amount they need to deposit* that will enable them to meet this goal in 6 yr.

The length of time $t$ (in years) required for a principal $P$ to grow to an amount $A$ at a given interest rate $r$ is given by $t = \frac{1}{r}\ln\left(\frac{A}{P}\right)$.

37. **Investment growth:** A small business is planning to build a new $350,000 facility in 8 yr. If they deposit $200,000 in an account that pays 5% interest compounded continuously, (a) will they have enough for the new facility in 8 yr? (b) If not, what amount should be invested on these terms to meet the goal?

38. **Investment growth:** After the twins were born, Sasan deposited $25,000 in an account paying 7.5% compounded continuously, with the goal of having $120,000 available for their college education 20 yr later. (a) Will Sasan meet the 20-yr goal? (b) If not, what amount should be invested on these terms to meet the goal?

If a periodic payment $\mathcal{P}$ is deposited $n$ times per year, with annual interest rate $r$ also compounded $n$ times per year for $t$ yr, the future value of the account is given by $A = \frac{\mathcal{P}}{R}[(1 + R)^N - 1]$, where $R = \frac{r}{n}$ and $N = nt$ (if the rate is 9% compounded monthly for 6 yr, $R = \frac{0.09}{12} = 0.0075$ and $N = (12)(6) = 72$).

39. **Saving for a rainy day:** How long would it take Jasmine to save $10,000 if she deposits $90/month at an annual rate of 7.5% compounded monthly?

**40. Saving for a sunny day:** What quarterly investment amount is required to ensure that Larry can save $4700 in 4 yr at an annual rate of 8.5% compounded quarterly?

**41. Saving for college:** At the birth of their first child, Latasha and Terrance opened an annuity account and have been depositing $50/month in the account ever since. If the account is now worth $30,000 and the interest on the account is 6.6% compounded monthly, how old is the child?

**42. Saving for a bequest:** When Cherie (Brandon's first granddaughter) was born, he purchased an annuity account for her and stipulated that she should receive the funds (in trust, if necessary) upon his death. The quarterly annuity payments were $250 and interest on the account was 7.6% compounded quarterly. The account balance of $17,500 was recently given to Cherie. How much longer did Brandon live?

**43. Saving for a down payment:** Tae-Hon is tired of renting and decides that within the next 5 yr he must save $22,500 for the down payment on a home. He finds an investment company that offers 9% interest compounded monthly and begins depositing $250 each month in the account. (a) Is this monthly amount sufficient to help him meet his 5-yr goal? (b) If not, find the *minimum amount he needs to deposit each month* that will enable him to meet his goal in 5 yr.

**44. Saving to open a business:** Madeline feels trapped in her current job and decides to save $75,000 over the next 7 yr to open up a Harley-Davidson franchise. To this end, she invests $145 every week in an account paying $7\frac{1}{2}$% interest compounded weekly. (a) Is this weekly amount sufficient to help her meet the 7-yr goal? (b) If not, find the *minimum amount she needs to deposit each week* that will enable her to meet this goal in 7 yr.

## ▶ WORKING WITH FORMULAS

**Solve for the indicated unknowns.**

**45.** $A = P + Prt$
  **a.** solve for $t$
  **b.** solve for $P$

**46.** $A = P(1 + r)^t$
  **a.** solve for $t$
  **b.** solve for $r$

**47.** $A = P\left(1 + \dfrac{r}{n}\right)^{nt}$
  **a.** solve for $r$
  **b.** solve for $t$

**48.** $A = Pe^{rt}$
  **a.** solve for $P$
  **b.** solve for $r$

**49.** $Q(t) = Q_0 e^{rt}$
  **a.** solve for $Q_0$
  **b.** solve for $t$

**50.** $P = \dfrac{AR}{(1 + R)^N - 1}$
  **a.** solve for $A$
  **b.** solve for $N$

**51. Amount of a mortgage payment:** $\mathcal{P} = \dfrac{PR}{1 - (1 + R)^{-N}}$

The mortgage payment required to pay off (or amortize) a loan is given by the formula shown, where $\mathcal{P}$ is the payment amount, $P$ is the original amount of the loan, $t$ is the time in years, $r$ is the annual interest rate, $n$ is the number of payments per year, $R = \frac{r}{n}$, and $N = nt$. Find the *monthly payment* required to amortize a $125,000 home, if the interest rate is 5.5% per year and the home is financed over 30 yr.

 **52. Time required to amortize a mortgage:** $t = 16.71 \ln\left(\dfrac{x}{x - 1000}\right), x > 1000.$

The number of years needed to amortize (pay off) a mortgage depends on the amount of the regular monthly payment. The formula shown approximates the years $t$ required to pay off a $200,000 mortgage at 6% interest, based on a monthly payment of $x$ dollars.

  **a.** Use the TABLE feature of your calculator to find the payment required to pay off this mortgage in 30 yr, and the amount of interest paid ($30 \times 12 = 360$ payments).

  **b.** Find the payment required to pay off this mortgage in 20 yr, and the amount of interest that would be paid. How much interest was saved by making a higher payment?

  **c.** Repeat part (b) for a complete payoff in 15 yr.

► **APPLICATIONS**

53. **Bacterial growth:** As part of a lab experiment, Luamata needs to grow a culture of 200,000 bacteria, which are known to double in number every 12 hr. Find (a) the growth rate $r$ and (b) how many hours it takes for the culture to produce the 200,000 bacteria if he begins with 1000 bacteria.

54. **Rabbit populations:** After the wolf population was decimated due to overhunting, the rabbit population in the Boluhti Game Reserve began to double every 6 months. Find (a) the growth rate $r$ and (b) find the number of months required for the population to reach 2500 if there were an estimated 120 rabbits initially.

55. **Population growth:** Between 2000 and 2010, the population of San Antonio, Texas, grew from 1,145,000 to 1,327,000. Assuming exponential growth, find (a) the year the population broke the one million mark, (b) the year the population will exceed two million, and (c) the population in 2025.

56. **Moore's law:** Introduced in 2000, the Pentium 4 processor contained 42 million transistors. According to Moore's law, this number should  double every 2 yr. If Moore's law continues to hold, as it has for the last $40+$ yr, estimate (a) the number of transistors on the original 1993 Pentium processor, (b) the number of transistors on the 2006 Core 2 Duo processor, and (c) the year in which the number of transistors per processor will exceed 10 billion.

57. **Iodine-131, radioactive decay:** The radioactive element iodine-131 has a half-life of 8 days and is often used to help diagnose patients with thyroid problems. If a certain thyroid procedure requires 0.5 g and is scheduled to take place in 3 days, what is the minimum amount that must be on hand now (to the nearest hundredth of a gram)?

58. **Sodium-24, radioactive decay:** The radioactive element sodium-24 has a half-life of 15 hr and is used to help locate obstructions in blood flow. If the procedure requires 0.75 g and is scheduled to take place in 2 days (48 hr), what minimum amount must be on hand *now* (to the nearest hundredth of a gram)?

59. **Americium-241, radioactive decay:** The radioactive element americium-241 has a half-life of 432 yr and although extremely small amounts are used (about 0.0002 g), it is the most vital component of standard household smoke detectors. How many years will it take a 10-g mass of americium-241 to decay to 2.7 g?

60. **Carbon-14, radioactive decay:** Carbon-14 is a radioactive compound that occurs naturally in all living organisms, with the amount in the organism constantly renewed. After death, no new carbon-14 is acquired and the amount in the organism begins to decay exponentially. If the half-life of carbon-14 is 5730 yr, how old is a mummy having only 45% of the normal amount of carbon-14?

61. **Dating the Lascaux Cave dwellers:** Bits of charcoal from Lascaux Cave (home of the prehistoric Lascaux Cave Paintings) were used to estimate that the fire had burned some 17,255 yr ago. What percent of the original amount of carbon-14 remained in the bits of charcoal?

62. **Dating Stonehenge:** Using organic fragments found near Stonehenge (England), scientists were able to determine that the organism that produced the  fragments lived about 3925 yr ago. What percent of the original amount of carbon-14 remained in the organism?

63. **Forgotten knowledge:** A college education is one of the largest investments many individuals will make in their lifetime.  Unfortunately, much of this education is quickly forgotten. Suppose that without review, your knowledge of the Civil War can be modeled by the function $p(w) = 10 + 90e^{-0.027w}$, where $p(w)$ is the percentage of facts remembered $w$ weeks after your first exam on the topic. Without any refresher: (a) What percent will you remember 12 weeks later on the final exam? (b) What percent will you remember a year later? (c) How long until you have forgotten all but 25% of what you once knew?

64. **Spreading rumors:** As the popularity of social networking has grown over recent years, so has the speed at which news and gossip spread. Suppose the spread of a false "class is canceled" rumor is modeled by the function $n(m) = 50 - 49e^{-0.02m}$, where $n(m)$ models the number of your classmates who have heard the rumor $m$ min after it was first announced. (a) How many students will receive the message within the first 10 min? If there are 60 students in the class, (b) how long will it take the rumor to spread to half the class? (c) If the class starts 3 hr after the initial post, how many will not have received the message before class?

► EXTENDING THE CONCEPT

**65.** Many claim that inheritance taxes were put in place simply to prevent a massive accumulation of wealth by a select few. Suppose that in 1890, your great-grandfather deposited $10,000 in an account paying 6.2% compounded continuously. If the account were to pass to you untaxed, what would it be worth in 2010? Do some research on the inheritance tax laws in your state. In particular, what amounts can be inherited untaxed (i.e., before the inheritance tax kicks in)?

**66.** Suppose your wealthy, but eccentric grandfather offers you the following choice: (1) $1,000,000 cash or (2) the total from placing 1 penny on the first dark square of a chessboard, 2 pennies on the second dark square, 4 on the third, and so on, each time doubling the amount on the previous square until all 32 dark squares are covered. Which offer is worth more?

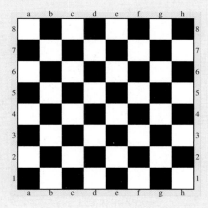

► MAINTAINING YOUR SKILLS

**67.** (2.5/3.1) Name the toolbox functions that are (a) one-to-one, (b) even, (c) increasing for $x \in \mathbb{R}$, and (d) asymptotically linear.

**68.** (R.4) In an effort to boost tourism, a trolley car is being built to carry sightseers from a strip mall to the top of Mt. Vernon, 1580-m high. Approximately how long will the trolley cables be?

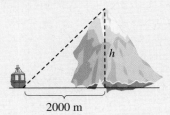

2000 m

**69.** (4.3) A polynomial with real coefficients is known to have the zeroes $x = 3$, $x = -1$, and $x = 1 + 2i$. Find the equation of the polynomial, given it has degree 4 and a $y$-intercept of $(0, -15)$.

**70.** (2.4) Is the following relation a function? If not, state how the definition of a function is violated.

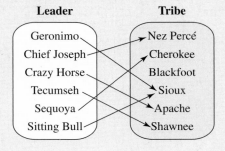

# MAKING CONNECTIONS

## Making Connections: Graphically, Symbolically, Numerically, and Verbally

Eight graphs (a) through (h) are given. Match the characteristics or equations shown in 1 through 16 to one of the eight graphs.

**(a)** 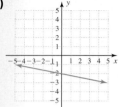    **(b)** 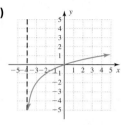    **(c)** 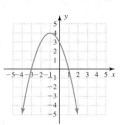    **(d)**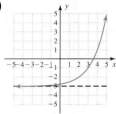

**(e)** 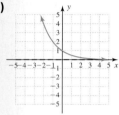    **(f)** 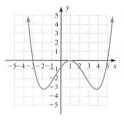    **(g)** 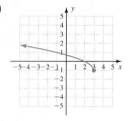    **(h)**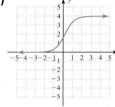

**1.** ___ $y = -\dfrac{1}{5}x - 2$

**2.** ___ domain: $x \in (-\infty, 3]$

**3.** ___ as $x \to \infty, y \to 0$

**4.** ___ $y = \log_2(x + 4) - 2$

**5.** ___ $y = -(x + 1)^2 + 4$

**6.** ___ as $x \to \infty, y \to 4$

**7.** ___ $f(0) = 1, f(-2) = 4$

**8.** ___ $f(x)\uparrow$ for $x \in (-\infty, \infty)$

**9.** ___ range: $y \in (-\infty, \infty)$

**10.** ___ $y = \dfrac{4}{1 + 1.5e^{-2x}}$

**11.** ___ $y = \sqrt{3 - x} - 1$

**12.** ___ $y = \dfrac{1}{20}(x + 3)(x - 1)^2(x - 5)$

**13.** ___ axis of symmetry: $x = -1$

**14.** ___ $y = 2^{-x}$

**15.** ___ $y = 2^{x-2} - 3$

**16.** ___ $f(x) \le 0$ for $x \in [-3, 5]$

## SUMMARY AND CONCEPT REVIEW

### SECTION 5.1   One-to-One and Inverse Functions

#### KEY CONCEPTS

- A function is one-to-one if each element of the range corresponds to a unique element of the domain.
- If every horizontal line intersects the graph of a function in at most one point, the function is one-to-one.
- If $f$ is a one-to-one function with ordered pairs $(a, b)$, then the inverse of $f$ exists and is that one-to-one function $f^{-1}$ with ordered pairs of the form $(b, a)$.
- The range of $f$ becomes the domain of $f^{-1}$, and the domain of $f$ becomes the range of $f^{-1}$.
- To find $f^{-1}$ using the algebraic method:
  1. Replace $f(x)$ with $y$.     2. Interchange $x$ and $y$.     3. Solve the equation for $y$.     4. Replace $y$ with $f^{-1}(x)$.
- If $f$ is a one-to-one function, the inverse $f^{-1}$ exists and satisfies both $(f \circ f^{-1})(x) = x$ and $(f^{-1} \circ f)(x) = x$.
- The graphs of $f$ and $f^{-1}$ are symmetric about the identity function $y = x$.

#### EXERCISES

Determine whether the functions given are one-to-one by noting the function family to which each belongs and mentally picturing the shape of the graph.

**1.** $h(x) = -|x - 2| + 3$       **2.** $p(x) = 2x^2 + 7$       **3.** $s(x) = \sqrt{x - 1} + 5$

Find the inverse of each function given. Then show using composition that your inverse function is correct. State any necessary restrictions.

**4.** $f(x) = -3x + 2$       **5.** $f(x) = x^2 - 2, x \geq 0$       **6.** $f(x) = \sqrt{x - 1}$

Determine the domain and range for each function whose graph is given, and use this information to state the domain and range of the inverse function. Then estimate the location of three points on the graph, and use these and the line $y = x$ to graph $f^{-1}(x)$ on the same grid.

**7.**
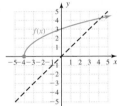

**8.**
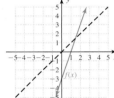

**9.**
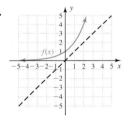

**10. Fines for overdue material:** Some libraries have set fees and penalties to discourage patrons from holding borrowed materials for an extended period. Suppose the fine for overdue DVDs is given by the function $f(t) = 0.15t + 2$, where $f(t)$ is the amount of the fine $t$ days after it is due. (a) What is the fine for keeping a DVD seven extra days? (b) Find $f^{-1}(t)$, then input your answer from part (a) and comment on the result. (c) If a fine of \$3.80 was assessed, how many days was the DVD overdue?

### SECTION 5.2   Exponential Functions

#### KEY CONCEPTS

- An exponential function is defined as $f(x) = b^x$, where $b > 0$, $b \neq 1$, and $x$ is any real number.
- The natural exponential function is $f(x) = e^x$, where $e \approx 2.71828182846$.
- For exponential functions, we have
  - one-to-one function   • $y$-intercept $(0, 1)$   • domain: $x \in \mathbb{R}$   • range: $y \in (0, \infty)$
  - increasing if $b > 1$   • decreasing if $0 < b < 1$   • asymptotic to $x$-axis
- If an equation can be written with like bases on each side, we solve it using the uniqueness property: If $b^m = b^n$, then $m = n$ (equal bases imply equal exponents).

#### EXERCISES

Graph each function using *transformations of the basic function*, then strategically plot a few points to check your work and round out the graph. Draw and label the asymptote.

**11.** $f(x) = 2^x + 3$       **12.** $g(x) = 2^{-x} - 1$       **13.** $h(x) = -e^{x+1} - 2$

Solve using the uniqueness property.

**14.** $3^{2x-1} = 27$  **15.** $4^x = \frac{1}{16}$  **16.** $e^x \cdot e^{x+1} = e^6$

**17.** A ballast machine is purchased new for \$142,000 by the AT & SF Railroad. The machine loses 15% of its value each year and must be replaced when its value drops below \$20,000. How many years will the machine be in service?

## SECTION 5.3   Logarithms and Logarithmic Functions

### KEY CONCEPTS

- A logarithm is an exponent. For $x, b > 0$, and $b \neq 1$, the expression $\log_b x$ represents the exponent that goes on base $b$ to obtain $x$: If $y = \log_b x$, then $b^y = x \Rightarrow b^{\log_b x} = x$ (by substitution).
- The equations $x = b^y$ and $y = \log_b x$ are equivalent. We say $x = b^y$ is the *exponential* form and $y = \log_b x$ is the *logarithmic* form of the equation.
- A logarithmic *function* is defined as $f(x) = \log_b x$, where $x, b > 0$, and $b \neq 1$.
  - $y = \log_{10} x = \log x$ is called the *common* logarithmic function.
  - $y = \log_e x = \ln x$ is called the *natural* logarithmic function.
- For logarithmic functions, we have
  - one-to-one function
  - $x$-intercept $(1, 0)$
  - domain: $x \in (0, \infty)$
  - range: $y \in \mathbb{R}$
  - increasing if $b > 1$
  - decreasing if $0 < b < 1$
  - asymptotic to $y$-axis

### EXERCISES

Write each expression in *exponential* form.

**18.** $\log_3 9 = 2$  **19.** $\log_5 \frac{1}{125} = -3$  **20.** $\ln 43 \approx 3.7612$

Write each expression in *logarithmic* form.

**21.** $5^2 = 25$  **22.** $e^{-0.25} \approx 0.7788$  **23.** $3^4 = 81$

Find the value of each expression without using a calculator.

**24.** $\log_2 32$  **25.** $\ln \frac{1}{e}$  **26.** $\log_9 3$

Graph each function using *transformations of the basic function,* then strategically plot a few points to check your work and round out the graph. Draw and label the asymptote.

**27.** $f(x) = \log_2 x$  **28.** $g(x) = \log_2(x + 3)$  **29.** $h(x) = 2 + \ln(x - 1)$

Find the domain of the following functions.

**30.** $g(x) = \log \sqrt{2x + 3}$  **31.** $f(x) = \ln(x^2 - 6x)$

**32.** The magnitude of an earthquake is given by $M(I) = \log\left(\frac{I}{I_0}\right)$, where $I$ is the intensity and $I_0$ is the reference intensity. (a) Find $M(I)$ given $I = 62{,}000I_0$ and (b) find the intensity $I$ given $M(I) = 7.3$.

## SECTION 5.4   Properties of Logarithms

### KEY CONCEPTS

- The basic definition of a logarithm gives rise to the following properties: For any base $b > 0$, $b \neq 1$,
  **1.** $\log_b b = 1$  **2.** $\log_b 1 = 0$  **3.** $\log_b b^x = x$  **4.** $b^{\log_b x} = x$

- Since a logarithm is an exponent, they have properties that parallel those of exponents.

| **Product Property** | **Quotient Property** | **Power Property** |
|---|---|---|
| like base and multiplication, add exponents: | like base and division, subtract exponents: | exponent raised to a power, multiply exponents: |
| $\log_b(MN) = \log_b M + \log_b N$ | $\log_b\left(\dfrac{M}{N}\right) = \log_b M - \log_b N$ | $\log_b M^P = p\log_b M$ |

- To evaluate logarithms with bases other than 10 or $e$, use the change-of-base formula:

$$\log_b M = \frac{\log M}{\log b} = \frac{\ln M}{\ln b}$$

- If an equation can be written with like bases on each side, we solve it using the uniqueness property: if $\log_b m = \log_b n$, then $m = n$ (equal bases imply equal arguments).

# EXERCISES

**33.** Solve each equation by applying fundamental properties.

   **a.** $\ln x = 32$     **b.** $\log x = 2.38$     **c.** $e^x = 9.8$     **d.** $10^x = \sqrt{7}$

**34.** Solve each equation. Write answers in exact form and in approximate form to four decimal places.

   **a.** $15 = 7 + 2e^{0.5x}$     **b.** $10^{0.2x} = 19$     **c.** $-2\log(3x) + 1 = -5$     **d.** $-2\ln x + 1 = 6.5$

**35.** Use the product or quotient property of logarithms to write each sum or difference as a single term.

   **a.** $\ln 7 + \ln 6$     **b.** $\log_9 2 + \log_9 15$     **c.** $\ln(x+3) - \ln(x-1)$     **d.** $\log x + \log(x+1)$

**36.** Use the power property of logarithms to rewrite each term as a product.

   **a.** $\log_5 9^2$     **b.** $\log_7 4^2$     **c.** $\ln 5^{2x-1}$     **d.** $\ln 10^{3x+2}$

**37.** Use the properties of logarithms to write the following expressions as sums or differences of simple logarithmic terms.

   **a.** $\ln(x\sqrt[4]{y})$     **b.** $\ln(\sqrt[3]{pq})$     **c.** $\log\left(\dfrac{\sqrt[3]{x^5y^4}}{\sqrt{x^5y^3}}\right)$     **d.** $\log\left(\dfrac{4\sqrt[3]{p^5q^4}}{\sqrt{p^3q^2}}\right)$

**38.** Evaluate using the change-of-base formula. Answer in exact form and approximate form to thousandths.

   **a.** $\log_6 45$     **b.** $\log_3 128$     **c.** $\log_2 124$     **d.** $\log_5 0.42$

## SECTION 5.5  Solving Exponential and Logarithmic Equations

### KEY CONCEPTS

- For a general base $b$, isolate the exponential term, apply the base-$b$ logarithm, then solve for $x$ using algebra and, if necessary, the change-of-base formula.
- If a logarithmic equation has a constant term, isolate all logarithmic terms to one side and consolidate using the product or quotient properties, then use the exponential form and algebra to solve.
- If a logarithmic equation has only logarithmic terms, consolidate logarithmic terms using the product or quotient properties, then use the uniqueness property and algebra to solve.

### EXERCISES

Solve each equation. Answer in both exact form and approximate form.

**39.** $2^x = 7$     **40.** $3^{x+1} = 5$     **41.** $e^{x-2} = 3^{-x}$

**42.** $\ln(x+1) = 2$     **43.** $\log x + \log(x-3) = 1$     **44.** $\log_{25}(x+2) - \log_{25}(x-3) = \frac{1}{2}$

**45.** The rate of decay for radioactive material is related to its half-life by the formula $r = \frac{\ln 2}{h}$, where $h$ represents the half-life of the material and $r$ is the rate of decay expressed as a decimal. Radon-222 has a half-life of approximately 3.9 days. (a) Find its rate of decay to the nearest hundredth of a percent. (b) Find the half-life of thorium-234 if its rate of decay is 2.89% per day.

**46.** The *barometric equation* $H = (30T + 8000)\ln(\frac{P_0}{P})$ relates the altitude $H$ to atmospheric pressure $P$, where $P_0 = 76$ cmHg. Find the atmospheric pressure at the summit of Mount Pico de Orizaba (Mexico), whose summit is at 5657 m. Assume the temperature at the summit is $T = 12°C$.

## SECTION 5.6  Applications from Business, Finance, and Science

### KEY CONCEPTS

- Simple interest: $I = Prt$; $P$ is the principal, $r$ is the interest rate per year, and $t$ is the time in years.
- Amount in an account after $t$ years: $A = P + Prt$ or $A = P(1 + rt)$.
- Interest compounded $n$ times per year: $A = P(1 + \frac{r}{n})^{nt}$; $P$ is the principal, $r$ is the interest rate per year, $t$ is the time in years, and $n$ is the number of times per year interest is compounded.
- Interest compounded continuously: $A = Pe^{rt}$; $P$ is the principal, $r$ is the interest rate per year, and $t$ is the time in years.
- If a loan or savings plan calls for a regular schedule of deposits, the plan is called an annuity.
- For periodic payment $\mathcal{P}$, deposited or paid $n$ times per year, at annual interest rate $r$, with interest compounded or calculated $n$ times per year for $t$ yr, $R = \frac{r}{n}$, and $N = nt$:
  - The accumulated value of the account is $A = \frac{\mathcal{P}}{R}[(1 + R)^N - 1]$.
  - The payment required to meet a future goal is $\mathcal{P} = \frac{AR}{(1 + R)^N - 1}$
  - The payment required to amortize an amount $P$ is $\mathcal{P} = \frac{PR}{1 - (1 + R)^{-N}}$.
- The general formulas for exponential growth and decay are $Q(t) = Q_0 e^{rt}$ and $Q(t) = Q_0 e^{-rt}$ respectively.

## EXERCISES

Solve each application.

**47.** Jeffery borrows $600.00 from his dad, who decides it's best to charge him interest. Three months later Jeff repays the loan plus interest, a total of $627.75. What was the annual simple interest rate on the loan?

**48.** To save money for her first car, Cheryl invests the $7500 she inherited in an account paying 7.8% interest compounded monthly. She hopes to buy the car in 6 yr and needs $12,000. Is this possible?

**49.** To save up for the vacation of a lifetime, Al-Harwi decides to save $15,000 over the next 4 yr. For this purpose he invests $260 every month in an account paying $7\frac{1}{2}$% interest compounded monthly. (a) Is this monthly amount sufficient to meet the 4-yr goal? (b) If not, find the *minimum amount he needs to deposit each month* that will enable him to meet this goal in 4 yr.

**50.** Eighty prairie dogs are released in a wilderness area in an effort to repopulate the species. Five years later a statistical survey reveals the population has reached 1250 dogs. Assuming the growth was exponential, approximate the growth rate to the nearest tenth of a percent.

# PRACTICE TEST

**1.** Write the expression $\log_3 81 = 4$ in exponential form.

**2.** Write the expression $25^{1/2} = 5$ in logarithmic form.

**3.** Write the expression $\log_b(\frac{\sqrt{x^5}y^3}{z})$ as a sum or difference of simple logarithmic terms.

**4.** Write the expression $\log_b m + \frac{3}{2}\log_b n - \frac{1}{2}\log_b p$ as a single logarithm.

Solve for $x$ using the uniqueness property.

**5.** $5^{x-7} = 125$          **6.** $2 \cdot 4^{3x} = \dfrac{8^x}{16}$

Given $\log_a 3 \approx 0.48$ and $\log_a 5 \approx 1.72$, evaluate the following without the use of a calculator:

**7.** $\log_a 45$          **8.** $\log_a 0.6$

Graph using transformations of the parent function.

**9.** $g(x) = -2^{x-1} + 3$          **10.** $h(x) = \log_2(x - 2) + 1$

**11.** Use the change-of-base formula to evaluate. Verify results using a calculator.

  **a.** $\log_3 100$          **b.** $\log_6 0.235$

**12.** State the domain and range of $f(x) = (x - 2)^2 - 3$ and determine if $f$ is a one-to-one function. If so, find its inverse. If not, restrict the domain of $f$ to create a one-to-one function, then find the inverse of this new function, including the domain and range.

Solve each equation.

**13.** $3^{x-1} = 89$          **14.** $\log_5 x + \log_5(x + 4) = 1$

**15.** A copier is purchased new for $8000. The machine loses 18% of its value each year and must be replaced when its value drops below $3000. How many years will the machine be in service?

**16.** In 1957, scientist Stanley Stevens proposed a mathematical model that attempted to compare the actual strength of a physical stimulus with the human perception (dead-reckoning) of its strength. The model has been

widely applied in comparisons of weight, sound, pressure, and other areas. If $M$ represents the measured strength of the stimulus, and $P$ the human perception of its strength, **Stevens' law** can be written as $\log P = \log k + \alpha \log M$, where $\alpha$ and $k$ are constants determined by the type of stimulation applied. In a controlled experiment, subjects were given a known amount of weight to lift, then asked to select an unmarked weight they felt was equal to half the known weight. (a) Solve the equation for $P$ and (b) use the result to determine what was perceived to be half of a 40-lb weight (assume $k = 0.89$ and $\alpha = 0.95$).

**17.** The number of ounces of unrefined platinum drawn from a mine is modeled by $Q(t) = -2600 + 1900 \ln t$, where $Q(t)$ represents the number of ounces mined in $t$ months. How many months did it take for the number of ounces mined to exceed 3000?

**18.** Jacob decides to save $4000 over the next 5 yr so that he can present his wife with a new diamond ring for their 20th anniversary. He invests $50 every month in an account paying $8\frac{1}{4}$% interest compounded monthly. (a) Is this amount sufficient to meet the 5-yr goal? (b) If not, find the *minimum amount he needs to save monthly* that will enable him to meet this goal.

**19.** Chaucer is a typical Welsh Corgi puppy. During his first year of life, his weight very closely follows the model $W(t) = 6.79 \ln t - 11.97$, where $W(t)$ is his weight in pounds after $t$ weeks and $8 \leq t \leq 52$.

  **a.** How much will Chaucer weigh when he is 26 weeks old (to the nearest one-tenth pound)?

  **b.** To the nearest week, how old is Chaucer when he weighs 12 lb?

**20.** The number of years $d$ required for an investment to double in value can be approximated by what is called the **rule of 72**, which is $d = \frac{72}{r}$, where $r$ is the annual *percent* interest rate. If a deposit is compounded monthly at an annual interest rate of 2%, find the doubling time using (a) the formula from Section 5.6 and (b) the rule of 72.

# CALCULATOR EXPLORATION AND DISCOVERY

## I. Solving Exponential Equations Graphically

In Section 5.2, we showed that the exponential function $f(x) = b^x$ was defined for all real numbers. This is important because it establishes that equations like $2^x = 7$ must have a solution, even if $x$ is not rational. In fact, since $2^2 = 4$ and $2^3 = 8$, the following inequalities indicate the solution must be between 2 and 3.

$$4 < 7 < 8 \qquad \text{7 is between 4 and 8}$$
$$2^2 < 2^x < 2^3 \qquad \text{replace 4 with } 2^2 \text{ and 8 with } 2^3$$
$$2 < x < 3 \qquad x \text{ must be between 2 and 3}$$

Although we have now developed an inverse for exponential functions, we are still unable to solve many equations involving exponentials and logarithms in exact form. We can, however, get a very close approximation using a graphing calculator. For the equation $2^x = 7$, enter $Y_1 = 2^x$ and $Y_2 = 7$ on the ⬤ screen. Then press ⬤ **6** to graph both functions (see Figure 5.37). To find the point of intersection, press ⬤ ⬤ (**CALC**), select option **5:intersect** and press ⬤ *three* times (to identify the intersecting functions and bypass "Guess"). The $x$- and $y$-coordinates of the point of intersection will appear at the bottom of the screen, with the $x$-coordinate being the solution. As you can see, $x$ is indeed $\log_2 7 = \frac{\ln 7}{\ln 2} \approx 2.8073549$. Solve the following equations graphically. Adjust the viewing window as needed.

**Figure 5.37**

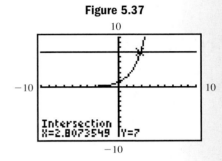

**Exercise 1:** $3^x = 22 - x$

**Exercise 2:** $2^x = \frac{3}{x}$

**Exercise 3:** $xe^{x-1} = 9$

**Exercise 4:** $e^{0.5x} = 0.1x^3$

## II. Investigating Logistic Equations

As we saw in Section 5.5, logistic models have the form $P(t) = \frac{c}{1 + ae^{-bt}}$, where $a$, $b$, and $c$ are positive constants and $P(t)$ represents the population at time $t$. For populations modeled by a logistic curve (sometimes called an "S" curve), growth is very rapid at first (like an exponential function), but this growth begins to slow down and level off due to various factors. Here, we investigate the effects that $a$ and $c$ have on the resulting graph.

1. Investigating $a$: From our earlier observation, as $t$ becomes larger and larger, the term $ae^{-bt}$ becomes smaller and smaller (approaching 0) because it is a decreasing function. If we allow that the term eventually becomes so small it can be disregarded, what remains is $P(t) = \frac{c}{1}$ or $c$. This is why $c$ is called the capacity constant and the population can get no larger than $c$. In Figure 5.38, the graph of $P(t) = \frac{1000}{1 + 50e^{-1x}}$ ($a = 50$, $b = 1$, and $c = 1000$) is shown using a lighter line, while the graph of $P(t) = \frac{750}{1 + 50e^{-1x}}$ ($a = 50$, $b = 1$, and $c = 750$), is given in bold.

**Figure 5.38**

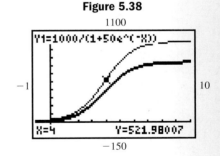

Also note that if $a$ is held constant, smaller values of $c$ cause the "interior" of the S curve to grow at a slower rate than larger values, a concept studied in some detail in a Calculus I class.

2. Investigating $c$: If $t = 0$, $ae^{-bt} = ae^0 = a$, and we note the ratio $P(0) = \frac{c}{1 + a}$ represents the *initial population*. This also means for constant values of $c$, larger values of $a$ make the ratio $\frac{c}{1 + a}$ smaller; while smaller values of $a$ make the ratio $\frac{c}{1 + a}$ larger. From this we conclude that $a$ primarily affects the initial population. In Figures 5.39 and 5.40 shown next, $P(t) = \frac{1000}{1 + 50e^{-1x}}$ (from 1) is graphed using a lighter line, while the graph of $P(t) = \frac{1000}{1 + 5e^{-1x}}$ ($a = 5$) and $P(t) = \frac{1000}{1 + 500e^{-1x}}$ ($a = 500$) are shown in bold.

**Figure 5.39**

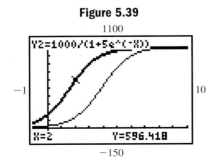

**Figure 5.40**

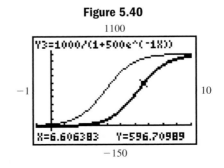

Note that changes in $a$ appear to have no effect on the rate of growth in the interior of the S curve.

The following exercises are based on the population of an ant colony, modeled by the logistic function $P(t) = \frac{2500}{1 + 25e^{-0.5x}}$. Respond to Exercises 5 through 9 without the use of a calculator.

**Exercise 5:** Identify the values of $a$, $b$, and $c$ for this logistics curve.

**Exercise 6:** What was the approximate initial population of the colony?

**Exercise 7:** Which gives a larger initial population: (a) $c = 2500$ and $a = 25$ or (b) $c = 3000$ and $a = 15$?

**Exercise 8:** What is the maximum population capacity for this colony?

**Exercise 9:** Which causes a slower population growth: (a) $c = 2000$ and $a = 25$ or (b) $c = 3000$ and $a = 25$?

**Exercise 10:** Verify your responses to Exercises 6 through 9 using a graphing calculator.

# STRENGTHENING CORE SKILLS

## The HerdBurn Scale—What's Hot and What's Not

The human mouth can easily distinguish between heat levels (the burning sensation) when eating foods "spiced" with various peppers. The level of "heat" is generally given in *Scoville units*, which is a measure of the element capsaicin that causes the burn. Sweet bell peppers and others have no capsaicin and a Scoville rating of 0, while red habanero peppers have a Scoville rating near 500,000. Although inedible, laboratory grades of capsaicin can have a Scoville rating near 16,000,000! This range of values makes a unit scale impractical for common use, and a logarithmic scale once again becomes more desirable. Using the newly developed *HerdBurn scale (hb)*, we have the following measures of "heat" for well known peppers of various types.

### The HerdBurn Scale

| HerdBurn Units (hb) | Chili Pepper | General Sensation | Caustic Power |
|---|---|---|---|
| 0 | sweet banana peppers | not sensed | none |
| 1 | cherry peppers | delicate warmth | none |
| 2 | pepperoncini peppers | strong warmth | none |
| 2.5 | Sonora peppers | slight burn | some reaction |
| 3 | ancho peppers | moderate burn | fanning the mouth |
| 3.5 | jalapeño peppers | strong burn | eyes water |
| 4 | hidalgo peppers | sizzling burn | pain threshold |
| 4.5 | cayenne peppers | scorching burn | painful |
| 5 | Bahamian peppers | blistering burn | very painful |
| 5.5 | habanero peppers | ruthless burn | intense pain |
| 6 | naga jolokia peppers | merciless burn | debilitating |
| 6.5 | military grade pepper spray | *inedible* | incapacitating |
| 7 | laboratory grade capsaicin | *inedible* | ruinous |
| 7.2 | pure capsaicin | *inedible* | deadly |

Similar to working with decibel levels or the Richter scale, we compare how many times hotter one pepper is than another by recognizing the values given are powers of 10. For example, a red habanero (5.7 hb) is about two times as hot as an orange habanero (5.4 hb): $\frac{10^{5.7}}{10^{5.4}} = 10^{0.3} \approx 2$, but nearly 100 times hotter than a red jalapeño pepper (3.7 hb): $\frac{10^{5.7}}{10^{3.7}} = 10^2 = 100$. Use this information to complete the following exercises.

**Exercise 1:** The "heat" in a rocotillo pepper measures about 3.4 on the HerdBurn scale, while a Jamaican hot pepper measures near 5.5. How many times hotter is the Jamaican pepper?

**Exercise 2:** A naga jolokia (6.0 hb) pepper is about 63 times as hot as a serrano pepper. What is the HerdBurn number for a serrano pepper?

## CUMULATIVE REVIEW CHAPTERS R–5

Use the quadratic formula to solve for $x$.

1. $x^2 - 4x + 53 = 0$

2. $6x^2 + 19x = 36$

3. Use substitution to show that $4 + 5i$ is a zero of $f(x) = x^2 - 8x + 41$.

4. Graph using transformations of a basic function: $y = 2\sqrt{x + 2} - 3$.

5. Find $(f \circ g)(x)$ and $(g \circ f)(x)$ and comment on what you notice: $f(x) = x^3 - 2$; $g(x) = \sqrt[3]{x + 2}$.

6. State the domain of $h(x)$ in interval notation:
$$h(x) = \frac{\sqrt{x + 3}}{x^2 + 6x + 8}.$$

7. According to the 2008 *National Vital Statistics Report* (Vol. 59, No. 1, page 64) there were 5877 sets of triplets born in the United States in 2008, and 6740 sets of triplets born in 1999. Assuming the relationship (year, sets of triplets) is linear: (a) find an equation of the line, (b) explain the meaning of the slope in this context, and (c) use the equation to estimate the number of sets born in 2005, and to project the number of sets that will be born in 2015 if this trend continues.

8. State the following geometric formulas:
   a. area of a circle    c. perimeter of a rectangle
   b. Pythagorean theorem    d. area of a trapezoid

9. Graph the following piecewise-defined function and state its domain, range, and intervals where it is increasing and decreasing.
$$h(x) = \begin{cases} -4, & -10 \le x < -2 \\ -x^2, & -2 \le x < 3 \\ 3x - 18, & x \ge 3 \end{cases}$$

10. Solve the inequality and write the solution in interval notation: $\frac{2x + 1}{x - 3} \ge 0$.

11. Use the rational roots theorem to find all zeroes of $f(x) = x^4 - 3x^3 - 12x^2 + 52x - 48$.

12. Given $f(c) = \frac{9}{5}c + 32$, find $k$, where $k = f(25)$. Then find the inverse function using the algebraic method, and verify that $f^{-1}(k) = 25$.

13. Solve the formula $V = \frac{1}{2}\pi ab^2$ (the volume of a paraboloid) for the variable $b$.

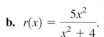

14. Use the *Guidelines for Graphing* to graph
   a. $p(x) = x^3 - 4x^2 + x + 6$.    b. $r(x) = \dfrac{5x^2}{x^2 + 4}$.

15. For $f(x) = \frac{2x + 3}{5}$, (a) find $f^{-1}$, (b) graph both functions and verify they are symmetric about the line $y = x$, and (c) show they are inverses using composition.

16. Solve for $x$: $10 = -2e^{-0.05x} + 25$.

17. Solve for $x$: $\ln(x + 3) + \ln(x - 2) = \ln 24$.

18. Once in orbit, satellites are often powered by radioactive isotopes. From the natural process of radioactive decay, the power output declines over a period of time. Suppose the power output is modeled by the function $p(t) = 50e^{-0.002t}$, where $p(t)$ is the power output in watts, $t$ days after the satellite has been put into service. (a) Approximately how much power remains 6 *months* later (assume 1 month $\approx$ 30.5 days)? (b) How many *years* until only one-fourth of the original power remains?

19. Simon and Christine own a sport wagon and a minivan. The sport wagon has a power curve that is closely modeled by $H(r) = 123 \ln r - 897$, where $H(r)$ is the horsepower at $r$ rpm, with $2200 \le r \le 5600$. The power curve for the minivan is $h(r) = 193 \ln r - 1464$, for $2600 < r \le 5800$.

   a. How much horsepower is generated by each engine at 3000 rpm?

   b. At what rpm are the engines generating the same horsepower?

   c. If Christine wants the maximum horsepower available, which vehicle should she drive? What is the maximum horsepower?

20. Wilson's disease is a hereditary disease that causes the body to retain copper. Radioactive copper, $^{64}$Cu, has been used extensively to study and understand this disease. $^{64}$Cu has a relatively short half-life of 12.7 hr. How many hours will it take for a 5-g mass of $^{64}$Cu to decay to 1 g?

21. Find the equation of the linear function shown.

| Exercise 21 | Exercise 22 |
|---|---|
|  | 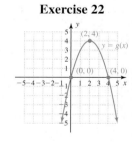 |

22. Find the equation of the quadratic function shown.

23. State the end-behavior, intercepts, domain, and range for (a) the linear function in Exercise 21 and (b) the quadratic function in Exercise 22.

24. Use the graphs given in Exercises 21 and 22 to find (a) $(f + g)(2)$, (b) $(g - f)(1)$, (c) $(g \cdot f)(4)$, and (d) $(\frac{f}{g})(0)$.

25. Draw the graph of a function $f$ with the following properties: (i) the domain is $[-3, 3]$, (ii) $f(-3) = f(0) = 0$, (iii) $y = -1$ is a local min at $x = -1$, (iv) $f(x)\downarrow$ for $x \in (-3, -1.5)$ and $f(x)\uparrow$ for $x \in (-1.5, 0)$, and (v) $f$ is symmetric about the origin.

# 6

# Systems of Equations and Inequalities

## CHAPTER CONNECTIONS

At the turn of the century, there was an explosion in the number of handheld electronic devices available to consumers. One device in particular, the popular smartphone, experienced a phenomenal growth in demand. With high demand and a large market, competition between manufacturers and suppliers is often fierce, with each fighting to earn and hold a share of the market. One significant factor in who gains the largest share is the price charged for the phone, with suppliers willing to supply more at a greater price, and consumers willing to buy more at a lesser price. Determining where the price will stabilize is an important component in the economics of *"supply and demand."* This application occurs as Exercise 82 in Section 6.1.

**Check out these other real-world connections:**

► Appropriate Measurements in Dietetics (Section 6.1, Exercise 70)

► Allocating Winnings to Different Investments (Section 6.2, Exercise 43)

► Market Pricing for Organic Produce (Section 6.3, Exercise 64)

► Minimizing Shipping Costs (Section 6.4, Exercise 68)

**LEARNING OBJECTIVES**

*In Section 6.1 you will see how we can:*

- ☐ **A.** Verify ordered pair solutions
- ☐ **B.** Solve linear systems by graphing
- ☐ **C.** Solve linear systems by substitution
- ☐ **D.** Solve linear systems by elimination
- ☐ **E.** Recognize inconsistent systems and dependent systems
- ☐ **F.** Use a system of equations to model and solve applications

In earlier chapters, we used linear equations in two variables to model a number of real-world situations. Graphing these equations gave us a visual image of how the variables were related, and helped us better understand this relationship. In many applications, two different measures of the independent variable must be considered simultaneously, leading to a **system of two linear equations in two unknowns.** Here, a graphical presentation once again supports a better understanding, as we explore systems and their many applications.

## A. Solutions to a System of Equations

A **system of equations** is a set of two or more equations for which a common solution is sought. Systems are widely used to model and solve applications when the information given enables the relationship between variables to be stated in different ways. For example, consider an amusement park that brought in $3100 in revenue by charging $9.00 for adults and $5.00 for children, while selling 500 tickets. Using $a$ for the number of adults and $c$ for the number of children, we could write one equation modeling the number of tickets sold: $a + c = 500$, and a second modeling the amount of revenue brought in: $9a + 5c = 3100$. To show that we're considering both equations simultaneously, a large "left brace" is used and the result is called a **system of two equations in two variables:**

$$\begin{cases} a + c = 500 & \text{number of tickets} \\ 9a + 5c = 3100 & \text{amount of revenue} \end{cases}$$

We note that both equations are linear and will have different slope values, so their graphs must intersect at some point. Since every point on a line satisfies the equation of that line, this point of intersection must satisfy *both* equations simultaneously and is the solution to the system. The figure that accompanies Example 1 shows the point of intersecion for this system is (150, 350).

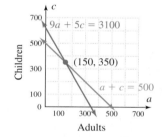

---

**EXAMPLE 1** ► **Verifying Solutions to a System**

Verify that (150, 350) is a solution to $\begin{cases} a + c = 500 \\ 9a + 5c = 3100 \end{cases}$.

**Solution** ► Substitute 150 for $a$ and 350 for $c$ in each equation.

$$\begin{array}{ll} a + c = 500 & \text{first equation} \\ (150) + (350) = 500 & \\ 500 = 500 \checkmark & \end{array} \qquad \begin{array}{ll} 9a + 5c = 3100 & \text{second equation} \\ 9(150) + 5(350) = 3100 & \\ 3100 = 3100 \checkmark & \end{array}$$

Since (150, 350) satisfies both equations, it is the solution to the system and we find the park sold 150 adult tickets and 350 tickets for children.

☑ **A.** You've just seen how we can verify ordered pair solutions

Now try Exercises 5 through 8 ►

---

## B. Solving Systems Graphically

To **solve a system of equations** means we apply various methods in an attempt to find ordered pair solutions. As Example 1 suggests, one method for finding solutions is to graph the system. Any method for graphing the lines can be employed, but to keep important concepts fresh, the slope-intercept method is used here.

---

**EXAMPLE 2** ▶   **Solving a System Graphically**

Solve the system by graphing: $\begin{cases} x + 2y = 4 \\ y = 3x - 5 \end{cases}$.

**Solution** ▶   To graph the first equation, the intercept method seems most convenient. Substituting 0 for $x$, we obtain $y = 2$ and the point $(0, 2)$. Substituting 0 for $y$ yields $x = 4$ and the point $(4, 0)$. Using these points we obtain the blue line shown. The second equation is in slope-intercept form, so we plot the $y$-intercept at $(0, -5)$ and use $\frac{\Delta y}{\Delta x} = \frac{3}{1}$ (up 3, over 1) to locate a second point at $(1, -2)$. These points give us the line shown in red. The point of intersection appears to be $(2, 1)$, and checking these values in both equations gives

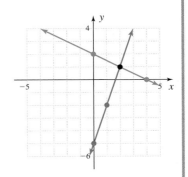

| | | |
|---|---|---|
| $x + 2y = 4$ | original equations | $y = 3x - 5$ |
| $2 + 2(1) = 4$ | substitute<br>2 for $x$ and 1 for $y$ | $1 = 3(2) - 5$ |
| $2 + 2 = 4$ | simplify | $1 = 6 - 5$ |
| $4 = 4$ ✓ | result checks | $1 = 1$ ✓ |

☑ **B. You've just seen how we can solve linear systems by graphing**

This verifies that $(2, 1)$ is the solution to the system.

**Now try Exercises 9 through 26** ▶

---

In Example 2, solving the first equation for $y$ yields $y = -\frac{1}{2}x + 2$. Since the slope of this line is $m = -\frac{1}{2}$ (from $y = mx + b$), while the slope of the second line is $m = 3$, the lines cannot be parallel and so must intersect at a single point.

## C. Solving Systems by Substitution

While a graphical approach best illustrates *why* the solution must be an ordered pair, it does have one obvious drawback—noninteger solutions are difficult to spot. The ordered pair $\left(\frac{1}{2}, -2\right)$ is the solution to $\begin{cases} 6x + y = 1 \\ y = 2x - 3 \end{cases}$, but this would be difficult to "pinpoint" as a precise location on a hand-drawn graph. To overcome this limitation, we next consider an algebraic method known as **substitution,** which works very well when $x$ or $y$ in either equation has a coefficient of 1. The method involves converting a system of two equations in two variables into a single equation in one variable by using an appropriate substitution. For $\begin{cases} 6x + y = 1 \\ y = 2x - 3 \end{cases}$, the second equation says "$y$ is 3 less than twice $x$." We reason that *all* points on this line are related this way, *including the point where this line intersects the other.* For this reason, we can substitute $2x - 3$ for $y$ *in the first equation*, obtaining a single equation in $x$.

If neither equation allows an immediate substitution, we first solve for one of the variables, either $x$ or $y$, and *then* substitute. The method is summarized here, and can actually be used with either like variables or like variable *expressions*. **See Exercises 37 through 40.**

> **Solving Systems Using Substitution**
>
> 1. Solve one of the equations for $x$ in terms of $y$ or $y$ in terms of $x$.
> 2. Substitute for the appropriate variable in the *other* equation and solve for the variable that remains.
> 3. Substitute the value from Step 2 into either of the original equations and solve for the other unknown.
> 4. Write the answer as an ordered pair and check the solution in both original equations.

**EXAMPLE 3** ▶ **Solving a System Using Substitution**

Solve using substitution: $\begin{cases} 6x + y = 1 \\ y = 2x - 3 \end{cases}$.

**Solution** ▶ Since $y = 2x - 3$, we can replace $y$ with $2x - 3$ in the first equation.

$$6x + y = 1 \qquad \text{first equation}$$
$$6x + (2x - 3) = 1 \qquad \text{substitute } 2x - 3 \text{ for } y$$
$$8x - 3 = 1 \qquad \text{combine like terms}$$
$$8x = 4 \qquad \text{add 3}$$
$$x = \frac{4}{8} \qquad \text{divide by 8}$$
$$= \frac{1}{2} \qquad \text{lowest terms}$$

The $x$-coordinate is $\frac{1}{2}$. To find the $y$-coordinate, we substitute $\frac{1}{2}$ for $x$ into either of the original equations, a process known as **back-substitution.** Substituting in the second equation gives

$$y = 2x - 3 \qquad \text{second equation}$$
$$= 2\left(\frac{1}{2}\right) - 3 \qquad \text{substitute } \tfrac{1}{2} \text{ for } x$$
$$= 1 - 3 \qquad \text{multiply}$$
$$= -2 \qquad \text{subtract}$$

To verify that $\left(\frac{1}{2}, -2\right)$ is the solution, we substitute $\frac{1}{2}$ for $x$ and $-2$ for $y$ in both equations.

$$6x + y = 1 \qquad \text{original equations} \qquad y = 2x - 3$$
$$6\left(\frac{1}{2}\right) + (-2) = 1 \qquad \substack{\text{substitute} \\ \tfrac{1}{2} \text{ for } x \text{ and } -2 \text{ for } y} \qquad -2 = 2\left(\frac{1}{2}\right) - 3$$
$$3 + (-2) = 1 \qquad \text{simplify} \qquad -2 = 1 - 3$$
$$1 = 1 ✓ \qquad \text{result checks} \qquad -2 = -2 ✓$$

☑ **C.** You've just seen how we can solve linear systems by substitution

This shows $\left(\frac{1}{2}, -2\right)$ is indeed the solution to the system.

**Now try Exercises 27 through 36** ▶

## D. Solving Systems Using Elimination

Now consider the system $\begin{cases} -2x + 5y = 13 \\ 2x - 3y = -7 \end{cases}$, where solving for any one of the variables will result in fractional coefficients. The substitution method can still be used, but often the **elimination method** is more efficient. The method takes its name from what happens when you add certain equations in a system (by adding the like terms from each). If the coefficients of either $x$ or $y$ are additive inverses—they sum to zero and are *eliminated*. For the system shown, "adding the equations" produces $2y = 6$, giving $y = 3$, then $x = 1$ using back-substitution.

$$\begin{cases} -2x +5y = 13 \\ \underline{2x -3y = -7} \end{cases} \quad \text{add like terms from each equation}$$
$$\begin{array}{rcl} 2y &=& 6 \quad \text{sum} \\ y &=& 3 \quad \text{divide by 2} \end{array}$$

If neither of the like-variable terms sum to zero, we can multiply one or both equations by a nonzero constant to "match up" the coefficients, so an elimination will take place. In doing so, we create an **equivalent system of equations,** meaning one that has the same solution as the original system. For $\begin{cases} 7x - 4y = 16 \\ -3x + 2y = -6 \end{cases}$, multiplying the second equation by 2 produces $\begin{cases} 7x - 4y = 16 \\ -6x + 4y = -12 \end{cases}$, and we note the $y$-terms now have a sum of zero. After adding the equations we see that $x = 4$.

$$\begin{cases} 7x -4y = 16 \\ \underline{-6x +4y = -12} \end{cases} \quad \text{add like terms from each equation}$$
$$\begin{array}{rcl} x &=& 4 \quad \text{sum} \end{array}$$

Using back-substitution and the first equation (substituting 4 for $x$) gives

$$\begin{array}{rcll} 7x - 4y &=& 16 & \text{first equation} \\ 7(4) - 4y &=& 16 & \text{substitute 4 for } x \\ 28 - 4y &=& 16 & \text{multiply} \\ -4y &=& -12 & \text{subtract 28} \\ y &=& 3 & \text{divide by } -4 \end{array}$$

and we see the solution is (4, 3).

Note the three systems produced are equivalent, and you can check that all have (4, 3) as a solution.

**1.** $\begin{cases} 7x - 4y = 16 \\ -3x + 2y = -6 \end{cases}$ **2.** $\begin{cases} 7x - 4y = 16 \\ -6x + 4y = -12 \end{cases}$ **3.** $\begin{cases} 7x - 4y = 16 \\ x = 4 \end{cases}$

In summary,

### Operations That Produce an Equivalent System

1. Changing the order of the equations.
2. Replacing an equation with a nonzero constant multiple of that equation.
3. Replacing an equation with the sum of two equations from the system.

Before beginning a solution using elimination, check to make sure the equations are written in the **standard form** $Ax + By = C$, so that like terms will appear above/below each other. Throughout this chapter, we will use R1 to represent the equation in *row 1* of the system, R2 to represent the equation in *row 2*, and so on. These designations are used to help describe and document the steps being used to solve a system, as in Example 4 where 2R1 + R2 indicates the equation in the first row has been multiplied by two, with the result added to the second equation.

**EXAMPLE 4** ▶ **Solving a System by Elimination**

Solve using elimination: $\begin{cases} 2x - 3y = 7 \\ 6y + 5x = 4 \end{cases}$.

**Solution** ▶ The second equation is not in standard form, so we rewrite the system as $\begin{cases} 2x - 3y = 7 \\ 5x + 6y = 4 \end{cases}$. If we "added the equations" now, we would get $7x + 3y = 11$, with neither variable eliminated. However, if we multiply the first equation by 2, we obtain $4x - 6y = 7$ and the $y$-coefficients will then sum to zero. This results in an equation with $x$ as the only unknown.

$$\begin{array}{r} 2R1 \\ + \\ R2 \end{array} \begin{cases} 4x - 6y = 14 \\ 5x + 6y = \underline{\phantom{0}4} \end{cases} \quad \text{add}$$
$$\begin{array}{r} 9x + 0y = 18 \quad \text{sum} \\ 9x = 18 \\ x = 2 \quad \text{solve for } x \end{array}$$

Substituting 2 for $x$ back into either of the original equations yields $y = -1$. The ordered pair solution is $(2, -1)$. Check by substituting 2 for $x$ and $-1$ for $y$ in both equations.

**Now try Exercises 41 through 50** ▶

The elimination method is summarized here. If either equation has fractional or decimal coefficients, we can "clear" them using an appropriate constant multiplier.

### Solving Systems Using Elimination

1. Write each equation in standard form: $Ax + By = C$.
2. Multiply one or both equations by a constant that will create coefficients of $x$ (or $y$) that are additive inverses.
3. Combine the two equations using vertical addition and solve for the variable that remains.
4. Substitute the value from Step 3 into either of the original equations and solve for the other unknown.
5. Write the answer as an ordered pair and check the solution in both original equations.

**WORTHY OF NOTE**

As the elimination method involves adding two equations, it is sometimes referred to as the *addition method* for solving systems.

**EXAMPLE 5** ▶  **Solving a System Using Elimination**

Solve using elimination: $\begin{cases} \frac{5}{2}x - 3y = 1 \\ 3x - 4y = 6 \end{cases}$.

**Solution** ▶  Multiplying the first equation by 2 (2R1) will clear the fraction and make the system easier to solve.

$$\frac{5}{2}x - 3y = 1 \qquad \text{original equation}$$

$$2\left(\frac{5}{2}x\right) - 2(3y) = 2(1) \qquad \text{multiply by 2}$$

$$5x - 6y = 2 \qquad \text{result}$$

The new system is $\begin{cases} 5x - 6y = 2 \\ 3x - 4y = 6 \end{cases}$.

In this case, we're unable to create additive inverses using a single integer multiplier. Instead, we multiply each equation by a constant that will "match up" one pair of variable terms, and allow an elimination to take place. Note that the $x$-terms can be eliminated if we use 3R1 + (−5R2).

$$\begin{array}{r} 3R1 \\ + \\ -5R2 \end{array} \begin{cases} 15x - 18y = 6 \\ -15x + 20y = -30 \end{cases} \quad \text{add}$$
$$\overline{\hphantom{-5R2}\quad 0x + 2y = -24} \quad \text{sum}$$
$$y = -12 \qquad \text{solve for } y$$

Substituting $y = -12$ in either of the original equations will give the corresponding value of $x$. Here, we'll use the second equation.

$$3x - 4y = 6 \qquad \text{second equation}$$
$$3x - 4(-12) = 6 \qquad \text{substitute } -12 \text{ for } y$$
$$3x + 48 = 6 \qquad \text{multiply}$$
$$3x = -42 \qquad \text{subtract 48}$$
$$x = -14 \qquad \text{divide by 3}$$

☑ **D.** You've just seen how we can solve linear systems by elimination

Use back-substitution to verify the solution to this system is $(-14, -12)$.

**Now try Exercises 51 through 56** ▶

**CAUTION** ▶  Be sure to multiply *all* terms (on both sides) of an equation when using a constant multiplier. Also, note that for Example 5, we could have eliminated the *y*-terms using 2R1−3R2.

## E. Inconsistent and Dependent Systems

A system having *at least one* solution is called a **consistent system.** As seen in Example 2, if the lines have different slopes, they intersect at a single point and the system has exactly one solution. Here, the lines are *independent* of each other and the system is called an **independent system.** If the lines have equal slopes *and* the same *y*-intercept, they are identical or **coincident lines.** Since one is right atop the other, they *intersect at all points,* and the system has an infinite number of solutions. Here, one line *depends* on the other and the system is called a **dependent system.** Using substitution or elimination on a dependent system results in the elimination of all variable terms and leaves a statement that is *always true,* such as $0 = 0$ or some other simple identity.

**EXAMPLE 6** ▶    **Solving a Dependent System**

Solve using elimination: $\begin{cases} 3x + 4y = 12 \\ 6x = 24 - 8y \end{cases}$.

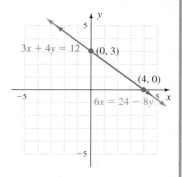

**Solution** ▶    Writing the system in standard form gives
$\begin{cases} 3x + 4y = 12 \\ 6x + 8y = 24 \end{cases}$.
Using $-2R1 + R2$, we can eliminate the variable $x$:

$$\begin{array}{r} -2R1 \\ + \\ R2 \end{array} \left\{ \begin{array}{r} -6x - 8y = -24 \\ 6x + 8y = 24 \end{array} \right.$$

$$\underline{\phantom{-2R1 \ \ }}$$

sum    $\quad 0x + 0y = \quad 0 \qquad$ add

$\qquad\qquad\quad 0 \ = \quad\ 0 \qquad$ variables are eliminated

true statement (identity)

Although we didn't expect it, both variables were eliminated and the final statement is always true ($0 = 0$). This indicates the system is dependent, meaning any ordered pair $(x, y)$ that satisfies one of the equations, will also satisfy the other. Writing both equations in slope-intercept form verifies their slopes and intercepts are equal, and that they represent the same line.

$$\begin{cases} 3x + 4y = 12 \\ 6x + 8y = 24 \end{cases} \longrightarrow \begin{cases} 4y = -3x + 12 \\ 8y = -6x + 24 \end{cases} \longrightarrow \begin{cases} y = -\dfrac{3}{4}x + 3 \\ y = -\dfrac{3}{4}x + 3 \end{cases}$$

The solutions of a dependent system are often written in set notation as the set of ordered pairs $(x, y)$, where $y$ is a specified function of $x$. Here the solution would be $\{(x, y) | y = -\frac{3}{4}x + 3\}$. Using an ordered pair with an arbitrary variable, called a **parameter,** is also common: $(p, \frac{-3p}{4} + 3)$, for any real number $p$. A few possible solutions would be $(0, 3)$ for $p = 0$, $(4, 0)$ for $p = 4$, and $(-2, \frac{9}{2})$ for $p = -2$.

**Now try Exercises 57 through 66** ▶

Finally, if the lines have equal slopes and *different y-intercepts,* they are parallel and the system will have no solution. A system with no solutions is called an **inconsistent system.** An "inconsistent system" produces an "inconsistent answer," such as $12 = 0$ or some other false statement when substitution or elimination is applied. In other words, all variable terms are once again eliminated, but the remaining statement is *false.* A summary of the three possibilities is shown in Figure 6.1 for arbitrary slope $m$ and y-intercept $(0, b)$.

**Figure 6.1**

| Consistent and Independent $m_1 \neq m_2$ | Consistent and Dependent $m_1 = m_2, b_1 = b_2$ | Inconsistent $m_1 = m_2, b_1 \neq b_2$ |
|---|---|---|
| One point in common | All points in common | No points in common |

☑ **E.** You've just seen how we can recognize inconsistent systems and dependent systems

### F. Systems and Modeling

In previous chapters, we solved numerous real-world applications by writing all given relationships in terms of a single variable. Many situations are easier to model using a system of equations with each relationship modeled independently using *two* variables. We begin here with a **mixture** application. Although they appear in many different forms (coin problems, metal alloys, investments, merchandising, and so on), mixture problems all have a similar theme. Generally one equation is related to *quantity* (how much of each item is being combined) and one equation is related to *value* (what is the value of each item being combined).

---

**EXAMPLE 7** ▶  **Solving a Mixture Application**

A jeweler is commissioned to create a piece of artwork that will weigh 14 oz and consist of 75% gold. She has on hand two alloys that are 60% and 80% gold respectively. How much of each should she use?

**Solution** ▶ Let $x$ represent ounces of the 60% alloy and $y$ represent ounces of the 80% alloy. The first equation must be $x + y = 14$, since the piece of art must weigh exactly 14 oz (this is the *quantity* equation). The $x$ ounces are 60% gold, the $y$ ounces are 80% gold, and the 14 oz will be 75% gold. This gives the *value* equation:

$0.6x + 0.8y = 0.75(14)$. The system is $\begin{cases} x + y = 14 \\ 6x + 8y = 105 \end{cases}$ (after clearing decimals).

Solving for $y$ in the first equation gives $y = 14 - x$. Substituting $14 - x$ for $y$ in the second equation gives

$$
\begin{aligned}
6x + 8y &= 105 && \text{second equation} \\
6x + 8(14 - x) &= 105 && \text{substitute } 14 - x \text{ for } y \\
6x + 112 - 8x &= 105 && \text{distribute} \\
-2x + 112 &= 105 && \text{simplify} \\
x &= \frac{7}{2} && \text{solve for } x
\end{aligned}
$$

Substituting $\frac{7}{2}$ for $x$ in the first equation gives $y = \frac{21}{2}$. She should use 3.5 oz of the 60% alloy and 10.5 oz of the 80% alloy.

> **WORTHY OF NOTE**
>
> As an estimation tool, note that if equal amounts of the 60% and 80% alloys were used (7 oz each), the result would be a 70% alloy (halfway in between). Since a 75% alloy is needed, more of the 80% gold must be used.

**Now try Exercises 69 through 74** ▶

---

A second application of systems involves uniform motion (*distance = rate · time*), and explores concepts of great importance to the navigation of ships and airplanes. As a simple illustration, if you've ever walked at your normal rate $r$ on the "moving walkways" at an airport, you likely noticed an increase in your total speed. This is because the resulting speed combines your walking rate $r$ with the speed $w$ of the walkway: *total speed = r + w*. If you walk in the opposite direction of the walkway, your total speed is much slower, as now *total speed = r − w*.

This same phenomenon is observed when an airplane is flying with or against the wind, or a ship is sailing with or against the current.

---

**EXAMPLE 8** ▶  **Solving an Application of Systems—Uniform Motion**

An airplane flying due south from St. Louis, Missouri, to Baton Rouge, Louisiana, uses a strong, steady tailwind to complete the trip in only 2.5 hr. On the return trip, the same wind slows the flight and it takes 3 hr to get back. If the flight distance between these cities is 912 km, what is the cruising speed of the airplane (speed with no wind)? How fast is the wind blowing in kilometers per hour (kph)?

**Solution** ▶ Let $r$ represent the rate of the plane and $w$ the rate of the wind. Since $D = RT$, the flight to Baton Rouge can be modeled by $912 = (r + w)(2.5)$, and the return flight by $912 = (r - w)(3)$. This produces the system $\begin{cases} 912 = 2.5r + 2.5w \\ 912 = \ 3r - \ 3w \end{cases}$.

Dividing R1 by 2.5 and R2 by 3 to simplify the system produces the following sequence:

$$\begin{matrix} \frac{R1}{2.5} \\ \frac{R2}{3} \end{matrix} \begin{cases} 912 = 2.5r + 2.5w \\ 912 = \ 3r - \ 3w \end{cases} \rightarrow \begin{cases} 364.8 = r + w \\ 304.0 = r - w \end{cases}$$

Using the new rows, R1 + R2 gives $668.8 = 2r$, showing $334.4 = r$. The speed of the plane is 334.4 kph. Substituting $334.4$ for $r$ in the second equation, we have:

$$\begin{aligned} 912 &= 3r - 3w & &\text{equation} \\ 912 &= 3(334.4) - 3w & &\text{substitute} \\ 912 &= 1003.2 - 3w & &\text{multiply} \\ -91.2 &= -3w & &\text{subtract 1003.2} \\ 30.4 &= w & &\text{divide by } -3 \end{aligned}$$

The speed of the wind is 30.4 kph. Verify that the same answer ($w = 30.4$) would have been obtained by simply using the new rows and R1 − R2.

**Now try Exercises 75 through 78** ▶

Systems of equations also play a significant role in *cost-based pricing* in the business world. The costs involved in running a business can broadly be understood as either a **fixed cost $k$** or a **variable cost $v$**. Fixed costs might include the monthly rent paid for facilities, which remains the same regardless of how many items are produced and sold. Variable costs would include the cost of materials needed to produce the item, which depends on the number of items made. The total cost can then be modeled by $C(x) = vx + k$ for $x$ number of items. Once a **selling price $p$** has been determined, the revenue equation is simply $R(x) = px$ (price times number of items sold). We can now set up and solve a system of equations that will determine how many items must be sold to break even, performing what is called a **break-even analysis** where $C(x) = R(x)$.

**EXAMPLE 9** ▶ **Solving an Application of Systems—Break-Even Analysis**

In home businesses that produce items to sell on eBay®, fixed costs are easily determined by rent and utilities, and variable costs by the price of materials needed to produce the item. Karen's home business makes large decorative candles for all occasions. The cost of materials is $3.50 per candle, and her rent and utilities average $900 per month. If her candles sell for $9.50, how many candles must be sold each month to break even?

**Solution ▶** Let $x$ represent the number of candles sold. Her total cost is $C(x) = 3.5x + 900$ (variable cost plus fixed cost), and projected revenue is $R(x) = 9.5x$. This gives the system $\begin{cases} C(x) = 3.5x + 900 \\ R(x) = 9.5x \end{cases}$. To break even, *Cost* = *Revenue*, which gives

$$\begin{aligned} 9.5x &= 3.5x + 900 && \text{Cost = Revenue} \\ 6x &= 900 && \text{subtract } 3.5x \\ x &= 150 && \text{divide by 6} \end{aligned}$$

The analysis shows that Karen must sell 150 candles each month to break even.

**Now try Exercises 79 through 82 ▶**

**WORTHY OF NOTE**

There are limitations to this model, and this interplay can be affected by the number of available consumers, production limits, "shelf-life" issues, and so on, but at any given moment in the life cycle of a product, consumer demand responds to price in this way. Producers also respond to price in a very predictable way.

In a "free-market" economy, also referred to as a "supply-and-demand" economy, there are naturally occurring forces that invariably come into play if no outside forces act on the producers (suppliers) and consumers (demanders). Generally speaking, the higher the price of an item, the lower the demand. A good advertising campaign can increase the demand, but the increasing demand brings an increase in price, which moderates the demand—and so it goes until a balance is reached. These free-market forces ebb and flow until **market equilibrium** occurs, at the specific price where the supply and demand are equal.

In Exercises 79 through 82, the equation models were artificially constructed to yield a "nice" solution. In actual practice, the equations and coefficients are not so "well behaved" and are based on the collection and interpretation of real data. While market analysts have sophisticated programs and numerous models to help develop these equations, here we'll use our experience with regression to develop the supply-and-demand curves.

**EXAMPLE 10 ▶** **Using Technology to Find Market Equilibrium**

A manufacturer of smartphones has hired a consulting firm to do market research on their "next-generation" phone. Over a 10-wk period, the firm collected the data shown for the smartphone market (data include smartphones sold and expected to sell).

| Price (Dollars) | Supply (Inventory) | Demand (Purchases) |
|---|---|---|
| 214.20 | 8,125 | 15,250 |
| 171.00 | 9,875 | 12,375 |
| 129.60 | 16,500 | 10,000 |
| 104.40 | 16,875 | 9,875 |
| 216.00 | 8,625 | 17,500 |
| 183.60 | 9,500 | 15,000 |
| 154.80 | 11,050 | 11,900 |
| 93.60 | 17,250 | 7,625 |
| 149.40 | 13,250 | 9,750 |
| 136.80 | 16,000 | 10,760 |

**a.** Use a graphing calculator to simultaneously display the demand and supply scatterplots.

**b.** Calculate a line of best fit for each and graph them with the scatterplots (identify each curve).

**c.** Find the equilibrium point.

**Solution ▶** **a.** Begin by clearing all lists. This can be done manually, or by pressing 2nd + **(MEM)** and selecting option **4:ClrAllLists** (the command appears on the home screen). Pressing ENTER will execute the command, and the word DONE will appear. Carefully input price in L1, supply in L2, and demand in L3 (see Figure 6.2). With the window settings given in Figure 6.3, pressing GRAPH will display the price/demand and price/supply scatterplots shown. If this

**Figure 6.2**

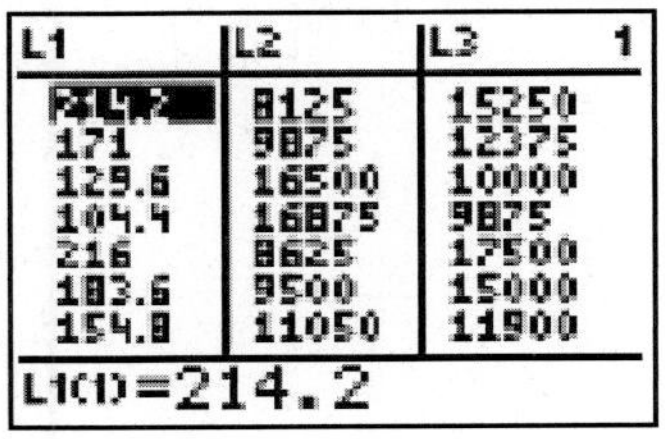


**Figure 6.3**

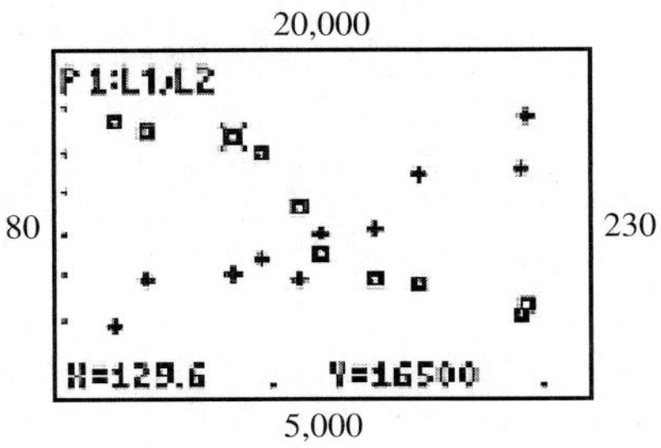

is not the case, use  (STAT PLOT) to be sure that "On" is highlighted in Plot1 and Plot2, and that Plot1 uses L1 and L2, while Plot2 uses L1 and L3 (Figure 6.4). Note we've chosen a different mark to indicate the data points for Plot2.

b. Calculate the linear regression equation for L1 and L2 (supply), and paste it in $Y_1$: **LinReg (ax + b) L1, L2, $Y_1$** . Next, calculate the linear regression for L1 and L3 (demand) and paste it in $Y_2$: **LinReg (ax + b) L1, L3, $Y_2$** (recall that $Y_1$ and $Y_2$ are accessed using the VARS key). The resulting equations and graphs are shown in Figures 6.5 and 6.6.

**Figure 6.4**

**Figure 6.5**

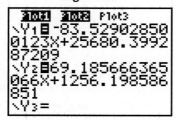

**Figure 6.6**

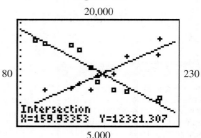

c. Once again we use (2nd) (TRACE) (CALC) 5:intersect to find the equilibrium point, which is approximately (159.93, 12,321). Supply and demand for this smartphone model are approximately equal at a price of $159.93, with 12,321 smartphones bought and sold.

☑ **F. You've just seen how we can use a system of equations to model and solve applications**

**Now try Exercises 83 and 84 ▶**

Other interesting applications can be found in the Exercise Set. **See Exercises 85 through 88.**

## 6.1 EXERCISES

### ▶ CONCEPTS AND VOCABULARY

**Fill in the blank with the appropriate word or phrase. Carefully reread the section, if necessary.**

1. Systems that have no solution are called _____ systems.

2. Systems having at least one solution are called _____ systems.

3. The given systems are equivalent. How do we obtain the second system from the first?
$$\begin{cases} \dfrac{2}{3}x + \dfrac{1}{2}y = \dfrac{5}{3} \\ 0.2x + 0.4y = 1 \end{cases} \quad \begin{cases} 4x + 3y = 10 \\ 2x + 4y = 10 \end{cases}$$

4. For the system shown, which solution method would be more efficient, substitution or elimination? Discuss/Explain why.
$$\begin{cases} 2x + 5y = 8 \\ 3x + 4y = 5 \end{cases}$$

## ▶ DEVELOPING YOUR SKILLS

**Use back-substitution to determine if the ordered pair given is a solution.**

5. $\begin{cases} 3x + y = 11 \\ -5x + y = -13 \end{cases}; (3, 2)$    6. $\begin{cases} 3x + 7y = -4 \\ 7x + 8y = -21 \end{cases}; (-6, 2)$    7. $\begin{cases} 4x - 3y = 7 \\ 5x + 6y = 12 \end{cases}; \left(2, \dfrac{1}{3}\right)$    8. $\begin{cases} 8x + y = 7 \\ 4x + 3y = 11 \end{cases}; \left(\dfrac{1}{2}, 3\right)$

**Solve each system by *graphing* manually. Check results using back-substitution.**

9. $\begin{cases} 3x + 2y = 12 \\ x - y = -1 \end{cases}$    10. $\begin{cases} 5x - 2y = 10 \\ -x + y = 1 \end{cases}$    11. $\begin{cases} y = -2x + 1 \\ 3x - y = -6 \end{cases}$    12. $\begin{cases} 4x + y = 12 \\ y = 3x - 2 \end{cases}$

13. $\begin{cases} -2x + 3y = 9 \\ 2y + x = -1 \end{cases}$    14. $\begin{cases} 3x + 4y = 8 \\ x = 2y + 6 \end{cases}$

**Given the graph shown, determine which equation(s) have the indicated point as a solution. If the point satisfies more than one equation, write the system for which it is a solution.**

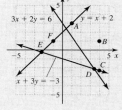

15. $A$           16. $B$           17. $C$

18. $D$           19. $E$           20. $F$

**Determine the number of solutions for each system by writing the equations in slope-intercept form. Do not solve the system.**

21. $\begin{cases} -2x + y = 3 \\ y - 8 = 3x \end{cases}$    22. $\begin{cases} y + 5 = 4x \\ -\dfrac{1}{2}x + y = 2 \end{cases}$    23. $\begin{cases} 6x + 3y = 4.5 \\ 2x + y = 7 \end{cases}$    24. $\begin{cases} 4x - 3y = -6 \\ 1.5y - 2x = 3 \end{cases}$

25. $\begin{cases} 7x - 4y = 24 \\ 4x + 3y = 15 \end{cases}$    26. $\begin{cases} 3x - 4y = 20 \\ 5x + 2y = -6 \end{cases}$

**Solve each system using *substitution*. Write solutions as an ordered pair.**

27. $\begin{cases} 2x + y = 10 \\ y = 3x \end{cases}$    28. $\begin{cases} y = 3x \\ -2x + y = 3 \end{cases}$    29. $\begin{cases} y + 2x = 9 \\ y = x - 3 \end{cases}$    30. $\begin{cases} x - 3y = 5 \\ x = y + 3 \end{cases}$

31. $\begin{cases} x = 5y - 9 \\ x - 2y = -6 \end{cases}$    32. $\begin{cases} 4x - 5y = 7 \\ 2x - 5 = y \end{cases}$    33. $\begin{cases} 3x - 4y = 24 \\ 5x + y = 17 \end{cases}$    34. $\begin{cases} 3x + 2y = 19 \\ x - 4y = -3 \end{cases}$

35. $\begin{cases} 5x - 6y = 12 \\ x + 2y = 4 \end{cases}$    36. $\begin{cases} 2x + 5y = 11 \\ 8x - y = 2 \end{cases}$

**The substitution method can be used for like variables *or for like expressions*. Solve the following systems, *using the expression* common to both equations.**

37. $\begin{cases} 2x + 4y = 16 \\ x + 22 = 4y \end{cases}$    38. $\begin{cases} 8x = 3y + 26 \\ 8x - 5y = 38 \end{cases}$    39. $\begin{cases} 5x - 11y = 21 \\ 11y = 5 - 8x \end{cases}$    40. $\begin{cases} -6x = 5y - 16 \\ 5y - 6x = 4 \end{cases}$

**Solve using *elimination* and check results using back-substitution.**

41. $\begin{cases} 2x - 4y = 10 \\ 3x + 4y = 5 \end{cases}$    42. $\begin{cases} -x + 5y = 8 \\ x + 2y = 6 \end{cases}$    43. $\begin{cases} 4x - 3y = 1 \\ 3y = -5x - 19 \end{cases}$    44. $\begin{cases} 5y - 3x = -5 \\ 3x + 2y = 19 \end{cases}$

45. $\begin{cases} 3x + 4y = 5 \\ 5x - 2y = -9 \end{cases}$    46. $\begin{cases} -2x + 5y = -11 \\ 3x - 10y = 19 \end{cases}$    47. $\begin{cases} 2x - 6y = -8 \\ -5x + 3y = -16 \end{cases}$    48. $\begin{cases} 4x - 3y = 24 \\ 6x - 5y = 38 \end{cases}$

49. $\begin{cases} 2x = -3y + 17 \\ 4x - 5y = 12 \end{cases}$    50. $\begin{cases} 2y = 5x + 2 \\ -4x = 17 - 6y \end{cases}$    51. $\begin{cases} \dfrac{1}{3}x + 2y = 8 \\ 2x - 5y = -3 \end{cases}$    52. $\begin{cases} -5x + 3y = 20 \\ 3x - \dfrac{2}{5}y = -5 \end{cases}$

53. $\begin{cases} 0.5x + 0.4y = 0.2 \\ 0.3y = 1.3 + 0.2x \end{cases}$    54. $\begin{cases} 0.2x + 0.3y = 0.8 \\ 0.3x + 0.4y = 1.3 \end{cases}$    55. $\begin{cases} -\dfrac{1}{6}u + \dfrac{1}{4}v = 4 \\ \dfrac{1}{2}u - \dfrac{2}{3}v = -11 \end{cases}$    56. $\begin{cases} \dfrac{3}{4}x + \dfrac{1}{3}y = -2 \\ \dfrac{3}{2}x + \dfrac{1}{5}y = 3 \end{cases}$

**Solve using any method and identify the system as consistent, inconsistent, or dependent.**

57. $\begin{cases} 7a + b = -25 \\ 2a - 5b = 14 \end{cases}$

58. $\begin{cases} -2m + 3n = -1 \\ 5m - 6n = 4 \end{cases}$

59. $\begin{cases} 2a = 2 - 3b \\ 6b + 4a = 7 \end{cases}$

60. $\begin{cases} 3p - 6q = -15 \\ 4p + 20 = 8q \end{cases}$

61. $\begin{cases} 6x - 22 = -y \\ 3x + \frac{1}{2}y = 11 \end{cases}$

62. $\begin{cases} 1.2x + 0.4y = 5 \\ 0.5y = -1.5x + 2 \end{cases}$

63. $\begin{cases} -10x + 35y = -5 \\ y = 0.25x \end{cases}$

64. $\begin{cases} 2x + 3y = 4 \\ x = -2.5y \end{cases}$

65. $\begin{cases} 0.2y = 0.3x + 4 \\ 0.6x - 0.4y = -1 \end{cases}$

66. $\begin{cases} 15 - 5y = -9x \\ -3x + \frac{5}{3}y = 5 \end{cases}$

## ▶ WORKING WITH FORMULAS

67. **Uniform motion with current:** $\begin{cases} (R + C)T_1 = D_1 \\ (R - C)T_2 = D_2 \end{cases}$

The formula shown can be used to solve uniform motion problems involving a *current,* where $D$ represents distance traveled, $R$ is the rate of the object with no current, $C$ is the speed of the current, and $T$ is the time. Chan-Li rows 9 mi up river (against the current) in 3 hr. It only took him 1 hr to row 5 mi downstream (with the current). How fast was the current? How fast can he row in still water?

68. **Fahrenheit and Celsius temperatures:** $\begin{cases} y = \frac{9}{5}x + 32 & °F \\ y = \frac{5}{9}(x - 32) & °C \end{cases}$

Many people are familiar with temperature measurement in degrees Celsius and degrees Fahrenheit, but few realize that the equations are linear and there is one temperature at which the two scales agree. Solve the system using the method of your choice and find this temperature.

## ▶ APPLICATIONS

Solve each application by modeling the situation described with a linear system.

### Mixture

69. **Theater productions:** At a recent production of *A Comedy of Errors,* the Community Theater brought in a total of $30,495 in revenue. If adult tickets were $9 and children's tickets were $6.50, how many tickets of each type were sold if 3800 tickets in all were sold?

70. **Milkfat requirements:** A dietician needs to mix 10 gal of milk that is $2\frac{1}{2}\%$ milkfat for the day's rounds. He has some milk that is 4% milkfat and some that is $1\frac{1}{2}\%$ milkfat. How much of each should be used?

71. **Filling the family cars:** Cherokee just filled both of the family vehicles at a service station. The total cost for 20 gal of regular unleaded and 17 gal of premium unleaded was $144.89. The premium gas was $0.10 more per gallon than the regular gas. Find the price per gallon for each type of gasoline.

72. **Household cleaners:** As a cleaning agent, a solution that is 24% vinegar is often used. How much pure (100%) vinegar and 5% vinegar must be mixed to obtain 50 oz of a 24% solution?

73. **Saving money:** Bryan has been doing odd jobs around the house, trying to earn enough money to buy a new Dirt-Surfer©. He saves all quarters and dimes in his piggy bank, while he places all nickels and pennies in a drawer to spend. So far, he has 225 coins in the piggy bank, worth a total of $45.00. How many of the coins are quarters? How many are dimes?

74. **Coin investments:** In 1990, Molly attended a coin auction and purchased some rare "Seated Liberty" fifty-cent pieces, and a number of very rare two-cent pieces from the Civil War Era. If she bought 47 coins with a face value of $10.06, how many of each denomination did she buy?

### Uniform Motion

75. **Canoeing on a stream:** On a recent camping trip, it took Molly and Sharon 2 hr to row 4 mi upstream from the drop in point to the campsite. After a leisurely weekend of camping, fishing, and relaxation, they rowed back downstream to the drop in point in just 30 min. Use this information to find (a) the speed of the current and (b) the speed Sharon and Molly would be rowing in still water.

76. **Taking a luxury cruise:** A luxury ship is taking a Caribbean cruise from Caracas, Venezuela, to Belize City on the Yucatan Peninsula, a distance of 1435 mi. En route they encounter the Caribbean Current, which flows to the northwest, parallel to the coastline. From Caracas to the Belize coast, the trip took 70 hr. After a few days of fun in the sun, the ship leaves for Caracas, with the return trip taking 82 hr. Use this information to find (a) the speed of the Caribbean Current and (b) the cruising speed of the ship.

77. **Airport walkways:** As part of an algebra field trip, Jason takes his class to the airport to use their moving walkways for a demonstration. The class measures the longest walkway, which turns out to be 256 ft long. Using a stopwatch, Jason shows it takes him just 32 sec

to complete the walk going in the same direction as the walkway. Walking in a direction opposite the walkway, it takes him 320 sec—10 times as long! The next day in class, Jason hands out a two-question quiz: (1) What was the speed of the walkway in feet per second? (2) What is my (Jason's) normal walking speed? Create the answer key for this quiz.

78. **Racing pigeons:** The American Racing Pigeon Union often sponsors opportunities for owners to fly their birds in friendly competitions. During a recent competition, Steve's birds were liberated in Topeka, Kansas, and headed almost due north to their loft in Sioux Falls, South Dakota, a distance of 308 mi. During the flight, they encountered a steady wind from the north and the trip took 4.4 hr. The next month, Steve took his birds to a competition in Grand Forks, North Dakota, with the birds heading almost due south to home, also a distance of 308 mi. This time the birds were aided by the same wind from the north, and the trip took only 3.5 hr. Use this information to find (a) the racing speed of Steve's birds and (b) the speed of the wind.

**Break-Even Analysis**

79. **Lawn service:** Dave and his sons run a lawn service, which includes mowing, edging, trimming, and aerating lawns. His fixed cost includes insurance, his salary, and monthly payments on equipment, and amounts to $4000/mo. The variable costs include gas, oil, hourly wages for his employees, and miscellaneous expenses, which run about $75 per lawn. The average charge for full-service lawn care is $115 per visit. Do a break-even analysis to determine (a) how many lawns Dave must service each month to break even and (b) the revenue required to break even.

80. **Production of mini-microwave ovens:** Due to high market demand, a manufacturer decides to introduce a new line of mini-microwave ovens for personal and office use. By using existing factory space and retraining some employees, fixed costs are estimated at $8400/mo. The components to assemble and test each microwave are expected to run $45 per unit. If market research shows consumers are willing to pay at least $69 for this product, find (a) how many units must be made and sold each month to break even and (b) the revenue required to break even.

**Market Equilibrium**

81. **Farm commodities:** One area where the law of supply and demand is clearly at work is farm commodities. Suppose that for $x$ billion bushels of soybeans, supply is modeled by $y = 1.5x + 3$, where $y$ is the current market price (in dollars per bushel). The related demand equation might be $y = -2.20x + 12$. (a) How many billion bushels will be supplied and demanded at a market price of $5.40? Is supply less than demand? (b) How many billion bushels will be supplied and demanded at a market price of $7.05? Is demand less than supply? (c) Find the equilibrium point for this market.

82. **Electronic media:** Competition in the smartphone market is often fierce—with suppliers fighting to earn and hold market shares. Suppose that for $x$ million smartphones sold, supply is modeled by $y = 3.8x + 40$, where $y$ is the current market price (in dollars). The related demand equation might be $y = -1.5x + 220$. (a) How many million smartphones will be supplied and demanded at a market price of $155? Is supply less than demand? (b) How many million smartphones will be supplied and demanded at a market price of $180? Is demand less than supply? (c) Find the equilibrium point for this market.

 83. **Pricing wakeboards:** A water sports company that manufactures high-end wakeboards has hired an outside consulting firm to do some market research on their wakeboard. This consulting firm collected the following supply-and-demand data for this and comparable wakeboards. Find the equilibrium point.

| Average Price (in U.S. dollars) | Available Inventory | Quantity Demanded |
| --- | --- | --- |
| 424.85 | 232 | 175 |
| 445.25 | 247 | 166 |
| 389.55 | 215 | 291 |
| 349.98 | 201 | 391 |
| 402.22 | 226 | 218 |
| 413.87 | 222 | 200 |
| 481.73 | 251 | 139 |
| 419.45 | 235 | 177 |
| 397.05 | 219 | 220 |
| 361.90 | 212 | 317 |

 84. **Pricing pet care products:** A metal shop that manufactures pens for pet rabbits has collected some data on sales and production. The following table shows the supply-and-demand data for these pens. Find the equilibrium point for this market.

| Average Price (in U.S. dollars) | Production (Supply) | Quantity Sold (Demand) |
| --- | --- | --- |
| 22.99 | 7 | 12 |
| 21.49 | 6 | 14 |
| 23.99 | 7 | 11 |
| 26.99 | 11 | 9 |
| 25.99 | 10 | 8 |
| 27.99 | 13 | 8 |
| 24.49 | 9 | 10 |
| 26.49 | 11 | 9 |

**Descriptive Translation**

**85. Important dates in U.S. history:** If you sum the year that the Declaration of Independence was signed and the year that the Civil War ended, you get 3641. There are 89 yr that separate the two events. What year was the Declaration signed? What year did the Civil War end?

**86. Architectural wonders:**
When it was first constructed in 1889, the Eiffel Tower in Paris, France, was the tallest structure in the world. In 2009, the Burj Khalifa skyscraper in Dubai, United Arab Emirates, became the world's tallest structure. The skyscraper is 235 ft less than

three times the height of the Eiffel Tower, and the sum of their heights is 3701 ft. How tall is each tower?

**87. Pacific islands land area:** In the South Pacific, the island nations of Tahiti and Tonga have a combined land area of 692 mi$^2$. Tahiti's land area is 112 mi$^2$ more than Tonga's. What is the land area of each island group?

**88. Card games:** On a cold winter night, in the lobby of a beautiful hotel in Sante Fe, New Mexico, Marc and Klay just barely beat John and Steve in a close game of Trumps. If the sum of the team scores was 990 points, and there was a 12-point margin of victory, what was the final score?

### ▶ EXTENDING THE CONCEPT

**89.** Federal income tax reform has been a hot political topic for many years. Suppose tax plan A calls for a flat tax of 20% tax on all income (no deductions or loopholes). Tax plan B requires taxpayers to pay $5000 plus 10% of all income. For what income level do both plans require the same tax?

**90.** Suppose a certain amount of money was invested at 6% per year, and another amount at 8.5% per year, with a total return of $1250. If the amounts invested at each rate were switched, the yearly income would have been $1375. To the nearest whole dollar, how much was invested at each rate?

**Given any two points, the equation of a line through these points can be found using a system of equations. While there are certainly more efficient methods, using a system here will show how we can find equations for polynomials of higher degree. The key is to note that each point will yield an equation of the form $y = mx + b$. For instance, the points (3, 6) and (−2, −4) yield the system** $\begin{cases} 6 = 3m + b \\ -4 = -2m + b \end{cases}$.

**91.** Use a system of equations to find the equation of the line containing the points (2, 7) and (−4, −5).

**92.** Use a system of equations to find the equation of the line containing the points (9, −1) and (−3, 7).

### ▶ MAINTAINING YOUR SKILLS

**93. (4.3)** Use the rational zeroes theorem to write the polynomial in completely factored form: $3x^4 - 19x^3 + 15x^2 + 27x - 10$.

**94. (3.1)** Use transformations of the toolbox function $f(x) = |x|$ to sketch the graph of $F(x) = -|x + 3| - 2$.

**95. (4.6)** Graph $y = x^2 - 6x - 16$ and state the interval where $f(x) \leq 0$.

**96. (5.5)** Solve for $x$ (rounded to the nearest thousandth): $33 = 77.5e^{-0.0052x} - 8.37$.

## LEARNING OBJECTIVES

*In Section 6.2 you will see how we can:*

☐ **A.** Visualize a solution in three dimensions

☐ **B.** Check ordered triple solutions

☐ **C.** Solve linear systems in three variables

☐ **D.** Recognize inconsistent and dependent systems

☐ **E.** Use a system of three equations in three variables to solve applications

The transition to systems of three equations in three variables requires a fair amount of "visual gymnastics" along with good organizational skills. Although the techniques used are identical and similar results are obtained, the third equation and variable give us more to track, and we must work more carefully toward the solution.

## A. Visualizing Solutions in Three Dimensions

The solution to an equation in one variable is the single number that satisfies the equation. For $x + 1 = 3$, the solution is $x = 2$ and its graph is a single *point* on the number line, a **one-dimensional graph.** The solution to an equation in two variables, such as $x + y = 3$, is any ordered pair $(x, y)$ that satisfies the equation. When we graph this solution set, the result is a *line* on the $xy$-plane, a **two-dimensional graph.** The solutions to an equation in three variables, such as $x + y + z = 6$, are the **ordered triples** $(x, y, z)$ that satisfy the equation. When we graph this solution set, the result is a **plane** in **space,** a **three-dimensional graph.** Recall a plane is a flat surface having infinite length and width, but no depth. We can graph this plane using the intercept method and the result is shown in Figure 6.7. For graphs in three dimensions, we often picture the $xy$-plane parallel to the ground (with the $y$-axis pointing to the right) and $z$ as the **vertical axis.** To find an additional point on this plane, we use any three numbers whose sum is 6, such as $(2, 3, 1)$. Move 2 units along the $x$-axis, 3 units parallel to the $y$-axis, and 1 unit parallel to the $z$-axis, as shown in Figure 6.8.

> **WORTHY OF NOTE**
>
> We can visualize the location of a point in space by considering a large rectangular box 2 ft long $\times$ 3 ft wide $\times$ 1 ft tall, placed snugly in the corner of a room. The floor is the $xy$-plane, one wall is the $xz$-plane, and the other wall is the $yz$-plane. The $z$-axis is formed where the two walls meet and the corner of the room is the origin $(0, 0, 0)$. To find the corner of the box located at $(2, 3, 1)$, first locate the point $(2, 3)$ in the $xy$-plane (the floor), then move up 1 ft.

**Figure 6.7**

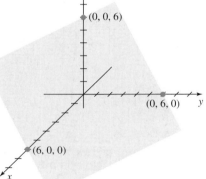

**Figure 6.8**

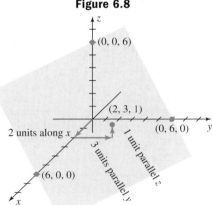

---

**EXAMPLE 1** ▶ **Finding Solutions to an Equation in Three Variables**

Use a guess-and-check method to find four additional points on the plane determined by $x + y + z = 6$.

**Solution** ▶ We can begin by letting $x = 0$, then use any combination of $y$ and $z$ that sum to 6. Two examples are $(0, 2, 4)$ and $(0, 5, 1)$. We could also select any two values for $x$ and $y$, then determine a value for $z$ that results in a sum of 6. Two examples are $(-2, 9, -1)$ and $(8, -3, 1)$.

☑ **A.** You've just seen how we can visualize a solution in three dimensions

Now try Exercises 5 through 8 ▶

## B. Solutions to a System of Three Equations in Three Variables

When solving a system of three equations in three variables, remember each equation represents a plane in space. These planes can intersect in various ways, creating different possibilities for a solution set (see Figures 6.9 through 6.12). The system could have a **unique** solution $(a, b, c)$, if the planes intersect at a single point (Figure 6.9) (the point satisfies all three equations simultaneously). If the planes intersect in a line (Figure 6.10), the system is **linearly dependent** and there is an infinite number of solutions. Unlike the two-dimensional case, the equation of a line in three dimensions is somewhat complex, and the coordinates of all points on this line are usually *represented* by a specialized ordered triple, which we use to state the solution set. If the planes intersect at all points, the system has **coincident dependence** (see Figure 6.11). This indicates the equations of the system differ by only a constant multiple—they are all "disguised forms" of the *same equation*. The solution set is any ordered triple $(a, b, c)$ satisfying this equation. Finally, the system may be **inconsistent** and have no solutions. This can happen a number of different ways, most notably if the planes intersect or are parallel as shown in Figure 6.12 (other possibilities are discussed in the exercises). In the case of "no solutions," an ordered triple may satisfy none of the equations, only one of the equations, or only two of the equations, but not all three equations.

| **Figure 6.9** | **Figure 6.10** | **Figure 6.11** | **Figure 6.12** |
|---|---|---|---|
| 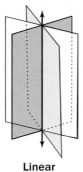 | 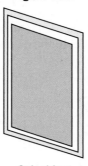 | 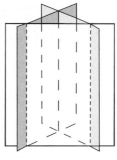 | 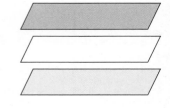 |
| Independent (Unique solution) | Linear dependence (Infinitely many solutions) | Coincident dependence (Infinitely many solutions) | Inconsistent (No solutions) |

---

**EXAMPLE 2** ▶ **Determining If an Ordered Triple Is a Solution**

Determine if the ordered triple $(1, -2, 3)$ is a solution to the systems shown.

**a.** $\begin{cases} x + 4y - z = -10 \\ 2x + 5y + 8z = 4 \\ x - 2y - 3z = -4 \end{cases}$
**b.** $\begin{cases} 3x + 2y - z = -4 \\ 2x - 3y - 2z = 2 \\ x - y + 2z = 9 \end{cases}$

**Solution** ▶ Substitute 1 for $x$, $-2$ for $y$, and 3 for $z$ in the first system.

**a.** $\begin{cases} x + 4y - z = -10 \\ 2x + 5y + 8z = 4 \\ x - 2y - 3z = -4 \end{cases} \rightarrow \begin{cases} (1) + 4(-2) - (3) = -10 \\ 2(1) + 5(-2) + 8(3) = 4 \\ (1) - 2(-2) - 3(3) = -4 \end{cases} \rightarrow \begin{cases} -10 = -10 \text{ true} \\ 16 = 4 \text{ false} \\ -4 = -4 \text{ true} \end{cases}$

No, the ordered triple $(1, -2, 3)$ is not a solution to the first system. Now use the same substitutions in the second system.

**b.** $\begin{cases} 3x + 2y - z = -4 \\ 2x - 3y - 2z = 2 \\ x - y + 2z = 9 \end{cases} \rightarrow \begin{cases} 3(1) + 2(-2) - (3) = -4 \\ 2(1) - 3(-2) - 2(3) = 2 \\ (1) - (-2) + 2(3) = 9 \end{cases} \rightarrow \begin{cases} -4 = -4 \text{ true} \\ 2 = 2 \text{ true} \\ 9 = 9 \text{ true} \end{cases}$

☑ **B.** You've just seen how we can check ordered triple solutions

The ordered triple $(1, -2, 3)$ is a solution to the second system only.

**Now try Exercises 9 and 10** ▶

## C. Solving Systems of Three Equations in Three Variables Using Elimination

From Section 6.1, we know that two systems of equations are **equivalent** if they have the same solution set. The systems

$$\begin{cases} 2x + y - 2z = -7 \\ x + y + z = -1 \\ -2y - z = -3 \end{cases} \quad \text{and} \quad \begin{cases} 2x + y - 2z = -7 \\ y + 4z = 5 \\ z = 1 \end{cases}$$

are equivalent, as both have the unique solution $(-3, 1, 1)$. In addition, it is evident that the second system can be solved more easily, since R2 and R3 have fewer variables than the first system. In the simpler system, mentally substituting 1 for $z$ into R2 immediately gives $y = 1$, and these values can be back-substituted into the first equation to find that $x = -3$. This observation guides us to a general approach for solving larger systems—we would like to *eliminate variables in the second and third equations, until we obtain an equivalent system that can easily be solved by back-substitution.* To begin, let's review the three operations that "transform" a given system, and produce an equivalent system.

### Operations That Produce an Equivalent System

1. Changing the order of the equations.
2. Replacing an equation by a nonzero constant multiple of that equation.
3. Replacing an equation with the sum of two equations from the system.

Building on the ideas from Section 6.1, we develop the following approach for solving a system of three equations in three variables.

### Solving a System of Three Equations in Three Variables

1. Write each equation in standard form: $Ax + By + Cz = D$.
2. If the "$x$" term in any equation has a coefficient of 1, interchange equations (if necessary) so this equation becomes R1.
3. Use the $x$-term in R1 to eliminate the $x$-terms from R2 and R3. The original R1, with the new R2 and R3, form an equivalent system that contains a smaller "subsystem" of two equations in two variables.
4. Solve the subsystem for either $y$ or $z$ and keep the result as the new R3.
5. Find the value of the remaining variables using back substitution.

We'll begin by solving the system $\begin{cases} 2x + y - 2z = -7 \\ x + y + z = -1 \\ -2y - z = -3 \end{cases}$ using the elimination method and the procedure outlined. In Example 3, the notation $-2R1 + R2 \rightarrow R2$ indicates the equation in row 1 has been multiplied by $-2$ and added to the equation in row 2, with the result placed in the system as the new row 2.

---

**EXAMPLE 3** ▶  **Solving a System of Three Equations in Three Variables**

Solve using elimination: $\begin{cases} 2x + y - 2z = -7 \\ x + y + z = -1 \\ -2y - z = -3 \end{cases}$.

**Solution** ▶

1. The system is in standard form.
2. If the $x$-term in any equation has a coefficient of 1, interchange equations so this equation becomes R1.

$$\begin{cases} 2x + y - 2z = -7 \\ x + y + z = -1 \\ -2y - z = -3 \end{cases} \xrightarrow{\text{R2} \leftrightarrow \text{R1}} \begin{cases} x + y + z = -1 \\ 2x + y - 2z = -7 \\ -2y - z = -3 \end{cases}$$

3. Use R1 to eliminate the $x$-term in R2 and R3. Since R3 has no $x$-term, the only elimination needed is the $x$-term from R2. Using $-2$R1 $+$ R2 will eliminate this term:

$$\begin{array}{rl} -2\text{R1} & -2x - 2y - 2z = \phantom{-}2 \\ + & \\ \underline{\text{R2}} & \underline{\phantom{-}2x + \phantom{-}y - 2z = -7} \\ & 0x - 1y - 4z = -5 \quad \text{sum} \\ & y + 4z = \phantom{-}5 \quad \text{simplify} \end{array}$$

The new R2 is $y + 4z = 5$. The original R1 and R3, along with the new R2 form an equivalent system that contains a smaller **subsystem.**

$$\begin{cases} x + y + z = -1 \\ 2x + y - 2z = -7 \\ -2y - z = -3 \end{cases} \xrightarrow{-2\text{R1} + \text{R2} \to \text{R2}} \begin{cases} x + y + z = -1 \\ y + 4z = \phantom{-}5 \\ -2y - z = -3 \end{cases} \begin{array}{l} \text{new} \\ \text{equivalent} \\ \text{system} \end{array}$$

4. Solve the subsystem for either $y$ or $z$, and keep the result as a *new* R3. We choose to eliminate $y$ using $2$R2 $+$ R3:

$$\begin{array}{rl} 2\text{R2} & 2y + 8z = \phantom{-}10 \\ + & \\ \underline{\text{R3}} & \underline{-2y - \phantom{8}z = -3} \\ & 0y + 7z = \phantom{-}7 \quad \text{sum} \\ & z = \phantom{-}1 \quad \text{simplify} \end{array}$$

The new R3 is $z = 1$.

$$\begin{cases} x + y + z = -1 \\ y + 4z = \phantom{-}5 \\ -2y - z = -3 \end{cases} \xrightarrow{2\text{R2} + \text{R3} \to \text{R3}} \begin{cases} x + y + z = -1 \\ y + 4z = \phantom{-}5 \\ z = \phantom{-}1 \end{cases} \begin{array}{l} \text{new} \\ \text{equivalent} \\ \text{system} \end{array}$$

The new R3, along with the original R1 and R2 from Step 3, form an equivalent system that can be solved using back-substitution. Substituting 1 for $z$ in R2 yields $y = 1$. Substituting 1 for $z$ and 1 for $y$ in R1 yields $x = -3$. The solution is $(-3, 1, 1)$. Check this result using back-substitution.

**Now try Exercises 11 through 14** ▶

This method for solving systems is called Gaussian Elimination (Carl Friedrich Gauss, 1777–1855), and we are said to be triangularizing the system.

While not absolutely needed for the elimination process, there are two reasons for wanting the coefficient of $x$ to be "1" in R1. First, it makes the elimination method more efficient since we can more easily see what to use as a multiplier. Second, it lays the foundation for developing other methods of solving larger systems. If no equation has an $x$-coefficient of 1, we simply use the $y$- or $z$-variable instead (see Example 7). Since solutions to larger systems generally are worked out in stages, we will sometimes track the transformations used by writing them *between* the original system and the equivalent system, rather than to the left as we did in Section 6.1.

Here is an additional example illustrating the elimination process, but in *abbreviated form*. Verify the calculations indicated using a separate sheet.

---

**EXAMPLE 4** ▶ **Solving a System of Three Equations in Three Variables**

Solve using elimination: $\begin{cases} -5y + 2x - z = -8 \\ -x + 3z + 2y = 13 \\ -z + 3y + x = 5 \end{cases}$.

**Solution** ▶ **1.** Write the equations in standard form: $\begin{cases} 2x - 5y - z = -8 \\ -x + 2y + 3z = 13 \\ x + 3y - z = 5 \end{cases}$

**2.** $\begin{cases} 2x - 5y - z = -8 \\ -x + 2y + 3z = 13 \\ x + 3y - z = 5 \end{cases}$ $\xrightarrow{\text{R3} \leftrightarrow \text{R1}}$ $\begin{cases} x + 3y - z = 5 \\ -x + 2y + 3z = 13 \\ 2x - 5y - z = -8 \end{cases}$  equivalent system

**3.** Using R1 + R2 will eliminate the *x*-term from R2, yielding $5y + 2z = 18$.
Using $-2$R1 + R3 eliminates the *x*-term from R3, yielding $-11y + z = -18$.

$\begin{cases} x + 3y - z = 5 \\ -x + 2y + 3z = 13 \\ 2x - 5y - z = -8 \end{cases}$ $\xrightarrow[\;-2\text{R1} + \text{R3} \to \text{R3}\;]{\text{R1} + \text{R2} \to \text{R2}}$ $\begin{cases} x + 3y - z = 5 \\ 5y + 2z = 18 \\ -11y + z = -18 \end{cases}$  equivalent system

**4.** Using $-2$R3 + R2 will eliminate *z* from the subsystem, leaving $27y = 54$.

$\begin{cases} x + 3y - z = 5 \\ 5y + 2z = 18 \\ -11y + z = -18 \end{cases}$ $\xrightarrow{-2\text{R3} + \text{R2} \to \text{R3}}$ $\begin{cases} x + 3y - z = 5 \\ 5y + 2z = 18 \\ 27y = 54 \end{cases}$  equivalent system

Solving for *y* in R3 yields: $\dfrac{\cancel{27}y}{\cancel{27}} = \dfrac{54}{27}$, showing $y = 2$. Substituting 2 for *y* in R2 gives,

$5(2) + 2z = 18$    substitute 2 for *y*
$10 + 2z = 18$    simplify
$2z = 8$    subtract 10
$z = 4$    divide by 2

Substituting 2 for *y* and 4 for *z* in R1 gives,

$x + 3(2) - 4 = 5$    substitute 2 for *y*, 4 for *z*
$x + 2 = 5$    simplify
$x = 3$    subtract 2

☑ **C.** You've just seen how we can solve linear systems in three variables

The solution is $(3, 2, 4)$, which can be checked using back-substitution.

**Now try Exercises 15 through 18** ▶

---

## D. Inconsistent and Dependent Systems

As mentioned, it is possible for larger systems to have no solutions or an infinite number of solutions. As with our work in Section 6.1, an inconsistent system (no solutions) will produce inconsistent results, ending with a statement such as $0 = -3$ or some other contradiction.

**EXAMPLE 5** ▶ **Attempting to Solve an Inconsistent System**

Solve using elimination: $\begin{cases} 2x + y - 3z = -3 \\ 3x - 2y + 4z = 2 \\ 4x + 2y - 6z = -7 \end{cases}$.

**Solution** ▶ 1. This system has no equation where the coefficient of $x$ is 1.

2. We can still use R1 to begin the solution process, but this time we'll use the variable $y$ since it *does* have coefficient 1.

Using 2R1 + R2 eliminates the $y$-term from R2, leaving $7x - 2z = -4$. But using $-2$R1 + R3 to eliminate the $y$-term from R3 results in a contradiction:

$$
\begin{array}{rl}
\text{2R1} & 4x + 2y - 6z = -6 \\
+ & \\
\underline{\text{R2}} & \underline{3x - 2y + 4z = \phantom{-}2} \\
& 7x \phantom{- 2y} - 2z = -4
\end{array}
\qquad
\begin{array}{rl}
\text{-2R1} & -4x - 2y + 6z = \phantom{-}6 \\
+ & \\
\underline{\text{R3}} & \underline{4x + 2y - 6z = -7} \\
& 0x + 0y + 0z = -1 \\
& \phantom{0x + 0y + 0}0 = -1 \quad \text{contradiction}
\end{array}
$$

We conclude the system is inconsistent. The answer is the empty set $\varnothing$, and we need work no further.

**Now try Exercises 19 and 20** ▶

Unlike our work with systems having only two variables, systems in three variables can have two forms of dependence—*linear dependence* (Figure 6.10) or *coincident dependence* (Figure 6.11). To help understand linear dependence, consider a system of two equations in three variables: $\begin{cases} -2x + 3y - z = 5 \\ x - 3y + 2z = -1 \end{cases}$. Each of these equations represents a plane, and unless the planes are parallel, their intersection will be a line. As in Section 6.1, we can state solutions to a dependent system using set notation with two of the variables written in terms of the third, or as an ordered triple using a parameter. The relationships named can then be used to generate specific solutions to the system.

Systems with two equations and two variables or three equations and three variables are called **square systems,** meaning there are exactly as many equations as there are variables. A system of linear equations cannot have a unique solution unless there are at least as many equations as there are variables in the system.

**EXAMPLE 6** ▶ **Solving a Dependent System**

Solve using elimination: $\begin{cases} 2x + 5y - 3z = -5 \\ -x - 5y + 2z = 1 \end{cases}$.

**Solution** ▶ We immediately note that R1 + R2 eliminates the $y$-term from R2, yielding $x - z = -4$ and the new system $\begin{cases} 2x + 5y - 3z = -5 \\ x \phantom{+ 5y} - z = -4 \end{cases}$. This means $(x, y, z)$ will satisfy both equations only when $x = z - 4$. Since $x$ is written in terms of $z$, we substitute $z - 4$ for $x$ *in either* of the original equations to find how $y$ is related to $z$. Using R2 in the original system we have:

$$
\begin{array}{rl}
-x - 5y + 2z = 1 & \text{row 2} \\
-(z - 4) - 5y + 2z = 1 & \text{substitute } z - 4 \text{ for } x \\
4 - 5y + z = 1 & \text{combine like terms} \\
-5y + z = -3 & \text{subtract 4} \\
-5y = -z - 3 & \text{subtract } z \\
y = \frac{1}{5}z + \frac{3}{5} & \text{solve for } y
\end{array}
$$

In set notation, the solution is $\{(x, y, z,) \mid x = z - 4, y = \frac{1}{5}z + \frac{3}{5}, z \in \mathbb{R}\}$. Randomly choosing $z = -3, 2$, and 7, the solutions would be $(-7, 0, -3)$, $(-2, 1, 2)$, and $(3, 2, 7)$ respectively. Verify that these satisfy both equations. Using $p$ as a parameter, the solution could also be written $(p - 4, \frac{1}{5}p + \frac{3}{5}, p)$ in parameterized form.

**Now try Exercises 21 through 24** ▶

The system in Example 6 was nonsquare, and we knew ahead of time the system would be dependent or inconsistent. The system in Example 7 *is* square, but only by applying the elimination process can we determine the nature of its solution(s).

---

**EXAMPLE 7** ▶ **Solving a Dependent System**

Solve using elimination: $\begin{cases} 3x - 2y + z = -1 \\ 2x + y - z = 5 \\ 10x - 2y = 8 \end{cases}$.

**Solution** ▶ This system has no equation where the coefficient of $x$ is 1. We will still use R1, noting that R1 + R2 eliminates the $z$-term from R2, yielding $5x - y = 4$.

$\begin{cases} 3x - 2y + z = -1 \\ 2x + y - z = 5 \\ 10x - 2y = 8 \end{cases}$ $\quad \dfrac{R1 + R2 \rightarrow R2}{R3 \rightarrow R3} \quad$ $\begin{cases} 3x - 2y + z = -1 \\ 5x - y = 4 \\ 10x - 2y = 8 \end{cases}$

We next solve the subsystem. Using $-2R2 + R3$ eliminates the $y$-term in R3, but also all other terms:

$$\begin{array}{lrl} -2R2 & -10x + 2y = -8 \\ + & \\ \underline{R3} & \underline{10x - 2y = 8} \\ & 0x + 0y = 0 & \text{sum} \\ & 0 = 0 & \text{result} \end{array}$$

Since R3 is the same as 2R2, the system is equivalent to $\begin{cases} 3x - 2y + z = -1 \\ 5x - y = 4 \end{cases}$.

We can solve for $y$ in R2 to write $y$ in terms of $x$: $y = 5x - 4$. Substituting $5x - 4$ for $y$ in R1 enables us to also write $z$ in terms of $x$:

$$\begin{array}{ll} 3x - 2y + z = -1 & \text{R1} \\ 3x - 2(5x - 4) + z = -1 & \text{substitute } 5x - 4 \text{ for } y \\ 3x - 10x + 8 + z = -1 & \text{distribute} \\ -7x + z = -9 & \text{simplify} \\ z = 7x - 9 & \text{solve for } z \end{array}$$

The solution set is $\{(x, y, z) \mid x \in \mathbb{R}, y = 5x - 4, z = 7x - 9\}$. Three of the infinite number of solutions are $(0, -4, -9)$ for $x = 0$, $(2, 6, 5)$ for $x = 2$, and $(-1, -9, -16)$ for $x = -1$. Verify these triples satisfy all three equations. Again using the parameter $p$, the solution could be written as $(p, 5p - 4, 7p - 9)$.

☑ **D.** You've just seen how we can recognize inconsistent and dependent systems

**Now try Exercises 25 through 28** ▶

For **coincident dependence** the equations in a system differ by only a constant multiple. After applying the elimination process, all variables are eliminated from the other equations, leaving statements that are always true (such as $2 = 2$ or some other true statement). For additional practice solving various kinds of systems, **see Exercises 29 through 40.**

## E. Applications

Applications of larger systems are simply an extension of our work with systems of two equations in two variables. Once again, the applications come in a variety of forms and from many fields. In the world of business and finance, systems can be used to diversify investments or spread out liabilities, a financial strategy hinted at in Example 8.

**EXAMPLE 8** ▶  **Modeling the Finances of a Business**

A small business borrowed \$225,000 from three different lenders to expand their product line. The interest rates were 5%, 6%, and 7%. Find how much was borrowed at each rate if the annual interest came to \$13,000 and twice as much was borrowed at the 5% rate than was borrowed at the 7% rate.

**Solution** ▶  Let $x$, $y$, and $z$ represent the amounts borrowed at 5%, 6%, and 7% respectively. This means our first equation is $x + y + z = 225$ (in thousands). The second equation is determined by the total interest paid, which was \$13,000: $0.05x + 0.06y + 0.07z = 13$. The third is found by carefully reading the problem. "*twice as much was borrowed at the 5% rate than was borrowed at the 7% rate,*" *or $x = 2z$.*

These equations form the system: $\begin{cases} x + y + z = 225 \\ 0.05x + 0.06y + 0.07z = 13. \\ x = 2z \end{cases}$

Written in standard form we have:

$$\begin{cases} x + y + z = 225 & \text{R1} \\ 5x + 6y + 7z = 1300 & \text{R2} \quad \text{multiply R2 by 100} \\ x \quad\quad - 2z = 0 & \text{R3} \end{cases}$$

Using $-5\text{R1} + \text{R2}$ will eliminate the $x$ term in R2, while $-\text{R1} + \text{R3}$ will eliminate the $x$-term in R3.

$$\begin{array}{ll} -5\text{R1} & -5x - 5y - 5z = -1125 \\ + & \\ \text{R2} & \underline{5x + 6y + 7z = \quad 1300} \\ & \quad\quad y + 2z = \quad 175 \end{array} \qquad \begin{array}{ll} -\text{R1} & -x - y - \ z = -225 \\ + & \\ \text{R3} & \underline{x \quad\quad - 2z = \quad\quad 0} \\ & \quad -y - 3z = -225 \end{array}$$

The new R2 is $y + 2z = 175$, and the new R3 (after multiplying by $-1$) is $y + 3z = 225$, yielding the equivalent system $\begin{cases} x + y + z = 225 \\ \quad\quad y + 2z = 175. \\ \quad\quad y + 3z = 225 \end{cases}$

Solving the $2 \times 2$ subsystem using $-\text{R2} + \text{R3}$ yields $z = 50$. Back-substitution shows $y = 75$ and $x = 100$, yielding the solution $(100, 75, 50)$. This means \$50,000 was borrowed at the 7% rate, \$75,000 was borrowed at 6%, and \$100,000 at 5%.

☑ **E. You've just seen how we can use a system of three equations in three variables to solve applications**

Now try Exercises 43 through 52 ▶

## 6.2 EXERCISES

### ▶ CONCEPTS AND VOCABULARY

**Fill in the blank with the appropriate word or phrase. Carefully reread the section, if necessary.**

**1.** The solution to an equation in three variables is an ordered _____.

**2.** The graph of the solutions to linear equations in three variables is a(n) _____.

**3.** Find a value of $z$ that makes the ordered triple $(2, -5, z)$ a solution to $2x + y + z = 4$. Discuss/Explain how this is accomplished.

**4.** Explain the difference between linear dependence and coincident dependence, and describe how the equations are related.

### ▶ DEVELOPING YOUR SKILLS

**Find four ordered triples that satisfy each equation.**

**5.** $x + 2y + z = 9$

**6.** $3x + y - z = 8$

**7.** $-x + y + 2z = -6$

**8.** $2x - y + 3z = -12$

**Determine if the given ordered triples are solutions to the system. If not a solution, identify which equation(s) are not satisfied.**

**9.** $\begin{cases} x + y - 2z = -1 \\ 4x - y + 3z = 3 \\ 3x + 2y - z = 4 \end{cases}$ ; $\begin{aligned} (0, 3, 2) \\ (-3, 4, 1) \end{aligned}$

**10.** $\begin{cases} 2x + 3y + z = 9 \\ 5x - 2y - z = -32; \\ x - y - 2z = -13 \end{cases}$ $\begin{aligned} (-4, 5, 2) \\ (5, -4, 11) \end{aligned}$

**Solve each system using elimination and back-substitution.**

**11.** $\begin{cases} x - y - 2z = -10 \\ x \quad\ - z = 1 \\ \quad\quad z = 4 \end{cases}$

**12.** $\begin{cases} x + y + 2z = -1 \\ 4x - y \quad\ = 3 \\ 3x \quad\quad = 6 \end{cases}$

**13.** $\begin{cases} x + 3y + 2z = 16 \\ -2y + 3z = 1 \\ 8y - 13z = -7 \end{cases}$

**14.** $\begin{cases} -x + y + 5z = 1 \\ 4x + y \quad\ = 1 \\ -3x - 2y \quad\ = 8 \end{cases}$

**15.** $\begin{cases} -x + y + 2z = -10 \\ x + y - z = 7 \\ 2x + y + z = 5 \end{cases}$

**16.** $\begin{cases} x + y - 2z = -1 \\ 4x - y + 3z = 3 \\ 3x + 2y - z = 4 \end{cases}$

**17.** $\begin{cases} 2x - 3y + 2z = 0 \\ 3x - 4y + z = -20 \\ x + 2y - z = 16 \end{cases}$

**18.** $\begin{cases} 3x - y + z = 6 \\ 2x + 2y - z = 5 \\ 2x - y + z = 5 \end{cases}$

**Solve using the elimination method. If a system is inconsistent or dependent, so state. For systems with linear dependence, write solutions in set notation and as an ordered triple in terms of a parameter.**

**19.** $\begin{cases} 3x + y + 2z = 3 \\ x - 2y + 3z = 1 \\ 4x - 8y + 12z = 7 \end{cases}$

**20.** $\begin{cases} 2x - y + 3z = 8 \\ 3x - 4y + z = 4 \\ -4x + 2y - 6z = 5 \end{cases}$

**21.** $\begin{cases} 4x + y + 3z = 8 \\ x - 2y + 3z = 2 \end{cases}$

**22.** $\begin{cases} 4x - y + 2z = 9 \\ 3x + y + 5z = 5 \end{cases}$

**23.** $\begin{cases} 6x - 3y + 7z = 2 \\ 3x - 4y + z = 6 \end{cases}$

**24.** $\begin{cases} 2x - 4y + 5z = -2 \\ 3x - 2y + 3z = 7 \end{cases}$

**Solve using elimination. If the system is linearly dependent, state the general solution in terms of a parameter. Then find four specific ordered triples that satisfy the system (solutions will vary).**

**25.** $\begin{cases} 3x - 4y + 5z = 5 \\ -x + 2y - 3z = -3 \\ 3x - 2y + z = 1 \end{cases}$

**26.** $\begin{cases} 5x - 3y + 2z = 4 \\ -9x + 5y - 4z = -12 \\ -3x + y - 2z = -12 \end{cases}$

**27.** $\begin{cases} x + 2y - 3z = 1 \\ 3x + 5y - 8z = 7 \\ x + y - 2z = 5 \end{cases}$

**28.** $\begin{cases} -2x + 3y - 5z = 3 \\ 5x - 7y + 12z = -8 \\ x - y + 2z = -2 \end{cases}$

**Solve using the elimination method. If a system is inconsistent or dependent, so state. For systems with linear dependence, write the answer in terms of a parameter. For coincident dependence, state the solution in set notation.**

**29.** $\begin{cases} 4x + 2y - 8z = 24 \\ -x - 0.5y + 2z = -6 \\ 2x + y - 4z = 12 \end{cases}$

**30.** $\begin{cases} 2x - 5y - 4z = 6 \\ x - 2.5y - 2z = 3 \\ -3x + 7.5y + 6z = -9 \end{cases}$

**31.** $\begin{cases} x - 2y + 2z = 6 \\ 2x - 6y + 3z = 13 \\ 3x + 4y - z = -11 \end{cases}$

**32.** $\begin{cases} 4x - 5y - 6z = 5 \\ 2x - 3y + 3z = 0 \\ x + 2y - 3z = 5 \end{cases}$

**33.** $\begin{cases} x - 5y - 4z = 3 \\ 2x - 9y - 7z = 2 \\ 3x - 14y - 11z = 5 \end{cases}$ **34.** $\begin{cases} 2x + 3y - 5z = 4 \\ x + y - 2z = 3 \\ x + 3y - 4z = -1 \end{cases}$

**35.** $\begin{cases} \frac{1}{6}x + \frac{1}{3}y - \frac{1}{2}z = 2 \\ \frac{3}{4}x - \frac{1}{3}y + \frac{1}{2}z = 9 \\ \frac{1}{2}x - y + \frac{1}{2}z = 2 \end{cases}$ **36.** $\begin{cases} \frac{x}{2} + \frac{y}{3} - \frac{z}{2} = 2 \\ \frac{2x}{3} - y - z = 8 \\ \frac{x}{6} + 2y + \frac{3z}{2} = 6 \end{cases}$

Some applications of systems lead to systems similar to those that follow. Solve using elimination.

**37.** $\begin{cases} -2A - B - 3C = 21 \\ B - C = 1 \\ A + B = -4 \end{cases}$ **38.** $\begin{cases} A - 2B = 5 \\ B + 3C = 7 \\ 2A - B - C = 1 \end{cases}$

**39.** $\begin{cases} A + 2C = 2 \\ 2A - 3B = 1 \\ 3A + 6B - 8C = 1 \end{cases}$ **40.** $\begin{cases} -A + 3B + 2C = 11 \\ 2B + C = 9 \\ B + 2C = 8 \end{cases}$

## ▶ WORKING WITH FORMULAS

**41. Dimensions of a rectangular solid:**
$$\begin{cases} 2w + 2h = P_1 \\ 2l + 2w = P_2 \\ 2l + 2h = P_3 \end{cases}$$

$P_2 = 16$ cm (top)
$P_3 = 18$ cm (front)
$P_1 = 14$ cm (side)
$h$ $l$ $w$

Using the formula shown, the dimensions of a rectangular solid can be found if the perimeters of the three distinct faces are known. Find the dimensions of the solid shown.

**42. Distance from a point $(x, y, z)$ to the plane**
$$Ax + By + Cz = D: \left| \frac{Ax + By + Cz - D}{\sqrt{A^2 + B^2 + C^2}} \right|$$

The perpendicular distance from a given point $(x, y, z)$ to the plane defined by $Ax + By + Cz = D$ is given by the formula shown. Consider the plane given in Figure 6.7 $(x + y + z = 6)$. What is the distance from this plane to the point $(3, 4, 5)$?

## ▶ APPLICATIONS

Solve the following applications by setting up and solving a system of three equations in three variables.

**Investment/Finance and Simple Interest Problems**

**43. Investing the winnings:** After winning $280,000 in the lottery, Maurika decided to place the money in three different investments: a certificate of deposit paying 4%, a money market certificate paying 5%, and some Aa bonds paying 7%. After 1 yr she earned $15,400 in interest. Find how much was invested at each rate if $20,000 more was invested at 7% than at 5%.

**44. Purchase at auction:** At an auction, a wealthy collector paid $7,000,000 for three paintings: a Monet, a Picasso, and a van Gogh. The Monet cost $800,000 more than the Picasso. The price of the van Gogh was $200,000 more than twice the price of the Monet. What was the price of each painting?

**Descriptive Translation**

**45. Major wars:** The United States has fought three major wars in modern times: World War II, the Korean War, and the Vietnam War. If you sum the years that each conflict ended, the result is 5871. The Vietnam War ended 20 years after the Korean War and 28 years after World War II. In what year did each end?

**46. Animal gestation periods:** The average gestation period (in days) of an elephant, rhinoceros, and camel sum to 1520 days. The gestation period of a rhino is 58 days longer than that of a camel. Twice the camel's gestation period decreased by 162 gives the gestation period of an elephant. What is the gestation period of each?

**47. Moments in U.S. history:** If you sum the year the Declaration of Independence was signed, the year the 13th Amendment to the Constitution abolished slavery, and the year the Civil Rights Act was signed, the total would be 5605. Ninety-nine years separate the 13th Amendment and the Civil Rights Act. The Civil Rights Act was signed 188 years after the Declaration of Independence. What year was each signed?

**48. Aviary wingspan:** If you combine the wingspan of the California Condor, the Wandering Albatross (see photo), and the prehistoric Quetzalcoatlus, you get an astonishing 18.6 m (over 60 ft). If the wingspan of the Quetzalcoatlus is equal to five times that of the  Wandering Albatross minus twice that of the California Condor, and six times the wingspan of the Condor is equal to five times the wingspan of the Albatross, what is the wingspan of each?

## Mixtures

**49. Chemical mixtures:** A chemist mixes three different solutions with concentrations of 20%, 30%, and 45% glucose to obtain 10 L of a 38% glucose solution. If the amount of 30% solution used is 1 L more than twice the amount of 20% solution used, find the amount of each solution used.

**50. Value of gold coins:** As part of a promotion, a local bank invites its customers to view a large sack full of $5, $10, and $20 gold pieces, promising to give the sack to the first person able to state the number of coins for each denomination. Customers are told there are exactly 250 coins, with a total face value of $1875. If there are also seven times as many $5 gold pieces as $20 gold pieces, how many of each denomination are there?

## Nutrition

**51. Industrial food production:** Acampana Soups is creating a new sausage and shrimp gumbo that contains three different types of fat: saturated, monounsaturated, and polyunsaturated. As a new member of the "Heart Healthy" menu, this soup must contain only 2.8 g of total fat per serving. Head chef Yev Kasem demands that the recipe provides exactly twice as much saturated fat as polyunsaturated fat. At the same time, the lead nutrition expert Florencia requires the amount of saturated fat in a serving to be 0.4 g less than the *combined* amount of unsaturated fats. How many grams of each type of fat will a serving of this soup contain?

**52. Geriatric nutrition:** The dietician at McKnight Place must create a balanced diet for Fred, consisting of a daily total of 1600 calories. For his successful rehabilitation, exact quantities of complex carbohydrates, fat, and protein must provide these calories. In this diet, carbohydrates provide 160 more calories than fat and protein together, while fat provides 1.25 times more calories than protein alone. How many calories should each nutrient provide on a daily basis?

## ▶ EXTENDING THE CONCEPT

Just as any two points determine a unique line, three noncollinear points determine a unique parabola (as long as no two have the same first coordinate). Each point $(x, y)$ will give an equation of the form $y = ax^2 + bx + c$, and we can create a $3 \times 3$ system that can be solved using elimination. For instance, the point $(2, 41)$ gives the equation $41 = a(2)^2 + 2b + c$.

**53. Height of a soccer ball:** One second after being kicked, a soccer ball is 26 ft high. After 2 sec, the ball is 41 ft high, and after 6 sec the ball is 1 ft above the ground. Use the ordered pairs (time, height) to find a function $h(t)$ modeling the height of the ball after $t$ sec, then use the equation to find (a) the maximum height of the kick, and (b) the height of the ball after 5.4 sec.

**54. Height of an arrow:** An archer is out in a large field testing a new bow. Pulling back the bow to near its breaking point, the archer lets the arrow fly. Suppose that 1 sec after release the arrow was 184 ft high, 4 sec later it was 600 ft high, and after 12 sec the arrow was 96 ft high. Use the ordered pairs (time, height) to find a function $h(t)$ modeling the height of the arrow after $t$ sec, then use the equation to (a) find the maximum height of the shot and (b) determine how long the arrow was airborne.

**55.** The system $\begin{cases} x - 2y - z = 2 \\ x - 2y + kz = 5 \\ 2x - 4y + 4z = 10 \end{cases}$ is dependent if $k = $ _____, and inconsistent otherwise.

**56.** One form of the equation of a circle is $x^2 + y^2 + Dx + Ey + F = 0$. Use a system to find the equation of the circle through the points $(2, -1)$, $(4, -3)$, and $(2, -5)$.

## ▶ MAINTAINING YOUR SKILLS

**57.** (4.6) If $p(x) = 2x^2 - x - 3$, in what interval(s) is $p(x) \leq 0$?

**58.** (4.4) Graph the polynomial defined by $f(x) = x^4 - 5x^2 + 4$.

**59.** (5.5) Solve the logarithmic equation: $\log(x + 2) + \log x = \log 3$

**60.** (2.5) Analyze the graph of $g$ shown. Clearly state the domain and range, zeroes, intervals where $g(x) > 0$, intervals where $g(x) < 0$, local maximums or minimums, and intervals where $g$ is increasing or decreasing. Assume the grid is scaled in units and estimate endpoints to tenths.

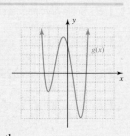

## MID-CHAPTER CHECK

1. Solve by graphing the system. State whether the system is consistent, inconsistent, or dependent.
$$\begin{cases} x - 3y = -2 \\ 2x + y = 3 \end{cases}$$

2. Solve the system using elimination. State whether the system is consistent, inconsistent, or dependent.
$$\begin{cases} x - 3y = -4 \\ 2x + y = 13 \end{cases}$$

3. Solve using a system of linear equations and any method you choose: How many ounces of a 40% acid should be mixed with 10 oz of a 64% acid to obtain a 48% acid solution?

4. Determine whether the ordered triple is a solution to the system. If not, identify which equation(s) are not satisfied.
$$\begin{cases} 5x + 2y - 4z = 22 \\ 2x - 3y + z = -1 \quad (2, 0, -3) \\ 3x - 6y + z = 2 \end{cases}$$

5. The system given is a dependent system. Without solving, state why.
$$\begin{cases} x + 2y - 3z = 3 \\ 2x + 4y - 6z = 6 \\ x - 2y + 5z = -1 \end{cases}$$

Solve each system of equations:

6. $$\begin{cases} x + 2y - 3z = -4 \\ 2y + z = 7 \\ 5y - 2z = 4 \end{cases}$$

7. $$\begin{cases} 2x + 3y - 4z = -4 \\ x - 2y + z = 0 \\ -3x - 2y + 2z = -1 \end{cases}$$

8. Solve the following system and write the solution as an ordered triple in terms of the parameter $p$.
$$\begin{cases} 2x - y + z = 1 \\ -5x + 2y - 3z = 2 \end{cases}$$

9. If you add Mozart's age when he wrote his first symphony, the age of American chess player Paul Morphy when he began dominating the international chess scene, and the age of Blaise Pascal when he formulated his well-known *Essai pour les coniques* (Essay on Conics), the sum is 37. At the time of each event, Paul Morphy's age was 3 yr less than twice Mozart's, and Pascal was 3 yr older than Morphy. Use a system to find the age of each.

10. The *William Tell Overture* (Gioachino Rossini, 1829) is one of the most famous, and best-loved overtures known. It is played in four movements: a prelude, the storm (often used in animations with great clashes of thunder and a driving rain), the sunrise (actually, *A call to the dairy cows…*), and the finale (better known as the *Lone Ranger* theme song). The prelude takes 2.75 min. Depending on how fast the finale is played, the total playing time is about 11 min. The playing time for the prelude and finale is 1 min longer than the playing time of the storm and the sunrise. Also, the playtime of the storm plus twice the playtime of the sunrise is 1 min longer than twice the finale. Use a system to find the playtime for each movement.

---

## 6.3  Nonlinear Systems of Equations and Inequalities

### LEARNING OBJECTIVES

*In Section 6.3 you will see how we can:*

☐ **A.** Visualize possible solutions

☐ **B.** Solve nonlinear systems using substitution

☐ **C.** Solve nonlinear systems using elimination

☐ **D.** Solve nonlinear systems of inequalities

☐ **E.** Solve applications of nonlinear systems

Equations where the variables have exponents other than 1 or that are transcendental (like logarithmic and exponential equations) are all *nonlinear* equations. A nonlinear system of equations has at least one nonlinear equation, and these systems occur in a great variety.

### A.  Possible Solutions for a Nonlinear System

When solving nonlinear systems, it is often helpful to *visualize* the graph of each equation in the system. This can help determine the number of possible intersections and further assist the solution process.

**EXAMPLE 1** ▶ **Sketching Graphs to Visualize the Number of Possible Solutions**

Identify each equation in the system as that of a line, parabola, circle, or one of the toolbox functions. Then determine the number of solutions possible by considering the different ways the graphs might intersect. Finally, solve the system by graphing.

$$\begin{cases} x^2 + y^2 = 25 \\ x - y = 1 \end{cases}$$

**Solution** ▶ The first equation contains a sum of second-degree terms with equal coefficients, which we recognize as the equation of a circle. The second equation is obviously linear. This means the system may have no solution, one solution, or two solutions, as shown in Figure 6.13. The graph of the system is shown in Figure 6.14 and consists of a line with slope $m = 1$ and $y$-intercept $(0, -1)$, with a circle of radius $r = 5$, centered at $(0, 0)$. The two points of intersection appear to be $(-3, -4)$ and $(4, 3)$. After checking these in the original equations we find that both are solutions to the system.

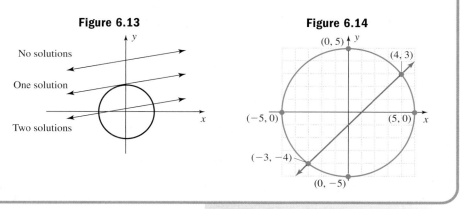

**Figure 6.13**

No solutions

One solution

Two solutions

**Figure 6.14**

$(0, 5)$ $(4, 3)$ $(-5, 0)$ $(5, 0)$ $(-3, -4)$ $(0, -5)$

☑ **A.** You've just seen how we can visualize possible solutions

Now try Exercises 5 through 10 ▶

## B. Solving Nonlinear Systems by Substitution

Since manually graphing nonlinear systems at best offers an *estimate* for the solution (points of intersection may not have rational values), we more often turn to algebraic methods or the use of a graphing calculator. Recall the substitution method involves solving one of the equations for a variable or expression that can be substituted in the other equation, in order to eliminate one of the variables. As with our study of linear systems, be aware that some nonlinear systems have no solutions.

**EXAMPLE 2A** ▶ **Solving a Nonlinear System Using Substitution**

Solve the system $\begin{cases} y = x^2 - 2x - 3 \\ 2x - y = 7 \end{cases}$.

**Solution** ▶ The first equation is that of a parabola. The second equation is linear. Since the first equation is already written with $y$ in terms of $x$, we can substitute $x^2 - 2x - 3$ for $y$ in the second equation and solve.

$$\begin{aligned} 2x - y &= 7 & \text{second equation} \\ 2x - (x^2 - 2x - 3) &= 7 & \text{substitute } x^2 - 2x - 3 \text{ for } y \\ 2x - x^2 + 2x + 3 &= 7 & \text{distribute} \\ -x^2 + 4x + 3 &= 7 & \text{combine like terms} \\ x^2 - 4x + 4 &= 0 & \text{set equal to zero} \\ (x - 2)^2 &= 0 & \text{factor} \end{aligned}$$

We find that $x = 2$ is a repeated root. Since the second equation is simpler than the first, we substitute 2 for $x$ in this equation and find $y = -3$. The ordered pair $(2, -3)$ checks in both equations and the system has only one (repeated) solution at $(2, -3)$.

**EXAMPLE 2B** ▶ **Solving a System Using Substitution**

Solve the system $\begin{cases} y = x^2 + 4x - 12 \\ -(x - 2)^2 = y \end{cases}$.

**Solution** ▶ Expanding the binomial square in the second equation gives $y = -(x^2 - 4x + 4)$, then $y = -x^2 + 4x - 4$ after simplification.

The system then becomes: $\begin{cases} y = \phantom{-}x^2 + 4x - 12 \\ y = -x^2 + 4x - \phantom{1}4 \end{cases}$. Substituting $-x^2 + 4x - 4$ for $y$ in the first equation gives the following:

$$
\begin{aligned}
y &= x^2 + 4x - 12 && \text{first equation} \\
-x^2 + 4x - 4 &= x^2 + 4x - 12 && \text{substitute } -x^2 + 4x - 4 \text{ for } y \\
0 &= 2x^2 - 8 && \text{add } x^2 \text{ and 4; subtract } 4x \\
8 &= 2x^2 && \text{isolate squared term} \\
4 &= x^2 && \text{divide by 2} \\
x = 2 \quad &\text{or} \quad x = -2 && \text{result}
\end{aligned}
$$

Substituting 2 for $x$ in the first equation gives

$$
\begin{aligned}
y &= x^2 + 4x - 12 && \text{first equation} \\
&= (2)^2 + 4(2) - 12 && \text{substitute 2 for } x \\
&= 4 + 8 - 12 && \text{simplify} \\
&= 0 && \text{result}
\end{aligned}
$$

☑ **B.** You've just seen how we can solve nonlinear systems using substitution

One solution is $(2, 0)$. Substituting $-2$ for $x$ shows the second solution is $(-2, -16)$. Both solutions can be verified using back-substitution.

**Now try Exercises 11 through 18** ▶

## C. Solving Nonlinear Systems by Elimination

When both equations in the system have second-degree terms with like variables, it is generally easier to use the elimination method, rather than substitution.

**EXAMPLE 3** ▶ **Solving a Nonlinear System Using Elimination or Graphing Technology**

Solve the system $\begin{cases} y - \frac{1}{2}x^2 = -3 \\ x^2 + y^2 = 41 \end{cases}$.

**Algebraic Solution** ▶ The first equation can be rewritten as $y = \frac{1}{2}x^2 - 3$ and is a parabola opening upward with vertex $(0, -3)$. The second equation represents a circle with center at

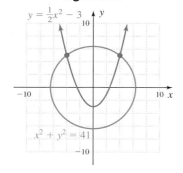

**Figure 6.15**

$y = \frac{1}{2}x^2 - 3$

$x^2 + y^2 = 41$

(0, 0) and radius $r = \sqrt{41} \approx 6.4$. Mentally visualizing these graphs indicates there will be two solutions (see Figure 6.15). After writing the system with $x$- and $y$-terms in the same order, we find that using $2R1 + R2$ will eliminate the variable $x$. From $2(-\frac{1}{2}x^2 + y) = 2(-3)$ we have

$$
\begin{array}{ll}
2R1 \quad -x^2 + 2y = -6 & \text{rewrite first equation; multiply by 2} \\
+ & \\
\underline{R2 \quad\quad x^2 + y^2 = 41} & \text{second equation} \\
\quad\quad\quad\quad y^2 + 2y = 35 & \text{add}
\end{array}
$$

To find solutions, we can set the equation equal to zero and factor, or use the quadratic formula if needed.

$$
\begin{array}{ll}
y^2 + 2y - 35 = 0 & \text{standard form} \\
(y + 7)(y - 5) = 0 & \text{factored form} \\
y = -7 \text{ or } y = 5 & \text{result}
\end{array}
$$

Since the radius of the circle is less than 7, the solution $y = -7$ leads to nonreal solutions. Using $y = 5$ in the second equation gives the following:

$$
\begin{array}{ll}
x^2 + y^2 = 41 & \text{equation 2} \\
x^2 + 5^2 = 41 & \text{substitute 5 for } y \\
x^2 + 25 = 41 & 5^2 = 25 \\
x^2 = 16 & \text{subtract 25} \\
x = \pm 4 & \text{square root property}
\end{array}
$$

The real number solutions are $(-4, 5)$ and $(4, 5)$.

> **WORTHY OF NOTE**
>
> To find the nonreal solutions we would simply continue by substituting $-7$ for $y$ in the second equation. This gives:
>
> $$x^2 + y^2 = 41$$
> $$x^2 + (-7)^2 = 41$$
> $$x^2 + 49 = 41$$
> $$x^2 = -8$$
> $$x = \pm 2i\sqrt{2}$$
>
> and the nonreal solutions would be $(2i\sqrt{2}, -7)$ and $(-2i\sqrt{2}, -7)$.

**Graphical Solution** ▶

Enter the equation $y = \frac{1}{2}x^2 - 3$ as $Y_1$ on the ⬭ screen. Solving $x^2 + y^2 = 41$ for $y$ gives the equations $y = \sqrt{41 - x^2}$ and $y = -\sqrt{41 - x^2}$, which we enter as $Y_2$ and $Y_3$. Using a square window and the ⬭ ⬭ (**CALC**) **5:intersect** feature as before, we obtain the graph shown in Figure 6.16, which shows the parabola and the circle intersect at $(4, 5)$. Since both the parabola and circle are symmetric to the $y$-axis, a second solution must be at $(-4, 5)$. These solutions check in both equations.

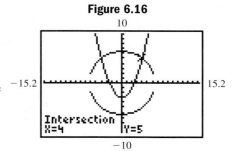

**Figure 6.16**

Now try Exercises 19 through 24 ▶

Nonlinear systems may involve other relations as well, including radical, rational, logarithmic, or exponential functions. These can also be solved using substitution.

**EXAMPLE 4** ▶ **Solving a System of Logarithmic Equations**

Solve the system using the method of your choice: $\begin{cases} y = -\log(x + 7) + 2 \\ y = \log(x + 4) + 1 \end{cases}$.

**Solution** ▶ Since both equations have $y$ written in terms of $x$, substitution appears to be the most convenient choice. The result is a logarithmic equation, which we can solve using the techniques from Chapter 5.

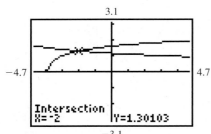

3.1

−4.7 ⊢ 4.7

Intersection
X=-2 | Y=1.30103

−3.1

$$\log(x + 4) + 1 = -\log(x + 7) + 2 \quad \text{substitute } \log(x + 4) + 1 \text{ for } y \text{ in first equation}$$
$$\log(x + 4) + \log(x + 7) = 1 \quad \text{add } \log(x + 7); \text{ subtract 1}$$
$$\log(x + 4)(x + 7) = 1 \quad \text{product property of logarithms}$$
$$(x + 4)(x + 7) = 10^1 \quad \text{exponential form}$$
$$x^2 + 11x + 18 = 0 \quad \text{multiply and write in standard form}$$
$$(x + 9)(x + 2) = 0 \quad \text{factor}$$
$$x + 9 = 0 \quad \text{or} \quad x + 2 = 0 \quad \text{zero factor theorem}$$
$$x = -9 \quad \text{or} \quad x = -2 \quad \text{possible solutions}$$

By inspection, we see that $x = -9$ is extraneous, since $\log(-9 + 4)$ and $-\log(-9 + 7)$ are not real numbers. Substituting $-2$ for $x$ in the second equation we find one form of the (exact) solution is $(-2, \log 2 + 1)$. If we substitute $-2$ for $x$ in the first equation the exact solution is $(-2, -\log 5 + 2)$. Using a calculator we can verify the ordered pairs are equivalent and approximately equal to $(-2, 1.3)$. A graphical check is shown in the margin.

☑ **C.** You've just seen how we can solve nonlinear systems using elimination

**Now try Exercises 25 through 36** ▶

For practice solving more complex systems using a graphing calculator, **see Exercises 37 through 42.**

**Figure 6.17**

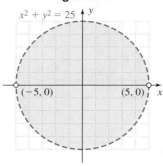

$x^2 + y^2 = 25$

$(-5, 0)$ $(5, 0)$

## D. Solving Systems of Nonlinear Inequalities

Nonlinear *inequalities* can be solved by graphing the boundary given by the related equation, and checking the regions that result using a test point. For example, the inequality $x^2 + y^2 < 25$ is solved by first graphing $x^2 + y^2 = 25$, which is a circle with radius 5 centered at $(0, 0)$. We then decide whether the boundary is included or excluded. In this case it is not included as indicated by the dashed boundary. We then use a test point from either "outside" or "inside" the region formed. The test point $(0, 0)$ results in a true statement since $0^2 + 0^2 < 25$, so the inside of the circle is shaded to indicate the solution region (Figure 6.17). For a *system* of nonlinear inequalities, we identify regions where the solution sets for both inequalities overlap, paying special attention to points of intersection.

**EXAMPLE 5** ▶ **Solving Systems of Nonlinear Inequalities**

Solve the system $\begin{cases} x^2 + y^2 < 25 \\ 2y - x \geq 5 \end{cases}$ by graphing.

**Solution** ▶ We recognize the first inequality as a circle with radius 5, with the solution region in the interior. The second inequality is linear and after solving for $x$ we'll use substitution to find points of intersection (if they exist). From $2y - x = 5$, we obtain $x = 2y - 5$.

$$x^2 + y^2 = 25 \quad \text{given}$$
$$(2y - 5)^2 + y^2 = 25 \quad \text{substitute } 2y - 5 \text{ for } x$$
$$4y^2 - 20y + 25 + y^2 = 25 \quad \text{expand}$$
$$5y^2 - 20y + 25 = 25 \quad \text{simplify}$$
$$y^2 - 4y = 0 \quad \text{subtract 25; divide by 5}$$
$$y(y - 4) = 0 \quad \text{factor}$$
$$y = 0 \quad \text{or} \quad y = 4 \quad \text{result}$$

Back-substitution shows the graphs intersect at $(-5, 0)$ and $(3, 4)$. Graphing a line through these points and using $(0, 0)$ as a test point shows the upper half-plane is the solution region for the linear inequality [$2(0) - 0 \geq 5$ is *false*]. The overlapping (solution) region for *both* inequalities is the circular section shown in purple. Note the points of intersection are graphed using "open dots" (see Figure 6.18), since points on the graph of the circle are excluded from the solution set.

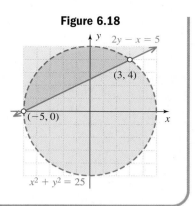

**Figure 6.18**

☑ **D.** You've just seen how we can solve nonlinear systems of inequalities

**Now try Exercises 43 through 50** ▶

## E. Applications of Nonlinear Systems

In the business world, a fast growing company can often reduce the average price of its products using what are called the **economies of scale.** These would include the ability to buy necessary materials in larger quantities, integrating new technology into the production process, and other means. However, there are also countering forces called the **diseconomies of scale,** which may include the need to hire additional employees, rent more production space, and the like. Companies often use what is called a **break-even analysis,** to determine the production level at which these forces stabilize.

**EXAMPLE 6** ▶ **Solving an Application of Nonlinear Systems**

Suppose the cost to produce a new and inexpensive shoe made from molded plastic is modeled by the function $C(x) = x^2 - 5x + 22$, where $C(x)$ represents the cost to produce $x$ thousand of these shoes. Similarly, revenue from the sales of these shoes is modeled by $R(x) = -x^2 + 10x$. Use a break-even analysis to find the quantity of sales that will cause the company to break even.

**Solution** ▶ Essentially we are asked to solve the system formed by the two equations:
$$\begin{cases} C(x) = x^2 - 5x + 22 \\ R(x) = -x^2 + 10x \end{cases}$$ Since we want to know the point where the company breaks even, we set $C(x) = R(x)$ and solve.

$$
\begin{aligned}
C(x) &= R(x) & &\text{break-even equation} \\
x^2 - 5x + 22 &= -x^2 + 10x & &\text{substitute for } C(x) \text{ and } R(x) \\
2x^2 - 15x + 22 &= 0 & &\text{set equal to zero} \\
(2x - 11)(x - 2) &= 0 & &\text{factored form} \\
x = \frac{11}{2} \text{ or } x &= 2 & &\text{result}
\end{aligned}
$$

With $x$ in thousands, it appears the company will break even if either 2000 shoes or 5500 shoes are made and sold.

**Now try Exercises 53 and 54** ▶

There are actually a large number of significant applications that involve nonlinear systems. Solutions to many of these are difficult to solve manually, and it's here that a skillful use of technology displays its ultimate value. One such application involves the manufacture of cardboard boxes (a billion dollar industry) used to ship goods and commodities all over the globe. Designing a box with the desired volume and needed dimensions often involves the use of a nonlinear system.

**EXAMPLE 7** ▶   **Manufacturing Cardboard Boxes**

In order to transport seedlings from the nursery to the market, a manufacturing engineer designs an open box with a square bottom and four sides of equal height. For the most efficient fit, the volume of the box must be 900 in³ and use 465 in² of cardboard stock. Find the dimensions of the box.

**Figure 6.19**

$y$ in.          $x$ in.

$x$ in.

**Solution** ▶   After carefully drawing a diagram (see Figure 6.19), we let $x$ represent the length and width of the square bottom, and $y$ represent the height. The volume of the box is then represented by $x^2y = 900$ in³ ($V = LWH$), and the surface area by $x^2 + 4xy = 465$ ($SA = S^2 + 4\,SH$). The resulting system is $\begin{cases} x^2y = 900 \\ x^2 + 4xy = 465 \end{cases}$. Solving for $y$ in the first equation, we obtain $y = \frac{900}{x^2}$ and can now substitute this result into the second equation to obtain an equation in $x$ alone:

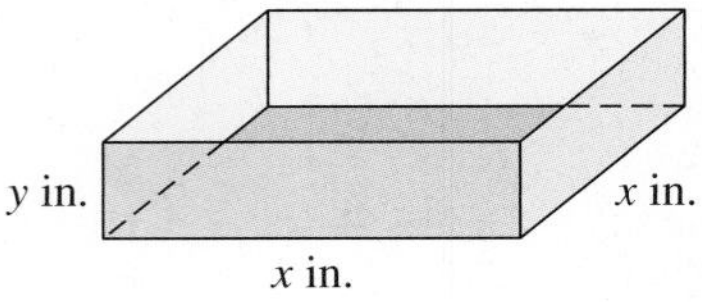

$$x^2 + 4xy = 465 \qquad \text{second equation}$$
$$x^2 + 4x\left(\frac{900}{x^2}\right) = 465 \qquad \text{substitute } \tfrac{900}{x^2} \text{ for } y$$
$$x^2 + \frac{3600}{x} = 465 \qquad \text{simplify}$$
$$x^3 + 3600 = 465x \qquad \text{multiply by } x \text{ (clear denominator)}$$
$$x^3 - 465x + 3600 = 0 \qquad \text{set equal to zero}$$

The zeroes of this cubic equation will give the dimensions of the base. A comprehensive graph of this function (using window size $[-30, 30]$ for $x$ and $[-2000, 7500]$ for $y$) shows there are three zeroes, but one of them is negative and is discounted in this context (Figure 6.20).

Using the ⌨ 2nd ⌨ TRACE (CALC) **2:zero** option, we find the positive zeroes are $x \approx 9.7$ (shown) and $x = 15$. Substituting 9.7 for $x$ in the first equation,

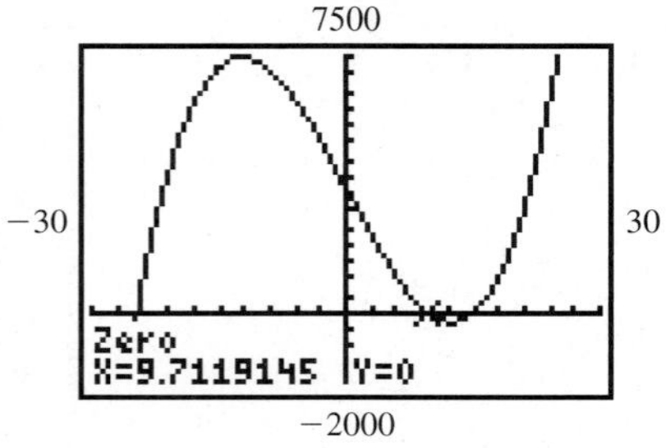

$$x^2y = 900 \qquad \text{first equation}$$
$$(9.7)^2y \approx 900 \qquad \text{substitute 9.7 for } x$$
$$(94.09)y \approx 900 \qquad \text{square 9.7}$$
$$y \approx 9.6 \qquad \text{divide}$$

**Figure 6.20**

7500

$-30$          30

Zero
X=9.7119145  Y=0

$-2000$

yields $y \approx 9.6$ in. for the height of the box, which does not seem practical for transporting seedlings. Substituting 15 for $x$ gives $y = 4$ in. as the height, which seems more reasonable. The dimensions of the box will most likely be 15 in. × 15 in. × 4 in.

☑ **E.** You've just seen how we can solve applications of nonlinear systems

**Now try Exercises 55 and 56** ▶

There are a number of other, interesting applications in the Exercise Set. **See Exercises 57 through 64.**

## 6.3 EXERCISES

### ▶ CONCEPTS AND VOCABULARY

**Fill in the blanks with the appropriate word or phrase. Carefully reread the section, if necessary.**

1. The solution to a system of nonlinear inequalities is a(n) _____ of the plane where the _____ for each individual inequality overlap.

2. When both equations in the system have at least one _____ -degree term, it is generally easier to use the _____ method to find a solution.

3. Discuss/Explain how the number of solutions for each system shown can be found by simply inspecting the equations. How many solutions does each have?

   **a.** $\begin{cases} y = |x| - 1 \\ x^2 + y^2 = 9 \end{cases}$   **b.** $\begin{cases} y = x^2 + 5 \\ x^2 + y^2 = 4 \end{cases}$

4. Discuss/Explain the reason(s) why $(-8, 6)$ and $(6, 8)$ are not solutions to the system shown, even though the line intersects the circle at these points:

   $\begin{cases} -x + 7y < 50 \\ x^2 + y^2 \leq 100 \end{cases}$.

### ▶ DEVELOPING YOUR SKILLS

**Identify each equation in the system as that of a line, parabola, circle, or absolute value function, then solve the system by graphing.**

5. $\begin{cases} x^2 + y = 6 \\ x + y = 4 \end{cases}$   6. $\begin{cases} -x + y = 4 \\ x^2 + y^2 = 16 \end{cases}$

7. $\begin{cases} y^2 + x^2 = 100 \\ y = |x - 2| \end{cases}$   8. $\begin{cases} x^2 + y^2 = 25 \\ x^2 + y = 13 \end{cases}$

9. $\begin{cases} -(x - 1)^2 + 2 = y \\ y - x^2 = -3 \end{cases}$   10. $\begin{cases} y - 4 = -x^2 \\ y = -|x - 1| + 3 \end{cases}$

**Solve using substitution. In Exercises 15 and 16, solve for $x^2$ or $y^2$ and use the result as a substitution.**

11. $\begin{cases} x^2 + y^2 = 25 \\ y - x = 1 \end{cases}$   12. $\begin{cases} x + 7y = 50 \\ x^2 + y^2 = 100 \end{cases}$

13. $\begin{cases} x^2 + y = 9 \\ -2x + y = 1 \end{cases}$   14. $\begin{cases} x^2 - y = 8 \\ x + y = 4 \end{cases}$

15. $\begin{cases} x^2 + y = 13 \\ x^2 + y^2 = 25 \end{cases}$   16. $\begin{cases} y^2 + (x - 3)^2 = 25 \\ y^2 + (x + 1)^2 = 9 \end{cases}$

17. $\begin{cases} y + x^2 = 6x + 1 \\ y = (x - 1)^2 \end{cases}$   18. $\begin{cases} x^2 + y^2 = 16 \\ y - x = -4 \end{cases}$

**Solve each system using elimination.**

19. $\begin{cases} x^2 + y^2 = 25 \\ \frac{1}{4}x^2 + y = 1 \end{cases}$   20. $\begin{cases} y - \frac{1}{2}x^2 = -1 \\ x^2 + y^2 = 65 \end{cases}$

21. $\begin{cases} x^2 + y^2 = 25 \\ y = x - 1 \end{cases}$   22. $\begin{cases} y + x^2 = 8x \\ y + 34 = (x + 4)^2 \end{cases}$

23. $\begin{cases} x^2 + y^2 = 65 \\ y = 3x + 25 \end{cases}$   24. $\begin{cases} y - 2x = 5 \\ x^2 + y^2 = 85 \end{cases}$

**Solve using the method of your choice.**

25. $\begin{cases} y - 5 = \log x \\ y = 6 - \log(x - 3) \end{cases}$   26. $\begin{cases} y = \log(x + 4) + 1 \\ y - 2 = -\log(x + 7) \end{cases}$

27. $\begin{cases} y = \ln(x^2) + 1 \\ y - 1 = \ln(x + 12) \end{cases}$   28. $\begin{cases} \log(x + 1.1) = y + 3 \\ y + 4 = \log(x^2) \end{cases}$

29. $\begin{cases} y - 9 = e^{2x} \\ 3 = y - 7e^x \end{cases}$   30. $\begin{cases} y - 2e^{2x} = 5 \\ y - 1 = 6e^x \end{cases}$

31. $\begin{cases} y = 4^{x+3} \\ y - 2^{x^2+3x} = 0 \end{cases}$   32. $\begin{cases} y - 3^{x^2+2x} = 0 \\ y = 9^{x+2} \end{cases}$

33. $\begin{cases} x^3 - y = 2x \\ y - 5x = -6 \end{cases}$   34. $\begin{cases} y - x^3 = -2 \\ y + 4 = 3x \end{cases}$

35. $\begin{cases} x^2 - 6x = y - 4 \\ y - 2x = -8 \end{cases}$   36. $\begin{cases} y + x = -2 \\ y + 4x = x^2 \end{cases}$

**Solve each system using a graphing calculator. Round solutions to hundredths (as needed).**

37. $\begin{cases} x^2 + y^2 = 34 \\ y^2 + (x - 3)^2 = 25 \end{cases}$   38. $\begin{cases} 5x^2 + 5y^2 = 40 \\ y + 2x = x^2 - 6 \end{cases}$

39. $\begin{cases} y = 2^x - 3 \\ y + 2x^2 = 9 \end{cases}$   40. $\begin{cases} y = -2\log(x + 8) \\ y + x^3 = 4x - 2 \end{cases}$

41. $\begin{cases} y = \dfrac{1}{(x - 3)^2} + 2 \\ (x - 3)^2 + y^2 = 10 \end{cases}$   42. $\begin{cases} y^2 + x^2 = 5 \\ y = \dfrac{1}{x - 1} - 2 \end{cases}$

**Solve each system of inequalities.**

43. $\begin{cases} y - x^2 \geq 1 \\ x + y \leq 3 \end{cases}$   44. $\begin{cases} x^2 + y^2 \leq 25 \\ x + 2y \leq 5 \end{cases}$

45. $\begin{cases} x^2 + y^2 > 16 \\ x^2 + y^2 \leq 64 \end{cases}$   46. $\begin{cases} y + 4 \geq x^2 \\ x^2 + y^2 \leq 34 \end{cases}$

47. $\begin{cases} y - x^2 \leq -16 \\ y^2 + x^2 < 9 \end{cases}$   48. $\begin{cases} x^2 + y^2 \leq 16 \\ x + 2y > 10 \end{cases}$

49. $\begin{cases} y^2 + x^2 \leq 25 \\ |x| - 1 > -y \end{cases}$   50. $\begin{cases} y^2 + x^2 \leq 4 \\ x + y < 4 \end{cases}$

## ▶ WORKING WITH FORMULAS

**51. Tunnel clearance:** $\begin{cases} h = \sqrt{r^2 - d^2} \\ d = k \end{cases}$

The maximum rectangular clearance allowed by a circular tunnel can be found using the formula shown, where $x^2 + y^2 = r^2$ models the tunnel's circular cross section and $h$ is the height of the tunnel at a distance $d$ from the center. If $r = 50$ ft, find the maximum clearance at distances of $k = 20$, 30, and 40 ft from center.

**52. Manufacturing cylindrical vents:** $\begin{cases} A = 2\pi rh \\ V = \pi r^2 h \end{cases}$

In the manufacture of cylindrical vents, a rectangular piece of sheet metal is rolled, riveted, and sealed to form the vent. The radius and height required to form a vent with a specified volume, using a piece of sheet metal with a given area, can be found by solving the system shown. Use the system to find the radius and height if the volume required is 4071 cm$^3$ and the area of the rectangular piece is 2714 cm$^2$.

## ▶ APPLICATIONS

**Solve the following applications of economies of scale.**

**53. World's most inexpensive car:** Early in 2008, the Tata Company (India) unveiled the new Tata Nano, the world's most inexpensive car. With its low price and 54 mpg, the car may prove to be very popular. *Assume* the cost to produce these cars is modeled by the function $C(x) = 2.5x^2 - 120x + 3500$, where $C(x)$ represents the cost to produce $x$-thousand cars. Suppose the revenue from the sale of these cars is modeled by $R(x) = -2x^2 + 180x - 500$. Use a break-even analysis to find the quantity of sales (to the nearest hundred) that will cause the company to break even.

**54. Document reproduction:** In a world of technology, document reproduction has become a billion dollar business. With very stiff competition, the price of a single black and white copy has varied greatly in recent years. Suppose the cost to produce these copies is modeled by the function $C(x) = 0.1x^2 - 1.2x + 7$, where $C(x)$ represents the cost to produce $x$ hundred thousand copies. If the revenue from the sale of these copies is modeled by $R(x) = -0.1x^2 + 1.8x - 2$, use a break-even analysis to find the number of copies (to the nearest thousand) that will cause the company to break even.

**Build the system of nonlinear equations needed to solve the following manufacturing applications, then solve.**

**55. Dimensions of a pool:** A homeowner wants to build a new swimming pool in her backyard. Due to size limitations, she decides to build a square pool with a flat bottom. If the volume of the pool must be 2000 ft$^3$ and the tile surface for the sides and bottom total 800 ft$^2$, what will be the final dimensions of the pool?

**56. Box manufacturing:** A cardboard box manufacturer receives an order for a large number of boxes for wrapping gifts during the Christmas season. The boxes are to have a square top and bottom, with a volume of 6750 cm$^3$ and a surface area of 2700 cm$^2$. Find the dimensions of the box fitting these requirements.

**Solve using a system of nonlinear equations.**

**57. Dimensions of a flag:** A large American flag has an area of 85 ft$^2$ and a perimeter of 37 ft. Find the dimensions of the flag.

**58. Dimensions of a sail:** The sail on a boat is a right triangle with a perimeter of 36 ft and a hypotenuse of 15 ft. Find the height and width of the sail.

**59. Dimensions of a tract:** The area of a rectangular tract of land is 45 km$^2$. The length of a diagonal is $\sqrt{106}$ km. Find the dimensions of the tract.

**60. Dimensions of a deck:** A rectangular deck has an area of 192 ft$^2$ and the length of the diagonal is 20 ft. Find the dimensions of the deck.

**61. Dimensions of a trailer:** The surface area of a closed rectangular trailer with square ends is 928 ft$^2$. If the sum of all edges of the trailer is 164 ft, find its dimensions.

**62. Dimensions of a cylindrical tank:** The surface area of a closed cylindrical tank is $192\pi$ m$^2$. Find the dimensions of the tank if the volume is $320\pi$ m$^3$ and the radius is as small as possible.

**63. Supply and demand:** Suppose the monthly demand $D$ (in ten-thousands of gallons) for a new synthetic oil is related to the price $P$ in dollars by the equation $10P^2 + 6D = 144$. For the price $P$, assume the amount $D$ that manufacturers are willing to supply is modeled by $8P^2 - 8P - 4D = 12$. What is the minimum price at which manufacturers are willing to begin supplying the oil? Use this information to create a system of nonlinear equations, then solve the system to find the equilibrium point.

**64. Supply and demand:** The weekly demand $D$ for organically grown carrots (in thousands of pounds) is related to the price per pound $P$ by the equation $8P^2 + 4D = 84$. At this market price, the amount that growers are willing to supply is modeled by the equation $8P^2 + 6P - 2D = 48$. What is the minimum price at which growers are willing to supply the organically grown carrots? Use this information to create a system of nonlinear equations, then solve the system to find the equilibrium point.

▶ **EXTENDING THE CONCEPT**

**65.** The area of a vertical parabolic segment is given by $A = \frac{2}{3}BH$, where $B$ is the length of the horizontal base of the segment and $H$ is the height from the base to the vertex. Investigate how this formula can be used to find the *area* of the solution region for the general system of inequalities shown.
$$\begin{cases} y \geq x^2 - bx + c \\ y \leq c + bx - x^2 \end{cases}$$

**66.** Find the area of the trapezoid formed by joining the points where the parabola $y = \frac{1}{2}x^2 - 26$ and the circle $x^2 + y^2 = 100$ intersect.

**67.** A rectangular fish tank has a bottom and four sides made out of glass. Use a system of equations to help find the dimensions of the tank if the height is 18 in., the surface area is 4806 in$^2$, the tank must hold 108 gal (1 gal = 231 in$^3$), and all three dimensions are integers.

**68.** Draw sketches showing the different ways each pair of relations can intersect and give zero, one, two, three, and/or four points of intersection. If a given number of intersections is not possible, so state.

    **a.** circle and line
    **b.** parabola and line
    **c.** circle and parabola
    **d.** circle and absolute value function
    **e.** absolute value function and line
    **f.** absolute value function and parabola

▶ **MAINTAINING YOUR SKILLS**

**69.** (1.5/1.6) Solve by factoring:
    **a.** $2x^2 + 5x - 63 = 0$
    **b.** $4x^2 - 121 = 0$
    **c.** $2x^3 - 3x^2 - 8x + 12 = 0$

**70.** (1.5/1.6) Solve each equation:
    **a.** $3x^2 + 4x - 12 = 0$
    **b.** $\sqrt{3x + 1} - \sqrt{2x} = 1$
    **c.** $\dfrac{1}{x + 2} + \dfrac{3}{x^2 + 5x + 6} = \dfrac{2}{x + 3}$

**71.** (6.2) Solve using any method. As an investment for retirement, Donovan bought three properties for a total of $250,000. Ten years later, the first property had doubled in value, the second property had tripled in value, and the third property was worth $10,000 less than when he bought it, for a total current value of $485,000. Find the original purchase price if he paid $20,000 more for the first property than he did for the second.

**72.** (2.3) In 2009, a small business purchased a copier for $4500. In 2012, the value of the copier had decreased to $3300. Assuming the depreciation is linear: (a) find the average rate of change and discuss its meaning in this context; (b) find the depreciation equation (assume 2009 corresponds to year 0); and (c) use the equation to predict the copier's value in 2016. (d) If the copier is traded in for a new model when its value is less than $700, how long will the company use this copier?

In this section, we'll build on many of the ideas from Section 6.3, with a more direct focus on systems of linear inequalities. While systems of equations have an unlimited number of applications, there are many situations that can only be modeled using an inequality. For example, decisions in business and industry are often based on a large number of limitations or constraints, with many different ways these constraints can be satisfied.

## A. Linear Inequalities in Two Variables

A linear equation in two variables is any equation that can be written in the form $Ax + By = C$, where $A$ and $B$ are real numbers, not simultaneously equal to zero. A **linear inequality** in two variables is similarly defined, with the " $=$ " sign replaced by the " $<$ ," " $>$ ," " $\leq$ ," " $\geq$ ", or " $\neq$ " symbol:

$$Ax + By < C \qquad Ax + By > C \qquad Ax + By \neq C$$
$$Ax + By \leq C \qquad Ax + By \geq C$$

Solving a linear inequality in two variables has many similarities with the one variable case. For one variable, we graph a *boundary point* on a number line, decide whether the endpoint is *included* or *excluded,* and *shade the appropriate half-line.* For $x + 1 \leq 3$, we have the solution $x \leq 2$ with the endpoint (boundary point) included and the line shaded to the left (Figure 6.21):

**Figure 6.21**

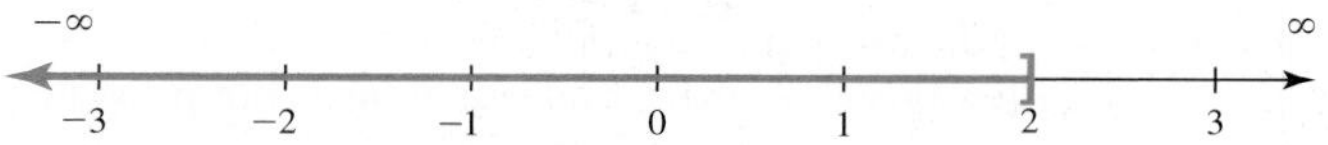

$-\infty$ ⟵————————————————————→ $\infty$

$-3 \quad -2 \quad -1 \quad 0 \quad 1 \quad 2 \quad 3$

Interval notation: $x \in (-\infty, 2]$

**Figure 6.22**

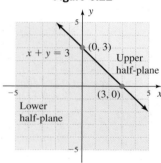

For linear inequalities in two variables, we graph a *boundary line,* decide whether the boundary line is *included* or *excluded,* and *shade the appropriate half-plane.* For $x + y \leq 3$, the boundary line $x + y = 3$ is graphed in Figure 6.22. Note it divides the coordinate plane into two regions called **half-planes,** and it forms the **boundary** between the two regions. If the boundary is **included** in the solution set, we graph it using a *solid line.* If the boundary is **excluded,** a *dashed line* is used. Recall that solutions to a linear equation are ordered pairs that make the equation true. We use a similar idea to find or verify solutions to linear inequalities. If any one point in a half-plane makes the inequality true, all points in that half-plane will satisfy the inequality.

---

**EXAMPLE 1** ▶ **Checking Solutions to an Inequality in Two Variables**

Determine whether the given ordered pairs are solutions to $-x + 2y \leq 2$:

**a.** $(4, -3)$      **b.** $(-2, 1)$      **c.** $(-4, -1)$

**Solution** ▶ **a.** Substitute 4 for $x$ and $-3$ for $y$:    $-(4) + 2(-3) \leq 2$    substitute 4 for $x$, $-3$ for $y$

$-10 \leq 2$    true

$(4, -3)$ is a solution.

**b.** Substitute $-2$ for $x$ and 1 for $y$:    $-(-2) + 2(1) \leq 2$    substitute $-2$ for $x$, 1 for $y$

$4 \leq 2$    false

$(-2, 1)$ is not a solution.

**c.** Substitute $-4$ for $x$ and $-1$ for $y$: $-(-4) + 2(-1) \leq 2$    substitute $-4$ for $x$, $-1$ for $y$

$2 \leq 2$    true

$(-4, -1)$ is a solution.

**Now try Exercises 5 through 8** ▶

Earlier we graphed linear equations by plotting a small number of ordered pairs or by solving for *y* and using the slope-intercept method. The line represented all ordered pairs that made the equation true, meaning *the left-hand expression was equal to the right-hand expression.* To graph linear inequalities, we reason that if the line represents all ordered pairs that make the expressions *equal,* then any point *not on that line* must make the expressions *unequal*—either greater than or less than. These ordered pair solutions must lie in one of the half-planes formed by the line, which we shade to indicate the **solution region.** Note this implies the boundary line for any inequality *is determined by the related equation,* created by temporarily replacing the inequality symbol with an "=" sign.

**EXAMPLE 2** ▶ **Solving an Inequality in Two Variables**

Solve the inequality $-x + 2y \leq 2$.

**Solution** ▶ The related equation and boundary line is $-x + 2y = 2$. Since the inequality is inclusive (less than *or equal to*), we graph a solid line. Using the intercepts, we graph the line through $(0, 1)$ and $(-2, 0)$ shown in Figure 6.23. To determine the solution region and which side to shade, we select $(0, 0)$ as a test point, which results in a true statement: $-(0) + 2(0) \leq 2$✓. Since $(0, 0)$ is in the "lower" half-plane, we shade this side of the boundary (see Figure 6.24).

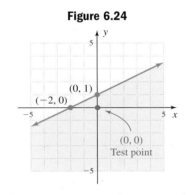

**Figure 6.23**

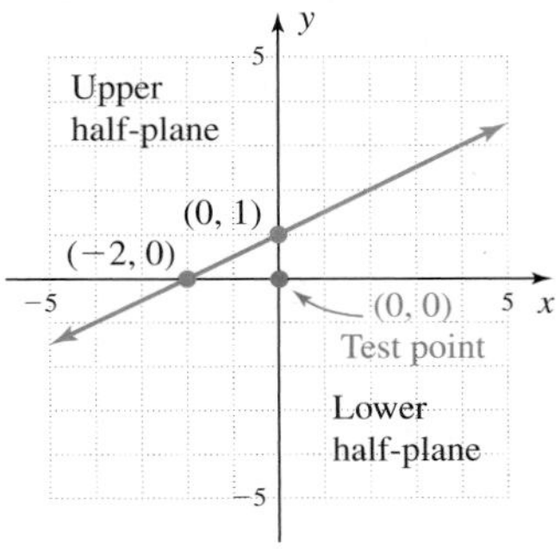

Upper half-plane

$(-2, 0)$  $(0, 1)$

$(0, 0)$ Test point

Lower half-plane

**Figure 6.24**

$(-2, 0)$  $(0, 1)$

$(0, 0)$ Test point

**✓ A.** You've just seen how we can solve a linear inequality in two variables

**Now try Exercises 9 through 12** ▶

The same solution would be obtained if we first solved for *y* and graphed the boundary line using the slope-intercept method. However, using the slope-intercept method offers a distinct advantage—test points are no longer necessary since solutions to "less than" inequalities will always appear *below* the (nonvertical) boundary line and solutions to "greater than" inequalities appear *above* the line. Written in slope-intercept form, the inequality from Example 2 is $y \leq \frac{1}{2}x + 1$. Note that $(0, 0)$ still results in a true statement, but the "less than or equal to" symbol now indicates directly that solutions will be found in the lower half-plane. These observations lead to our general approach for solving linear inequalities:

**Solving a Linear Inequality**

1. Graph the boundary line by solving for *y* and using the slope-intercept form.
   - Use a solid line if the boundary is included in the solution set.
   - Use a dashed line if the boundary is excluded from the solution set.
2. For "greater than" inequalities shade the upper half-plane. For "less than" inequalities shade the lower half-plane.

## B. Solving Systems of Linear Inequalities

To solve a **system of inequalities,** we apply the procedure outlined above to all inequalities in the system, and note the ordered pairs that satisfy *all inequalities simultaneously.* In other words, we find *the intersection of all solution regions* (where they overlap), which then represents the solution for the system. In the case of vertical boundary lines, the designations *"above"* or *"below" the line* cannot be applied, and instead we simply note that for any vertical line $x = k$, points with $x$-coordinates larger than $k$ will occur to the right.

**EXAMPLE 3** ▶ **Solving a System of Linear Inequalities**

Solve the system of inequalities: $\begin{cases} 2x + y \geq 4 \\ x - y < 2 \end{cases}$.

**Solution** ▶ Solving for $y$, we obtain $y \geq -2x + 4$ and $y > x - 2$. The line $y = -2x + 4$ will be a solid boundary line (included), while $y = x - 2$ will be dashed (not included). Both inequalities are "greater than" and so we shade the upper half-plane for each. The regions overlap and form the solution region (the lavender region shown). This sequence of events is illustrated here:

Shade above $y = -2x + 4$ (in blue)

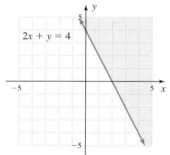

Shade above $y = x - 2$ (in pink)

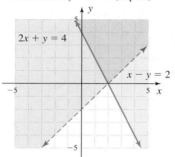

Overlapping region

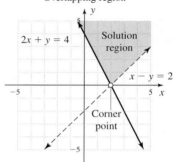

The solutions are all ordered pairs found in this region and its included boundaries. To verify the result, test the point $(2, 3)$ from inside the region, and $(5, -2)$ from outside the region.

$(2, 3)$: $\begin{cases} 2(2) + 3 \geq 4 \\ 2 - 3 < 2 \end{cases}$     $\begin{cases} 7 \geq 4 \text{ True} \\ -1 < 2 \text{ True} \end{cases}$

$(5, -2)$: $\begin{cases} 2(5) + (-2) \geq 4 \\ 5 - (-2) < 2 \end{cases}$     $\begin{cases} 8 \geq 4 \text{ True} \\ 7 < 2 \text{ False} \end{cases}$

Since $(2, 3)$ satisfies *both* equations, it is a solution to the system. Since $(5, -2)$ does not satisfy *both* equations, it is not a solution. The point $(2, 0)$ is not a solution since it does not satisfy $x - y < 2$ ($2 - 0 < 2$ is false), and is on the dashed (excluded) line.

 **B. You've just seen how we can solve a system of linear inequalities**

Now try Exercises 13 through 42 ▶

**CAUTION** ▶ When working with inequalities, keep in mind that multiplying or dividng by a negative value will reverse the relationships involved, and the inequality symbol must likewise be reversed to maintain a true statement.

For future reference, the point of intersection $(2, 0)$ is called a **corner point** or **vertex** of the solution region. If the point of intersection is not easily found from the graph, we can find it by solving a linear system using the two lines. For Example 3, the system is

$$\begin{cases} 2x + y = 4 \\ \quad x - y = 2 \end{cases}$$

and solving by elimination gives $3x = 6$, $x = 2$, and $(2, 0)$ as the point of intersection.

## C. Applications of Systems of Linear Inequalities

Systems of inequalities give us a way to model the decision-making process when certain *constraints* must be satisfied. A constraint is a fact or consideration that somehow limits or governs possible solutions, like the number of acres a farmer plants—which may be limited by time, size of land, government regulations, and so on.

**EXAMPLE 4** ▶  **Investment Planning**

As part of their retirement planning, James and Lily decide to invest up to $30,000 in two separate investment vehicles. The first is a bond issue paying 5% and the second is a money market certificate paying 3%. A financial adviser suggests they invest at least $10,000 in the certificate and not more than $15,000 in bonds.

  **a.** What various amounts can be invested in each?

  **b.** Is an investment of $5 thousand in bonds and $25 thousand in a money market certificate possible?

**Solution** ▶  **a.** Consider the ordered pairs $(B, C)$, where $B$ represents the money invested in bonds and $C$ the money invested in the certificate. Since they plan to invest no more than $30,000, the investment constraint would be $B + C \leq 30$ (in thousands). Following the adviser's recommendations, the constraints on each investment would be $B \leq 15$ and $C \geq 10$. Since they cannot invest less than zero dollars, the last two constraints are $B \geq 0$ and $C \geq 0$ (note that $C \geq 0$ is redundant, since we already have $C \geq 10$).

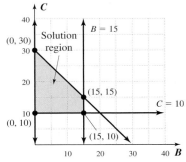

$$\begin{cases} B + C \leq 30 \\ B \leq 15 \\ C \geq 10 \\ B \geq 0 \end{cases}$$

The graph of $B + C = 30$ is a line with negative slope through the points $(0, 30)$ and $(30, 0)$. The graph of $B = 15$ is a vertical line through $(15, 0)$, and the graph of $C = 10$ is a horizontal line through $(0, 10)$. The resulting system is shown in the figure. The vertical boundary line at $B = 15$ has shading to the left (less than) and the horizontal boundary line at $C = 10$ has shading above (greater than). We now see the solution region is a quadrilateral with vertices at $(0, 10)$, $(0, 30)$, $(15, 10)$, and $(15, 15)$, as shown. Any ordered pair in this region or on its boundaries would represent an investment of the form (money in bonds, money in CDs) → $(B, C)$, and would satisfy all constraints in the system.

  **b.** Yes, since the ordered pair $(5, 25)$ satisfies all three inequalities (and is on an included boundary of the region).

✓ **C.** You've just seen how we can solve applications using a system of linear inequalities

**Now try Exercises 53 through 56** ▶

For Example 4, a natural follow-up question would be, "What combination of (money in bonds, money in CDs) would offer the greatest return?" This would depend on the interest being paid on each investment, and introduces us to a study of **linear programming,** which follows Example 5.

Systems of linear inequalities often involve various constraints on the variables, and may involve more than one nonvertical, nonhorizontal boundary line.

---

**EXAMPLE 5** ▶  **Planning a Wedding**

After carefully considering the cost and size of her wedding, Clarissa has determined she can afford to invite no more than 100 people (family and guests) to her wedding. To satisfy tradition and propriety, Clarissa feels that between 30 and 60 of the invitations must go to her immediate and extended family. Due to her extensive network of friends and business contacts, she wants to invite no more than 60 but no less than 20 other guests.

**a.** What possibilities will satisfy these constraints?

**b.** If she invites 25 family members, can she invite 75 guests?

**Solution** ▶  **a.** For ordered pairs of the form (family, guests) $\rightarrow (f, g)$, the first equation is $g + f \leq 100$, since no more than 100 invitations will be given out. We also know Clarissa will invite between 30 and 60 family members: $(30 \leq f \leq 60)$, and between 20 and 60 other guests: $(20 \leq g \leq 60)$. The system of inequalities is

$$\begin{cases} g + f \leq 100 \\ 30 \leq f \leq 60 \\ 20 \leq g \leq 60 \end{cases}$$

The graph of $g + f = 100$ is a line with negative slope through the points $(0, 100)$ and $(100, 0)$, with shading beneath this line. For $30 \leq f \leq 60$, we have a pair of horizontal lines through $x = 30$ and $x = 60$ with shading between them. Similarly, the graph of $20 \leq g \leq 60$ is a pair of vertical lines with shading between them. The resulting graph is shown here, with the possible number of invitations for each group drawn from the lavender solution region.

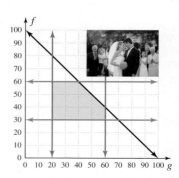

**b.** No. Even though she is planning on no more than 100 guests, inviting 25 family members and 75 guests does not satisfy the third inequality.

**Now try Exercises 57 through 60** ▶

---

## D. Linear Programming

To become as profitable as possible, corporations look for ways to maximize their revenue and minimize their costs, while keeping up with delivery schedules and product demand. To operate at peak efficiency, plant managers must find ways to maximize productivity, while considering employee welfare, union agreements, and other factors. Problems where the goal is to **maximize** or **minimize** the value of a given quantity under certain **constraints** or restrictions are called programming problems. The quantity we seek to maximize or minimize is called the **objective function.** For situations where *linear* programming is used, the objective function is given as a linear function in two variables and is denoted $f(x, y)$. A function in two variables is evaluated in much the same way as a single variable function. To evaluate $f(x, y) = 2x + 3y$ at the point $(4, 5)$, we substitute 4 for $x$ and 5 for $y$: $f(4, 5) = 2(4) + 3(5) = 23$.

**EXAMPLE 6** ▶ **Determining Maximum Values**

Of the points given, determine which ordered pair returns a maximum value for $f(x, y) = 5x + 4y$: (0, 6), (5, 0), (0, 0), or (4, 2).

Solution ▶ Organizing our work in table form gives

| Given Points | Objective Function $f(x, y) = 5x + 4y$ |
|---|---|
| (0, 6) | $f(0, 6) = 5(0) + 4(6) = 24$ |
| (5, 0) | $f(5, 0) = 5(5) + 4(0) = 25$ |
| (0, 0) | $f(0, 0) = 5(0) + 4(0) = 0$ |
| (4, 2) | $f(4, 2) = 5(4) + 4(2) = 28$ |

Of the points given, the function $f(x, y) = 5x + 4y$ is maximized at (4, 2).

Now try Exercises 43 through 46 ▶

When the objective is stated as a linear function in two variables and the constraints are expressed as a system of linear inequalities, we have what is called a **linear programming** problem. The systems of inequalities solved earlier produced solution regions that were either **bounded** (as in Example 4) or **unbounded** (as in Example 3). We interpret the word *bounded* to mean we can enclose the solution region within a circle of appropriate size. If we cannot draw a circle around the region because it extends indefinitely in some direction, the region is said to be *unbounded*. In this study, we will consider only situations that produce bounded solution regions. The regions we study will also be **convex,** meaning that for any two points in the enclosed region, the line segment between them is also in the region (Figure 6.25). Under these conditions, it can be shown that the maximum or minimum values *must occur at one of the corner points of the solution region,* also called the **feasible region.**

**Figure 6.25**

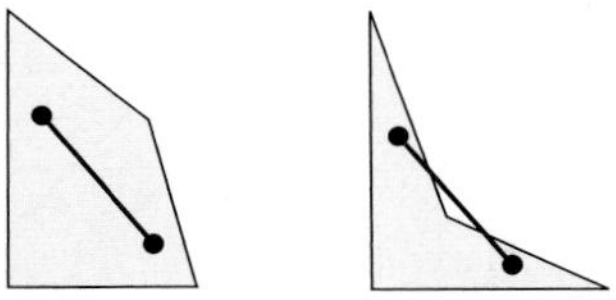

Convex          Not convex

**EXAMPLE 7** ▶ **Finding the Maximum of an Objective Function**

Find the maximum value of the objective function $f(x, y) = 2x + y$ given the

constraints shown: $\begin{cases} x + y \le 4 \\ 3x + y \le 6 \\ x \ge 0 \\ y \ge 0 \end{cases}$.

Solution ▶ Begin by noting that the solutions must be in QI, since $x \ge 0$ and $y \ge 0$. Graph the boundary lines $y = -x + 4$ and $y = -3x + 6$, shading the lower half-plane in each case since they are "less than" inequalities. This produces the feasible region shown in lavender. There are four corner points to this region: (0, 0), (0, 4), (2, 0), and (1, 3). Three of these points are intercepts and can be found quickly. The point (1, 3) was found by solving the

system $\begin{cases} x + y = 4 \\ 3x + y = 6 \end{cases}$. Knowing that the objective

function will be maximized at one of the corner points, we test them in the objective function, using a table to organize our work.

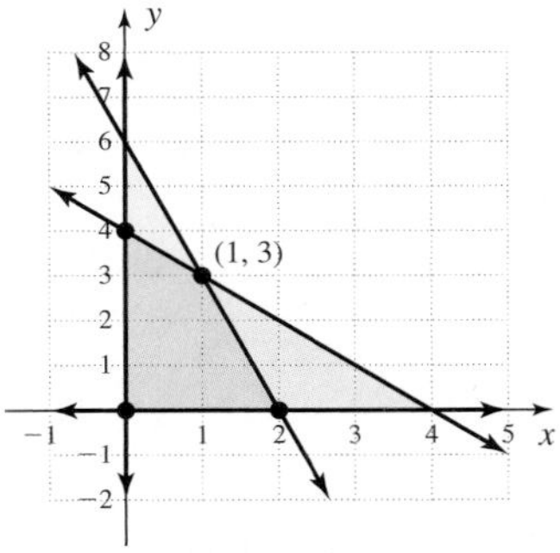

| Corner Points | Objective Function $f(x, y) = 2x + y$ |
|---|---|
| $(0, 0)$ | $f(0, 0) = 2(0) + (0) = 0$ |
| $(0, 4)$ | $f(0, 4) = 2(0) + (4) = 4$ |
| $(2, 0)$ | $f(2, 0) = 2(2) + (0) = 4$ |
| $(1, 3)$ | $f(1, 3) = 2(1) + (3) = 5$ |

Under the constraints given, the objective function $f(x, y) = 2x + y$ is maximized at $(1, 3)$.

**Now try Exercises 47 through 50** ▶

To help understand why solutions must occur at a vertex, note the objective function $f(x, y)$ is maximized using only ordered pairs $(x, y)$ from the feasible region. If we let $K$ represent this maximum value, the function from Example 7 becomes $K = 2x + y$ or $y = -2x + K$, which is a line with slope $-2$ and $y$-intercept $K$. The table in Example 7 suggests that $K$ should range from 0 to 5 and graphing $y = -2x + K$ for $K = 1$, $K = 3$, and $K = 5$ produces the family of parallel lines shown in Figure 6.26. Note that values of $K$ larger than 5 will cause the line to miss the solution region altogether, and the maximum value of 5 occurs where the line intersects the feasible region at the vertex $(1, 3)$. These observations lead to the following principles, which we offer without a formal proof.

**Figure 6.26**

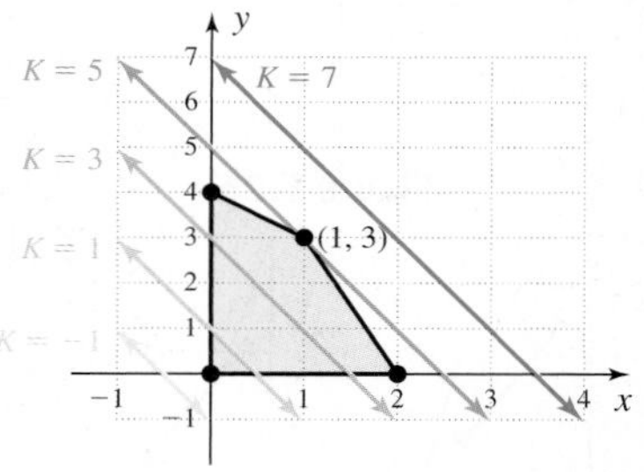

### Linear Programming Solutions

1. If the feasible region is convex and bounded, a maximum and a minimum value exist.
2. If a unique solution exists, it will occur at a vertex of the feasible region.
3. If more than one solution exists, they must occur at adjacent vertices and at every point on the boundary line between them.
4. If the feasible region is unbounded, the application may have no solutions.

Solving linear programming problems depends in large part on two things: (1) identifying the **objective** and the **decision variables** (what each variable represents in context), and (2) using the decision variables to write the *objective function* and **constraint inequalities.** This brings us to our five-step approach for solving linear programming applications.

### Solving Linear Programming Applications

1. Identify the main objective and the decision variables (descriptive variables may help), then write the objective function in terms of these variables.
2. Organize all information in a table, with the *decision variables* and *constraints* heading up the columns, and their *components* leading each row.
3. Complete the table using the information given, and write the constraint inequalities using the decision variables, constraints, and the domain.
4. Graph the constraint inequalities, determine the feasible region, and identify all corner points.
5. Test these points in the objective function to determine the optimal solution(s).

**EXAMPLE 8** ▶    Solving an Application of Linear Programming

The owner of a snack food business wants to create two nut mixes for the holiday season. The regular mix will have 14 oz of peanuts and 4 oz of cashews, while the deluxe mix will have 12 oz of peanuts and 6 oz of cashews. The owner estimates he will make a profit of $3 on the regular mixes and $4 on the deluxe mixes. How many of each should be made in order to maximize profit, if only 840 oz of peanuts and 348 oz of cashews are available?

**Solution** ▶    Our *objective* is to maximize profit, and the *decision variables* could be $r$ to represent the number of regular mixes sold, and $d$ for the number of deluxe mixes. This gives $P(r, d) = \$3r + \$4d$ as our *objective function*. The information is organized in Table 6.1, using the variables $r$, $d$, and the constraints to head each column. Since the mixes are composed of peanuts and cashews, these lead the rows in the table.

**Table 6.1**

$$P(r, d) = \$3r \quad + \quad \$4d$$
$$\qquad\qquad\qquad \downarrow \qquad\qquad\quad \downarrow$$

|  | Number of Regular Mixes, $r$ | Number of Deluxe Mixes, $d$ | *Maximum* Number of Ounces Available |
|---|---|---|---|
| Ounces of **Peanuts** | 14 | 12 | 840 |
| Ounces of **Cashews** | 4 | 6 | 348 |
| **Profit** (in Dollars) | 3 | 4 | |

After filling in the appropriate values, reading the table from left to right along the "peanut" row and the "cashew" row gives the constraint inequalities $14r + 12d \leq 840$ and $4r + 6d \leq 348$. Realizing we won't be making negative numbers of mixes, the remaining constraints are $r \geq 0$ and $d \geq 0$. The complete system is

$$\begin{cases} 14r + 12d \leq 840 \\ 4r + 6d \leq 348 \\ r \geq 0 \\ d \geq 0 \end{cases}$$

Note once again that the solutions must be in QI, since $r \geq 0$ and $d \geq 0$. Graphing the first two inequalities using slope-intercept form gives $d \leq -\frac{7}{6}r + 70$ and $d \leq -\frac{2}{3}r + 58$ producing the feasible region shown in lavender. The four corner points are $(0, 0)$, $(60, 0)$, $(0, 58)$, and $(24, 42)$. Three of these points are intercepts while the point $(24, 42)$ was found by solving the system $\begin{cases} 14r + 12d = 840 \\ 4r + 6d = 348 \end{cases}$.

Knowing the solution must occur at one of these points, we test them in the objective function (Table 6.2).

**Table 6.2**

| Corner Points | Objective Function $P(r, d) = \$3r + \$4d$ |
|---|---|
| $(0, 0)$ | $P(0, 0) = \$3(0) + \$4(0) = \$0$ |
| $(60, 0)$ | $P(60, 0) = \$3(60) + \$4(0) = \$180$ |
| $(0, 58)$ | $P(0, 58) = \$3(0) + \$4(58) = \$232$ |
| $(24, 42)$ | $P(24, 42) = \$3(24) + \$4(42) = \$240$ |

A maximum profit of $240 is made if 24 boxes of the regular mix and 42 boxes of the deluxe mix are made and sold.

**Now try Exercises 61 through 66** ▶

Linear programming can also be used to minimize an objective function, as in Example 9.

---

**EXAMPLE 9 ▶**    **Minimizing Costs Using Linear Programming**

A beverage producer operating in Missouri needs to minimize shipping costs from its two primary plants in Kansas City (KC) and St. Louis (STL). All wholesale orders within the state are shipped from one of these plants. An outlet in Macon orders 200 cases of soft drinks on the same day an order for 240 cases comes from Springfield. The plant in KC has 300 cases ready to ship and the plant in STL has 200 cases. The cost of shipping each case to Macon is $0.50 from KC, and $0.70 from STL. The cost of shipping each case to Springfield is $0.60 from KC, and $0.65 from STL. How many cases should be shipped from each warehouse to minimize costs?

**Solution ▶**    Our *objective* is to minimize costs, which depends on the number of cases shipped from each plant to each destination. To begin we use the following assignments:

$$A \rightarrow \text{cases shipped from KC to Macon}$$
$$B \rightarrow \text{cases shipped from KC to Springfield}$$
$$C \rightarrow \text{cases shipped from STL to Macon}$$
$$D \rightarrow \text{cases shipped from STL to Springfield}$$

From this information, the equation for total cost $T$ is

$$T = 0.5A + 0.6B + 0.7C + 0.65D,$$

an equation in *four* variables. To make the cost equation more manageable, note that since Macon ordered 200 cases, $A + C = 200$. Similarly, Springfield ordered 240 cases, so $B + D = 240$. After solving for $C$ and $D$ respectively these equations enable us to substitute for $C$ and $D$, resulting in an equation with just two variables. For $C = 200 - A$ and $D = 240 - B$ we have

$$T(A, B) = 0.5A + 0.6B + 0.7(200 - A) + 0.65(240 - B)$$
$$= 0.5A + 0.6B + 140 - 0.7A + 156 - 0.65B$$
$$= 296 - 0.2A - 0.05B$$

The constraints involving the KC plant are $A + B \leq 300$ with $A \geq 0, B \geq 0$. The constraints for the STL plant are $C + D \leq 200$ with $C \geq 0, D \geq 0$. To write the system in terms of $A$ and $B$ alone, note the number of cases ordered will be equal to the number of cases shipped:

$$\text{Macon} + \text{Springfield} = (A + B) + (C + D)$$

Since Macon ordered 200 and Springfied ordered 240 cases, we have

$$200 + 240 = (A + B) + (C + D)$$

Using the constraints for the St. Louis plant $(C + D \leq 200)$ gives

$$200 + 240 \leq A + B + 200$$
$$240 \leq A + B$$

Finally, since Kansas City isn't going to ship more cases than were ordered, we have $A \leq 200$ and $B \leq 240$.

Combining the new STL constraints with those from KC produces the following system and solution. All points of intersection were read from the graph or located using the related system of equations.

$$\begin{cases} A + B \le 300 \\ A + B \ge 240 \\ A \le 200 \\ B \le 240 \\ A \ge 0 \\ B \ge 0 \end{cases}$$

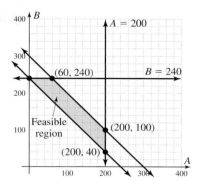

To find the minimum cost, we check each vertex in the objective function.

| Corner Points | Objective Function $T(A, B) = 296 - 0.2A - 0.05B$ |
|---|---|
| $(0, 240)$ | $P(0, 240) = 296 - 0.2(0) - 0.05(240) = \$284$ |
| $(60, 240)$ | $P(60, 240) = 296 - 0.2(60) - 0.05(240) = \$272$ |
| $(200, 100)$ | $P(200, 100) = 296 - 0.2(200) - 0.05(100) = \$251$ |
| $(200, 40)$ | $P(200, 40) = 296 - 0.2(200) - 0.05(40) = \$254$ |

A minimum cost of \$251 occurs when $A = 200$ and $B = 100$, meaning the producer should ship the following quantities:

$A \rightarrow$ cases shipped from KC to Macon $= 200$
$B \rightarrow$ cases shipped from KC to Springfield $= 100$
$C \rightarrow$ cases shipped from STL to Macon $= 0$ $(C = 200 - A)$
$D \rightarrow$ cases shipped from STL to Springfield $= 140$ $(D = 240 - B)$

 **D.** You've just seen how we can solve applications using linear programming

Now try Exercises 67 and 68 ▶

## 6.4 EXERCISES

### ▶ CONCEPTS AND VOCABULARY

**Fill in the blank with the appropriate word or phrase. Carefully reread the section, if necessary.**

**1.** In linear programming, if $(x, y)$ is a unique solution, it must be a(n) _____ of the feasible region.

**2.** For the line $y = mx + b$ drawn in the coordinate plane, solutions to $y > mx + b$ are found in the region _____ the line.

**3.** Suppose two boundary lines in a system of linear inequalities intersect, but the point of intersection is not a vertex of the feasible region. Describe how this is possible.

**4.** Describe the conditions necessary for a linear programming problem to have multiple solutions. (*Hint:* Consider the diagram in Figure 6.26, and the slope of the line from the objective function.)

## ▶ DEVELOPING YOUR SKILLS

**Determine whether the ordered pairs given are solutions to the inequality shown.**

**5.** $2x + y > 3$; $(0, 0)$, $(3, -5)$, $(-3, -4)$, $(-3, 9)$

**6.** $3x - y > 5$; $(0, 0)$, $(4, -1)$, $(-1, -5)$, $(1, -2)$

**7.** $4x - 2y \leq -8$; $(0, 0)$, $(-3, 5)$, $(-3, -2)$, $(-1, 1)$

**8.** $3x + 5y \geq 15$; $(0, 0)$, $(3, 5)$, $(-1, 6)$, $(7, -3)$

**Solve the linear inequalities by shading the appropriate half-plane.**

**9.** $x + 2y < 8$

**10.** $x - 3y > 6$

**11.** $2x - 3y \geq 9$

**12.** $4x + 5y \geq 15$

**Determine whether the ordered pairs given are solutions to the accompanying system.**

**13.** $\begin{cases} 5y - x \geq 10 \\ 5y + 2x \leq -5 \end{cases}$;

$(-2, 1)$, $(-5, -4)$, $(-6, 2)$, $(-8, 2.2)$

**14.** $\begin{cases} 8y + 7x \geq 56 \\ 3y - 4x \geq -12 \\ y \geq 4 \end{cases}$;

$(1, 5)$, $(4, 6)$, $(8, 5)$, $(5, 3)$

**Solve each system of inequalities by graphing the solution region. Verify the solution using a test point.**

**15.** $\begin{cases} x + 2y \geq 1 \\ 2x - y \leq -2 \end{cases}$

**16.** $\begin{cases} -x + 5y < 5 \\ x + 2y \geq 1 \end{cases}$

**17.** $\begin{cases} 3x + y > 4 \\ x > 2y \end{cases}$

**18.** $\begin{cases} 3x \leq 2y \\ y \geq 4x + 3 \end{cases}$

**19.** $\begin{cases} 2x + y < 4 \\ 2y > 3x + 6 \end{cases}$

**20.** $\begin{cases} x - 2y < -7 \\ 2x + y > 5 \end{cases}$

**21.** $\begin{cases} x > -3y - 2 \\ x + 3y \leq 6 \end{cases}$

**22.** $\begin{cases} 2x - 5y < 15 \\ 3x - 2y > 6 \end{cases}$

**23.** $\begin{cases} 5x + 4y \geq 20 \\ x - 1 \geq y \end{cases}$

**24.** $\begin{cases} 10x - 4y \leq 20 \\ 5x - 2y > -1 \end{cases}$

**25.** $\begin{cases} y \leq \dfrac{3}{2}x \\ 4y \geq 6x - 12 \end{cases}$

**26.** $\begin{cases} 3x + 4y > 12 \\ y < \dfrac{2}{3}x \end{cases}$

**27.** $\begin{cases} x + y \leq 9 \\ 5x + 4y \leq 40 \\ x \geq 1 \\ y \geq 2 \end{cases}$

**28.** $\begin{cases} x + y \leq 8 \\ 3x + 5y \leq 30 \\ x \geq 3 \\ y \geq 1 \end{cases}$

**29.** $\begin{cases} 3x + 4y \geq 24 \\ 5x + 3y \geq 30 \\ x \leq 7 \\ y \leq 7 \end{cases}$

**30.** $\begin{cases} 2x + 5y \geq 20 \\ 3x + 2y \geq 18 \\ x \leq 8 \\ y \leq 7 \end{cases}$

**31.** $\begin{cases} y \leq x + 3 \\ x + 2y \leq 4 \\ x \geq 0, y \geq 0 \end{cases}$

**32.** $\begin{cases} 4y < 3x + 12 \\ y \leq x + 1 \\ x \geq 0, y \geq 0 \end{cases}$

**33.** $\begin{cases} 2x + 3y \leq 18 \\ 2x + y \leq 10 \\ x \geq 0, y \geq 0 \end{cases}$

**34.** $\begin{cases} 8x + 5y \leq 40 \\ x + y \leq 7 \\ x \geq 0, y \geq 0 \end{cases}$

**35.** $\begin{cases} y + 2x < 8 \\ y + x < 6 \\ x \geq 0 \\ y \geq 0 \end{cases}$

**36.** $\begin{cases} x + 2y < 10 \\ x + y < 7 \\ x \geq 0 \\ y \geq 0 \end{cases}$

**37.** $\begin{cases} -2x - y > -8 \\ -x - 2y < -7 \\ x \geq 0 \\ y \geq 0 \end{cases}$

**38.** $\begin{cases} y + 2x \geq 10 \\ 2y + x \leq 11 \\ x \geq 0 \\ y \geq 0 \end{cases}$

**Use the equations given to write the system of linear inequalities represented by each graph.**

**39.**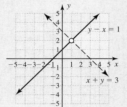

**40.**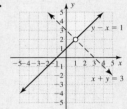

**41.**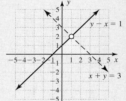

**42.**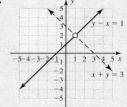

**Determine which of the ordered pairs given produces the maximum value of $f(x, y)$.**

**43.** $f(x, y) = 12x + 10y$; $(0, 0)$, $(0, 8.5)$, $(7, 0)$, $(5, 3)$

**44.** $f(x, y) = 50x + 45y$; $(0, 0)$, $(0, 21)$, $(15, 0)$, $(7.5, 12.5)$

**Determine which of the ordered pairs given produces the minimum value of $f(x, y)$.**

**45.** $f(x, y) = 8x + 15y$; $(0, 20)$, $(35, 0)$, $(5, 15)$, $(12, 11)$

**46.** $f(x, y) = 75x + 80y$; $(0, 9)$, $(10, 0)$, $(4, 5)$, $(5, 4)$

For Exercises 47 and 48, find the *maximum* value of the objective function $f(x, y) = 8x + 5y$ and where this value occurs, given the constraints shown.

**47.** $\begin{cases} x + 2y \leq 6 \\ 3x + y \leq 8 \\ x \geq 0 \\ y \geq 0 \end{cases}$     **48.** $\begin{cases} 2x + y \leq 7 \\ x + 2y \leq 5 \\ x \geq 0 \\ y \geq 0 \end{cases}$

For Exercises 49 and 50, find the *minimum* value of the objective function $f(x, y) = 36x + 40y$ and where this value occurs, given the constraints shown.

**49.** $\begin{cases} 3x + 2y \geq 18 \\ 3x + 4y \geq 24 \\ x \geq 0 \\ y \geq 0 \end{cases}$     **50.** $\begin{cases} 2x + y \geq 10 \\ x + 4y \geq 3 \\ x \geq 2 \\ y \geq 0 \end{cases}$

## ▶ WORKING WITH FORMULAS

### Area Formulas

**51.** The area of a triangle is usually given as $A = \frac{1}{2}BH$, where $B$ and $H$ represent the base and height respectively. The area of a rectangle can be stated as $A = BH$. If the base of both a triangle and rectangle is equal to 20 in., what are the possible values for $H$ if the triangle must have an area *greater than* 50 in$^2$ and the rectangle must have an area *less than* 200 in$^2$?

### Volume Formulas

**52.** The volume of a cone is $V = \frac{1}{3}\pi r^2 h$, where $r$ is the radius of the base and $h$ is the height. The volume of a cylinder is $V = \pi r^2 h$. If the radius of both a cone and cylinder is equal to 10 cm, what are the possible values for $h$ if the cone must have a volume *greater than* 200 cm$^3$ and the cylinder must have a volume *less than* 850 cm$^3$?

## ▶ APPLICATIONS

Write a system of linear inequalities that models the information given, then solve.

**53. Losing weight:** The number of calories needed to remain healthy while maintaining or losing body weight depends on age, activity levels, current weight, and other factors. For males ages 18 to 35 with a slightly active lifestyle, the relationship between current weight and calories needed can be approximated by the system and graph shown. (a) Will a slightly active 25-yr-old male weighing 180 lb lose weight on a daily diet of 3000 calories? (b) Verify that Mitchell (point $M$) will likely be losing weight over time (point $M$ is a solution to the system). (c) Verify that Jaymes (point $J$) will likely *not* be losing weight over time (point $J$ is not a solution to the system).

$\begin{cases} c \leq 14.4w + 536 \\ c \geq 14.4w - 215 \\ 135 \leq w \leq 185 \end{cases}$

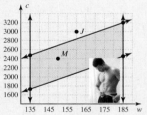

**54. Target heart rates:** At many health clubs, trainers use what is called a target heart rate to assist people wanting to do aerobic exercises (exercise resulting in 70% to 80% of a maximum heart rate). Typically, the target rate $t$ depends on the age $a$ of the individual, with the relationship modeled by the system and graph shown. (a) According to the system and graph, is a 50-yr-old

$\begin{cases} t \leq -0.7a + 174 \\ t \geq -0.6a + 152 \\ 20 \leq a \leq 70 \end{cases}$

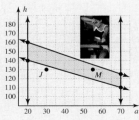

person with an exercise heart rate of 110 performing aerobic exercise? (b) Verify that Jessie (point $J$) is not yet at the point of doing aerobic exercise (point $J$ is not a solution to the system). (c) Verify that Mackenzie (point $M$) is performing aerobic exercise (point $M$ is a solution to the system).

**55. Gifts to grandchildren:** Grandpa Augustus is considering how to divide some or all of a $50,000 gift between his two grandchildren, Julius and Anthony. After weighing their respective positions in life and family responsibilities, he decides he must bequeath at least $20,000 to Julius, but no more than $25,000 to Anthony. Determine the possible ways that Grandpa can divide the $50,000.

**56. Guns versus butter:** Every year, governments around the world have to make the decision as to how much of their revenue must be spent on national defense and domestic improvements (guns versus butter). Suppose total expenditures for these two needs could not exceed $120 billion, and a government decides they need to spend at least $42 billion on butter and no more than $80 billion on defense. Determine the possible amounts that can go toward each need.

**57. Family time versus personal time:** As a single parent, Conner is struggling to get ahead at work while raising children and keeping his sanity. After a careful look at his schedule, he finds there are at most 420 min during the week that he isn't working, sleeping, cleaning, or paying bills. Out of love and fairness, he decides to devote

between 150 and 350 min to games and activities with his children. To keep himself mentally and physically healthy, he feels he should use at least 60 min but no more than 180 min exercising or interacting with other adults. (a) Write the system of inequalities that models this situation. (b) What possibilities will satisfy these constraints? (c) If he spends 300 min with his children this week, will he be able to exercise at the health club for 150 min?

**58. Setting priorities:** As newlyweds, Yvonne and Gary are trying to decide how much of their "extra income" they should put in long-term savings and how much they can use for discretionary spending (movies, dates, travel, etc.). After setting a budget, they find that their extra income will not be more than $500 each month. After talking to a financial counselor, they decide to place

at least $150 but not more than $350 in savings. To have fun and keep the love alive, Gary wants to spend at least $100 each month. While Yvonne okays that, she sets a maximum amount of $300 for this entertainment. (a) Write the system of inequalities that models this situation. (b) What possibilities will satisfy these constraints? (c) If Yvonne and Gary place only $175 in savings, what is the maximum amount they could spend on entertainment and still satisfy the system?

**59. Snack nut mixes:** The owner of Natalie's Nut Shoppe wants to sell two new combinations of macadamia nuts and pecans, a standard mix and a high quality mix. To begin the process,

Natalie purchases 480 oz of pecans and 420 oz of macadamia nuts. For the pecans, the standard mix will have 6 oz and the quality mix will have 4 oz. As for the macadamia nuts, the standard mix will have 4 oz, while the quality mix will have 6 oz. (a) Using $s$ for the standard mix and $q$ for the quality mix, write an inequality that models how the 480 oz of pecans could be used, and another inequality that models how the 420 oz of macadamia nuts could be used. (b) What other constraints should be used for this model? (c) Write the final result as a system and graph the solution region.

**60. Feeding construction workers:** Sammy's Snack Wagon travels to various construction sites, offering a variety of sandwiches for the

workers. His two most popular sandwiches, the "Big-n-Meaty" and the "Large-n-Luscious," are a combination of ham and salami. To make his sandwiches for the week, Sammy purchases 240 oz of salami and 300 oz of ham. For the salami, the Big-n-Meaty will use 4 oz and the Large-n-Luscious will use 2 oz. As for the ham, the Big-n-Meaty will use 3 oz, while the Large-n-Luscious will use 4 oz. (a) Using $b$ for the Big-n-Meaty sandwich and $l$ for the Large-n-Luscious, write an inequality that models how the 240 oz of salami could be used, and another inequality that models how the 300 oz of ham could be used. (b) What other constraints should be used for this model? (c) Write the final result as a system and graph the solution region.

**Solve the following applications of linear programming.**

**61. Land/crop allocation:** A farmer has 500 acres of land to plant corn and soybeans. During the last few years, market prices have been stable and the farmer anticipates a profit of $900 per acre on the corn harvest and $800 per acre on the soybeans. The farmer must take into account the time it takes to plant and harvest each crop, which is 3 hr/acre for corn and 2 hr/acre for soybeans. If the farmer has at most 1300 hr to plant, care for, and harvest each crop, how many acres of each crop should be planted in order to maximize profits?

**62. Coffee blends:** The owner of a coffee shop has decided to introduce two new blends of coffee in order to attract new customers—a *Deluxe Blend* and a *Savory Blend*. Each pound of the deluxe blend contains 30% Colombian and 20% Sumatran coffee, while each pound of the savory blend contains 35% Colombian and 15% Sumatran coffee (the remainder of each is made up of cheap and plentiful domestic varieties). The profit on the deluxe blend will be $1.25/lb, while the profit on the savory blend will be $1.40/lb. How many pounds of each should the owner make in order to maximize profit, if only 455 lb of Colombian coffee and 250 lb of Sumatran coffee are currently available?

**63. Manufacturing screws:** A machine shop manufactures two types of screws—sheet metal screws and wood screws, using three different machines. Machine Moe can make a sheet metal screw in 20 sec and a wood screw in 5 sec. Machine Larry can make a sheet metal screw in 5 sec and a wood screw in 20 sec. Machine Curly, the newest machine (nyuk, nyuk) can make a sheet metal screw in 15 sec and a wood screw in 15 sec. (Machine Shemp kept breaking down and was pulled from operation.) Each machine can operate for only 3 hr each day before shutting down for maintenance. If sheet metal screws sell for 10 cents and wood screws sell for 12 cents, how many of each type should the machines be programmed to make in order to maximize revenue? (*Hint:* Standardize time units.)

64. **Hauling hazardous waste:** A waste disposal company is contracted to haul away some hazardous waste material. A full container of liquid waste weighs 800 lb and has a volume of 20 ft$^3$. A full container of solid waste weighs 600 lb and has a volume of 30 ft$^3$. The trucks used can carry at most 10 tons (20,000 lb) and have a carrying volume of 800 ft$^3$. If the trucking company makes \$300 for disposing of liquid waste and \$400 for disposing of solid waste, what is the maximum revenue per truck that can be generated?

65. **Maximizing profit—food service:** JW and JP are starting up a fast-food restaurant specializing in peanut butter and jelly sandwiches. Independent research has determined the two most popular sandwiches will be the Commoner (smooth peanut butter and grape jelly), and the Clubhouse (three slices of bread). A Commoner uses 2 oz of peanut butter and 3 oz of jelly. The Clubhouse uses 4 oz of peanut butter and 5 oz of jelly. The Commoner will be priced at \$2.00, and a Clubhouse at \$3.50. If the restaurant has 250 oz of smooth peanut butter and 345 oz of grape jelly on hand for opening day, how many of each should they make and sell to maximize revenue?

66. **Maximizing profit—construction materials:** Mooney and Sons produces and sells two varieties of concrete mixes. The mixes are packaged in 50-lb bags. Type A is appropriate for finish work, and contains 20 lb of cement and 30 lb of sand. Type B is appropriate for foundation and footing work, and contains 10 lb of cement and 20 lb of sand. The remaining weight comes from a cheap gravel aggregate. The profit on type A is \$1.20/bag, while the profit on type B is \$0.90/bag. How many bags of each should the company make to maximize profit, if 2750 lb of cement and 4500 lb of sand are currently available?

67. **Minimizing transportation costs:** Robert's Las Vegas Tours needs to drive 375 people and 19,450 lb of luggage from Salt Lake City, Utah, to Las Vegas, Nevada, and can charter buses from two companies. The buses from company X carry 45 passengers and 2750 lb of luggage at a cost of \$1250 per trip. Company Y offers buses that carry 60 passengers and 2800 lb of luggage at a cost of \$1350 per trip. How many buses should be chartered from each company in order for Robert to minimize the cost?

68. **Minimizing shipping costs:** An oil company is trying to minimize shipping costs from its two primary refineries in Tulsa, Oklahoma, and Houston, Texas. All orders within the region are shipped from one of these two refineries. An order for 220,000 gal comes in from a location in Colorado, and another for 250,000 gal from a location in Mississippi. The Tulsa refinery has 320,000 gal ready to ship, while the Houston refinery has 240,000 gal. The cost of transporting each gallon to Colorado is \$0.05 from Tulsa and \$0.075 from Houston. The cost of transporting each gallon to Mississippi is \$0.06 from Tulsa and \$0.065 from Houston. How many gallons should be distributed from each refinery to minimize the cost of filling both orders?

▶ **EXTENDING THE CONCEPT**

69. Graph the feasible region formed by the system
$$\begin{cases} x + y \le 6 \\ 0 \le x \le 4. \\ 0 \le y \le 5 \end{cases}$$

(a) Select random points on the boundary line formed by the first constraint and evaluate the objective function $f(x, y) = 5x + 5y$. At what point(s) will this function be maximized? (b) Describe the characteristic shared by the objective function and this constraint that leads to the presence of multiple optimal solutions.

70. Find the maximum value of the objective function $f(x, y) = 4x + 15y$ given the constraints
$$\begin{cases} 2x + 5y \le 24 \\ 3x + 4y \le 29 \\ x + 6y \le 26 \\ x \ge 0 \\ y \ge 0 \end{cases}$$

▶ **MAINTAINING YOUR SKILLS**

71. (1.6) Find all solutions (real and nonreal) by factoring: $x^3 - 5x^2 + 3x - 15 = 0$.

72. (4.6) Solve the rational inequality. Write your answer in interval notation. $\frac{x + 2}{x^2 - 9} > 0$

73. (3.3) The resistance to current flow in copper wire varies directly with its length and inversely with the square of its diameter. A wire 8 m long with a 0.004-m diameter has a resistance of 1500 Ω. Find the resistance in a wire of like material that is 2.7 m long with a 0.005-m diameter.

74. (5.5) Solve for $x$: $-350 = 211e^{-0.025x} - 450$. Round solutions to hundredths as needed.

## MAKING CONNECTIONS

## Making Connections: Graphically, Symbolically, Numerically, and Verbally

**Match the characteristics shown in 1 through 16 to one or more of the eight graphs given in (a) through (h).**

**(a)**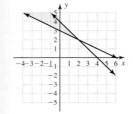

**(b)**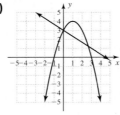

**(c)**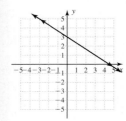

**(d)**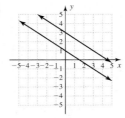

**(e)**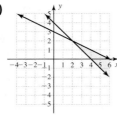

**(f)**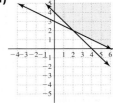

**(g)**

**(h)**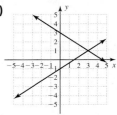

**1.** _____ $\begin{cases} x + y \leq 4 \\ x + 2y \leq 6 \end{cases}$

**2.** _____ $\begin{cases} 2x + 3y = 9 \\ x + \dfrac{3}{2}y = \dfrac{9}{2} \end{cases}$

**3.** _____ $m_1 = m_2, b_1 \neq b_2$

**4.** _____ $(0, 0)$ is a solution

**5.** _____ $\begin{cases} 2x + 3y = 9 \\ -2x + 3y = -3 \end{cases}$

**6.** _____ $\begin{cases} x + y \leq 4 \\ x + 2y \geq 6 \end{cases}$

**7.** _____ $(-1, 4)$ is a solution

**8.** _____ nonlinear system

**9.** _____ $\begin{cases} 2x + 3y = 9 \\ 2x + 3y = 3 \end{cases}$

**10.** _____ $\begin{cases} x + y \geq 4 \\ x + 2y \geq 6 \end{cases}$

**11.** _____ consistent, dependent

**12.** _____ inconsistent system

**13.** _____ $\begin{cases} x + y \geq 4 \\ x + 2y \leq 6 \end{cases}$

**14.** _____ $\begin{cases} 2x + 3y = 9 \\ x^2 + y = 2x + 3 \end{cases}$

**15.** _____ exactly two solutions

**16.** _____ consistent, independent

**64. Classical music:** 90.7 FM—*The Home of the Classics*—is having their annual fund drive. Being a loyal listener, Mitchell decides that for the next 3 days he will donate money according to the number of times his favorite composers come on the air: $3 for every piece by Mozart (*M*), $2.50 for every piece by Beethoven (*B*), and $2 for every piece by Vivaldi (*V*). This information is displayed in matrix *D*. The number of pieces he heard from each composer is displayed in matrix *C*. Compute the product *DC* and discuss what each entry of the product matrix represents.

$$\begin{array}{c} \quad\text{Mon. Tue. Wed.} \\ \begin{array}{c} M \\ B \\ V \end{array} \left[\begin{array}{ccc} 4 & 3 & 5 \\ 3 & 2 & 4 \\ 2 & 3 & 3 \end{array}\right] = C \end{array}$$

$$\begin{array}{ccc} M & B & V \\ [3 & 2.5 & 2] = D \end{array}$$

**65. Pizza and salad:** The science and math departments of a local college are at a presemester retreat, and decide to have pizza, salads, and soft drinks for lunch. The quantity of food ordered by each department is shown in matrix *Q*. The cost of the food item at each restaurant is shown in matrix *C* using the published prices from three popular restaurants: Pizza Home (PH), Papa Jeff's (PJ), and Dynamos (D).

**a.** What is the total cost to the math department if the food is ordered from Pizza Home?

**b.** What is the total cost to the science department if the food is ordered from Papa Jeff's?

**c.** Compute the product *QC* and discuss the meaning of each entry in the product matrix.

$$\begin{array}{c} \quad\text{Pizza Salad Drink} \\ \begin{array}{c} \text{Science} \\ \text{Math} \end{array} \left[\begin{array}{ccc} 8 & 12 & 20 \\ 10 & 8 & 18 \end{array}\right] = Q \end{array}$$

$$\begin{array}{c} \quad\text{PH} \quad \text{PJ} \quad\ \ \text{D} \\ \begin{array}{c} \text{Pizza} \\ \text{Salad} \\ \text{Drink} \end{array} \left[\begin{array}{ccc} 8 & 7.5 & 10 \\ 1.5 & 1.75 & 2 \\ 0.90 & 1 & 0.75 \end{array}\right] = C \end{array}$$

**66. Manufacturing pool tables:** Cue Ball Incorporated makes pool tables for home, commercial, and professional use. The amount of time required to pack, load, and install each type is summarized in matrix *T*, with all times in hours. The cost of these components in dollars per hour is summarized in matrix *C* for two of its warehouses, one on the West Coast and the other in the Midwest.

**a.** What is the cost to package, load, and install a commercial pool table from the coastal warehouse?

**b.** What is the cost to package, load, and install a commercial pool table from the warehouse in the Midwest?

**c.** Compute the product *TC* and discuss the meaning of each entry in the product matrix.

$$\begin{array}{c} \qquad\quad\text{Pack} \ \ \text{Load} \ \ \text{Install} \\ \begin{array}{c} \text{Home} \\ \text{Comm} \\ \text{Prof} \end{array} \left[\begin{array}{ccc} 1 & 0.2 & 1.5 \\ 1.5 & 0.5 & 2.2 \\ 1.75 & 0.75 & 2.5 \end{array}\right] = T \end{array}$$

$$\begin{array}{c} \qquad\quad\text{Coast} \ \ \text{Midwest} \\ \begin{array}{c} \text{Pack} \\ \text{Load} \\ \text{Install} \end{array} \left[\begin{array}{cc} 10 & 8 \\ 12 & 10.5 \\ 13.5 & 12.5 \end{array}\right] = C \end{array}$$

**67. Joining a club:** The likelihood a student joins a particular club depends on their class standing. This information is stored in matrix *C*. The number of males and females from each class that are projected to join a club is stored in matrix *J*. Compute the product *JC* and use the result to answer the following:

**a.** Approximately how many females joined the chess club?

**b.** Approximately how many males joined the writing club?

**c.** What does the entry $p_{13}$ of the product matrix tell us?

$$\begin{array}{c} \qquad\quad\text{Fresh Soph Junior} \\ \begin{array}{c} \text{Female} \\ \text{Male} \end{array} \left[\begin{array}{ccc} 25 & 18 & 21 \\ 22 & 19 & 18 \end{array}\right] = J \end{array}$$

$$\begin{array}{c} \qquad\quad\text{Spanish Chess Writing} \\ \begin{array}{c} \text{Fresh} \\ \text{Soph} \\ \text{Junior} \end{array} \left[\begin{array}{ccc} 0.6 & 0.1 & 0.3 \\ 0.5 & 0.2 & 0.3 \\ 0.4 & 0.2 & 0.4 \end{array}\right] = C \end{array}$$

**68. Designer shirts:** The SweatShirt Shoppe sells three types of designs on its products: stenciled (S), embossed (E), and applique (A). The quantity of each size sold is shown in matrix *Q*. The retail price of each sweatshirt depends on its size and whether it was finished by hand or machine. Retail prices are shown in matrix *C*. Assuming all stock is sold, compute the product *QC* and use the result to answer the following.

**a.** How much revenue was generated by the large sweatshirts?

**b.** How much revenue was generated by the extra-large sweatshirts?

**c.** What does the entry $p_{11}$ of the product matrix *QC* tell us?

$$\begin{array}{c} \qquad\quad\text{S} \quad\ \text{E} \quad\ \text{A} \\ \begin{array}{c} \text{med} \\ \text{large} \\ \text{x-large} \end{array} \left[\begin{array}{ccc} 30 & 30 & 15 \\ 60 & 50 & 20 \\ 50 & 40 & 30 \end{array}\right] = Q \end{array}$$

$$\begin{array}{c} \qquad\quad\text{Hand Machine} \\ \begin{array}{c} \text{S} \\ \text{E} \\ \text{A} \end{array} \left[\begin{array}{cc} 40 & 25 \\ 60 & 40 \\ 90 & 60 \end{array}\right] = C \end{array}$$

**69. Rabbits and foxes:**
In Sherwood Forest, the
rabbit and fox populations
are related by the matrix
equation

$$\begin{bmatrix} r_{t+1} \\ f_{t+1} \end{bmatrix} = \begin{bmatrix} 1.2 & -0.4 \\ 0.2 & 0.9 \end{bmatrix} \begin{bmatrix} r_t \\ f_t \end{bmatrix},$$ where $r_t$ and $f_t$ represent

the rabbit and fox populations at the beginning of year $t$
and $r_{t+1}$ and $f_{t+1}$ give the populations at the beginning of
the following year. (a) Find the populations of each
animal at the beginning of years 2 and 3, if the current

populations are $\begin{bmatrix} r_1 \\ f_1 \end{bmatrix} = \begin{bmatrix} 1000 \\ 100 \end{bmatrix}$. (b) Repeat part (a) with

initial populations of $\begin{bmatrix} r_1 \\ f_1 \end{bmatrix} = \begin{bmatrix} 200 \\ 100 \end{bmatrix}$.

**70. Sharks and lobsters:** Along a certain stretch of the
Baja California coast, the red rock lobster and great white
shark populations are related by the matrix equation

$$\begin{bmatrix} l_{t+1} \\ s_{t+1} \end{bmatrix} = \begin{bmatrix} 1.15 & -0.2 \\ 0.01 & 0.80 \end{bmatrix} \begin{bmatrix} l_t \\ s_t \end{bmatrix},$$ where $l_t$ and $s_t$ represent

the lobster and shark populations at the beginning of year $t$
and $l_{t+1}$ and $s_{t+1}$ give the populations at the beginning of
the following year. (a) Find the populations of each
animal 5 yr from now, if the most recent survey

places the current populations at $\begin{bmatrix} l_0 \\ s_0 \end{bmatrix} = \begin{bmatrix} 3000 \\ 50 \end{bmatrix}$.

(b) Repeat part (a) with the predicted future populations

of $\begin{bmatrix} l_0 \\ s_0 \end{bmatrix} = \begin{bmatrix} 100 \\ 100 \end{bmatrix}$.

**71. Bicycle inventory:** Century
Cycling is in the business of
high-end road bicycles. Use
matrix multiplication and the
table to determine the retail and
wholesale values of their current
inventory.

| Model | Number in Stock | Retail Value | Wholesale Value |
|---|---|---|---|
| Comp | 20 | $1200 | $ 900 |
| Sport | 10 | $1500 | $1250 |
| Pro | 5 | $2500 | $2000 |

**72. Currency conversion:** Katia returns home from her North
American backpacking adventure with $125 in pesos (MXN),
$20 U.S. (USD), and
$15 Canadian (CAD). After
resting and resupplying,
she plans to join her friends
for a weekend in Tokyo. Use
matrix multiplication and the
conversions given in the table
to determine the value of her
money in euros (EUR)
and yen (JPY).

| Currency | Value in EUR | Value in JPY |
|---|---|---|
| MXN | 0.06 | 12 |
| USD | 0.71 | 79 |
| CAD | 0.74 | 83 |

▶ **EXTENDING THE CONCEPT**

**73.** For the matrix $A$ shown, use your calculator to
compute $A^2$, $A^3$, $A^4$, and $A^5$. Do you notice a pattern?
Try to write a "matrix formula" for $A^n$, where $n$ is a
positive integer, then use your formula to find $A^6$.
Check results using a calculator.

$$A = \begin{bmatrix} 1 & 0 & 1 \\ 1 & 1 & 1 \\ 1 & 0 & 1 \end{bmatrix}$$

**74.** The matrix $M = \begin{bmatrix} 2 & 1 \\ -3 & -2 \end{bmatrix}$ has some very interesting

properties. Compute the powers $M^2$, $M^3$, $M^4$, and $M^5$,
then discuss what you find. Try to find/create another
$2 \times 2$ matrix that has similar properties.

**75.** (a) Compute the value of $A^2$ and $B^2$ for

$$A = \begin{bmatrix} \dfrac{3}{5} & -\dfrac{4}{5} \\ -\dfrac{4}{5} & -\dfrac{3}{5} \end{bmatrix} \text{ and } B = \begin{bmatrix} -\dfrac{12}{13} & \dfrac{5}{13} \\ \dfrac{5}{13} & \dfrac{12}{13} \end{bmatrix} \text{ to show that}$$

each is a square root of $I_2 = \begin{bmatrix} 1 & 0 \\ 0 & 1 \end{bmatrix}$ (i.e., $A^2 = B^2 = I_2$).

After noting the pattern of signs, along with the fact that
3-4-5 and 5-12-13 form Pythagorean triples, construct
two of your own square roots for $I_2$.

## ▶ MAINTAINING YOUR SKILLS

**76.** (4.6) Given $f(x) = x^4 - 10x^2 + 9$, solve $f(x) \geq 0$.

**77.** (6.2) Solve the system using elimination.

$$\begin{cases} x + 2y - z = 3 \\ -2x - y + 3z = -5 \\ 5x + 3y - 2z = 2 \end{cases}$$

**78.** (4.2) Find the quotient using synthetic division, then check using multiplication.

$$\frac{x^3 - 9x + 10}{x - 2}$$

**79.** (5.4) Evaluate using the change-of-base formula, then check using exponentiation.

$$\log_2 21$$

---

# MID-CHAPTER CHECK

**State the size of each matrix and identify the entry in the second row, third column.**

**1.** $A = \begin{bmatrix} 0.4 & 1.1 & 0.2 \\ -0.2 & 0.1 & -0.9 \\ 0.7 & 0.4 & 0.8 \end{bmatrix}$

**2.** $B = \begin{bmatrix} -2 & 1 & \frac{1}{2} & 5 \\ 4 & \frac{3}{4} & 0 & -3 \end{bmatrix}$

**Write each system in matrix form and solve using row operations to triangularize the matrix. If the system is linearly dependent, write the solution using a parameter.**

**3.** $\begin{cases} 2x + 3y = -5 \\ -5x - 4y = 2 \end{cases}$

**4.** $\begin{cases} -x + y - 5z = 23 \\ 2x + 4y - z = 9 \\ 3x - 5y + z = 1 \end{cases}$

**5.** $\begin{cases} x + y - 3z = -11 \\ 4x - y - 2z = -4 \\ 3x - 2y + z = 7 \end{cases}$

**6.** For matrices $A$ and $B$ given, compute:

$$A = \begin{bmatrix} -3 & -2 \\ 5 & 4 \end{bmatrix}, B = \begin{bmatrix} 10 & 15 \\ -30 & -5 \end{bmatrix}$$

   **a.** $A - B$     **b.** $\frac{2}{5}B$     **c.** $5A + B$

 **7.** For matrices $C$ and $D$ given, use a calculator to find:

$$C = \begin{bmatrix} -0.2 & 0 & 0.2 \\ 0.4 & 0.8 & 0 \\ 0.1 & -0.2 & -0.1 \end{bmatrix}, D = \begin{bmatrix} 5 & 2.5 & 10 \\ -2.5 & 0 & -5 \\ 10 & 2.5 & 10 \end{bmatrix}$$

   **a.** $C + \frac{1}{5}D$     **b.** $-0.6D$     **c.** $CD$

**8.** For the matrices $A$, $B$, $C$, and $D$ given, compute the products indicated (if possible):

$$A = \begin{bmatrix} 4 & -1 \\ 0 & -5 \end{bmatrix} \qquad B = \begin{bmatrix} 6 & -2 \\ 0 & 1 \\ 4 & 7 \end{bmatrix}$$

$$C = \begin{bmatrix} 4 & -8 & -3 \\ -1 & 0 & 1 \end{bmatrix} \qquad D = \begin{bmatrix} 2 & 0 & -6 \\ -1 & -3 & 0 \\ 1 & 5 & -4 \end{bmatrix}$$

   **a.** $AC$    **b.** $-2CD$    **c.** $BA$    **d.** $CB - 4A$

**9.** Create a system of equations to model this exercise, then write the system in matrix form and solve. The campus bookstore offers both new and used texts to students. In a recent biology class with 24 students, 14 bought used texts and 10 bought new texts, with the class as a whole paying $2370. Of the 6 premed students in class, 2 bought used texts, and 4 bought new texts, with the group paying a total of $660. How much does a used text cost? How much does a new text cost?

**10.** Table $A$ shown gives the number and type of extended warranties sold to individual car owners and to business fleets. Table $B$ shows the promotions offered to those making the purchase. Write the entries of each table in matrix form and compute the product matrix $P = AB$, then state what each entry of the product matrix represents.

**Table A**

| Extended Warranties | 80,000 mi | 100,000 mi | 120,000 mi |
|---|---|---|---|
| Individuals | 30 | 25 | 10 |
| Businesses | 20 | 12 | 5 |

**Table B**

| Promotions | Rebate | Free AAA Membership |
|---|---|---|
| 80,000 mi | $ 50 | 1 yr |
| 100,000 mi | $ 75 | 2 yr |
| 120,000 mi | $100 | 3 yr |

## LEARNING OBJECTIVES

*In Section 7.3 you will see how we can:*

- ☐ **A.** Recognize the identity matrix for multiplication
- ☐ **B.** Find the inverse of a square matrix
- ☐ **C.** Solve systems using matrix equations
- ☐ **D.** Use determinants to find whether a matrix is invertible

While using matrices and row operations offers a degree of efficiency in solving systems, we are still required to solve for each variable *individually*. Using matrix multiplication we can actually rewrite a given system as a single *matrix equation,* in which the values of the variables are computed *simultaneously*. As with other kinds of equations, the use of identities and inverses is involved, which we now develop in the context of matrices.

## A. Multiplication and Identity Matrices

From the properties of real numbers, 1 is the identity for multiplication since $n \cdot 1 = 1 \cdot n = n$. A similar identity exists for matrix multiplication. Consider the $2 \times 2$ matrix $A = \begin{bmatrix} 1 & 4 \\ -2 & 3 \end{bmatrix}$. While matrix multiplication is not *generally* commutative, if we can find a matrix $B$ where $AB = BA = A$, then $B$ is a prime candidate for the identity matrix, which is denoted $I$. For the products $AB$ and $BA$ to be possible and have the same order as $A$, we note $B$ must also be a $2 \times 2$ matrix. Using the arbitrary matrix $B = \begin{bmatrix} a & b \\ c & d \end{bmatrix}$, we have the following.

---

**EXAMPLE 1A** ▶ **Solving $AB = A$ to Find the Identity Matrix**

For $\begin{bmatrix} 1 & 4 \\ -2 & 3 \end{bmatrix} \begin{bmatrix} a & b \\ c & d \end{bmatrix} = \begin{bmatrix} 1 & 4 \\ -2 & 3 \end{bmatrix}$, use matrix multiplication, the equality of matrices, and systems of equations to find the value of $a$, $b$, $c$, and $d$.

**Solution** ▶ The product on the left gives $\begin{bmatrix} \mathbf{a + 4c} & b + 4d \\ -2a + 3c & -2b + 3d \end{bmatrix} = \begin{bmatrix} \mathbf{1} & 4 \\ -2 & 3 \end{bmatrix}$.

Since corresponding entries must be equal (shown by matching colors), we can find $a$, $b$, $c$, and $d$ by solving the systems $\begin{cases} \mathbf{a + 4c = 1} \\ -2a + 3c = -2 \end{cases}$ and $\begin{cases} b + 4d = 4 \\ -2b + 3d = 3 \end{cases}$. For the first system, $2R1 + R2$ shows $a = 1$ and $c = 0$. Using $2R1 + R2$ for the second shows $b = 0$ and $d = 1$. It appears $\begin{bmatrix} 1 & 0 \\ 0 & 1 \end{bmatrix}$ is a candidate for the identity matrix.

---

Before we name $B$ as the identity matrix, we must show that $AB = BA = A$.

---

**EXAMPLE 1B** ▶ **Verifying $AB = BA = A$**

Given $A = \begin{bmatrix} 1 & 4 \\ -2 & 3 \end{bmatrix}$ and $B = \begin{bmatrix} 1 & 0 \\ 0 & 1 \end{bmatrix}$, determine if $AB = A$ and $BA = A$.

**Solution** ▶ $AB = \begin{bmatrix} 1 & 4 \\ -2 & 3 \end{bmatrix} \begin{bmatrix} 1 & 0 \\ 0 & 1 \end{bmatrix} = \begin{bmatrix} 1(1) + 4(0) & 1(0) + 4(1) \\ -2(1) + 3(0) & -2(0) + 3(1) \end{bmatrix} = \begin{bmatrix} 1 & 4 \\ -2 & 3 \end{bmatrix} = A \checkmark$

$BA = \begin{bmatrix} 1 & 0 \\ 0 & 1 \end{bmatrix} \begin{bmatrix} 1 & 4 \\ -2 & 3 \end{bmatrix} = \begin{bmatrix} 1(1) + 0(-2) & 1(4) + 0(3) \\ 0(1) + 1(-2) & 0(4) + 1(3) \end{bmatrix} = \begin{bmatrix} 1 & 4 \\ -2 & 3 \end{bmatrix} = A \checkmark$

☑ **A. You've just seen how we can recognize the identity matrix for multiplication**

Since $AB = A = BA$, $B$ is the identity matrix $I$.

---

Now try Exercises 5 through 8 ▶

By replacing the entries of $A = \begin{bmatrix} 1 & 4 \\ -2 & -3 \end{bmatrix}$ with those of the general matrix $\begin{bmatrix} a_{11} & a_{12} \\ a_{21} & a_{22} \end{bmatrix}$, we can show that $I_2 = \begin{bmatrix} 1 & 0 \\ 0 & 1 \end{bmatrix}$ is the identity for *all* $2 \times 2$ matrices. In considering the identity for larger matrices, we find that only *square matrices* can have inverses, since $AI = IA$ requires the multiplication must be possible in both directions. (This is commonly referred to as *multiplication from the right* and *multiplication from the left*.) Using the same procedure as before we can show $\begin{bmatrix} 1 & 0 & 0 \\ 0 & 1 & 0 \\ 0 & 0 & 1 \end{bmatrix}$ is the identity for $3 \times 3$ matrices (denoted $I_3$). The $n \times n$ identity matrix $I_n$ is unique and consists of 1's down the main diagonal and 0's for all other entries.

As in Section 7.2, a graphing calculator can be used to investigate operations on matrices and matrix properties. For the $3 \times 3$ matrix $A = \begin{bmatrix} 2 & 5 & 1 \\ 4 & -1 & 1 \\ 0 & 3 & -2 \end{bmatrix}$ and $I_3 = \begin{bmatrix} 1 & 0 & 0 \\ 0 & 1 & 0 \\ 0 & 0 & 1 \end{bmatrix}$, a calculator will confirm that $AI_3 = A = I_3 A$. Carefully enter $A$ into your calculator as matrix $A$, and $I_3$ as matrix $B$. Figure 7.6 shows $AB = A$ and $BA = A$. **See Exercises 9 through 12.**

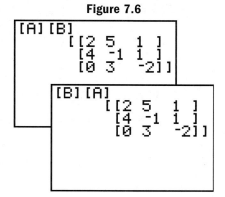

**Figure 7.6**

## B. The Inverse of a Matrix

Again from the properties of real numbers, we know the multiplicative inverse for $a$ is $a^{-1} = \dfrac{1}{a}$ $(a \neq 0)$, since the products $a \cdot a^{-1}$ and $a^{-1} \cdot a$ yield the identity 1. To show that a similar inverse exists for matrices, consider the square matrix $A = \begin{bmatrix} 6 & 5 \\ 2 & 2 \end{bmatrix}$ and an arbitrary matrix $B = \begin{bmatrix} a & b \\ c & d \end{bmatrix}$. If we can find a matrix $B$, where $AB = BA = I$, then $B$ is a prime candidate for the inverse matrix of $A$, which is denoted $A^{-1}$. In an attempt to find such a matrix $B$, we proceed as in Examples 1A and 1B.

**EXAMPLE 2A** ▶ **Solving $AB = I$ to find $A^{-1}$**

For $\begin{bmatrix} 6 & 5 \\ 2 & 2 \end{bmatrix} \begin{bmatrix} a & b \\ c & d \end{bmatrix} = \begin{bmatrix} 1 & 0 \\ 0 & 1 \end{bmatrix}$, use matrix multiplication, the equality of matrices, and systems of equations to find the entries of $B$.

**Solution** ▶ The product on the left gives $\begin{bmatrix} \mathbf{6a + 5c} & 6b + 5d \\ 2a + 2c & 2b + 2d \end{bmatrix} = \begin{bmatrix} \mathbf{1} & 0 \\ 0 & 1 \end{bmatrix}$. Since corresponding entries must be equal (shown by matching colors), we find the values of $a$, $b$, $c$, and $d$ by solving the systems $\begin{cases} \mathbf{6a + 5c = 1} \\ 2a + 2c = 0 \end{cases}$ and $\begin{cases} 6b + 5d = 0 \\ 2b + 2d = 1 \end{cases}$. Using $-3R2 + R1$ for the first system shows $a = 1$ and $c = -1$, while $-3R2 + R1$ for the second shows $b = -2.5$ and $d = 3$. Matrix $B = \begin{bmatrix} a & b \\ c & d \end{bmatrix} = \begin{bmatrix} 1 & -2.5 \\ -1 & 3 \end{bmatrix}$ is the prime candidate for $A^{-1}$.

It may have occurred to you that the two systems used in Example 2A have *exactly the same coefficients*. In matrix form, these systems are $\begin{bmatrix} 6 & 5 & | & 1 \\ 2 & 2 & | & 0 \end{bmatrix}$ and $\begin{bmatrix} 6 & 5 & | & 0 \\ 2 & 2 & | & 1 \end{bmatrix}$. Instead of solving these two systems separately, we can solve them *simultaneously* by simply using the coefficient matrix augmented with the identity matrix: $\begin{bmatrix} 6 & 5 & | & 1 & 0 \\ 2 & 2 & | & 0 & 1 \end{bmatrix}$. Row operations are then used to transform the given matrix into the identity: $\begin{bmatrix} 1 & 0 & | & e & f \\ 0 & 1 & | & g & h \end{bmatrix}$, in a sense "preserving" the inverse operations needed. The inverse matrix is then $\begin{bmatrix} e & f \\ g & h \end{bmatrix}$.

---

**EXAMPLE 2B** ▶ **Finding a 2 × 2 Inverse Using the Augmented Matrix**

Use the augmented matrix method to find $A^{-1}$ for $A = \begin{bmatrix} 6 & 5 \\ 2 & 2 \end{bmatrix}$. Then check by verifying $AA^{-1} = A^{-1}A = I$.

**Solution** ▶ First we augment the given matrix with the identity: $\begin{bmatrix} 6 & 5 & | & 1 & 0 \\ 2 & 2 & | & 0 & 1 \end{bmatrix}$. Then we use row operations to transform $A$ into the identity matrix.

$$\begin{bmatrix} 6 & 5 & | & 1 & 0 \\ 2 & 2 & | & 0 & 1 \end{bmatrix}$$

next: $-3R2 + R1 \rightarrow R2$

$$\begin{array}{rrrr} -3R2 & -6 & -6 & 0 & -3 \\ +R1 & 6 & 5 & 1 & 0 \\ \hline \text{New R2} & 0 & -1 & 1 & -3 \end{array}$$

$$\begin{bmatrix} 6 & 5 & | & 1 & 0 \\ 0 & -1 & | & 1 & -3 \end{bmatrix}$$

next: $5R2 + R1 \rightarrow R1$

$$\begin{array}{rrrr} 5R2 & 0 & -5 & 5 & -15 \\ +R1 & 6 & 5 & 1 & 0 \\ \hline \text{New R1} & 6 & 0 & 6 & -15 \end{array}$$

$$\begin{bmatrix} 6 & 0 & | & 6 & -15 \\ 0 & -1 & | & 1 & -3 \end{bmatrix} \qquad \begin{bmatrix} 1 & 0 & | & 1 & -2.5 \\ 0 & 1 & | & -1 & 3 \end{bmatrix}$$

next: $\dfrac{R1}{6} \rightarrow R1, \; -1R2 \rightarrow R2$

As in Example 2A, a prime candidate for $A^{-1}$ is $\begin{bmatrix} 1 & -2.5 \\ -1 & 3 \end{bmatrix}$. To determine if $A^{-1}$ has truly been found, we check to see if multiplication from the right and multiplication from the left yields the identity matrix $I$:

**Check:**

$$AA^{-1} = \begin{bmatrix} 6 & 5 \\ 2 & 2 \end{bmatrix}\begin{bmatrix} 1 & -2.5 \\ -1 & 3 \end{bmatrix} = \begin{bmatrix} 6(1) + 5(-1) & 6(-2.5) + 5(3) \\ 2(1) + 2(-1) & 2(-2.5) + 2(3) \end{bmatrix} = \begin{bmatrix} 1 & 0 \\ 0 & 1 \end{bmatrix} = I \checkmark$$

$$A^{-1}A = \begin{bmatrix} 1 & -2.5 \\ -1 & 3 \end{bmatrix}\begin{bmatrix} 6 & 5 \\ 2 & 2 \end{bmatrix} = \begin{bmatrix} 1(6) + (-2.5)(2) & 1(5) + (-2.5)(2) \\ -1(6) + 3(2) & -1(5) + 3(2) \end{bmatrix} = \begin{bmatrix} 1 & 0 \\ 0 & 1 \end{bmatrix} = I \checkmark$$

☑ **B.** You've just seen how we can find the inverse of a square matrix

Since $AA^{-1} = I = A^{-1}A$, we have indeed found our inverse.

**Now try Exercises 13 through 20** ▶

These observations guide us to the following definition of an inverse matrix.

> ### The Inverse of a Matrix
>
> Given an $n \times n$ matrix $A$, if there exists an $n \times n$ matrix $A^{-1}$ such that $AA^{-1} = A^{-1}A = I_n$, then $A^{-1}$ is the inverse of matrix $A$.

As mentioned earlier, only square matrices have inverses, but not every square matrix has an inverse. If an inverse exists, the matrix is said to be **invertible.** For $2 \times 2$ matrices that are invertible, a simple formula exists for computing the inverse. The formula is derived in the *Strengthening Core Skills* feature at the end of this chapter.

> ### The Inverse of a 2 × 2 Matrix
>
> If $A = \begin{bmatrix} a & b \\ c & d \end{bmatrix}$, then $A^{-1} = \dfrac{1}{ad - bc}\begin{bmatrix} d & -b \\ -c & a \end{bmatrix}$ provided $ad - bc \neq 0$.

To "test" the formula, again consider the matrix $A = \begin{bmatrix} 6 & 5 \\ 2 & 2 \end{bmatrix}$, where $a = 6$, $b = 5$, $c = 2$, and $d = 2$:

$$A^{-1} = \frac{1}{(6)(2) - (5)(2)}\begin{bmatrix} 2 & -5 \\ -2 & 6 \end{bmatrix} = \frac{1}{2}\begin{bmatrix} 2 & -5 \\ -2 & 6 \end{bmatrix} = \begin{bmatrix} 1 & -2.5 \\ -1 & 3 \end{bmatrix} \checkmark$$

**See Exercises 67 through 70** for more practice with this formula.

Almost without exception, real-world applications involve much larger matrices, with entries that are not integer-valued. Although the augmented matrix method from Example 2 can be extended to find the inverses of larger matrices, the process becomes very tedious and too time consuming to be useful. The process for a $3 \times 3$ matrix is discussed in the *Strengthening Core Skills* feature at the end of this chapter. But for practical reasons, we will rely on a calculator to produce these larger inverse matrices. This is done by (1) carefully entering a square matrix $A$ into the calculator, (2) returning to the home screen, and (3) calling up matrix $A$ and pressing 🔳 🔳 to find $A^{-1}$ (see Figure 7.7). **See Exercises 21 through 24.**

**Figure 7.7**

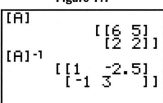

### C. Solving Systems Using Matrix Equations

One reason matrix multiplication has its row $\times$ column definition is to assist in writing a linear system of equations as a single matrix equation. The equation consists of the matrix of constants $B$ on the right, and a product of the coefficient matrix $A$ with the matrix of variables $X$ on the left: $AX = B$. For $\begin{cases} x + 4y - z = 10 \\ 2x + 5y - 3z = 7 \\ 8x + y - 2z = 11 \end{cases}$, the matrix

$$\underset{A}{\begin{bmatrix} 1 & 4 & -1 \\ 2 & 5 & -3 \\ 8 & 1 & -2 \end{bmatrix}} \underset{X}{\begin{bmatrix} x \\ y \\ z \end{bmatrix}} = \underset{B}{\begin{bmatrix} 10 \\ 7 \\ 11 \end{bmatrix}}$$

equation is . Note that computing the product on the left will yield the original system.

Once written as a matrix equation, the system can be solved using an inverse matrix and the following sequence. If $A$ represents the matrix of coefficients, $X$ the matrix of variables, $B$ the matrix of constants, and $I$ the appropriate identity, the sequence is

$$AX = B \qquad \text{matrix equation}$$
$$A^{-1}(AX) = A^{-1}B \qquad \text{multiply from the left by the inverse of } A$$
$$(A^{-1}A)X = A^{-1}B \qquad \text{associative property}$$
$$IX = A^{-1}B \qquad A^{-1}A = I$$
$$X = A^{-1}B \qquad IX = X$$

These steps illustrate why the method works. In actual practice, after carefully entering the matrices, only the last step is used when solving matrix equations with technology. Once matrix $A$ is entered, the calculator will automatically *find* and *use* $A^{-1}$ as we enter $A^{-1}B$.

---

**EXAMPLE 3** ▶ **Using Technology to Solve a Matrix Equation**

Use a calculator and a matrix equation to solve the system
$$\begin{cases} x + 4y - z = 10 \\ 2x + 5y - 3z = 7. \\ 8x + y - 2z = 11 \end{cases}$$

**Solution** ▶ As before, the matrix equation is

$$\overset{A}{\begin{bmatrix} 1 & 4 & -1 \\ 2 & 5 & -3 \\ 8 & 1 & -2 \end{bmatrix}} \overset{X}{\begin{bmatrix} x \\ y \\ z \end{bmatrix}} = \overset{B}{\begin{bmatrix} 10 \\ 7 \\ 11 \end{bmatrix}}.$$

Carefully enter (and double-check) the matrix of coefficients as matrix $A$ in your calculator, and the matrix of constants as matrix $B$. The product $A^{-1}B$ shows the solution is $x = 2, y = 3, z = 4$ (see Figure 7.8). Verify by substitution.

**Figure 7.8**

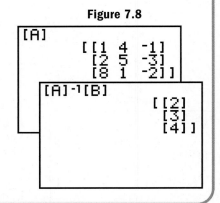

☑ **C.** You've just seen how we can solve systems using matrix equations

**Now try Exercises 25 through 30** ▶

---

The matrix equation method does have a few shortcomings. Consider the system whose corresponding matrix equation is $\begin{bmatrix} 4 & -10 \\ -2 & 5 \end{bmatrix} \begin{bmatrix} x \\ y \end{bmatrix} = \begin{bmatrix} -8 \\ 13 \end{bmatrix}$. After entering the matrix of coefficients $A$ and matrix of constants $B$, attempting to compute $A^{-1}B$ results in the error message shown in Figure 7.9. The calculator is unable to return a solution due to something called a **"singular matrix."** To investigate further, we attempt to find $A^{-1}$ for $\begin{bmatrix} 4 & -10 \\ -2 & 5 \end{bmatrix}$ using the formula for a $2 \times 2$ matrix. With $a = 4$, $b = -10, c = -2$, and $d = 5$, we have

$$A^{-1} = \frac{1}{ad - bc}\begin{bmatrix} d & -b \\ -c & a \end{bmatrix} = \frac{1}{(4)(5) - (-10)(-2)}\begin{bmatrix} 5 & 10 \\ 2 & 4 \end{bmatrix} = \frac{1}{0}\begin{bmatrix} 5 & 10 \\ 2 & 4 \end{bmatrix}$$

**Figure 7.9**

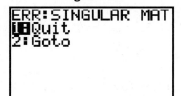

Since division by zero is undefined, we conclude that matrix $A$ has no inverse. Such matrices are said to be **singular** or **noninvertible.** Solving systems using matrix equations is only possible when the matrix of coefficients is **nonsingular. See Exercises 31 through 42.**

## D.  Determinants and Singular Matrices

As a practical matter, it becomes important to know ahead of time whether a particular matrix has an inverse. To help with this, we introduce one additional operation on a square matrix, that of calculating its **determinant.** For a $1 \times 1$ matrix the determinant is the entry itself. For a $2 \times 2$ matrix $A = \begin{bmatrix} a_{11} & a_{12} \\ a_{21} & a_{22} \end{bmatrix}$, the determinant of $A$, written as $\det(A)$ or denoted with vertical bars as $|A|$, is computed as *a difference of diagonal products* beginning with the upper-left entry:

$$\det(A) = \begin{vmatrix} a_{11} & a_{12} \\ a_{21} & a_{22} \end{vmatrix} = a_{11}a_{22} - a_{21}a_{12}$$

2nd diagonal product

1st diagonal product

### The Determinant of a 2 × 2 Matrix

Given any $2 \times 2$ matrix $A = \begin{bmatrix} a_{11} & a_{12} \\ a_{21} & a_{22} \end{bmatrix}$,

$$\det(A) = |A| = a_{11}a_{22} - a_{21}a_{12}$$

**EXAMPLE 4** ▶ **Calculating Determinants**

Compute the determinant of each matrix.

**a.** $B = \begin{bmatrix} 3 & 2 \\ 1 & -6 \end{bmatrix}$   **b.** $C = \begin{bmatrix} 5 & 2 & 1 \\ -1 & -3 & 4 \end{bmatrix}$   **c.** $D = \begin{bmatrix} 4 & -10 \\ -2 & 5 \end{bmatrix}$

Solution ▶   **a.** $\det(B) = \begin{vmatrix} 3 & 2 \\ 1 & -6 \end{vmatrix} = (3)(-6) - (1)(2) = -20$

**b.** Determinants are only defined for square matrices.

**c.** $\det(D) = \begin{vmatrix} 4 & -10 \\ -2 & 5 \end{vmatrix} = (4)(5) - (-2)(-10) = 20 - 20 = 0$

Now try Exercises 43 through 46 ▶

Notice from Example 4(c), the determinant of $\begin{bmatrix} 4 & -10 \\ -2 & 5 \end{bmatrix}$ is zero, and this is the same matrix we earlier found had no inverse. This observation can be extended to larger matrices and offers the connection we seek between a given matrix, its inverse, and matrix equations.

### Singular Matrices

If $A$ is a square matrix and $\det(A) = 0$, the inverse matrix *does not exist* and $A$ is said to be *singular* or *noninvertible.*

In summary, inverses exist only for square matrices, but not every square matrix has an inverse. If the determinant of a square matrix is zero, an inverse does not exist and the method of matrix equations cannot be used to solve the system.

To use the determinant test for a $3 \times 3$ system, we need to compute a $3 \times 3$ determinant. At first glance, our experience with $2 \times 2$ determinants appears to be of little help. However, every entry in a $3 \times 3$ matrix is associated with a smaller $2 \times 2$ matrix, formed by *deleting the row and column* of that entry and using the entries that remain. These $2 \times 2$'s are called the **associated minor matrices** or simply the **minors.** Using a general matrix of coefficients, we'll identify the minors associated with the entries in the first row.

**WORTHY OF NOTE**

For the determinant of a general $n \times n$ matrix using cofactors, see Appendix II.

$$\begin{bmatrix} a_{11} & a_{12} & a_{13} \\ a_{21} & a_{22} & a_{23} \\ a_{31} & a_{32} & a_{33} \end{bmatrix} \qquad \begin{bmatrix} a_{11} & a_{12} & a_{13} \\ a_{21} & a_{22} & a_{23} \\ a_{31} & a_{32} & a_{33} \end{bmatrix} \qquad \begin{bmatrix} a_{11} & a_{12} & a_{13} \\ a_{21} & a_{22} & a_{23} \\ a_{31} & a_{32} & a_{33} \end{bmatrix}$$

| **Entry: $a_{11}$** **associated minor** | **Entry: $a_{12}$** **associated minor** | **Entry: $a_{13}$** **associated minor** |
|---|---|---|
| $\begin{bmatrix} a_{22} & a_{23} \\ a_{32} & a_{33} \end{bmatrix}$ | $\begin{bmatrix} a_{21} & a_{23} \\ a_{31} & a_{33} \end{bmatrix}$ | $\begin{bmatrix} a_{21} & a_{22} \\ a_{31} & a_{32} \end{bmatrix}$ |

To illustrate, consider the system shown, and (1) form the matrix of coefficients, (2) identify the minor matrices associated with the entries in the first row, and (3) compute the determinant of each *minor*.

$$\begin{cases} 2x + 3y - z = 1 \\ x - 4y + 2z = -3 \\ 3x + y = -1 \end{cases} \qquad \textbf{(1) Matrix of coefficients } \begin{bmatrix} 2 & 3 & -1 \\ 1 & -4 & 2 \\ 3 & 1 & 0 \end{bmatrix}$$

**(2)** $\begin{bmatrix} 2 & 3 & 1 \\ 1 & -4 & 2 \\ 3 & 1 & 0 \end{bmatrix} \qquad \begin{bmatrix} 2 & 3 & -1 \\ 1 & -4 & 2 \\ 3 & 1 & 0 \end{bmatrix} \qquad \begin{bmatrix} 2 & 3 & -1 \\ 1 & -4 & 2 \\ 3 & 1 & 0 \end{bmatrix}$

| **Entry $a_{11}$: 2** **associated minor** | **Entry $a_{12}$: 3** **associated minor** | **Entry $a_{13}$: $-1$** **associated minor** |
|---|---|---|
| $\begin{bmatrix} -4 & 2 \\ 1 & 0 \end{bmatrix}$ | $\begin{bmatrix} 1 & 2 \\ 3 & 0 \end{bmatrix}$ | $\begin{bmatrix} 1 & -4 \\ 3 & 1 \end{bmatrix}$ |
| **(3) Determinant of minor** | **Determinant of minor** | **Determinant of minor** |
| $\begin{vmatrix} -4 & 2 \\ 1 & 0 \end{vmatrix} = -2$ | $\begin{vmatrix} 1 & 2 \\ 3 & 0 \end{vmatrix} = -6$ | $\begin{vmatrix} 1 & -4 \\ 3 & 1 \end{vmatrix} = 13$ |

For computing a $3 \times 3$ determinant, we illustrate a technique called **expansion by minors.**

**The Determinant of a $3 \times 3$ Matrix—Expansion by Minors**

For the matrix $A$ shown, $\det(A)$ is the unique number computed as follows:

**matrix $A$**

$$\begin{bmatrix} a_{11} & a_{12} & a_{13} \\ a_{21} & a_{22} & a_{23} \\ a_{31} & a_{32} & a_{33} \end{bmatrix}$$

1. Select any row or column and form the product of each entry with its minor matrix. The illustration here uses the entries in row 1:

$$\det(A) = +a_{11}\begin{vmatrix} a_{22} & a_{23} \\ a_{32} & a_{33} \end{vmatrix} - a_{12}\begin{vmatrix} a_{21} & a_{23} \\ a_{31} & a_{33} \end{vmatrix} + a_{13}\begin{vmatrix} a_{21} & a_{22} \\ a_{31} & a_{32} \end{vmatrix}$$

**Sign Chart**

$$\begin{bmatrix} + & - & + \\ - & + & - \\ + & - & + \end{bmatrix}$$

2. The *signs used between terms* of the expansion depend on the row or column chosen, according to the *sign chart* shown.

The determinant of a matrix is unique and *any* row or column can be used. For this reason, it's helpful to select the row or column having the most zero, integer, positive, and/or small entries.

**EXAMPLE 5** ▶ **Calculating a 3 × 3 Determinant**

Compute the determinant of $A = \begin{bmatrix} 2 & 1 & -3 \\ 1 & -1 & 0 \\ -2 & 1 & 4 \end{bmatrix}$.

**Solution** ▶ Since the second row has the "smallest" entries as well as a zero entry, we compute the determinant using this row. According to the sign chart, the signs of the terms will be negative–positive–negative, giving

$$\det(A) = -(1)\begin{vmatrix} 1 & -3 \\ 1 & 4 \end{vmatrix} + (-1)\begin{vmatrix} 2 & -3 \\ -2 & 4 \end{vmatrix} - (0)\begin{vmatrix} 2 & 1 \\ -2 & 1 \end{vmatrix}$$
$$= -1(4 + 3) + (-1)(8 - 6) - (0)(2 + 2)$$
$$= \quad -7 \quad + \quad (-2) \quad - \quad 0$$
$$= -9$$

The value of $\det(A)$ is $-9$.

Now try Exercises 47 through 50 ▶

Try computing the determinant of $A$ two more times, using a different row or column each time. Since the determinant is unique, you should obtain the same result.

There are actually other alternatives for computing a 3 × 3 determinant. The first is called **determinants by column rotation,** and takes advantage of patterns generated from the expansion of minors. This method is applied to the matrix shown, which uses alphabetical entries for simplicity.

$$\det\begin{bmatrix} a & b & c \\ d & e & f \\ g & h & i \end{bmatrix} \begin{matrix} = a(ei - fh) - b(di - fg) + c(dh - eg) \\ = aei - afh - bdi + bfg + cdh - ceg \\ = aei + bfg + cdh - afh - bdi - ceg \end{matrix}$$    expansion using R1   distribute   rewrite result

Although history is unsure of who should be credited, notice that if you repeat the first two columns to the right of the given matrix ("rotation of columns"), identical products are obtained using the six diagonals formed—three in the downward direction using addition, three in the upward direction using subtraction.

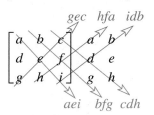

Adding the products in **blue** (regardless of sign) and subtracting the products in **red** (regardless of sign) gives the determinant. This method is more efficient than expansion by minors, *but can only be used for* 3 × 3 *matrices!*

**EXAMPLE 6** ▶ **Calculating det(A) Using Column Rotation**

Use the column rotation method to find the determinant of $A = \begin{bmatrix} 1 & 5 & 3 \\ -2 & -8 & 0 \\ -3 & -11 & 1 \end{bmatrix}$.

**Solution** ▶ Rotate columns 1 and 2 to the right, and compute the diagonal products.

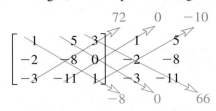

Adding the products in **blue** (regardless of sign) and subtracting the products in **red** (regardless of sign) shows $\det(A) = -4$:

$$-8 + 0 + 66 - 72 - 0 - (-10) = -4.$$

**Now try Exercises 51 through 54** ▶

**Figure 7.10**

The final method is presented in the *Extending the Concept* feature of the Exercise Set, and shows that if certain conditions are met, the determinant of a matrix can be found using its triangularized form.

As with the operations studied in Section 7.2, the process of computing a determinant becomes very cumbersome for larger matrices, or those with rational or radical entries. Most graphing calculators are programmed to handle these computations easily. After accessing the matrix menu (2nd $x^{-1}$), calculating a determinant is the first option under the **MATH** submenu (Figure 7.10). The calculator results for det([A]) as defined in Example 6 is also shown in Figure 7.10. **See Exercises 59 through 62.**

**EXAMPLE 7** ▶ **Solving a System after Verifying A Is Invertible**

Given the system shown here, (1) form the matrix equation $AX = B$; (2) compute the determinant of the coefficient matrix and determine if you can proceed; and (3) if so, solve the system using a matrix equation.

$$\begin{cases} 2x + y - 3z = 11 \\ x - y \quad\quad = 1 \\ -2x + y + 4z = -8 \end{cases}$$

**Solution** ▶ **1.** Form the matrix equation $AX = B$:

$$\begin{bmatrix} 2 & 1 & -3 \\ 1 & -1 & 0 \\ -2 & 1 & 4 \end{bmatrix} \begin{bmatrix} x \\ y \\ z \end{bmatrix} = \begin{bmatrix} 11 \\ 1 \\ -8 \end{bmatrix}$$

**2.** Enter the matrices $A$ and $B$ into the calculator. Since $\det(A)$ is nonzero (from Example 5 and Figure 7.11), we can proceed.

**3.** Enter $A^{-1}B$ on the home screen and press ▭ (Figure 7.12).

**Figure 7.11**    **Figure 7.12**

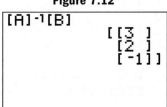

The solution is the ordered triple $(3, 2, -1)$.

**Now try Exercises 63 through 66** ▶

We close this section with an application involving a $4 \times 4$ system. There is a large variety of additional applications in the Exercise Set.

**EXAMPLE 8** ▶    **Solving an Application Using Technology and Matrix Equations**

A local theater sells four sizes of soft drinks: 32 oz @ $4.50, 24 oz @ $3.80, 16 oz @ $3.00, and 12 oz @ $2.40/each. As part of a "free guest pass" promotion, the manager asks employees to try and determine the number of each size sold, given the following: (1) the total revenue from soft drinks was $1439.60; (2) there were 9096 oz of soft drink sold; (3) there were 394 soft drinks sold; and (4) the number of 24-oz and 12-oz drinks sold was 12 more than the number of 32-oz and 16-oz drinks sold. Write a system of equations that models this information, then solve the system using a matrix equation.

**Solution** ▶    If we let $x$, $l$, $m$, and $s$ represent the number of 32-oz, 24-oz, 16-oz, and 12-oz soft drinks sold, the following system is produced:

$$\begin{array}{rl}\text{revenue:} & \left\{\begin{array}{l} 4.5x + 3.8l + 3m + 2.4s = 1439.6 \\ 32x + 24l + 16m + 12s = 9096 \\ x + l + m + s = 394 \\ l + s = x + m + 12 \end{array}\right. \\ \text{ounces sold:} \\ \text{quantity sold:} \\ \text{amounts sold:} \end{array}$$

When written as a matrix equation the system becomes:

$$\begin{bmatrix} 4.5 & 3.8 & 3 & 2.4 \\ 32 & 24 & 16 & 12 \\ 1 & 1 & 1 & 1 \\ -1 & 1 & -1 & 1 \end{bmatrix} \begin{bmatrix} x \\ l \\ m \\ s \end{bmatrix} = \begin{bmatrix} 1439.6 \\ 9096 \\ 394 \\ 12 \end{bmatrix}$$

To solve, carefully enter the matrix of coefficients as matrix $A$ (see Figure 7.13), and the matrix of constants as matrix $B$, then [since $\det(A) \neq 0$] compute $A^{-1}B = X$. This gives a solution of $(x, l, m, s) = (112, 151, 79, 52)$ (Figure 7.14). The theater sold 112 32-oz drinks, 151 24-oz drinks, 79 16-oz drinks, and 52 12-oz drinks.

**Figure 7.13**        **Figure 7.14**

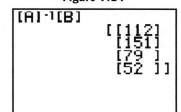

✓ **D.** You've just seen how we can use determinants to find whether a matrix is invertible

Now try Exercises 71 through 88 ▶

## 7.3 EXERCISES

▶ CONCEPTS AND VOCABULARY

**Fill in the blank with the appropriate word or phrase. Carefully reread the section, if necessary.**

1. The $n \times n$ identity matrix $I_n$ consists of 1's down the _____ and _____ for all other entries.

2. If the determinant of a matrix is zero, the matrix is said to be _____ or _____, meaning no inverse exists.

3. Explain why inverses exist only for square matrices, then discuss why some square matrices do not have an inverse. Illustrate each point with an example.

4. What is the connection between the determinant of a $2 \times 2$ matrix and the formula for finding its inverse? Use the connection to create a $2 \times 2$ matrix that is invertible, and another that is not.

▶ DEVELOPING YOUR SKILLS

**Use matrix multiplication, equality of matrices, and the arbitrary matrix given to show that** $\begin{bmatrix} a & b \\ c & d \end{bmatrix} = \begin{bmatrix} 1 & 0 \\ 0 & 1 \end{bmatrix}$.

5. $A = \begin{bmatrix} 2 & 5 \\ -3 & -7 \end{bmatrix} \begin{bmatrix} a & b \\ c & d \end{bmatrix} = \begin{bmatrix} 2 & 5 \\ -3 & -7 \end{bmatrix}$

6. $A = \begin{bmatrix} 9 & -7 \\ -5 & 4 \end{bmatrix} \begin{bmatrix} a & b \\ c & d \end{bmatrix} = \begin{bmatrix} 9 & -7 \\ -5 & 4 \end{bmatrix}$

7. $A = \begin{bmatrix} a & b \\ c & d \end{bmatrix} \begin{bmatrix} 0.4 & 0.6 \\ 0.3 & 0.2 \end{bmatrix} = \begin{bmatrix} 0.4 & 0.6 \\ 0.3 & 0.2 \end{bmatrix}$

8. $A = \begin{bmatrix} a & b \\ c & d \end{bmatrix} \begin{bmatrix} \frac{1}{2} & \frac{1}{4} \\ \frac{1}{3} & \frac{1}{8} \end{bmatrix} = \begin{bmatrix} \frac{1}{2} & \frac{1}{4} \\ \frac{1}{3} & \frac{1}{8} \end{bmatrix}$

**Show $AI = IA = A$ for each matrix given.**

9. $\begin{bmatrix} -3 & 8 \\ -4 & 10 \end{bmatrix}$

10. $\begin{bmatrix} 0.5 & -0.2 \\ -0.7 & 0.3 \end{bmatrix}$

11. $\begin{bmatrix} -4 & 1 & 6 \\ 9 & 5 & 3 \\ 0 & -2 & 1 \end{bmatrix}$

12. $\begin{bmatrix} 9 & 1 & 3 & -1 \\ 2 & 0 & -5 & 3 \\ 4 & 6 & 1 & 0 \\ 0 & -2 & 4 & 1 \end{bmatrix}$

**Find the inverse of each $2 \times 2$ matrix using matrix multiplication, equality of matrices, and a system of equations.**

13. $\begin{bmatrix} 5 & -4 \\ 2 & 2 \end{bmatrix}$

14. $\begin{bmatrix} 1 & -5 \\ 0 & -4 \end{bmatrix}$

**Find the inverse of each matrix by augmenting the identity matrix and using row operations.**

15. $\begin{bmatrix} 1 & -3 \\ 4 & -10 \end{bmatrix}$

16. $\begin{bmatrix} -2 & 0.4 \\ 1 & 0.8 \end{bmatrix}$

**Demonstrate that $B = A^{-1}$, by showing $AB = BA = I$. Do not use a calculator.**

17. $A = \begin{bmatrix} 1 & 5 \\ -2 & -9 \end{bmatrix}$

$B = \begin{bmatrix} -9 & -5 \\ 2 & 1 \end{bmatrix}$

18. $A = \begin{bmatrix} -2 & -6 \\ 4 & 11 \end{bmatrix}$

$B = \begin{bmatrix} 5.5 & 3 \\ -2 & -1 \end{bmatrix}$

19. $A = \begin{bmatrix} 4 & -5 \\ 0 & 2 \end{bmatrix}$

$B = \begin{bmatrix} \frac{1}{4} & \frac{5}{8} \\ 0 & \frac{1}{2} \end{bmatrix}$

20. $A = \begin{bmatrix} -2 & 5 \\ 3 & -4 \end{bmatrix}$

$B = \begin{bmatrix} \frac{4}{7} & \frac{5}{7} \\ \frac{3}{7} & \frac{2}{7} \end{bmatrix}$

**Use a calculator to find $A^{-1} = B$, then confirm the inverse by showing $AB = BA = I$.**

21. $A = \begin{bmatrix} -2 & 3 & 1 \\ 5 & 2 & 4 \\ 2 & 0 & -1 \end{bmatrix}$

22. $A = \begin{bmatrix} 0.5 & 0.2 & 0.1 \\ 0 & 0.3 & 0.6 \\ 1 & 0.4 & -0.3 \end{bmatrix}$

23. $A = \begin{bmatrix} -7 & 5 & -3 \\ 1 & 9 & 0 \\ 2 & -2 & -5 \end{bmatrix}$

24. $A = \dfrac{1}{12} \begin{bmatrix} 12 & -6 & 3 & 0 \\ 0 & -4 & 8 & -12 \\ 12 & -12 & 0 & 0 \\ 0 & 12 & 0 & -12 \end{bmatrix}$

**Write each system in the form of a matrix equation. Do not solve.**

**25.** $\begin{cases} 2x - 3y = 9 \\ -5x + 7y = 8 \end{cases}$

**26.** $\begin{cases} 0.5x - 0.6y = \quad\;\; 0.6 \\ -0.7x + 0.4y = -0.375 \end{cases}$

**27.** $\begin{cases} x + 2y - z = 1 \\ x \qquad\;\; + z = 3 \\ 2x - \;\; y + z = 3 \end{cases}$

**28.** $\begin{cases} 2x - \;\; 3y - 2z = \;\; 4 \\ \frac{1}{4}x - \frac{2}{5}y + \frac{3}{4}z = \frac{-1}{3} \\ -2x + 1.3y - 3z = \;\; 5 \end{cases}$

**29.** $\begin{cases} -2w + \;\; x - 4y + 5z = -3 \\ 2w - 5x + \;\; y - 3z = \;\; 4 \\ -3w + \;\; x + 6y + \;\; z = \;\; 1 \\ w + 4x - 5y + \;\; z = -9 \end{cases}$

**30.** $\begin{cases} 1.5w + 2.1x - 0.4y + \quad\; z = 1 \\ 0.2w - 2.6x + \quad\; y \qquad\quad = 5.8 \\ \qquad\qquad\; 3.2x \qquad\;\; + z = 2.7 \\ 1.6w + \;\; 4x - \;\; 5y + 2.6z = -1.8 \end{cases}$

 **Write each system as a matrix equation and solve (if possible) using inverse matrices and your calculator. If the coefficient matrix is singular, write *no solution using matrix equations*.**

**31.** $\begin{cases} 0.05x - 3.2y = -15.8 \\ 0.02x + 2.4y = \;\; 12.08 \end{cases}$

**32.** $\begin{cases} 0.3x + 1.1y = \quad\; 3.5 \\ -0.5x - 2.9y = -10.1 \end{cases}$

**33.** $\begin{cases} \frac{-1}{6}u + \frac{1}{4}v = \quad\; 1 \\ \frac{1}{2}u - \frac{2}{3}v = -2 \end{cases}$

**34.** $\begin{cases} \sqrt{2}a + \sqrt{3}b = \sqrt{5} \\ \sqrt{6}a + \quad 3b = \sqrt{7} \end{cases}$

**35.** $\begin{cases} \frac{-1}{8}a + \frac{3}{5}b = \frac{5}{6} \\ \frac{5}{16}a - \frac{3}{2}b = \frac{-4}{5} \end{cases}$

**36.** $\begin{cases} 3\sqrt{2}a + 2\sqrt{3}b = 12 \\ 5\sqrt{2}a - 3\sqrt{3}b = \;\; 1 \end{cases}$

**37.** $\begin{cases} 0.2x - 1.6y + \;\; 2z = -1.9 \\ -0.4x - \quad y + 0.6z = \;\; -1 \\ 0.8x + 3.2y - 0.4z = \;\; 0.2 \end{cases}$

**38.** $\begin{cases} 1.7x + 2.3y - \quad 2z = 41.5 \\ 1.4x - 0.9y + 1.6z = -10 \\ -0.8x + 1.8y - 0.5z = 16.5 \end{cases}$

**39.** $\begin{cases} x - \quad 2y + \;\; 2z = \;\; 6 \\ 2x - 1.5y + 1.8z = \;\; 2.8 \\ \frac{-2}{3}x + \;\; \frac{1}{2}y - \;\; \frac{3}{5}z = -\frac{11}{30} \end{cases}$

**40.** $\begin{cases} 4x - \quad 5y - 6z = 5 \\ \frac{1}{8}x - \frac{3}{5}y + \frac{5}{4}z = \frac{-2}{3} \\ -0.5x + 2.4y - 5z = 5 \end{cases}$

**41.** $\begin{cases} -2w + \;\; 3x - \;\; 4y + \;\; 5z = \;\; -3 \\ 0.2w - 2.6x + \quad y - 0.4z = \;\; 2.4 \\ -3w + 3.2x + 2.8y + \quad z = \;\; 6.1 \\ 1.6w + \;\; 4x - \;\; 5y + 2.6z = -9.8 \end{cases}$

**42.** $\begin{cases} 2w - \;\; 5x + \;\; 3y - \;\; 4z = 7 \\ 1.6w \qquad\quad + 4.2y - 1.8z = 5.4 \\ 3w + 6.7x - \;\; 9y + \;\; 4z = -8.5 \\ \qquad\quad 0.7x \qquad\;\; - 0.9z = 0.9 \end{cases}$

**Compute the determinant of each matrix and state whether an inverse matrix exists. Do not use a calculator.**

**43.** $\begin{bmatrix} 4 & -7 \\ 3 & -5 \end{bmatrix}$        **44.** $\begin{bmatrix} 0.6 & 0.3 \\ 0.4 & 0.5 \end{bmatrix}$

**45.** $\begin{bmatrix} 1.2 & -0.8 \\ 0.3 & -0.2 \end{bmatrix}$      **46.** $\begin{bmatrix} -2 & 6 \\ -3 & 9 \end{bmatrix}$

**Compute the determinant of each matrix without using a calculator. If the determinant is zero, write *singular matrix*.**

**47.** $A = \begin{bmatrix} 1 & 0 & -2 \\ 0 & -1 & -1 \\ 2 & 1 & -4 \end{bmatrix}$   **48.** $B = \begin{bmatrix} -2 & 2 & 1 \\ 0 & -1 & 2 \\ 4 & -4 & 0 \end{bmatrix}$

**49.** $C = \begin{bmatrix} -2 & 3 & 4 \\ 0 & 6 & 2 \\ 1 & -1.5 & -2 \end{bmatrix}$

**50.** $D = \begin{bmatrix} 1 & 2 & -0.8 \\ 2.5 & 5 & -2 \\ 3 & 0 & -2.5 \end{bmatrix}$

**Compute the determinant of each matrix using the column rotation method.**

**51.** $\begin{bmatrix} 2 & -3 & 1 \\ 4 & -1 & 5 \\ 1 & 0 & -2 \end{bmatrix}$    **52.** $\begin{bmatrix} -3 & 2 & 4 \\ -1 & -2 & 0 \\ 3 & 1 & 5 \end{bmatrix}$

**53.** $\begin{bmatrix} 1 & -1 & 2 \\ 3 & -2 & 4 \\ 4 & 3 & 1 \end{bmatrix}$    **54.** $\begin{bmatrix} 5 & 6 & 2 \\ -2 & 1 & -2 \\ 3 & 4 & -1 \end{bmatrix}$

**Use determinants to solve the following equations.**

**55.** $\begin{vmatrix} -x & -9 \\ -2 & 3-x \end{vmatrix} = 0$    **56.** $\begin{vmatrix} 2-x & 3 \\ 6 & 5-x \end{vmatrix} = 0$

**57.** $\begin{vmatrix} 1-x & -9 & 0 \\ 0 & -2-x & 1 \\ 0 & 0 & -x \end{vmatrix} = 0$

**58.** $\begin{vmatrix} 4-x & 0 & 0 \\ 4 & 3-x & 0 \\ -5 & 3 & -1-x \end{vmatrix} = 0$

Use a calculator to compute the determinant of each matrix. If the determinant is zero, write *singular matrix*. If the determinant is nonzero, find $A^{-1}$ and store the result as matrix $B$ ( **STO▸** **2nd** **x⁻¹** 2: [B] **ENTER** ). Then verify the inverse by showing $AB = BA = I$.

59. $A = \begin{bmatrix} 1 & 0 & 3 & -4 \\ 2 & 5 & 0 & 1 \\ 8 & 15 & 6 & -5 \\ 0 & 8 & -4 & 1 \end{bmatrix}$

60. $B = \begin{bmatrix} 2 & 1 & -3 & 5 \\ -3 & 0 & 4 & -2 \\ 1 & 5 & 0 & -3 \\ 0 & -12 & -2 & 0 \end{bmatrix}$

61. $N = \begin{bmatrix} 0 & -2 & 0 & 3 \\ -1 & 1 & -2 & 0 \\ 1 & 2 & -2 & -1 \\ 1 & -1 & 1 & 1 \end{bmatrix}$

62. $M = \begin{bmatrix} 1 & 2 & 1 & 1 \\ 0 & 1 & -3 & 2 \\ -1 & 0 & 2 & -3 \\ 2 & -1 & 1 & 4 \end{bmatrix}$

 For each system shown, (1) form the matrix equation $AX = B$; (2) compute the determinant of the coefficient matrix and determine if you can proceed; and (3) if possible, solve the system using the matrix equation.

63. $\begin{cases} x - 2y + 2z = 7 \\ 2x + 2y - z = 5 \\ 3x - y + z = 6 \end{cases}$

64. $\begin{cases} 2x - 3y - 2z = 7 \\ x - y + 2z = -5 \\ 3x + 2y - z = 11 \end{cases}$

65. $\begin{cases} x - 3y + 4z = -1 \\ 4x - y + 5z = 7 \\ 3x + 2y + z = -3 \end{cases}$

66. $\begin{cases} 5x - 2y + z = 1 \\ 3x - 4y + 9z = -2 \\ 4x - 3y + 5z = 6 \end{cases}$

## ▶ WORKING WITH FORMULAS

The inverse of a 2 × 2 matrix: $A = \begin{bmatrix} a & b \\ c & d \end{bmatrix} \Rightarrow A^{-1} = \dfrac{1}{ad - bc} \begin{bmatrix} d & -b \\ -c & a \end{bmatrix}$

The inverse of a 2 × 2 matrix can be found using the formula shown, as long as $ad - bc \neq 0$. Use the formula to find inverses for the matrices here (if possible), then verify by showing $AA^{-1} = A^{-1}A = I$.

67. $A = \begin{bmatrix} 3 & -5 \\ 2 & 1 \end{bmatrix}$

68. $A = \begin{bmatrix} 2 & 3 \\ -5 & -4 \end{bmatrix}$

69. $A = \begin{bmatrix} 0.3 & -0.4 \\ -0.6 & 0.8 \end{bmatrix}$

70. $A = \begin{bmatrix} 0.2 & 0.3 \\ -0.4 & -0.6 \end{bmatrix}$

## ▶ APPLICATIONS

Solve each application using a matrix equation.

 **Descriptive Translation**

71. **Convenience store sales:** The local Moto-Mart sells four different sizes of Slushies—behemoth, 60 oz @ $2.59; gargantuan, 48 oz @ $2.29; mammoth, 36 oz @ $1.99; and jumbo, 24 oz @ $1.59. As part of a promotion, the owner offers free gas to any customer who can tell how many of each size were sold last week, given the following: (1) the total revenue for the Slushies was $402.29, (2) 7884 oz were sold, (3) 191 Slushies were sold, and (4) the number of behemoth Slushies sold was one more than the number of jumbo. How many of each size were sold?

72. **Cartoon characters:** In America, four of the most beloved cartoon characters are Foghorn Leghorn, Elmer Fudd, Bugs Bunny, and Tweety Bird. Suppose that Bugs Bunny is four times as tall as Tweety Bird. Elmer Fudd is as tall as the combined height of Bugs Bunny and Tweety Bird. Foghorn Leghorn is 20 cm taller than the combined height of Elmer Fudd and Tweety Bird. The combined height of all four characters is 500 cm. How tall is each one?

73. **Rolling Stones music:** One of the most prolific and popular rock-and-roll bands of all time is the Rolling Stones. Four of their many great hits include: *Jumpin' Jack Flash, Tumbling Dice, You Can't Always Get What You Want,* and *Wild Horses.* The total playing time of all four songs is 20.75 min. The combined playing time of *Jumpin' Jack Flash* and *Tumbling Dice* equals that of *You Can't Always Get What You Want. Wild Horses* is 2 min longer than *Jumpin' Jack Flash,* and *You Can't Always Get What You Want* is twice as long as *Tumbling Dice.* Find the playing time of each song.

74. **Mozart's arias:** Mozart wrote some of vocal music's most memorable arias in his operas, including *Tamino's Aria, Papageno's Aria,* the *Champagne Aria,* and the *Catalogue Aria.* The total playing time of all four arias is 14.3 min. *Papageno's Aria* is 3 min shorter than the *Catalogue Aria.* The *Champagne Aria* is 2.7 min shorter than *Tamino's Aria.* The combined time of *Tamino's Aria* and *Papageno's Aria* is five times that of the *Champagne Aria.* Find the playing time of all four arias.

## Manufacturing

**75. Resource allocation:** Time Pieces Inc. manufactures four different types of grandfather clocks. Each clock requires these four stages: (1) assembly, (2) installing the clockworks, (3) inspection and testing, and (4) packaging for delivery. The number of hours required for each stage is shown in the table for each of the four clock types. At the end of a busy week, the owner determines that personnel on the assembly line worked for 262 hr, the installation crews for 160 hr, the testing department for 29 hr, and the packaging department for 68 hr. How many clocks of each type were made?

| Dept. | Clock A | Clock B | Clock C | Clock D |
|---|---|---|---|---|
| Assemble | 2.2 | 2.5 | 2.75 | 3 |
| Install | 1.2 | 1.4 | 1.8 | 2 |
| Test | 0.2 | 0.25 | 0.3 | 0.5 |
| Pack | 0.5 | 0.55 | 0.75 | 1.0 |

**76. Resource allocation:** Figurines Inc. makes and sells four sizes of metal figurines, mostly historical figures and celebrities. Each figurine goes through four stages of development: (1) casting, (2) trimming, (3) polishing, and (4) painting. The number of hours required for each stage is shown in the table for each of the four sizes. At the end of a busy week, the manager finds that the casting department put in 62 hr, and the trimming department worked for 93.5 hr, with the polishing and painting departments logging 138 hr and 358 hr respectively. How many figurines of each type were made?

| Dept. | Small | Medium | Large | X-Large |
|---|---|---|---|---|
| Casting | 0.5 | 0.6 | 0.75 | 1 |
| Trimming | 0.8 | 0.9 | 1.1 | 1.5 |
| Polishing | 1.2 | 1.4 | 1.7 | 2 |
| Painting | 2.5 | 3.5 | 4.5 | 6 |

**77. Thermal conductivity:** In lab experiments designed to measure the heat conductivity of a square metal plate of uniform density, the edges are held at four different (constant) temperatures. The *mean-value principle* from physics tells us that the temperature at a given point $p_i$ on the plate is equal to the average temperature of nearby points. Use this information to form a system of four equations in four variables, and determine the temperature at interior points $p_1, p_2, p_3,$ and $p_4$ on the plate shown. (*Hint:* Use the temperature of the four points closest to each.)

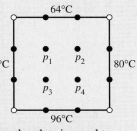

**78. Thermal conductivity:** Repeat Exercise 77 if (a) the temperatures at the top and bottom of the plate were *increased* by 10°, with the temperatures at the left and right edges *decreased* by 10° (what do you notice?); (b) the temperature at the top and left were *decreased* by 10°, with the temperatures at the bottom and right held at their original temperature.

## Curve Fitting

**79. Quadratic fit:** Use a matrix equation to find a quadratic function of the form $y = ax^2 + bx + c$ such that $(-4, -5), (0, -5),$ and $(2, 7)$ are on the graph of the function.

**80. Quadratic fit:** Use a matrix equation to find a quadratic function of the form $f(x) = ax^2 + bx + c$ matching the graph given.

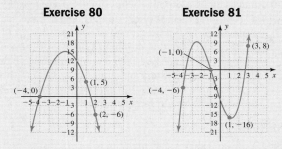

**81. Cubic fit:** Use a matrix equation to find a cubic function of the form $g(x) = ax^3 + bx^2 + cx + d$ matching the graph given.

**82. Cubic fit:** Use a matrix equation to find a cubic function of the form $y = ax^3 + bx^2 + cx + d$ such that $(-2, 5), (0, 1), (2, -3),$ and $(3, 25)$ are on the graph of the function.

## Investing

**83. Wise investing:** Morgan received an $800 gift from her grandfather, and showing wisdom beyond her years, decided to place the money in a certificate of deposit (CD) and a money market fund (MM). At the time, CDs were earning 3.5% and MMs were earning 2.5%. At the end of 1 yr, she cashed both in and received a total of $824.50. How much was deposited in each?

**84. Baseball cards:** Gary has a passion for baseball, which includes a collection of rare baseball cards. His most prized cards feature Willie Mays (1953 Topps) and Mickey Mantle (1959 Topps). The Willie Mays card has appreciated 28% and the Mickey Mantle card 25% since he purchased them, and together they are now worth $17,100. If he paid a total of $13,507.50 at auction for both cards, what was the original price of each?

**85. Retirement planning:** Using payroll deduction, Jeanette was able to put aside $4800 last year for her impending retirement. Last year, her company retirement fund paid 4.2% and her mutual funds returned 5.75%, but her stock fund actually lost 2.5% in value. If her net gain for the year was $104.25 and $300 more was placed in stocks than in mutual funds, how much was placed in each investment vehicle?

**86. Charitable giving:** The hyperbolic funnels seen at many shopping malls are primarily used by nonprofit organizations to raise funds for worthy causes. A coin is launched down a ramp into the funnel and seemingly makes endless circuits before finally disappearing down a "black hole" (the collection bin). During one such collection, the bin was found to hold $112.89, and 1450 coins consisting of pennies, nickels, dimes, and quarters. How many of each denomination were there, if the number of quarters and dimes was equal to the number of nickels, and the number of pennies was twice the number of quarters?

## Nutrition

**87. Animal diets:** A zoo dietitian needs to create a specialized diet that regulates an animal's intake of fat, carbohydrates, and protein. The table given shows three different foods and the amount of these nutrients (in grams) that each ounce of food provides. How many ounces of each should the dietitian recommend

to supply 20 g of fat, 30 g of carbohydrates, and 44 g of protein?

| Nutrient | Food I | Food II | Food III |
|----------|--------|---------|----------|
| Fat      | 2      | 4       | 3        |
| Carb.    | 4      | 2       | 5        |
| Protein  | 5      | 6       | 7        |

**88. Training diet:** A dietitian is designing a prerace meal for one of her clients, an elite swimmer. She knows he'll need roughly 24 g of fat, 244 g of carbohydrates, and 40 g of protein. The table given shows three different foods and the amount of these nutrients (in grams) that each ounce of food provides. How many ounces of each should she recommend?

| Nutrient | Food I | Food II | Food III |
|----------|--------|---------|----------|
| Fat      | 2      | 5       | 0        |
| Carb.    | 10     | 15      | 18       |
| Protein  | 2      | 10      | 0.75     |

## ▶ EXTENDING THE CONCEPT

**89.** Suppose $A = \begin{bmatrix} a & b \\ c & d \end{bmatrix}$ and $B = \begin{bmatrix} e & f \\ g & h \end{bmatrix}$ are invertible matrices and $k \neq 1$ is a nonzero constant. Verify the following properties: (a) $|AB| = |A||B|$, (b) $|A^{-1}| = |A|^{-1}$, and (c) $|kA| \neq k|A|$. Note these properties actually hold for *all* square matrices.

**90.** Another alternative for finding determinants uses the triangularized form of a matrix and is offered without proof: *If nonsingular matrix A is written in triangularized form without scaling or reordering any row ($R_i \leftrightarrow R_j$ and $kR_i \rightarrow R_i$ are not permitted, but $kR_i + R_j \rightarrow R_j$ is), then det(A) is equal to the product of resulting diagonal entries.* Compute the determinant of each matrix using this method. Be careful not to interchange rows or replace any row by a multiple of itself in the process.

a. $\begin{bmatrix} 1 & -2 & 3 \\ -4 & 5 & -6 \\ 2 & 5 & 3 \end{bmatrix}$    b. $\begin{bmatrix} 2 & 5 & -1 \\ -2 & -3 & 4 \\ 4 & 6 & 5 \end{bmatrix}$    c. $\begin{bmatrix} -2 & 4 & 1 \\ 5 & 7 & -2 \\ 3 & -8 & -1 \end{bmatrix}$    d. $\begin{bmatrix} 3 & -1 & 4 \\ 0 & -2 & 6 \\ -2 & 1 & -3 \end{bmatrix}$

▶ **MAINTAINING YOUR SKILLS**

**91.** (4.3) Solve using the rational zeroes theorem:
$x^3 - 7x^2 = -36$.

**92.** (3.1/5.3) Match each equation to its related graph. Justify your answers.

_____ $y = \log_2 (x - 2)$     _____ $y = \log_2 x - 2$
**a.**                       **b.**

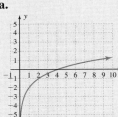
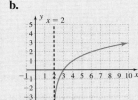

**93.** (1.3) Solve the absolute value inequality:
$-3|2x + 5| - 7 \le -19$.

**94.** (3.3) A coin collector believes that the value of a coin varies inversely with the number of coins still in circulation. If 4 million coins are in circulation, the coin has a value of $25. (a) Find the variation equation and (b) determine how many coins are in circulation if the value of the coin is $6.25.

---

## 7.4   Applications of Matrices and Determinants: Cramer's Rule, Partial Fractions, and More

**LEARNING OBJECTIVES**

*In Section 7.4 you will see how we can:*

☐ **A.** Solve a system using determinants and Cramer's rule

☐ **B.** Decompose a rational expression into partial fractions

☐ **C.** Use determinants in applications involving geometry in the coordinate plane

Besides solving systems, matrices can be used to accomplish such diverse things as finding the volume of a three-dimensional solid to checking whether solutions to a differential equation are linearly independent.

### A. Solving Systems Using Determinants and Cramer's Rule

In addition to identifying singular matrices, determinants can actually be used to *develop a formula* for the solution of a system. Consider the following solution to a *general* 2 × 2 system, which parallels the solution to a specific 2 × 2 system. With a view toward a solution involving determinants, the coefficients of $x$ are written as $a_{11}$ and $a_{21}$ in the general system, and the coefficients of $y$ are $a_{12}$ and $a_{22}$.

<table>
<tr><td align="center">**Specific System**</td><td align="center">**General System**</td></tr>
<tr><td align="center">$\begin{cases} 2x + 5y = 9 \\ 3x + 4y = 10 \end{cases}$</td><td align="center">$\begin{cases} a_{11}x + a_{12}y = c_1 \\ a_{21}x + a_{22}y = c_2 \end{cases}$</td></tr>
<tr><td align="center">eliminate the *x*-term<br>$-3R1 + 2R2$</td><td align="center">eliminate the *x*-term<br>$-a_{21}R1 + a_{11}R2$</td></tr>
</table>

sums to zero                                  sums to zero

$$\begin{aligned} -3 \cdot 2x - 3 \cdot 5y &= -3 \cdot 9 \\ 2 \cdot 3x + 2 \cdot 4y &= 2 \cdot 10 \\ \hline 2 \cdot 4y - 3 \cdot 5y &= 2 \cdot 10 - 3 \cdot 9 \end{aligned}$$

$$\begin{aligned} -a_{21}a_{11}x - a_{21}a_{12}y &= -a_{21}c_1 \\ a_{11}a_{21}x + a_{11}a_{22}y &= a_{11}c_2 \\ \hline a_{11}a_{22}y - a_{21}a_{12}y &= a_{11}c_2 - a_{21}c_1 \end{aligned}$$

Notice the $x$-terms sum to zero in both systems. We are deliberately leaving the solution on the left unsimplified to show the pattern developing on the right. Next we solve for $y$.

| **Factor out $y$ and solve:** | **Factor out $y$ and solve:** |
|---|---|

$$(2 \cdot 4 - 3 \cdot 5)y = 2 \cdot 10 - 3 \cdot 9$$

$$(a_{11}a_{22} - a_{21}a_{12})y = a_{11}c_2 - a_{21}c_1$$

divide $\quad y = \dfrac{2 \cdot 10 - 3 \cdot 9}{2 \cdot 4 - 3 \cdot 5}$

divide $\quad y = \dfrac{a_{11}c_2 - a_{21}c_1}{a_{11}a_{22} - a_{21}a_{12}}$

On the left we find $y = \frac{-7}{-7} = 1$ and back-substitution shows $x = 2$. But more important, on the right we obtain a formula for the $y$-value of *any* $2 \times 2$ system: $y = \frac{a_{11}c_2 - a_{21}c_1}{a_{11}a_{22} - a_{21}a_{12}}$. If we had chosen to solve for $x$, the solution would be $x = \frac{a_{22}c_1 - a_{12}c_2}{a_{11}a_{22} - a_{21}a_{12}}$. Note these formulas are defined only if $a_{11}a_{22} - a_{21}a_{12} \neq 0$.

You may have already noticed, but this denominator is the *determinant of the coefficients matrix* $\begin{bmatrix} a_{11} & a_{12} \\ a_{21} & a_{22} \end{bmatrix}$ from the previous section! Since the numerator is also a difference of two products, we investigate the possibility that it too can be expressed as a determinant. Working backward, we're able to reconstruct the numerator for $x$ in determinant form as $\begin{vmatrix} c_1 & a_{12} \\ c_2 & a_{22} \end{vmatrix}$, where it is apparent this determinant was formed by *replacing the coefficients of the x-variables with the constant terms.*

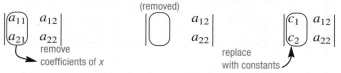

It is also apparent the numerator for $y$ can be also written in determinant form as $\begin{vmatrix} a_{11} & c_1 \\ a_{21} & c_2 \end{vmatrix}$, or the determinant formed by *replacing the coefficients of the y-variables with the constant terms*:

If we use the notation $D_y$ for this determinant, $D_x$ for the determinant where $x$ coefficients were replaced by the constants, and $D$ as the determinant for the coefficients matrix—the solutions can be written as shown next, with the result known as **Cramer's rule.**

---

**Cramer's Rule for 2 × 2 Systems**

Given a $2 \times 2$ system of linear equations

$$\begin{cases} a_{11}x + a_{12}y = c_1 \\ a_{21}x + a_{22}y = c_2 \end{cases}$$

the solution (if one exists) is the ordered pair $(x, y)$, where

$$x = \frac{D_x}{D} = \frac{\begin{vmatrix} c_1 & a_{12} \\ c_2 & a_{22} \end{vmatrix}}{\begin{vmatrix} a_{11} & a_{12} \\ a_{21} & a_{22} \end{vmatrix}} \quad \text{and} \quad y = \frac{D_y}{D} = \frac{\begin{vmatrix} a_{11} & c_1 \\ a_{21} & c_2 \end{vmatrix}}{\begin{vmatrix} a_{11} & a_{12} \\ a_{21} & a_{22} \end{vmatrix}}$$

provided $D \neq 0$.

**EXAMPLE 1** ▶ **Solving a System Using Cramer's Rule**

Use Cramer's rule to solve the system $\begin{cases} 2x - 5y = 9 \\ -3x + 4y = -10 \end{cases}$.

**Solution** ▶ Begin by finding the value of $D$, $D_x$, and $D_y$.

$$D = \begin{vmatrix} 2 & -5 \\ -3 & 4 \end{vmatrix} = (2)(4) - (-3)(-5) = -7$$

$$D_x = \begin{vmatrix} 9 & -5 \\ -10 & 4 \end{vmatrix} = (9)(4) - (-10)(-5) = -14$$

$$D_y = \begin{vmatrix} 2 & 9 \\ -3 & -10 \end{vmatrix} = (2)(-10) - (-3)(9) = 7$$

This gives $x = \frac{D_x}{D} = \frac{-14}{-7} = 2$ and $y = \frac{D_y}{D} = \frac{7}{-7} = -1$. The solution is $(2, -1)$. Check by substituting these values into the original equations.

**Now try Exercises 5 through 12** ▶

Regardless of the method used to solve a system, always be aware that a system could be consistent and independent (unique solution), consistent and dependent (infinitely many solutions), or inconsistent (no solution). The system $\begin{cases} y - 2x = -3 \\ 4x + 6 = 2y \end{cases}$ yields $\begin{cases} -2x + y = -3 \\ 4x - 2y = -6 \end{cases}$ in standard form, with $D = \begin{vmatrix} -2 & 1 \\ 4 & -2 \end{vmatrix} = (-2)(-2) - (4)(1) = 0$. Since $\det(D) = 0$, Cramer's rule cannot be applied, and the system is either inconsistent or dependent. To find out which, we write the equations in function form (solve for $y$). The result is $\begin{cases} y = 2x - 3 \\ y = 2x + 3 \end{cases}$, showing the system consists of two parallel lines and has no solutions.

## Cramer's Rule for 3 × 3 Systems

Cramer's rule can be extended to a 3 × 3 system of linear equations, using the same pattern as for 2 × 2 systems. Given the general 3 × 3 system

$$\begin{cases} a_{11}x + a_{12}y + a_{13}z = c_1 \\ a_{21}x + a_{22}y + a_{23}z = c_2 \\ a_{31}x + a_{32}y + a_{33}z = c_3 \end{cases}$$

the solutions are $x = \frac{D_x}{D}$, $y = \frac{D_y}{D}$, and $z = \frac{D_z}{D}$, where $D_x$, $D_y$, and $D_z$ are again formed by replacing the coefficients of the indicated variable with the constants, and $D$ is the determinant of the coefficient matrix ($D \neq 0$).

### Cramer's Rule Applied to 3 × 3 Systems

Given a 3 × 3 system of linear equations

$$\begin{cases} a_{11}x + a_{12}y + a_{13}z = c_1 \\ a_{21}x + a_{22}y + a_{23}z = c_2 \\ a_{31}x + a_{32}y + a_{33}z = c_3 \end{cases}$$

the solution (if one exists) is the ordered triple $(x, y, z)$, where

$$x = \frac{D_x}{D} = \frac{\begin{vmatrix} c_1 & a_{12} & a_{13} \\ c_2 & a_{22} & a_{23} \\ c_3 & a_{32} & a_{33} \end{vmatrix}}{\begin{vmatrix} a_{11} & a_{12} & a_{13} \\ a_{21} & a_{22} & a_{23} \\ a_{31} & a_{32} & a_{33} \end{vmatrix}} \qquad y = \frac{D_y}{D} = \frac{\begin{vmatrix} a_{11} & c_1 & a_{13} \\ a_{21} & c_2 & a_{23} \\ a_{31} & c_3 & a_{33} \end{vmatrix}}{\begin{vmatrix} a_{11} & a_{12} & a_{13} \\ a_{21} & a_{22} & a_{23} \\ a_{31} & a_{32} & a_{33} \end{vmatrix}}$$

$$z = \frac{D_z}{D} = \frac{\begin{vmatrix} a_{11} & a_{12} & c_1 \\ a_{21} & a_{22} & c_2 \\ a_{31} & a_{32} & c_3 \end{vmatrix}}{\begin{vmatrix} a_{11} & a_{12} & a_{13} \\ a_{21} & a_{22} & a_{23} \\ a_{31} & a_{32} & a_{33} \end{vmatrix}},$$

provided $D \neq 0$.

**EXAMPLE 2** ▶ **Solving a 3 × 3 System Using Cramer's Rule**

Solve using Cramer's rule: $\begin{cases} x - 2y + 3z = -1 \\ -2x + y - 5z = 1. \\ 3x + 3y + 4z = 2 \end{cases}$

**Solution** ▶ Begin by computing the determinant of the coefficient matrix, to ensure that Cramer's rule can be applied. Using the third row, we have

$$D = \begin{vmatrix} 1 & -2 & 3 \\ -2 & 1 & -5 \\ 3 & 3 & 4 \end{vmatrix} = +3\begin{vmatrix} -2 & 3 \\ 1 & -5 \end{vmatrix} - 3\begin{vmatrix} 1 & 3 \\ -2 & -5 \end{vmatrix} + 4\begin{vmatrix} 1 & -2 \\ -2 & 1 \end{vmatrix}$$

$$= 3(7) - 3(1) + 4(-3) = 6$$

Since $D \neq 0$ we continue, electing to compute the remaining determinants using a calculator.

$$D_x = \begin{vmatrix} -1 & -2 & 3 \\ 1 & 1 & -5 \\ 2 & 3 & 4 \end{vmatrix} = 12, \; D_y = \begin{vmatrix} 1 & -1 & 3 \\ -2 & 1 & -5 \\ 3 & 2 & 4 \end{vmatrix} = 0, \; D_z = \begin{vmatrix} 1 & -2 & -1 \\ -2 & 1 & 1 \\ 3 & 3 & 2 \end{vmatrix} = -6$$

The solution is $x = \frac{D_x}{D} = \frac{12}{6} = 2$, $y = \frac{D_y}{D} = \frac{0}{6} = 0$, and $z = \frac{D_z}{D} = \frac{-6}{6} = -1$, or $(2, 0, -1)$ in triple form. Check this solution in the original equations.

 **A.** You've just seen how we can solve a system using determinants and Cramer's rule

Now try Exercises 13 through 22 ▶

## B. Rational Expressions and Partial Fractions

Recall that a rational expression is one of the form $\frac{P(x)}{Q(x)}$, where $P$ and $Q$ are polynomials and $Q(x) \neq 0$. The addition of rational expressions is widely taught in courses prior to college algebra, and involves combining two rational expressions into a single term using a common denominator. In some applications of higher mathematics, we seek to reverse this process and *decompose* a rational expression into a sum of its **partial fractions.** To begin, we make the following observations:

1.  Consider the sum $\frac{7}{x+2} + \frac{5}{x-3}$, noting both terms are proper fractions (the degree of the numerator is less than the degree of the denominator) and have distinct linear denominators.

$$\frac{7}{x+2} + \frac{5}{x-3} = \frac{7(x-3)}{(x+2)(x-3)} + \frac{5(x+2)}{(x-3)(x+2)} \qquad \text{common denominator}$$

$$= \frac{7(x-3) + 5(x+2)}{(x+2)(x-3)} \qquad \text{combine numerators}$$

$$= \frac{12x - 11}{(x+2)(x-3)} \qquad \text{result}$$

Assuming we didn't have the original sum to look at, reversing the process would require us to begin with a **decomposition template** such as

$$\frac{12x - 11}{(x+2)(x-3)} = \frac{A}{x+2} + \frac{B}{x-3}$$

and solve for the *constants A and B*. We know the numerators must be constants, else the partial fraction(s) would be improper while the original expression is not.

2.  Consider the sum $\frac{3}{x-1} + \frac{5}{x^2 - 2x + 1}$, again noting both terms are proper fractions.

$$\frac{3}{x-1} + \frac{5}{x^2 - 2x + 1} = \frac{3}{x-1} + \frac{5}{(x-1)(x-1)} \qquad \text{factor denominators}$$

$$= \frac{3(x-1)}{(x-1)(x-1)} + \frac{5}{(x-1)(x-1)} \qquad \text{common denominator}$$

$$= \frac{(3x-3) + 5}{(x-1)(x-1)} \qquad \text{combine numerators}$$

$$= \frac{3x + 2}{(x-1)^2} \qquad \text{result}$$

Note that while the new denominator is the repeated factor $(x-1)^2$, *both* $(x-1)$ and $(x-1)^2$ were denominators in the original sum. Assuming we didn't know the original sum, reversing the process would require us to begin with the template

$$\frac{3x + 2}{(x-1)^2} = \frac{A}{x-1} + \frac{B}{(x-1)^2}$$

and solve for the constants $A$ and $B$. As with observation 1, we know the numerator of the first term must be constant. While the second term would still be a proper fraction if the numerator were linear (degree 1), the denominator is a *repeated* linear factor and using a single constant in the numerator of *all such fractions* will ensure we obtain unique values for $A$ and $B$. In the end, for any repeated linear factor $(ax + b)^n$ in the original denominator, terms of the form

$$\frac{A_1}{ax+b} + \frac{A_2}{(ax+b)^2} + \cdots + \frac{A_{n-1}}{(ax+b)^{n-1}} + \frac{A_n}{(ax+b)^n}$$

must appear in the decomposition template, although some of these numerators may turn out to be zero.

**EXAMPLE 3** ▶   **Writing the Decomposition Template for Distinct and Repeated Linear Factors**

Write the decomposition template for

**a.** $\dfrac{x - 8}{2x^2 + 5x + 3}$    **b.** $\dfrac{x + 1}{x^2 - 6x + 9}$

**Solution** ▶   **a.** Factoring the denominator gives $\dfrac{x - 8}{(2x + 3)(x + 1)}$. With two distinct linear factors in the denominator, the decomposition template is

$$\frac{x - 8}{(2x + 3)(x + 1)} = \frac{A}{2x + 3} + \frac{B}{x + 1} \quad \text{decomposition template}$$

**b.** After factoring we have $\dfrac{x + 1}{(x - 3)^2}$, and the denominator is a repeated linear factor. Using our previous observations the template would be

$$\frac{x + 1}{(x - 3)^2} = \frac{A}{x - 3} + \frac{B}{(x - 3)^2} \quad \text{decomposition template}$$

**Now try Exercises 23 through 28** ▶

When both distinct and repeated linear factors are present in the denominator, the decomposition template maintains the elements illustrated in both observations 1 and 2.

**EXAMPLE 4** ▶   **Writing the Decomposition Template for Distinct and Repeated Linear Factors**

Write the decomposition template for $\dfrac{x^2 - 4x - 15}{x^3 - 2x^2 + x}$.

**Solution** ▶   Factoring the denominator gives $\dfrac{x^2 - 4x - 15}{x(x^2 - 2x + 1)}$, or $\dfrac{x^2 - 4x - 15}{x(x - 1)^2}$ after factoring completely. With a distinct linear factor of $x$, and the repeated linear factor $(x - 1)^2$, the decomposition template becomes

$$\frac{x^2 - 4x - 15}{x(x - 1)^2} = \frac{A}{x} + \frac{B}{x - 1} + \frac{C}{(x - 1)^2} \quad \text{decomposition template}$$

**Now try Exercises 29 and 30** ▶

To continue our observations,

**3.** Consider the sum $\dfrac{4}{x} + \dfrac{2x + 3}{x^2 + 1}$, noting the denominator of the first term is linear, while the denominator of the second is an irreducible quadratic (meaning it has only nonreal zeroes).

$$\frac{4}{x} + \frac{2x + 3}{x^2 + 1} = \frac{4(x^2 + 1)}{x(x^2 + 1)} + \frac{(2x + 3)x}{(x^2 + 1)x} \quad \text{find common denominator}$$

$$= \frac{(4x^2 + 4) + (2x^2 + 3x)}{x(x^2 + 1)} \quad \text{combine numerators}$$

$$= \frac{6x^2 + 3x + 4}{x(x^2 + 1)} \quad \text{result}$$

Here, reversing the process would require us to begin with the template

$$\frac{6x^2 + 3x + 4}{x(x^2 + 1)} = \frac{A}{x} + \frac{Bx + C}{x^2 + 1},$$

allowing that the numerator of the second term might be linear since the denominator is quadratic but *not due to a repeated linear factor*.

4. Finally, consider the sum $\frac{2x+1}{x^2+3} + \frac{x-2}{(x^2+3)^2}$, where the denominator of the first term is an irreducible quadratic and the second has *the same factor* with multiplicity two.

$$\frac{2x+1}{x^2+3} + \frac{x-2}{(x^2+3)^2} = \frac{(2x+1)(x^2+3)}{(x^2+3)(x^2+3)} + \frac{x-2}{(x^2+3)(x^2+3)} \qquad \text{common denominator}$$

$$= \frac{(2x^3+x^2+6x+3)+(x-2)}{(x^2+3)(x^2+3)} \qquad \text{combine numerators}$$

$$= \frac{2x^3+x^2+7x+1}{(x^2+3)^2} \qquad \text{result after simplifying}$$

Reversing the process would require us to begin with the template

$$\frac{2x^3+x^2+7x+1}{(x^2+3)^2} = \frac{Ax+B}{x^2+3} + \frac{Cx+D}{(x^2+3)^2}$$

allowing that the numerator of either term might be nonconstant for the reasons in observation 3. Similar to our reasoning in observation 2, all powers of a repeated quadratic factor must be present in the template.

When both distinct and repeated factors are present in the denominator, the decomposition template maintains the essential elements determined by observations 1 through 4. Using these observations, we can formulate a general approach to the decomposition template.

> **WORTHY OF NOTE**
>
> Note that the second term in the decomposition template would still be a proper fraction if the numerator were quadratic or cubic, but since the denominator is a *repeated* quadratic factor, using only a linear form ensures we obtain unique values for all coefficients.

---

### Decomposition Template for Rational Expressions

For a proper rational expression $\frac{P(x)}{Q(x)}$ in lowest terms …

1. Factor $Q$ completely into linear and irreducible quadratic factors.
2. For the linear factors, each distinct factor and each power of a repeated factor must appear in the decomposition template with a constant numerator.
3. For the irreducible quadratic factors, each distinct factor and each power of a repeated factor must appear in the decomposition template with a linear numerator.

If the rational expression is improper (the degree of $P$ is greater than or equal to the degree of $Q$), first find the quotient and remainder using polynomial division, then apply the steps above to the remainder. Only the remainder portion need be decomposed into partial fractions.

---

**EXAMPLE 5** ▶  **Writing the Decomposition Template for Linear and Quadratic Factors**

Write the decomposition template for

a. $\dfrac{x^2+10x+1}{(x+1)(x^2+x+3)}$    b. $\dfrac{x^2}{(x^2+2)^3}$

**Solution** ▶  a. One factor of the denominator is a distinct linear factor, and the other is an irreducible quadratic [since the discriminant in the quadratic formula is negative: $D = (1)^2 - 4(1)(3) < 0$]. The decomposition template is

$$\frac{x^2+10x+1}{(x+1)(x^2+x+3)} = \frac{A}{x+1} + \frac{Bx+C}{x^2+x+3} \qquad \text{decomposition template}$$

**b.** The denominator consists of a repeated, irreducible quadratic factor. Using our previous observations the template would be

$$\frac{x^2}{(x^2 + 2)^3} = \frac{Ax + B}{x^2 + 2} + \frac{Cx + D}{(x^2 + 2)^2} + \frac{Ex + F}{(x^2 + 2)^3} \qquad \text{decomposition template}$$

**Now try Exercises 31 and 32 ▶**

Once the template is obtained, we multiply both sides of the equation by the factored form of the original denominator and simplify. The resulting equation is an identity—a true statement for all real numbers $x$. And in many cases, the constants $A$, $B$, $C$, and so on can be identified using a choice of **convenient values** for $x$, as in Example 6.

**EXAMPLE 6 ▶** **Decomposing a Rational Expression with Distinct Linear Factors**

Decompose the expression $\frac{4x + 11}{x^2 + 7x + 10}$ into partial fractions.

**Solution ▶** Factoring the denominator gives $\frac{4x + 11}{(x + 5)(x + 2)}$ with two distinct linear factors. The required template is

$$\frac{4x + 11}{(x + 5)(x + 2)} = \frac{A}{x + 5} + \frac{B}{x + 2} \qquad \text{decomposition template}$$

Multiplying both sides by $(x + 5)(x + 2)$ clears all denominators and yields

$$4x + 11 = A(x + 2) + B(x + 5) \qquad \text{clear denominators}$$

Since the equation must be true for all $x$, using $x = -5$ will *conveniently* eliminate the term with $B$, and enable us to solve for $A$ directly:

$$\begin{aligned}
4(-5) + 11 &= A(-5 + 2) + B(-5 + 5) &&\text{substitute } -5 \text{ for } x \\
-20 + 11 &= -3A + B(0) &&\text{simplify} \\
-9 &= -3A &&\text{term with } B \text{ is eliminated} \\
3 &= A &&\text{solve for } A
\end{aligned}$$

To find $B$, we repeat this procedure, using an $x$-value that *conveniently* eliminates the term with $A$, namely, $x = -2$.

$$\begin{aligned}
4x + 11 &= A(x + 2) + B(x + 5) &&\text{original equation} \\
4(-2) + 11 &= A(-2 + 2) + B(-2 + 5) &&\text{substitute } -2 \text{ for } x \\
-8 + 11 &= A(0) + 3B &&\text{simplify} \\
3 &= 3B &&\text{term with } A \text{ is eliminated} \\
1 &= B &&\text{solve for } B
\end{aligned}$$

With $A = 3$ and $B = 1$, the complete decomposition is

$$\frac{4x + 11}{(x + 5)(x + 2)} = \frac{3}{x + 5} + \frac{1}{x + 2}$$

which can be checked by adding the rational expressions on the right.

**Now try Exercises 33 through 38 ▶**

**EXAMPLE 7** ▶ **Decomposing a Rational Expression with Repeated Linear Factors**

Decompose the expression $\frac{9}{(x + 5)(x^2 + 7x + 10)}$ into partial fractions.

**Solution** ▶ Factoring the denominator gives $\frac{9}{(x + 5)(x + 2)(x + 5)} = \frac{9}{(x + 2)(x + 5)^2}$
(one distinct linear factor, one repeated linear factor). The decomposition template is $\frac{9}{(x + 2)(x + 5)^2} = \frac{A}{x + 2} + \frac{B}{x + 5} + \frac{C}{(x + 5)^2}$. Multiplying both sides by $(x + 2)(x + 5)^2$ clears all denominators and yields

$$9 = A(x + 5)^2 + B(x + 2)(x + 5) + C(x + 2).$$

Using $x = -5$ will eliminate the terms with $A$ and $B$, giving

| | |
|---|---|
| $9 = A(-5 + 5)^2 + B(-5 + 2)(-5 + 5) + C(-5 + 2)$ | substitute $-5$ for $x$ |
| $9 = A(0) + B(-3)(0) - 3C$ | simplify |
| $9 = -3C$ | terms with $A$ and $B$ are eliminated |
| $-3 = C$ | solve for $C$ |

Using $x = -2$ will eliminate the terms with $B$ and $C$, and we have

| | |
|---|---|
| $9 = A(x + 5)^2 + B(x + 2)(x + 5) + C(x + 2)$ | original equation |
| $9 = A(-2 + 5)^2 + B(-2 + 2)(-2 + 5) + C(-2 + 2)$ | substitute $-2$ for $x$ |
| $9 = A(3)^2 + B(0)(3) + C(0)$ | simplify |
| $9 = 9A$ | terms with $B$ and $C$ are eliminated |
| $1 = A$ | solve for $A$ |

To find $B$, we substitute $A = 1$ and $C = -3$ into the original equation, *with any value of $x$ that does not eliminate $B$.* For efficiency, we'll often use $x = 0$ or $x = 1$.

| | |
|---|---|
| $9 = A(x + 5)^2 + B(x + 2)(x + 5) + C(x + 2)$ | original equation |
| $9 = 1(0 + 5)^2 + B(0 + 2)(0 + 5) - 3(0 + 2)$ | substitute 1 for $A$, $-3$ for $C$, 0 for $x$ |
| $9 = 25 + 10B - 6$ | simplify |
| $-1 = B$ | solve for $B$ |

With $A = 1$, $B = -1$, and $C = -3$ the complete decomposition is

$$\frac{9}{(x + 2)(x + 5)^2} = \frac{1}{x + 2} + \frac{-1}{x + 5} + \frac{-3}{(x + 5)^2}$$

$$= \frac{1}{x + 2} - \frac{1}{x + 5} - \frac{3}{(x + 5)^2}$$

**Now try Exercises 39 and 40** ▶

As an alternative to using convenient values, a system of equations can be set up by multiplying out the right-hand side (after clearing fractions) and equating coefficients of the terms with like degrees.

**EXAMPLE 8** ▶ **Decomposing a Rational Expression with Linear and Quadratic Factors**

Decompose the given expression into partial fractions: $\frac{3x^2 - x - 11}{x^3 - 3x^2 + 4x - 12}$.

**Solution** ▶ A careful inspection indicates the denominator will factor by grouping, giving $x^3 - 3x^2 + 4x - 12 = x^2(x - 3) + 4(x - 3) = (x - 3)(x^2 + 4)$. With one linear factor and one irreducible quadratic factor, the required template is

$$\frac{3x^2 - x - 11}{(x - 3)(x^2 + 4)} = \frac{A}{x - 3} + \frac{Bx + C}{x^2 + 4} \qquad \text{decomposition template}$$

$$3x^2 - x - 11 = A(x^2 + 4) + (Bx + C)(x - 3) \qquad \begin{array}{l}\text{multiply by } (x - 3)(x^2 + 4) \\ \text{(clear denominators)}\end{array}$$

$$= Ax^2 + 4A + Bx^2 - 3Bx + Cx - 3C \qquad \text{distribute/F-O-I-L}$$

$$= (A + B)x^2 + (C - 3B)x + 4A - 3C \qquad \text{group and factor}$$

For the left side to equal the right, we must equate coefficients of terms with like degree: $A + B = 3$ (for $x^2$), $C - 3B = -1$ (for $x$), and $4A - 3C = -11$ (for the constants). This gives the $3 \times 3$ system $\begin{bmatrix} 1 & 1 & 0 & | & 3 \\ 0 & -3 & 1 & | & -1 \\ 4 & 0 & -3 & | & -11 \end{bmatrix}$ in matrix form (verify this). Using the coefficient matrix we find that $D = 13$, and we complete the solution using Cramer's rule.

$$D_A = \begin{vmatrix} 3 & 1 & 0 \\ -1 & -3 & 1 \\ -11 & 0 & -3 \end{vmatrix} = 13, \quad D_B = \begin{vmatrix} 1 & 3 & 0 \\ 0 & -1 & 1 \\ 4 & -11 & -3 \end{vmatrix} = 26, \quad D_C = \begin{vmatrix} 1 & 1 & 3 \\ 0 & -3 & -1 \\ 4 & 0 & -11 \end{vmatrix} = 65$$

The result is $A = \frac{13}{13} = 1$, $B = \frac{26}{13} = 2$, and $C = \frac{65}{13} = 5$, giving the decomposition

$$\frac{3x^2 - x - 11}{(x - 3)(x^2 + 4)} = \frac{1}{x - 3} + \frac{2x + 5}{x^2 + 4}.$$

☑ **B. You've just seen how we can decompose a rational expression into partial fractions**

**Now try Exercises 41 through 46 ▶**

Although the last example could be solved using the "convenient values" method (try $x = 3$, $x = 0$, and $x = 1$, for instance), a system of equations is sometimes our *only* option. Also, if the decomposition template produces a large or cumbersome system, a graphing calculator can assist the solution process using a matrix equation feature. **See Exercises 47 and 48.**

As a final reminder, if the degree of the numerator is *greater than or equal to* the degree of the denominator, divide using long division and apply the preceding methods to the remainder polynomial. For instance, you can check that $\frac{3x^3 + 6x^2 + 5x - 7}{x^2 + 2x + 1} = 3x + \frac{2x - 7}{(x + 1)^2}$, and decomposing the remainder polynomial gives a final result of $3x + \frac{2}{x + 1} - \frac{9}{(x + 1)^2}$. **See Exercises 49 and 50.**

## C. Determinants, Geometry, and the Coordinate Plane

As mentioned in the introduction, the use of determinants extends far beyond solving systems of equations. Here, we'll demonstrate how determinants can be used to find the area of a triangle whose vertices are given as three points in the coordinate plane.

**The Area of a Triangle in the *xy*-Plane**

Given a triangle with vertices at $(x_1, y_1)$, $(x_2, y_2)$, and $(x_3, y_3)$,

$$\text{Area} = \frac{1}{2}|\det(T)|, \text{ where } T = \begin{bmatrix} x_1 & y_1 & 1 \\ x_2 & y_2 & 1 \\ x_3 & y_3 & 1 \end{bmatrix}$$

**EXAMPLE 9** ▶ **Finding the Area of a Triangle Using Determinants**

Find the area of a triangle with vertices at $(3, 1)$, $(-2, 3)$, and $(1, 7)$ (see Figure 7.15).

**Solution** ▶ Begin by forming matrix $T$ and computing $\det(T)$ (see Figure 7.16):

**Figure 7.15**

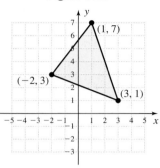

$$\det(T) = \begin{vmatrix} x_1 & y_1 & 1 \\ x_2 & y_2 & 1 \\ x_3 & y_3 & 1 \end{vmatrix} = \begin{vmatrix} 3 & 1 & 1 \\ -2 & 3 & 1 \\ 1 & 7 & 1 \end{vmatrix}$$

$$= 3(3 - 7) - 1(-2 - 1) + 1(-14 - 3)$$
$$= -12 + 3 + (-17)$$
$$= -26$$

Compute the area: $A = \frac{1}{2}|\det(T)|$
$$= \frac{1}{2}|-26|$$
$$= 13$$

**Figure 7.16**
$T = [A]$

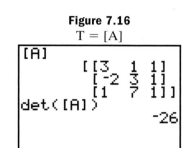

The area of this triangle is 13 units$^2$.

**Now try Exercises 53 through 58 ▶**

☑ **C. You've just seen how we can use determinants in applications involving geometry in the coordinate plane**

As an extension of this formula, what if the three points were collinear? After a moment, it may occur to you that the formula would give an area of 0 units$^2$, since no triangle could be formed. This gives rise to a **test for collinear points.**

**Test for Collinear Points**

Three points $(x_1, y_1)$, $(x_2, y_2)$, and $(x_3, y_3)$ are collinear if

$$\det(T) = \begin{vmatrix} x_1 & y_1 & 1 \\ x_2 & y_2 & 1 \\ x_3 & y_3 & 1 \end{vmatrix} = 0.$$

**See Exercises 59 through 62.** There are a variety of additional applications in the Exercise Set. **See Exercises 63 through 68.**

## 7.4 EXERCISES

▶ **CONCEPTS AND VOCABULARY**

**Fill in the blank with the appropriate word or phrase. Carefully reread the section, if necessary.**

**1.** The determinant $\begin{vmatrix} a_{11} & a_{12} \\ a_{21} & a_{22} \end{vmatrix}$ is evaluated as: _____.

**2.** The three points $(x_1, y_1)$, $(x_2, y_2)$, and $(x_3, y_3)$ are collinear if $|T| = \begin{vmatrix} x_1 & y_1 & 1 \\ x_2 & y_2 & 1 \\ x_3 & y_3 & 1 \end{vmatrix}$ has a value of ____.

**3.** Discuss/Explain the process of writing $\frac{8x-3}{x^2-x}$ as a sum of partial fractions.

**4.** Discuss/Explain why Cramer's rule cannot be applied if $D = 0$. Use an example to illustrate.

## ▶ DEVELOPING YOUR SKILLS

Solve each system using Cramer's rule, if possible. Do not use a calculator.

5. $\begin{cases} 2x + 5y = 7 \\ -3x + 4y = 1 \end{cases}$

6. $\begin{cases} -x + 5y = 12 \\ 3x - 2y = -8 \end{cases}$

7. $\begin{cases} 4x + y = -11 \\ 3x - 5y = -60 \end{cases}$

8. $\begin{cases} x = -2y - 11 \\ y = 2x - 13 \end{cases}$

9. $\begin{cases} \dfrac{x}{8} + \dfrac{y}{4} = 1 \\ \dfrac{y}{5} = \dfrac{x}{2} + 6 \end{cases}$

10. $\begin{cases} \dfrac{2}{3}x - \dfrac{3}{8}y = \dfrac{7}{5} \\ \dfrac{5}{6}x + \dfrac{3}{4}y = \dfrac{11}{10} \end{cases}$

11. $\begin{cases} 0.6x - 0.3y = 8 \\ 0.8x - 0.4y = -3 \end{cases}$

12. $\begin{cases} -2.5x + 6y = -1.5 \\ 0.5x - 1.2y = 3.6 \end{cases}$

The two systems given in Exercises 13 and 14 are identical except for the third equation. For the first system given, (a) write the determinants $D$, $D_x$, $D_y$, and $D_z$ then (b) determine if a solution using Cramer's rule is possible by computing $|D|$ without the use of a calculator (do not solve the system). Then (c) compute $|D|$ for the second system and try to determine how the equations in the second system are related.

13. $\begin{cases} 4x - y + 2z = -5 \\ -3x + 2y - z = 8, \\ x - 5y + 3z = -3 \end{cases}$ $\begin{cases} 4x - y + 2z = -5 \\ -3x + 2y - z = 8 \\ x + y + z = 3 \end{cases}$

14. $\begin{cases} 2x + 3z = -2 \\ -x + 5y + z = 12, \\ 3x - 2y + z = -8 \end{cases}$ $\begin{cases} 2x + 3z = -2 \\ -x + 5y + z = 12 \\ x + 5y + 4z = 10 \end{cases}$

Solve each system using Cramer's rule, if possible. Verify computations using a graphing calculator.

15. $\begin{cases} x + 2y + 5z = 10 \\ 3x + 4y - z = 10 \\ x - y - z = -2 \end{cases}$

16. $\begin{cases} x + 3y + 5z = 6 \\ 2x - 4y + 6z = 14 \\ 9x - 6y + 3z = 3 \end{cases}$

17. $\begin{cases} 4x - 6y - 8z = 1 \\ -2x + 3y + 4z = 2 \\ 5y + z = -3 \end{cases}$

18. $\begin{cases} x - 3y - 5z = 2 \\ -2x - y + 2z = 4 \\ 3x - 2y - 7z = -5 \end{cases}$

19. $\begin{cases} y + 2z = 1 \\ 4x - 5y + 8z = -8 \\ 8x - 9z = 9 \end{cases}$

20. $\begin{cases} x + 2y + 5z = 10 \\ 3x - z = 8 \\ -y - z = -3 \end{cases}$

21. $\begin{cases} w + 2x - 3y = -8 \\ x - 3y + 5z = -22 \\ 4w - 5x = 5 \\ -y + 3z = -11 \end{cases}$

22. $\begin{cases} w - 2x + 3y - z = 11 \\ 3w - 2y + 6z = -13 \\ 2x + 4y - 5z = 16 \\ 3x - 4z = 5 \end{cases}$

## ▶ DECOMPOSITION OF RATIONAL EXPRESSIONS

Exercises 23 through 32 are designed solely to reinforce the various possibilities for decomposing a rational expression. All are proper fractions whose denominators are completely factored. Set up the decomposition template using appropriate numerators, but *do not solve*.

23. $\dfrac{3x + 2}{(x + 3)(x - 2)}$

24. $\dfrac{-4x + 1}{(x - 2)(x - 5)}$

25. $\dfrac{3x^2 - 2x + 5}{(x - 1)(x + 2)(x - 3)}$

26. $\dfrac{-2x^2 + 3x - 4}{(x + 3)(x + 1)(x - 2)}$

27. $\dfrac{x^2 + 5}{x(x - 3)(x + 1)}$

28. $\dfrac{x^2 - 7}{(x + 4)(x - 2)x}$

29. $\dfrac{x^2 + x - 1}{x^2(x + 2)}$

30. $\dfrac{x^2 - 3x + 5}{(x - 3)(x + 2)^2}$

31. $\dfrac{x^3 + 3x - 2}{(x + 1)(x^2 + 2)^2}$

32. $\dfrac{2x^3 + 3x^2 - 4x + 1}{x(x^2 + 3)^2}$

Decompose each rational expression into partial fractions.

33. $\dfrac{4 - x}{x^2 + x}$

34. $\dfrac{3x + 13}{x^2 + 5x + 6}$

35. $\dfrac{2x - 27}{2x^2 + x - 15}$

36. $\dfrac{-11x + 6}{5x^2 - 4x - 12}$

37. $\dfrac{8x^2 - 3x - 7}{x^3 - x}$

38. $\dfrac{x^2 + 24x - 12}{x^3 - 4x}$

39. $\dfrac{3x^2 + 7x - 1}{x^3 + 2x^2 + x}$

40. $\dfrac{-2x^2 - 7x + 28}{x^3 - 4x^2 + 4x}$

41. $\dfrac{3x^2 + 10x + 4}{8 - x^3}$

42. $\dfrac{3x^2 + 4x - 1}{x^3 - 1}$

43. $\dfrac{6x^2 + x + 13}{x^3 + 2x^2 + 3x + 6}$

44. $\dfrac{2x^2 - 14x - 7}{x^3 - 2x^2 + 5x - 10}$

**45.** $\dfrac{x^4 - x^2 - 2x + 1}{x^5 + 2x^3 + x}$

**46.** $\dfrac{-3x^4 + 13x^2 + x - 12}{x^5 + 4x^3 + 4x}$

 **47.** $\dfrac{x^3 - 17x^2 + 76x - 98}{(x^2 - 6x + 9)(x^2 - 2x - 3)}$

**48.** $\dfrac{16x^3 - 66x^2 + 98x - 54}{(2x^2 - 3x)(4x^2 - 12x + 9)}$

**49.** $\dfrac{x^4 - 4x^3 - 7x^2 - 28x - 47}{x^2 - 5x - 6}$

**50.** $\dfrac{x^4 + x^3 - 8x^2 + 6x + 20}{x^3 - 4x}$

## ▶ WORKING WITH FORMULAS

**51. Equation of a line:** $\begin{vmatrix} x & y & 1 \\ x_1 & y_1 & 1 \\ x_2 & y_2 & 1 \end{vmatrix} = 0$

The determinant shown can be used to find the equation of the line passing through the points $(x_1, y_1)$ and $(x_2, y_2)$.

(a) Show that the formula agrees with the point-slope formula $y - y_1 = m(x - x_1)$, where $m = \dfrac{y_2 - y_1}{x_2 - x_1}$. Then

(b) use the formula given to find the equation of the line passing through the points $(-2, 3)$ and $(1, -5)$.

**52. Area of a Norman window:** $A = \begin{vmatrix} L & r^2 \\ -\dfrac{\pi}{2} & W \end{vmatrix}$

The determinant shown can be used to find the area of a Norman window (rectangle + half-circle) with length $L$, width $W$, and radius $r = \frac{1}{2}W$. (a) Show that the formula agrees with area found using basic geometry formulas. Then (b) use the formula given to find the area of the window.

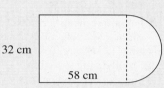

32 cm

58 cm

## ▶ APPLICATIONS

**Geometric Applications**

**Find the area of the triangle with the vertices given. Assume units are in centimeters.**

**53.** $(2, 1), (3, 7),$ and $(5, 3)$

**54.** $(-2, 3), (-3, -4),$ and $(-6, 1)$

**Find the area of the parallelogram with vertices given. Assume units are in feet.**

**55.** $(-4, 2), (-6, -1), (3, -1),$ and $(5, 2)$

**56.** $(-5, -6), (5, 0), (5, 4),$ and $(-5, -2)$

**The volume of a triangular pyramid is given by the formula** $V = \frac{1}{3}Bh$**, where *B* represents the area of the triangular base and *h* is the height of the pyramid. Find the volume of a triangular pyramid whose height is given and whose base has the coordinates shown. Assume units are in meters.**

**57.** $h = 6$ m; vertices $(3, 5), (-4, 2),$ and $(-1, 6)$

**58.** $h = 7.5$ m; vertices $(-2, 3), (-3, -4),$ and $(-6, 1)$

**Determine if the following points are collinear.**

**59.** $(1, 5), (-2, -1),$ and $(4, 11)$

**60.** $(1, 1), (3, -5),$ and $(-2, 9)$

**61.** $(-2.5, 5.2), (1.2, -5.6),$ and $(2.2, -8.5)$

**62.** $(-0.5, 2.55), (-2.8, 1.63),$ and $(3, 3.95)$

**Write a linear system that models each application. Then solve using Cramer's rule.**

**63. Return on investments:** If \$15,000 is invested at a certain interest rate and \$25,000 is invested at another interest rate, the total return was \$2900. If the investments were reversed, the return would be \$2700. What was the interest rate paid on each investment?

**64. Cost of fruit:** Many years ago, 2 lb of apples, 2 lb of kiwi, and 10 lb of pears cost \$3.26. Three pounds of apples, 2 lb of kiwi, and 7 lb of pears cost \$2.98. Two pounds of apples, 3 lb of kiwi, and 6 lb of pears cost \$2.89. Find the cost of a pound of each fruit.

**65. Forces on trusses:** If we consider a very simple truss in the form of an equilateral triangle, the forces exerted along the rafters of the truss by a weight at the apex can be modeled by a 2 × 2 system of linear equations. If a 180-lb carpenter is working at the center of this truss, the forces along each rafter can be modeled by the system shown. Find the force along each rafter.

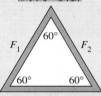

$$\begin{cases} \dfrac{\sqrt{3}}{2}(F_1 + F_2) = 180 \\ F_1 - F_2 = 0 \end{cases}$$

**66. Dietary research for pets:** As part of a research project, a college student is mixing a special diet for pet mice from two available sources. The diet must offer exactly 22.8 g of protein and 5 g of fat. Given the protein and fat values for the food sources shown, how much of each should be used?

|  | Source 1 | Source 2 |
|---|---|---|
| protein value | 0.18 | 0.24 |
| fat value | 0.06 | 0.04 |

**67. High-altitude weather research:** A high-altitude weather balloon carrying a heavy payload has suddenly ruptured and is plummeting back to Earth. An onboard altimeter transmits its height in feet every few seconds. For data of the form (time in seconds, height in feet), three of the readings are (5, 9600), (10, 8400), and (15, 6400). (a) Use this data to find an equation of the form $h = at^2 + bt + c$ that models the height of the balloon at any time $t$. (b) At what height did the balloon rupture? (c) What is the altitude of the balloon after 20 sec? (d) How many seconds until the payload hits the ground?

**68. Manufacturing surfboards:** Australian Waterglide is a manufacturer of custom surfboards for beginners, recreational surfers, and competitive surfers. For each board, production is handled in three stages: forming, fiberglass, and finishing. The number of hours required for each stage is given in the table. If the company has 80 labor hours per week available for forming, 152 hr available for fiberglass, and 145 hr available for finishing, how many boards of each type should be made?

|  | Beginner | Recreational | Competition |
|---|---|---|---|
| forming | 3 | 4 | 5 |
| fiberglass | 4.5 | 8.5 | 9 |
| finishing | 5.5 | 7.5 | 8 |

▶ **EXTENDING THE CONCEPT**

**69.** Find the area of the pentagon whose vertices are: $(-5, -5)$, $(5, -5)$, $(8, 6)$, $(-8, 6)$, and $(0, 12.5)$.

**70.** The polynomial form for the equation of a circle is $x^2 + y^2 + Dx + Ey + F = 0$. Find the equation of the circle that contains the points $(-1, 7)$, $(2, 8)$, and $(5, -1)$.

▶ **MAINTAINING YOUR SKILLS**

**71. (4.4)** Graph the polynomial using information about end-behavior, intercepts, and midinterval points: $f(x) = x^3 - 2x^2 - 7x + 6$.

**72. (3.1)** Which is the graph (left or right) of $g(x) = -|x + 1| + 3$? Justify your answer.

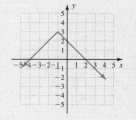

**73. (5.3/5.5)** Solve the equation $3^{2x-1} = 9^{2-x}$ two ways: first using logarithms, then by equating the bases and using the uniqueness property of exponents.

**74. (4.4)** Which is the graph (left or right) of a degree 3 polynomial? Justify your answer.

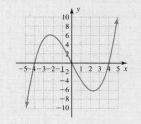

## MAKING CONNECTIONS

## Making Connections: Graphically, Symbolically, Numerically, and Verbally

Eight matrices $A$ through $H$ are given. Use a graphing calculator to help match the characteristics or operations indicated in 1 through 16 to one of the eight matrices. In some cases, the response requires two matrices.

$$A = \begin{bmatrix} 3 & -2 \\ 1 & 4 \end{bmatrix} \qquad B = \begin{bmatrix} -2 & 3 \\ 2 & -4 \end{bmatrix} \qquad C = \begin{bmatrix} 4 & 0 & -2 \\ 1 & -3 & 5 \end{bmatrix} \qquad D = \begin{bmatrix} -2 & 5 \\ 1 & 4 \\ 0 & 3 \end{bmatrix}$$

$$E = \begin{bmatrix} 0 & 3 & 2 \\ -2 & 4 & -1 \\ 1 & 5 & -6 \end{bmatrix} \qquad F = \begin{bmatrix} 1 & 0 & 0 \\ 0 & 1 & 0 \\ 0 & 0 & 1 \end{bmatrix} \qquad G = \begin{bmatrix} 1 & 3 & 0 \\ -3 & 4 & -1 \\ -2 & 7 & -1 \end{bmatrix} \qquad H = \begin{bmatrix} 3 & -1 & 1 & 0 \\ -5 & 2 & 0 & 1 \end{bmatrix}$$

**1.** \_\_\_\_ $3 \times 3$, noninvertible

**2.** \_\_\_\_ determinant is 1

**3.** \_\_\_\_ entry $a_{3,2}$ is 3

**4.** \_\_\_\_ sum is $\begin{bmatrix} 1 & 1 \\ 3 & 0 \end{bmatrix}$

**5.** \_\_\_\_ determinant is 14

**6.** \_\_\_\_ matrix squared is $\begin{bmatrix} 10 & -18 \\ -12 & 22 \end{bmatrix}$

**7.** \_\_\_\_ matrix inverse is $\begin{bmatrix} -2 & -1.5 \\ -1 & -1 \end{bmatrix}$

**8.** \_\_\_\_ entry $a_{3,1}$ is $-2$

**9.** \_\_\_\_ product is $\begin{bmatrix} -8 & 14 \\ -5 & 8 \end{bmatrix}$

**10.** \_\_\_\_ product is $\begin{bmatrix} -3 & -15 & 29 \\ 8 & -12 & 18 \\ 3 & -9 & 15 \end{bmatrix}$

**11.** \_\_\_\_ determinant is $-67$

**12.** \_\_\_\_ determinant is 0

**13.** \_\_\_\_ $3 \times 2$ matrix

**14.** \_\_\_\_ $2 \times 3$ matrix

**15.** \_\_\_\_ augmented matrix

**16.** \_\_\_\_ identity matrix

## SUMMARY AND CONCEPT REVIEW

## SECTION 7.1    Solving Linear Systems Using Matrices and Row Operations

### KEY CONCEPTS

- A *matrix* is a rectangular arrangement of numbers. An $m \times n$ matrix has $m$ rows and $n$ columns.
- An *augmented matrix* is derived from a system of linear equations by augmenting the *coefficient matrix* with the *matrix of constants*.
- An inconsistent system has no solutions and will yield a contradictory statement such as $0 = 1$.
  A dependent system has infinitely many solutions and will yield an identity statement such as $0 = 0$.

### EXERCISES

**1.** Write an example of the following matrices:

    **a.** $2 \times 3$    **b.** $3 \times 2$    **c.** $3 \times 3$, in triangular form

Solve by triangularizing the augmented matrix. If the system is linearly dependent, state the answer using a parameter. Use a calculator for Exercise 7.

**2.** $\begin{cases} x - 2y = 6 \\ 4x - 3y = 4 \end{cases}$

**3.** $\begin{cases} x - 2y + 2z = 7 \\ 2x + 2y - z = 5 \\ 3x - y + z = 6 \end{cases}$

**4.** $\begin{cases} 2x - y + 2z = -1 \\ x + 2y + 2z = -3 \\ 3x - 4y + 2z = 1 \end{cases}$

**5.** $\begin{cases} 5x - 2y + 4z = -3 \\ -2x + 5y + z = 1 \\ 6x + 6y + 10z = 4 \end{cases}$

**6.** $\begin{cases} 2x + 2y - 6z = -4 \\ -3x - 3y + 9z = 6 \\ 4x + 4y - 12z = -8 \end{cases}$

**7.** $\begin{cases} 2w + x + 2y - 3z = -19 \\ w - 2x - y + 4z = 15 \\ x + 2y - z = 1 \\ 3w - 2x - 5z = -60 \end{cases}$

## SECTION 7.2   The Algebra of Matrices

### KEY CONCEPTS

- The entries of a matrix are denoted $a_{ij}$, where $i$ gives the row and $j$ gives the column of its location.
- The sum or difference of two matrices of equal order is found by combining corresponding entries: $A + B = [a_{ij} + b_{ij}]$.
- To perform *scalar multiplication,* take the product of the constant with each entry in the matrix, forming a new matrix of like size. For constant $k$ and matrix $A$: $kA = [ka_{ij}]$.
- *Matrix multiplication* is performed by multiplying a row in the first matrix by a column in the second. For an $m \times n$ matrix $A = [a_{ij}]$ and an $s \times t$ matrix $B = [b_{ij}]$, $AB$ is possible if $n = s$. The result will be an $m \times t$ matrix $P = [p_{ij}]$, where $p_{ij}$ is the product of the $i$th row of $A$ with the $j$th column of $B$.

### EXERCISES

Compute the operations indicated below (if possible), using the following matrices.

$$A = \begin{bmatrix} \frac{-1}{4} & \frac{-3}{4} \\ \frac{-1}{8} & \frac{-7}{8} \end{bmatrix} \qquad B = \begin{bmatrix} -7 & 6 \\ 1 & -2 \end{bmatrix} \qquad C = \begin{bmatrix} -1 & 3 & 4 \\ 5 & -2 & 0 \\ 6 & -3 & 2 \end{bmatrix} \qquad D = \begin{bmatrix} 2 & -3 & 0 \\ 0.5 & 1 & -1 \\ 4 & 0.1 & 5 \end{bmatrix}$$

**8.** $A + B$  
**9.** $B - A$  
**10.** $C - B$  
**11.** $8A$  
**12.** $BA$

**13.** $C + D$  
**14.** $D - C$  
**15.** $BC$  
**16.** $-4D$  
**17.** $CD$

## SECTION 7.3   Solving Linear Systems Using Matrix Equations

### KEY CONCEPTS

- Any $n \times n$ system of equations can be written as a matrix equation and solved (if a unique solution exists) using an inverse matrix. The system $\begin{cases} 2x + 3y = 7 \\ x - 4y = -2 \end{cases}$ is written as $\begin{bmatrix} 2 & 3 \\ 1 & -4 \end{bmatrix}\begin{bmatrix} x \\ y \end{bmatrix} = \begin{bmatrix} 7 \\ -2 \end{bmatrix}$.
- Every square matrix has a real number associated with it, called its *determinant.* For the $2 \times 2$ matrix $A = \begin{bmatrix} a_{11} & a_{12} \\ a_{21} & a_{22} \end{bmatrix}$, $\det(A) = a_{11}a_{22} - a_{21}a_{12}$.
- If the determinant of a matrix is zero, the matrix is said to be *singular* or *noninvertible.* If the coefficient matrix of a matrix equation is noninvertible, the system is either inconsistent or dependent.

### EXERCISES

Complete Exercises 18 through 20 using the following matrices:

$$A = \begin{bmatrix} 1 & 0 \\ 0 & 1 \end{bmatrix} \qquad B = \begin{bmatrix} 0.2 & 0.2 \\ -0.6 & 0.4 \end{bmatrix} \qquad C = \begin{bmatrix} 2 & -1 \\ 3 & 1 \end{bmatrix} \qquad D = \begin{bmatrix} 10 & -6 \\ -15 & 9 \end{bmatrix}$$

**18.** Exactly one of the matrices given is singular. Compute each determinant to identify it.

**19.** Show that $AB = BA = B$. What can you conclude about matrix $A$?

**20.** Show that $BC = CB = I$. What can you conclude about matrix $C$?

Use a graphing calculator to complete Exercises 21 through 23, using the matrices given:

$$E = \begin{bmatrix} 1 & -2 & 3 \\ -2 & 1 & -5 \\ -1 & -1 & -2 \end{bmatrix} \qquad F = \begin{bmatrix} 1 & -1 & 1 \\ 0 & 1 & 0 \\ -2 & 1 & -1 \end{bmatrix} \qquad G = \begin{bmatrix} -1 & 0 & -1 \\ 0 & 1 & 0 \\ 2 & 1 & 1 \end{bmatrix}$$

**21.** Exactly one of the matrices is singular. Compute each determinant to identify it.

**22.** Compute the products $FG$ and $GF$. What can you conclude about matrix $G$?

**23.** Verify that $EG \ne GE$ and $EF \ne FE$. What can you conclude?

Solve manually using a matrix equation.

**24.** $\begin{cases} 2x - 5y = 14 \\ -3y + 4x = -14 \end{cases}$

Solve using a matrix equation and your calculator.

 **25.** $\begin{cases} 0.5x - 2.2y + 3z = -8 \\ -0.6x - y + 2z = -7.2 \\ x + 1.5y - 0.2z = 2.6 \end{cases}$

## SECTION 7.4   Applications of Matrices and Determinants: Cramer's Rule, Partial Fractions, and More

### KEY CONCEPTS

- Cramer's rule uses a ratio of determinants to solve systems of equations (if solutions exist).
- Determinants can be used to find the area of a triangle in the plane if the vertices of the triangle are known, and as a test to see if three points are collinear.

### EXERCISES

Solve using Cramer's rule. Use a graphing calculator for Exercise 28.

**26.** $\begin{cases} 5x + 6y = 8 \\ 10x - 2y = -9 \end{cases}$

**27.** $\begin{cases} 2x + y - z = -1 \\ x - 2y + z = 5 \\ 3x - y + 2z = 8 \end{cases}$

**28.** $\begin{cases} 2x + y = -2 \\ -x + y + 5z = 12 \\ 3x - 2y + z = -8 \end{cases}$

**29.** Find the area of a triangle whose vertices have the coordinates $(6, 1)$, $(-1, -6)$, and $(-6, 2)$.

**30.** Find the partial fractions decomposition for $\dfrac{7x^2 - 5x + 17}{x^3 - 2x^2 + 3x - 6}$.

## PRACTICE TEST

**Solve each system by triangularizing the augmented matrix and using back-substitution.**

**1.** $\begin{cases} 3x + 8y = -5 \\ x + 10y = 2 \end{cases}$

**2.** $\begin{cases} 3x - y + 5z = 1 \\ 3x + y + 4z = 4 \\ x + y + z = \frac{7}{3} \end{cases}$

**3.** $\begin{cases} 4x - 5y - 6z = 5 \\ 2x - 3y + 3z = 0 \\ x + 2y - 3z = 5 \end{cases}$

**4.** Given matrices $A$ and $B$, compute:   **a** $A - B$   **b.** $\dfrac{2}{5}B$   **c.** $AB$   **d.** $A^{-1}$   **e.** $|A|$

$$A = \begin{bmatrix} -3 & -2 \\ 5 & 4 \end{bmatrix} \qquad B = \begin{bmatrix} 3 & 3 \\ -3 & -5 \end{bmatrix}$$

**5.** Given matrices $C$ and $D$, use a calculator to find:
   **a.** $C - D$  **b.** $-0.6D$  **c.** $DC$  **d.** $D^{-1}$  **e.** $|D|$

$$C = \begin{bmatrix} 0.5 & 0 & 0.2 \\ 0.4 & -0.5 & 0 \\ 0.1 & -0.4 & -0.1 \end{bmatrix}$$

$$D = \begin{bmatrix} 0.5 & 0.1 & 0.2 \\ -0.1 & 0.1 & 0 \\ 0.3 & 0.4 & 0.8 \end{bmatrix}$$

**6.** Use matrices to find three different solutions of the dependent system:

$$\begin{cases} 2x - y + z = 4 \\ 3x - 2y + 4z = 9 \\ x - 2y + 8z = 11 \end{cases}$$

**7.** Solve using Cramer's rule: $\begin{cases} 2x - 3y = 2 \\ x - 6y = -2 \end{cases}$

**8.** Solve using a calculator and Cramer's rule:

$$\begin{cases} 2x + 3y + z = 3 \\ x - 2y - z = 4 \\ x - y - 2z = -1 \end{cases}$$

**9.** Solve using a matrix equation:

$$\begin{cases} 2x - 5y = 11 \\ 4x + 7y = 4 \end{cases}$$

**10.** Solve using a matrix equation and your calculator:

$$\begin{cases} x - 2y + 2z = 7 \\ 2x + 2y - z = 5 \\ 3x - y + z = 6 \end{cases}$$

**11.** Use the equality of matrices to write and solve a system that gives values of $x$, $y$, and $z$ so that $A = B$.

$$A = \begin{bmatrix} 2x + y & 3 \\ x + z & 3x + 2z \end{bmatrix}$$

$$B = \begin{bmatrix} z - 1 & 3 \\ 2y + 5 & y + 8 \end{bmatrix}$$

**12.** Given matrix $X$ is a solution to $AX = B$ for the matrix $A$ given, find matrix $B$.

$$X = \begin{bmatrix} -1 \\ \frac{-3}{2} \\ 2 \end{bmatrix} \quad A = \begin{bmatrix} 1 & -2 & 2 \\ 2 & -6 & 3 \\ 3 & 4 & -1 \end{bmatrix}$$

**13.** Use matrices and a graphing calculator to determine which three of the following four points are collinear: $(-1, 4), (1, 3), (2, 1), (4, -1)$.

**14.** A farmer plants a triangular field with wheat. The first vertex of the triangular field is 1 mi east and 1 mi north of his house. The second vertex is 3 mi east and 1 mi south of his house. The third vertex is 1 mi west and 2 mi south of his house. What is the area of the field?

**15.** For $A = \begin{bmatrix} r & 2 \\ 3 & s \end{bmatrix}$ and $A^2 = \begin{bmatrix} 10 & -2 \\ -3 & 7 \end{bmatrix}$ given, find $r$ and $s$.

**Create a system of equations to model each exercise, then solve using any matrix method.**

**16.** Dr. Brown and Dr. Stamper graduate from medical school with $155,000 worth of student loans. Due to her state's tuition reimbursement plan, Dr. Brown owes one fourth of what Dr. Stamper owes. How much does each doctor owe?

**17.** Justin is rehabbing two old houses simultaneously. Last week, he spent a total of 23 hr working on the houses, with one house receiving 8 more hours of his attention than the other. How many hours did he spend on each house?

**18.** In his first month as assistant principal of Washington High School, Mr. Johnson gave out 20 detentions. They were either for 1 day, 2 days, or 5 days. He recorded a total of 38 days of detention served. He also noted that there were twice as many 2-day detentions as 5-day detentions. How many of each type of detention did Mr. Johnson give out?

**19.** The city of Cherrywood has approved a $1,800,000 plan to renovate its historic commercial district. The money will be coming from three separate sources. The first is a federal program that charges a low 2% interest annually. The second is a municipal bond offering that will cost 5% annually. The third is a standard loan from a neighborhood bank, but it will cost 8.5% annually. In the first year, the city will not make any repayment on these loans and will accrue $94,500 more debt. The federal program and bank loan together are responsible for $29,500 of this interest. How much money was originally provided by each source?

**20.** Decompose the expression into partial fractions:

$$\frac{4x^2 - 4x + 3}{x^3 - 27}.$$

## CALCULATOR EXPLORATION AND DISCOVERY

## I. Solving Systems with Technology and Cramer's Rule

In Section 7.4, we saw that one interesting application of matrices is Cramer's rule. You may have noticed that when technology is used with Cramer's rule, the chances of making an error are fairly high, as we need to input the entries for numerous matrices. However, as we mentioned in the chapter introduction, one of the advantages of matrices is that they *are easily programmable,* and we can actually write a very simple program that will make Cramer's rule a more efficient method.

To begin, press the **PRGM** key, and then the right arrow ◯ twice and ▭ to create a name for our program. At the prompt, we'll enter **CRAMER2**. As we write the program, note that most of the needed commands (**ClrHome, Disp, Pause, Prompt, Stop**) are located in the submenus of the **PRGM** key. The exceptions are the ▭ = ▭ sign found under the **TEST** menu (accessed using the **2nd** **MATH** keys), the arrows " → " used to indicate the **STO▸** key, and the convert-to-fraction command " ▸**Frac**" available as option 1 in the **MATH** menu.

The program shown takes the coefficients and constants of a 2 × 2 linear system, and returns the ordered pair solution in the form of $x = h$ and $y = k$. Even with minimal programming experience, commands can be very intuitive and reading through the program will help you identify that Cramer's rule is being used.

| | |
|---|---|
| **ClrHome** | **(CE–BF)/(AE–BD)→X** |
| **Disp "2×2 SYSTEMS"** | **(AF–DC)/(AE–BD)→Y** |
| **Pause** | **ClrHome** |
| **Disp "AX+BY = C"** | **Disp "THE SOLUTION IS"** |
| **Disp "DX+EY = F"** | **Disp ""** |
| **Disp ""** | **Disp "X="** |
| **Disp "ENTER THE VALUES"** | **Disp X ▶ Frac** |
| **Disp "FOR A,B,C,D,E,F"** | **Disp "Y="** |
| **Disp ""** | **Disp Y ▶ Frac** |
| **Prompt A,B,C,D,E,F** | **Stop** |

**Exercise 1:** Use the program to check the answers to Exercises 1 and 7 of the Practice Test.

**Exercise 2:** Create 2 × 2 systems of your own that are (a) consistent and independent, (b) consistent and dependent, and (c) inconsistent. Then verify results using the program.

**Exercise 3:** Use the box on page 608 of Section 7.4 to write a similar program for 3 × 3 systems. Call the program CRAMER3, and repeat parts (a), (b), and (c) from Exercise 2 with a 3 × 3 system.

## II. Solving Systems with Matrices, Row Operations, and Technology

Graphing calculators also offer a very efficient way to solve systems using matrices. Once the system has been written in matrix form, it can easily be entered and solved by asking the calculator to instantly perform the row operations needed. Pressing **2nd** **x⁻¹** (**MATRIX**) gives a screen similar to the one shown in Figure 7.17, where we begin by selecting the **EDIT** option (push the right arrow ◯ twice). Pressing ▭ places you on a screen where you can EDIT matrix *A*, changing the size as needed. Using the 3 × 4 system and matrix below,

$$\begin{cases} x + \ y + \ z = -1 \\ 2x + \ y - 2z = -7 \\ \ - 2y - \ z = -3 \end{cases} \qquad \begin{bmatrix} 1 & 1 & 1 & \vdots & -1 \\ 2 & 1 & -2 & \vdots & -7 \\ 0 & -2 & -1 & \vdots & -3 \end{bmatrix}$$

we press 3 and ▭, then 4 and ▭, giving the screen shown in Figure 7.18. The dash marks to the right indicate that there is a fourth column that cannot be seen, but that comes into view as you enter the elements of the matrix. Begin entering the first row of the matrix which has entries {1, 1, 1, −1}. Press ▭ after each entry and the cursor automatically goes to the next position in the matrix. After entering the second row {2, 1, −2, −7} and the third row {0, −2, −1, −3}, the completed matrix should look like the one shown in Figure 7.19 (the matrix is currently shifted to the right, showing the fourth column). To rewrite this matrix in reduced row-echelon form (**rref**) we now return to the home screen by pressing **2nd** **MODE** (QUIT), then press the **CLEAR** key for a clean screen. To access the **rref** function, press **2nd**

**Figure 7.17**

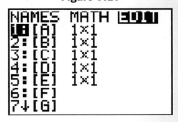

**Figure 7.18**              **Figure 7.19**              **Figure 7.20**

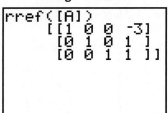

 (**MATRIX**) and select the **MATH** option, then scroll upward (or downward) until you get to **B:rref.** Pressing ⬤ places this function on the home screen, where we must tell it to perform the **rref** operation on matrix [A]. Press ⬤ ⬤ (**MATRIX**) to select a matrix (notice that matrix **NAMES** is automatically highlighted). Press ⬤ to select matrix [A] as the object of the **rref** function. After pressing ⬤ the calculator quickly computes the reduced row-echelon form and displays it on the screen as in Figure 7.20. The solution is easily read as $x = -3$, $y = 1$, and $z = 1$, which can be confirmed by substituting these values back into the original system. Use these ideas to complete the following.

**Exercise 4:** Use this method to solve the $2 \times 2$ system from Exercise 30 in Section 7.1.

**Exercise 5:** Use this method to solve the $3 \times 3$ system from Exercise 32 in Section 7.1.

## STRENGTHENING CORE SKILLS

### Augmented Matrices and Matrix Inverses

The formula for finding the inverse of a $2 \times 2$ matrix has its roots in the more general method of computing the inverse of an $n \times n$ matrix. This involves augmenting a square matrix $M$ with its corresponding identity $I_n$ on the right (forming an $n \times 2n$ matrix), and using row operations to *transform $M$ into the identity*. In some sense, as the original matrix is transformed, the "identity part" keeps track of the operations we used to convert $M$ and we can use the results to "get back home," so to speak. We'll illustrate by performing row operations similar to those used in Example 2B from Section 7.3, but instead on the general matrix $\begin{bmatrix} a & b \\ c & d \end{bmatrix}$.

$$\left[\begin{array}{cc|cc} a & b & 1 & 0 \\ c & d & 0 & 1 \end{array}\right]$$

next: $-cR1 + aR2 \to R2$

$$\begin{array}{r} -cR1 \\ +aR2 \\ \hline \text{New R2} \end{array} : \begin{array}{cccc} -ac & -bc & -c & 0 \\ ac & ad & 0 & a \\ \hline 0 & ad-bc & -c & a \end{array}$$

$$\left[\begin{array}{cc|cc} a & b & 1 & 0 \\ 0 & ad-bc & -c & a \end{array}\right]$$

next: $-bR2 + (ad-bc)R1 \to R1$

$$\begin{array}{r} -bR2 \\ +(ad-bc)R1 \\ \hline \text{New R1} \end{array} : \begin{array}{cccc} 0 & -b(ad-bc) & bc & -ab \\ a(ad-bc) & b(ad-bc) & ad-bc & 0 \\ \hline a(ad-bc) & 0 & ad & -ab \end{array}$$

$$\left[\begin{array}{cc|cc} a(ad-bc) & 0 & ad & -ab \\ 0 & ad-bc & -c & a \end{array}\right]$$

next: $\frac{R1}{a(ad-bc)} \to R1, \frac{R2}{ad-bc} \to R2$

$$\begin{array}{c} \dfrac{R1}{a(ad-bc)}: \\[2ex] \dfrac{R2}{ad-bc}: \end{array} \begin{array}{cccc} \dfrac{a(ad-bc)}{a(ad-bc)} & 0 & \dfrac{ad}{a(ad-bc)} & \dfrac{-ab}{a(ad-bc)} \\[2ex] 0 & \dfrac{ad-bc}{ad-bc} & \dfrac{-c}{ad-bc} & \dfrac{a}{ad-bc} \end{array}$$

$$\left[\begin{array}{cc|cc} 1 & 0 & \frac{d}{ad-bc} & \frac{-b}{ad-bc} \\[1ex] 0 & 1 & \frac{-c}{ad-bc} & \frac{a}{ad-bc} \end{array}\right]$$

This shows $A^{-1} = \begin{bmatrix} \frac{d}{ad-bc} & \frac{-b}{ad-bc} \\[1ex] \frac{-c}{ad-bc} & \frac{a}{ad-bc} \end{bmatrix}$ and factoring out $\frac{1}{ad-bc}$ produces the familiar formula.

As you might imagine, attempting this on a general $3 \times 3$ matrix is problematic at best, and instead we simply apply the augmented matrix method to find $A^{-1}$ for the $3 \times 3$ matrix shown in blue.

$$\left[\begin{array}{ccc|ccc} 2 & 1 & 0 & 1 & 0 & 0 \\ -1 & 3 & -2 & 0 & 1 & 0 \\ 3 & -1 & 2 & 0 & 0 & 1 \end{array}\right]$$

next: R1 + 2R2 → R2, −3R1 + 2R3 → R3

| R1 | 2 | 1 | 0 | 1 | 0 | 0 |
|---|---|---|---|---|---|---|
| + 2R2 | −2 | 6 | −4 | 0 | 2 | 0 |
| New R2 | 0 | 7 | −4 | 1 | 2 | 0 |

| −3R1 | −6 | −3 | 0 | −3 | 0 | 0 |
|---|---|---|---|---|---|---|
| + 2R3 | 6 | −2 | 4 | 0 | 0 | 2 |
| New R3 | 0 | −5 | 4 | −3 | 0 | 2 |

$$\left[\begin{array}{ccc|ccc} 2 & 1 & 0 & 1 & 0 & 0 \\ 0 & 7 & -4 & 1 & 2 & 0 \\ 0 & -5 & 4 & -3 & 0 & 2 \end{array}\right]$$

next: −1R2 + 7R1 → R1, 5R2 + 7R3 → R3

| −1R2 | 0 | −7 | 4 | −1 | −2 | 0 |
|---|---|---|---|---|---|---|
| + 7R1 | 14 | 7 | 0 | 7 | 0 | 0 |
| New R1 | 14 | 0 | 4 | 6 | −2 | 0 |

| 5R2 | 0 | 35 | −20 | 5 | 10 | 0 |
|---|---|---|---|---|---|---|
| + 7R3 | 0 | −35 | 28 | −21 | 0 | 14 |
| New R3 | 0 | 0 | 8 | −16 | 10 | 14 |

$$\left[\begin{array}{ccc|ccc} 14 & 0 & 4 & 6 & -2 & 0 \\ 0 & 7 & -4 & 1 & 2 & 0 \\ 0 & 0 & 8 & -16 & 10 & 14 \end{array}\right]$$

next: −1R3 + 2R1 → R1, R3 + 2R2 → R2

| −1R3 | 0 | 0 | −8 | 16 | −10 | −14 |
|---|---|---|---|---|---|---|
| + 2R1 | 28 | 0 | 8 | 12 | −4 | 0 |
| New R1 | 28 | 0 | 0 | 28 | −14 | −14 |

| R3 | 0 | 0 | 8 | −16 | 10 | 14 |
|---|---|---|---|---|---|---|
| + 2R2 | 0 | 14 | −8 | 2 | 4 | 0 |
| New R2 | 0 | 14 | 0 | −14 | 14 | 14 |

$$\left[\begin{array}{ccc|ccc} 28 & 0 & 0 & 28 & -14 & -14 \\ 0 & 14 & 0 & -14 & 14 & 14 \\ 0 & 0 & 8 & -16 & 10 & 14 \end{array}\right] \quad \left[\begin{array}{ccc|ccc} 1 & 0 & 0 & 1 & -0.5 & -0.5 \\ 0 & 1 & 0 & -1 & 1 & 1 \\ 0 & 0 & 1 & -2 & 1.25 & 1.75 \end{array}\right]$$

next: $\dfrac{R1}{28} \to R1, \dfrac{R2}{14} \to R2, \dfrac{R3}{8} \to R3$

To verify, we show $AA^{-1} = I$: $\left[\begin{array}{ccc} 2 & 1 & 0 \\ -1 & 3 & -2 \\ 3 & -1 & 2 \end{array}\right]\left[\begin{array}{ccc} 1 & -0.5 & -0.5 \\ -1 & 1 & 1 \\ -2 & 1.25 & 1.75 \end{array}\right] = \left[\begin{array}{ccc} 1 & 0 & 0 \\ 0 & 1 & 0 \\ 0 & 0 & 1 \end{array}\right]$ ✓ ($A^{-1}A = I$ also checks).

**Exercise 1:** Use the preceding inverse and a matrix equation to solve the system

$$\begin{cases} 2x + y = -2 \\ -x + 3y - 2z = -15. \\ 3x - y + 2z = 9 \end{cases}$$

# CUMULATIVE REVIEW CHAPTERS R–7

1. Perform the operations indicated.
   a. $(3 - 2i)(2 + i)$
   b. $(5 - 3i)^2$
   c. $\dfrac{8 - i}{2 + i}$
   d. $i^{49}$

2. Solve $S = 2\pi rh + 2\pi r^2$ for $h$.

**Solve the following equations.**

3. $2x - 4(3x + 1) = 5 - 4x$

4. $\dfrac{x + 6}{x + 2} - \dfrac{1}{x} = \dfrac{12}{x^2 + 2x}$

5. $9x^2 + 1 = 6x$

6. $\sqrt{2x + 11} - x = 6$

7. Find an equation of the line that passes through the points $(2, -2)$ and $(-3, 5)$.

8. Find an equation of the line perpendicular to the line shown, with the same $y$-intercept.

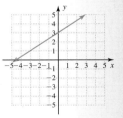

9. Given $f(x) = 3 - 4x - x^2$ and $g(x) = |4 - \sqrt{x + 2}|$, find
   a. $f(3)$
   b. $g(23)$
   c. $f(-2)$
   d. $g(-3)$

**Graph the following by using transformations of the parent function.**

**10.** $y = (x - 3)^2$

**11.** $y = \dfrac{1}{2}|x| - 3$

**12.** $y = -(x + 2)^3$

**13.** $y = \sqrt{-x} - 2$

**Solve the following inequalities. Express your answer in interval notation.**

**14.** $2(x - 2) + 3 \le 8$

**15.** $x^2 + 5 > 6x$

**16.** A long-distance calling card advertises one of the lowest rates available: 3¢ per minute after a 15¢ connection fee. After reading the fine print, you discover the cost of a call is modeled by the ceiling function $c(m) = 9\lceil \frac{1}{3}m \rceil + 15$, where $m$ is the length of the call in minutes and $c$ is the cost of the call in cents. Graph this function and explain why this card may not be as inexpensive as advertised.

**Name interval(s) where the following functions are increasing, decreasing, or constant. Write answers using interval notation.**

**17.** $y = f(x)$

**18.** $g(x) = \dfrac{1}{(x + 3)} + 3$

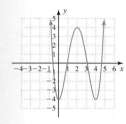

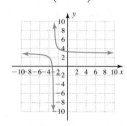

**19.** Compute the quotient $(4y^2 - 3y + 10) \div (2y + 1)$.

**20.** Use the remainder theorem to find $P(-3)$ for $P(x) = x^3 + 4x^2 - 15$.

**21.** Find a cubic polynomial with real coefficients having the roots $x = -2$ and $x = 3 - i$. Assume a leading coefficient of 1.

**22.** State the end-behavior and $y$-intercept of the following functions.
   **a.** $f(x) = 3x^3 - 16$
   **b.** $g(x) = -5x^6 + 2x^3 - 3x + 7$
   **c.** $h(x) = 10 - 2x^5$

**23.** Solve the system of equations using any method.
$$\begin{cases} 5x + 6y = -30 \\ -3x + 2y = 4 \end{cases}$$

**24.** Solve the system of inequalities by graphing.
$$\begin{cases} y > x^2 - 1 \\ x^2 + y^2 \le 13 \end{cases}$$

**25.** Simon Legré, a notorious French criminal, is planning to embezzle money from the Prefect of Police, and believes he can remove 12% of their $5,000,000 budget without being detected. He plans to use his ill-gotten gain to buy a legitimate business, then purchase a condominium and delivery truck for the business. The business will cost $20,000 more than twice the delivery truck, and the condominium will cost $200,000 more than the truck and business combined. How much will each cost?

# Analytic Geometry and the Conic Sections

## CHAPTER OUTLINE

## CHAPTER CONNECTIONS

In 1609, Johannes Kepler (1571–1630) first published his laws of planetary motion, irrefutably challenging the long-held beliefs of Aristotle and Ptolemy that the Earth was the center of the universe. His efforts were also instrumental in establishing that a planet's orbit was elliptical rather than circular, and in explaining why the speed of a planet seems to vary in its orbit. Building on his work, Sir Isaac Newton (1643–1727) would later show Kepler's laws were a natural result of gravitational attraction between heavenly bodies. Their work laid the foundations of modern space exploration. Some of these applications appear in Exercises 71 through 76 of Section 8.2.

**Check out these other real-world connections:**

▶ Designing an Elliptical Garden with Fountains (Section 8.2, Exercise 64)

▶ The Design of a Lithotripter for Treating Kidney Stones with Shockwaves (Section 8.2, Exercise 66)

▶ Locating a Ship Using Radar (Section 8.3, Exercise 75)

▶ Parabolic Shape of a Solar Furnace (Section 8.4, Exercise 76)

## LEARNING OBJECTIVES

*In Section 8.1 you will see how we can:*

☐ **A.** Verify theorems from basic geometry involving the distance between two points

☐ **B.** Verify that points $(x, y)$ are an equal distance from a given point and a given line

☐ **C.** Use the defining characteristics of a conic section to find its equation

Generally speaking, **analytical geometry** is a study of geometry using the tools of algebra and a coordinate system. These tools include the midpoint and distance formulas; the algebra of parallel, perpendicular, and intersecting lines; and other tools that help establish geometric concepts. In this section, we'll use these tools to verify certain relationships, then use these relationships to introduce a family of curves known as the **conic sections.**

## A. Verifying Relationships from Plane Geometry

For the most part, the algebraic tools used in this study were introduced in previous chapters. As the midpoint and distance formulas play a central role, they are restated here for convenience.

---

### Algebraic Tools Used in Analytical Geometry

Given points $P_1 = (x_1, y_1)$ and $P_2 = (x_2, y_2)$ in the $xy$-plane:

**Midpoint Formula**

The midpoint of line segment $P_1P_2$ is

$$(x, y) = \left( \frac{x_1 + x_2}{2}, \frac{y_1 + y_2}{2} \right)$$

**Distance Formula**

The distance from $P_1$ to $P_2$ is

$$d = \sqrt{(x_2 - x_1)^2 + (y_2 - y_1)^2}$$

---

These formulas can be used to verify the conclusion of many theorems from Euclidean geometry, while providing important links to an understanding of the conic sections.

**EXAMPLE 1** ▶ **Verifying a Theorem from Basic Geometry**

A theorem from basic geometry states: *The midpoint of the hypotenuse of a right triangle is an equal distance from all three vertices.* Verify this statement for the right triangle formed by $(-4, -2)$, $(4, -2)$, and $(4, 4)$.

**Solution** ▶

After the plotting points and drawing a triangle, we note the hypotenuse has endpoints $(-4, -2)$ and $(4, 4)$, with midpoint $\left( \frac{4 + (-4)}{2}, \frac{4 + (-2)}{2} \right) = (0, 1)$. Using the distance formula to find the distance from $(0, 1)$ to $(4, 4)$ gives

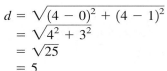

$$d = \sqrt{(4 - 0)^2 + (4 - 1)^2}$$
$$= \sqrt{4^2 + 3^2}$$
$$= \sqrt{25}$$
$$= 5$$

From the definition of midpoint, $(0, 1)$ is also 5 units from $(-4, -2)$.

Checking the distance from $(0, 1)$ to the vertex $(4, -2)$ gives

$$d = \sqrt{(4 - 0)^2 + (-2 - 1)^2}$$
$$= \sqrt{4^2 + (-3)^2}$$
$$= \sqrt{25}$$
$$= 5$$

It appears the midpoint of the hypotenuse is indeed an equal distance from all three vertices (see the figure).

**Now try Exercises 5 through 8** ▶

Recall from Section 2.1 that a circle is the set of all points that are an equal distance (called the radius) from a given point (called the center). If all three vertices of a triangle lie on the circumference of a circle, we say the circle **circumscribes** the triangle. Based on our earlier work, it appears we could also state the theorem in Example 1 as *For any circle in the xy-plane whose center (h, k) is the midpoint of the hypotenuse L of a right triangle, the circle defined by* $(x - h)^2 + (y - k)^2 = (\frac{L}{2})^2$, *circumscribes the triangle.* The circle and triangle from Example 1 illustrate this theorem in Figure 8.1, where the equation of the circle is $(x - 0)^2 + (y - 1)^2 = (\frac{10}{2})^2$. **See Exercises 9 through 14.**

Figure 8.1

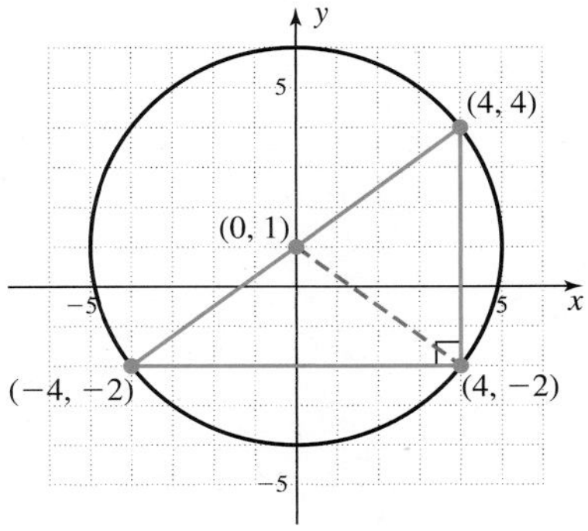

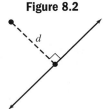

**A.** You've just seen how we can verify theorems from basic geometry involving the distance between two points

Figure 8.2

## B. The Distance between a Point and a Line

In a study of analytical geometry, we are also interested in the distance $d$ between a point and a *line*. This is always defined as the **perpendicular distance,** or the length of a line segment perpendicular to the given line, with the given point and the point of intersection as endpoints (see Figure 8.2).

**EXAMPLE 2 ▶ Locating Points That Are an Equal Distance from a Given Point and Line**

In Figure 8.3, the origin (0, 0) is seen to be an equal distance from the point (0, 2) and the line $y = -2$. Show that the following points are also an equal distance from (0, 2) and $y = -2$:

**a.** $(2, \frac{1}{2})$ **b.** (4, 2) **c.** (8, 8)

**Solution ▶** Since the given line is horizontal, the perpendicular distance from the line to each point can be found by vertically counting the units. It remains to show that this is also the distance from the given point to (0, 2) (see Figure 8.4).

**a.** The distance from $(2, \frac{1}{2})$ to $y = -2$ is **2.5 units**. The distance from $(2, \frac{1}{2})$ to (0, 2) is

$$\begin{aligned} d &= \sqrt{(0 - 2)^2 + (2 - 0.5)^2} \\ &= \sqrt{(-2)^2 + 1.5^2} \\ &= \sqrt{6.25} \\ &= 2.5 \checkmark \end{aligned}$$

Figure 8.3

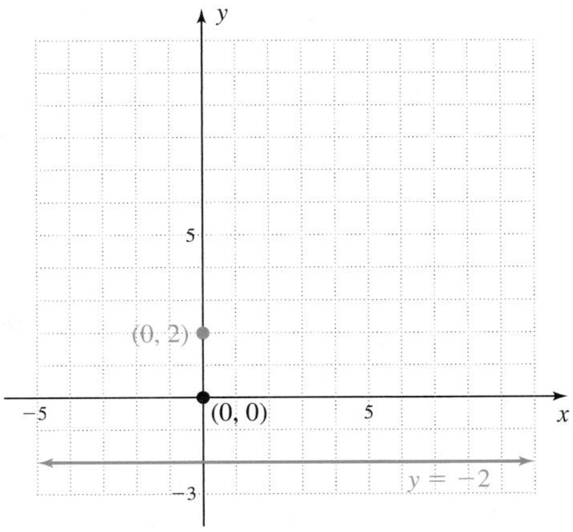

**b.** The distance from (4, 2) to $y = -2$ is **4 units**. The distance from (4, 2) to (0, 2) is

$$\begin{aligned} d &= \sqrt{(0 - 4)^2 + (2 - 2)^2} \\ &= \sqrt{(-4)^2 + 0^2} \\ &= \sqrt{16} \\ &= 4 \checkmark \end{aligned}$$

Figure 8.4

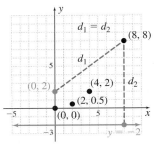

**c.** The distance $d_2$ from (8, 8) to $y = -2$ is **10 units**. The distance $d_1$ from (8, 8) to (0, 2) is

$$\begin{aligned} d_1 &= \sqrt{(0 - 8)^2 + (2 - 8)^2} \\ &= \sqrt{(-8)^2 + (-6)^2} \\ &= \sqrt{100} \\ &= 10 \checkmark \end{aligned}$$

**B.** You've just seen how we can verify that points $(x, y)$ are an equal distance from a given point and a given line

Now try Exercises 17 through 20 ▶

**Figure 8.5**

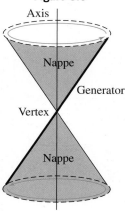

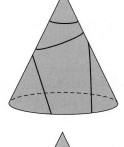

Circle

Ellipse

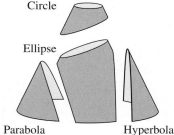

Parabola          Hyperbola

## C. Characteristics of the Conic Sections

Examples 1 and 2 bring us one step closer to the wider application of these ideas in a study of the conic sections. But before the connection is clearly made, we'll introduce some background on this family of curves. In common use, a cone might bring to mind the conical paper cups found at a water cooler. The point of the cone is called the **vertex** and the sheet of paper forming the sides is called a **nappe.** In mathematical terms, a cone has two nappes, formed by rotating a nonvertical line (called the generator), about a vertical line (called the axis), at their point of intersection—the vertex (see Figure 8.5). The conic sections are so named because all curves in the family can be formed by a *section* of the *cone,* or more precisely the intersection of a plane and a cone. Figure 8.6 shows that if the plane does not go through the vertex, the intersection will produce a circle, ellipse, parabola, or hyperbola.

**Figure 8.6**

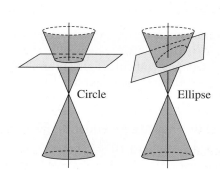
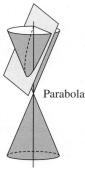
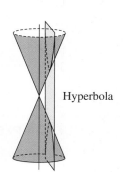

Circle          Ellipse          Parabola          Hyperbola

If the plane *does* go through the vertex, the result is a single point, a single line (if the plane contains the generator), or a pair of intersecting lines (if the plane contains the axis).

The connection we seek to make is that each conic section can be defined in terms of the distance between points in the plane, as in Example 1, or the distance between a given point and a line, as in Example 2. In Example 1, we noted the points $(-4, -2)$, $(4, -2)$, and $(4, 4)$ were all on a circle of radius 5 with center $(0, 1)$, in line with the analytic definition of a circle: *A circle is the set of all points that are an equal distance (called the radius) from a given point (called the center).*

In Example 2, you may have noticed that the points seemed to form the right branch of a parabola (see Figure 8.7), and in fact, this example illustrates the analytic definition of a parabola: *A parabola is the set of all points that are an equal distance from a given point (called the **focus**), and a given line (called the **directrix**).*

The focus and directrix are not actually part of the graph, they are simply used to locate points on the graph. For this reason all foci (plural of focus) will be represented by a "✳" symbol rather than a point.

**Figure 8.7**

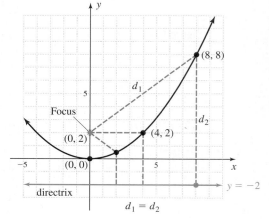

**EXAMPLE 3** ▶  **Finding an Equation for All Points That Form a Certain Parabola**

With Example 2 as a pattern, use the analytic definition to find a formula (equation) for the set of all points that form the parabola.

**Solution** ▶  Use the ordered pair $(x, y)$ to represent an arbitrary point on the parabola. Since any point on the line $y = -2$ has coordinates $(x, -2)$, we set the distance from $(x, -2)$ to $(x, y)$ equal to the distance from $(0, 2)$ to $(x, y)$. The result is

$$\sqrt{(x - x)^2 + [y - (-2)]^2} = \sqrt{(x - 0)^2 + (y - 2)^2} \qquad \text{distances are equal}$$

$$\sqrt{(y + 2)^2} = \sqrt{x^2 + (y - 2)^2} \qquad \text{simplify}$$

$$(y + 2)^2 = x^2 + (y - 2)^2 \qquad \text{power property}$$

$$y^2 + 4y + 4 = x^2 + y^2 - 4y + 4 \qquad \text{expand binomials}$$

$$8y = x^2 \qquad \text{simplify}$$

$$y = \frac{1}{8}x^2 \qquad \text{result}$$

All points satisfying these conditions are on the parabola defined by $y = \frac{1}{8}x^2$.

**Now try Exercises 21 and 22** ▶

   At this point, it seems reasonable to ask what happens when the distance from the focus to $(x, y)$ is *less than* the distance from the directrix to $(x, y)$. For example, what if the distance is only two-thirds as long? As you might guess, the result is one of the other conic sections, in this case an ellipse. If the distance from the focus to a point $(x, y)$ is *greater than* the distance from the directrix to $(x, y)$, one branch of a hyperbola is formed. While we will defer a development of their general equations until later in the chapter, the following diagrams serve to illustrate this relationship for the ellipse, and show why we refer to the conic sections as a *family of curves*. In Figure 8.8, the line segment from the focus to each point on the graph (shown in blue), is exactly two-thirds the length of the line segment from the directrix to the same point (shown in red). Note the graph of these points forms the right half of an ellipse. In Figure 8.9, the lines and points forming the first half are hidden to more clearly show the remaining points that form the complete graph.

**Figure 8.8**

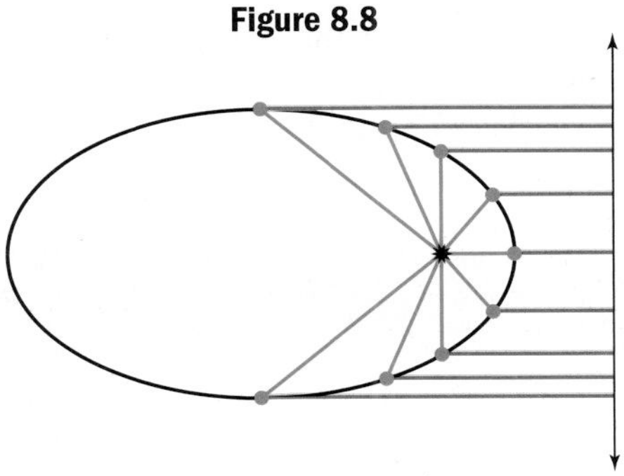

**Figure 8.9**

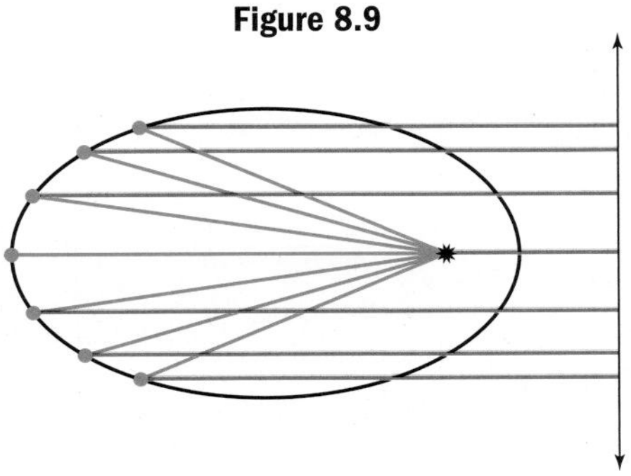

**EXAMPLE 4** ▶  **Finding an Equation for All Points That Form a Certain Ellipse**

Suppose we arbitrarily select the point $(1, 0)$ as a focus and the (vertical) line $x = 4$ as the directrix. Use these to find an equation for the set of all points where the distance from the focus to a point $(x, y)$ is $\frac{1}{2}$ the distance from the directrix to $(x, y)$.

**Solution** ▶  Since any point on the line $x = 4$ has coordinates $(4, y)$, we have:

Distance from $(1, 0)$ to $(x, y) = \dfrac{1}{2}$ [distance from $(4, y)$ to $(x, y)$]   in words

$$\sqrt{(x-1)^2 + [y-0]^2} = \frac{1}{2}\sqrt{(x-4)^2 + (y-y)^2}$$   resulting equation

$$\sqrt{(x-1)^2 + y^2} = \frac{1}{2}\sqrt{(x-4)^2}$$   simplify

$$(x-1)^2 + y^2 = \frac{1}{4}(x-4)^2$$   power property

$$x^2 - 2x + 1 + y^2 = \frac{1}{4}(x^2 - 8x + 16)$$   expand binomials

$$x^2 - 2x + 1 + y^2 = \frac{1}{4}x^2 - 2x + 4$$   distribute

$$\frac{3}{4}x^2 + y^2 = 3$$   simplify: $1x^2 - \frac{1}{4}x^2 = \frac{3}{4}x^2$

$$3x^2 + 4y^2 = 12$$   polynomial form

All points satisfying these conditions are on the ellipse defined by $3x^2 + 4y^2 = 12$.

☑ **C.** You've just seen how we can use the defining characteristics of a conic section to find its equation

**Now try Exercises 23 and 24** ▶

Actually, any given ellipse has two foci (see Figure 8.10) and the equation from Example 4 could also have been developed using the left focus (with the directrix also on the left). This symmetrical relationship leads us to an *alternative definition* for the ellipse, which we will explore further in Section 8.2:

> *For foci $f_1$ and $f_2$, an ellipse is the set of all points $(x, y)$ where the sum of the distances from $f_1$ to $(x, y)$ and $f_2$ to $(x, y)$ is constant.*

See Figure 8.11 and **Exercises 25 and 26.** Both the focus/directrix definition and the "two foci" definition have merit, and simply tend to call out different characteristics and applications of the ellipse. The hyperbola also has a focus/directrix definition and a "two foci" definition. **See Exercises 27 and 28.**

**Figure 8.10**

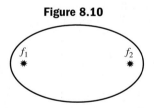

**Figure 8.11**

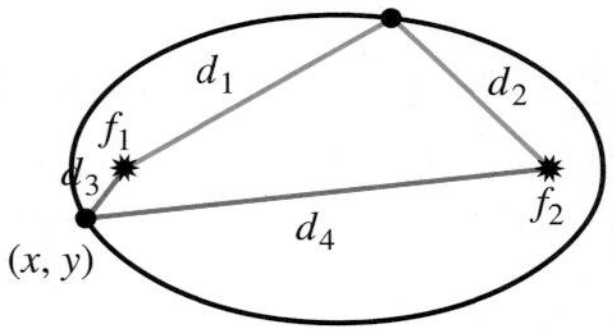

$d_1 + d_2 = d_3 + d_4$

## 8.1 EXERCISES

### ▶ CONCEPTS AND VOCABULARY

**Fill in the blank with the appropriate word or phrase. Carefully reread the section, if necessary.**

**1.** The distance between a point and a line always refers to the _____ distance.

**2.** The conic sections are formed by the intersection of a(n) _____ and a(n) _____.

**3.** If a plane intersects a cone at its vertex, the result is a(n) _____, a line, or a pair of _____ lines.

**4.** A circle is defined relative to an equal distance between two _____. A parabola is defined relative to an equal distance between a(n) _____ and a(n) _____.

## ▶ DEVELOPING YOUR SKILLS

**The three points given form a right triangle. Find the midpoint of the hypotenuse and verify that the midpoint is an equal distance from all three vertices.**

**5.** $P_1 = (-5, 2)$
$P_2 = (1, 2)$
$P_3 = (-5, -6)$

**6.** $P_1 = (3, 2)$
$P_2 = (3, 14)$
$P_3 = (8, 2)$

**7.** $P_1 = (10, -21)$
$P_2 = (-6, -9)$
$P_3 = (3, 3)$

**8.** $P_1 = (6, -6)$
$P_2 = (-12, 18)$
$P_3 = (20, 42)$

**9.** Find an equation of the circle that circumscribes the triangle in Exercise 5.

**10.** Find an equation of the circle that circumscribes the triangle in Exercise 6.

**11.** Find an equation of the circle that circumscribes the triangle in Exercise 7.

**12.** Find an equation of the circle that circumscribes the triangle in Exercise 8.

**13.** Of the following six points, four are an equal distance from the point $A(2, 3)$ and two are not. (a) Identify which four, and (b) find any two additional points that are this same (nonvertical, nonhorizontal) distance from $(2, 3)$:

$B(7, 15)$    $C(-10, 8)$    $D(9, 14)$    $E(-3, -9)$
$F(5, 4 + 3\sqrt{10})$    $G(2 - 2\sqrt{30}, 10)$

**14.** Of the following six points, four are an equal distance from the point $P(-1, 4)$ and two are not. (a) Identify which four, and (b) find any two additional points that are the same (nonvertical, nonhorizontal) distance from $(-1, 4)$.

$Q(-9, 10)$    $R(5, 12)$    $S(-7, 11)$    $T(4, 4 + 5\sqrt{3})$
$U(-1 + 4\sqrt{6}, 6)$    $V(-7, 4 + \sqrt{51})$

## ▶ WORKING WITH FORMULAS

**The Perpendicular Distance from a Point to a Line:** $d = \frac{|Ax_1 + By_1 + C|}{\sqrt{A^2 + B^2}}$. **The perpendicular distance from a point $(x_1, y_1)$ to a given line can be found using the formula shown, where $Ax + By + C = 0$ is the equation of the line in standard form.**

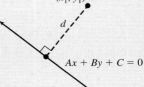

**15.** Use the formula to verify that $P(-6, 2)$ and $Q(6, 4)$ are an equal distance from the line $y = -\frac{1}{2}x + 3$.

**16.** Find the value(s) for $y$ that ensure $(1, y)$ is this same distance from $y = -\frac{1}{2}x + 3$.

## ▶ APPLICATIONS

**17.** Of the following four points, three are an equal distance from the point $A(0, 1)$ and the line $y = -1$. (a) Identify which three, and (b) find any two additional points that satisfy these conditions.

$B(-6, 9)$    $C(4, 4)$    $D(-2\sqrt{2}, 6)$    $E(4\sqrt{2}, 8)$

**18.** Of the following four points, three are an equal distance from the point $P(2, 4)$ and the line $y = -4$. (a) Identify which three, and (b) find any two additional points that satisfy these conditions.

$Q(-10, 9)$    $R(2 + 4\sqrt{2}, 3)$    $S(10, 4)$
$T(2 - 4\sqrt{5}, 5)$

**19.** Consider the fixed *point* $(0, -4)$ and the fixed *line* $y = 4$. Verify that the distance from each point given to $(0, -4)$ is equal to the distance from the point to the line $y = 4$.

$A(4, -1)$    $B\left(10, -\frac{25}{4}\right)$    $C(4\sqrt{2}, -2)$
$D(8\sqrt{5}, -20)$

**20.** Consider the fixed *point* $(0, -2)$ and the fixed *line* $y = 2$. Verify that the distance from each point given to

$(0, -2)$, is equal to the distance from the point to the line $y = 2$.

$P(12, -18)$    $Q\left(6, -\frac{9}{2}\right)$    $R(4\sqrt{5}, -10)$
$S(4\sqrt{6}, -12)$

**21.** The points from Exercise 19 are on the graph of a parabola. Find an equation of the parabola and verify the given points satisfy the equation.

**22.** The points from Exercise 20 are on the graph of a parabola. Find an equation of the parabola and verify the given points satisfy the equation.

**23.** Using $(0, -2)$ as the focus and the horizontal line $y = -8$ as the directrix, find an equation for the set of all points $(x, y)$ where the distance from the focus to $(x, y)$ is one-half the distance from the directrix to $(x, y)$.

**24.** Using $(4, 0)$ as the focus and the vertical line $x = 9$ as the directrix, find an equation for the set of all points $(x, y)$ where the distance from the focus to $(x, y)$ is two-thirds the distance from the directrix to $(x, y)$.

**25.** From Exercise 23, verify the points $(-3, 2)$ and $(\sqrt{12}, 0)$ are on the ellipse defined by $4x^2 + 3y^2 = 48$. Then verify that $d_1 + d_2 = d_3 + d_4$.

**Exercise 25**

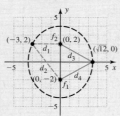

**26.** From Exercise 24, verify the points $\left(4, \frac{10}{3}\right)$ and $(-3, -\sqrt{15})$ are on the ellipse defined by $5x^2 + 9y^2 = 180$. Then verify that $d_1 + d_2 = d_3 + d_4$.

**Exercise 26**

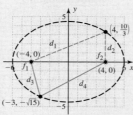

**27.** From the focus/directrix definition of a hyperbola: *If the distance from the focus to a point $(x, y)$ is **greater than** the distance from the directrix to $(x, y)$, one branch of a hyperbola is formed.* Using $(2, 0)$ as the focus and the vertical line $x = \frac{1}{2}$ as the directrix, find an equation for the set of all points $(x, y)$ where the distance from the focus to $(x, y)$, is twice the distance from the directrix to $(x, y)$.

**28.** From the "two foci" definition of a hyperbola: *For foci $f_1$ and $f_2$, a hyperbola is the set of all points $(x, y)$ where the difference of the distances from $f_1$ to $(x, y)$ and $f_2$ to $(x, y)$ is constant.* Verify the points $(2, 3)$ and $(-3, -2\sqrt{6})$ are on the graph of the hyperbola from Exercise 27. Then verify $d_1 - d_2 = d_3 - d_4$.

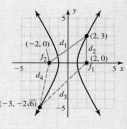

▶ **EXTENDING THE CONCEPT**

**29. Properties of a circle:** A theorem from elementary geometry states: *If a radius is perpendicular to a chord, it bisects the chord.* Verify this is true for the circle, radii, and chords shown.

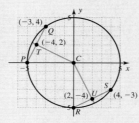

**30.** Verify that points $C(-2, 3)$ and $D(2\sqrt{2}, \sqrt{6})$ are points on the ellipse with foci at $A(-2, 0)$ and $B(2, 0)$, by verifying $d(AC) + d(BC) = d(AD) + d(BD)$. The expression that results has the form $\sqrt{U + V} + \sqrt{U - V}$, which prior to the common use of technology, had to be simplified using the formula $\sqrt{U + V} + \sqrt{U - V} = \sqrt{a + \sqrt{b}}$, where $a = 2U$ and $b = 4(U^2 - V^2)$. Use this relationship to simplify the equation above.

▶ **MAINTAINING YOUR SKILLS**

**31.** (5.6) $5000 is deposited at 4% compounded continuously. How many years will it take for the account to exceed $8000?

**32.** (4.3) Use the rational zeroes theorem and other tools to factor $f(x)$ and sketch its graph: $f(x) = x^4 - 3x^3 - 3x^2 + 11x - 6$.

**33.** (5.5) Solve for $x$ in both exact and approximate form:
   **a.** $5 = \dfrac{10}{1 + 9e^{-0.5x}}$   **b.** $345 = 5e^{0.4x} + 75$

**34.** (4.5) Sketch a complete graph of $h(x) = \frac{x^2 - 9}{x^2 - 4}$. Clearly label all intercepts and asymptotes.

## ▶ APPLICATIONS

**63. Decorative fireplaces:** A bricklayer intends to build an elliptical fireplace 3 ft high and 8 ft wide, with two glass doors that open at the middle. The hinges to these doors are to be screwed onto a spine that is perpendicular to the hearth and goes through the foci of the ellipse. How far from center will the spines be located? How tall will each spine be?

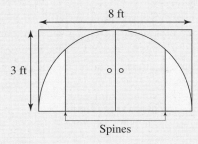

8 ft

3 ft

Spines

**64. Decorative gardens:** A retired math teacher decides to present her husband with a beautiful elliptical garden to help celebrate their 50th anniversary. The ellipse is to be 8 m long and 5 m across, with decorative fountains located at the foci. How far from the center of the ellipse should the fountains be located (round to the nearest 100th of a meter)? How far apart are the fountains?

**65. Attracting attention to art:** A gallery owner asks a student from the university to design an exhibit that will highlight an ancient sculpture. The student decides to create an elliptical showroom with reflective walls, with a rotating laser light on a stand at one focus, and the sculpture placed at the other focus. If the elliptical room is 24 ft long and 16 ft wide, how far from the center of the ellipse should the stands be located (round to the nearest 10th of a foot)? How far apart are the stands?

**66. Medical procedures:** *Lithotripsy* is a noninvasive medical procedure that is used to break up kidney and bladder stones in the body. A machine called a *lithotripter* uses its three-dimensional semielliptical shape and the properties of an ellipse to concentrate shock waves generated at one focus, on a kidney stone located at the other focus (see diagram—not drawn to scale). If the lithotripter has a length (semimajor axis) of 16 cm and a radius (semiminor axis) of 10 cm, how far from the vertex should a kidney stone be located for the best result?

**Exercise 66**

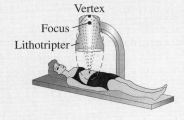

Vertex

Focus

Lithotripter

**67. Elliptical arches:** In some situations, bridges are built using uniform elliptical archways as shown in the figure given. Find the equation of a central ellipse forming each arch if it has a total width of 30 ft and a maximum center height (above level ground) of 8 ft. What is the height of a point 9 ft to the right of the center of each arch?

**Exercise 67**

8 ft

60 ft

**68. Elliptical arches:** An elliptical arch bridge is built across a one-lane highway. The arch is 20 ft across and has a maximum center height of 12 ft. Will a farm truck hauling a load 10 ft wide with a clearance height of 11 ft be able to go under the bridge without damage? (*Hint:* See Exercise 67.)

**69. Plumbing:** A properly vented home enables water to run freely throughout its plumbing system, while helping to prevent sewage gases from entering the home. Find the equation of the elliptical hole cut in a roof in order to allow a 3-in. vent pipe to exit, if the roof has a slope of $\frac{4}{3}$.

**70. Light projection:** Standing a short distance from a wall, Kymani's flashlight projects a circle of radius 30 cm. When holding the flashlight at an angle, a vertical ellipse 100 cm long is formed, with the focus 10 cm from the vertex. Find the equation of the central circle and ellipse, and the area of the wall that each illuminates.

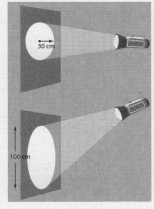

30 cm

100 cm

As a planet orbits around the Sun, it traces out an ellipse with the Sun at one *focus* of the ellipse. Use this information and the graphs provided to complete Exercises 71 through 76.

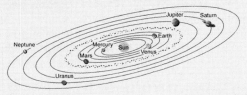

**71. Orbit of Mercury:** The approximate orbit of the planet Mercury is shown in the figure given. Find an equation that models this orbit.

**Exercise 71**

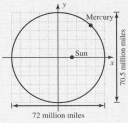

Mercury

Sun

70.5 million miles

72 million miles

72. **Orbit of Pluto:** The approximate orbit of the Kuiper object Pluto is shown in the figure given. Find an equation that models this orbit.

**Exercise 72**

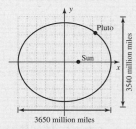

73. **Planetary orbits:** The maximum distance from the planet Mars to the Sun is about 156 million miles, while the minimum distance is about 128 million miles. Use this information to find the lengths of the semimajor and semiminor axes, rounded to the nearest million. If Mars has an orbital velocity of 54,000 miles per hour (1.296 million miles per day), how many days does it take Mars to orbit the Sun?

74. **Planetary orbits:** The maximum distance from the planet Saturn to the Sun is about 940 million miles, while the minimum distance is about 840 million miles. Use this information to find the lengths of the semimajor and semiminor axes, rounded to the nearest million. If Saturn has an orbital velocity of 21,650 miles per hour (about 0.52 million miles per day), how many days does it take Saturn to orbit the Sun? (*Hint:* Use the formula from Exercise 62.)

75. **Orbital velocity of Earth:** The minimum distance from the Earth to the Sun is about 91 million mi, maximum distance is 95 million mi, and it completes one orbit in about 365 days. Use this information and the formula from Exercise 62 to find Earth's orbital speed around the Sun in miles per hour.

76. **Orbital velocity of Jupiter:** The planet Jupiter has a perihelion of 460 million mi, an aphelion of 508 million mi, and completes one orbit in about 4329 days. Use this information and the formula from Exercise 62 to find Jupiter's orbital speed around the Sun in miles per hour.

77. **Area of a race track:** Suppose the *Toronado 500* is a car race that is run on an elliptical track. The track is bounded by two ellipses with equations of $4x^2 + 9y^2 = 900$ and $9x^2 + 25y^2 = 900$, where $x$ and $y$ are in hundreds of yards. Use the formula given in Exercise 61 to find the area of the race track.

78. **Area of a border:** The tablecloth for a large oval table is elliptical in shape. It is designed with two concentric ellipses (one within the other) as shown in the figure. The equation of the outer ellipse is $9x^2 + 25y^2 = 225$, and the equation of the inner ellipse is $4x^2 + 16y^2 = 64$ with $x$ and $y$ in feet. Use the formula given in Exercise 61 to find the area of the border of the tablecloth.

**Exercise 78**

79. **Whispering galleries:** Due to their unique properties, ellipses are used in the construction of *whispering galleries* like those in St. Paul's Cathedral (London) and Statuary Hall in the U.S. Capitol. Suppose a whispering gallery was built using the equation shown, with the dimensions in feet. (a) How tall is the ceiling at its highest point? (b) How wide is the gallery vertex to vertex? (c) How far from the base of the doors at either end, should a young couple stand so that one can clearly hear the other whispering, "I love you."?

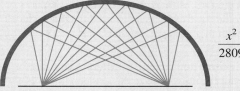

$$\frac{x^2}{2809} + \frac{y^2}{2025} = 1$$

80. While an elliptical billiard table has little practical value, it offers an excellent illustration of elliptical properties. A ball placed at one focus and hit with the cue stick from any angle, will hit the cushion and immediately rebound to the other focus and continue through each focus until coming to rest. Suppose one such table was constructed using the equation $\frac{x^2}{9} + \frac{y^2}{4} = 1$ as a model, with the dimensions in feet. (a) How far apart are the vertices? (b) How far apart are the foci?

▶ **EXTENDING THE CONCEPT**

81. For $6x^2 + 36x + 3y^2 - 24y + 74 = -28$, does the equation appear to be that of a circle, ellipse, or parabola? Write the equation in factored form. What do you notice? What can you say about the graph of this equation?

82. Algebraically verify that for a horizontal ellipse ($a > b$) $\frac{x^2}{a^2} + \frac{y^2}{b^2} = 1$, the length of the focal chord is $\frac{2a^2}{b}$.

▶ **MAINTAINING YOUR SKILLS**

83. (5.4) Evaluate the expression $\log_3 20$.

84. (1.4) Compute the product and quotient of: $z_1 = 2\sqrt{3} + 2i\sqrt{3}$; $z_2 = 5\sqrt{3} - 5i$

85. (1.3) Solve the absolute value inequality (a) graphically and (b) analytically:

$$-2|x - 3| + 10 > 4.$$

86. (3.3) The resistance $R$ to current flow in an electrical wire varies directly with the length $L$ of the wire and inversely with the square of its diameter $d$. (a) Write the variation equation; (b) find the constant of variation if a wire 2 m long with diameter $d = 0.005$ m has a resistance of 240 ohms ($\Omega$); and (c) find the resistance in a similar wire 3 m long and 0.006 m in diameter.

## MID-CHAPTER CHECK

**Sketch the graph of each conic section.**

1. $(x - 4)^2 + (y + 3)^2 = 9$
2. $x^2 + y^2 - 10x + 4y + 4 = 0$
3. $\dfrac{(x - 2)^2}{16} + \dfrac{(y + 3)^2}{1} = 1$
4. $9x^2 + 4y^2 + 18x - 24y + 9 = 0$
5. $\dfrac{(x + 3)^2}{9} + \dfrac{(y - 4)^2}{4} = 1$
6. $9x^2 + 16y^2 - 36x + 96y + 36 = 0$
7. Find the equation for all points located an equal distance from the point $(0, 3)$ and the line $y = -3$.

8. Find the equation of each relation and state its domain and range.

   **a.** 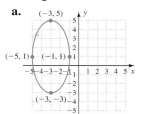     **b.**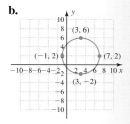

9. Find the equation of the ellipse (in standard form) having foci at $(0, 13)$ and $(0, -13)$, with a minor axis of length 10 units.

10. Find the equation of the ellipse (in standard form) if the vertices are $(-4, 0)$ and $(4, 0)$ and the distance between the foci is $4\sqrt{3}$ units.

---

## 8.3 | The Hyperbola

### LEARNING OBJECTIVES

*In Section 8.3 you will see how we can:*

☐ **A.** Use the equation of a hyperbola to graph central and noncentral hyperbolas

☐ **B.** Distinguish between the equations of circles, ellipses, and hyperbolas

☐ **C.** Locate the foci of a hyperbola and use the foci and other features to write its equation

☐ **D.** Solve applications involving foci

As seen in Section 8.1 (see Figure 8.21), a hyperbola is a conic section formed by a plane that cuts both nappes of a right circular cone. A hyperbola has two symmetric parts called **branches,** which open in opposite directions. Although the branches appear to resemble parabolas, we will soon discover they are a very different curve.

**Figure 8.21**

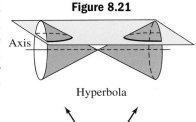

### A. The Equation of a Hyperbola

In Section 8.2 we noted that for the equation $Ax^2 + By^2 = F$ $(A, B, F, > 0)$, if $A = B$ the equation is that of a circle, and if $A \neq B$ the equation represents an ellipse. Both cases contain a *sum* of second-degree terms. Perhaps driven by curiosity, we might wonder what happens if the equation has a *difference* of second-degree terms. Consider the equation $9x^2 - 16y^2 = 144$. It appears the graph will be centered at $(0, 0)$ since no shifts are applied ($h$ and $k$ are both zero). Using the intercept method to graph this equation reveals an entirely new curve, called a *hyperbola*.

---

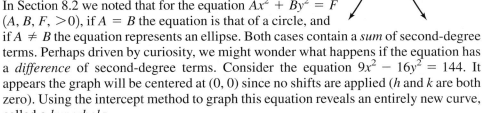

**EXAMPLE 1** ▶   **Graphing a Central Hyperbola**

Graph the equation $9x^2 - 16y^2 = 144$ using intercepts and additional points as needed.

**Solution** ▶   Substituting 0 for $x$ to find the $y$-intercepts gives

$$9x^2 - 16y^2 = 144 \quad \text{given}$$
$$9(0)^2 - 16y^2 = 144 \quad \text{substitute 0 for } x$$
$$-16y^2 = 144 \quad \text{simplify}$$
$$y^2 = -9 \quad \text{divide by } -16$$

Since $y^2$ can never be negative, we conclude that the graph has *no y-intercepts.* Substituting 0 for $y$ to find the $x$-intercepts gives

$$9x^2 - 16y^2 = 144 \qquad \text{given}$$
$$9x^2 - 16(0)^2 = 144 \qquad \text{substitute 0 for } y$$
$$9x^2 = 144 \qquad \text{simplify}$$
$$x^2 = 16 \qquad \text{divide by 9}$$
$$x = \sqrt{16} \quad \text{and} \quad x = -\sqrt{16} \qquad \text{square root property}$$
$$x = 4 \quad \text{and} \quad x = -4 \qquad \text{simplify}$$
$$(4, 0) \quad \text{and} \quad (-4, 0) \qquad x\text{-intercepts}$$

Knowing the graph has no $y$-intercepts, we select inputs greater than 4 and less than $-4$ to help sketch the graph. Using $x = 5$ and $x = -5$ yields

$$9x^2 - 16y^2 = 144 \qquad \text{given} \qquad\qquad 9x^2 - 16y^2 = 144$$
$$9(5)^2 - 16y^2 = 144 \qquad \text{substitute for } x \qquad 9(-5)^2 - 16y^2 = 144$$
$$9(25) - 16y^2 = 144 \qquad 5^2 = (-5)^2 = 25 \qquad 9(25) - 16y^2 = 144$$
$$225 - 16y^2 = 144 \qquad \text{simplify} \qquad\qquad 225 - 16y^2 = 144$$
$$-16y^2 = -81 \qquad \text{subtract 225} \qquad -16y^2 = -81$$
$$y^2 = \frac{81}{16} \qquad \text{divide by } -16 \qquad\qquad y^2 = \frac{81}{16}$$
$$y = \frac{9}{4} \quad y = -\frac{9}{4} \qquad \text{square root property} \qquad y = \frac{9}{4} \quad y = -\frac{9}{4}$$
$$y = 2.25 \quad y = -2.25 \qquad \text{decimal form} \qquad y = 2.25 \quad y = -2.25$$
$$(5, 2.25) \quad (5, -2.25) \qquad \text{ordered pairs} \qquad (-5, 2.25) \quad (-5, -2.25)$$

Plotting these points and connecting them with a smooth curve, while *knowing there are no y-intercepts,* produces the graph in the figure. The point at the origin (in blue) is not a part of the graph, and is given only to indicate the "center" of the hyperbola. The points $(-4, 0)$ and $(4, 0)$ are called **vertices,** and the **center** of the hyperbola is always the point halfway between them.

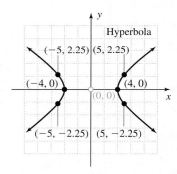

**Now try Exercises 5 through 16** ▶

Since the hyperbola in Example 1 crosses a horizontal line of symmetry, it is referred to as a **horizontal hyperbola.** If the center is at the origin, we have a **central hyperbola.** The line passing through the center and both vertices is called the **transverse axis** (vertices are always on the transverse axis), and the line passing through the center and perpendicular to this axis is called the **conjugate axis** (see Figure 8.22).

In Example 1, the coefficient of $x^2$ was positive and we were subtracting $16y^2$: $9x^2 - 16y^2 = 144$. The result was a horizontal hyperbola. If the $y^2$-term is positive and we subtract the term containing $x^2$, the result is a **vertical hyperbola** (Figure 8.23).

**Figure 8.22**

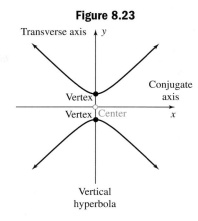

Horizontal
hyperbola

**Figure 8.23**

Vertical
hyperbola

**EXAMPLE 2** ▶ **Identifying the Axes, Vertices, and Center of a Hyperbola from Its Graph**

For the hyperbola shown, state the location of the
vertices and the equation of the transverse axis.
Then identify the location of the center and the
equation of the conjugate axis.

**Solution** ▶ By inspection we locate the vertices at (0, 0) and
(0, 4). The equation of the transverse axis is
$x = 0$. The center is halfway between the vertices
at (0, 2), meaning the equation of the conjugate
axis is $y = 2$.

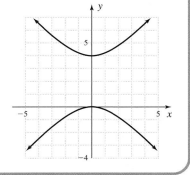

**Now try Exercises 17 through 20** ▶

### Standard Form

As with the ellipse, the polynomial form of the equation is helpful for *identifying*
hyperbolas, but not very helpful when it comes to *graphing* a hyperbola (since we still
must go through the laborious process of finding additional points). For graphing,
standard form is once again preferred. Consider the hyperbola $9x^2 - 16y^2 = 144$
from Example 1. To write the equation in standard form, we divide by 144 and obtain
$\frac{x^2}{4^2} - \frac{y^2}{3^2} = 1$. By comparing the standard form to the graph, we note $a = 4$ represents
the distance from center to vertices, similar to the way we used $a$ previously. But since
the graph has no $y$-intercepts, what could $b = 3$ represent? The answer lies in the fact that
branches of a hyperbola are **linearly asymptotic,** meaning they will approach and
become very close to imaginary lines that can be used to sketch the graph. For a central
hyperbola, the slopes of the asymptotic lines are given by the ratios $\frac{b}{a}$ and $-\frac{b}{a}$, with the
related equations being $y = \frac{b}{a}x$ and $y = -\frac{b}{a}x$. The graph from Example 1 is repeated
in Figure 8.24, with the asymptotes drawn. For a clearer understanding of how the
equations for the asymptotes were determined, **see Exercise 77.**

A second method of drawing the asymptotes involves drawing a **central rectangle** with dimensions 2*a* by 2*b*, as shown in Figure 8.25. The asymptotes will be the *extended diagonals* of this rectangle. This brings us to the equation of a hyperbola in standard form.

**Figure 8.24**

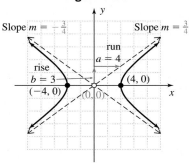

Slope method

**Figure 8.25**

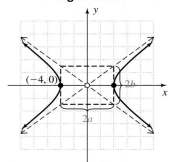

Central rectangle method

---

### The Equation of a Hyperbola in Standard Form

The equation

$$\frac{(x - h)^2}{a^2} - \frac{(y - k)^2}{b^2} = 1$$

represents a *horizontal* hyperbola with center $(h, k)$
- *transverse* axis $y = k$
- *conjugate* axis $x = h$
- $a$ gives the distance from center to vertices.

The equation

$$\frac{(y - k)^2}{b^2} - \frac{(x - h)^2}{a^2} = 1$$

represents a *vertical* hyperbola with center $(h, k)$
- *transverse* axis $x = h$
- *conjugate* axis $y = k$
- $b$ gives the distance from center to vertices.

- Asymptotes can be drawn by starting at $(h, k)$ and using slopes $m = \pm\frac{b}{a}$.

---

**EXAMPLE 3** ▶    **Graphing a Hyperbola Using Its Equation in Standard Form**

Sketch the graph of $16(x - 2)^2 - 9(y - 1)^2 = 144$, and label the center, vertices, and asymptotes.

**Solution** ▶    Begin by noting a difference of the second-degree terms, with the $x^2$-term occurring first. This means we'll be graphing a horizontal hyperbola whose center is at $(2, 1)$. Continue by writing the equation in standard form.

$$16(x - 2)^2 - 9(y - 1)^2 = 144 \qquad \text{given equation}$$

$$\frac{16(x - 2)^2}{144} - \frac{9(y - 1)^2}{144} = \frac{144}{144} \qquad \text{divide by 144}$$

$$\frac{(x - 2)^2}{9} - \frac{(y - 1)^2}{16} = 1 \qquad \text{simplify}$$

$$\frac{(x - 2)^2}{3^2} - \frac{(y - 1)^2}{4^2} = 1 \qquad \text{write denominators in squared form}$$

Since $a = 3$ the vertices are a horizontal distance of 3 units from the center $(2, 1)$, giving $(2 + 3, 1) = (5, 1)$ and $(2 - 3, 1) = (-1, 1)$. After plotting the center and vertices, we can begin at the center and count off slopes of $m = \pm\frac{b}{a} = \pm\frac{4}{3}$, or draw a rectangle centered at $(2, 1)$ with dimensions $2(3) = 6$ (horizontal dimension) by $2(4) = 8$ (vertical dimension) to sketch the asymptotes. The complete graph is shown here.

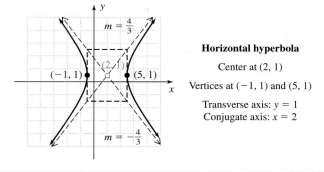

**Horizontal hyperbola**

Center at $(2, 1)$

Vertices at $(-1, 1)$ and $(5, 1)$

Transverse axis: $y = 1$
Conjugate axis: $x = 2$

Now try Exercises 21 through 34 ▶

If the hyperbola in Example 3 were a central hyperbola, the equations of the asymptotes (or the point-slope formula) would be $y = \frac{4}{3}x$ and $y = -\frac{4}{3}x$. But the center of this graph has been shifted 2 units right and 1 unit up. Using our knowledge of transformations, the equations for the asymptotes of the shifted hyperbola must be

$$(y - 1) = +\frac{4}{3}(x - 2) \text{ and } (y - 1) = -\frac{4}{3}(x - 2)$$

### Polynomial Form

If the equation is given as a polynomial in expanded form, complete the square in $x$ and $y$, then write the equation in standard form.

**EXAMPLE 4 ▶ Graphing a Hyperbola by Completing the Square**

Graph the equation $9y^2 - x^2 + 54y + 4x + 68 = 0$ by completing the square. Label the center and vertices, and sketch the asymptotes.

**Solution ▶** Since the $y^2$-term occurs first, we assume the equation represents a vertical hyperbola, but wait for the factored form to be sure (**see Exercise 79**).

$$9y^2 - x^2 + 54y + 4x + 68 = 0 \qquad \text{given}$$

$$9y^2 + 54y - x^2 + 4x = -68 \qquad \text{collect like-variable terms; subtract 68}$$

$$9(y^2 + 6y + \underline{\phantom{xx}}) - 1(x^2 - 4x + \underline{\phantom{xx}}) = -68 \qquad \text{factor out 9 from } y\text{-terms and } -1 \text{ from } x\text{-terms}$$

$$\underbrace{9(y^2 + 6y + 9)}_{\text{adds } 9(9) = 81} - \underbrace{1(x^2 - 4x + 4)}_{\text{adds } -1(4) = -4} = -68 + 81 + (-4) \qquad \begin{array}{l}\text{complete the square}\\ \text{add } 81 + (-4) \text{ to right}\end{array}$$

$$9(y + 3)^2 - 1(x - 2)^2 = 9 \qquad \text{factor}$$

$$\frac{(y + 3)^2}{1} - \frac{(x - 2)^2}{9} = 1 \qquad \text{divide by 9 (standard form)}$$

$$\frac{(y + 3)^2}{1^2} - \frac{(x - 2)^2}{3^2} = 1 \qquad \text{write denominators in squared form}$$

The center of the hyperbola is $(2, -3)$ with $a = 3$, $b = 1$, and a transverse axis of $x = 2$. The vertices are at $(2, -3 + 1) = (2, -2)$ and $(2, -3 - 1) = (2, -4)$. After plotting the center and vertices, we draw a rectangle centered at $(2, -3)$ with a horizontal "width" of $2(3) = 6$ and a vertical "length" of $2(1) = 2$ to sketch the asymptotes. The completed graph is given in Figure 8.26.

**Figure 8.26**

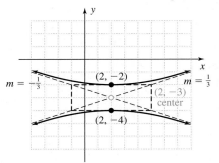

Vertical hyperbola

Center at $(2, -3)$
Vertices at $(2, -2)$ and $(2, -4)$

Transverse axis: $x = 2$
Conjugate axis: $y = -3$

✓ **A.** You've just seen how we can use the equation of a hyperbola to graph central and noncentral hyperbolas

Now try Exercises 35 through 38 ▶

## B. Distinguishing between the Equations of Circles, Ellipses, and Hyperbolas

So far we've explored numerous graphs of circles, ellipses, and hyperbolas. In Example 5 we'll attempt to identify a given conic section from its equation alone (without graphing the equation). As you've seen, the corresponding equations have unique characteristics that can help distinguish one from the other.

**EXAMPLE 5** ▶  **Identifying a Conic Section from Its Equation**

Identify each equation as that of a circle, ellipse, or hyperbola. Justify your choice and name the center, but do not draw the graphs.

**a.** $y^2 = 36 + 9x^2$          **b.** $4x^2 = 16 - 4y^2$
**c.** $x^2 = 225 - 25y^2$       **d.** $25x^2 = 100 + 4y^2$
**e.** $3(x - 2)^2 + 4(y + 3)^2 = 12$    **f.** $4(x + 5)^2 = 36 + 9(y - 4)^2$

**Solution** ▶  **a.** Writing the equation as $y^2 - 9x^2 = 36$ shows $h = 0$ and $k = 0$. Since the equation contains a difference of second-degree terms, it is the equation of a (vertical) hyperbola ($A$ and $B$ have opposite signs). The center is at $(0, 0)$.

**b.** Rewriting the equation as $4x^2 + 4y^2 = 16$ and dividing by 4 gives $x^2 + y^2 = 4$. The equation represents a circle ($A = B$) of radius 2, with the center at $(0, 0)$.

**c.** Writing the equation as $x^2 + 25y^2 = 225$ we note a sum of second-degree terms with unequal coefficients. The equation is that of an ellipse ($A \neq B$), with the center at $(0, 0)$.

**d.** Rewriting the equation as $25x^2 - 4y^2 = 100$ we note the equation contains a difference of second-degree terms. The equation represents a central (horizontal) hyperbola ($A$ and $B$ have opposite signs), whose center is at $(0, 0)$.

**e.** The equation is in factored form and contains a sum of second-degree terms with unequal coefficients. This is the equation of an ellipse ($A \neq B$) with the center at $(2, -3)$.

**f.** Rewriting the equation as $4(x + 5)^2 - 9(y - 4)^2 = 36$ we note a difference of second-degree terms. The equation represents a horizontal hyperbola ($A$ and $B$ have opposite signs) with center $(-5, 4)$.

✓ **B.** You've just seen how we can distinguish between the equations of circles, ellipses, and hyperbolas

Now try Exercises 39 through 50 ▶

## C. The Foci of a Hyperbola

Like the ellipse, the foci of a hyperbola play an important part in their application. A long distance radio navigation system (called LORAN for short), can be used to determine the location of ships and airplanes and is based on the characteristics of a hyperbola (**see Exercises 75 and 76**). Hyperbolic mirrors are also used in some telescopes, and have the property that a beam of light directed at one focus will be reflected to the second focus. To understand and appreciate these applications, we use the analytic definition of a hyperbola:

> ### Definition of a Hyperbola
>
> Given two fixed points $f_1$ and $f_2$ in a plane, a hyperbola is the set of all points $(x, y)$ such that the distance $d_1$ from $f_1$ to $(x, y)$ and the distance $d_2$ from $f_2$ to $(x, y)$, satisfy the equation
>
> $$|d_1 - d_2| = k.$$
>
> In other words, the difference of these two distances is a positive constant. The fixed points $f_1$ and $f_2$ are called the foci of the hyperbola.

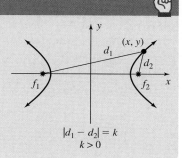

$$|d_1 - d_2| = k$$
$$k > 0$$

**Figure 8.27**

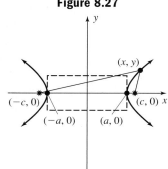

As with the analytic definition of the ellipse, it can be shown that the constant $k$ is again equal to $2a$ (for horizontal hyperbolas). To find the equation of a hyperbola in terms of $a$ and $b$, we use an approach similar to that of the ellipse (**see Appendix III**), and the result is identical to that seen earlier: $\frac{x^2}{a^2} - \frac{y^2}{b^2} = 1$, where $b^2 = c^2 - a^2$ (see Figure 8.27).

We now have the ability to *find the foci of any hyperbola*—and can use this information in many significant applications. Since the location of the foci play such an important role, it is best to remember the relationship as $c^2 = a^2 + b^2$ (called the **foci formula** for hyperbolas), noting that for a hyperbola, $c > a$ and $c > b$. Be sure to note that for ellipses, the foci formula is $c^2 = |a^2 - b^2|$ since $a > c$ (horizontal ellipses) or $b > c$ (vertical ellipses).

---

**EXAMPLE 6** ▶ **Graphing a Hyperbola and Identifying Its Foci by Completing the Square**

For the hyperbola defined by $7x^2 - 9y^2 - 14x + 72y - 200 = 0$, find the coordinates of the center, vertices, foci, and the dimensions of the central rectangle. Then sketch the graph, including the asymptotes.

**Solution** ▶

$$7x^2 - 9y^2 - 14x + 72y - 200 = 0 \qquad \text{given}$$
$$7x^2 - 14x - 9y^2 + 72y = 200 \qquad \text{group terms; add 200}$$
$$7(x^2 - 2x + \underline{\quad}) - 9(y^2 - 8y + \underline{\quad}) = 200 \qquad \text{factor out leading coefficients}$$
$$7(x^2 - 2x + 1) - 9(y^2 - 8y + 16) = 200 + 7 + (-144)$$

add $7 + (-144)$ to right-hand side

adds $7(1) = 7$        adds $-9(16) = -144$

$$7(x - 1)^2 - 9(y - 4)^2 = 63 \qquad \text{factored form}$$
$$\frac{(x - 1)^2}{9} - \frac{(y - 4)^2}{7} = 1 \qquad \text{divide by 63 and simplify}$$
$$\frac{(x - 1)^2}{3^2} - \frac{(y - 4)^2}{(\sqrt{7})^2} = 1 \qquad \text{write denominators in squared form}$$

This is a horizontal hyperbola with $a = 3$ and $b = \sqrt{7}$. The center is at $(1, 4)$, with vertices $(1 - 3, 4)$ and $(1 + 3, 4)$. Using the foci formula $c^2 = a^2 + b^2$ yields $c^2 = 9 + 7 = 16$, showing the foci are $(1 - 4, 4)$ and $(1 + 4, 4)$ (4 units from center).

The central rectangle is $2(3) = 6$ by $2\sqrt{7} \approx 5.29$. Drawing the rectangle and sketching the asymptotes results in the graph shown.

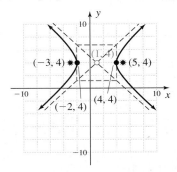

**Horizontal hyperbola**

Center at $(1, 4)$
Vertices at $(-2, 4)$ and $(4, 4)$

Transverse axis: $y = 4$
Conjugate axis: $x = 1$
Location of foci: $(-3, 4)$ and $(5, 4)$

☑ **C.** You've just seen how we can locate the foci of a hyperbola and use the foci and other features to write its equation

**Now try Exercises 51 through 60** ▶

The focal chord for a horizontal hyperbola is a vertical line segment through the focus and parallel to the conjugate axis, with endpoints on the hyperbola. Similar to the focal chord of an ellipse, we can use its length to find additional points on the graph of the hyperbola. The total length is once again $L = \frac{2b^2}{a}$ (for a horizontal hyperbola), meaning the distance from the foci to the graph (along the focal chord) is $\frac{b^2}{a}$. For Example 6, $a = 3$ and $b^2 = 7$, so the vertical distance from focus to graph (in either direction) is $\frac{7}{3} = 2.\overline{3}$. From the left focus $(-3, 4)$, we can now graph the additional points $(-3, 4 - 2.\overline{3}) = (-3, 1.\overline{6})$, and $(-3, 4 + 2.\overline{3}) = (-3, 6.\overline{3})$. From the right focus $(5, 4)$, we obtain $(5, 1.\overline{6})$ and $(5, 6.\overline{3})$. Graphical verification is provided in Figure 8.28. Also **see Exercise 70.**

As with the ellipse, if any two of the values for $a$, $b$, and $c$ are known, the relationship between them can be used to construct the equation of the hyperbola. **See Exercises 61 through 68.**

**Figure 8.28**

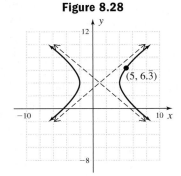

## D. Applications Involving Foci

Applications involving the foci of a conic section can take many forms. As before, only partial information about the hyperbola may be available, and we'll determine a solution by manipulating a given equation or constructing an equation from given facts.

**EXAMPLE 7** ▶ **Applying the Properties of a Hyperbola—The Path of a Comet**

In rare cases, comets traveling near the Sun are not captured by its gravity, but are slung around it in a hyperbolic path with the Sun at one focus. If the path illustrated by the graph shown is modeled by the equation $2116x^2 - 400y^2 = 846{,}400$, how close did the comet get to the Sun? Assume units are in millions of miles and round to the nearest million.

**Solution** ▶ We are essentially asked to find the distance between a vertex and focus. Begin by writing the equation in standard form:

$2116x^2 - 400y^2 = 846{,}400$    given

$\dfrac{x^2}{400} - \dfrac{y^2}{2116} = 1$    divide by 846,400

$\dfrac{x^2}{20^2} - \dfrac{y^2}{46^2} = 1$    write denominators in squared form

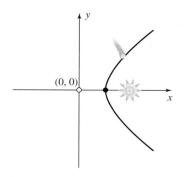

This is a horizontal hyperbola with $a = 20$ and $b = 46$. Use the foci formula to find $c^2$ and $c$.

$$c^2 = a^2 + b^2$$
$$c^2 = 400 + 2116$$
$$c^2 = 2516$$
$$c \approx -50 \text{ and } c \approx 50$$

Since $a = 20$ and $|c| \approx 50$, the comet came within about $50 - 20 = 30$ million miles of the Sun.

**Now try Exercises 71 through 74 ▶**

**EXAMPLE 8 ▶    Applying the Properties of a Hyperbola—The Location of a Storm**

Two amateur meteorologists, living 4 km apart (4000 m), see a storm approaching. The one farthest from the storm hears a loud clap of thunder 9 sec after the one nearest. Assuming the speed of sound is 340 m/sec, determine an equation that models possible locations for the storm at this time.

**Solution ▶**    Let $M_1$ represent the meteorologist nearest the storm and $M_2$ the farthest. Since $M_2$ heard the thunder 9 sec after $M_1$, $M_2$ must be $9 \cdot 340 = 3060$ m farther away from the storm $S$. In other words, from our definition of a hyperbola, we have $|d_1 - d_2| = 3060$. Let's place the information on a coordinate grid. For convenience, we'll use the straight line distance between $M_1$ and $M_2$ as the $x$-axis, with the origin an equal distance from each. With the constant difference equal to 3060, we have $2a = 3060$, $a = 1530$ and $\frac{x^2}{1530^2} - \frac{y^2}{b^2} = 1$. Since the meteorologists live 4000 m apart, we also have $2c = 4000$ and $c = 2000$. Using $c^2 = a^2 + b^2$, we can now find the value of $b$: $2000^2 = 1530^2 + b^2$ or $b^2 = (2000)^2 - (1530)^2 = 1{,}659{,}100 \approx 1288^2$. The equation that models possible locations of the storm is $\frac{x^2}{1530^2} - \frac{y^2}{1288^2} \approx 1$.

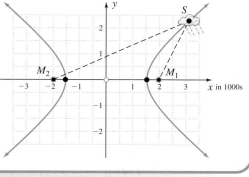

☑ **D.** You've just seen how we can solve applications involving foci

**Now try Exercises 75 and 76 ▶**

## 8.3 EXERCISES

### ▶ CONCEPTS AND VOCABULARY

**Fill in the blank with the appropriate word or phrase. Carefully reread the section, if necessary.**

**1.** The line that passes through the vertices of a hyperbola is called the _____ axis.

**2.** The center of a hyperbola is located _____ between the foci.

**3.** The conjugate axis is _____ to the _____ axis and contains the _____ of the hyperbola.

**4.** The center of the hyperbola defined by
$25(x - 2)^2 - 16(y - 3)^2 = 400$ is at _____.

### ▶ DEVELOPING YOUR SKILLS

**Graph each hyperbola. Label the center, vertices, and any additional points used.**

**5.** $\dfrac{x^2}{9} - \dfrac{y^2}{4} = 1$

**6.** $\dfrac{x^2}{16} - \dfrac{y^2}{9} = 1$

**7.** $\dfrac{x^2}{49} - \dfrac{y^2}{16} = 1$

**8.** $\dfrac{x^2}{25} - \dfrac{y^2}{9} = 1$

**9.** $\dfrac{x^2}{36} - \dfrac{y^2}{16} = 1$

**10.** $\dfrac{x^2}{81} - \dfrac{y^2}{16} = 1$

**11.** $\dfrac{y^2}{9} - \dfrac{x^2}{1} = 1$

**12.** $\dfrac{y^2}{1} - \dfrac{x^2}{4} = 1$

**13.** $\dfrac{y^2}{9} - \dfrac{x^2}{9} = 1$

**14.** $\dfrac{y^2}{4} - \dfrac{x^2}{4} = 1$

**15.** $\dfrac{y^2}{36} - \dfrac{x^2}{25} = 1$

**16.** $\dfrac{y^2}{16} - \dfrac{x^2}{4} = 1$

**For the graphs given, state the location of the vertices and the equation of the transverse axis. Then identify the location of the center and the equation of the conjugate axis. Note the scale used.**

**17.**
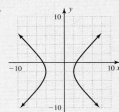

**18.**
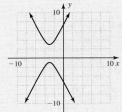

**19.**
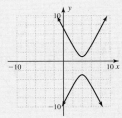

**20.**
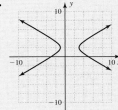

**Sketch a complete graph of each equation, including the asymptotes. Be sure to identify the center and vertices.**

**21.** $\dfrac{(y + 1)^2}{4} - \dfrac{x^2}{25} = 1$

**22.** $\dfrac{y^2}{4} - \dfrac{(x - 2)^2}{9} = 1$

**23.** $\dfrac{(x - 3)^2}{36} - \dfrac{(y + 2)^2}{49} = 1$

**24.** $\dfrac{(x - 2)^2}{9} - \dfrac{(y - 1)^2}{4} = 1$

**25.** $\dfrac{(y + 1)^2}{7} - \dfrac{(x + 5)^2}{9} = 1$

**26.** $\dfrac{(y - 3)^2}{16} - \dfrac{(x + 2)^2}{5} = 1$

**27.** $16x^2 - 9y^2 = 144$

**28.** $16x^2 - 25y^2 = 400$

**29.** $9y^2 - 4x^2 = 36$

**30.** $25y^2 - 4x^2 = 100$

**31.** $(x - 2)^2 - 4(y + 1)^2 = 16$

**32.** $9(x + 1)^2 - (y - 3)^2 = 81$

**33.** $2(y + 3)^2 - 5(x - 1)^2 = 50$

**34.** $9(y - 4)^2 - 5(x - 3)^2 = 45$

**Graph each hyperbola by writing the equation in standard form. Label the center and vertices, and sketch the asymptotes.**

**35.** $4x^2 - y^2 + 40x - 4y + 60 = 0$

**36.** $x^2 - 4y^2 - 12x - 16y + 16 = 0$

**37.** $4y^2 - x^2 - 24y - 4x + 28 = 0$

**38.** $-9x^2 + 4y^2 - 18x - 24y - 9 = 0$

**Classify each equation as that of a circle, ellipse, or hyperbola. Justify your response (assume all are nondegenerate).**

**39.** $-4x^2 - 4y^2 = -24$

**40.** $9y^2 = -4x^2 + 36$

**41.** $x^2 + y^2 = 2x + 4y + 4$

**42.** $x^2 = y^2 + 6y - 7$

**43.** $2x^2 - 4y^2 = 8$

**44.** $36x^2 + 25y^2 = 900$

**45.** $x^2 + 5 = 2y^2$

**46.** $x + y^2 = 3x^2 + 9$

**47.** $2x^2 = -2y^2 + x + 20$

**48.** $2y^2 + 3 = 6x^2 + 8$

**49.** $16x^2 + 5y^2 - 3x + 4y = 538$

**50.** $9x^2 + 9y^2 - 9x + 12y + 4 = 0$

**Use the definition of a hyperbola to find the distance between the vertices and the dimensions of the rectangle centered at (0, 0). Figures are not drawn to scale. Note that Exercises 53 and 54 are *vertical hyperbolas*.**

**51.**
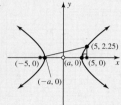

**52.**
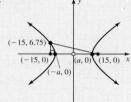

**53.**
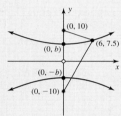

**54.**
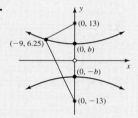

**Write each equation in standard form to find and list the coordinates of the (a) center, (b) vertices, (c) foci, and (d) dimensions of the central rectangle. Then (e) sketch the graph, including the asymptotes.**

**55.** $4x^2 - 9y^2 - 24x + 72y - 144 = 0$

**56.** $4x^2 - 36y^2 - 40x + 144y - 188 = 0$

**57.** $4y^2 - 16x^2 - 24y - 28 = 0$

**58.** $4y^2 - 81x^2 - 162x - 405 = 0$

**59.** $9x^2 - 3y^2 - 54x - 12y + 33 = 0$

**60.** $10x^2 + 60x - 5y^2 + 20y - 20 = 0$

**Find the equation of the hyperbola (in standard form) that satisfies the following conditions:**

**61.** vertices at $(-6, 0)$ and $(6, 0)$; foci at $(-8, 0)$ and $(8, 0)$

**62.** vertices at $(-4, 0)$ and $(4, 0)$; foci at $(-6, 0)$ and $(6, 0)$

**63.** foci at $(-2, -3\sqrt{2})$ and $(-2, 3\sqrt{2})$; length of conjugate axis: 6 units

**64.** foci at $(-5, 2)$ and $(7, 2)$; length of conjugate axis: 8 units

**Use the characteristics of a hyperbola and the graph given to write the related equation and state the location of the foci (65 and 66) or the dimensions of the central rectangle (67 and 68).**

**65.**
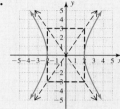

**66.**
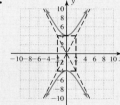

**67.**
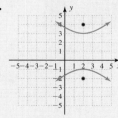

**68.**
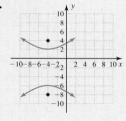

## ▶ WORKING WITH FORMULAS

**69. Equation of a semi-hyperbola:** $y = \sqrt{\dfrac{36 - 4x^2}{-9}}$

The "upper half" of a certain hyperbola is given by the equation shown. (a) Simplify the radicand, (b) state the domain of the expression, and (c) graph the upper half of the hyperbola. (d) What is the equation for the "lower half" of this hyperbola?

**70. Focal chord of a horizontal hyperbola:** $L = \dfrac{2b^2}{a}$

The focal chords of a hyperbola are line segments containing points $f_1$ and $f_2$, parallel to the conjugate axis, with endpoints on the hyperbola (see graph). The length of the chord is given by the formula shown. Use the formula to find the length of the focal chord for the hyperbola indicated, then compare the calculated value with the length estimated from the given graph:

$$\frac{(x-2)^2}{4} - \frac{(y-1)^2}{5} = 1.$$

**Exercise 70**
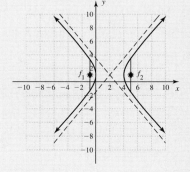

## ▶ APPLICATIONS

**71. Stunt pilots:** At an air show, a stunt plane dives along a hyperbolic path whose vertex is directly over the grandstands. If the plane's flight path can be modeled by the hyperbola $25y^2 - 1600x^2 = 40,000$, what is the minimum altitude of the plane as it passes over the stands? Assume $x$ and $y$ are in yards.

**72. Flying clubs:** To test their skill as pilots, the members of a flight club attempt to drop sandbags on a target placed in an open field, by diving along a hyperbolic path whose vertex is directly over the target area. If the flight path of the plane flown by the club's president is modeled by $9y^2 - 16x^2 = 14,400$, what is the minimum altitude of her plane as it passes over the target? Assume $x$ and $y$ are in feet.

**73. Charged particles:** It has been shown that when like particles with a common charge are hurled at each other, they deflect and travel along paths that are hyperbolic. Suppose the paths of two such particles is modeled by the hyperbola $x^2 - 9y^2 = 36$. What is the minimum distance between the paths of the particles as they approach each other? Assume $x$ and $y$ are in microns.

**74. Nuclear cooling towers:** The natural draft cooling towers for nuclear power stations are called

*hyperboloids of one sheet.* The perpendicular cross sections of these hyperboloids form two branches of a hyperbola. Suppose the central cross section of one such tower is modeled by the hyperbola $1600x^2 - 400(y - 50)^2 = 640,000$. What is the minimum distance between the sides of the tower? Assume $x$ and $y$ are in feet.

**75. Locating a ship using radar:** Under certain conditions, the properties of a hyperbola can be used to help locate the position of a ship. Suppose two radio stations are located 100 km apart along a straight shoreline. A ship is sailing parallel to the shore and is 60 km out to sea. The ship sends out a distress call that is picked up by the closer station in 0.4 milliseconds (msec—one-thousandth of a second), while it takes 0.5 msec to reach the station that is farther away. Radio waves travel about 300 km/msec. Use this information to find the equation of a hyperbola that will help you find the location of the ship, then find the coordinates of the ship.

**76. Locating a plane using radar:** Two radio stations are located 80 km apart along a straight shoreline, when a "mayday" call (a plea for immediate help) is received from a plane that is about to ditch in the ocean. The plane was flying at low altitude, parallel to the shoreline, and 20 km out when it ran into trouble. The plane's distress call is picked up by the closer station in 0.1 msec, while it takes 0.3 msec to reach the other. Use this information to construct the equation of a hyperbola that will help you find the location of the ditched plane, then find the coordinates of the plane. Also see Exercise 75.

## ▶ EXTENDING THE CONCEPT

**77.** For a greater understanding as to *why* the branches of a hyperbola are asymptotic, solve $\frac{x^2}{a^2} - \frac{y^2}{b^2} = 1$ for $y$, then consider what happens as $x \to \infty$ (note that $x^2 - k \approx x^2$ for large $x$).

**78.** Which has a greater area: (a) The central rectangle of the hyperbola given by $(x - 5)^2 - (y + 4)^2 = 57$, (b) the circle given by $(x - 5)^2 + (y + 4)^2 = 57$, or (c) the ellipse given by $9(x - 5)^2 + 10(y + 4)^2 = 570$?

**79.** It is possible for the plane to intersect only the vertex of the cone or to be tangent to the sides. These are called **degenerate cases** of a conic section. Many times we're unable to tell if the equation represents a degenerate case until it's written in standard form. Write the following equations in standard form and comment.

    **a.** $4x^2 - 32x - y^2 + 4y + 60 = 0$            **b.** $x^2 - 4x + 5y^2 - 40y + 84 = 0$

## ▶ MAINTAINING YOUR SKILLS

**80. (3.4)** Graph the piecewise-defined function:
$$f(x) = \begin{cases} 4 - x^2, & -2 \le x < 3 \\ 5, & x \ge 3 \end{cases}$$

**81. (4.2)** Use the remainder theorem to determine if $x = 2$ is a zero of $g(x) = x^5 - 5x^4 + 4x^3 + 16x^2 - 32x + 16$. If yes, find its multiplicity.

**82. (1.4)** The number $z = 1 + i\sqrt{2}$ is a solution to two out of the three equations given. Which two?

    **a.** $x^4 + 4 = 0$         **c.** $x^2 - 2x + 3 = 0$

    **b.** $x^3 - 6x^2 + 11x - 12 = 0$

**83. (6.4)** A government-approved company is licensed to haul toxic waste. Each container of solid waste weighs 800 lb and has a volume of 100 ft$^3$. Each container of liquid waste weighs 1000 lb and is 60 ft$^3$ in volume. The revenue from hauling solid waste is $300 per container, while the revenue from liquid waste is $350 per container. The truck used by this company has a weight capacity of 39.8 tons and a volume capacity of 6960 ft$^3$. What combination of solid and liquid waste containers will produce the maximum revenue?

### LEARNING OBJECTIVES

*In Section 8.4 you will see how we can:*

☐ **A.** Graph parabolas with a horizontal axis of symmetry

☐ **B.** Identify and use the focus-directrix form of the equation of a parabola

☐ **C.** Solve nonlinear systems involving the conic sections

☐ **D.** Solve applications of the analytic parabola

In previous coursework, you likely learned that the graph of a quadratic function was a parabola. Parabolas are actually the fourth and final member of the family of conic sections, and as we saw in Section 8.1, the graph can be obtained by observing the intersection of a plane and a cone. If the plane is parallel to the generator of the cone (shown as a dark line in Figure 8.29), the intersection of the plane with one nappe forms a parabola. In this section we develop the general equation of a parabola from its analytic definition, opening a new realm of applications that extends far beyond those involving only zeroes and extreme values.

**Figure 8.29**

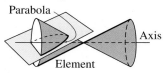

### A. Parabolas with a Horizontal Axis

An introductory study of parabolas generally involves those with a vertical axis, defined by the equation $y = ax^2 + bx + c$. Unlike the previous conic sections, this equation has *only one second-degree (squared) term* and defines a function. As a review, the primary characteristics are listed here and illustrated in Figure 8.30. **See Exercises 5 through 10.**

**Figure 8.30**

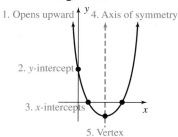

1. Opens upward
2. *y*-intercept
3. *x*-intercepts
4. Axis of symmetry
5. Vertex

**Vertical Parabolas**

For a second-degree equation of the form $y = ax^2 + bx + c$, the graph is a vertical parabola with these characteristics:

1. opens upward if $a > 0$, downward if $a < 0$.
2. $y$-intercept: $(0, c)$ (substitute 0 for $x$)
3. $x$-intercept(s): substitute 0 for $y$ and solve.
4. axis of symmetry: $x = \frac{-b}{2a}$
5. vertex $(h, k)$: $\left(\frac{-b}{2a}, c - \frac{b^2}{4a}\right)$

### Horizontal Parabolas

Similar to our study of horizontal and vertical hyperbolas, the graph of a parabola can open *to the right or left,* as well as up or down. After interchanging the variables $x$ and $y$ in the standard equation, we obtain the parabola $x = ay^2 + by + c$, noting the resulting graph will be a reflection about the line $y = x$. Here, the axis of symmetry is a horizontal line and factoring or the quadratic formula is used to find the *y-intercepts* (if they exist). Note that although the graph is still a parabola—*it is not the graph of a function.*

**Horizontal Parabolas**

For a second-degree equation of the form $x = ay^2 + by + c$, the graph is a horizontal parabola with these characteristics:

1. opens right if $a > 0$, left if $a < 0$.
2. $x$-intercept: $(c, 0)$ (substitute 0 for $y$)
3. $y$-intercept(s): substitute 0 for $x$ and solve.
4. axis of symmetry: $y = \frac{-b}{2a}$
5. vertex $(h, k)$: $\left(c - \frac{b^2}{4a}, \frac{-b}{2a}\right)$

**EXAMPLE 1** ▶ **Graphing a Horizontal Parabola**

Graph the relation whose equation is $x = y^2 + 3y - 4$, then state the domain and range of the relation.

**Solution** ▶ Since the equation has a single squared term in $y$, the graph will be a horizontal parabola.

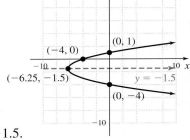

1. Since $a > 0$, the parabola opens to the right.
2. Substituting 0 for $y$, we find the $x$-intercept is $(-4, 0)$.
3. After factoring, we find the $y$-intercepts are $(0, -4)$ and $(0, 1)$.
4. The axis of symmetry is $y = -\frac{b}{2a} = -\frac{3}{2}$ or $-1.5$.
5. For the vertex we have $c - \frac{b^2}{4a} = -4 - \frac{3^2}{4} = -6.25$, and $-\frac{b}{2a} = -1.5$. The coordinates of the vertex are $(-6.25, -1.5)$.

Using horizontal and vertical boundary lines we find the domain for this relation is $x \in [-6.25, \infty)$ and the range is $y \in (-\infty, \infty)$. The graph is shown in the figure.

**Now try Exercises 11 through 16** ▶

As with the vertical parabola, the equation of a horizontal parabola can be written as a transformation: $x = a(y \pm k)^2 \pm h$ by completing the square. Note that in this case, the vertical shift is $k$ units *opposite the sign,* with a horizontal shift of $h$ units in the same direction as the sign.

**EXAMPLE 2** ▶ **Graphing a Horizontal Parabola by Completing the Square**

Graph by completing the square: $x = -2y^2 - 8y - 9$, then state the domain and range.

**Solution** ▶ Using the original equation, we note the graph will be a horizontal parabola opening to the left $(a < 0)$ and have an $x$-intercept of $(-9, 0)$. Completing the square gives $x = -2(y^2 + 4y + 4) - 9 + 8$, so the equation of this parabola in shifted form is

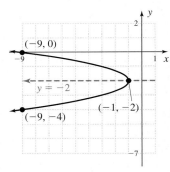

$$x = -2(y + 2)^2 - 1$$

The vertex is at $(-1, -2)$ and $y = -2$ is the axis of symmetry. This means there are no $y$-intercepts, a fact that comes to light when we attempt to solve the equation after substituting 0 for $x$:

$$-2(y + 2)^2 - 1 = 0 \qquad \text{substitute 0 for } x$$

$$(y + 2)^2 = -\frac{1}{2} \qquad \text{no real roots}$$

Using symmetry, the point $(-9, -4)$ is also on the graph. After plotting these points we obtain the graph shown. From the graph, the domain is $x \in (-\infty, -1]$ and the range is $y \in \mathbb{R}$.

☑ **A.** You've just seen how we can graph parabolas with a horizontal axis of symmetry

**Now try Exercises 17 through 30** ▶

## B. The Focus-Directrix Form of the Equation of a Parabola

As with the ellipse and hyperbola, many significant applications of the parabola rely on its analytical definition rather than its algebraic form. From the construction of radio telescopes to the manufacture of flashlights, the location of the focus of a parabola is critical. To understand these and other applications, we use the analytic definition of a parabola first introduced in Section 8.1.

### Definition of a Parabola

Given a fixed point $f$ and fixed line $D$ in the plane, a parabola is the set of all points $(x, y)$ such that the distance from $f$ to $(x, y)$ is equal to the distance from $D$ to $(x, y)$. The fixed point $f$ is the **focus** of the parabola, and the fixed line is the **directrix**.

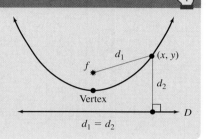

**WORTHY OF NOTE**

For the analytic parabola, we use $p$ to designate the focus since $c$ is so commonly used as the constant term in $y = ax^2 + bx + c$.

The general equation of a parabola can be obtained by combining this definition with the distance formula. With no loss of generality, we can assume the parabola shown in the definition box is oriented in the plane with the vertex at $(0, 0)$ and the focus at $(0, p)$. As the diagram in Figure 8.31 indicates, this gives the directrix an equation of $y = -p$ with all points on $D$ having coordinates of $(x, -p)$. Using $d_1 = d_2$ the distance formula yields

**Figure 8.31**

$$\sqrt{(x - 0)^2 + (y - p)^2} = \sqrt{(x - x)^2 + (y + p)^2} \quad \text{from the definition}$$
$$(x - 0)^2 + (y - p)^2 = (x - x)^2 + (y + p)^2 \quad \text{square both sides}$$
$$x^2 + y^2 - 2py + p^2 = 0 + y^2 + 2py + p^2 \quad \text{simplify; expand binomials}$$
$$x^2 - 2py = 2py \quad \text{subtract } p^2 \text{ and } y^2$$
$$x^2 = 4py \quad \text{isolate } x^2$$

The resulting equation is called the **focus-directrix form** of a *vertical parabola* with center at $(0, 0)$. If we had begun by orienting the parabola so it opened to the right, we would have obtained the equation of a *horizontal parabola* with center $(0, 0)$: $y^2 = 4px$. If the vertex $(h, k)$ is not at the origin, we use our understanding of transformations to write a more general equation.

### The Equation of a Parabola in Focus-Directrix Form

| **Vertical Parabola** | **Horizontal Parabola** |
|---|---|
| $(x - h)^2 = 4p(y - k)$ | $(y - k)^2 = 4p(x - h)$ |
| focus: $(h, k + p)$ | focus: $(h + p, k)$ |
| directrix: $y = k - p$ | directrix: $x = h - p$ |
| If $p > 0$, opens upward. | If $p > 0$, opens to the right. |
| If $p < 0$, opens downward. | If $p < 0$, opens to the left. |

*For a parabola, note there is only one second-degree term.*

**EXAMPLE 3** ▶ **Locating the Focus and Directrix of a Parabola**

Find the vertex, focus, and directrix for the parabola defined by $x^2 = -12y$. Then sketch the graph, including the focus and directrix.

**Solution** ▶ Since the $x$-term is squared and no shifts have been applied, the graph will be a vertical parabola with a vertex of $(0, 0)$. Use a direct comparison between the given equation and the focus-directrix form to determine the value of $p$:

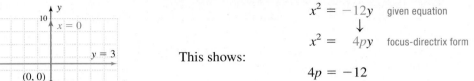

$$x^2 = -12y \quad \text{given equation}$$
$$\downarrow$$
$$x^2 = \phantom{-}4py \quad \text{focus-directrix form}$$

This shows:

$$4p = -12$$
$$p = -3$$

Since $p < 0$, the parabola opens downward, with the focus at $(0, -3)$ and a directrix of $y = 3$. To complete the graph we need a few additional points. Since $(\pm 6)^2 = 36$ is divisible by 12, we use inputs of $x = 6$ and $x = -6$ to get the points $(6, -3)$ and $(-6, -3)$. Note the axis of symmetry is $x = 0$. The graph is shown.

**Now try Exercises 31 through 38** ▶

**Figure 8.32**

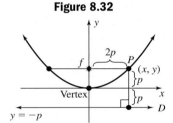

As an alternative to calculating additional points to sketch the graph, we can use what is called the **focal chord** of the parabola. Similar to the ellipse and hyperbola, the focal chord is the line segment that contains the focus, is parallel to the directrix, and has its endpoints on the graph. Using the definition of a parabola and the diagram in Figure 8.32, we note the vertical distance from $(x, y)$ to the directrix is $2p$. This means a line segment parallel to the directrix from the focus to the graph will also have a length of $|2p|$, and the focal chord of any parabola has a total length of $|4p|$. Note that in Example 3, the points we happened to choose were actually the endpoints of the focal chord.

**EXAMPLE 4** ▶ **Locating the Focus and Directrix of a Parabola**

Find the vertex, focus, and directrix for the parabola whose equation is given, then sketch the graph, including the focus, focal chord, and directrix: $x^2 - 6x + 12y - 15 = 0$.

**Solution** ▶ Since only the $x$-term is squared, the graph will be a vertical parabola. To find the end-behavior, vertex, focus, and directrix, we complete the square in $x$ and use a direct comparison between the shifted form and the focus-directrix form:

$$x^2 - 6x + 12y - 15 = 0 \quad \text{given equation}$$
$$x^2 - 6x + \underline{\phantom{9}} = -12y + 15 \quad \text{complete the square in } x$$
$$x^2 - 6x + 9 = -12y + 24 \quad \text{add 9}$$
$$(x - 3)^2 = -12(y - 2) \quad \text{factor}$$

Notice the parabola has been shifted 3 units right and 2 up, so *all features of the parabola will likewise be shifted*. Since we have $4p = -12$ (the coefficient of the linear term), we know $p = -3$ ($p < 0$) and the parabola opens downward. If the parabola were in standard position, the vertex would be at $(0, 0)$, the focus at $(0, -3)$ and the directrix a horizontal line at $y = 3$. But since the parabola is shifted 3 right and 2 up, we add 3 to all $x$-values and 2 to all $y$-values to locate the features of the shifted parabola. The vertex is at $(0 + 3, 0 + 2) = (3, 2)$. The focus is $(0 + 3, -3 + 2) = (3, -1)$ and the directrix is $y = 3 + 2 = 5$. Finally, the horizontal distance from the focus to the graph is $|2p| = 6$ units (since $|4p| = 12$), giving us the additional points $(3 - 6, -1) = (-3, -1)$ and $(3 + 6, -1) = (9, -1)$ as endpoints of the focal chord. The graph is shown.

**Now try Exercises 39 through 48** ▶

In many cases, we need to construct the equation of the parabola when only partial information is known, as illustrated in Example 5.

**EXAMPLE 5** ▶　**Constructing the Equation of a Parabola**

Find the equation of the parabola with vertex $(4, 4)$ and focus $(1, 4)$. Then graph the parabola using the equation and focal chord.

**Solution** ▶　As the vertex and focus are on a horizontal line, we have a horizontal parabola with general equation $(y \pm k)^2 = 4p(x \pm h)$. The distance from vertex to focus is 3 units. With the focus to the left of the vertex, the parabola opens left so $p = -3$, and $4p = -12$. The vertex is shifted 4 units right and 4 units up from $(0, 0)$, showing $h = 4$ and $k = 4$, and the equation of the parabola must be $(y - 4)^2 = -12(x - 4)$, with directrix $x = 7$. Using the focal chord, the vertical distance from $(1, 4)$ to the graph is $|2p| = |2(-3)| = 6$, giving points $(1, 10)$ and $(1, -2)$. The graph is shown.

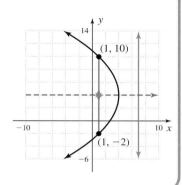

☑ **B. You've just seen how we can identify and use the focus-directrix form of the equation of a parabola**

Now try Exercises 49 through 62 ▶

## C. Nonlinear Systems and the Conic Sections

Similar to our work with nonlinear systems in Section 6.3, the graphing, substitution, or elimination method can still be used when the system involves a conic section. When both equations in the system have at least one second degree term, it is generally easier to use the elimination method.

**EXAMPLE 6** ▶　**Solving a System of Nonlinear Equations**

Solve the system using elimination: $\begin{cases} 2y^2 - 5x^2 = 13 \\ 3x^2 + 4y^2 = 39 \end{cases}$

**Solution** ▶　The first equation represents a vertical and central hyperbola, while the second represents a horizontal and central ellipse. After writing the system with the $x$- and $y$-terms in the same order, we obtain $\begin{cases} -5x^2 + 2y^2 = 13 \\ 3x^2 + 4y^2 = 39 \end{cases}$. Using $-2R1 + R2$ will eliminate the $y$-term.

$$\begin{array}{rl} 10x^2 - 4y^2 = -26 & \quad -2R1 \\ \underline{3x^2 + 4y^2 = \phantom{-}39} & \quad +R2 \\ 13x^2 \phantom{- 4y^2} = \phantom{-}13 & \quad \text{sum} \\ x^2 = 1 & \quad \text{divide by 13} \\ x = -1 \quad \text{or} \quad x = 1 & \quad \text{square root property} \end{array}$$

Substituting $x = 1$ and $x = -1$ into the second equation we obtain

$$\begin{aligned} 3(1)^2 + 4y^2 &= 39 & \qquad 3(-1)^2 + 4y^2 &= 39 \\ 3 + 4y^2 &= 39 & 3 + 4y^2 &= 39 \\ 4y^2 &= 36 & 4y^2 &= 36 \\ y^2 &= 9 & y^2 &= 9 \\ y = -3 \quad \text{or} \quad y &= 3 & y = -3 \quad \text{or} \quad y &= 3 \end{aligned}$$

Since $-1$ and $1$ each generated *two outputs,* there are a total of four ordered pair solutions: $(1, -3)$, $(1, 3)$, $(-1, -3)$, and $(-1, 3)$. The graph is shown and supports our results.

☑ **C. You've just seen how we can solve nonlinear systems involving the conic sections**

Now try Exercises 63 through 68 ▶

## D. Applications of the Analytic Parabola

Here is just one of the many ways the analytic definition of a parabola can be applied. There are several others in the Exercise Set. Many applications use the property that light or sound coming in parallel to the axis of a parabola will be reflected to the focus.

**EXAMPLE 7** ▶ **Locating the Focus of a Parabolic Receiver**

The diagram shows the cross section of a radio antenna dish. Engineers have located a point on the cross section that is 0.75 m above and 6 m to the right of the vertex. At what coordinates should the engineers build the focus of the antenna?

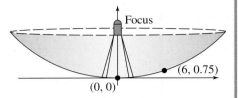

**Solution** ▶ By inspection we see this is a vertical parabola with center at $(0, 0)$. This means its equation must be of the form $x^2 = 4py$. Because we know $(6, 0.75)$ is a point on this graph, we can substitute $(6, 0.75)$ in this equation and solve for $p$:

$$x^2 = 4py \qquad \text{equation for vertical parabola, vertex at } (0, 0)$$
$$(6)^2 = 4p(0.75) \qquad \text{substitute 6 for } x \text{ and 0.75 for } y$$
$$36 = 3p \qquad \text{simplify}$$
$$p = 12 \qquad \text{result}$$

With $p = 12$, we see that the focus must be located at $(0, 12)$, or 12 m directly above the vertex.

☑ **D.** You've just seen how we can solve applications of the analytic parabola

**Now try Exercises 71 through 78** ▶

Note that in many cases, the focus of a parabolic dish may be above the rim of the dish.

## 8.4 EXERCISES

### ▶ CONCEPTS AND VOCABULARY

**Fill in the blank with the appropriate word or phrase. Carefully reread the section, if necessary.**

1. The equation $x = ay^2 + by + c$ is that of a(n) _____ parabola, opening to the _____ if $a > 0$ and to the left if _____.

2. Given $y^2 = 4px$, the focus is at _____ and the equation of the directrix is _____.

3. Given $x^2 = -16y$, the value of $p$ is _____ and the coordinates of the focus are _____.

4. Discuss/Explain how to find the vertex, directrix, and focus from the equation $(x - h)^2 = 4p(y - k)$.

### ▶ DEVELOPING YOUR SKILLS

**Sketch the graph using the $x$- and $y$-intercepts (if they exist) and the vertex of the parabola. Also state the domain and range.**

5. $y = x^2 - 2x - 3$   6. $y = x^2 + 6x + 5$

7. $y = -2x^2 + 8x + 10$   8. $y = -3x^2 - 12x + 15$

9. $y = 2x^2 + 5x - 7$   10. $y = 2x^2 - 7x + 3$

**Find the $x$- and $y$-intercepts (if they exist) and the vertex of the graph. Then sketch the graph using symmetry and state the relation's domain and range.**

11. $x = y^2 - 2y - 3$   12. $x = y^2 - 4y - 12$

13. $x = -y^2 + 6y + 7$   14. $x = -y^2 + 8y - 12$

15. $x = -y^2 + 8y - 16$   16. $x = -y^2 + 6y - 9$

**Sketch the graph of each relation using shifts of a basic function, then state the domain and range of the relation.**

17. $y = (x - 2)^2 + 3$

18. $y = (x + 2)^2 - 4$

19. $x = 2(y - 3)^2 + 1$

20. $x = -2(y + 3)^2 - 5$

21. $x = y^2 - 6y$

22. $x = y^2 - 8y$

23. $x = y^2 - 4$

24. $x = y^2 - 9$

25. $x = -y^2 + 2y - 1$

26. $x = -y^2 + 4y - 4$

27. $x = y^2 - 10y + 4$

28. $x = y^2 + 12y - 5$

29. $x = 3 - 8y - 2y^2$

30. $x = 2 - 12y + 3y^2$

**Find the vertex, focus, and directrix for the parabolas defined by the equations given, then use this information to sketch a complete graph. For Exercises 39 through 48, also include the focal chord.**

31. $x^2 = 8y$

32. $x^2 = 16y$

33. $x^2 = -24y$

34. $x^2 = -20y$

35. $y^2 = 18x$

36. $y^2 = 20x$

37. $y^2 = -10x$

38. $y^2 = -14x$

39. $x^2 - 8x - 8y + 16 = 0$

40. $x^2 - 10x - 12y + 25 = 0$

41. $x^2 - 14x - 24y + 1 = 0$

42. $x^2 - 10x - 12y + 1 = 0$

43. $3x^2 - 24x - 12y + 12 = 0$

44. $2x^2 - 8x - 16y - 24 = 0$

45. $y^2 - 6y + 4x + 1 = 0$

46. $y^2 - 2y + 8x + 9 = 0$

47. $2y^2 - 20y + 8x + 2 = 0$

48. $3y^2 - 18y + 12x + 3 = 0$

**Find the equation of the parabola in standard form that satisfies the conditions given.**

49. focus: $(0, 2)$
    directrix: $y = -2$

50. focus: $(-3, 0)$
    directrix: $x = 3$

51. focus: $(0, -5)$
    directrix: $y = 5$

52. focus: $(5, 0)$
    directrix: $x = -5$

53. vertex: $(2, -2)$
    focus: $(-1, -2)$

54. vertex: $(4, 1)$
    focus: $(1, 1)$

55. vertex: $(4, -7)$
    focus: $(4, -4)$

56. vertex: $(-3, -4)$
    focus: $(-3, -1)$

57. focus: $(3, 4)$
    directrix: $y = 0$

58. focus: $(-1, 2)$
    directrix: $x = -5$

**For the graphs in Exercises 59 through 62, only two of the following four features are displayed: vertex, focus, directrix, and endpoints of the focal chord. Find the remaining two features and the equation of the parabola.**

59.

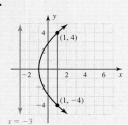

60.

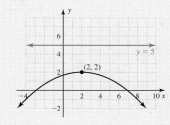

61.

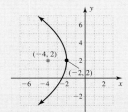

62.

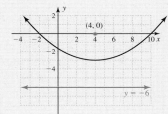

**Solve using substitution or elimination, then verify your solutions using back-substitution.**

63. $\begin{cases} x^2 + y^2 = 25 \\ 2x^2 - 3y^2 = 5 \end{cases}$

64. $\begin{cases} y^2 - x^2 = 12 \\ x^2 + y^2 = 20 \end{cases}$

65. $\begin{cases} x^2 - y = 4 \\ y^2 - x^2 = 16 \end{cases}$

66. $\begin{cases} 2x^2 - 3y^2 = 38 \\ x^2 + 5y = 35 \end{cases}$

67. $\begin{cases} 5x^2 - 2y^2 = 75 \\ 2x^2 + 3y^2 = 125 \end{cases}$

68. $\begin{cases} 3x^2 - 7y^2 = 20 \\ 4x^2 + 9y^2 = 45 \end{cases}$

▶ **WORKING WITH FORMULAS**

69. **The area of a right parabolic segment:** $A = \frac{2}{3}ab$

A *right parabolic segment* is formed by cutting the parabola with a line perpendicular to its axis of symmetry. The area of this segment is given by the formula shown, where $b$ is the length of the chord cutting the parabola and $a$ is the perpendicular distance from the vertex to this chord. What is the area of the parabolic segment shown in the figure?

 70. **The arc length of a right parabolic segment:**

$$L = \frac{1}{2}\sqrt{b^2 + 16a^2} + \frac{b^2}{8a}\ln\left(\frac{4a + \sqrt{b^2 + 16a^2}}{b}\right)$$

To find the length $L$ of the arc $ABC$ shown, we use the formula given where "ln" represents the natural log function. Suppose a baseball thrown from centerfield reaches a maximum height of 20 ft and traverses an arc length of 340 ft. Will the ball reach the catcher 325 ft away without bouncing?

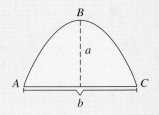

► **APPLICATIONS**

**71. Parabolic car headlights:** The cross section of a typical car headlight can be modeled by an equation similar to $25x = 16y^2$, where $x$ and $y$ are in inches and $x \in [0, 4]$. Use this information to graph the relation for the indicated domain.

**72. Parabolic flashlights:** The cross section of a typical flashlight reflector can be modeled by an equation similar to $4x = y^2$, where $x$ and $y$ are in centimeters and $x \in [0, 2.25]$. Use this information to graph the relation for the indicated domain.

**73. Parabolic sound receivers:** Sound technicians at professional sports events often use parabolic receivers as they move along the sidelines. If a cross section of the receiver is modeled by the equation $y^2 = 54x$, and is 36 in. in *diameter,* how deep is the parabolic receiver? What is the location of the focus?

**74. Parabolic sound receivers:** Private investigators will often use a smaller and less expensive parabolic receiver (see Exercise 73) to gather information for their clients. If a cross section of the receiver is modeled by the equation $y^2 = 24x$, and the receiver is 12 in. in *diameter,* how deep is the parabolic dish? What is the location of the focus?

**75. Parabolic radio wave receivers:** The program known as S.E.T.I. (Search for Extra-Terrestrial Intelligence) involves a group of scientists using radio telescopes

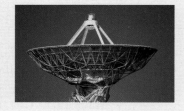

to look for radio signals from possible intelligent species in outer space. The radio telescopes are actually parabolic dishes that vary in size from a few feet to hundreds of feet in diameter. If a particular radio telescope is 100 ft in diameter and has a cross section modeled by the equation $x^2 = 167y$, how deep is the parabolic dish? What is the location of the focus?

**76. Solar furnace:** Another form of technology that uses a parabolic dish is called a solar furnace. In general, the rays of the Sun are reflected by the dish and concentrated at the focus, producing extremely high temperatures. Suppose the dish of one of these parabolic reflectors has a 30-ft diameter and a cross section modeled by the equation $x^2 = 50y$. How deep is the parabolic dish? What is the location of the focus?

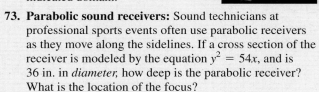

**77. Commercial flashlights:** The reflector of a large, commercial flashlight has the shape of a parabolic dish, with a diameter of 10 cm and a depth of 5 cm. What equation will the engineers and technicians use for the manufacture of the dish? How far from the vertex (the lowest point of the dish) will the bulb be placed?

**78. Industrial spotlights:** The reflector of an industrial spotlight has the shape of a parabolic dish with a diameter of 120 cm. What is the depth of the dish if the correct placement of the bulb is 11.25 cm above the vertex (the lowest point of the dish)? What equation will the engineers and technicians use for the manufacture of the dish?

► **EXTENDING THE CONCEPT**

**79.** In a study of quadratic graphs from the equation $y = ax^2 + bx + c$, no mention is made of a parabola's focus and directrix. Generally, when $a \geq 1$, the focus of a parabola is very near its vertex. Complete the square of the function $y = 2x^2 - 8x + 12$ and write the result in the form $(x - h)^2 = 4p(y - k)$. What is the value of $p$? What are the coordinates of the vertex?

**80.** Like the ellipse and hyperbola, the focal chord of a parabola (also called the **latus rectum**) can be used to help sketch its graph. From our earlier work, we know the endpoints of the focal chord are $|2p|$ units from the focus. Write the equation $-12y + 15 = x^2 - 6x$ in the form $(x \pm h)^2 = 4p(y \pm k)$, and use the endpoints of the focal chord to help graph the parabola.

► **MAINTAINING YOUR SKILLS**

**81. (6.2)** Construct a system of three equations in three variables using the equation $y = ax^2 + bx + c$ and the points $(-3, 3)$, $(0, 6)$, and $(1, -1)$. Then solve the system to find the equation of the parabola containing these points.

**82. (1.5/1.6)** Find all roots (real and nonreal) of the equation $x^6 - 64 = 0$. (*Hint:* Begin by factoring as the difference of two perfect squares.)

**83. (2.5)** What are the characteristics of an *even* function? What are the characteristics of an *odd* function?

**84. (5.5)** Solve each equation and write solutions in both exact and approximate form: (a) $3 \ln x + 5 = 17$ (b) $2e^{5x} - 7 = 11$

## MAKING CONNECTIONS

## Making Connections: Graphically, Symbolically, Numerically, and Verbally

Eight graphs (a) through (h) are given. Match the characteristics shown in 1 through 16 to one of the eight graphs.

**(a)** 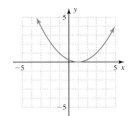    **(b)** 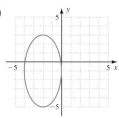    **(c)** 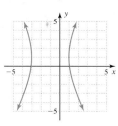    **(d)**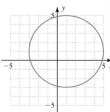

**(e)** 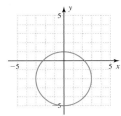    **(f)** 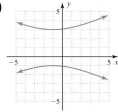    **(g)** 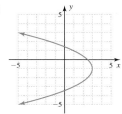    **(h)**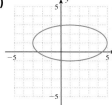

**1.** ____ $(x - 1)^2 + (y - 1)^2 = 16$    **2.** ____ $y = \frac{1}{4}(x - 1)^2$    **3.** ____ foci at $(-1, 1 \pm \sqrt{5})$

**4.** ____ domain: $x \in [-4, 0]$, range: $y \in [-5, 3]$    **5.** ____ $x = -\frac{1}{2}(y + 1)^2 + 3$    **6.** ____ domain: $x \in [-3, 5]$, range: $y \in [-3, 5]$

**7.** ____ $4(x + 1)^2 - (y - 1)^2 = 16$    **8.** ____ $x^2 + (y - 2)^2 = 9$    **9.** ____ vertices at $(-3, 1)$ and $(5, 1)$

**10.** ____ $4(y - 1)^2 - (x + 1)^2 = 16$    **11.** ____ center at $(0, -2)$    **12.** ____ focus at $(1, 1)$

**13.** ____ $4(x + 2)^2 + (y + 1)^2 = 16$    **14.** ____ $(x - 1)^2 + 4(y - 1)^2 = 16$    **15.** ____ axis of symmetry: $y = -1$

**16.** ____ transverse axis $y = 1$

## SUMMARY AND CONCEPT REVIEW

### SECTION 8.1   A Brief Introduction to Analytical Geometry

#### KEY CONCEPTS

- The midpoint and distance formulas play important roles in the study of analytical geometry:

$$\text{midpoint: } (x, y) = \left(\frac{x_1 + x_2}{2}, \frac{y_1 + y_2}{2}\right) \quad \text{distance: } d = \sqrt{(x_2 - x_1)^2 + (y_2 - y_1)^2}$$

- Using these tools, we can verify or construct relationships between points, lines, and curves in the plane; verify properties of geometric figures; prove theorems from Euclidean geometry; and construct relationships that define the conic sections.

#### EXERCISES

**1.** Verify the closed figure with vertices $(-3, -4)$, $(-5, 4)$, $(3, 6)$, and $(5, -2)$ is a square.

**2.** Find the equation of the circle that circumscribes the square in Exercise 1.

**3.** A theorem from Euclidean geometry states: *If any two points are equidistant from the endpoints of a line segment, they are on the perpendicular bisector of the segment.* Determine if the line through $(-3, 6)$ and $(6, -9)$ is the perpendicular bisector of the segment through $(-5, -2)$ and $(5, 4)$.

**4.** Four points are given here. Verify that the distance from each point to the line $y = -1$ is the same as the distance from the given point to the fixed point $(0, 1)$: $(-6, 9)$, $(-2, 1)$, $(4, 4)$, and $(8, 16)$.

## SECTION 8.2    The Circle and the Ellipse

### KEY CONCEPTS

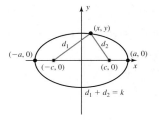

- The equation of a circle centered at $(h, k)$ with radius $r$ is $(x - h)^2 + (y - k)^2 = r^2$.
- The equation of an ellipse in standard form is $\frac{(x - h)^2}{a^2} + \frac{(y - k)^2}{b^2} = 1$. The center is at $(h, k)$, with horizontal distance $a$ and vertical distance $b$ from center to graph.
- Given two fixed points $f_1$ and $f_2$ in a plane (called the foci), an ellipse is the set of all points $(x, y)$ such that the distance from the first focus to $(x, y)$, plus the distance from the second focus to $(x, y)$, remains constant.
- For an ellipse, the distance from center to vertex is *greater than* the distance $c$ from center to one focus. To find the foci, we use $c^2 = |a^2 - b^2|$.

### EXERCISES

Sketch the graph of each equation in Exercises 5 through 9.

**5.** $x^2 + y^2 = 16$   **6.** $x^2 + 4y^2 = 36$   **7.** $x^2 + y^2 + 6x + 4y + 12 = 0$   **8.** $\dfrac{(x + 3)^2}{16} + \dfrac{(y - 2)^2}{9} = 1$

**9.** Find the equation of the ellipse with (a) vertices at $(\pm 13, 0)$ and foci at $(\pm 12, 0)$; (b) foci at $(0, \pm 16)$ with major axis of length 40 units.

**10.** Write the equation in standard form and sketch the graph, noting all of the characteristic features of the ellipse. $4x^2 + 25y^2 - 16x - 50y - 59 = 0$

## SECTION 8.3    The Hyperbola

### KEY CONCEPTS

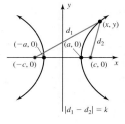

- The equation of a *horizontal* hyperbola in standard form is $\frac{(x - h)^2}{a^2} - \frac{(y - k)^2}{b^2} = 1$. The center is at $(h, k)$ with horizontal distance $a$ from center to vertices, and vertical distance $b$ from center to the central rectangle.
- Given two fixed points $f_1$ and $f_2$ in a plane (called the foci), a hyperbola is the set of all points $(x, y)$ such that the distance from one focus to point $(x, y)$, less the distance from the other focus to $(x, y)$, remains a positive constant: $|d_1 - d_2| = k$.
- For a hyperbola, the distance from center to one vertex is *less than* the distance from center to the focus $c$. To find the foci we have $c^2 = a^2 + b^2$.

### EXERCISES

Sketch the graph of each equation, indicating the center, vertices, and asymptotes.

**11.** $4y^2 - 25x^2 = 100$     **12.** $\dfrac{(x + 2)^2}{9} - \dfrac{(y - 1)^2}{4} = 1$

**13.** $9y^2 - x^2 - 18y - 72 = 0$     **14.** $x^2 - 4y^2 - 12x - 8y + 16 = 0$

**15.** Find the equation of the hyperbola with (a) vertices at $(\pm 15, 0)$, foci at $(\pm 17, 0)$, and (b) foci at $(0, \pm 5)$ with vertical dimension of central rectangle 8 units.

**16.** Write the equation in standard form and sketch the graph, noting all of the characteristic features of the hyperbola. $4x^2 - 9y^2 - 40x + 36y + 28 = 0$

## SECTION 8.4    The Analytic Parabola; More on Nonlinear Systems

### KEY CONCEPTS

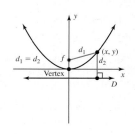

- Horizontal parabolas have equations of the form $x = ay^2 + by + c$; $a \neq 0$.
- A horizontal parabola will open to the right if $a > 0$, and to the left if $a < 0$. The axis of symmetry is $y = \frac{-b}{2a}$, with the vertex $(h, k)$ found by writing the equation in shifted form: $(x - h) = a(y - k)^2$.
- Given a fixed point $f$ (called the focus) and fixed line $D$ (called the directrix) in the plane, a parabola is the set of all points $(x, y)$ such that the distance from $f$ to $(x, y)$ is equal to the distance from $(x, y)$ to line $D$.
- The equation $x^2 = 4py$ describes a vertical parabola, opening upward if $p > 0$, and downward if $p < 0$.

- The equation $y^2 = 4px$ describes a horizontal parabola, opening to the right if $p > 0$, and to the left if $p < 0$.
- $p$ is the distance from the vertex to the focus (or from the vertex to the directrix).
- The focal chord of a parabola is a line segment that contains the focus and is parallel the directrix, with its endpoints on the graph. It has a total length of $|4p|$.

## EXERCISES

Find the vertex and $x$- and $y$-intercepts if they exist. Then sketch the graph using symmetry or by completing the square and using a transformation.

**17.** $x = y^2 - 4$           **18.** $x = y^2 + y - 6$

**21.** Identify the conic sections in the system, then solve.

$$\begin{cases} (x + 7)^2 + y^2 = 20 \\ y^2 - 7 = x \end{cases}$$

Find the vertex, focus, and directrix for each parabola. Then sketch the graph using this information and the focal chord. Also graph the directrix.

**19.** $x^2 = -20y$           **20.** $x^2 - 8x - 8y + 16 = 0$

---

## PRACTICE TEST

**By inspection only (no graphing, completing the square, etc.), match each equation to its correct description.**

**1.** $x^2 + y^2 - 6x + 4y + 9 = 0$ _____

**2.** $4y^2 + x^2 - 4x + 8y + 20 = 0$ _____

**3.** $y - x^2 - 4x + 20 = 0$ _____

**4.** $x^2 - 4y^2 - 4x + 12y + 20 = 0$ _____

   **a.** Parabola    **b.** Hyperbola    **c.** Circle    **d.** Ellipse

**Graph each conic section, and label the center, vertices, foci, focal chords, asymptotes, and other important features where applicable.**

**5.** $(x - 4)^2 + (y + 3)^2 = 9$

**6.** $\dfrac{(x - 2)^2}{16} + \dfrac{(y + 3)^2}{1} = 1$

**7.** $\dfrac{(x + 3)^2}{9} - \dfrac{(y - 4)^2}{4} = 1$

**8.** $x^2 + y^2 - 10x + 4y + 4 = 0$

**9.** $9x^2 + 4y^2 + 18x - 24y + 9 = 0$

**10.** $9x^2 - 4y^2 + 18x - 24y - 63 = 0$

**11.** $x = (y + 3)^2 - 2$

**12.** $y^2 - 6y - 12x - 15 = 0$

**Solve each nonlinear system using any method.**

**13. a.** $\begin{cases} 4x^2 - y^2 = 16 \\ y - x = 2 \end{cases}$   **b.** $\begin{cases} 2y^2 - x^2 = 4 \\ x^2 + y^2 = 8 \end{cases}$

**14.** A support bracket on the frame of a large ship is a steel right triangle with a hypotenuse of 25 ft and a perimeter of 60 ft. Find the lengths of the other sides using a system of nonlinear equations.

**15.** Find an equation for the circle whose center is at $(-2, 5)$ and whose graph goes through the point $(0, 3)$.

**16.** Find the equation of the ellipse (in standard form) with vertices at $(\pm 4, 0)$ and foci at $(\pm 2, 0)$. Then determine where this ellipse and the circle $x^2 + y^2 = 13$ intersect.

**17.** The orbit of Halley's Comet around the Sun is elliptical, with the Sun at one focus. When the orbit is expressed as a central ellipse on the coordinate grid, its equation is $\dfrac{x^2}{17.8^2} + \dfrac{y^2}{4.8^2} = 1$, with $a$ and $b$ in astronomical units. Use this information to find the distance from the Sun at its farthest point, and the distance from the Sun at its closest point.

**Determine the equation of each relation and state its domain and range. For the parabola and the ellipse, also give the location of the foci.**

**18.**

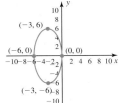

**19.**

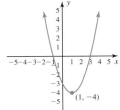

**20.**

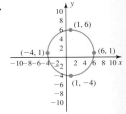

## CALCULATOR EXPLORATION AND DISCOVERY

### I. The Graph of a Circle

When using a graphing calculator to study circles, it is important to keep two things in mind. First, we must modify the equation of the circle before it can be graphed using this technology. Second, most standard viewing windows have the $x$- and $y$-values preset at $[-10, 10]$ even though the calculator screen is not square. This tends to compress the $y$-values and give a skewed image of the graph. Consider the *relation* $x^2 + y^2 = 25$, which we know is the equation of a circle centered at $(0, 0)$ with radius $r = 5$. To enable the calculator to graph this relation, we must define it in two pieces by solving for $y$:

$$x^2 + y^2 = 25 \qquad \text{original equation}$$
$$y^2 = 25 - x^2 \qquad \text{isolate } y^2$$
$$y = \pm\sqrt{25 - x^2} \qquad \text{solve for } y$$

Note that we can separate this result into two parts, enabling the calculator to draw the circle: $Y_1 = \sqrt{25 - x^2}$ gives the "upper half" of the circle, and $Y_2 = -\sqrt{25 - x^2}$ gives the "lower half." Enter these on the ⬭ screen (note that $Y_2 = -Y_1$ can be used instead of reentering the entire expression: ⬭ ⬭ ⬭). But if we graph $Y_1$ and $Y_2$ on the standard screen, the result appears more oval than circular (Figure 8.33). One way to fix this is to use the ⬭ **5:ZSquare** option, which forces the same scale to be used on both axes (see Figure 8.34). Although it is a much improved graph, the circle does not appear "closed" as the calculator lacks sufficient pixels to show the proper curvature. A second alternative is to manually set a "friendly" window. Using Xmin $= -9.4$, Xmax $= 9.4$, Ymin $= -6.2$, and Ymax $= 6.2$ will generate a better graph, which we can use to study the relation more closely. Note that we can jump between the upper and lower halves of the circle using the up ⬭ or down ⬭ arrows.

**Figure 8.33**

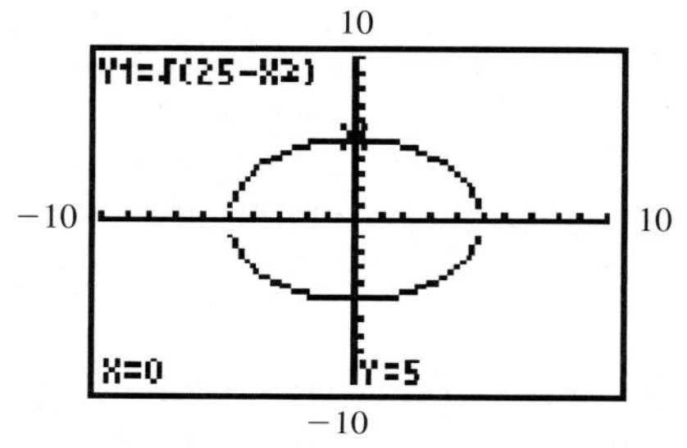

**Figure 8.34**

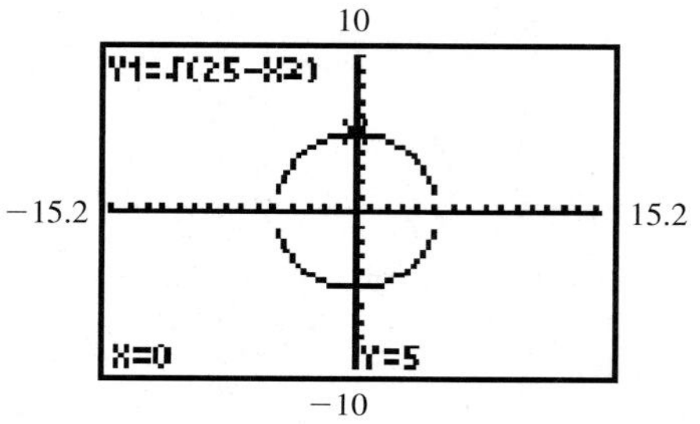

**Exercise 1:** Graph the circle defined by $x^2 + y^2 = 36$ using a friendly window, then use the ⬭ feature to find the value of $y$ when $x = 3.6$. Now find the value of $y$ when $x = 4.8$. Explain why the values seem "interchangeable."

**Exercise 2:** Graph the circle defined by $(x - 3)^2 + y^2 = 16$ using a friendly window, then use the ⬭ feature to find the value of the $y$-intercepts. Show you get the same intercept by computation.

### II. Elongation and Eccentricity

Technically speaking, a circle is an ellipse with both foci at the center. As the distance between foci increases, the ellipse becomes more elongated. We saw other instances of elongation in stretches and compressions of parabolic graphs, and in hyperbolic graphs where the asymptotic slopes varied depending on the values $a$ and $b$. The measure used to quantify this elongation is called the *eccentricity $e$*, and is determined by the ratio $e = \frac{c}{a}$. For this *Exploration and Discovery*, we'll use the **repeat graph** feature of a graphing calculator to explore the eccentricity of the graph of a conic. The "repeat graph" feature enables you to graph a family of curves by enclosing the values of a parameter in braces "{ }." For instance, entering $\{-2, -1, 0, 1, 2\}X + 3$ as $Y_1$ on the ⬭ screen will automatically graph these five lines:

$$y = -2x + 3, \, y = -1x + 3, \, y = 0x + 3, \, y = 1x + 3, \text{ and } y = 2x + 3$$

We'll use this feature to graph a family of ellipses, observing the result and calculating the eccentricity for each curve in the family. The standard form is $\frac{x^2}{a^2} + \frac{y^2}{b^2} = 1$, which we'll solve for $y$ and enter as $Y_1$ and $Y_2$. After simplification the result is $y = \pm b\sqrt{1 - \frac{x^2}{a^2}}$, but for this investigation we'll use the constant $b = 2$ and vary the parameter $a$ using the values $a = 2, 4, 6,$ and $8$. The result is $y = 2\sqrt{1 - \frac{x^2}{\{2^2, 4^2, 6^2, 8^2\}}}$. Note from Figure 8.35 that we've set $Y_2 = -Y_1$ to graph the lower half of the ellipse. Using the "friendly window" shown (Figure 8.36) gives the graphs displayed in Figure 8.37, where we see the ellipse is increasingly elongated in the horizontal direction (note when $a = 2$ the result is a circle).

**Figure 8.35**

**Figure 8.36**

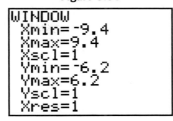

**Figure 8.37**

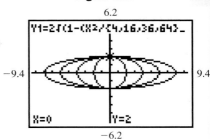

Using $a = 2, 4, 6,$ and $8$ with $b = 2$ in the foci formula $c = \sqrt{a^2 - b^2}$ gives $c = 0, 2\sqrt{3}, 4\sqrt{2},$ and $2\sqrt{15}$ respectively, with these eccentricities: $e = \frac{c}{a} = 0, \frac{\sqrt{3}}{2}, \frac{2\sqrt{2}}{3},$ and $\frac{\sqrt{15}}{4}$. While difficult to see in radical form, we find that the eccentricity of an ellipse always satisfies the inequality $0 \le e < 1$. To two decimal places, the values are $e = 0, 0.87, 0.94,$ and $0.97$ respectively.

As a final note, it's interesting how the $e = \frac{c}{a}$ definition of eccentricity relates to our everyday use of the word "eccentric." A normal or "noneccentric" person is thought to be well-rounded, and sure enough $e = 0$ produces a well-rounded figure—a circle. A person who is highly eccentric is thought to be far from the norm, deviating greatly from the center, and greater values of $e$ produce very elongated ellipses.

**Exercise 3:** Perform a similar exploration using a family of *hyperbolas*. What do you notice about the eccentricity?

**Exercise 4:** Perform a similar exploration using a family of *parabolas*. What do you notice about the eccentricity?

## STRENGTHENING CORE SKILLS

### More on Completing the Square

From our work so far in Chapter 8, we realize the process of *completing the square* has much greater use than simply as a tool for working with quadratic equations. It is a valuable tool in the application of the conic sections, as well as other areas. The purpose of this *Strengthening Core Skills* is to reinforce the ability and confidence needed to apply the process correctly. This is important because in some cases the values of $a$ and $b$ are rational or irrational numbers. No matter what the context,

1. **The process begins *with a coefficient of 1*.** For $20x^2 + 120x + 27y^2 - 54y + 192 = 0$, we recognize the equation of an ellipse, since the coefficients of the squared terms are positive and unequal. To study or graph this ellipse, we'll use the standard form to identify the values of $a$, $b$, and $c$. Grouping the like-variable terms gives

$$(20x^2 + 120x \quad) + (27y^2 - 54y \quad) + 192 = 0$$

and to complete the square, we factor out the lead coefficient of each group (to get a coefficient of 1):

$$20(x^2 + 6x \quad) + 27(y^2 - 2y \quad) + 192 = 0$$

Subtracting 192 from both sides brings us to the fundamental step for completing the square.

2. **The quantity $[\frac{1}{2} \text{ (linear coefficient)}]^2$ will complete a trinomial square.** For this example we obtain $[\frac{1}{2}(6)]^2 = 9$ for $x$, and $[\frac{1}{2}(-2)]^2 = 1$ for $y$, with these numbers inserted in the appropriate group:

$$20(x^2 + 6x + 9) + 27(y^2 - 2y + 1) = -192 + \underline{\qquad} \qquad \text{complete the square}$$

Due to the distributive property, we have in effect added $20 \cdot 9 = 180$ and $27 \cdot 1 = 27$ (for a total of 207) to the left side of the equation:

$$\underset{\substack{\text{adds } 20 \cdot 9 = 180 \\ \text{to left side}}}{20(x^2 + 6x + 9)} + \underset{\substack{\text{adds } 27 \cdot 1 = 27 \\ \text{to left side}}}{27(y^2 - 2y + 1)} = -192 + \underline{\qquad}$$

This brings us to the final step.

3. **Keep the equation in balance.** Since the left side was increased by 207, we also increase the right side by 207.

$$\underset{\substack{\text{adds } 20 \cdot 9 = 180 \\ \text{to left side}}}{20(x^2 + 6x + 9)} + \underset{\substack{\text{adds } 27 \cdot 1 = 27 \\ \text{to left side}}}{27(y^2 - 2y + 1)} = -192 + \underset{\substack{\text{add } 180 + 27 = 207 \\ \text{to right side}}}{207}$$

The quantities in parentheses factor, giving $20(x + 3)^2 + 27(y - 1)^2 = 15$. We then divide by 15 and simplify, obtaining the standard form $\frac{4(x + 3)^2}{3} + \frac{9(y - 1)^2}{5} = 1$. Note the coefficient of each binomial square is not 1, even after setting the equation equal to 1. Practice completing the square using these exercises.

**Exercise 1:** $100x^2 - 400x + 18y^2 - 108y + 554 = 0$

**Exercise 2:** $28x^2 - 56x + 48y^2 + 192y + 195 = 0$

## CUMULATIVE REVIEW CHAPTERS R–8

**Solve each equation.**

1. $x^3 - 2x^2 + 4x - 8 = 0$

2. $2|n + 4| + 3 = 13$

3. $\sqrt{x - 3} + 5 = x$

4. $x^{\frac{3}{2}} + 8 = 0$

5. $x^2 - 6x + 13 = 0$

6. $4 \cdot 2^{x+1} = \frac{1}{8}$

7. $3^{x-2} = 7$

8. $\ln x = 2$

9. $\log x + \log(x - 3) = 1$

**Graph each relation. Include vertices, x- and y-intercepts, asymptotes, and other features.**

10. $y = \frac{2}{3}x + 2$

11. $y = |x - 2| + 3$

12. $y = \frac{1}{x - 1} + 2$

13. $y = \sqrt{x - 3} + 1$

14. **a.** $g(x) = (x - 3)(x + 1)(x + 4)$

   **b.** $f(x) = x^4 + x^3 - 13x^2 - x + 12$

15. $h(x) = \frac{x - 2}{x^2 - 9}$

16. $q(x) = 2^x + 3$

17. $f(x) = \log_2(x + 1)$

18. $x = y^2 + 4y + 7$

19. $x^2 + y^2 + 10x - 4y + 20 = 0$

20. $4(x - 1)^2 - 36(y + 2)^2 = 144$

21. Determine the following for the indicated graph (write all answers in interval notation): (a) domain, (b) range, (c) interval(s) where $f(x)$ is increasing or decreasing, (d) location of any maximum or minimum value(s), (e) interval(s) where $f(x)$ is positive and negative.

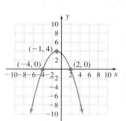

**Solve each system of equations.**

22. $\begin{cases} 4x + 3y = 13 \\ -9y + 5z = 19 \\ x - 4z = -4 \end{cases}$

23. $\begin{cases} x^2 + y^2 = 25 \\ 64x^2 + 12y^2 = 768 \end{cases}$

24. If a person invests $5000 at 9% compounded quarterly, how long would it take for the money to grow to $12,000?

25. A radiator contains 10 L of liquid that is 40% antifreeze. How much should be drained off and replaced with pure antifreeze for a 60% mixture?

**Solve each equation using the approach specified.**

26. $\frac{1}{8}x^3 - 4x + 3 = \frac{1}{4}x^2 - 5$; Rational roots theorem

27. $|x + 4| = 8 - |x|$; Graphing

28. $e^{2x} - 3e^x = 4$; u-substitution

29. $3^{x-1} = 2^{2-x}$; Logarithms

30. $\frac{x + 3}{x - 4} \geq 3$; Interval tests

# Sequences and Series; Counting and Probability

## CHAPTER OUTLINE

## CHAPTER CONNECTIONS

Accurately modeling the size of an endangered population is a critical step in preserving the species. Without such estimates, there would be no way of measuring how successful efforts to rebuild the population have been. Once an accurate population count is made, wildlife experts can use this initial value along with the natural reproduction rate, relocation data, and other factors to predict the size of the population in future years. In Chapter 9, we'll see how sequences and series can be used to model endangered populations, even when such limited information is available. This application appears as Exercise 76 in Section 9.1.

**Check out these other real-world connections:**

▶ Blue-Book Values
(Section 9.1, Exercise 73)
▶ Scheduling Classes
(Section 9.4, Exercises 23–26)
▶ Disarming Explosives
(Section 9.5, Exercise 56)
▶ Pollution Testing
(Section 9.6, Exercise 38)

## LEARNING OBJECTIVES

*In Section 9.1 you will see how we can:*

☐ **A.** Write out the terms of a sequence given the general or *n*th term

☐ **B.** Work with recursive sequences and sequences involving a factorial

☐ **C.** Find the partial sum of a series

☐ **D.** Use summation notation to write and evaluate series

☐ **E.** Use sequences to solve applications

A *sequence* can be thought of as a pattern of numbers listed in a prescribed order. A *series* is the sum of the numbers in a sequence. Sequences and series come in countless varieties, and we'll introduce some general forms here. In following sections we'll focus on two special types: arithmetic and geometric sequences. These are used in a number of different fields, with a wide variety of significant applications.

## A. Finding the Terms of a Sequence Given the General Term

Suppose you had $10,000 to invest, and decided to place the money in government bonds that guarantee an annual return of 7%. From our work in Chapter 5, we know the amount of money in the account after $x$ years can be modeled by the function $f(x) = 10,000(1.07)^x$. If you reinvest your earnings each year, the amount in the account would be (rounded to the nearest dollar):

| Year: | $f(1)$ | $f(2)$ | $f(3)$ | $f(4)$ | $f(5)\ldots$ |
|---|---|---|---|---|---|
| | ↓ | ↓ | ↓ | ↓ | ↓ |
| Value: | $10,700 | $11,449 | $12,250 | $13,108 | $14,026\ldots$ |

Note the relationship (year, value) is a function that pairs 1 with $10,700, 2 with $11,449, 3 with $12,250, and so on. This is an example of a **sequence.** To distinguish sequences from other algebraic functions, we commonly name the functions $a$ instead of $f$, use the variable $n$ instead of $x$, and employ a subscript notation. The function $f(x) = 10,000(1.07)^x$ would then be written $a_n = 10,000(1.07)^n$. Using this notation $a_1 = 10,700$, $a_2 = 11,449$, and so on.

The values $a_1, a_2, a_3, a_4, \ldots$ are called the **terms** of the sequence, while the subscripted variables 1, 2, 3, 4, ... are called the indices (plural for **index**) of the terms. If the account were closed after a certain number of years (for example, after the fifth year) we have a **finite sequence.** If we let the investment grow indefinitely, the result is called an **infinite sequence.** The expression $a_n$ that defines the sequence is called the **general** or **nth term** and the terms immediately preceding it are called the $(n - 1)$st term, the $(n - 2)$nd term, and so on.

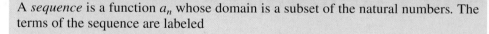

### Sequences

A *sequence* is a function $a_n$ whose domain is a subset of the natural numbers. The terms of the sequence are labeled

$$a_1, a_2, a_3, \ldots, a_{n-1}, a_n, a_{n+1}, \ldots$$

where $a_n$ represents an arbitrary "interior" term of the sequence.

---

**EXAMPLE 1A** ▶ **Computing Specified Terms of a Sequence**

For $a_n = \frac{n+1}{n^2}$, find $a_1, a_3, a_6,$ and $a_7$.

**Solution** ▶

$$a_1 = \frac{1+1}{1^2} = 2 \qquad a_3 = \frac{3+1}{3^2} = \frac{4}{9}$$

$$a_6 = \frac{6+1}{6^2} = \frac{7}{36} \qquad a_7 = \frac{7+1}{7^2} = \frac{8}{49}$$

---

**EXAMPLE 1B** ▶ | **Computing the First Few Terms of a Sequence**

Write the first four terms of the sequence $a_n = (-1)^n 2^n$ as a list.

**Solution** ▶

$a_1 = (-1)^1 2^1 = -2$        $a_2 = (-1)^2 2^2 = 4$
$a_3 = (-1)^3 2^3 = -8$        $a_4 = (-1)^4 2^4 = 16$

The sequence can be written $-2, 4, -8, 16, \ldots$, or more generally as $-2, 4, -8, 16, \ldots, (-1)^n 2^n, \ldots$ to show how each term was generated.

☑ **A.** You've just seen how we can write out the terms of a sequence given the general or $n$th term

**Now try Exercises 5 through 18** ▶

When the terms of a sequence *alternate in sign* as in Example 1B, we call it an **alternating sequence**.

## B. Recursive Sequences and Factorial Notation

Sometimes the formula defining a sequence uses the preceding term or terms to generate those that follow. These are called **recursive sequences** and are particularly useful in writing computer programs. Because of how they are defined, recursive sequences must give an inaugural term or **seed element(s)** to begin the recursion process.

Perhaps the most famous recursive sequence is associated with the work of Leonardo of Pisa (A.D. 1180–1250), better known to history as *Fibonacci*. In the Fibonacci sequence, each successive term is the sum of the previous two, beginning with 1, 1, … .

**EXAMPLE 2** ▶ | **Computing the Terms of a Recursive Sequence**

Write out the first eight terms of the Fibonacci sequence, which is defined by $c_1 = 1$, $c_2 = 1$, and $c_n = c_{n-1} + c_{n-2}$.

**Solution** ▶ | The first two terms are given, so we begin with $n = 3$.

$$
\begin{array}{lll}
c_3 = c_{3-1} + c_{3-2} & c_4 = c_{4-1} + c_{4-2} & c_5 = c_{5-1} + c_{5-2} \\
\;\;\; = c_2 + c_1 & \;\;\; = c_3 + c_2 & \;\;\; = c_4 + c_3 \\
\;\;\; = 1 + 1 & \;\;\; = 2 + 1 & \;\;\; = 3 + 2 \\
\;\;\; = 2 & \;\;\; = 3 & \;\;\; = 5
\end{array}
$$

At this point we can use the fact that each successive term is simply the sum of the preceding two, and find that $c_6 = 5 + 3 = 8$, $c_7 = 8 + 5 = 13$, and $c_8 = 13 + 8 = 21$. The first eight terms are 1, 1, 2, 3, 5, 8, 13, and 21.

**Now try Exercises 19 through 24** ▶

**WORTHY OF NOTE**

One application of the Fibonacci sequence involves the Fibonacci spiral, found in the growth of many ferns and the spiral shell of many mollusks.

Some sequences may involve the computation of a **factorial**, which is the product of a given natural number with all those that precede it. The expression **5!** is read, "five factorial," and is evaluated as: $5! = 5 \cdot 4 \cdot 3 \cdot 2 \cdot 1 = 120$.

**Factorials**

For any natural number $n$,
$$n! = n \cdot (n - 1) \cdot (n - 2) \cdots \cdots 3 \cdot 2 \cdot 1$$

Rewriting a factorial in equivalent forms often makes it easier to simplify certain expressions. For example, we can rewrite 5! as $5 \cdot 4!$ or $5! = 5 \cdot 4 \cdot 3!$, and so on. Consider Example 3.

**EXAMPLE 3** ▶ **Simplifying Expressions Using Factorial Notation**

Simplify by writing the numerator in an equivalent form.

**a.** $\dfrac{9!}{7!}$      **b.** $\dfrac{11!}{8!3!}$      **c.** $\dfrac{10!}{3!5!2!}$

Solution ▶

**a.** $\dfrac{9!}{7!} = \dfrac{9 \cdot 8 \cdot \cancel{7!}}{\cancel{7!}}$    **b.** $\dfrac{11!}{8!3!} = \dfrac{11 \cdot 10 \cdot 9 \cdot \cancel{8!}}{\cancel{8!}3!}$    **c.** $\dfrac{10!}{3!5!2!} = \dfrac{10 \cdot 9 \cdot 8 \cdot 7 \cdot 6 \cdot \cancel{5!}}{3!\cancel{5!}2!}$

$= 9 \cdot 8$        $= \dfrac{990}{6}$        $= \dfrac{10 \cdot 9 \cdot 8 \cdot 7 \cdot \cancel{6}}{\cancel{6} \cdot 2}$

$= 72$        $= 165$        $= \dfrac{5040}{2} = 2520$

> **WORTHY OF NOTE**
>
> Most calculators have a factorial option or key. On many calculator models it is located in a submenu of the **MATH** menu: **MATH** ◀ PRB 4: !.

Now try Exercises 25 through 30 ▶

---

**EXAMPLE 4** ▶ **Computing a Specified Term from a Sequence Defined Using Factorials**

Find and simplify the third term of each sequence.

**a.** $a_n = \dfrac{n!}{2^n}$      **b.** $b_n = \dfrac{(-1)^n(2n-1)!}{n!}$

Solution ▶

**a.** $a_3 = \dfrac{3!}{2^3}$      **b.** $b_3 = \dfrac{(-1)^3[2(3)-1]!}{3!}$

$= \dfrac{6}{8} = \dfrac{3}{4}$      $= \dfrac{(-1)(5!)}{3!} = \dfrac{(-1)(5 \cdot 4 \cdot \cancel{3!})}{\cancel{3!}}$

$= -20$

☑ **B.** You've just seen how we can work with recursive sequences and sequences involving a factorial

Now try Exercises 31 through 36 ▶

## C. Series and Partial Sums

Sometimes the terms of a sequence are dictated by context rather than a formula. Consider the stacking of large pipes in a storage yard. If there are 10 pipes in the bottom row, then 9 pipes, then 8, and so on (see Figure 9.1), how many pipes are in the stack if there is a single pipe at the top? The sequence generated is 10, 9, 8, … , 3, 2, 1 and to answer the question we would have to *compute the sum of all terms in the sequence.* When the terms of a finite sequence are added, the result is called a **finite series.**

**Figure 9.1**

> **Finite Series**
>
> Given the sequence $a_1, a_2, a_3, a_4, \ldots , a_n$, the sum of the terms is called a **finite series** or **partial sum** and is denoted $S_n$:
>
> $$S_n = a_1 + a_2 + a_3 + \cdots + a_{n-1} + a_n$$

---

**EXAMPLE 5** ▶ **Computing a Partial Sum**

Given $a_n = 2n - 1$, find the value of (a) $S_4$ and (b) $S_7$.

Solution ▶ Since we eventually need the sum of the first seven terms, begin by writing out these terms: 1, 3, 5, 7, 9, 11, and 13.

**a.** $S_4 = a_1 + a_2 + a_3 + a_4$      **b.** $S_7 = a_1 + a_2 + a_3 + a_4 + a_5 + a_6 + a_7$

$= 1 + 3 + 5 + 7$      $= 1 + 3 + 5 + 7 + 9 + 11 + 13$

$= 16$      $= 49$

☑ **C.** You've just seen how we can find the partial sum of a series

Now try Exercises 37 through 42 ▶

## D. Summation Notation

When the general term of a sequence is known, the Greek letter *sigma* $\Sigma$ can be used to write the related series in a more compact form. For instance, to indicate the sum of the first four terms of $a_n = 3n + 2$, we write $\sum_{i=1}^{4}(3i + 2)$ with this notation indicating we are to compute the sum of all terms generated as $i$ cycles from 1 through 4. This result is called **summation** or **sigma notation** and the letter $i$ is called the **index of summation**. The letters $j$, $k$, $l$, and $m$ are also used as indices (plural for index), and the summation need not start at 1.

---

**EXAMPLE 6** ▶  **Computing a Partial Sum**

Compute each sum:

    **a.** $\sum_{i=1}^{4}(3i + 2)$      **b.** $\sum_{j=1}^{5}\frac{1}{j}$      **c.** $\sum_{k=3}^{6}(-1)^k k^2$

**Solution** ▶   **a.** $\sum_{i=1}^{4}(3i + 2) = (3 \cdot 1 + 2) + (3 \cdot 2 + 2) + (3 \cdot 3 + 2) + (3 \cdot 4 + 2)$

$$= 5 + 8 + 11 + 14 = 38$$

    **b.** $\sum_{j=1}^{5}\frac{1}{j} = \frac{1}{1} + \frac{1}{2} + \frac{1}{3} + \frac{1}{4} + \frac{1}{5}$

$$= \frac{60}{60} + \frac{30}{60} + \frac{20}{60} + \frac{15}{60} + \frac{12}{60} = \frac{137}{60}$$

    **c.** $\sum_{k=3}^{6}(-1)^k k^2 = (-1)^3 \cdot 3^2 + (-1)^4 \cdot 4^2 + (-1)^5 \cdot 5^2 + (-1)^6 \cdot 6^2$

$$= (-9) + 16 + (-25) + 36 = 18$$

**Now try Exercises 43 through 54** ▶

---

> **WORTHY OF NOTE**
>
> The finite series given in Example 6(b) is a partial sum for the **harmonic series** $\sum_{j=1}^{\infty}\frac{1}{j} = 1 + \frac{1}{2} + \frac{1}{3} + \frac{1}{4} + \frac{1}{5} + \cdots$.
> This series, so named because of its connection to harmonics in music, is a special case of several important series studied in calculus.

If a definite pattern is noted in a given series expansion, this process can be reversed, with the expanded form being expressed in summation notation using the *n*th term.

---

**EXAMPLE 7** ▶  **Writing a Sum in Sigma Notation**

Write each of the following sums in summation (sigma) notation.

    **a.** $-2 + 4 - 8 + 16 - 32$      **b.** $12 + 15 + 18 + 21 + \cdots$

**Solution** ▶   **a.** The series has five terms and each term is a power of 2 with alternating signs. The general term is $a_n = (-2)^n$, and the series is $\sum_{n=1}^{5}(-2)^n$.

    **b.** The raised ellipsis "$\cdots$" indicates the sum continues indefinitely. Since the terms are consecutive multiples of 3, we identify the general term as $a_n = 3n$, while noting the series starts at $n = 4$ (instead of $n = 1$). Since the sum continues indefinitely, we use the infinity symbol $\infty$ as the "ending" value in sigma notation. The series is $\sum_{n=4}^{\infty} 3n$.

> **WORTHY OF NOTE**
>
> By varying the formula for the general term and/or where the sum begins, more than one acceptable form is possible.
> For Example 7(b), $\sum_{k=1}^{\infty}(9 + 3k)$ also works.

**Now try Exercises 55 through 64** ▶

Since the commutative and associative laws hold for the addition of real numbers, summations have the following properties:

> ### Properties of Summation
>
> Given any real number $c$ and natural number $n$,
>
> (I) $\displaystyle\sum_{i=1}^{n} c = cn$
>
> If you add a constant $c$ "$n$" times the result is $cn$.
>
> (II) $\displaystyle\sum_{i=1}^{n} ca_i = c\sum_{i=1}^{n} a_i$
>
> A constant can be factored out of a sum.
>
> (III) $\displaystyle\sum_{i=1}^{n} (a_i \pm b_i) = \sum_{i=1}^{n} a_i \pm \sum_{i=1}^{n} b_i$
>
> A summation can be decomposed into two (or more) simpler summations.
>
> (IV) $\displaystyle\sum_{i=1}^{n} a_i = \sum_{i=1}^{m} a_i + \sum_{i=m+1}^{n} a_i; \; 1 \le m < n$
>
> A summation is cumulative and can be written as a sum of smaller parts.

The verification of property II depends solely on the distributive property.

**Proof:** $\displaystyle\sum_{i=1}^{n} ca_i = ca_1 + ca_2 + ca_3 + \cdots + ca_n$   expand sum

$\qquad = c(a_1 + a_2 + a_3 + \cdots + a_n)$   factor out $c$

$\qquad = c\displaystyle\sum_{i=1}^{n} a_i$   write series in summation form

The verifications of properties III and IV simply use the commutative and associative properties of addition. You are asked to prove property III in **Exercise 79.**

---

**EXAMPLE 8** ▶ **Computing a Sum Using Summation Properties**

Recompute the sum $\displaystyle\sum_{i=1}^{4} (3i + 2)$ from Example 6(a) using summation properties.

**Solution** ▶ $\displaystyle\sum_{i=1}^{4} (3i + 2) = \sum_{i=1}^{4} 3i + \sum_{i=1}^{4} 2$   property III

$\qquad = 3\displaystyle\sum_{i=1}^{4} i + \sum_{i=1}^{4} 2$   property II

$\qquad = 3(10) + 2(4)$   $1 + 2 + 3 + 4 = 10$; property I

$\qquad = 38$   result

☑ **D.** You've just seen how we can use summation notation to write and evaluate series

**Now try Exercises 65 through 68** ▶

## E. Applications of Sequences

To solve applications of sequences, begin by identifying where the sequence begins (the initial term), then write out the first few terms to help identify the $n$th term.

**EXAMPLE 9** ▶  Solving an Application—Accumulation of Stock

Hydra already owned 1420 shares of stock when her company began offering employees the opportunity to purchase 175 discounted shares per year. If she made no purchases other than these discounted shares each year, how many shares will she have 9 yr later? If this continued for the 25 yr she will work for the company, how many shares will she have at retirement?

**Solution** ▶  To begin, it helps to simply write out the first few terms of the sequence. Since she already had 1420 shares before the company made this offer, we let $a_0 = 1420$ be the inaugural term. After 1 yr, she owns $a_1 = 1420 + 175 = 1595$ shares. The next few terms are $a_2 = 1770$, $a_3 = 1945$, $a_4 = 2120$, and so on. This supports a general term of $a_n = 1420 + 175n$.

| After 9 years | After 25 years |
|---|---|
| $a_9 = 1420 + 175(9)$ | $a_{25} = 1420 + 175(25)$ |
| $= 2995$ | $= 5795$ |

After 9 yr she would have 2995 shares. Upon retirement she would own 5795 shares of company stock.

 **E.** You've just seen how we can use sequences to solve applications

**Now try Exercises 71 through 78** ▶

## 9.1 EXERCISES

▶ **CONCEPTS AND VOCABULARY**

**Fill in the blank with the appropriate word or phrase. Carefully reread the section, if necessary.**

1. A sequence is a(n) _____ of numbers listed in a specific _____.

2. A series is the _____ of the numbers from a given sequence.

3. Describe the characteristics of a recursive sequence and give one example.

4. Describe the characteristics of an alternating sequence and give one example.

▶ **DEVELOPING YOUR SKILLS**

**Find the first four terms of each sequence.**

5. $a_n = 2n - 1$

6. $a_n = 3n^2 - 3$

7. $a_n = (-1)^n n$

8. $a_n = \dfrac{n}{n + 1}$

9. $a_n = \left(\dfrac{1}{2}\right)^n$

10. $a_n = \dfrac{1}{n}$

11. $a_n = \dfrac{(-1)^n}{n(n + 1)}$

12. $a_n = (-1)^n 2^n$

**Find the indicated term for each sequence. Use fractions if possible.**

13. $a_n = n^2 - 2$; $a_9$

14. $a_n = \dfrac{(-1)^{n+1}}{n}$; $a_5$

15. $a_n = \dfrac{(-1)^{n+1}}{2n - 1}$; $a_5$

16. $a_n = 2\left(\dfrac{1}{2}\right)^{n-1}$; $a_7$

17. $a_n = \left(1 + \dfrac{1}{n}\right)^n$; $a_{10}$

18. $a_n = \dfrac{1}{n(2n + 1)}$; $a_4$

**Find the first five terms of each recursive sequence.**

19. $\begin{cases} a_1 = 2 \\ a_n = 5a_{n-1} - 3 \end{cases}$

20. $\begin{cases} a_1 = 3 \\ a_n = 2a_{n-1} - 3 \end{cases}$

21. $\begin{cases} a_1 = -1 \\ a_n = (a_{n-1})^2 + 3 \end{cases}$

22. $\begin{cases} a_1 = -2 \\ a_n = a_{n-1} - 16 \end{cases}$

23. $\begin{cases} c_1 = 64, c_2 = 32 \\ c_n = \dfrac{c_{n-2} - c_{n-1}}{2} \end{cases}$

24. $\begin{cases} c_1 = 1, c_2 = 2 \\ c_n = c_{n-1} + (c_{n-2})^2 \end{cases}$

**Simplify each factorial expression.**

25. $\dfrac{8!}{5!}$

26. $\dfrac{12!}{10!}$

27. $\dfrac{9!}{7!2!}$

28. $\dfrac{6!}{3!3!}$

29. $\dfrac{8!}{2!6!}$

30. $\dfrac{10!}{3!7!}$

**Write out the first four terms in each sequence.**

**31.** $a_n = \dfrac{(-1)^n}{(n+1)!}$     **32.** $a_n = \dfrac{n!}{(n+3)!}$

**33.** $a_n = \dfrac{(n+1)!}{(3n)!}$     **34.** $a_n = \dfrac{(n+3)!}{(2n)!}$

**35.** $a_n = \dfrac{n^n}{n!}$     **36.** $a_n = \dfrac{2^n}{n!}$

**Find the indicated partial sum for each sequence.**

**37.** $a_n = n; S_5$     **38.** $a_n = n^2; S_7$

**39.** $a_n = 2n - 1; S_8$     **40.** $a_n = 3n - 1; S_6$

**41.** $a_n = \dfrac{1}{n}; S_5$     **42.** $a_n = \dfrac{n}{n+1}; S_4$

**Expand and evaluate each series.**

**43.** $\displaystyle\sum_{i=1}^{4} (3i - 5)$    **44.** $\displaystyle\sum_{i=1}^{5} (2i - 3)$    **45.** $\displaystyle\sum_{k=1}^{5} (2k^2 - 3)$

**46.** $\displaystyle\sum_{k=1}^{5} (k^2 + 1)$    **47.** $\displaystyle\sum_{k=1}^{7} (-1)^k k$    **48.** $\displaystyle\sum_{k=1}^{5} (-1)^k 2^k$

**49.** $\displaystyle\sum_{i=1}^{4} \dfrac{i^2}{2}$    **50.** $\displaystyle\sum_{i=2}^{4} i^2$    **51.** $\displaystyle\sum_{j=3}^{7} 2j$

**52.** $\displaystyle\sum_{j=3}^{7} \dfrac{j}{2^j}$    **53.** $\displaystyle\sum_{k=3}^{8} \dfrac{(-1)^k}{k(k-2)}$    **54.** $\displaystyle\sum_{k=2}^{6} \dfrac{(-1)^{k+1}}{k^2 - 1}$

**Write each sum using sigma notation.**

**55. a.** $4 + 8 + 12 + 16 + 20$
    **b.** $5 + 10 + 15 + 20 + 25$

**56. a.** $-1 + 4 - 9 + 16 - 25 + 36$
    **b.** $1 - 8 + 27 - 64 + 125 - 216$

**57. a.** $1 + 3 + 5 + 7 + 9 + 11 + \cdots$
    **b.** $1 + \dfrac{1}{2} + \dfrac{1}{4} + \dfrac{1}{8} + \dfrac{1}{16} + \dfrac{1}{32} + \cdots$

**58. a.** $0.1 + 0.01 + 0.001 + 0.0001 + \cdots$
    **b.** $1 + \dfrac{1}{2} + \dfrac{1}{6} + \dfrac{1}{24} + \dfrac{1}{120} + \dfrac{1}{720} + \cdots$

**For the given general term $a_n$, write the indicated sum using sigma notation.**

**59.** $a_n = n + 3; S_5$

**60.** $a_n = \dfrac{n^2 + 1}{n + 1}; S_4$

**61.** $a_n = \dfrac{n^2}{3}$; third partial sum

**62.** $a_n = 2n - 1$; sixth partial sum

**63.** $a_n = \dfrac{n}{2^n}$; sum for $n = 3$ to $7$

**64.** $a_n = n^2$; sum for $n = 2$ to $6$

**Compute each sum by applying properties of summation.**

**65.** $\displaystyle\sum_{i=1}^{5} (4i - 5)$     **66.** $\displaystyle\sum_{i=1}^{6} (3 + 2i)$

**67.** $\displaystyle\sum_{k=1}^{4} (3k^2 + k)$     **68.** $\displaystyle\sum_{k=1}^{4} (2k^3 + 5)$

▶ **WORKING WITH FORMULAS**

Triangular numbers: $a_n = \dfrac{n(n+1)}{2}$

The number of dots in the pattern shown represents a sequence of numbers aptly called the triangular numbers. The $n$th term of this sequence $a_n$ can be found using the formula given.

**69.** (a) Find $a_6$ using the formula and verify by drawing the next figure in the sequence. (b) Solve the given formula for $n$ in terms of $a_n$. Then (c) use your formula from part (b) to determine if 5050 is a triangular number. If so, which one?

**70.** (a) By rearranging the dots in $a_4$ and $a_5$, show *graphically* that $a_4 + a_5 = 5^2$. Then (b) show *algebraically* that this relationship holds at any position in the sequence by verifying $a_{n-1} + a_n = n^2$.

▶ **APPLICATIONS**

Use the information given in each exercise to determine the $n$th term $a_n$ for the sequence described. Then use the $n$th term to list the specified number of terms.

**Exercise 72**

**71. Wage increases:** Latisha gets \$7.25 an hour for filling candy machines for Archtown Vending. Each year she receives a \$0.50 hourly raise. List Latisha's hourly wage for the first 5 yr. How much will she make in the fifth year if she works 8 hr/day for 240 working days?

**72. Average birth weight:** The average birth weight of a certain animal species is 900 g, with the baby gaining 125 g each day for the first 10 days. List the infant's weight for the first 10 days. How much does the infant weigh on the 10th day?

**73. Blue-book value:** Steve's car has a blue-book value of $6000. Each year it loses 20% of its value. List the value of Steve's car for the next 5 yr.

**74. Effects of inflation:** Suppose inflation will average 4% for the next 5 yr. List the growing cost (year by year) of a DVD that costs $15 right now.

**75. Stocking a lake:** A local fishery stocks a large lake with 1500 bass and then adds an additional 100 mature bass per month until the lake nears maximum capacity. If the bass population grows at a rate of 5%/month through natural reproduction, the number of bass in the pond after $n$ months is given by the recursive sequence $b_0 = 1500$, $b_n = 1.05b_{n-1} + 100$. How many bass will be in the lake after 6 months?

**76. Species preservation:** The Interior Department introduces 50 wolves into a large wildlife area in an effort to preserve the species. Each year about 12 additional adult wolves are added from capture and relocation programs. If the wolf population grows at a rate of 10%/yr through natural reproduction, the number of wolves in the area after $n$ years is given by the recursive sequence $w_0 = 50$, $w_n = 1.10w_{n-1} + 12$. How many wolves are in the wildlife area after 6 years?

**77.** Handicaps are used in golf (and other sports) to estimate how far a player is from par (average). On a standard golf course, handicaps are computed according to the recurrence relation $a_n = \frac{(n-1)a_{n-1} + (x_n - 72)}{n}$, where $a_n$ is the handicap after playing $n$ rounds with $x_n$ strokes taken in the last round. Suppose your first six rounds after joining the local country club are (in order): 92, 87, 89, 93, 81, and 89. Compute the terms of the sequence formed by updating your handicap after each round. After all terms are found, round each to the nearest whole point.

**78.** In a typical "90% of 210" bowling league, handicaps are computed according to the recurrence relation $a_n = \frac{(n-1)a_{n-1} + 0.9(210 - x_n)}{n}$, where $a_n$ is the handicap after bowling $n$ games with a score of $x_n$ in the last game bowled. Suppose your first six scores after joining the league are (in order): 142, 149, 133, 167, 162, and 173. Compute the terms of the sequence formed by updating your handicap after each game. After all terms are found, round each to the nearest whole point.

▶ **EXTENDING THE CONCEPT**

**79.** Verify that a summation may be decomposed into two (or more) simpler sequences. That is, verify that the following statement is true:

$$\sum_{i=1}^{n}(a_i \pm b_i) = \sum_{i=1}^{n}a_i \pm \sum_{i=1}^{n}b_i.$$

**Surprisingly, some of the most celebrated numbers in mathematics can be represented or approximated by series expansion. Use your calculator to find the partial sums for $n = 10$, $n = 25$, and $n = 50$ for the summations given, and attempt to name the number the summation approximates:**

**80.** $\displaystyle\sum_{k=1}^{n} \frac{1}{2^k}$

**81.** $\displaystyle\sum_{k=0}^{n} \frac{1}{k!}$

**82.** $\displaystyle\sum_{k=0}^{n} \frac{4(-1)^k}{2k + 1}$

▶ **MAINTAINING YOUR SKILLS**

**83. (5.3)** Write $\log_3 \frac{1}{81} = -x$ in exponential form, then solve by equating bases.

**84. (2.4)** Set up the difference quotient for $f(x) = \sqrt{x}$, then rationalize the numerator.

**85. (8.4)** Solve the nonlinear system. $\begin{cases} x^2 + y^2 = 9 \\ 9y^2 - 4x^2 = 16 \end{cases}$

**86. (7.3)** Solve the system using a matrix equation.
$$\begin{cases} 25x + y - 2z = -14 \\ 2x - y + z = 40 \\ -7x + 3y - z = -13 \end{cases}$$

## 9.2 Arithmetic Sequences

**LEARNING OBJECTIVES**

*In Section 9.2 you will see how we can:*

- ☐ **A.** Identify an arithmetic sequence and its common difference
- ☐ **B.** Find the $n$th term of an arithmetic sequence
- ☐ **C.** Find the $n$th partial sum of an arithmetic sequence
- ☐ **D.** Solve applications involving arithmetic sequences

Similar to the way polynomials fall into certain groups or families (linear, quadratic, cubic, etc.), sequences and series with common characteristics are likewise grouped. In this section, we focus on arithmetic sequences, which behave much like the linear functions studied in Chapter 2.

### A. Identifying an Arithmetic Sequence and Finding the Common Difference

An **arithmetic sequence** is one where each successive term is found by adding a fixed constant to the preceding term. For instance 3, 7, 11, 15, … is an arithmetic sequence, since adding 4 to any given term produces the next term. This also means if you take the difference of any two consecutive terms, the result will be 4 and in fact, 4 is called the **common difference** $d$ for this sequence. Using the notation developed earlier, we can write $d = a_{n+1} - a_n$, where $a_n$ represents any term of the sequence and $a_{n+1}$ represents the term that follows $a_n$.

> **Arithmetic Sequences**
>
> Given a sequence $a_1, a_2, a_3, \ldots, a_n, a_{n+1}, \ldots$, where $n \in \mathbb{N}$,
> if there exists a common difference $d$ such that $a_{n+1} - a_n = d$ for all $n$,
> then the sequence is an *arithmetic sequence*.

The difference of successive terms can be rewritten as $a_{n+1} = a_n + d$ (for $n \geq 1$) to highlight that each term is found by adding $d$ to the previous term.

---

**EXAMPLE 1** ▶  **Identifying an Arithmetic Sequence**

Determine if the given sequence is arithmetic. If yes, name the common difference. If not, try to determine the pattern that forms the sequence.

  **a.** 2, 5, 8, 11, …     **b.** $\frac{1}{2}, \frac{5}{6}, \frac{13}{12}, \frac{77}{60}, \frac{29}{20}, \ldots$

**Solution** ▶  **a.** Begin by looking for a common difference $d = a_{n+1} - a_n$. Checking each pair of consecutive terms we have

$$5 - 2 = 3, \qquad 8 - 5 = 3, \qquad 11 - 8 = 3, \quad \text{and so on.}$$

This is an arithmetic sequence with common difference $d = 3$.

**b.** Checking each pair of consecutive terms yields

$$\frac{5}{6} - \frac{1}{2} = \frac{5}{6} - \frac{3}{6} \qquad\qquad \frac{13}{12} - \frac{5}{6} = \frac{13}{12} - \frac{10}{12} \qquad\qquad \frac{77}{60} - \frac{13}{12} = \frac{77}{60} - \frac{65}{60}$$

$$= \frac{2}{6} = \frac{1}{3} \qquad\qquad\qquad = \frac{3}{12} = \frac{1}{4} \qquad\qquad\qquad = \frac{12}{60} = \frac{1}{5}$$

Since the difference is not constant, this is not an arithmetic sequence. It appears the sequence is formed by adding $\frac{1}{n+1}$ to each previous term.

**Now try Exercises 5 through 16** ▶

**EXAMPLE 2 ▶**    **Writing the First Few Terms of an Arithmetic Sequence**

Write the first five terms of the arithmetic sequence, given the first term $a_1$ and the common difference $d$.

   **a.** $a_1 = 12$ and $d = -4$          **b.** $a_1 = \frac{1}{2}$ and $d = \frac{1}{3}$

**Solution ▶**    **a.** $a_1 = 12$ and $d = -4$. Starting at $a_1 = 12$, add $-4$ to each new term to generate the sequence.

$$a_1 = 12$$
$$a_2 = 12 + (-4) = 8$$
$$a_3 = 8 + (-4) = 4$$
$$a_4 = 4 + (-4) = 0$$
$$a_5 = 0 + (-4) = -4$$

The first five terms of this sequence are 12, 8, 4, 0, and $-4$.

   **b.** $a_1 = \frac{1}{2}$ and $d = \frac{1}{3}$. Starting at $a_1 = \frac{1}{2}$ and adding $\frac{1}{3}$ to each new term will generate the sequence: $\frac{1}{2}, \frac{5}{6}, \frac{7}{6}, \frac{3}{2}, \frac{11}{6}\ldots$ . Note that since the common denominator is 6, terms of the sequence can quickly be found by adding $\frac{1}{3} = \frac{2}{6}$ to the previous term and reducing if possible.

☑ **A. You've just seen how we can identify an arithmetic sequence and its common difference**

**Now try Exercises 17 through 24 ▶**

## B. Finding the *n*th Term of an Arithmetic Sequence

If the values $a_1$ and $d$ from an arithmetic sequence are known, we could generate the terms of the sequence by adding *multiples of d to the first term*, instead of repeatedly adding $d$ to new terms. For example, we can generate the sequence 3, 8, 13, 18, 23 by adding multiples of 5 to the first term $a_1 = 3$:

| | |
|---|---|
| $3 = 3 + (0)5$ | $a_1 = a_1 + 0d$ |
| $8 = 3 + (1)5$ | $a_2 = a_1 + 1d$ |
| $13 = 3 + (2)5$ | $a_3 = a_1 + 2d$ |
| $18 = 3 + (3)5$ | $a_4 = a_1 + 3d$ |
| $23 = 3 + (4)5$ | $a_5 = a_1 + 4d$ |

current term ⟶    initial term    ⟵ coefficient of common difference

It's helpful to note the coefficient of $d$ is 1 less than the subscript of the current term (as shown): $5 - 1 = 4$. This observation leads us to a formula for the *n*th term.

### The *n*th Term of an Arithmetic Sequence

The *n*th term of an *arithmetic sequence* is given by

$$a_n = a_1 + (n - 1)d,$$

where $d$ is the common difference.

**EXAMPLE 3** ▶ **Finding a Specified Term in an Arithmetic Sequence**

Find the 24th term of the sequence 0.1, 0.4, 0.7, 1, ... .

Solution ▶ Instead of creating all terms up to the 24th, we determine the constant $d$ and use the $n$th term formula. By inspection we note $a_1 = 0.1$ and $d = 0.4 - 0.1 = 0.3$.

$$
\begin{aligned}
a_n &= a_1 + (n - 1)d && \text{$n$th term formula} \\
&= 0.1 + (n - 1)0.3 && \text{substitute 0.1 for $a_1$ and 0.3 for $d$} \\
&= 0.1 + 0.3n - 0.3 && \text{distribute 0.3} \\
&= 0.3n - 0.2 && \text{simplify}
\end{aligned}
$$

To find the 24th term we substitute 24 for $n$:

$$
\begin{aligned}
a_{24} &= 0.3(24) - 0.2 && \text{substitute 24 for $n$} \\
&= 7 && \text{result}
\end{aligned}
$$

**Now try Exercises 25 through 36** ▶

**EXAMPLE 4** ▶ **Determining the Number of Terms in an Arithmetic Sequence**

Find the number of terms in the arithmetic sequence 2, $-5$, $-12$, $-19$, ..., $-411$.

Solution ▶ By inspection we see that $a_1 = 2$ and $d = (-5) - 2 = -7$. As before,

$$
\begin{aligned}
a_n &= a_1 + (n - 1)d && \text{$n$th term formula} \\
&= 2 + (n - 1)(-7) && \text{substitute 2 for $a_1$ and $-7$ for $d$} \\
&= 2 - 7n + 7 && \text{distribute $-7$} \\
&= -7n + 9 && \text{simplify}
\end{aligned}
$$

Although we don't know the number of terms in the sequence, we *do* know the last term is $-411$. Substituting $-411$ for $a_n$ gives

$$
\begin{aligned}
-411 &= -7n + 9 && \text{substitute $-411$ for $a_n$} \\
60 &= n && \text{solve for $n$}
\end{aligned}
$$

There are 60 terms in this sequence.

**Now try Exercises 37 through 44** ▶

Note that in both Examples 3 and 4, the $n$th term had the form of a linear function ($y = mx + b$) after simplifying: $a_n = 0.3n - 0.2$ and $a_n = -7n + 9$. This is a characteristic of arithmetic sequences, with the common difference $d$ corresponding to the slope $m$. This means the graph of an arithmetic sequence will always be a set of discrete points that lie on a straight line.

If the first term $a_1$ is unknown but another term $a_k$ is given, the $n$th term can be written

$$a_n = a_k + (n - k)d$$

(the subscript of the term $a_k$ and coefficient of $d$ sum to $n$).

**EXAMPLE 5** ▶ **Finding the First Term of an Arithmetic Sequence**

Given an arithmetic sequence where $a_6 = 0.55$ and $a_{13} = 0.9$, find the common difference $d$ and the value of $a_1$.

Solution ▶ At first it seems that not enough information is given, but recall we can express $a_{13}$ as the sum of any earlier term and the appropriate multiple of $d$. Since $a_6$ is known, we write $a_{13} = a_6 + 7d$ (note $13 = 6 + 7$ as required).

$$a_{13} = a_6 + 7d \qquad \text{$a_1$ is unknown}$$
$$0.9 = 0.55 + 7d \qquad \text{substitute 0.9 for $a_{13}$ and 0.55 for $a_6$}$$
$$0.35 = 7d \qquad \text{subtract 0.55}$$
$$d = 0.05 \qquad \text{solve for $d$}$$

Having found $d$, we can now solve for $a_1$.

$$a_{13} = a_1 + 12d \qquad \text{$n$th term formula for $n = 13$}$$
$$0.9 = a_1 + 12(0.05) \qquad \text{substitute 0.9 for $a_{13}$ and 0.05 for $d$}$$
$$0.9 = a_1 + 0.6 \qquad \text{simplify}$$
$$a_1 = 0.3 \qquad \text{solve for $a_1$}$$

☑ **B. You've just seen how we can find the $n$th term of an arithmetic sequence**

The first term is $a_1 = 0.3$ and the common difference is $d = 0.05$.

**Now try Exercises 45 through 50** ▶

## C. Finding the $n$th Partial Sum of an Arithmetic Sequence

Using sequences and series to solve applications often requires computing the sum of a given number of terms. To develop and understand the approach used, consider the sum of the first 10 natural numbers. Using $S_{10}$ to represent this sum, we have $S_{10} = 1 + 2 + 3 + 4 + 5 + 6 + 7 + 8 + 9 + 10$. We could just use brute force, but if we rewrite the sum a second time *in reverse order,* then add it to the first, we find that each column adds to 11.

| $S_{10} =$ | 1 | + | 2 | + | 3 | + | 4 | + | 5 | + | 6 | + | 7 | + | 8 | + | 9 | + | 10 |
|---|---|---|---|---|---|---|---|---|---|---|---|---|---|---|---|---|---|---|---|
| $S_{10} =$ | 10 | + | 9 | + | 8 | + | 7 | + | 6 | + | 5 | + | 4 | + | 3 | + | 2 | + | 1 |
| $2S_{10} =$ | 11 | + | 11 | + | 11 | + | 11 | + | 11 | + | 11 | + | 11 | + | 11 | + | 11 | + | 11 |

Since there are 10 columns, the total is $11 \times 10 = 110$. Because this is twice the actual sum, we find that $2S_{10} = 110$, and so $S_{10} = 55$.

Now consider the sequence $a_1, a_2, a_3, a_4, \ldots, a_n$ with common difference $d$. Use $S_n$ to represent the sum of the first $n$ terms. Then write the original series, and the series in reverse order underneath. Since one row increases at the same rate the other decreases, the sum of each column remains constant, and for simplicity's sake we choose $a_1 + a_n$ to represent this sum.

| $S_n =$ | $a_1$ | + | $a_2$ | + | $a_3$ | $+ \cdots +$ | $a_{n-2}$ | + | $a_{n-1}$ | + | $a_n$ |
|---|---|---|---|---|---|---|---|---|---|---|---|
| $S_n =$ | $a_n$ | + | $a_{n-1}$ | + | $a_{n-2}$ | $+ \cdots +$ | $a_3$ | + | $a_2$ | + | $a_1$ |
| $2S_n =$ | $(a_1 + a_n)$ | + | $(a_1 + a_n)$ | + | $(a_1 + a_n)$ | $+ \cdots +$ | $(a_1 + a_n)$ | + | $(a_1 + a_n)$ | + | $(a_1 + a_n)$ |

To understand why each column adds to $a_1 + a_n$, consider the sum in the second column: $a_2 + a_{n-1}$. From $a_2 = a_1 + d$ and $a_{n-1} = a_n - d$, we obtain $a_2 + a_{n-1} = (a_1 + d) + (a_n - d)$ by adding the equations, which gives a result of $a_1 + a_n$. Since there are $n$ columns, we end up with $2S_n = n(a_1 + a_n)$, and solving for $S_n$ gives the formula for the first $n$ terms of an arithmetic sequence.

> **The _n_th Partial Sum of an Arithmetic Sequence**
>
> Given an arithmetic sequence with first term $a_1$, the _n_th partial sum is given by
>
> $$S_n = \frac{n}{2}(a_1 + a_n).$$
>
> In words: The sum of an arithmetic sequence is one-half the number of terms times the sum of the first and last terms.

**EXAMPLE 6 ▶  Computing the Sum of an Arithmetic Sequence**

Use the summation formula to find the sum of the first 100 positive odd integers:

$$\sum_{k=1}^{100}(2k - 1).$$

**Solution ▶**  The initial terms of the sequence are $1, 3, 5, \ldots$ and we note $a_1 = 1$, $d = 2$, and $n = 100$. To use the sum formula, we need the value of $a_n = a_{100}$. The _n_th term formula shows $a_{100} = a_1 + (100 - 1)d = 1 + (99)2$, so $a_{100} = 199$.

$$S_n = \frac{n}{2}(a_1 + a_n) \qquad \text{sum formula}$$

$$S_{100} = \frac{100}{2}(a_1 + a_{100}) \qquad \text{substitute 100 for } n$$

$$= \frac{100}{2}(1 + 199) \qquad \text{substitute 1 for } a_1, 199 \text{ for } a_{100}$$

$$= 10,000 \qquad \text{result}$$

☑ **C.** You've just seen how we can find the _n_th partial sum of an arithmetic sequence

The sum of the first 100 positive odd integers is 10,000.

Now try Exercises 51 through 64 ▶

## D. Applications Involving Arithmetic Sequences and Series

In the evolution of certain plants and shelled animals, sequences and series seem to have been one of nature's favorite tools. The spirals found on many ferns and flowers are excellent examples of sequences in nature, as are the size of the chambers in nautilus shells (see Figures 9.2 and 9.3). Sequences and series also provide a good mathematical model for a variety of other situations as well.

**Figure 9.2**                    **Figure 9.3**

spiral fern                        nautilus

**EXAMPLE 7** ▶    **Solving an Application of Arithmetic Sequences: Seating Capacity**

Cox Auditorium is an amphitheater that has 40 seats in the first row, 42 in the second row, 44 in the third, and so on. If the auditorium's seating capacity is 8550, how many rows are there in the auditorium?

**Solution** ▶    The number of seats in each row gives the terms of an arithmetic sequence with $a_1 = 40$ and $d = 2$. The seating capacity is the sum of this arithmetic sequence and is known to be $S_n = 8550$. To find the number of rows, we might attempt to solve for $n$ in the partial sum formula $S_n = \frac{n}{2}(a_1 + a_n)$, but this formula depends on two unknowns: $n$ and $a_n$. By substituting the $n$th term formula into the partial sum formula, we're able to relate the sum directly to $n$ and bypass $a_n$ altogether.

$$S_n = \frac{n}{2}(a_1 + a_n) \qquad \text{sum formula}$$

$$= \frac{n}{2}(a_1 + [a_1 + (n-1)d]) \quad \text{substitute } a_1 + (n-1)d \text{ for } a_n$$

$$= \frac{n}{2}[2a_1 + (n-1)d] \qquad \text{simplify}$$

Substituting 40 for $a_1$, 2 for $d$, and 8550 for $S_n$ gives

$$8550 = \frac{n}{2}[2 \cdot 40 + (n-1) \cdot 2] \qquad \text{substitute 40 for } a_1\text{, 2 for } d\text{, and 8550 for } S_n$$

$$= 40n + n(n-1) \qquad \text{distribute } \tfrac{n}{2}$$
$$= n^2 + 39n \qquad \text{distribute } n\text{; collect like terms}$$
$$0 = n^2 + 39n - 8550 \qquad \text{subtract 8550}$$
$$= (n + 114)(n - 75) \qquad \text{factor}$$

Since $n = -114$ does not fit the context, the result is $n = 75$. The auditorium has 75 rows [with $a_1 + (n-1)d = 40 + (75-1) \cdot 2 = 188$ seats in the last row].

 **D.** You've just seen how we can solve applications involving arithmetic sequences

**Now try Exercises 67 through 72** ▶

---

## 9.2 EXERCISES

### ▶ CONCEPTS AND VOCABULARY

**Fill in the blank with the appropriate word or phrase. Carefully reread the section, if necessary.**

1. The formula for the $n$th partial sum of an arithmetic sequence is $S_n = $ _____, where $a_n$ is the _____ term.

2. The $n$th term formula for an arithmetic sequence is $a_n = $ _____, where $a_1$ is the _____ term and $d$ is the _____ _____.

3. Describe how the formula for the $n$th partial sum was derived, and illustrate its application using a sequence from the Exercise Set.

4. Describe the similarities and differences between linear functions and arithmetic sequences. Illustrate by discussing the inputs, outputs, and graphs of $f(x) = 2x + 3$ and $a_n = 2n + 3$.

## ▶ DEVELOPING YOUR SKILLS

**Determine if the sequence given is arithmetic. If yes, name the common difference. If not, try to determine the pattern that forms the sequence.**

5. $-5, -2, 1, 4, 7, 10, \ldots$

6. $1, -2, -5, -8, -11, -14, \ldots$

7. $0.5, 3, 5.5, 8, 10.5, \ldots$

8. $1.2, 3.5, 5.8, 8.1, 10.4, \ldots$

9. $2, 3, 5, 7, 11, 13, 17, \ldots$

10. $1, 4, 8, 13, 19, 26, 34, \ldots$

11. $\frac{1}{24}, \frac{1}{12}, \frac{1}{8}, \frac{1}{6}, \frac{5}{24}, \ldots$

12. $\frac{1}{12}, \frac{1}{15}, \frac{1}{20}, \frac{1}{30}, \frac{1}{60}, \ldots$

13. $1, 4, 9, 16, 25, 36, \ldots$

14. $-125, -64, -27, -8, -1, \ldots$

15. $\pi, \frac{5\pi}{6}, \frac{2\pi}{3}, \frac{\pi}{2}, \frac{\pi}{3}, \frac{\pi}{6}, \ldots$

16. $\pi, \frac{7\pi}{8}, \frac{3\pi}{4}, \frac{5\pi}{8}, \frac{\pi}{2}, \ldots$

**Write the first four terms of the arithmetic sequence.**

17. $a_1 = 2, d = 3$

18. $a_1 = 7, d = -2$

19. $a_1 = 0.3, d = 0.03$

20. $a_1 = 0.5, d = 0.25$

21. $a_1 = \frac{3}{2}, d = \frac{1}{2}$

22. $a_1 = \frac{3}{4}, d = -\frac{1}{8}$

23. $a_1 = -2, d = -3$

24. $a_1 = -4, d = -4$

**Identify the first term and the common difference. Then write the expression for the general term $a_n$ and use it to find the 6th, 10th, and 12th terms of the sequence.**

25. $2, 7, 12, 17, \ldots$

26. $7, 4, 1, -2, -5, \ldots$

27. $5.10, 5.25, 5.40, \ldots$

28. $9.75, 9.40, 9.05, \ldots$

29. $\frac{3}{2}, \frac{9}{4}, 3, \frac{15}{4}, \ldots$

30. $\frac{5}{7}, \frac{3}{14}, -\frac{2}{7}, -\frac{11}{14}, \ldots$

**Find the indicated term using the information given.**

31. $a_1 = 5, d = 4$; find $a_{15}$

32. $a_1 = 9, d = -2$; find $a_{17}$

33. $a_1 = \frac{3}{2}, d = -\frac{1}{12}$; find $a_7$

34. $a_1 = \frac{12}{25}, d = -\frac{1}{10}$; find $a_9$

35. $a_1 = -0.025, d = 0.05$; find $a_{50}$

36. $a_1 = 3.125, d = -0.25$; find $a_{20}$

**Find the number of terms in each sequence.**

37. $a_1 = 2, a_n = -22, d = -3$

38. $a_1 = 4, a_n = 42, d = 2$

39. $a_1 = 0.4, a_n = 10.9, d = 0.25$

40. $a_1 = -0.3, a_n = -36, d = -2.1$

41. $-3, -0.5, 2, 4.5, 7, \ldots, 47$

42. $-3.4, -1.1, 1.2, 3.5, \ldots, 38$

43. $\frac{1}{12}, \frac{1}{8}, \frac{1}{6}, \frac{5}{24}, \frac{1}{4}, \ldots, \frac{9}{8}$

44. $\frac{1}{12}, \frac{1}{15}, \frac{1}{20}, \frac{1}{30}, \ldots, -\frac{1}{4}$

**Find the common difference $d$ and the value of $a_1$ using the information given.**

45. $a_3 = 7, a_7 = 19$

46. $a_5 = -17, a_{11} = -2$

47. $a_2 = 1.025, a_{26} = 10.025$

48. $a_6 = -12.9, a_{30} = 1.5$

49. $a_{10} = \frac{13}{18}, a_{24} = \frac{27}{2}$

50. $a_4 = \frac{5}{4}, a_8 = \frac{9}{4}$

**Evaluate each sum. For Exercises 55 and 56, use the summation properties from Section 9.1.**

51. $\sum_{n=1}^{30} (3n - 4)$

52. $\sum_{n=1}^{29} (4n - 1)$

53. $\sum_{n=1}^{37} \left(\frac{3}{4}n + 2\right)$

54. $\sum_{n=1}^{20} \left(\frac{5}{2}n - 3\right)$

55. $\sum_{n=4}^{15} (3 - 5n)$

56. $\sum_{n=7}^{20} (7 - 2n)$

**Express the total area of the shaded rectangles in terms of a series written in sigma notation. Then evaluate the sum.**

57.

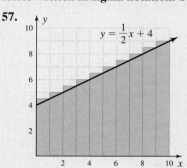

58.

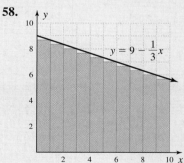

**Compute the partial sums indicated.**

59. The partial sum $S_{15}$ for the series
$-12 + (-9.5) + (-7) + (-4.5) + \cdots$

60. The partial sum $S_{20}$ for the series $\frac{9}{2} + \frac{7}{2} + \frac{5}{2} + \frac{3}{2} + \cdots$

61. The partial sum $S_{30}$ for the series
$0.003 + 0.173 + 0.343 + 0.513 + \cdots$

62. The partial sum $S_{50}$ for the series
$(-2) + (-7) + (-12) + (-17) + \cdots$

63. The partial sum $S_{20}$ for the series
$\sqrt{2} + 2\sqrt{2} + 3\sqrt{2} + 4\sqrt{2} + \cdots$

64. The partial sum $S_{10}$ for the series
$12\sqrt{3} + 10\sqrt{3} + 8\sqrt{3} + 6\sqrt{3} + \cdots$

## ▶ WORKING WITH FORMULAS

**65. Sum of the first $n$ natural numbers:** $S_n = \dfrac{n(n + 1)}{2}$

The sum of the first $n$ natural numbers can be found using the formula shown, where $n$ represents the number of terms in the sum. Verify the formula by adding the first six natural numbers by hand, and then evaluating $S_6$. Then find the sum of the first 75 natural numbers.

**66. Sum of the squares of the first $n$ natural numbers:**

$$S_n = \dfrac{n(n + 1)(2n + 1)}{6}$$

The sum of the squares of the first $n$ natural numbers can be found using the formula shown, where $n$ represents the number of terms in the sum. Verify the formula by adding the squares of the first six natural numbers by hand, and then evaluating $S_6$. Then find the sum of the squares of the first 20 natural numbers.

## ▶ APPLICATIONS

**67. Temperature fluctuation:** At 5 P.M. in Coldwater, the temperature was a chilly 36°F. If the temperature decreased by 3°F every half-hour for the next 7 hr, at what time did the temperature hit 0°F?

**68. Arc of a baby swing:** When Mackenzie's baby swing is started, the first swing (one way) is a 30-in. arc. As the swing slows down, each successive arc is 1.5 in. less than the previous one. Find the length of the tenth swing, and how far Mackenzie has traveled during the 10 swings.

**69. Computer animations:** The animation on a new computer game initially allows the hero of the game to jump a (screen) distance of 10 in. over booby traps and obstacles. Each successive jump is limited to $\frac{3}{4}$ in. less than the previous one. Find the length of the seventh jump, and the total distance covered after seven jumps.

**70. Seating capacity:** The Fox Theater creates a "theater in the round" when it shows any of Shakespeare's plays. The first row has 80 seats, the second row has 88, the third row has 96, and so on. How many seats are in the 10th row? If there is room for 25 rows, how many chairs will be needed to set up the theater?

**71. Sales goals:** *At the time that I was newly hired, 100 sales per month I required. Each following month—the last plus 20 more, as I work for the goal of top sales score. When 2500 sales are thusly made, it's Tahiti and piña coladas in the shade.* How many sales were made by this person in the seventh month? What were the total sales after the 12th month? Was the goal of 2500 total sales met?

**72. Bequests to charity:** *At the time our mother left this Earth, she gave $9000 to her children of birth. This we kept and each year added $3000 more, as a lasting memorial from the children she bore. When $42,000 is thusly attained, all goes to charity that her memory be maintained.* What was the balance in the sixth year? In what year was the goal of $42,000 met?

## ▶ EXTENDING THE CONCEPT

**73.** From a study of numerical analysis, a function is known to be linear if its "first differences" (differences between successive outputs) are constant. Likewise, a function is known to be quadratic if its "second differences" (differences between successive first differences) are constant. Use this information to determine if the following sets of output come from a linear or quadratic function:

   **a.** 19, 11.8, 4.6, −2.6, −9.8, −17, −24.2, …

   **b.** −10.31, −10.94, −11.99, −13.46, −15.35, …

**74.** From elementary geometry it is known that the interior angles of a triangle sum to 180°, the interior angles of a quadrilateral sum to 360°, the interior angles of a pentagon sum to 540°, and so on. Use this pattern to develop a formula for the sum of the interior angles of an $n$-sided polygon. The interior angles of a decagon (10 sides) sum to how many degrees?

▶ **MAINTAINING YOUR SKILLS**

**75.** (5.5) Solve for $t$: $2530 = 500e^{0.45t}$

**76.** (4.1) Graph by completing the square. Label all important features: $y = x^2 - 2x - 3$.

**77.** (2.3) In 2000, the deer population was 972. By 2005 it had grown to 1217. Assuming the growth is linear, find the function that models this data and use it to estimate the deer population in 2008.

**78.** (3.3) Suppose $y$ varies inversely with $x$ and directly with $w$. If $y = 14$ when $x = 15$ and $w = 52.5$, find the value of $y$ when $x = 32$ and $w = 208$.

---

## 9.3  Geometric Sequences

### LEARNING OBJECTIVES

*In Section 9.3 you will see how we can:*

☐ **A.** Identify a geometric sequence and its common ratio

☐ **B.** Find the $n$th term of a geometric sequence

☐ **C.** Find the $n$th partial sum of a geometric sequence

☐ **D.** Find the sum of an infinite geometric series

☐ **E.** Solve applications involving geometric sequences and series

Recall that arithmetic sequences are those where each term is found by *adding* a constant value to the preceding term. In this section, we consider geometric sequences, which behave much like the exponential functions studied in Chapter 5.

### A.  Geometric Sequences

A **geometric sequence** is one where each successive term is found by multiplying the preceding term by a fixed constant. Consider growth of a bacteria population, where a single cell splits in two every hour over a 24-hr period. Beginning with a single bacterium ($a_0 = 1$), after 1 hr there are 2, after 2 hr there are 4, and so on. Writing the number of bacteria as a sequence we have:

| hours: | $a_1$ | $a_2$ | $a_3$ | $a_4$ | $a_5$ | … |
|---|---|---|---|---|---|---|
| | ↓ | ↓ | ↓ | ↓ | ↓ | |
| bacteria: | 2 | 4 | 8 | 16 | 32 | … |

The sequence 2, 4, 8, 16, 32, … is a geometric sequence since each term is found by multiplying the previous term by the constant factor 2. This also means that the ratio of any two consecutive terms must be 2 and in fact, 2 is called the **common ratio $r$** for this sequence. Using the notation from Section 9.1 we can write $r = \frac{a_{n+1}}{a_n}$, where $a_n$ represents any term of the sequence and $a_{n+1}$ represents the term that follows $a_n$.

> **Geometric Sequences**
>
> Given a sequence $a_1, a_2, a_3, \ldots, a_n, a_{n+1}, \ldots$, where $n \in \mathbb{N}$,
>
> if there exists a common ratio $r$ such that $\frac{a_{n+1}}{a_n} = r$ for all $n$,
>
> then the sequence is a *geometric sequence*.

The ratio of successive terms can be rewritten as $a_{n+1} = a_n r$ (for $n \geq 1$) to highlight that each term is found by multiplying the preceding term by $r$.

**EXAMPLE 1** ▶ **Identifying a Geometric Series**

Determine if the given sequence is geometric. If yes, name the common ratio. If not, try to determine the pattern that forms the sequence.

**a.** $1, 0.5, 0.25, 0.125, \ldots$   **b.** $\frac{1}{7}, \frac{2}{7}, \frac{6}{7}, \frac{24}{7}, \frac{120}{7}, \ldots$

**Solution** ▶ Apply the definition to check for a common ratio $r = \frac{a_{n+1}}{a_n}$.

    **a.** For $1, 0.5, 0.25, 0.125, \ldots$, the ratio of consecutive terms gives

$$\frac{0.5}{1} = 0.5, \qquad \frac{0.25}{0.5} = 0.5, \qquad \frac{0.125}{0.25} = 0.5, \qquad \text{and so on.}$$

This is a geometric sequence with common ratio $r = 0.5$.

    **b.** For $\frac{1}{7}, \frac{2}{7}, \frac{6}{7}, \frac{24}{7}, \frac{120}{7}, \ldots$, we have:

$$\frac{2}{7} \div \frac{1}{7} = \frac{2}{7} \cdot \frac{7}{1} \qquad \frac{6}{7} \div \frac{2}{7} = \frac{6}{7} \cdot \frac{7}{2} \qquad \frac{24}{7} \div \frac{6}{7} = \frac{24}{7} \cdot \frac{7}{6} \quad \text{and so on.}$$
$$= 2 \qquad\qquad\qquad = 3 \qquad\qquad\qquad = 4$$

Since the ratio is not constant, this is not a geometric sequence. The sequence appears to be formed by dividing $n!$ by 7: $a_n = \frac{n!}{7}$.

**Now try Exercises 5 through 18 ▶**

---

**EXAMPLE 2** ▶ **Writing the First Few Terms of a Geometric Sequence**

Write the first five terms of the geometric sequence, given the first term $a_1 = -16$ and the common ratio $r = 0.25$.

**Solution** ▶ Starting at $a_1 = -16$, multiply each term by 0.25 to generate the sequence.

$a_1 = -16$
$\qquad a_2 = -16(0.25) = -4$
$\qquad\qquad a_3 = -4(0.25) = -1$
$\qquad\qquad\qquad a_4 = -1(0.25) = -0.25$
$\qquad\qquad\qquad\qquad a_5 = -0.25(0.25) = -0.0625$

☑ **A.** You've just seen how we can identify a geometric sequence and its common ratio

The first five terms of this sequence are $-16, -4, -1, -0.25,$ and $-0.0625$.

**Now try Exercises 19 through 26 ▶**

---

**B. Finding the *n*th Term of a Geometric Sequence**

If the values $a_1$ and $r$ from a geometric sequence are known, we could generate the terms of the sequence by applying *additional factors of r to the first term,* instead of repeatedly multiplying new terms by $r$. If $a_1 = 3$ and $r = 2$, we simply begin at $a_1$, and continue applying additional factors of $r$ for each successive term.

$$3 = 3 \cdot 2^0 \qquad a_1 = a_1 r^0$$
$$6 = 3 \cdot 2^1 \qquad a_2 = a_1 r^1$$
$$12 = 3 \cdot 2^2 \qquad a_3 = a_1 r^2$$
$$24 = 3 \cdot 2^3 \qquad a_4 = a_1 r^3$$
$$48 = 3 \cdot 2^4 \qquad a_5 = a_1 r^4$$

current term ⟶    initial term    exponent on common ratio

From this pattern, we note the exponent on $r$ is always 1 less than the subscript of the current term: $5 - 1 = 4$, which leads us to the formula for the *n*th term of a geometric sequence.

> ### The nth Term of a Geometric Sequence
>
> The nth term of a *geometric sequence* is given by
>
> $$a_n = a_1 r^{n-1},$$
>
> where $r$ is the common ratio.

---

**EXAMPLE 3** ▶  **Finding a Specific Term in a Geometric Sequence**

Identify the common ratio $r$, and use it to write the expression for the $n$th term. Then find the 10th term of the sequence: $3, -6, 12, -24, \ldots$.

**Solution** ▶  By inspection we note that $a_1 = 3$ and $r = \frac{-6}{3} = -2$. This gives

$$a_n = a_1 r^{n-1} \qquad \text{\textit{n}th term formula}$$
$$= 3(-2)^{n-1} \qquad \text{substitute 3 for } a_1 \text{ and } -2 \text{ for } r$$

To find the 10th term we substitute $n = 10$:

$$a_{10} = 3(-2)^{10-1} \qquad \text{substitute 10 for } n$$
$$= 3(-2)^9 = -1536 \qquad \text{simplify}$$

**Now try Exercises 27 through 38** ▶

---

**EXAMPLE 4** ▶  **Determining the Number of Terms in a Geometric Sequence**

Find the number of terms in the geometric sequence $4, 2, 1, \ldots, \frac{1}{64}$.

**Solution** ▶  Observing that $a_1 = 4$ and $r = \frac{2}{4} = \frac{1}{2}$, we have

$$a_n = a_1 r^{n-1} \qquad \text{\textit{n}th term formula}$$
$$= 4\left(\frac{1}{2}\right)^{n-1} \qquad \text{substitute 4 for } a_1 \text{ and } \tfrac{1}{2} \text{ for } r$$

Although we don't know the number of terms in the sequence, we *do* know the last term is $\frac{1}{64}$. Substituting $a_n = \frac{1}{64}$ gives

$$\frac{1}{64} = 4\left(\frac{1}{2}\right)^{n-1} \qquad \text{substitute } \tfrac{1}{64} \text{ for } a_n$$
$$\frac{1}{256} = \left(\frac{1}{2}\right)^{n-1} \qquad \text{divide by 4 (multiply by } \tfrac{1}{4}\text{)}$$

From our work in Chapter 5, we attempt to write both sides as exponentials with a like base, or apply logarithms. Since $256 = 2^8$, we equate bases.

$$\left(\frac{1}{2}\right)^8 = \left(\frac{1}{2}\right)^{n-1} \qquad \text{write } \tfrac{1}{256} \text{ as } \left(\tfrac{1}{2}\right)^8$$
$$8 = n - 1 \qquad \text{like bases imply like exponents}$$
$$9 = n \qquad \text{solve for } n$$

This shows there are nine terms in the sequence.

**Now try Exercises 39 through 46** ▶

---

**WORTHY OF NOTE**

The sequence from Example 3 is an *alternating sequence,* and the exponential nature of its graph can be seen by taking a look at the graph of $b_n = |a_n| = |3(-2)^{n-1}|$.

Note that in both Examples 3 and 4, the $n$th term had the form of an exponential function $(y = a \cdot b^n)$ after simplifying: $a_n = 3(-2)^{n-1}$ and $a_n = 4(\frac{1}{2})^{n-1}$. This is in fact a characteristic of geometric sequences, with the common ratio $r$ corresponding to the base $b$. This means the graph of a geometric sequence will always be a set of discrete points that lie on an exponential graph when $r > 0$, or oscillate between exponential graphs $(y = \pm a \cdot r^x)$ when $r < 0$ (see *Worthy of Note*).

If the first term $a_1$ is unknown but another term $a_k$ is given, the $n$th term can be written

$$a_n = a_k r^{n-k}$$

(the subscript on the term $a_k$ and the exponent on $r$ sum to $n$).

---

**EXAMPLE 5** ▶ **Finding the First Term of a Geometric Sequence**

Given a geometric sequence where $a_4 = 0.075$ and $a_7 = 0.009375$, find the common ratio $r$ and the value of $a_1$.

**Solution** ▶ Since $a_1$ is not known, we express $a_7$ as the product of a known term and the appropriate number of common ratios: $a_7 = a_4 r^3$ ($7 = 4 + 3$, as required).

$$
\begin{aligned}
a_7 &= a_4 \cdot r^3 && \text{$a_1$ is unknown} \\
0.009375 &= 0.075 r^3 && \text{substitute 0.009375 for $a_7$ and 0.075 for $a_4$} \\
0.125 &= r^3 && \text{divide by 0.075} \\
r &= 0.5 && \text{solve for $r$}
\end{aligned}
$$

Having found $r$, we can now solve for $a_1$

$$
\begin{aligned}
a_7 &= a_1 r^6 && \text{$n$th term formula} \\
0.009375 &= a_1 (0.5)^6 && \text{substitute 0.009375 for $a_7$ and 0.5 for $r$} \\
0.009375 &= a_1 (0.015625) && \text{simplify} \\
a_1 &= 0.6 && \text{solve for $a_1$}
\end{aligned}
$$

The first term is $a_1 = 0.6$ and the common ratio is $r = 0.5$.

☑ **B.** You've just seen how we can find the $n$th term of a geometric sequence

Now try Exercises 47 through 52 ▶

---

## C. Finding the $n$th Partial Sum of a Geometric Sequence

As with arithmetic series, applications of geometric series often involve computing sums of consecutive terms. We can adapt the method for finding the sum of an arithmetic sequence to develop a formula for finding the sum of a geometric sequence.

For the $n$th partial sum, we have $S_n = a_1 + a_1 r + a_1 r^2 + a_1 r^3 + \cdots + a_1 r^{n-1}$. If we multiply $S_n$ by $-r$ then add the original series, the "interior terms" cancel.

$$
\begin{array}{lcccccccc}
S_n & = & a_1 & + a_1 r & + a_1 r^2 & + \cdots & + a_1 r^{n-2} & + a_1 r^{n-1} & \\
-rS_n & = & & - a_1 r & - a_1 r^2 & - \cdots & - a_1 r^{n-2} & - a_1 r^{n-1} & - a_1 r^n \\
\hline
S_n - rS_n & = & a_1 & + 0 & + 0 & + \cdots + & 0 & + 0 & - a_1 r^n
\end{array}
$$

We then have $S_n - rS_n = a_1 - a_1 r^n$, and can now solve for $S_n$:

$$
\begin{aligned}
S_n(1 - r) &= a_1(1 - r^n) && \text{factor out $S_n$ and $a_1$} \\
S_n &= \frac{a_1(1 - r^n)}{1 - r} && \text{solve for $S_n$ (divide by $1 - r$)}
\end{aligned}
$$

The result is a formula for the $n$th partial sum of a geometric sequence.

---

### The $n$th Partial Sum of a Geometric Sequence

Given a geometric sequence with first term $a_1$ and common ratio $r$, the $n$th partial sum is

$$S_n = \frac{a_1 - a_1 r^n}{1 - r} = \frac{a_1(1 - r^n)}{1 - r},\ r \neq 1$$

In words: The sum of the first $n$ terms of a geometric sequence is the difference of the first and $(n + 1)$st term, divided by 1 minus the common ratio.

**EXAMPLE 6** ▶ **Computing the Sum of a Geometric Sequence**

Use the summation formula to find the sum of the first 9 powers of 3:

$$\sum_{i=1}^{9} 3^i.$$

Solution ▶ The initial terms of this series are $3 + 9 + 27 + \cdots$, and we note $a_1 = 3$, $r = 3$, and $n = 9$. We could find the first nine terms and add, but using the partial sum formula is much faster and gives

$$S_n = \frac{a_1(1 - r^n)}{1 - r} \qquad \text{sum formula}$$

$$S_9 = \frac{3(1 - 3^9)}{1 - 3} \qquad \text{substitute 3 for } a_1, 9 \text{ for } n, \text{ and 3 for } r$$

$$= \frac{3(-19{,}682)}{-2} \qquad \text{simplify}$$

$$= 29{,}523 \qquad \text{result}$$

☑ **C.** You've just seen how we can find the *n*th partial sum of a geometric sequence

**Now try Exercises 53 through 72** ▶

## D. The Sum of an Infinite Geometric Series

To this point we've considered only partial sums of a series. While it is impossible to add an infinite number of terms, some of these "infinite sums" appear to have what is called a **limiting value.** The sum appears to get ever closer to this value—much like the asymptotic behavior of some graphs. We will define the sum of this **infinite series** to be this limiting value, if it exists. Consider the illustration in Figure 9.4, where a standard sheet of typing paper is cut in half. One of the halves is again cut in half and the process is continued indefinitely, as shown. Notice the "halves" create an infinite sequence $\frac{1}{2}, \frac{1}{4}, \frac{1}{8}, \frac{1}{16}, \frac{1}{32}, \ldots$ with $a_1 = \frac{1}{2}$ and $r = \frac{1}{2}$. The corresponding infinite series is $\frac{1}{2} + \frac{1}{4} + \frac{1}{8} + \frac{1}{16} + \frac{1}{32} + \cdots + \left(\frac{1}{2}\right)^n + \cdots$.

**Figure 9.4**

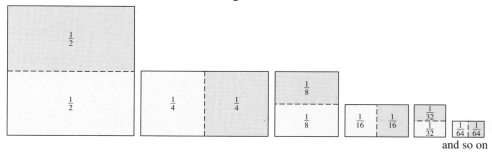

and so on

**Figure 9.5**

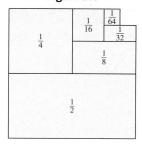

If we arrange the halves from each stage as shown in Figure 9.5, we would be rebuilding the original sheet of paper. As we add more and more of these halves together, we get closer and closer to the size of the original sheet. We gain an intuitive sense that this series must add to 1, because the *pieces* of the original sheet of paper must total 1 whole sheet. To explore this idea further, consider what happens to the size of the missing corner, $\left(\frac{1}{2}\right)^n$, as $n$ becomes large.

$$n = 4: \left(\frac{1}{2}\right)^4 \approx 0.06 \qquad n = 8: \left(\frac{1}{2}\right)^8 \approx 0.004 \qquad n = 12: \left(\frac{1}{2}\right)^{12} \approx 0.0002$$

Although a definitive proof is left to a future course, it seems clear that the size of the missing corner is becoming very close to zero as $n$ becomes large. In fact, the same can be shown for any $|r| < 1$: as $n \to \infty$, $r^n \to 0$. This suggests that as $n \to \infty$, $S_n = \frac{a_1 - a_1 r^n}{1 - r} \to \frac{a_1 - 0}{1 - r}$ and the limiting value (represented by $S_\infty$) is $S_\infty = \frac{a_1}{1 - r}$ for $|r| < 1$.

**WORTHY OF NOTE**

The formula for the sum of an infinite geometric series can also be derived by noting that
$$S_\infty = a_1 + a_1 r + a_1 r^2 + \cdots$$
can be rewritten as
$$S_\infty = a_1 + r(a_1 + a_1 r + a_1 r^2 + \cdots)$$
$$= a_1 + r S_\infty.$$
If $|r| < 1$, we then have
$$S_\infty - r S_\infty = a_1$$
$$(1 - r)S_\infty = a_1$$
$$S_\infty = \frac{a_1}{1 - r}.$$

### Infinite Geometric Series

Given a geometric sequence with first term $a_1$ and $|r| < 1$, the sum of the related infinite series is given by

$$S_\infty = \frac{a_1}{1 - r}$$

If $|r| > 1$, no finite sum exists.

**EXAMPLE 7 ▶  Computing an Infinite Sum**

Find the limiting value of each infinite geometric series (if it exists).

   **a.** $1 + 2 + 4 + 8 + \cdots$   **b.** $3 + 2 + \frac{4}{3} + \frac{8}{9} + \cdots$

   **c.** $0.185 + 0.000185 + 0.000000185 + \cdots$

**Solution ▶**  Begin by determining if the infinite series is geometric with $|r| < 1$. If so, use $S_\infty = \frac{a_1}{1 - r}$.

   **a.** Since $r = 2$ (by inspection), a finite sum does not exist.

   **b.** Using the ratio of consecutive terms we find $r = \frac{2}{3}$ and the infinite sum exists. With $a_1 = 3$, we have

$$S_\infty = \frac{3}{1 - \frac{2}{3}} = \frac{3}{\frac{1}{3}} = 9$$

   **c.** This series is equivalent to the repeating decimal $0.185185185\ldots = 0.\overline{185}$. The common ratio is $r = \frac{0.000185}{0.185} = 0.001$ and the infinite sum exists:

$$S_\infty = \frac{0.185}{1 - 0.001} = \frac{5}{27}$$

☑ **D.** You've just seen how we can find the sum of an infinite geometric series

Now try Exercises 73 through 86 ▶

## E.  Applications Involving Geometric Sequences and Series

Here are a few of the ways these ideas can be put to use.

**EXAMPLE 8 ▶  Solving an Application of Geometric Sequences: Pendulums**

A pendulum is any object attached to a fixed point and allowed to swing freely under the influence of gravity. Suppose each swing (in one direction) is 90% the length of the previous one. Gradually the swings become shorter and shorter and at some point the pendulum will appear to have stopped (although *theoretically* it never does).

   **a.** How far does the pendulum travel on its eighth swing, if the first was 2 m?

   **b.** What is the total distance traveled by the pendulum for these eight swings?

   **c.** How many swings until the length of each swing falls below 0.5 m?

   **d.** What total distance does the pendulum travel before coming to rest?

**Solution** ▶

**a.** The lengths of each swing form the terms of a geometric sequence with $a_1 = 2$ and $r = 0.9$. The first few terms are 2, 1.8, 1.62, 1.458, and so on. For the 8th term we have:

$$a_n = a_1 r^{n-1} \qquad \text{$n$th term formula}$$
$$a_8 = 2(0.9)^{8-1} \qquad \text{substitute 8 for $n$, 2 for $a_1$, and 0.9 for $r$}$$
$$\approx 0.957 \qquad \text{result}$$

The pendulum travels about 0.957 m on its 8th swing.

**b.** For the total distance traveled after eight swings, we compute the value of $S_8$.

$$S_n = \frac{a_1(1 - r^n)}{1 - r} \qquad \text{$n$th partial sum formula}$$
$$S_8 = \frac{2(1 - 0.9^8)}{1 - 0.9} \qquad \text{substitute 2 for $a_1$, 0.9 for $r$, and 8 for $n$}$$
$$\approx 11.4 \qquad \text{result}$$

The pendulum has traveled about 11.4 m by the end of the 8th swing.

**c.** To find the number of swings until the length of each swing is less than 0.5 m, we solve for $n$ in the equation $0.5 = 2(0.9)^{n-1}$. This yields

$$0.25 = 0.9^{n-1} \qquad \text{divide by 2}$$
$$\ln 0.25 = \ln 0.9^{n-1} \qquad \text{take the natural log of both sides}$$
$$\ln 0.25 = (n - 1)\ln 0.9 \qquad \text{apply power property}$$
$$\frac{\ln 0.25}{\ln 0.9} + 1 = n \qquad \text{solve for $n$ (exact form)}$$
$$14.16 \approx n \qquad \text{solve for $n$ (approximate form)}$$

After the 14th swing, each successive swing will be less than 0.5 m.

**d.** For the total distance traveled before coming to rest, we consider the related infinite geometric series, with $a_1 = 2$ and $r = 0.9$.

$$S_\infty = \frac{a_1}{1 - r} \qquad \text{infinite sum formula}$$
$$S_\infty = \frac{2}{1 - 0.9} \qquad \text{substitute 2 for $a_1$ and 0.9 for $r$}$$
$$= 20 \qquad \text{result}$$

The pendulum would travel 20 m before coming to rest.

**Now try Exercises 89 and 90** ▶

As mentioned in Section 9.1, sometimes the sequence or series for a particular application will use the preliminary or inaugural term $a_0$, as when an initial amount of money is deposited before any interest is earned, or the efficiency of a new machine after purchase—prior to any wear and tear.

**EXAMPLE 9** ▶ **Equipment Efficiency—Furniture Manufacturing**

The manufacturing of mass-produced furniture requires robotic machines to drill numerous holes for the bolts used in the assembly process. When new, the drill bits are capable of drilling through hardwood at a rate of 6 cm/sec. As the bit becomes worn, it loses 4% of its drilling speed per day.

**a.** How fast can the bit drill after a 5-day workweek?

**b.** When the drilling speed falls below 3.6 cm/sec, the bit must be replaced. After how many days must the bit be replaced?

**Solution** ▶ The efficiency of a new drill bit (prior to use) is given as $a_0 = 6$ cm/sec. Since the bit *loses* $4\% = 0.04$ of its efficiency per day, it maintains $96\% = 0.96$ of its efficiency, showing that after 1 day of use $a_1 = 0.96(6) = 5.76$. This means the $n$th term formula will be $a_n = 5.76(0.96)^{n-1}$.

**a.** At the end of day 5 we have

$$a_5 = 5.76(0.96)^{5-1} \qquad \text{substitute 5 for } n$$
$$= 5.76(0.96)^4 \qquad \text{simplify}$$
$$\approx 4.9 \qquad \text{result}$$

After 5 days, the bit can drill through the hardwood at about 4.9 cm/sec.

**b.** To find the number of days until the efficiency falls below 3.6 cm/sec, we replace $a_n$ with 3.6 and solve for $n$.

$$3.6 = 5.76(0.96)^{n-1} \qquad \text{substitute 3.6 for } a_n$$
$$0.625 = 0.96^{n-1} \qquad \text{divide by 5.76}$$
$$\ln 0.625 = \ln 0.96^{n-1} \qquad \text{take the natural log of both sides}$$
$$\ln 0.625 = (n-1)\ln 0.96 \qquad \text{apply power property}$$
$$\frac{\ln 0.625}{\ln 0.96} = n - 1 \qquad \text{divide by ln 0.96}$$
$$\frac{\ln 0.625}{\ln 0.96} + 1 = n \qquad \text{solve for } n \text{ (exact form)}$$
$$12.5 \approx n \qquad \text{solve for } n \text{ (approximate form)}$$

☑ **E. You've just seen how we can solve application problems involving geometric sequences and series**

The drill bit must be replaced after 12 full days of use.

**Now try Exercises 91 through 104** ▶

## 9.3 EXERCISES

▶ **CONCEPTS AND VOCABULARY**

**Fill in the blank with the appropriate word or phrase. Carefully reread the section, if necessary.**

**1.** The $n$th term of a geometric sequence is given by $a_n = $ _____, for any $n \geq 1$.

**2.** For the geometric sequence $a_1, a_1r, a_1r^2, \dots$, the $n$th partial sum is given by $S_n = $ _____.

**3.** Describe the similarities and differences between exponential functions and geometric sequences. Illustrate by discussing the inputs, outputs, and graphs of $f(x) = 8(\frac{1}{2})^x$ and $a_n = 8(\frac{1}{2})^n$.

**4.** Describe the difference(s) between an arithmetic and a geometric sequence. How can a student prevent confusion between the formulas?

▶ **DEVELOPING YOUR SKILLS**

**Determine if the sequence given is geometric. If yes, name the common ratio. If not, try to determine the pattern that forms the sequence.**

**5.** 4, 8, 16, 32, …

**6.** 2, 6, 18, 54, 162, …

**7.** 3, −6, 12, −24, 48, …

**8.** 128, −32, 8, −2, …

**9.** 2, 5, 10, 17, 26, …

**10.** −13, −9, −5, −1, 3, …

**11.** 3, 0.3, 0.03, 0.003, …

**12.** 12, 0.12, 0.0012, 0.000012, …

**13.** $\frac{1}{2}, \frac{1}{4}, \frac{1}{8}, \frac{1}{16}, \dots$

**14.** $\frac{2}{3}, \frac{4}{9}, \frac{8}{27}, \frac{16}{81}, \dots$

**15.** $3, \dfrac{12}{x}, \dfrac{48}{x^2}, \dfrac{192}{x^3}, \ldots$

**16.** $5, \dfrac{10}{a}, \dfrac{20}{a^2}, \dfrac{40}{a^3}, \ldots$

**17.** $240, 120, 40, 10, 2, \ldots$

**18.** $-120, -60, -20, -5, -1, \ldots$

**Write the first four terms of the sequence.**

**19.** $a_1 = 5, r = 2$

**20.** $a_1 = 2, r = -4$

**21.** $a_1 = -6, r = -\frac{1}{2}$

**22.** $a_1 = \frac{2}{3}, r = \frac{1}{5}$

**23.** $a_1 = 4, r = \sqrt{3}$

**24.** $a_1 = \sqrt{5}, r = \sqrt{5}$

**25.** $a_1 = 0.1, r = 0.1$

**26.** $a_1 = 0.024, r = 0.01$

**Write the expression for the $n$th term, then find the indicated term for each sequence.**

**27.** $a_1 = -24, r = \frac{1}{2}$; find $a_7$

**28.** $a_1 = 48, r = -\frac{1}{3}$; find $a_6$

**29.** $a_1 = -\frac{1}{20}, r = -5$; find $a_4$

**30.** $a_1 = \frac{3}{20}, r = 4$; find $a_5$

**31.** $a_1 = 2, r = \sqrt{2}$; find $a_7$

**32.** $a_1 = \sqrt{3}, r = \sqrt{3}$; find $a_8$

**Identify $a_1$ and $r$, then write the expression for the $n$th term $a_n = a_1 r^{n-1}$ and use it to find $a_6$, $a_{10}$, and $a_{12}$.**

**33.** $\frac{1}{27}, -\frac{1}{9}, \frac{1}{3}, -1, 3, \ldots$

**34.** $-\frac{7}{8}, \frac{7}{4}, -\frac{7}{2}, 7, -14, \ldots$

**35.** $729, 243, 81, 27, 9, \ldots$

**36.** $625, 125, 25, 5, 1, \ldots$

**37.** $0.2, 0.08, 0.032, 0.0128, \ldots$

**38.** $0.5, -0.35, 0.245, -0.1715, \ldots$

**Find the number of terms in each sequence.**

**39.** $a_1 = 9, a_n = 729, r = 3$

**40.** $a_1 = 1, a_n = -128, r = -2$

**41.** $a_1 = 16, a_n = \frac{1}{64}, r = \frac{1}{2}$

**42.** $a_1 = 4, a_n = \frac{1}{512}, r = \frac{1}{2}$

**43.** $2, -6, 18, -54, \ldots, -4374$

**44.** $3, -6, 12, -24, \ldots, -6144$

**45.** $\frac{3}{8}, -\frac{3}{4}, \frac{3}{2}, -3, \ldots, 96$

**46.** $-\frac{5}{27}, \frac{5}{9}, -\frac{5}{3}, -5, \ldots, -135$

**Find the common ratio $r$ and the value of $a_1$ using the information given (assume $r > 0$).**

**47.** $a_3 = 324, a_7 = 64$

**48.** $a_5 = 6, a_9 = 486$

**49.** $a_4 = \frac{4}{9}, a_8 = \frac{9}{4}$

**50.** $a_2 = \frac{16}{81}, a_5 = \frac{2}{3}$

**51.** $a_4 = \frac{32}{3}, a_8 = 54$

**52.** $a_3 = \frac{16}{25}, a_7 = 25$

**Find the partial sum indicated.**

**53.** $a_1 = 8, r = -2$; find $S_{12}$

**54.** $a_1 = 96, r = \frac{1}{3}$; find $S_5$

**55.** $a_1 = 8, r = \frac{3}{2}$; find $S_7$

**56.** $a_1 = -1, r = -\frac{3}{2}$; find $S_{10}$

**57.** $2 + 6 + 18 + \cdots$; find $S_6$

**58.** $16 - 8 + 4 - \cdots$; find $S_8$

**59.** $\frac{4}{3} + \frac{2}{9} + \frac{1}{27} + \cdots$; find $S_9$

**60.** $\frac{1}{18} - \frac{1}{6} + \frac{1}{2} - \cdots$; find $S_7$

**Find the partial sum indicated. For Exercises 65 and 66, use the summation properties from Section 9.1.**

**61.** $\displaystyle\sum_{j=1}^{5} 4^j$

**62.** $\displaystyle\sum_{k=1}^{10} 2^k$

**63.** $\displaystyle\sum_{k=1}^{8} 5\left(\frac{2}{3}\right)^{k-1}$

**64.** $\displaystyle\sum_{j=1}^{7} 3\left(\frac{1}{5}\right)^{j-1}$

**65.** $\displaystyle\sum_{i=4}^{10} 9\left(-\frac{1}{2}\right)^{i-1}$

**66.** $\displaystyle\sum_{i=3}^{8} 5\left(-\frac{1}{4}\right)^{i-1}$

**Find the indicated partial sum using the information given. Write all results in simplest form.**

**67.** $a_2 = -5, a_5 = \frac{1}{25}$; find $S_5$

**68.** $a_3 = 1, a_6 = -27$; find $S_6$

**69.** $a_4 = \frac{1}{3}, a_7 = \frac{9}{64}$; find $S_6$

**70.** $a_2 = \frac{16}{81}, a_5 = \frac{2}{3}$; find $S_8$

**71.** $a_3 = 2\sqrt{2}, a_6 = 8$; find $S_7$

**72.** $a_2 = 3, a_5 = 9\sqrt{3}$; find $S_7$

**Determine whether the infinite geometric series has a finite sum. If so, find the limiting value.**

**73.** $9 + 3 + 1 + \cdots$

**74.** $3 + 6 + 12 + 24 + \cdots$

**75.** $4 + 8 + 16 + 32 + \cdots$

**76.** $25 + 10 + 4 + \frac{8}{5} + \cdots$

**77.** $6 + 3 + \frac{3}{2} + \frac{3}{4} + \cdots$

**78.** $6 - 3 + \frac{3}{2} - \frac{3}{4} + \cdots$

**79.** $0.3 + 0.03 + 0.003 + \cdots$

**80.** $0.63 + 0.0063 + 0.000063 + \cdots$

**81.** $\displaystyle\sum_{k=1}^{\infty} \frac{3}{4}\left(\frac{2}{3}\right)^k$

**82.** $\displaystyle\sum_{i=1}^{\infty} 5\left(\frac{1}{2}\right)^i$

**83.** $\displaystyle\sum_{j=1}^{\infty} 9\left(-\frac{5}{4}\right)^j$

**84.** $\displaystyle\sum_{k=1}^{\infty} 12\left(\frac{4}{3}\right)^k$

**Express the total area of the shaded rectangles in terms of a series written in sigma notation. Then evaluate the sum.**

**85.**

**86.**

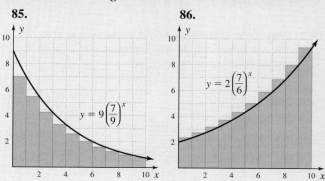

## ▶ WORKING WITH FORMULAS

**87. Sum of the cubes of the first $n$ natural numbers:**

$$S_n = \frac{n^2(n+1)^2}{4}$$

The sum of the cubes of the first $n$ natural numbers can be found using the formula shown, where $n$ represents the number of terms in the sum. Verify the formula by adding the cubes of the first eight natural numbers by hand, and then evaluating $S_8$.

**88. Student loan payment:** $A_n = P(1 + r)^n$

If $P$ dollars is borrowed at an annual interest rate $r$ with interest compounded annually, the amount of money to be paid back after $n$ yr is given by the indicated formula. Find the total amount of money that the student must repay to clear the loan, if $8000 is borrowed at 4.5% interest and the loan is paid back in 10 yr.

## ▶ APPLICATIONS

**Write the $n$th term formula for each application, then solve.**

**89. Pendulum movement:** On each swing, a pendulum travels only 80% as far as it did on the previous swing. If the first swing is 24 ft, how far does the pendulum travel on the 7th swing? What total distance is traveled before the pendulum comes to rest?

**90. Tire swings:** Ernesto is swinging to and fro on his backyard tire swing. With each swing, Ernesto travels 75% as far as he did on the previous swing. If the first arc (or swing) is 30 ft, find the distance Ernesto travels on the 5th arc. What total distance will he travel before coming to rest?

**Identify the inaugural term and write the $n$th term formula for each application, then solve.**

**91. Depreciation—automobiles:** A certain new SUV depreciates in value about 20% per year. If the SUV is purchased for $46,000, how much is it worth 4 yr later? How many years until its value is less than $5000?

**92. Depreciation—business equipment:** A new photocopier under heavy use will depreciate about 25% per year. If the copier is purchased for $7000, how much is it worth 4 yr later? How many years until its value is less than $1246?

**93. Equipment aging—industrial oil pumps:** Tests have shown that the pumping power of a heavy-duty oil pump decreases by 3% per month. If the pump can move 160 gallons per minute (gpm) new, how many gpm can the pump move 8 months later? If the pumping rate falls below 118 gpm, the pump must be replaced. How many months until this pump is replaced?

**94. Equipment aging—lumber production:** At the local mill, a certain type of saw blade can saw approximately 2 log-ft/sec when it is new. As time goes on, the blade becomes worn and loses 6% of its cutting speed each week. How many log-ft/sec can the saw blade cut after 6 wk? If the cutting speed falls below 1.2 log-ft/sec, the blade must be replaced. During what week of operation will this blade be replaced?

**95. Population growth—United States:** In April 2000, the population of the United States was approximately 281 million. If the population was growing at a rate of 0.96% per year, what was the population in April 2010?

**96. Population growth—space colony:** The population of the Zeta Colony on Mars is 1000 people. Determine the population of the colony 20 yr from now, if the population is growing at a constant rate of 5% per year.

**97. Creating a vacuum:** To create a vacuum, a hand pump is used to remove the air from an air-tight cube with a volume of 462 in³. With each stroke of the pump, two-fifths of the air that remains in the cube is removed. How much air remains inside after the 5th stroke? How many strokes are required to remove all but 12.9 in³ of the air?

**98. Atmospheric pressure:** In 1654, scientist Otto von Guericke performed his famous demonstration of atmospheric pressure and the strength of a vacuum in front of Emperor Ferdinand III of Hungary. After joining two hemispheres with mating rims, he used a vacuum pump to remove all of the air from the sphere formed. He then attached a team of 15 horses to each hemisphere and despite their efforts, they could not pull the hemispheres apart. If the sphere held a volume of 4200 in³ of air and one-tenth of the remaining air was removed with each stroke of the pump, how much air was still in the sphere after the 11th stroke? How many strokes were required to remove 85% of the air?

**99. Treating swimming pools:** In preparation for the summer swim season, chlorine is added to swimming pools to control algae and bacteria. However, careful measurements must be taken as levels above 5 ppm (parts per million) can be highly irritating to the eyes and throat, while levels below 1 ppm will be ineffective (3.0 to 3.5 ppm is ideal). In addition, the water must be treated daily since within a 24-hr period, about 25% of the chlorine will dissipate into the air. If the chlorine level in a swimming pool is 8 ppm after its initial treatment, how many days should the County Pool Supervisor wait before opening it up to the public? If left untreated, how many days until the chlorine level drops below 1 ppm?

100. **Venting landfill gases:** The gases created from the decomposition of waste in landfills must be carefully managed, as their release can cause terrible odors, harm the landfill structure, damage vegetation, or even cause an explosion. Suppose the accumulated volume of gas is 50,000 $ft^3$, and civil engineers are able to vent 2.5% of this gas into the atmosphere daily. What volume of gas remains after 21 days? How many days until the volume of gas drops below 10,000 $ft^3$?

101. **Population growth—bacteria:** A biologist finds that the population of a certain type of bacteria doubles *each half-hour.* If an initial culture has 50 bacteria, what is the population after 5 hr? How long will it take for the number of bacteria to reach 204,800?

102. **Population growth—boom towns:** Suppose the population of a "boom town" in the old west doubled *every 2 months* after gold was discovered. If the initial population was 219, what was the population 8 months later? How many months until the population exceeded 28,000?

103. **Elastic rebound—super balls:** Megan discovers that a rubber ball dropped from a height of 2 m rebounds four-fifths of the distance it has previously fallen. How high does it rebound on the 7th bounce? How far does the ball travel before coming to rest?

104. **Elastic rebound—computer animation:** The screen saver on my laptop is programmed to send a colored ball vertically down the middle of the screen so that it rebounds 95% of the distance it last traversed. If the ball always begins at the top and the screen is 36 cm tall, how high does the ball bounce on its 8th rebound? How far does the ball travel before coming to rest?

▶ **EXTENDING THE CONCEPT**

105. A standard piece of typing paper is approximately 0.001 in. thick. Suppose you were able to fold this piece of paper in half 26 times. How thick would the result be? As tall as a hare, as tall as a hen, as tall as a horse, as tall as a house, or taller than a high-rise? Find the actual height by computing the 27th term of a geometric sequence. Discuss what you find.

106. Consider the following situation. A person is hired at a salary of $40,000/yr, with a guaranteed raise of $1750/yr. At the same time, inflation is running about

4% per year. How many years until this person's salary is overtaken and eaten up by the actual cost of living?

107. Verify the following statements.
   a. If $a_1, a_2, a_3, \ldots, a_n$ is a geometric sequence with $r$ and $a_1$ greater than zero, then $\log a_1, \log a_2, \log a_3, \ldots, \log a_n$ is an arithmetic sequence.
   b. If $a_1, a_2, a_3, \ldots, a_n$ is an arithmetic sequence, then $10^{a_1}, 10^{a_2}, \ldots, 10^{a_n}$ is a geometric sequence.

▶ **MAINTAINING YOUR SKILLS**

108. (2.5) Find the zeroes of $f$ using the quadratic formula: $f(x) = x^2 + 5x + 9$.

109. (1.6) Solve for $x$: $\dfrac{3}{x^2 - 3x - 10} - \dfrac{4}{x - 5} = \dfrac{1}{x + 2}$

110. (4.5) Graph the rational function: $h(x) = \dfrac{x^2}{x - 1}$

111. (5.6) Given the logistic function shown, find $p(50)$, $p(75)$, $p(100)$, and $p(150)$:

$$p(t) = \frac{4200}{1 + 10e^{-0.055t}}$$

## MID-CHAPTER CHECK

**In Exercises 1 through 3, the $n$th term is given. Write the first three terms of each sequence and find $a_9$.**

1. $a_n = 7n - 4$
2. $a_n = n^2 + 3$
3. $a_n = (-1)^n(2n - 1)$

4. Evaluate the sum $\displaystyle\sum_{n=1}^{4} 3^{n+1}$.

5. Rewrite using sigma notation.
   $1 + 4 + 7 + 10 + 13 + 16$

**Match each formula to its correct description.**

**6.** $S_n = \dfrac{n(a_1 + a_n)}{2}$

**7.** $a_n = a_1 r^{n-1}$

**8.** $S_\infty = \dfrac{a_1}{1 - r}$

**9.** $a_n = a_1 + (n-1)d$

**10.** $S_n = \dfrac{a_1(1 - r^n)}{1 - r}$

    **a.** sum of an infinite geometric series (with $|r| < 1$)

    **b.** $n$th term formula for an arithmetic series

    **c.** sum of a finite geometric series

    **d.** summation formula for an arithmetic series

    **e.** $n$th term formula for a geometric series

**11.** Identify $a_1$ and the common difference $d$. Then find an expression for the general term $a_n$.

    **a.** 2, 5, 8, 11, ...

    **b.** $\frac{3}{2}, \frac{9}{4}, 3, \frac{15}{4}, \dots$

**Find the number of terms in each series, then find the sum.**

**12.** $2 + 5 + 8 + 11 + \cdots + 74$

**13.** $\frac{1}{2} + \frac{3}{2} + \frac{5}{2} + \frac{7}{2} + \cdots + \frac{31}{2}$

**14.** For an arithmetic series, $a_3 = -8$ and $a_7 = 4$. Find $S_{10}$.

**15.** For a geometric series, $a_3 = -81$ and $a_6 = 3$. Find $S_{10}$.

**16.** Identify $a_1$ and the common ratio $r$. Then find an expression for the general term $a_n$.

    **a.** 2, 6, 18, 54, ...

    **b.** $\frac{1}{2}, \frac{1}{4}, \frac{1}{8}, \frac{1}{16}, \dots$

**17.** Find the number of terms in the series, then compute the sum. $\frac{1}{54} + \frac{1}{18} + \frac{1}{6} + \cdots + \frac{81}{2}$

**18.** Find the infinite sum (if it exists).

    $-49 + (-7) + (-1) + \left(-\frac{1}{7}\right) + \cdots$

**19.** Barrels of toxic waste are stacked at a storage facility in pyramid form, with 60 barrels in the first row, 59 in the second row, and so on, until there are 10 barrels in the top row. How many barrels are in the storage facility?

**20.** As part of a conditioning regimen, a drill sergeant orders her platoon to do 25 continuous standing broad jumps. The best of these recruits was able to jump 96% of the distance from the previous jump, with a first jump distance of 8 ft. Use a sequence/series to determine the distance the recruit jumped on the 15th try, and the total distance traveled by the recruit after all 25 jumps.

---

## 9.4    Counting Techniques

**LEARNING OBJECTIVES**

*In Section 9.4 you will see how we can:*

☐ **A.** Count possibilities using lists and tree diagrams

☐ **B.** Count possibilities using the fundamental principle of counting

☐ **C.** Quick-count distinguishable permutations

☐ **D.** Quick-count nondistinguishable permutations

☐ **E.** Quick-count using combinations

How long would it take to estimate the number of fans sitting shoulder-to-shoulder at a sold-out basketball game? Well, it depends. You could actually begin counting 1, 2, 3, 4, 5, ... , which would take a very long time, or you could try to simplify the process by counting the number of fans in the first row and multiplying by the number of rows. Techniques for "quick-counting" the objects in a large set play an important role in a study of probability.

### A. Counting by Listing and Tree Diagrams

Consider the simple spinner shown in Figure 9.6, which is divided into three equal parts. What are the different possible outcomes for two spins, spin 1 followed by spin 2? We might begin by organizing the possibilities using a **tree diagram.** As the name implies, each choice or possibility appears as the branch of a tree, with the total possibilities being equal to the number of paths from the beginning point to the end of a branch. Figure 9.7 shows how the spinner exercise would appear (for two spins). Moving from top to bottom we can trace nine possible paths: *AA, AB, AC, BA, BB, BC, CA, CB,* and *CC.*

**Figure 9.6**

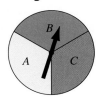

**Figure 9.7**

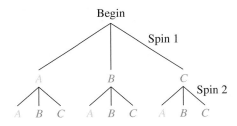

**EXAMPLE 1** ▶  **Listing Possibilities Using a Tree Diagram**

A basketball player is fouled and awarded three free throws. Let H represent the possibility of a hit (basket is made), and M the possibility of a miss. Determine the possible outcomes for the three shots using a tree diagram.

**Solution** ▶  Each shot has two possibilities, hit (H) or miss (M), so the tree will branch in two directions at each level. As illustrated in the figure, there are a total of eight possibilities: HHH, HHM, HMH, HMM, MHH, MHM, MMH, and MMM.

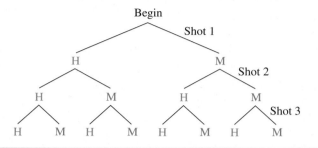

**Now try Exercises 5 through 8** ▶

To assist our discussion, an **experiment** is any task that can be done repeatedly and has a well-defined set of possible outcomes. Each repetition of the experiment is called a **trial**. A **sample outcome** is any *single* outcome of a trial, while the **sample space** is the set of *all* possible outcomes.

In our first illustration, the *experiment* was spinning a spinner with *three sample outcomes* (A, B, or C), and after *two trials* (spin 1 and spin 2), there were *nine* elements in the *sample space*. Note that after the first trial, we had three possibilities (A, B, and C). For two trials we have $3^2 = 9$ possibilities, while three trials would yield a sample space with $3^3 = 27$ possibilities. In general, for $N$ equally likely outcomes we have

**A "Quick-Counting" Formula for a Sample Space**

If an experiment has $N$ sample outcomes that are equally likely and the experiment is repeated $t$ times, the number of elements in the sample space is

$$N^t$$

**EXAMPLE 2** ▶  **Counting the Outcomes in a Sample Space**

Many locks have the digits 0 through 39 arranged along a circular dial. Opening the lock requires stopping at a sequence of three numbers within this range, going counterclockwise to the first number, clockwise to the second, and counterclockwise to the third. How many three-number sequences are possible?

**Solution** ▶  In this experiment, there are three trials ($t = 3$), each with 40 possible outcomes ($N = 40$). The quick-counting formula gives $40^3 = 64,000$ possible sequences.

 **A.** You've just seen how we can count possibilities using lists and tree diagrams

**Now try Exercises 9 and 10** ▶

## B. Fundamental Principle of Counting

The number of possible outcomes may differ depending on how the event is defined. For instance, some security systems, license plates, and telephone numbers exclude certain numbers. As an example, phone numbers cannot begin with 0 or 1 because these are reserved for operator assistance, long distance, and international calls. Constructing a three-digit area code is like filling in three blanks $\underline{\phantom{xx}}\,\underline{\phantom{xx}}\,\underline{\phantom{xx}}$ with three digits.

Since the area code must start with a number between 2 and 9, there are eight choices for the first blank. Since there are 10 choices for the second digit and 10 choices for the third, there are $8 \cdot 10 \cdot 10 = 800$ possibilities.

---

**EXAMPLE 3** ▶ **Counting Possibilities for a Four-Digit Security Code**

A digital security system requires that you enter a four-digit PIN (personal identification number), using only the digits 1 through 9. How many codes are possible if

   **a.** Repetition of digits is allowed?

   **b.** Repetition is not allowed?

   **c.** The first digit must be even and repetitions are not allowed?

**Solution** ▶  **a.** Consider filling in the four blanks $\underline{\phantom{xx}}\,\underline{\phantom{xx}}\,\underline{\phantom{xx}}\,\underline{\phantom{xx}}$ with the number of ways the digit can be chosen. If repetition is allowed, the experiment is similar to that of Example 2 and there are $N^t = 9^4 = 6561$ possible PINs.

   **b.** If repetition is not allowed, there are only eight possible choices for the second digit of the PIN, then seven for the third, and six for the fourth. The number of possible PIN numbers decreases to $9 \cdot 8 \cdot 7 \cdot 6 = 3024$.

   **c.** There are four choices for the first digit (2, 4, 6, 8). Once this choice has been made there are eight choices for the second digit, seven for the third, and six for the last: $4 \cdot 8 \cdot 7 \cdot 6 = 1344$ possible codes.

**Now try Exercises 11 through 14** ▶

---

Given *any* experiment involving a sequence of tasks, if the first task can be completed in $p$ possible ways, the second task has $q$ possibilities, and the third task has $r$ possibilities, a tree diagram will show that the number of possibilities for $task_1$–$task_2$–$task_3$ is $p \cdot q \cdot r$. This situation is simply a generalization of the previous quick-counting formula. Even though the examples we've considered to this point have varied a great deal, this idea was fundamental to counting all possibilities in a sample space and is, in fact, known as the **fundamental principle of counting (FPC)**.

**Fundamental Principle of Counting (Applied to Three Tasks)**

Given any experiment with three defined tasks, if there are $p$ possibilities for the first task, $q$ possibilities for the second, and $r$ possibilities for the third, the total number of ways the experiment can be completed is

$$p \cdot q \cdot r$$

This fundamental principle can be extended to include any number of tasks, and can be applied in many different ways. **See Exercises 15 through 18.**

**EXAMPLE 4** ▶ **Counting Possibilities for Seating Arrangements**

Adrienne, Bob, Carol, Dax, Earlene, and Fabian bought tickets to see *The Marriage of Figaro*. Assuming they sat together in a row of six seats, how many different seating arrangements are possible if

**a.** Bob and Carol are sweethearts and must sit together?

**b.** Bob and Carol are enemies and must not sit together?

**Solution** ▶ **a.** Since a restriction has been placed on the seating arrangement, it will help to divide the experiment into a sequence of tasks—*task 1:* they sit together; *task 2:* either Bob is on the left or Bob is on the right; and *task 3:* the other four are seated. Bob and Carol can sit together in five different ways, as shown in Figure 9.8, so there are five possibilities for task 1. There are two ways they can be side-by-side: Bob on the left and Carol on the right, as shown, or Carol on the left and Bob on the right. The remaining four people can be seated randomly, so task 3 has $4 \cdot 3 \cdot 2 \cdot 1 = 4! = 24$ possibilities. Under these conditions they can be seated $5 \cdot 2 \cdot 4! = 240$ ways.

**Figure 9.8**

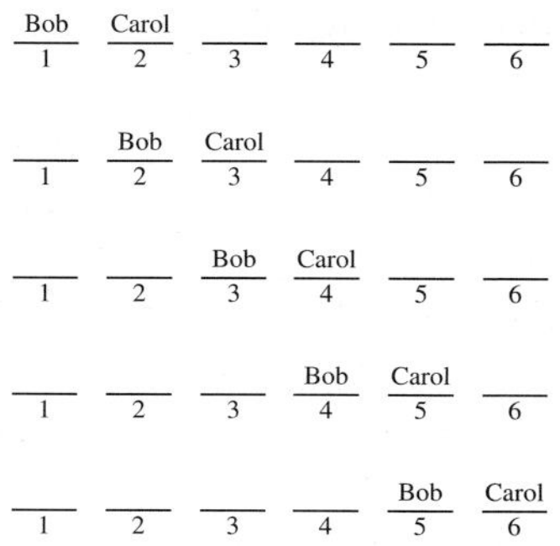

**b.** This is similar to part (a), but now we have to count the number of ways they can be separated by *at least one seat*—*task 1:* Bob and Carol are in nonadjacent seats; *task 2:* either Bob is on the left or Bob is on the right; and *task 3:* the other four are seated. For tasks 1 and 2, be careful to note there is no multiplication involved, just simple counting. If Bob sits in seat 1 (to the left of Carol), there are four nonadjacent seats on the right. If Bob sits in seat 2, there are three nonadjacent seats on the right. With Bob in seat 3, there are two nonadjacent seats to his right. Similar reasoning for the remaining seats shows there are $2(4 + 3 + 2 + 1) = 20$ possibilities for Bob and Carol not sitting together (the possibilities are doubled because Bob could also sit to the right of Carol). Multiplying by the number of ways the other four can be seated (task 3) gives $20 \cdot 4! = 480$ possible seating arrangements. We could also reason that since there are $6! = 720$ random seating arrangements and 240 of them consist of Bob and Carol sitting together, the remaining $720 - 240 = 480$ must consist of Bob and Carol *not* sitting together. More will be said about this type of reasoning in Section 9.5.

☑ **B.** You've just seen how we can count possibilities using the fundamental principle of counting

**Now try Exercises 19 through 26** ▶

## C. Distinguishable Permutations

In the game of Scrabble® (Milton Bradley), players attempt to form words by rearranging letters. Suppose a player has the letters P, S, T, and O at the end of the game. These letters could be rearranged or *permuted* to form the words POTS, SPOT, TOPS, OPTS, POST, or STOP. These arrangements, obtained by changing an existing order, are called **distinguishable permutations.**

Example 4 considered the various ways six people could fill six seats. But what if there were fewer seats than people? By the FPC, with six people and four seats there could be $6 \cdot 5 \cdot 4 \cdot 3 = 360$ different arrangements, with six people and three seats there are $6 \cdot 5 \cdot 4 = 120$ different arrangements, and so on. Generally, for n people and r seats, the first r factors of n! will be used. The notation and formula for *distinguishable permutations of n objects taken r at a time* is $_nP_r = \frac{n!}{(n-r)!}$. By defining $0! = 1$, the formula includes the case where all n objects are selected, which of course results in $_nP_n = \frac{n!}{(n-n)!} = \frac{n!}{0!} = \frac{n!}{1} = n!$.

### Distinguishable Permutations: Unique Elements

If $r$ objects are selected from a set containing $n$ unique elements and placed in an ordered arrangement, the number of distinguishable permutations is

$$_nP_r = \frac{n!}{(n-r)!} \quad \text{or} \quad _nP_r = n(n-1)(n-2)\cdots(n-r+1)$$

---

**EXAMPLE 5** ▶ **Computing a Permutation**

Compute each value of $_nP_r$.

    **a.** $_7P_4$           **b.** $_{10}P_3$

**Solution** ▶ Begin by evaluating each expression using the formula $_nP_r = \frac{n!}{(n-r)!}$, noting the third line (in bold) gives the first $r$ factors of $n!$.

    **a.** $_7P_4 = \dfrac{7!}{(7-4)!}$                    **b.** $_{10}P_3 = \dfrac{10!}{(10-3)!}$

               $= \dfrac{7\cdot6\cdot5\cdot4\cdot3!}{3!}$                        $= \dfrac{10\cdot9\cdot8\cdot7!}{7!}$

               $= \mathbf{7\cdot6\cdot5\cdot4}$                           $= \mathbf{10\cdot9\cdot8}$

               $= 840$                                  $= 720$

**Now try Exercises 27 through 34** ▶

---

**EXAMPLE 6** ▶ **Counting the Possibilities for Finishing a Race**

As part of a sorority's initiation process, the nine new inductees must participate in a barefoot and blindfolded 1-mi race. Assuming there are no ties, how many first-through fifth-place finishes are possible if it is well known that Mediocre Mary will finish fifth and Lightning Louise will finish first?

**Solution** ▶ To help understand the situation, we can diagram the possibilities for finishing first through fifth. Since Louise will finish first, this slot can be filled in only one way, by Louise herself. The same goes for Mary and her fifth-place finish:

                Louise    _____   _____   _____   Mary
                 1st        2nd        3rd        4th        5th

The remaining three slots can be filled in $_7P_3 = 7\cdot6\cdot5$ different ways, indicating that under these conditions, there are $1\cdot7\cdot6\cdot5\cdot1 = 210$ different ways to finish.

☑ **C.** You've just seen how we can quick-count distinguishable permutations

**Now try Exercises 35 through 40** ▶

---

## D. Nondistinguishable Permutations

As the name implies, certain permutations are nondistinguishable, meaning you cannot tell one from another. Such is the case when the original set contains elements or outcomes that are identical. Consider a family with four children, Megan, Morgan, Michael, and Mitchell, who are at the photo studio for a family picture. Michael and Mitchell are identical twins and cannot be told apart. In how many ways can they be lined up for the picture? Since this is an ordered arrangement of four children taken from a group of four, there are $4! = 24$ ways to line them up. A few of them are

| | | | | | | | |
|---|---|---|---|---|---|---|---|
| Megan | Morgan | Michael | Mitchell | Megan | Morgan | Mitchell | Michael |
| Megan | Michael | Morgan | Mitchell | Megan | Mitchell | Morgan | Michael |
| Michael | Megan | Morgan | Mitchell | Mitchell | Megan | Morgan | Michael |

But of these six arrangements, half will appear to be the same picture, since the difference between Michael and Mitchell cannot be distinguished. In fact, of the 24 total permutations, every picture where Michael and Mitchell have switched places will be nondistinguishable. To find the *distinguishable* permutations, we need to take the total permutations (4!) *and divide by 2!, the number of ways the twins can be permuted*: $\frac{4!}{2!} = \frac{24}{2} = 12$ distinguishable pictures.

These ideas can be generalized and stated in the following way.

### Distinguishable Permutations: Nonunique Elements

In a set containing $n$ elements where one element is repeated $p$ times, another is repeated $q$ times, and another is repeated $r$ times $(p + q + r = n)$, the number of distinguishable permutations is

$$\frac{n!}{p!q!r!}$$

The idea can be extended to include any number of repeated elements.

**EXAMPLE 7** ▶ **Counting Nondistinguishable Permutations**

A Scrabble player starts the game with the seven letters S, A, O, O, T, T, and T in her rack. How many distinguishable arrangements can be formed as she attempts to play a word?

**Solution** ▶ Essentially the exercise asks for the number of distinguishable permutations of the seven letters, given T is repeated three times and O is repeated twice (for S and A, 1! = 1 can be disregarded). There are $\frac{7!}{3!2!1!1!} = \frac{7!}{3!2!} = 420$ distinguishable permutations.

☑ **D.** You've just seen how we can quick-count nondistinguishable permutations

**Now try Exercises 41 through 52** ▶

### E. Combinations

Similar to nondistinguishable permutations, there are other times the total number of permutations must be reduced to quick-count the elements of a desired subset. Consider a vending machine that offers a variety of 40¢ candies. If you have a quarter (Q), dime (D), and nickel (N), the machine wouldn't care about the order the coins are deposited. Even though QDN, QND, DQN, DNQ, NQD, and NDQ give the $_3P_3 = 6$ possible permutations, the machine considers them as equal and will vend your snack. Using sets, this is similar to saying the set $A = \{X, Y, Z\}$ has only one subset with three elements, since $\{X, Z, Y\}$, $\{Y, X, Z\}$, $\{Y, Z, X\}$, and so on, all represent the same set. Similarly, there are six two-letter permutations of $X$, $Y$, and $Z$ $(_3P_2 = 6)$: $XY$, $XZ$, $YX$, $YZ$, $ZX$, and $ZY$, but only three two-letter subsets $_3P_1 = 3$: $\{X, Y\}$, $\{X, Z\}$ and $\{Y, Z\}$. When groupings having the same elements (but perhaps arranged differently) are considered identical, the result is called a **combination** and is denoted $_nC_r$. Since the $r$ objects can be selected in $r!$ ways, we divide $_nP_r$ by $r!$ to "quick-count" the number of possibilities: $_nC_r = \frac{_nP_r}{r!}$, which can be thought of as *the first r factors of n!, divided by r!*. By substituting $\frac{n!}{(n-r)!}$ for $_nP_r$ in this formula, we find an alternative method for computing $_nC_r$ is $\frac{n!}{r!(n-r)!}$. Take special note that when order is unimportant, the result is a *combination*, not a permutation.

**WORTHY OF NOTE**

In Example 7, if a Scrabble player is able to play all seven letters in one turn, he or she "bingos" and is awarded 50 extra points. The player in Example 7 did just that. Can you determine what word was played?

> ### Combinations
>
> The number of combinations of $n$ objects taken $r$ at a time is given by
>
> $$_nC_r = \frac{_nP_r}{r!} \quad \text{or} \quad _nC_r = \frac{n!}{r!(n-r)!}$$

**EXAMPLE 8 ▶**  **Computing Combinations Using a Formula**

Compute each value of $_nC_r$.

    **a.** $_7C_4$         **b.** $_8C_3$         **c.** $_5C_2$

**Solution ▶**    **a.** $_7C_4 = \dfrac{7 \cdot 6 \cdot 5 \cdot 4}{4!}$      **b.** $_8C_3 = \dfrac{8 \cdot 7 \cdot 6}{3!}$      **c.** $_5C_2 = \dfrac{5 \cdot 4}{2!}$

             $= 35$                    $= 56$                $= 10$

Now try Exercises 53 through 62 ▶

**EXAMPLE 9 ▶**  **Applications of Combinations—Lottery Results**

A small city is getting ready to draw five Ping-Pong balls of the nine they have numbered 1 through 9 to determine the winner(s) for its annual raffle. If a ticket holder has the same five numbers, they win. In how many ways can the winning numbers be drawn?

**Solution ▶**    Since the winning numbers can be drawn in any order, we have a combination of 9 things taken 5 at a time. The five numbers can be drawn in
$_9C_5 = \frac{9 \cdot 8 \cdot 7 \cdot 6 \cdot 5}{5!} = 126$ ways.

Now try Exercises 63 and 64 ▶

Somewhat surprisingly, there are many situations where the order things are listed is not important. Such situations include

- The formation of committees, if the order people volunteer is unimportant
- Card games with a standard deck, if the order cards are dealt is unimportant
- Playing BINGO, if the order the winning numbers are called is unimportant

When the order in which people or objects are selected from a group is unimportant, the number of possibilities is a *combination,* not a permutation.

Another way to tell the difference between permutations and combinations is the following memory device: *Permutations* have *Priority* or *Precedence;* in other words, the *Position* of each element matters. By contrast, a *Combination* is like a *Committee* of *Colleagues* or *Collection* of *Commoners;* all members have equal rank. For permutations, *a-b-c* is different from *b-a-c.* For combinations, *a-b-c* is the same as *b-a-c.*

**EXAMPLE 10 ▶**  **Applications of Quick-Counting—Committees and Governance**

The Sociology Department of Lakeside Community College has 12 dedicated faculty members. (a) In how many ways can a three-member textbook selection committee be formed? (b) If the department is in need of a Department Chair, Curriculum Chair, and Technology Chair, in how many ways can the positions be filled?

**Solution** ▶    **a.** Since textbook selection depends on a *Committee* of *Colleagues*, the order members are chosen is not important. This is a *Combination* of 12 people taken 3 at a time, and there are $_{12}C_3 = 220$ ways the committee can be formed.

**b.** Since those selected will have *Position* or *Priority*, this is a *Permutation* of 12 people taken 3 at a time, giving $_{12}P_3 = 1320$ ways the positions can be filled.

 **E.** You've just seen how we can quick-count using combinations

**Now try Exercises 65 through 76** ▶

The Exercise Set contains a wide variety of additional applications. **See Exercises 79 through 105.**

## 9.4 EXERCISES

### ▶ CONCEPTS AND VOCABULARY

**Fill in the blank with the appropriate word or phrase. Carefully reread the section, if necessary.**

**1.** A(n) _____ is any task that can be repeated and has a well-defined set of possible _____.

**2.** If some elements of a group are identical, certain rearrangements are identical and the result is a(n) _____ permutation.

**3.** A three-digit number is formed from digits 1 to 9. Explain how forming the number with repetition differs from forming it without repetition.

**4.** Discuss/Explain the difference between a permutation and a combination. Try to think of new ways to help remember the distinction.

### ▶ DEVELOPING YOUR SKILLS

**5.** For the spinner shown here, (a) draw a tree diagram illustrating all possible outcomes for two spins and (b) create an ordered list showing all possible outcomes for two spins.

**6.** For the fair coin shown here, (a) draw a tree diagram illustrating all possible outcomes for four flips and (b) create an ordered list showing the possible outcomes for four flips.

**7.** A fair coin is flipped five times. If you extend the tree diagram from Exercise 6, how many possibilities are there?

**8.** A spinner has the two equally likely outcomes *A* or *B* and is spun four times. How is this experiment related to the one in Exercise 6? How many possibilities are there?

**9.** An inexpensive lock uses the numbers 0 to 24 with a three-number sequence. How many different sequences are possible?

**10.** Grades at a local college consist of A, B, C, D, F, and W. If four classes are taken, how many different report cards are possible?

**License plates:** In a certain (English-speaking) country, license plates for automobiles consist of two letters followed by one of four symbols (■, ◆, ○, or ●), followed by three digits. How many license plates are possible if

**11.** repetition is allowed?

**12.** repetition is not allowed?

**13.** A remote access door opener requires a five-digit (1–9) sequence. How many sequences are possible if (a) repetition is allowed? (b) repetition is not allowed?

**14.** An instructor is qualified to teach Math 020, 030, 140, and 160. How many different four-course schedules are possible if (a) repetition is allowed? (b) repetition is not allowed?

**Use the fundamental principle of counting and other quick-counting techniques to respond.**

15. **Menu items:** At Joe's Diner, the manager is offering a dinner special that consists of one choice of entree (chicken, beef, soy meat, or pork), two vegetable servings (corn, carrots, green beans, peas, broccoli, or okra), and one choice of pasta, rice, or potatoes. How many different meals are possible?

16. **Getting dressed:** A frugal businessman has five shirts, seven ties, four pairs of dress pants, and three pairs of dress shoes. Assuming that all possible arrangements are appealing, how many different shirt-tie-pants-shoes outfits are possible?

17. **Number combinations:** How many four-digit numbers can be formed using the even digits 0, 2, 4, 6, 8, if (a) no repetitions are allowed; (b) repetitions are allowed; (c) repetitions are not allowed and the number must be less than 6000 and divisible by 10?

18. **Number combinations:** If I was born in March, April, or May, after the 19th but before the 30th, and after 1949 but before 1981, how many different MM–DD–YYYY dates are possible for my birthday?

**Seating arrangements:** William, Xayden, York, and Zelda decide to sit together at the movies. How many ways can they be seated if

19. they sit in random order?

20. York must sit next to Zelda?

21. York and Zelda must be on the outside?

22. William must have the only aisle seat?

**Course schedule:** A college student is trying to set her schedule for the next semester and is planning to take five classes: English, art, math, fitness, and science. How many different schedules are possible if

23. the classes can be taken in any order?

24. she wants her science class to immediately follow her math class?

25. she wants her English class to be first and her fitness class to be last?

26. she can't decide on the best order and simply takes the classes in alphabetical order?

**Find the value of $_nP_r$ in two ways: (a) compute $r$ factors of $n!$ and (b) use the formula $_nP_r = \frac{n!}{(n-r)!}$.**

27. $_{10}P_3$         28. $_{12}P_2$         29. $_9P_4$

30. $_5P_3$          31. $_8P_7$          32. $_8P_1$

**Determine the number of three-letter permutations of the letters given, then use an organized list to write them all out. How many of them are actually words or common names?**

33. T, R, and A          34. P, M, and A

35. The regional manager for an office supply store needs to replace the manager and assistant manager at the downtown store. In how many ways can this be done if she selects the personnel from a group of 10 qualified applicants?

36. The local chapter of Mu Alpha Theta will soon be electing a president, vice-president, and treasurer. In how many ways can the positions be filled if the chapter has 15 members?

37. The local school board is going to select a principal, vice-principal, and assistant vice-principal from a pool of eight qualified candidates. In how many ways can this be done?

38. From a pool of 32 applicants, a board of directors must select a president, vice-president, labor relations liaison, and a director of personnel for the company's day-to-day operations. Assuming all applicants are qualified and willing to take on any of these positions, how many ways can this be done?

39. A hugely popular chess tournament now has six finalists. Assuming there are no ties, (a) in how many ways can the finalists place in the final round? (b) In how many ways can they finish first, second, and third? (c) In how many ways can they finish if it's sure that Roberta Fischer is going to win the tournament and that Geraldine Kasparov will come in sixth?

40. A field of 10 horses has just left the paddock area and is heading for the gate. Assuming there are no ties in the big race, (a) in how many ways can the horses place in the race? (b) In how many ways can they finish in the win, place, or show positions? (c) In how many ways can they finish if it's sure that John Henry is going to win, Seattle Slew will come in second, and either Dumb Luck or Calamity Jane will come in tenth?

**Assuming all multiple births are identical and the children cannot be told apart, how many distinguishable photographs can be taken of a family of six, if they stand in a single row and there is**

41. one set of twins

42. one set of triplets

43. one set of twins and one set of triplets

44. one set of quadruplets

45. How many distinguishable numbers can be made by rearranging the digits of 105,001?

46. How many distinguishable numbers can be made by rearranging the digits in the palindrome 1,234,321?

**How many distinguishable permutations can be formed from the letters of the given word?**

**47.** logic          **48.** leave

**49.** lotto          **50.** levee

**A Scrabble player (see Example 7) has the six letters shown remaining in her rack. How many distinguishable, six-letter permutations can be formed? (If all six letters are played, what was the word?)**

**51.** A, A, A, N, N, B          **52.** D, D, D, N, A, E

**Find the value of $_nC_r$ in two ways: (a) using $_nC_r = \frac{_nP_r}{r!}$ ($r$ factors of $n!$ over $r!$) and (b) using $_nC_r = \frac{n!}{r!(n-r)!}$.**

**53.** $_9C_4$          **54.** $_{10}C_3$          **55.** $_8C_5$

**56.** $_6C_3$          **57.** $_6C_6$          **58.** $_6C_0$

**Verify that each pair of combinations is equal.**

**59.** $_9C_4, _9C_5$          **60.** $_{10}C_3, _{10}C_7$

**61.** $_8C_5, _8C_3$          **62.** $_7C_2, _7C_5$

**63.** A platoon leader needs to send four soldiers to do some reconnaissance work. There are 12 soldiers in the platoon and each soldier is assigned a number between 1 and 12. The numbers 1 through 12 are placed in a helmet and drawn randomly. If a soldier's number is drawn, then that soldier goes on the mission. In how many ways can the reconnaissance team be chosen?

**64.** Seven colored balls (red, indigo, violet, yellow, green, blue, and orange) are placed in a bag and three are then withdrawn. In how many ways can the three balls be drawn?

**65.** When the company's switchboard operators went on strike, the company president asked for three volunteers from the managerial ranks to temporarily take their place. In how many ways can the three volunteers "step forward," if there are 14 managers and assistant managers in all?

**66.** Becky has identified 12 books she wants to read this year and decides to take four with her to read while on vacation. She chooses *Pastwatch* by Orson Scott Card for sure, then decides to randomly choose any three of the remaining books. In how many ways can she select the four books she'll end up taking?

**67.** A new garage band has built up their repertoire to 10 excellent songs that really rock. Next month they'll be playing in a *Battle of the Bands* contest, with the winner getting some guaranteed gigs at the city's most popular hot spots. In how many ways can the band select 5 of their 10 songs to play at the contest?

**68.** Pierre de Guirré is an award-winning chef and has just developed 12 delectable, new main-course recipes for his restaurant. In how many ways can he select three of the recipes to be entered in an international culinary competition?

**For each exercise, determine the most appropriate tool for obtaining a solution, then solve. Some exercises can be completed using more than one method.**

**69.** In how many ways can eight second-grade children line up for lunch?

**70.** If you flip a fair coin five times, how many different outcomes are possible?

**71.** Eight sprinters are competing for the gold, silver, and bronze medals. In how many ways can the medals be awarded?

**72.** Motorcycle license plates are made using two letters followed by three numbers. How many plates can be made if repetition of letters (only) is allowed?

**73.** A committee of five students is chosen from a class of 20 to attend a seminar. How many different ways can this be done?

**74.** If onions, cheese, pickles, and tomatoes are available to dress a hamburger, how many different hamburgers can be made?

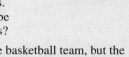

**75.** A caterer offers eight kinds of fruit to make various fruit trays. How many different trays can be made using four different fruits?

**76.** Eighteen females try out for the basketball team, but the coach can only place 15 on her roster. How many different teams can be formed?

## ▶ WORKING WITH FORMULAS

**77. Stirling's Formula: $n! \approx \sqrt{2\pi n}\left(\dfrac{n}{e}\right)^n$**

Values of $n!$ grow very quickly as $n$ gets larger (13! is already in the billions). For some applications, scientists find it useful to use the approximation for $n!$ shown, called Stirling's Formula.

**a.** Compute the value of 7! on your calculator, then use Stirling's Formula with $n = 7$. By what percent does the approximate value differ from the true value?

**b.** Compute the value of 10! on your calculator, then use Stirling's Formula with $n = 10$. By what percent does the approximate value differ from the true value?

**78. Factorial formulas: For whole numbers $k$ and $n$, with $k < n$, $\dfrac{n!}{(n-k)!} = n(n-1)(n-2)\cdots(n-k+1)$**

**a.** Verify the formula for $n = 7$ and $k = 5$.

**b.** Verify the formula for $n = 9$ and $k = 6$.

▶ **APPLICATIONS**

**79. Yahtzee:** In the game of "Yahtzee"® (Milton Bradley) five dice are rolled simultaneously on the first turn in an attempt to obtain various arrangements (worth various point values). How many different arrangements are possible?

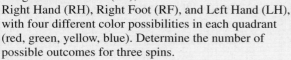

**80. Twister:** In the game of "Twister"® (Milton Bradley) a simple spinner is divided into four quadrants designated Left Foot (LF), Right Hand (RH), Right Foot (RF), and Left Hand (LH), with four different color possibilities in each quadrant (red, green, yellow, blue). Determine the number of possible outcomes for three spins.

**81. Clue:** In the game of "Clue"® (Parker Brothers) a crime is committed in one of nine rooms, with one of six implements, by one of six people. In how many different ways can the crime be committed?

**Phone numbers:** In North America, phone numbers have 10 digits: a three-digit area code, a three-digit exchange number, and the four final digits that make each phone number unique. Neither area codes nor exchange numbers can start with 0 or 1. Prior to 1994 the second digit of the area code *had to be* a 0 or 1. Sixteen area codes are reserved for special services (such as 911 and 411). In 1994, the last area code was used up and the rules were changed to allow the digits 2 through 9 as the middle digit in area codes.

**82.** How many different area codes were possible prior to 1994?

**83.** How many different exchange numbers were possible prior to 1994?

**84.** How many different phone numbers were possible *prior to* 1994?

**85.** How many different phone numbers were possible *after* 1994?

**Aircraft N-Numbers:** In the United States, private aircraft are identified by an "N-Number," which is generally the letter "N" followed by five characters and includes these restrictions: (1) the N-Number can consist of five digits, four digits followed by one letter, or three digits followed by two letters; (2) the first digit cannot be a zero; (3) to avoid confusion with the numbers zero and one, the letters O and I cannot be used; and (4) repetition of digits and letters is allowed. How many unique N-Numbers can be formed

**86.** that have four digits and one letter?

**87.** that have three digits and two letters?

**88.** that have five digits?

**89.** that have three digits and two letters with no repetitions of any kind allowed?

**Seating arrangements:** Eight people would like to be seated. Assuming some will have to stand, in how many ways can the seats be filled if the number of seats available is

**90.** eight

**91.** five

**92.** three

**93.** one

**Seating arrangements:** In how many different ways can eight people (six students and two teachers) sit in a row of eight seats if

**94.** the teachers must sit on the ends

**95.** the teachers must sit together

**Television station programming:** A television station needs to fill eight half-hour slots for its Tuesday evening schedule with eight programs. In how many ways can this be done if

**96.** there are no constraints

**97.** *The Office* must have the 8:00 P.M. slot

**98.** *The Office* must have the 8:00 P.M. slot and *30 Rock* must be shown at 6:00 P.M.

**99.** *The Office* can be aired at 7:00 or 9:00 P.M. and *30 Rock* can be aired at 6:00 or 8:00 P.M.

**Scholarship awards:** Fifteen students at Roosevelt Community College have applied for six scholarships. How many ways can the awards be given if

**100.** there are six different awards given to six different students

**101.** there are six identical awards given to six different students

**Committee composition:** The local city council has 10 members and is trying to decide if they want to be governed by a committee of three people or by a president, vice-president, and secretary.

**102.** If they are to be governed by committee, how many unique committees can be formed?

**103.** How many different president, vice-president, and secretary possibilities are there?

**104. Team rosters:** A soccer team has three goalies, eight defensive players, and eight forwards on its roster. How many different starting line-ups can be formed (one goalie, three defensive players, and three forwards)?

**105. e-mail addresses:** A business wants to standardize the e-mail addresses of its employees. To make them easier to remember and use, they consist of two letters and two digits (followed by @esmtb.com), with zero being excluded from use as the first digit and no repetition of letters or digits allowed. Will this provide enough unique addresses for their 53,000 employees worldwide?

▶ **EXTENDING THE CONCEPT**

**Tic-Tac-Toe:** In the game *Tic-Tac-Toe,* players alternately write an "X" or an "O" in one of nine squares on a 3 × 3 grid. If either player gets three in a row horizontally, vertically, or diagonally, that player wins. If all nine squares are played with neither person winning, the game is a draw. Assuming "X" always goes first,

**106.** How many different "ending boards" are possible if the game ends after five plays?

**107.** How many different "ending boards" are possible if the game ends after six plays?

▶ **MAINTAINING YOUR SKILLS**

**108.** (6.4) Solve the given system of linear inequalities by graphing. Shade the feasible region.

$$\begin{cases} 2x + y < 6 \\ x + 2y < 6 \\ x \geq 0 \\ y \geq 0 \end{cases}$$

**109.** (9.2) For the series $1 + 5 + 9 + 13 + \cdots + 197$, state the *n*th term formula. Then find the 35th term and the sum of the first 35 terms.

**110.** (7.2/7.3) Given matrices *A* and *B* shown, use a calculator to find $A + B, AB,$ and $A^{-1}$.

$$A = \begin{bmatrix} 1 & 0 & 3 \\ -2 & 5 & 1 \\ 2 & 1 & 4 \end{bmatrix} \quad B = \begin{bmatrix} 0.5 & 0.2 & -7 \\ -9 & 0.1 & 8 \\ 1.2 & 0 & 6 \end{bmatrix}$$

**111.** (8.3) Graph the hyperbola that is defined by

$$\frac{(x - 2)^2}{4} - \frac{(y + 3)^2}{9} = 1.$$

# 9.5 | Introduction to Probability

**LEARNING OBJECTIVES**

*In Section 9.5 you will see how we can:*

☐ **A.** Define an event on a sample space
☐ **B.** Compute elementary probabilities
☐ **C.** Use certain properties of probability
☐ **D.** Compute probabilities using quick-counting techniques
☐ **E.** Compute probabilities involving nonexclusive events

There are few areas of mathematics that give us a better view of the world than **probability** and **statistics.** Unlike statistics, which seeks to analyze and interpret data, probability (for our purposes) attempts to use observations and data to make statements concerning the likelihood of future events. Such predictions of what *might* happen have found widespread application in such diverse fields as politics, manufacturing, gambling, opinion polls, product reliability, and many others. In this section, we develop the basic elements of probability.

## A. Defining an Event

In Section 9.4 we defined the following terms: experiment and sample outcome. Flipping a coin twice in succession is an *experiment,* and two sample outcomes are HH and HT. An **event *E*** is *any designated set of sample outcomes,* and is a subset of the sample space. One event might be $E_1$: (two heads occur), another possibility is $E_2$: (at least one tail occurs).

**EXAMPLE 1** ▶ **Stating a Sample Space and Defining an Event**

Consider the experiment of rolling one standard, six-sided die (plural is dice). State the sample space *S* and define any two events relative to *S*.

**Solution** ▶ *S* is the set of all possible outcomes, so $S = \{1, 2, 3, 4, 5, 6\}$. Two possible events are $E_1$: (a 5 is rolled) and $E_2$: (an even number is rolled).

☑ **A.** You've just seen how we can define an event on a sample space

**Now try Exercises 5 through 8** ▶

## B. Elementary Probability

When rolling a die, we know the result can be any of the six equally likely outcomes in the sample space, so the chance of $E_1$:(a five is rolled) is $\frac{1}{6}$. Since three of the elements in $S$ are even numbers, the chance of $E_2$:(an even number is rolled) is $\frac{3}{6} = \frac{1}{2}$. This suggests the following definition.

> ### The Probability of an Event $E$
>
> Given $S$ is a sample space of equally likely events and $E$ is an event relative to $S$, the probability of $E$, written $P(E)$, is computed as
>
> $$P(E) = \frac{n(E)}{n(S)},$$
>
> where $n(E)$ represents the number of elements in $E$ and $n(S)$ represents the number of elements in $S$.

> **WORTHY OF NOTE**
>
> Our study of probability will involve only those sample spaces with events that are equally likely.

A standard deck of playing cards consists of 52 cards divided in four groups or *suits*. There are 13 hearts (♥), 13 diamonds (♦), 13 spades (♠), and 13 clubs (♣). As you can see in Figure 9.9, each of the 13 cards in a suit is labeled A, 2, 3, 4, 5, 6, 7, 8, 9, 10, J, Q, and K. Also notice that 26 of the cards are red (hearts and diamonds), 26 are black (spades and clubs), and 12 of the cards are "face cards" (J, Q, K of each suit).

**Figure 9.9**

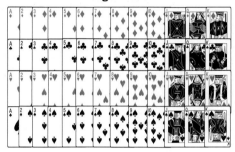

---

**EXAMPLE 2** ▶ **Stating a Sample Space and the Probability of a Single Outcome**

A single card is drawn from a well-shuffled deck. Define $S$ and state the probability of any single outcome. Then define $E$ as *a King is drawn* and find $P(E)$.

**Solution** ▶ Sample space: $S = \{\text{the 52 cards}\}$. There are 52 equally likely outcomes, so the probability of any one outcome is $\frac{1}{52}$. Since $S$ has four Kings, $P(E) = \frac{n(E)}{n(S)} = \frac{4}{52} = \frac{1}{13}$ or about 0.077.

**Now try Exercises 9 through 12** ▶

---

**EXAMPLE 3** ▶ **Stating a Sample Space and the Probability of a Single Outcome**

A family of five has two girls and three boys whose ages are 21, 19, 15, 13, and 9. One is to be selected randomly. Find the probability a teenager is chosen.

**Solution** ▶ The sample space is $S = \{9, 13, 15, 19, 21\}$. Three of the five are teenagers, meaning the probability is $\frac{3}{5}$, 0.6, or 60%.

☑ **B.** You've just seen how we can compute elementary probabilities

**Now try Exercises 13 and 14** ▶

## C. Properties of Probability

A study of probability necessarily includes recognizing some basic and fundamental properties. For example, when a fair coin is flipped, what is $P(E)$ if $E$ is defined as *a head or tails is flipped*? The event $E$ will occur 100% of the time, since a head or a tail are the only possibilities. In symbols we write $P$(outcome is in the sample space) or simply $P(S) = 1$ (100%).

What percent of the time will an event *not* in the sample space occur? Since the coin has only the two sides (heads or tails), the probability of flipping something else is zero. In symbols, $P$(outcome is not in sample space) $= 0$ or simply $P(\sim S) = 0$.

> **WORTHY OF NOTE**
>
> In probability studies, the tilde "~" acts as a negation symbol. For any event $E$, $\sim E$ means the event does not occur.

### Properties of Probability

Given sample space $S$ and any event $E$ defined relative to $S$:

    **1.** $P(S) = 1$         **2.** $P(\sim S) = 0$         **3.** $0 \leq P(E) \leq 1$

---

**EXAMPLE 4 ▶**   **Determining the Probability of an Event**

A game is played using a spinner like the one shown. Determine the probability of the following events:

$E_1$: A nine is spun.
$E_2$: An integer greater than 0 and less than 9 is spun.

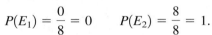

**Solution ▶**   The sample space consists of eight equally likely outcomes.

$$P(E_1) = \frac{0}{8} = 0 \qquad P(E_2) = \frac{8}{8} = 1.$$

Technically, $E_1$: A nine is spun is not an "event," since it is not in the sample space and cannot occur, while $E_2$ contains the entire sample space and must occur.

**Now try Exercises 15 and 16 ▶**

Because we know $P(S) = 1$, the probabilities of all single events defined on the sample space must sum to 1. For the experiment of rolling a fair die, the sample space has six (equally likely) outcomes. Note that $P(1) = P(2) = P(3) = P(4) = P(5) = P(6) = \frac{1}{6}$, and $\frac{1}{6} + \frac{1}{6} + \frac{1}{6} + \frac{1}{6} + \frac{1}{6} + \frac{1}{6} = 1$.

### Probability and Sample Outcomes

Given a sample space $S$ with $n$ sample outcomes $s_1, s_2, s_3, \ldots, s_n$:

$$\sum_{i=1}^{n} P(s_i) = P(s_1) + P(s_2) + P(s_3) + \cdots + P(s_n) = 1$$

The **complement** of an event $E$ is the set of sample outcomes in $S$ not in $E$. Symbolically, $\sim E$ is the complement of $E$.

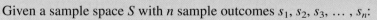

### Probability and Complementary Events

Given sample space $S$ and any event $E$ defined relative to $S$, the complement of $E$, written $\sim E$, is the set of all outcomes not in $E$ and:

    **1.** $P(\sim E) = 1 - P(E)$         **2.** $P(E) + P(\sim E) = 1$

**EXAMPLE 5** ▶   **Stating a Probability Using Complements**

Use complementary events to answer the following questions:

**a.** A single card is drawn from a well-shuffled deck. What is the probability that it is not a diamond?

**b.** A single letter is picked at random from the letters in the word "divisibility." What is the probability it is not an "i"?

**Solution** ▶   **a.** Since there are 13 diamonds in a standard 52-card deck:
$P(\sim D) = 1 - P(D) = 1 - \frac{13}{52} = \frac{39}{52} = 0.75$.

**b.** Of the 12 letters in d-i-v-i-s-i-b-i-l-i-t-y, 5 are "i's." This means $P(\sim i) = 1 - P(i)$, or $1 - \frac{5}{12} = \frac{7}{12}$. The probability of choosing a letter other than "i" is $0.58\overline{3}$.

> **WORTHY OF NOTE**
>
> Probabilities can be written in fraction form, decimal form, or as a percent. For $P(\sim D)$ from Example 5(a), the probability could be written $\frac{3}{4}$, 0.75, or 75%.

**Now try Exercises 17 through 20** ▶

---

**EXAMPLE 6** ▶   **Stating a Probability Using Complements**

Inter-Island Waterways has just opened hydrofoil service between several islands. The hydrofoil is powered by two engines, one forward and one aft, and will operate if either of its two engines is functioning. Due to testing and past experience, the company knows the probability of the aft engine failing is $P(\text{aft engine fails}) = 0.05$, the probability of the forward engine failing is $P(\text{forward engine fails}) = 0.03$, and the probability that both fail is $P(\text{both engines simultaneously fail}) = 0.012$. What is the probability the hydrofoil completes its next trip?

**Solution** ▶   Although the answer may *seem* complicated, note that $P(\text{trip is completed})$ and $P(\text{both engines simultaneously fail})$ are complements.

$$P(\text{trip is completed}) = 1 - P(\text{both engines simultaneously fail})$$
$$= 1 - 0.012$$
$$= 0.988$$

There is close to a 99% probability the trip will be completed.

**Now try Exercises 21 and 22** ▶

The chart in Figure 9.10 shows all 36 possible outcomes (the sample space) from the experiment of rolling two fair dice.

**Figure 9.10**

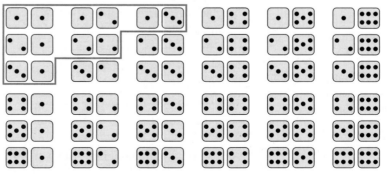

**EXAMPLE 7** ▶ **Stating a Probability Using Complements**

Two fair dice are rolled. What is the probability the sum of both dice is greater than or equal to 5, $P(\text{sum} \geq 5)$?

**Solution** ▶ See Figure 9.10. For $P(\text{sum} \geq 5)$ it may be easier to use complements as there are far fewer possibilities: $P(\text{sum} \geq 5) = 1 - P(\text{sum} < 5)$, which gives

$$1 - \frac{6}{36} = 1 - \frac{1}{6} = \frac{5}{6} = 0.8\overline{3}.$$

☑ **C.** You've just seen how we can use certain properties of probability

Now try Exercises 23 and 24 ▶

## D. Probability and Quick-Counting

Quick-counting techniques were introduced earlier to help count the number of elements in a large or more complex sample space, and the number of sample outcomes in an event.

**EXAMPLE 8A** ▶ **Stating a Probability Using Combinations**

Five cards are drawn from a shuffled 52-card deck. Calculate the probability of $E_1$:(*all five cards are face cards*) or $E_2$:(*all five cards are hearts*).

**Solution** ▶ The sample space for both events consists of all five-card groups that can be formed from the 52-card deck, which gives $_{52}C_5$ total possibilities. For $E_1$ we are to select five face cards from the 12 that are available (three from each of the four suits), for $_{12}C_5$ different combinations. The probability of five face cards is $P(E_1) = \frac{n(E_1)}{n(S)} = \frac{_{12}C_5}{_{52}C_5}$, which gives $\frac{792}{2,598,960} \approx 0.0003$. For $E_2$ we are to select five hearts from the 13 available giving $_{13}C_5$ different possibilities. The probability of five hearts is $P(E_2) = \frac{n(E_2)}{n(S)} = \frac{_{13}C_5}{_{52}C_5}$, which is $\frac{1287}{2,598,960} \approx 0.0005$.

> **WORTHY OF NOTE**
>
> It seems reasonable that the probability of 5 hearts is slightly higher, as 13 of the 52 cards are hearts, while only 12 are face cards.

**EXAMPLE 8B** ▶ **Stating a Probability Using Combinations and the Fundamental Principle of Counting**

Of the 42 seniors at Jacoby High School, 23 are female and 19 are male. A group of five students is to be selected at random to attend a conference in Reno, Nevada. What is the probability the group will have exactly three females?

**Solution** ▶ The sample space consists of all five-person groups that can be formed from the 42 seniors, which gives $_{42}C_5$ total possibilities. The event consists of selecting 3 females from the 23 available ($_{23}C_3$) and 2 males from the 19 available ($_{19}C_2$). Using the fundamental principle of counting $n(E) = (_{23}C_3) \cdot (_{19}C_2)$ and the probability the group has 3 females is $\frac{n(E)}{n(S)} = \frac{(_{23}C_3) \cdot (_{19}C_2)}{_{42}C_5}$, which gives $\frac{302,841}{850,668} \approx 0.356$. There is approximately a 35.6% probability the group will have exactly 3 females.

☑ **D.** You've just seen how we can compute probabilities using quick-counting techniques

Now try Exercises 25 through 28 ▶

While the fundamentals of probability are usually introduced using dice, cards, and very basic applications, this should not take away from its true power and utility. We use probability to help us to analyze things quantitatively so that important decisions can be made. In manufacturing, the probability a product becomes defective during its warranty period is crucial to the company's financial stability.

Health organizations rely heavily on probability to plan against the anticipated spread of an illness through a given population. In these and many other situations, the sample spaces are very large, making the use of technology an integral part of probability studies.

### E. Probability and Nonexclusive Events

Sometimes the way events are defined causes them to share sample outcomes. Using a standard deck of playing cards once again, if we define the events $E_1$:(a club is drawn) and $E_2$:(a face card is drawn), they share the outcomes J♣, Q♣, and K♣ as shown in Figure 9.11. This overlapping region is the intersection of the events, or $E_1 \cap E_2$. If we hastily compute $n(E_1 \cup E_2)$ as $n(E_1) + n(E_2)$, this intersecting region gets counted *twice*!

**Figure 9.11**

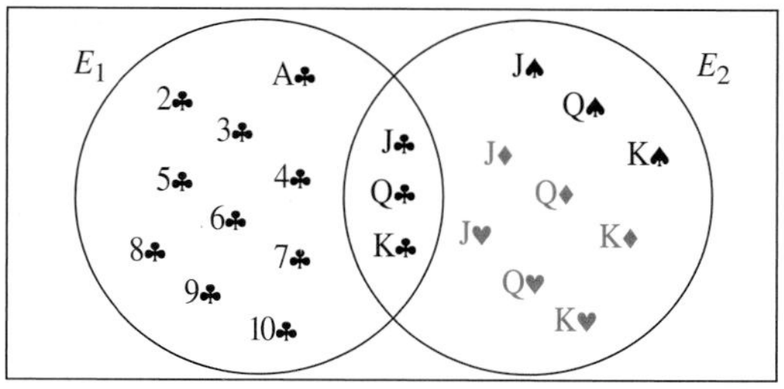

> **WORTHY OF NOTE**
>
> This can be verified by simply counting the elements involved: $n(E_1) = 13$ and $n(E_2) = 12$ so $n(E_1) + n(E_2) = 25$. However, there are only 22 possibilities in $E_1 \cup E_2$—the J♣, Q♣, and K♣ got counted twice.

In cases where the events are **nonexclusive** (not mutually exclusive), we maintain the correct count by subtracting one of the two intersections, obtaining $n(E_1 \cup E_2) = n(E_1) + n(E_2) - n(E_1 \cap E_2)$. This leads to the following calculation for the probability of nonexclusive events:

$$P(E_1 \cup E_2) = \frac{n(E_1) + n(E_2) - n(E_1 \cap E_2)}{n(S)} \quad \text{definition of probability}$$

$$= \frac{n(E_1)}{n(S)} + \frac{n(E_1)}{n(S)} - \frac{n(E_1 \cap E_2)}{n(S)} \quad \text{property of rational expressions}$$

$$= P(E_1) + P(E_2) - P(E_1 \cap E_2) \quad \text{definition of probability}$$

---

### Union Property for Probability

Given sample space $S$ and events $E_1$ and $E_2$ defined relative to $S$, the probability of $E_1$ *or* $E_2$ is given by

$$P(E_1 \cup E_2) = P(E_1) + P(E_2) - P(E_1 \cap E_2)$$

---

**EXAMPLE 9A** ▶   **Stating the Probability of Nonexclusive Events**

What is the probability that a club or a face card is drawn from a standard deck of 52 well-shuffled cards?

Solution ▶   As before, define the events $E_1$:(a club is drawn) and $E_2$:(a face card is drawn). Since there are 13 clubs and 12 face cards, $P(E_1) = \frac{13}{52}$ and $P(E_2) = \frac{12}{52}$. But three of the face cards are clubs, so $P(E_1 \cap E_2) = \frac{3}{52}$. This leads to

$$P(E_1 \cup E_2) = P(E_1) + P(E_2) - P(E_1 \cap E_2) \quad \text{union property for probability}$$

$$= \frac{13}{52} + \frac{12}{52} - \frac{3}{52} \quad \text{substitute}$$

$$= \frac{22}{52} \approx 0.423 \quad \text{combine terms}$$

There is about a 42% probability that a club or face card is drawn.

**EXAMPLE 9B** ▶  **Stating the Probability of Nonexclusive Events**

A survey of 100 voters was taken to gather information on critical issues and the demographic information collected is shown in the table. One out of the 100 voters is to be drawn at random to be interviewed on the 5 P.M. News. What is the probability the person is a woman ($W$) or a Republican ($R$)?

|  | Women | Men | Totals |
|---|---|---|---|
| Republican | 17 | 20 | 37 |
| Democrat | 22 | 17 | 39 |
| Independent | 8 | 7 | 15 |
| Green Party | 4 | 1 | 5 |
| Tax Reform | 2 | 2 | 4 |
| Totals | 53 | 47 | 100 |

**Solution** ▶  Since there are 53 women and 37 Republicans, $P(W) = 0.53$ and $P(R) = 0.37$. The table shows 17 people are both female and Republican so $P(W \cap R) = 0.17$.

$$P(W \cup R) = P(W) + P(R) - P(W \cap R) \quad \text{union property for probability}$$
$$= 0.53 + 0.37 - 0.17 \qquad\qquad \text{substitute}$$
$$= 0.73 \qquad\qquad\qquad\qquad \text{combine}$$

There is a 73% probability the person is a woman or a Republican.

 **E.** You've just seen how we can compute probabilities involving nonexclusive events

**Now try Exercises 29 through 42** ▶

Two events that have no common outcomes are called **mutually exclusive** events (one excludes the other and vice versa). For example, in rolling one die, $E_1$:(*a 2 is rolled*) and $E_2$:(*an odd number is rolled*) are mutually exclusive, since 2 is not an odd number. For the probability of $E_3$:(*a 2 is rolled* or *an odd number is rolled*), we note that $P(E_1 \cap E_2) = 0$ and the previous formula simply reduces to $P(E_1) + P(E_2)$. **See Exercises 43 and 44.** There is a large variety of additional applications in the Exercise Set. **See Exercises 47 through 62.**

## 9.5 EXERCISES

### ▶ CONCEPTS AND VOCABULARY

**Fill in the blank with the appropriate word or phrase. Carefully reread the section, if necessary.**

1. Given a sample space $S$ and an event $E$ defined relative to $S$: _____ $\leq P(E) \leq$ _____, $P(S) =$ _____, and $P(\sim S) =$ _____.

2. The _____ of an event $E$ is the set of sample outcomes in $S$ which are not contained in $E$.

3. Discuss/Explain the difference between mutually exclusive events and nonexclusive events. Give an example of each.

4. A single die is rolled. With no calculations, explain why the probability of rolling an even number is greater than rolling a number greater than four.

### ▶ DEVELOPING YOUR SKILLS

**State the sample space $S$ and the probability of a single outcome. Then define any two events $E$ relative to $S$ (many answers possible).**

5. Two fair coins are flipped.

6. The simple spinner shown is spun.

**Exercise 6**

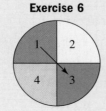

7. The head coaches for six little league teams (the Patriots, Cougars, Angels, Sharks, Eagles, and Stars) have gathered to discuss new changes in the rule book. One of them is randomly chosen to ask the first question.

8. Experts on the planets Mercury, Venus, Mars, Jupiter, Saturn, Uranus, Neptune, and the Kuiper object (formerly known as the planet) Pluto have gathered at a space exploration conference. One group of experts is selected at random to speak first.

**Find $P(E)$ for the events defined.**

**9.** Nine index cards numbered 1 through 9 are shuffled and placed in an envelope, then one of the cards is randomly drawn. Define event $E$ as *the number drawn is even*.

**10.** Eight flash cards used for studying basic geometric shapes are shuffled and one of the cards is drawn at random. The eight cards include information on circles, squares, rectangles, kites, trapezoids, parallelograms, pentagons, and triangles. Define event $E$ as *a quadrilateral is drawn.*

**11.** One card is drawn at random from a standard deck of 52 cards. What is the probability of
  **a.** drawing a Jack
  **b.** drawing a spade
  **c.** drawing a black card
  **d.** drawing a red three

**12.** Pinochle is a card game played with a deck of 48 cards consisting of 2 Aces, 2 Kings, 2 Queens, 2 Jacks, 2 Tens, and 2 Nines in each of the four standard suits [hearts (♥), diamonds (♦), spades (♠), and clubs (♣)]. If one card is drawn at random from this deck, what is the probability of
  **a.** drawing an Ace
  **b.** drawing a club
  **c.** drawing a red card
  **d.** drawing a face card (Jack, Queen, King)

**13.** A group of finalists on a game show consists of three males and five females. Hank has a score of 520 points, with Harry and Hester having 490 and 475 points respectively. Madeline has 532 points, with Mackenzie, Morgan, Maggie, and Melanie having 495, 480, 472, and 470 points respectively. One of the contestants is randomly selected to start the final round. Define $E_1$ as *Hester is chosen*, $E_2$ as *a female is chosen,* and $E_3$ as *a contestant with fewer than 500 points is chosen.* Find the probability of each event.

**14.** Soccer coach Maddox needs to fill the last spot on his starting roster for the opening day of the season and has to choose between three forwards and five defenders. The forwards have jersey numbers 5, 12, and 17, while the defenders have jersey numbers 7, 10, 11, 14, and 18. Define $E_1$ as *a forward is chosen,* $E_2$ as *a defender is chosen,* and $E_3$ as *a player whose jersey number is greater than 10 is chosen.* Find the probability of each event.

**15.** A game is played using a spinner like the one shown. For each spin,
  **a.** What is the probability the arrow lands in a shaded region?
  **b.** What is the probability your spin is less than 5?
  **c.** What is the probability you spin a 2?
  **d.** What is the probability the arrow points to prime number?

**16.** A game is played using a spinner like the one shown here. For each spin,
  **a.** What is the probability the arrow lands in a purple region?
  **b.** What is the probability your spin is greater than 2?
  **c.** What is the probability the arrow lands in a shaded region?
  **d.** What is the probability you spin a 5?

**Use the complementary events to complete Exercises 17 through 20.**

**17.** One card is drawn from a standard deck of 52. What is the probability it is not a club?

**18.** A corporation will be moving its offices to Los Angeles, Miami, Atlanta, Dallas, or Phoenix. If the site is randomly selected, what is the probability Dallas is not chosen?

**19.** A single digit is randomly selected from among the digits of 10!. What is the probability the digit is not a 2?

**20.** Four standard dice are rolled. What is the probability the sum is less than 24?

**21.** A large manufacturing plant can remain at full production as long as one of its two generators is functioning. Due to past experience and the age difference between the systems, the plant manager estimates the probability of the main generator failing is 0.05, the probability of the secondary generator failing is 0.01, and the probability of both failing is 0.009. What is the probability the plant remains in full production today?

**22.** A fire station gets an emergency call from a shopping mall in the mid-afternoon. From a study of traffic patterns, Chief Nozawa knows the probability the  most direct route is clogged with traffic is 0.07, while the probability of the secondary route being clogged is 0.05. The probability both are clogged is 0.02. What is the probability they can respond to the call unimpeded using one of these routes?

**23.** Two fair dice are rolled (see Figure 9.10). What is the probability of
  **a.** a sum less than four
  **b.** a sum less than eleven
  **c.** the sum is not nine
  **d.** a roll is not a "double" (both dice the same)

**"Double-six" dominos is a game played with the 28 numbered tiles shown in the diagram.**

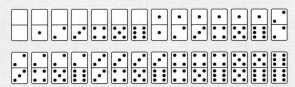

24. The 28 dominos are placed in a bag, shuffled, and then one domino is randomly drawn. What is the probability the total number of dots on the domino
    a. is three or less
    b. is greater than three
    c. does not have a blank half
    d. is not a "double" (both sides the same)

25. Five cards are drawn from a well-shuffled, standard deck of 52 cards. Which has the greater probability: (a) all five cards are red or (b) all five cards are numbered cards? How much greater?

26. Five cards are drawn from a well-shuffled pinochle deck of 48 cards (see Exercise 12). Which has the greater probability: (a) all five cards are face cards (King, Queen, or Jack) or (b) all five cards are black? How much greater?

27. A dietetics class has 24 students. Of these, 9 are vegetarians and 15 are not. The instructor receives enough funding to send six students to a conference. If the students are selected randomly, what is the probability the group will have
    a. exactly two vegetarians
    b. exactly four nonvegetarians
    c. at least three vegetarians

28. A large law firm has a support staff of 15 employees: six paralegals and nine legal assistants. Due to recent changes in the law, the firm wants to send five of them to a forum on the new changes. If the selection is done randomly, what is the probability the group will have
    a. exactly three paralegals
    b. exactly two legal assistants
    c. at least two paralegals

**Find the probability indicated using the information given.**

29. Given $P(E_1) = 0.7$, $P(E_2) = 0.5$, and $P(E_1 \cap E_2) = 0.3$, compute $P(E_1 \cup E_2)$.

30. Given $P(E_1) = 0.6$, $P(E_2) = 0.3$, and $P(E_1 \cap E_2) = 0.2$, compute $P(E_1 \cup E_2)$.

31. Given $P(E_1) = \frac{3}{8}$, $P(E_2) = \frac{3}{4}$, and $P(E_1 \cup E_2) = \frac{5}{6}$, compute $P(E_1 \cap E_2)$.

32. Given $P(E_1) = \frac{1}{2}$, $P(E_2) = \frac{3}{5}$, and $P(E_1 \cup E_2) = \frac{17}{20}$, compute $P(E_1 \cap E_2)$.

33. Given $P(E_1 \cup E_2) = 0.72$, $P(E_2) = 0.56$, and $P(E_1 \cap E_2) = 0.43$, compute $P(E_1)$.

34. Given $P(E_1 \cup E_2) = 0.85$, $P(E_1) = 0.4$, and $P(E_1 \cap E_2) = 0.21$, compute $P(E_2)$.

35. Two fair dice are rolled. What is the probability the sum of the dice is
    a. a multiple of 3 and an odd number
    b. greater than 5 with a 3 on at least one die
    c. an even number and a number greater than 9
    d. an odd number and a number less than 10

36. *Eight Ball* is a game played on a pool table with 15 balls numbered 1 through 15 and a cue ball that is solid white. Of the 15 numbered balls, 8 are a solid (nonwhite) color and numbered 1 through 8. The other seven are striped balls numbered 9 through 15. The fifteen numbered balls (no cueball) are placed in a large bowl and mixed, then one is drawn out. What is the probability of drawing

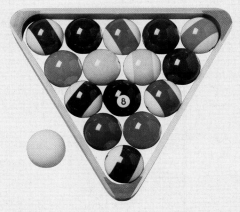

    a. the eight ball
    b. a number greater than fifteen
    c. an even number
    d. a multiple of three
    e. a solid color and an even number
    f. a striped ball and an odd number
    g. an even number and a number divisible by 3
    h. an odd number and a number divisible by 4

37. A survey of 50 veterans was taken to gather information on their service career and what life is like out of the military. A breakdown of those surveyed is shown in the table. One out of the 50 will be selected at random for an interview and a biographical sketch. What is the probability the person chosen is

| | Women | Men | Totals |
|---|---|---|---|
| Private | 6 | 9 | 15 |
| Corporal | 10 | 8 | 18 |
| Sergeant | 4 | 5 | 9 |
| Lieutenant | 2 | 1 | 3 |
| Captain | 2 | 3 | 5 |
| Totals | 24 | 26 | 50 |

    a. a woman and a sergeant
    b. a man and a private
    c. a private and a sergeant
    d. a woman and an officer
    e. a person in the military

**38.** Referring to Exercise 37, what is the probability the person chosen is
   **a.** a woman or a sergeant
   **b.** a man or a private
   **c.** a woman or a man
   **d.** a woman or an officer
   **e.** a captain or a lieutenant

**A computer is asked to randomly generate a three-digit number. What is the probability the**

**39.** tens digit is odd or the ones digit is even

**40.** first digit is prime and the number is a multiple of 10

**A computer is asked to randomly generate a four-digit number. What is the probability the number is**

**41.** at least 4000 or a multiple of 5

**42.** less than 7000 and an odd number

**43.** Two fair dice are rolled. What is the probability of
   **a.** boxcars (sum of 12) or snake eyes (sum of 2)
   **b.** a sum of 7 or a sum of 11
   **c.** an even-numbered sum or a prime sum
   **d.** an odd-numbered sum or a sum that is a multiple of 4
   **e.** a sum of 15 or a multiple of 12
   **f.** a sum that is a prime number

**44.** Suppose all 16 balls from a game of pool (see Exercise 36) are placed in a large leather bag and mixed, then one is drawn out. Consider the cue ball as "0." What is the probability of drawing
   **a.** a striped ball          **b.** a solid-colored ball
   **c.** a polka-dotted ball     **d.** the cue ball
   **e.** the cue ball or the eight ball
   **f.** a striped ball or a number less than five
   **g.** a solid color or a number greater than 12
   **h.** an odd number or a number divisible by 4

## ▶ WORKING WITH FORMULAS

**Expected value:** $E = \left(\begin{array}{c}\textbf{probability}\\\textbf{of win}\end{array}\right)\left(\begin{array}{c}\textbf{money}\\\textbf{won}\end{array}\right) - \left(\begin{array}{c}\textbf{probability}\\\textbf{of loss}\end{array}\right)\left(\begin{array}{c}\textbf{money}\\\textbf{lost}\end{array}\right)$

**When playing games of chance, the expected value $E$ of your bet is given by the formula shown. If the game is fair, your expected value is zero. Unfortunately, most casino games favor "the house" and your expected value is negative.**

**45.** In roulette, there are 38 numbers on the wheel: the odd numbers 1, 3, 5, ... , 35 in red; the even numbers 2, 4, 6, ... , 36 in black; and the numbers 0 and 00 in green. In a "straight up" bet, money is placed on any one particular number. If the number is rolled, the payout is 35 times the size of the bet. Otherwise, the bet is lost. What is the expected value on a $10 straight-up bet on lucky number 7? Is a straight-up bet fair?

**46.** In a color bet, money is placed on either red or black. If a number is rolled in the chosen color, the payout matches the size of the bet. Otherwise, the bet is lost. What is the expected value for a $10 bet on red? Is a color bet better or worse in the long run than a straight-up bet (see Exercise 45)?

## ▶ APPLICATIONS

**47.** To improve customer service, a company tracks the number of minutes a caller is "on hold" and waiting for a customer service representative. The table shows the probability that a caller will wait $m$ minutes. Based on the table, what is the probability a caller waits
   **a.** at least 2 min        **b.** less than 2 min
   **c.** 4 min or less         **d.** over 4 min
   **e.** less than 2 or more than 4 min
   **f.** 3 or more min

| Wait Time (minutes $m$) | Probability |
|---|---|
| 0 | 0.07 |
| $0 < m < 1$ | 0.28 |
| $1 \leq m < 2$ | 0.32 |
| $2 \leq m < 3$ | 0.25 |
| $3 \leq m < 4$ | 0.08 |

**48.** To study the impact of technology on American families, a researcher first determines the probability that a family has $n$ computers at home. Based on the table, what is the probability a home
   **a.** has at least one computer
   **b.** has two or more computers
   **c.** has fewer than four computers
   **d.** has five computers
   **e.** has one, two, or three computers
   **f.** does not have two computers

| Number of Computers | Probability |
|---|---|
| 0 | 9% |
| 1 | 51% |
| 2 | 28% |
| 3 | 9% |
| 4 | 3% |

**Jolene is an experienced markswoman and is able to hit a 10 in. by 20 in. target 100% of the time at a range of 100 yd. Assuming the probability she hits a target is proportional to its area, what is the probability she hits the shaded portions shown?**

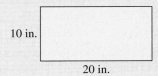

10 in.

20 in.

**49. a.**

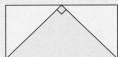

**b.**

**50. a.**

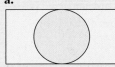

**b.**    |←—10 in.—→|

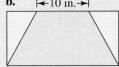

**51.** A circular dartboard has a total radius of 8 in., with circular bands that are 2 in. wide, as shown. You are skilled enough to hit this board 100% of the time so you always score at least two points each time you throw a dart. Assuming the probabilities are related to area, on the next dart that you throw what is the probability you

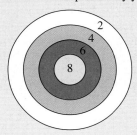

2
4
6
8

**a.** score at least a 4?    **b.** score at least a 6?
**c.** hit the bull's-eye?    **d.** score exactly 4 points?

**52.** Three red balls, six blue balls, and four white balls are placed in a bag. What is the probability the first ball you draw out is

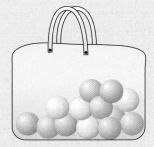

**a.** red                **b.** blue
**c.** not white          **d.** purple
**e.** red or white       **f.** red and white

**53.** Three red balls, six blue balls, and four white balls are placed in a bag, then two are drawn out and placed in a rack. What is the probability the balls drawn are
**a.** first red, second blue
**b.** first blue, second red
**c.** both white
**d.** first blue, second not red
**e.** first white, second not blue
**f.** first not red, second not blue

**54.** Due to inevitable oversights and human fault, errors randomly occur in a first draft manuscript. Suppose colleague A finds 90% of all existing errors, while colleague B (working independently) only spots 85% of the errors. If the two catch 98% of all errors by working together, (a) what percent of the errors were found by both? (b) What percent was missed by both?

**55.** Your instructor surprises you with a True/False quiz for which you are totally unprepared and must guess randomly. What is the probability you pass the quiz with an 80% or better if there are
**a.** three questions
**b.** four questions
**c.** five questions

**56.** A robot is sent out to disarm a timed explosive device by randomly changing some switches from a neutral position to a *positive flow* or *negative flow* position. The problem is, the switches are independent and unmarked, and it is unknown which direction is positive and which direction is negative. The bomb is harmless if a majority of the switches yield a positive flow. All switches must be thrown. What is the probability the device is disarmed if there are
**a.** three switches
**b.** four switches
**c.** five switches

**57.** A survey of 100 retirees was taken to gather information concerning how they viewed the Vietnam War back in the early 1970s. A breakdown of those surveyed is shown in the table. One out of the hundred will be selected at random for a personal, taped interview. What is the probability the person chosen had a
**a.** career of any kind and opposed the war
**b.** medical career and supported the war
**c.** military career and opposed the war
**d.** legal or business career and opposed the war
**e.** academic or medical career and supported the war

| Career    | Support | Opposed | Total |
|-----------|---------|---------|-------|
| Military  | 9       | 3       | 12    |
| Medical   | 8       | 16      | 24    |
| Legal     | 15      | 12      | 27    |
| Business  | 18      | 6       | 24    |
| Academics | 3       | 10      | 13    |
| Totals    | 53      | 47      | 100   |

**58.** Referring to Exercise 57, what is the probability the person chosen

   **a.** had a career of any kind or opposed the war

   **b.** had a medical career or supported the war

   **c.** supported the war or had a military career

   **d.** had a medical or a legal career

   **e.** supported or opposed the war

**59.** The Board of Directors for a large hospital has 15 members. There are six doctors of nephrology (kidneys), five doctors of gastroenterology (stomach and intestines), and four doctors of endocrinology (hormones and glands). Eight of them will be selected to visit the nation's premier hospitals on a 3-week, expenses-paid tour. What is the probability the group of eight selected consists of exactly

   **a.** four nephrologists and four gastroenterologists

   **b.** three endocrinologists and five nephrologists

**60.** A support group for hodophobics (an irrational fear of travel) has 32 members. There are 15 aviophobics (fear of air travel), eight siderodrophobics (fear of train travel), and nine thalassophobics (fear of ocean travel) in the group. Twelve of them will be randomly selected to participate in a new therapy. What is the probability the group of 12 selected consists of exactly

   **a.** two aviophobics, six siderodrophobics, and four thalassophobics

   **b.** five thalassophobics, four aviophobics, and three siderodrophobics

**61.** A trained chimpanzee is given a box containing eight wooden cubes with the letters p, a, r, a, l, l, e, l printed on them (one letter per block). Assuming the chimp can't read or spell, what is the probability he draws the eight blocks in order and actually forms the word "parallel"?

**62.** A number is called a "perfect number" if the sum of its proper factors is equal to the number itself. Six is the first perfect number since the sum of its proper factors is six: $1 + 2 + 3 = 6$. Twenty-eight is the second since: $1 + 2 + 4 + 7 + 14 = 28$. A young child is given a box containing eight wooden blocks with the following numbers printed on them (one per block): four 3's, two 5's, one 0, and one 6. What is the probability she draws the eight blocks in order and forms the fifth perfect number: 33,550,336?

▶ **EXTENDING THE CONCEPT**

**63.** The function $f(x) = (\frac{1}{2})^x$ gives the probability that $x$ number of flips will all result in heads (or tails). Compute the probability that 20 flips results in *20 heads in a row,* then use the Internet or some other resource to find the probability of winning a state lottery. Which is more likely to happen (which has the greater probability)? Were you surprised?

**64.** Consider the 210 discrete points found in the first and second quadrants where $-10 \leq x \leq 10$, $1 \leq y \leq 10$, and $x$ and $y$ are integers. The coordinates of each point are written on a slip of paper and placed in a box. One of the slips is then randomly drawn. What is the probability the point $(x, y)$ drawn

   **a.** is on the graph of $y = |x|$

   **b.** is on the graph of $y = 2|x|$

   **c.** is on the graph of $y = 0.5|x|$

   **d.** has $y$-coordinate $y > -2$

   **e.** has $x$-coordinate $x \leq 5$

   **f.** is between the branches of $y = x^2$

▶ **MAINTAINING YOUR SKILLS**

**65.** (7.1) Solve the system using matrices and row reduction: $\begin{cases} x - 2y + 3z = 10 \\ 2x + y - z = 18 \\ 3x - 2y + z = 26 \end{cases}$

**66.** (5.4) Complete the following logarithmic properties:

   $\log_b b =$ _____      $\log_b 1 =$ _____

   $\log_b b^n =$ _____      $b^{\log_b n} =$ _____

**67.** (4.6) Solve the inequality by graphing the function and labeling the appropriate interval(s):

   $$\frac{x^2 - 1}{x} \geq 0.$$

**68.** (9.3) A rubber ball is dropped from a height of 25 ft onto a hard surface. With each bounce, it rebounds 60% of the height from which it last fell. Use sequences/series to find (a) the height of the sixth bounce, (b) the total distance traveled up to the sixth bounce, and (c) the distance the ball will travel before coming to rest.

**The Binomial Theorem**

## LEARNING OBJECTIVES

*In Section 9.6 you will see how we can:*

- ☐ **A.** Use Pascal's triangle to find $(a + b)^n$
- ☐ **B.** Find binomial coefficients using $\binom{n}{k}$ notation
- ☐ **C.** Use the binomial theorem to find $(a + b)^n$
- ☐ **D.** Find a specific term of a binomial expansion
- ☐ **E.** Solve applications of binomial powers

Strictly speaking, a binomial is a polynomial with two terms. This limits us to terms with real number coefficients and whole number powers on variables. In this section, we will loosely regard a binomial as the sum or difference of *any* two terms. Hence $3x^2 - y^4$, $\sqrt{x} + 4$, $x + \frac{1}{x}$, and $-\frac{1}{2} + \frac{\sqrt{3}}{2}i$ are all "binomials." Our goal is to develop an ability to raise a binomial to any natural number power, with the results having important applications in genetics, probability, polynomial theory, and other areas. The tool used for this purpose is called the *binomial theorem*.

## A. Binomial Powers and Pascal's Triangle

Much of our mathematical understanding comes from a study of patterns. One area where the study of patterns has been particularly fruitful is **Pascal's triangle** (Figure 9.12), named after the French scientist Blaise Pascal (although the triangle was well known before his time). It begins with a 1 at the vertex of the triangle, with 1's extending diagonally downward to the left and right as shown. The entries on the interior of the triangle are found by adding the two entries directly above and to the left and right of each new position.

**Figure 9.12**

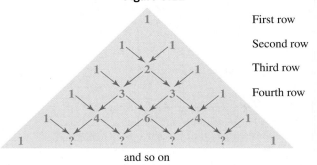

First row
Second row
Third row
Fourth row

and so on

There are a variety of patterns hidden within the triangle. In this section, we'll use the *horizontal rows* of the triangle to help us raise a binomial to various powers. To begin, recall that $(a + b)^0 = 1$ and $(a + b)^1 = 1a + 1b$ (unit coefficients are included for emphasis). In our earlier work, we saw that a binomial square (a binomial raised to the second power) always followed the pattern $(a + b)^2 = 1a^2 + 2ab + 1b^2$. Observe the overall pattern that is developing as we include $(a + b)^3$:

$$(a + b)^0 = \qquad\qquad 1 \qquad\qquad \text{row 1}$$
$$(a + b)^1 = \qquad\qquad 1a + 1b \qquad\qquad \text{row 2}$$
$$(a + b)^2 = \qquad 1a^2 + 2ab + 1b^2 \qquad \text{row 3}$$
$$(a + b)^3 = 1a^3 + 3a^2b + 3ab^2 + 1b^3 \quad \text{row 4}$$

Apparently the coefficients of $(a + b)^n$ will occur in row $n + 1$ of Pascal's triangle. Also observe that in each term of the expansion, the exponent of the first term $a$ *decreases by 1* as the exponent on the second term $b$ *increases by 1,* keeping the degree of each term constant (recall the degree of a term with more than one variable is the sum of the exponents).

$$1a^3b^0 + 3a^2b^1 + 3a^1b^2 + 1a^0b^3$$

$$\underset{\text{degree 3}}{3 + 0} \quad \underset{\text{degree 3}}{2 + 1} \quad \underset{\text{degree 3}}{1 + 2} \quad \underset{\text{degree 3}}{0 + 3}$$

These observations help us to quickly expand a binomial power.

**EXAMPLE 1** ▶ **Expanding a Binomial Using Pascal's Triangle**

Use Pascal's triangle and the patterns noted to expand $\left(x + \frac{1}{2}\right)^4$.

**Solution** ▶ Working step-by-step we have

1. The coefficients will be in the fifth row of Pascal's triangle.

$$1 \quad 4 \quad 6 \quad 4 \quad 1$$

2. The exponents on $x$ begin at 4 and *decrease,* while the exponents on $\frac{1}{2}$ begin at 0 and *increase.*

$$1x^4\left(\frac{1}{2}\right)^0 + 4x^3\left(\frac{1}{2}\right)^1 + 6x^2\left(\frac{1}{2}\right)^2 + 4x^1\left(\frac{1}{2}\right)^3 + 1x^0\left(\frac{1}{2}\right)^4$$

3. Simplify each term.

The result is $x^4 + 2x^3 + \frac{3}{2}x^2 + \frac{1}{2}x + \frac{1}{16}$.

Now try Exercises 5 through 8 ▶

If the exercise involves a difference rather than a sum, we simply rewrite the expression using algebraic addition and proceed as before.

**EXAMPLE 2** ▶ **Raising a Complex Number to a Power Using Pascal's Triangle**

Use Pascal's triangle and the patterns noted to compute $(3 - 2i)^5$.

**Solution** ▶ Begin by rewriting $(3 - 2i)^5$ as $[3 + (-2i)]^5$.

1. The coefficients will be in the sixth row of Pascal's triangle.

$$1 \quad 5 \quad 10 \quad 10 \quad 5 \quad 1$$

2. The exponents on 3 begin at 5 and *decrease,* while the exponents on $(-2i)$ begin at 0 and *increase.*

$$1(3^5)(-2i)^0 + 5(3^4)(-2i)^1 + 10(3^3)(-2i)^2 + 10(3^2)(-2i)^3 + 5(3^1)(-2i)^4 + 1(3^0)(-2i)^5$$

3. Simplify each term.

$$243 - 810i - 1080 + 720i + 240 - 32i$$

**☑ A.** You've just seen how we can use Pascal's triangle to find $(a + b)^n$

The result is $-597 - 122i$.

Now try Exercises 9 and 10 ▶

---

**Expanding Binomial Powers $(a + b)^n$**

1. The coefficients will be in row $n + 1$ of Pascal's triangle.
2. The exponents on *the first term* begin at $n$ and *decrease,* while the exponents on *the second term* begin at 0 and *increase.*
3. For any binomial difference $(a - b)^n$, rewrite the base as $[a + (-b)]^n$ using algebraic addition and proceed as before, then simplify each term.

## B. Binomial Coefficients and Factorials

Pascal's triangle can easily be used to find the coefficients of $(a + b)^n$, as long as the exponent is relatively small. If we needed to expand $(a + b)^{25}$, writing out the first 26 rows of the triangle would be rather tedious. To overcome this limitation, we introduce a *formula* that enables us to find the coefficients of any binomial expansion.

> **The Binomial Coefficients**
>
> For whole numbers $n$ and $r$ where $n \geq r$, the expression $\binom{n}{r}$, read "$n$ choose $r$," is called a **binomial coefficient** and is evaluated as:
>
> $$\binom{n}{r} = \frac{n!}{r!(n - r)!}$$

Notice the formula for determining binomial coefficients is identical to that for $_nC_r$, and it turns out the coefficients are actually found using a combination, with the new notation used primarily as a convenience. In Example 1, we found the coefficients of $(a + b)^4$ using the fifth row of Pascal's triangle. In Example 3, these coefficients are found using the formula for binomial coefficients.

**EXAMPLE 3** ▶ **Computing Binomial Coefficients**

Evaluate $\binom{n}{r} = \frac{n!}{r!(n - r)!}$ as indicated:

**a.** $\binom{4}{1}$  **b.** $\binom{4}{2}$  **c.** $\binom{4}{3}$

**Solution** ▶ **a.** $\binom{4}{1} = \frac{4!}{1!(4 - 1)!} = \frac{4 \cdot 3!}{1!3!} = 4$

**b.** $\binom{4}{2} = \frac{4!}{2!(4 - 2)!} = \frac{4 \cdot 3 \cdot 2!}{2!2!} = \frac{4 \cdot 3}{2} = 6$

**c.** $\binom{4}{3} = \frac{4!}{3!(4 - 3)!} = \frac{4 \cdot 3!}{3!1!} = 4$

**Now try Exercises 11 through 14** ▶

Note $\binom{4}{1} = 4$, $\binom{4}{2} = 6$ and $\binom{4}{3} = 4$ give the *interior entries* in the fifth row of Pascal's triangle: 1 4 6 4 1. For consistency and symmetry, we define $0! = 1$, which enables the formula to generate all entries of the triangle, including the 1's.

$$\binom{4}{0} = \frac{4!}{0!(4 - 0)!} = \frac{4!}{1 \cdot 4!} = 1$$

$$\binom{4}{4} = \frac{4}{4!(4 - 4)!} = \frac{4!}{4! \cdot 0!} = \frac{4!}{4! \cdot 1} = 1$$

The formula for $\binom{n}{r}$ with $0 \leq r \leq n$ now gives all coefficients in the $(n + 1)$st row. For $n = 5$, we have

$$\binom{5}{0} \quad \binom{5}{1} \quad \binom{5}{2} \quad \binom{5}{3} \quad \binom{5}{4} \quad \binom{5}{5}$$

$$1 \qquad 5 \qquad 10 \qquad 10 \qquad 5 \qquad 1$$

**EXAMPLE 4** ▶ **Computing Binomial Coefficients**

Compute the binomial coefficients:

**a.** $\binom{9}{0}$     **b.** $\binom{9}{1}$     **c.** $\binom{6}{5}$     **d.** $\binom{6}{6}$

Solution ▶ **a.** $\binom{9}{0} = \dfrac{9!}{0!(9-0)!} = \dfrac{9!}{9!} = 1$     **b.** $\binom{9}{1} = \dfrac{9!}{1!(9-1)!} = \dfrac{9!}{8!} = 9$

☑ **B.** You've just seen how we can find binomial coefficients using $\binom{n}{k}$ notation

**c.** $\binom{6}{5} = \dfrac{6!}{5!(6-5)!} = \dfrac{6!}{5!} = 6$     **d.** $\binom{6}{6} = \dfrac{6!}{6!(6-6)!} = \dfrac{6!}{6!} = 1$

**Now try Exercises 15 and 16** ▶

As mentioned, the formulas for $\binom{n}{r}$ and $_nC_r$ yield like results for given values of $n$ and $r$. For future use, it will help to commit the general results from Example 4 to memory: $\binom{n}{0} = 1$, $\binom{n}{1} = n$, $\binom{n}{n-1} = n$, and $\binom{n}{n} = 1$. By exploiting the symmetry in Pascal's triangle, we can often half the number of computations needed by using the identity $\binom{n}{n-r} = \binom{n}{r}$.

## C. The Binomial Theorem

Using $\binom{n}{r}$ notation and the observations made regarding binomial powers, we can now state the **binomial theorem.**

### Binomial Theorem

For any binomial $(a + b)$ and natural number $n$,

$$(a+b)^n = \binom{n}{0}a^n b^0 + \binom{n}{1}a^{n-1}b^1 + \binom{n}{2}a^{n-2}b^2 + \cdots + \binom{n}{n-1}a^1 b^{n-1} + \binom{n}{n}a^0 b^n$$

The theorem can also be stated in summation form as

$$(a+b)^n = \sum_{r=0}^{n} \binom{n}{r}a^{n-r}b^r$$

The expansion actually looks overly impressive in this form, and it helps to summarize the process in words, as we did earlier. The exponents on the first term $a$ begin at $n$ and decrease, while the exponents on the second term $b$ begin at 0 and increase, keeping the degree of each term constant. The $\binom{n}{r}$ notation simply gives the coefficients of each term. As a final note, observe that the $r$ in $\binom{n}{r}$ gives the exponent on $b$.

**EXAMPLE 5** ▶ **Expanding a Binomial Using the Binomial Theorem**

Expand $(a + b)^6$ using the binomial theorem.

Solution ▶ $(a+b)^6 = \binom{6}{0}a^6 b^0 + \binom{6}{1}a^5 b^1 + \binom{6}{2}a^4 b^2 + \binom{6}{3}a^3 b^3 + \binom{6}{4}a^2 b^4 + \binom{6}{5}a^1 b^5 + \binom{6}{6}a^0 b^6$

$= 1a^6 + 6a^5 b + \dfrac{6!}{2!4!}a^4 b^2 + \dfrac{6!}{3!3!}a^3 b^3 + \dfrac{6!}{4!2!}a^2 b^4 + 6ab^5 + 1b^6$

$= a^6 + 6a^5 b + 15a^4 b^2 + 20a^3 b^3 + 15a^2 b^4 + 6ab^5 + b^6$

**Now try Exercises 17 through 24** ▶

**EXAMPLE 6** ▶ **Using the Binomial Theorem to Find the Initial Terms of an Expansion**

Find the first three terms of $(2x + y^2)^{10}$.

**Solution** ▶ Use the binomial theorem with $a = 2x$, $b = y^2$, and $n = 10$.

$$(2x + y^2)^{10} = \binom{10}{0}(2x)^{10}(y^2)^0 + \binom{10}{1}(2x)^9(y^2)^1 + \binom{10}{2}(2x)^8(y^2)^2 + \cdots \quad \text{first three terms}$$

$$= (1)1024x^{10} + (10)512x^9y^2 + \frac{10!}{2!8!}256x^8y^4 + \cdots \quad \binom{10}{0} = 1, \binom{10}{1} = 10$$

$$= 1024x^{10} + 5120x^9y^2 + (45)256x^8y^4 + \cdots \quad \frac{10!}{2!8!} = 45$$

$$= 1024x^{10} + 5120x^9y^2 + 11{,}520x^8y^4 + \cdots \quad \text{result}$$

☑ **C.** You've just seen how we can use the binomial theorem to find $(a + b)^n$

**Now try Exercises 25 through 28** ▶

## D. Finding a Specific Term of the Binomial Expansion

In some applications of the binomial theorem, our main interest is a *specific term* of the expansion, rather than the expansion as a whole. To find a specified term, it helps to consider that the expansion of $(a + b)^n$ has $n + 1$ terms: $(a + b)^0$ has one term, $(a + b)^1$ has two terms, $(a + b)^2$ has three terms, and so on. Because the notation $\binom{n}{r}$ always begins at $r = 0$ for the first term, the value of $r$ will be *1 less than the term we are seeking*. In other words, for the seventh term of $(a + b)^9$, we use $r = 6$.

**The kth Term of a Binomial Expansion**

For the binomial expansion $(a + b)^n$, the $k$th term is given by

$$\binom{n}{r}a^{n-r}b^r, \text{ where } r = k - 1.$$

**EXAMPLE 7** ▶ **Finding a Specific Term of a Binomial Expansion**

Find the eighth term in the expansion of $(x + 2y)^{12}$.

**Solution** ▶ By comparing $(x + 2y)^{12}$ to $(a + b)^n$ we have $a = x$, $b = 2y$, and $n = 12$. Since we want the eighth term, $k = 8$ and $r = 7$. The eighth term of the expansion is

$$\binom{12}{7}x^5(2y)^7 = \frac{12!}{7!5!}128x^5y^7 \quad 2^7 = 128$$

$$= (792)(128x^5y^7) \quad \binom{12}{7} = 792$$

$$= 101{,}376x^5y^7 \quad \text{result}$$

☑ **D.** You've just seen how we can find a specific term of a binomial expansion

**Now try Exercises 29 through 34** ▶

## E. Applications

One application of the binomial theorem involves a **binomial experiment** and **binomial probability.** For binomial probabilities, the following must be true: (1) the experiment must have only two possible outcomes, typically called success and failure, and (2) if the experiment has $n$ trials, the probability of success must be constant for all $n$ trials. If the probability of success for each trial is $p$, the formula $\binom{n}{k}(1 - p)^{n-k}p^k$ gives the probability that exactly $k$ trials will be successful.

### Binomial Probability

Given a binomial experiment with $n$ trials, where the probability for success in each trial is $p$, the probability that exactly $k$ trials are successful is given by

$$\binom{n}{k}(1-p)^{n-k}p^{k}.$$

---

**EXAMPLE 8** ▶ **Applying the Binomial Theorem—Binomial Probability**

Tamika has a free-throw shooting average of 85%. On the last play of the game, with her team behind by three points, she is fouled at the three-point line, and is awarded two additional free throws via technical fouls on the opposing coach (for a total of five free-throws). What is the probability she makes *at least three* (meaning they at least tie the game)?

**Solution** ▶ Here we have $p = 0.85$, $1 - p = 0.15$, and $n = 5$. The key idea is to recognize the phrase *at least three* means "3 or 4 or 5." So $P(\text{at least } 3) = P(3 \cup 4 \cup 5)$.

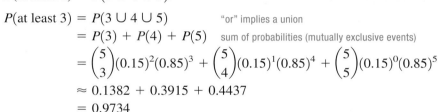

$$P(\text{at least } 3) = P(3 \cup 4 \cup 5) \qquad \text{"or" implies a union}$$
$$= P(3) + P(4) + P(5) \qquad \text{sum of probabilities (mutually exclusive events)}$$
$$= \binom{5}{3}(0.15)^2(0.85)^3 + \binom{5}{4}(0.15)^1(0.85)^4 + \binom{5}{5}(0.15)^0(0.85)^5$$
$$\approx 0.1382 + 0.3915 + 0.4437$$
$$= 0.9734$$

Tamika's team has an excellent chance ($\approx 97.3\%$) of at least tying the game.

 **E. You've just seen how we can solve applications of binomial powers**

**Now try Exercises 37 through 40** ▶

---

## 9.6 EXERCISES

▶ **CONCEPTS AND VOCABULARY**

**Fill in the blank with the appropriate word or phrase. Carefully reread the section, if necessary.**

**1.** In all terms in the expanded form of $(a + b)^n$, the exponents on $a$ and $b$ must sum to _____.

**2.** In a binomial experiment with $n$ trials each with probability $p$ of success, the probability of exactly $k$ successes is given by the formula _____.

**3.** Discuss why the expansion of $(a + b)^n$ has $n + 1$ terms.

**4.** For any defined binomial experiment, discuss the relationships between the phrases, "exactly $k$ successes," "more than $k$ successes," and "at least $k$ successes."

# Appendix I

# Geometry Review

**LEARNING OBJECTIVES**

*In Section A.I you will review how to:*

- ☐ **A.** Find the perimeter and area of common geometric figures
- ☐ **B.** Compute the volume of common geometric solids
- ☐ **C.** Apply similar triangles to find missing sides of triangles
- ☐ **D.** Solve applications using basic geometric properties

Developing the ability to use mathematics as a descriptive tool is a major goal of this text. Without a solid understanding of basic geometry, this goal would be difficult to achieve—as many of the tasks we perform daily are based on decisions regarding size, measurement, configuration, and the like.

## A. Perimeter and Area Formulas

Basic geometry plays an important role in the application of mathematics. For your convenience, the most common formulas are collected in Table A.I.1, and a focused effort should be made to commit them to memory. Note that some of the formulas use **subscripted variables,** or a variable with a small case number to the lower right ($s_1$, $s_2$, and so on). To help understand the table, we quickly review some fundamental terms and their meaning. A **plane** is the infinite extension of length and width along a flat surface. **Perimeter** is the distance around a two-dimensional figure, or a closed figure that lies in a plane. Many times these figures are **polygons,** or closed figures composed of line segments. The general name for a four-sided polygon is a **quadrilateral.** A **right angle** is an angle measuring 90°. A quadrilateral with four right angles is called a **rectangle.** **Area** is a measure of the amount of surface covered by a plane figure, with the measurement given in **square units.**

**Table A.I.1**

| | Definitions and Diagrams | Perimeter Formula (linear units or *units*) | Area Formula (square units or *units²*) |
|---|---|---|---|
| triangle | a three-sided polygon | $P = s_1 + s_2 + s_3$ | $A = \dfrac{1}{2}bh$ |
| rectangle | a quadrilateral with four right angles | $P = 2L + 2W$ | $A = LW$ |
| square | a rectangle with four equal sides | $P = 4s$ | $A = s^2$ |
| trapezoid | a quadrilateral with one pair of parallel sides (called bases $b_1$ and $b_2$) | $P = s_1 + s_2 + s_3 + s_4$ | $A = \dfrac{b_1 + b_2}{2} \cdot h$ |
| circle | the set of all points lying in a plane that are an equal distance (called the radius $r$) from a given point (called the center $C$) | $C = 2\pi r$ or $C = \pi d$ | $A = \pi r^2$ |

The formulas $C = \pi d$ and $C = 2\pi r$ both use the symbol "$\pi$," which represents the ratio of a circle's circumference to its diameter. We will use a two decimal approximation in calculations done by hand: $\pi \approx 3.14$. On many calculators, $\pi$ is the ⓝ function to the ⬤ key and produces a much better approximation (see Figure A.I.1). When using a calculator, we most often use all displayed digits and round only the final answer to the desired level of accuracy.

**Figure A.I.1**

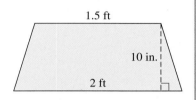

π | 3.141592654

If a problem or application uses a formula, begin by stating the formula rather than by immediately making any substitutions. This will help to prevent many careless errors. For Example 1, recall that a trapezoid is a quadrilateral with two parallel sides.

---

**EXAMPLE 1** ▶  **Finding the Area of a Window**

A basement window is shaped like an isosceles trapezoid (base angles equal, opposite sides equal in length), with a height of 10 in. and bases of 1.5 ft and 2 ft. What is the area of the glass in the window?

1.5 ft

10 in.

2 ft

**Solution** ▶  Before applying the area formula, all measures must use the same unit. In inches we have 1.5 ft = 18 in. and 2 ft = 24 in.

$$A = \frac{a + b}{2} \cdot h \qquad \text{given formula}$$

$$= \left(\frac{18 \text{ in.} + 24 \text{ in.}}{2}\right)(10 \text{ in.}) \qquad \text{substitute 10 in. for } h, \text{ 18 in. for } b_1 \text{ and 24 in. for } b_2$$

$$= \left(\frac{42 \text{ in.}}{2}\right)(10 \text{ in.}) \qquad \text{simplify}$$

$$= (21 \text{ in.})(10 \text{ in.})$$

$$= 210 \text{ in}^2 \qquad \text{result}$$

The area of the glass in the window is $210 \text{ in}^2$.

**WORTHY OF NOTE**

In actual practice, most calculations are done without using the units of measure, with the correct units supplied in the final answer. When like units do occur in an exercise, they are treated just as the numeric factors. If they are part of a product, we write the units with an appropriate exponent as in Example 1. If the like units occur in the numerator and denominator, they "cancel."

**Now try Exercises 1 through 12** ▶

## Composite Figures

The largest part of geometric applications, whether in art, construction, or architecture, involves **composite figures,** or figures that combine basic shapes. In many cases we are able to **partition** or break the figure into more common shapes using an **auxiliary line,** or a dashed line drawn to highlight certain features of the diagram. When computing a perimeter, we use only the exposed, outer edges, much as a soldier would guard the base camp by marching along the outer edge—the perimeter. For composite figures, it's helpful to verbally describe the situation given, creating a verbal model that can easily be translated into an equation model.

**EXAMPLE 2** ▶  **Determining the Perimeter and Area of a Composite Figure**

Find the perimeter and area of the composite Figure A.I.2. Use $\pi \approx 3.14$.

**Figure A.I.2**

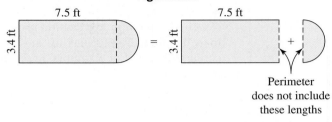

7.5 ft

3.4 ft

Solution ▶  To compute the perimeter, use only the exposed (outer) edges as shown in Figure A.I.3.

**Figure A.I.3**

7.5 ft

3.4 ft

=

7.5 ft

3.4 ft

+

Perimeter does not include these lengths

- Perimeter = three sides of rectangle + one-half circle    verbal model

$$P = 2L + W + \frac{\pi d}{2}$$    formula model

$$\approx 2(7.5) + 3.4 + \frac{(3.14)(3.4)}{2}$$    substitute 7.5 for L, 3.4 for W, 3.4 for d, and 3.14 for $\pi$

$$= 15 + 3.4 + 5.338$$    simplify

$$= 23.738$$    result

The perimeter of the figure is about 23.7 ft.

- Total Area = area of rectangle + one-half area of circle    verbal model

$$A = LW + \frac{\pi r^2}{2}$$    formula model

$$\approx (7.5)(3.4) + \frac{(3.14)(1.7)^2}{2}$$    substitute 1.7 for r (d/2)

$$= 25.5 + 4.5373$$    simplify

$$= 30.0373$$    result

The area of the figure is about 30.0 ft².

☑ **A.** You've just reviewed how to find the perimeter and area of common geometric figures

**Now try Exercises 13 through 24** ▶

## B.  Volume Formulas

**Volume** is a measure of the amount of space occupied by a three-dimensional object and is measured in **cubic units.** Some of the more common formulas are given in Table A.I.2.

**Table A.I.2**

| | Definitions and Diagrams | | Volume Formula (cubic units or units$^3$) |
|---|---|---|---|
| rectangular solid | a six-sided solid figure with opposite faces congruent and adjacent faces meeting at right angles 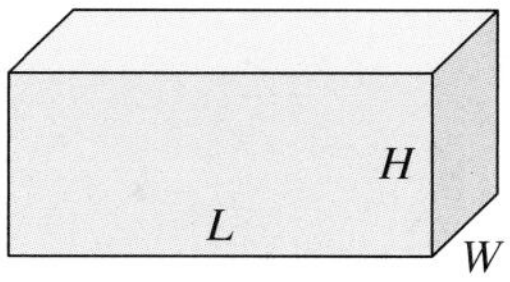 | | $V = LWH$ |
| cube | a rectangular solid with six congruent, square faces 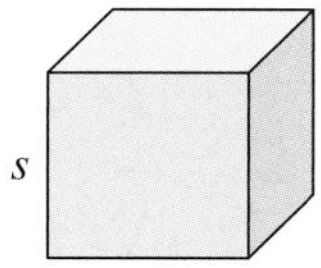 | | $V = s^3$ |
| sphere | the set of all points an equal distance (called the radius) from a given point (called the center) 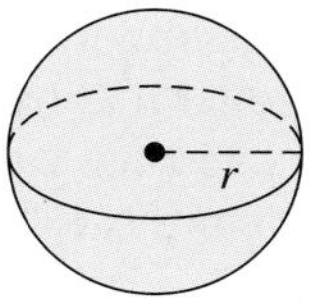 | | $V = \dfrac{4}{3}\pi r^3$ |
| right circular cylinder | union of all line segments connecting two congruent circles in parallel planes, meeting each at a right angle 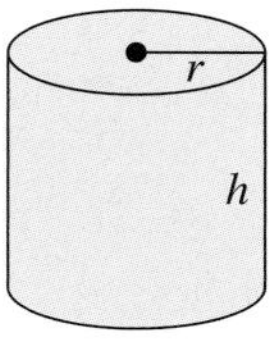 | | $V = \pi r^2 h$ |
| right circular cone | union of all line segments connecting a given point (vertex) to a given circle (base) and whose altitude meets the center of the base at a right angle 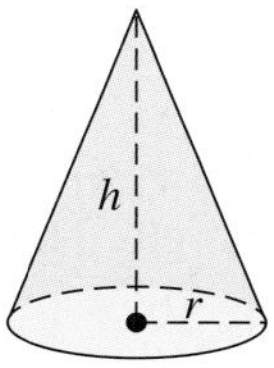 | | $V = \dfrac{1}{3}\pi r^2 h$ |
| right square pyramid | union of all line segments connecting a given point (vertex) to a given square (base) and whose altitude meets the center of the base at a right angle 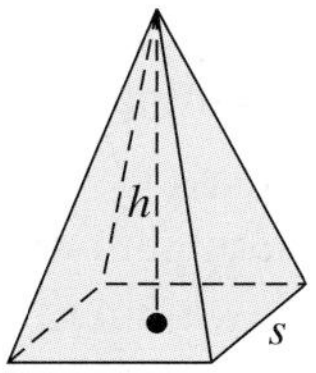 | | $V = \dfrac{1}{3}s^2 h$ |

**EXAMPLE 3** ▶ **Determining the Volume of a Composite Figure**

Sand is being loaded onto light trucks for delivery to a glass manufacturer. As the sand overflows the truck bed, which is 3 ft deep, it forms a right square pyramid (see figure). How many cubic feet of sand are there at the moment the pyramid is 4 ft high, if the truck bed is 6 ft by 6 ft?

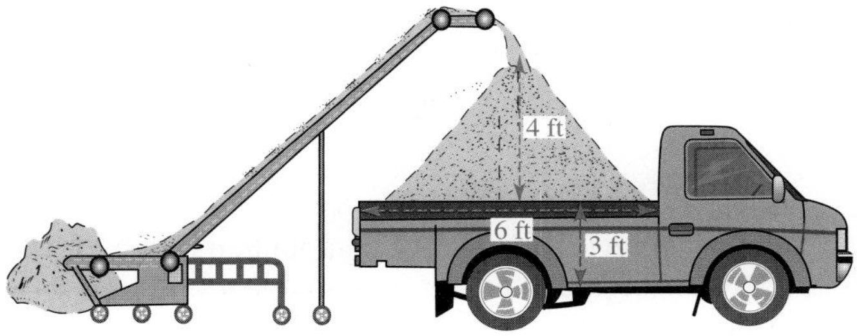

**WORTHY OF NOTE**

It is again worth noting that units of measure are treated as though they were numeric factors. For the volume of the truck bed in Example 3: (6 ft)(6 ft)(3 ft) = 108 ft$^3$. This concept is an important part of the unit conversions often used in the application of mathematics.

**Solution** ▶   Total Volume = volume of truck + volume of pyramid   verbal model

$$V = LWH + \frac{1}{3}s^2h$$   formula model

$$= (6)(6)(3) + \frac{1}{3}(6)^2(4)$$   substitute 6 for *L*, 6 for *W*, 3 for *H*; 6 for *s* and 4 for *h*

$$= 108 + 48$$   simplify

$$= 156$$   result

There are about 156 ft³ of sand in the load.

☑ **B.** You've just reviewed how to compute the volume of common geometric figures

Now try Exercises 25 and 26 ▶

## C.  Similar Triangles

Another important geometric relationship is that of **similar triangles.** Two triangles are similar if corresponding angles are equal. The angles are usually named with capital letters and the side opposite each angle is named using the related lowercase letter (see Figure A.I.4). Similar triangles have the following useful properties:

**Figure A.I.4**

Similar Triangles

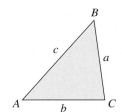
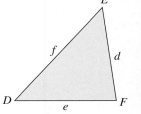

Given △*ABC* and △*DEF* as shown.
If ∠*A* = ∠*D*, ∠*B* = ∠*E* and ∠*C* = ∠*F*,
then △*ABC* and △*DEF* are *similar triangles* and
corresponding sides are in proportion.

$$\frac{a}{d} = \frac{b}{e} \qquad \frac{b}{e} = \frac{c}{f} \qquad \frac{c}{f} = \frac{a}{d}$$

The phrase *corresponding sides* means sides that are in the same relative position in each triangle. This property enables us to find the length of a missing side by setting up and solving a proportion. One important application of similar triangles involves ramps or other inclined planes.

**EXAMPLE 4** ▶   **Determining the Height of a Mountain**

A cyclist's GPS cyclometer shows an elevation gain of 79.2 ft for the first $\frac{1}{4}$ mi of a mountain ascent. If the road grade remains constant over the 15-mi climb, how tall is the mountain?

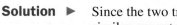

**Solution** ▶   Since the two triangles in question are similar, we set up and solve a proportion. Let *h* represent the height of the mountain (see figure).

$$\frac{\text{elevation gain so far}}{\text{distance ridden so far}} = \frac{\text{total elevation gain}}{\text{total distance ridden}}$$   corresponding sides are in proportion

$$\frac{79.2}{0.25} = \frac{h}{15}$$   substitute known values

$$1188 = 0.25h$$   clear denominators

$$4752 = h$$   solve for *h*

☑ **C.** You've just reviewed how to apply similar triangles to find missing sides of triangles

The mountain is 4752 ft tall.

Now try Exercises 29 and 30 ▶

## D. Applications of Basic Geometric Properties

Cubes and rectangular boxes are part of a larger family of solids called **prisms**. A prism is any **solid figure** with bases of the same size and shape (the top and bottom are both called **bases**). A **right prism** has sides which are perpendicular to the bases, with all cross sections congruent. The volume of a right prism is the area of its base times its height. This is easily seen in the case of a rectangular box: $V = LWH = (LW)H = (\text{area of base}) \cdot H$.

### Volume of a Right Prism

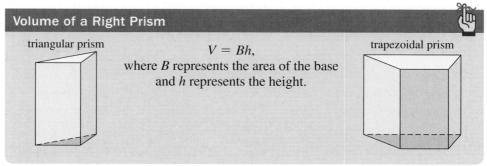

triangular prism

$V = Bh,$
where $B$ represents the area of the base
and $h$ represents the height.

trapezoidal prism

---

**EXAMPLE 5** ▶ **Determining the Volume of a Right Prism**

The feeding trough at a large cattle ranch is a right prism with trapezoidal bases (see figure). The trough is 1.4 ft deep and 38 ft long. If the bases of the trapezoid are 2 ft and 3 ft, how many loads of feed are needed to fill the trough if each load is 54 ft³?

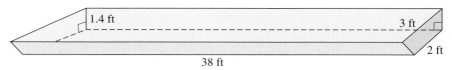

1.4 ft                                         3 ft
                                               2 ft
38 ft

**Solution** ▶ Notice the trough is a right trapezoidal prism.

**a.** Find the area of the trapezoidal bases:

$$A = \frac{b_1 + b_2}{2} \cdot h \qquad \text{area of a trapezoid}$$

$$= \left(\frac{3 + 2}{2}\right)(1.4) \qquad \text{substitute 1.4 for } h, \text{3 for } b_1, \text{and 2 for } b_2$$

$$= 3.5 \text{ ft}^2 \qquad \text{result}$$

The area of each base is 3.5 ft².

**b.** Compute the volume of the trough (the right prism).

$$V = Bh \qquad \text{volume = area} \cdot \text{height}$$

$$= (3.5)(38) \qquad \text{substitute 3.5 for } B, \text{38 for } h$$

$$= 133 \text{ ft}^3 \qquad \text{result}$$

The volume of the trough is 133 ft³.

**c.** Find how many 54-ft³ loads are needed for 133 ft³ (divide).

$$\text{loads} = \frac{133 \text{ ft}^3}{54 \text{ ft}^3}$$

$$\approx 2.463$$

Approximately 2.5 loads of feed are needed to fill the trough.

☑ **D.** You've just reviewed how to solve applications involving basic geometric properties

Now try Exercises 31 and 32 ▶

Various other applications of geometry can be found in **Exercises 33 through 46.**

## A.I EXERCISES

▶ **DEVELOPING YOUR SKILLS**

**Compute the area of each trapezoid using the dimensions given.**

**1.**

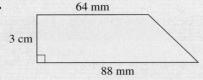

64 mm

3 cm

88 mm

**2.**

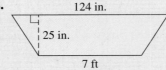

124 in.

25 in.

7 ft

**Use a ruler to estimate the area of each trapezoid using the actual measurements.**

**3.** In square millimeters,

**4.** In square inches,

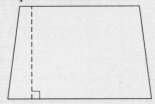

**Graph paper can be an excellent aid in understanding why perimeter must be measured in linear units, while area is measured in square units. Suppose the graph paper shown has squares that are 1 cm by 1 cm. Use the grid to answer each question by counting and by computation.**

**5.** What are the perimeter and area of the rectangle shown?

**6.** What are the perimeter and area of the triangle shown?

**Exercises 5 and 6**

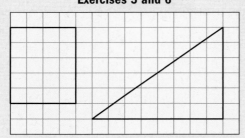

**7.** What are the perimeter and area of the trapezoid shown?

**8.** What are the perimeter and area of the triangle shown?

**Exercises 7 and 8**

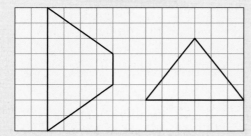

**9.** What is the outer circumference of a circular gear with a radius (center to teeth) of 2.5 cm? Round to tenths.

**10.** If the working depth of the gear (height of gear teeth) from Exercise 9 is 0.3 cm, what is the inner area of the gear?

**11.** Find the missing length, then compute the area and perimeter.

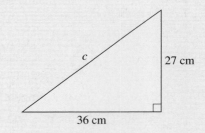

$c$

27 cm

36 cm

**12.** Find the missing length, then compute the area and perimeter.

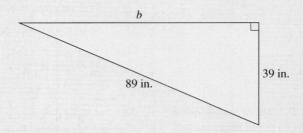

$b$

39 in.

89 in.

**In Exercises 13–20, a composite figure is shown. Construct an appropriate formula and use it to find the perimeter, area, or volume as indicated.**

**13.** Perimeter

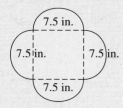

7.5 in.

7.5 in.      7.5 in.

7.5 in.

**14.** Perimeter

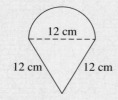

12 cm

12 cm      12 cm

**15.** Area

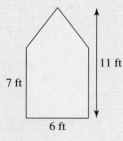

11 ft

7 ft

6 ft

**16.** Area

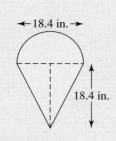

←18.4 in.→

18.4 in.

**17.** Area

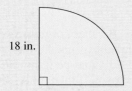

18 in.

**18.** Area

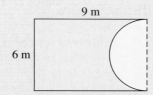

9 m

6 m

**19.** Determine the outer perimeter and total area of the track shown.

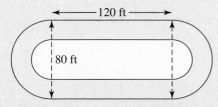

←——120 ft——→

80 ft

**20.** Find the perimeter and area of the skirt pattern shown.

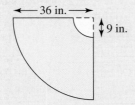

←— 36 in. —→

9 in.

**21.** How many square inches of paper are needed to cover the kite?

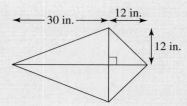

←— 30 in. —→   12 in.

12 in.

**22.** Find the area of the mainsail and the area of the jib sail of the sailboat.

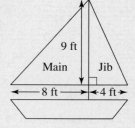

9 ft

Main      Jib

←—— 8 ft ——→←4 ft→

**23.** Find the length of the missing side.

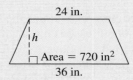

24 cm

Perimeter = 72 cm      W

**24.** Find the height of the trapezoid.

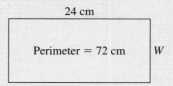

24 in.

h

Area = 720 in²

36 in.

**25.** Find the composite volume.

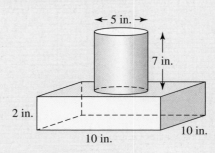

←— 5 in. →

7 in.

2 in.

10 in.

10 in.

**26.** Find the composite volume.

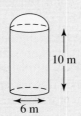

10 m

6 m

**27. Matching:** Place the correct letter in the corresponding blank without using any reference material.

**A.** perimeter of a square          **B.** perimeter of a rectangle          **C.** perimeter of a triangle

**D.** area of a triangle          **E.** circumference of a circle          **F.** volume of a cube

**G.** area of a square          **H.** perimeter of a trapezoid          **I.** volume of a rectangular solid

**J.** area of a circle          **K.** area of a rectangle          **L.** area of a trapezoid

_____ $LW$                           _____ $s_1 + s_2 + s_3$                           _____ $\pi r^2$

_____ $4s$                             _____ $LWH$                              _____ $\dfrac{1}{2}bh$

_____ $s_1 + s_2 + s_3 + s_4$          _____ $2\pi r$                         _____ $s^2$

_____ $2L + 2W$                     _____ $s^3$                           _____ $A = \dfrac{b_1 + b_2}{2} \cdot h$

**28. Surface Area of a Rectangular Box:** $SA = 2(LW + LH + WH)$ The surface area $SA$ of a rectangular box is found by summing the areas of all six sides. Find the surface area of the top and bottom if the box is 15 in. tall, 8 in. wide, and has a surface area of 378 in$^2$.

**Exercise 28**

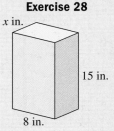

$x$ in.

15 in.

8 in.

## ▶ APPLICATIONS (ROUND TO TENTHS AS NEEDED)

**29. Similar triangles:** To estimate the height of a flagpole, Mitchell reasons his shadow must be proportional to the shadow of the flagpole due to similar triangles. Mitchell is 4 feet tall and his shadow measures 7.5 ft. The shadow of the flagpole measures 60 ft. How high is the pole?

**Exercise 29**

**30. Similar triangles:** A triangular image measuring 8 in. × 9 in. × 10 in. is projected on a screen using an overhead projector. If the smallest side of the projected image is 2 ft, what are the other dimensions of the projected image?

**31. Volume of a prism:** Using a sheet of canvas that is 10 ft by 16 ft, a simple tent is made using a long pole through the middle and pegging the sides into the ground as shown. If the tent is 4 ft high at the apex, what is the volume of space contained in the tent?

**Exercise 31**

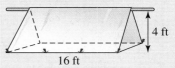

4 ft

16 ft

**32. Volume of a prism:** The feeding trough at a pig farm is a right prism with bases that are isosceles triangles (two equal sides). The trough is 3 ft wide, 2 ft deep, and 12 ft long. What is its volume? If each load is 7 ft$^3$, how many loads of slop are needed to fill the trough?

**Exercise 32**

3 ft

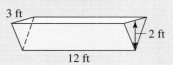

2 ft

12 ft

**33. Length of a cable:** A cellular relay tower is secured by cables which are clamped 21.5 m up the tower and anchored in the ground 9 m from its base. If 30 cm lengths are needed to secure the cable at each end, how long are the cables? Round to hundredths of a meter.

**Exercise 33**

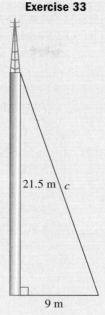

21.5 m $\ c$

9 m

**34. Wind power:** Seeking to reduce energy consumption, many shipping companies are harnessing wind power using specialized kites. If such a kite is flying 150 m ahead of the ship and 216 m of cable has been let out, what is the height $h$ of the kite's bridle point $B$ (see illustration)?

**Exercise 34**

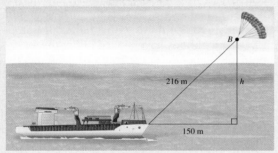

$B$

216 m

$h$

150 m

**35. Most economical purchase:** Missy's Famous Pizza Emporium is running a special—one large for $8.99 or two mediums for $13.49. If a large pizza has a 14-in. diameter and a medium pizza has a 12-in. diameter, which is the better buy (least expensive per square inch)?

**36. Most economical purchase:** A large can of peaches costs 79¢ and is 15 cm high with a radius of 6 cm. A small can of peaches costs 39¢ and is 15 cm high with a radius of 4 cm. Which is the better buy (least expensive per cubic cm)?

**37. Volume of a jacuzzi:** A certain Jacuzzi tub is 84 in. long, 60 in. wide, and 30 in. deep. How many gallons of water will it take to fill this tub? (*Hint:* One cubic foot is approximately equal to 7.48 gal.)

**38. Volume of a birdbath:** I am trying to fill an outdoor birdbath, and all I can find is a plastic dish-pan 7 in. high, with a 12 in. by 15 in. base. If the birdbath holds 15 gal, how many trips will have to be made?

**Exercise 38**

**39. Paving a walkway:** Current plans call for building a circular fountain 6 m in diameter, surrounded by a circular walkway 1.5 m wide. (a) What is the approximate area of the walkway? (b) If the concrete for the walkway is 6 cm deep, what volume of cement must be used? (c) If the cement costs $125/m³ with a 7% sales tax, what is the cost of the materials for this walkway?

**Exercise 39**

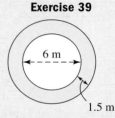

6 m

1.5 m

**40. Paving a driveway:** A driveway and turn-about has the dimensions shown in the figure. (a) What is the area of the driveway? (b) If the concrete was poured to a depth of 4 in., what volume of cement was used to construct the driveway? (c) If the cement cost $3.50 ft³ and a 5.65% tax was paid, what was the total cost of the materials used to complete the driveway?

**Exercise 40**

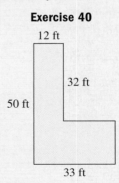

12 ft

32 ft

50 ft

33 ft

**41. Cost of drywall:** After the studs are up, the wall shown in the figure must be covered in drywall. (a) How many square feet of drywall are needed? (b) If drywall is sold only in 4 ft-by-8 ft sheets, about how many sheets are required? (c) If drywall costs $8.25 per sheet and a 5.75% tax must be paid, what is the total cost of the material?

**Exercise 41**

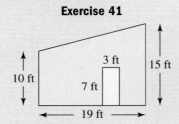

**42. Cost of baseboards:** The dimensions for the living room/dining room of a home are shown. (a) How many feet and inches of molding are needed for the baseboards around the perimeter of the room? (b) If the molding is only sold in 8-ft lengths, how many are needed? (c) If the molding costs $1.74 per foot and sales tax is 5.75%, what is the total cost of the baseboards?

**Exercise 42**

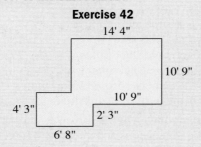

**43. Dimensions of an index card:** A popular-sized index card has a length that is 1 in. less than twice its width. If the card has an area of 15 in$^2$, find the length and width.

**44. Dimensions of a ruler:** The plastic ruler that Albert uses for graphing lines has a length that is 1 cm more than seven times its width. If the ruler has an area of 30 cm$^2$, find its length and width.

**45. Tracking an oil leak:** At an oil storage facility, one of the tanks has a leaky nozzle. If the tank shown was full to begin with, how many gallons have been lost at the moment the height of the oil in the tank is 24 ft?

**Exercise 45**

**46. Tracking a water leak:** A cylindrical water tank has developed a slow leak. If the tank is standing vertically and is 4 ft tall with a radius of 9 in., how many gallons have leaked out at the moment the height of the water in the tank is 3 ft?

# Appendix II

# More on Matrices

## Reduced Row-Echelon Form

A matrix is in reduced row-echelon form if it satisfies the following conditions:

**1.** All null rows (zeroes for all entries) occur at the bottom of the matrix.
**2.** The first nonzero entry of any row must be a 1.
**3.** For any two consecutive, nonzero rows, the leading 1 in the higher row is to the left of the leading 1 in the lower row.
**4.** Every column with a leading 1 has zeroes for all other entries in the column.

Matrices A through D are in reduced row-echelon form.

$$A = \begin{bmatrix} 0 & 1 & 0 & 0 & 0 & 5 \\ 0 & 0 & 0 & 1 & 0 & 3 \\ 0 & 0 & 0 & 0 & 1 & 2 \end{bmatrix} \quad B = \begin{bmatrix} 1 & 0 & 0 & 5 \\ 0 & 1 & 0 & 3 \\ 0 & 0 & 0 & 0 \end{bmatrix} \quad C = \begin{bmatrix} 1 & 0 & 0 & 5 \\ 0 & 1 & 3 & -2 \\ 0 & 0 & 0 & 0 \end{bmatrix} \quad D = \begin{bmatrix} 1 & 0 & 5 & 0 \\ 0 & 1 & 2 & 0 \\ 0 & 0 & 0 & 1 \end{bmatrix}$$

Where *Gaussian elimination* places a matrix in *row-echelon form* (satisfying the first three conditions), *Gauss-Jordan elimination* places a matrix in *reduced row-echelon form*. To obtain this form, continue applying row operations to the matrix until the fourth condition above is also satisfied. For a $3 \times 3$ system having a unique solution, the diagonal entries of the coefficient matrix will be 1's, with 0's for all other entries. To illustrate, we'll extend Example 4 from Section 7.1 until reduced row-echelon form is obtained.

---

**EXAMPLE 4** ▶ **Solving Systems Using the Augmented Matrix and Gauss-Jordan Elimination**

Solve using Gauss-Jordan elimination: $\begin{cases} 2x + y - 2z = -7 \\ x + y + z = -1 \\ -2y - z = -3 \end{cases}$

**Solution** ▶

$\begin{cases} 2x + y - 2z = -7 \\ x + y + z = -1 \\ -2y - z = -3 \end{cases}$ matrix form → $\begin{bmatrix} 2 & 1 & -2 & -7 \\ 1 & 1 & 1 & -1 \\ 0 & -2 & -1 & -3 \end{bmatrix}$ R1 ↔ R2 $\begin{bmatrix} 1 & 1 & 1 & -1 \\ 2 & 1 & -2 & -7 \\ 0 & -2 & -1 & -3 \end{bmatrix}$

$\begin{bmatrix} 1 & 1 & 1 & -1 \\ 2 & 1 & -2 & -7 \\ 0 & -2 & -1 & -3 \end{bmatrix}$ $-2R1 + R2 \rightarrow R2$ $\begin{bmatrix} 1 & 1 & 1 & -1 \\ 0 & -1 & -4 & -5 \\ 0 & -2 & -1 & -3 \end{bmatrix}$ $-1R2 \rightarrow R2$ $\begin{bmatrix} 1 & 1 & 1 & -1 \\ 0 & 1 & 4 & 5 \\ 0 & -2 & -1 & -3 \end{bmatrix}$

$\begin{bmatrix} 1 & 1 & 1 & -1 \\ 0 & 1 & 4 & 5 \\ 0 & -2 & -1 & -3 \end{bmatrix}$ $2R2 + R3 \rightarrow R3$ $\begin{bmatrix} 1 & 1 & 1 & -1 \\ 0 & 1 & 4 & 5 \\ 0 & 0 & 7 & 7 \end{bmatrix}$ $\dfrac{R3}{7} \rightarrow R3$ $\begin{bmatrix} 1 & 1 & 1 & -1 \\ 0 & 1 & 4 & 5 \\ 0 & 0 & 1 & 1 \end{bmatrix}$

$\begin{bmatrix} 1 & 1 & 1 & -1 \\ 0 & 1 & 4 & 5 \\ 0 & 0 & 1 & 1 \end{bmatrix}$ $-R2 + R1 \rightarrow R1$ $\begin{bmatrix} 1 & 0 & -3 & -6 \\ 0 & 1 & 4 & 5 \\ 0 & 0 & 1 & 1 \end{bmatrix}$ $\begin{matrix} 3R3 + R1 \rightarrow R1 \\ -4R3 + R2 \rightarrow R2 \end{matrix}$ $\begin{bmatrix} 1 & 0 & 0 & -3 \\ 0 & 1 & 0 & 1 \\ 0 & 0 & 1 & 1 \end{bmatrix}$

The final matrix is in reduced row-echelon form with solution $(-3, 1, 1)$ just as in Section 7.1.

---

## The Determinant of a General Matrix

To compute the determinant of a general square matrix, we introduce the idea of a **cofactor.** For an $n \times n$ matrix $A$, $A_{ij} = (-1)^{i+j}|M_{ij}|$ is the cofactor of matrix element $a_{ij}$, where $|M_{ij}|$ represents the determinant of the corresponding minor matrix. Note that $i + j$ is the sum of the row and column of the entry, and if this sum is even, $(-1)^{i+j} = 1$, while if the sum is odd, $(-1)^{i+j} = -1$. (This is how the sign table for a $3 \times 3$ determinant was generated.) To compute the determinant of an $n \times n$ matrix, multiply each element in any row or column by its cofactor and add. The result is a tier-like process in which the determinant of a larger matrix requires computing the determinant of smaller matrices. In the case of a $4 \times 4$ matrix, each of the minor matrices will be size $3 \times 3$, whose determinant then requires the computation of other $2 \times 2$ determinants. In the following illustration, two of the entries in the first row are zero for convenience. For

$$A = \begin{bmatrix} -2 & 0 & 3 & 0 \\ 1 & 2 & 0 & -2 \\ 3 & -1 & 4 & 1 \\ 0 & -3 & 2 & 1 \end{bmatrix},$$

$$\text{we have } \det(A) = -2 \cdot (-1)^{1+1} \begin{vmatrix} 2 & 0 & -2 \\ -1 & 4 & 1 \\ -3 & 2 & 1 \end{vmatrix} + (3) \cdot (-1)^{1+3} \begin{vmatrix} 1 & 2 & -2 \\ 3 & -1 & 1 \\ 0 & -3 & 1 \end{vmatrix}$$

Computing the first $3 \times 3$ determinant gives $-16$; the second $3 \times 3$ determinant is 14. This gives

$$\det(A) = -2(-16) + 3(14)$$
$$= 74$$

# Appendix III

# Deriving the Equation of a Conic

## The Equation of an Ellipse

In Section 8.2, the equation $\sqrt{(x + c)^2 + y^2} + \sqrt{(x - c)^2 + y^2} = 2a$ was developed using the distance formula and the definition of an ellipse. To find the standard form of the equation, we treat this result as a radical equation, isolating one of the radicals and squaring both sides.

$$\sqrt{(x + c)^2 + y^2} = 2a - \sqrt{(x - c)^2 + y^2} \qquad \text{isolate one radical}$$
$$(x + c)^2 + y^2 = 4a^2 - 4a\sqrt{(x - c)^2 + y^2} + (x - c)^2 + y^2 \qquad \text{square both sides}$$

We continue by simplifying the equation, isolating the remaining radical, and squaring again.

$$x^2 + 2cx + c^2 + y^2 = 4a^2 - 4a\sqrt{(x - c)^2 + y^2} + x^2 - 2cx + c^2 + y^2 \qquad \text{expand binomials}$$
$$4cx = 4a^2 - 4a\sqrt{(x - c)^2 + y^2} \qquad \text{simplify}$$
$$a\sqrt{(x - c)^2 + y^2} = a^2 - cx \qquad \text{isolate radical; divide by 4}$$
$$a^2[(x - c)^2 + y^2] = a^4 - 2a^2cx + c^2x^2 \qquad \text{square both sides}$$
$$a^2x^2 - 2a^2cx + a^2c^2 + a^2y^2 = a^4 - 2a^2cx + c^2x^2 \qquad \text{expand and distribute } a^2 \text{ on left}$$
$$a^2x^2 - c^2x^2 + a^2y^2 = a^4 - a^2c^2 \qquad \text{add } 2a^2cx \text{ and rewrite equation}$$
$$x^2(a^2 - c^2) + a^2y^2 = a^2(a^2 - c^2) \qquad \text{factor}$$
$$\frac{x^2}{a^2} + \frac{y^2}{a^2 - c^2} = 1 \qquad \text{divide by } a^2(a^2 - c^2)$$

Since $a > c$, we know $a^2 > c^2$ and $a^2 - c^2 > 0$. For convenience, let $b^2 = a^2 - c^2$ (it also follows that $a^2 > b^2$ and $a > b$, since $c > 0$). Substituting $b^2$ for $a^2 - c^2$ we obtain the standard form of the equation of an ellipse (major axis horizontal, since we stipulated $a > b$): $\dfrac{x^2}{a^2} + \dfrac{y^2}{b^2} = 1$. Note once again the $x$-intercepts are $(\pm a, 0)$, while the $y$-intercepts are $(0, \pm b)$. For the foci, $c^2 = |a^2 - b^2|$ to allow for a major axis that may be vertical.

## The Equation of a Hyperbola

Similar to the development of the equation of an ellipse, the equation $\sqrt{(x + c)^2 + y^2} - \sqrt{(x - c)^2 + y^2} = 2a$ can be developed using the distance formula and the definition of a hyperbola. To find the standard form of this equation, we apply the same procedures as before.

$$\sqrt{(x + c)^2 + y^2} = 2a + \sqrt{(x - c)^2 + y^2} \qquad \text{isolate one radical}$$

$$(x + c)^2 + y^2 = 4a^2 + 4a\sqrt{(x - c)^2 + y^2} + (x - c)^2 + y^2 \qquad \text{square both sides}$$

$$x^2 + 2cx + c^2 + y^2 = 4a^2 + 4a\sqrt{(x - c)^2 + y^2} + x^2 - 2cx + c^2 + y^2 \qquad \text{expand binomials}$$

$$4cx = 4a^2 + 4a\sqrt{(x - c)^2 + y^2} \qquad \text{simplify}$$

$$cx - a^2 = a\sqrt{(x - c)^2 + y^2} \qquad \text{isolate radical; divide by 4}$$

$$c^2x^2 - 2a^2cx + a^4 = a^2[(x - c)^2 + y^2] \qquad \text{square both sides}$$

$$c^2x^2 - 2a^2cx + a^4 = a^2x^2 - 2a^2cx + a^2c^2 + a^2y^2 \qquad \text{expand and distribute } a^2 \text{ on the right}$$

$$c^2x^2 - a^2x^2 - a^2y^2 = a^2c^2 - a^4 \qquad \text{add } 2a^2cx \text{ and rewrite equation}$$

$$x^2(c^2 - a^2) - a^2y^2 = a^2(c^2 - a^2) \qquad \text{factor}$$

$$\frac{x^2}{a^2} - \frac{y^2}{c^2 - a^2} = 1 \qquad \text{divide by } a^2(c^2 - a^2)$$

From the definition of a hyperbola we have $0 < a < c$, showing $c^2 > a^2$ and $c^2 - a^2 > 0$. For convenience, let $b^2 = c^2 - a^2$ and substitute to obtain the standard form of the equation of a hyperbola (transverse axis horizontal): $\dfrac{x^2}{a^2} - \dfrac{y^2}{b^2} = 1$. Note the $x$-intercepts are $(0, \pm a)$ and there are no $y$-intercepts. For the foci, $c^2 = a^2 + b^2$.

## The Asymptotes of a Central Hyperbola

From our work in Section 8.3, a central hyperbola with a horizontal axis will have asymptotes at $y = \pm \dfrac{b}{a}x$. To understand why, recall that for asymptotic behavior we investigate what happens to the relation for large values of $x$, meaning as $|x| \to \infty$. Starting with $\dfrac{x^2}{a^2} - \dfrac{y^2}{b^2} = 1$, we have

$$b^2x^2 - a^2y^2 = a^2b^2 \qquad \text{clear denominators}$$

$$a^2y^2 = b^2x^2 - a^2b^2 \qquad \text{isolate term with } y$$

$$a^2y^2 = b^2x^2\left(1 - \frac{a^2}{x^2}\right) \qquad \text{factor out } b^2x^2 \text{ from right side}$$

$$y^2 = \frac{b^2}{a^2}x^2\left(1 - \frac{a^2}{x^2}\right) \qquad \text{divide by } a^2$$

$$y = \pm\frac{b}{a}x\sqrt{1 - \frac{a^2}{x^2}} \qquad \text{take square roots of both sides}$$

As $|x| \to \infty$, $\dfrac{a^2}{x^2} \to 0$, and we find that for large values of $x$, $y \approx \pm\dfrac{b}{a}x$.

# Appendix IV

# Proof Positive—A Selection of Proofs from College Algebra

## Proof from Chapter 2

### Perpendicularity

*If two lines are perpendicular, their slopes have a product of $-1$.*

### Proof of Perpendicularity

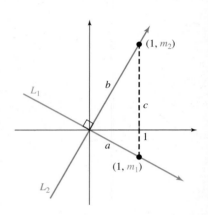

Assume that lines $L_1$ and $L_2$, with slopes $m_1$ and $m_2$ respectively, intersect in the plane at right angles. If we consider the point of intersection to be $(0, 0)$, the lines have equations of $y = m_1 x$ and $y = m_2 x$, with the point $(1, m_1)$ on $L_1$ and the point $(1, m_2)$ on $L_2$ (see figure). We now consider the right triangle formed by sides $a$, $b$, and $c$ shown. Using the distance formula with the points $(0, 0)$ and $(1, m_1)$, we have $a^2 = 1^2 + m_1^2$. Likewise, $b^2 = 1^2 + m_2^2$ and $c^2 = (1 - 1)^2 + (m_1 - m_2)^2$. From $a^2 + b^2 = c^2$, we have

$$(1^2 + m_1^2) + (1^2 + m_2^2) = (1 - 1)^2 + (m_1 - m_2)^2 \qquad \text{Pythagorean theorem}$$
$$2 + m_1^2 + m_2^2 = m_1^2 - 2m_1 m_2 + m_2^2 \qquad \text{combine terms, square binomial}$$
$$2 = -2m_1 m_2 \qquad \text{subtract } m_1^2, m_2^2$$
$$-1 = m_1 m_2 \ \checkmark \qquad \text{divide by } -2$$

Note that $m_1 m_2 = -1$ also implies that $m_1 = -\dfrac{1}{m_2}$, showing their slopes must be negative reciprocals.

## Proofs from Chapter 4

### The Remainder Theorem

*If a polynomial $p(x)$ is divided by $x - c$ using synthetic division, the remainder is equal to $p(c)$.*

### Proof of the Remainder Theorem

From our previous work, any number $c$ used in synthetic division will occur as the factor $x - c$ when written as (quotient)(divisor) + remainder: $p(x) = (x - c)q(x) + r$. Here, $q(x)$ represents the quotient polynomial and $r$ is a constant. Evaluating $p(c)$ gives

$$p(x) = (x - c)q(x) + r$$
$$p(c) = (c - c)q(c) + r$$
$$= 0 \cdot q(c) + r$$
$$= r \ \checkmark$$

### The Factor Theorem

*Given a polynomial $p(x)$,*
*(1) if $p(c) = 0$, then $x - c$ is a factor of $p(x)$, and*
*(2) if $x - c$ is a factor of $p(x)$, then $p(c) = 0$.*

**Proof of the Factor Theorem**

1. Consider a polynomial $p$ written in the form $p(x) = (x - c)q(x) + r$. From the remainder theorem we know $p(c) = r$, and substituting $p(c)$ for $r$ in the equation shown gives

$$p(x) = (x - c)q(x) + p(c)$$

If $p(c) = 0$, then $x - c$ is a factor of $p(x)$:

$$p(x) = (x - c)q(x) \checkmark$$

2. The steps from part 1 can be reversed, since any factor $x - c$ of $p(x)$, can be written in the form $p(x) = (x - c)q(x)$. Evaluating at $x = c$ produces a result of zero:

$$p(c) = (c - c)q(x)$$
$$= 0 \checkmark$$

## Complex Conjugates Theorem

*Given $p(x)$ is a polynomial with real coefficients, complex solutions must occur in conjugate pairs. If $a + bi$ is a solution, then $a - bi$ must also be a solution.*

To prove this for polynomials of degree $n > 2$, we let $z_1 = a + bi$ and $z_2 = c + di$ be complex numbers. Then $\bar{z}_1 = a - bi$, and $\bar{z}_2 = c - di$ represent their conjugates, and we can observe the following properties:

1. The conjugate of a sum is equal to the sum of the conjugates.

| sum: $z_1 + z_2$ | | sum of conjugates: $\bar{z}_1 + \bar{z}_2$ |
|---|---|---|
| $(a + bi) + (c + di)$ | | $(a - bi) + (c - di)$ |
| $(a + c) + (b + d)i$ | $\rightarrow$ conjugate of sum $\rightarrow$ | $(a + c) - (b + d)i \checkmark$ |

2. The conjugate of a product is equal to the product of the conjugates.

| product: $z_1 \cdot z_2$ | | product of conjugates: $\bar{z}_1 \cdot \bar{z}_2$ |
|---|---|---|
| $(a + bi) \cdot (c + di)$ | | $(a - bi) \cdot (c - di)$ |
| $ac + adi + bci + bdi^2$ | | $ac - adi - cbi + bdi^2$ |
| $(ac - bd) + (ad + bc)i$ | $\rightarrow$ conjugate of product $\rightarrow$ | $(ac - bd) - (ad + bc)i \checkmark$ |

Since polynomials involve only sums and products, and the complex conjugate of any real number is the number itself, we have the following:

**Proof of the Complex Conjugates Theorem**

Given polynomial $p(x) = a_n x^n + a_{n-1}x^{n-1} + \cdots + a_1 x + a_0$, where $a_n, a_{n-1}, \ldots, a_1, a_0$ are real numbers and $z = a + bi$ is a zero of $p$, we must show that $\bar{z} = a - bi$ is also a zero.

$$a_n z^n + a_{n-1}z^{n-1} + \cdots + a_1 z + a_0 = p(z) \qquad \text{evaluate } p(x) \text{ at } z$$
$$a_n z^n + a_{n-1}z^{n-1} + \cdots + a_1 z + a_0 = 0 \qquad p(z) = 0 \text{ given}$$
$$\overline{a_n z^n + a_{n-1}z^{n-1} + \cdots + a_1 z + a_0} = \bar{0} \qquad \text{conjugate both sides}$$
$$\overline{a_n z^n} + \overline{a_{n-1}z^{n-1}} + \cdots + \overline{a_1 z} + \overline{a_0} = \bar{0} \qquad \text{property 1}$$
$$\bar{a}_n(\bar{z}^n) + \bar{a}_{n-1}(\bar{z}^{n-1}) + \cdots + \bar{a}_1(\bar{z}) + \bar{a}_0 = \bar{0} \qquad \text{property 2}$$
$$a_n(\bar{z}^n) + a_{n-1}(\bar{z}^{n-1}) + \cdots + a_1(\bar{z}) + a_0 = 0 \qquad \text{conjugate of a real number is the number}$$
$$p(\bar{z}) = 0 \quad \checkmark \quad \text{result}$$

An immediate and useful result of this theorem is that any polynomial of odd degree must have at least one real root.

## Linear Factorization Theorem

*If $p(x)$ is a complex polynomial of degree $n \geq 1$, then $p$ has exactly $n$ linear factors and can be written in the form $p(x) = a_n(x - c_1)(x - c_2) \cdot \cdots \cdot (x - c_n)$, where $a_n \neq 0$ and $c_1, c_2, \ldots, c_n$ are complex numbers. Some factors may have multiplicities greater than 1 ($c_1, c_2, \ldots, c_n$ are not necessarily distinct).*

## Proof of the Linear Factorization Theorem

Given $p(x) = a_n x^n + a_{n-1} x^{n-1} + \cdots + a_1 x + a_0$ is a complex polynomial, the Fundamental Theorem of Algebra establishes that $p(x)$ has a least one complex zero, call it $c_1$. The factor theorem stipulates $x - c_1$ must be a factor of $P$, giving

$$p(x) = (x - c_1)q_1(x)$$

where $q_1(x)$ is a complex polynomial of degree $n - 1$.

Since $q_1(x)$ is a complex polynomial in its own right, it too must also have a complex zero, call it $c_2$. Then $x - c_2$ must be a factor of $q_1(x)$, giving

$$p(x) = (x - c_1)(x - c_2)q_2(x)$$

where $q_2(x)$ is a complex polynomial of degree $n - 2$.

Repeating this rationale $n$ times will cause $p(x)$ to be rewritten in the form

$$p(x) = (x - c_1)(x - c_2) \cdot \cdots \cdot (x - c_n)q_n(x)$$

where $q_n(x)$ has a degree of $n - n = 0$, a nonzero constant typically called $a_n$.

The result is $p(x) = a_n(x - c_1)(x - c_2) \cdot \cdots \cdot (x - c_n)$, and the proof is complete.


## Proofs from Chapter 5

### The Product Property of Logarithms

*Given $M$, $N$, and $b \neq 1$ are positive real numbers, $\log_b(MN) = \log_b M + \log_b N$.*

**Proof of the Product Property**

For $P = \log_b M$ and $Q = \log_b N$, we have $b^P = M$ and $b^Q = N$ in exponential form. It follows that

$$
\begin{aligned}
\log_b(MN) &= \log_b(b^P b^Q) && \text{substitute } b^P \text{ for } M \text{ and } b^Q \text{ for } N \\
&= \log_b(b^{P+Q}) && \text{properties of exponents} \\
&= P + Q && \text{log Property III} \\
&= \log_b M + \log_b N && \text{substitute } \log_b M \text{ for } P \text{ and } \log_b N \text{ for } Q
\end{aligned}
$$

### The Quotient Property of Logarithms

*Given $M$, $N$, and $b \neq 1$ are positive real numbers, $\log_b\left(\dfrac{M}{N}\right) = \log_b M - \log_b N$.*

**Proof of the Quotient Property**

For $P = \log_b M$ and $Q = \log_b N$, we have $b^P = M$ and $b^Q = N$ in exponential form. It follows that

$$
\begin{aligned}
\log_b\left(\frac{M}{N}\right) &= \log_b\left(\frac{b^P}{b^Q}\right) && \text{substitute } b^P \text{ for } M \text{ and } b^Q \text{ for } N \\
&= \log_b(b^{P-Q}) && \text{properties of exponents} \\
&= P - Q && \text{log Property III} \\
&= \log_b M - \log_b N && \text{substitute } \log_b M \text{ for } P \text{ and } \log_b N \text{ for } Q
\end{aligned}
$$

### The Power Property of Logarithms

*Given $M$, $N$, and $b \neq 1$ are positive real numbers and $x$ is any real number, $\log_b M^x = x \log_b M$.*

**Proof of the Power Property**

For $P = \log_b M$, we have $b^P = M$ in exponential form. It follows that

$$
\begin{aligned}
\log_b(M)^x &= \log_b(b^P)^x && \text{substitute } b^P \text{ for } M \\
&= \log_b(b^{Px}) && \text{properties of exponents} \\
&= Px && \text{log Property III} \\
&= (\log_b M)x && \text{substitute } \log_b M \text{ for } P \\
&= x \log_b M && \text{rewrite factors}
\end{aligned}
$$

# Student Answer Appendix

## CHAPTER R
### Exercises R.1, pp. 10–12

**1.** repeating, terminating, irrational **3.** positive; negative; 7; $-7$; principal **5. a.** $\{1, 2, 3, 4, 5\}$ **b.** $\{\ \}$ **7.** True **9.** True **11.** True

**13.** $1.\overline{3}$

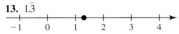

**15.** $2.\overline{5}$

**17. a.** 40,000 acres
**b.** ten 3's (0.333 333 333 3, may vary by calculator model)

**19.** $\approx 2.65$

**21.** $\approx 1.73$

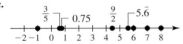

**23. a. i.** $\{8, 7, 6\}$ **ii.** $\{8, 7, 6\}$ **iii.** $\{-1, 8, 7, 6\}$
**iv.** $\{-1, 8, 0.75, \frac{9}{2}, 5.\overline{6}, 7, \frac{3}{5}, 6\}$ **v.** $\{\ \}$
**vi.** $\{-1, 8, 0.75, \frac{9}{2}, 5.\overline{6}, 7, \frac{3}{5}, 6\}$ **b.** $\{-1, \frac{3}{5}, 0.75, \frac{9}{2}, 5.\overline{6}, 6, 7, 8\}$
**c.**

$$\frac{3}{5} \quad 0.75 \quad \frac{9}{2} \quad 5.\overline{6}$$

**25. a. i.** $\{\sqrt{49}, 2, 6, 4\}$ **ii.** $\{\sqrt{49}, 2, 6, 0, 4\}$
**iii.** $\{-5, \sqrt{49}, 2, -3, 6, -1, 0, 4\}$ **iv.** $\{-5, \sqrt{49}, 2, -3, 6, -1, 0, 4\}$
**v.** $\{\sqrt{3}, \pi\}$ **vi.** $\{-5, \sqrt{49}, 2, -3, 6, -1, \sqrt{3}, 0, 4, \pi\}$
**b.** $\{-5, -3, -1, 0, \sqrt{3}, 2, \pi, 4, 6, \sqrt{49}\}$
**c.**

$$\sqrt{3} \quad \pi \quad \sqrt{49}$$

**27.** False; not all real numbers are irrational.
**29.** False; not all rational numbers are integers.
**31.** False; $\sqrt{25} = 5$ is not irrational.
**33.** c; IV **35.** a; VI **37.** d; III
**39.** Let $a$ represent Kylie's age: $a \geq 6$ yr.
**41.** Let $n$ represent the number of incorrect words: $n \leq 2$ incorrect.
**43.** $x \geq \dfrac{3}{2}$ **45.** $x < 2$ **47.** $5 \leq w < 32$ **49.** 2.75 **51.** $-4$

**53.** $\dfrac{1}{2}$ **55.** $\dfrac{3}{4}$ **57.** $|-7.5 - 2.5|, |2.5 - (-7.5)|, 10$ **59.** $-8, 2$

**61.** negative **63.** $-n$ **65.** undefined, since $12 \div 0 = k$ implies $k \cdot 0 = 12$
**67.** undefined, since $7 \div 0 = k$ implies $k \cdot 0 = 7$
**69. a.** positive **b.** negative **c.** negative **d.** negative

**71.** $-\dfrac{11}{6}$ **73.** $-2$ **75.** $9^2 = 81$ is closest **77.** 7 **79.** $-2.185$ **81.** $4\frac{1}{3}$

**83.** $-\dfrac{29}{12}$ or $-2\frac{5}{12}$ **85.** 0 **87.** $-5$ **89.** $-\dfrac{1}{10}$ **91.** $-\dfrac{7}{8}$ **93.** $-4$ **95.** $\dfrac{-11}{12}$

**97.** 64 **99.** 4489.70 **101.** 3 **103.** $D \approx 4.6$ cm **105.** 32°F
**107.** 179°F **109.** about 68.13 cm **111.** Tsu Ch'ung-chih: $\frac{355}{113}$
**113.** positive

### Exercises R.2, pp. 18–21

**1.** coefficient **3.** Answers will vary.
**5.** two; 3 and $-5$ **7.** two; 2 and $\frac{1}{4}$ **9.** three; $-2$, 1, and $-5$

**11.** one; $-1$ **13.** $n - 7$ **15.** $n + 4$ **17.** $(n - 5)^2$ **19.** $2n - 13$
**21.** $\frac{2}{3}n - 5$ **23.** $3(n + 5) - 7$ **25.** Let $w$ represent the width in meters. Then $2w$ represents twice the width and $2w - 3$ represents three meters less than twice the width. **27.** Let $b$ represent the speed of the bus. Then $b + 15$ represents 15 mph more than the speed of the bus.
**29.** $h = b + 150$ **31.** $L = 2W + 20$ **33.** $M = 2.5N$
**35.** $T = 12.50g + 50$ **37.** 14 **39.** 19 **41.** 0 **43.** 16
**45.** 51 **47.** 2 **49.** $-\frac{41}{5}$ **51.** 24

**53.**

| $x$ | Output |
|---|---|
| $-3$ | 14 |
| $-2$ | 6 |
| $-1$ | 0 |
| 0 | $-4$ |
| 1 | $-6$ |
| 2 | $-6$ |
| 3 | $-4$ |

$-1$ gives an output of 0.

**55.**

| $x$ | Output |
|---|---|
| $-3$ | $-18$ |
| $-2$ | $-15$ |
| $-1$ | $-12$ |
| 0 | $-9$ |
| 1 | $-6$ |
| 2 | $-3$ |
| 3 | 0 |

3 gives an output of 0.

**57.**

| $x$ | Output |
|---|---|
| $-3$ | $-5$ |
| $-2$ | 8 |
| $-1$ | 9 |
| 0 | 4 |
| 1 | $-1$ |
| 2 | 0 |
| 3 | 13 |

2 gives an output of 0.

**59. a.** $7 + (-5) = 2$ **b.** $n + (-2)$ **c.** $a + (-4.2) + 13.6 = a + 9.4$
**d.** $x + 7 - 7 = x$ **61. a.** 3.2 **b.** $\frac{5}{6}$ **63.** $-5x + 13$
**65.** $-\frac{2}{15}p + 6$ **67.** $-2a$ **69.** $\frac{17}{12}x$ **71.** $-2a^2 + 2a$ **73.** $6x^2 - 3x$
**75.** $2a + 3b + 2c$ **77.** $\frac{29}{8}n + \frac{38}{5}$ **79.** $7a^2 - 13a - 5$ **81.** 10 ohms
**83. a.** $t = \frac{1}{2}j$ **b.** $t = 275$ mph **85. a.** $L = 2W + 3$ **b.** 107 ft
**87.** $t = c + 31$; 44¢ **89.** $C = 25t + 43.50$; $81
**91. a.** positive odd integer

### Exercises R.3, pp. 32–34

**1.** power **3. a.** cannot be simplified, unlike terms
**b.** can be simplified, like bases **5.** $14n^7$ **7.** $-12p^5q^4$ **9.** $a^{14}b^7$

**11.** $216p^3q^6$ **13.** $32.768h^3k^6$ **15.** $\dfrac{p^2}{4q^2}$ **17.** $49c^{14}d^4$ **19.** $\frac{9}{16}x^6y^2$

**21. a.** $V = 27x^6$ **b.** $1728$ units$^3$ **23.** $3w^3$ **25.** $-3ab$

**27.** $\dfrac{27}{8}$ **29.** $2h^3$ **31.** $\dfrac{-1}{8}$ **33.** $-8$ **35.** $\dfrac{4p^8}{q^6}$ **37.** $\dfrac{8x^6}{27y^9}$ **39.** $\dfrac{25m^4n^6}{4r^8}$

**41.** $\dfrac{3p^2}{-4q^2}$ **43.** $\dfrac{5}{3h^7}$ **45.** $\dfrac{1}{a^3}$ **47.** $\dfrac{a^{12}}{b^4c^8}$ **49.** $\dfrac{-12}{5x^4}$ **51.** 2 **53.** $\frac{7}{10}$

**55.** $\frac{13}{9}$ **57.** $-4$ **59.** $6.97 \times 10^9$ **61.** $0.000\ 000\ 006\ 5$
**63.** about 185 gal/day **65.** polynomial, none of these, degree 3
**67.** nonpolynomial because exponents are not whole numbers, NA, NA
**69.** polynomial, binomial, degree 3
**71.** $-w^3 - 3w^2 + 7w + 8.2; -1$ **73.** $c^3 + 2c^2 - 3c + 6; 1$
**75.** $\frac{-2}{3}x^2 + 12; \frac{-2}{3}$ **77.** $3p^3 - 3p^2 - 12$ **79.** $7.85b^2 - 0.6b - 1.9$
**81.** $\frac{1}{4}x^2 - 8x + 6$ **83.** $q^6 + q^5 - q^4 + 2q^3 - q^2 - 2q$
**85.** $-3x^3 + 3x^2 + 18x$ **87.** $3r^2 - 11r + 10$ **89.** $x^3 - 27$
**91.** $b^3 - b^2 - 34b - 56$ **93.** $21v^2 - 47v + 20$ **95.** $9 - m^2$

**97.** $x^2 + \frac{3}{4}x + \frac{1}{8}$  **99.** $m^2 - \frac{9}{16}$  **101.** $6x^2 + 11xy - 10y^2$
**103.** $12c^2 + 23cd + 5d^2$  **105.** $4m + 3; 16m^2 - 9$
**107.** $7x + 10; 49x^2 - 100$  **109.** $6 - 5k; 36 - 25k^2$
**111.** $x - \sqrt{6}; x^2 - 6$  **113.** $x^2 + 8x + 16$  **115.** $16g^2 + 24g + 9$
**117.** $16 - 8\sqrt{x} + x$  **119.** $xy + 2x - 3y - 6$
**121.** $k^3 + 3k^2 - 28k - 60$  **123. a.** 340 mg, 292.5 mg
**b.** Less, amount is decreasing.  **c.** after 5 hr **125.** $F = kPQd^{-2}$
**127.** $5x^{-3} + 3x^{-2} + 2x^{-1} + 4$  **129.** $15  **131.** 6

## Exercises R.4, pp. 43–46

**1.** $(16^{\frac{1}{4}})^3$  **3.** Answers will vary.  **5. a.** $|9| = 9$ **b.** $|-10| = 10$
**7. a.** $7|p|$  **b.** $|x - 3|$  **c.** $9m^2$  **d.** $|x - 3|$  **9. a.** 4  **b.** $-6x$  **c.** $6z^4$
**d.** $\frac{v}{-2}$  **11. a.** 2  **b.** not a real number  **c.** $3x^2$  **d.** $-3x$  **e.** $k - 3$
**f.** $|h + 2|$  **13. a.** $-5$  **b.** $-3|n^3|$  **c.** not real number  **d.** $\frac{7|v^5|}{6}$
**15. a.** 4  **b.** $\frac{64}{125}$  **c.** $\frac{125}{8}$  **d.** $\frac{9p^4}{4q^2}$  **17. a.** $-1728$
**b.** not a real number  **c.** $\frac{1}{9}$  **d.** $\frac{-256}{81x^4}$  **19. a.** $\frac{32n^{10}}{p^2}$  **b.** $\frac{1}{2y^{\frac{1}{4}}}$
**21. a.** $3m\sqrt{2}$  **b.** $10pq^2\sqrt[3]{q}$  **c.** $\frac{3}{2}mn\sqrt[3]{n^2}$  **d.** $4pq^3\sqrt{2p}$
**e.** $-3 + \sqrt{7}$  **f.** $\frac{9}{2} - \sqrt{2}$  **23. a.** $15a^2$  **b.** $-4b\sqrt{b}$  **c.** $\frac{x^4\sqrt{y}}{3}$
**d.** $3u^2v\sqrt[3]{v}$  **25. a.** $2m^2$  **b.** $3n$  **c.** $\frac{3\sqrt{5}}{4x}$  **d.** $\frac{18\sqrt[3]{3}}{z^3}$
**27. a.** $2x^2y^3$  **b.** $x^2\sqrt[4]{x}$  **c.** $\sqrt[12]{b}$  **d.** $\frac{1}{\sqrt[6]{6}} = \frac{\sqrt[6]{6^5}}{6}$  **e.** $b^{\frac{3}{4}}$
**29. a.** $9\sqrt{2}$  **b.** $14\sqrt{3}$  **c.** $16\sqrt{2m}$  **d.** $-5\sqrt{7p}$
**31. a.** $-x\sqrt[3]{2x}$  **b.** $5\sqrt{5} - \sqrt{3x}$
**c.** $3x\sqrt{7x} - 8\sqrt{3}$  **33. a.** 98  **b.** $\sqrt{15} + \sqrt{21}$
**c.** $n^2 - 5$  **d.** $39 - 12\sqrt{3}$
**35. a.** $-19$  **b.** $15\sqrt{3} - 15\sqrt{2}$
**c.** $42 + 34\sqrt{3}$
**37.** verified  **39.** verified  **41. a.** $\frac{\sqrt{3}}{2}$  **b.** $\frac{2\sqrt{15x}}{9x^2}$  **c.** $\frac{3\sqrt{6b}}{10b}$
**d.** $\frac{\sqrt[3]{2p^2}}{2p}$  **e.** $\frac{5\sqrt[3]{a^2}}{a}$  **43. a.** $-12 + 4\sqrt{11}; 1.27$  **b.** $\frac{6\sqrt{x} + 6\sqrt{2}}{x - 2}$
**45. a.** $6\sqrt{2} - 5\sqrt{3} \approx -0.17$
**b.** $-4 - \frac{10\sqrt{2}}{3} \approx -8.71$  **47.** 8.33 ft
**49. a.** $8\sqrt{10}$ m  **b.** about 25.3 m  **51. a.** 365.02 days  **b.** 688.69 days
**c.** 87.91 days  **53.** 9,491,274 m or about 9,491 km
**55. a.** 36 mph  **b.** 46.5 mph  **57.** $12\pi\sqrt{34} \approx 219.82$ m$^2$
**59. a.** $(x + \sqrt{5})(x - \sqrt{5})$  **b.** $(n + \sqrt{19})(n - \sqrt{19})$
**61.** re-order the exponents, 3  **63.** $\frac{1}{\sqrt{x + h} + \sqrt{x}}$
**65.** $x \in [1, 2) \cup (2, \infty)$  **67. a.** $\sqrt{6 + 2\sqrt{7}}$  **b.** verified

## Exercises R.5, pp. 55–58

**1.** binomial; conjugate  **3.** Answers will vary.
**5. a.** $-17(x^2 - 3)$  **b.** $7b(3b^2 - 2b + 8)$  **c.** $-3a^2(a^2 + 2a - 3)$
**7. a.** $(a + 2)(2a + 3)$  **b.** $(b^2 + 3)(3b + 2)$  **c.** $(n + 7)(4m - 11)$
**9. a.** $(3q + 2)(3q^2 + 5)$  **b.** $(h - 12)(h^4 - 3)$  **c.** $(k^2 - 7)(k^3 - 5)$
**11. a.** $-1(p - 7)(p + 2)$  **b.** prime  **c.** $(n - 4)(n - 5)$
**13. a.** $(3p + 2)(p - 5)$  **b.** $(4q - 5)(q + 3)$  **c.** $(5u + 3)(2u - 5)$
**15. a.** $(2s + 5)(2s - 5)$  **b.** $(3x + 7)(3x - 7)$  **c.** $2(5x + 6)(5x - 6)$
**d.** $(11h + 12)(11h - 12)$  **e.** $(b + \sqrt{5})(b - \sqrt{5})$  **17. a.** $(a - 3)^2$
**b.** $(b + 5)^2$  **c.** $(2m - 5)^2$  **d.** $(3n - 7)^2$

**19. a.** $(2p - 3)(4p^2 + 6p + 9)$  **b.** $(m + \frac{1}{2})(m^2 - \frac{1}{2}m + \frac{1}{4})$
**c.** $(g - 0.3)(g^2 + 0.3g + 0.09)$  **d.** $-2t(t - 3)(t^2 + 3t + 9)$
**21. a.** $(x + 3)(x - 3)(x + 1)(x - 1)$  **b.** $(x^2 + 9)(x^2 + 4)$
**c.** $(x - 2)(x^2 + 2x + 4)(x + 1)(x^2 - x + 1)$
**23. a.** $(n + 1)(n - 1)$  **b.** $(n - 1)(n^2 + n + 1)$
**c.** $(n + 1)(n^2 - n + 1)$  **d.** $7x(2x + 1)(2x - 1)$
**25.** $(a + 5)(a + 2)$  **27.** $2(x - 2)(x - 10)$  **29.** $-1(3m + 8)(3m - 8)$
**31.** $(r - 3)(r - 6)$  **33.** $(2h + 3)(h + 2)$  **35.** $(3k - 4)^2$
**37.** $-3x(2x - 7)(x - 3)$  **39.** $4m(m + 5)(m - 2)$  **41.** $(a + 5)(a - 12)$
**43.** $(2x - 5)(4x^2 + 10x + 25)$  **45.** prime  **47.** $(x - 5)(x + 3)(x - 3)$
**49. a.** H  **b.** E  **c.** C  **d.** F  **e.** B  **f.** A  **g.** I  **h.** D  **i.** G
**51.** $2\pi r(r + h)$, $7000\pi$ cm$^2$; 21,991 cm$^2$
**53.** $V = \frac{1}{3}\pi h(R + r)(R - r)$; $6\pi$ cm$^3$; 18.8 cm$^3$
**55.** $V = x(x + 5)(x + 3)$ **a.** 3 in. **b.** 5 in. **c.** $V = 24(29)(27) = 18,792$ in$^3$
**57.** $L = L_0\sqrt{\left(1 + \frac{v}{c}\right)\left(1 - \frac{v}{c}\right)}$, $L = 12\sqrt{(1 + 0.75)(1 - 0.75)}$
$= 3\sqrt{7}$ in. $\approx 7.94$ in.
**59. a.** $V = 4x^3 - 4xy^2$  **b.** $V = 4x(x + y)(x - y)$  **c.** 220 in$^3$
**61. a.** $\frac{1}{8}(4x^4 + x^3 - 6x^2 + 32)$  **b.** $\frac{1}{18}(12b^5 - 3b^3 + 8b^2 - 18)$
**63.** $-12$  **65.** $5x^2(x - 4)(x - 5)(x + 1)$
**67.** $7c(c^2 - 5)^{-\frac{1}{2}}(2c + 5)(2c - 5)$
**69.** $(b^2 - 1)^{-\frac{1}{2}}(5b + 2)(b - 2)(b - 8)$  **71.** $H = -33$

## Exercises R.6, pp. 65–68

**1.** common denominator  **3.** F; numerator should be $-1$
**5. a.** $\frac{1}{3}$  **b.** $\frac{x + 3}{2x(x - 2)}$  **7. a.** $\frac{r + 5}{r + 3}$  **b.** simplified
**9. a.** $-1$  **b.** $-1$  **11. a.** $-3ab^9$  **b.** $\frac{x + 3}{9}$  **c.** $-1(y + 3)$  **d.** $\frac{-1}{m}$
**13. a.** $\frac{2n + 3}{n}$  **b.** $\frac{3x + 5}{2x + 3}$  **c.** $x + 2$  **d.** $n - 2$  **15.** $\frac{(a - 2)(a + 1)}{(a + 3)(a + 2)}$
**17.** 1  **19.** $\frac{(p - 4)^2}{p^2}$  **21.** $\frac{-15}{4}$  **23.** $\frac{3}{2}$  **25.** $\frac{y}{x}$
**27.** $\frac{m}{m - 4}$  **29.** $\frac{y + 3}{3y(y + 4)}$  **31.** $\frac{x + 0.3}{x - 0.2}$  **33.** $\frac{n + \frac{1}{5}}{n + \frac{2}{3}}$
**35.** $\frac{3(a^2 + 3a + 9)}{2}$  **37.** $\frac{2n + 1}{n}$  **39.** $\frac{3 + 20x}{8x^2}$  **41.** $\frac{14y - x}{8x^2y^4}$
**43.** $\frac{2}{p + 6}$  **45.** $\frac{-3m - 16}{(m + 4)(m - 4)}$  **47.** $\frac{-5m + 37}{m - 7}$
**49.** $\frac{-y + 11}{(y + 6)(y - 5)}$  **51.** $\frac{1}{x - 1}$  **53.** $\frac{-2a - 21}{3a(a + 4)}$
**55.** $\frac{y^2 + 26y - 1}{(5y + 1)(y + 3)(y - 2)}$  **57. a.** $\frac{1}{p^2} - \frac{5}{p}; \frac{1 - 5p}{p^2}$
**b.** $\frac{1}{x^2} + \frac{2}{x^3}; \frac{x + 2}{x^3}$  **59.** $\frac{4a}{a + 20}$  **61.** $p - 1$
**63.** $\frac{-2}{y + 31}$  **65. a.** $\frac{1 + \frac{3}{m}}{1 - \frac{3}{m}}; \frac{m + 3}{m - 3}$  **b.** $\frac{1 + \frac{2}{x^2}}{1 - \frac{2}{x^2}}; \frac{x^2 + 2}{x^2 - 2}$
**67.** $\frac{f_1 + f_2}{f_1 f_2}$  **69.** $\frac{-a}{x(x + h)}$  **71.** $\frac{-(2x + h)}{2x^2(x + h)^2}$

**73. a.** $300 million; $2550 million  **75.** Price rises rapidly for first four days, then begins a gradual decrease. Yes, on the 35th day of trading.

| P | $\dfrac{450P}{100 - P}$ |
|---|---|
| 40 | 300 |
| 60 | 675 |
| 80 | 1800 |
| 90 | 4050 |
| 93 | 5979 |
| 95 | 8550 |
| 98 | 22050 |
| 100 | ERROR |

| Day | Price |
|---|---|
| 0 | 10 |
| 1 | 16.67 |
| 2 | 32.76 |
| 3 | 47.40 |
| 4 | 53.51 |
| 5 | 52.86 |
| 6 | 49.25 |
| 7 | 44.91 |
| 8 | 40.75 |
| 9 | 37.03 |
| 10 | 33.81 |

**77.** $t = 8$ weeks  **79.** $66 million dollars

**81. b.** $20 \cdot n \div 10 \cdot n = 2n^2$, all others equal 2  **83. b.** $\dfrac{2x + 3y}{5}$

## Practice Test, pp. 71–72

**1. a.** True  **b.** True  **c.** False, $\sqrt{2}$ cannot be expressed as a ratio of integers.  **d.** True  **2. a.** False; parentheses first  **b.** False; undefined  **c.** True  **d.** False; $-2x + 6$  **3. a.** 11  **b.** $-5$  **c.** not a real number  **d.** 20  **4. a.** $\frac{9}{8}$  **b.** $\frac{-7}{6}$  **c.** 0.5  **d.** $-4.6$  **5. a.** $\frac{28}{3}$  **b.** 0.9  **c.** 4  **d.** $-7$  **6.** $\approx 4439.28$  **7. a.** 0  **b.** undefined  **8. a.** 3; $-2, 6, 5$  **b.** 2; $\frac{1}{3}, 1$  **9. a.** $-13$  **b.** $\approx 7.29$  **10. a.** $x^3 - (2x - 9)$

**b.** $2n - 3\left(\dfrac{n}{2}\right)^2$  **11. a.** Let $r$ represent Earth's radius. Then $11r - 119$ represents Jupiter's radius.  **b.** Let $e$ represent this year's earnings. Then $4e + 1.2$ million represents last year's earnings.  **12. a.** $9v^2 + 3v - 7$  **b.** $-7b + 8$  **c.** $x^2 + 6x$  **13. a.** $(3x + 4)(3x - 4)$  **b.** $v(2v - 3)^2$  **c.** $(x + 5)(x + 3)(x - 3)$  **14. a.** $(x - 3)(x^2 + 3x + 9)$

**b.** $3a(2c + 3b)(4c^2 - 6bc + 9b^2)$  **15. a.** $5b^3$  **b.** $4a^{12}b^{12}$  **c.** $\dfrac{m^6}{8n^3}$

**d.** $\dfrac{25}{4}p^2q^2$  **16. a.** $-4ab$  **b.** $6.4 \times 10^{-2} = 0.064$  **c.** $\dfrac{a^{12}}{b^4c^8}$  **d.** $-6$

**17. a.** $9x^4 - 25y^2$  **b.** $4a^2 + 12ab + 9b^2$  **18. a.** $7a^4 - 5a^3 + 8a^2 - 3a - 18$  **b.** $-7x^4 + 4x^2 + 5x$  **19. a.** (v)  **b.** (iv)  **c.** (i)  **d.** (ii)  **e.** (iii)

**20. a.** $-1$  **b.** $\dfrac{2 + n}{2 - n}$  **c.** $x - 3$  **d.** $\dfrac{x - 5}{3x - 2}$  **e.** $\dfrac{x - 5}{3x + 1}$

**f.** $\dfrac{3(m + 7)}{5(m + 4)(m - 3)}$  **21. a.** $|x + 11|$  **b.** $\dfrac{-2}{3v}$  **c.** $\dfrac{64}{125}$  **d.** $-\dfrac{1}{2} + \dfrac{\sqrt{2}}{2}$

**e.** $11\sqrt{10}$  **f.** $x^2 - 5$  **g.** $\dfrac{\sqrt{10x}}{5|x|}$  **h.** $2(\sqrt{6} + \sqrt{2})$

**22.** $-0.5x^2 + 10x + 1200$;  **a.** 10 decreases of $0.50 or $5.00  **b.** Maximum revenue is $1250.  **23.** 58 cm

**24. a.** $V = 3.6 \times 10^9$ m$^3$  **b.** Weight: $2(3.6 \times 10^9) = 7.2 \times 10^9$ g or $7.2 \times 10^6$ kg (over 15,873,000 lb)

**25. a.** $d = \dfrac{\sqrt{(L + D)(L - D)}}{2}$  **b.** $d = 2.5$ m

# CHAPTER 1

## Exercises 1.1, pp. 82–85

**1.** identity; unknown; contradiction; unknown  **3.** Answers will vary.

**5.** $x = 3$  **7.** $b = \dfrac{6}{5}$  **9.** $b = -15$  **11.** $m = -\dfrac{27}{4}$  **13.** $x = 12$

**15.** $x = 12$  **17.** $p = -56$  **19.** $a = -3.6$  **21.** $v = -0.5$

**23.** $n = \dfrac{20}{21}$  **25.** $p = \dfrac{12}{5}$  **27.** contradiction; $x \in \{\ \}$

**29.** conditional; $n = -\dfrac{11}{10}$  **31.** identity; $x \in \mathbb{R}$

**33.** $C = \dfrac{P}{1 + M}$  **35.** $r = \dfrac{C}{2\pi}$  **37.** $L = \dfrac{4V}{\pi d^2}$  **39.** $n = \dfrac{2S_n}{a_1 + a_n}$

**41.** $P = \dfrac{2(S - B)}{S}$  **43.** $y = \dfrac{-A}{B}x + \dfrac{C}{B}$  **45.** $y = \dfrac{-20}{9}x + \dfrac{16}{3}$

**47.** $y = \dfrac{-4}{5}x - 5$  **49.** $a = 3, b = 2, c = -19; x = -7$

**51.** $a = 7, b = -13, c = -27; x = -2$  **53. a.** $8 + 6 - 12 = 2$✓  **b.** $6 + 8 - 12 = 2$✓  **c.** $V = E - F + 2$  **d.** 20  **55.** 12 in.  **57.** $57,971  **59.** 10.5 m  **61.** 56 in.  **63.** 4.6 in.  **65.** 11: 30 A.M.  **67.** 15 mi  **69.** 12 lb  **71.** 16 lb  **73.** Answers will vary.  **75.** 69  **77.** $-3$  **79. a.** $(2x + 3)(2x - 3)$  **b.** $(x - 3)(x^2 + 3x + 9)$

## Exercises 1.2, pp. 93–95

**1.** set; number line; interval  **3.** Answers will vary.  **5.** $w \geq 45$  **7.** $250 < T < 450$

**9.**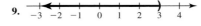

**11.** (number line from $-4$ to $6$)

**13.** (number line from $-4$ to $6$)

**15.** (number line from $-4$ to $7$)

**17.** $\{x | x \geq -2\}$; $[-2, \infty)$  **19.** $\{x | -2 \leq x \leq 1\}$; $[-2, 1]$

**21.** $\{a | a \geq 2\}$; (number line from $-3$ to $5$) $; a \in [2, \infty)$

**23.** $\{n | n \geq 1\}$; (number line from $0$ to $3$) $; n \in [1, \infty)$

**25.** $\{x | x < \frac{-32}{5}\}$; (number line from $-10$ to $0$) $; x \in (-\infty, \frac{-32}{5})$

**27.** $x \in \{\ \}$  **29.** $x \in \mathbb{R}$  **31.** $p \in \mathbb{R}$

**33.** $\{2\}; \{-3, -2, -1, 0, 1, 2, 3, 4, 6, 8\}$

**35.** $\{\ \}; \{-3, -2, -1, 0, 1, 2, 3, 4, 5, 6, 7\}$

**37.** $\{4, 6\}; \{2, 4, 5, 6, 7, 8\}$

**39.** $x \in (-\infty, -2) \cup (1, \infty)$; (number line from $-4$ to $3$)

**41.** $x \in [-2, 5)$; (number line from $-3$ to $6$)

**43.** no solution

**45.** $x \in (-\infty, \infty)$; (number line from $-4$ to $4$)

**47.** $x \in [-5, 0]$; (number line from $-6$ to $1$)

**49.** $x \in \left(-\frac{1}{3}, -\frac{1}{4}\right)$; (number line from $-1$ to $1$)

**51.** $x \in (-\infty, \infty)$; (number line from $-4$ to $4$)

**53.** $x \in [-4, 1)$;

**55.** $x \in [-1.4, 0.8]$;

**57.** $x \in [-16, 8)$;

**59.** $m \in (-\infty, 0) \cup (0, \infty)$    **61.** $y \in (-\infty, -7) \cup (-7, \infty)$
**63.** $a \in (-\infty, \frac{1}{2}) \cup (\frac{1}{2}, \infty)$    **65.** $x \in (-\infty, 4) \cup (4, \infty)$
**67.** $x \in [2, \infty)$    **69.** $n \in [4, \infty)$    **71.** $b \in \left[\frac{4}{3}, \infty\right)$

**73.** $y \in [-\infty, 2)$    **75. a.** $W = \dfrac{BH^2}{704}$    **b.** $B \geq 30.2$

**77.** $\frac{82 + 76 + 65 + 71 + x}{5} \geq 75$; at least 81%
**79.** $\frac{1125 + 850 + 625 + 400 + x}{5} \geq 1000$; at least $2000
**81.** $0 < 20w < 150; 0 < w < 7.5$ m
**83.** $45 < \frac{9}{5}C + 32 < 85; 7.2° < C < 29.4°$
**85.** $4.5h + 20 < 6h + 11$; more than 6 hr
**87. a.** $54.69    **b.** $45.00    **c.** at least 8.9 mi closer
**89. a.** $\geq$    **b.** $<$    **c.** $>$    **d.** $<$
**91.** Since $a < b, a - b < 0$ and all but the first inequality should be changed to "greater than" inequalities. The final statement is false since a perfect square must be nonnegative.
**93.** $\dfrac{17}{18}x - 5$    **95.** $m = \dfrac{-5}{24}$

## Exercises 1.3, pp. 101–103

**1.** isolate    **3.** no solution; Answers will vary.    **5.** $\{-4, 6\}$
**7.** $\{-3.35, 0.85\}$    **9.** $\{-\frac{8}{7}, 2\}$    **11.** $\{-\frac{1}{2}, \frac{1}{2}\}$    **13.** $\{-5\}$    **15.** $\{\ \}$
**17.** $\{-10, -6\}$    **19.** $\{3.5, 11.5\}$    **21.** $\{-1.6, 1.6\}$    **23.** $\{-3, -\frac{1}{2}\}$
**25.** $(-5, -3)$    **27.** $\{\ \}$    **29.** $\left[\frac{8}{3}, \frac{14}{3}\right]$    **31.** $(-\infty, \infty)$    **33.** $\left(-1, \frac{3}{5}\right)$
**35.** $\left[-\frac{7}{4}, 0\right]$    **37.** $(-\infty, -10) \cup (4, \infty)$    **39.** $(-\infty, -3] \cup [3, \infty)$
**41.** $\left(-\infty, -\frac{7}{3}\right] \cup \left[\frac{7}{3}, \infty\right)$    **43.** $\left(-\infty, \frac{5}{7}\right) \cup \left(\frac{5}{7}, \infty\right)$
**45.** $(-\infty, -\frac{7}{15}) \cup (1, \infty)$    **47.** $\left\{\frac{5}{2}\right\}$    **49.** $\{\ \}$    **51.** $(-\infty, \infty)$
**53. a.** $d - L \leq x \leq d + L$    **b.** $45 \leq d \leq 51$ in.
**55.** in feet: [32,500, 37,600]; yes    **57.** in feet: $d < 210$ or $d > 578$
**59. a.** $|s - 37.58| \leq 3.35$    **b.** [34.23, 40.93]
**61. a.** $|180 - 72t| < 90$    **b.** Picks up station at 1:15 P.M., loses reception at 3:45 P.M.    **63. a.** $|d - 42.7| < 0.03$    **b.** $|d - 73.78| < 1.01$
**c.** $|d - 57.150| < 0.127$    **d.** $|d - 217.105| < 1.205$
**e.** golf: $t \approx 0.0014$    **65. a.** $x = 4$    **b.** $x \in [\frac{4}{3}, 4]$    **c.** $x = 0$
**d.** $x \in (-\infty, \frac{3}{5}]$    **e.** $x \in \{\ \}$
**67. a.** Answers will vary.    **b.** $A = 0$ or $B = 0$
**69.** $3x(2x + 5)(3x - 4)$    **71.** $\dfrac{-3 + \sqrt{3}}{6} \approx -0.21$

## Mid-Chapter Check, p. 104

**1. a.** $r = -9$    **b.** $x = -6$    **c.** identity; $m \in \mathbb{R}$    **d.** $y = \dfrac{50}{13}$
**e.** contradiction; $j \in \{\ \}$    **f.** $x = 5.5$    **2.** $v_0 = \dfrac{H + 16t^2}{t}$
**3.** $y = \dfrac{S - 2\pi x^2}{\pi x^2}$
**4. a.** $x \geq 1$ or $x \leq -2$

**b.** $16 < x \leq 19$

**5. a.** $x \in \left(-\infty, \frac{5}{2}\right) \cup \left(\frac{5}{2}, \infty\right)$    **b.** $x \in \left(-\infty, \frac{17}{6}\right]$
**6. a.** $\{-4, 14\}$    **b.** $\{\ \}$    **7. a.** $q \in (-8, 0)$    **b.** $x = -6$
**8. a.** $d \in (-\infty, 0] \cup [4, \infty)$    **b.** $y \in \left(-\infty, -\frac{19}{2}\right) \cup \left(\frac{23}{2}, \infty\right)$
**c.** $k \in (-\infty, \infty)$    **9.** 1 hr, 20 min    **10.** $w \in [8, 26]$; yes

## Exercises 1.4, pp. 111–114

**1.** $3 - 2i$    **3.** (b) is correct.    **5. a.** $12i$    **b.** $7i$    **c.** $3\sqrt{3}$    **d.** $6\sqrt{2}$
**7. a.** $-3i\sqrt{2}$    **b.** $-5i\sqrt{2}$    **c.** $15i$    **d.** $6i$    **9. a.** $i\sqrt{19}$    **b.** $i\sqrt{31}$
**c.** $\dfrac{2\sqrt{3}}{5}i$    **d.** $\dfrac{3\sqrt{2}}{8}i$    **11. a.** $1 + i$    **b.** $2 + i\sqrt{3}$    **13. a.** $4 + 2i$
**b.** $2 - i\sqrt{2}$    **15. a.** $5 + 0i; a = 5, b = 0$    **b.** $0 + 3i; a = 0, b = 3$
**17. a.** $0 + 18i; a = 0, b = 18$    **b.** $0 + \dfrac{\sqrt{2}}{2}i; a = 0, b = \dfrac{\sqrt{2}}{2}$
**19. a.** $4 + 5i\sqrt{2}; a = 4, b = 5\sqrt{2}$    **b.** $-5 + 3i\sqrt{3}; a = -5, b = 3\sqrt{3}$
**21. a.** $\dfrac{7}{4} + \dfrac{7\sqrt{2}}{8}i; a = \dfrac{7}{4}, b = \dfrac{7\sqrt{2}}{8}$    **b.** $\dfrac{1}{2} + \dfrac{\sqrt{10}}{2}i; a = \dfrac{1}{2}, b = \dfrac{\sqrt{10}}{2}$
**23. a.** $19 + i$    **b.** $2 - 4i$    **c.** $9 + 10i\sqrt{3}$    **25. a.** $-3 + 2i$
**b.** $8$    **c.** $2 - 8i$    **27. a.** $2.7 + 0.2i$    **b.** $15 + \dfrac{1}{12}i$    **c.** $-2 - \dfrac{1}{8}i$
**29. a.** $15 + 0i$    **b.** $16 + 0i$    **31. a.** $-21 - 35i$    **b.** $-42 - 18i$
**33. a.** $-8 - 14i$    **b.** $-7 + 17i$    **35. a.** $-5 + 12i$    **b.** $-7 - 24i$
**37. a.** $-21 - 20i$    **b.** $7 + 6i\sqrt{2}$    **39. a.** $4 - 5i; 41$
**b.** $3 + i\sqrt{2}; 11$    **41. a.** $-7i; 49$    **b.** $\frac{1}{2} + \frac{2}{3}i; \frac{25}{36}$    **43.** no    **45.** yes
**47.** yes    **49.** no    **51.** yes    **53. a.** $1$    **b.** $-1$    **c.** $-i$    **d.** $i$
**55. a.** $0 + \dfrac{2}{7}i$    **b.** $\dfrac{2}{5} + \dfrac{4}{5}i$    **57. a.** $1 - \dfrac{3}{4}i$    **b.** $-\dfrac{1}{2} + 0i$
**59. a.** $\dfrac{21}{13} - \dfrac{14}{13}i$    **b.** $\dfrac{-10}{13} - \dfrac{15}{13}i$    **61. a.** $\sqrt{13}$    **b.** $5$    **c.** $\sqrt{11}$
**63.** $A + B = 10\checkmark$    $AB = 40\checkmark$    **65.** $(7 - 5i)\ \Omega$    **67.** $(25 + 5i)$ V
**69.** $\left(\dfrac{7}{4} + i\right)\Omega$    **71.** $-8 - 6i$    **73. a.** $3 + 4i$    **b.** $3 - 2i$
**c.** $\dfrac{3\sqrt{2}}{2} + \dfrac{\sqrt{2}}{2}i$    **75. a.** $(-\infty, 8) \cup (8, \infty)$    **b.** $(-\infty, 0]$    **c.** $(\frac{-3}{2}, \infty]$
**d.** $(-\infty, -2) \cup (-2, 2) \cup (2, \infty)$    **77.** John

## Exercises 1.5, pp. 124–127

**1.** descending; 0    **3.** GCF factoring: $x = 0, x = \dfrac{5}{4}$
**5.** $a = -1, b = 2, c = -15$    **7.** not quadratic (degree 1)
**9.** $a = \frac{1}{4}, b = -6, c = 0$    **11.** $a = 2, b = 0, c = 7$
**13.** not quadratic (degree 3)    **15.** $a = 1, b = -1, c = -5$
**17.** $x = 5$ or $x = -3$    **19.** $m = 4$    **21.** $p = 0$ or $p = 2$
**23.** $h = 0$ or $h = \frac{-1}{2}$    **25.** $a = 3$ or $a = -3$    **27.** $g = -9$
**29.** $m = -5$ or $m = -3$ or $m = 3$    **31.** $c = -3$ or $c = 15$
**33.** $r = 8$ or $r = -3$    **35.** $t = -13$ or $t = 2$
**37.** $x = 5$ or $x = -3$    **39.** $w = -\frac{1}{2}$ or $w = 3$    **41.** $m = \pm 4$
**43.** $y = \pm 2\sqrt{7}; y \approx \pm 5.29$    **45.** no real solutions
**47.** $x = \pm\frac{\sqrt{21}}{4}; x \approx \pm 1.15$    **49.** $n = 9; n = -3$
**51.** $w = -5 \pm \sqrt{3}; w \approx -3.27$ or $w \approx -6.73$    **53.** no real solutions
**55.** $m = 2 \pm \frac{3\sqrt{2}}{7}; m \approx 2.61$ or $m \approx 1.39$    **57.** 9; $(x + 3)^2$
**59.** $\frac{9}{4}; (n + \frac{3}{2})^2$    **61.** $\frac{1}{9}; (p + \frac{1}{3})^2$    **63.** $x = -1; x = -5$
**65.** $p = 3 \pm \sqrt{6}; p \approx 5.45$ or $p \approx 0.55$
**67.** $p = -3 \pm \sqrt{5}; p \approx -0.76$ or $p \approx -5.24$
**69.** $m = \frac{-3}{2} \pm \frac{\sqrt{13}}{2}; m \approx 0.30$ or $m \approx -3.30$
**71.** $n = \frac{5}{2} \pm \frac{3\sqrt{5}}{2}; n \approx 5.85$ or $n \approx -0.85$

**73.** $x = \frac{1}{2}$ or $x = -4$; $x = 0.5$ or $x = -4$
**75.** $n = 3$ or $n = \frac{-3}{2}$; $n = 3$ or $n = -1.5$
**77.** $p = \frac{3}{8} \pm \frac{\sqrt{41}}{8}$; $p \approx 1.18$ or $p \approx -0.43$
**79.** $m = \frac{7}{2} \pm \frac{\sqrt{33}}{2}$; $m \approx 6.37$ or $m \approx 0.63$
**81.** $p = \frac{-1}{3}$ or $p = 0$; $p \approx -0.33$ or $p = 0$
**83.** $w = -3 \pm \sqrt{10}$; $w \approx -6.16$ or $w \approx 0.16$
**85.** $w = \frac{-1}{2}$ or $w = \frac{2}{3}$; $w = -0.5$ or $w \approx 0.67$
**87.** $a = \frac{1}{2} \pm \frac{\sqrt{2}}{2}$; $a \approx -0.21$ or $a \approx 1.21$
**89.** $m = \pm\frac{5}{2}$; $m = -2.5$ or $m = 2.5$
**91.** $n = \frac{1}{3} \pm \frac{\sqrt{10}}{3}$; $n \approx -0.72$ or $n \approx 1.39$
**93.** $m = \frac{-2}{3}$ or $m = 3$; $m \approx -0.67$ or $m = 3$
**95.** $x = \frac{24}{25} \pm \frac{3\sqrt{1006}}{50}$; $x \approx -0.94$ or $x \approx 2.86$
**97.** $a = -3 \pm \frac{3\sqrt{2}}{2}$; $a \approx -5.12$ or $a \approx -0.88$
**99.** $x = 1 \pm \frac{\sqrt{6}}{2}i$; $x \approx 1 \pm 1.22i$    **101.** $a = \frac{5}{6} \pm \frac{\sqrt{47}}{6}i$; $a \approx 0.83 \pm 1.14i$
**103.** $a = \frac{1}{6} \pm \frac{\sqrt{23}}{6}i$; $a \approx 0.17 \pm 0.80i$    **105.** two rational; factorable
**107.** two nonreal    **109.** two rational; factorable    **111.** two nonreal
**113.** two irrational    **115.** one repeated; factorable    **117.** $x = \frac{3}{2} \pm \frac{1}{2}i$
**119.** $x = -\frac{1}{2} \pm \frac{\sqrt{3}}{2}i$    **121.** $x = \frac{5}{4} \pm \frac{3\sqrt{7}}{4}i$    **123.** $t = \frac{v \pm \sqrt{v^2 - 64h}}{32}$
**125.** 29.6 sec    **127.** 13.9 ft or 13 ft, 11 in.    **129.** 20,000 printers
**131.** 16 in.    **133.** 2014    **135.** 110 yd by 45 yd    **137.** Answers will vary.
**139. a.** $z = -2i$ or $z = 11i$    **b.** $z = \frac{-13}{2}i$ or $z = 2i$
**c.** $z = -9 + 3i$ or $z = 1 + 3i$    **141. a.** $P = 2L + 2W, A = LW$
**b.** $P = 2\pi r, A = \pi r^2$    **c.** $P = c + h + b_1 + b_2, A = \frac{1}{2}h(b_1 + b_2)$
**d.** $P = a + b + c, A = \frac{1}{2}bh$    **143.** 700 $30 tickets; 200 $20 tickets

## Exercises 1.6, pp. 137–140

**1.** zero product    **3.** Answers will vary.
**5.** $x = -2, x = 0, x = 11$    **7.** $x = -3, x = 0, x = \frac{2}{3}$
**9.** $x = -\frac{3}{2}, x = 0, x = 3$    **11.** $x = -7, x = -3, x = 3$
**13.** $x = -5, x = 2, x = 5$    **15.** $x = 0, x = 2, x = -1 \pm i\sqrt{3}$
**17.** $x = \pm 2, x = 5$    **19.** $x = 3, x = \pm 2i$    **21.** $x = \pm\sqrt{5}, x = 6$
**23.** $x = 0, x = 7, x = \pm 2i$    **25.** $x = \pm 4, x = \pm 4i$
**27.** $x = \pm\sqrt{2}, x = \pm 1, x = \pm i$    **29.** $x = \pm 1, x = 2, x = -1 \pm i\sqrt{3}$
**31.** $x = -\frac{1}{2} \pm \frac{i\sqrt{3}}{2}, x = \frac{1}{2} \pm \frac{i\sqrt{3}}{2}, x = \pm 1$    **33.** $x = 1$    **35.** $a = \frac{3}{2}$
**37.** $y = 12$    **39.** $x = 3; x = 7$ is extraneous    **41.** $n = 7$
**43.** $a = -1, a = -8$    **45.** $f = \frac{f_1 f_2}{f_1 + f_2}$    **47.** $r = \frac{E - IR}{I}$ or $\frac{E}{I} - R$
**49.** $T_2 = \frac{T_1 P_2 V_2}{P_1 V_1}$    **51.** $r = \frac{A - P}{Pt}$    **53. a.** $x = \frac{14}{3}$
**b.** $x = 8; x = 1$ is extraneous    **55. a.** $m = 3$    **b.** $x = 5$    **c.** $m = -64$
**d.** $x = -16$    **57. a.** $x = 25$    **b.** $x = 7; x = -2$ is extraneous
**c.** $x = 2, x = 18$    **d.** $x = 6; x = 0$ is extraneous    **59.** $x = -32$
**61.** $x = 9$    **63.** $x = -32, x = 22$    **65.** $x = -27, x = 125$
**67.** $x = \pm 2, x = \pm 5$    **69.** $b = -2, b = -1, b = 4, b = 5$
**71.** $x = -1, x = \frac{1}{4}$    **73.** $x = \pm\frac{1}{3}, x = \pm\frac{1}{2}$    **75.** $x = -4, x = 45$
**77.** $x = -6; x = \frac{-74}{9}$ is extraneous    **79. a.** $h = \sqrt{\left(\frac{S}{\pi r}\right)^2 - r^2}$
**b.** $S = 12\pi\sqrt{34}$ m$^2$    **81.** $x = \pm 3, x = -2$    **83.** 2, 4, 6 or $-2, 0, 2$
**85.** $r = 3$ m; $r = 0$ m and $r = -12$ m do not fit the context
**87.** 20 min    **89.** $v = 6$ mph    **91.** $P = 80\%$    **93. a.** 36 million mi
**b.** 67 million mi    **c.** 93 million mi    **d.** 142 million mi    **e.** 484 million mi
**f.** 887 million mi    **95.** Answers will vary.
**97. a.** $x = 6$ (since $7 - 6 > 0$, but $7 - 9 < 0$)
**b.** $x = 9$ (since $9 - 8 > 0$, but $6 - 8 < 0$)
**c.** $x = 6$ and $x = 9$ (since $9 - 3 > 0$, and $6 - 3 > 0$)
**d.** neither (since $6 - 12 < 0$ and $9 - 12 < 0$)    **99.** 1.7 hr    **101.** $3\frac{1}{2}$

## Making Connections, p. 140

**1.** b    **3.** c    **5.** b    **7.** d    **9.** g    **11.** h    **13.** c    **15.** c, e

## Summary and Concept Review, pp. 141–144

**1. a.** no    **b.** yes    **c.** yes    **2.** $b = 6$    **3.** $n \in \mathbb{R}$    **4.** $m = -1$
**5.** $x = \frac{1}{6}$    **6.** $p \in \{\}$    **7.** $g = 10$    **8.** $h = \frac{V}{\pi r^2}$    **9.** $L = \frac{P - 2W}{2}$
**10.** $x = \frac{c - b}{a}$    **11.** $y = \frac{2}{3}x - 2$    **12.** 8 gal    **13.** $12 + \frac{9}{8}\pi$ ft$^2 \approx 15.5$ ft$^2$
**14.** $\frac{2}{3}$ hr = 40 min    **15.** $a \geq 35$    **16.** $a < 2$    **17.** $s \leq 65$
**18.** $c \geq 1200$    **19.** $(5, \infty)$    **20.** $(-10, \infty)$    **21.** $(-\infty, 2]$
**22.** $(-9, 9]$    **23.** $(-6, \infty)$    **24.** $(-\infty, -1.6) \cup (2.3, \infty)$
**25. a.** $(-\infty, 3) \cup (3, \infty)$    **b.** $\left(-\infty, \frac{3}{2}\right) \cup \left(\frac{3}{2}, \infty\right)$    **c.** $[-5, \infty)$
**d.** $(-\infty, 6]$    **26.** $x \geq 96\%$    **27.** $\{-4, 10\}$    **28.** $\{-7, 3\}$    **29.** $\{-5, 8\}$
**30.** $\{-4, -1\}$    **31.** $(-\infty, -6) \cup (2, \infty)$    **32.** $[4, 32]$    **33.** $\{\}$    **34.** $\{\}$
**35.** $(-\infty, \infty)$    **36.** $[-2, 6]$    **37.** $(-\infty, -2] \cup [\frac{10}{3}, \infty)$
**38. a.** $|r - 2.5| \leq 1.7$    **b.** highest: 4.2 in., lowest: 0.8 in.    **39.** $6i\sqrt{2}$
**40.** $24i\sqrt{3}$    **41.** $-2 + i\sqrt{2}$    **42.** $3i\sqrt{2}$    **43.** $i$    **44.** $21 + 20i$
**45.** $-2 + i$    **46.** $-5 + 7i$    **47.** 13    **48.** $-20 - 12i$    **49.** verified
**50.** verified    **51. a.** $2x^2 + 3 = 0; a = 2, b = 0, c = 3$    **b.** not quadratic
**c.** $x^2 - 8x - 99 = 0; a = 1, b = -8, c = -99$
**d.** $x^2 + 16 = 0; a = 1, b = 0, c = 16$    **52. a.** $x = 5$ or $x = -2$
**b.** $x = -5$ or $x = 5$    **c.** $x = -\frac{1}{2}$ or $x = \frac{2}{3}$    **d.** $x = -\frac{5}{3}$ or $x = 3$
**53. a.** $x = \pm 3$    **b.** $x = 2 \pm \sqrt{5}$    **c.** $x = \pm i\sqrt{5}$    **d.** $x = \pm 5$
**54. a.** $x = 3$ or $x = -5$    **b.** $x = -8$ or $x = 2$
**c.** $x = 1 \pm \frac{\sqrt{10}}{2}$; $x \approx 2.58$ or $x \approx -0.58$    **d.** $x = 2$ or $x = \frac{1}{3}$
**55. a.** $x = \frac{3 \pm \sqrt{2}}{2}$; $x \approx 2.21$ or $x \approx 0.79$
**b.** $x = -4 \pm \sqrt{21}$; $x \approx -8.58$ or $x \approx 0.58$
**c.** $x = 2 \pm i\sqrt{5}$; $x \approx 2.00 \pm 2.24i$    **d.** $x = \frac{3}{2} \pm \frac{1}{2}i$
**56. a.** 1 sec    **b.** 244 ft    **c.** 8 sec    **57.** 11 in. $\times$ 17 in.
**58.** $x = \pm\sqrt{3}, x = 7$    **59.** $x = -2, x = 0, x = \frac{1}{3}$
**60.** $x = 0, x = 2, x = -1 \pm i\sqrt{3}$    **61.** $x = \pm\frac{1}{2}, x = \pm\frac{1}{2}i$    **62.** $x = \frac{-1}{2}$
**63.** $h = -\frac{5}{3}, h = 2$    **64.** $n = 13; n = -2$ is extraneous
**65.** $x = -3, x = 3$    **66.** $x = -4; x = 5$
**67.** $x = -1; x = 7$ is extraneous    **68.** $x = \frac{5}{2}$    **69.** $x = -5.8, x = 5$
**70.** $x = -2, x = -1, x = 4, x = 5$
**71.** $x = -3, x = 3, x = -i\sqrt{2}, x = i\sqrt{2}$
**72.** steam 60 mph, diesel 80 mph    **73.** 16 sec

## Practice Test, pp. 144–145

**1. a.** $x = 27$    **b.** $x = 2$    **c.** $C = \frac{P}{1 + k}$    **d.** $x = -4, x = -1$
**2.** 30 gal    **3. a.** $x > -30$    **b.** $-5 \leq x < 4$    **c.** $x \in \mathbb{R}$
**d.** $2 \leq x \leq 4$    **e.** $x < -4$ or $x > 2$    **4.** $S \geq 177$
**5.** $-\frac{4}{3} + \frac{i\sqrt{5}}{3}$    **6.** $-i$    **7. a.** 1    **b.** $i\sqrt{3}$    **c.** 1    **8.** $13 + i$    **9.** $-\frac{3}{2} + \frac{3}{2}i$
**10.** $(2 - 3i)^2 - 4(2 - 3i) + 13 = 0; -5 - 12i - 8 + 12i + 13 = 0; 0 = 0$ ✓
**11. a.** $z = -3, z = 10$    **b.** $x = \frac{2}{3}, x = 6$    **12. a.** $x = \pm 5$
**b.** $x = 1 \pm i\sqrt{3}$    **13. a.** $x = 5 \pm \frac{\sqrt{2}}{2}$    **b.** $x = \frac{5}{4} + \frac{i\sqrt{7}}{4}$
**14. a.** $x = \frac{3 \pm \sqrt{3}}{3}$    **b.** $x = 1 \pm 3i$    **15. a.** $P = -x^2 + 120x - 2000$
**b.** 10,000    **16. a.** $t = 5$ (May)    **b.** $t = 9$ (Sept.)    **c.** July; $3000 more
**17.** $x = 0, x = \pm\frac{1}{3}$    **18.** $x = -2, x = \pm\frac{3}{2}$
**19.** $x = 6$ ($x = -2$ is extraneous)    **20.** $x = -\frac{3}{2}, x = 2$
**21.** $x = 28$    **22.** $x = 16$ ($x = 4$ is extraneous)    **23.** $x = -11, x = 5$
**24.** $x = \pm 1, x = \pm 4$    **25. a.** $F \approx 64.8$ g    **b.** $W \approx 256$ g

## Calculator Exploration and Discovery, pp. 145–146

**1.** 16 lb for 71; 6 lb for 72
**2.** $x \in (-\infty, -4] \cup [2, \infty)$
**3.** $x \in [-5, 1]$
**4.** $x \in \mathbb{R}$

## Strengthening Core Skills p. 146

**Exercise 1:** $x = -3$ or $x = 7$
**Exercise 2:** $x \in [-5, 3]$
**Exercise 3:** $x \in (-\infty, -1] \cup [4, \infty)$

# CHAPTER 2
## Exercises 2.1, pp. 157–161

**1.** first, second    **3.** independent, output

**5.**
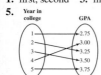

$D = \{1, 2, 3, 4, 5\}$
$R = \{2.75, 3.00, 3.25, 3.50, 3.75\}$

**7.** $D = \{1, 3, 5, 7, 9\}; R = \{2, 4, 6, 8, 10\}$

**9.** $D = \{4, -1, 2, -3\}; R = \{0, 5, 4, 2, 3\}$

**11.** $D: x \in \mathbb{R},$    **13.** $D: x \geq -2,$    **15.** $D: x \in \mathbb{R}$
$\;\;\;\;R: y \in \mathbb{R}$    $\;\;\;\;\;\;R: y \in \mathbb{R}$    $\;\;\;\;\;R: y \geq -1$

| x | y |
|---|---|
| −6 | 5 |
| −3 | 3 |
| 0 | 1 |
| 3 | −1 |
| 6 | −3 |
| 8 | $\frac{-13}{3}$ |

| x | y |
|---|---|
| −2 | 0 |
| 0 | 2, −2 |
| 1 | 3, −3 |
| 3 | 5, −5 |
| 6 | 8, −8 |
| 7 | 9, −9 |

| x | y |
|---|---|
| −3 | 8 |
| −2 | 3 |
| 0 | −1 |
| 2 | 3 |
| 3 | 8 |
| 4 | 15 |

**17.** $D: -5 \leq x \leq 5$    **19.** $D: x \geq 1$    **21.** $D: x \in \mathbb{R}$
$\;\;\;\;\;R: 0 \leq y \leq 5$    $\;\;\;\;\;R: y \in \mathbb{R}$    $\;\;\;\;\;R: y \in \mathbb{R}$

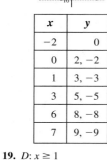

| x | y |
|---|---|
| −4 | 3 |
| −3 | 4 |
| 0 | 5 |
| 2 | $\sqrt{21}$ |
| 3 | 4 |
| 4 | 3 |

| x | y |
|---|---|
| 10 | 3, −3 |
| 5 | 2, −2 |
| 4 | $\sqrt{3}, -\sqrt{3}$ |
| 2 | 1, −1 |
| 1.25 | 0.5, −0.5 |
| 1 | 0 |

| x | y |
|---|---|
| −9 | −2 |
| −2 | −1 |
| −1 | 0 |
| 0 | 1 |
| 4 | $\sqrt[3]{5}$ |
| 7 | 2 |

**23.** $(3, 1)$    **25.** $(-0.7, -0.3)$    **27.** $(\frac{1}{20}, \frac{1}{24})$    **29.** $(0, -1)$    **31.** $(-1, 0)$

**33.** $2\sqrt{34}$    **35.** 10    **37.** right triangle    **39.** not a right triangle

**41.** right triangle

**43.** $x^2 + y^2 = 9$    **45.** $(x - 5)^2 + y^2 = 3$

**47.** $(x - 4)^2 + (y + 3)^2 = 4$    **49.** $(x + 7)^2 + (y + 4)^2 = 7$

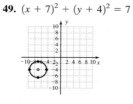

**51.** $(x - 1)^2 + (y + 2)^2 = 9$    **53.** $(x - 4)^2 + (y - 5)^2 = 12$

**55.** $(x - 7)^2 + (y - 1)^2 = 100$    **57.** $(x - 3)^2 + (y - 4)^2 = 41$

**59.** $(x - 5)^2 + (y - 4)^2 = 9$    **61.** $x^2 + y^2 = 25$

**63.** $(2, 3), r = 2, x \in [0, 4], y \in [1, 5]$

**65.** $(-1, 2), r = 2\sqrt{3}, x \in [-1 - 2\sqrt{3}, -1 + 2\sqrt{3}],$
$\;\;\;\;\;y \in [2 - 2\sqrt{3}, 2 + 2\sqrt{3}]$

**67.** $(-4, 0), r = 9, x \in [-13, 5], y \in [-9, 9]$

**69.** $(x - 5)^2 + (y - 6)^2 = 57, (5, 6), r = \sqrt{57}$

**71.** $(x - 5)^2 + (y + 2)^2 = 25, (5, -2), r = 5$

**73.** $x^2 + (y + 3)^2 = 14, (0, -3), r = \sqrt{14}$

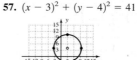

**75.** $(x + 2)^2 + (y + 5)^2 = 11, (-2, -5), r = \sqrt{11}$

**77.** $(x + 7)^2 + y^2 = 37, (-7, 0), r = \sqrt{37}$

**79.** $(x - 3)^2 + (y + 5)^2 = 32, (3, -5), r = 4\sqrt{2}$

**81. a.** $y = x^2 - 6x$  **83. b.** $x^2 + (y - 3)^2 = 36$
**85. f.** $(x - 1)^2 + (y + 2)^2 = 49$  **87. j.** $6x + y = x^2 + 9$
**89.** $A = \frac{1}{2}(8) + 22 - 1 = 25$ units$^2$
**91. a.** $(x - 5)^2 + (y - 12)^2 = 625$  **b.** no
**93.** Red: $(x - 2)^2 + (y - 2)^2 = 4$;
Blue: $(x - 2)^2 + y^2 = 16$;
Area blue $= 12\pi$ units$^2$
**95.**

No, distance between
centers is less
than sum of radii.

**97. a.** $(x - 3)^2 + (y - 5)^2 = 62^2$  **b.** raccoon: 61 m, opossum: 65 m
**c.** the raccoon  **99.** Answers will vary.
**101.** Let $s$ represent the length of the sides of the square. From the given
information, the length of diagonal $d$ is $s\sqrt{2}$. Since $d = 2r$, we have
$2r = s\sqrt{2}$ and solving for $s$ gives $s = r\sqrt{2}$. The area of a square is $A = s^2$
so $A = (r\sqrt{2})^2 = 2r^2$. Solving $A = 2r^2$ for $r$ gives $r = \sqrt{\frac{A}{2}}$ (since $r > 0$).
**103. a.** $40\frac{1}{3}$  **b.** $x^4$  **c.** $\frac{1}{5}$  **d.** 9  **e.** $4m^6n^2$  **f.** 6  **105.** $x = -3, 9$

## Exercises 2.2, pp. 171–175

**1.** 0, 0  **3.** yes, $m_1 \neq m_2$; no, $m_1 \cdot m_2 \neq -1$
**5.**

| $x$ | $y$ |
|---|---|
| $-6$ | 6 |
| $-3$ | 4 |
| 0 | 2 |
| 3 | 0 |

**7.**

| $x$ | $y$ |
|---|---|
| $-2$ | 1 |
| 0 | 4 |
| 2 | 7 |
| 4 | 10 |

**9.** $-0.5 = \frac{3}{2}(-3) + 4$
$-0.5 = -\frac{9}{2} + 4$
$-0.5 = -0.5$ ✓
$\frac{19}{4} = \frac{3}{2}(\frac{1}{2}) + 4$
$\frac{19}{4} = \frac{3}{4} + 4$
$\frac{19}{4} = \frac{19}{4}$ ✓

**11.**

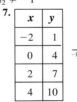

**13.**

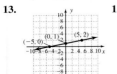

**15.**

**17.**

**19.**

**21.**

**23.**

**25.**

**27.**

**29.**

**31.** $m = 1$;
$(2, 4)$ and $(1, 3)$

**33.** $m = \frac{4}{3}$;
$(7, -1)$ and $(1, -9)$

**35.** $m = \frac{-15}{4}$;
$(5, -23)$ and $(-7, 22)$

**37.** $m = \frac{-4}{7}$;
$(-10, 10)$ and $(11, -2)$

**39.** $m = \frac{-3}{8}$, $(\frac{-1}{4}, \frac{1}{4})$, $(\frac{7}{4}, -\frac{1}{2})$

**41. a.** $m = 125$, cost increased \$125,000/1000 ft$^2$  **b.** \$375,000
**43. a.** $m = 12$, 12 m$^3$ dumped per garbage truck  **b.** 83 trucks
**45. a.** $m = \frac{23}{6}$, a person weighs 23 lb more for each additional 6 in. in
height  **b.** $\approx 3.8$
**47.** In inches: $(0, -6)$ and $(576, -18)$: $m = \frac{-1}{48}$. The sewer line is 1 in.
deeper for each 48 in. in length.
**49.** $m = 0$ (line is horizontal)  **51.** slope is undefined (line is vertical)
$(2, -5), (-7, -5)$       $(-5, 1), (-5, -6)$

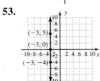

**53.**

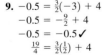

**55.**

**57.** $L_1: x = 2$; $L_2: y = 4$; point of intersection $(2, 4)$
**59. a.** For any two points chosen $m = 0$, indicating
there has been no increase or decrease in the number
of supreme court justices.  **b.** For any two points
chosen $m = \frac{1}{10}$, which indicates that over the last 5
decades, one nonwhite or nonmale justice has been
added to the court every 10 yr.

**61.** parallel   **63.** neither   **65.** parallel   **67.** right triangle
**69.** right triangle   **71.** not a right triangle
**73. a.** 78.2 yr   **b.** $T = \frac{L - 73.7}{0.15}$   **c.** 2015
**75.** $v = -1250t + 8500$   **a.** \$3500   **b.** 5 yr
**77.** $h = -3t + 300$   **a.** 273 in.   **b.** 20 months
**79. a.** \$7360   **b.** 2014   **81. a.** 24.1   **b.** 2009
**83.** Yes they will meet, the two roads are not parallel: $\frac{38}{12} \neq \frac{30}{9.5}$.
**85. a.** (1999, 6.5), (2005, 44.5); $\frac{\Delta s}{\Delta t} = \frac{19}{3} = \frac{6.3}{1}$: screen time for kids
8 to 18 yrs is increasing at a rate of over 6 hr/yr.   **b.** The rate of
change $\frac{\Delta s}{\Delta t} = \frac{6.3}{1}$ gives $(2005 + 5, 44.5 + 5 \cdot 6.3) \rightarrow (2010, 76.1\overline{6})$; just over
76 hr/week   **87.** $a = -6$   **89. a.** 142   **b.** $-83$   **c.** 9   **d.** $\frac{27}{2}$
**91. a.** $10\sqrt{5}$   **b.** 4   **93. a.** $(x - 3)(x + 2)(x - 2)$
**b.** $(x - 24)(x + 1)$   **c.** $(x - 5)(x^2 + 5x + 25)$

## Exercises 2.3, pp. 184–187

**1.** $\frac{-7}{4}$, (0, 3)   **3.** Answers will vary.
**5.** $y = \frac{-4}{5}x + 2$   **7.** $y = 2x + 7$   **9.** $y = \frac{-5}{3}x - 5$

| x | y |
|----|----|
| $-5$ | 6 |
| $-2$ | $\frac{18}{5}$ |
| 0 | 2 |
| 1 | $\frac{6}{5}$ |
| 3 | $\frac{-2}{5}$ |

| x | y |
|----|----|
| $-5$ | $-3$ |
| $-2$ | 3 |
| 0 | 7 |
| 1 | 9 |
| 3 | 13 |

| x | y |
|----|----|
| $-5$ | $\frac{10}{3}$ |
| $-2$ | $\frac{-5}{3}$ |
| 0 | $-5$ |
| 1 | $\frac{-20}{3}$ |
| 3 | $-10$ |

**11.** $y = 2x - 3$: 2, $-3$   **13.** $y = \frac{5}{-3}x - 7$: $\frac{5}{-3}$, $-7$
**15.** $y = \frac{-35}{6}x - 4$: $\frac{-35}{6}$, $-4$
**17.**    **19.**    **21.**

**23.** $\frac{-3}{4}$, $y = \frac{-3}{4}x + 3$; The coefficient of $x$ is the slope and the constant is
the $y$-intercept.   **25.** $\frac{2}{5}$, $y = \frac{2}{5}x - 2$; The coefficient of $x$ is the slope and
the constant is the $y$-intercept.   **27.** $\frac{4}{5}$, $y = \frac{4}{5}x + 3$; The coefficient of $x$ is
the slope and the constant is the $y$-intercept.
**29.** $y = \frac{-2}{3}x + 2$, $f(x) = \frac{-2}{3}x + 2$, $m = \frac{-2}{3}$, $y$-intercept (0, 2)
**31.** $y = \frac{-5}{4}x + 5$, $f(x) = \frac{-5}{4}x + 5$, $m = \frac{-5}{4}$, $y$-intercept (0, 5)
**33.** $y = \frac{1}{3}x$, $f(x) = \frac{1}{3}x$, $m = \frac{1}{3}$, $y$-intercept (0, 0)
**35.** $y = \frac{-3}{4}x + 3$, $f(x) = \frac{-3}{4}x + 3$, $m = \frac{-3}{4}$, $y$-intercept (0, 3)
**37.** $y = \frac{2}{3}x + 1$   **39.** $y = -2x - 3$   **41.** $y = \frac{-3}{2}x - 4$
**43.** $y = 2x - 13$   **45.** $y = 250x + 500$

**47.**    **49.**    **51.**

**53.**    **55.** $y = -\frac{3}{5}x + 4$   **57.** $y = \frac{2}{3}x - 5$

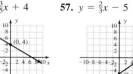

**59.** $y = \frac{2}{5}x + 4$   **61.** $y = \frac{-5}{3}x + 7$   **63.** $y = \frac{-12}{5}x - \frac{29}{5}$
**65.** $y = 5$   **67.** perpendicular   **69.** neither   **71.** parallel
**73. a.** $y = \frac{4}{9}x + \frac{31}{9}$   **b.** $y = \frac{-9}{4}x + \frac{3}{4}$   **75. a.** $y = \frac{-1}{2}x - 2$
**b.** $y = 2x - 2$

**77.** $y + 5 = 2(x - 2)$

**79.** $y + 4 = \frac{3}{8}(x - 3)$

**81.** $y + 3.1 = 0.5(x - 1.8)$

**83.** $y - 2 = \frac{6}{5}(x - 4)$; For each 5000 additional sales, income rises
\$6000.   **85.** $y - 100 = \frac{-20}{1}(x - 0.5)$; For every hour of television, a
student's final grade falls 20%.   **87.** $y - 10 = \frac{35}{2}(x - \frac{1}{2})$; Every 2 in. of
rainfall increases the number of cattle raised per acre by 35.   **89.** C   **91.** A
**93.** B   **95.** D   **97.** $m = \frac{-a}{b}$, $y$-intercept $= \frac{c}{b}$   **a.** $m = \frac{-3}{4}$, $y$-intercept
(0, 2)   **b.** $m = \frac{-2}{5}$, $y$-intercept (0, $-3$)   **c.** $m = \frac{5}{6}$, $y$-intercept (0, 2)
**d.** $m = \frac{5}{3}$, $y$-intercept (0, 3)   **99. a.** As the temperature increases 5°C, the
velocity of sound waves increases 3 m/sec. At a temperature of 0°C, the
velocity is 331 m/sec.   **b.** 343 m/sec   **c.** 50°C   **101. a.** value depends
on time: $V = 5t + 190$   **b.** $\frac{\Delta V}{\Delta t} = \frac{5}{1}$, the coin gains \$5.00 in value each
year; $5(0) + 190 = \$190$ is the purchase price   **c.** $5(8) + 190 = 230$, the
penny will be worth \$230.00   **d.** 16 yr   **103. a.** number of Internet
users depends on time: $U = 11t + 132$   **b.** $\frac{\Delta U}{\Delta t} = \frac{11}{1}$, the number of users
grows by 11 million every year   **c.** $11(10) + 132 = 242$, there will be
about 242 million users in 2010   **d.** $U = 300$ gives $300 = 11t + 132$,
with $t \approx 15.3$; in the year 2015   **105. a.** elevation depends on driving
distance: $E = -365d + 11,200$   **b.** $\frac{\Delta E}{\Delta d} = \frac{-365}{1}$, elevation decreases 365 ft
each mile; $-365(0) + 11,200 = 11,200$ ft, tunnel elevation
**c.** $-365(8) + 11,200 = 8280$, the elevation would be 8280 ft
**d.** $\frac{\Delta E}{\Delta d} = \frac{-365}{5280} \approx -0.069$ shows the grade is about 6.9%
**107. a.** $y = \frac{-3}{2}x + 6$   **b.** $a = 3, b = 2, x = 6$, and $y = -7$ gives
$(3)(6) + (2)(-7) = k$, with $4 = k$. The equation is $3x + 2y = 4$,
or $y = \frac{-3}{2}x + 2$; verified   **c.** $a = 3, b = 2, x = 6$, and $y = -7$ gives
$(2)(6) - (3)(-7) = k$, with $33 = k$. The equation is $2x - 3y = 33$, or
$y = \frac{2}{3}x - 11$; verified   **109. 1.** d.   **2.** a.   **3.** c.   **4.** b.   **5.** f.   **6.** h.
**111. a.** 9   **b.** $9|x|$   **113.** 113.10 yd$^2$

## Mid-Chapter Check, p. 188

**1.**

| x | y |
|----|----|
| $-6$ | 69 |
| $-4$ | 29 |
| $-2$ | 5 |
| 0 | $-3$ |
| 2 | 5 |
| 4 | 29 |
| 6 | 69 |

**2.** 11 mi. away at (0.2, 0.6)
**3.**  center $(-5, 12)$; radius 13   **4.**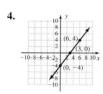

**5.** $m = \frac{-18}{7}$   **6.** positive, loss is decreasing (profit is increasing); $m = \frac{3}{2}$,
yes; $\frac{1.5}{1}$. Each year Data.com's loss decreases by 1.5 million.

**7. a.** $E(x) = 7.5x + 950$ **b.** $1100, $1175, $1250
**c.**

**d.** 47 snowboards

**8.** $L_1$: $x = -3$, $m$ undefined; $L_2$: $y = 2$, $m = 0$
**9. a.** $V = -3000t + 30,000$ **b.** $15,000 **c.** 2019
**10. a.** for $3x + 4y = 12$: (4, 0), (0, 3); for $4x - 3y = 12$: (3, 0), (0, -4)
**b.** for $3x + 4y = 12$: $m = \frac{-3}{4}$; for $4x - 3y = 12$: $m = \frac{4}{3}$; perpendicular

## Exercises 2.4, pp. 197–202

**1.** first **3.** Answers will vary. **5.** function **7.** Not a function; Shaq is paired with two heights. **9.** The summer temperatures "hit" 106° in two different years and we cannot be sure which year is intended.
**11.** Not a function; 4 is paired with 2, 7, and $-5$. **13.** function
**15.** function **17.** Not a function; $-2$ is paired with 3 and $-4$.
**19.** function **21.** function **23.** Not a function; 0 is paired with 4 and $-4$.
**25.** function **27.** Not a function; 4 is paired with $-1$ and 1.
**29.** function
**31.**

**33.**

function          function

**35.** function, $x \in [-4, 5]$, $y \in [-2, 3]$
**37.** function, $x \in [-4, \infty)$, $y \in [-4, \infty)$
**39.** function, $x \in [-4, 4]$, $y \in [-5, -1]$
**41.** function, $x \in (-\infty, \infty)$, $y \in (-\infty, \infty)$
**43.** not a function, $x \in [-3, 5]$, $y \in [-3, 3]$
**45.** not a function, $x \in (-\infty, 3]$, $y \in (-\infty, \infty)$
**47.** $x \in (-\infty, 5) \cup (5, \infty)$ **49.** $a \in [\frac{-5}{3}, \infty)$
**51.** $x \in (-\infty, -5) \cup (-5, 5) \cup (5, \infty)$
**53.** $v \in (-\infty, -3\sqrt{2}) \cup (-3\sqrt{2}, 3\sqrt{2}) \cup (3\sqrt{2}, \infty)$
**55.** $x \in (-\infty, \infty)$ **57.** $n \in (-\infty, \infty)$
**59.** $x \in (-\infty, \infty)$ **61.** $x \in (-\infty, -2) \cup (-2, 5) \cup (5, \infty)$
**63.** $x \in [2, \frac{5}{2}) \cup (\frac{5}{2}, \infty)$ **65.** $x \in (-4, \infty)$ **67.** $x \in (3, \infty)$
**69.** $x \in (\frac{7}{3}, \infty)$ **71.** $0, \frac{18}{5}, c + 3, \frac{c + 7}{2}$ **73.** $204, \frac{12}{5}, 20c^2 - 8c$,
$5c^2 + 6c + 1$ **75.** $4, -18, \frac{4}{a}, \frac{12}{a - 2}$ **77.** $5, -5, -5$ if
$a < 0$ or 5 if $a > 0$, 5 if $a > 2$ or $-5$ if $a < 2$
**79.** $8\pi, 3\pi, 4\pi c, 2\pi(c + 3)$ **81.** $16\pi, \frac{9}{4}\pi, 4\pi c^2, (c^2 + 6c + 9)\pi$
**83.** $3, \sqrt{2}, \sqrt{6a - 1}, \sqrt{2a - 3}$ **85.** $\frac{14}{5}, \frac{7}{9}, \frac{27a^2 - 5}{9a^2}, \frac{3a^2 - 6a - 2}{a^2 - 2a + 1}$
**87. a.** $D$: $\{-1, 0, 1, 2, 3, 4, 5\}$ **b.** $R$: $\{-2, -1, 0, 1, 2, 3, 4\}$
**c.** 1 **d.** $-1$ **89. a.** $D$: $[-5, 5]$ **b.** $y \in [-3, 4]$ **c.** $-2$ **d.** $-4$ and 0
**91. a.** $D$: $[-3, \infty)$ **b.** $y \in (-\infty, 4]$ **c.** 2 **d.** $-2$ and 2 **93. a.** 186.5 lb
**b.** 37 lb **95. a.** (3, 183), (5, 241), (7, 299), (9, 357), (11, 415); yes
**b.** $473 **c.** 2014 **97. a.** $N(g) = 2.5g$ **b.** $g \in [0, 5]$; $N \in [0, 12.5]$
**99. a.** $[0, \infty)$ **b.** about 2356 units$^3$ **c.** 800 units$^3$
**101. a.** $c(t) = 42.50t + 50$ **b.** $156.25 **c.** 5 hr
**d.** $t \in [0, 10.6]$; $c \in [0, 500]$ **103. a.** Grass height depends on time:
$H(t) = 2.1t + 1.9$. **b.** $H(3) = 8.2$, the grass will be 8.2 cm tall.
**c.** $H(t) = 14.5, 14.5 = 2.1t + 1.9$ gives $t = 6$. He can avoid mowing the yard for 6 weeks. **105. a.** yes, passes vertical line test **b.** between 35 and 40 **c.** 2013
**107.** negative outputs become positive

**109.** $(x - 4)^2 + (y + 1)^2 = 25$

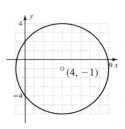

**111.** $x = \frac{7}{3}, -1$

## Exercises 2.5, pp. 217–223

**1.** even, $y$-axis, odd, origin **3.** Answers will vary.
**5.**

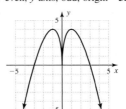

**7.** even **9.** even

**11.**

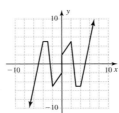

**13.** odd **15.** not odd

**17.** neither **19.** odd **21.** neither **23.** odd **25.** $x \in [-1, 1] \cup [3, \infty]$
**27.** $x \in (-\infty, -1) \cup (-1, 1) \cup [1, \infty)$ **29.** $x \in \{-2\} \cup [2, \infty)$
**31.** $x \in (3, \infty)$ **33.** $V(x)$↑: $x \in (-3, 1) \cup (4, 6)$;
$V(x)$↓: $x \in (-\infty, -3) \cup (1, 4)$; constant: none **35.** $f(x)$↑: $x \in (1, 4)$;
$f(x)$↓: $x \in (-2, 1) \cup (4, \infty)$; constant: $x \in (-\infty, -2)$
**37. a.** $p(x)$↑: $x \in (-\infty, \infty)$; $p(x)$↓: none **b.** down, up
**39. a.** $f(x)$↑: $x \in (-3, 0) \cup (3, \infty)$; $f(x)$↓: $x \in (-\infty, -3) \cup (0, 3)$
**b.** up, up **41. a.** $x \in (-\infty, \infty)$; $y \in (-\infty, 5]$ **b.** $x = 1, 3$
**c.** $H(x) \geq 0$: $x \in [1, 3]$; $H(x) \leq 0$: $x \in (-\infty, 1] \cup [3, \infty)$
**d.** $H(x)$↑: $x \in (-\infty, 2)$; $H(x)$↓: $x \in (2, \infty)$ **e.** local max: $y = 5$ at (2, 5)
**43. a.** $x \in (-\infty, \infty)$; $y \in (-\infty, \infty)$ **b.** $x = -1, 5$
**c.** $g(x) \geq 0$: $x \in [-1, \infty)$; $g(x) \leq 0$: $x \in (-\infty, -1] \cup \{3.5\}$
**d.** $g(x)$↑: $x \in (-\infty, 1) \cup (5, \infty)$; $g(x)$↓: $x \in (1, 5)$ **e.** local max: $y = 6$
at (1, 6); local min: $y = 0$ at (5, 0) **45. a.** $x \in [-4, \infty)$; $y \in (-\infty, 3]$
**b.** $x = -4, 2$ **c.** $Y_1 \geq 0$: $x \in [-4, 2]$; $Y_1 \leq 0$: $x \in [2, \infty]$
**d.** $Y_1$↑: $x \in (-4, -2)$; $Y_1$↓: $x \in (-2, \infty)$ **e.** local max: $y = 3$
at $(-2, 3)$; endpoint min $y = 0$ at $(-4, 0)$
**47. a.** $x \in (-\infty, \infty)$, $y \in (-\infty, \infty)$
**b.** $x = -4$ **c.** $p(x) \geq 0$: $x \in [-4, \infty)$; $p(x) \leq 0$: $x \in (-\infty, -4]$
**d.** $p(x)$↑: $x \in (-\infty, -3) \cup (-3, \infty)$; $p(x)$↓: never decreasing **e.** local
max: none; local min: none **49.** 0 **51.** 0.5 **53.** $-1.6$ **55.** $-0.1$
**57. a.** 48 ft/sec **b.** 32 ft/sec **c.** 16 ft/sec **d.** $-32$ ft/sec
**59. a.** 0 m/sec **b.** 0 m/sec **c.** 4.9 m/sec **d.** $-4.9$ m/sec
**61.** Answers will vary. **63.** Answers will vary.

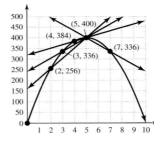 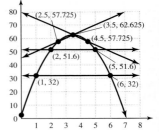

**65.** $\frac{\Delta y}{\Delta x} = 39$   **67.** $\frac{\Delta y}{\Delta x} = 1$   **69.** $\frac{\Delta F}{\Delta m} = 9.8$   **71.** $\frac{\Delta A}{\Delta r} \approx 37.70$
**73. a.** $\frac{\Delta g}{\Delta x} = 2x + 2 + h$   **b.** $\frac{\Delta g}{\Delta x} = -3.9$   **c.** $\frac{\Delta g}{\Delta x} = 3.01$
**d.** The rates of change have opposite sign, with the
secant line to the left being slightly more steep.

**75. a.** $\frac{\Delta g}{\Delta x} = 3x^2 + 3xh + h^2$   **b.** $\frac{\Delta g}{\Delta x} = 12.61$   **c.** $\frac{\Delta g}{\Delta x} \approx 0.49$
**d.** Both lines have a positive slope, but the line at
$x = -2$ is much steeper.

**77. a.** $\frac{\Delta j}{\Delta x} = \frac{-2x - h}{x^2(x + h)^2}$   **b.** $[0.50, 0.51]$: $\frac{\Delta j}{\Delta x} \approx -15.5$
**c.** $[1.50, 1.51]$: $\frac{\Delta j}{\Delta x} \approx -0.6$   **d.** Answers will vary.
**79. a.** $\frac{\Delta g}{\Delta x} = 3x^2 + 3xh + h^2$   **b.** $[-2.01, -2.00]$: $\frac{\Delta g}{\Delta x} \approx 12.1$
**c.** $[0.40, 0.41]$: $\frac{\Delta g}{\Delta x} \approx 0.5$   **d.** Answers will vary.   **81.** 0   **83.** $2x + h$
**85.** $3x^2 + 3xh + h^2$   **87.** $-\frac{2x + h}{x^2(x + h)^2}$   **89.** 0   **91.** $2x + h$
**93.** $6x^2 + 6xh + 2h^2$   **95.** $\frac{3(2x + h)}{x^2(x + h)^2}$   **97.** $y = \sin x$:   **a.** $y \in [-1, 1]$
**b.** $(-360, 0), (-180, 0), (0, 0), (180, 0), (360, 0)$
**c.** $f(x)\uparrow$: $x \in (-360, -270) \cup (-90, 90) \cup (270, 360)$; $f(x)\downarrow$:
$x \in (-270, -90) \cup (90, 270)$   **d.** min: $(-90, -1)$ and $(270, -1)$,
max: $(-270, 1)$ and $(90, 1)$   **e.** odd;
$y = \cos x$:   **a.** $y \in [-1, 1]$   **b.** $(-270, 0), (-90, 0), (90, 0), (270, 0)$
**c.** $f(x)\uparrow$: $x \in (-180, 0) \cup (180, 360)$; $f(x)\downarrow$: $x \in (-360, -180) \cup (0, 180)$
**d.** min: $(-180, -1)$ and $(180, -1)$, max: $(-360, 1)$, $(0, 1)$, and $(360, 1)$
**e.** even   **99. a.** $x \in [0, 260]$, $y \in [0, 80]$   **b.** 80 ft   **c.** 120 ft   **d.** yes
**e.** $(0, 120)$   **f.** $(120, 260)$   **101. a.** $x \in (-\infty, \infty)$; $y \in [-1, \infty)$
**b.** $(-1, 0), (1, 0)$   **c.** $f(x) \geq 0$: $x \in (-\infty, -1] \cup [1, \infty)$;
$f(x) < 0$: $x \in (-1, 1)$   **d.** $f(x)\uparrow$: $x \in (0, \infty)$, $f(x)\downarrow$: $x \in (-\infty, 0)$
**e.** up/up   **f.** min: $(0, -1)$
**103.** zeroes: $(-8, 0), (-4, 0), (0, 0), (4, 0)$;
min: $(-10, -6), (-2, -1), (4, 0)$; max: $(-6, 2), (2, 2)$

**105. a.** $\frac{\Delta \text{weight}}{\Delta \text{time}} = \frac{50}{1}$, positive, 50 g are gained each week   **b.** 25th to 29th:
$\frac{\Delta w}{\Delta t} = \frac{50}{1}$; 32nd to 36th: $\frac{\Delta w}{\Delta t} = \frac{250}{1}$; the weight gain is five times greater in the
later weeks.   **107. a.** 176 ft   **b.** 320 ft   **c.** 144 ft/sec   **d.** $-144$ ft/sec;
The arrow is returning to Earth.   **109. a.** 17.89 ft/sec; 25.30 ft/sec
**b.** 30.98 ft/sec; 35.78 ft/sec   **c.** between 5 and 10   **d.** 1.48 (ft/sec)/ft,
0.96 (ft/sec)/ft; For low heights, small increases in drop height result in
dramatic increases in impact velocity.
**111. a.** March: $\frac{\Delta d}{\Delta t} \approx 15$, June: $\frac{\Delta d}{\Delta t} \approx 3$, 5 times faster
**b.** decreasing $(\frac{\Delta d}{\Delta t} < 0)$; 5000 units/month   **113. a.** verified   **b.** verified
**115.** Answers will vary.   **117.** $x = -2, x = 10$   **119. a.** $\frac{12}{4 - x^2}$   **b.** $\frac{9}{4 - x^2}$

## Exercises 2.6, pp. 232–236

**1.** linear   **3.** positive
**5. a.**

**b.** positive

**7. a.**

**b.** linear   **c.** positive

**9. a.**

**b.** positive   **c.** strong

**11. a.** (A) (D) (C) (B)
**b.**

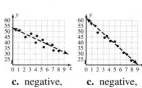

**c.** positive,   **c.** positive,   **c.** negative,   **c.** negative,
**d.** $m \approx 3.8$   **d.** $m \approx 4.2$   **d.** $m \approx -2.4$   **d.** $m \approx -4.6$
**13. a.** linear   **b.** positive   **c.** strong   **d.** $m \approx 4.2$   **15. a.** nonlinear
**b.** positive   **c.** NA   **d.** NA   **17. a.** nonlinear   **b.** negative   **c.** NA
**d.** NA

**19. a.**

**b.** positive
**c.** $f(x) = 2.4x + 62.3$,
$f(5) = 74.3(74,300)$,
$f(21) = 112.7 (112,700)$

**21.** Using $(5, 7.6)$ and $(36, 39.7)$: $y = 1.04x + 2.42$; GDP in 2010 is
predicted to be just over 44,000.
**23. a.**

| $x$ | $y$ |
|---|---|
| 1 | 6.28 |
| 2 | 12.57 |
| 3 | 18.85 |
| 4 | 25.13 |
| 5 | 31.42 |
| 6 | 37.70 |

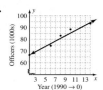

**b.** positive, larger radius $\Rightarrow$ larger area   **c.** perfect correlation
**d.** $m = 2\pi$

**25. a.**

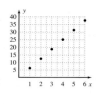

**b.** linear   **c.** positive
**d.** $y = 0.96x + 1.55$, 63.95 in.

**27. a.**

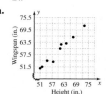

**b.** linear   **c.** positive   **d.** $y \approx 1.45x + 3.18$; about 24,900 stores
**29. a.**

**b.** linear   **c.** positive
**d.** $y = 0.203x + 73.32$,
about 96.9 in.

**31. a.**

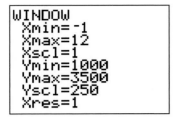

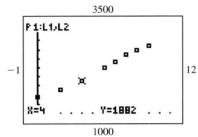

**b.**

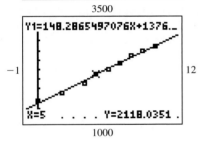

**c.** $Y_1(15) \approx 3600.9$, over 3600 billion
**d.** for $4000 \approx 148.29x + 1376.60$; $x \approx 17.7$, in 2017
**33. a.** $r^2 \approx 0.985$   **b.** $r^2 \approx 0.985$   **c.** they are almost identical; context, pattern of scatterplot, anticipated growth, etc.   **35.** No. Except for the endpoints of the domain, one $x$ is mapped to two $y$'s.   **37.** $r = \frac{A - P}{Pt}$

## Making Connections, p. 236

**1.** d   **3.** a   **5.** b   **7.** c   **9.** f   **11.** d   **13.** f   **15.** a

## Summary and Concept Review, pp. 236–240

**1.** $x \in \{-7, -4, 0, 3, 5\}$ $y \in \{-2, 0, 1, 3, 8\}$

**2.**

| $x$ | $y$ |
|-----|-----|
| $-5$ | 0 |
| $-4$ | 3 |
| $-2$ | $\sqrt{21} \approx 4.58$ |
| 0 | 5 |
| 2 | $\sqrt{21} \approx 4.58$ |
| 4 | 3 |
| 5 | 0 |

$x \in [-5, 5]$   $y \in [0, 5]$

**3.**      $(0, -5), (-3, 0)$

**4.** 65 mi   **5.** $\left(\frac{5}{2}, -3\right)$

**6.** 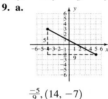    **7.**

**8.** $(x + 1.5)^2 + (y - 2)^2 = 6.25$

**9. a.**     **b.**

$\frac{-5}{9}, (14, -7)$          $\frac{1}{3}, (0, 3)$

**10. a.** parallel   **b.** perpendicular

**11. a.**     **b.**

**12. a.**     **b.**

**13. a.** vertical
    **b.** horizontal
    **c.** neither

**14.** $m = \frac{2}{3}$, $y$-intercept $(0, 2)$; When the rodent population increases by 3000, the hawk population increases by 200.
**15. a.** $y = \frac{-4}{3}x + 4$, $m = \frac{-4}{3}$, $y$-intercept $(0, 4)$
**b.** $y = \frac{5}{3}x - 5$, $m = \frac{5}{3}$, $y$-intercept $(0, -5)$

**16. a.**     **b.**

**17. a.**  **b.**

**18.** $y = 5$, $x = -2$; $y = 5$  **19.** $y = \frac{-3}{4}x + \frac{11}{4}$  **20.** $f(x) = \frac{4}{3}x$

**21. a.** $m = \frac{2}{5}$, $y$-intercept $(0, 2)$  **b.** $y = \frac{2}{5}x + 2$  **c.** When the rabbit population increases by 500, the wolf population increases by 200.

**22. a.** $y - 90 = \frac{-15}{2}(x - 2)$  **b.** $f(x) = \frac{-15}{2}x + 105$
**c.** $(14, 0)$, $(0, 105)$  **d.** $f(20) = -45$, $x = 12$

**23. a.** $x \in [\frac{-5}{4}, \infty)$  **b.** $x \in (-\infty, -2) \cup (-2, 3) \cup (3, \infty)$

**24.** $14$; $\frac{26}{9}$; $18a^2 - 9a$  **25.** It is a function.

**26. I. a.** $D = \{-1, 0, 1, 2, 3, 4, 5\}$, $R = \{-2, -1, 0, 1, 2, 3, 4\}$  **b.** 1  **c.** 2
**II. a.** $x \in (-\infty, \infty)$, $y \in (-\infty, \infty)$  **b.** $-1$  **c.** 3
**III. a.** $x \in [-3, \infty)$, $y \in [-4, \infty)$  **b.** $-1$  **c.** $-3$ or 3

**27.** $D: x \in (-\infty, \infty)$, $R: y \in [-5, \infty)$, $f(x)\uparrow: x \in (2, \infty)$,
$f(x)\downarrow: x \in (-\infty, 2)$, $f(x) > 0: x \in (-\infty, -1) \cup (5, \infty)$,
$f(x) < 0: x \in (-1, 5)$

**28.** $D: x \in [-3, \infty)$, $R: y \in (-\infty, 0]$, $f(x)\uparrow:$ none, $f(x)\downarrow: x \in (-3, \infty)$,
$f(x) > 0:$ none, $f(x) < 0: x \in (-3, \infty)$

**29.** $D: x \in (-\infty, \infty)$, $R: y \in (-\infty, \infty)$, $f(x)\uparrow: x \in (-\infty, -3) \cup (1, \infty)$,
$f(x)\downarrow: x \in (-3, 1)$, $f(x) > 0: x \in (-5, -1) \cup (4, \infty)$,
$f(x) < 0: x \in (-\infty, -5) \cup [-1, 4)$

**30. a.** odd  **b.** even  **c.** neither  **d.** odd

**31.**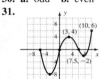
zeroes: $(-6, 0)$, $(0, 0)$
$(6, 0)$ $(9, 0)$
min: $(-3, -8)$,
$(-7.5, -2)$
max: $(-6, 0)$, $(3, 4)$

**32. a.** 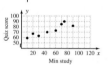  **b.** linear  **c.** positive

**33. a.** $f(x) = 0.35x + 56.10$  **b.** 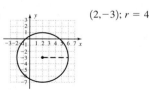  **c.** strong

**34. a.** $f(60) = 77.1$; over 77%  **b.** about 97 min

## Practice Test, pp. 240–241

**1.** $a$ and $c$ are nonfunctions, they do not pass the vertical line test
**2.** neither
**3.** 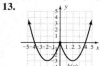  **4.**   $(2, -3)$; $r = 4$

**5.** $V = \frac{20}{3}t + \frac{20}{3}$, $66\frac{2}{3}$ mph  **6.** $y = \frac{-6}{5}x + \frac{2}{5}$
**7. a.** $(7.5, 1.5)$  **b.** $\approx 61.27$ mi  **8.** $L_1: x = -3$, $L_2: y = 4$
**9. a.** $x \in \{-4, -2, 0, 2, 4, 6\}$; $y \in \{-2, -1, 0, 1, 2, 3\}$
**b.** $x \in [-2, 6]$; $y \in [1, 4]$  **10. a.** 300  **b.** 30  **c.** $W(h) = \frac{25}{2}h$
**d.** Wages are $12.50/hr.  **e.** $h \in [0, 40]$; $w \in [0, 500]$  **11. a.** $\frac{7}{2}$
**b.** $\frac{-a^2 - 6a - 7}{a^2 + 6a + 9}$  **12. a.** $D: x \in [-4, \infty)$, $R: y \in [-3, \infty)$  **b.** $f(-1) \approx 2.2$
**c.** $f(x) < 0: x \in [-4, -3)$; $f(x) > 0: x \in (-3, \infty)$
**d.** $f(x)\uparrow: x \in [-4, \infty)$; never decreasing or constant
**13.**   **14.**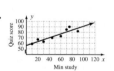

**15.**

| $x$ | $f(x)$ |
|-----|--------|
| $-5$ | 9 |
| $-3$ | $3\sqrt{7} \approx 7.9$ |
| $-1$ | $3\sqrt{5} \approx 6.7$ |
| 0 | 6 |
| 1 | $3\sqrt{3} \approx 5.2$ |
| 3 | 3 |
| 5 | undefined |

**16. a.** $D: x \in [0, \infty)$; $R: y \in (-\infty, 50]$  **b.** down on the right  **c.** endpoint max of 50 at $(0, 50)$, local min of 12 at $(23, 12)$, local max of 33 at $(57, 33)$

**17. a.** $\frac{\Delta r}{\Delta t} = 16t + 8h$  **b.** 24; Over this period, the rabbit population was growing at an average rate of 24 rabbits per year.  **18. a.** $\frac{\Delta \text{sales}}{\Delta \text{time}} = \frac{15.55}{1}$
**b.** Sales are increasing at a rate of 15.55 million phones per year.  **c.** 2009: about 34.4 million sales, 2012: about 81.05 million sales, 2015: about 127.7 million sales

**19. a.** 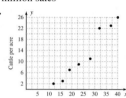  **b.** linear  **c.** positive

**20. a.** $f(x) = 0.91x - 10.78$  **b.** $f(50) \approx 35$  **c.** strong

## Calculator Exploration and Discovery, pp. 241–242

**1.** $(-1.5, 0)$, $(0, 1)$  **2.** $(11, 0)$, $(0, -15.8)$  **3.** $x \approx -2.87$, $x \approx 0.87$, min: $y = -7$ at $(-1, -7)$, no max  **4.** $x \approx -1.88$, $x \approx 0.35$, $x \approx 1.53$, max: $y = 3$ at $(-1, 3)$, min: $y = -1$ at $(1, -1)$  **5.** $x \approx 1.35$, $x \approx 6.65$, min: $y = -7$ at $(4, -7)$, no max  **6.** $x = -2$, $x = 2$, min: $y = 0$ at $(2, 0)$, max: $y \approx 9.48$ at $(-0.67, 9.48)$  **7.** $x = -2$, $x \approx -0.41$, $x = 0$, $x \approx 2.41$, min: $y \approx -3.20$ at $(-1.47, -3.20)$, min: $y \approx -9.51$ at $(1.67, -9.51)$, max: $y \approx 0.20$ at $(-0.20, 0.20)$  **8.** $x = -4$, $x = 0$, min: $y = -3.08$ at $(-2.67, -3.08)$

## Strengthening Core Skills, p. 243

**1. a.** $\frac{1}{3}$, increasing  **b.** $y - 5 = \frac{1}{3}(x - 0)$,
$y = \frac{1}{3}x + 5$
**c.** $(0, 5)$, $(-15, 0)$

**2. a.** $\frac{-7}{3}$, decreasing  **b.** $y - 9 = \frac{-7}{3}(x - 0)$,
$y = \frac{-7}{3}x + 9$
**c.** $(0, 9)$, $(\frac{27}{7}, 0)$

**3. a.** $\frac{1}{2}$, increasing  **b.** $y - 2 = \frac{1}{2}(x - 3)$,
$y = \frac{1}{2}x + \frac{1}{2}$
**c.** $(0, \frac{1}{2})$, $(-1, 0)$

**4. a.** $\frac{3}{4}$, increasing  **b.** $y + 4 = \frac{3}{4}(x + 5)$,
$y = \frac{3}{4}x - \frac{1}{4}$
**c.** $(0, \frac{-1}{4})$, $(\frac{1}{3}, 0)$

**5. a.** $\frac{-3}{4}$, decreasing　**b.** $y - 5 = \frac{-3}{4}(x + 2)$,
$y = \frac{-3}{4}x + \frac{7}{2}$　**c.** $(0, \frac{7}{2}), (\frac{14}{3}, 0)$

**6. a.** $\frac{-1}{2}$, decreasing　**b.** $y + 7 = \frac{-1}{2}(x - 2)$,
$y = \frac{-1}{2}x - 6$　**c.** $(0, -6), (-12, 0)$

**7. a.** slope is undefined; none　**b.** $x = -3$
**c.** $(-3, 0)$; no $y$-intercept

**8. a.** $m = 0$; constant　**b.** $y = 0x + 8$
**c.** $(0, 8)$; no $x$-intercept

### Cumulative Review Chapters R–2, p. 244

**1.** $x^2 + 2$　**2.** $-2 < x < 8$　**3.** 29.45 cm

**4.** $r = \dfrac{-\pi h + \sqrt{\pi^2 h^2 + 2\pi A}}{2\pi}$　**5.** $x = 1$　**6.** $\frac{4}{9}$　**7. a.** $\frac{-1}{3}$　**b** $\frac{3}{5}$

**8. a.**　　　　　**b.**　　　　　**9.** $y = \frac{1}{2}x + \frac{7}{2}$

**10.**　　　　$x^2 - 2x + 26 = 0$
　　　　$(1 + 5i)^2 - 2(1 + 5i) + 26 = 0$
　　　$1 + 10i - 25 - 2 - 10i + 26 = 0$
　　　　　　　　　　$0 = 0$

**11.** $\frac{\Delta y}{\Delta x} = -3$　**12.**

**13. a.** $D: x \in [-7, \infty), R: x \in [-6, \infty)$
**b.** $f(-3) = 3, f(-1) = 0, f(1) = -3, f(3) = -6$
**c.** $x = -5, x = -1, x = 5$
**d.** $f(x) < 0$ for $x \in [-7, -5) \cup (-1, 5)$
　　$f(x) > 0$ for $x \in [-5, -1) \cup (5, \infty)$
**e.** $f(x) \uparrow$ for $x \in (-7, -3) \cup (3, \infty)$
　　$f(x) \downarrow$ for $x \in (-3, 3)$
**f.** local max of $y = 3$ at $x = -3$, local min of $y = -6$ at $x = 3$
**14. a.** $f(x)$　**b.** $g(x)$　**15. a.** $\frac{x - 7}{(x - 5)(x + 2)}$　**b.** $\frac{b^2 - 4ac}{4a^2}$
**16. a.** $\frac{-5}{2} + \frac{3\sqrt{2}}{2}$　**b.** $\frac{\sqrt{2}}{2}$　**17.** $y = \frac{-2}{3}x + 2$
**18.** not a function; One element of the domain (Raphael) is paired with two elements of the range.
**19.** $x = -5 \pm \frac{\sqrt{2}}{2}; x \approx -5.707, x \approx -4.293$　**20.** $x = -5 \pm \frac{i\sqrt{2}}{2}$
**21.** $W = 31$ cm, $L = 47$ cm　**22. a.** $-21 + 20i$　**b.** $\frac{-3}{5} - \frac{4}{5}i$
**23. a.** $x = \frac{-4}{3}, \frac{5}{2}$　**b.** $x = -5, -\sqrt{3}, \sqrt{3}$
**24.** $m_1 = \frac{1}{2} \Rightarrow m_2 = -2$, so $y = -2x + 4$
**25.** $P = 5 + 10 + \sqrt{97}$; no, since $5^2 + (\sqrt{97})^2 \neq 10^2$

## CHAPTER 3
### Exercises 3.1, pp. 255–259

**1.** $(-5, -9)$, upward　**3.** Answers will vary.　**5. a.** squaring　**b.** vertex $(-3, -4)$　**c.** up/up　**d.** $(-5, 0), (-1, 0), (0, 5)$　**e.** min of $-4$ at $(-3, -4)$　**f.** $x \in (-\infty, \infty); y \in [-4, \infty)$　**g.** positive for $x \in (-\infty, -5) \cup (-1, \infty)$; negative for $x \in (-5, -1)$　**h.** increasing for $x \in (-3, \infty)$; decreasing for $x \in (-\infty, -3)$　**7. a.** absolute value　**b.** vertex $(2, 3)$　**c.** down/down　**d.** $(0, -3), (1, 0), (3, 0)$　**e.** max of 3 at $(2, 3)$　**f.** $x \in (-\infty, \infty)$; $y \in (-\infty, 3]$　**g.** positive for $x \in (1, 3)$; negative for $(-\infty, 1) \cup (3, \infty)$　**h.** increasing for $x \in (-\infty, 2)$; decreasing for $x \in (2, \infty)$　**9. a.** cube root　**b.** inflection point $(0, 1)$　**c.** up/down　**d.** $(0, 1), (1, 0)$　**e.** no max or min　**f.** $x \in (-\infty, \infty); y \in (-\infty, \infty)$　**g.** positive for $x \in (-\infty, 1)$; negative for $x \in (1, \infty)$　**h.** decreasing for $x \in (-\infty, \infty)$　**11. a.** square root　**b.** initial point $(-4, -2)$　**c.** up on right　**d.** $(-3, 0), (0, 2)$　**e.** min of $-2$ at $(-4, -2)$　**f.** $x \in [-4, \infty); y \in [-2, \infty)$　**g.** positive for $x \in (-3, \infty)$; negative for $x \in [-4, -3)$　**h.** increasing for $x \in (-4, \infty)$　**13. a.** cubing　**b.** inflection point $(-1, -1)$　**c.** up/down　**d.** $(-2, 0), (0, -2)$　**e.** no max or min　**f.** $x \in (-\infty, \infty); y \in (-\infty, \infty)$　**g.** positive for $x \in (-\infty, -2)$; negative for $x \in (-2, \infty)$　**h.** decreasing for $x \in (-\infty, \infty)$

**15.**　　　　　　　**17.**

**19.**　　　　　　　**21.**

　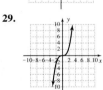

**23.**　　　　　　　**25.**

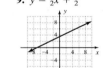　

**27.**　　　　　　　**29.**

**31.** g　**33.** i　**35.** e　**37.** j　**39.** l　**41.** c
**43.** left 2, down 1　　**45.** left 3, reflected across $x$-axis, down 2　　**47.** left 3, down 1

　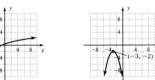　

**49.** left 1, down 2　　**51.** left 3, reflected across $x$-axis, down 2　　**53.** left 1, reflected across $x$-axis, stretched vertically, down 3

**55.** left 2, reflected across x-axis, compressed vertically, down 1

**57.** reflected across x-axis, right 3, compressed vertically, down 1

**59.** left 1, reflected across x = −1, reflected across x-axis, stretched vertically, up 3

**61.** left 2, compressed vertically, down 3

**63. a.**

**b.**

**c.**

**d.**

**65. a.**

**b.**

**c.**

**d.**

**67. a.** (−3, 0), (0, 8), (1, 0)    **b.** (3, 0), (0, −12), (−1, 0)
**c.** (−5, 0), (−2, 16), (−1, 0)    **d.** (−2, 5), (1, 1), (2, 5)
**69. a.** $x \in [2, \infty)$; $y \in (-\infty, 12)$    **b.** $x \in (-\infty, -2]$; $y \in (-3, \infty)$
**c.** $x \in [1, \infty)$; $y \in (-\infty, 6)$    **d.** $x \in [4, \infty)$; $y \in (-\infty, 6)$
**71. a.** down/up    **b.** increasing: $x \in (-\infty, 1) \cup (7, \infty)$;
decreasing: $x \in (1, 7)$    **c.** max at (1, 2); min at (7, −10)
**73.** $f(x) = -(x - 2)^2$    **75.** $p(x) = 1.5\sqrt{x + 3}$    **77.** $f(x) = \frac{4}{5}|x + 4|$
**79. a.**

**b.** about 65 in³, $V \approx 65.4$ in³; yes
**c.** $r = \sqrt[3]{\frac{3}{4\pi}V}$   **d.** 6 in.

**81. a.** compressed vertically

**b.** 2.25 sec; yes

**83. a.** compressed vertically

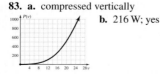

**b.** 216 W; yes

**85. a.** vertical stretch by a factor of 2    **b.** 12.5 ft; yes

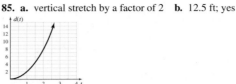

**87. a.**

**b.**

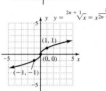

**c.**

**d.**

**89. a.** horizontal 3; vertical 9    **b.** horizontal 2; vertical 8    **c.** horizontal 5; vertical 5    **d.** horizontal 27; vertical 3
**91. a.** shifted left 3    **b.** shifted up 3    **c.** shifted up $f(3) = 9$

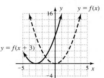

**d.** The graph of $y = f(x + h)$ is the graph of $y = f(x)$ shifted left $h$ units. The graph of $y = f(x) + h$ is the graph of $y = f(x)$ shifted up $h$ units. The graph of $y = f(x) + f(h)$ is the graph of $y = f(x)$ shifted up $f(h)$ units.
**93.** Any points in Quadrants III and IV will reflect across the x-axis and move to Quadrants I and II.

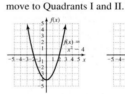

**95.** $x = -5$    **97.** $f(x)\uparrow: x \in (4, \infty); f(x)\downarrow: x \in (-\infty, 4)$

## Exercises 3.2, pp. 270–274

**1.** vertical, $y = 2$    **3.** Answers will vary.
**5. a.** as $x \to -\infty, y \to 2$      **7. a.** as $x \to -\infty, y \to 1$
as $x \to \infty, y \to 2$            as $x \to \infty, y \to 1$
**b.** as $x \to 1^-, y \to -\infty$      **b.** $y = 1$
as $x \to 1^+, y \to \infty$         **c.** as $x \to -2^-, y \to \infty$
as $x \to -2^+, y \to \infty$
**9.** $\to -2$    **11.** $\to -\infty$    **13.** $-1, \pm\infty$
**15.** down 1, $x \in (-\infty, 0) \cup (0, \infty)$, $y \in (-\infty, -1) \cup (-1, \infty)$

**17.** left 2, $x \in (-\infty, -2) \cup (-2, \infty)$, $y \in (-\infty, 0) \cup (0, \infty)$

**19.** right 2, reflected across $x$-axis,
$x \in (-\infty, 2) \cup (2, \infty), y \in (-\infty, 0) \cup (0, \infty)$

**21.** left 2, down 1,
$x \in (-\infty, -2) \cup (-2, \infty), y \in (-\infty, -1) \cup (-1, \infty)$

**23.** right 1, $x \in (-\infty, 1) \cup (1, \infty), y \in (0, \infty)$

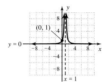

**25.** left 2, reflected across $x$-axis,
$x \in (-\infty, -2) \cup (-2, \infty), y \in (-\infty, 0)$

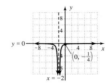

**27.** down 2, $x \in (-\infty, 0) \cup (0, \infty), y \in (-2, \infty)$

**29.** left 2, up 1, $x \in (-\infty, -2) \cup (-2, \infty)$,
$y \in (1, \infty)$

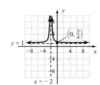

**31.** reciprocal square, $S(x) = \dfrac{1}{(x + 1)^2} - 2$

**33.** reciprocal function, $Q(x) = \dfrac{1}{x + 1} - 2$

**35.** reciprocal square, $v(x) = \dfrac{1}{(x + 2)^2} - 5$

**37.** $g(x)$ increases faster $(3 > 2)$    **a.** $x = 0$ and $1$
**b.** $(-\infty, 0) \cup (0, 1)$    **c.** $(1, \infty)$

**39.** $f(x)$ increases faster $(4 > 2)$    **a.** $x = -1, 0,$ and $1$
**b.** $(-1, 0) \cup (0, 1)$    **c.** $(-\infty, -1) \cup (1, \infty)$

**41.** $g(x)$ increases faster $\left(\dfrac{4}{5} > \dfrac{2}{3}\right)$    **a.** $x = -1, 0,$ and $1$
**b.** $(-1, 0) \cup (0, 1)$    **c.** $(-\infty, -1) \cup (1, \infty)$

**43.** $g(x)$ increases faster $\left(\dfrac{1}{3} > \dfrac{1}{6}\right)$    **a.** $x = 0$ and $1$    **b.** $(0, 1)$    **c.** $(1, \infty)$

**45.** $g(x)$ increases faster $\left(\dfrac{5}{4} > \dfrac{2}{3}\right)$    **a.** $x = 0$ and $1$    **b.** $(0, 1)$    **c.** $(1, \infty)$

**47.** $[0, \infty)$    **49.** $(-\infty, \infty)$    **51.** $(-\infty, \infty)$
**53.** defined: b, c, d; undefined: a    **55.** defined: a, b, d; undefined: c
**57.** $F$ is the graph of $f$ shifted left 1 unit and down 2; verified

**59.** $P$ is the graph of $p$ shifted right 2 units and reflected across the $x$-axis; verified
**61. a.** $F$ becomes very small.    **b.** $y = \dfrac{1}{x^2}$    **c.** $m_2 = \dfrac{d^2 F}{k m_1}$
**63. a.** It decreases: $D(1) = 75, D(3) = 25,$ and $D(5) = 15$.
**b.** It approaches 0.    **c.** as $p \to 0, D \to \infty$

**65. a.** It decreases: $I(5) = 100, I(10) = 25,$ and $I(15) \approx 11.1$.
**b.** toward the light source    **c.** as $d \to 0, l \to \infty$

**67. a.** \$20,000, \$80,000, \$320,000; Cost increases dramatically.
**b.**

**c.** as $p \to 100, C \to \infty$
**69. a.** 253 ft/sec (about 172 mph) **b.** approx. 791 ft
**71. a.** size 10    **b.** approx. 7 ft, 1 in.
**73. a.**    **b.** $P = 32.251 w^{0.246}$
**c.** about 63 days
**d.** about 6.9 kg

**75. a.**    **b.** $S = 1.687 a^{0.386}$
**c.** about 33 species
**d.** about 37,200 $\text{mi}^2$

**77.** The area is always 1 $\text{unit}^2$. The area is always $\frac{1}{x}$ $\text{units}^2$.
**79.** $y = \frac{-2}{3}x + 5$,    **81.** $c = \sqrt{\dfrac{E}{m}}$

## Exercises 3.3, pp. 280–284

**1.** decreases    **3.** Answers will vary.    **5.** $d = kt$    **7.** $F = ka$
**9.** $y = 0.025x$

| $x$ | $y$ |
|-----|-----|
| 500 | 12.5 |
| 650 | 16.25 |
| 750 | 18.75 |

**11.** $P \approx 0.44d$; 26.6 psi; $k \approx 0.44$ psi/ft (or $\approx 0.44$ psi increase in pressure for each 1 ft increase in depth)
**13. a.** $S = \frac{192}{47}h$    **b.** 330 stairs
**c.** $S = 331$; yes

**15.** 32 oz   **17.** $\frac{1}{16}$ T   **19.** 62.5 lb/ft³   **21.** $V = ks^3$   **23.** $P = kc^2$

**25.** $p = 1.28q^4$

| q | p |
|---|---|
| 1.5 | 6.48 |
| 2.5 | 50 |
| 10 | 12,800 |

**27. a.** Area varies directly as a side squared.   **b.** $A = ks^2$

**c.**    **d.**

| s | A |
|---|---|
| 0 | 0 |
| 5 | 150 |
| 10 | 600 |
| 15 | 1350 |
| 20 | 2400 |
| 25 | 3750 |
| 30 | 5400 |

**e.** $k = 6$; $A = 6s^2$; 55,303,776 m²

**29. a.** Distance varies directly as time squared.   **b.** $D = kt^2$

**c.** 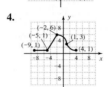   **d.**

| t | D |
|---|---|
| 1 | 16 |
| 1.5 | 36 |
| 2 | 64 |
| 2.5 | 100 |
| 3 | 144 |
| 3.5 | 196 |
| 4 | 256 |

**e.** $k = 16$; $d = 16t^2$; about 3.5 sec; 121 ft

**31.** $F = \frac{k}{d^2}$   **33.** $S = \frac{k}{L}$

**35.** $Y = \dfrac{12,321}{Z^2}$

| Z | Y |
|---|---|
| 37 | 9 |
| 74 | 2.25 |
| 111 | 1 |

**37.** $w = \dfrac{3,072,000,000}{r^2}$; 48 kg   **39.** $l = krt$   **41.** $A = kh(B + b)$

**43.** $R = \frac{kL}{A}$

**45.** $C = \frac{6.75R}{S^2}$

| R | S | C |
|---|---|---|
| 120 | 6 | 22.5 |
| 200 | 12.5 | 8.64 |
| 350 | 15 | 10.5 |

**47.** $E = 0.5mv^2$; 800 J   **49.** c   **51.** d   **53. a.** Force varies jointly with the square of the velocity and the cross-sectional area.   **b.** $k \approx 0.0004$
**c.** $F \approx 2.7$ lb   **55.** $T = \frac{48}{V}$; 32 volunteers   **57.** $M = \frac{1}{6}E$; $\approx 41.7$ kg
**59.** $D = 21.6\sqrt{S}$; $\approx 144.9$ ft   **61.** $C = 8.5LD$; $76.50
**63.** $C \approx (4.4 \times 10^{-4})\frac{P_1P_2}{d^2}$; about 222 calls   **65. a.** about 23.39 cm³
**b.** about 191%   **67. a.** $M = kwh^2(\frac{1}{L})$   **b.** 180 lb

**69.** $S \approx \frac{0.0182r}{g}$; about 30 mph   **71. a.** $6.67 \times 10^{-7}$ N   **b.** $667 \times 10^{-15}$ N
**73. a.** $y_2 = (\frac{x_2}{x_1})y_1$   **b.** $M_2 = (\frac{E_2}{E_1})M_1 = (\frac{170}{140}) 53.2 \approx 64.6$
**75.** $x = 0, -4, -2$
**77.**

## Mid-Chapter Check, p. 285

**1.** $g(x) = \sqrt{x + 4} + 2$
**2. a.** cubic   **b.** up on the left, down on the right; inflection point: (2, 2); $x$-int: (4, 0); $y$-int: (0, 5)   **c.** $D: x \in (-\infty, \infty)$; $R: y \in (-\infty, \infty)$   **d.** $k = 1$
**3.**

 $q(x)$ is a reflection of $p(x)$ across the $x$-axis, and $r(x)$ is the same as $q(x)$, but compressed by a factor of $\frac{1}{2}$.

**4.**

**5. a.** $\infty$   **b.** 2   **6. a.** $D: x \in [0, \infty)$; $R: y \in [0, \infty)$
**b.** $D: x \in (-\infty, \infty)$; $R: y \in (-\infty, \infty)$   **c.** $D: x \in (0, \infty)$; $R: y \in (0, \infty)$
**d.** $D: x \in (-\infty, 0) \cup (0, \infty)$; $R: y \in (-\infty, 0) \cup (0, \infty)$   **7. a.** 50   **b.** 60
**c.** 6 hr   **d.** 16 hr   **e.** no   **8.** 20 lb   **9.** 22.8
**10.** $W = 4x^3\sqrt[3]{y}$

| W | x | y |
|---|---|---|
| 8 | 1 | 8 |
| 324 | 3 | 27 |
| 128 | 2 | 64 |

## Exercises 3.4, pp. 293–297

**1.** smooth   **3.** Each piece must be continuous on the corresponding interval, and the function values at the endpoints of each interval must be equal. Answers will vary.
**5. a.** $f(x) = \begin{cases} x^2 - 6x + 10, & 0 \le x \le 5 \\ \frac{3}{2}x - \frac{5}{2}, & 5 < x \le 9 \end{cases}$   **b.** $x \in [0, 9)$ $y \in [1, 11]$
**7.** $x \in (2, 4] \cup (6, \infty)$; $y \in (-\infty, 3) \cup [4, 10)$
**9.** $-2, 2, \frac{1}{2}, 0, 2.999, 5$   **11.** 5, 5, 0, −4, 5, 11
**13.** $D: x \in [-2, \infty)$; $R: y \in [-4, \infty)$

**15.** $D: x \in (-\infty, 9)$; $R: y \in [2, \infty)$

**17.** $D: x \in (-\infty, \infty); R: y \in [0, \infty)$

**19.** $D: x \in (-\infty, \infty); R: y \in (-\infty, 3) \cup (3, \infty)$

**21.** discontinuity at $x = -3$, redefine $f(x) = -6$ at $x = -3; c = -6$

**23.** discontinuity at $x = 1$, redefine $f(x) = 3$ at $x = 1; c = 3$

**25.** $(4, 11); c = \frac{9}{4}$  **27.** $(0, 5); c = 2$

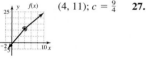 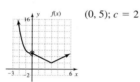

**29.** $f(x) = \begin{cases} \frac{1}{2}x - 1, & -4 \le x < 2 \\ 3x - 6, & x \ge 2 \end{cases}$

**31.** $p(x) = \begin{cases} (x + 1)^2 - 4, & x \le 1 \\ x + 1, & x > 1 \end{cases}$

**33.** $f(x) = \begin{cases} |x + 3| - 2, & x < 1 \\ x^2 - 5, & x \ge 1 \end{cases}$

**35.** $p(x) = \begin{cases} \sqrt{-x - 1} - 3, & x < -1 \\ -2x, & -1 \le x \le 1 \\ -\sqrt{x - 1} + 3, & x > 1 \end{cases}$

**37.** Graph is discontinuous at $x = 0$; $f(x) = 1$ for $x > 0$; $f(x) = -1$ for $x < 0$.

**39. a.** $S(t) = \begin{cases} -t^2 + 6t, & 0 \le t \le 5 \\ 5, & t > 5 \end{cases}$  **b.** $S(t) \in [0, 9]$

**41. a.**

| Year (0 → 1950) | Percent |
|---|---|
| 5 | 7.33 |
| 15 | 14.13 |
| 25 | 14.93 |
| 35 | 22.65 |
| 45 | 41.55 |
| 55 | 51.20 |
| 65 | 51.20 |

**b.** Each piece gives a slightly different value due to rounding of coefficients in each model. At $t = 30$, we use the "first" piece: $P(30) = 13.08$.

**43. a.** $C(h) = \begin{cases} 0.09h, & 0 \le h \le 1000 \\ 0.18h - 90, & h > 1000 \end{cases}$

**b.** $C(1200) = \$126$

**45. a.** $C(t) = \begin{cases} 0.75t, & 0 \le t \le 25 \\ 1.5t - 18.75, & t > 25 \end{cases}$

**b.** $C(45) = \$48.75$

**47. a.** $S(t) = \begin{cases} -1.4t^2 + 31.6t + 155, & 0 \le t \le 12 \\ 2.6t^2 - 81.5t + 944, & 12 < t \le 29 \end{cases}$

**b.** \$526.5 billion, \$834 billion, \$1271.5 billion

**49. a.** $c(m) = \begin{cases} 3.3m, & 0 \le m \le 30 \\ 7m - 111, & m > 30 \end{cases}$

**b.** \$2.11

**51. a.** $C(a) = \begin{cases} 0, & a < 2 \\ 2, & 2 \le a < 13 \\ 5, & 13 \le a < 20 \\ 7, & 20 \le a < 65 \\ 5, & a \ge 65 \end{cases}$

**b.** \$38

**53. a.** $C(w - 1) = 17\lceil w - 1 \rceil + 88$  **b.** $0 < w \le 13$  **c.** 88¢
**d.** 173¢  **e.** 173¢  **f.** 173¢  **g.** 190¢

**55. a.** yes

**b.** $h(x) = \begin{cases} 5, & x \le -3 \\ -2x - 1, & -3 < x < 2 \\ -5, & x \ge 2 \end{cases}$

**57.** no; $f(x)$ has a removable discontinuity at $x = -2$; $g(x)$ has a nonremovable discontinuity at $x = -2$;

**59. a.** **b.**

**c.** $g(x) = \begin{cases} 3(x + 2)^2 - 11, & -4 \le x \le -1 \\ 10, & -1 < x < 1 \\ -6x + 19, & x \ge 1 \end{cases}$

**61.** $x = -7, x = 4$  **63.** $y = \frac{4}{3}x - 2$

## Exercises 3.5, pp. 308–314

**1.** $f(5) \cdot g(5), (f \cdot g)(5)$    **3.** The domain of $h$ is the empty set, since $f$ and $g$ do not intersect.    **5. a.** $x \in \mathbb{R}$    **b.** $f(-2) - g(-2) = 13$
**7. a.** $h(x) = x^2 - 6x - 3$    **b.** $h(-2) = 13$    **c.** They are identical.
**9. a.** $x \in [3, \infty)$    **b.** $h(x) = \sqrt{x-3} + 2x^3 - 54$
**c.** $h(4) = 75$, 2 is not in the domain of $h$.
**11. a.** $x \in [-5, 3]$    **b.** $r(x) = \sqrt{x+5} + \sqrt{3-x}$
**c.** $r(2) = \sqrt{7} + 1$, 4 is not in the domain of $r$.
**13. a.** $x \in [-4, \infty)$    **b.** $h(x) = \sqrt{x+4}(2x+3)$
**c.** $h(-4) = 0, h(21) = 225$
**15. a.** $x \in [-1, 7]$    **b.** $r(x) = \sqrt{-x^2 + 6x + 7}$
**c.** 15 is not in the domain of $r$, $r(3) = 4$.
**17. a.** $x \in (-\infty, 5) \cup (5, \infty)$    **b.** $h(x) = \dfrac{x+1}{x-5}, x \ne 5$
**19. a.** $x \in (-\infty, -4) \cup (-4, \infty)$    **b.** $h(x) = x - 4, x \ne -4$
**21. a.** $x \in (-\infty, -4) \cup (-4, \infty)$    **b.** $h(x) = x^2 - 2, x \ne -4$
**23. a.** $x \in (-\infty, 1) \cup (1, \infty)$    **b.** $h(x) = x^2 - 6x, x \ne 1$
**25. a.** $x \in (-\infty, 3)$    **b.** $r(x) = \dfrac{1-x}{\sqrt{3-x}}$
**c.** 6 is not in the domain of $r$, $r(-6) = \dfrac{7}{3}$.
**27. a.** $x \in \left(-\dfrac{13}{2}, \infty\right)$    **b.** $r(x) = \dfrac{x^2 - 36}{\sqrt{2x+13}}$    **c.** $r(6) = 0, r(-6) = 0$
**29. a.** $h(x) = \dfrac{2x+4}{x-3}$    **b.** $x \in (-\infty, 3) \cup (3, \infty)$    **c.** $x \ne -2, x \ne 0$
**31. a.** $h(x) = \dfrac{x^2 - 4x - 12}{x^2 - 4x + 3}$    **b.** $x \in (-\infty, 1) \cup (1, 3) \cup (3, \infty)$
**c.** $x \ne -1, x \ne -2, x \ne 2$
**33.** sum: $3x + 1, x \in (-\infty, \infty)$; difference: $x + 5, x \in (-\infty, \infty)$;
product: $2x^2 - x - 6, x \in (-\infty, \infty)$;
quotient: $\dfrac{2x+3}{x-2}, x \in (-\infty, 2) \cup (2, \infty)$
**35.** sum: $x^2 + 3x - 4, x \in (-\infty, \infty)$; difference: $x^2 + x - 2$,
$x \in (-\infty, \infty)$; product: $x^3 + x^2 - 5x + 3, x \in (-\infty, \infty)$;
quotient: $x + 3, x \in (-\infty, 1) \cup (1, \infty)$
**37.** sum: $x + 2 + \sqrt{x+6}, x \in [-6, \infty)$; difference: $x + 2 - \sqrt{x-6}$,
$x \in [-6, \infty)$; product: $(x+2)\sqrt{x+6}, x \in [-6, \infty)$;
quotient: $\dfrac{x+2}{\sqrt{x+6}}, x \in (-6, \infty)$
**39.** sum: $\dfrac{7x-11}{(x-3)(x+2)}, x \in (-\infty, -2) \cup (-2, 3) \cup (3, \infty)$;
difference: $\dfrac{-3x+19}{(x-3)(x+2)}, x \in (-\infty, -2) \cup (-2, 3) \cup (3, \infty)$;
product: $\dfrac{10}{(x-3)(x+2)}, x \in (-\infty, -2) \cup (-2, 3) \cup (3, \infty)$;
quotient: $\dfrac{2x+4}{5(x-3)}, x \in (-\infty, -2) \cup (-2, 3) \cup (3, \infty)$
**41.** $0, 0, 4a^2 - 10a - 14, a^2 - 9a$
**43. a.** $h(x) = \sqrt{2x-2}$    **b.** $H(x) = 2\sqrt{x+3} - 5$
**c.** $D$ of $h(x)$: $x \in [1, \infty)$; $D$ of $H(x)$: $x \in [-3, \infty)$
**45. a.** $h(x) = \sqrt{3x+1}$    **b.** $H(x) = 3\sqrt{x-3} + 4$
**c.** $D$ of $h(x)$: $x \in [-\tfrac{1}{3}, \infty)$; $D$ of $H(x)$: $x \in [3, \infty)$
**47. a.** $h(x) = x^2 + x - 2$    **b.** $H(x) = x^2 - 3x + 2$
**c.** $D$ of $h(x)$: $x \in (-\infty, \infty)$; $D$ of $H(x)$: $x \in (-\infty, \infty)$
**49. a.** $h(x) = x^2 + 7x + 8$    **b.** $H(x) = x^2 + x - 1$
**c.** $D$ of $h(x)$: $x \in (-\infty, \infty)$; $D$ of $H(x)$: $x \in (-\infty, \infty)$
**51. a.** $h(x) = |-3x + 1| - 5$    **b.** $H(x) = -3|x| + 16$
**c.** $D$ of $h(x)$: $x \in (-\infty, \infty)$; $D$ of $H(x)$: $x \in (-\infty, \infty)$
**53. a.** $(f \circ g)(x)$: For $g(x)$ to be defined, $x \ne 0$.

For $f[g(x)] = \dfrac{2g(x)}{g(x) + 3}, g(x) \ne -3$ so $x \ne -\dfrac{5}{3}$.

domain: all real numbers except $x = -3$ and $x = -\dfrac{5}{3}$

**b.** $(g \circ f)(x)$: For $f(x)$ to be defined, $x \ne -3$.

For $g[f(x)] = \dfrac{5}{f(x)}, f(x) \ne 0$ so $x \ne 0$.

domain: all real numbers except $x = -3$ and $x = 0$

**c.** $(f \circ g)(x) = \dfrac{10}{5+3x}; (g \circ f)(x) = \dfrac{5x+15}{2x}$; The domain of a composition cannot always be determined from the composed form.
**55. a.** $(f \circ g)(x)$: For $g(x)$ to be defined, $x \ne 5$.

For $f[g(x)] = \dfrac{4}{g(x)}, g(x) \ne 0$ and $g(x)$ is never zero.

domain: $\{x | x \ne 5\}$

**b.** $(g \circ f)(x)$: For $f(x)$ to be defined, $x \ne 0$.

For $g[f(x)] = \dfrac{1}{f(x) - 5}, f(x) \ne 5$ so $x \ne \dfrac{4}{5}$.

domain: all real numbers except $x = 0$ and $x = \dfrac{4}{5}$

**c.** $(f \circ g)(x) = 4x - 20; (g \circ f)(x) = \dfrac{x}{4-5x}$; The domain of a composition cannot always be determined from the composed form.
**57. a.** 41    **b.** 41    **59.** $f(x) = 5x^8$    **61.** $g(x) = 2x + 3$
**63.** $f(x) = x^3 - 5, g(x) = \sqrt{x-2} + 1$
**65.** $j(x) = 2x^2 + 16x + 29; k(x) = 4x^2 + 28x + 48$
**67. a.** $4x + 9$    **b.** $\frac{x-9}{4}$    **c.** $x$    **d.** $x$
**69. a.** 6000    **b.** 3000    **c.** 8000    **d.** $C(9) - T(9); 4000$
**71. a.** \$1 billion    **b.** \$5 billion    **c.** 2003, 2007, 2010
**d.** $t \in (2000, 2003) \cup (2007, 2010)$    **e.** $t \in (2003, 2007)$
**f.** $R(5) - C(5); \$4$ billion
**73. a.** 4    **b.** 0    **c.** 2    **d.** 3    **e.** $-\frac{1}{3}$    **f.** 6    **g.** $-3$    **h.** 1    **i.** 1
**j.** undefined    **k.** 0.5    **l.** 2    **75.** $h(x) = -\frac{2}{3}x + 4$    **77.** $h(x) = 4x - x^2$
**79. a.** $A = 2\pi r(20 + r); f(r) = 2\pi r, g(r) = 20 + r$
**b.** $A(5) = 250\pi$ units$^2$
**81. a.** $C(x) = 28,000x + 108,000, R(x) = 40,000x$,
$P(x) = 12,000x - 108,000$    **b.** Nine boats must be sold.    **c.** \$72,000
**83. a.** $P(n) = 11.45n - 0.1n^2$    **b.** \$123    **c.** \$327    **d.** $C(115) > R(115)$
**85. a.** $T(d) = \dfrac{\sqrt{d}}{4} + \dfrac{d}{1116}$    **b.** 4 sec    **c.** about 494 ft
**87. a.** $h(x) = x - 2.5$    **b.** 10.5    **c.** 15    **89. a.** $A(t) = 4\pi t^2$    **b.** 14,400$\pi$ m$^2$
**91. a.** $L(0) = 500$ lions and $H(500) = 400$ hyenas
**b.** $H[L(x)] = 400 + 0.0075x, (H \circ L)(16,000) = 520$ hyenas    **c.** prior to an increase of 30,000    **93.** Answers will vary.
**95. a.** $(-\infty, -2) \cup (-2, 2) \cup (2, \infty)$    **b.** verified
**c.** $(-\infty, -2) \cup (-2, -1] \cup [1, 2) \cup (2, \infty)$; Answers will vary.

| $x$ | $h(x)$ | $(f \circ g)(x)$ |
|---|---|---|
| $-3$ | $\frac{1}{5}$ | $\frac{1}{5}$ |
| $-2$ | undefined | undefined |
| $-1$ | $-\frac{1}{3}$ | $-\frac{1}{3}$ |
| $0$ | $-\frac{1}{4}$ | undefined |
| $1$ | $-\frac{1}{3}$ | $-\frac{1}{3}$ |
| $2$ | undefined | undefined |
| $3$ | $\frac{1}{5}$ | $\frac{1}{5}$ |

**97. a.**    **b.**    **c.**

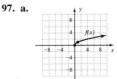

**99.** $y = -\frac{3}{2}x$

## Exercises 3.6, pp. 321–325

**1.** maximum, $y = 12$　**3.** Answers will vary.
**5.** $d(x) = \sqrt{(x-3)^2 + x^4}$　**7.** $P(w) = \frac{2w^2 + 100}{w}$　**9.** $A(x) = 0.5x^5$
**11.** $A(x) = 4x - x^3$　**13. a.** $A(x) = 2x\sqrt{9 - x^2}$
**b.** $P(x) = 4x + 2\sqrt{9 - x^2}$　**15.** $A(x) = \frac{-5}{12}x^2 + 5x, 0 < x < 12$
**17.** $A(x) = 4x\sqrt{25 - x^2}$　**19.** $V(r) = 3\pi r^2(8 - r)$
**21.** (1) f, h, m; (2) a, g, k, o; (3) f, r; (4) i, n, s; (5) a, g; (6) b, d;
(7) e, l; (8) j, n, p, s; (9) q; (10) c, t
**23. a.** $h = \frac{2A}{b}$　**b.** $h = \frac{3V}{\pi r^2}$　**c.** $h = \sqrt{\left(\frac{SA}{\pi r}\right)^2 - r^2}$　**d.** $h = \frac{SA - 2lw}{2l + 2w}$
**25.** about $2.12 \times 4.24$ units; $A = 9$ units$^2$　**27.** $(-1, 3)$; $d \approx 6.71$ units
**29.** $4 \times 7.5$ cm; $A = 30$ cm$^2$　**31.** about $16.97 \times 16.97$ in.; $A = 288$ in$^2$
**33.** $V \approx 302$ in$^3$; base radius 4 in., height 6 in.
**35. a.** 7.99 cm from the end used for the triangular earring ($A \approx 3.07$ cm$^2$)
**b.** 8.48 cm from the end used for the triangular earring ($A \approx 6.12$ cm$^2$)
**37. a.** $S = 2\pi r^2 + 2\pi rh$　**b.** $S(r) = 2\pi r^2 + \frac{25,410}{r}$
**c.** $r \approx 12.6$ in., $h \approx 25.3$ in.　**39. a.** $T = \frac{500 - x}{7} + \frac{\sqrt{x^2 + 200^2}}{4}$
**b.** exit the woods at $x \approx 139$ yd, leaving 361 yd to run
**41. a.** $V = \frac{75\pi h^2}{h - 30}$　**b.** base radius: 21.2 cm; height: 60 cm, $V \approx 28,274$ cm$^3$
**43.** $x = 4 \pm \sqrt{5}$　**45.** 43.5 thousand (43,500) units would be sold.

## Making Connections, p. 325

**1.** g　**3.** a　**5.** d　**7.** a　**9.** b　**11.** h　**13.** f　**15.** f

## Summary and Concept Review, pp. 326–329

**1.** squaring function　**a.** up on left/up on the right　**b.** $x$-intercept:
$(-4, 0)$, $(0, 0)$; $y$-intercept: $(0, 0)$　**c.** vertex $(-2, -4)$
**d.** $x \in (-\infty, \infty)$, $y \in [-4, \infty)$
**2.** square root function　**a.** down on the right　**b.** $x$-intercept: $(0, 0)$;
$y$-intercept: $(0, 0)$　**c.** initial point $(-1, 2)$;　**d.** $x \in [-1, \infty)$, $y \in (-\infty, 2]$
**3.** cubing function　**a.** down on left/up on the right　**b.** $x$-intercept(s):
$(2, 0)$; $y$-intercept: $(0, -2)$　**c.** inflection point: $(1, -1)$
**d.** $x \in (-\infty, \infty)$, $y \in (-\infty, \infty)$
**4.** absolute value function　**a.** down on left/down on the right
**b.** $x$-intercepts: $(-1, 0)$, $(3, 0)$; $y$-intercept: $(0, 1)$　**c.** vertex: $(1, 2)$
**d.** $x \in (-\infty, \infty)$, $y \in (-\infty, 2]$
**5.** cube root function　**a.** up on left, down on right
**b.** $x$-intercept: $(1, 0)$; $y$-intercept: $(0, 1)$　**c.** inflection point: $(1, 0)$
**d.** $x \in (-\infty, \infty)$, $y \in (-\infty, \infty)$

**6.** quadratic

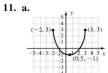

**7.** absolute value

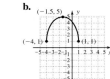

**8.** cubic

**9.** square root

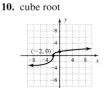

**10.** cube root

**11. a.**

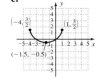

**b.**

**c.**

**12.**

**13.**

**14. a.** $\approx$\$32,143; \$75,000; \$175,000;
\$675,000; cost increases dramatically
**c.** as $p \to 100$, $C \to \infty$

**b.**

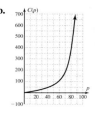

**15.**

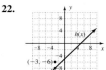

Domain of $f(x)$ is
$(-\infty, \infty)$; domain of
$g(x)$ and $h(x)$ is $[0, \infty)$.

**16. a.** 88.4 hr　**b.** 570 km
**17.** $k = 17.5$; $y = 17.5\sqrt[3]{x}$　**18.** $k = 0.72$; $z = \frac{0.72v}{w^2}$

| $x$ | $y$ |
|-----|-----|
| 216 | 105 |
| 0.343 | 12.25 |
| 729 | 157.5 |

| $v$ | $w$ | $z$ |
|-----|-----|-----|
| 196 | 7 | 2.88 |
| 38.75 | 1.25 | 17.856 |
| 24 | 0.6 | 48 |

**19.** $t = 160$　**20.** 4.5 sec
**21. a.** $f(x) = \begin{cases} 5, & x \le -4 \\ -x + 1, & -4 < x \le 3 \\ 3\sqrt{x - 3} - 1, & x > 3 \end{cases}$　**b.** $R: y \in [-2, \infty)$

**22.**

$D: x \in (-\infty, \infty)$,
$R: y \in (-\infty, -8) \cup (-8, \infty)$,
discontinuity at $x = -3$;
define $h(x) = -8$ at $x = -3$

**23.** $-4, -4, -4.5, -4.99, 3\sqrt{3} - 9, 3\sqrt{3.5} - 9$

**24.** $D: x \in (-\infty, \infty)$;
$R: y \in [-4, \infty)$

**25.** $C(x) = \begin{cases} 20x, & x \le 2 \\ 30x - 20, & 2 < x \le 4 \\ 40x - 60, & x > 4 \end{cases}$
For 5 hr the total cost is \$140.

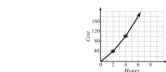

**26.** $a^2 + 7a - 2$　**27.** 147　**28.** $x \in (-\infty, \frac{2}{3}) \cup (\frac{2}{3}, \infty)$
**29.** $4x^2 + 8x - 3$　**30.** 99　**31.** $x$; $x$　**32.** $f(x) = \sqrt{x} + 1$; $g(x) = 3x - 2$
**33.** $f(x) = x^2 - 3x - 10$; $g(x) = x^{\frac{1}{3}}$　**34. a.** 4　**b.** 6　**c.** $\frac{-1}{5}$　**d.** 14
**e.** 0　**f.** 7　**35. a.** $P(x) = 84.95n - (-0.002n^2 + 20n + 30,000) =$
$0.002n^2 + 64.95n - 30,000$　**b.** $-$\$3700　**c.** \$344,750　**d.** 456
**36.** $A(t) = \pi(2t + 3)^2$　**37. a.** $A(W) = 300W - W^2$　**b.** $W = L = 150$ ft
**c.** 22,500 ft$^2$　**38. a.** $V = b^3 + (\frac{1}{3})b^2h$　**b.** $V(b) = 2b^3$
**c.** $b = 21$ cm, $h = 63$ cm

## Practice Test, pp. 329–330

**1. I. a.** square root   **b.** $x \in [-4, \infty), y \in [-3, \infty)$   **c.** $(-2, 0), (0, 1)$
**d.** up on right   **e.** $x \in (-2, \infty)$; $x \in [-4, -2)$
**II. a.** cubic   **b.** $x \in (-\infty, \infty), y \in (-\infty, \infty)$   **c.** $(2, 0), (0, -1)$
**d.** down on left, up on right   **e.** $x \in (2, \infty)$; $x \in (-\infty, 2)$
**III. a.** absolute value   **b.** $x \in (-\infty, \infty), y \in (-\infty, 4]$
**c.** $(-1, 0), (3, 0), (0, 2)$   **d.** down on left, down on right   **e.** $x \in (-1, 3)$;
$x \in (-\infty, -1) \cup (3, \infty)$
**IV. a.** quadratic   **b.** $x \in (-\infty, \infty)$; $y \in [-5.5, \infty)$
**c.** $(0, 0), (5, 0), (0, 0)$   **d.** up on left, up on right
**e.** $x \in (-\infty, 0) \cup (5, \infty)$; $x \in (0, 5)$

**2.**     **3.**     **4.**

**5. a.** $(-\infty, \infty)$   **b.** $[0, \infty)$   **c.** $[0, \infty)$   **6.** VA: $x = -2$; HA: $y = -1$
**7. a.**

**b.** $S(t) = 17.27t^{2.50}$   **c.** 3.05 mm   **d.** 0.95 sec   **8.** $a = 1, b = 4$
**9. a.** $4, -4, \frac{25}{4}$     **b.**

**10. a.** $W(t) = \begin{cases} (t-4)^2 + 209, & 0 \le t < 4 \\ -2t + 217, & 4 \le t \le 8 \\ 201, & t > 8 \end{cases}$   **b.** 225 lb   **c.** 205 lb

**11.** 1617 kW·h/yr   **12.** $M = kd^3(\frac{1}{p^2})$, approx. $2.2788 \times 10^8$   **13.** 520 lb
**14. a.** $P(x) = 200x - 20{,}000$   **b.** $\overline{P}(x) = 200 - \frac{20{,}000}{x}$
**c.** −$200/item; On a average, the company lost $200 for each of the first
50 items produced.   **d.** $100/item; On average, the company made $100
for each of the first 200 items produced.   **e.** The average profit becomes
closer and closer to the marginal profit of $200 per item.
**15. a.** $(f \cdot g)(x) = (x - 1)\sqrt{2 - x}$; $x \in (-\infty, 2]$
**b.** $(\frac{f}{g})(x) = \frac{\sqrt{2-x}}{x-1}$; $x \in (-\infty, 1) \cup (1, 2]$
**c.** $(\frac{g}{f})(x) = \frac{x-1}{\sqrt{2-x}}$; $x \in (-\infty, 2)$
**16.** $3x + 1$; $x \in [\frac{1}{3}, \infty]$   **17. a.** $V(t) = \frac{4}{3}\pi(\sqrt{t})^3$   **b.** $36\pi$ in³
**18.** max: $y = 8$ at $x = -2$; min: $y = -7$ at $x \approx -5.87$ and $y = -7$ at $x \approx 1.87$
**19. a.** $V(h) = 480\pi h$   **b.** about 22,619.5 ft³   **20. a.** 4   **b.** −6
**c.** 8   **d.** 2.5

## Calculator Exploration and Discovery, pp. 331–332

**1.** shifted right 3 units; Answers will vary.
**2.** shifted right 3 units; Answers will vary.
**3.** They are approaching 4; not defined
**4.** $Y_1 = 4, Y_2$ shows an error; calculator is *rounding* to 4; no

## Strengthening Core Skills, pp. 332–333

**1.** $D: x \in (-\infty, \infty)$,   $R: y \in (-\infty, 7]$
**2.** $D: x \in [-1, 7]$,   $R: y \in [-6, 9]$
**3.** $D: x \in (-\infty, 9]$,   $R: y \in [-3, \infty)$
**4.** $D: x \in (-\infty, \infty)$,   $R: y \in (-\infty, 9]$

## Cumulative Review Chapters R–3, pp. 333–334

**1.** $R = \frac{R_1 R_2}{R_1 + R_2}$   **3. a.** $(x - 1)(x^2 + x + 1)$   **b.** $(x - 3)(x + 2)(x - 2)$
**5.** all reals   **7.** verified   **9.** $y = \frac{11}{60}x + \frac{1009}{60}$; 39 min, driving time
increases 11 min every 60 days   **11. a.** $y = 3x - 32$ (Answers will vary.)
**b.** month 11   **13. a.** $x$   **b.** $x$
**15.**

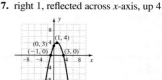

**17.** $X = 63$
**19. a.** $f(4) = -1, g(2) = 4, (f \circ g)(2) = -1$
**b.** $g(4) = 0, f(8) = 4, (g \circ f)(8) = 0$
**c.** $(f \cdot g)(0) = (-2)(4) = -8, (\frac{g}{f})(0) = \frac{4}{-2} = -2$
**d.** $(f + g)(1) = -3 + 5 = 2, (g - f)(9) = 2 - 2 = 0$
**21. a.** all real numbers   **b.** $x \ge 0$
**23.** $x \in (-\frac{11}{2}, 4)$   **25. a.** false: $x^2 + 6x + 9$   **b.** false: $25x^{4}y^{6}$
**c.** false: −9   **d.** false: undefined   **e.** false: $\frac{1}{9}$   **f.** false: 17

# CHAPTER 4

## Exercises 4.1, pp. 344–348

**1.** extreme, vertex   **3.** Answers will vary.
**5.** left 2, down 9           **7.** right 1, reflected across $x$-axis, up 4

**9.** left 1, stretched vertically,       **11.** right 2, reflected across $x$-axis,
down 8                         stretched vertically, up 15

**13.** right $\frac{5}{2}$, down $\frac{17}{4}$         **15.** right $\frac{9}{8}$, stretched vertically,
                                     down $\frac{49}{16}$

**17.** left 1, down 7           **19.** right 2, reflected across
                              $x$-axis, up 6

**21.** right $\frac{3}{2}$, stretched vertically, down 6

**23.** left 3, compressed vertically, down $\frac{19}{2}$

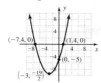

**25.** $y = 1(x - 2)^2 - 1$   **27.** $y = -1(x + 2)^2 + 4$
**29.** $y = -\frac{3}{2}(x + 2)^2 + 3$   **31. i.** $x = -3 \pm \sqrt{5}$   **ii.** $x = 4 \pm \sqrt{3}$
**iii.** $x = -4 \pm \frac{\sqrt{14}}{2}$   **iv.** $x = 2 \pm \sqrt{2}$   **v.** $t = -2.7, t = 1.3$
**vi.** $t = -1.4, t = 2.6$   **33. a.** $(0, -66,000)$; When no cars are produced, there is a daily loss of $66,000.   **b.** $(20, 0), (330, 0)$; No profit will be made if either 20 or less, or 330 or more cars are produced.   **c.** 175   **d.** $240,250
**35. a.** $2   **b.** $44   **c.** $8800   **d.** $23; $44,100   **37.** 6000; $3200
**39. a.** $h(t) = -16t^2 + 240t + 544$   **b.** 544 ft; That is when the fuel is exhausted.   **c.** 1344 ft   **d.** 1344 ft   **e.** It is coming back down.
**f.** 1444 ft   **g.** 17 sec   **41. a.** 14.4 ft   **b.** 41 ft   **c.** 48.02 ft   **d.** 90 ft
**43. a.** 25 ft   **b.** approx. 3.43 sec   **c.** 67.25 ft
**45. a.** 2500 ft$^2$; 50 ft × 50 ft   **b.** 5000 ft$^2$; 50 ft × 100 ft
**47. a.** approx. 29.5 in. wide by 18.7 in. long   **b.** approx. 930 in$^2$
**49. a.** $1.25 each; $781.25   **b.** about $0.85 each
**51. a.** $R = -43.07t^2 + 976.53t - 126.8$   **b.** 4598   **c.** $t = 8.26$ days (early on the ninth day)   **d.** about 5408 participants
**53.** Answers will vary.   **55. a.** verified   **b.** verified
**57.** $m = \frac{4}{3}$, $y$-intercept $(0, 3)$   **59.** $g(x) = -\sqrt[3]{x + 1} - 3$

### Exercises 4.2, pp. 356–359

**1.** $P(c)$, remainder   **3.** Answers will vary.
**5.** $x^3 - 5x^2 - 4x + 23 = (x - 2)(x^2 - 3x - 10) + 3$
**7.** $2x^3 + 5x^2 + 4x + 17 = (x + 3)(2x^2 - x + 7) - 4$
**9.** $x^3 - 8x^2 + 11x + 20 = (x - 5)(x^2 - 3x - 4) + 0$
**11. a.** $\dfrac{2x^2 - 5x - 3}{x - 3} = (2x + 1) + \dfrac{0}{x - 3}$
**b.** $2x^2 - 5x - 3 = (x - 3)(2x + 1) + 0$
**13. a.** $\dfrac{x^3 - 3x^2 - 14x - 8}{x + 2} = (x^2 - 5x - 4) + \dfrac{0}{x + 2}$
**b.** $x^3 - 3x^2 - 14x - 8 = (x + 2)(x^2 - 5x - 4) + 0$
**15. a.** $\dfrac{x^3 - 5x^2 - 4x + 23}{x - 2} = (x^2 - 3x - 10) + \dfrac{3}{x - 2}$
**b.** $x^3 - 5x^2 - 4x + 23 = (x - 2)(x^2 - 3x - 10) + 3$
**17. a.** $\dfrac{2x^3 - 5x^2 - 11x - 17}{x - 4} = (2x^2 + 3x + 1) + \dfrac{-13}{x - 4}$
**b.** $2x^3 - 5x^2 - 11x - 17 = (x - 4)(2x^2 + 3x + 1) - 13$
**19.** $x^3 + 5x^2 + 7 = (x + 1)(x^2 + 4x - 4) + 11$
**21.** $x^3 - 13x - 12 = (x - 4)(x^2 + 4x + 3) + 0$
**23.** $3x^3 - 8x + 12 = (x - 1)(3x^2 + 3x - 5) + 7$
**25.** $n^3 + 27 = (n + 3)(n^2 - 3n + 9) + 0$
**27.** $x^4 + 3x^3 - 16x - 8 = (x - 2)(x^3 + 5x^2 + 10x + 4) + 0$
**29.** $\dfrac{2x^3 + 7x^2 - x + 26}{x^2 + 3} = (2x + 7) + \dfrac{-7x + 5}{x^2 + 3}$
**31.** $\dfrac{x^4 - 5x^2 - 4x + 7}{x^2 - 1} = (x^2 - 4) + \dfrac{-4x + 3}{x^2 - 1}$
**33. a.** $-30$   **b.** 12   **35. a.** $-1$   **b.** 3   **37. a.** 31   **b.** 0
**39. a.** $-10$   **b.** 0

**41.**
$$-3 \,\big|\; 1 \quad 2 \quad -5 \quad -6$$
$$\underline{\phantom{1}\;\; -3 \quad 3 \quad 6}$$
$$\phantom{-3\,|}\; 1 \; -1 \; -2 \quad 0$$

**43.**
$$2 \,\big|\; 1 \quad 0 \quad -7 \quad 6$$
$$\underline{\phantom{1}\;\; 2 \quad 4 \quad -6}$$
$$\phantom{2\,|}\; 1 \quad 2 \; -3 \quad 0$$

**45.**
$$\tfrac{2}{3} \,\big|\; 9 \quad 18 \quad -4 \quad -8$$
$$\underline{\phantom{9}\;\; 6 \quad 16 \quad 8}$$
$$\phantom{\tfrac{2}{3}\,|}\; 9 \quad 24 \quad 12 \quad 0$$

**47. a.** yes   **b.** yes   **49. a.** no   **b.** yes   **51. a.** yes   **b.** yes
**53.** $P(x) = (x + 2)(x - 3)(x + 5)$; $P(x) = x^3 + 4x^2 - 11x - 30$
**55.** $P(x) = (x + 2)(x - \sqrt{3})(x + \sqrt{3})$; $P(x) = x^3 + 2x^2 - 3x - 6$
**57.** $P(x) = (x + 5)(x - 2\sqrt{3})(x + 2\sqrt{3})$; $P(x) = x^3 + 5x^2 - 12x - 60$
**59.** $P(x) = (x - 1)(x + 2)(x - \sqrt{10})(x + \sqrt{10})$;
$P(x) = x^4 + x^3 - 12x^2 - 10x + 20$   **61.** $P(x) = (x + 2)(x - 3)(x - 4)$
**63.** $p(x) = (x + 3)^2(x - 3)(x - 1)$   **65.** $f(x) = 2(x - \frac{3}{2})(x + 2)(x + 5)$
**67.** $p(x) = (x + 3)(x - 3)^2$   **69.** $p(x) = (x - 2)^3$
**71.** $V(x) = (24 - 2x)(18 - 2x)x$; $V(x) = 4x^3 - 84x^2 + 432x$
**73. a.** week 10; 22.5 thousand   **b.** one week before closing; 36 thousand
**c.** week 9   **75. a.** 198 ft$^3$   **b.** 2 ft   **c.** about 7 ft
**77. a.** $C(2) = \$175.00$   **b.** $C(7) = \$110$; the cost is higher in February, perhaps due to the cost of heating the house during the winter.
**79.** $k = 10$   **81.** $k = -3$   **83.** $\dfrac{x^3 + x^2 + 9x + 6}{x + 3} = x^2 - 2x + 15 - \dfrac{39}{x + 3}$;
$92 = 92$, way cool!   **85. a.** $P(3i) = 0$   **b.** $P(-2i) = 0$
**c.** $P(1 + 2i) = 0$   **d.** $P(1 + 3i) = 0$   **87.** $w = 0, w = 1$
**89.** $\dfrac{\Delta y}{\Delta x} = \dfrac{-1.9}{10}$

### Exercises 4.3, pp. 371–374

**1.** $n$, complex   **3.** b; 4 is not a factor of 6   **5. a.** yes   **b.** yes
**7.** There is at least one zero in the intervals $(-2, -1)$, $(0, 1)$, and $(1, 2)$

| $x$ | $y$ |
|---|---|
| $-4$ | $-51$ |
| $-3$ | $-17$ |
| $-2$ | $-1$ |
| $-1$ | 3 |
| 0 | 1 |
| 1 | $-1$ |
| 2 | 3 |
| 3 | 19 |
| 4 | 53 |

**9.** $\{\pm 1, \pm 15, \pm 3, \pm 5, \pm\frac{1}{4}, \pm\frac{15}{4}, \pm\frac{3}{4}, \pm\frac{5}{4}, \pm\frac{1}{2}, \pm\frac{15}{2}, \pm\frac{3}{2}, \pm\frac{5}{2}\}$
**11.** $\{\pm 1, \pm 15, \pm 3, \pm 5, \pm\frac{1}{2}, \pm\frac{15}{2}, \pm\frac{3}{2}, \pm\frac{5}{2}\}$
**13.** $\{\pm 1, \pm 28, \pm 2, \pm 14, \pm 4, \pm 7, \pm\frac{1}{6}, \pm\frac{14}{3}, \pm\frac{1}{3}, \pm\frac{7}{3}, \pm\frac{2}{3}, \pm\frac{7}{6}, \pm\frac{1}{2}, \pm\frac{7}{2}, \pm\frac{28}{3}, \pm\frac{4}{3}\}$
**15.** $\{\pm 1, \pm 3, \pm\frac{1}{32}, \pm\frac{1}{2}, \pm\frac{1}{16}, \pm\frac{1}{4}, \pm\frac{1}{8}, \pm\frac{3}{32}, \pm\frac{3}{2}, \pm\frac{3}{16}, \pm\frac{3}{4}, \pm\frac{3}{8}\}$
**17.** $(x + 4)(x - 1)(x - 3)$, $x = -4, 1, 3$
**19.** $(x + 3)(x + 2)(x - 5)$, $x = -3, -2, 5$
**21.** $(x + 3)(x - 1)(x - 4)$, $x = -3, 1, 4$
**23.** $(x + 2)(x - 3)(x - 5)$, $x = -2, 3, 5$
**25.** $(x + 7)(x + 2)(x + 1)(x - 3)$, $x = -7, -2, -1, 3$
**27.** $(2x + 3)(2x - 1)(x - 1)$; $x = -\frac{3}{2}, \frac{1}{2}, 1$
**29.** $(2x + 3)^2(x - 1)$; $x = -\frac{3}{2}, 1$
**31.** $(x + 2)(x - 1)(2x - 5)$; $x = -2, 1, \frac{5}{2}$
**33.** $(x + 1)(2x + 1)(x - \sqrt{5})(x + \sqrt{5})$; $x = -1, -\frac{1}{2}, \sqrt{5}, -\sqrt{5}$
**35.** $r(x) = (x + 2)(3x - 2)(x + \sqrt{2})(x - \sqrt{2})$; $x = -2, \frac{2}{3}, -\sqrt{2}, \sqrt{2}$
**37. a.** $\{\pm 1, \pm 12, \pm 2, \pm 6, \pm 3, \pm 4\}$
**b.**

| Possible positive zeroes | Possible negative zeroes | Possible nonreal zeroes | Total zeroes |
|---|---|---|---|
| 2 | 1 | 0 | 3 |
| 0 | 1 | 2 | 3 |

**c.** $x = -3, 1, 4$

**39. a.** $\{\pm1, \pm8, \pm2, \pm4\}$

**b.**

| Possible positive zeroes | Possible negative zeroes | Possible nonreal zeroes | Total zeroes |
|---|---|---|---|
| 1 | 3 | 0 | 4 |
| 1 | 1 | 2 | 4 |

**c.** $x = -1, -2$ (repeated zero), 2

**41. a.** $\{\pm1, \pm60, \pm2, \pm30, \pm3, \pm20, \pm4, \pm15, \pm5, \pm12, \pm6, \pm10\}$

**b.**

| Possible positive zeroes | Possible negative zeroes | Possible nonreal zeroes | Total zeroes |
|---|---|---|---|
| 2 | 0 | 2 | 4 |
| 0 | 2 | 2 | 4 |
| 2 | 2 | 0 | 4 |
| 0 | 0 | 4 | 4 |

**c.** $x = -3, -2, 2, 5$

**43. a.** $\{\pm1, \pm80, \pm2, \pm40, \pm4, \pm20, \pm5, \pm16, \pm8, \pm10, \pm\frac{1}{2}, \pm\frac{5}{2}\}$

**b.**

| Possible positive zeroes | Possible negative zeroes | Possible nonreal zeroes | Total zeroes |
|---|---|---|---|
| 2 | 0 | 2 | 4 |
| 0 | 2 | 2 | 4 |
| 2 | 2 | 0 | 4 |
| 0 | 0 | 4 | 4 |

**c.** $x = -4, -\frac{5}{2}, 2$ (repeated zero)

**45.** $P(x) = (x + 2)(x - 2)(x + 3i)(x - 3i)$
$x = -2, x = 2, x = 3i, x = -3i$

**47.** $Q(x) = (x + 2)(x - 2)(x + 2i)(x - 2i)$
$x = -2, x = 2, x = 2i, x = -2i$

**49.** $P(x) = (x + 1)(x + 1)(x - 1); x = -1, x = -1, x = 1$

**51.** $Q(x) = (x - 5)(x + 5)(x - 5); x = 5, x = -5, x = 5$

**53.** $(x - 5)^3(x + 9)^2; x = 5$, multiplicity 3; $x = -9$, multiplicity 2

**55.** $(x - 7)^2(x + 2)^2(x + 7); x = 7$, multiplicity 2; $x = -2$, multiplicity 2; $x = -7$, multiplicity 1

**57.** $P(x) = x^3 - 3x^2 + 4x - 12$    **59.** $P(x) = x^4 - x^3 - x^2 - x - 2$

**61.** $P(x) = x^4 - 6x^3 + 13x^2 - 24x + 36$

**63.** $P(x) = x^4 + 2x^2 + 8x + 5$

**65.** $(x - 4)(2x - 3)(2x + 3); x = 4, \frac{3}{2}, -\frac{3}{2}$

**67.** $(x - 2)(x + 12)(4x^2 + 3); x = 2, -12, \frac{\sqrt{3}}{2}i, -\frac{\sqrt{3}}{2}i$

**69.** $x = 1, 2, 4, -2$    **71.** $x = 1, 2, 3, \frac{-3}{2}$    **73.** $x = -2, \frac{-3}{2}, 4$

**75.** $x = 3, -1, \frac{5}{3}$    **77.** $x = -3, 4, -3, \pm\sqrt{2}$

**79.** $x = 1, 2, -3, \pm i\sqrt{7}$    **81.** $x = -1, \frac{3}{2}, \pm i\sqrt{3}$

**83. a.** $A(1) \approx 11.6$ million km$^2$; $A(10) \approx 6.0$ million km$^2$
**b.** month 9 (September); about 5.6 million km$^2$
$x$: [0, 12], $y$: [0, 20]

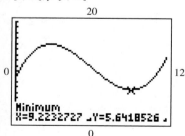

**85. a.** $A(1) = 980$, the altitude was $2000 + 980 = 2980$ m
$A(7) = -64$, the altitude was $2000 - 64 = 1936$ m
$A(13) = 980$, the altitude was $2000 + 980 = 2980$ m
**b.** $A(d) = -(d + 1)(d - 6)(d - 8)(d - 15)$,
three times, at $d = 6, d = 8$, and $d = 15$.

**87.** The zeroes of $f$ are $x = -2, x = 1$, and $x = 4$; yes

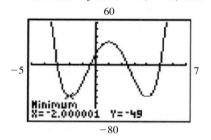

**89. a.** $4\,\text{cm} \times 4\,\text{cm} \times 4\,\text{cm}$    **b.** $5\,\text{cm} \times 5\,\text{cm} \times 5\,\text{cm}$
**91.** length 10 in., width 5 in., height 3 in.    **93.** 1994, 1998, 2002; about 5 yr
**95. a.** 8.97 m, 11.29 m, 12.05 m, 12.94 m    **b.** 9.7 m; +3.7
**97. a.** yes   **b.** no   **c.** about 14.88    **99. a.** $x = -\sqrt{7}, \sqrt{7}$
**b.** $x = -2\sqrt{3}, 2\sqrt{3}$   **c.** $x = -3\sqrt{2}, 3\sqrt{2}$
**101. a.** $C(z) = (z - 4i)(z + 3)(z - 2)$
**b.** $C(z) = (z - 9i)(z + 4)(z + 1)$   **c.** $C(z) = (z - 2 + i)(z - 3i)(z + i)$
**d.** $C(z) = (z - 2 + 3i)(z - 5i)(z + 2i)$    **103. a.** $w = 150$ ft, $l = 300$
**b.** $A = 15,000$ ft$^2$    **105.** $r(x) = 2\sqrt{x + 4} - 2$

## Mid-Chapter Check, pp. 374–375

**1. a.** $x^3 + 8x^2 + 7x - 14 = (x^2 + 6x - 5)(x + 2) - 4$
**b.** $\dfrac{x^3 + 8x^2 + 7x - 14}{x + 2} = x^2 + 6x - 5 - \dfrac{4}{x + 2}$
**2.** $f(x) = (2x + 3)(x + 1)(x - 1)(x - 2)$    **3.** $f(-2) = 7$
**4.** $f(x) = x^3 - 2x + 4$    **5.** $g(2) = -8$ and $g(3) = 5$ have opposite signs
**6.** $f(x) = (x - 2)(x + 1)(x + 2)(x + 4)$
**7.** $x = -2, x = 1, x = -1 \pm 3i$    **8. a.** $a > 0$, minimum
**b.** A minimum value of $y = -11$ occurs at $x = 4$.    **9. a.** $2100
**b.** A maximum revenue of $44,100 occurs after two price decreases of
$200 each.    **10. a.** degree 4; three turning points    **b.** 2 sec
**c.** $A(t) = (t - 1)^2(t - 3)(t - 5), A(t) = t^4 - 10t^3 + 32t^2 - 38t + 15$
$A(2) = 3$; altitude is 300 ft above hard-deck, $A(4) = -9$; altitude is 900 ft
below hard-deck

## Exercises 4.4, pp. 387–391

**1.** zero, $m$    **3.** Answers will vary.    **5.** polynomial, degree 5
**7.** not a polynomial, sharp turns    **9.** polynomial, degree 4
**11.** up/down    **13.** down/down    **15.** down/up; $(0, -2)$
**17.** down/down; $(0, -6)$    **19.** up/down; $(0, -6)$
**21. a.** even   **b.** $-3$ odd, $-1$ even, 3 odd   **c.** $f(x) = (x + 3)$
$(x + 1)^2(x - 3)$, deg 4   **d.** $x \in \mathbb{R}, y \in [-9, \infty)$
**23. a.** even   **b.** $-3$ odd, $-1$ odd, 2 odd, 4 odd
**c.** $f(x) = -(x + 3)(x + 1)(x - 2)(x - 4)$, deg 4
**d.** $x \in \mathbb{R}, y \in (-\infty, 25]$    **25. a.** odd   **b.** $-1$ even, 3 odd
**c.** $f(x) = -(x + 1)^2(x - 3)$, deg 3   **d.** $x \in \mathbb{R}, y \in \mathbb{R}$
**27.** degree 6; up/up; $(0, -12)$    **29.** degree 5; up/down; $(0, -24)$
**31.** degree 6; up/up; $(0, -192)$    **33.** degree 5; up/down; $(0, 2)$
**35.** b    **37.** e    **39.** c

**41.**
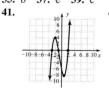

**43.**

**45.**

**47.**
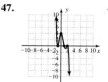

**49.**
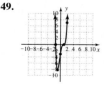

**51.**
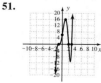

**53.**  **55.**  **57.**

**59.**  **61.**  **63.**

**65.**  **67.**

**69.** $P(x) = \frac{1}{6}(x + 4)(x - 1)(x - 3)$, $P(x) = \frac{1}{6}(x^3 - 13x + 12)$
**71. a.** $(-3)^2 + (-1)^2 + (2)^2 + (4)^2 = (-2)^2 - 2(-13)$
$9 + 1 + 4 + 16 = 4 + 26$
$30 = 30$ ✓
**b.** $(x + 3)(x + 1)(x - 2)(x - 4) = x^4 - 2x^3 - 13x^2 + 14x + 24$ ✓
**73. a.** 280 vehicles above average, 216 vehicles below average, 154 vehicles above average **b.** 6:00 A.M. ($t = 0$), 10:00 A.M. ($t = 4$), 3:00 P.M. ($t = 9$), 6:00 P.M. ($t = 12$)
**c.** max: about 300 vehicles above average at 7:30 A.M.; min: about 220 vehicles below average at 12 noon

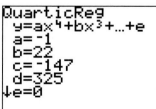

**75. a.** 3 **b.** 5 **c.** $B(x) = \frac{1}{4}x(x - 4)(x - 9)$, $-\$80,000$
**77. a.**

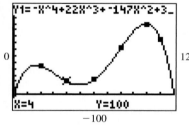

, quartic;

**b.** $t \approx 1.7$ (7:42 A.M.), 227 vehicles; $t \approx 9.9$ (3:54 P.M.), 551 vehicles
**c.** $t \approx 7.93$ (1:56 P.M.) and $t \approx 11.27$ (5:16 P.M.)

**79. a.** $f(x) \to \infty, f(x) \to -\infty$ **b.** $g(x) \to \infty, g(x) \to \infty$; $x^4 \geq 0$ for all $x$
**81.** verified **83.** $h(x) = \dfrac{1 - 2x}{x^2}$, $D : x \in \{x | x \neq 0\}$; $H(x) = \dfrac{1}{x^2 - 2x}$, $D : x \in \{x | x \neq 0, x \neq 2\}$ **85. a.** $x = 2$ **b.** $x = 8$ **c.** $x = 4, x = -6$

## Exercises 4.5, pp. 402–406

**1.** as $x \to -\infty, y \to 2$ **3.** horizontal, vertical; Answers will vary.
**5.** $x = 3; x \in (-\infty, 3) \cup (3, \infty)$
**7.** $x = \frac{-5}{2}, x = 1; x \in (-\infty, -\frac{5}{2}) \cup (-\frac{5}{2}, 1) \cup (1, \infty)$

**9.** No V.A., $x \in (-\infty, \infty)$ **11.** $x = 3$, yes; $x = -2$, yes **13.** $x = 3$, no
**15.** $x = 2$, yes; $x = -2$, no **17. a.** $y = 0$ **b.** crosses at $(\frac{3}{2}, 0)$
**19. a.** $y = 4$ **b.** crosses at $(-\frac{21}{4}, 4)$ **21. a.** $y = 3$ **b.** does not cross
**23.** $q(x) = 0$, $r(x) = 8x$ directly; the graph will cross the horizontal asymptote at $x = 0$. **25.** $q(x) = 2$, $r(x) = -8x - 8$; the graph will cross the horizontal asymptote at $x = 1$. **27.** $(0, 0)$ cross, $(3, 0)$ cross; $(0, 0)$
**29.** $(-4, 0)$ cross; $(0, 4)$ **31.** $(0, 0)$ cross, $(3, 0)$ bounce; $(0, 0)$

**33.**  **35.**  **37.**

**39.**  **41.**  **43.**

**45.**

**47.** $f(x) = \dfrac{(x - 4)(x + 1)}{(x + 2)(x - 3)}$ **49.** $f(x) = \dfrac{5x}{(x + 3)^2(x - 3)}$

**51.**  **53.**  **55.**

**57.**  **59.**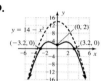

**61.**  **a.** It is impossible to remove 100% of the pollutants. **b.** $250 thousand, $277.8 thousand **c.** 90%

**63. a.** 5 hr; about 0.28 **b.** $-0.019, -0.005$; As the number of hours increases, the rate of change decreases. **c.** $h \to \infty, C \to 0$; horizontal asymptote **65. a.** $20,000, $80,000, $320,000; cost increases dramatically
**b.**

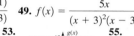

**c.** as $p \to 100^-, C \to \infty$

**67. a.** 2; 10 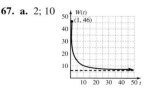 **b.** 14; 20 **c.** On average, 6 words will be remembered for life.

**69. a.** $0.75x + 0.25(40) = (x + 40)C$     **b.** 35%; 62.5%; 160 gal
$\frac{3}{4}x + 10 = (x + 40)C$, solve for $C$     **c.** 160 gal; 200 gal
     **d.** 70%; 75%

**71. a.** 5     **b.** 18     **c.** The horizontal asymptote at $y = 95$ means her average grade will approach 95 as the number of tests taken increases; no     **d.** 6
**73. a.** $225; $175     **b.** 2000 heaters     **c.** 4000 heaters     **d.** The horizontal asymptote at $y = 125$ means the average cost approaches $125 as monthly production gets very large. Due to limitations on production (maximum of 5000 heaters) the average cost will never fall below $A(5000) = 135$.     **75. a.** $A(x) = \frac{4x^2 + 53x + 250}{x}$; $x = 0$, $y = 4x + 53$
**b.** cost: $307, $372, $445; avg. cost: $307, $186, $148.33

**c.**

**d.** 8; $116.25; verified     **77. a.** $S = \frac{\pi r^3 + 2V}{r}$     **b.** $r \approx 3.1$ in., $h \approx 3$ in.
**79.** $y = \frac{-4}{3}x - \frac{1}{3}$     **81.** $\frac{-16}{3}, \frac{3}{4}$; $(3x + 16)(4x - 3) = 0$

## Exercises 4.6, pp. 415–418

**1.** vertical, multiplicity     **3.** Answers will vary.     **5.** $x \in (0, 4)$
**7.** $x \in (-\infty, -5] \cup [1, \infty)$     **9.** $x \in (-1, \frac{7}{2})$     **11.** $x \in [-\sqrt{7}, \sqrt{7}]$
**13.** $x \in (-\infty, -\frac{5}{3}] \cup [1, \infty)$     **15.** $x \in (-\infty, \infty)$     **17.** { }
**19.** $x \in (-\infty, 5) \cup (5, \infty)$     **21.** { }     **23.** $x \in (-3, 1)$
**25.** $x \in (-\infty, \frac{-3}{2}] \cup [2, \infty)$     **27.** $x \in (-\infty, \infty)$
**29.** $x \in (-\infty, -5] \cup [5, \infty)$     **31.** $x \in (-\infty, 0] \cup [5, \infty)$     **33.** a
**35.** b     **37.** $x \in (-3, 5)$     **39.** $x \in [4, \infty) \cup \{-1\}$
**41.** $x \in (-\infty, -2] \cup \{2\} \cup [4, \infty)$     **43.** $x \in (-2 - \sqrt{3}, -2 + \sqrt{3})$
**45.** $x \in (-\infty, -3] \cup [5, \infty)$     **47.** $x \in [-3, 0] \cup [3, \infty)$
**49.** $x \in (-3, 1) \cup (2, \infty)$     **51.** $x \in (-\infty, -3) \cup (-1, 1) \cup (3, \infty)$
**53.** $x \in (-\infty, -2) \cup (-2, 1) \cup (3, \infty)$     **55.** $x \in (-\infty, -2) \cup (2, 3)$
**57.** $x \in [-3, 2)$     **59.** $x \in (-\infty, -2) \cup (-2, -1)$
**61.** $x \in (-\infty, -2) \cup [2, 3)$     **63.** $x \in (-\infty, -5) \cup (0, 1) \cup (2, \infty)$
**65.** $x \in (-4, -2] \cup (1, 2] \cup (3, \infty)$     **67.** $x \in (-\infty, -2] \cup (0, 2)$
**69.** $x \in (-\infty, -17) \cup (-2, 1) \cup (7, \infty)$     **71.** $x \in (-3, \frac{-7}{4}] \cup (2, \infty)$
**73.** $x \in (-2, \infty)$     **75.** $(-\infty, -3) \cup (3, \infty)$     **77.** d
**79.** b     **81. a.** $x \in (-\infty, 3) \cup (3, \infty)$     **b.** $x \in (-\infty, 3)$     **c.** $x \in (3, \infty)$
**d.** $x \in (-\infty, \frac{3}{2}) \cup (3, \infty)$     **83.** $d(x) = k(x^3 - 192x + 1024)$
**a.** $x \in (5, 8]$     **b.** $320k$ units     **c.** $x \in [0, 3)$     **d.** 2 ft     **85. a.** verified
**b.** horizontal: $r_2 = 20$, as $r_1$ increases, $r_2$ decreases to maintain $R = 40$; vertical: $r_1 = 20$, as $r_1$ decreases, $r_2$ increases to maintain $R = 40$
**c.** $r_1 \in (20, 40)$     **87.** $R(t) = 0.01t^2 + 0.1t + 30$     **a.** $[0°, 30°)$
**b.** $(20°, \infty)$     **c.** $(50°, \infty)$     **89.** $F(x) > 0$ for $x \in (-2, 1) \cup (4, \infty)$
**91.** $x(x + 2)(x - 1)^2 > 0$; $\frac{x(x + 2)}{(x - 1)^2} > 0$

**93.**

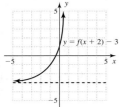
$y = f(x + 2) - 3$

**95.** $F(x) = \begin{cases} f(x) & x \neq -4 \\ -6 & x = -4 \end{cases}$

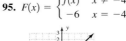
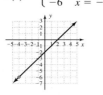

## Making Connections, p. 418

**1.** e     **3.** b     **5.** a     **7.** h     **9.** d     **11.** d, f     **13.** g     **15.** c

## Summary and Concept Review, pp. 419–422

**1.**

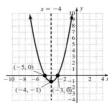

**2.**

**3.**

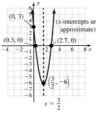
(x-intercepts are approximate)

**4. a.** 0 ft     **b.** 108 ft     **c.** 2.25 sec     **d.** 144 ft; $t = 3$ sec
**5.** $q(x) = x^2 + 6x + 7$; $r = 8$     **6.** $q(x) = x + 1$; $r = 3x - 4$
**7.** $(x + 4)(x + 1)(x - 3)$

**8.** $\underline{-7|}$  1    9    13   −10
$\qquad\qquad\quad -7\ \ -14\qquad 7$
$\qquad\qquad 1\quad 2\quad -1\quad \boxed{-3}$
$h(-7) = -3$

**9.** $\frac{1}{2}|$  4    8   −3   −1
$\qquad\qquad\ \ 2\quad\ 5\quad\ 1$
$\qquad\quad 4\ \ \ 10\quad 2\quad \boxed{0}$
Since $r = 0$, $\frac{1}{2}$ is a root and $x - \frac{1}{2}$ is a factor.

**10.** $3i|$  1       −2       9      −18
$\qquad\qquad\qquad 3i\quad -9 - 6i\quad 18$
$\qquad\quad 1\ \ -2 + 3i\quad -6i\qquad \boxed{0}$
Since $r = 0$, $3i$ is a zero.
**11.** $P(x) = x^3 - x^2 - 5x + 5$     **12. a.** $C(0) = 350$ customers     **b.** more at 2 P.M.; 170     **c.** busier at 1 P.M.; $760 > 710$
**13.** $\{\pm 1, \pm 10, \pm 2, \pm 5, \pm \frac{1}{2}, \pm \frac{5}{2}, \pm \frac{1}{4}, \pm \frac{5}{4}\}$     **14.** $x = -\frac{1}{2}, 2, \frac{5}{2}$
**15.** $P(x) = (2x + 3)(x - 4)(x + 1)$     **16.** only $\pm 1, \pm 3$ are possibilities, none give a remainder of zero     **17.** [1, 2], [4, 5]; verified     **18.** one sign change for $g(x) \to 1$ positive zero; three sign changes for $g(-x) \to 3$ or 1 negative zeroes; 1 positive, 3 negative, 0 complex, or 1 positive, 1 negative, 2 complex; 1 positive, 1 negative, 2 complex, verified
**19.** degree 5; up/down; $(0, -4)$     **20.** degree 4; up/up; $(0, 8)$

**21.**

**22.**

**23.**

**24. a.** even     **b.** $x = -2$, odd; $x = -1$, even; $x = 1$, odd     **c.** deg 6: $P(x) = (x + 2)(x + 1)^2(x - 1)^3$     **25. a.** $\{x | x \in \mathbb{R}; x \neq -1, 4\}$
**b.** HA: $y = 1$; VA: $x = -1, x = 4$     **c.** $V(0) = \frac{9}{4}$ (y-intercept); $x = -3, 3$ (x-intercepts)     **d.** $V(1) = \frac{4}{3}$     **26.** no—even multiplicity; yes—odd multiplicity

**27.**

**28.**

**29.**     **30.**

**31.** $V(x) = \frac{x^2 - x - 12}{x^2 - x - 6}$; $V(0) = 2$    **32. a.** $y = 15$;
as $|x| \to \infty$ $A(x) \to 15^+$. As production increases, average cost decreases and approaches 15.    **b.** $x > 2000$

**33. a.**     **b.** about 2450 favors    **c.** about $2.90 each.

**34.** factored form $(x + 4)(x - 1)(x - 2) > 0$

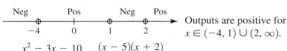

Outputs are positive for $x \in (-4, 1) \cup (2, \infty)$.

**35.** $\dfrac{x^2 - 3x - 10}{x - 2} = \dfrac{(x - 5)(x + 2)}{x - 2} \geq 0$

Outputs are positive or zero for $x \in [-2, 2) \cup [5, \infty)$.

**36.** $\dfrac{(x + 2)(x - 1)}{x(x - 2)} \leq 0$

Outputs are negative or zero for $x \in [-2, 0) \cup [1, 2)$.

### Practice Test, pp. 422–423

**1. a.** $f(x) = -(x - 5)^2 + 9$    **b.** $g(x) = \frac{1}{2}(x + 4)^2 + 8$

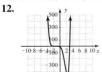 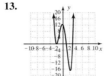

**2.** $(-2, 0)$, $y = 2x^2 + 4x$    **3. a.** 40 ft, 48 ft    **b.** 49 ft    **c.** 14 sec

**4.** $x - 5 + \dfrac{14x + 3}{x^2 + 2x + 1}$    **5.** $x^2 + 2x - 9 + \dfrac{-2}{x + 2}$

**6.** $\underline{-3}|$   1   0   −15   −10   24
       −3   9   18   −24
    1   −3   −6   8   |0   $r = 0$ ✓

**7.** −1    **8.** $2x^3 - 4x^2 + 18x - 36$
**9.** $Q(x) = (x - 2)^2(x - 1)^2(x + 1)$; 2 mult 2, 1 mult 2, −1 mult 1
**10. a.** $\pm 1, \pm 18, \pm 2, \pm 9, \pm 3, \pm 6$    **b.** 1 positive zero; 3 or 1 negative zeroes; 2 or 0 complex zeroes    **c.** $C(x) = (x + 2)(x - 1)(x - 3i)(x + 3i)$
**11. a.** 2002, 2004, 2008    **b.** 4 yr    **c.** deficit of $7.5 million

**12.**     **13.**     **14.**

**15. a.** removal of 100% of the contaminants    **b.** $500,000; $3,000,000; dramatic increase    **c.** 88%

**16. a.**     **b.**

**17.** 800    **18. a.** $x \in (-\infty, -3] \cup [-1, 4]$    **b.** $x \in (-\infty, -4) \cup (0, 2)$
**19. a.**

**b.** $h = -\sqrt[3]{55}$; no    **c.** 28.6%; 29.6%    **d.** $\approx 11.7$ hr    **e.** 4 hr; 43.7%
**f.** The amount of the chemical in the bloodstream becomes negligible.
**20.** $V(x) = \dfrac{x^2 + x - 6}{x^2 - 2x - 3}$; $V(0) = 2$

### Calculator Exploration and Discovery, pp. 423–424

**1.** They do not affect the solution set.    **2.** They do not affect the solution set.
**3.** $(-2, -4)$    **4.** $(-1, -4)$    **5.** $(-1, 3)$    **6.** $(-3, -4)$ and $(2, 1)$

### Strengthening Core Skills, pp. 424–425

**1.** $x \in (-\infty, 3]$    **2.** $x \in (-2, -1) \cup (2, \infty)$
**3.** $x \in (-\infty, -4) \cup (1, 3)$    **4.** $x \in [-2, \infty)$
**5.** $x \in (-\infty, -2) \cup (2, \infty)$    **6.** $x \in [-3, 1] \cup [3, \infty)$

### Cumulative Review Chapters R–4 pp. 425–426

**1.** $h = \dfrac{A - 2\pi r^2}{2\pi r}$    **3. a.** $(2x - 3)^2$    **b.** $(x - 1)^2(x + 2)$
**5.** no solution    **7.** verified    **9. a.** strength depends on time:
$s(t) = 0.5t + 124.5$    **b.** 154.5 lb    **c.** 151 days    **11. a.** height depends on time in months: $h(t) \approx 45.2t + 7.5$ (answers will vary)    **b.** 685.5 ft
**c.** about 36 months    **13.** −30.5    **15. a.** degree 5, five zeroes (including real, nonreal, and repeated zeroes)    **b.** $g(x)$ has no sign changes, no positive zeroes    **c.** $g(-x)$ has one sign change, one negative zero    **d.** $\{\pm 1, \pm 3\}$
**e.** 1 and 3 (no positive roots)    **f.** $x = -1$    **17. a.** −4    **b.** 4
**19.**

**21.** 0.18637; 2.86    **23. a.** $-5 + 12i$    **b.** $1 - i$    **c.** 41    **d.** $0 - i$
**25. a.** $Y_6$    **b.** $Y_7$    **c.** $Y_4$    **d.** $Y_3$    **e.** $Y_5$    **f.** $Y_2$

## CHAPTER 5

### Exercises 5.1, pp. 435–438

**1.** second, one    **3.** False; Answers will vary.    **5.** one-to-one
**7.** one-to-one    **9.** not one-to-one, fails horizontal line test: $x = -3$, $x = -0.5$, and $x = 2$ are paired with $y = 0$    **11.** not a function
**13.** one-to-one    **15.** not one-to-one, $y = 1$ is paired with $x = -6$ and $x = 8$    **17.** one-to-one    **19.** not one-to-one; $h(x) < 3$, corresponds to two $x$-values    **21.** one-to-one    **23.** not one-to-one; $y = 3$ corresponds to more than one $x$-value    **25.** $f^{-1} = \{(1, -2), (4, -1), (5, 0), (9, 2), (15, 5)\}$
**27.** $v^{-1} = \{(3, -4), (2, -3), (1, 0), (0, 5), (-1, 12), (-2, 21), (-3, 32)\}$
**29.** $f^{-1}(x) = x - 5$    **31.** $p^{-1}(x) = \dfrac{-5}{4}x$    **33.** $f^{-1}(x) = \dfrac{x - 3}{4}$
**35.** $t(x) = x^3 + 4$    **37.** $x \in \mathbb{R}, y \in \mathbb{R}; f^{-1}(x) = x^3 + 2, x \in \mathbb{R}, y \in \mathbb{R}$; verified    **39.** $x \in \mathbb{R}, y \in \mathbb{R}; f^{-1}(x) = \sqrt[3]{x} - 1, x \in \mathbb{R}, y \in \mathbb{R}$; verified
**41.** $x \neq -2, y \neq 0; f^{-1}(x) = \dfrac{8}{x} - 2, x \neq 0, y \neq -2$; verified
**43.** $x \neq -1, y \neq 1; f^{-1}(x) = \dfrac{x}{1 - x}, x \neq 1, y \neq -1$; verified
**45. a.** $x \geq -5, y \geq 0$    **b.** $f^{-1}(x) = \sqrt{x} - 5, x \geq 0, y \geq -5$
**47. a.** $x > 3, y > 0$    **b.** $v^{-1}(x) = \sqrt{\dfrac{8}{x}} + 3, x > 0, y > 3$
**49. a.** $x \geq -4, y \geq -2$    **b.** $p^{-1}(x) = \sqrt{x + 2} - 4, x \geq -2, y \geq -4$
**51.** $(f \circ g)(x) = x, (g \circ f)(x) = x$    **53.** $(f \circ g)(x) = x, (g \circ f)(x) = x$
**55.** $(f \circ g)(x) = x, (g \circ f)(x) = x$    **57.** $f^{-1}(x) = 2x + 5$
**59.** $f^{-1}(x) = 2x + 6$    **61.** $f^{-1}(x) = \dfrac{x^3 - 1}{2}$    **63.** $f^{-1}(x) = 2\sqrt[3]{x} + 1$

**65.** $D: x \geq -\frac{2}{3}, R: y \geq 0; f^{-1}(x) = \frac{x^2 - 2}{3}, D: x \geq 0, R: y \geq -\frac{2}{3}$

**67.** $D: x \geq 3, R: y \geq 0; p^{-1}(x) = \frac{x^2}{4} + 3, D: x \geq 0, R: y \geq 3$

**69.** $D: x \geq 0, R: y \geq 3; v^{-1}(x) = \sqrt{x - 3}, D: x \geq 3, R: y \geq 0$

**71.**
$D: x \in [0, \infty), R: y \in [-2, \infty);$
$D: x \in [-2, \infty), R: y \in [0, \infty)$

**73.**
$D: x \in (0, \infty), R: y \in (-\infty, \infty);$
$D: x \in (-\infty, \infty), R: y \in (0, \infty)$

**75.**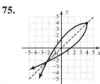
$D: x \in (-\infty, 4], R: y \in (-\infty, 4];$
$D: x \in (-\infty, 4], R: y \in (-\infty, 4]$

**77. a.** $f^{-1}(x) = \frac{x-1}{2}$  **b.** $(-3, -5), (0, 1),$ and $(1, 3)$ are on the graph of $f; (-5, -3), (1, 0),$ and $(3, 1)$ are on the graph of $f^{-1}$   **c.** verified
**d.**

**79. a.** $h^{-1}(x) = \frac{x}{1-x}$  **b.** $(0, 0), (1, \frac{1}{2}),$ and $(2, \frac{2}{3})$ are on the graph of $h;$ $(0, 0), (\frac{1}{2}, 1),$ and $(\frac{2}{3}, 2)$ are on the graph of $h^{-1}$   **c.** verified
**d.**

**81. a.** 31.5 cm  **b.** The result is 80 cm. It gives the distance of the projector from the screen.  **83. a.** $-63.5°F$  **b.** $T^{-1}(x) = \frac{-2}{7}(x - 59);$ independent: temperature, dependent: altitude  **c.** 22,000 ft

**85. a.** 144 ft  **b.** $d^{-1}(x) = \frac{\sqrt{x}}{4};$ independent: distance fallen, dependent: time fallen  **c.** 7 sec  **87. a.** 56,520 ft³  **b.** $V^{-1}(x) = \sqrt[3]{\frac{12x}{\pi}};$ independent: volume, dependent: height  **c.** 18 ft

**89. a.** $f^{-1}(x) = \frac{dx - b}{-cx + a}$  **b.** $f^{-1}(x) = \frac{x}{-x + 2};$ The results are the same.  **91.** d  **93. a.** $P = 2l + 2w$  **b.** $A = \pi r^2$  **c.** $V = \pi r^2 h$
**d.** $V = \frac{1}{3}\pi r^2 h$  **e.** $C = 2\pi r$  **f.** $A = \frac{1}{2}bh$  **g.** $A = \frac{a + b}{2} \cdot h$
**h.** $V = \frac{4}{3}\pi r^3$  **i.** $a^2 + b^2 = c^2$  **95.** $\approx 0.472, \approx 0.365;$ rate of change is greater in $[1, 2]$ due to shape of the graph.

## Exercises 5.2, pp. 447–450

**1.** real, numbers; $(0, \infty); \to 0$   **3.** False; For $b < 1$ and $x_2 > x_1, b^{x_2} < b^{x_1},$ so function is decreasing.  **5.** 16, 2, 8, 11.036
**7.** $1, \frac{1}{64}, \frac{1}{4}, 64$

**9.**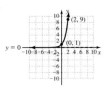
increasing

**11.**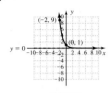
decreasing

**13.** $y = 3^x;$ up 2

**15.** $y = 3^x;$ left 3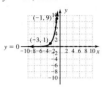

**17.** $y = 3^x;$ reflect across $y$-axis

**19.** $y = (\frac{1}{3})^x;$ up 1

**21.** $y = (\frac{1}{3})^x;$ right 2
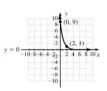

**23.** e  **25.** a  **27.** b  **29.** 2.718282  **31.** 7.389056  **33.** 4.113250

**35.**   **37.** 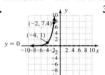  **39.**

**41.** 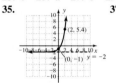  **43.**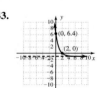

**45.** 3  **47.** $\frac{3}{2}$  **49.** $-\frac{1}{3}$  **51.** 4  **53.** $-3$  **55.** 3  **57.** 2  **59.** $-2$
**61.** 2  **63.** 3  **65.** $x \approx 2.8$  **67.** $x \approx 3.2$
**69. a.** 1732, 3000, 5196, 9000  **b.** yes  **c.** as $t \to \infty, P \to \infty$
**d.**

**71. a.** $100,000  **b.** 3 yr  **73. a.** $\approx$86,806  **b.** 3 yr
**75. a.** $40 million  **b.** 7 yr  **77.** No, they will have to wait about 5 min.
**79.** 32% transparent  **81.** 17% transparent  **83.** $\approx$25,526
**85. a.** 8 g  **b.** 48 min

**87.** exponential, $y \approx 346.35(0.94)^x$

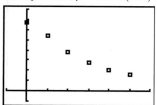

    **a.** 88,100   **b.** 43,100   **c.** 2021

**89.** exponential, $y \approx 103.83(1.0595)^x$

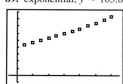

    **a.** 220   **b.** The 22nd note, or F#
    **c.** frequency doubles; yes

**91.** $\frac{1}{5}$   **93.** 75   **95.** $9.5 \times 10^{-7}$; Answers will vary.
**97.** $5; \frac{-7}{9}; 2a^2 - 3a; 2a^2 + 4ah + 2h^2 - 3a - 3h$
**99. a.** no solution   **b.** $\{-5, 6\}$

## Exercises 5.3, pp. 460–463

**1.** real; numbers, $(0, \infty)$; $-\infty$   **3.** 5; Answers will vary.   **5.** $2^3 = 8$
**7.** $7^x = \frac{1}{7}$   **9.** $9^0 = 1$   **11.** $8^{\frac{1}{3}} = y$   **13.** $2^1 = 2$   **15.** $x^2 = 49$
**17.** $10^2 = 100$   **19.** $e^y \approx 54.598$   **21.** $\log_4 64 = 3$   **23.** $\log_3 \frac{1}{9} = x$
**25.** $0 = \ln 1$   **27.** $\log_{\frac{1}{3}} y = -3$   **29.** $\log 1000 = 3$   **31.** $\log \frac{1}{100} = x$
**33.** $\log_4 8 = \frac{3}{2}$   **35.** $\log_{\frac{1}{8}} \frac{1}{y} = \frac{-3}{-2}$   **37.** 1   **39.** 3   **41.** 1   **43.** $\frac{1}{2}$   **45.** $-2$
**47.** $-2$   **49.** $\frac{2}{3}$   **51.** 1.6990   **53.** 0.4700   **55.** 5.4161   **57.** 0.7841
**59.** shift up 3               **61.** shift left 1

**63.** shift left 1,           **65.** right 2, stretch vertically
    reflect across $x$-axis          by a factor of 3

**67.** reflect about the $x$-axis and shift up 2

**69.** II   **71.** VI   **73.** V   **75.** $x \in (3, \infty)$   **77.** $x \in (-\infty, -1) \cup (3, \infty)$
**79.** $x \in (\frac{3}{2}, \infty)$   **81.** $x \in (-3, 3)$   **83.** pH $\approx 4.1$; acid
**85. a.** $\approx 4.7$   **b.** $\approx 7.9$   **87.** about 400 times   **89.** about 3.2 times
**91. a.** $\approx 2.4$   **b.** $\approx 1.2$   **93. a.** 20 dB   **b.** 120 dB
**95.** about 500 times   **97.** about 3160 times   **99.** 6194 m
**101. a.** about 5434 m   **b.** 4000 m   **103. a.** 2225 items   **b.** 2732 items
**c.** \$117,000   **105. a.** about 58.6 cfm   **b.** about 1605 ft²   **107. a.** 95%
**b.** 67%   **109.** Answers will vary.   **a.** 0 dB   **b.** 90 dB   **c.** 15 dB
**d.** 120 dB   **e.** 100 dB   **f.** 140 dB   **111. a.** $\frac{-2}{3}$   **b.** $\frac{-3}{2}$   **c.** $\frac{-5}{2}$

**113.** $D: x \in \mathbb{R}, R: y \in \mathbb{R}$

**115.** $x \in (-\infty, -5); f(x) = (x + 5)(x - 4)^2 = x^3 - 3x^2 - 24x + 80$

## Mid-Chapter Check, pp. 463–464

**1. a.** $\frac{2}{3} = \log_{27} 9$   **b.** $\frac{5}{4} = \log_{81} 243$   **2. a.** $8^{\frac{5}{3}} = 32$   **b.** $1296^{0.25} = 6$
**3. a.** $x = 5$   **b.** $b = \frac{5}{4}$   **4. a.** $x = 3$   **b.** $b = 5$   **5. a.** \$71,191.41
**b.** 6 yr   **6.** $F(x) = 4 \cdot 5^{x-3} + 2$   **7.** $f^{-1}(x) = (x - 1)^2 + 3$;
$D: x \in [1, \infty); R: y \in [3, \infty)$; verified   **8. a.** $\log 2 \approx 0.301$; verified
**b.** $4 \approx \ln 54.598$; verified   **9. a.** $10^{2.845} \approx 700$; verified
**b.** $e^{1.4} \approx 4.0552$; verified   **10.** $\approx 7.9$ times more intense

## Exercises 5.4, pp. 470–473

**1.** quotient; uniqueness   **3.** Answers will vary; Yes,
$1.5663025 = 1.5663025$   **5.** $x = e^{3.4} \approx 29.964$   **7.** $x = 10^{\frac{1}{4}} \approx 1.778$
**9.** $x = \ln 9.025 \approx 2.200$   **11.** $x = \log 18.197 \approx 1.260$
**13.** $x = \ln \frac{65}{4} + 2, x \approx 4.7881$   **15.** $x = \log(78) - 5, x \approx -3.1079$
**17.** $x = -\frac{\ln 2.32}{0.75}, x \approx -1.1221$   **19.** $x = e^{\frac{8}{3}} - 4, x \approx 10.3919$
**21.** $x = 5 - 10^{1.25}, x \approx -12.7828$   **23.** $x = \frac{e^{0.4} - 5}{2}, x \approx -1.7541$
**25.** $\log_2 42$   **27.** $\ln(2x^2 - 14x)$   **29.** $\log(x^2 - 1)$   **31.** $\log_3 4$
**33.** $\log\left(\frac{x}{x+1}\right)$   **35.** $\ln\left(\frac{x-5}{x}\right)$   **37.** $\ln(x - 2)$   **39.** $\log_5(x - 2)$
**41.** $\log_3\left(\frac{x^2 - x - 6}{x + 6}\right)$   **43.** $\ln\left(\frac{3z}{y^2}\right)$   **45.** $\log(x^4 - x^3)$
**47.** $\log\left(\frac{x^2}{\sqrt{x^2 + 1}}\right)$   **49.** $\log_3\left(\frac{p^2}{16q^4}\right)$   **51.** $(x + 2)\log 8$   **53.** $(2x - 1)\ln 5$
**55.** $\frac{1}{2}\log 22$   **57.** $4\log_5 3$   **59.** $3\log a + \log b$   **61.** $\ln x + \frac{1}{4}\ln y$
**63.** $2\ln x - \ln y$   **65.** $\frac{1}{2}\log(x - 2) - \frac{1}{2}\log x$
**67.** $\log_2 3 + 2\log_2 x + 4\log_2 y + \frac{1}{2}\log_2 z$
**69.** $\frac{1}{2}\log_3(x + 2) - \frac{1}{2}\log_3(x - 2) - \frac{1}{2}\log_3(x + 3)$
**71.** $\frac{\ln 60}{\ln 7}$; 2.104076884   **73.** $\frac{\ln 152}{\ln 5}$; 3.121512475
**75.** $\frac{\log 1.73205}{\log 3}$; 0.499999576   **77.** $\frac{\log 0.125}{\log 0.5}$; 3
**79.** $f(x) = \frac{\log(x)}{\log(3)}; f(5) \approx 1.4650; f(15) \approx 2.4650; f(45) \approx 3.4650$;
outputs increase by 1; $f(3^3 \cdot 5) \approx 4.4650$
**81.** $h(x) = \frac{\log(x)}{\log(9)}; h(2) \approx 0.3155; h(4) \approx 0.6309; h(8) \approx 0.9464$;
outputs are multiples of 0.3155; $h(2^4) \approx 4(0.3155) \approx 1.2619$
**83.** verified   **85. a.** $N = AX^{-m}$   **b.** $\approx 3500$ people   **87.** no, pH $\approx 7.32$
**89.** No, pH $\approx 5.9$ and the soil must be treated further.
**91.** $600^{601} > 601^{600}; \frac{1}{99^{100}} < \frac{1}{100^{99}}$
**93. a.** $f(x + 1) = m(x + 1) + b = m + (mx + b) = m + f(x)$
**b.** $g(x + 1) = a \cdot b^{x+1} = b \cdot (a \cdot b^x) = b \cdot g(x)$
**c.** For linear functions, increasing the input by one *adds* the previous
output to $m$. For exponential functions, increasing the input by one
*multiplies* the previous output by $b$.   **95.** zeroes at $x = 3$ and $x = -2$;
HA: $y = 1$, VA $x = -1, x = 1$   **97. a.** $S = C/6 + 3$   **b.** 15-in. shelves

## Exercises 5.5, pp. 480–482

**1.** combine, exponential   **3.** False; Answers will vary.   **5.** $x = 32$
**7.** $x = 20$; $-5$ is extraneous   **9.** $x = 2$; $-\frac{5}{2}$ is extraneous   **11.** $x = 0$
**13.** $x = \frac{5}{2}$   **15.** $x = \frac{2}{3}$   **17.** $x = \frac{-19}{9}$   **19.** $x = \frac{e^2 - 63}{9}$; $x \approx -6.1790$
**21.** $x = 2$; $-9$ is extraneous   **23.** no solution
**25.** $x = 2 + \sqrt{3}$; $2 - \sqrt{3}$ is extraneous

**27.** $x = \dfrac{\ln 231}{\ln 7}; x \approx 2.7968$   **29.** $x = \dfrac{\ln 128{,}967}{3 \ln 5}; x \approx 2.4371$

**31.** $x = \dfrac{\ln 2}{\ln 3 - \ln 2}; x \approx 1.7095$   **33.** $x = \frac{\ln 9 - \ln 5}{2 \ln 5 - \ln 9}; x \approx 0.5753$

**35.** $x = \dfrac{200 \ln 2}{3}; x \approx 46.2098$   **37.** $x \approx -4.815, x \approx 102.084$

**39.** $x \approx 2.013, x \approx 3.608$   **41.** $t = \dfrac{\ln\!\left(\dfrac{\frac{C}{P} - 1}{a}\right)}{-k}; t \approx 55$

**43. a.** 30 fish   **b.** about 37 months   **45.** about 3.2 cmHg   **47.** about 50 min

**49.** $15,641   **51. a.** 6 hr   **b.** 18.0%   **53.** $M_f = 52.76$ tons

**55. a.** 26 planes   **b.** 9 days   **57.** $x = \ln\!\left(\frac{3}{2}\right) \approx 1.60944$

**59. a.** $f^{-1}(x) = 10^{x-1} + 2; D: x \in (-\infty, \infty); R: y \in (2, \infty)$
**b.** $f^{-1}(x) = \ln(x + 2) - 5; D: x \in (-2, \infty); R: y \in (-\infty, \infty)$
**61. a.** d   **b.** e   **c.** b   **d.** f   **e.** a   **f.** c
**63. a.** $x \in [-\frac{3}{2}, \infty), y \in [0, \infty)$   **b.** $x \in (-\infty, \infty), y \in [-3, \infty)$
**65.** 13.5 tons

## Exercises 5.6, pp. 491–495

**1.** Compound   **3.** Answers will vary.   **5.** $4896   **7.** 250%   **9.** $2152.47
**11.** 5.25 yr   **13.** 80%   **15.** 4 yr   **17.** 16 yr   **19.** $7561.33   **21.** about
5 yr   **23.** 7.5 yr   **25.** no   **27. a.** no   **b.** 9.12%   **29.** 7.9 yr   **31.** 7.5 yr
**33. a.** no   **b.** 9.4%   **35. a.** no   **b.** approx. 13,609 euros   **37. a.** no
**b.** $234,612.02   **39.** about 7 yr   **41.** 22 yr   **43. a.** no   **b.** $298.31

**45. a.** $t = \dfrac{A - P}{Pr}$   **b.** $P = \dfrac{A}{1 + rt}$   **47. a.** $r = n\!\left(\sqrt[nt]{\dfrac{A}{P}} - 1\right)$

**b.** $t = \dfrac{\ln\!\left(\dfrac{A}{P}\right)}{n \ln\!\left(1 + \dfrac{r}{n}\right)}$   **49. a.** $Q_0 = \dfrac{Q(t)}{e^{rt}}$   **b.** $t = \dfrac{\ln\!\left[\dfrac{Q(t)}{Q_0}\right]}{r}$

**51.** $709.74   **53. a.** about 5.8%   **b.** about 92 hr   **55. a.** 1990   **b.** 2037
**c.** about 1,656,000   **57.** 0.65 g   **59.** about 816 yr   **61.** about 12.4%
**63. a.** 75%   **b.** 32%   **c.** 66 weeks   **65.** $17,027,502.21
**67. a.** $f(x) = x^3, f(x) = x, f(x) = \sqrt{x}, f(x) = \sqrt[3]{x}, f(x) = \frac{1}{x}$

**b.** $f(x) = |x|, f(x) = x^2, f(x) = \dfrac{1}{x^2}$   **c.** $f(x) = x, f(x) = x^3, f(x) = \sqrt{x},$

$f(x) = \sqrt[3]{x}$   **d.** $f(x) = \dfrac{1}{x}, f(x) = \dfrac{1}{x^2}$

**69.** $P(x) = x^4 - 4x^3 + 6x^2 - 4x - 15$

## Making Connections, p. 496

**1.** a   **3.** e   **5.** c   **7.** e   **9.** b   **11.** g   **13.** c   **15.** d

## Summary and Concept Review, pp. 497–500

**1.** no   **2.** no   **3.** yes   **4.** $f^{-1}(x) = \dfrac{x - 2}{-3}$   **5.** $f^{-1}(x) = \sqrt{x + 2}$

**6.** $f^{-1}(x) = x^2 + 1; x \geq 0$   **7.** $f(x)$: $D: x \in [-4, \infty), R: y \in [0, \infty)$;
$f^{-1}(x)$: $D: x \in [0, \infty), R: y \in [-4, \infty)$
**8.** $f(x)$: $D: x \in (-\infty, \infty), R: y \in (-\infty, \infty)$;
$f^{-1}(x)$: $D: (-\infty, \infty), R: y \in (-\infty, \infty)$   **9.** $f(x)$: $D: x \in (-\infty, \infty)$,
$R: y \in (0, \infty); f^{-1}(x)$: $D: x \in (0, \infty), R: y \in (-\infty, \infty)$
**10. a.** $3.05   **b.** $f^{-1}(t) = \frac{t - 2}{0.15}, f^{-1}(3.05) = 7$   **c.** 12 days
**11.**   **12.**   **13.**

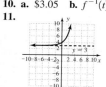      

**14.** 2   **15.** $-2$   **16.** $\frac{5}{2}$   **17.** 12.1 yr   **18.** $3^2 = 9$   **19.** $5^{-3} = \frac{1}{125}$
**20.** $e^{3.7612} \approx 43$   **21.** $\log_5 25 = 2$   **22.** $\ln 0.7788 \approx -0.25$
**23.** $\log_3 81 = 4$   **24.** 5   **25.** $-1$   **26.** $\frac{1}{2}$

**27.**   **28.**   **29.**

**30.** $x \in (-\frac{3}{2}, \infty)$   **31.** $x \in (-\infty, 0) \cup (6, \infty)$   **32. a.** $\approx 4.8$
**b.** $\approx 20{,}000{,}000I_0$   **33. a.** $x = e^{32}$   **b.** $x = 10^{2.38}$   **c.** $x = \ln 9.8$

**d.** $x = \dfrac{1}{2} \log 7$   **34. a.** $x = \dfrac{\ln 4}{0.5}, x \approx 2.7726$   **b.** $x = \dfrac{\log 19}{0.2}, x \approx 6.3938$

**c.** $x = \dfrac{10^3}{3}, x \approx 333.3333$   **d.** $x = e^{-2.75}, x \approx 0.0639$

**35. a.** $\ln 42$   **b.** $\log_9 30$   **c.** $\ln\!\left(\frac{x + 3}{x - 1}\right)$   **d.** $\log(x^2 + x)$
**36. a.** $2 \log_5 9$   **b.** $2 \log_7 4$   **c.** $(2x - 1)\ln 5$   **d.** $(3x + 2)\ln 10$
**37. a.** $\ln x + \frac{1}{4} \ln y$   **b.** $\frac{1}{3} \ln p + \ln q$
**c.** $\frac{5}{3} \log x + \frac{4}{3} \log y - \frac{5}{2} \log x - \frac{3}{2} \log y$

**d.** $\log 4 + \frac{5}{3} \log p + \frac{4}{3} \log q - \frac{3}{2} \log p - \log q$   **38. a.** $\dfrac{\log 45}{\log 6} \approx 2.215$

**b.** $\dfrac{\log 128}{\log 3} \approx 4.417$   **c.** $\dfrac{\ln 124}{\ln 2} \approx 6.954$   **d.** $\dfrac{\ln 0.42}{\ln 5} \approx -0.539$

**39.** $x = \frac{\ln 7}{\ln 2}, x \approx 2.8074$   **40.** $x = \frac{\ln 5}{\ln 3} - 1, x \approx 0.4650$
**41.** $x = \frac{2}{1 + \ln 3}, x \approx 0.9530$   **42.** $x = e^2 - 1, x \approx 6.3891$
**43.** $x = 5; -2$ is extraneous   **44.** $x = 4.25$   **45. a.** 17.77%   **b.** 23.98 days
**46.** 38.6 cmHg   **47.** 18.5%   **48.** Almost, she needs $42.15 more.
**49. a.** no   **b.** $268.93   **50.** 55.0%

## Practice Test, p. 500

**1.** $3^4 = 81$   **2.** $\log_{25} 5 = \frac{1}{2}$   **3.** $\frac{5}{2} \log_b x + 3 \log_b y - \log_b z$

**4.** $\log_b \dfrac{m\sqrt{n^3}}{\sqrt{p}}$   **5.** $x = 10$   **6.** $x = \dfrac{-5}{3}$   **7.** 2.68   **8.** $-1.24$

**9.**   **10.**

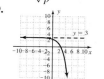   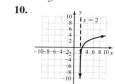

**11. a.** 4.19   **b.** $-0.81$   **12.** $f$ is a parabola (hence not one-to-one),
$x \in \mathbb{R}, y \in [-3, \infty)$; vertex is at $(2, -3)$; so restricted domain could be
$x \in [2, \infty)$ to create a one-to-one function;
$f^{-1}(x) = \sqrt{x + 3} + 2, x \in [-3, \infty), y \in [2, \infty)$.
**13.** $x = 1 + \frac{\ln 89}{\ln 3}, x \approx 5.0857$   **14.** $x = 1; x = -5$ is extraneous
**15.** $\approx 5$ yr   **16. a.** $P = kM^\alpha$   **b.** $P \approx 15.3$ lb $(M = 20)$
**17.** 19.1 months   **18. a.** no   **b.** $54.09   **19 a.** 10.2 lb   **b.** 34 weeks
**20. a.** 34.7 yr   **b.** 36 yr

## Calculator Exploration and Discovery pp. 501–502

**1.** $x \approx 2.7$   **2.** $x \approx 1.3$   **3.** $x \approx 2.3$   **4.** $x \approx 4.8, x \approx 7.4$
**5.** $a = 25, b = 0.5, c = 2500$   **6.** $\frac{2500}{26} \approx 96$ ants   **7.** b
**8.** 2500   **9.** a   **10.** verified

## Strengthening Core Skills p. 503

**1.** about 126 times hotter   **2.** about 4.2 hb

## Cumulative Review Chapters R–5, p. 504

**1.** $x = 2 \pm 7i$   **3.** $(4 + 5i)^2 - 8(4 + 5i) + 41 = 0$
$\qquad\qquad -9 + 40i - 32 - 40i + 41 = 0$
$\qquad\qquad\qquad\qquad\qquad\qquad 0 = 0 \checkmark$
**5.** $f(g(x)) = x; g(f(x)) = x$; They are inverse functions.
**7. a.** $T(t) = 6644 - 96t$ (2008 → year 8)
**b.** $\frac{\Delta T}{\Delta t} \approx \frac{-96}{1}$, triplet births decreased (on average) by 96 per year.
**c.** $T(5) \approx 6165$ sets of triplets, $T(15) \approx 5206$ sets of triplets

**9.**

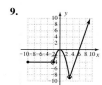

$D: x \in [-10, \infty), R: y \in [-9, \infty)$
$h(x)\!\uparrow: x \in (-2, 0) \cup (3, \infty)\ h(x)\!\downarrow: x \in (0, 3)$

**11.** $x = 3, x = 2$ (multiplicity 2), $x = -4$   **13.** $\sqrt{\frac{2V}{\pi a}} = b$

**15. a.** $f^{-1}(x) = \dfrac{5x - 3}{2}$   **b.**

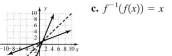

**c.** $f^{-1}(f(x)) = x$

**17.** $x = 5; x = -6$ is an extraneous root   **19. a.** about 88 hp for sport wagon, about 81 hp for minivan   **b.** $\approx 3294$ rpm   **c.** minivan, 208 hp at 5800 rpm   **21.** $f(x) = \dfrac{-3}{2}x + 3$   **23. a.** up/down; (0, 3), (2, 0); $D = \mathbb{R}; R = \mathbb{R}$   **b.** down/down; (0, 0), (4, 0); $D = \mathbb{R}; R = (-\infty, 4]$
**25.** Answers will vary.

# CHAPTER 6
## Exercises 6.1, pp. 516–520

**1.** inconsistent   **3.** Multiply the first equation by 6 and the second equation by 10.   **5.** yes   **7.** yes

**9.**

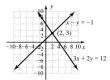

(2, 3)   **11.**

$(-1, 3)$

**13.**

$(-3, 1)$

**15.** $y = x + 2$   **17.** $x + 3y = -3$   **19.** $\begin{cases} y = x + 2 \\ x + 3y = -3 \end{cases}$
**21.** $y = 2x + 3, y = 3x + 8, m_1 \neq m_2$: one solution
**23.** $y = -2x + 1.5, y = -2x + 7, m_1 = m_2, b_1 \neq b_2$: no solutions
**25.** $y = \dfrac{7}{4}x - 6, y = -\dfrac{4}{3}x + 5, m_1 \neq m_2$: one solution
**27.** (2, 6)   **29.** (4, 1)   **31.** $(-4, 1)$   **33.** $(4, -3)$   **35.** $(3, \frac{1}{2})$
**37.** $(-2, 5)$   **39.** $(2, -1)$   **41.** $(3, -1)$   **43.** $(-2, -3)$   **45.** $(-1, 2)$
**47.** (5, 3)   **49.** $(\frac{11}{2}, 2)$   **51.** (6, 3)   **53.** $(-2, 3)$   **55.** $(-6, 12)$
**57.** $(-3, -4)$; consistent/independent   **59.** $\varnothing$; inconsistent
**61.** $\{(x, y)|y = -6x + 22\}$; consistent/dependent   **63.** (4, 1); consistent/independent   **65.** $\varnothing$; inconsistent   **67.** 1 mph, 4 mph
**69.** 2318 adult tickets; 1482 child tickets   **71.** premium: \$3.97, regular: \$3.87   **73.** 150 quarters, 75 dimes   **75. a.** 3 mph
**b.** 5 mph   **77. a.** 3.6 ft/sec   **b.** 4.4 ft/sec   **79. a.** 100 lawns/mo
**b.** \$11,500/mo   **81. a.** 1.6 billion bu, 3 billion bu; yes   **b.** 2.7 billion bu, 2.25 billion bu; yes   **c.** \$6.65, 2.43 billion bu   **83.** about 227 boards at \$410 per board   **85.** 1776; 1865   **87.** Tahiti: 402 mi$^2$, Tonga: 290 mi$^2$
**89.** \$50,000   **91.** $y = 2x + 3$   **93.** $(x - 5)(x - 2)(x + 1)(3x - 1)$
**95.** $x \in [-2, 8]$

## Exercises 6.2, pp. 529–531

**1.** triple   **3.** $z = 5$; Answers will vary.   **5.** Answers will vary.
**7.** Answers will vary.   **9.** yes, no; R2 and R3   **11.** (5, 7, 4)
**13.** $(-2, 4, 3)$   **15.** $(4, 0, -3)$   **17.** (5, 12, 13)   **19.** no solution; inconsistent   **21.** $\{(x, y, z)|x \in \mathbb{R}, y = 2 - x, z = 2 - x\}$; $(p, 2 - p, 2 - p)$, other solutions possible
**23.** $\left\{(x, y, z)\middle| x = -\dfrac{5}{3}z - \dfrac{2}{3}, y = -z - 2, z \in \mathbb{R}\right\}; \left(-\dfrac{5}{3}p - \dfrac{2}{3}, -p - 2, p\right)$, other solutions possible   **25.** $(p, 2p, p + 1)$   **27.** $(p + 9, p - 4, p)$
**29.** $\left\{(x, y, z)\middle| x + \dfrac{1}{2}y - 2z = 6\right\}$   **31.** $(-1, \frac{-3}{2}, 2)$
**33.** $(-p - 17, -p - 4, p)$   **35.** (12, 6, 4)   **37.** $(1, -5, -6)$
**39.** $(1, \frac{1}{3}, \frac{1}{2})$   **41.** (5 cm, 3 cm, 4 cm)   **43.** \$80,000 at 4%; \$90,000 at 5%; \$110,000 at 7%   **45.** World War II, 1945; Korean, 1953; Vietnam, 1973
**47.** Declaration of Independence, 1776; 13th Amendment, 1865; Civil Rights Act, 1964   **49.** 1 L 20% solution; 3 L 30% solution; 6 L 45% solution
**51.** saturated: 1.2 g; monounsaturated: 1.0 g; polyunsaturated: 0.6 g
**53.** $h(t) = -5t^2 + 30t + 1$;   **a.** 46 ft   **b.** 17.2 ft   **55.** 2   **57.** $\left[-1, \frac{3}{2}\right]$
**59.** $x = 1$

## Mid-Chapter Check, p. 532

**1.** (1, 1); consistent   **2.** (5, 3); consistent   **3.** 20 oz   **4.** no; R2 and R3
**5.** 2R1 = R2   **6.** (1, 2, 3)   **7.** (1, 2, 3)   **8.** $(p, p - 5, -p - 4)$
**9.** Morphy: 13; Mozart: 8; Pascal: 16   **10.** prelude: 2.75 min; storm: 2.5 min; sunrise: 2.5 min; finale: 3.25 min

## Exercises 6.3, pp. 539–541

**1.** region, solutions   **3. a.** two solutions   **b.** no solutions
**5.** parabola, line;   **7.** circle, absolute value;
   $(-1, 5), (2, 2)$      $(-6, 8), (8, 6)$

**9.**

parabola, parabola; $(-1, -2), (2, 1)$

**11.** $(-4, -3), (3, 4)$   **13.** (2, 5), $(-4, -7)$   **15.** $(-3, 4), (-4, -3)$, (3, 4), $(4, -3)$   **17.** (0, 1), (4, 9)   **19.** $(4, -3), (-4, -3)$
**21.** (4, 3), $(-3, -4)$   **23.** $(-8, 1), (-7, 4)$   **25.** $(5, \log 5 + 5)$
**27.** $(-3, \ln 9 + 1), (4, \ln 16 + 1)$   **29.** (0, 10), (ln 6, 45)
**31.** $(-3, 1), (2, 1024)$   **33.** $(-3, -21), (1, -1), (2, 4)$
**35.** $(2, -4), (6, 4)$   **37.** $(3, 5), (3, -5)$   **39.** $(-2.43, -2.81), (2, 1)$
**41.** (0.72, 2.19), (2, 3), (4, 3), (5.28, 2.19)

**43.**

**45.**

**47.** no solution

**49.**

**51.** $h \approx 45.8$ ft; $h = 40$ ft; $h = 30$ ft   **53.** The company breaks even if either 18,400 or 48,200 cars are sold.   **55.** $\begin{cases} x^2 y = 2000 \\ x^2 + 4xy = 800 \end{cases}$; approx.

(12.4, 13) or (20, 5); The pool will likely have the dimensions 20 ft by 20 ft by 5 ft.   **57.** 8.5 ft $\times$ 10 ft   **59.** 5 km $\times$ 9 km   **61.** 8 ft $\times$ 8 ft $\times$ 25 ft

**63.** \$1.83; $\begin{cases} 10P^2 + 6D = 144 \\ 8P^2 - 8P - 4D = 12 \end{cases}$; 90,000 gal at \$3/gal

**65.** Answers will vary.   **67.** 18 in. by 18 in. by 77 in.

**69. a.** $x = -7, x = \frac{9}{2}$   **b.** $x = \frac{-11}{2}, x = \frac{11}{2}$   **c.** $x = 2, x = -2, x = \frac{3}{2}$

**71.** \$95,000, \$75,000, \$80,000

## Exercises 6.4, pp. 551–555

**1.** vertex   **3.** The feasible region may be bordered by three or more oblique lines, with two of them intersecting outside and away from the feasible region.   **5.** no, no, no, no   **7.** no, yes, yes, no

**9.**    **11.**    **13.** no, no, no, yes

**15.**    **17.**    **19.**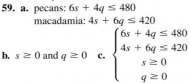

**21.** 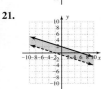   **23.**    **25.**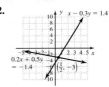

**27.**    **29.**    **31.**

**33.** 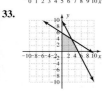   **35.**  (2, 4)   **37.**  (3, 2)

**39.** $\begin{cases} y - x \le 1 \\ x + y > 3 \end{cases}$   **41.** $\begin{cases} y - x \le 1 \\ x + y < 3 \\ y \ge 0 \end{cases}$   **43.** (5, 3)   **45.** (12, 11)

**47.** 26 at (2, 2)   **49.** 264 at (4, 3)   **51.** $5 < H < 10$   **53. a.** Yes, (180, 3000) is in the solution region.   **b.** Verified, the point (180, 3000) satisfies all inequalities in the system.   **c.** Verified, the point (160, 3000) does not satisfy the first inequality.

**55.**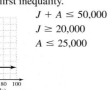
$$J + A \le 50,000$$
$$J \ge 20,000$$
$$A \le 25,000$$

**57. a.** Using $c$ for time with children and $p$ for personal time, the system of inequalities is
$$\begin{cases} c + p \le 420 \\ 150 \le c \le 350. \\ 60 \le p \le 180 \end{cases}$$

**b.** The solution region (all possibilities) is shown in the graph.   **c.** No. Although (300, 150) satisfies the second and third inequalities, $300 + 150 > 420$ and is not a possible solution.

**59. a.** pecans: $6s + 4q \le 480$
    macadamia: $4s + 6q \le 420$
**b.** $s \ge 0$ and $q \ge 0$   **c.** $\begin{cases} 6s + 4q \le 480 \\ 4s + 6q \le 420 \\ s \ge 0 \\ q \ge 0 \end{cases}$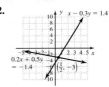

**61.** 300 acres of corn; 200 acres of soybeans   **63.** 240 sheet metal screws; 480 wood screws   **65.** 65 Commoners, 30 Clubhouses

**67.** 3 buses from company X; 4 buses from company Y

**69. a.** any point on $x + y \le 6$ for $1 \le x \le 4$   **b.** The objective function and the constraint both have the same slope and are parallel.

**71.** $x = 5, \pm i\sqrt{3}$   **73.** 324 $\Omega$

## Making Connections p. 556

**1.** c   **3.** g   **5.** d   **7.** a   **9.** g   **11.** e   **13.** f   **15.** b

## Summary and Concept Review, pp. 557–558

**1.** 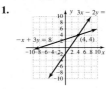   **2.**

**3.**

**4.** no solution; inconsistent   **5.** $(5, -1)$; consistent   **6.** $(7, 2)$; consistent
**7.** $(3, -1)$; consistent   **8.** $\left(\frac{11}{4}, \frac{-1}{6}\right)$; consistent   **9.** \$1.20   **10.** $(0, 3, 2)$
**11.** $(1, 1, 1)$   **12.** no solution; inconsistent   **13.** 1530 quarters, 1180 dimes, 710 nickels   **14.** circle, line; $(4, 3), (-3, -4)$   **15.** parabola, line; $(3, -2)$
**16.** parabola, circle; $(\sqrt{3}, 2), (-\sqrt{3}, 2)$
**17.** circle, parabola; $(1, 3), (-1, 3)$

**18.** 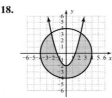   **19.**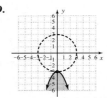

parabola, circle

note the open circle showing noninclusion at $(0, -3)$; circle, parabola

**20.** 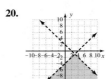   **21.**    **22.**

**23.**  Maximum of 270 occurs at both $(0, 6)$ and $(3, 4)$ on the line $y = \frac{-2}{3}x + 6$, and all points on this line in between these endpoints.

**24.** 50 cows, 425 chickens

## Practice Test, p. 559

**1.** 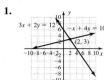 $(2, 3)$ **2.** $\left(\frac{2}{5}, \frac{-4}{5}\right)$ **3.** $(-3, 2)$ **4.** $(2, -1, 4)$

**5.** $\{(x, y, z) \,|\, x = 2z - 1, y = 5z - 6, z \in \mathbb{R}\}$ **6.** $a = 5, b = 2$
**7.** 21.59 cm by 35.56 cm **8.** Canary Islands: 3000 mi$^2$; the Azores: 890 mi$^2$ **9.** corn 25¢, beans 20¢, peas 29¢ **10.** \$15,000 at 7%, \$8000 at 5%, \$7000 at 9%

**11.**  **12.** maximum of $P = 400$ at $(8, 0)$

**13.** 30 plain; 20 deluxe **14.** $(-1 - \sqrt{7}, 1 - \sqrt{7}), (-1 + \sqrt{7}, 1 + \sqrt{7})$
**15.** $(\sqrt{3}, 1), (-\sqrt{3}, 1)$ **16.** 15 ft, 20 ft
**17.**

**18. a.** $Y_2 \approx -0.765x + 105.468$, $Y_1 \approx 0.261x - 15.794$
**b.** 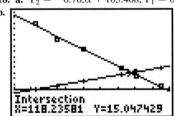 **c.** Equilibrium occurs when approx 150,000 cards are sold at a price near \$1.18.

**19.**  **20.** Answers will vary. Possible solution:
$$\begin{cases} x^2 + y^2 > 1 \\ x^2 + y^2 < 4 \\ x > 0, y < 0 \end{cases}$$

## Calculator Exploration and Discovery, p. 560

**Exercise 1:** $(-1, 4)$ **Exercise 2:** $\left(\frac{1}{2}, -\frac{2}{3}\right)$

**Exercise 3:** **Exercise 4:** **Exercise 5:**
 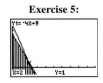

## Strengthening Core Skills, p. 561

**Exercise 1:** $(3, -1, 6)$ **Exercise 2:** $\left(-2, \frac{1}{2}, 1\right)$

## Cumulative Review Chapters R–6, p. 562

**1.**  **3.**  **5.**

**7. a.** $D: x \in (-\infty, \infty)$ **b.** $R: y \in (-\infty, 4]$ **c.** $f(x)\!\uparrow: x \in (-\infty, -1)$; $f(x)\!\downarrow: x \in (-1, \infty)$ **d.** max: $(-1, 4)$ **e.** $f(x) > 0: x \in (-4, 2)$; $f(x) < 0: x \in (-\infty, -4) \cup (2, \infty)$ **f.** $\frac{\Delta y}{\Delta x} = \frac{7}{4}$ **9.** $x = 9$, $\pm i\sqrt{2}$
**11.** $3x^2 + 5$ **13.** 1 **15.** $(x - 2)(x + 2)(x - 3 - 3\sqrt{2})(x - 3 + 3\sqrt{2})$
**17.** $x \in \left(-\infty, \frac{-11}{2}\right] \cup (-3, \infty)$ **19.** $\approx 9.7$ yr

**21.**  **23.** 80 arrows, 25 balls, 15 bats

**25.** $x = 5, y \approx 4.7$ **27.** $f(-2)$ is undefined, $f(2) = 3$, $f(x) = 0$ for $x = -1$ and $x = 4$ **29.** $x = \dfrac{-100 \ln\left(\frac{4}{143}\right)}{9}, x \approx 39.74$

## CHAPTER 7
### Exercises 7.1, pp. 572–576

**1.** 2, 3, 1 **3.** Multiply R1 by $-2$ and add that result to R2. This sum will be the new R2. **5.** $3 \times 2$; 5.8 **7.** $4 \times 3$; $-1$
**9.** $\begin{bmatrix} 1 & 2 & -1 & | & 1 \\ 1 & 0 & 1 & | & 3 \\ 2 & -1 & 1 & | & 3 \end{bmatrix}$; diagonal entries 1, 0, 1
**11.** $\begin{cases} x + 4y = 5 \\ \quad\quad y = \frac{1}{2} \end{cases} \rightarrow \left(3, \frac{1}{2}\right)$ **13.** $\begin{cases} x + 2y - z = 0 \\ \quad\quad y + 2z = 2 \\ \quad\quad\quad\quad z = 3 \end{cases} \rightarrow (11, -4, 3)$
**15.** $\begin{cases} x + 3y - 4z = 29 \\ \quad\quad y - \frac{3}{2}z = \frac{21}{2} \\ \quad\quad\quad\quad z = 3 \end{cases} \rightarrow (-4, 15, 3)$
**17.** $\begin{bmatrix} 1 & -6 & | & -2 \\ 0 & -28 & | & -6 \end{bmatrix}$ **19.** $\begin{bmatrix} 1 & -3 & 3 & | & 2 \\ 0 & 23 & -12 & | & -15 \\ -2 & 1 & 0 & | & 4 \end{bmatrix}$
**21.** $\begin{bmatrix} 3 & 1 & 1 & | & 8 \\ 0 & -3 & -3 & | & -6 \\ 0 & -10 & -13 & | & 34 \end{bmatrix}$ **23.** $2R1 + R2 \rightarrow R2$, $-3R1 + R3 \rightarrow R3$
**25.** $-5R1 + R2 \rightarrow R2$, $4R1 + R3 \rightarrow R3$ **27.** $\left(\frac{9}{14}, \frac{11}{7}\right)$
**29.** $(20, 10)$ **31.** $(1, 6, 9)$ **33.** $\left(\frac{3}{5}, \frac{-4}{5}\right)$ **35.** $(1, 1, 2)$ **37.** $(1, 1, 1)$
**39.** $\left(-1, \frac{-3}{2}, 2\right)$ **41.** linear dependence; $(p - 4, -2p + 8, p)$
**43.** coincident dependence; $\{(x, y, z) \,|\, 3x - 4y + 2z = -2\}$
**45.** inconsistent; no solution **47.** linear dependence; $\left(-\frac{3}{5}p + \frac{12}{5}, -\frac{1}{5}p + \frac{4}{5}, p\right)$ **49.** inconsistent; no solution **51. a.** \$30 **b.** price \$22.50, quantity 1500 items **c.** \$33,750 **53.** LA to STL, 1600 mi; STL to CIN, 310 mi; CIN to NY, 570 mi **55.** Moe 90, Larry 45, Curly 30 **57.** 15 m, 36 m, 39 m **59.** \$2000 at 5%; \$3000 at 7%; \$5000 at 9% **61. a.** $w_1 \approx 14.1$ lb and $w_2 \approx 19.3$ lb **b.** $w_2 \approx 13.7$ lb and $w_3 \approx 7.1$ lb **63. a.** intersection 1: 220, intersection 2: 370, and intersection 3: 345 **b.** intersection 1: 200, intersection 2: 350, and intersection 3: 325 **65.** $x = 84°$; $y = 25°$
**67.** $x^3 + x - 10$; $x^3 - x - 6$; $x^4 - 2x^3 - 8x + 16$; $x^2 + 2x + 4$
**69.** $C > 40,000$ in the year 2018 ($t \approx 13.4$)

## Exercises 7.2, pp. 584–589

**1.** order, $a_{ij}$; $b_{ij}$   **3.** Answers will vary.   **5.** $2 \times 2$, $a_{12} = -3$, $a_{21} = 5$
**7.** $2 \times 3$, $a_{12} = -3$, $a_{23} = 6$, $a_{22} = 5$   **9.** $3 \times 3$, $a_{12} = 1$, $a_{23} = 1$, $a_{31} = 5$
**11.** true   **13.** conditional; $a = -4$, $b = 3$, $c = -2$

**15.** $\begin{bmatrix} 10 & 0 \\ 0 & 10 \end{bmatrix}$   **17.** different orders, sum not possible

**19.** $\begin{bmatrix} \frac{263}{32} & -\frac{19}{8} \\ -\frac{85}{16} & \frac{35}{16} \end{bmatrix}$   **21.** $\begin{bmatrix} -\frac{1}{2} & \frac{13}{8} & -\frac{1}{4} \\ \frac{1}{4} & \frac{5}{2} & -\frac{21}{8} \\ -\frac{31}{8} & -\frac{9}{4} & \frac{7}{2} \end{bmatrix}$   **23.** $\begin{bmatrix} 20 & -15 \\ -25 & -10 \end{bmatrix}$

**25.** $\begin{bmatrix} \frac{-5}{2} & -1 & 0 \\ 0 & \frac{-7}{2} & 1 \\ 2 & \frac{3}{2} & -6 \end{bmatrix}$   **27.** $\begin{bmatrix} 1 & -2 & 0 \\ 0 & -1 & 2 \\ 4 & 3 & -6 \end{bmatrix}$   **29.** $\begin{bmatrix} 1 & 0 \\ 0 & 1 \end{bmatrix}$

**31.** matrix mult. not possible   **33.** $\begin{bmatrix} 12 & -24 & 90 \\ -6 & 15 & -57 \end{bmatrix}$

**35.** $\begin{bmatrix} 79 & -30 \\ -50 & 19 \end{bmatrix}$   **37.** $\begin{bmatrix} 42 & 18 & -60 \\ -12 & -42 & 36 \end{bmatrix}$   **39.** $\begin{bmatrix} 1 & 0 \\ 0 & 1 \end{bmatrix}$

**41.** $\begin{bmatrix} 1 & 0 & 0 \\ 0 & 1 & 0 \\ 0 & 0 & 1 \end{bmatrix}$   **43.** $\begin{bmatrix} \frac{-3}{19} & \frac{4}{57} \\ \frac{1}{19} & \frac{5}{57} \end{bmatrix}$   **45.** $\begin{bmatrix} 0 & \frac{3}{4} & \frac{1}{4} \\ \frac{-1}{2} & \frac{3}{8} & \frac{1}{8} \\ \frac{-1}{4} & \frac{11}{16} & \frac{1}{16} \end{bmatrix}$

**47.** $\begin{bmatrix} 1.75 & 2.5 \\ 7.5 & 13 \end{bmatrix}$   **49.** $\begin{bmatrix} -4 & 28 & 4 \\ -8 & 17 & 3 \end{bmatrix}$

**51.** $\begin{bmatrix} 2x - 4 & -2y - 4 & -2x + 3y \\ 10 - y & 4x - 5 & 0 \end{bmatrix}$   **53.** $\begin{bmatrix} 12y^2 + 4y & 6y^2 - 4y \\ 4y^2 + 30y & -10 \end{bmatrix}$

**55.** verified   **57.** verified

**59.** $\begin{bmatrix} \text{Perimeter} \\ \text{Area} \end{bmatrix} = \begin{bmatrix} 2L + 2W \\ LW \end{bmatrix} \Rightarrow \begin{cases} P = 2L + 2W \checkmark \\ A = LW \end{cases}$

**61. a.**

$V = D\begin{array}{c} \\ \text{S} \\ \text{D} \\ \text{P} \end{array}\begin{bmatrix} 3820 & 1960 \\ 2460 & 1240 \\ 1540 & 920 \end{bmatrix}$ (T, S)     $M = D\begin{array}{c} \\ \text{S} \\ \text{D} \\ \text{P} \end{array}\begin{bmatrix} 4220 & 2960 \\ 2960 & 3240 \\ 1640 & 820 \end{bmatrix}$ (T, S)

**b.** 3900 more by Minsk
**c.**
$V = \begin{bmatrix} 3972.8 & 2038.4 \\ 2558.4 & 1289.6 \\ 1601.6 & 956.8 \end{bmatrix}$   **d.** $\begin{bmatrix} 8361.6 & 5116.8 \\ 5636.8 & 4659.2 \\ 3307.2 & 1809.6 \end{bmatrix}$

$M = \begin{bmatrix} 4388.8 & 3078.4 \\ 3078.4 & 3369.6 \\ 1705.6 & 852.8 \end{bmatrix}$

**63.** $[22{,}000 \; 19{,}000 \; 23{,}500 \; 14{,}000]$; total profit N: \$22,000, S: \$19,000, E: \$23,500, W: \$14,000

**65. a.** \$108.20   **b.** \$101
**c.** $\begin{bmatrix} 100 & 101 & 119 \\ 108.2 & 107 & 129.5 \end{bmatrix}$ First row, total cost for science from each restaurant; second row, total cost for math from each restaurant.

**67.** $\begin{bmatrix} 32.4 & 10.3 & 21.3 \\ 29.9 & 9.6 & 19.5 \end{bmatrix}$ **a.** 10   **b.** 20   **c.** $p_{13}$ gives the approximate number of females expected to join the writing club.

**69. a.** year 2: 1160 rabbits and 290 foxes; year 3: 1276 rabbits and 493 foxes **b.** year 2: 200 rabbits and 130 foxes; year 3: 188 rabbits and 157 foxes
**71.** retail \$51,500; wholesale \$40,500

**73.** $\begin{bmatrix} 2^{n-1} & 0 & 2^{n-1} \\ 2^n - 1 & 1 & 2^n - 1 \\ 2^{n-1} & 0 & 2^{n-1} \end{bmatrix}$   **75. a.** verified   **b.** Answers will vary.

**77.** $(-1, 1, -2)$   **79.** $\approx 4.39$

## Mid-Chapter Check p. 589

**1.** $3 \times 3$, $-0.9$   **2.** $2 \times 4$, $0$   **3.** $(2, -3)$   **4.** $(2, 0, -5)$
**5.** $(p - 3, 2p - 8, p)$   **6. a.** $\begin{bmatrix} -13 & -17 \\ 35 & 9 \end{bmatrix}$   **b.** $\begin{bmatrix} 4 & 6 \\ -12 & -2 \end{bmatrix}$

**c.** $\begin{bmatrix} -5 & 5 \\ -5 & 15 \end{bmatrix}$   **7. a.** $\begin{bmatrix} 0.8 & 0.5 & 2.2 \\ -0.1 & 0.8 & -1 \\ 2.1 & 0.3 & 1.9 \end{bmatrix}$   **b.** $\begin{bmatrix} -3 & -1.5 & -6 \\ 1.5 & 0 & 3 \\ -6 & -1.5 & -6 \end{bmatrix}$

**c.** $\begin{bmatrix} 1 & 0 & 0 \\ 0 & 1 & 0 \\ 0 & 0 & 1 \end{bmatrix}$   **8. a.** $\begin{bmatrix} 17 & -32 & -13 \\ 5 & 0 & -5 \end{bmatrix}$   **b.** $\begin{bmatrix} -26 & -18 & 24 \\ 2 & -10 & -4 \end{bmatrix}$

**c.** $\begin{bmatrix} 24 & 4 \\ 0 & -5 \\ 16 & -39 \end{bmatrix}$   **d.** $\begin{bmatrix} -4 & -33 \\ -2 & 29 \end{bmatrix}$   **9.** used: \$80, new: \$125

**10.** $\begin{bmatrix} 4375 & 110 \\ 2400 & 59 \end{bmatrix}$; $p_{11}$: total rebates paid to individuals, $p_{21}$: total rebates paid to business, $p_{12}$: free AAA years given to individuals, $p_{22}$: free AAA years given to business

## Exercises 7.3, pp. 600–605

**1.** diagonal, zeroes   **3.** Answers will vary.   **5.** verified   **7.** verified
**9.** verified   **11.** verified   **13.** $\begin{bmatrix} \frac{1}{9} & \frac{2}{9} \\ \frac{-1}{9} & \frac{5}{18} \end{bmatrix}$   **15.** $\begin{bmatrix} -5 & 1.5 \\ -2 & 0.5 \end{bmatrix}$   **17.** verified

**19.** verified   **21.** $\begin{bmatrix} \frac{-2}{39} & \frac{1}{13} & \frac{10}{39} \\ \frac{1}{3} & 0 & \frac{1}{3} \\ \frac{-4}{39} & \frac{2}{13} & \frac{-19}{39} \end{bmatrix}$   **23.** $\begin{bmatrix} \frac{-9}{80} & \frac{31}{400} & \frac{27}{400} \\ \frac{1}{80} & \frac{41}{400} & \frac{-3}{400} \\ \frac{-1}{20} & \frac{-1}{100} & \frac{-17}{100} \end{bmatrix}$

**25.** $\begin{bmatrix} 2 & -3 \\ -5 & 7 \end{bmatrix}\begin{bmatrix} x \\ y \end{bmatrix} = \begin{bmatrix} 9 \\ 8 \end{bmatrix}$   **27.** $\begin{bmatrix} 1 & 2 & -1 \\ 1 & 0 & 1 \\ 2 & -1 & 1 \end{bmatrix}\begin{bmatrix} x \\ y \\ z \end{bmatrix} = \begin{bmatrix} 1 \\ 3 \\ 3 \end{bmatrix}$

**29.** $\begin{bmatrix} -2 & 1 & -4 & 5 \\ 2 & -5 & 1 & -3 \\ -3 & 1 & 6 & 1 \\ 1 & 4 & -5 & 1 \end{bmatrix}\begin{bmatrix} w \\ x \\ y \\ z \end{bmatrix} = \begin{bmatrix} -3 \\ 4 \\ 1 \\ -9 \end{bmatrix}$

**31.** $(4, 5)$   **33.** $(12, 12)$   **35.** no solution using matrix equations
**37.** $(1.5, -0.5, -1.5)$   **39.** no solution using matrix equations
**41.** $(-1, -0.5, 1.5, 0.5)$   **43.** 1, yes   **45.** 0, no   **47.** 1   **49.** singular matrix   **51.** $-34$   **53.** 7   **55.** $x = -3, 6$   **57.** $x = -2, 0, 1$
**59.** singular matrix

**61.** $\begin{bmatrix} 1 & -2 & 1 & -2 \\ -5 & 9 & -3 & 12 \\ -3 & 5 & -2 & 7 \\ -3 & 6 & -2 & 8 \end{bmatrix}$ verified   **63.** $\det(A) = -5$; $(1, 6, 9)$

**65.** $\det(A) = 0$   **67.** $A^{-1} = \begin{bmatrix} \frac{1}{13} & \frac{5}{13} \\ \frac{-2}{13} & \frac{3}{13} \end{bmatrix}$ verified   **69.** singular

**71.** 31 behemoth, 52 gargantuan, 78 mammoth, 30 jumbo
**73.** *Jumpin' Jack Flash:* 3.75 min
    *Tumbling Dice:* 3.75 min
    *You Can't Always Get What You Want:* 7.5 min
    *Wild Horses:* 5.75 min
**75.** 30 of clock A; 20 of clock B; 40 of clock C; 12 of clock D
**77.** $p_1 = 72.25°$, $p_2 = 74.75°$, $p_3 = 80.25°$, $p_4 = 82.75°$
**79.** $y = x^2 + 4x - 5$   **81.** $g(x) = x^3 + 2x^2 - 9x - 10$
**83.** \$450 in the CD, \$350 in the MM   **85.** \$1500 in retirement fund, \$1500 in mutual fund, \$1800 in stock fund   **87.** 2 oz Food I, 1 oz Food II, 4 oz Food III

**89. a.** $|AB| = \left| \begin{bmatrix} a & b \\ c & d \end{bmatrix}\begin{bmatrix} e & f \\ g & h \end{bmatrix} \right| = \begin{vmatrix} ae + bg & af + bh \\ ce + dg & cf + dh \end{vmatrix}$
$= (ae + bg)(cf + dh) - (af + bh)(ce + dg)$
$= \cancel{aeef} + adeh + bcfg + \cancel{bdgh} - \cancel{aeef} - adfg - bceh - \cancel{bdgh}$
$= adeh - adfg - bceh + bcfg = ad(eh - fg) - bc(eh - fg)$
$= (ad - bc)(eh - fg) = \begin{vmatrix} a & b \\ c & d \end{vmatrix}\begin{vmatrix} e & f \\ g & h \end{vmatrix} = |A||B| \checkmark$

**b.** From (a), $|A||A^{-1}| = |AA^{-1}| = |I_2| = \begin{vmatrix} 1 & 0 \\ 0 & 1 \end{vmatrix} = 1 \cdot 1 - 0 \cdot 0 = 1.$

Dividing the first and last quantities (in bold) by the nonzero number $|A|$ gives $|A^{-1}| = |A|^{-1}$. ✓

**c.** $|kA| = \left| k \begin{bmatrix} a & b \\ c & d \end{bmatrix} \right| = \begin{vmatrix} ka & kb \\ kc & kd \end{vmatrix} = (ka)(kd) - (kb)(kc) =$

$k^2(ad - bc) \neq k(ad - bc) = k \begin{vmatrix} a & b \\ c & d \end{vmatrix} = k|A|.$

**91.** $x = 6, x = 3, x = -2$  **93.** $x \in (-\infty, -\frac{9}{2}] \cup [-\frac{1}{2}, \infty)$

## Exercises 7.4 pp. 615–618

**1.** $a_{11}a_{22} - a_{21}a_{12}$  **3.** Answers will vary.

**5.** $(1, 1)$  **7.** $(-5, 9)$

**9.** $\left( \frac{-26}{3}, \frac{25}{3} \right)$  **11.** not possible with Cramer's rule

**13. a.** $D = \begin{vmatrix} 4 & -1 & 2 \\ -3 & 2 & -1 \\ 1 & -5 & 3 \end{vmatrix}$  $D_x = \begin{vmatrix} -5 & -1 & 2 \\ 8 & 2 & -1 \\ -3 & -5 & 3 \end{vmatrix}$

$D_y = \begin{vmatrix} 4 & -5 & 2 \\ -3 & 8 & -1 \\ 1 & -3 & 3 \end{vmatrix}$  $D_z = \begin{vmatrix} 4 & -1 & -5 \\ -3 & 2 & 8 \\ 1 & -5 & -3 \end{vmatrix}$

**b.** $|D| = 22$, solutions possible
**c.** $|D| = 0$, Cramer's rule cannot be used: coefficients R1 + R2 = R3
**15.** $(1, 2, 1)$  **17.** not possible with Cramer's rule

**19.** $\left( \frac{3}{4}, \frac{5}{3}, \frac{-1}{3} \right)$  **21.** $(0, -1, 2, -3)$  **23.** $\frac{A}{x + 3} + \frac{B}{x - 2}$

**25.** $\frac{A}{x - 1} + \frac{B}{x + 2} + \frac{C}{x - 3}$  **27.** $\frac{A}{x} + \frac{B}{x - 3} + \frac{C}{x + 1}$

**29.** $\frac{A}{x} + \frac{B}{x^2} + \frac{C}{x + 2}$  **31.** $\frac{A}{x + 1} + \frac{Bx + C}{x^2 + 2} + \frac{Dx + E}{(x^2 + 2)^2}$

**33.** $\frac{4}{x} - \frac{5}{x + 1}$  **35.** $\frac{-4}{2x - 5} + \frac{3}{x + 3}$  **37.** $\frac{7}{x} + \frac{2}{x + 1} - \frac{1}{x - 1}$

**39.** $\frac{-1}{x} + \frac{4}{x + 1} + \frac{5}{(x + 1)^2}$  **41.** $\frac{3}{2 - x} - \frac{4}{x^2 + 2x + 4}$

**43.** $\frac{5}{x + 2} + \frac{x - 1}{x^2 + 3}$  **45.** $\frac{1}{x} - \frac{3x + 2}{(x^2 + 1)^2}$

**47.** $\frac{3}{x + 1} - \frac{2}{x - 3} + \frac{1}{(x - 3)^3}$  **49.** $x^2 + x + 4 - \frac{5}{x - 6} + \frac{3}{x + 1}$

**51. a.** $\begin{vmatrix} x & y & 1 \\ x_1 & y_1 & 1 \\ x_2 & y_2 & 1 \end{vmatrix} = 0$

$x \begin{vmatrix} y_1 & 1 \\ y_2 & 1 \end{vmatrix} - y \begin{vmatrix} x_1 & 1 \\ x_2 & 1 \end{vmatrix} + 1 \begin{vmatrix} x_1 & y_1 \\ x_2 & y_2 \end{vmatrix} = 0$

$x(y_1 - y_2) - y(x_1 - x_2) + (x_1 y_2 - x_2 y_1) = 0$

$(x_2 - x_1)(y - y_1) - (y_2 - y_1)(x - x_1) = 0$

$(x_2 - x_1)(y - y_1) = (y_2 - y_1)(x - x_1)$

$y - y_1 = \frac{y_2 - y_1}{x_2 - x_1}(x - x_1)$

$y - y_1 = m(x - x_1)$, where $m = \frac{y_2 - y_1}{x_2 - x_1}$ ✓

**b.** $y = \frac{-8}{3}x - \frac{7}{3}$  **53.** 8 cm$^2$  **55.** 27 ft$^2$  **57.** 19 m$^3$  **59.** yes  **61.** no

**63.** $\begin{cases} 15{,}000x + 25{,}000y = 2900 \\ 25{,}000x + 15{,}000y = 2700 \end{cases}$; 6%, 8%

**65.** $F_1 = F_2 = 60\sqrt{3} \approx 103.9$ lb

**67. a.** $h = 10{,}000 - 16t^2$  **b.** 10,000 ft  **c.** 3600 ft  **d.** 25 sec

**69.** $A = 195$ units$^2$

**71.**

**73.** $3^{2x - 1} = 3^{4 - 2x}; x = 1.25$

## Making Connections, p. 619

**1.** $G$  **3.** $D$  **5.** $A$  **7.** $B$  **9.** $CD$  **11.** $E$  **13.** $D$  **15.** $H$

## Summary and Concept Review, pp. 619–621

**1.** Answers will vary.  **2.** $(-2, -4)$  **3.** $(1, 6, 9)$
**4.** $\left( -\frac{6}{5}p - 1, -\frac{2}{5}p - 1, p \right)$  **5.** no solution
**6.** $\{(x, y, z) | x + y - 3z = -2\}$  **7.** $(-2, 7, 1, 8)$
**8.** $\begin{bmatrix} -7.25 & 5.25 \\ 0.875 & -2.875 \end{bmatrix}$  **9.** $\begin{bmatrix} -6.75 & 6.75 \\ 1.125 & -1.125 \end{bmatrix}$  **10.** not possible

**11.** $\begin{bmatrix} -2 & -6 \\ -1 & -7 \end{bmatrix}$  **12.** $\begin{bmatrix} 1 & 0 \\ 0 & 1 \end{bmatrix}$  **13.** $\begin{bmatrix} 1 & 0 & 4 \\ 5.5 & -1 & -1 \\ 10 & -2.9 & 7 \end{bmatrix}$

**14.** $\begin{bmatrix} 3 & -6 & -4 \\ -4.5 & 3 & -1 \\ -2 & 3.1 & 3 \end{bmatrix}$  **15.** not possible  **16.** $\begin{bmatrix} -8 & 12 & 0 \\ -2 & -4 & 4 \\ -16 & -0.4 & -20 \end{bmatrix}$

**17.** $\begin{bmatrix} 15.5 & 6.4 & 17 \\ 9 & -17 & 2 \\ 18.5 & 20.8 & 13 \end{bmatrix}$  **18.** $D$  **19.** It's an identity matrix.
**20.** It's the inverse of $B$.  **21.** $E$  **22.** It is the inverse of $F$.
**23.** Matrix multiplication is not generally commutative.  **24.** $(-8, -6)$

**25.** $(2, 0, -3)$  **26.** $\left( \frac{-19}{35}, \frac{25}{14} \right)$  **27.** $(1, -1, 2)$  **28.** $\left( \frac{-37}{19}, \frac{36}{19}, \frac{31}{19} \right)$

**29.** $\frac{91}{2}$ units$^2$  **30.** $\frac{5}{x - 2} + \frac{2x - 1}{x^2 + 3}$

## Practice Test, pp. 621–622

**1.** $\left( -3, \frac{1}{2} \right)$  **2.** $\left( -3p + \frac{16}{3}, p, 2p - 3 \right)$  **3.** $\left( 2, 1, \frac{-1}{3} \right)$

**4. a.** $\begin{bmatrix} -6 & -5 \\ 8 & 9 \end{bmatrix}$  **b.** $\begin{bmatrix} 1.2 & 1.2 \\ -1.2 & -2 \end{bmatrix}$  **c.** $\begin{bmatrix} -3 & 1 \\ 3 & -5 \end{bmatrix}$  **d.** $\begin{bmatrix} -2 & -1 \\ 2.5 & 1.5 \end{bmatrix}$

**e.** $-2$

**5. a.** $\begin{bmatrix} 0 & -0.1 & 0 \\ 0.5 & -0.6 & 0 \\ -0.2 & -0.8 & -0.9 \end{bmatrix}$  **b.** $\begin{bmatrix} -0.3 & -0.06 & -0.12 \\ 0.06 & -0.06 & 0 \\ -0.18 & -0.24 & -0.48 \end{bmatrix}$

**c.** $\begin{bmatrix} 0.31 & -0.13 & 0.08 \\ -0.01 & -0.05 & -0.02 \\ 0.39 & -0.52 & -0.02 \end{bmatrix}$  **d.** $\begin{bmatrix} \frac{40}{17} & 0 & \frac{-10}{17} \\ \frac{40}{17} & 10 & \frac{-10}{17} \\ \frac{-35}{17} & -5 & \frac{30}{17} \end{bmatrix}$  **e.** $\frac{17}{500}$

**6.** $(-1, -6, 0), (1, -1, 1), (3, 4, 2)$, answers vary as $(2p - 1, 5p - 6, p)$

**7.** $\left( 2, \frac{2}{3} \right)$  **8.** $(3, -2, 3)$  **9.** $\left( \frac{97}{34}, \frac{-18}{17} \right)$  **10.** $(1, 6, 9)$

**11.** $x = 1, y = -1, z = 2$  **12.** $B = \begin{bmatrix} 6 \\ 13 \\ -11 \end{bmatrix}$

**13.** $(-1, 4), (2, 1), (4, -1)$  **14.** 5 mi$^2$  **15.** $r = -2, s = 1$
**16.** Dr. Brown owes \$31,000; Dr. Stamper owes \$124,000.
**17.** 7.5 hr, 15.5 hr  **18.** 11 one-day, 6 two-day, 3 five-day
**19.** federal program: \$200,000; municipal bonds: \$1,300,000;

bank loan: \$300,000  **20.** $\frac{1}{x - 3} + \frac{3x + 2}{x^2 + 3x + 9}$

## Calculator Exploration and Discovery, pp. 623–624

**1.** $(3, \frac{1}{2}), (2, \frac{2}{3})$  **2.** Answers will vary.  **3.** Answers will vary.
**4.** $(10, 12)$  **5.** $(2, 1, -3)$

## Strengthening Core Skills, pp. 624–625

**1.** $(1, -4, 1)$

## Cumulative Review Chapters R–7, pp. 625–626

**1. a.** $8 - i$  **b.** $16 - 30i$  **c.** $3 - 2i$  **d.** $i$  **3.** $x = \frac{-3}{2}$  **5.** $x = \frac{1}{3}$

**7.** $y = \dfrac{-7}{5}x + \dfrac{4}{5}$   **9. a.** $-18$   **b.** 1   **c.** 7   **d.** $-3$ is not in the domain

**11.**     **13.**

**15.** $(-\infty, 1) \cup (5, \infty)$

**17.** $f(x)\uparrow: x \in (0, 2) \cup (4, \infty), f(x)\downarrow: x \in (-\infty, 0) \cup (2, 4)$, constant: none

**19.** $2y - \dfrac{5}{2} + \dfrac{25}{2(2y + 1)}$   **21.** $x^3 - 4x^2 - 2x + 20$   **23.** $\left(-3, \dfrac{-5}{2}\right)$

**25.** condominium: \$400,000; delivery truck: \$60,000; business: \$140,000

# CHAPTER 8

## Exercises 8.1, pp. 632–634

**1.** perpendicular   **3.** point, intersecting   **5.** $(-2, -2)$; verified
**7.** $\left(\frac{13}{2}, -9\right)$; verified   **9.** $(x + 2)^2 + (y + 2)^2 = 5^2$
**11.** $\left(x - \frac{13}{2}\right)^2 + (y + 9)^2 = \left(\frac{25}{2}\right)^2$   **13. a.** $d = 13; B, C, E, G$   **b.** Answers
will vary.   **15.** verified, $d = \frac{8\sqrt{5}}{5}$   **17. a.** $B, C, E$   **b.** Answers will vary.
**19.** verified   **21.** $y = -\frac{1}{16}x^2$; verified   **23.** $4x^2 + 3y^2 = 48$
**25.** verified; verified   **27.** $3x^2 - y^2 = 3$   **29.** verified   **31.** about 12 yr
**33. a.** $x = 2 \ln 9, x \approx 4.39$   **b.** $x = \frac{5}{2} \ln 54, x \approx 9.97$

## Exercises 8.2, pp. 643–646

**1.** $c^2 = |a^2 - b^2|$   **3.** $2a, 2b$
**5.** $x^2 + y^2 = 49$       **7.** $(x - 5)^2 + y^2 = 3$
**9.** $(x - 1)^2 + (y - 5)^2 = 25$    **11.** $(x - 5)^2 + (y - 7)^2 = 49$
**13.** $(x - 6)^2 + (y - 5)^2 = 9$    **15.** $(x - 2)^2 + (y + 5)^2 = 25$
   center: $(6, 5), r = 3$         center: $(2, -5), r = 5$

            (graph for 15)

**17.** $(x + 3)^2 + y^2 = 14$    **19.**
   center: $(-3, 0), r = \sqrt{14}$

**21.**         **23.**

**25. a.** $\dfrac{x^2}{16} + \dfrac{y^2}{4} = 1, (0, 0), a = 4, b = 2$
  **b.** $(-4, 0), (4, 0), (0, -2), (0, 2)$
  **c.**

**27. a.** $\dfrac{x^2}{9} + \dfrac{y^2}{16} = 1, (0, 0), a = 3, b = 4$
  **b.** $(0, -4), (0, 4), (-3, 0), (3, 0)$
  **c.**

**29. a.** $\dfrac{x^2}{5} + \dfrac{y^2}{2} = 1, (0, 0), a = \sqrt{5}, b = \sqrt{2}$
  **b.** $(-\sqrt{5}, 0), (\sqrt{5}, 0), (0, -\sqrt{2}), (0, \sqrt{2})$
  **c.**

**31.** ellipse       **33.** circle

**35.** ellipse

**37.** $x^2 + \dfrac{(y + 3)^2}{4} = 1$    **39.** $\dfrac{(x + 2)^2}{16} + \dfrac{(y - 1)^2}{4} = 1$

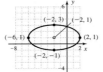    (graph for 39)

     $D: x \in [-1, 1], R: y \in [-5, -1]$      $D: x \in [-6, 2], R: y \in [-1, 3]$

**41.** $\dfrac{(x - 3)^2}{4} + \dfrac{(y + 5)^2}{10} = 1$    **43.** $\dfrac{(x - 3)^2}{25} + \dfrac{(y + 2)^2}{10} = 1$

   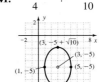

$D: x \in [1, 5],$
$R: y \in [-5 - \sqrt{10}, -5 + \sqrt{10}]$

$D: x \in [-2, 8],$
$R: y \in [-2 - \sqrt{10}, -2 + \sqrt{10}]$

**45. a.** $(2, 1)$   **b.** $(-3, 1)$ and $(7, 1)$   **c.** $(2 - \sqrt{21}, 1)$ and $(2 + \sqrt{21}, 1)$
  **d.** $(2, 3)$ and $(2, -1)$   **e.**

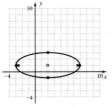

**47. a.** $(4, -3)$   **b.** $(4, 2)$ and $(4, -8)$   **c.** $(4, 0)$ and $(4, -6)$   **d.** $(0, -3)$
and $(8, -3)$   **e.**

**49. a.** $(-2, -2)$   **b.** $(-5, -2)$ and $(1, -2)$
**c.** $(-2 + \sqrt{3}, -2)$ and $(-2 - \sqrt{3}, -2)$   **d.** $(-2, -2 + \sqrt{6})$ and
$(-2, -2 - \sqrt{6})$   **e.**

**51. a.** $\dfrac{x^2}{289^2} + \dfrac{y^2}{136^2} = 1$

**b.** $\dfrac{2(136)^2}{289} = 128$ units
$(\pm 255, \pm 64)$
$\dfrac{255^2}{289^2} + \dfrac{64^2}{136^2} = 1$ ✓

**53. a.** $\dfrac{(x-3)^2}{3^2} + \dfrac{(y+2)^2}{5^2} = 1$

**b.** $\dfrac{2(3)^2}{5} = 3.6$ units
$(1.2, 2), (4.8, 2),$
$(1.2, -6), (4.8, -6);$
verified

**55. a.** $\dfrac{(x-2)^2}{5^2} + \dfrac{(y-1)^2}{2^2} = 1$

**b.** $\dfrac{2(2)^2}{5} = 1.6$ units
$(-2.6, 1.8), (-2.6, 0.2),$
$(6.6, 1.8), (6.6, 0.2);$
verified

**57.** $\dfrac{x^2}{16} + \dfrac{y^2}{9} = 1, (\pm\sqrt{7}, 0)$

**59.** $\dfrac{(x+3)^2}{4} + \dfrac{(y+1)^2}{16} = 1, (-3, -1 \pm 2\sqrt{3})$   **61.** $A = 12\pi$ units$^2$

**63.** $\sqrt{7} \approx 2.65$ ft, $2.25$ ft   **65.** $8.9$ ft, $17.9$ ft   **67.** $\dfrac{x^2}{15^2} + \dfrac{y^2}{8^2} = 1; 6.4$ ft

**69.** $\dfrac{x^2}{6.25} + \dfrac{y^2}{2.25} = 1$   **71.** $\dfrac{x^2}{36^2} + \dfrac{y^2}{(35.25)^2} = 1$   **73.** $a \approx 142$ million
miles, $b \approx 141$ million miles, orbit time $\approx 686$ days   **75.** about 66,697 mph
**77.** $90{,}000\pi$ yd$^2$   **79. a.** $45$ ft   **b.** $106$ ft   **c.** $25$ ft   **81.** ellipse, since
squared terms are positive and $A \neq B$; $6(x+3)^2 + 3(y-4)^2 = 0$; the
constant term becomes zero; the graph is the single point $(-3, 4)$
**83.** $\dfrac{\log 20}{\log 3} \approx 2.73$
**85. a.** $x \in (0, 6)$       **b.** $-2|x-3| + 10 > 4$

$|x - 3| < 3$
$x - 3 > -3$ or $x - 3 < 3$
$x > 0$ or $x < 6$
$0 < x < 6$

## Mid-Chapter Check, p. 647

**1.** 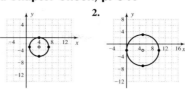   **2.**    **3.**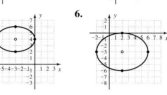

**4.** 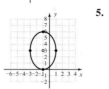   **5.**    **6.**

**7.** $y = \frac{1}{12}x^2$

**8. a.** $\dfrac{(x+3)^2}{4} + \dfrac{(y-1)^2}{16} = 1; D: x \in [-5, -1], R: y \in [-3, 5]$

**b.** $(x-3)^2 + (y-2)^2 = 16; D: x \in [-1, 7], R: y \in [-2, 6]$

**9.** $\frac{x^2}{25} + \frac{y^2}{194} = 1$   **10.** $\frac{x^2}{16} + \frac{y^2}{4} = 1$

## Exercises 8.3, pp. 656–658

**1.** transverse   **3.** perpendicular, transverse, center

**5.** 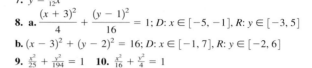   **7.**   **9.**

**11.** 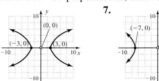   **13.**   **15.**

**17.** $(-4, -2), (2, -2), y = -2, (-1, -2); x = -1$
**19.** $(4, 1), (4, -3), x = 4, (4, -1); y = -1$

**21.** 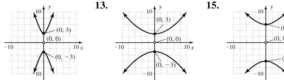   **23.**   **25.**

**27.** 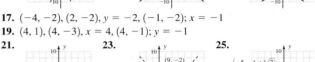   **29.**   **31.**

**33.**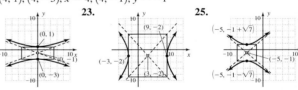

**35.** $\dfrac{(x+5)^2}{9} - \dfrac{(y+2)^2}{36} = 1$   **37.** $\dfrac{(y-3)^2}{1} - \dfrac{(x+2)^2}{4} = 1$

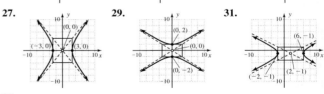

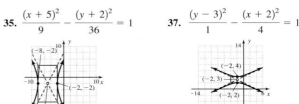

**39.** circle; $A = B$   **41.** circle; $A = B$   **43.** hyperbola; $A, B$ opposite signs
**45.** hyperbola; $A, B$ opposite signs   **47.** circle; $A = B$
**49.** ellipse; $A \neq B$   **51.** 8, $2a = 8, 2b = 6$   **53.** 12, $2a = 16, 2b = 12$
**55.** $\dfrac{(x - 3)^2}{9} - \dfrac{(y - 4)^2}{4} = 1$   **a.** $(3, 4)$   **b.** $(0, 4)$ and $(6, 4)$
**c.** $(3 - \sqrt{13}, 4)$ and $(3 + \sqrt{13}, 4)$   **d.** $2a = 6, 2b = 4$
**e.**

**57.** $\dfrac{(y - 3)^2}{16} - \dfrac{x^2}{4} = 1$   **a.** $(0, 3)$   **b.** $(0, 7), (0, -1)$
**c.** $(0, 3 + 2\sqrt{5}), (0, 3 - 2\sqrt{5})$   **d.** $2a = 4, 2b = 8$
**e.**

**59.** $\dfrac{(x - 3)^2}{4} - \dfrac{(y + 2)^2}{12} = 1$   **a.** $(3, -2)$   **b.** $(1, -2)$ and $(5, -2)$
**c.** $(-1, -2)$ and $(7, -2)$   **d.** $2a = 4, 2b = 4\sqrt{3}$
**e.**

**61.** $\dfrac{x^2}{36} - \dfrac{y^2}{28} = 1$   **63.** $\dfrac{y^2}{9} - \dfrac{(x + 2)^2}{9} = 1$   **65.** $\dfrac{x^2}{4} - \dfrac{y^2}{9} = 1, (\pm\sqrt{13}, 0)$
**67.** $\dfrac{(y - 1)^2}{4} - \dfrac{(x - 2)^2}{5} = 1$, 4 by $2\sqrt{5}$
**69.** **a.** $y = \frac{2}{3}\sqrt{x^2 - 9}$   **b.** $x \in (-\infty, -3] \cup [3, \infty)$
**c.**
    **d.** $y = \dfrac{-2}{3}\sqrt{x^2 - 9}$

**71.** 40 yd   **73.** 12 microns   **75.** $\dfrac{x^2}{225} - \dfrac{y^2}{2275} = 1$; about $(24.1, 60)$ or
$(-24.1, 60)$   **77.** $y = \pm\sqrt{\dfrac{b^2}{a^2}x^2 - b^2}$; as $x \to \infty, y \to \pm\sqrt{\dfrac{b^2}{a^2}x^2} = \pm\dfrac{b}{a}x$
**79.** **a.** $\dfrac{(x - 4)^2}{\frac{1}{4}} - (y - 2)^2 = 0$   **b.** $(x - 2)^2 + \dfrac{(y - 4)^2}{\frac{1}{5}} = 0$
**81.** yes; 3   **83.** 42 solid, 46 liquid

## Exercises 8.4, pp. 664–666

**1.** horizontal, right, $a < 0$   **3.** $-4, (0, -4)$
**5.** $x \in (-\infty, \infty), y \in [-4, \infty)$    **7.** $x \in (-\infty, \infty), y \in (-\infty, 18]$

**9.** $x \in (-\infty, \infty), y \in [-10.125, \infty)$   **11.** $x \in [-4, \infty), y \in (-\infty, \infty)$
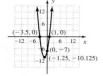   

**13.** $x \in (-\infty, 16], y \in (-\infty, \infty)$    **15.** $x \in (-\infty, 0], y \in (-\infty, \infty)$

**17.** $x \in (-\infty, \infty), y \in [3, \infty)$    **19.** $x \in [1, \infty), y \in (-\infty, \infty)$

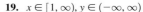

**21.** $x \in [-9, \infty), y \in (-\infty, \infty)$    **23.** $x \in [-4, \infty), y \in (-\infty, \infty)$

**25.** $x \in (-\infty, 0], y \in (-\infty, \infty)$    **27.** $x \in [-21, \infty), y \in (-\infty, \infty)$

**29.** $x \in (-\infty, 11], y \in (-\infty, \infty)$
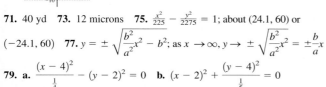

**31.**    **33.**    **35.**
 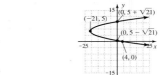

**37.**    **39.**    **41.**
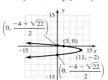

**43.**    **45.**    **47.**
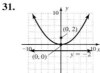 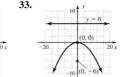 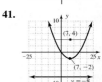
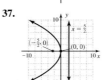 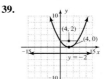 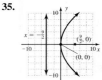
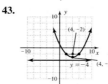 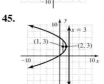 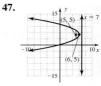

**49.** $x^2 = 8y$   **51.** $x^2 = -20y$   **53.** $(y + 2)^2 = -12(x - 2)$
**55.** $(x - 4)^2 = 12(y + 7)$   **57.** $(x - 3)^2 = 8(y - 2)$   **59.** $y^2 = 8(x + 1)$;
vertex $(-1, 0)$; focus $(1, 0)$   **61.** $(y - 2)^2 = -8(x + 2)$; directrix: $x = 0$;
endpoints $(-4, 6)$ and $(-4, -2)$   **63.** $(4, 3), (4, -3), (-4, 3), (-4, -3)$;
verified   **65.** $(3, 5), (-3, 5), (0, -4)$; verified
**67.** $(5, 5), (5, -5), (-5, 5), (-5, -5)$; verified   **69.** 16 units$^2$

**71.**

**73.** 6 in.; $(13.5, 0)$   **75.** $\approx 14.97$ ft; $(0, 41.75)$

**77.** $y^2 = 5x$ or $x^2 = 5y$; 1.25 cm   **79.** $(x - 2)^2 = \frac{1}{2}(y - 4)$; $p = \frac{1}{8}$; $(2, 4)$
**81.** $y = -2x^2 - 5x + 6$   **83.** symmetric to the $y$-axis, $f(-x) = f(x)$; symmetric to the origin, $f(-x) = -f(x)$

## Making Connections, p. 667

**1.** d   **3.** f   **5.** g   **7.** c   **9.** h   **11.** e   **13.** b   **15.** g

## Summary and Concept Review, pp. 667–669

**1.** verified (segments are perpendicular and equal length)
**2.** $x^2 + (y - 1)^2 = 34$   **3.** yes   **4.** verified
**5.**    **6.**

**7.** 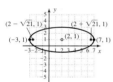   **8.**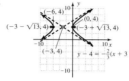

**9. a.** $\dfrac{x^2}{169} + \dfrac{y^2}{25} = 1$   **b.** $\dfrac{x^2}{144} + \dfrac{y^2}{400} = 1$

**10.** $\dfrac{(x - 2)^2}{25} + \dfrac{(y - 1)^2}{4} = 1$   **11.**

**12.**    **13.** 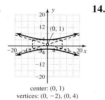   **14.**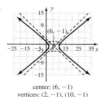

center: $(-2, 1)$
vertices: $(-5, 1), (1, 1)$

center: $(0, 1)$
vertices: $(0, -2), (0, 4)$

center: $(6, -1)$
vertices: $(2, -1), (10, -1)$

**15. a.** $\dfrac{x^2}{225} - \dfrac{y^2}{64} = 1$   **b.** $\dfrac{y^2}{16} - \dfrac{x^2}{9} = 1$

**16.** $\dfrac{(x - 5)^2}{9} - \dfrac{(y - 2)^2}{4} = 1$   **17.**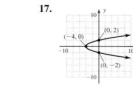

**18.**    **19.**    **20.**

**21.** circle, parabola; $(-3, 2), (-3, -2)$

## Practice Test, p. 669

**1.** c   **2.** d   **3.** a   **4.** b
**5.** 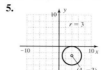   **6.**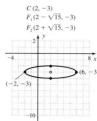

$C(2, -3)$
$F_1(2 - \sqrt{15}, -3)$
$F_2(2 + \sqrt{15}, -3)$

**7.**    **8.**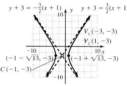

**9.**    **10.**

**11.**    **12.**

**13. a.** $\left(\dfrac{10}{3}, \dfrac{16}{3}\right), (-2, 0)$   **b.** $(2, 2), (-2, 2), (-2, -2), (2, -2)$

**14.** 15 ft, 20 ft   **15.** $(x + 2)^2 + (y - 5)^2 = 8$
**16.** $\dfrac{x^2}{16} + \dfrac{y^2}{12} = 1$; $(2, 3), (2, -3), (-2, -3), (-2, 3)$
**17.** 34.9 AU or about 3244 million miles; 0.7 AU or about 65 million miles
**18.** $y = (x - 1)^2 - 4$; $D: x \in (-\infty, \infty)$, $R: y \in [-4, \infty)$; focus: $\left(1, \dfrac{-15}{4}\right)$
**19.** $(x - 1)^2 + (y - 1)^2 = 25$; $D: x \in [-4, 6]$, $R: y \in [-4, 6]$
**20.** $\dfrac{(x + 3)^2}{9} + \dfrac{y^2}{36} = 1$; $D: x \in [-6, 0]$, $R: y \in [-6, 6]$;
foci: $(-3, -3\sqrt{3}), (-3, 3\sqrt{3})$

## Calculator Exploration and Discovery, pp. 670–671

**Exercise 1:** $y = \pm 4.8$; $y = \pm 3.6$; Answers will vary.
**Exercise 2:** $(0, \pm 2.6457513), (0, \pm \sqrt{7})$
**Exercise 3:** $e > 1$
**Exercise 4:** $e = 1$

## Strengthening Core Skills, pp. 671–672

**Exercise 1:** $\dfrac{25(x - 2)^2}{2} + \dfrac{9(y - 3)^2}{4} = 1$

**Exercise 2:** $\dfrac{28(x - 1)^2}{25} + \dfrac{48(y + 2)^2}{25} = 1$

## Cumulative Review Chapters R–8, p. 672

**1.** $x = 2, x = \pm 2i$   **3.** $x = 7; x = 4$ is extraneous   **5.** $x = 3 \pm 2i$
**7.** $x = 2 + \dfrac{\ln 7}{\ln 3}$   **9.** $x = 5; x = -2$ is extraneous

**11.**    **13.**    **15.**

**17.**    **19.**

**21. a.** $x \in (-\infty, \infty)$   **b.** $y \in (-\infty, 4]$
**c.** $f(x)\uparrow: x \in (-\infty, -1), f(x)\downarrow: x \in (-1, \infty)$   **d.** max: $(-1, 4)$
**e.** $f(x) > 0: x \in (-4, 2), f(x) < 0: x \in (-\infty, -4) \cup (2, \infty)$
**23.** $(3, 4), (3, -4), (-3, 4), (-3, -4)$   **25.** $3\frac{1}{3}$ L   **27.** $x = -6, x = 2$
**29.** $x \approx 1.387 \left( x = \dfrac{2 \ln 2 + \ln 3}{\ln 2 + \ln 3} \text{ or } \dfrac{\ln 12}{\ln 6} \right)$

## CHAPTER 9
### Exercises 9.1, pp. 679–681

**1.** pattern, order   **3.** formula defining the sequence uses the preceding term(s); Answers will vary.   **5.** 1, 3, 5, 7   **7.** $-1, 2, -3, 4$
**9.** $\dfrac{1}{2}, \dfrac{1}{4}, \dfrac{1}{8}, \dfrac{1}{16}$   **11.** $\dfrac{-1}{2}, \dfrac{1}{6}, \dfrac{-1}{12}, \dfrac{1}{20}$
**13.** 79   **15.** $\frac{1}{9}$   **17.** approx. 2.6   **19.** 2, 7, 32, 157, 782
**21.** $-1, 4, 19, 364, 132,499$   **23.** 64, 32, 16, 8, 4   **25.** 336   **27.** 36
**29.** 28   **31.** $\dfrac{-1}{2}, \dfrac{1}{6}, \dfrac{-1}{24}, \dfrac{1}{120}$   **33.** $\frac{1}{3}, \frac{1}{120}, \frac{1}{15,120}, \frac{1}{3,991,680}$   **35.** $1, 2, \frac{9}{2}, \frac{32}{3}$
**37.** 15   **39.** 64   **41.** $\frac{137}{60}$   **43.** $-2 + 1 + 4 + 7 = 10$
**45.** $-1 + 5 + 15 + 29 + 47 = 95$
**47.** $-1 + 2 + (-3) + 4 + (-5) + 6 + (-7) = -4$
**49.** $0.5 + 2 + 4.5 + 8 = 15$
**51.** $6 + 8 + 10 + 12 + 14 = 50$
**53.** $-\dfrac{1}{3} + \dfrac{1}{8} + \left(-\dfrac{1}{15}\right) + \dfrac{1}{24} + \left(-\dfrac{1}{35}\right) + \dfrac{1}{48} = -\dfrac{27}{112}$
**55. a.** $\displaystyle\sum_{n=1}^{5} 4n$   **b.** $\displaystyle\sum_{n=1}^{5} 5n$   **57. a.** $\displaystyle\sum_{k=1}^{\infty}(2k-1)$   **b.** $\displaystyle\sum_{k=1}^{\infty}\frac{1}{k^2}$
**59.** $\displaystyle\sum_{n=1}^{5}(n+3)$   **61.** $\displaystyle\sum_{n=1}^{3}\frac{n^2}{3}$   **63.** $\displaystyle\sum_{n=3}^{7}\frac{n}{2^n}$   **65.** 35   **67.** 100
**69. a.** $a_6 = 21$,   **b.** $n = \dfrac{\sqrt{8a_n + 1} - 1}{2}$   **c.** yes, the 100th
**71.** $a_n = 6.75 + 0.5n$; \$7.25, \$7.75, \$8.25, \$8.75, \$9.25; \$17,760
**73.** $a_n = 6000(0.8)^{n-1}$; \$6000, \$4800, \$3840, \$3072, \$2457.60, \$1966.08
**75.** $\approx 2690$   **77.** 20, 18, 17, 18, 16, 17
**79.** $\displaystyle\sum_{i=1}^{n}(a_i \pm b_i) = (a_1 \pm b_1) + (a_2 \pm b_2) + (a_3 \pm b_3) + \cdots + (a_n \pm b_n)$
$= a_1 + a_2 + a_3 + \cdots + a_n \pm b_1 \pm b_2 \pm b_3 \pm \cdots \pm b_n$
$= (a_1 + a_2 + a_3 + \cdots + a_n) \pm (b_1 + b_2 + b_3 + \cdots + b_n)$
$= \displaystyle\sum_{i=1}^{n}a_i \pm \sum_{i=1}^{n}b_i$
**81.** approaches $e$   **83.** $3^{-x} = \frac{1}{81}; x = 4$
**85.** $(\sqrt{5}, -2), (\sqrt{5}, 2), (-\sqrt{5}, -2), (-\sqrt{5}, 2)$

### Exercises 9.2, pp. 687–690

**1.** $\dfrac{n(a_1 + a_n)}{2}$, $n$th   **3.** Answers will vary.   **5.** arithmetic; $d = 3$
**7.** arithmetic; $d = 2.5$   **9.** not arithmetic; all prime   **11.** arithmetic; $d = \frac{1}{24}$
**13.** not arithmetic; $a_n = n^2$   **15.** arithmetic; $d = \frac{-\pi}{6}$   **17.** 2, 5, 8, 11
**19.** 0.3, 0.33, 0.36, 0.39   **21.** $\frac{3}{2}, 2, \frac{5}{2}, 3$   **23.** $-2, -5, -8, -11$
**25.** $a_1 = 2, d = 5, a_n = 5n - 3, a_6 = 27, a_{10} = 47, a_{12} = 57$

**27.** $a_1 = 5.10, d = 0.15$,
$a_n = 0.15n + 4.95, a_6 = 5.85$,
$a_{10} = 6.45, a_{12} = 6.75$
**29.** $a_1 = \frac{3}{2}, d = \frac{3}{4}, a_n = \frac{3}{4}n + \frac{3}{4}$,
$a_6 = \frac{21}{4}, a_{10} = \frac{33}{4}, a_{12} = \frac{39}{4}$
**31.** 61   **33.** 1   **35.** 2.425   **37.** 9   **39.** 43   **41.** 21   **43.** 26
**45.** $d = 3, a_1 = 1$   **47.** $d = 0.375, a_1 = 0.65$   **49.** $d = \frac{115}{126}, a_1 = \frac{-472}{63}$
**51.** 1275   **53.** 601.25   **55.** $-534$   **57.** $\displaystyle\sum_{n=1}^{10}\left(\frac{1}{2}n + 4\right) = 67.5 \text{ unit}^2$
**59.** 82.5   **61.** 74.04   **63.** $210\sqrt{2}$   **65.** $S_6 = 21; S_{75} = 2850$
**67.** at 11 P.M.   **69.** $5\frac{1}{2}$ in; $54\frac{1}{4}$ in.   **71.** $a_7 = 220; a_{12} = 2520$; yes
**73. a.** linear   **b.** quadratic   **75.** $t \approx 3.6$   **77.** $f(x) = 49x + 972; 1364$

### Exercises 9.3, pp. 697–700

**1.** $a_1 r^{n-1}$   **3.** Answers will vary.   **5.** geometric; $r = 2$
**7.** geometric; $r = -2$   **9.** not geometric; $a_n = n^2 + 1$
**11.** geometric; $r = 0.1$   **13.** geometric; $r = \frac{1}{2}$   **15.** geometric; $r = \frac{4}{x}$
**17.** not geometric; $a_n = \dfrac{240}{n!}$   **19.** 5, 10, 20, 40   **21.** $-6, 3, \frac{-3}{2}, \frac{3}{4}$
**23.** $4, 4\sqrt{3}, 12, 12\sqrt{3}$   **25.** 0.1, 0.01, 0.001, 0.0001
**27.** $a_n = -24\left(\dfrac{1}{2}\right)^{n-1}; a_7 = -\dfrac{3}{8}$   **29.** $a_n = -\dfrac{1}{20}(-5)^{n-1}; a_4 = \dfrac{25}{4}$
**31.** $a_n = 2(\sqrt{2})^{n-1} = (\sqrt{2})^{n+1}; a_7 = 16$
**33.** $a_1 = \frac{1}{27}, r = -3; a_n = \frac{1}{27}(-3)^{n-1}; a_6 = -9, a_{10} = -729$,
$a_{12} = -6561$
**35.** $a_1 = 729, r = \frac{1}{3}; a_n = 729(\frac{1}{3})^{n-1}; a_6 = 3, a_{10} = \frac{1}{27}, a_{12} = \frac{1}{243}$
**37.** $a_1 = 0.2, r = 0.4; a_n = 0.2(0.4)^{n-1}$;
$a_6 = 0.002048, a_{10} = 0.0000524288, a_{12} = 0.000008388608$
**39.** 5   **41.** 11   **43.** 8   **45.** 9   **47.** $r = \frac{2}{3}, a_1 = 729$
**49.** $r = \frac{3}{2}, a_1 = \frac{32}{243}$   **51.** $r = \frac{3}{2}, a_1 = \frac{256}{81}$   **53.** $-10,920$
**55.** $\dfrac{2059}{8} = 257.375$   **57.** 728   **59.** $\approx 1.60$   **61.** 1364   **63.** $\dfrac{31,525}{2187}$
**65.** $-\dfrac{387}{512}$   **67.** $\frac{521}{25}$   **69.** $\frac{3367}{1296}$   **71.** $14 + 15\sqrt{2}$   **73.** $\frac{27}{7}$   **75.** no
**77.** 12   **79.** $\frac{1}{3}$   **81.** $\frac{3}{2}$   **83.** No finite sum exists.
**85.** $\displaystyle\sum_{n=1}^{10}9\left(\dfrac{7}{9}\right)^n \approx 28.95 \text{ unit}^2$   **87.** 1296
**89.** $a_n = 24(0.8)^{n-1}; a_7 \approx 6.3$ ft; $S_\infty = 120$ ft
**91.** $a_0 = 46,000; a_n = 36,800(0.8)^{n-1}; a_4 \approx \$18,841.60$; 10 yr
**93.** $a_0 = 160; a_n = 155.2(0.97)^{n-1}; a_8 \approx 125.4$ gpm; 10 months
**95.** $a_0 = 281; a_n = 283.7(1.0096)^{n-1}; a_{10} \approx 309$ million
**97.** $a_0 = 462; a_n = 277.2\left(\dfrac{3}{5}\right)^{n-1}; a_5 \approx 35.9 \text{ in}^3$; 7 strokes
**99.** $a_0 = 8, a_n = 6(0.75)^{n-1}$; 2 days; 8 days
**101.** $a_0 = 50; a_n = 100(2)^{n-1}; a_{10} = 51,200$ bacteria; 12 half-hours (6 hr)
**103.** $a_0 = 2; a_n = 1.6\left(\dfrac{4}{5}\right)^{n-1}; a_7 \approx 0.42$ m;
total distance $= a_0 + 2S_\infty = 18$ m
**105.** about 67,109 in. This is almost 1.06 mi.
**107. a.** For an arithmetic sequence, the difference $d = a_k - a_{k-1}$ must be constant. For $a_k = a_1 r^{k-1}$ and $a_{k-1} = a_1 r^{k-2}$, we have
$d = \log(a_1 r^{k-1}) - \log(a_1 r^{k-2})$
$= \log(a_1) + \log(r^{k-1}) - [\log(a_1) + \log(r^{k-2})]$
$= \log(a_1) + (k-1)\log(r) - \log(a_1) - (k-2)\log(r)$
$= \log(r)[(k-1) - (k-2)] = \log(r) ✓$

**b.** For a geometric sequence, the ratio $r = \dfrac{a_k}{a_{k-1}}$ must be constant.
For $a_k = a_1 + (k-1)d$ and $a_{k-1} = a_1 + (k-2)d$, we have
$r = \dfrac{10^{a_1 + (k-1)d}}{10^{a_1 + (k-2)d}} = \dfrac{10^{a_1}10^{(k-1)d}}{10^{a_1}10^{(k-2)d}} = \dfrac{10^{(k-1)d}}{10^{(k-2)d}}$
$= 10^{(k-1)d - (k-2)d} = 10^{d[(k-1)d - (k-2)]} = 10^d ✓$

**109.** $x = 0$   **111.** $p(50) \approx 2562.1, p(75) \approx 3615.6, p(100) \approx 4035.1$,
$p(150) \approx 4189.1$

## Mid-Chapter Check, pp. 700–701

**1.** $3, 10, 17, a_9 = 59$    **2.** $4, 7, 12, a_9 = 84$    **3.** $-1, 3, -5, a_9 = -17$
**4.** 360    **5.** $\sum_{k=1}^{6}(3k-2)$    **6.** d    **7.** e    **8.** a    **9.** b    **10.** c
**11. a.** $a_1 = 2, d = 3; a_n = 3n - 1$    **b.** $a_1 = \frac{3}{2}, d = \frac{3}{4}; a_n = \frac{3}{4}n + \frac{3}{4}$
**12.** $n = 25; S_{25} = 950$    **13.** $n = 16; S_{16} = 128$    **14.** $S_{10} = -5$
**15.** $S_{10} = \dfrac{-14{,}762}{27}$    **16. a.** $a_1 = 2, r = 3; a_n = 2(3)^{n-1}$
**b.** $a_1 = \frac{1}{2}, r = \frac{1}{2}; a_n = (\frac{1}{2})^n$    **17.** $n = 8; S_8 = \frac{1640}{27}$    **18.** $\frac{-343}{6}$    **19.** 1785
**20.** $\approx 4.5$ ft; $\approx 127.9$ ft

## Exercises 9.4, pp. 708–712

**1.** experiment, outcomes    **3.** Answers will vary.
**5. a.** 16 possible

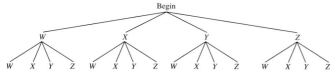

**b.** *WW, WX, WY, WZ, XW, XX, XY, XZ, YW, YX, YY, YZ, ZW, ZX, ZY, ZZ*
**7.** 32    **9.** 15,625    **11.** 2,704,000    **13. a.** 59,049    **b.** 15,120
**15.** 360 if double veggies are not allowed, 432 if double veggies are
allowed    **17. a.** 96    **b.** 500    **c.** 12    **19.** 24    **21.** 4    **23.** 120
**25.** 6    **27.** 720    **29.** 3024    **31.** 40,320    **33.** 6; 3    **35.** 90    **37.** 336
**39. a.** 720    **b.** 120    **c.** 24    **41.** 360    **43.** 60    **45.** 60    **47.** 120
**49.** 30    **51.** 60, BANANA    **53.** 126    **55.** 56    **57.** 1    **59.** verified
**61.** verified    **63.** 495    **65.** 364    **67.** 252    **69.** $8! = 40{,}320$
**71.** $_8P_3 = 336$    **73.** $_{20}C_5 = 15{,}504$    **75.** $_8C_4 = 70$    **77. a.** $\approx 1.2\%$
**b.** $\approx 0.83\%$    **79.** 7776    **81.** 324    **83.** 800    **85.** 6,272,000,000
**87.** 518,400    **89.** 357,696    **91.** 6720    **93.** 8    **95.** 10,080    **97.** 5040
**99.** 2880    **101.** 5005    **103.** 720    **105.** 52,650, no
**107.** $8 \cdot (_6C_3) - 6 \cdot 2 = 148$    **109.** $a_n = 4n - 3; 137; 2415$
**111.**

## Exercises 9.5, pp. 718–723

**1.** $0, 1, 1, 0$    **3.** Answers will vary.    **5.** $S = \{HH, HT, TH, TT\}, \frac{1}{4}$
**7.** $S = \{$coach of Patriots, Cougars, Angels, Sharks, Eagles, Stars$\}, \frac{1}{6}$
**9.** $P(E) = \frac{4}{9}$    **11. a.** $\frac{1}{13}$    **b.** $\frac{1}{4}$    **c.** $\frac{1}{2}$    **d.** $\frac{1}{26}$
**13.** $P(E_1) = \frac{1}{8}, P(E_2) = \frac{5}{8}, P(E_3) = \frac{3}{4}$    **15. a.** $\frac{3}{4}$    **b.** 1    **c.** $\frac{1}{4}$    **d.** $\frac{1}{2}$
**17.** $\frac{3}{4}$    **19.** $\frac{6}{7}$    **21.** 0.991    **23. a.** $\frac{1}{12}$    **b.** $\frac{11}{12}$    **c.** $\frac{8}{9}$    **d.** $\frac{5}{6}$
**25.** b, about 12%    **27. a.** 0.3651    **b.** 0.3651    **c.** 0.3969    **29.** 0.9
**31.** $\frac{7}{24}$    **33.** 0.59    **35. a.** $\frac{1}{6}$    **b.** $\frac{7}{36}$    **c.** $\frac{1}{9}$    **d.** $\frac{4}{9}$    **37. a.** $\frac{2}{25}$    **b.** $\frac{9}{50}$
**c.** 0    **d.** $\frac{2}{25}$    **e.** 1    **39.** $\frac{3}{4}$    **41.** $\frac{11}{15}$    **43. a.** $\frac{1}{18}$    **b.** $\frac{2}{9}$    **c.** $\frac{8}{9}$    **d.** $\frac{5}{9}$    **e.** $\frac{1}{36}$
**f.** $\frac{5}{12}$    **45.** \$0.53 loss; no    **47. a.** 0.33    **b.** 0.67    **c.** 1    **d.** 0    **e.** 0.67
**f.** 0.08    **49. a.** $\frac{1}{2}$    **b.** $\frac{1}{8}$    **51. a.** $\frac{9}{16}$    **b.** $\frac{1}{4}$    **c.** $\frac{1}{16}$    **d.** $\frac{5}{16}$    **53. a.** $\frac{3}{26}$
**b.** $\frac{3}{26}$    **c.** $\frac{1}{13}$    **d.** $\frac{9}{26}$    **e.** $\frac{2}{13}$    **f.** $\frac{11}{26}$    **55. a.** $\frac{1}{8}$    **b.** $\frac{1}{16}$    **c.** $\frac{3}{16}$    **57. a.** $\frac{47}{100}$
**b.** $\frac{2}{25}$    **c.** $\frac{3}{100}$    **d.** $\frac{9}{50}$    **e.** $\frac{11}{100}$    **59. a.** $\frac{5}{429}$    **b.** $\frac{8}{2145}$    **61.** $\frac{1}{3360}$
**63.** $\frac{1}{1{,}048{,}576}$; answers will vary; 20 heads in a row.    **65.** $(9, 1, 1)$
**67.** $x \in [-1, 0) \cup [1, \infty)$

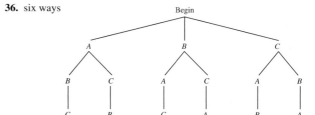

## Exercises 9.6, pp. 729–731

**1.** $n$    **3.** Answers will vary.
**5.** $x^5 + 5x^4y + 10x^3y^2 + 10x^2y^3 + 5xy^4 + y^5$
**7.** $16x^4 + 96x^3 + 216x^2 + 216x + 81$    **9.** $41 + 38i$    **11.** 35
**13.** 1140    **15.** 1    **17.** $c^5 + 5c^4d + 10c^3d^2 + 10c^2d^3 + 5cd^4 + d^5$
**19.** $a^6 - 6a^5b + 15a^4b^2 - 20a^3b^3 + 15a^2b^4 - 6ab^5 + b^6$
**21.** $16x^4 - 96x^3 + 216x^2 - 216x + 81$    **23.** $-11 + 2i$
**25.** $x^9 + 18x^8y + 144x^7y^2$    **27.** $v^{24} - 6v^{22}w + \frac{33}{2}v^{20}w^2$    **29.** $35x^4y^3$
**31.** $1792p^2$    **33.** $264x^2y^{10}$    **35.** $\approx 0.25$    **37. a.** $\approx 17.8\%$    **b.** $\approx 23.0\%$
**39. a.** $\approx 0.89\%$    **b.** $\approx 7.9\%$    **c.** $\approx 99.0\%$    **d.** $\approx 61.0\%$    **41.** $2^{n-1}, 2048$
**43.**

$f(3) = 1$

**45.** $g(x) > 0: x \in (-2, 0) \cup (3, \infty)$

## Making Connections, p. 731

**1.** c    **3.** b    **5.** d    **7.** b, d, f    **9.** d    **11.** a    **13.** e    **15.** g

## Summary and Concept Review, pp. 732–735

**1.** $1, 6, 11, 16; a_{10} = 46$    **2.** $1, \frac{3}{5}, \frac{2}{5}, \frac{5}{17}; a_{10} = \frac{11}{101}$
**3.** $a_n = n^4; a_6 = 1296$    **4.** $a_n = 3n - 20; a_6 = -2$
**5.** $\frac{255}{256}$    **6.** $-112$    **7.** 140    **8.** 35    **9.** $4, -8, 8, \frac{-16}{3}, \frac{8}{3}$    **10.** $\frac{1}{2}, \frac{3}{4}, \frac{5}{4}, \frac{9}{4}, \frac{17}{4}$
**11.** $\sum_{i=1}^{7}(i^2 + 3i - 2); 210$    **12. a.** about 134 hawks    **b.** 8 yr
**13.** $a_n = 3n - 1; 119$    **14.** $a_n = -2n + 5; -65$
**15.** 740    **16.** 1335    **17.** 630    **18.** $-11.25$    **19.** 875    **20.** 3240
**21.** 7.55 m    **22.** 32    **23.** 2401    **24.** 3645    **25.** 6560    **26.** $\frac{819}{512}$
**27.** 10.75    **28.** $\frac{50}{9}$    **29.** 4    **30.** does not exist    **31.** does not exist
**32.** 5    **33.** $\frac{63{,}050}{6561}$
**34.** $a_0 = 1225, a_1 \approx 1311, a_n = 1311(1.07)^{n-1}; a_{15} \approx 3380; S_{15} \approx 32{,}944$
**35.** $a_0 = 121{,}500, a_1 = 81{,}000, a_n = 81{,}000(\frac{2}{3})^{n-1}; a_7 \approx 7111$ ft$^3$
**36.** six ways

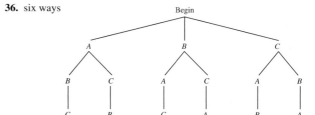

**37. a.** 720    **b.** 1000    **38.** 24    **39.** 220    **40. a.** 5040    **b.** 840    **c.** 35
**41. a.** 720    **b.** 120    **c.** 24    **42.** 3360    **43. a.** 220    **b.** 1320
**44.** $\frac{4}{13}$    **45.** $\frac{3}{13}$    **46.** $\frac{5}{6}$    **47.** $\frac{7}{24}$    **48.** $\frac{175}{396}$    **49. a.** 0.608    **b.** 0.392
**c.** 1    **d.** 0    **e.** 0.928    **f.** 0.178    **50. a.** 21    **b.** 56
**51. a.** $x^4 - 4x^3y + 6x^2y^2 - 4xy^3 + y^4$    **b.** $41 - 38i$
**52. a.** $a^8 + 8\sqrt{3}a^7 + 84a^6 + 168\sqrt{3}a^5$
**b.** $78{,}125a^7 + 218{,}750a^6b + 262{,}500a^5b^2 + 175{,}000a^4b^3$
**53. a.** $280x^4y^3$    **b.** $-64{,}064a^5b^9$    **54. a.** about 93.3%    **b.** about 62.4%

## Practice Test, pp. 735–736

**1. a.** $\frac{1}{2}, \frac{4}{5}, 1, \frac{8}{7}; a_8 = \frac{16}{11}$    **b.** $3, 6, 15, 45; a_8 = 14{,}175$
**c.** $3, 2\sqrt{2}, \sqrt{7}, \sqrt{6}; a_8 = \sqrt{2}$    **2. a.** 165    **b.** $\frac{311}{420}$    **c.** $\frac{-2343}{512}$    **d.** 7

**3. a.** $a_1 = 7, d = -3, a_n = 10 - 3n$
**b.** $a_1 = -8, d = 2, a_n = 2n - 10$   **c.** $a_1 = 4, r = -2, a_n = 4(-2)^{n-1}$
**d.** $a_1 = 10, r = \frac{2}{5}, a_n = 10(\frac{2}{5})^{n-1}$   **4. a.** 199   **b.** 9   **c.** $\frac{3}{4}$   **d.** 6
**5. a.** 1712   **b.** 2183   **c.** 2188   **d.** 12   **6. a.** $\approx 8.82$ ft   **b.** $\approx 72.4$ ft
**7.** $6756.57   **8.** 900,900
**9. a.**

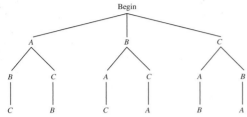

**b.** $ABC, ACB, BAC, BCA, CAB, CBA$
**10.** 302,400   **11.** $\frac{1}{3}$   **12.** 302,400
**13. a.** $x^4 - 8x^3y + 24x^2y^2 - 32xy^3 + 16y^4$   **b.** $-4$
**14. a.** $x^{10} + 10\sqrt{2}\,x^9 + 90x^8$   **b.** $a^8 - 16a^7b^3 + 112a^6b^6$
**15.** 0.989   **16. a.** 0.08   **b.** 0.92   **c.** 1   **d.** 0   **e.** 0.95   **f.** 0.03
**17. a.** $\frac{59}{100}$   **b.** $\frac{53}{100}$   **c.** $\frac{13}{100}$   **d.** $\frac{47}{100}$   **18. a.** 0.1875   **b.** 0.589
**c.** 0.4015   **d.** 0.2945   **e.** 0.4110   **f.** 0.2055   **19. a.** about 27.9%
**b.** about 97.6%   **20. a.** 0.8075   **b.** 0.0075   **c.** 0.9925

## Calculator Exploration and Discovery, pp. 736–737

**Exercise 1:** $S_5 = 0.6875, S_{10} = 0.666015625, S_{20} = 0.6666660309$;
It seems the limiting value is $S_\infty = \frac{2}{3}$. $S_{50} \approx 0.6666666667$ and
$S_{100} \approx 0.6666666667$ further support this conjecture.
**Exercise 2:** $S_5 \approx 0.8333333333, S_{10} \approx 0.9090909091$,
$S_{20} \approx 0.9523809524$; It seems the limiting value is $S_\infty = 1$.
$S_{50} \approx 0.9803921569$ and $S_{100} \approx 0.9900990099$ further support this
conjecture.
**Exercise 3:** $S_5 \approx 2.708333333, S_{10} \approx 2.718281526$,
$S_{20} \approx 2.718281828$; It seems the limiting value is $S_\infty = e$.
$S_{50} \approx 2.718281828$ and $S_{100} \approx 2.718281828$ further support this conjecture.
**Exercise 4: a.** 36   **b.** 84   **c.** 126   **d.** 126
**Exercise 5:** 1,221,759   **Exercise 6:** 20

## Strengthening Core Skills, pp. 737–738

**Exercise 1:** $71,500

## Cumulative Review Chapters R–9, pp. 738–739

**1. a.** 23 cards are assembled each hour.   **b.** $y = 23x - 155$
**c.** 184 cards   **d.** $\approx 6:45$ A.M.   **3.** $x = \dfrac{-5 \pm \sqrt{109}}{6}$; $x \approx 0.91$; $x \approx -2.57$
**5.** $Y = \dfrac{kVW}{X}$; $Y = \dfrac{3VW}{2X}$   **7.** verified; reflections across $y = x$
**9. a.** $4x + 2h - 3$   **b.** $\dfrac{-1}{(x + h - 2)(x - 2)}$

**11.**

**13. a.** $x^3 = 125$   **b.** $e^5 = 2x - 1$   **15. a.** $x \approx 3.19$   **b.** $x = 334$
**17.** (5, 10, 15)   **19.** $\dfrac{(x + 3)^2}{16} + \dfrac{(y - 3)^2}{4} = 1$; $(-3, 3)$; $(-7, 3)$, $(1, 3)$;
$(-3 - 2\sqrt{3}, 3), (-3 + 2\sqrt{3}, 3)$   **21.** 1333   **23. a.** $\approx 7.0\%$
**b.** $\approx 91.9\%$   **c.** $\approx 98.9\%$   **d.** $\binom{12}{0}(0.04)^{12}(0.96)^0$; virtually nil
**25.** $\begin{cases} 9a + 3b + c = 56 \\ 25a + 5b + c = 126 \\ 81a + 9b + c = 98 \end{cases}$; $y = -7x^2 + 91x - 154$   **a.** mid-June
**b.** $\approx 142$   **c.** 98   **d.** November to February

# APPENDIX

## APPENDIX A.I pp. A-7–A-11

**1.** $A = 2280$ mm$^2$ or 22.8 cm$^2$   **3.** $A \approx 950$ mm$^2$
**5.** $P = 18$ cm           **7.** $P = 20$ cm           **9.** $C \approx 15.7$ cm
    $A = 20$ cm$^2$              $A = 20$ cm$^2$
**11.** $c = 45$ cm                          **13.** $P = 2(\pi d) = 15\pi \approx 47.1$ in.
    $P = 108$ cm
    $A = 486$ cm$^2$
**15.** $A = \dfrac{bh}{2} + LW = 54$ ft$^2$       **17.** $A = \dfrac{\pi r^2}{4} = 81\pi \approx 254.5$ in$^2$
**19.** $P = 2L + \pi d = 240 + 80\pi \approx 491.3$ ft
    $A = LW + \pi r^2 = 9600 + 1600\pi \approx 14,626.5$ ft$^2$
**21.** 504 in$^2$   **23.** $W = 12$ cm
**25.** $V = LWH + \pi r^2 h = 200 + 43.75\pi \approx 337.4$ in$^3$
**27.** K: $LW$                    C: $s_1 + s_2 + s_3$           J: $\pi r^2$

    A: $4s$                       I: $LWH$                      D: $\dfrac{1}{2}bh$

    H: $s_1 + s_2 + s_3 + s_4$    E: $2\pi r$                   G: $s^2$

    B: $2L + 2W$                  F: $s^3$                      L: $A = \dfrac{b_1 + b_2}{2} \cdot h$
**29.** 32 ft   **31.** 192 ft$^3$   **33.** 23.91 m
**35.** one large at about 5.8¢/in$^2$
**37.** about 654.5 gal   **39. a.** 35.3 m$^2$   **b.** 2.1 m$^3$   **c.** about $280.88
**41. a.** 216.5 ft$^2$   **b.** about 7   **c.** about $61.07   **43.** 5 in. × 3 in.
**45.** $2500\pi \approx 7854$ ft$^3$
    about 58,752 gal

# Photo Credits

## Chapter R

pg. 1: © Corbis/RF; pg. 12/top: © Comstock Images/Alamy/RF; pg. 12/middle: © Ryan McVay/Getty Images/RF; pg. 12/bottom: Courtesy Amanda Wilson; pg. 19/left: © Photodisc/Getty Images/RF; pg. 19/right: © Getty Images/Steve Allen/RF; pg. 20: © Royalty-Free/Corbis/RF; pg. 34: © Ingram Publishing; pg. 45: © David Frazier Photolibrary, Inc./RF; pg. 46: © Glen Allison/Getty Images/RF; pg. 72: © PhotoLink/Getty Images/RF.

## Chapter 1

pg. 73: © Corbis/RF; pg. 84/right: © Stocktrek Images, Inc./Alamy/RF; pg. 84/left: © Authors Image/Alamy/RF; pg. 103: © PhotoLink/Getty Images/RF; pg. 135: © Getty Images/Digital Vision; pg. 139/top: © Brand X Pictures/Punchstock/RF; pg. 139/bottom: © Photodisc Collection/Getty Images/RF; pg. 144: Courtesy Jeremy Coffelt.

## Chapter 2

pg. 147: © The McGraw-Hill Companies, Inc.; pg. 148/left: © PhotoAlto/RF; pg. 148/middle: © Brand X Pictures/Punchstock/RF; pg. 148/right: © Lars Niki/RF; pg. 170: © Epoxydude/Getty Images/RF; pg. 174: © Imagery Majestic/Cutcaster/RF; pg. 186: © Stockbyte/Punchstock/RF; pg. 188: © Royalty-Free/Corbis/RF; pg. 201/left: © Gabriela Medina/Blend Images LLC/RF; pg. 201/top right: © Ariel Skelley/Blend Images LLC/RF; pg. 201/bottom right: Courtesy Twitter; pg. 202: © Sieda Preis/Getty Images/RF; pg. 204: © The McGraw-Hill Companies, Inc./ Ken Cavanagh Photographer; pg. 211: © Ingram Publishing/RF; pg. 212: © Ingram Publishing/ SuperStock/RF; pg. 215: © The McGraw-Hill Companies, Inc./Barry Barker, photographer; pg. 216: © Royalty-Free/Corbis; pg. 222: © Steve Cole/Getty Images/RF; pg. 223: © Charles Smith/Corbis/RF; pg. 231: © Lisa F. Young/Alamy/RF; pg. 234/left: © Tom Grill/ Corbis/RF; pg. 234/top right: © The McGraw-Hill Companies, Inc./Jill Braaten, photographer; pg. 234/ bottom right: © The McGraw-Hill Companies, Inc./RF; pg. 241: © Lourens Smak/Alamy/RF.

## Chapter 3

pg. 245: © Richard Carey/Alamy/RF; pg. 254: © Keith Leighton/Alamy/RF; pg. 258: © Paul Edmondson/Getty Images/RF; pg. 264: Courtesy Jeremy Coffelt; pg. 268/top: © Scenics of America/PhotoLink/Getty Images/RF; pg. 268/bottom: © Brand X Pictures/ PunchStock/RF; pg. 275: © Brand X Pictures/ PunchStock/RF; pg. 279: © Royalty-Free/Corbis; pg. 285: © The McGraw-Hill Companies, Inc./John Flournoy, photographer; pg. 286: © Allan and Sandy Carey/Getty Images/RF; pg. 292: © Brand X Pictures/ PunchStock/RF; pg. 296: Courtesy John Coburn; pg. 301: © U.S. Fish & Wildlife Service/Tracy Brooks/ RF; pg. 306/inset: © Library of Congress Prints &

Photographs Division (LC-USZ62-60242)/RF; pg. 306: © Stocktrek/age photostock/RF; pg. 306: © Chad Baker/Photodisc/Getty Images/RF; pg. 307: © Royalty-Free/Corbis/RF; pg. 312: © Digital Vision/PunchStock/RF; pg. 313/left: © Goodshot/ PunchStock/RF; pg. 313/right: © Royalty-Free/Corbis; pg. 324/left: © Ingram Publishing/Fotosearch/RF; pg. 324/top middle: © Thinkstock images/Jupiter Images/RF; pg. 324/right: © Purestock/Superstock/ RF; pg. 324/middle bottom: © RubberBall Productions/RF.

## Chapter 4

pg. 335: © Blend Images/Getty Images/RF; pg. 343: © Photodisc Collection/Getty Images/RF; pg. 347: © Corbis/Royalty-Free; pg. 356: © Adalberto Rios Szalay/Sexto Sol/Getty Images/RF; pg. 358: © Royalty-Free/Corbis; pg. 373: © Royalty-Free/ Corbis; pg. 405: © Royalty-Free/Corbis; pg. 414: © Olivier Renck/Getty Images; pg. 417: © Royalty-Free/Corbis.

## Chapter 5

pg. 427: © Brand X Pictures/Jupiter Images/RF; pg. 437: © Brand X Pictures/RF; pg. 444: © The McGraw-Hill Companies, Inc.; pg. 448/left: © Geostock/Getty Images/RF; pg. 448/right: © Lawrence M. Sawyer/Getty Images/RF; pg. 458: © W.C. Mendenhall/USGS/RF; pg. 459: © Lars Niki/RF; pg. 462/top: © Corbis/RF; pg. 462/bottom: © Medioimages/Superstock/RF; pg. 472/top left: © Ingram Publishing/age Fotostock/RF; pg. 472/ bottom left: © Keith Brofsky/Getty Images/RF; pg. 472/top right: © The McGraw-Hill Companies, Inc./Andrew Resek, photographer; pg. 472/bottom right: © McGraw-Hill Higher Education, Inc./Carlyn Iverson, photographer; pg. 481/left: © Kent Knudson/ PhotoLink/Getty Images/RF; pg. 481/right: © StockTrek/ Getty Images/RF; pg. 489: © Tom Brakefield/Getty Images/RF; pg. 491: © Comstock Images/Getty Images/RF; pg. 493: © Brand X Pictures/RF; pg. 494/left: © Photographer's Choice/Getty Images/ RF; pg. 494/top right: © Photographer's Choice/Getty Images/RF; pg. 494/bottom right: Library of Congress Prints and Photographs Division/RF; pg. 500: © CMCD/Getty Images/RF; pg. 503: © John A. Rizzo/Getty Images/RF.

## Chapter 6

pg. 505: Coutesy Ashley Zellmer; pg. 506: © Hill Street Studios/Blend Images LLC/RF; pg. 514: © Creatas/PunchStock/RF; pg. 519: © Lourens Smak/Alamy/RF; pg. 520: Courtesy Julia Zellmer; pg. 530: © Creatas/Punchstock/RF; pg. 538: © Ryan McVay/Getty Images/RF; pg. 540: © The McGraw-Hill Companies, Inc.; pg. 546: © Ariel Skelley/Blend Images LLC/RF; pg. 553/top left: © Jose Luis Pelaez Inc./Blend Images/LLC/RF; pg. 553/bottom left:

© Getty Images/RF; pg. 553/right: © Todd Wright/ Blend Images LLC/RF; pg. 554/top left: © PhotosIndia .com/Getty Images/RF; pg. 554/bottom left: © Burke/ Triolo/Getty Images/RF; pg. 554/right: © Burke/ Triolo/Brand X Pictures/RF; pg. 557: © F. Schussler/ PhotoLink/Getty Images/RF.

## Chapter 7

pg. 563: © Comstock/JupiterImages/RF; pg. 571: © Photographer's Choice/Getty Images/RF; pg. 584: © Blend Images/SuperStock/RF; pg. 587: © David Papazian/Beatworks/Corbis/RF; pg. 588/left: © Digital Vision/PunchStock/RF; pg. 588/top right: © DreamPictures/Blend Images LLC/RF; pg. 588/bottom right: © ImageState/ Alamy/RF; pg. 599: © Fuse/Getty Images/RF; pg. 603/left: © Ingram Publishing/Alamy/RF; pg. 603: © Mark Steinmetz/RF; pg. 604: © liquidlibrary/PictureQuest/RF; pg. 618/left: © Ingram Publishing/SuperStock/RF; pg. 618/top right: © Image Source/Getty Images/RF; pg. 618/bottom right: © Stockbyte/Getty Images/RF.

## Chapter 8

pg. 627: © Stocktrek/Alamy/RF; pg. 654: © Brand X Pictures/Punchstock/RF; pg. 655: © Digital Vision/Getty Images/RF; pg. 658: © H. Wiesenhofer/ PhotoLink/Getty Images/RF; pg. 666/top left: © Jim Wehtje/Getty Images/RF; pg. 666/bottom left: © Edmond Van Hoorick/Getty Images/RF; pg. 666/ right: © Creatas/PunchStock/RF.

## Chapter 9

pg. 673: © Getty Images/Digital Vision/RF; pg. 675: © Ingram Publishing/Alamy/RF; pg. 680: © Brand X Pictures/PunchStock/RF; pg. 681/top: © Brand X Pictures/PunchStock/RF; pg. 681/ bottom: © ColorBlind Images/Blend Images LLC/RF; pg. 686/left: © IT Stock/age fotostock/ RF; pg. 686/right: © Leslie Richard Jacobs/ Corbis; pg. 689/right: © Royalty-Free/Corbis; pg. 689/left: © Dynamic Graphics/JupiterImages/ RF; pg. 696: © Stockbyte/PunchStock/RF; pg. 699: © Anderson Ross/Getty Images/RF; pg. 700: © John Lund/Drew Kelly/Blend Images LLC/RF; pg. 702: © Steve Sant/RF; pg. 708: © Brand X Pictures/PunchStock/RF; pg. 709/top: © Royalty-Free/Corbis/RF; pg. 709/bottom: © Nancy R. Cohen/Getty Images/RF; pg. 710/left: © Tomi/PhotoLink/Getty Images/RF; pg. 710/right: © Burke/Triolo/Brand X Pictures; pg. 711: © Ingram Publishing/RF; pg. 712: © Brand X Pictures/RF; pg. 719: © Kris Legg/Alamy/RF; pg. 720: © Royalty-Free/Corbis; pg. 729: © Corbis/ SuperStock/RF.

## Appendix

A-10: © Sandra Ivany/Brand X Pictures/Getty Images/RF.

# Subject Index